Springer Monographs in Mathematics

Springer
Berlin
Heidelberg
New York
Barcelona
Budapest
Hong Kong
London
Milan
Paris
Santa Clara
Singapore
Tokyo

Victor Ivrii

Microlocal Analysis and Precise Spectral Asymptotics

Springer

Victor Ivrii
Department of Mathematics
University of Toronto
100, St. George Street
Toronto, Ontario M5S 1A1
Canada

Library of Congress Cataloging-in-Publication Data

Ivrii, Victor, 1949-
Microlocal analysis and precise spectral asymptotics / Victor Ivrii.
p. cm. -- (Springer monographs in mathematics)
Includes bibliographical references and index.
ISBN 3-540-62780-4 (hardcover : alk. paper)
1. Microlocal analysis. 2. Spectral theory (Mathematics)
3. Eigenvalues. 4. Asymptotic expansions. I. Title. II. Series.
QA299.6.I87 1998
515'.7242--dc21 97-49029
CIP

Mathematics Subject Classification (1991): 35P, 35J, 35L, 58G, 81Q

ISBN 3-540-62780-4 Springer-Verlag Berlin Heidelberg New York

Typesetting by the author using LateX
SPIN 10521749 41/3143-5 4 3 2 1 0 – Printed on acid-free paper

Preface

The problem of spectral asymptotics, in particular the problem of the asymptotic distribution of eigenvalues, is one of the central problems in the spectral theory of partial differential operators; moreover, it is very important for the general theory of partial differential operators.

I started working in this domain in 1979 after R. Seeley found a remainder estimate of the same order as the then hypothetical second term for the Laplacian in domains with boundary, and M. Shubin and B. M. Levitan suggested that I should try to prove Weyl's conjecture. During the past fifteen years I have not left the topic, although I had such intentions in 1985 when the methods I invented seemed to fail to provide further progress and only a couple of not very exciting problems remained to be solved. However, at that time I made the step toward local semiclassical spectral asymptotics and rescaling, and new horizons opened.

So I can say that this book is the result of fifteen years of work in the theory of spectral asymptotics. However, it would be more accurate to introduce it as the result of eight years of great effort, since during the last six years I wrote and rewrote preliminary versions of this book from the initial version to the final. Nine preprints were published from the total volume of about 900 pages of the preliminary version. Almost all the results presented here were obtained by myself (sometimes in cooperation) and almost all of them have been published only in short notes or in preprints.

Many of my colleagues influenced this book. Fifteen years ago B. M. Levitan and M. Shubin drew my attention to the theory of spectral asymptotics. Numerous discussions with M. S. Agranovich, M. S. Birman, S. Levendorskii, V. Petkov, G. Rozenblioum, Yu. Safarov, M. Z. Solomyak, D. Vassiliev, A. Volovoi and later with L. Boutet de Monvel, Y. Colin de Verdiere, A. Grigis, C. Gerard, B. Helffer, L. Hörmander, Ya. Kannai, B. and R. Lascars, D. Robert, I. M. Sigal, J. Sjöstrand, A. Sobolev, S. Vugalter and G. M. Zhislin were very stimulating and useful. I am very grateful to all the graduate students at the École Polytechnique and the University of Toronto who took my courses; their interest was very stimulating as well! Certain results were obtained in collaboration with my graduate students S. Fedorova (now Fainshtein), E. Filippov and A. Kachalkina, and later results were obtained in cooperation with I. M. Sigal and A. Sobolev. I would like to express my sincere and deep gratitude to all of them.

I started this work in Magnitogorsk, continued it at the École Polytechnique where I spent two years (1990–1992), and finished it at the University of Toronto. I am very grateful to all my colleagues in these places and at the University of Bologna, the Weizmann Institute of Science, the Mittag-Leffler Institute, l'Université Paris-7, l'Université Paris-Sud and l'Université Paris-Nord where I spent a few months each. I am especially grateful to those at the École Polytechnique who taught me to use Macintosh computers and TeX. J.-M. Bony occupies a very special place in this list of gratitudes.

I am very grateful to Ed Doolittle who did a wonderful editorial job.

Finally, I am very grateful to my wife Olga for her understanding, approval, patience and support without which I would have been unable to realize even the first stage of my project.

Victor Ivrii, Toronto, 1997.

Table of Contents

Part III. Estimates of the Spectrum

Part IV. Asymptotics of Spectra

0. Introduction

0.1 Problems to Study

The problem of spectral asymptotics, in particular the problem of the asymptotic distribution of eigenvalues, is one of the central problems of the spectral theory of partial differential operators, and is very important for the general theory of partial differential operators. Apart from applications in quantum mechanics, radio physics, continuum media mechanics (elasticity, hydrodynamics, theory of shells), etc., there are also applications to mathematics itself and deep though non-obvious links with differential geometry, dynamical systems theory and ergodic theory; even the term "spectral geometry" has arisen. All these circumstances make this topic very attractive for a mathematician.

The problem originated in 1911 when H. Weyl published a paper devoted to eigenvalue asymptotics for the Laplace operator in a bounded domain with regular boundary. After this article a huge number of different papers devoted to spectral asymptotics were published among the authors of which were numerous prominent mathematicians. The theory developed in two directions: on one hand the theory was extended to consider more general operators, boundary conditions and geometrical domains; on the other hand the theory was improved and more accurate remainder estimates were derived. In the latter extensions the links with differential geometry, dynamical systems theory and ergodic theory appeared. Even the theory of eigenvalue asymptotics for the Laplace (or Laplace-Beltrami) operator has a long, dramatic and still unfinished history. Apart from asymptotics with respect to the spectral parameter, asymptotics with respect to other parameters appeared; the most important among them are (in my opinion) semiclassical asymptotics, i.e., asymptotics with respect to a small parameter h (the Planck constant in physics) tending to $+0$. For a long time these asymptotics were in the shadows: most attention was paid to the eigenvalue asymptotics for operators on compact manifolds (with or without boundary). The results which had been obtained in that case were then reproved for operators in $\mathbb{R}^d$ such as the Schrödinger operator $-h^2\Delta + V(x)$ with fixed $h > 0$ and with $V(x) \to +\infty$ as $|x| \to \infty$; less attention was paid to semiclassical asymptotics (i.e., asymptotics of eigenvalues less than some fixed level λ as $h \to +0$). The asymptotics of small negative eigenvalues were considered in the case of fixed h and $V(x)$ decreasing at infinity as $|x|^{2m}$ with $m \in (-1, 0)$; under reasonable conditions in this case the discrete spectrum of the operator has accumulation point -0 and the essential spectrum coincides with $[0, +\infty)$. The result of the development of the theory was that at a certain time there existed four parallel (though not equally developed and not equally respected)[1)] theories, the statements in each of which had to be proven separately.

[1)] If the achievements in the first theory were broadly known by mathematicians, the other results were familiar to the narrow circle of PDE-folk and mathematical physicists many of whom (including myself, it is a real shame for me) even thought that only the first topic is really decent.

This pluralism is now over (at least in my papers) because each of the other results are easily derived from *Local Semiclassical Spectral Asymptotics* (*LSSA* in what follows).[2)]

Namely semiclassical spectral asymptotics are the main object of this book and all the other results are obtained as applications.

0.2 Methods

In his papers H. Weyl applied the *variational* method (Dirichlet-Neumann bracketing) he invented; later this method was improved in various directions by many outstanding mathematicians and mathematical physicists (R. Courant, M. Birman, and others). This method is simple and works in general situations (even for non-semibounded operators) and seems to be the best method if one does not care about precise remainder estimates.

Later *Tauberian* methods appeared (first due to T. Carleman). The idea is to consider some function $F(A,t)$ of an operator and an auxiliary parameter t. Then

$$\operatorname{Tr} F(A,t) = \int F(\tau,t)d_\tau \operatorname{Tr} E(\tau)$$

where $E(\tau)$ is the spectral projector. One then tries to construct the left-hand expression by means of PDE theory and then recover $\operatorname{Tr} E(\tau)$ with some error by means of an appropriate Tauberian theorem. The choice of the function $F(A,t)$ is very important here. Taking $F = \exp(-tA)$ (for a positive operator A), $F = (t-A)^{-1}$ and $F = A^t$ one obtains respectively the *heat equation, resolvent and ζ-function methods*, which are usually simple in their PDE part but difficult in the Tauberian part and almost always fail to provide a decent remainder estimate.

On the other hand, the *hyperbolic operator method* due to Levitan and Avvakumovič is difficult in its PDE part but is almost trivial in the Tauberian part and provides accurate remainder estimates. By this method all the asymptotics with the most accurate remainder estimates were obtained. This method is based on the function $U(t) = F(A,t) = \exp(itA)$ which satisfies $(D_t - A)U = 0$, $U(0) = I$, and therefore the Schwartz kernel $u(x,y,t)$ of the operator is the fundamental solution to the Cauchy problem (or the initial-boundary value problem) for the operator $D_t - A$ (where $D_t = -i\partial_t$, etc.). The Schwartz kernel is connected to the eigenvalue counting function of the operator A by the formula

$$\int u(x,x,t)dx = \int \exp it\lambda \, d_\lambda N(\lambda). \tag{0.1}$$

In the case of a matrix operator A, $u(x,y,t)$ is a matrix-valued function and should be replaced by its trace in the left-hand expression; here and below $N(\lambda)$ is a number of eigenvalues of A less than λ (here we only consider operators which are semibounded from below with discrete spectra). Then by means of the inverse Fourier transform we can recover $N(\lambda)$ provided we have constructed $u(x,y,t)$ by means of the methods of partial differential operator theory. In fact we are never able (excluding some very special

[2)] I started to study semiclassical spectral asymptotics at the time when I foolishly assumed that the first theory was completely finished and I had nothing to do; so I decided to extend the results to semiclassical asymptotics before leaving the area. When I told my first results to M. Z. Solomyak, he asked if those results could be deduced from the results of the classical theory by the Birman-Schwinger principle. A quick check gave a very surprising answer: using classical asymptotics one can prove semiclassical asymptotics only under very unnatural restrictions; but using semiclassical asymptotics one can extend classical asymptotics to more general operators and problems.

and rare cases) to construct $u(x, y, t)$ precisely and for all the values $t \in \mathbb{R}$. Usually (now we assume that A is an elliptic first-order pseudo-differential operator) the fundamental solution is constructed approximately (modulo smooth functions) for t belonging to some interval $[-T, T]$ with $T > 0$. As a consequence we obtain, modulo $O(\lambda^{-K})$ with any arbitrarily chosen K, an expression for

$$F_{t\to\tau}\chi_T(t)\int u(x,x,t)dx = \int \hat{\chi}_T(\tau-\lambda)d_\lambda N(\lambda) \tag{0.2}$$

where χ is a fixed smooth function supported in $[-1,1]$, $\chi_T(t) = \chi(\frac{t}{T})$ and the hat as well as $F_{t\to\tau}$ mean the Fourier transform. Then if we know the left-hand expression we are able, by the Tauberian theorem due to Hörmander, to recover $N(\lambda)$ approximately by the formula

$$N(\lambda) = \int_{-\infty}^{\lambda} \bigl(F_{t\to\tau}\chi_T(t)\sigma(t)\bigr)(\tau)d\tau + O\bigl(\lambda^{d-1}\bigr) \tag{0.3}$$

where d is the dimension of the domain and

$$\sigma(t) = \int u(x,x,t)dx. \tag{0.4}$$

The explicit construction of $u(x, x, t)$ in this situation leads us to the formula

$$N(\lambda) = \kappa_0\lambda^d + O\bigl(\lambda^{d-1}\bigr) \tag{0.5}$$

with leading coefficient

$$\kappa_0 = (2\pi)^{-d}\int_{a(x,\xi)<1} dxd\xi \tag{0.6}$$

where $a(x,\xi)$ is the principal symbol of the operator A.

We see that the crucial step in this approach is the construction of the fundamental solution. This construction by means of Fourier integral operators is now standard and well known provided we consider scalar operators or operators with constant multiplicities of the eigenvalues of the principal symbol. We construct $u(x, y, t)$ on a compact set K contained in the interior of our domain X (and T depends on the distance between K and ∂X). If one of those assumptions is violated then the construction is much more sophisticated and is possible only under some very restrictive conditions. In the presence of the boundary (but only in the case of constant multiplicities of the eigenvalues of the principal symbol) this construction was realized in papers due to R. Seeley, D. Vassiliev and R. Melrose. But it is possible to avoid all these troubles by means of another approach suggested by V. Ivrii [7–9] (see also [Hörmander 1] and [Melrose 1]) based on the investigation of the propagation of singularities for $u(x, y, t)$ and the construction of an "approximation" (in a rather exotic sense) for the distribution which leads to an approximation in a reasonable sense for $\sigma(t)$ for $|t| \le T$ with appropriate $T > 0$. For the h-pseudo-differential operators considered in this book this approach is essentially simpler and more transparent because there is a selected parameter h; this case will be discussed below. This method allows us to prove the asymptotics (0.3) for an arbitrary self-adjoint m-th order elliptic operator with $m > 0$ and the spectral parameter λ^m on a compact manifold with or without boundary (in the former case the boundary conditions are also supposed to be elliptic), scalar or matrix, semibounded from below or not semibounded at all (in this case $N(\lambda)$ is replaced by $N^{\pm}(\lambda)$ which is the number of eigenvalues lying between 0 and $\pm\lambda^m$); the formula for κ_0 should be changed if necessary.

The two-term asymptotics

$$N(\lambda) = \kappa_0 \lambda^d + \kappa_1 \lambda^{d-1} + o(\lambda^{d-1}) \tag{0.7}$$

suggested by H. Weyl (who also gave a formula for κ_1) fails to be true without some additional condition. Surely this asymptotic expansion is wrong for $d = 1$ and for the Laplace-Beltrami operator on the sphere $\mathbb{S}^d$; this fact is associated with the high multiplicities of its eigenvalues. Moreover, it remains wrong in the case when this Laplace-Beltrami operator is perturbed by a potential or even by a symmetric first-order operator with small coefficients; in these cases all the eigenvalues of high multiplicities will be separated into narrow eigenvalue clusters separated by lacunas. On the other hand, under conditions of a global nature the asymptotics (0.7) have their place. For a scalar operator on a compact manifold without boundary the relevant condition is *"The measure of the {set of all the points of the cotangent bundle periodic with respect to the Hamiltonian flow generated by the principal symbol} equals 0."* The condition is more complicated for matrix operators. For a scalar second-order operator on a compact manifold with boundary one need only consider trajectories transversal to the boundary which reflect according the laws of geometrical optics. There are some points of the cotangent bundle through which infinitely long trajectories do not pass but the measure of these dead-end points vanishes and we need not take them into account. However, for higher-order operators as well as for matrix operators the trajectories reflected from the boundary can branch and in this case it is necessary to follow every branch. This circumstance guides us to a situation which is essentially more complicated: the measure of the set of all these dead-end points does not necessarily vanish, so the following additional condition appears: *"the measure of {the set of all dead-end points} equals* 0;" this condition is not automatically fulfilled now.

Let us clarify, for scalar first-order operators on a manifold without boundary, the link between the asymptotics (0.7) and periodic Hamiltonian trajectories. It is well known that singularities of the solutions of hyperbolic equations propagate along Hamiltonian trajectories. This fact guides us to the conclusion that the singular support $\sigma(t)$ is contained in the set of all the periods of the Hamiltonian trajectories including $t = 0$ (this is the so-called *Poisson relation*); in particular, $t = 0$ is an isolated point of the singular support (this fact remains true in the more general situation). Hence if there is no periodic trajectory with period not exceeding T then we know that in the interval $[-T, T]$ the distribution $\sigma(t)$ is singular only at 0 and hence we know $\sigma(t)$ in this interval modulo a smooth function. The Tauberian theorem permits us to obtain that the remainder in asymptotics (0.7) does not exceed $C\lambda^{d-1}T^{-1} + O(\lambda^{d-2})$ with a constant C which does not depend on T; however "O" here is not necessarily uniform with respect to T. Hence if $t = 0$ were the unique period (surely I know that it is impossible!) then we could choose T and obtain the remainder estimate $o(\lambda^{d-1})$. In the general (realistic) case one should consider a partition of unity by means of two pseudo-differential operators for every chosen T such that the support of the first operator contains no periodic point with period not exceeding T and the measure of the support of the symbol of the second operator is less than ϵ with arbitrarily chosen $\epsilon > 0$. Applying the Tauberian theorem to each term

$$N_j(\lambda) = \int (Q_j e)(x, x, \lambda)\, dx$$

in $N(\lambda)$ we obtain the remainder estimate $C\lambda^{d-1}T^{-1} + O(\lambda^{d-2})$ for $j = 1$ and $C\varepsilon\lambda^{d-1} + O(\lambda^{d-2})$ for $j = 2$, and these estimates again imply (0.7). Here and in what follows $e(x, y, \lambda)$ is the Schwartz kernel of the spectral projector. Moreover, under certain more restrictive conditions on the Hamiltonian flow one can improve the remainder estimate in (0.7) up to $O(\lambda^{d-1}\log^{-1}\lambda)$ or even $O(\lambda^{d-1-\delta})$ with a small exponent $\delta > 0$.

It appeared recently that the presence of periodic trajectories and moreover the presence of eigenvalue clusters allow two-term asymptotics of the form

$$N(\lambda) = \kappa_0\lambda^d + f(\lambda)\lambda^{d-1} + o(\lambda^{d-1}) \tag{0.8}$$

with an explicitly calculable function $f(\lambda)$ which is bounded and oscillating as $\lambda \to +\infty$ with the characteristic "period" of oscillations $\asymp 1$. In particular, this fact permits us to obtain the asymptotic distribution of eigenvalues inside clusters. In addition, under some assumptions including the assumption that all the trajectories are periodic, one can obtain the asymptotics (0.8) with the remainder estimate $O(\lambda^{d-2})$!

It is well known that in a large number of cases the Weylian formula fails in its standard form. In these cases it is necessary either to remove from the domain of integration some part of the phase space or to separate variables x and ξ to two parts: $x = (x', x'')$, $\xi = (\xi', \xi'')$, and only consider the variables (x', ξ') as Weylian, which means that one should consider the operator in question as a (partial) differential operator with respect to x' with operator-valued coefficients and apply the Weylian procedure to this operator. In the more general case one should divide the phase space into a few parts one of which should be removed from consideration altogether while in the other parts the "Weylization" (preceded by a certain transform) should be made only with respect to certain variables.

The previous remarks are not a survey, even an incomplete one. Their purpose is only to motivate this book in general and the following part of this introduction in particular. Nor are the comments in each chapter complete. As the best surveys I recommend the books of M. S. Birman and M. Z. Solomyak [1] and G. Rozenblioum, M. Z. Solomyak and M. Shubin [1].

It also should be noted that though this book demands from the reader only a good and detailed knowledge of the theory of pseudo-differential operators and a familiarity with the theory of Fourier integral operators (see, e.g., the books of L. Hörmander [1], F. Treves [1], M. Taylor [1] and M. Shubin [1]); an elementary familiarity with the theory of spectral asymptotics (see the same books) is also very desirable.

It should be noted that this book has few intersections with the books of S. Levendorskii [1] and Yu. Safarov and D. Vassiliev [1] which are also devoted to spectral asymptotics. I recommend both of those books, especially the second one.

Finally, I would like to mention that there are some problems in this book. Two stars denote problems the solutions of which are not known to me. One star denotes problems the solutions of which are known to me but which cannot be published here because of lack of space and which also cannot be published elsewhere by me because of lack of time due to my efforts on *this* book. I welcome readers to solve these problems either by the method suggested or by some other method and to publish their solutions. The unstarred problems I consider rather trivial exercises.

The structure of this book and the numbering of statements and formulas is clear from the table of contents; I mention only that the lowest-order subsections are not reflected in the numbering of statements, formulas, etc.

0.3 What is in the Book? Chapters 1–3

Let us now discuss the contents and principal ideas of this book.

There are no spectral asymptotics yet in Part I, "Semiclassical Microlocal Analysis"; we develop the apparatus of semiclassical microlocal analysis here. **Chapter 1**, "Introduction to semiclassical microlocal analysis," contains the necessary survey of the theory of the h-pseudo-differential operators (in *section 1.1*), the functional calculus of these operators (in *section 1.4*), the theory of the h-Fourier integral operators (in *section 1.2*) and the theory of singularities (oscillation front sets and related topics, in *section 1.3*). There are no new non-trivial results in this chapter but there is a rather obvious but very important new

Definition 0.1. A family of functions u depending on a small parameter $h \in [0, h_0]$ (and maybe depending on some other parameters of an arbitrary nature), tempered (i.e., with $\|u\| \leq Ch^{-n}$ with some constant C and exponent n where here and below $\|.\|$ means the L^2-norm) is *negligible in the box* (parallelepiped) $\Pi = \{(x,\xi) : |x_j - \bar{x}_j| \leq \gamma_j, |\xi_j - \bar{\xi}_j| \leq \rho_j \ \forall j = 1, \dots, d\}$ if there exists a function $\phi \in C_0^\infty(\mathbb{R}^{2d})$ equal to 1 in the unit cube $\{(x,\xi) : |x_j| \leq 1, |\xi_j| \leq 1 \quad \forall j = 1, \dots d\}$ such that $\|\phi_\Pi^w(x, hD)u\| \leq Ch^s$ where $\phi_\Pi(x,\xi) = \phi(\gamma^{-1}(x - \bar{x}), \rho^{-1}(\xi - \bar{\xi}))$; note that the exponent s enters into the definition of negligibility. ◊

In the standard definition leading to the notion of an *oscillation front set* one takes a *fixed* box Π (i.e., with constant $(\bar{x}, \bar{\xi})$ and $(\gamma_1, \dots, \rho_d)$); this definition is too rough for our needs and we admit variable (i.e., depending on parameters) $(\bar{x}, \bar{\xi})$ and $\gamma_1, \dots, \rho_d$ satisfying the *uncertainty principle*

$$\gamma_j \rho_j \geq h^{1-\delta} \qquad \forall j = 1 \dots d \tag{0.9}$$

with some arbitrarily small fixed exponent $\delta > 0$. An important special case is $\gamma_1 = \dots = \rho_d \geq h^{\frac{1}{2}(1-\delta)}$.

Chapter 2 is devoted to the propagation of singularities (or more precisely of the negligibility) of solutions in the interior of the domain. The key definition of the *microhyperbolicity of an operator P at* the given *point z* of the cotangent bundle *in the direction $T \in T_z T^*X$* appears in *section 2.1*, and the main theorem concerning propagation of singularities is proven there; the theorem is formulated in terms of real-valued functions ϕ_k such that the operator P under consideration is microhyperbolic at every point $z \in \Omega$ in the directions of the Hamiltonian fields generated by ϕ_k; here and in what follows $\Omega \subset T^*X$ is a phase domain under consideration. This theorem is proven by means of energy estimates. We then apply this theorem to the operator $P = hD_t - A^w(x, hD_x, h)$ and to the analysis of the fundamental solution $u(x,y,t)$ of the Cauchy problem for this operator, establishing the finite speed of propagation of singularities (in the domain $\Omega \subset T^*X$). Under the microhyperbolicity condition for the operator A at the energy level λ we prove the statement that the singularities of $u(x,y,t)$ leave the diagonal $\{x = y, \xi = \eta\}$; this statement yields the negligibility of

$$\sigma_Q(t) = \int (Q_x)(x,x,t)dx$$

in boxes of the form $\{(t,\tau) : h^{1-\delta} \leq \pm t \leq T_0, |\tau - \lambda| \leq \epsilon_0\}$ with small enough positive constants T_0 and ϵ_0 and with arbitrarily small exponent $\delta > 0$; here and in what follows Q is a cutting h-pseudo-differential operator supported in Ω. The results of section 2.1 are reformulated in *section 2.2* in terms of the generalized bicharacteristics of P which, in the case of constant multiplicities of the eigenvalues $\alpha_j(x,\xi)$ of the symbol $a(x,\xi) = A(x,\xi,0)$, are the standard bicharacteristics of the symbols $\tau - \alpha_j(x,\xi)$. In *section 2.3* the classical theorem on the propagation of singularities along bicharacteristics for operators with constant multiplicities of the characteristics is proven in a more general and refined form. Namely for the operator $P = hD_t - A^w(x, hD_x, h)$ with self-adjoint operator $A^w(x, hD_x, h)$ and with constant multiplicities of the eigenvalues $\alpha_j(x,\xi)$ of the symbol $a(x,\xi)$ we replace the usually used negligibility in a small fixed neighborhood by negligibility in the γ-neighborhood of a Hamiltonian trajectory with $\gamma \geq h^{\frac{1}{2}(1-\delta)}$; it is assumed that this trajectory is not too long (i.e., $T \leq ch^{-\delta'}$ with a small enough exponent $\delta' > 0$); the symbol $A(x, hD_x, h)$ and its derivatives up to large enough order K and the elements of the Jacobi matrix of the corresponding Hamiltonian flow are also assumed to be not too large (i.e., their absolute values or norms are also less than $ch^{-\delta'}$). The proof is based on the Heisenberg equation. In what follows we sometimes return to the investigation of the propagation of singularities.

In **Chapter 3** we realize approximately the same program for operators near the boundary on which boundary conditions are given. In *section 3.1* we give the key definition of the *microhyperbolicity of the problem* (P, B) *at the point* $z \in T^*\partial X$ *in the multidirection* $\mathcal{T} = (T', \zeta_1, \dots, \zeta_n) \in T_z T^*\partial X \times \mathbb{R}^n$ where n is the number of the different characteristic points z_k of the operator P lying over z. Then we prove the main theorem concerning propagation of singularities; this theorem is formulated in terms of real-valued functions $\phi_k(x, \xi')$ coinciding at $\{x_1 = 0\}$ and such that the problem (P, B) is microhyperbolic in the domain $\Omega \subset T^*\partial X$ under consideration in the multidirection $(H_\phi, \partial_{x_1}\phi_1, \dots, \partial_{x_1}\phi_n)$, where we assume that $X = \{x_1 = 0\}$ locally and $x' = (x_2, \dots, x_d)$, $\phi = \phi(0, x, \xi')$. Then in *section 3.2* we deduce the same corollaries as in section 2.1 (concerning the finite speed of the propagation of singularities for the operator $P = hD_t - A^w(x, hD_x, h)$ and the negligibility of $\sigma_Q(t)$ in the same boxes as before provided (A, B) is microhyperbolic at the level λ) from this theorem. We discuss some geometrical aspects of the propagation of singularities; however the general results are not very precise. Finally, *section 3.3* is devoted to the propagation of singularities along "long trajectories"; here we consider only h-differential second-order operators and assume that at the point where a trajectory meets the boundary and reflects the trajectory is transversal enough to the boundary (with angle greater than $h^{\delta'}$); moreover, we assume that a certain non-degeneracy condition on the boundary operator is also fulfilled at such reflection points. It should be noted that the number of reflection points is not too large (less than $h^{-\delta''}$ with exponent $\delta'' > 0$ depending on δ' and tending to $+0$ as δ' tends to $+0$).

0.4 What is in the Book? Chapters 4–5

The goal of Part II, "Local and Microlocal Semiclassical Spectral Asymptotics," is to obtain asymptotics as $h \to +0$ for

$$\mathcal{E}_Q(\lambda_1, \lambda_2) = \int (Q_x e)(x, x, \lambda_1, \lambda_2)\, dx$$

where

$$e(x, y, \lambda_1, \lambda_2) = e(x, y, \lambda)|_{\lambda=\lambda_1}^{\lambda=\lambda_2}$$

and either $Q = Q^w(x, hD_x)$ is an h-pseudo-differential operator with compactly supported symbol or $Q = \psi(x)$ is a compactly supported function; we speak about *microlocal spectral asymptotics* and *local spectral asymptotics* respectively. Here and below the subscript "x" means that the corresponding operator acts with respect to x, etc.

In **Chapter 4** we consider the case when $\operatorname{supp} Q$ is disjoint from the boundary of the domain X. *Section 4.1* contains a preliminary analysis and the results obtained there are rather trivial. First of all microlocal estimates $\|\mathcal{E}_Q(\lambda_1, \lambda_2)\| \le Ch^{-d}$ are obtained in the general case and estimates $\|\mathcal{E}_Q(\lambda_1, \lambda_2)\| \le Ch^s$ are obtained in the case when the interval (λ_1, λ_2) is *classically forbidden* in a neighborhood of $\operatorname{supp} Q$ (i.e., the interval is disjoint from the spectrum of the symbol $A(x, \xi, h)$ for the indicated (x, ξ)); one can weaken this condition provided the multiplicities of the eigenvalues of the symbol $a(x, \xi)$ are constant. We obtain similar local estimates for operators elliptic in the standard sense.

Then in *section 4.2* we apply the *method of successive approximations* to $u(x, y, t)$; we take an unperturbed operator $\bar{A} = A(y, hD_x, 0)$ with symbol depending only on ξ. This method turns out to be applicable in the rather short time interval $\{|t| \le h^{\frac{1}{2}+\delta}\}$ with arbitrary small exponent $\delta > 0$. This immediately yields in *section 4.3* complete semiclassical asymptotics for the spectral means

$$\int \phi(\frac{1}{L}(\lambda - \bar{\lambda}))d_\lambda \mathcal{E}_Q(\lambda_1, \lambda) \tag{0.10}$$

with any $L \ge h^{\frac{1}{2}-\delta}$ and with $\phi \in C_0^K(\mathbb{R})$; moreover, one can take $L \ge h^{1-\delta}$ provided the eigenvalues of $a(x, \xi)$ have constant multiplicities. This yields complete asymptotics for $\mathcal{E}_Q(\lambda_1, \lambda_2)$ in the case when both levels λ_1 and λ_2 are classically forbidden but the whole interval (λ_1, λ_2) is not; in all these cases the leading term is $\kappa_0 h^{-d}$. Then we consider the main case when $\bar{\lambda}$ is not a classically forbidden level and A is microhyperbolic at this level. Combining the results of section 4.2 with the assertion from section 2.1 concerning the negligibility of $\sigma_Q(t)$ we obtain complete asymptotics of $\sigma_Q(t)$ in the zone $\{(t, \tau) : |t| \le T_0, |\tau - \bar{\lambda}| \le \epsilon_0\}$ with small positive constants T_0 and ϵ_0. For $|\tau - \bar{\lambda}| \le \epsilon_0$ we get asymptotics in the interval $[-h^{\frac{1}{2}+\delta}, h^{\frac{1}{2}+\delta}]$. It was proven in section 2.1 that $\sigma(t)$ is negligible in $[-T_0, -h^{1-\delta}] \cup [h^{1-\delta}, T_0]$, and these intervals overlap! We immediately get asymptotics of (0.10) for $L \ge h^{1-\delta}$. Moreover, the Tauberian method yields the estimate

$$|\mathcal{E}_Q(\lambda_1, \lambda_2) - \kappa_{0Q}(\lambda_1, \lambda_2)h^{-d}| \le Ch^{1-d} \tag{0.11}$$

for $\lambda_1, \lambda_2 \in (\bar{\lambda} - \frac{1}{2}\epsilon_0, \bar{\lambda} + \frac{1}{2}\epsilon_0)$ and miscellaneous related asymptotics in which the possible smallness of $\operatorname{supp} Q$ is taken into account. The same method also yields asymptotics of the Riesz means of order $\vartheta > 0$ with remainder estimate $Ch^{1-d+\vartheta}$. However, all these asymptotics are due to the simple Tauberian estimate

$$|\mathcal{E}_Q(\lambda_1, \lambda_2)| \le Ch^{1-d} \qquad \forall \lambda_1, \lambda_2 \in [\bar{\tau} - \epsilon_0, \bar{\tau} + \epsilon_0] : |\lambda_1 - \lambda_2| \le h \tag{0.12}$$

and asymptotics of (0.10) with $L \ge h$ and remainder estimate $Ch^{1-d}\left(\frac{h}{L}\right)^s$ with an arbitrarily large exponent s. Actually estimates of this type seem to be the most important for internal needs.

Applying the results of section 2.3 under a condition forbidding periodic trajectories of length not exceeding T we obtain complete asymptotics of $\sigma_Q(t)$ in the zone indicated above with T_0 replaced by T; then the same arguments yield the estimate

$$|\mathcal{E}_Q(\lambda_1, \lambda_2) - \kappa_{0Q}(\lambda_1, \lambda_2)h^{-d} - \kappa_{1Q}(\lambda_1, \lambda_2)h^{1-d}| \le C\varepsilon h^{1-d} + Ch^{1-d+\delta'} \tag{0.13}$$

with $\epsilon = T^{-1}$ and small enough exponent $\delta' > 0$; under some more general and realistic conditions we obtain an estimate with a more complicated expression for ε. Those adjustments are also achieved for Riesz means (the first term in the remainder estimate is $Ch^{1-d+\vartheta}T^{-1-\vartheta}$) and for expression (0.10) with $L \ge hT^{-1}$ (the first term of the remainder estimate is $Ch^{1-d}T^{-1}\left(\frac{h}{LT}\right)^s$).

In *section 4.4* we analyze operators with constant multiplicities of the eigenvalues of the principal symbol. By means of *an admissible partition of unity and dilatations with respect to x with a change of h on every element of this partition* we replace the microhyperbolicity condition by another much less restrictive condition (which is void for Riesz means). The method applied here is very important and will be used very often in this book. We call it the *rescaling method* or *partition-dilatation method* in what follows. To get the best possible results one should combine this method with a refined analysis of the propagation of singularities along trajectories which are long *after* rescaling but are usually of bounded length or even short *before*: time and energy are rescaled automatically.

Considering $a(x,\xi)$ as the symbol of an m-th order differential operator and assuming that it is elliptic in the standard sense we turn from microlocal to local asymptotics.

Section 4.5 may be summarized simply: for the Schrödinger and Dirac operators in dimension $d \geq 2$ one can simply drop the microhyperbolicity condition. Here we apply the same rescaling method as before with some simplifications and modifications.

In *section 4.6* we extend the above results to families of commuting operators.

Finally, *section 4.7* has a more special character. Here we consider scalar operators with periodic Hamiltonian flow of the principal symbol, or some weak perturbation of an operator of this type; there automatically appears an additional parameter $\eta \in [h^n, h^\delta]$ with arbitrarily large n and arbitrarily small $\delta > 0$ (this parameter equals h in the "normal" situation); for $\eta \leq \epsilon h$ with a small constant $\epsilon > 0$ there arise spectral zones of length $\leq C\eta$ separated by lacunary zones of length $\asymp h$ in the semiclassical approximation to the spectrum; it is proven that $\mathcal{E}_Q(\lambda_1,\lambda_2) = O(h^s)$ with arbitrarily large s provided the interval (λ_1,λ_2) lies in a lacunary zone; in the case when λ_1 and λ_2 belong to different lacunary zones we obtain complete non-Weylian asymptotics for $\mathcal{E}_Q(\lambda_1,\lambda_2)$. Turning to the long-time propagation of singularities (with $|t| \leq \epsilon\eta^{-1}$) we prove that singularities propagate along Hamiltonian trajectories of the symbol $a(x,\xi)$, drifting in the direction of the Hamiltonian field generated by an auxiliary symbol $b(x,\xi)$ with drift speed $\asymp \eta$. Now turning to $\mathcal{E}_Q(\lambda_1,\lambda_2)$ we finally derive, under certain conditions, the estimate

$$(0.14) \quad \bigl|\mathcal{E}_Q(\lambda_1,\lambda_2) - \kappa_{0Q}(\lambda_1,\lambda_2)h^{-d} - \kappa_{1Q}(\lambda_1,\lambda_2) - f_Q(\lambda_1,\lambda_2)h^{1-d}\bigr| \\ \leq C(\eta+h)h^{1-d}$$

where the function f is explicitly calculated and quickly oscillating. It arises from singularities of $\sigma_Q(t)$ different from $t=0$. Nevertheless the last term in the asymptotics is of the order of the second one and its variation on the interval of length $\asymp h$ is of the same order $\asymp h^{1-d}$ as the variation of the first term.[3)] For $\eta \geq h^{1-\delta}$ we obtain asymptotics (0.13) with $\varepsilon = \eta$ under certain conditions.

I analyze Chapter 4 so thoroughly because its results and methods are highly important for the parts of the book which follow. The scheme of the arguments in the main stream of the chapter is the following: *construction (by means of the method of successive approximations) of $u(x,y,t)$ and hence $\sigma_Q(t)$ in the short time-interval $\{|t| \leq T_1\}$—step-by-step extension of this interval by different methods—application of the Tauberian method;* this scheme will be used repeatedly. An important general idea is that in the theory of spectral asymptotics the step-by-step approach with improvement is much better than more bold attempts to do everything in one shot.

It should be mentioned that we do not use the results of Chapter 3 in Chapter 4 and that the results of sections 4.1–4.3 are obtained for standard operators as well as for operators with operator-valued symbols; however, the results for operators with operator-valued symbols are applied only in sections 12.1–12.4.

Chapter 5 is essentially parallel to Chapter 4 without sections 4.4, 4.6–4.7, and the content of each of *sections 5.1–5.4* is similar to the content of the respective sections 4.1–4.5. We underline only the following "defects":

(i) We now consider only standard operators, not operators with operator-valued symbols;

(ii) We do not weaken the microhyperbolicity condition for operators with constant multiplicities of the eigenvalues of the principal symbol; we do it only for the Schrödinger operator in section 5.4.

[3)] As it should be because the decrease of the last term along the spectral gap should compensate for the increase of the first term; along the cluster both terms grow and if $\eta \ll h$ then the last term grows more quickly.

(iii) In section 5.4 we analyze only the Schrödinger operator, not the Dirac operator.

It also should be noted that if the interval $[\lambda_1, \lambda_2]$ is classically forbidden for the symbol $a(x,\xi)$ but is not forbidden for the problem $\big(a(x', D_t, \xi'), b(x', D_t, \xi')\big)$ on the semiaxis $\mathbb{R}^+ \ni t$ (as above we assume that $X = \{x_1 > 0\}$ locally) then in the above estimates and asymptotics one should replace d by $d-1$.

0.5 What is in the Book? Chapters 6–7

In **Chapters 6, 7** we consider respectively the Schrödinger and Dirac operators with strong magnetic field; now apart from the semiclassical parameter h we have also a large parameter μ (a coupling constant) responsible for the interaction of the particle with the magnetic field. We consider only dimensions 2 and 3 and assume that the the unit ball under consideration is entirely contained in the domain X and that the intensity of the magnetic field nowhere vanishes in this ball. The analysis in these two chapters is essentially parallel. In *sections 6.1, 7.1* a preliminary analysis is done and in particular $(\mathrm{tr})e(x,x,\lambda)$ is calculated explicitly in the case when the metric coefficients and the scalar potential are constant and the vector potential is linear (so the intensity of the magnetic field is also constant) and $X = \mathbb{R}^d$; these formulas are good approximations for the general case which we consider. For the Schrödinger (but not for the Dirac!) operator we prove the estimates $|\mathcal{E}_\psi(0)| \le ch^{-d}$ in the general case and $|\mathcal{E}_\psi(0)| \le \mu^{-s}$ for $\mu h \ge C_0$ (we take $\lambda_1 = -\infty$ for the Schrödinger operator); hence for the Schrödinger operator (but not for the Dirac operator!) we assume in what follows that $\mu h \le c$. In *sections 6.2, 7.2* we reduce our operators microlocally to canonical forms which are essentially different in the cases $d = 2, 3$; we obtain, modulo $O(\mu^{-s})$, $\mu^{-1}h$-pseudo-differential operators with respect to $x' = x_1$ for $d = 2$ and $(\mu^{-1}h, h)$-pseudo-differential operators with respect to $x' = (x_1, x_2)$ for $d = 3$, and x_d enters in these forms only through the harmonic oscillator. Based on this reduction, in *sections 6.3, 7.3* we analyze the propagation of singularities with respect to x' for $\mu \le h^{1-\delta}$ and under some restrictions we prove that on the intervals $[-T, -T_0]$ and $[T_0, T]$ the distribution $\sigma_\psi(t)$ is negligible, where either $T_0 \asymp \mu^{-1}$ or $T_0 \asymp h^{1-\delta}$, while on the interval $[-2T_0, 2T_0]$ for $\sigma_\psi(t)$ the standard formulas of section 4.2 hold; $T \asymp 1$ for $d = 3$ and $T \asymp \mu$ or larger in certain cases for $d = 2$. We always assume that there is an appropriate cutting near the energy level 0 with respect to the dual variable τ. Applying the Tauberian method we obtain asymptotics for $\mathcal{E}_\psi(0)$ for the Schrödinger operator (for the Dirac operator the situation is more complicated) with principal part $\asymp h^{-d}$ and remainder term $O(h^{-2})$ for $d = 3$ and $O(\mu^{-1}h^{-1})$ or even better for $d = 2$. We also analyze less restrictive conditions in these and the following sections and obtain asymptotics in these cases with the same or worse remainder estimates. So, at this time *the canonical form is applied only for the study of the long-time propagation of singularities in order to extend the time interval and improve the remainder estimate in the Weyl-type asymptotics.*

In *sections 6.4, 7.4* the case $\mu \ge h^\delta$ is considered. Then the remainder term $O(\mu^{-s})$ in the microlocal canonical form is negligible and after decomposition with respect to Hermitian functions of $\sqrt{\mu h^{-1}}x_d$ we obtain families of $\mu^{-1}h$-pseudo-differential operators with respect to $x' = x_1$ for $d = 2$ and $(\mu^{-1}h, h)$-pseudo-differential operators with respect to $x' = (x_1, x_2)$ for $d = 3$. These operators depend on the number n of the Hermitian function; in fact this is completely true only for the Schrödinger operator (for the Dirac operator the situation is more complicated and certain additional arguments are necessary). Applying the results of Chapter 4 to these families of operators, we obtain asymptotics for $\mathcal{E}_\psi(0)$ for the Schrödinger operator under certain conditions (for the Dirac operator the situation is more complicated again) with the same remainder estimates as in sections 6.3, 7.3. So, here *the canonical form is used fully*. These asymptotics are not completely effective (i.e., one can calculate some terms only "in principle" but not "in fact").

The conditions of this section and those of section 6.3 (or section 7.3 for the Dirac operator) overlap and under all of them both asymptotics are valid. These asymptotics are of different types but comparing them we can remove some extra terms and obtain asymptotics which are completely effective even if the conditions of one of these sections (6.3 or 6.4 for the Schrödinger operator, 7.3 or 7.4 for the Dirac operator) fail to hold. These two ideas—*application of the canonical form for the analysis of propagation of singularities and the extension of the interval for* $\sigma_\psi(t)$ and *comparison of the asymptotics obtained by different methods*—seem to be very useful and we use them heavily in the chapters of the book which follow.

In *section 7.5* we consider the Dirac operator in the case $\mu h \geq c$; moreover for $d = 2$ only $\mu^{-1}h$ is assumed to be a small parameter (so h is not necessarily small here). In this case it is necessary to consider only one number $n = 0$ and we obtain asymptotics with principal part $\asymp \mu h^{1-d}$ and remainder estimate $O(\mu h^{-1})$ for $d = 3$ and $O(1)$ for $d = 2$.

For $d = 3$ in *sections 6.5, 7.6* we improve the remainder estimates slightly under certain conditions on the magnetic lines along which the singularities propagate in this case. Moreover, we discuss the case $d = 2$ in which one should replace magnetic lines by drift lines in the propagation theorems.

In *sections 6.6, 7.7* we consider the case $d = 2$. It is possible that lacunas appear in the semiclassical approximation to the spectrum in this case. For the lacunary values of the spectral parameters we obtain complete semiclassical spectral asymptotics.

Finally, in *sections 6.7, 7.8* we apply the method of partition of unity and dilatations (mentioned in the discussion of section 4.3) and weaken the conditions of the theorems proven in the previous sections.

It should be noted that Chapters 6 and 7 do not use Chapter 5 or the "operator-valued" results of Chapter 4.

0.6 What is in the Book? Chapters 8–9

The character of Part III, "Estimates of the Spectrum," is essentially different from the character of the previous parts. There are only a few micro-analytical considerations in this part but there are many functional-analytical arguments. The substance of these arguments is the following: in Part II we considered $\mathcal{E}_\psi(0)$ for a function ψ supported in $B(0, \frac{1}{2})$ (where $B(y, r)$ is the ball with radius r and center y in $\mathbb{R}^d$) and all conditions (apart from the self-adjointness condition and conditions associated with Hamiltonian trajectories) were assumed to be fulfilled only in $B(0, 1)$. We had asymptotics with remainder estimates with constants depending only on the constants used in these conditions. Hence those estimates yield a similar estimate in the case in which ψ is supported in $B(y, \frac{1}{2}\gamma(y))$ and it is assumed that after the dilatation (an affine transformation in $\mathbb{R}^d$) transforming $B(y, \gamma(y))$ into $B(0, 1)$ and multiplication by $\varrho^{-1}(y)$ for the operator in question the conditions of Part II will be fulfilled with the parameter h replaced by a new parameter $h_{\text{eff}} = h\rho(y)^{-1}\gamma(y)^{-1}$ (moreover, in Chapters 6 and 7 a parameter μ_{eff} also appears). One should assume here that $h_{\text{eff}} \leq 1$ (i.e., that $\rho(y)\gamma(y) \geq h$) and one can rewrite the conditions mentioned above for the operator before dilatation. Let us assume that $\gamma(y) > 0$ and $|\nabla\gamma(y)| \leq 1$; then one can take a γ-admissible partition of unity and under the condition $\rho(y)\gamma(y) \geq h$ one can sum the derived estimates and obtain an estimate of the form[4)]

$$\left|N^-(0) - \mathcal{N}^-(0)\right| \le C\mathcal{R}_1$$

with the same (Weylian or non-Weylian) principal part $\mathcal{N}^-(0)$ as before and with some remainder estimate $\mathcal{R}_1$. However there are many important cases in which this condition $\rho(y)\gamma(y) \ge h$ is fulfilled only in a subdomain X' of the domain X; then summing with respect to a partition of unity in X' we obtain estimates only for $\mathcal{E}_\psi(0)$ (but not for $N^-(0)$) with an admissible function ψ such that $\operatorname{supp}\psi \subset X'$ and $\psi = 1$ outside of $X^{IV} = \{\rho(y)\gamma(y) \le \frac{3}{2}h\}$. Thus, in order to obtain an estimate for $N^-(0) = \mathcal{E}_1(0)$, we need to derive an estimate for $\mathcal{E}_{1-\psi}(0)$.

On the other hand, if we had $X = X'' = \{\rho(y)\gamma(y) \le 2h\}$ then we very often would have some estimates for $N^-(0)$ of a completely different (variational) origin which have the form $N^-(0) \le C\mathcal{R}_2$. In this book we almost exclusively apply the Rozenblioum variational estimate, but one can use other estimates of this type. It appears that in the general case one can prove these estimates together and derive the estimate

$$\left|N^-(0) - \mathcal{N}^-(0)\right| \le C\bigl(\mathcal{R}_1 + \mathcal{R}_2 + \mathcal{R}_{12}\bigr)$$

where $\mathcal{N}^-(0)$ and $\mathcal{R}_1$ (associated with $\mathcal{E}_\psi(0)$) are exactly the same as before in the zone X', the estimate $\mathcal{R}_2$ is derived for a self-adjoint operator A' given in the domain X'' where A' coincides with A in the zone $\{\rho(y)\gamma(y) \le \frac{7}{4}h\}$ (and the boundary conditions also coincide at the common part of the boundary), and $\mathcal{R}_{12}$ is an additional term depending on an auxiliary construction. The proper construction using the *Birman-Schwinger principle* with an appropriate auxiliary function makes this term less than $C\bigl(\mathcal{R}_1 + \mathcal{R}_2\bigr)$. This construction is done in **Chapter 8**. Namely, in *section 8.1* the principal theorems are formulated, in *section 8.2* we prove them and in *section 8.3* we adjust them for Riesz means. Finally, in *section 8.4* we reformulate them for quadratic forms. Here we have two important ideas: *dilatation with multiplication and a change of parameters (rescaling)* (we mentioned this idea in the discussion of section 4.4) and *combining the asymptotics obtained in Part II with estimates of variational origin which makes it possible to consider a domain containing not only a semiclassical zone X' but also a singular zone X''*. This task is done in Chapter 8 for semibounded operators.

In **Chapter 9** we consider the non-semibounded case and our analysis here is essentially parallel to the analysis of Chapter 8; a very important role is played by the following form of the Birman-Schwinger principle:

$$\mathbf{N}(A;[\lambda_1,\lambda_2)) = \mathbf{N}(A - \bar{t}J;[\lambda_1,\lambda_2)) - \mathbf{N}(-J^{-1}(A-\lambda I);[0,\bar{t}))\Big|_{\lambda=\lambda_1}^{\lambda=\lambda_2}$$

which applies (under some hypotheses) in the case of a positively defined matrix-valued function J and for $\bar{t} > 0$ (a similar form applies to the case $\bar{t} < 0$) where $\mathbf{N}(A;[\lambda_1,\lambda_2))$ is the number of eigenvalues of the operator A in the interval $[\lambda_1,\lambda_2)$ (provided $\operatorname{Spec}_{\mathrm{ess}}(A) \cap [\lambda_1,\lambda_2) = \emptyset$; otherwise $\mathbf{N}(A;[\lambda_1,\lambda_2)) = \infty$) and $\mathbf{N}(-J^{-1}(A-\lambda I);[0,\bar{t}))$ is the number of eigenvalues of the spectral problem

$$\bigl((A-\lambda I) + tJ\bigr)u = 0$$

in the interval $[0,\bar{t})$. One can iterate this formula, at each step replacing the operator A by $A - t_1J_1 - \dots - t_{k-1}J_{k-1}$, the number $\bar{t}$ by t_k and the function J by J_k with $k = 1,\dots,n$; here $t_1,\dots,t_n$ *are not necessarily of the same sign*. As a result we obtain the following more complicated formula

---

[4)] At this time we discuss only the case of Chapter 8 when the operator is semibounded from below.

$$\begin{aligned}\mathbf{N}(A;[\lambda_1,\lambda_2]) =&\mathbf{N}(A-t_1J_1-\dots-t_{k-1}J_{k-1};[\lambda_1,\lambda_2])-\\&\mathbf{N}(-J_1^{-1}(A-\lambda I);[0,t_1))-\dots\\&\mathbf{N}(-J_n^{-1}(A-\lambda I-t_1J_1-\dots-t_{n-1}J_{n-1});[0,t_n))\Big|_{\lambda=\lambda_1}^{\lambda=\lambda_2}\end{aligned}$$

to the right-hand expression of which we apply arguments similar to those discussed above (in the discussion of Chapter 8). Actually, we multiply the last equality by $\varphi_1(t_1)\dots\varphi_n(t_n)$ with appropriate functions φ_k and integrate on $(t_1,\dots,t_n)\in\mathbb{R}^n$ (this rather technical trick is useful).

It should be noted that the reader who is interested only in semibounded operators has no need to read Chapter 9. It should also be noted that Chapters 8 and 9 would be absolutely trivial if we assumed that the singular zone is empty; however in this case we would obtain a much more restricted series of results in Chapters 10 and 11 and we would not be ready for Chapter 12 psychologically.

0.7 What is in the Book? Chapters 10–11

In Part IV, "Asymptotics of Spectra," we return to spectral asymptotics now based on Parts II and III. The essence of Chapters 10 and 11 is very simple: *we substitute an operator A depending on certain parameters into the assertions of Chapters 8, 9 (and even Chapters 4–7 after partition of unity, dilatation, multiplication and change of parameters), check that the conditions of these assertions are uniformly fulfilled, and as a result we obtain asymptotics with respect to these parameters.* The arguments of these two chapters are in terms of "arithmetics," dynamical systems theory (when we check conditions associated with the Hamiltonian trajectories) and rather simple functional analysis (when we check conditions associated with the singular zone). So *having started with (micro)local semiclassical asymptotics we come to asymptotics with respect to quite different parameters (e.g., the spectral parameter).* Actually, semiclassical asymptotics are also treated but our results are not the same as before and one should distinguish the actual semiclassical parameter h and the effective semiclassical parameter h_{eff}. The latter depends on the point and arises always, no matter what parameters were in the original problem.

So **Chapter 10** contains, in *section 10.1*, another treatment of semiclassical asymptotics, but now we are interested in $\mathbf{N}_h(\lambda_1,\lambda_2)$ and in situations when there are singularities, exits to infinity (or even the domain $\mathbb{R}^d$) and other non-regular phenomena. In *sections 10.2–10.4* we investigate asymptotics with respect to a large spectral parameter. In section 10.2 we consider cases in which a domain may have singularities on the boundary (edges, vertices, etc.), cusps (i.e., exits to infinity) which are not too massive and also cases in which coefficients may have singularities which are not too strong (a good example is the Schrödinger operator with potential similar to $\pm|x|^{2m}$ near the origin with $m\in(-1,0)$). In section 10.3 stronger singularities of coefficients are allowed but a *coercivity* condition is assumed (a good example is the Schrödinger operator with potential similar to $+|x|^{2m}$ near the origin with $m\le-1$). In section 10.4 we consider the situation in which a domain is "essentially non-bounded" or even coincides with $\mathbb{R}^d$; e.g., we consider the Schrödinger operator in $\mathbb{R}^d$ with potential similar to $+|x|^{2m}$ near infinity with $m>0$. However we are able to consider certain situations only when the potential does not tend to $+\infty$ along some directions (along these directions, $V(x)=0$ or even $V(x)$ tends to $-\infty$, for example) but the arising "canyons" are assumed to be narrow enough to provide the standard asymptotics. In *section 10.5* we analyze eigenvalues tending to -0; e.g., we consider the Schrödinger operator in $\mathbb{R}^d$ with potential similar to $-|x|^{2m}$ near infinity with $m\in(-1,0)$; for this operator the essential spectrum is $[0, +\infty)$ and $(-\infty, 0)$ contains the discrete spectrum tending to -0, the asymptotics of which is the object of our interest. Finally, in *section 10.6* we consider asymptotics with several parameters: when we have different semiclassical parameters with respect to different variables or when we have semiclassical and spectral parameters at the same time. In Appendix H the distribution of the spectra of negative-order pseudo-differential operators on compact closed manifolds is derived.

Chapter 11 is devoted to the distribution of spectra for Schrödinger and Dirac operators with strong magnetic field. The idea is exactly the same as in Chapter 10 but now we consider the case when we apply the results of Chapters 6 and 7 after the partition-dilatation-multiplication procedure. So we now have two local effective parameters: h_{eff} and μ_{eff}, and on elements with $\mu_{\mathsf{eff}} \gg 1$ we refer to the results of Chapters 6 and 7 while on elements with $\mu_{\mathsf{eff}} \le \mathrm{const}$ we refer to the results of Chapters 4 and 5. In *sections 11.1, 11.2* we consider the case of a fixed spectral parameter in dimensions 2 and 3 respectively. In *sections 11.3–11.5* we are interested in asymptotics containing the spectral parameter (and usually only this parameter). Some results appear similar to the results of Chapter 10 but these assertions have quite different conditions and quite different orders of the leading terms and remainder estimates are derived. Moreover, there are examples of a quite different sort: the Schrödinger and Dirac operators in $\mathbb{R}^2$ with constant metric coefficients, vanishing scalar potential and constant non-vanishing magnetic field have pure point spectra of infinite multiplicity. If we perturb these operators by a scalar potential tending to 0 at $-\infty$, then every time this eigenvalue of infinite multiplicity decomposes to a sequence of eigenvalues of finite multiplicities tending to a point of the essential spectrum (which is an eigenvalue of the unperturbed operator) and we are interested in the asymptotic distribution of eigenvalues in each sequence. Another typical "2-dimensional magnetic" example is the Dirac operator in $\mathbb{R}^2$ with discrete spectrum tending to both infinities and maybe to some finite point. In general in sections 11.3–11.5 each of these four operators (2- and 3-dimensional Schrödinger and Dirac operators) appear to be essentially distinct.

0.8 What is in the Book? Chapter 12

The character of **Chapter 12** is quite different. In this chapter we improve and generalize some results of Chapter 10, but the results of Chapters 4 and 5 are insufficient for our purposes in this case and we are forced to repeat some arguments of Chapters 4 and even 2 in certain specific situations. Basically we apply the ideas discussed above. The asymptotics obtained are more or less non-Weylian.

In *section 12.1* elliptic operators in domains with thick cusps are discussed. If we treat $\mathbf{N}(\tau)$ then a cusp is thick if the area of its boundary is infinite, and the standard Weyl asymptotics with remainder estimate $O\bigl(\tau^{\frac{1}{m}(d-1)}\bigr)$ fail to hold. Moreover, if the volume of the cusp is infinite then $\mathbf{N}(\tau) = O\bigl(\tau^{\frac{d}{m}}\bigr)$ is no longer true. For Riesz means the notion of thickness depends on the order of the mean. Let us consider the Dirichlet Laplacian for simplicity. Making a change of variables which transforms the cusp to the cylinder $\mathbb{R}^{d'} \times \Omega$ with $\Omega \Subset \mathbb{R}^{d''}$, one can consider the transformed operator as a Schrödinger operator in $\mathbb{R}^{d'}$ with operator valued coefficients. The auxiliary space is $\mathbb{H} = L^2(\Omega)$. Applying the results of Chapter 4 concerning operators with operator valued symbols we obtain non-Weylian asymptotics. However, these asymptotics are not very effective in the sense that the principal part depends on spectral asymptotics for the family of the auxiliary operators in $\mathbb{H}$. On the other hand, one can study the propagation of singularities for the transformed operator treated as an operator with an operator valued symbol. We then obtain some results for the original operator as well. Namely, we obtain that for a not very remote cusp in Weylian asymptotics one can take take much larger T (depending on τ, $|x'|$) and drastically improve the remainder estimate. These asymptotics are not very effective either. Under a stabilization condition in the remote cusp one can replace an auxiliary operator by its limit value. So, we get two different asymptotics both of which are not effective. However, combining them we obtain much more effective asymptotics: in the best possible case one has only one auxiliary operator instead of a family. Thus, the ideas are basically the same as in Chapter 6: we take two different approaches (on one hand we apply the theory of operators with operator valued symbols to derive spectral asymptotics, and on the other we apply the operator-valued theory only in order to get better propagation results) and then we combine the asymptotics obtained.

In *section 12.2* we consider the class of elliptic operators in $\mathbb{R}^d$. A good example of such operators is the Schrödinger operator with potential V failing to tend to $+\infty$ along directions $x \in \Lambda$. So, the domain $\{x, V(x) \le \tau\}$ contains cusps and if the canyons are not narrow and steep enough the Weylian asymptotics with standard remainder estimate fail. Let us assume that Λ is a smooth d'-dimensional manifold and let us transform it to $\mathbb{R}^{d'}$. Then the ideas of section 12.1 yield results of the same type (heuristically one can say that we apply the results of the previous section but the cusp depends on τ).

In *section 12.3* maximally hypoelliptic operators on a compact manifold are treated. We assume that the operator uniformly degenerates on the symplectic conic manifold Λ. Microlocally $\Lambda = \{x'' = \xi'' = 0\}$. The same basic ideas as above are applied. We treat the operator in question as $A(x', D')$ in $L^2(\mathbb{R}^{d'}, \mathbb{H})$ with $\mathbb{H} = \mathbb{R}^{d''}$.

In *section 12.4* we consider semiclassical asymptotics of spectral Riesz means for operators with singularities at disjoint points. In this case the asymptotics of $\mathbf{N}(h)$ are Weylian as are the asymptotics of $\mathbf{N}_\vartheta(h)$ with $\vartheta < \bar{\vartheta}$ where ϑ is the order of the Riesz mean. So, we are interested in the case $\vartheta \ge \bar{\vartheta}$. One very early idea was to use the equality

$$\operatorname{Tr}\bigl(E^1_\vartheta(\tau) - E^0_\vartheta(\tau)\bigr) = \text{const} \int_0^1 \operatorname{Tr}(A^1 - A^0)E^t_{\vartheta-1}(\tau)d\tau$$

where $A^t = tA^1 + (1-t)A^0$ and the superscript $E^t(\tau)$ is related to A^t. Then taking appropriate partitions and applying the standard Weylian approximation to some elements and this formula (or a more sophisticated and general form) to the other elements we obtain the final answer as a sum of the Weylian expression and the term(s) associated with the model operator A^0. We pick A^0 as an operator with positively homogeneous coefficients and get a more or less simple expression for the non-Weylian term(s). Actually this idea is adjusted in different ways, one of which is an investigation of the very long range (in the local scale) propagation of singularities.

In *section 12.5* we perform a similar task for the Schrödinger operator with strong magnetic field.

Finally, in *section 12.6*, three-dimensional Schrödinger and Dirac operators with constant magnetic field and quickly decreasing electric potential are considered. We obtain asymptotics of eigenvalues near the boundary of the essential spectrum. If the potential decreases quickly enough, the results of Chapters 6 and 7 fail to apply so we use the theory of spectral asymptotics for operators with operator-valued symbols: if the magnetic field is directed along x_3 we introduce an auxiliary space $\mathbb{H} = L^2(\mathbb{R}_{x_3})$ and consider x' as the main variable.

0.9 How LSSA are Applied to Classical Asymptotics

Finally I would like to discuss three examples which illustrate Chapter 10 and show how LSSA yield asymptotics with respect to the spectral parameter.

(i) Let X be a compact Riemannian manifold without boundary and let Δ be a (positive) Laplace-Beltrami operator. Let us consider the Schrödinger operator $A_h = h^2\Delta + V(x)$; it is well known that

$$\mathbf{N}^-(A_h) = \kappa_0 h^{-d} + O(h^{1-d}) \tag{0.15}$$

with Weylian constant κ_0 provided 0 is not a critical value of $V(x)$.[5] The Birman-Schwinger principle implies that $\mathbf{N}^-(A_h) = \mathbf{N}(h^{-2})$ where $\mathbf{N}(\lambda)$ is the number of eigenvalues $\mu \in [0, \lambda)$ of the spectral problem $(\Delta + \mu V(x))u = 0$ (counting their eigenvalues). These two equalities immediately yield the asymptotics

$$\mathbf{N}(\lambda) = \kappa_0 \lambda^{\frac{d}{2}} + O(\lambda^{\frac{1}{2}(d-1)}) \tag{0.16}$$

provided 0 is not a critical value of $V(x)$ (the condition which was claimed not to be necessary). Try to obtain this result directly in the case when $V(x)$ vanishes at some point! It is striking that nobody had observed this non-trivial new result which trivially follows from two well known facts.

(ii) Let us consider the Schrödinger operator $A = \Delta + V(x)$ in $\mathbb{R}^d$ where Δ is the (positive) Laplacian and the real-valued potential $V(x)$ satisfies the following conditions:

$$|D^\alpha V| \le \langle x\rangle^{2m-\alpha} \qquad \forall \alpha : |\alpha| \le K, \tag{0.17}$$

$$V \ge \epsilon_0 |x|^{2m} \qquad \text{for } x : |x| \ge c \tag{0.18}$$

where $\langle x\rangle = (|x|^2 + 1)^{\frac{1}{2}}$ here and in what follows. Let us take a γ-admissible partition of unity with $\gamma(x) = \frac{1}{4}\langle x\rangle$. If we make the dilatation transforming $B(\bar{x}, \gamma(\bar{x}))$ to $B(0,1)$ and multiply $A - \lambda$ by λ^{-1} we obtain a Schrödinger operator for $|x| \le C\lambda^{\frac{1}{2m}}$ in $B(0,1)$ with standard restrictions on the potential and with $h = \gamma(\bar{x})^{-1}\lambda^{-\frac{1}{2}}$. Let us apply a result of Chapter 4;[6] then we obtain Weylian asymptotics for spatial means of $e(x,x,\lambda)$ in $B(\bar{x}, \gamma(\bar{x}))$ with remainder estimate $O(h^{1-d}) = O(\lambda^{\frac{1}{2}(d-1)}\gamma(\bar{x})^{d-1})$; there is no additional here factor because $e(x,x,\lambda)$ is a density but not a function. Summing with respect to the partition of unity we obtain the remainder estimate $O(\lambda^{(d-1)l})$ with $l = \frac{1}{2m}(1+m)$. On the other hand, let us consider the ball $B(\bar{x}, \gamma(\bar{x}))$ with $|x| \ge C\lambda^{\frac{1}{2m}}$. After dilatation and multiplication of $A-\lambda$ by $\rho^{-2}(\bar{x})$ with $\rho(\bar{x}) = \langle x\rangle^m$ we obtain a Schrödinger operator with standard restrictions on the potential $v(x) \ge \epsilon$ and with $h = \gamma(\bar{x})^{-1}\rho(\bar{x})^{-1}$; we then derive the estimate $|e(x,x,\lambda)| \le Ch^s\gamma(\bar{x})^{-d}$ with an arbitrary exponent s and hence the contribution of the domain $\{x, |x| \ge C\lambda^{\frac{1}{2m}}\}$ to the remainder estimate is $O(\lambda^{-s})$ with another, also arbitrarily large, exponent s. We also obtain the asymptotics

$$\mathbf{N}(\lambda) = \mathcal{N}(\lambda) + O(\lambda^{(d-1)l}) \tag{0.19}$$

with

$$\mathcal{N}(\lambda) = \kappa_0 \int (\lambda - V(x))_+^{\frac{d}{2}} dx \tag{0.20}$$

as $\lambda \to +\infty$ and $\mathcal{N}(\lambda) \asymp \lambda^{dl}$.

(iii) Let us consider the Schrödinger operator in $\mathbb{R}^d$ with potential $V(x)$ satisfying (0.17) with $m \in (-1, 0)$ and now let $\lambda < 0$ be a small parameter. Then for the same

---

5) It is shown in this book that one can reject this condition for $d \ge 2$ and for manifolds with boundary.

6) In which case we are able not to assume that V is non-degenerate, i.e., $|\nabla V| \ge \epsilon_0 |x|^{2m-1}$ for $|x| \ge c$.

ρ and γ as before the dilatation of $B(\bar{x}, \gamma(\bar{x}))$ into $B(0,1)$ and multiplication of $A - \lambda$ by $\rho(\bar{x})^{-2}$ give a Schrödinger operator with standard restrictions on the potential and with $h = \gamma(\bar{x})^{-1}\rho(\bar{x})^{-1}$ provided $|x| \le C|\lambda|^{\frac{1}{2m}}$. Hence for the spatial mean of $e(x,x,\lambda)$ in $B(\bar{x}, \gamma(\bar{x}))$ the Weylian formula holds with remainder estimate $O(h^{1-d}) = O(\rho(\bar{x})^{d-1}\gamma(\bar{x})^{d-1})$. On the other hand, for $|x| \ge C|\lambda|^{\frac{1}{2m}}$ dilatation and multiplication by $|\lambda|^{-1}$ give a Schrödinger operator with standard restrictions on the potential $v(x) \ge \epsilon$ and with $h = \gamma(\bar{x})^{-1}|\lambda|^{-\frac{1}{2}}$. Hence the estimate $|e(x,x,\lambda)| \le Ch^s\gamma(\bar{x})^{-d}$ again holds. Summing with respect to a partition of unity we obtain asymptotics (0.19)–(0.20) as $\lambda \to -0$ with the same l as in (i); moreover $\mathcal{N}(\lambda) \asymp |\lambda|^{dl}$ provided

$$V \le -\epsilon |x|^{2m} \qquad \text{for } x : |x| \ge c \tag{0.21}$$

in some non-empty open cone in $\mathbb{R}^d$.

0.10 Paths

For those interested in the basics of the theory I recommend the following sequence of chapters: 1 $\longrightarrow$ 2 $\longrightarrow$ 4 $\longrightarrow$ 10 and then 1 $\longrightarrow$ 2 $\longrightarrow$ 4 $\longrightarrow$ 8 $\longrightarrow$ 10; in addition, I suggest considering only the Schrödinger operator in Chapter 10 and only the standard (not the operator-valued) case in Chapters 2 and 4. Those interested in domains with boundary should read Chapters 3 and 5. Those interested in magnetic field theory should read Chapters 6 (and maybe 7) and 11 (but Chapters 3 and 5 are not necessary). Those interested in the results of Chapter 12 should consider references to every section.

Part I

Semiclassical Microlocal Analysis

1. Introduction to Semiclassical Microlocal Analysis

In this chapter I give a survey of the theory of h-pseudo-differential operators (in section 1.1), h-Fourier integral operators (in section 1.2), the notion of oscillation front sets and related topics (in section 1.3) and a functional calculus for h-pseudo-differential operators (in section 1.4). Proofs are mostly absent since all the results of this chapter are either trivial consequences of the results proven in [Hörmander 1] or can be proven by trivial modifications of the arguments used in that book.

1.1 h-Pseudo-Differential Operators

1.1.1 Quantization, symbols and pseudo-differential operators

The most convenient starting definition of h-pseudo-differential operators is a definition based on double symbols defined on $\mathbb{R}^d \times \mathbb{R}^d \times \mathbb{R}^d$. Let $\mathbb{H}$, $\mathbb{H}'$, etc., be auxiliary Hilbert spaces (finite dimensional for beginners), $\mathcal{L}(\mathbb{H}', \mathbb{H})$ the space of linear bounded operators from $\mathbb{H}$ to $\mathbb{H}'$ equipped with the uniform operator norm and $a(x, \xi, y)$ a function on $\mathbb{R}^d \times \mathbb{R}^d \times \mathbb{R}^d$. Then under appropriate conditions on a, the formula

$$(Au)(x) = (2\pi)^{-d} \iint a(x, y, h\xi) e^{i\langle x-y, \xi\rangle} u(y)\, dy d\xi \tag{1.1.1}$$

defines an operator a from $\mathcal{S}(\mathbb{R}^d, \mathbb{H})$ to $\mathcal{S}'(\mathbb{R}^d, \mathbb{H}')$ which will be denoted by $a(\overset{3}{x}, \overset{1}{x}, \overset{2}{hD})$.

In particular, if a only depends on x and y we obtain the operator of multiplication by $a(x, x)$; if a only depends on x and ξ we obtain the operator $a(\overset{2}{x}, \overset{1}{hD})$ which can be rewritten in the form

$$(Au)(x) = \int a(x, h\xi) e^{i\langle x, \xi\rangle} \hat{u}(y)\, dy \tag{1.1.2}$$

where $\hat{u} = F_{x\to\xi} u$ is the Fourier transform of u; if a only depends on y and ξ we obtain the operator $a(\overset{1}{x}, \overset{2}{hD})$ which can be rewritten in the form

$$\widehat{Au}(\xi) = (2\pi)^{-d} \int a(y, h\xi) e^{-i\langle y, \xi\rangle} u(y)\, dy; \tag{1.1.3}$$

if a only depends on $\frac{1}{2}(x+y)$ and ξ we obtain the operator $a^{\mathrm{w}}(x, hD)$:

$$(Au)(x) = (2\pi)^{-d} \iint a(\frac{x+y}{2}, h\xi) e^{i\langle x-y, \xi\rangle} u(y)\, dy d\xi; \tag{1.1.4}$$

finally, if a depends only on ξ we obtain a multiplicator $a(hD)$:

$$\widehat{Au}(\xi) = a(h\xi)\hat{u}(\xi). \tag{1.1.5}$$

Definitions (1.1.1)–(1.1.4) can be rewritten in the form

$$A = \iiint \hat{a}(\alpha, \alpha', \beta) e^{i\langle\alpha,x\rangle} e^{i\langle\beta,hD\rangle} e^{i\langle\alpha',x\rangle}\, d\alpha d\alpha' d\beta$$

$$A = \iint \hat{a}(\alpha, \beta) e^{i\langle\alpha,x\rangle} e^{i\langle\beta,hD\rangle}\, d\alpha d\beta$$

$$A = \iint \hat{a}(\alpha, \beta) e^{i\langle\beta,hD\rangle} e^{i\langle\alpha,x\rangle}\, d\alpha d\beta$$

$$A = \iint \hat{a}(\alpha, \beta) e^{i\langle\beta,hD\rangle + i\langle\alpha,x\rangle}\, d\alpha d\beta$$

respectively where $\hat{a}$ is the complete Fourier transform of a and the following useful relations hold:

$$e^{-\frac{ih}{2}\langle\alpha,\beta\rangle} e^{i\langle\alpha,x\rangle} e^{i\langle\beta,hD\rangle} = e^{\frac{ih}{2}\langle\alpha,\beta\rangle} e^{i\langle\beta,hD\rangle} e^{i\langle\alpha,x\rangle} = e^{i\langle\alpha,x\rangle + i\langle\beta,hD\rangle}. \tag{1.1.6}$$

The transitions from the symbol $a(x,\xi)$ to the operators $A = a(\overset{2}{x}, \overset{1}{hD})$, $A = a(\overset{1}{x}, \overset{2}{hD})$ and $a^{\mathrm{w}}(x, hD)$ are called *qp-*, *pq-* and *Weyl (symmetric) quantizations of* a respectively; conversely, a is called the *qp-*, *pq-* and *Weyl symbol of* A respectively. In what follows "quantization," "operator" and "symbol" mean any quantization, operator or symbol if the type is not indicated. We sometimes write $A = \mathrm{Op}(a)$ and $a = \mathrm{symb}(A)$, and super- or subscripts (if they are indicated) give the type. In principle, according to Chapter 18.4 of [Hörmander 1], the different types of quantization are unnecessary; namely an operator $a = a(\overset{3}{x}, \overset{1}{x}, \overset{2}{hD})$ can be rewritten in each form $a_{(1)}(\overset{2}{x}, \overset{1}{hD})$, $a_{(2)}(\overset{1}{x}, \overset{2}{hD})$, $a^{w}_{(3)}(x, hD)$ where the symbols $a_{(j)}$ are given by the formulas

$$a_{(1)}(x,\xi) = \Gamma_x e^{ih\langle D_y, D_\xi\rangle} a(x,y,\xi), \tag{1.1.7$_1$}$$

$$a_{(2)}(x,\xi) = \Gamma_x e^{-ih\langle D_x, D_\xi\rangle} a(x,y,\xi), \tag{1.1.7$_2$}$$

$$a_{(3)}(x,\xi) = \Gamma_x e^{\frac{ih}{2}\langle D_y - D_x, D_\xi\rangle} a(x,y,\xi), \tag{1.1.7$_3$}$$

where Γ_x is the operator of restriction to the diagonal $\{x = y\}$. Moreover, it is not necessary to introduce operators with multiple symbols:

$$A = \int \cdots \int \hat{a}(\alpha_{(1)}, \ldots, \alpha_{(n+1)}, \beta_{(1)}, \ldots, \beta_{(n)}) \times \\ e^{i\langle\alpha_{(1)},x\rangle} e^{i\langle\beta_{(1)},hD\rangle} \ldots e^{i\langle\beta_{(n)},hD\rangle} e^{i\langle\alpha_{(n+1)},x\rangle} d\alpha_{(1)} \ldots d\alpha_{(n+1)} d\beta_{(1)} \ldots d\beta_{(n)} \tag{1.1.8}$$

where $a(x_{(1)}, \ldots, x_{(n+1)}, \xi_{(1)}, \ldots, \xi_{(n)})$ is a multiple symbol and $x_{(j)}, \xi_{(j)}, \alpha_{(j)}, \beta_{(j)} \in \mathbb{R}^d$.

The following simple assertion is useful:

Lemma 1.1.1. *Let $b(x,\xi)$ be a linear scalar symbol. Then for an arbitrary symbol a and any type of quantization*

$$[\mathrm{Op}(a), \mathrm{Op}(b)] = -ih\, \mathrm{Op}\{a, b\} \tag{1.1.9}$$

where $[.,.]$ is the commutator and $\{.,.\}$ are Poisson brackets:

$$\{a,b\} = \sum_j ((\partial_{\xi_j} a)(\partial_{x_j} b) - (\partial_{x_j} a)(\partial_{\xi_j} b)).$$

Moreover, for Weyl quantization

$$\mathrm{Op}^{\mathrm{w}}(a)\,\mathrm{Op}^{\mathrm{w}}(b) + \mathrm{Op}^{\mathrm{w}}(b)\,\mathrm{Op}^{\mathrm{w}}(a) = 2\,\mathrm{Op}^{\mathrm{w}}(ab) \tag{1.1.10}$$

for a linear scalar symbol b, and (1.1.9) holds even for a quadratic scalar symbol b.

Theorem 18.4.10 and formula (18.4.8) of [Hörmander 1] imply the following

Proposition 1.1.2. *Suppose*

$$(1.1.11)\quad |D_x^\alpha D_y^\beta D_\xi^\nu a| \le L\boldsymbol{\rho}^{-\nu}\boldsymbol{\gamma}^{-\alpha-\beta}$$
$$\forall \alpha,\beta,\nu : |\alpha|+|\beta|+|\nu| \le 2N+K \quad \forall (x,y,\xi)\in \mathbb{R}^d\times\mathbb{R}^d\times\mathbb{R}^d$$

where $\boldsymbol{\rho}=(\rho_1,\dots,\rho_d)$, $\boldsymbol{\gamma}=(\gamma_1,\dots,\gamma_d)$ *are vectors with positive coordinates and* $|.|$ *means the norm in* $\mathcal{L}(\mathbb{H}',\mathbb{H})$; *we omit the subscript indicating the norm if no confusion arises. Then for* $h\in(0,h_0]$ *with* $h_0=\min_{1\le j\le d}\rho_j\gamma_j$,

$$|D_x^\alpha D_\xi^\nu a_{(j)}| \le CL\boldsymbol{\rho}^{-\nu}\boldsymbol{\gamma}^{-\alpha} \qquad \forall \alpha,\nu : |\alpha|+|\nu|\le 2N \quad \forall (x,\xi)\in\mathbb{R}^d\times\mathbb{R}^d$$

where here and below N *is arbitrary large and* $C=C(d,N)$ *and* $K=K(d)$ *are made as large as necessary.*

For every $M\le N$,

$$|D_x^\alpha D_\xi^\nu (a_{(j)}-a_{(j,M)})| \le CL\rho^{-\nu}\gamma^{-\alpha}(\frac{h}{h_0})^{M+1}$$
$$\forall \alpha,\nu : |\alpha|+|\nu|\le 2(N-M) \quad \forall (x,\xi)\in\mathbb{R}^d\times\mathbb{R}^d$$

where

$$(1.1.12)_1 \qquad a_{(1,M)}(x,\xi) = \Gamma \sum_{0\le n\le M}\frac{1}{n!}(ih\langle D_y,D_\xi\rangle)^n a(x,y,\xi),$$

$$(1.1.12)_2 \qquad a_{(2,M)}(x,\xi) = \Gamma \sum_{0\le n\le M}\frac{1}{n!}(-ih\langle D_x,D_\xi\rangle)^n a(x,y,\xi),$$

$$(1.1.12)_3 \qquad a_{(3,M)}(x,\xi) = \Gamma \sum_{0\le n\le M}\frac{1}{n!}(\frac{1}{2}ih\langle D_y-D_x,D_\xi\rangle)^n a(x,y,\xi).$$

1.1.2 Estimates

Theorem 18.6.2 of [Hörmander 1] implies the following.

Theorem 1.1.3. *Let* (1.1.11) *be satisfied with* $N=0$. *Then for* $h\in(0,h_0]$ *the* $\mathcal{L}(L^2(\mathbb{R}^d,\mathbb{H}'),L^2(\mathbb{R}^d,\mathbb{H}))$*-norm of* $a(\overset{3}{x},\overset{1}{x},\overset{2}{hD})$ *does not exceed* CL.

Remark 1.1.4. *We will use h-pseudo-differential operators with symbols satisfying (1.1.11). Making the dilatation* $x_{j,\text{new}}=x_{j,\text{old}}/\gamma_j$ *they can be reduced to symbols satisfying (1.1.11) with* $\boldsymbol{\gamma}=\mathbf{1}=(1,\dots,1)$. *The choice of h is also arbitrary. Therefore the only essential data is the vector* $\mathbf{h}=(h_1,\dots,h_d)$ *with* $h_j=h/\rho_j\gamma_j$. *The operator* $a(x,hD)$ *can always be reduced to the operator* $a'(x,\mathbf{h}D)$ *with symbol* a' *satisfying (1.1.11) with* $\boldsymbol{\rho}=\boldsymbol{\gamma}=\mathbf{1}$ *and* $\mathbf{h}D=(h_1D_1,\dots,h_dD_d)$. ◊

Remark 1.1.5. *Note that the adjoint to the operator* $a(\overset{3}{x},\overset{1}{x},\overset{2}{hD})$ *is the operator* $a^\dagger(\overset{1}{x},\overset{3}{x},\overset{2}{hD})$ *where* † *means Hermitian conjugation: if* $a\in\mathcal{L}(\mathbb{H}',\mathbb{H})$ *then* $a^\dagger\in\mathcal{L}(\mathbb{H},\mathbb{H}')$. *In particular, the adjoint to an operator with qp-, pq- or Weyl symbol* $a(x,\xi)$ *is the operator with pq-, qp- or Weyl symbol* $a^\dagger(x,\xi)$, *and for* $\mathbb{H}'=\mathbb{H}$ *an operator is self-adjoint if and only if its Weyl symbol is Hermitian. This fact is one of the reasons why Weyl quantization is the most important.* ◊

Theorem 1.1.2 and theorem 18.5.4 of [Hörmander 1] imply

Theorem 1.1.6. *Let a and b be $\mathcal{L}(\mathbb{H}'', \mathbb{H}')$- and $\mathcal{L}(\mathbb{H}', \mathbb{H})$-valued symbols such that*

$$|D_x^\alpha D_\xi^\nu a| \le L\rho^{-\nu}\gamma^{-\alpha} \quad \forall \alpha,\nu : |\alpha|+|\nu| \le 2N+K \quad \forall (x,\xi)\in \mathbb{R}^d\times\mathbb{R}^d \tag{1.1.13}$$

with $L = L_a$, and the same inequalities hold for b with $L = L_b$. Let $a(x,hD)$, $b(x,hD)$ and $(a\#_M)(x,hD)$ be quantizations of the symbols a, b and $a\#_M b$ where $M \le N$ and here and below

$$a\#_M b = \sum_{0\le|\alpha|\le M} \frac{(ih)^{|\alpha|}}{\alpha!}(D_\xi^\alpha a)(D_x^\alpha b), \tag{1.1.14$_1$}$$

$$a\#_M b = \sum_{0\le|\alpha|\le M} -\frac{(ih)^{|\alpha|}}{\alpha!}(D_x^\alpha a)(D_\xi^\alpha b), \tag{1.1.14$_2$}$$

$$a\#_M b = \sum_{0\le|\alpha|+|\beta|\le M} \frac{(ih)^{|\alpha|-|\beta|}}{\alpha!\beta!}(D_\xi^\alpha D_x^\beta a)(D_x^\alpha D_\xi^\beta b) \tag{1.1.14$_3$}$$

for qp-, pq- and Weyl quantization respectively.

Then the $\mathcal{L}(L^2(\mathbb{R}^d,\mathbb{H}''), L^2(\mathbb{R}^d,\mathbb{H}))$-norm of the operator

$$a(x,hD)b(x,hD) - (a\#b_M)(x,hD)$$

does not exceed $C(h/h_0)^{M+1}L_aL_b$.

By the methods of [Hörmander 1] the following statement can be proven easily.

Proposition 1.1.7. *Let the symbols a and b satisfy (1.1.11) with $L = L_a$, $\rho = \rho_a$, $\gamma = \gamma_a$ and $L = L_b$, $\rho = \rho_b$, $\gamma = \gamma_b$ respectively and let $\rho_{a,j}\gamma_{a,j} \ge h$, $\rho_{b,j}\gamma_{b,j} \ge h$ $\forall j$.*

(i) Suppose either a or b is a scalar symbol. Then

$$\|[\mathrm{Op}(a),\mathrm{Op}(b)]\| \le CL_aL_bh', \quad h' = h\max_{1\le j\le d}\Bigl(\frac{1}{\rho_{a,j}\gamma_{b,j}} + \frac{1}{\rho_{b,j}\gamma_{a,j}}\Bigr). \tag{1.1.15}$$

(ii) Let a and b be supported in the boxes $\Pi_a = \{|x_j - x_{a,j}| \le \gamma_{a,j}, |\xi_j - \xi_{a,j}| \le \rho_{a,j}\}$ and $\Pi_b = \{|x_j - x_{b,j}| \le \gamma_{b,j}, |\xi_j - \xi_{b,j}| \le \rho_{b,j}\}$ respectively. Assume that

$$\sum_{1\le j\le d}\bigl(|x_j-y_j|\gamma_j^{-1} + |\xi_j-\eta_j|\rho_j^{-1}\bigr) \ge \epsilon \qquad \forall (x,\xi)\in\Pi_a \quad \forall (y,\eta)\in\Pi_b$$

with $\rho_j = \max(\rho_{a,j},\rho_{b,j})$, $\gamma_j = \max(\gamma_{a,j},\gamma_{b,j})$. Then

$$\|\mathrm{Op}(a)\,\mathrm{Op}(b)\| \le CL_aL_bh'^N$$

with h' defined in (i) and with arbitrarily large N.

The following assertion will be necessary:

Theorem 1.1.8. *Let $a = a(x,y,\xi)$ be an $\mathbb{H}$-valued[1)] symbol such that*

$$|D_x^\alpha D_y^\beta D_\xi^\nu a| \le L\rho^{-\nu}\gamma^{-\alpha-\beta}\Phi(\xi)^{-m}\phi(x)^{-l} \tag{1.1.16}$$
$$\forall \alpha,\beta,\nu : |\alpha|+|\beta|+|\nu| \le 2N+K$$
$$\forall (x,y,\xi)\in\mathbb{R}^d\times\mathbb{R}^d\times\mathbb{R}^d$$

---

1) Therefore the corresponding operator acts from $L^2(\mathbb{R}^d)$ to $L^2(\mathbb{R}^d,\mathbb{H})$.

where $\phi(x) = (1 + \sum_j x_j^2/\bar\gamma_j^2)^{\frac12}$, $\Phi(\xi) = (1 + \sum_j \xi_j^2/\bar\rho_j^2)^{\frac12}$, *or such that similar estimates are satisfied with* x *replaced by* y *in the right-hand expression, or such that* $a(x, y, \xi) = a(\frac12(x+y), \xi)$ *with* $a(x,\xi)$ *satisfying* (1.1.16). *Moreover, assume that* $\rho_j > 0$, $\gamma_j > 0$ *and* $\rho_j\gamma_j \ge h$, $\bar\rho_j\gamma_j \ge h$, $\rho_j\bar\gamma_j \ge h$ *for* $j = 1, \dots, d$. *Recall that* ρ_j *and* γ_j *are components of* ρ *and* γ *respectively.*

Let

$$\mathcal{K}_A(x,y) = (2\pi)^{-d}\int e^{i\langle x-y,\xi\rangle} a(x,y,h\xi)d\xi$$

be the Schwartz kernel of the operator $a(\overset{3}{x},\overset{1}{x},h\overset{2}{D})$.

Then

$$\int |\mathcal{K}_A(x,y)|^2 dy \le CL^2\bar\rho_1\dots\bar\rho_d h^{-d} \qquad \forall x\in\mathbb{R}^d \tag{1.1.17}$$

provided $m > \frac{d}{2}$, $l = 0$ *and*

$$\iint |\mathcal{K}_A(x,y)|^2 dxdy \le CL^2\bar\rho_1\dots\bar\rho_j\bar\gamma_1\dots\bar\gamma_d h^d \tag{1.1.18}$$

provided $m > \frac{d}{2}$, $l > \frac{d}{2}$.

Proof. The conditions on $\gamma_j, \bar\gamma_j, \rho_j$ and $\bar\rho_j$ imply that the weight functions $\Phi(\xi)^m$ and $\phi(x)^l$ are tempered with respect to the metric

$$\sum_j \gamma_j^{-2}dx_j^2 + h^2\rho_j^{-2}d\xi_j^2$$

and hence theorem 18.5.5 of [Hörmander 1] implies that the operator $\Phi(hD)^m A^*A\Phi(hD)^m$ is a scalar h-pseudo-differential operator with symbol satisfying (1.1.13) with L_a replaced by CL^2; hence it is a bounded operator from $L^2(\mathbb{R}^d)$ to $L^2(\mathbb{R}^d)$ and its norm does not exceed CL^2. The Sobolev embedding theorem then implies that the norm of the operator $A^*A\Phi(hD)^m$ from $L^2(\mathbb{R}^d)$ to $L^\infty(\mathbb{R}^d)$ does not exceed $CL^2\bar\rho_1^{\frac12}\dots\bar\rho_d^{\frac12}h^{-\frac{d}{2}}$ and hence

$$\int |\mathcal{K}_{A^*A\Phi(hD)^m}(x,y)|^2 dy \le CL^4\bar\rho_1\dots\bar\rho_d h^{-d} \qquad \forall x\in\mathbb{R}^d.$$

But the left-hand expression is equal to

$$\int |\Phi(hD_y)^m\mathcal{K}_{A^*A}(x,y)|^2 dy$$

and the same arguments imply that

$$|\mathcal{K}_{A^*A}(x,y)| \le CL^2\bar\rho_1\dots\bar\rho_d h^{-d} \qquad \forall x,y\in\mathbb{R}^d. \tag{1.1.19}$$

On the other hand,

$$\mathcal{K}_{A^*A}(x,x) = \int \mathcal{K}_A^\dagger(x,y)\mathcal{K}_A(x,y)dy = \int |\mathcal{K}_A(x,y)|^2 dy \tag{1.1.20}$$

since the range of $\mathcal{K}_A(x,y)$ is a matrix-column and this equality and (1.1.19) imply (1.1.17). Moreover, (1.1.19) holds with A replaced by $A\phi(x)^l$ and hence the inequality for A remains true if we integrate with respect to $x\in\mathbb{R}^d$ provided $l < \frac{d}{2}$. This inequality and (1.1.20) imply (1.1.18). □

Corollary 1.1.9. *(i) Let $a(x,y,\xi)$ be an $\mathcal{L}(\mathbb{H}',\mathbb{H})$-valued symbol satisfying* (1.1.16) *(in the original form or with x replaced by y in the right-hand expression) or $a(x,y,\xi) = a(\frac{1}{2}(x+y),\xi)$ where $a(x,\xi)$ satisfies* (1.1.16). *Let $\mathcal{K}_A(x,y)$ be the same as before. Then*

$$\int |\mathcal{K}_A(x,y)w|^2 dy \le CL^2\bar\rho_1\dots\bar\rho_d h^{-d}|w|^2 \quad \forall x\in\mathbb{R}^d \quad \forall w\in\mathbb{H} \tag{1.1.21}$$

provided $m>\frac{d}{2}$, $l=0$ and

$$\iint |\mathcal{K}_A(x,y)w|^2 dxdy \le CL^2\bar\rho_1\dots\bar\rho_d\bar\gamma_1\dots\bar\gamma_d h^{-d}|w|^2 \quad \forall w\in\mathbb{H} \tag{1.1.22}$$

provided $m>\frac{d}{2}$, $l>\frac{d}{2}$;

(ii) Moreover, estimates (1.1.17) *and* (1.1.18) *remain true provided $\mathbb{H}$ is a finite-dimensional space with C depending on $\mathrm{D}=\dim\mathbb{H}<\infty$.*

(iii) Let A, B be h-pseudo-differential operators with $\mathcal{L}(\mathbb{H}''',\mathbb{H}'')$- and $\mathcal{L}(\mathbb{H}'',\mathbb{H}')$-valued symbols respectively satisfying (1.1.16) *with $L=L_a$ and $L=L_b$ respectively and let E be an operator from $L^2(\mathbb{R}^d,\mathbb{H}')$ to $L^2(\mathbb{R}^d,\mathbb{H}'')$ with the norm L. Then the same arguments as before yield*

$$|\mathcal{K}_{AEB}(x,y)| \le CLL_aL_b\bar\rho_1\dots\bar\rho_d h^{-d} \qquad \forall x,y\in\mathbb{R}^d \tag{1.1.23}$$

provided $m>\frac{d}{2}$, $l=0$ and

$$\int |\mathcal{K}_{AEB}(x,x)|dx \le CL\bar\rho_1\dots\bar\rho_d\bar\gamma_1\dots\bar\gamma_d h^{-d} \tag{1.1.24}$$

provided $m>\frac{d}{2}$, $l>\frac{d}{2}$.

(iv) All the conclusions of theorem 1.1.8 and this corollary remain true with constants independent of $\bar x$, $\bar\xi$ if we replace $\Phi(\xi)$ and $\phi(x)$ by $\Phi(\xi-\bar\xi)$ and $\phi(x-\bar x)$ respectively, with arbitrary $\bar x,\bar\xi\in\mathbb{R}^d$; moreover in (iii) one can take different $\bar x$, $\bar\xi$ for a and b.

Note that the shift $x\to x-\bar x$, $\xi\to\bar\xi$ is equivalent to the transformation $A\in T^*AT$ with the operator T: $(Tu)(x)=e^{-i\langle x,\bar\xi\rangle}u(x+\bar x)$ which preserves all the L^p-norms.

Theorems 18.6.4 and 18.6.8 of [Hörmander 1] imply the following.

Theorem 1.1.10 (Gårding inequality). *(i) Let a be an $\mathcal{L}(\mathbb{H},\mathbb{H})$-valued Hermitian symbol satisfying* (1.1.13) *with $N=0$ such that*

$$\langle a(x,\xi)w,w\rangle \ge 0 \qquad \forall (x,\xi)\in\mathbb{R}^d\times\mathbb{R}^d \quad \forall w\in\mathbb{H} \tag{1.1.25}$$

where here $\langle .,.\rangle$ means the inner product in $\mathbb{H}$. Then for $h\in(0,h_0]$

$$(a^{\mathrm{w}}(x,hD)u,u) \ge -CL\frac{h}{h_0}\|u\|^2 \qquad \forall u\in L^2(\mathbb{R}^d,\mathbb{H}) \tag{1.1.26}$$

where here $(.,.)$ and $\|.\|$ mean the inner product and norm in $L^2(\mathbb{R}^d,\mathbb{H})$ and $h_0=\min_{1\le j\le d}\rho_j\gamma_j$.

(ii) In the scalar case (i.e., for $\mathbb{H}=\mathbb{C}$) (1.1.25) *means exactly that*

$$a(x,\xi)\ge 0 \tag{1.1.25$'$}$$

and the stronger inequality

$$(a^{w}(x,hD)u,u) \ge -CL(\frac{h}{h_0})^2\|u\|^2 \qquad \forall u\in L^2(\mathbb{R}^d) \tag{1.1.26*}$$

holds.

Remark 1.1.11. *(i) In [Hörmander 1] it was assumed that inequalities (1.1.11) and (1.1.13) are satisfied for every α, β, ν with constants depending on α, β, ν; however my "improvements" are only formal: they follow immediately from either the corresponding statements in that book or their proofs. Moreover, there were statements about continuity in the corresponding topology in that book.*

(ii) Theorem 1.1.10(ii) remains true provided only that the principal symbol a_0 is scalar.

(iii) A slightly worse estimate holds if a_0 is scalar and

$$(1.1.18)^{**} \qquad a_0(x,\xi) \ge 0, \qquad a^s(x,\xi) + \frac{1}{2}\operatorname{tr}^+ F_{a_0}(x,\xi) \ge 0 \qquad \forall (x,\xi);$$

a_0 and the subprincipal symbol a^s are introduced in the following subsection, F_{a_0} is the skew-Hessian of a_0 and $\operatorname{tr}^+ F = \frac{1}{2}\sum |f|_j$ where f_j are all the eigenvalues of F. Certain results of this type were recently obtained even for matrix-valued a_0 (but the definition of $\operatorname{tr}^+ F$ is more complicated). However, when estimates of this type are needed, they will be derived in a better form (in special situations), so I do not discuss these results further here. ◊

1.1.3 Symbol classes

Let us introduce the classes $\mathrm{S}_{h,\rho,\gamma,N}(\mathbb{R}^d \times \mathbb{R}^d, \mathbb{H}', \mathbb{H})$ and $\mathrm{S}^{\mathrm{double}}_{h,\rho,\gamma,N}(\mathbb{R}^d \times \mathbb{R}^d \times \mathbb{R}^d, \mathbb{H}', \mathbb{H})$ of ordinary and double symbols of the form

$$(1.1.27) \qquad a = \sum_{0 \le n \le N} a_n h^n$$

where a_n are $\mathcal{L}(\mathbb{H}', \mathbb{H})$-valued symbols such that

$$(1.1.28) \quad |D_x^\alpha D_\xi^\nu a_n| \le L \rho^{-\nu} \gamma^{-\alpha} h_0^{-n} \qquad \forall \alpha, \nu : |\alpha| + |\nu| \le 2(N-n) + K \quad \forall (x,\xi) \in \mathbb{R}^d \times \mathbb{R}^d,$$

$$(1.1.29) \quad |D_x^\alpha D_y^\beta D_\xi^\nu a_n| \le L \rho^{-\nu} \gamma^{-\alpha-\beta} h_0^{-n} \qquad \forall \alpha, \beta, \nu : |\alpha| + |\beta| + |\nu| \le 2(N-n) + K \quad \forall (x,y,\xi) \in \mathbb{R}^d \times \mathbb{R}^d \times \mathbb{R}^d$$

respectively for ordinary and double symbols; recall that $h_0 = \min_j \rho_j \gamma_j$.

Let us introduce norms in these symbol spaces: $|\!|\!| a |\!|\!|$ is the least number L for which all the above inequalities hold.

For $a \in \mathrm{S}^{\mathrm{double}}_{h,\rho,\gamma,N}$ define $a_{(j,n)}$ by the new formula

$$a_{(j,n)} = \sum_{0 \le n \le N} a_{n(j,N-n)'} h^n$$

and for $a, b \in \mathrm{S}_{h,\rho,\gamma,N}$ define $a \#_N b$ by the new formula

$$a \#_N b = \sum_{0 \le n+m \le N} a_m \#'_{N-m-n} b_n h^{m+n}$$

where in both formulas the prime means the old definition of these operations. These operations depend on the type of quantization. One can easily see that the operation $\#_N$ is associative.

In what follows in the case $\boldsymbol{\rho} = \rho\mathbf{1}$ with $\mathbf{1} = (1, \dots, 1)$, $\boldsymbol{\rho}$ is replaced by ρ in the notation and $\boldsymbol{\rho} = \mathbf{1}$ is omitted entirely; and the same convention is used for $\boldsymbol{\gamma}$. For $a \in \mathrm{S}_{h,\boldsymbol{\rho},\boldsymbol{\gamma},N}$ and $A = a^w(x, hD, h)$, $a^0 = a_0$ and $a^s = a_1$ are called the *principal symbol* and *subprincipal symbol* of A respectively. In fact the principal symbol does not depend on the type of quantization and the subprincipal symbol is given by the formulas

$$a^s = a_1 \pm \frac{i}{2}\langle D_x, D_\xi\rangle a_0 \tag{1.1.30}$$

for *qp*- and *pq*-quantization respectively. It is easy to show that

$$(a\#b)_0 = a_0 b_0, \tag{1.1.31}$$

$$(a\#b)^s = a_0 b^s + a^s b_0 - \frac{i}{2}\{a_0, b_0\} \tag{1.1.32}$$

where

$$\{a, b\} = \sum_j \bigl((D_{x_j} a)(D_{\xi_j} b) - (D_{\xi_j} a)(D_{x_j} b)\bigr) \tag{1.1.33}$$

are *Poisson brackets.*

Theorems 1.1.3 and 1.1.6 imply

Theorem 1.1.12. *(i) Suppose $a \in \mathrm{S}^{\mathrm{double}}_{h,\boldsymbol{\rho},\boldsymbol{\gamma},N}(\mathbb{R}^d \times \mathbb{R}^d \times \mathbb{R}^d, \mathbb{H}', \mathbb{H})$. Then for $h \in (0, h_0]$, $a_{(j,N)} \in \mathrm{S}_{h,\boldsymbol{\rho},\boldsymbol{\gamma},N}(\mathbb{R}^d \times \mathbb{R}^d, \mathbb{H}', \mathbb{H})$ and*

$$\interleave a_{(j,N)} \interleave_S \le C \interleave a \interleave_{S^{\mathrm{double}}}, \tag{1.1.34}$$

$$\| a_{(j,N)}(x, hD, h) - a(\overset{3}{x}, \overset{1}{x}, \overset{2}{hD}) \| \le C\Bigl(\frac{h}{h_0}\Bigr)^{N+1} \interleave a \interleave_{S^{\mathrm{double}}} \tag{1.1.35}$$

where $\|.\|$ means the $\mathcal{L}(L^2(\mathbb{R}^d, \mathbb{H}'), L^2(\mathbb{R}^d, \mathbb{H}))$-norm;

(ii) Let $a, b \in \mathrm{S}_{h,\boldsymbol{\rho},\boldsymbol{\gamma},N}$ be $\mathcal{L}(\mathbb{H}'', \mathbb{H}')$- and $\mathcal{L}(\mathbb{H}', \mathbb{H})$-valued symbols. Then for $h \in (0, h_0]$,

$$\interleave a \#_N b \interleave_S \le C \interleave a \interleave_S \cdot \interleave b \interleave_S \tag{1.1.36}$$

and

$$\| a(x, hD, h) b(x, hD, h) - (a \#_N b)(x, hD, h) \| \le C\Bigl(\frac{h}{h_0}\Bigr)^{N+1} \interleave a \interleave_S \cdot \interleave b \interleave_S . \tag{1.1.37}$$

Applying the standard parametrix construction we obtain the results of the following section.

1.1.4 Parametrices

The following statement is standard:

Theorem 1.1.13. *Suppose $a \in \mathrm{S}_{h,\rho,\gamma,N}(\mathbb{R}^d \times \mathbb{R}^d, \mathbb{H}', \mathbb{H})$. Assume that $\|a\| \le c$ and at every point $(x,\xi) \in \Omega \subset \mathbb{R}^d \times \mathbb{R}^d$ its principal symbol a_0 is invertible and*

$$|a_0(x,\xi)^{-1}| \le c. \tag{1.1.38}$$

Then there exists a symbol $b \in \mathrm{S}_{h,\rho,\gamma,N}(\mathbb{R}^d \times \mathbb{R}^d, \mathbb{H}, \mathbb{H}')$ with $\|b\| \le C = C(d,N,c)$ such that

$$a \#_N b = I, \quad b \#_N a = I \quad \text{in} \quad \Omega.$$

This symbol is defined uniquely in Ω and $b_n(x,\xi)$ depends only on $D_x^\alpha D_\xi^\nu a_k$ with $|\alpha| \le n - k, |\nu| \le n - k$ calculated at (x,ξ). In particular, $b_0 = a_0^{-1}$ and if $a_0 = I, a_1 = \dots = a_n = 0$ in Ω then $b_0 = I, b_1 = \dots = b_n = 0$ in Ω.

For Weyl quantization one can easily prove that $a^\dagger \#_N a$ is always a Hermitian symbol. Using this fact it is easy to prove

Theorem 1.1.14. *Let $a \in \mathrm{S}_{h,\rho,\gamma,N}(\mathbb{R}^d \times \mathbb{R}^d, \mathbb{H}, \mathbb{H})$ be a Hermitian symbol such that $\|a\| \le c$ and*

$$\langle a_0(x,\xi)w, w\rangle \ge \epsilon_0 |w|^2 \qquad \forall (x,\xi) \in \Omega \quad \forall w \in \mathbb{H}. \tag{1.1.39}$$

Then there exists a Hermitian symbol $b \in \mathrm{S}_{h,\rho,\gamma,N}(\mathbb{R}^d \times \mathbb{R}^d, \mathbb{H}, \mathbb{H})$ with $\|b\| \le C = C(d,N,c,\epsilon_0)$ such that $b_0 = a_0^{\frac{1}{2}}$ and

$$b \#_N b = a$$

in Ω. This symbol is defined uniquely in Ω and $b_n(x,\xi)$ depends only on $D_x^\alpha D_\xi^\nu a_k$ with $|\alpha| \le n - k, |\nu| \le n - k$ calculated at (x,ξ). In particular, $b_0 = a_0^{\frac{1}{2}}$ and if $a_0 = I, a_1 = \dots = a_n = 0$ in Ω then $b_0 = I, b_1 = \dots = b_n = 0$ in Ω.

1.1.5 Alternative symbol classes

The classes of h-pseudo-differential operators introduced in subsection 1.1.3 are completely sufficient for our purposes, but they are not invariant with respect to non-linear changes of variable even if $\boldsymbol{\rho} = \rho\mathbf{1}$ and $\boldsymbol{\gamma} = \gamma\mathbf{1}$ with $\rho, \gamma \in (0,1]$ (which is assumed in what follows), so I now describe invariant classes of operators. Replace (1.1.11) and (1.1.13) by

$$|D_x^\alpha D_y^\beta D_\xi^\nu a| \le L\rho^{-|\nu|}\gamma^{-|\alpha|-|\beta|}\langle\xi\rangle^{m-|\nu|} \qquad \forall \alpha,\beta,\nu : |\alpha|+|\beta|+|\nu| \le 2N+K \quad \forall (x,y,\xi) \in \mathbb{R}^d \times \mathbb{R}^d \times \mathbb{R}^d \tag{1.1.11$'$}$$

$$|D_x^\alpha D_\xi^\nu a| \le L\rho^{-|\nu|}\gamma^{-|\alpha|}\langle\xi\rangle^{m-|\nu|} \qquad \forall \alpha,\nu : |\alpha|+|\nu| \le 2N+K \quad \forall (x,\xi) \in \mathbb{R}^d \times \mathbb{R}^d \tag{1.1.13$'$}$$

where $\langle\xi\rangle = (1+|\xi|^2)^{\frac{1}{2}}$ and $m \in \mathbb{R}$.

Then the results of Chapters 18.4–18.6 of [Hörmander 1] imply that proposition 1.1.2 remains true with the additional factors $\langle\xi\rangle^{m-|\nu|}$ and $\langle\xi\rangle^{m-|\nu|-M}$ in the first and the second inequalities in the conclusion respectively. Moreover, theorems 1.1.3 and 1.1.6 remain true provided $m \le 0$ and $m_a + m_b \le 0$ respectively, where the obvious notation is used and K and C depend on arbitrary n such that $m_* \le n$ for all the exponents m_*.

Theorem 1.1.10 remains true provided $m \le 1$ in the general case and $m \le 2$ in the scalar case.

Let us introduce the classes $\tilde{S}_{h,\rho,\gamma,N}(\mathbb{R}^d \times \mathbb{R}^d, \mathbb{H}', \mathbb{H})$ and $\tilde{S}^{\mathrm{double}}_{h,\rho,\gamma,N}(\mathbb{R}^d \times \mathbb{R}^d \times \mathbb{R}^d, \mathbb{H}', \mathbb{H})$ of ordinary and double symbols of the form (1.1.27) where the right-hand expressions of (1.1.28) and (1.1.29) now have the additional factor $\langle\xi\rangle^{m-|\nu|-n}$. Let us introduce the corresponding norms $|\!|\!|.|\!|\!|_{\tilde{S}}$ by the same method as before (but with modified inequalities) and let us define the operations $a \mapsto a_{(j,N)}$ and $\#_N$ in the same way as before. Then theorems 1.1.12–1.1.14 remain true with the following modifications:

(i) In theorem 1.1.13 we now assume that $|a_0(x,\xi)^{-1}| \le c\langle\xi\rangle^{-m}$ and obtain $b \in \tilde{S}^{-m}$;

(ii) In theorem 1.1.9 we now assume that

$$\langle a_0(x,\xi)w, w\rangle \ge \epsilon_0 \langle\xi\rangle^m |w|^2 \tag{1.1.36$'$}$$

and obtain $b \in \tilde{S}^{\frac{m}{2}}$.

Repeating the proof of theorem 18.1.17 of [Hörmander 1] one can easily prove the following

Theorem 1.1.15. *Suppose $a \in \tilde{S}_{h,\rho,\gamma,N}(\mathbb{R}^d \times \mathbb{R}^d, \mathbb{H}', \mathbb{H})$ with $0 < \gamma \le \rho \le 1, h \in (0, h_0]$. Let G and G' be open domains in $\mathbb{R}^d$ and let $\varkappa : G \to G'$ be a diffeomorphism such that*

$$|D^\alpha \varkappa(x)| \le c \qquad \forall \alpha : |\alpha| \le K' \quad \forall x \in G \tag{1.1.40}$$

and the same inequalities hold in G' for the inverse diffeomorphism $\varkappa^{-1}$ where $K' = K'(d,m,n,K)$ and $K = K(d,m,n)$. Then there exists a symbol $a_\varkappa \in \tilde{S}_{h,\rho,\gamma,N}(\mathbb{R}^d \times \mathbb{R}^d, \mathbb{H}', \mathbb{H})$ such that $|\!|\!|a_\varkappa|\!|\!|_{\tilde{S}} \le C|\!|\!|a|\!|\!|_{\tilde{S}}$ and for every pair of functions $f, g \in C_0^K(G)$ the following estimate holds:

$$\|\Phi(hD)^k(\varkappa^{*-1} f a(x,hD,h) g \varkappa^* - f' a_\varkappa(x,hD,h) g')\Phi(hD)^l\| \le C'(\frac{h}{h_0})^{N+1} |\!|\!|a|\!|\!| \tag{1.1.41}$$

provided $l \le n, k \le n, l + m + k \le N+1$ where n is arbitrary, $\varkappa^*$ is the induced map $\varkappa^* u = u \circ \varkappa$, $f' = f \circ \varkappa^{-1}, g' = g \circ \varkappa^{-1}$ and $C = C(d,m,n,N,c)$, $C' = C'(d,m,n,N,c,c')$.

Moreover, in T^*G' the symbol $a_\varkappa$ is defined uniquely:

$$a_{\varkappa n} = \sum_{0 \le r \le n} (\ell_{n-r} a_r) \circ \Psi^{-1} \tag{1.1.42}$$

where ℓ_k are linear $2k$-th degree differential operators depending on $\varkappa$ and on the type of quantization, $\ell_0 = I$, $\ell_k 1 = 0$ for $k \ge 1$ and $\Psi : T^*G \to T^*G'$ is the symplectomorphism induced by $\varkappa$:

$$\Psi(x,\xi) = (\varkappa(x), {}^t\varkappa'^{-1}(x)\xi); \tag{1.1.43}$$

$\varkappa'(x)$ is the Jacobi matrix of $\varkappa$ at x and ^{t} means matrix transposition.

Remark 1.1.16. *According to formula (18.1.30) of [Hörmander 1],*

$$a_\varkappa(x,\xi) \sim \Gamma \sum_\alpha \frac{1}{\alpha!} a^{(\alpha)}(x, {}^t\varkappa'(x)\xi)(hD_y)^\alpha \exp(ih^{-1}\langle(\varkappa(y) - \varkappa(x) - \varkappa'(x)(y-x), \xi\rangle) \tag{1.1.44}$$

for qp-quantization where $a^{(\alpha)} = \partial_\xi^\alpha a$. Hence for this method of quantization

(1.1.45)
$$\ell_k = \sum_{k\le|\nu|\le 2k} \ell_{k\nu}(x,\xi)\partial_\xi^\nu$$

and for the other two types of quantization

(1.1.46)
$$\ell_k = \sum_{k\le|\nu|,|\nu|+|\alpha|\le 2k} \ell_{k\nu\alpha}(x,\xi)\partial_\xi^\nu\partial_x^\alpha;$$

moreover, we prove below that for Weyl quantization, ℓ_k are $(2k-1)$-th degree operators for $k>0$. ◊

1.1.6 Classes of pseudo-differential operators

Let us introduce the following classes of operators: $A\in \boldsymbol{\Psi}_{h,\rho,\gamma,N}(\mathbb{R}^d,\mathbb{H}',\mathbb{H})$ iff there exits a symbol $a\in \mathrm{S}_{h,\rho,\gamma,N}(\mathbb{R}^d\times\mathbb{R}^d,\mathbb{H}',\mathbb{H})$ such that $A-\mathrm{Op}(a)$ is a negligible operator in the sense that its operator norm $\|.\|$ does not exceed Lh^{-N-1} [2)] in which case $|\!|\!|A|\!|\!| = \inf_a(|\!|\!|a|\!|\!| + h^{-N-1}\|A-\mathrm{Op}(a)\|)$.

One can introduce the classes $\tilde{\boldsymbol{\Psi}}^m_{h,\rho,\gamma,N}(\mathbb{R}^d,\mathbb{H}',\mathbb{H})$ in a similar way.[3)]

1.2 Lagrangian Distributions and Fourier Integral Operators

1.2.1 Lagrangian distributions and manifolds

We start with the theory of Lagrangian distributions. Let $\mathbb{H}$ be an auxiliary Hilbert space as before.

Definition 1.2.1. An $\mathbb{H}$-*valued Lagrangian distribution* is a distribution of the form

(1.2.1)
$$u(x) = (2\pi h)^{-\frac{r}{2}}\int a(x,\theta)e^{ih^{-1}\phi(x,\theta)}\,d\theta$$

where the *phase function* ϕ and *amplitude* a satisfy the following conditions:

(1.2.2) ϕ is real-valued and belongs to $C^M(\bar G)$ where $G \Subset \mathbb{R}^d\times\mathbb{R}^r$, i.e., $G\subset\{(x,\xi): |x|\le c, |\xi|\le c\}$, $\bar G$ means the closure of G and

$$|D_x^\alpha D_\theta^\nu \phi|\le c \qquad \forall\alpha,\nu: |\alpha|\le M, |\nu|\le M \quad \forall(x,\theta)\in G$$

for some c,

(1.2.3) ϕ is *non-degenerate*, i.e., if at some point $(x,\theta)\in\bar G$, $\phi'_\theta(x,\theta)=0$ then at this point the rank of the $(d+r)\times r$ matrix $(\phi''_{\theta x},\phi''_{\theta\theta})$ equals r, i.e., if $|\phi'_\theta(x,\theta)|\le\epsilon$, then the matrix $(\phi''_{\theta x},\phi''_{\theta\theta})$ has a minor (determinant) of dimension r, the absolute value of which is not less than ϵ with a small enough constant $\epsilon>0$,

(1.2.4) $a(x,\theta)$ is an $\mathbb{H}$-valued symbol supported in G (i.e., $\mathrm{dist}(\mathrm{supp}\,a,\partial G)\ge\epsilon$) such that

$$|D_x^\alpha D_\theta^\nu a|\le L \qquad \forall\alpha,\theta: |\alpha|\le M, |\nu|\le M \quad \forall(x,\theta);$$

more generally, we consider amplitudes of the form (1.1.27) where a_n satisfy the previous estimate for $|\alpha|+|\nu|\le K+2(N-n)$; the class of these amplitudes we denote by $\mathrm{S}^0_{h,N}(G,\mathbb{H})$ and the least L for which these estimates hold we denote by $|\!|\!|a|\!|\!|$. ◊

2) There is a minor difference between negligible operators and operators of the form $A=\mathrm{Op}(a)$ with symbols a negligible in the sense that $|\!|\!|a|\!|\!|\le Lh^{N+1}$.

3) Surely, the notion of a negligible operator should be modified: now the operator A is negligible if $\|D^\alpha AD^\beta\|\le Lh^{N+1-|\alpha|-|\beta|}$ $\forall\alpha,\beta: |\alpha|+|\beta|\le N+1$.

Let us introduce the *Lagrangian manifold generated by* ϕ:

(1.2.5) $$\Lambda_\phi = \{(x, \phi'_x(x,\theta)) : (x,\theta) \in \Sigma_\phi\} \subset T^*\mathbb{R}^d$$

where

(1.2.6) $$\Sigma_\phi = \{(x,\theta) \in G, \phi'_\theta(x,\theta) = 0\}.$$

Then $\Lambda = \Lambda_\phi$ and Σ_ϕ are respectively C^{M-2} and C^{M-1} manifolds of dimension d and Λ_ϕ is a Lagrangian manifold (i.e., $\dim(\Lambda) = d$ and the restriction of the symplectic form $\sigma = \sum_j dx_j \wedge d\xi_j$ to Λ equals 0) and the map $\iota : \Sigma_\phi \ni (x,\theta) \to (x, \phi'_x(x,\theta)) \in \Lambda_\phi$ is a local diffeomorphism. It is easy to prove that

(1.2.7) $$r - \operatorname{rank} \phi''_{\theta\theta}(x,\theta) = d - \operatorname{rank} \pi'_{\Lambda,x}(\iota(x,\theta)),$$

where $\pi_{\Lambda,x} : \Lambda \ni (x,\xi) \to x \in \mathbb{R}^d$ and $\pi'_{\Lambda,x}$ is its differential.

For the justification of all our future calculations we need

Proposition 1.2.2. *Suppose that conditions* (1.2.2)–(1.2.4) *are satisfied with* $M = K = K_0(d,r)$, $N = 0$. *Then the* $L^2(\mathbb{R}^d, \mathbb{H})$*-norm of* u *does not exceed* $C\|a\|$ *with* $C = C(d,r,c,\epsilon)$.

The proof of this proposition follows from the arguments of Chapter 21.1 of [Hörmander 1]. Since $(hD)^\alpha u$ is again a distribution of the same form we obtain that its $L^2(\mathbb{R}^d, \mathbb{H})$-norm also does not exceed $C\|a\|$ with $C = C(d,r,k,c,\epsilon)$ provided $|\alpha| \le k$ and $M = K = K_0 + k$ in the conditions.

Then the stationary phase method implies

Proposition 1.2.3. *Let* ι *be a one-to-one map, and let* $\pi_{\Lambda,x}$ *be a diffeomorphism (i.e., let* $\pi_{\Lambda,x}$ *be a one-to-one map with* $\operatorname{rank} \pi'_{\Lambda,x} = d$ *at every point of* Λ*; then* $|\det \pi'_{\Lambda,x}| \ge \epsilon$ *at every point of* Λ*). Then for* $K = K(d,r)$ *and* $M = M(d,r,N)$, *a distribution* u *of the form* (1.2.1) *can be rewritten in the following form:*

(1.2.8) $$u(x) = \tilde{a}(x) e^{ih^{-1}\tilde{\phi}(x)} + U(x)$$

where

(1.2.9) $$\tilde{\phi}(x) = \phi(\iota^{-1}\pi'^{-1}_{\Lambda,x}(x)),$$

$\tilde{a} \in \mathrm{S}^0_{h,N}(\pi'_{\Lambda,x}\iota G)$ *with* $K = K(d,0)$, $\|\tilde{a}\| \le C\|a\|$ *where the norms are calculated in the corresponding classes,*

(1.2.10) $$\tilde{a}_0(x) = e^{\frac{i\pi}{4}\operatorname{sgn}(\phi''_{\theta\theta})} |\det(\phi''_{\theta\theta})|^{-\frac{1}{2}} a_0 \big|_{(x,\theta) = \iota^{-1}\pi'^{-1}_{\Lambda,x}(x)},$$

sgn *means the signature of the symmetric matrix (i.e., the difference between the numbers of positive and negative eigenvalues) and the* $L^2(\mathbb{R}^d, \mathbb{H})$*-norm of* U *does not exceed* $Ch^{N+1}\|a\|$ *with* $C = C(d,r,N,c,\epsilon)$*; moreover the same is true for* $(hD)^\alpha U$ *provided* $|\alpha| \le k$ *and we add* k *to* K *and* M *and take* C *also depending on* k.[4)]

Remark. *It is easy to see that* $\Lambda = \{(x, \tilde{\phi}'_x(x)) : x \in \pi'_{\Lambda,x}\iota G\}$ *in this case.* ◊

Definition 1.2.4. A point $z \in \Lambda$ is called *regular* provided

$$\operatorname{rank} \pi'_{\Lambda,x}(z) = d.$$

◊

4) Such functions we call (k,N)-negligible or simply negligible.

It is well known [Maslov 1] that if ι is a one-to-one map and z_1 and z_2 are regular points then

$$\text{(1.2.11)} \qquad \operatorname{sgn} \phi''_{\theta\theta}\Big|_{\iota^{-1}z_1}^{\iota^{-1}z_2} = 2\,\mathrm{ind_M}\,\gamma(z_1, z_2)$$

where $\mathrm{ind_M}$ is the Maslov index and $\gamma(z_1, z_2)$ is any curve from z_1 to z_2.

So proposition 1.2.3 states that if all points of a Lagrangian manifold are regular then one can take $r = 0$ in (1.2.1) (we assume that ι is a one-to-one map and that K and M are as above).

Let us consider the case when $\operatorname{rank} \pi'_{\Lambda,x}(z) = k$ with $0 \le k \le d-1$. Then there exists a partition of variables $x = (x', x'') \in \mathbb{R}^k \times \mathbb{R}^{d-k}$ (preceded by a change of enumeration of the coordinates if necessary) such that if $\xi = (\xi', \xi'')$ is the corresponding partition of the dual variables in $T^*_x\mathbb{R}^d$ then the following condition is satisfied at z:

$$\text{(1.2.12)} \qquad \operatorname{rank} \tilde{\pi}'_{\Lambda} = d$$

where $\tilde{\pi}_\Lambda : \Lambda \ni (x, \xi) \to (x', \xi'')$ and the prime means the differential (i.e., the absolute value of the determinant of this matrix is not less than ϵ).

It is then easy to prove that the $r^* \times r^*$ matrix

$$\Phi = \begin{pmatrix} \phi''_{\theta\theta} & \phi''_{\theta x''} \\ \phi''_{x''\theta} & \phi''_{x''x''} \end{pmatrix}$$

with $r^* = r + d - k$ is non-degenerate. Let us consider the function

$$v(x', \xi'') = (2\pi h)^{\frac{k-d}{2}} \int u(x) e^{ih^{-1}\langle x'', \xi''\rangle}\, d\xi''$$

which is a partial h^{-1}-Fourier transform according to Maslov's terminology. Then v is a distribution of the form (1.2.1) with $r, x, \theta, a(x,\theta)$ and $\phi(x,\theta)$ replaced by $r^*, y = (x', \xi'')$, $\theta^* = (\theta, x'')$, $a^*(y, \theta^*) = a(x, \theta)$ and $\phi^*(y, \theta^*) = \phi(x, \theta) - \langle x'', \xi''\rangle$ respectively; surely the new amplitude is not compactly supported with respect to ξ'' but it can easily be improved by multiplying it by a cutting function $\zeta(\xi'')$ with a fixed $\zeta \in C_0^\infty(\mathbb{R}^{d-k})$ supported in the ball of appropriate radius R and equal to 1 in the ball of radius $\frac{1}{2}R$; it is easy to prove by the stationary phase method that the error function V as well as $\langle \xi \rangle^k V$ are negligible. Applying proposition 1.2.3 one can rewrite v in the form (1.2.8) with phase function

$$S(x', \xi'') = \phi(\iota^{-1}\tilde{\pi}_\Lambda^{-1}(x', \xi'')) - \langle x'', \xi''\rangle$$

with x'' calculated at the point $x = \pi_{\Lambda,x}\tilde{\pi}_\Lambda^{-1}(x', \xi'')$ instead of the phase function $\tilde{\phi}$. Hence we obtain

Proposition 1.2.5. *Suppose* (1.2.2)–(1.2.4) *are satisfied with* $K = K(d, r)$, $M = M(d, r, N)$; *let* ι *be a one-to-one map and for all* $z \in \Lambda_\phi$ *suppose condition* (1.2.12) *is fulfilled. Then one can rewrite (modulo negligible functions) a distribution* u *of the form* (1.2.1) *in the same form with* θ *replaced by* ξ'', *with phase function*

$$\text{(1.2.13)} \qquad \psi(x, \xi'') = S(x, \xi'') + \langle x'', \xi''\rangle$$

and with a new amplitude b *satisfying the same inequality for amplitude norms, with the same constants here and in the negligibility condition as before.*

One can easily prove that

(1.2.14) $$\Lambda_\phi = \Lambda_\psi = \{(x,\xi) : x'' = -S'_{\xi''}(x',\xi''), \xi' = S'_{x'}(x'',\xi'')\}.$$

Also any distribution of the form (1.2.1) can be rewritten modulo negligible functions as the sum of distributions of the same form with $r \le d$; recall that amplitudes can be chosen compactly supported.

Remark 1.2.6. *The arguments used in the proofs of propositions 1.2.3 and 1.2.5 imply that the terms in the decomposition of the amplitude $b(x',\xi'')$ are defined by the formula*

(1.2.15) $$b_n = \sum_{0\le j\le n} (\ell_{n-j}a_j)\circ\iota^{-1}\tilde{\pi}_\Lambda^{-1}$$

where ℓ_j are $2j$-th order linear differential operators depending on ϕ. In particular,

(1.2.16) $$b_0 = e^{\frac{i\pi}{4}\operatorname{sgn}\Phi}|\det\Phi|^{-\frac12} a_0\circ\iota^{-1}\tilde{\pi}_\Lambda^{-1}.$$

This implies that classes of distributions defined by different phase functions coincide modulo negligible functions provided these functions generate the same Lagrangian manifold and we take appropriate K and M.

On the other hand, it is well known [Hörmander 1], [Maslov 1] that if Λ is a Lagrangian manifold and if condition (1.2.12) is fulfilled at $z\in\Lambda$ then Λ is generated in a neighborhood of z by the phase function (1.2.13) with S defined uniquely modulo an additive constant.

So we obtain that for a compact subset Λ_0 of the Lagrangian C^M manifold Λ, the spaces $\mathbf{I}_{h,N,k}(\mathbb{R}^d,\Lambda_0,\mathbb{H})$ of Lagrangian distributions are well defined. Here we assume that in the ϵ-neighborhood of each point $z\in\Lambda_0$, the Lagrangian manifold coincides with $\{l_j(x,\xi)=0\quad \forall j=1,\dots,d\}$ with all the derivatives up to order M of the functions l_j not exceeding c and with the absolute value of some d-dimensional minor of the $d\times 2d$ matrix $(l_{jx_k}, l_{j\xi_k})_{j,k=1,\dots,d}$ greater than or equal to ϵ; then locally after an appropriate partition of coordinates in $\mathbb{R}^d\times\mathbb{R}^d$, Λ is defined by (1.2.14) with appropriate function S. $\diamondsuit$

Let us consider the density δ_Σ on the manifold $\Sigma=\Sigma_\phi$ which is the push-forward of the Dirac measure δ in $\mathbb{R}^r$ under the map $(x,\theta)\to\phi'_\theta(x,\theta)$; then

$$\delta_\Sigma = \frac{|dz|}{\left|\frac{D(\lambda,\phi'_{\theta'})}{D(x,\theta)}\right|}$$

where $z=(z_1,\dots,z_d)$ is an arbitrary local coordinate system on Λ continued as C^1 functions into a neighborhood of Λ and $|dz|$ is the Lebesgue measure on Λ. Then (1.2.16) implies that

$$a_0\delta_\Sigma^{\frac12}\circ\iota^{-1} = \tilde{a}_0\delta_{\tilde\Sigma}^{\frac12}\circ\tilde\iota^{-1}$$

provided that in a neighborhood of z the manifold Λ is generated by two phase functions ϕ and $\tilde\phi$ with coinciding r, $\operatorname{sgn}\Phi$ and with coinciding values on Λ [5] and a_0 and $\tilde a_0$ are the leading terms of the symbols a and $\tilde a$, and (a,ϕ) and $(\tilde a,\tilde\phi)$ define the same Lagrangian distribution (modulo negligible functions).

[5] The values of the functions are well defined on Λ since Λ and Σ are identified by means of ι; it is easy to prove that two phase functions with coinciding r, $\operatorname{sgn}\Phi$ and Λ coincide at Λ provided they coincide at z.

More generally, (1.2.16) implies that the semidensity

$$a^0 = e^{\frac{i\pi}{4}\operatorname{sgn}\Phi} a_0 \delta_\Sigma^{\frac{1}{2}} \circ \iota^{-1}$$

does not depend (modulo a factor i^p with $p \in \mathbb{Z}$) on the choice of ϕ generating Λ where here

$$\Phi = \begin{pmatrix} \phi''_{\theta\theta} & \phi''_{\theta x} \\ \phi''_{x\theta} & \phi''_{xx} \end{pmatrix}$$

is a $(d+r) \times (d+r)$ matrix; one can prove (see, e.g., [Hörmander 1]) that the difference $\operatorname{sgn}\Phi - \operatorname{sgn}\tilde{\Phi}$ is the same for three choices of Φ and $\tilde{\Phi}$: this, the previous and $\Phi = (\phi''_{xx})$.

So modulo a factor i^p on Λ there is a well defined $\mathbb{H}$-valued semidensity a^0 which is called the *principal symbol of the Lagrangian distribution* u; surely, the coordinate system is now assumed to be fixed but in section 25.1 of [Hörmander 1] it is proven that if u changes as an $\mathbb{H}$-valued semidensity in $\mathbb{R}^d$ then a_0 also changes as an $\mathbb{H}$-valued semidensity in Λ.

Locally on Λ a branch of i^p can be chosen uniquely but globally this is possible only if

$$\text{(1.2.17)} \qquad \operatorname{ind}_{\mathrm{M}} \gamma \equiv 0 \pmod{\mathbb{Z}}$$

for every closed curve γ on Λ (more precisely, for every class $[\gamma]$ of homotopic closed curves since Maslov indices of homotopic closed curves coincide).

Since $d\phi = \langle \xi, dx \rangle$ on Λ [6] it follows that for a global unique continuous definition of a^0 it is necessary that

$$\text{(1.2.18)} \qquad \int_\gamma \langle \xi, dx \rangle = 0$$

for every closed curve γ on Λ (more precisely, for every class $[\gamma]$ of homotopic closed curves since these integrals for homotopic closed curves coincide): for a Lagrangian manifold the exterior form $\langle \xi, dx \rangle|_\Lambda$ is closed.[7]

On the other hand, under quantization conditions (1.2.17) and (1.2.18) the symbol a^0 can be chosen globally uniquely for u and conversely starting from a^0 one can define $u \in \mathbf{I}_{h,N,k}(\mathbb{R}^d, \Lambda, \mathbb{H})$ with principal symbol a^0. The isomorphic map $a^0 \mapsto u$ is called the *Maslov canonical operator*.

The same arguments imply that if we do not want to consider all the values of $h \in (0, h_0]$ but only a certain sequence tending to $+0$ then we can replace conditions (1.2.17) and (1.2.18) by one quantization condition

$$\text{(1.2.19)} \qquad \frac{1}{4} \operatorname{ind}_{\mathrm{M}} [\gamma] + (2\pi h)^{-1} \int_{[\gamma]} \langle \xi, dx \rangle \equiv 0 \pmod{\mathbb{Z}}$$

for every class $[\gamma]$ of homotopic closed curves and for every h from this sequence.

In conclusion we give some other properties of Lagrangian distributions which can also be proven by the arguments of section 25.1 of [Hörmander 1].

Theorem 1.2.7. *Suppose u is a distribution of the form* (1.2.1) *and ϕ satisfies conditions* (1.2.2)–(1.2.3) *with $M = M(d,r,k,N)$, $a \in \mathrm{S}^0_{h,N}(G, \mathbb{H})$ with $K = K(d,r,k)$. Let us assume that for every $n \le s$ the amplitude a_n has a $(2s - 2n + 1)$-th order zero at*

---

[6] In fact, $d\phi = \langle \phi'_x, dx \rangle + \langle \phi'_\theta, d\theta \rangle = \langle \xi, dx \rangle$.

[7] In fact, $d\langle \xi, dx \rangle = -\sigma$.

Σ_ϕ. Then there exists an amplitude $a' \in \mathrm{S}^0_{h,N}(G,\mathbb{H})$ with $a'_n = 0$ for $n \le s$ such that $\|a'\| \le C\|a\|$ and a' defines u modulo negligible functions.

The terms in the decomposition of a' are given by the formula

$$a'_n = \sum_{0\le j\le n-s-1} \ell_{n-s-1-j} a_j$$

where ℓ_j are $2j$-th order linear differential operators.

Theorem 1.2.8. *Suppose u is a distribution of the form (1.2.1) and ϕ satisfies conditions (1.2.2)–(1.2.3) with $M = M(d,r,k,N)$, $a \in \mathrm{S}^0_{h,N}(G,\mathbb{H})$ with $K = K(d,r,k)$. Let $P \in \mathrm{S}_{h,N}(\mathbb{R}^d \times \mathbb{R}^d, \mathbb{H}', \mathbb{H})$. Then there exists a symbol $b \in \mathrm{S}^0_{h,N}(G,\mathbb{H}')$ such that $\|b\| \le C\|a\| \cdot \|P\|$ (we recall that all norms are calculated in the corresponding classes) and b defines a Lagrangian distribution v with the same phase function coinciding modulo negligible functions with $P(x,hD,h)u$. More precisely, the $L^2(\mathbb{R}^d,\mathbb{H}')$-norm of $(hD)^\alpha(P(x,hD,h)u - v)$ does not exceed $C\|a\| \cdot \|P\|$ for $|\alpha| \le k$. Moreover,*

$$b_0(x,\theta) = P_0(x,\phi'_x(x,\theta))a_0(x,\theta) \tag{1.2.20}$$

and

$$b^0(x,\xi) = P^0(x,\xi)a^0(x,\xi) \quad \text{on } \Lambda. \tag{1.2.20$'$}$$

Finally, if $P^0(x,\xi) = 0$ on Λ then $v = hw$ where w is also a Lagrangian distribution with principal symbol

$$c^0 = -i\{P_0, a^0\} + P^s a^0 \quad \text{on } \Lambda \tag{1.2.21}$$

where the Poisson bracket does not depend on the smooth continuation of a^0 from Λ.

Remark. *Obviously all these assertions remain true if we replace $\mathbb{H}$-valued distributions and symbols (amplitudes) by $\mathcal{L}(\mathbb{H},\mathbb{H}_0)$-valued ones where $\mathbb{H}_0$ is also an auxiliary Hilbert space.* ◊

1.2.2 Fourier integral operators

Let us now proceed to the theory of h-Fourier integral operators. Let Ω_X and Ω_Y be domains in T^*X and T^*Y respectively where $X = \mathbb{R}^d \ni x$ and $Y = \mathbb{R}^d \ni y$ are equipped with symplectic forms $\sigma_X = \sum_j dx_j \wedge d\xi_j$ and $\sigma_Y = \sum_j dy_j \wedge d\eta_j$ respectively. Let $\Psi : \Omega_Y \to \Omega_X$ be a diffeomorphism such that

$$|D^\alpha_{y,\eta}\Psi(y,\eta)| \le c \qquad \forall \alpha : |\alpha| \le M \quad \forall (y,\eta) \in \Omega_Y \tag{1.2.22}$$

and the same inequalities remain true for the inverse diffeomorphism Ψ^{-1} and $(x,\xi) \in \Omega_X$. Then the following assertions are equivalent:

(i) Ψ is a symplectomorphism, i.e., Ψ preserves the symplectic structure ($\Psi_*\sigma_Y = \sigma_X$) or (what is the same) Ψ preserves Poisson brackets ($\{f\circ\Psi, g\circ\Psi\} = \{f,g\}\circ\Psi$ for all functions f,g on Ω_X; one need only check $f,g \in \{x_1,\dots,x_d,\xi_1,\dots,\xi_d\}$).

(ii) $\Lambda = \mathrm{graph}(\Psi)$ is a Lagrangian manifold with respect to the symplectic form $\sigma = \sigma_X - \sigma_Y$ or (what is the same) $\Lambda^T = \{(x,\xi,y,\eta) : (x,\xi,y,-\eta) \in \Lambda\}$ is a Lagrangian manifold with respect to the symplectic form $\sigma^T = \sigma_X + \sigma_Y$.

Definition 1.2.9. Let $\Psi : \Omega_Y \to \Omega_X$ be a symplectomorphism, Λ its graph, Λ_0 a compact subset of Λ with $\mathrm{dist}(\Lambda_0, \Lambda\setminus\Lambda_0) \ge \epsilon > 0$. A *Fourier integral operator* is an operator A with Schwartz kernel $\mathcal{K}_A(x,y)$ such that $(2\pi h)^{d/2}\mathcal{K}_A(x,y) \in \mathbf{I}_{h,N,k}(\mathbb{R}^d \times \mathbb{R}^d, \Lambda_0^T, \mathcal{L}(\mathbb{H}',\mathbb{H}))$. ◊

So, if $\phi(x, y, \theta)$ is a phase function satisfying (1.2.2)–(1.2.3) with (x, y) instead of x and generating Λ^T then

$$\mathcal{K}_A(x,y) = (2\pi h)^{-\frac{r+d}{2}} \int a(x,y,\theta) e^{ih^{-1}\phi(x,y,\theta)} d\theta \tag{1.2.23}$$

with $a \in \mathrm{S}_{h,N}(G_0, \mathcal{L}(\mathbb{H}', \mathbb{H}))$ where G_0 is an admissible domain in $\mathbb{R}^d \times \mathbb{R}^d \times \mathbb{R}^r$.

Then locally

$$\Lambda = \{(x, \phi'_x(x,y,\theta), y, -\phi'_y(x,y,\theta)) : \phi'_\theta(x,y,\theta) = 0\}. \tag{1.2.24}$$

Example 1.2.10. *(i) If the differential of the map* $\Lambda \ni (x,\xi,y,\eta) \to (x,\eta)$ *is non-degenerate then we can take* $\phi(x,y,\theta) = S(x,\eta) - \langle y, \eta\rangle$ *with* $\theta = \eta$ *in which case*

$$\mathcal{K}_A(x,y) = (2\pi h)^{-\frac{d}{2}} \int a(x,y,\eta) e^{ih^{-1}(S(x,\eta) - \langle y,\eta\rangle)} d\eta$$

and $\Lambda = \{(x, S'_x(x,\eta), S'_\eta(x,\eta), \eta)\}$.

(ii) In particular, for $S(x,\eta) = \langle y, \eta\rangle$ *we obtain an h-pseudo-differential operator with compactly supported symbol.*

(iii) The map $\Psi : (y,\eta) = (y', y', \eta', \eta') \to (y', \eta', \eta', -y'')$ *with* $y' \in \mathbb{R}^k, y' \in \mathbb{R}^{d-k}$ *corresponds to the phase function* $\phi(x,y,\eta') = \langle x' - y', \eta'\rangle - \langle x'', y''\rangle$ *and for a symbol equal to 1 we obtain the operator* $F'' = h^{k-d} F_{x'' \to h^{-1}\xi''}$*, the partial h-Fourier transform:*

$$(F''u)(x) = (2\pi h)^{k-d} \int e^{-ih^{-1}\langle x'', y'\rangle} u(x', y'') dy''$$

(surely 1 is not a compactly supported function but we can save this by multiplying F'' *on the left or right by an h-pseudo-differential operator with compactly supported symbol (see theorem 1.2.12(iv) below)).*

(iv) Suppose $\mathcal{U}$ *and* $\mathcal{V}$ *are domains in* X *and* Y *respectively and* $\psi : \mathcal{V} \to \mathcal{U}$ *is a diffeomorphism; let us introduce the operator* $A : u \to P(fug \circ \psi^{-1})$ *with* $f, g \in C_0^M(\mathcal{V})$ *and an h-pseudo-differential operator* P *with compactly supported symbol. Then* A *corresponds to the symplectomorphism* $\Psi : (y,\eta) \to (\psi(y), {}^t\psi'^{-1}(y)\eta)$ *where the notation of section 1.1 is used.*

Egorov's theorem [Hörmander 1] states that every symplectomorphism is locally a composition of those described in (i), (iii) and (iv); moreover, microlocally (see the next subsection) every Fourier integral operator is a product of those described in (i), (iii) and (iv).

(v) The special case of a symplectomorphism described in (i) is a map $(y,\eta) \to (y, \eta - \psi')$ *with a real-valued scalar function* ψ*; the corresponding Fourier integral operator is of the form* $Pe^{ih^{-1}\psi}$ *(or* $e^{ih^{-1}\psi}P$*) where* P *is an h-pseudo-differential operator with compactly supported symbol.* ◊

By the arguments of section 25.2 of [Hörmander 1] it is easy to prove

Theorem 1.2.11. *Let* $\phi(x,y,\theta)$ *be a phase function which satisfies conditions* (1.2.2)–(1.2.3) *and suppose* Λ *is the Lagrangian manifold given by* (1.2.24). *Assume that*

(1.2.25) *At every point* $(x,\xi,y,\eta) \in \Lambda$ *the differentials of the maps* $\Pi_X : \Lambda \ni (x,\xi,y,\eta) \to (x,\xi)$ *and* $\Pi_Y : \Lambda \ni (x,\xi,y,\eta) \to (y,\eta)$ *have rank 2d or (which is the same) at every point* $(x,y,\theta) \in \Sigma_\phi$ *the matrix* $\begin{pmatrix} \phi''_{\theta\theta} & \phi''_{\theta}x \\ \phi''_{x\theta} & \phi''_{xx}\end{pmatrix}$ *is non-degenerate (i.e.,* $|\det \begin{pmatrix} \phi''_{\theta\theta} & \phi''_{\theta x} \\ \phi''_{x\theta} & \phi''_{xx}\end{pmatrix}| \ge \epsilon > 0 \quad \forall (x,y,\theta) \in \Sigma_\phi$*). Suppose* $a \in \mathrm{S}^0_{h,N}(G, \mathcal{L}(\mathbb{H}', \mathbb{H}))$ *with* $N = 0$, $M = K = K(d,r)$ *and* A *is an operator with Schwartz kernel given by* (1.2.23).

Then $\|A\| \le C\|a\|$ *where* $\|.\|$ *means the* $\mathcal{L}(L^2(\mathbb{R}^d,\mathbb{H}'),\mathbb{R}^d,\mathbb{H}))$*-norm and* $C = C(d,r,c,\epsilon)$.[8)]

Applying this theorem it is easy to prove that all the statements of proposition 1.2.5, remark 1.2.6, theorems 1.2.7 and 1.2.8 (with P staying either to the left or to the right of A) remain true for Fourier integral operators; one should only replace (1.2.1) by (1.2.22) and the $L^2(\mathbb{R}^d,\mathbb{H})$-norm by the $\mathcal{L}(L^2(\mathbb{R}^d,\mathbb{H}'),L^2(\mathbb{R}^d,\mathbb{H}))$-norm and in (1.2.20) one should replace $P_0(x,\phi'_x)$ by $P_0(y,-\phi'_y)$ in the case when P is to the right of A; formulas (1.2.20)′ and (1.2.21) should be changed in the same manner.

Moreover, we obtain

Theorem 1.2.12. *(i) Let A be a Fourier integral operator with phase function $\phi(x,y,\theta)$ and symbol $a(x,y,\theta)$. Then the adjoint operator A^* is also a Fourier integral operator with phase function $-\phi(y,x,\theta)$ and symbol $a^\dagger(y,x,\theta)$ and*

$$\Psi_{A^*} = \Psi_A^{-1} \tag{1.2.26}$$

where here and below Ψ_A is the symplectomorphism corresponding to A.

(ii) Let A and B be Fourier integral operators with phase functions $\phi_A(x,y,\theta')$ and $\phi_B(x,y,\theta'')$ and symbols $a(x,y,\theta')$ and $b(x,y,\theta'')$ respectively. Then AB is also a Fourier integral operator with phase function $\phi_A(x,z,\theta')+\phi_B(z,y,\theta'')$ and symbol $a(x,z,\theta')b(z,y,\theta'')$, where $\theta=(\theta',z,\theta'')\in\mathbb{R}^r$, $r=r'+d+r''$. Moreover, if ϕ_A and ϕ_B satisfy (1.2.2), (1.2.3) and (1.2.25) then the new phase function also satisfies those conditions (maybe with other constants c and ϵ) and

$$\Psi_{AB} = \Psi_A \circ \Psi_B. \tag{1.2.27}$$

Moreover, if $a\in \mathrm{S}^0_{h,N}(G'',\mathbb{H}'',\mathbb{H}')$ and $b\in \mathrm{S}^0_{h,N}(G',\mathbb{H}',\mathbb{H})$ respectively and we remove all terms with degree of h greater than N from the product ab then the $\mathcal{L}(L^2(\mathbb{R}^d,\mathbb{H}''),L^2(\mathbb{R}^d,\mathbb{H}))$-norm of the error does not exceed $C\|a\|\cdot\|b\|h^{N+1}$; moreover, footnote [8)] remains true.

(iii) Let A be a Fourier integral operator with phase function $\phi(x,y,\theta')$ which satisfies (1.2.2), (1.2.3) and (1.2.25) and with symbol $a(x,y,\theta')$ and let u be a Lagrangian distribution with phase function $\psi(x,\theta'')$ which satisfies (1.2.2) and (1.2.3) and with amplitude $b(x,\theta'')$. Then Au is a Lagrangian distribution with phase function $\psi'(x,\theta)=\phi(x,z,\theta')+\psi(x,\theta'')$ and amplitude $b(x,\theta)=a(x,z,\theta')b(z,\theta'')$, where $\theta=(\theta',z,\theta'')\in\mathbb{R}^r$, $r=r'+d+r''$. Moreover, $\psi'(x,\theta)$ satisfies (1.2.2) and (1.2.3) and

$$\Lambda_{Au} = \Psi_A(\Lambda_u); \tag{1.2.28}$$

here Λ_u is the Lagrangian manifold corresponding to the Lagrangian distribution u. Moreover, if $a\in \mathrm{S}^0_{h,N}(G'',\mathbb{H}',\mathbb{H})$ and $b\in \mathrm{S}^0_{h,N}(G',\mathbb{H})$ and we remove all terms with degree of h greater than N from the product ab then the $L^2(\mathbb{R}^d,\mathbb{H}')$-norm of the error does not exceed $C\|a\|\cdot\|b\|h^{N+1}$; moreover, the natural modification of footnote [8)] remains true.

(iv) Let A be a Fourier integral operator with phase function ϕ satisfying (1.2.2), (1.2.3) and (1.2.25) with $M=M(d,r,N)$, $a\in \mathrm{S}^0_{h,N}(G,\mathbb{H}',\mathbb{H})$ with $K=K(d,r)$. Moreover, suppose $p\in \mathrm{S}_{h,N}(\mathbb{R}^d\times\mathbb{R}^d,\mathbb{H}'',\mathbb{H}')$ and $q\in \mathrm{S}_{h,N}(\mathbb{R}^d\times\mathbb{R}^d,\mathbb{H},\mathbb{H}'')$. Then there exist symbols $b\in \mathrm{S}^0_{h,N}(G,\mathbb{H}'',\mathbb{H})$ and $b'\in \mathrm{S}^0_{h,N}(G,\mathbb{H}',\mathbb{H}')$ with $\|b\|\le C\|P\|\cdot\|a\|$, $\|b'\|\le C\|P\|\cdot\|a\|$ such that the $\mathcal{L}(L^2(\mathbb{R}^d,\mathbb{H}''),L^2(\mathbb{R}^d,\mathbb{H}))$-norm of the operator $p(x,hD,h)A-B$ and the $\mathcal{L}(L^2(\mathbb{R}^d,\mathbb{H}'),L^2(\mathbb{R}^d,\mathbb{H}''))$-norm of the operator $Aq(x,hD,h)-B'$ do not exceed $C\|P\|\cdot\|a\|h^{N+1}$ and, moreover, footnote [8)] remains true. Here B and B' are Fourier integral operators with phase function ϕ and symbols b and b' respectively.

[8)] Moreover, the same is true for the operator $(hD)^\alpha A(hD)^\beta$ provided $|\alpha|+|\beta|\le k$, M and K are replaced by $K+k$ and $M+k$ in the conditions and C now depends on k.

Moreover,

$$b_0(x,y,\theta) = p_0(x,\phi'_x(x,\theta))a_0(x,y,\theta),\qquad (1.2.29)$$
$$b'_0(x,y,\theta) = a_0(x,y,\theta)q_0(x,-\phi'_y(x,\theta))$$

and hence

$$b^0(x,\xi,y,\eta) = p_0(x,\xi)a^0(x,\xi,y,\eta),\qquad (1.2.29)'$$
$$b'^0(x,\xi,y,\eta) = a^0(x,\xi,y,\eta)q_0(y,\eta)$$

where the principal symbol of a Fourier integral operator equals the principal symbol of its Schwartz kernel at the ^{T}*-transposed point.*

Remark 1.2.13. *Similar formulas also hold for assertions (i) and (ii). Namely, in (ii)*

$$c^0(x,\xi,y,\eta) = a^0(x,\xi,z,\zeta)b^0(z,\zeta,y,\eta)\qquad (1.2.30)$$

for $(x,\xi) = \Psi_A(z,\zeta)$, $(z,\zeta) = \Psi_B(y,\eta)$ *and in (iii)*

$$d^0(x,\xi) = a^0(x,\xi,y,\eta)b^0(y,\eta)\qquad (1.2.30)'$$

for $(x,\xi) = \Psi_A(y,\eta)$ *where* c^0 *and* d^0 *are the principal symbols of* AB *and* Au *respectively.* ◊

In what follows we assume that the domains and ranges of all symbols and operators agree.

Corollary 1.2.14. *(i) Let* A *and* B *be Fourier integral operators with symplectomorphisms* Ψ *and* Ψ^{-1} *respectively and let* $p(x,hD,h)$ *be an* h*-pseudo-differential operator. Then modulo negligible operators* $q(x,hD,h) \equiv Bp(x,hD,h)A$ *is also an* h*-pseudo-differential operator and*

$$q_n = \sum_{0\le j\le n} \ell_{n-j}(p_j)\circ\Psi\qquad (1.2.31)$$

where ℓ_j *are* $2j$*-th degree linear differential operators of the form*

$$\ell_j(q) = \sum_{k,|\alpha|\le 2j} c_{jk\alpha}(D^\alpha_{x,\xi}q)c'_{jk\alpha}$$

with operator-valued coefficients $c_{jk\alpha}$ *and* $c'_{jk\alpha}$ *(for scalar* A *and* B *we obtain scalar differential operators); in particular,*

$$q_0(x,\xi) = b^0((x,\xi),\Psi(x,\xi))p_0(\Psi(x,\xi))a^0(\Psi(x,\xi),(x,\xi)).\qquad (1.2.32)$$

(ii) Hence if the operator A *is elliptic for* (x,ξ,y,η) *with* $(x,\xi) = \Psi_A(y,\eta)$ *and* $(y,\eta)\in\Omega'_Y \Subset \Omega_Y$*, i.e., if* $a^0(x,\xi,y,\eta)$ *is invertible and the norm of the inverse symbol does not exceed c as well as* $\|a\|$ *then there exists a Fourier integral operator* B *corresponding to the symplectomorphism* Ψ_A^{-1} *such that* $BA\equiv I$ *in a neighborhood of* Ω'_Y *(i.e., the equality is true modulo the sum of an* h*-pseudo-differential operator with symbol vanishing in this neighborhood and a negligible operator) and* $AB\equiv I$ *in the neighborhood of* $\Omega'_X = \Psi_A(\Omega_Y)$.

In particular, if A *is a scalar operator and* $P = p(x,hD,h)$ *is an* h*-pseudo-differential operator then in a neighborhood of* Ω_Y *for the principal symbol of* $Q\equiv BPA$ *the equality*

$$q_0 = p_0\circ\Psi\qquad (1.2.33)$$

holds.

Let us note that for a given symplectomorphism Ψ and domain Ω'_Y an elliptic symbol does not necessarily exist because of obstacles of a topological nature. However, such symbols certainly exist (even in the scalar case) provided quantization conditions (1.2.17) and (1.2.18) are fulfilled; in particular, such symbols exist provided Ω'_Y is simply connected. In this case A^*A is an h-pseudo-differential operator with Hermitian symbol and with positive definite principal symbol in the neighborhood of Ω'_Y. Then proposition 1.1.12 implies

Theorem 1.2.15. *Let $\Psi : \Omega_Y \to \Omega_X$ be a symplectomorphism satisfying condition (1.2.22) with graph satisfying (1.2.17) and (1.2.18). Suppose $\Omega'_Y \Subset \Omega_Y$ and let $\Omega'_X = \Psi(\Omega'_Y)$. Then there exists a scalar Fourier integral operator A corresponding to Ψ such that $A^*A \equiv I$ and $AA^* \equiv I$ in neighborhoods of Ω'_Y and Ω'_X respectively.*

One can easily rewrite all these assertions in the uniform form which we used before. Using a partition of unity in $\mathbb{R}^d \times \mathbb{R}^d$ and corollary 1.2.14 one can prove

Theorem 1.2.16. *Let A and B be scalar Fourier integral operators with symplectomorphisms Ψ and Ψ^{-1} respectively and let $p \in \mathrm{S}^0_{h,\rho,\rho,N}(G,\mathbb{H}',\mathbb{H})$, $G \Subset \mathbb{R}^d \times \mathbb{R}^d$ and $h' = h\rho^{-2} \in (0,1]$, $\rho \in (0,1]$. Then there exists $q \in \mathrm{S}^0_{h,\rho,\rho,N}(G',\mathbb{H}',\mathbb{H})$ with $G' = \Psi(G)$ such that $\|\!|q|\!\| \le C\|\!|p|\!\|$ and*

$$\|q(x,hD,h) - Bp(x,hD,h)A\| \le C\rho^{-4d}h'^{N+1}\|\!|p|\!\|$$

with $C = C(d,c,N,G,A,B)$, $K = K(d)$ and $M = M(d,N)$ in the conditions on Ψ, A, B. Moreover, footnote [8)] remains true.

Remark. *One can easily replace the factor ρ^{-4d} by the factor $(\rho^{-2d}\,\mathrm{mes}\,\mathrm{supp}\,P + 1)$ in this estimate. I do not know whether it is possible to replace the factor ρ^{-4d} by 1 here. However we have no need of these improvements.* ◊

Finally, the following statement is well known (see, e.g., [Hörmander 1]):

Theorem 1.2.17. *Let $A \in \boldsymbol{\Psi}_{h,N}(\mathbb{R}^d,\mathbb{H},\mathbb{H})$ and suppose its principal symbol a_0 is real-valued and scalar. Let $P \in \boldsymbol{\Psi}_{h,N}(\mathbb{R}^d,\mathbb{H},\mathbb{H})$ be an operator with compactly supported symbol. Then $e^{ith^{-1}A}P$ and $Pe^{ith^{-1}A}$ are Fourier integral operators (for any $t \in \mathbb{R}$) with symplectomorphism $\Psi = \Phi_t$ where $\Phi_t = \exp(tH_{a_0})$ is the Hamiltonian flow generated by the Hamiltonian field $-H_{a_0}$ where $H_{a_0} = (\partial_\xi a_0, -\partial_x a_0)$.*

In conclusion let us note that the notations for negligible functions and negligible operators do not agree completely: if $h^{d/2}\mathcal{K}_A$ is an (N,k)-negligible function then A is an $(N-\frac{d}{2},k)$-negligible operator; conversely, if A is an (N,k)-negligible operator then $h^{d/2}\mathcal{K}_A$ is an $(N,k-[\frac{d}{2}]-1]$-negligible function. However this fact should not lead to confusion.

1.2.3 Metaplectic operators

Let us consider operators with quadratic symbols:

$$(1.2.34)_1 \qquad A = \sigma(FZ,Z) = \frac{1}{2}\sum_{i,j}\bigl(\alpha_{ij}x_ix_j + \alpha'_{ij}(x_ihD_j + hD_ja_i) + \alpha''_{ij}h^2D_iD_j\bigr),$$

$$(1.2.35)_1 \qquad a = \sigma(Fz,z) = \frac{1}{2}\sum_{i,j}\bigl(\alpha a_{ij}x_ix_j + 2\alpha'_{ij}x_i\xi_j + \alpha''_{ij}\xi_i\xi_j\bigr)$$

where $\sigma(z,z') = \langle \xi, x'\rangle - \langle x, \xi'\rangle$ for $z=(x,\xi)$ and $z'=(x',\xi')$ is the skew-symmetric bilinear *symplectic* form, $Z=(x,hD)$ is a vector-valued operator from $H^1(\mathbb{R}^d)$ to $L^2(\mathbb{R}^d, \mathbb{C}^{2d})$, $F=\begin{pmatrix} -{}^t\alpha' & -\alpha'' \\ \alpha & \alpha' \end{pmatrix}$ is the *skew-Hessian* of a, $\alpha=(\alpha_{ij})$, $\alpha'=(\alpha'_{ij})$, $\alpha''=(\alpha''_{ij})$ are $d\times d$ matrices and (α), (α'') are assumed to be symmetric, and all the coefficients here and below are assumed to be real.

Moreover, let us consider operators with linear symbols:

$$A = \sigma(\ell, Z) = \sum_j (\alpha_i x_i + \alpha'_i hD_i), \tag{1.2.34$_{\frac{1}{2}}$}$$

$$a = \sigma(\ell, z) = \sum_j (\alpha_i x_i + \alpha'_i \xi_i), \tag{1.2.35$_{\frac{1}{2}}$}$$

where $\ell = (-\alpha, \alpha')$, $\alpha = (\alpha_1, \dots, \alpha_d)$, $\alpha' = (\alpha'_1, \dots, \alpha'_d)$ and the subscript means the genuine order of the operator in the context of the "correct" calculus.

One can easily see that

$$(Fz, z') = -\sigma(z, Fz'), \qquad \sigma(H_f, H_g) = \{f, g\}, \tag{1.2.36$_{1-2}$}$$

and

(1.2.37) $H_a = -Fz$, $H_a = -\ell$ for quadratic and linear symbols respectively.

Let A_1 and $\mathrm{A}_{\frac{1}{2}}$ be the spaces of operators with quadratic and linear (real-valued) symbols respectively and let $\mathrm{A}_0 = \mathbb{R}$. Then $\mathrm{A} = \mathrm{A}_1$ and $\tilde{\mathrm{A}} = \mathrm{A}_1 \oplus \mathrm{A}_{\frac{1}{2}} \oplus \mathrm{A}_0$ are Lie algebras with Lie operation $A, B \mapsto ih^{-1}[A,B]$ (moreover, $\tilde{\mathrm{A}}$ is a graded Lie algebra) and

$$\dim \mathrm{A} = d(2d+1), \qquad \dim \tilde{\mathrm{A}} = (d+1)(2d+1). \tag{1.2.38}$$

On the other hand, we have the natural isomorphisms $\mathrm{A}_j \simeq \mathrm{B}_j$, etc., where B_1 is the space of all real $2d\times 2d$ matrices satisfying $(12.4.36)_1$, $\mathrm{B}_{\frac{1}{2}} = \mathbb{R}^d$, $\mathrm{B}_0 = \mathbb{R}$ and we obtain the induced Lie operations $F, F' \mapsto -[F,F']$ and $(F,\ell,c),(F',\ell',c') \mapsto (-[F,F'], -Fl' + F'l, \sigma(l,l'))$ on B, $\tilde{\mathrm{B}}$ respectively.

The following statements are either well known or easily proven (see, e.g., [Leray 1]):

Proposition 1.2.18. *(i) If $A \in \tilde{\mathrm{A}}$ and $\mathrm{D}(A) = \{u \in L^2(\mathbb{R}^d), Au \in L^2(\mathbb{R}^d)\}$ then A is a self-adjoint operator.*

(ii) The sets G and $\tilde{\mathrm{G}}$ of all operators of the form $e^{ih^{-1}A}$ with $A \in \mathrm{A}$, $A \in \tilde{\mathrm{A}}$ are Lie groups and $\dim \mathrm{G} = d(2d+1)$, $\dim \tilde{\mathrm{G}} = (d+1)(2d+1)$.

(iii) The set $\mathcal{G}$ of all matrices of the form $\exp(-F)$ with $F \in \mathrm{B}$ coincides with the Lie group of all symplectic matrices (i.e., matrices S such that $\sigma(Fz, Fz') = \sigma(z, \zeta')$). Moreover, the set $\tilde{\mathcal{G}}$ of all transformations of the form $\exp(-\tilde{F})|_{\mathbb{R}^{2d}}$ with $\tilde{F} \in \tilde{\mathrm{B}}$ coincides with the Lie group of all symplectic affine transformations of $\mathbb{R}^{2d}$ (i.e., transformations of the form $z \to Sz + L$ with $S \in \mathcal{G}$, $L \in \mathbb{R}^{2d}$). For $\tilde{F} = (F, \ell, c)$ we have $S = exp(-F)$, $L = \int_0^1 \exp(-tF)\ell dt$. Moreover, $\dim \mathcal{G} = d(2d+1)$, $\dim \tilde{\mathcal{G}} = d(2d+3)$.

(iv) The isomorphisms $\mathrm{A} \simeq \mathrm{B}$ and $\tilde{\mathrm{A}} \simeq \tilde{\mathrm{B}}$ generate natural maps $\theta : \mathrm{G} \to \mathcal{G}$ where $\tilde{\theta} : \tilde{\mathrm{G}} \to \tilde{\mathcal{G}}$ and θ is two-sheeted covering and for any $g \in \tilde{\mathcal{G}}$, $\tilde{\theta}^{-1} g \simeq \mathbb{S}^1 \subset \mathbb{C}$.

(v) Suppose $G \in \mathrm{G}$ or $G \in \tilde{\mathrm{G}}$ and let P be an h-pseudo-differential operator with compactly supported symbol. Then PG and GP are Fourier integral operators with symplectomorphism $S = \theta(G)$ or $S = \tilde{\theta}(G)$. They can be defined by a quadratic phase function (with linear part in the second case). Moreover,

$$GZ = (SZ)G. \tag{1.2.39}$$

(vi) On the other hand, (1.2.39) *with linear (affine) transformation* S *yields that* $G \in \mathrm{G}$ $(G \in \tilde{\mathrm{G}})$ *and* $S = \theta(G)$ $(S = \tilde{\theta}(G)$ *respectively).*

Definition 1.2.19. Elements of G are *metaplectic operators.* Elements of $\tilde{\mathrm{G}}$ are *generalized metaplectic operators.* ◊

Example 1.2.20. *(i) Partial h-Fourier transforms (with appropriate numerical factor) are metaplectic operators. Operators of the form* $(Gu)(x) = |\det\beta|^{1/2}u(\beta x)$ *with non-degenerate matrix* β *are metaplectic operators. Operators of the form* $(Gu)(x) = e^{\frac{1}{2}ih^{-1}\langle Wx,x\rangle}u(x)$ *with symmetric matrix* W *are metaplectic operators. All metaplectic operators are products of operators of these three types.*

(ii) Operators of the form $(Gu)(x) = e^{ih^{-1}(\langle bx\rangle+\phi)}u(x-a)$ *are (all) generalized metaplectic operators with linear phase functions.*

(iii) All generalized metaplectic operators are products of operators of the form (i) and (ii). ◊

1.2.4 Jets

Lemma 1.1.1 yields the following

Proposition 1.2.21. *Let* $p_j(x,\xi)$ *with* $j = 1,\dots,s-1$ *be scalar linear functions on* T^*X*. Then*

$$(p_1\dots p_s)^w(x,hD) = \frac{1}{s!}\sum_{|J|=s} p^w_{j_1}(x,hD)\dots p^w_{j_s}(x,hD) \tag{1.2.40}$$

where the summation is taken over all the permutations J *(i.e., the right-hand expression is a symmetrized product).*

Moreover $(1.1.14)_3$ implies

Proposition 1.2.22. *Let* $p_j(x,\xi,h)$, $j = 1,\dots,s > 1$ *be scalar symbols and suppose* $p_{j0}(x,\xi)$ *vanish at* $(\bar{x},\bar{\xi}) \in T^*\mathbb{R}^d$*. Let* q *be the Weyl symbol of the symmetric product of* $p^w_1(x,hD,h),\dots,p^w_s(x,hD,h)$*. Then*

$$D^\alpha_{x,\xi}q_n(\bar{x},\bar{\xi}) = 0 \quad \forall n\geq 1, \alpha : |\alpha|+2n\leq s.$$

Remark. *These equalities remain true for other types of quantization only provided* $|\alpha|+2n<s$. ◊

Let $p_j(x,\xi,h)$ be the symbols described in proposition 1.2.22 and let $\tilde{p}_j(x,\xi,h)$ be Weyl symbols of $Bp^w_j(x,hD,h)A$ where A and B are Fourier integral operators corresponding to symplectomorphisms Ψ and Ψ^{-1} respectively such that $AB\equiv I$ and $BA\equiv I$ in $\Omega_X \ni (\bar{x},\bar{\xi})$ and $\Omega_Y \ni (\bar{y},\bar{\eta}) = \Psi^{-1}(\bar{x},\bar{\xi})$ respectively. Then $\tilde{p}_{j0} = p_{j0}\circ\Psi$ vanish at $(\bar{y},\bar{\eta})$. On the other hand,

$$\prod_{0\leq j\leq s}(Bp^w_j(x,hD,h)A) \equiv B(\prod_{0\leq j\leq s}p^w_j(x,hD,h))A$$

and the same is also true for the symmetric product. Applying corollary 1.2.14(i) and proposition 1.2.22 to both sides of the last equality we obtain that the degrees of the operators ℓ_j in (1.2.31) do not exceed $2j-1$ for $j\geq 1$. Moreover the general case of corollary 1.2.14(i) can be easily reduced to the case considered (when A and B are scalar and mutually inverse in the indicated sense) by replacing A and B by $A' = AR_1$ and $B' = R_2B$ respectively with h-pseudo-differential operators R_1 and R_2. So we obtain

Proposition 1.2.23. *For Weyl quantization the ℓ_j in* (1.2.31) *(and hence also in* (1.1.22)*) are $(2j-1)$-th degree operators for $j \geq 1$.*

It is easy to observe that this assertion fails for other types of quantization (it fails even for $j = 1$).

Definition 1.2.24. Let $P = p(x, hD, h)$ be an operator with symbol p such that

$$\partial_x^\alpha \partial_\xi^\beta p_n(\bar{x}, \bar{\xi}) = 0 \quad \forall \alpha, \beta, n : l|\alpha| + r|\beta| + n(l+r) < m \tag{1.2.41}$$

in $\mathbb{R}^d$ where $l > 0, r > 0, (\bar{x}, \bar{\xi}) \in T^*\mathbb{R}^d$ (without loss of generality one can assume that $l + r = 1$). Then the operator

$$q(z, hD_z, h) = \sum_{\alpha, \beta, n: l|\alpha| + r|\beta| + n(l+r) = m} \frac{1}{\alpha!\beta!} (\partial_x^\alpha \partial_\xi^\beta p_n)(\bar{x}, \bar{\xi}) l_{\alpha\beta}(z, hD_z) h^n$$

is called an *(l, r)-jet of P at the point $(\bar{x}, \bar{\xi})$;* here $l_{\alpha\beta}(z, \zeta) = z^\alpha \zeta^\beta$ and we use the same type of quantization for p and $z^\alpha \zeta^\beta$.

It is easy to show that the microlocalization does not depend on the type of quantization. One can also introduce anisotropic microlocalizations (with vector l and r); moreover it is possible to consider the case in which some components of l and r vanish, but we have no need of it.

It is easy to see that the product of operators corresponds to the product of microlocalizations; (l, r) are supposed to be the same and then the degrees (m) sum. Corollary 1.2.14, theorem 1.1.13 and proposition 1.2.23 imply

Proposition 1.2.25. *Let* (1.2.41) *be fulfilled. Then*

(i) If A and B are scalar Fourier integral operators corresponding to symplectomorphisms Ψ and Ψ^{-1} respectively such that $BA \equiv I$ on Ω_Y and $AB \equiv I$ on $\Omega_X \ni (\bar{x}, \bar{\xi})$ then for $l = r$ the operator $\tilde{P} = BPA$ satisfies (1.2.41) *at $(\bar{y}, \bar{\eta}) = \Psi(\bar{x}, \bar{\xi})$ and the jet of $\tilde{P}$ at $(\bar{y}, \bar{\eta})$ is connected to the the jet of P at $(\bar{x}, \bar{\xi})$ by the equality*

$$\tilde{q}(z, hD_z, h) = q(\Psi'_{(\bar{y}, \bar{\eta})}(z, hD_z), h) = G^* q(z, hD_z, h) G \tag{1.2.42}$$

where $\Psi'_{(\bar{y}, \bar{\eta})}$ is the differential of Ψ at $(\bar{y}, \bar{\eta})$ (so $\Psi'_{(\bar{y}, \bar{\eta})}$ is a linear symplectomorphism in $\mathbb{R}^d \times \mathbb{R}^d$) and G is the corresponding metaplectic operator.

(ii) This assertion remains true provided $l < r$ and $Au = u \circ \varkappa$, $Bu = u \circ \varkappa^{-1}$ where $\varkappa$ is a diffeomorphism.[9)]

1.3 Oscillation Front Sets and Related Topics

1.3.1 Principal definitions

In this book we consider various objects depending on a parameter (or parameters) $\mathbf{m} \in \mathbf{M}$ of an arbitrary nature (i.e., we consider families of the individual objects); one of these parameters is the semiclassical parameter $h \in (0, 1]$ [10)] and the exact nature of other parameters is as a rule unessential.[11)] Objects not depending on the parameters

[9)] In this case $(Gu)(x) = |\det \beta|^{\frac{1}{2}} u(\beta x)$ with $\beta = \varkappa'(\bar{y})$.

[10)] Surely, h can run over a certain subset of $(0, 1]$ with an accumulation point 0.

[11)] For example, for h-differential operators one can treat coefficients as parameters belonging to certain functional spaces. On the other hand, we can consider coefficients also depending on h via other parameters.

(i.e., individual objects) are called *fixed* while numbers depending on parameters are also called *parameters*.[12)] If B is a Banach space of individual objects then $\mathbf{B}$ is the Banach space of these objects uniformly bounded (in which case $\|u\|_{\mathbf{B}} = \sup_{\mathbf{m}\in\mathbf{M}} \|u\|_B$); if λ is a (numerical) parameter then $\lambda\mathbf{B}$ is the space of families of the form λu with $u \in \mathbf{B}$ and $\|v\|_{\lambda\mathbf{B}} = \|\lambda^{-1}u\|_{\mathbf{B}}$.

In what follows we speak of objects but mean families of these objects (this carelessness does not lead to confusion because we use the word *fixed* for individual objects).

As *admissible* we consider *tempered* functions, i.e., functions $u \in h^{-L}\mathbf{L}^2(\mathbb{R}^d, \mathbb{H})$ where $\mathbb{H}$, $\mathbb{H}'$, etc., are auxiliary Hilbert spaces. Sometimes we replace $\mathbb{H}$ in this definition by $\mathcal{L}(\mathbb{H}, \mathbb{H}')$ (in what follows we use plain letters instead of bold). Similarly as *admissible* we consider *tempered* operators, i.e., operators $A \in h^{-L}\boldsymbol{\mathcal{L}}(L^2(\mathbb{R}^d, \mathbb{H}), L^2(\mathbb{R}^{d'}, \mathbb{H}'))$.

Functions $u \in h^s\mathbf{L}^2(\mathbb{R}^d, \mathbb{H})$ or $\in h^s\mathbf{L}^2(\mathbb{R}^d, \mathcal{L}(\mathbb{H}, \mathbb{H}'))$ and operators $A \in h^s$ $\boldsymbol{\mathcal{L}}(L^2(\mathbb{R}^d, \mathbb{H}), L^2(\mathbb{R}^{d'}, \mathbb{H}'))$ are called *negligible*; in these cases we write $u \equiv 0$ and $A \equiv 0$ respectively. Exponents L and s enter into these definitions (so in fact we speak about L-admissible or s-negligible functions, etc.).

As an *admissible box* we consider a parallelepiped $\Pi = \{(x, \xi) : |x_j - \bar{x}_j| \le \gamma_j, |\xi_j - \bar{\xi}_j| \le \rho_j \ \forall j = 1, \dots, d\}$ where $(\bar{x}, \bar{\xi})$, $\rho = (\rho_1, \dots, \rho_d)$, $\gamma = (\gamma_1, \dots, \gamma_d)$ are parameters such that

$$\infty \ge \gamma_j > 0, \quad \infty \ge \rho_j > 0, \quad \rho_j\gamma_j \ge h^{1-\delta} \quad \forall j = 1, \dots, d \tag{1.3.1}$$

where here and in what follows δ is always an arbitrarily small positive exponent; so an admissible box is not necessarily fixed.

Definition 1.3.1. (i) Let u be an admissible function and Π an admissible box. Then u *is negligible in* Π ($u \equiv 0$ in Π) if there exists a symbol $\varphi \in \mathrm{S}_{h,\rho,\gamma,N}(\mathbb{R}^d \times \mathbb{R}^d, \mathbb{C})$ with $K = K(d, L, s, \delta)$ in the definition of the class such that $\varphi = I$ in Π and $\varphi(x, hD, h)u \equiv 0$ (this agrees with the definition given in the introduction).

(ii) Similarly, let A be an admissible operator and $\Pi = \Pi' \times \Pi''$ an admissible box. Then A is *negligible in* Π ($A \equiv 0$ in Π) if there exist symbols $\varphi' \in \mathrm{S}_{h,\rho',\gamma',N}(\mathbb{R}^{d'} \times \mathbb{R}^{d'}, \mathbb{C})$ and $\varphi'' \in \mathrm{S}_{h,\rho'',\gamma'',N}(\mathbb{R}^{d''} \times \mathbb{R}^{d''}, \mathbb{C})$ with $K = K(d, L, s, \delta)$ in the definition of the class such that $\varphi' = I$ in Π', $\varphi'' = I$ in Π'' and $\mathrm{Op}(\varphi'')A\,\mathrm{Op}(\varphi') \equiv 0$. ◇

Remark 1.3.2. *(i) Surely these definitions depend on exponents L and s; the results of section 1.1 imply that these definitions do not depend on the choice of K_1 large enough;*

(ii) Though all admissible (negligible) functions are Schwartz kernels of admissible (negligible) operators the converse assertion fails to be true. ◇

Definition 1.3.3. (i) Let u be an admissible function; let us introduce the set $\mathrm{OF}^s(u) \subset T^*\mathbb{R}^d$: $(\bar{x}, \bar{\xi}) \notin \mathrm{OF}^s(u)$ iff $u \equiv 0$ in some fixed box centered in $(\bar{x}, \bar{\xi})$ (so, the point $(\bar{x}, \bar{\xi})$ is also fixed);

(ii) Let A be an admissible operator; let us introduce $\mathrm{OF}^s(A) \subset T^*(\mathbb{R}^{d'} \times \mathbb{R}^{d''})$: $(\bar{x}, \bar{\xi}, \bar{y}, \bar{\eta}) \notin \mathrm{OF}^s(A)$ iff $A \equiv 0$ in some fixed box centered in $(\bar{x}, \bar{\xi}, \bar{y}, \bar{\eta})$.

$\mathrm{OF}^s(.)$ are called *oscillation front sets.* ◇

Remark 1.3.4. *The notion of oscillation front set is rather rough and it would be desirable to give a more refined and sophisticated definition also based on definition 1.3.1 and leading to no difficulty in the advanced development of the theory.*[13)] *However we simply use definition 1.3.1 instead of definition 1.3.3 if the latter is too rough for our needs.* ◇

---

12) There is no room for confusion: fixed numbers are called *constants* and some of these constants are called *exponents*.

13) One can refer to this hope as to Problem 1.1.1**. Maybe non-standard analysis will be useful?

1.3.2 Properties

The results of sections 1.1 and 1.2 imply

Theorem 1.3.5. *(i) Let* $p \in \mathrm{S}_{h,\rho,\gamma,N}(\mathbb{R}^d \times \mathbb{R}^d, \mathbb{H}', \mathbb{H})$ *with* $K = K(d,s,\delta)$*. Then* $P = p(x, hD, h)$ *is negligible in every box*

$$(1.3.2)\quad \Pi = \Pi' \times \Pi'' = \{(x,\xi) : |x_j - \bar{x}_j| \le \gamma'_j, |\xi_j - \bar{\xi}_j| \le \rho'_j \quad \forall j = 1,\dots,d\} \times \\ \{(y,\eta) : |y_j - \bar{y}_j| \le \gamma''_j, |\eta_j - \bar{\eta}_j| \le \rho''_j \quad \forall j = 1,\dots,d\}$$

provided

$$(1.3.3)\qquad \gamma'_j \asymp \gamma''_j, \quad \rho'_j \asymp \rho''_j, \quad \min(\rho'_j, \rho_j)\min(\gamma'_j, \gamma_j) \ge h^{1-\delta} \quad \forall j,$$

$$(1.3.4)\quad \sum_{1\le j\le d} \Bigl(\frac{1}{\gamma_j}(|x_j - y_j| + |x_j - z_j|) + \frac{1}{\rho_j}(|\xi_j - \eta_j| + |\xi_j - \zeta_j|)\Bigr) \ge \epsilon \\ \forall (x,\xi) \in \Pi', \quad \forall (y,\eta) \in \Pi'', \quad \forall (z,\zeta) \in \operatorname{supp} p;$$

recall that ϵ *is an arbitrary positive constant.*

(ii) Let A *be a Fourier integral operator corresponding to the symplectomorphism* Ψ *and let all the conditions on phase functions and symbols be fulfilled with* $M = K = K(d,r,s,\delta)$ *uniformly with respect to all the parameters. Then* A *is negligible in every box of the form* (1.3.2) *such that*

$$(1.3.5)\qquad \gamma'_j \asymp \gamma''_j \asymp \gamma, \quad \rho'_j \asymp \rho''_j \asymp \rho, \quad \gamma = \rho \ge \epsilon h^{\frac{1}{2}(1-\delta)},$$

$$(1.3.6)\quad \operatorname{dist}((x,\xi), \Psi(y,\eta)) + \operatorname{dist}((x,\xi), (z, \phi'_z(z,z',\theta)) \ge \epsilon\rho \\ \forall (x,\xi) \in \Pi', \quad \forall (y,\eta) \in \Pi'', \quad \forall (z,z',\theta) \in \operatorname{supp} a \cap \{|\phi'_\theta(z,z',\theta)| \le \epsilon\rho\}.$$

(iii) Let $Au = (fu)\circ \varkappa^{-1}$ *where* $\varkappa$ *is a* C^K *diffeomorphism from* $Y \subset \mathbb{R}^d$ *to* $X \subset \mathbb{R}^d$ *and* $f \in C_0^K(Y)$ *uniformly with respect to all the parameters,* $K = K(d,s,\delta)$*. Then* A *is negligible in every box of the form* (1.3.2) *such that*

$$(1.3.7)\qquad \gamma'_j \asymp \gamma''_j \asymp \gamma,\ \rho'_j \asymp \rho''_j \asymp \rho,\ \gamma \le \rho,\ \min(\gamma,1)\min(\rho,1) \ge h^{1-\delta},$$

$$(1.3.8)\quad \frac{1}{\gamma}(|x - \varkappa(y)| + |y - z|) + \frac{1}{\rho}|{}^t\varkappa'\xi - \eta| \ge \epsilon \\ \forall (x,\xi) \in \Pi', \quad \forall (y,\eta) \in \Pi'', \quad \forall z \in \operatorname{supp} f.$$

(iv) Let u *be a Lagrangian distribution with Lagrangian manifold* Λ *and let all the conditions on* Λ*, phase functions and amplitudes be fulfilled with* $M = K = K(d,r,s,\delta)$ *uniformly with respect to all the parameters. Then* u *is negligible in every box*

$$(1.3.9)\qquad \Pi = \{(x,\xi) : |x_j - \bar{x}_j| \le \gamma'_j, |\xi_j - \bar{\xi}_j| \le \rho'_j \quad \forall j = 1,\dots,d\}$$

provided

$$(1.3.10)\qquad \gamma'_j \asymp \gamma, \quad \rho'_j \asymp \rho \quad \forall j, \quad \gamma = \rho \ge \epsilon h^{\frac{1}{2}(1-\delta)},$$

$$\text{(1.3.11)}\quad \operatorname{dist}((x,\xi),(z,\phi'_z(z,\theta)))\ge\epsilon\rho \qquad \forall(x,\xi)\in\Pi\quad\forall(z,\theta)\in\operatorname{supp}a\cap\{|\phi'_\theta(z,\theta)|\le\epsilon\rho\}.$$

(v) Moreover, if $\Lambda=\{x'=\xi''=0\}$ where $x=(x';x'')=(x_1,\dots,x_{d'};\,x_{d'+1},\dots,x_d)$, etc., with $0\le d'\le d$ then u is negligible in every box of the form (1.3.9) *such that*

$$\text{(1.3.12)}\qquad \min(\rho_j,1)\min(\gamma_j,1)\ge\epsilon h^{1-\delta}\quad\forall j,$$

$$\text{(1.3.13)}\qquad \sum_{1\le j\le d'}\frac{1}{\gamma_j}|x_j|+\sum_{d'+1\le j\le d}\frac{1}{\rho_j}|\xi_j|\ge\epsilon.$$

Using assertion (v) one can easily find boxes in which the operators A: $(Au)(x')=(f(hD)u)(x',0)$ and B: $(Bu)(x')=\int f(x)u(x)|_{x=(x',x'')}dx''$ with compactly supported smooth function f are negligible (see also theorem 1.3.6(ix),(x)). It is also easy to prove that oscillation front sets have the following properties similar to properties of wave front sets:

Theorem 1.3.6. *(i) $\mathrm{OF}^s(.)$ is a closed set;*

(ii) If A and u are admissible with $L=0$ and either A or u is compactly supported[14] *then*

$$\text{(1.3.14)}\quad \mathrm{OF}^s(Au)\subset\mathrm{OF}^s(A)\circ\mathrm{OF}^s(u)= \{(x,\xi:\exists(y,\eta)\in\mathrm{OF}^s(u):(x,\xi,y,\eta)\in\mathrm{OF}^s(A)\};$$

(iii) If A and B are admissible with $L=0$ and either A or B is compactly supported then

$$\text{(1.3.15)}\quad \mathrm{OF}^s(AB)\subset\mathrm{OF}^s(A)\circ\mathrm{OF}^s(B)= \{(x,\xi,z,\zeta):\exists(y,\eta):(x,\xi,y,\eta)\in\mathrm{OF}^s(A),(y,\eta,z,\zeta)\in\mathrm{OF}^s(B)\};$$

(iv) If u and v are admissible with $L=0$ and $u\otimes v(x)=u(x')v(x'')$ in the notation of theorem 1.3.5(v) then

$$\text{(1.3.16)}\qquad \mathrm{OF}^s(u\otimes v)=\mathrm{OF}^s(u)\times\mathrm{OF}^s(v);$$

(v) If $P=p(x,hD,h)$ with $p\in\mathrm{S}_{h,\rho,\gamma,N}(\mathbb{R}^d\times\mathbb{R}^d,\mathbb{H}',\mathbb{H})$ with $K=K(d,s,\delta)$ and if

$$\min(1,\rho_j)\min(1,\gamma_j)\ge\epsilon h^{1-\delta}\qquad\forall j$$

then

$$\text{(1.3.17)}\qquad \mathrm{OF}^s(P)=\{(x,\xi,x,\xi):(x,\xi)\in\operatorname{supp}p\}$$

and for every admissible function u

$$\text{(1.3.18)}\qquad \mathrm{OF}^s(Pu)\subset\mathrm{OF}^s(u)\subset\mathrm{OF}^s(Pu)\cup\mathrm{Char}(P)$$

provided K also depends on L and $\mathrm{Char}(P)$ is the set of all the points (x,ξ) at which the symbol $p(x,\xi,0)$ is not invertible with the norm of the inverse symbol not exceeding C;

(vi) The following inclusions hold:

---

[14] The operator A is called compactly supported if for every fixed $K\Subset\mathbb{R}^{d'}$ there exists fixed $K'\Subset\mathbb{R}^{d''}$ such that $\operatorname{supp}u\subset K$ implies that $\operatorname{supp}Au\subset K'$, and if the adjoint operator A^* has the same property.

$$\mathrm{OF}^s(A) \subset \mathrm{OF}^s(\mathcal{K}_A)^T \tag{1.3.19}$$

and for admissible $\mathcal{K}_A$

$$\mathrm{OF}^s(\mathcal{K}_A)^T \subset \mathrm{OF}^{s+\frac{d}{4}}(A) \tag{1.3.20}$$

with $d = d' + d''$;

(vii) If u *is a Lagrangian distribution with Lagrangian manifold* Λ *then*

$$\mathrm{OF}^s(u) \subset \Lambda \tag{1.3.21}$$

provided $M = K = K(d, r, s)$;

(viii) The same inclusion is true for a Fourier integral operator with Λ *replaced by* Λ^T;

(ix) If $a \in \mathrm{S}_{h,N}(\mathbb{R}^d \times \mathbb{R}^d, \mathbb{H}', \mathbb{H})$ *with* $K = K(d, L, s)$, $\mathrm{supp}(a) \subset \{|\xi''| \le c\}$ *and* $v(x') = (a(x, hD, h)u)(x', 0)$ *then*

$$\mathrm{OF}^s(v) \subset \{(x', \xi') : \exists \xi'' : (x', 0, \xi', \xi'') \in \mathrm{OF}^{s+\frac{d''}{2}}(u)\} \tag{1.3.22}$$

in the notation of theorem 1.3.5(iv);

(x) If $a \in \mathrm{S}_{h,N}(\mathbb{R}^d \times \mathbb{R}^d, \mathbb{H}', \mathbb{H})$ *with* $K = K(d, L, s)$, $\mathrm{supp}\, a \subset \{|x''| \le c\}$ *and* $v(x') = \int_{\mathbb{R}^{d''}} (a(x, hD, h)u)(x', x'')dx''$ *then*

$$\mathrm{OF}^s(v) \subset \{(x', \xi') : \exists x'' : (x', x'', \xi', 0) \in \mathrm{OF}^s(u)\}. \tag{1.3.23}$$

Remark 1.3.7. *(i) If* $u \in L^2(X, \mathbb{H})$ *with* $X \subset \mathbb{R}^d$ *then definition 1.3.1(i) remains reasonable provided*

$$\Pi_{\epsilon x} = \{|x_j - \bar{x}_j| \le (1 + \epsilon)\gamma_j \quad \forall j\} \subset X$$

with some small positive constant ϵ*; in this case one should replace the cutting operator* $\varphi(x, hD, h)$ *by* $\varphi(x, hD, h)\psi(x)$ *where* $\psi = 1$ *in* Π, ψ *is supported in* $\Pi_{\epsilon x}$ *and belongs to the same symbol class. So definition 1.3.3(i) remains reasonable provided* $\mathrm{dist}(\bar{x}, \complement X) \ge \epsilon$;

(ii) If $A \in \mathcal{L}(L^2(Y, \mathbb{H}'), L^2(X, \mathbb{H}))$ *with* $X \subset \mathbb{R}^d$ *and* $Y \subset \mathbb{R}^{d'}$ *then definition 1.3.1(ii) remains reasonable provided* $\Pi'_{\epsilon x} \subset X$ *and* $\Pi''_{\epsilon y} \subset Y$*; one should only take cutting operators of the form* $(\mathrm{Op}(\varphi'))\psi'$ *and* $\psi'' \mathrm{Op}(\varphi'')$*. So definition 1.3.3(ii) remains reasonable provided* $\mathrm{dist}(\bar{x}, \complement X) \ge \epsilon$ *and* $\mathrm{dist}(\bar{y}, \complement Y) \ge \epsilon$.

(iii) Let $v \in L^2(X, \mathbb{H})$ *be a fixed function. Let* $u = u_h = \phi(h|D|)v$ *be a family of functions where* $\phi \in C_0^\infty(\mathbb{R}^+)$, $\phi \ne 0$*. Then* $(\mathrm{OF}(u))_{\mathrm{cone}} = \mathrm{WF}(v)$ *where* M_{cone} *is the cone hull (with respect to* ξ*) of the set* M *and* $\mathrm{WF}(v)$ *is the standard wave front set.* ◊

1.4 Functional Calculus of h-Pseudo-Differential Operators

The results of this section will not be used before section 4.7, so references to sections up to 4.7 are permissible. Here I construct two variants of functional calculus: for general operators and analytic functions and for Hermitian operators and smooth functions. I consider only functions of one variable but one can easily extend these calculi to consider functions of several variables (in which case one should assume that the operators commute).

1.4.1 Analytic theory

Let $A = a(x, hD, h)$ with $a \in \mathrm{S}_{h,N}(\mathbb{R}^d \times \mathbb{R}^d, \mathbb{H}, \mathbb{H})$ and let us consider the operator $f(A)$ where $f(\cdot)$ is an analytic function in the domain $G \supset \operatorname{Spec}(A)$. The standard definition is

$$f(A) = \frac{1}{2\pi i}\int_\Gamma (\tau - A)^{-1} f(\tau)d\tau \tag{1.4.1}$$

where here $\Gamma = \partial G'$, $\operatorname{Spec}(A) \subset G' \Subset G$ and Γ is oriented counter-clockwise.

The standard parametrix construction implies that

(1.4.2) $\operatorname{Spec}(A)$ is contained in the η-neighborhood of

$$\Sigma_A = \overline{\cup_{(x,\xi)} \operatorname{Spec}(a_0(x,\xi))} \qquad \Longleftarrow \qquad h \in (0, \bar{h}]$$

where $a_0(x,\xi) = a(x,\xi,0)$ and

$$\eta = \epsilon, \qquad \bar{h} = \epsilon_1 \tag{1.4.3}$$

where here and below $\epsilon > 0$ is an arbitrarily small constant, $\epsilon_1 = \epsilon_1(d, |\!|\!|a|\!|\!|_1, \epsilon) > 0$ is also a small constant and $|\!|\!|.|\!|\!|_\gamma$ means the $\mathrm{S}_{h,\gamma,\gamma,N}$-norm.

Moreover, in this case

$$\|(\tau - A)^{-1} - \operatorname{Op}(E_\tau)\| \le Ch^s, \tag{1.4.4}$$

and

$$|\!|\!|E_\tau|\!|\!|_1 \le C, \quad |\!|\!|E_\tau - (\tau - a_0)^{-1}|\!|\!|_1 \le Ch \tag{1.4.5}$$

provided

$$\tau \in \mathbb{C}, \quad \operatorname{dist}(\tau, \Sigma_A) \ge \eta \tag{1.4.6}$$

for an appropriate $\mathrm{S}_{h,\gamma,\gamma,N}$-valued function $E_\tau = E_\tau(x,\xi,h)$ analytic with respect to τ, where $C = C(d, N, s, |\!|\!|a|\!|\!|_1, \epsilon)$ and $K = K(d,s)$, $\gamma = 1$ in the definition of the symbol classes.

Hence we obtain

Proposition 1.4.1. *Let $A = a(x, hD, h)$ with $a \in \mathrm{S}_{h,N}(\mathbb{R}^d \times \mathbb{R}^d, \mathbb{H}, \mathbb{H})$ and let f be a function analytic in the domain $G \Supset G' \Supset \Sigma_A$. Moreover, let*

$$\Gamma = \partial G', \qquad \operatorname{dist}(\Gamma, \Sigma_A) \ge \eta, \tag{1.4.7}$$

and suppose (1.4.3) *holds. Then for $h \in (0, \bar{h}]$*

$$\|f(A) - \operatorname{Op}(F)\| \le CMh^s \tag{1.4.8}$$

and

$$|\!|\!|F|\!|\!|_1 \le CM, \qquad |\!|\!|F - f(a_0)|\!|\!|_1 \le CMh \tag{1.4.9}$$

for an appropriate $F \in \mathrm{S}_{h,N}(\mathbb{R}^d \times \mathbb{R}^d, \mathbb{H}, \mathbb{H})$ where

$$M = \|f\|_{L^1(\Gamma)}. \tag{1.4.10}$$

Let us now assume that

$$\|(\tau - a_0(x,\xi))^{-1}\| \le c\eta^{-p} \quad \forall \tau \in \mathbb{C} : \mathrm{dist}(\tau, \Sigma_A) \ge \eta \tag{1.4.11}$$

with a small parameter $\eta \in (0,1]$ and an exponent $p \ge 1$.

Remark 1.4.2. *Condition (1.4.11) is satisfied provided either*

$$\mathbb{H} = \mathbb{C}^r \otimes \mathbb{H}_0, \quad a_0(x,\xi) = b(x,\xi) \otimes I, \quad p = r \tag{1.4.12}$$

or

$$a_0(x,\xi)^\dagger = a_0(x,\xi) \quad \forall (x,\xi), \quad p = 1. \tag{1.4.13}$$

Then (1.4.2) remains true provided

$$\eta \ge h^\nu, \quad \bar{h} = \epsilon_1, \quad \nu = \frac{1}{2p} - \delta \tag{1.4.3$'$}$$

where the exponent $\delta > 0$ is arbitrarily small and K and all the constants depend on δ.◊

In this case (1.4.4) remains true and

$$\|E_\tau\|_\gamma \le C\gamma^{-1}, \qquad \|E_\tau - (\tau - a_0)^{-1}\|_\gamma \le Ch\gamma^{-3} \tag{1.4.5$'$}$$

where $\gamma = \eta^p$. Hence (1.4.8) also remains true and

$$\|F\|_\gamma \le CM\gamma^{-1}, \qquad \|F - f(a_0)\|_\gamma \le CMh\gamma^{-3}. \tag{1.4.9$'$}$$

So we have proven

Proposition 1.4.1$'$. *Suppose the conditions of proposition 1.4.1 are satisfied with* (1.4.3) *replaced by* (1.4.3)$'$ *and suppose* (1.4.11) *holds. Then* (1.4.8) *and* (1.4.9)$'$ *hold.*

In particular, let us assume that (1.4.13) holds; then $p = 1$, $\gamma = \eta$. Consider $f(\tau) = e^{it\tau}$ with a large parameter t; take $G' = \{\tau : |\mathrm{Re}\,\tau| \le c, |\mathrm{Im}\,\tau| \le \gamma\}$ with $\gamma = \langle t\rangle^{-1}$ where $\langle t\rangle = (t^2+1)^{1/2}$ and assume here that $\Sigma_A \subset [-\frac{1}{2}c, \frac{1}{2}c]$. We obtain that $M \asymp 1$ and hence

$$\|e^{itA} - \mathrm{Op}(F_t)\| \le Ch^s, \tag{1.4.14}$$

where

$$\|F_t\|_\gamma \le C\langle t\rangle, \qquad \|F_t - e^{ita_0}\|_\gamma \le Ch\langle t\rangle^3 \tag{1.4.15}$$

provided $t \in \mathbb{R}, |t| \le h^{(\delta-1)/2}, \gamma = \langle t\rangle^{-1}$.

Moreover, note that the terms $E_{\tau n}$ in the decomposition of the symbol E are sums of terms of the form

$$(\tau - a_0)^{-1} b_1 (\tau - a_0)^{-1} \dots b_{j-1} (\tau - a_0)^{-1} \tag{1.4.16}$$

with $j \le 2n+1$, $b_k = b_{njk} \in \mathrm{S}_{h,N}(\mathbb{R}^d \times \mathbb{R}^d, \mathbb{H}, \mathbb{H})$ uniformly with respect to all the parameters and the number of these terms does not exceed $K_0(n)$. Then we obtain that the terms F_{tn} in the decomposition of the symbol F_t are sums of terms of the form $g(t) * b_1 g(t) * \dots b_{j-1} G(t)$ where "$*$" means convolution and

$$g(t) = \frac{1}{2\pi i} \int_\Gamma (\tau - a_0)^{-1} e^{it\tau} d\tau = e^{ita_0};$$

hence F_{tn} is a sum of terms of the form

$$t^{j-1}\int\cdots\int_{\Delta_{j-1}} e^{itz_1a_0}b_1e^{itz_2a_0}\dots b_{j-1}e^{itz_ja_0}dz' \tag{1.4.17}$$

with $z_j = 1 - z_1 - \dots - z_{j-1}$, $dz' = dz_1\dots dz_{j-1}$, $\Delta_{j-1} = \{z' \in \mathbb{R}^{j-1} : z_k > 0 \quad \forall k, z_1 + \dots z_{j-1} < 1\}$. Recall that $j \le 2n+1$.

Note that for any first-order differentiation ∂ the following formula holds:

$$\partial e^{ita_0} = \int_0^t e^{it'a_0}(i\partial a_0)e^{i(t-t')a_0}dt'. \tag{1.4.18}$$

Applying this formula to (1.4.17) we easily obtain that

$$\|F_t\|_\gamma \le C, \qquad \|F_t - e^{ita_0}\|_\gamma \le Ch\gamma^{-2} \tag{1.4.15*}$$

provided $|t| \le h^{(\delta-1)/2}, \gamma = \langle t\rangle^{-1}$.

So we have improved estimates (1.4.15).

Note that for a scalar symbol a_0 the symbol F_t is a symbol of the form

$$e^{ita_0}\sum_{0\le n\le N,\, j\le 2n} b_{nj}t^j.$$

1.4.2 Self-adjoint theory

Now suppose $A^* = A$ and $f \in C^K(\mathbb{R})$ satisfies

$$\langle\tau\rangle^2|D_\tau^j f(\tau)| \le M\gamma^{-j} \quad \forall j \le K \qquad \text{with } \gamma \in [h^{\frac12(1-\delta)}, 1],\ \delta > 0. \tag{1.4.19}$$

Then $f(A)$ is given by the formula

$$f(A) = \int_{\mathbb{R}} \hat f(t)e^{itA}dt \tag{1.4.20}$$

where (1.4.19) implies that

$$|\hat f(t)| \le CM(1+\gamma|t|)^{-K}. \tag{1.4.21}$$

Consider integral (1.4.20) separately on $[-h^{(\delta'-1)/2}, h^{(\delta'-1)/2}]$ and on the remaining part of $\mathbb{R}$ where $\delta' > 0$ is a small enough exponent; then (1.4.21) implies that the operator norm of the second integral does not exceed CMh^s while in the first integral we can replace e^{itA} by $\mathrm{Op}(F_t)$ and then take the integral on $\mathbb{R}$ again; the operator norm of the error also does not exceed CMh^s. We then obtain the sum of terms of the form

$$h^n\int_{\mathbb{R}} \mathcal{J}_{nj}(\tau)f(\tau)d\tau \tag{1.4.22}$$

where $\mathcal{J}_{nj}(\tau) = \mathrm{Res}_{\mathbb{R}} J_{nj} = \frac{1}{2\pi i}(J_{nj}(\tau+i0) - J_{nj}(\tau-i0))$ and $J_{nj}(\tau)$ is an operator-valued function of the form (1.4.16). Hence $\mathcal{J}_{nj}(\tau)$ is a distribution supported in $\mathcal{C} = \{(\tau,x,\xi) : \tau \in \mathrm{Spec}(a_0(x,\xi))\}$; so $\mathrm{supp}\,\mathcal{J}_{nj} \subset \{|\tau| \le C_0\}$. Moreover $|J_{nj}(\tau)| \le C|\mathrm{Im}\,\tau|^{-j} \quad \forall\tau \in \mathbb{C}$ and hence

$$\left|\int J_{nj}(\tau \pm i0)f(\tau)d\tau\right| \le C\|f\|_{C^j} \tag{1.4.23}$$

for every $f \in C_0^j([-2c, 2c])$. Therefore (1.4.18) implies that the term (1.4.22) does not exceed $Ch^n\gamma^{-j}$. Moreover, applying $D_{x,\xi}^\alpha$ to (1.4.16) we obtain a function of the same type with $j + |\alpha|$ instead of j. So we obtain the first assertion of the following statement:

Proposition 1.4.3. *(i) If conditions* (1.4.13) *and* (1.4.19) *are satisfied then* (1.4.8) *and* (1.4.9)$'$ *remain true.*

(ii) Moreover, if both conditions (1.4.12) *and* (1.4.13) *are satisfied then*

$$\|\!|F|\!\|_\gamma \le CM, \qquad \|\!|F - f(a_0)|\!\|_\gamma \le CMh\gamma^{-2}; \tag{1.4.9$''$}$$

in this case C depends on r.

The proof of the second assertion is based on the fact that in this case

$$\int_{-2c}^{2c} |J_{nj}(\tau \pm i\eta)|d\tau \le C\eta^{1-j}$$

provided $j > 1$ and hence one can replace j by $j - 1$ in the right-hand expression in (1.4.23).

Remark 1.4.4. *Instead of (1.4.19) we can assume that $f_1 = \langle\tau\rangle^{-2l} f \in C_b^K$ and obtain proposition 1.4.3 with $M = \|f_1\|_{C_b^K}$ where*

$$C_b^K = \{g : \|g\|_K = \sum_{|\alpha|\le K} \sup|D^\alpha g| < \infty\};$$

surely in this case K and C also depend on l. In fact, $f(A) = (A^2 + I)^l f_1(A)$ where f_1 satisfies (1.4.19) with $\gamma = 1$. ◊

1.4.3 Subprincipal symbols

Let us consider the case of a scalar symbol a_0 and show the equality of subprincipal symbols:

$$f(A)^s = f'(a_0)a^s \tag{1.4.24}$$

where $f'(\tau) = df(\tau)/d\tau$. Our previous calculations show that it is sufficient to prove this formula for $f(\tau) = (\tau - z)^{-1}$ and moreover it is sufficient to consider the case $z = 0$ only. Let $B = \mathrm{Op}(b)$ be a parametrix of A; then $b_0 = a_0^{-1}$, $\{b_0, a_0\} = 0$ and hence $BA \equiv I$ implies that $b^s a_0 + b_0 a^s = 0$ and we obtain that $b^s = -a_0^{-2}a^s$ which coincides with (1.4.24) for this choice of f.

1.4.4 Functional calculus for other classes of PDO

Let us now assume that $a \in \bar{\mathrm{S}}^m_{h,N}(\mathbb{R}^d \times \mathbb{R}^d, \mathbb{H}, \mathbb{H})$ and (1.4.13) is satisfied. Moreover, let us assume that in the domain $\Omega \subset \{(x,\xi) : |\xi| \le c\}$ the symbol a coincides with the symbol $a' \in \mathrm{S}_{h,N}(\mathbb{R}^d \times \mathbb{R}^d, \mathbb{H}, \mathbb{H})$ which also satisfies (1.4.13). Moreover suppose $q' \in \mathrm{S}_{h,N}(\mathbb{R}^d \times \mathbb{R}^d, \mathbb{H}, \mathbb{H})$ is supported in $\Omega' \Subset \Omega$ and $\mathrm{dist}(\Omega', \complement\Omega) \ge \epsilon_0$. Then proposition 2.1.9 implies that for $|t| \le \epsilon h^{-1}$

$$\|(e^{itA} - e^{itA'})Q\| \le Ch^s \|\!|q|\!\|$$

where $A = \mathrm{Op}(a), A' = \mathrm{Op}(a'), Q = \mathrm{Op}(q)$ and ϵ, C depend on d, c, ϵ_0 and the norm of a and C depends on N, s and the norm of a' also. Then the construction used in subsection 1.4.2 yields

$$\|(f(A) - f(A'))Q\| \le Ch^s \|\!|q|\!\|.$$

So we constructed $f(A)$ modulo negligible operators outside of $\Omega' \times \mathbb{R}^{2d} \cup \mathbb{R}^{2d} \times \Omega'$.

Let us now assume that

$$|a_0(x,\xi)w| \geq \epsilon_0 |\xi|^m |w| \qquad \forall w \in \mathbb{H} \quad \forall (x,\xi) \in \mathbb{R}^{2d} : |\xi| \geq c \tag{1.4.25}$$

and let us consider $f \in C_0^K([-c_0, c_0])$; choose c_1 large enough and let $\operatorname{supp}(1-q) \subset \{|\xi| \geq \frac{1}{2}c\}$ and $\chi \in C_0^K([-1,1]), \chi_T(t) = \chi(\frac{t}{T})$ with small enough T. Then the standard parametrix construction implies that the operator $f(hD_t)\chi_T(t)\exp(ih^{-1}tA)(I-Q)$ is negligible. On the other hand, modulo negligible operators this operator equals

$$\sum_{0 \leq n \leq N} \frac{1}{n!}(-i\frac{h}{T})^n \chi_T^{(n)} f^{(n)}(hD_t)\exp(ih^{-1}tA)(I-Q) =$$

$$\sum_{0 \leq n \leq N} \frac{1}{n!}(-i\frac{h}{T})^n \chi_T^{(n)} f^{(n)}(A)\exp(ih^{-1}tA)(I-Q);$$

then taking $\chi = 1$ in $(-\frac{1}{2}, \frac{1}{2})$ and substituting $t = 0$ we obtain that $f(A)(I-Q) \equiv 0$ and hence $f(A) = f(A)Q \in \mathrm{S}_{h,K}$.

We do not need to extend this calculus.

Comments

I have sent the reader to the book [Hörmander 1] for details, proofs and advanced results. On the other hand, there are many remarkable properties of Weyl quantization related to the formula

$$\mathrm{Op}^{\mathrm{w}}(e^{i(\langle \alpha, x\rangle + \langle \beta, \xi \rangle)}) = e^{i(\langle \alpha, x\rangle + \langle \beta, hD_x\rangle)}$$

discussed in [Shubin 1]. Other types of symbols are also considered there.

2. Propagation of Singularities in the Interior of a Domain

In this chapter I consider microlocal propagation of singularities (i.e., oscillation front sets) in the interior of a domain. In general in this book I consider two types of microlocal operator, namely, microelliptic and microhyperbolic. For microelliptic operators there is no propagation of singularities at all and I prove the different variants of this fact which I need by means of either a parametrix construction or the Gårding inequality; sometimes I combine both methods. In this chapter I treat microhyperbolic operators. In section 2.1 I prove the main general theorem and certain of its corollaries which will be necessary for spectral asymptotics; this theorem is formulated in terms of auxiliary real-valued functions $\phi_j(x,\xi)$ with $j = 1,\dots,J$. Then in section 2.2 I reformulate this theorem in terms of bicharacteristics and generalized bicharacteristics. Finally, in section 2.3 I study the propagation of singularities for scalar and related operators for a large time interval by quite different methods.

In section 4.7 and in Chapters 6, 7 and 12 other very long-time propagation results will be obtained by the same and other methods.

2.1 General Theory

2.1.1 Principal theorem

Let us start with the key notion and key statement of this chapter:

Definition 2.1.1. Let $p \in \mathrm{S}_{h,0}(T^*\mathbb{R}^d, \mathbb{H}, \mathbb{H})$ be a Hermitian symbol; then p is *microhyperbolic at the point* $z = (x,\xi) \in T^*\mathbb{R}^d$ *in the direction* $\mathcal{T} = (\mathcal{T}_x, \mathcal{T}_\xi) \in T_z T^*\mathbb{R}^d$ iff

$$-\langle (\mathcal{T}p)(z)w, w\rangle \ge \epsilon_0 |w|^2 - c_1 |p(z)w|^2 \qquad \forall w \in \mathbb{H} \tag{2.1.1}$$

where $\mathcal{T}p = (\langle \mathcal{T}_x \partial_x\rangle + \langle \mathcal{T}_\xi, \partial_\xi\rangle)p$ and $\epsilon_0 > 0$ and c_1 are appropriate constants. $\Diamond$

We have the following principal

Theorem 2.1.2. *Suppose* $P \in \boldsymbol{\Psi}_{h,N}(\mathbb{R}^d, \mathbb{H}, \mathbb{H})$ *and assume that*

(2.1.1) $p(z) = P_0(z)$ is a Hermitian symbol for all z.

Let $\Omega \Subset T^*\mathbb{R}^d$ *and let* $\phi_1, \dots, \phi_J$ *be real-valued* C_b^N *functions in* $T^*\mathbb{R}^d$ *such that at every point* $z \in \Omega_\epsilon$ *the symbol* p *is microhyperbolic in the directions* $H_{\phi_1}, \dots, H_{\phi_J}$ *where here and below* M_ϵ *means the* ϵ*-neighborhood of the set* M *and* $H_\phi = \langle (\partial_\xi \phi), \partial_x\rangle - \langle (\partial_x \phi), \partial_\xi\rangle$ *is the Hamiltonian field generated by* ϕ. *Let* u *be tempered, i.e.,*

$$\|u\| \le h^{-l}, \tag{2.1.3}$$

and

$$\text{OF}^{s+1}(Pu) \cap \Omega \cap \{\phi_1 \le 0\} \cap \cdots \cap \{\phi_J \le 0\} = \emptyset, \tag{2.1.4}$$

$$\text{OF}^{s}(u) \cap \partial\Omega \cap \{\phi_1 \le 0\} \cap \cdots \cap \{\phi_J \le 0\} = \emptyset. \tag{2.1.5}$$

Then

$$\text{OF}^{s}(u) \cap \Omega \cap \{\phi_1 \le 0\} \cap \cdots \cap \{\phi_J \le 0\} = \emptyset. \tag{2.1.6}$$

Proof. Let $q_1, q_2, q_3 \in \mathrm{S}_{h,N}(\mathbb{R}^d, \mathbb{C})$ be arbitrary symbols such that $q_1 = 1$ in $\Omega_\epsilon \cap \{\phi_1 \le \epsilon\} \cap \cdots \cap \{\phi_J \le \epsilon\}$, $q_2 = 1$ in $(\partial\Omega)_\epsilon \cap \{\phi_1 \le \epsilon\} \cap \cdots \cap \{\phi_J \le \epsilon\}$ and $\operatorname{supp} q_3 \subset \Omega_{\frac{1}{2}\epsilon} \cap \{\phi_1 \le \frac{1}{2}\epsilon\} \cap \cdots \cap \{\phi_J \le \frac{1}{2}\epsilon\}$. Let $Q_k = \mathrm{Op}^{\mathrm{w}}(q_k)$ and $N = N(s,l)$ in the definition of the symbol classes, etc. One can easily see that in order to prove the theorem it is sufficient to prove that (2.1.3) and

$$\|Q_1 Pu\| \le h^{s+1}, \qquad \|Q_2 u\| \le h^s \tag{2.1.7$_{1,2}$}$$

yield

$$\|Q_3 u\| \le Ch^s \tag{2.1.8}$$

where $\|.\|$ means the $L^2(\mathbb{R}^d, \mathbb{H})$-norm and C depends on $d, s, l, \epsilon_0, c_1, \epsilon$,

$$c_2 = \|\!|P|\!\| + \sum_k \|\!|q_k|\!\| + \sum_j \|\phi_j\|_{C^N};$$

however, C neither depends on Ω nor even its diameter.

We prove (2.1.8) under the additional assumption

$$\|Q_1 u\| \le h^{s-\frac{1}{2}}; \tag{2.1.9}$$

obviously, the general case is reduced to this one by induction on s. Let $\chi \in C^\infty(\mathbb{R})$ be a fixed real-valued function such that

$$\operatorname{supp}\chi \subset (-\infty, 0],\ \chi > 0 \text{ on } (-\infty, 0),\ \chi' \ge c_3\chi \text{ and} \qquad \chi_1 = \chi^{\frac{1}{2}},\ \chi_2 = (-\chi')^{\frac{1}{2}} \in C^\infty(\mathbb{R}) \tag{2.1.10}$$

where the prime means the derivative and the constant c_3 can be chosen later (such a function χ exists for every c_3). We can replace ϕ_j by $\phi_j - \frac{1}{4}\epsilon$ and ϵ by $\frac{3}{4}\epsilon$; hence it is sufficient to prove (2.1.8) only for $Q_3 = \mathrm{Op}^{\mathrm{w}}(q_3)$ with

$$q_3(z) = \zeta(z) \prod_{1 \le k \le J} \chi(\phi_k(z))$$

where $\zeta \in C_0^N(\Omega_\epsilon)$, $0 \le \zeta \le 1$, $\zeta = 1$ in $\Omega_{\frac{2}{3}\epsilon}$; there exists such a function with C^N-norm not exceeding $C_1 = C_1(d, N, \epsilon)$.

Let us consider the identity

$$\begin{aligned} &- \operatorname{Re} i([P,Q]u,u) = \\ &\operatorname{Re} i((Q+Q^*)Pu,u) + \operatorname{Re} i((P^*-P)Qu,u). \end{aligned} \tag{2.1.11}$$

Let us take $Q = Q_3^2$; then $Q = Q^*$ and the absolute value of the first term in the right-hand expression in (2.1.11) does not exceed $\|Q_3 Pu\| \cdot \|Q_3 u\|$. Then (2.1.3) and (2.1.6) imply that $\|Q_3 Pu\| \le 2h^{s+1}$ (in fact, $q_1 = 1$ in the ϵ_1-neighborhood of $\operatorname{supp} q_3$ with $\epsilon_1 = \epsilon_1(d, \epsilon_0, c_1, J, \epsilon) > 0$). So the absolute value of the first term in the right-hand expression in (2.1.11) does not exceed $2h^{s+1}\|Q_3 u\|^2$. The second term in the right-hand expression in (2.1.11) equals

$$\operatorname{Re} i((P^* - P)Q_3 u, Q_3 u) + \operatorname{Re} i([P^* - P, Q_3]Q_3 u, u).$$

Then (2.1.2) yields that the absolute value of the first term does not exceed $C_2 h\|Q_3 u\|^2$ where C_2 does not depend on c_3; (2.1.2), (2.1.3) and (2.1.9) imply that the absolute value of the second term does not exceed $Ch^{s+\frac{3}{2}}\|Q_3 u\|^2$ where C depends on c_3. Thus (2.1.11) implies that

$$|-\operatorname{Re} ih^{-1}([P,Q]u,u)| \le C_2\|Q_3 u\|^2 + Ch^{2s}. \tag{2.1.12}$$

Let us note that the principal symbol of the operator $-ih^{-1}[P,Q]$ is equal to

$$\sum_{1\le j\le J} 2\{p,\phi_j\}r_j^2 - \{p,\zeta^2\}\prod_{1\le k\le J}\chi^2(\phi_k)$$

where

$$r_j = \zeta\chi_1(\phi_j)\chi_2(\phi_j)\prod_{k\ne j}\chi(\phi_k).$$

Since $q_2 = 1$ in the ϵ_1-neighborhood of the support of the second term we obtain from (2.1.3), (2.1.7), (2.1.12) and (2.1.9) that

$$|\sum_{1\le j\le J}(S_j R_j u, R_j u)| \le C_2\|Q_3 u\|^2 + Ch^{2s} \tag{2.1.13}$$

where $R_j = \operatorname{Op}^{\mathrm{w}}(r_j)$, $S_j = \operatorname{Op}^{\mathrm{w}}(\{p,\phi_j\})$.

The microhyperbolicity of p in the directions H_{ϕ_j} yields that there exists a symbol q_4 vanishing in the ϵ_1-neighborhood of the supports of r_j for all j such that $|\!|\!|q_4|\!|\!| \le C$ and the following inequalities are fulfilled:

$$\langle(\{p,\phi_j\} + cp^\dagger p + (q_4 - \epsilon_0)I)w, w\rangle \ge 0 \qquad \forall w\in\mathbb{H}\quad \forall z\in T^*\mathbb{R}^d\quad \forall j = 1,\dots,J. \tag{2.1.14}$$

Then passing from symbols to operators we obtain by means of the Gårding inequality (theorem 1.1.10(i)) that

$$\operatorname{Re}(S_j v, v) + c_1\|Pv\|^2 + \operatorname{Re}(Q_4 v, v) \ge (\epsilon_0 - Ch)\|v\|^2. \tag{2.1.15}$$

Let us take $v = R_j u$ here. Observe that (2.1.3), (2.1.6) and (2.1.9) yield that

$$\|Pv\| \le \|R_j Pu\| + \|[R_j, P]u\| \le Ch^s.$$

Moreover, observe that $|(Q_4 v, v)| \le Ch^{2s}$ because the supports of Q_4 and R_j are disjoint and that $\|v\| \le Ch^{s-\frac{1}{2}}$ by (2.1.9) and (2.1.3). Therefore

$$\operatorname{Re}(S_j R_j u, R_j u) \ge \epsilon_0\|R_j u\|^2 - Ch^{2s}. \tag{2.1.16}$$

This inequality and (2.1.13) yield that

$$\sum_{1\le j\le J}\|R_j u\| \le C_2\|Q_3 u\|^2 + Ch^{2s}. \tag{2.1.17}$$

Moreover, (2.1.10) yields that $q_3^2 \le \epsilon_3^{-1} r_j^2$ with $\epsilon_3 = c_3^{-1}$. Therefore the Gårding inequality yields that

$$\|Q_3 v\|^2 \le \epsilon_3 \|R_j v\|^2 + Ch\|v\|^2.$$

Let us take $v = Q_5 u$ here where $q_5 = 1$ in the ϵ_1-neighborhood of $\operatorname{supp} q_5$; then this estimate and (2.1.3) and (2.1.6) yield that

$$\|Q_3 u\|^2 \le c_3^{-1}\|R_j u\|^2 + Ch^{2s}.$$

This estimate, (2.1.6), the possibility of choosing ϵ_3 arbitrarily small and the fact that C_2 does not depend on c_3 (and ϵ_3) together imply (2.1.8). □

Remark 2.1.3. *(i) In fact, we could prove this theorem for $J = 1$ from which the general case follows;*

(ii) The theorem remains true if we assume that p is microhyperbolic in Ω_ϵ in the direction H_ψ with a real-valued function ψ with $\|\psi\|_{C^N} \le c_2$ and that p is non-strictly microhyperbolic in Ω_ϵ in the directions H_{ϕ_j} for $j = 1, \dots, J$; this means that for $\mathcal{T} = H_{\phi_j}$

$$-\langle(\mathcal{T}p)(z)w, w\rangle \ge -c_1 |p(z)w|^2 \qquad \forall w \in \mathbb{H};$$

we reduce this case to the one considered above replacing ϕ_j by $\phi_j + \varepsilon\psi$ with small enough $\varepsilon > 0$;

(iii) For fixed Ω and $\phi_1, \dots, \phi_J$ one can reformulate this theorem in terms of oscillation front sets: If (2.1.3) holds and if

$$\mathrm{OF}^{s+1}(Pu) \cap \bar\Omega \cap \{\phi_1 \le 0\} \cap \dots \cap \{\phi_J \le 0\} = \emptyset,$$
$$\mathrm{OF}^{s}(u) \cap (\partial\Omega) \cap \{\phi_1 \le 0\} \cap \dots \cap \{\phi_J \le 0\} = \emptyset$$

then

$$\mathrm{OF}^{s}(Pu) \cap \bar\Omega \cap \{\phi_1 \le 0\} \cap \dots \cap \{\phi_J \le 0\} = \emptyset$$

and here one can replace $\bar\Omega$ by Ω; this is the standard form of statements of this type. ◊

In what follows we need

Proposition 2.1.4. *Theorem 2.1.2 remains true if in the definition of c_2 and hence in the definition of C we replace $\|\!|P|\!\|$ by $\|\!|P - \bar P|\!\|$ with an arbitrary Hermitian matrix $\bar P \in \mathcal{L}(\mathbb{H}, \mathbb{H})$.*

Proof. Let $P = \bar P + P'$, $\|\!|P'|\!\| \le c_2 - 1$. Without loss of generality one can assume that $\mathbb{H} = \mathbb{H}_1 \otimes \mathbb{H}_2$, $\bar P \mathbb{H}_j \subset \mathbb{H}_j$ and $\|\bar P w_1\| \le c_2 \|w_1\|$, $\|\bar P w_2\| \ge c_2\|w_2\|$, $\forall w_j \in \mathbb{H}_j$. Then

$$P = \begin{pmatrix} P_1 & P_{12} \\ P_{21} & P_2 \end{pmatrix}$$

and the principal symbol of P_2 is invertible and the norm of the inverse symbol does not exceed 1; then its parametrix $R_2 \in \boldsymbol{\Psi}_{h,N}(\mathbb{R}^d, \mathbb{H}_2, \mathbb{H}_2)$ exists with $\|\!|R_2|\!\| \le C_0 = C_0(d, N, c, c_1, c_2)$. Thus the treatment of the propagation of singularities for P is reduced to that for $P'' = P_1 - P_{12}R_2P_{21} \in \boldsymbol{\Psi}_{h,N}(\mathbb{R}^d, \mathbb{H}_1, \mathbb{H}_1)$ with $\|\!|P''|\!\| \le C_0$. The principal symbol $p'' = p_1 - p_{12}p_2^{-1}p_{21}$ is Hermitian because $p_{12} = p_{21}^\dagger$. Moreover, the equality

$$\langle\{p'' \phi\}w_1, w_1\rangle = \langle\{p, \phi\}w, w\rangle$$

holds for $w = (w_1, w_2) = (w_1, -p_2^{-1}p_{21}w_1)$ and hence condition (2.1.2) for p implies the same condition for p'' with the same constants c_1 and c_2. □

2.1.2 Cauchy problem. I

Let us now consider the Cauchy problem for the operator

$$P = hD_t - A \tag{2.1.18}$$

with $A \in C([0,T], \boldsymbol{\Psi}_{h,N}(\mathbb{R}^d, \mathbb{H}, \mathbb{H}))$ where

(2.1.19) $a(t,x,\xi) = A_0(t,x,\xi)$ is a Hermitian symbol for all (t,x,ξ).

We assume that $\Omega \Subset \mathbb{R} \times T^*\mathbb{R}^d$ and $\phi_j \in C([0,T], C_b^N(T^*\mathbb{R}^d))$ are real-valued functions such that

$$\langle \{p, \phi_j\}(t,z)w, w\rangle \ge \epsilon_0 |w|^2 \quad \forall w \in \mathbb{H}. \tag{2.1.20}$$

Then we replace (2.1.11) by the identity

$$h\,\mathrm{Re}(Qu,u)_T - \mathrm{Re}\, i([P,Q]u,u)_{[0,T]} = \\ h\,\mathrm{Re}(Qu,u)_0 + \mathrm{Re}\, i((Q+Q^*)u,u)_{[0,T]} + \mathrm{Re}\, i((P-P^*)Qu,u)_{[0,T]}$$

where $(.,.)_{[0,T]}$ and $(.,.)_t$ are inner products in $L^2((0,T)\times\mathbb{R}^d, \mathbb{H})$ and in $L^2(\mathbb{R}^d, \mathbb{H})$ for the indicated t respectively; for the norms similar notation will be used. Repeating the proof of theorem 2.1.2 with the obvious modifications we obtain

Proposition 2.1.5. *Let conditions* (2.1.18) *and* (2.1.19) *be fulfilled and suppose* $\Omega \Subset \mathbb{R} \times T^*\mathbb{R}^d$. *Let* $\phi_j \in C([0,T], C^N(T^*\mathbb{R}^d))$ *be real-valued functions satisfying* (2.1.20) *in* Ω_ϵ. *Let* $q_1, q_2, q_3 \in C([0,T], \mathrm{S}_{h,N}(T^*\mathbb{R}^d, \mathbb{C}))$ *be symbols such that* $q_1 = 1$ *in* $\Omega_\epsilon \cap \{\phi_1 \le \epsilon\} \cap \cdots \cap \{\phi_J \le \epsilon\}$, $q_2 = 1$ *in* $(\partial\Omega)_\epsilon \cap \{\phi_1 \le \epsilon\} \cap \cdots \cap \{\phi_J \le \epsilon\}$ *and* $\operatorname{supp} q_3 \subset \Omega_{\frac{\epsilon}{2}} \cap \{\phi_1 \le \frac{\epsilon}{2}\} \cap \cdots \cap \{\phi_J \le \frac{\epsilon}{2}\}$. *Let* $Q_k = \mathrm{Op}^w(q_k)$ *and* $N = N(s)$ *in the definition of symbol classes, etc. Then*

$$\sup_{t\in[0,T]} \|u\|_t \le 1, \qquad \|Q_1 u\|_0 \le h^s,$$
$$\int_0^T (\|Q_1 Pu\|_t + h\|Q_2 u\|_t)dt \le h^{s+1}$$

imply that

$$\sup_{t\in[0,T]} \|Q_3 u\|_t \le h^s$$

with the same type of constant C *as in theorem 2.1.2.*[1] *This statement remains true if we replace* $\sup_{t\in[0,T]} \|\cdot\|_t$ *by* $\frac{1}{\sqrt{T}}\|\cdot\|_{[0,T]}$ *in the conditions and in the conclusion. Moreover, proposition 2.1.4 remains true.*[2]

Let us note that if

$$|\nabla_z a| \le c \qquad \forall (t,z) \in [0,T] \times T^*\mathbb{R}^d \tag{2.1.21}$$

then (2.1.20) holds for the function $\phi = t + \psi(z)$ provided $\|\nabla_z \psi\| \le c^{-1}(1-\epsilon)$. Hence we obtain

---

[1] Surely with $|\!|\!|P|\!|\!|$ replaced by $|\!|\!|A|\!|\!|$ in the definition of c_2.

[2] Surely with $|\!|\!|P - \bar{P}|\!|\!|$ replaced by $|\!|\!|A - \bar{A}|\!|\!|$ in the definition of c_2.

Corollary 2.1.6. *Let conditions (2.1.17), (2.1.19) and (2.1.21) be fulfilled. Then the statement of proposition 2.1.5 holds for* $\Omega = [0,T] \times T^*\mathbb{R}^d$, $q_k = q_k(z)$ $(k = 1,3)$, $q_2 = 0$ *provided*

$$\operatorname{dist}(\operatorname{supp}(1 - q_1), \operatorname{supp} q_3) \geq CT + \epsilon$$

with an arbitrarily small constant $\epsilon > 0$.

Let us introduce

Definition 2.1.7. Let $x = (x', x'') \in \mathbb{R}^{d'} \times \mathbb{R}^{d''}$ be a partition of variables, $d = d' + d''$. Let u be an admissible (in the previous sense) function. Let us introduce $\mathrm{OF}^{s\prime}(u) \subset T^*\mathbb{R}^{d'} \times \mathbb{R}^{d''}$: $(\bar{x}, \bar{\xi}') \notin \mathrm{OF}^{s\prime}(u)$ iff there exists a symbol $q(x,\xi') \in C(\mathbb{R}^{d''}, \mathrm{S}_{h,N}(T^*\mathbb{R}^{d'}, \mathbb{C}))$ such that $q(\bar{x}, \bar{\xi}') \neq 0$ and $\|q(x, hD')u\| \leq Ch^s$ where $\|.\|$ means the $L^2(\mathbb{R}^d, \mathbb{H})$-norm. ◊

Using the natural partition (t,x) and in the definition replacing the $L^2(\mathbb{R}^d, \mathbb{H})$-norm by the $L^2([0,T] \times \mathbb{R}^d, \mathbb{H})$-norm we obtain

Corollary 2.1.6′. *Under the assumptions of corollary 2.1.6*

$$\begin{aligned}\mathrm{OF}^{s\prime}(u) \subset \{(t,x,\xi) : \exists (t', y, \eta) \in \mathrm{OF}^{(s+1)\prime}(Pu) \text{ with } \quad & \pm t' \in [0, \pm t] \\ \text{and } \operatorname{dist}((x,\xi),(y,\eta)) \leq c|t - t'|\} \cup & \\ \{(t,x,\xi) : \exists (y,\eta) \in \mathrm{OF}^s((u|_{t=0}) \text{ with } \operatorname{dist}((x,\xi),(y,\eta)) \leq c|t|\};\end{aligned}$$

this assertion holds for intervals $[0,T]$, $[-T,0]$, $[-T,T]$; *in particular in the last case we obtain that*

$$\Pi' \mathrm{OF}^s(u)\big|_{t=0}) \subset \mathrm{OF}^{s\prime}(u)\big|_{t=0} \subset \mathrm{OF}^s(u\big|_{t=0}) \cup \mathrm{OF}^{(s+1)\prime}(Pu)\big|_{t=0}$$

where $\Pi' : (t, x, \tau, \xi) \to (t, x, \xi)$ *is a natural map.*

Corollary 2.1.6′ states the finite speed of propagation of singularities. The second of its assertions is also useful: it justifies the application of theorem 2.1.2 for the propagation of singularities of solutions to the Cauchy problem without loss in the indices.

2.1.3 Cauchy problem. II

Let us now consider the Cauchy problem for the operator

$$P = B^{-1} hD_t - A \tag{2.1.22}$$

where $A \in \Psi_{h,N}(\mathbb{R}^d, \mathbb{H}, \mathbb{H})$ and

(2.1.19)′ $a(x,\xi) = A_0(x,\xi)$ is a Hermitian symbol for all (x,ξ),

(2.1.23) B is a self-adjoint operator in $\mathbb{H}$ such that

$$\langle Bw, w\rangle \geq |w|^2 \qquad \forall w \in \mathrm{D}(B).$$

Let $u^\pm = \theta(\pm t)u$ where here and in what follows θ is the Heaviside function. Then

$$Pu^\pm = (Pu)^\pm \mp ih\delta(t)B^{-1}u_0, \qquad u_0 = \mathfrak{d}_0 u \tag{2.1.24}$$

where $\mathfrak{d}_0$ is the operator of restriction to $\{t = 0\}$. Then we easily obtain

Lemma 2.1.8. *Let conditions (2.1.22) and (2.1.23) be fulfilled and let $N = N(d,s)$. Then*

$$\mathrm{OF}^{s+1}(Pu^{\pm}) \subset \mathrm{OF}^{s+1}(Pu) \cap \{\pm t \geq 0\} \cup$$
$$\Pi_0^{-1}(\mathrm{OF}^{(s+1)\prime}(Pu)\big|_{t=0} \cup \mathrm{OF}^{s+\frac{1}{2}}(\eth_0 u))$$

where $\Pi_0 : (0,x,\tau,\xi) \to (x,\xi)$ is a natural map.

It is obvious that the function $\phi = t$ satisfies the microhyperbolicity condition in the domain $\Omega' \times \{|\tau| \leq 2c'\}$ if and only if

$$|B^{-\frac{1}{2}}w| + |a(x,\xi)w| \geq \epsilon_0|w| \qquad \forall w \in \mathbb{H} \quad \forall (x,\xi) \in \Omega'. \tag{2.1.25}$$

More precisely, the microhyperbolicity condition (for $\phi = t$) at some energy level $\tau \in [-2c', 2c']$ [3] yields (2.1.25), and (2.1.25) yields the microhyperbolicity condition at every level $\tau \in [-2c', 2c']$; without loss of generality we assume that $c' = c$ here. One can easily check that replacing $B^{-\frac{1}{2}}$ by $B^{-\delta}$ with another positive exponent δ we obtain exactly the same condition.

If conditions (2.1.25) and (2.1.21) are fulfilled then condition (2.1.1) is fulfilled for $\phi = t + \psi(x,\xi)$ with any real-valued function ψ such that $|\nabla_{x,\xi}\psi| \leq \epsilon_2$ with $\epsilon_2 = \epsilon_2(\epsilon_0, c) > 0$. This fact and the uniform version of lemma 2.1.8, theorem 2.1.2 and the fact that P and $\Phi(hD_t)$ commute imply

Proposition 2.1.9. *Let conditions (2.1.21)–(2.1.23), (2.1.19)′ and (2.1.25) be fulfilled and let $T \in [\epsilon, c_1], T' = (1+\epsilon)T$ and $Q_j = \mathrm{Op}(q_j)$ with $q_j \in \mathrm{S}_{h,N}(T^*\mathbb{R}, \mathbb{C})$ such that*

$$\mathrm{dist}(\complement\Omega', \mathrm{supp}(1-q_1), \mathrm{supp}\, q_2) \geq C_0 T \tag{2.1.26}$$

with $C_0 = C_0(c, \epsilon_0)$. Let $\chi \in C_0^N([-1,1])$ and $\Phi \in C_0^N([-c,c])$ be fixed functions, $\chi_T(t) = \chi(\frac{t}{T})$. Then

(i) The inequalities

$$\|u\|_{[-T',T']} \leq h^{-l}, \qquad \|Q_1 Pu\|_{[-T',T']} \leq h^{s+1},$$
$$\|B^{-\frac{1}{2}} Q_1 u\|_0 \leq h^{s+\frac{1}{2}}$$

imply that

$$\|\Phi(hD_t)\chi_T(t)Q_2 u\| \leq Ch^s$$

where C depends on d, s, l, c, ϵ_0, c_1, c_2, and ϵ, $c_2 = \|A - \bar{A}\| + \sum_k \|q_k\|$, $\bar{A}$ is an arbitrarily chosen (constant) Hermitian operator in $\mathbb{H}$ and $N = N(d,s,l)$ in the conditions;

(ii) In particular, for $\|u\| \leq 1$ we obtain that

$$\mathrm{OF}^s(u) \cap \{|t| \leq T, |\tau| \leq c, (x,\xi) \in \Omega'\}$$
$$\subset \{(t,x,\xi) : \exists (t', y, \eta) \in \mathrm{OF}^{s+1}(Pu)$$
$$\text{with } \pm t' \in [0, \pm t] \text{ and } \mathrm{dist}((x,\xi),(y,\eta)) \leq C_0|t-t'|\} \cup$$
$$\{(t,x,\xi) : \exists (y,\eta) \in \mathrm{OF}^{s+\frac{1}{2}}(B^{-\frac{1}{2}}u\big|_{t=0}) \cup \partial\Omega'$$
$$\text{with } \mathrm{dist}((x,\xi),(y,\eta)) \leq C_0|t|\}.$$

So we state here the finite speed of propagation of singularities for energy levels $\tau \in [-c, c]$.

---

[3] I.e., the standard microhyperbolicity condition for the operator $A - B^{-1}\tau I$.

2.1.4 Properties of fundamental solutions

Now let us treat the (approximate) fundamental solution to the Cauchy problem for operator (2.1.22). Let $U(t,x,y)$ be an $\mathcal{L}(\mathbb{H},\mathbb{H})$-valued distribution such that for an operator $\mathbf{U}: L^2(\mathbb{R}^d,\mathbb{H}) \to L^2(\mathbb{R}^d,\mathbb{H})$ with Schwartz kernel $U(t,x,y)$ the following estimates in the $\mathcal{L}(L^2(\mathbb{R}^{d+1},\mathbb{H}), L^2(\mathbb{R}^d,\mathbb{H}))$-norms are fulfilled:

$$\|\mathbf{U}\| \le M, \tag{2.1.27}$$

$$\|\bar{\chi}(t)\operatorname{Op}(\zeta)P\mathbf{U}\| \le Mh^{s+1} \tag{2.1.28}$$

and the estimate

$$\|\operatorname{Op}(\zeta)(\eth_0\mathbf{U} - I)\| \le Mh^{s+\frac{1}{2}} \tag{2.1.29}$$

in the $\mathcal{L}(L^2(\mathbb{R}^d,\mathbb{H}), L^2(\mathbb{R}^d,\mathbb{H}))$-norm is fulfilled where $\zeta = 1$ in $\Omega'_\epsilon \Subset T^*\mathbb{R}^d$, $\bar{\chi} = 1$ on $(-(1+\epsilon)T_0, (1+\epsilon)T_0)$, $\epsilon \le T_0 \le c_1$, $|\!|\!|\zeta|\!|\!| \le c_1$ and the constant M will be discussed later. Then proposition 2.1.9 yields

Proposition 2.1.10. *Let conditions* (2.1.21)–(2.1.23), (2.1.19)′, (2.1.25) *and* (2.1.27)–(2.1.29) *be fulfilled with* $M \ge 1$, $N = N(d,s)$. *Let* $\chi \in C_0^N([-1,1])$ *and* $\Phi \in C_0^N([-c,c])$ *be fixed functions and let* $Q_k = \operatorname{Op}(q_k)$ *where* $q_k \in \mathrm{S}_{h,N}(T^*\mathbb{R}^d,\mathbb{H},\mathbb{H})$ *satisfy the condition*

$$\operatorname{dist}(\operatorname{supp} q_1, \operatorname{supp} q_2 \cup \complement\Omega') \ge C_0 T \tag{2.1.30}$$

with $C_0 = C_0(c)$. *Then*

$$\|\Phi(hD_t)\chi_T(t)Q_1\mathbf{U}Q_2\| \le CMh^s.$$

This assertion, theorem 1.1.4 and corollary 1.1.5 yield

Corollary 2.1.11. *Let* $q_k \in \mathrm{S}_{h,N}(T^*\mathbb{R}^d,\mathbb{H},\mathbb{H})$ *and* $\operatorname{diam}(\operatorname{supp} q_k) \le c$, $k = 1,2$. *Then*

(i) If (2.1.27) *holds then for* $T \in [\epsilon, T_0]$ *the estimate*

$$\|F_{t\to h^{-1}\tau}\chi_T(t)Q_{1x}U\,{}^tQ_{2y}\| \le CMh^{-d} \qquad \forall\tau: |\tau| \le c \quad \forall x,y \in \mathbb{R}^d$$

holds where here and in what follows $({}^tQu) = (Q^*u^\dagger)^\dagger$, *"†" is a natural map from* $\mathbb{H}$ *to the dual space* $\mathbb{H}^\dagger$ *and inversely, and operators with respect to* y *are written to the right of the function. The latter is convenient because it agrees with matrix theory notation (but contradicts standard operator theory notation).*

(ii) In terms of proposition 2.1.10 the estimate

$$\|F_{t\to h^{-1}\tau}\chi_T(t)Q_{1x}U\,{}^tQ_{2y}\| \le CMh^s \qquad \forall\tau: |\tau| \le c \quad \forall x,y \in \mathbb{R}^d$$

holds.

Remark 2.1.12. *(i) In fact we do not use proposition 2.1.10 for the first assertion;*

(ii) If $Q = q(\overset{3}{x}, \overset{1}{x}, \overset{2}{hD})$ *then* ${}^tQ = q(\overset{1}{x}, \overset{3}{x}, \overset{2}{-hD})$.

(iii) C *is the same type of constant as before.* ◊

Let us assume that $-a$ is microhyperbolic in Ω' with respect to H_ϕ. Let us apply theorem 2.1.2 to $u^\pm$ and equation (2.1.24) and take $\phi_1 = \pm t$ and $\phi_2 = \pm(\phi - \epsilon_1 t)$ and replace P by $\pm P$ with a small enough constant ϵ_1 (one can check that these functions and operators satisfy the microhyperbolicity condition); we then obtain

Proposition 2.1.13. *Let conditions* (2.1.21)–(2.1.23), (2.1.19)′, (2.1.25) *and* (2.1.27)–(2.1.29) *be satisfied with* $M \ge 1$, $N = N(d,s)$. *Let* $\phi \in C_b^N(T^*\mathbb{R}^d)$ *be a real-valued function such that* $\|\phi\|_{C^N} \le c_1$ *and* $-a$ *is microhyperbolic in* Ω' *with respect to* H_ϕ.[4] *Let* $q_k \in \mathrm{S}_{h,N}(T^*\mathbb{R}^d, \mathbb{H}, \mathbb{H})$ *satisfy the conditions*

$$\pm\phi(x,\xi) \le \pm\phi(y,\eta) + \epsilon_1 T - \epsilon \qquad \forall (x,\xi) \in \operatorname{supp} q_1\ \forall (y,\eta) \in \operatorname{supp} q_2 \tag{2.1.31}$$

and

$$\operatorname{dist}(\operatorname{supp} q_1, \complement\Omega') \ge C_0 T + \epsilon \tag{2.1.32}$$

where $T \in [\epsilon, T_0]$ *is arbitrary and* $T_0 = T_0(c) > 0$ *is a small enough constant,* $\epsilon > 0$ *is an arbitrarily small constant. Moreover, let* $\chi \in C_0^N([\frac{1}{2}, 1])$ *and* $\Phi \in C_0^N((-2\epsilon_1, 2\epsilon_1))$ *be fixed functions. Then*

$$\|\Phi(hD_t)\chi_{\pm T}(t) Q_1 \mathbf{U} Q_2\| \le CMh^s.$$

Representing q_1 and q_2 as sums of functions with small enough supports and applying theorem 1.1.4 and corollary 1.1.5 we obtain

Corollary 2.1.14. *Let the conditions of proposition 2.1.13 be fulfilled and let* $\operatorname{diam}(\operatorname{supp} q_k) \le c$, $k = 1, 2$. *Then*

(*i*) *The estimate*

$$\|F_{t\to h^{-1}\tau}\chi_{\pm T}(t)\Gamma(Q_{1x}U^t Q_{2y})\| \le CMh^s \qquad \forall \tau : |\tau| \le \varepsilon_0 \tag{2.1.33}$$

holds where here and in what follows $(\Gamma_x u)(x) = u(x,x)$ *and* $(\Gamma u) = \int (\Gamma_x u)(x)dx$.

(*ii*) *Moreover, if* $\phi = \phi(x)$ *and* $\operatorname{supp} q_k \subset \{|\xi| \le c_1\}$ $\forall k = 1, 2$ *then the estimate*

$$\|F_{t\to h^{-1}\tau}\chi_{\pm T}(t)\Gamma_x(Q_{1x}U^t Q_{2y})\| \le CMh^s \qquad \forall \tau : |\tau| \le \varepsilon_0 \quad \forall x \in \mathbb{R}^d \tag{2.1.34}$$

holds.

It appears that this corollary easily implies a stronger version of itself:

Theorem 2.1.15. *Let conditions* (2.1.21)–(2.1.23), (2.1.19)′, (2.1.25) *and* (2.1.27)–(2.1.29) *be satisfied with* $M \ge 1$, $N = N(d,s)$. *Let* $\phi \in C_b^N(T^*\mathbb{R}^d)$ *be a real-valued function such that* $\|\phi\|_{C^N} \le c_1$ *and assume* $-a$ *is microhyperbolic in* Ω' *with respect to* H_ϕ. *Let* $q_k \in \mathrm{S}_{h,N}(T^*\mathbb{R}^d, \mathbb{H}, \mathbb{H})$ *satisfy condition* (2.1.32) *and let* $\operatorname{diam}(\operatorname{supp} q_k) \le c$, $k = 1, 2$. *Let* $\chi \in C_0^N([\frac{1}{2}, 1])$ *be a fixed function and* $T_0 = T_0(c, \epsilon_0) > 0$ *be a small enough constant. Then*

(*i*) *For* $T \in [h, T_0]$ *the estimate*

$$\|F_{t\to h^{-1}\tau}\chi_{\pm T}(t)\Gamma(Q_{1x}U^t Q_{2y})\| \le CMh^{1-d}\Bigl(\frac{h}{T}\Bigr)^s \qquad \forall \tau : |\tau| \le \varepsilon_0 \tag{2.1.35}$$

holds.

(*ii*) *Moreover, if* $\phi = \phi(x)$ *and* $\operatorname{supp} q_k \subset \{|\xi| \le c_1\}$ $\forall k = 1, 2$ *then the estimate*

$$\|F_{t\to h^{-1}\tau}\chi_{\pm T}(t)\Gamma_x(Q_{1x}U^t Q_{2y})\| \le CMh^{1-d}\Bigl(\frac{h}{T}\Bigr)^s \qquad \forall \tau : |\tau| \le \varepsilon_0 \quad \forall x \in \mathbb{R}^d \tag{2.1.36}$$

holds.

---

[4] Then p is also microhyperbolic in $\Omega = \{(t,x,\tau,\xi) : (x,\xi) \in \Omega', |\tau| \le 2\epsilon_1\}$ with respect to H_ϕ where $\epsilon_1 = \epsilon_1(\epsilon_0, c, c_1)$.

Proof. Let us note that we need to prove estimates (2.1.35) and (2.1.36) only provided $\operatorname{diam}(\operatorname{supp} q_k) \le \epsilon_2$, $k = 1,2$ with a small enough constant $\epsilon_2 = \epsilon_2(d,\epsilon_0,c,c_1) > 0$ (otherwise one can make a partition of unity). Then without loss of generality one can assume that $\phi = x_1$; in fact, we can reduce the case $\phi = \phi(x)$ to this one by a change of coordinate system (and (2.1.36) remains true) and we can reduce the most general case to this one by replacing U by $U' = \mathcal{F}_x U^t \mathcal{F}_y$ where $\mathcal{F}$ is a scalar Fourier integral operator such that $\mathcal{F}^*\mathcal{F} \equiv I$ in a neighborhood of $\operatorname{supp} q_1$, $\mathcal{F}\mathcal{F}^* \equiv I$ in a neighborhood of $\Psi_{\mathcal{F}}(\operatorname{supp} q_1)$ and $\phi = x_1 \circ \Psi_{\mathcal{F}}$ where $\Psi_{\mathcal{F}}$ is the corresponding symplectomorphism.

Let $\psi_j(x)$ $(j = 1,\dots,J)$ be real-valued functions such that $\sum_j \psi_j^2(x) = 1$ in a neighborhood of $\operatorname{supp} q_1$, $\operatorname{diam}(\operatorname{supp}\psi_j) \le r$, $|D_x^\alpha \psi_j| \le c_3 r^{-|\alpha|}$ $\forall\alpha : |\alpha| \le K$; $J = c_3(1+[\frac{1}{r}])^d$ and the multiplicity of the covering by $\operatorname{supp}\psi_j$ is less than c_3. Obviously, for $c_3 = c_3(d,c)$ such functions exist. Let us replace Q_1 and Q_2 by $\psi_j Q_1$ and $\psi_j Q_2$ respectively and make the rescaling (dilatation) $t' = \frac{t}{r}$, $x' = \frac{x}{r}$, $h' = \frac{h}{r}$ (in which case $h'D_{x'} = hD_x$ and $h'D_{t'} = hD_t$) with $r \in [h,1]$. Now we are in precisely the situation of corollary 2.1.14 and hence we obtain the estimate (2.1.34) in the new variables. Then in the old variables we obtain the estimate

$$\|F_{t\to h^{-1}\tau}\chi_{\pm rT_0}(t)\Gamma_{\tilde{x}}(\psi_j^2(x)Q_{1x}U^tQ_{2y})\| \le CMr^{1-d}\Big(\frac{h}{r}\Big)^s \qquad \forall\tau : |\tau| \le \varepsilon_0 \quad \forall x \in \mathbb{R}^d \quad \forall j = 1,\dots,J.$$

The factor $r^{-d} =$ in the right-hand expression appears because *the Schwartz kernel is a function with respect to x and a density with respect to y and therefore the restriction of the Schwartz kernel to the diagonal is also a density*. The factor r in the right-hand expression appears because $F_{t\to h^{-1}\tau} = rF_{t'\to h'^{-1}\tau}$. These two simple observations are very important.

Let us sum with respect to j and pick $r = \frac{T}{T_0}$. We get (2.1.36). □

Moreover, we need

Remark 2.1.16. *Let us in addition assume that*

$$\|B_1^{2j}\Phi(hD_t)\chi_T(t)\mathbf{U}(t)B_1^{2j}\| \le M \tag{2.1.37}$$

where $\Phi = 1$ at $(-c,c)$ and B_1 is an operator satisfying (2.1.23). Then interpolation arguments imply that every estimate obtained remains true if we multiply the left-hand expression by B_1^j on both sides and replace the s-degree of h or $\frac{h}{T}$ in the right-hand expression by the corresponding $\frac{s}{2}$-degree. ◊

2.1.5 Overdetermined systems

First of all, let us note that if in theorem 2.1.2 we know that

$$\|Q_1P_j'u\| \le h^s \qquad \forall j = 1,\dots,m \tag{2.1.38}$$

then one can weaken condition (2.1.1):

$$-\langle(Tp)(z)w,w\rangle \ge \epsilon_0|w|^2 - c_1|p(z)w|^2 - \sum_{1\le j\le m}|p_j'(z)w|^2 \quad \forall w \in \mathbb{H} \tag{2.1.1$'$}$$

where p_j are the principal symbols of P_j. In fact, this leads to the additional term $\sum_j C|q_1p_jw|^2$ in the left-hand expression of (2.1.14) and to the additional term $\sum_j C\|Q_1P_jv\|^2$ in the left-hand expression of (2.1.15). Other details of the proof do not change.

Now, let us consider the solution of the multi-time Cauchy problem

(2.1.39) $$hD_{t_j}U \equiv A_j\mathbf{U} \quad \text{in } \Omega \quad (j=1,\dots,m),$$

(2.1.40) $$\mathbf{U}\big|_{t_1=\dots=t_m=0} \equiv I$$

where

(2.1.41) $A_j = A_j(x,hD,h)$ commute modulo negligible operators.

Then proposition 2.1.13 and corollary 2.1.14 remain true with $t=(t_1,\dots,t_m)$, $\tau=(\tau_1,\dots,\tau_m)$ and with $\chi\in C_0^K([-1,1]^m)$. Moreover, proposition 2.1.13 remains true under microhyperbolicity condition (2.1.1)′ with $P=-A_\ell=\ell_1A+\dots+\ell_mA_m$, $P_j=A_j$ for some $\ell\in\mathbb{R}^m$ with $|\ell l=1$ and χ supported in the vicinity of 0.

Repeating the proof of theorem 2.1.15 (with $F_{t\to h^{-1}\tau}=r^mF_{t'\to h'^{-1}\tau}$) we obtain

Theorem 2.1.15′. *Let conditions* (2.1.21), (2.1.40), (2.1.19), (2.1.27) *and* (2.1.39)–(2.1.40) *be satisfied for* $j=1,\dots,m$ *with* $M\ge1$, $N=N(d,s)$. *Suppose* $\phi\in C_b^N(T^*\mathbb{R}^d)$ *is a real-valued function such that* $\|\phi\|_{C^N}\le c_1$ *and the microhyperbolicity condition* (2.1.1)′ *is fulfilled in* Ω' *with* $T=H_\phi$, $p=a_\ell=\ell_1a_1+\dots+\ell_ma_m$, $p'_j=a_j$. *Let* $q_k\in \mathrm{S}_{h,N}(T^*\mathbb{R}^d,\mathbb{H},\mathbb{H})$ *satisfy condition* (2.1.32) *and let* $\operatorname{diam}(\operatorname{supp} q_k)\le c$, $k=1,2$. *Let* $\chi\in C^N$ *be a fixed function supported in the neighborhood of* ℓ *and* $T_0=T_0(c,\epsilon_0)>0$ *be a small enough constant. Then*

(i) For $T\in[h,T_0]$ *the estimate*

(2.1.35)′ $$\|F_{t\to h^{-1}\tau}\chi_{\pm T}(t)\Gamma(Q_{1x}U^tQ_{2y})\|\le CMh^{m-d}\Big(\frac{h}{T}\Big)^s \qquad \forall\tau:|\tau|\le\varepsilon_0$$

holds for a sufficiently small constant $\epsilon_0=\epsilon_0(c,0)$.

(ii) Moreover, if $\phi=\phi(x)$ *and* $\operatorname{supp} q_k\subset\{|\xi|\le c_1\}$ $\forall k=1,2$ *then the estimate*

(2.1.36)′ $$\|F_{t\to h^{-1}\tau}\chi_{\pm T}(t)\Gamma_x(Q_{1x}U^tQ_{2y})\|\le CMh^{1m-d}\Big(\frac{h}{T}\Big)^s \qquad \forall\tau:|\tau|\le\varepsilon_0 \quad \forall x\in\mathbb{R}^d$$

holds.

(iii) If an appropriate function $\phi=\phi_\ell$ *exists for every* $\ell\in\mathbb{S}^{m-1}$ *then the corresponding estimate ((2.1.35′) or (2.1.36)′) holds for every function* $\chi\in C_0^K(B(0,1)\setminus B(0,\frac12))$.

2.2 Geometric Interpretation

2.2.1 Discussion of the microhyperbolicity condition

In this section I give a more geometrical version of theorem 2.1.2, the main theorem of the previous section. However I do not discuss various special cases of this principal theorem. The reader can find them in, e.g., [Ivrii 1]. I assume throughout this section that condition (2.1.2) holds (i.e., $P\in\Psi_{h,N}(\mathbb{R}^d,\mathbb{H},\mathbb{H})$ and that its principal symbol p is Hermitian), that $N=N(d,s)$, $\Omega\Subset T^*\mathbb{R}^d$ and

(2.2.1) p is microhyperbolic at every point $z=(x,\xi)\in\Omega$ in the direction H_{x_1}.

Let us introduce the closed invariant subspace $\mathbb{H}_\rho(p,z)\subset\mathbb{H}$ of the matrix[5)] $p(z)$ corresponding to the interval $[-\rho,\rho]$ of its spectrum where $\rho>0$ and $z\in\omega$. Let us define the sets

5) It is more convenient to use term "matrix" for an operator acting in the auxiliary spaces.

(2.2.2) $$\varGamma_\rho(p,z) = \{\mathcal{T} \in T_z\Omega \setminus 0 : -\langle(\mathcal{T}^0 p)(z)w, w\rangle \geq \rho|w|^2 \quad \forall w \in \mathbb{H}_\rho(p,z)\}$$

where $\mathcal{T}^0 = \mathcal{T}/|\mathcal{T}|$. Moreover, let us introduce the set

(2.2.3) $$\varGamma(p,z) = \bigcup_{\rho>0} \varGamma_\rho(p,z).$$

The following assertion is obvious:

Lemma 2.2.1. *(i) If $|\mathcal{T}| \leq c$ satisfies (2.1.1) then $\mathcal{T} \in \varGamma_\rho(p,z)$ with $\rho = \rho(\epsilon_0, c_1) > 0$. On the other hand, if $|\mathcal{T}| \geq \rho$ and $\mathcal{T} \in \varGamma_\rho(p,z)$ then (2.1.1) holds with $\epsilon_0 = \epsilon_0(\rho) > 0$ and $c_1 = c_1(\rho)$;*

(ii) $\varGamma_\rho(p,z)$ is a conical strictly convex respectively closed subset of $T_z\Omega\setminus 0$; moreover, $\{(z,\mathcal{T}) \in T\Omega : \mathcal{T} \in \varGamma_\rho(p,z)\}$ is a respectively closed subset of $T\Omega \setminus 0$ where 0 means the 0-section of this bundle. Moreover, let the index set $\mathbf{M}$ be a Hausdorff topological space and assume that for all $\mathbf{m} \in \mathbf{M}$ the symbol $p_\mathbf{m}$ satisfies (2.1.1) and (2.2.1) and that the map $\mathbf{M} \ni \mathbf{m} \to p_\mathbf{m} \in C^1_b(T^*\mathbb{R}^d)$ is continuous; then the set $\{(\mathbf{m}, z, \mathcal{T}) \in \mathbf{M} \times T\Omega : \mathcal{T} \in \varGamma_\rho(p,z)\}$ is a respectively closed subset of $\mathbf{M} \times (T\Omega \setminus 0)$.

(iii) For $\rho' < \rho$, $\varGamma_{\rho'}(p,z)$ contains the ϵ-neighborhood of $\varGamma_\rho(p,z) \cap \{|\mathcal{T}| = 1\}$ with $\epsilon > 0$ depending on $\epsilon_0, c_1, \rho, \rho'$. This assertion remains true if we replace $\varGamma_{\rho'}(p,z)$ by $\{(z,\mathcal{T}) \in T\Omega : \mathcal{T} \in \varGamma_\rho(p,z)\}$ and even by $\{(\mathbf{m}, z, \mathcal{T}) \in \mathbf{M}\times T\Omega : \mathcal{T} \in \varGamma_\rho(p,z)\}$ provided the index set is a metric space, and for all $\mathbf{m} \in \mathbf{M}$ the symbol $p_\mathbf{m}$ satisfies (2.1.1) and (2.2.1) and the map $\mathbf{M} \ni \mathbf{m} \to p_\mathbf{m} \in C^1_b(T^*\mathbb{R}^d)$ is uniformly continuous (surely in this case ϵ depends on this map also);

(iv) Hence $\varGamma(p,z)$ is a conical convex open subset of $T_z\Omega\setminus 0$; moreover, $\{(z,\mathcal{T}) \in T\Omega : \mathcal{T} \in \varGamma(p,z)\}$ is an open subset of $T\Omega\setminus 0$; moreover, $\{(\mathbf{m}, z, \mathcal{T}) \in \mathbf{M}\times T\Omega : \mathcal{T} \in \varGamma(p,z)\}$ is an open subset of $\mathbf{M} \times (T\Omega \setminus 0)$ provided the additional conditions at the end of (ii) are fulfilled;

(v) If $\bar{\mathcal{T}} \in \partial\varGamma(p,z)$ or $(\bar{z}, \bar{\mathcal{T}}) \in \partial\{(z,\mathcal{T}) \in T\Omega : \mathcal{T} \in \varGamma(p,z)\}$ and $\bar{\mathcal{T}} \neq 0$ then there exist sequences $\{\rho_n\}$ and $\{w_n \in \mathbb{H}_{\rho_n}(p,z)\}$ such that $|w_n| = 1$, $\rho_n \to +0$ and if w'_n is another sequence with the same properties then $-\langle(\bar{\mathcal{T}}p)(\bar{z})w'_n, w'_n\rangle \geq -\rho'_n$ and $|\langle(\bar{\mathcal{T}}p)(\bar{z})w_n, w'_n\rangle| \leq -\rho'_n$ where $\rho'_n \to +0$ and do not depend on the choice of w'_n.

Furthermore, let us introduce the subsets $\varGamma^\sigma_\rho(p,z)$ and $\varGamma^\sigma(p,z)$ where for any set $G \subset T^*\mathbb{R}^d$,

$$G^\sigma = \{\zeta \in T^*\mathbb{R}^d, \sigma(\zeta,\zeta') \leq 0 \quad \forall \zeta' \in G\}$$

where $\sigma(.,.)$ is a bilinear symplectic form. Surely, G^σ is a conical convex closed set in $T^*\mathbb{R}^d$.

Example 2.2.2. *(i) If p is elliptic at z, i.e., if*

(2.2.4) $$|p(z)w| \geq \epsilon|w| \qquad \forall w \in \mathbb{H}$$

then $\varGamma(p,z) = T^*\mathbb{R}^d \setminus 0$ and $\varGamma^\sigma(p,z) = 0$;

(ii) Suppose that for $z \in \Omega$ the symbol $p(z)$ is unitarily equivalent to

$$\begin{pmatrix} q(z)(\xi_1 - \lambda(z')) & 0 \\ 0 & p''(z) \end{pmatrix}, \qquad \mathbb{H} = \mathbb{H}' \oplus \mathbb{H}''$$

where the symbols q and p'' are bounded and Hermitian, λ is a real-valued function with $z' = (x, \xi')$, $\xi' = (\xi_2, \ldots, \xi_d)$ in this example and p'' satisfies (2.2.4) on $\mathbb{H}''$,

(2.2.5) $$\langle q(z)w, w\rangle \geq \epsilon|w|^2 \qquad \forall w \in \mathbb{H}'.$$

Then in a neighborhood of $\Sigma = \{\xi_1 = \lambda(z')\}$ all those symbols and functions (including the symbol of the operator effecting the unitary equivalence) can be chosen as smooth as p; moreover, for $z \in \Sigma$

$$\Gamma(p,z) = \{(\mathcal{T}',\eta_1) \in T^*\mathbb{R}^d : \eta_1 > \mathcal{T}'(\lambda)(z')\}$$

and

$$\Gamma^\sigma(p,z) = (H_{\xi_1-\lambda})_{\text{cone}} = \{tH_{\xi_1-\lambda}, t>0\};$$

(iii) Suppose that for $z \in \Omega$, $p(z)$ is unitarily equivalent to the symbol

$$\begin{pmatrix} p_1(z)\otimes I_1 & \dots & 0 & 0 \\ \vdots & \ddots & \vdots & 0 \\ 0 & \dots & p_n(z)\otimes I_n & 0 \\ 0 & \dots & 0 & p''(z) \end{pmatrix}$$

where $\mathbb{H} = \mathbb{H}_1' \oplus \dots \oplus \mathbb{H}_n' \oplus \mathbb{H}''$, $\mathbb{H}_j' = \mathbb{H}_j \otimes \mathbb{H}_j^0$, $\mathbb{H}_j$ are finite-dimensional spaces, $p_j(z)$ are matrices in $\mathbb{H}_j$ and I_j are the identity operators in $\mathbb{H}_j^0$. Let us assume that all the symbols are Hermitian and smooth, the symbol effecting the unitary equivalence is also smooth, and p'' satisfies (2.2.2) in $\mathbb{H}''$. Then

$$\Gamma(p,z) = \bigcap_{1\le j\le n} \Gamma(p_j,z)$$

and

$$\Gamma^\sigma(p,z) = \text{convex hull}(\bigcup_{1\le j\le n} \Gamma^\sigma(p_j,z)) = \oplus \sum_{1\le j\le n} \Gamma^\sigma(p_j,z)).$$

◊

Let us consider the *characteristic symbol*

(2.2.7) $$g(z) = \prod_{1\le j\le n} \det p_j(z).$$

Proposition 2.2.3. *Let g have a zero of order exactly r at $\bar{z} \in \Omega$.*
(i) Then

(2.2.8) $$(H_{x_1}^r g)(\bar{z}) \neq 0$$

and $\mathcal{T} \in \Gamma(p,\bar{z})$ if and only if

(2.2.9) $$(\mathcal{T}^j H_{x_1}^{r-j} g)(z^*)(H_{x_1}^r g)^{-1}(\bar{z}) > 0 \qquad \forall j = 1,\dots,r.$$

Moreover, $\mathcal{T} \in \overline{\Gamma(p,\bar{z})}$ if and only if the corresponding non-strict inequalities are fulfilled.
(ii) Moreover, $\Gamma(p,\bar{z})$ coincides with the connected component of $\{\zeta, g_{\bar{z}}(\zeta) \neq 0\}$ where

$$g_{\bar{z}}(\zeta) = \sum_{|\alpha|=r} \frac{1}{\alpha!}(\partial_z^\alpha g)(\bar{z})\zeta^\alpha$$

is a jet of g at $\bar{z}$. Moreover, $g_{\bar{z}}(\zeta)$ is a hyperbolic polynomial of ζ with respect to H_{x_1}.

Proof. Let us first consider the special case $r = \mathrm{D}_1 + \cdots + \mathrm{D}_n$ where $\mathrm{D}_j = \dim \mathbb{H}_j$. Then (2.2.1) easily implies that $p_1(\bar{z}) = \cdots = p_n(\bar{z}) = 0$ and (2.2.8) holds. In this case $\mathcal{T} \in \Gamma(p, z)$ iff $-(\mathcal{T}p)(\bar{z})$ is a positive definite matrix, i.e., iff $((-\mathcal{T} - tH_{x_1})p)(\bar{z})$ is a positive definite matrix for every $t \geq 0$ and that is obviously equivalent to (2.2.9). Moreover, statement (ii) is also obvious in this case.

On the other hand, if $r < \mathrm{D}_1 + \cdots + \mathrm{D}_n$ then in a neighborhood of $\bar{z}$ the finite-dimensional symbol

$$p' = \begin{pmatrix} p_1 & \dots & 0 \\ \vdots & \ddots & \vdots \\ 0 & \dots & p_n \end{pmatrix}$$

is unitarily equivalent to the matrix $\begin{pmatrix} p''' & 0 \\ 0 & p^{IV} \end{pmatrix}$ where p^{IV} satisfies (2.2.4) on the corresponding space and the dimension of the first component of the decomposition is r. It is obvious that condition (2.2.8) or (2.2.9) for $g = \det p'$ is fulfilled if and only if the same condition is fulfilled for $g''' = \det p'''$. Moreover, the jets of g and g''' in $\bar{z}$ coincide modulo a constant factor. Finally, it is obvious that the microhyperbolicity conditions for p and p''' are equivalent. □

2.2.2 Principal theorem

We need to prove the following version of theorem 2.1.2:

Proposition 2.2.4. *Suppose conditions* (2.1.2) *and* (2.2.1) *are fulfilled. Let* $\bar{z} = (\bar{x}, \bar{\xi}) \in \Omega$, $\bar{t} = \bar{x}_1 - \varepsilon\rho$ *where* $\rho \in (0, \varepsilon]$, $\varepsilon = \varepsilon(c, \epsilon_0, c_1, c_2) > 0$ *is a small enough constant and* $c_2 = \|p\|_{C^2}$. *Moreover, let*

$$(\bar{z} - \Gamma^\sigma_\rho(p, \bar{z})) \cap \{t \leq x_1 \leq \bar{x}_1\} \subset \Omega_{(\rho)} = \{z : \operatorname{dist}(z, \complement\Omega) \geq \rho\}, \tag{2.2.10}$$

$t \in [\bar{t}, \bar{x}_1]$, *let* u *be admissible and let*

$$\begin{aligned} \mathrm{OF}^{s+1}(Pu) \cap (\bar{z} - \Gamma^\sigma_\rho(p, \bar{z})) \cap \{t \leq x_1 \leq \bar{x}_1\} &= \emptyset, \\ \mathrm{OF}^{s}(u) \cap (\bar{z} - \Gamma^\sigma_\rho(p, \bar{z})) \cap \{x_1 = t\} &= \emptyset. \end{aligned}$$

Then

$$\mathrm{OF}^{s}(u) \cap (\bar{z} - \Gamma^\sigma_\rho(p, \bar{z})) \cap \{t \leq x_1 \leq \bar{x}_1\} = \emptyset.$$

Proof. Note that $\Gamma_\rho(p, z)$ contains the ϵ-neighborhood of H_{x_1} with $\epsilon = \epsilon(\epsilon_0, c, c_1, c_2)$ and hence $\Gamma^\sigma_\rho(p, z)$ is a convex proper cone contained in the cone $\{z = (x, \xi) : |x'| + |\xi| \leq Cx_1\}$ with an appropriate constant $C = C(\epsilon_0, c_1, c_2)$. Hence for an appropriate $\varepsilon = \varepsilon(\epsilon_0, c_1, c_2, \mu) > 0$

$$(\bar{z} - \Gamma^\sigma_\rho(p, \bar{z})) \cap \{t \leq x_1 \leq \bar{x}_1\}$$

is contained in the $\mu\rho$-neighborhood of $\bar{z}$ where $\mu > 0$ will be chosen later. Note that $\Gamma_\rho(p, z)$ strictly increases and $\Gamma^\sigma_\rho(p, z)$ strictly decreases as $\rho > 0$ decreases. Hence it is sufficient to prove this proposition for $\Gamma_\rho(p, \bar{z})$ replaced by some strictly convex cone Γ' such that $\Gamma_\rho(p, \bar{z}) \subset \Gamma' \subset \Gamma_{\frac{1}{2}\rho}(p, \bar{z})$ and $\partial\Gamma' \setminus 0$ is smooth and its Gaussian curvature is positive. Then the dual cone Γ'^σ has the same properties. Without loss of generality one can assume that $\bar{z} = 0$ and $\Gamma'^\sigma \ni (1, 0, \dots, 0)$; one can always arrange that by a linear symplectic transformation in $T^*\mathbb{R}^d$ preserving hyperplanes $\{x_1 = \text{const}\}$. But in this case $-\Gamma'^\sigma = \{x_1 + \psi(z') \le 0\}$ where ψ is a real-valued function, positively homogeneous of degree 1, positive and smooth for $z' \ne 0$. Moreover, $H_{x_1+\psi}(0) \in \Gamma_{\frac{1}{2}\rho}(p, 0)$ and hence for admissible $\mu = \mu(\epsilon_0, c_1, c_2)$ we obtain that $H_{x_1+\psi}(z) \in \Gamma_{\frac{1}{3}\rho}(p, 0)$ in the $\mu\rho$-neighborhood of 0 for $z' \ne 0$. Let us replace ψ by $\psi_\eta = \psi * \omega_\eta - c_3\eta$ where $\omega_\eta(z') = \eta^{2d-1}\omega(z'/\eta)$, $\omega \in C_0^\infty(\mathbb{R}^{2d-1})$, $\omega \ge 0$ and $\int_{\mathbb{R}^{2d-1}} \omega dz' = 1$. Then for admissible small $\eta > 0$ depending on ϵ_0, c_1, c_2 the convexity of all the cones yields that $H_{x_1+\psi_\eta}(z) \in \Gamma_{\frac{1}{4}\rho}(p, z)$ at every point z of the $\mu\rho$-neighborhood of 0; moreover, $x_1 + \psi_\eta < 0$ in $-\Gamma'^\sigma$ for admissible c_3 which does not depend on η. Finally, $\{x_1 + \psi_\eta < 0\}$ contracts to $-\Gamma'^\sigma$ as $\eta \to +0$. Then theorem 2.1.2 yields the proposition. □

Now let us give the key

Definition 2.2.5. (i) *A generalized bicharacteristic (generalized ρ-bicharacteristic)* of p is a Lipschitz curve $\{z = z(t)\}$ such that for a.e. t, $z(t)$ is differentiable and $\frac{dz}{dt} \in \Gamma^\sigma(p, z)$ ($\frac{dz}{dt} \in \Gamma^\sigma_\rho(p, z)$ respectively). Since $\Gamma^\sigma_\rho(p, z) \subset \{|z'| \le Cx_1\}$ then one can take x_1 as a parameter t (let us recall that $z = (x, \xi)$).

(ii) Let $\mathrm{K}^\pm(p, z, \Omega)$ ($\mathrm{K}^\pm_\rho(p, z, \Omega)$) be the union of all the generalized bicharacteristics (generalized ρ-bicharacteristics) of p lying in $\bar\Omega$ and going out from the point $z \in \Omega$ in the direction of increasing $\pm t$. ◊

Lemma 2.2.6. *The sets introduced above have the following properties:*

(i) $\mathrm{K}^\pm(p, z, \Omega) \subset \mathrm{K}^\pm_\rho(p, z, \Omega) \subset K^\pm_{\rho'}(p, z, \Omega)$ for $0 < \rho < \rho'$; moreover, $\mathrm{K}^\pm_{\rho'}(p, \bar z, \Omega)$ contains the $\varepsilon|x_1 - \bar x_1|$-neighborhood of every point $z = (x, \xi) \in \mathrm{K}^\pm_\rho(p, \bar z, \Omega)$ intersected with Ω where $\varepsilon = \varepsilon(\epsilon_0, c_1, c_2, \rho, \rho') > 0$;

(ii) $z' \in \mathrm{K}^\pm(p, z, \Omega)$ if and only if $z \in \mathrm{K}^\mp(p, z', \Omega)$;

(iii) If $z'' \in \mathrm{K}^\pm(p, z', \Omega)$ and $z' \in \mathrm{K}^\pm(p, z, \Omega)$ then $z'' \in \mathrm{K}^\pm(p, z, \Omega)$; assertions (ii) and (iii) also remain true for $\mathrm{K}^\pm_\rho$;

(iv) $\mathrm{K}^\pm(p, z, \Omega)$ and $\mathrm{K}^\pm_\rho(p, z, \Omega)$ are closed sets;

(v) $\mathrm{K}^\pm(p, z, \Omega) = \bigcap_{\rho>0} \mathrm{K}^\pm_\rho(p, z, \Omega)$;

(vi) If $z_n \to z$, $z'_n \to z'$ and $p_n \to p$ in C^2_b as $n \to +\infty$ and if $z'_n \in \mathrm{K}^\pm(p_n, z_n, \Omega)$ then $z' \in \mathrm{K}^\pm(p, z, \Omega)$.[6]

Proof. Assertions (i)–(iii) are obvious; we prove (v); (iv) and (vi) are proven in the same way. Let us consider $\mathrm{K}^+(p, z, \Omega)$. Then (i) implies that $\mathrm{K}^+(p, z, \Omega)$ is contained in $\bigcap_{\rho>0} \mathrm{K}_\rho(p, z, \Omega)$; let us prove the reverse inclusion. If $z' \in \bigcap_{\rho>0} \mathrm{K}^+_\rho(p, z, \Omega)$ then there exist a sequence $\rho_n \to +0$ and generalized ρ_n-bicharacteristics $z = z_n(t)$ such that $z_n(0) = z$, $z_n(t_n) = z'$; since one can take $t = x_1 + \text{const}$ then one can assume that $t_n = t_0$. Since $|\frac{dz_n(t)}{dt}|$ is uniformly bounded for all n and $t \in [0, t_0]$ then by the Ascoli-Arzelà theorem one can assume that $z_n(t) \to z(t)$ as $n \to +\infty$ uniformly with respect to $t \in [0, t_0]$ where $z = z(t)$ is a Lipschitz curve; since $z(t)$ is a function with bounded variation, by the Lebesgue theorem [Kolmogorov, Fomin 1] it is differentiable a.e. and we need only to prove that $\frac{dz}{dt} \in \Gamma^\sigma(p, z(t))$ for a.e. t. It is obvious that for every $\rho > 0$ there exist $\epsilon > 0$ and n_0 such that $z_n - z_n(t') \in \pm\Gamma^\sigma_\rho(p, z(t'))$ for all $n \ge n_0$ and $t \in [t' - \epsilon, t' + \epsilon] \cap [0, t_0]$ with $\pm(t - t') > 0$. Then after the limit transition we obtain that $z - z(t') \in \pm\Gamma^\sigma_\rho(p, z(t'))$ for every $\rho > 0$ and indicated t, t'. This yields that if $z(t)$ is differentiable at $t = t'$ then $\frac{dz}{dt} \in \Gamma^\sigma_\rho(p, z(t))$ for $t = t'$ and for all $\rho > 0$ and hence $\frac{dz}{dt} \in \Gamma^\sigma(p, z(t))$ for $t = t'$. □

Now we can prove the central theorem of this section:

[6] However, even for $p_n = p$ it is possible that the set $\mathrm{K}^\pm(p, z, \Omega)$ is wider than the set of such limits.

Theorem 2.2.7. *Suppose $\Omega \Subset T^*\mathbb{R}^d$ and conditions (2.1.2) and (2.2.1) are fulfilled. Moreover, let $\bar{z} \in \Omega$. Then if u is admissible and*

$$\mathrm{OF}^{s+1}(Pu) \cap \mathrm{K}^{\mp}(p, \bar{z}, \Omega) = \emptyset$$

and

$$\mathrm{OF}^{s}(u) \cap \mathrm{K}^{\mp}(p, \bar{z}, \Omega) \cap \partial\Omega = \emptyset$$

then

$$\mathrm{OF}^{s}(u) \cap \mathrm{K}^{\mp}(p, \bar{z}, \Omega) = \emptyset.$$

Proof. I will prove this statement with $\mathrm{K}^{\mp}_{\rho}$ instead of $\mathrm{K}^{\mp}$. Then assertions (v) and (i) of lemma 2.2.6 yield the statement in its original form. Let us consider K^{-}_{ρ}; replacing t, P by $-t, -P$ one can reduce the K^{+}_{ρ} case to this one.

It is easy to see that if $\epsilon > 0$ and $\rho' > 0$ are arbitrarily small and if $\varepsilon = \varepsilon(\epsilon_0, \rho, \rho', c, c_1, c_2) > 0$ is small enough then for $\bar{z} = (\bar{x}, \bar{\xi}) \in \Omega_{(\rho')}$ and $t = \bar{x}_1 - \varepsilon\rho$ the following inclusions hold:

$$\mathrm{K}^{-}_{(1+\epsilon)\rho}(p, \bar{z}, \Omega) \supset (\bar{z} - \Gamma^{\sigma}_{\rho}(p, \bar{z})) \cap \{x_1 \geq t\} \supset \mathrm{K}^{-}_{(1-\epsilon)\rho}(p, \bar{z}, \Omega) \cap \{x_1 \geq t\}.$$

Then the assertions of the theorem for the domain $\Omega_n = \{z \in \Omega, -c_4 + \varepsilon\rho n > x_1\}$ instead of Ω are easily proven for an appropriate constant c_4 by means of induction with respect to n: for $n = 0$ the assertion is trivial since $\Omega_0 = \emptyset$ for appropriate c_4, and the induction step is provided by proposition 2.2.4. Therefore this assertion holds for Ω also. □

Remark 2.2.8. *It is easy to see that under the assumptions of theorem 2.1.2 the function ϕ increases along the generalized ρ-bicharacteristics provided $\rho > 0$ is small enough; hence if $\phi(\bar{z}) < 0$ and $z \in \mathrm{K}^{-}_{\rho}(p, \bar{z}, \Omega)$ then $\phi(z) < 0$. Hence theorem 2.1.2 and this theorem are equivalent.* ◊

Example 2.2.9. *(i) If p satisfies (2.2.4) then $\mathrm{K}^{\pm}(p, z, \Omega) = z$;*

(ii) If in Ω

$$p(z) = \begin{pmatrix} g(z)q(z) & 0 \\ 0 & p''(z) \end{pmatrix},$$

$p''(z)$ and $q(z)$ satisfy (2.2.4) and (2.2.5) respectively on the corresponding subspaces, $g(z)$ is a real-valued scalar function with $dg \neq 0$, and if $g(\bar{z}) = 0$, then $\mathrm{K}^{\pm}(p, \bar{z}, \Omega)$ is an interval of the semibicharacteristic of g lying in $\bar{\Omega}$, i.e., of the curve $z = z(t)$ with

$$\frac{dz}{dt} = H_g(z), \quad z(0) = \bar{z} \qquad (\pm t \geq 0). \tag{2.2.11}$$

In particular, if $g = \tau - a(x, \xi)$ where a is a scalar real-valued symbol and (x, ξ) is replaced by (t, x, τ, ξ) we obtain the system

$$\frac{dz}{dt} = -H_a(z), \quad z = (x, \xi), \quad \tau = a(z) = \text{const.} \tag{2.2.12}$$

This case will be analyzed very carefully in the next section. ◊

2.3 Propagation of Singularities along Long Bicharacteristics

The results of the previous section imply the well known fact that singularities propagate along bicharacteristics for operators with scalar principal symbols. In this section we improve and generalize this result in two directions: namely, we treat large time intervals depending on h and we replace the description of the singularities by means of oscillation front sets with a more refined description related to boxes depending on h.

2.3.1 Evolution of operators

We start with the Cauchy problem for the operator

$$P = hD_t - A \tag{2.3.1}$$

where

(2.3.2) $A = A^w(x, hD_x, h)$ is an operator with a Hermitian Weyl symbol and hence A is a self-adjoint operator

and

(2.3.3) $a = A_0$ is a scalar real-valued symbol.

It is obvious that the investigation of the propagation of singularities of solutions to this Cauchy problem can be reduced to the treatment of the operator

$$Q^t = e^{-ih^{-1}tA} Q^0 e^{ih^{-1}tA} \tag{2.3.4}$$

where $Q^0 = \mathrm{Op}^{\mathrm{w}}(q^0)$ is also a pseudo-differential operator.[7] It is known ([Hörmander 1], [Treves 1], [Taylor 3], [Shubin 1], etc.) that for standard pseudo-differential operators A and Q and finite time t the *propagators* $e^{\pm ih^{-1}tA}$ are Fourier integral operators corresponding to the symplectomorphisms $\Phi_{\pm t}$ where $\Phi_t : z(0) \to z(t)$ and $z(t)$ is a solution of system (2.2.12). This implies that Q^t is also a pseudo-differential operator. However, we use these results only as a heuristic preliminary. Definition (2.3.4) implies that Q^t is a solution of the Cauchy problem

$$\partial_t Q^t = -ih^{-1}[A, Q^t], \tag{2.3.5}$$

$$Q^t\big|_{t=0} = Q^0. \tag{2.3.6}$$

Let us assume that Q^t is a pseudo-differential operator with Weyl symbol q^t where for now we treat symbols as formal power series; then the Cauchy problem (2.3.5)–(2.3.6) implies the similar Cauchy problem for symbols; namely, for a term q^t_n in the decomposition of q^t we obtain the problem

$$\partial_t q^t_n + \{a, q^t_n\} + i[a^s, q^t_n] = \\ \sum_{j+|\alpha|+|\beta|=n-k}^{0\le k\le n-1} \frac{1}{2^{|\alpha|+|\beta|}\alpha!\beta!}\Bigl((-i)^{|\alpha|-|\beta|} A^{(\alpha)}_{j(\beta)} q^{t(\beta)}_{k(\alpha)} - i^{|\alpha|-|\beta|} q^{t(\beta)}_{k(\alpha)} A^{(\alpha)}_{j(\beta)}\Bigr), \tag{2.3.7$_n$}$$

$$q^t_n\big|_{t=0} = q^0_n. \tag{2.3.8$_n$}$$

---

[7] Physicists would say that we use the Heisenberg representation.

Recall that $\{.,.\}$ are Poisson brackets, $[.,.]$ is the commutator in $\mathcal{L}(\mathbb{H},\mathbb{H})$, a^s is the subprincipal symbol of A (i.e., $a^s = A_1$), A_n are terms in the decomposition of the symbol of A and $b^{(\alpha)}_{(\beta)} = \partial_x^\beta \partial_\xi^\alpha b$.

Now we throw off our assumption that Q^t is an h-pseudo-differential operator. We needed this assumption *only* to derive problems $(2.3.7)_n$–$(2.3.8)_n$. So, let us consider this problem, where for now "symbol" means "function" rather than "symbol of an operator."

Let us start with the principal symbol q_0^t which satisfies the following *transport equation*

$$\partial_t q_0^t + \{a, q_0^t\} + i[a^s, q_0^t] = 0 \tag{2.3.7$_0$}$$

where $[.,.]$ means the matrix (i.e., operator in the auxiliary space $\mathbb{H}$) commutator. Moreover, if a^s is a scalar symbol we obtain the equation

$$\partial_t q_0^t + \{a, q_0^t\} = 0 \tag{2.3.7$'_0$}$$

which can be rewritten in the form

$$\partial_t (q_0^t \circ \Phi_{-t}) = 0 \tag{2.3.9$'_0$}$$

where $\Phi_t : T^*\mathbb{R}^d \to T^*\mathbb{R}^d$ is a symplectomorphism; namely, $\Phi_t : z(0) \to z(t)$ where $z(t)$ is a solution of the dynamical system

$$\frac{dz(t)}{dt} = -H_a(z(t)) \tag{2.3.10}$$

and hence

$$q_0^t = q_0^0 \circ \Phi_t. \tag{2.3.11$'_0$}$$

Definition 2.3.1. The family Φ_t is called the *Hamiltonian flow generated by the Hamiltonian a.* ◊

In the general case equation $(2.3.7)_0$ can be rewritten in the form

$$\partial_t \left(T_t^{-1}(q_0^t \circ \Phi_{-t})T_t\right) = 0 \tag{2.3.9$_0$}$$

where T_t is the differentiable matrix family in $\mathbb{H}$ satisfying the Cauchy problem

$$\partial_t T_t = -i(a^s \circ \Phi_{-t})T_t, \qquad T_0 = I. \tag{2.3.12}$$

These matrices are unitary since a^s is Hermitian. Therefore

$$q_0^t = (T_t q_0^0 T_t^\dagger) \circ \Phi_t. \tag{2.3.11$_0$}$$

One can rewrite this formula in the form

$$q_0^t = T_{t,t}(q_0^0 \circ \Phi_t)T_{t,t}^\dagger \tag{2.3.13$_0$}$$

where $T_{t,\tau}(z) = T_\tau(\Phi_t(z))$ is the family of matrices satisfying the Cauchy problem

$$\partial_\tau T_{t,\tau} = -i(a^s \circ \Phi_{t-\tau})T_{t,\tau}, \qquad T_{t,0} = I. \tag{2.3.14}$$

In order to use q_0^t and q_n^t we need to prove the following series of assertions:

Lemma 2.3.2. *Let $\Phi_t : \mathbb{R}^d \to \mathbb{R}^d$ be a flow defined by the dynamical system $\frac{dz}{dt} = F(z)$. Let $\bar{z} \in \mathbb{R}^d$, $\bar{t} > 0$ and $z(t) = \Phi_t(\bar{z})$. Let us assume that*

$$|(\partial_z^\alpha F)(z(t))| \le c\omega^{|\alpha|} \qquad \forall t \in [0, \bar{t}] \quad \forall \alpha : 1 \le |\alpha| \le r \tag{2.3.15}$$

and

$$|\Phi'_{t-t'}(z(t'))| \le c\omega \qquad \forall t, t' : 0 \le t' \le t \le \bar{t} \tag{2.3.16}$$

where Φ'_t is the differential of Φ_t (with respect to z) and $1 \le \omega$ is a parameter. Then

$$|(\partial_z^\alpha \Phi_t)(\bar{z})| \le C\omega\omega_1^{|\alpha|-1} \qquad \forall t \in [0, \bar{t}] \quad \forall \alpha : 1 \le |\alpha| \le r \tag{2.3.17}$$

where here and below $\omega_1 = (\bar{t} + 1)\omega^4$, $C = C(d, r, c)$.

Proof. Let us use induction with respect to $r \ge 1$; for $r = 1$ the estimate (2.3.17) coincides with (2.3.16). Let us assume that (2.3.16) holds for all α with $|\alpha| < r$. Let us consider $z(t, u) = \Phi_t(u)$. Differentiating the initial system r times with respect to u and setting $u = \bar{z}$ we obtain that

$$\frac{dz_k^{(\alpha)}}{dt} = \sum_{1 \le j \le d} F'_{kj}(z) z_j^{(\alpha)} + f_{k,\alpha} \tag{2.3.18}$$

where F_k and z_k are components of F and z respectively and $F'_{kj} = \partial_{z_j} F_k$, $z_k^{(\alpha)} = \partial_u^\alpha z\big|_{u=\bar{z}}$,

$$f_{k,\alpha} = \sum_{2 \le |I| \le r} \sum_{\substack{\alpha_1, \dots, \alpha_s : s = |I|, \\ |\alpha_i| \ge 1, \alpha_1 + \dots + \alpha_s = \alpha}} \bar{c}_{d, I, \alpha_1, \dots, \alpha_s} F_I(z(t)) z_{i_1}^{\alpha_1} \dots z_{i_s}^{\alpha_s},$$

$I = (i_1, \dots, i_s) \in \{1, \dots, d\}^s$, $F_I = \partial_{z_{i_1}} \dots \partial_{z_{i_s}}$, and $\bar{c}\dots$ are absolute constants. Then condition (2.3.15) and the induction assumption imply that

$$|f_{k,\alpha}| \le C\omega^4 \omega_1^{r-2}. \tag{2.3.19}$$

On the other hand, (2.3.18) and the equality $z_k^{(\alpha)} = 0$ for $t = 0$ and $|\alpha| \ge 2$ imply that

$$z^{(\alpha)}(t) = \int_0^t \Phi'_{t-t'}(z(t')) f_\alpha(t') dt'$$

where $z^{(\alpha)}$ and f_α are vectors with components $z_k^{(\alpha)}$ and $f_{k,\alpha}$ respectively; this formula, (2.3.16) and (2.3.19) imply (2.3.17) for $|\alpha| = r$ where the constant C is changed if necessary. □

Remark 2.3.3. *(i) For system (2.3.10) condition (2.3.15) means precisely that*

$$|a_{(\beta)}^{(\alpha)}(z(t))| \le c\omega^{|\alpha|+|\beta|-1} \qquad \forall t : [0, \bar{t}] \quad \forall \alpha, \beta : 2 \le |\alpha| + |\beta| \le r + 1. \tag{2.3.15$'$}$$

(ii) If condition (2.3.16) is fulfilled for $t' = 0$ for system (2.3.10) then the condition is fulfilled for all $t' \in [0, t]$ with ω replaced by ω^{2d}; in fact,

$$\Phi'_{t-t'}(z(t')) = \Phi'_t(z(0))(\Phi'_{t'}(z(0)))^{-1}$$

and since system (2.3.10) is Hamiltonian we obtain that $\det \Phi'_\tau = 1$ and $(\Phi'_\tau)^{-1}$ is a co-factor matrix. ◊

Lemma 2.3.4. *Let the conditions of lemma 2.3.2 be fulfilled and let a^s be a Hermitian symbol such that*

$$(2.3.20) \qquad |(\partial_z^\alpha a^s)(z(t))| \le c\omega^{|\alpha|+1} \qquad \forall t \in [0,\bar t] \quad \forall \alpha : 1 \le |\alpha| \le r$$

where here and below $|.|$ means the $\mathcal{L}(\mathbb{H},\mathbb{H})$-norm. Then

(i) The following estimates hold:

$$|(\partial_u^\alpha (a^s \circ \Phi_t))(u))|_{u=\bar z}| \le C\omega^2\omega_1^{|\alpha|-1} \qquad \forall t \in [0,\bar t] \quad \forall \alpha : 1 \le |\alpha| \le r.$$

(ii) If $T_{t,\tau}(z)$ is a unitary operator family defined by the Cauchy problem (2.3.14) then

$$(2.3.21) \qquad |(\partial_z^\alpha T_{t,\tau})(\bar z)| \le C\omega_1^{|\alpha|} \qquad \forall t,\tau : 0 \le \tau \le t \le \bar t \quad \forall \alpha : 1 \le |\alpha| \le r.$$

(iii) Moreover, if

$$|(\partial_z^\alpha q_0^0)(z(\bar t))| \le c\rho^{|\alpha|} \qquad \forall \alpha : 1 \le |\alpha| \le r$$

with $\rho \ge 1$ then

$$(2.3.22) \qquad |(\partial_z^\alpha (q_0^0 \circ \Phi_t))(z(\bar t - t))| \le C\rho_0^{|\alpha|} \quad \forall t \in [0,\bar t] \quad \forall \alpha : 1 \le |\alpha| \le r$$

and

$$(2.3.23) \qquad |(\partial_z^\alpha q_0^t)(z(\bar t - t))| \le C\rho_0^{|\alpha|} \quad \forall t \in [0,\bar t] \quad \forall \alpha : 1 \le |\alpha| \le r$$

where q_0^t is given by $(2.3.13)_0$ and $\rho_0 = \omega\bar\rho$, $\bar\rho = \rho + \omega(\bar t + 1)$.

Proof. Let us use the rule for differentiation of composite functions (see the proof of lemma 2.3.2). Then assertion (i) follows from lemma 2.3.2 and (iii) follows from (ii). Let us prove (ii) by means of induction with respect to $r \ge 0$; for $r = 0$ (2.3.21) obviously holds. Let us assume that (2.3.21) holds for all α with $|\alpha| \le r$. Differentiating (2.3.14) r times with respect to z (with $r \ge 1$) we obtain that

$$\partial_\tau T_{t,\tau}^{(\alpha)} = -i(a^s \circ \Phi_{t-\tau})T_{t,\tau}^{(\alpha)} + f_\alpha(t,\tau), \qquad T_{t,0}^{(\alpha)} = 0$$

where

$$f_\alpha(t,\tau) = -i \sum_{\beta \le \alpha, |\beta| \ge 1} \frac{1}{\alpha!(\alpha-\beta)!}(a^s \circ \Phi_{t-\tau})^{(\beta)} T_{t,\tau}^{(\alpha-\beta)}$$

and hence

$$T_{t,\tau}^{(\alpha)} = \int_0^\tau T_{t,\tau-\tau'} f_\alpha(t,\tau')d\tau';$$

the induction assumption easily yields that

$$|f_\alpha(t,\tau')| \le C(\bar t + 1)^{-1}\omega_1^r$$

and hence (2.3.21) holds for $|\alpha| = r$. □

Proposition 2.3.5. *Let Φ_t be the flow defined by system (2.3.10), $\bar z \in T^*\mathbb{R}^d$, $\bar t > 0$ and $z(t) = \Phi_t(\bar z)$. Let us assume that*

$$(2.3.24) \quad |A_{k(\beta)}^{(\alpha)}(z(t))| \le c\omega^{|\alpha|+|\beta|+2k-1}$$

$$\forall t \in [0,\bar t] \quad \forall k \le n \quad \forall \alpha,\beta : 2 \le |\alpha| + |\beta| + 2k \le r+1,$$

$$(2.3.25) \qquad |q^{0(\alpha)}_{k(\beta)}(z(\bar t))| \le c\rho^{|\alpha|+|\beta|+2k} \qquad \forall k \le n \quad \forall \alpha,\beta : 1 \le |\alpha|+|\beta|+2k \le r$$

where $\omega \ge 1, \rho \ge 1, 2n \le r$; moreover, let us assume that (2.3.16) is fulfilled. Let q^t_k be solutions of the recurrent Cauchy problems (2.3.7)–(2.3.8). Then

$$(2.3.26) \quad |q^{t(\alpha)}_{k(\beta)}(z(\bar t - t))| \le C\zeta_k \rho_k^{|\alpha|+|\beta|}$$
$$\forall t \in [0,\bar t] \quad \forall k \le n \quad \forall \alpha,\beta : 1 \le |\alpha|+|\beta|+2k \le r$$

where $\rho_k = \omega^{k+1}\bar\rho$, $\bar\rho = \rho + \omega(\bar t + 1)$, $\zeta_k = \omega^{k(k+3)}\bar\rho^{2k}(\bar t + 1)^k$.

Proof. Let us use induction with respect to k; for $k = 0$ (2.3.26) has been proven. Let us assume that (2.3.26) holds for $k < n$. Then it easy to prove that

$$|f^{(\alpha)}_{n(\beta)}(t, z(\bar t - t))| \le C\omega^2\zeta_{n-1}\rho_{n-1}^{|\alpha|+|\beta|+2}$$
$$\forall t \in [0,\bar t] \quad \forall \alpha,\beta : |\alpha|+|\beta|+2n \le r$$

where $f_n(t,z)$ is the right-hand expression in $(2.3.7)_n$. On the other hand, $(2.3.7)_n$ and $(2.3.13)_0$ imply that

$$(2.3.12)_n \qquad q^t_n = \int_0^t T_{t,\tau}(f_n(t-\tau,.)\circ \Phi_\tau)T^\dagger_{t,\tau}d\tau + T_{t,t}(q^0_n \circ \Phi_t)T^\dagger_{t,t}$$

and then lemma 2.3.4 implies (2.3.26) with $k = n$ provided $\rho_n \ge \omega\rho_{n-1}$, $\zeta_n \ge \omega^2\rho^2_{n-1}\zeta_{n-1}(\bar t + 1)$; these inequalities are provided by the definitions of ρ_k, ζ_k. □

Let us proceed from symbols to operators:

Theorem 2.3.6. *Suppose that $A \in \boldsymbol{\Psi}_{h,\gamma,\gamma,N}(\mathbb{R}^d, \mathbb{H}, \mathbb{H})$ satisfies conditions (2.3.2) and (2.3.3) and let Φ_t be the Hamiltonian flow generated by a, $q^0 \in S_{h,\gamma,\gamma,N}(T^*\mathbb{R}^d, \mathbb{H}, \mathbb{H})$. Let us assume that*

$$(2.3.27) \quad |A^{(\alpha)}_{k(\beta)}(\Phi_{-t}(z))| \le c\omega^{|\alpha|+|\beta|+2k}$$
$$\forall t \in [0,\bar t] \quad \forall z \in \operatorname{supp} q \quad \forall k,\alpha,\beta : 2 \le |\alpha|+|\beta|+2k \le 2N+K+1$$

and that

$$(2.3.16)'' \qquad |\Phi'_{-t}(z))| \le c\omega \qquad \forall t \in [0,\bar t] \quad \forall z \in \operatorname{supp} q.$$

Moreover, let us assume that

$$(2.3.28) \qquad \bar t \le \omega, \quad (\bar t + 1) \le \gamma h^{-\frac12+\delta}, \quad 1 \ge \rho = \frac{1}{\omega} \ge h^{\delta'}$$

where $\delta > 0$, s and $\delta' \le \delta'_0$ are arbitrary, $K = K(d)$, $N = N(d,s,\delta)$ are large enough and $\delta'(d,s,\delta)_0 > 0$ is small enough. Let $q^t = \sum_{n\le N} q^t_n h^n$ where q^t_n are solutions of the recurrent Cauchy problems $(2.3.7)_n$–$(2.3.8)_n$. Then for $t \in [0,\bar t]$,

(i) $q^t \in S_{h,\gamma',\gamma',N}(T^*\mathbb{R}^d, \mathbb{H}, \mathbb{H})$ with $\gamma' = \gamma h^{\delta''}$ where $\delta'' \to +0$ as $\delta' \to +0$ and

$$|\!|\!| q^t |\!|\!|_{\gamma'} \le C |\!|\!| q^0 |\!|\!|_\gamma$$

with a constant $C = C(d,s,\delta,\delta',c,c_1)$ where $c_1 = |\!|\!| A |\!|\!|_\gamma$;

(ii) For $Q^t = \operatorname{Op}^w(q^t)$ the estimate

$$(2.3.29) \qquad \|Q^t - e^{-ih^{-1}tA}Q^0e^{ih^{-1}tA}\| \le Ch^s |\!|\!| q^0 |\!|\!|_\gamma$$

holds where $\|.\|$ means the $\mathcal{L}(L^2(\mathbb{R}^d,\mathbb{H}), L^2(\mathbb{R}^d,\mathbb{H}))$-norm, i.e., $Q^t = e^{-ith^{-1}A}$ $Q^0e^{-ith^{-1}A} \equiv \operatorname{Op}^w(q^t)$ and hence $Q^t \in \boldsymbol{\Psi}_{h,\gamma',\gamma',N}(\mathbb{R}^d, \mathbb{H}, \mathbb{H})$.

Proof. Assertion (i) obviously follows from proposition 2.3.5 and remark 2.3.3. Then the construction of q^t and the conditions on γ, ω, $\bar{t}$ and N imply that equation (2.3.5) holds approximately:

$$\|\partial_t Q^t + ih^{-1}[A, Q^t]\| \le Ch^{s+1} |||q^0|||_\gamma$$

which is equivalent to

$$\|\partial_t (e^{ih^{-1}tA} Q^t e^{ih^1 tA})\| \le Ch^{s+1} |||q^0|||_\gamma$$

because the operator A is self-adjoint; this inequality combined with (2.3.6) (which is fulfilled because of the definitions of Q^t and q^t) and with the estimate $(\bar{t}+1)h \le c$ imply that

$$\|e^{ih^{-1}tA} Q^t e^{ih^1 tA} - Q^0\| \le Ch^s |||q^0|||_\gamma$$

which is equivalent to (2.3.29). □

2.3.2 Propagation of singularities

Theorem 2.3.6 immediately implies

Theorem 2.3.7. *Suppose $A \in \Psi_{h,\gamma,\gamma,N}(\mathbb{R}^d, \mathbb{H}, \mathbb{H})$ satisfies conditions (2.3.2) and (2.3.3) and let Φ_t be the Hamiltonian flow generated by a. Let $q, q' \in S_{h,\gamma,\gamma,N}(T^*\mathbb{R}^d, \mathbb{H}, \mathbb{H})$. Let us assume that* ***either***

(i) For all $z \in \operatorname{supp} q$, $t \in [0, \bar{t}]$ conditions (2.3.27) and (2.3.16)″ are fulfilled and

(2.3.30) $$\operatorname{dist}(z', \Phi_{-t}(z)) \ge \epsilon\gamma \qquad \forall z' \in \operatorname{supp} q'$$

or

(ii) The conditions of (i) are fulfilled with permuted q and q' and Φ_t replaced by Φ_{-t}.

Suppose condition (2.3.28) and all the conditions of theorem 2.3.6 on γ, ω, $\bar{t}$, N, K, δ' are fulfilled. Then for $Q = \operatorname{Op}(q)$ and $Q' = \operatorname{Op}(q')$ the inequality

(2.3.31) $$\|Q e^{ih^{-1}t^* A} Q'\| \le Ch^s |||q|||_\gamma \cdot |||q'|||_\gamma$$

holds.

Moreover, let $U(t, x, y)$ be the Schwartz kernel of $e^{ih^{-1}tA}$ and $\operatorname{supp} q \cup \operatorname{supp} q' \subset \{|\xi| \le c\}$. Then the estimate

(2.3.32) $$|Q_x U(t^*, ., .)^t Q'_y| \le C' h^s |||q|||_\gamma \cdot |||q'|||_\gamma \qquad \forall x, y \in \mathbb{R}^d$$

holds with $C' = C'(d, s, \delta, \delta', c, c_1, \epsilon)$.

We need to generalize this theorem:

Theorem 2.3.8. *Let $A \in \Psi_{h,\gamma,\gamma,N}(\mathbb{R}^d, \mathbb{H}, \mathbb{H})$ be a self-adjoint operator and let $q, q' \in S_{h,\gamma,\gamma,N}(T^*\mathbb{R}^d, \mathbb{H}, \mathbb{H})$ and $\operatorname{supp} q \cup \operatorname{supp} q' \subset \{|\xi| \le c\}$. Let $a = A_0$, $\tau_1, \tau_2 \in \mathbb{R}$, $\tau_1 < \tau_2$ and let $\lambda_1, \dots, \lambda_m$ be real-valued Lipschitz functions on $T^*\mathbb{R}^d$ such that*

(2.3.33) $$\operatorname{Spec} a(z) \cap [\tau_1, \tau_2] = \{\lambda_1(z), \dots, \lambda_m(z)\} \cap [\tau_1, \tau_2] \qquad \forall z \in \operatorname{supp} q.$$

Let $\Phi_{j,t}$ be the Hamiltonian flows generated by the symbols λ_j and let us assume that for all $z \in (\operatorname{supp} q)_\rho$ (where M_ρ means the ρ-neighborhood of M and $\rho = \frac{1}{\omega}$),

(2.3.34) $$\operatorname{Spec} a(\Phi_{j,-t}(z)) \cap [\lambda_j(z) - \rho, \lambda_j(z) + \rho] = \lambda_j(z)$$

and conditions (2.3.16)″ and (2.3.27) are fulfilled; moreover, let condition (2.3.27) be fulfilled for $|\alpha| + |\beta| = 1, k = 0$ *also. Finally, let us assume that for* $t = t^*$ *condition (2.3.30) is fulfilled. Let* $\chi \in C_0^N(\mathbb{R})$ *be supported in* $[t^* - \epsilon\rho\gamma, t^* + \epsilon\rho\gamma] \cap [0, \bar{t}]$ *where* $\varepsilon = \varepsilon(c, \epsilon)$ *is small enough. Let the conditions of theorem 2.3.6 on* γ, ω, $\bar{t}$, N, K, δ' *be fulfilled,* $Q = \mathrm{Op}(q)$ *and* $Q' = \mathrm{Op}(q')$. *Let* $U(t,x,y)$ *be the Schwartz kernel of* $e^{ih^{-1}tA}$. *Then*

(i) *The estimate*

$$(2.3.35)\quad |F_{t\to h^{-1}\tau}\chi(t)Q_x U(t,.,.)^t Q'_y\| \le C' h^s |\!|\!| q |\!|\!|_\gamma \cdot |\!|\!| q' |\!|\!|_\gamma \cdot \|\chi|_{C^N} \qquad \forall x,y \in \mathbb{R}^d \quad \forall \tau \in [\tau_1 + \epsilon'\rho, \tau_2 - \epsilon'\rho]$$

holds where $\varepsilon' > 0$ *is arbitrary and* C' *also depends on* ϵ' *and* m.

(ii) *This assertion remains true if we permute* q *and* q' *and replace* $\Phi_{j,t}$ *by* $\Phi_{-j,t}$ *in the conditions; this case reduces to the previous because* $U^\dagger(t,x,y) = U(-t,y,x)$; *this equality follows from the self-adjointness of* A.

In order to prove this theorem we prove a series of auxiliary assertions. Let us note that without loss of generality one can assume that $\gamma \le \bar{\gamma} = \varepsilon\rho^2$ with a small enough constant $\varepsilon = \varepsilon(c,\epsilon) > 0$ and that q is supported in the $\frac{\gamma}{2}$-neighborhood of some point $\bar{z}$. Let Ω be the γ-neighborhood of $\bar{z}$ and $\Omega_j(t) = \{\Phi_{j,-t}(z), z \in \Omega\}$ and $t \in [0,\bar{t}]$. Then (2.3.16)″ yields that $\Omega_j(t)$ is contained in the $C_0\omega\gamma$-neighborhood of $\Phi_{j,-t}(\bar{z})$ and contains the $\varepsilon'\gamma\omega^{-1}$-neighborhood of $\{\Phi_{j,-t}(z), z \in \operatorname{supp} q\}$.

Lemma 2.3.9. *Suppose the conditions of theorem 2.3.8 are fulfilled. Then there exist unitary symbols* $r(z), r_1(t,z), \dots, r_m(t,z)$ *defined in* $\Omega, \Omega_1(t), \dots, \Omega_m(t)$ *respectively such that in* Ω,

$$(2.3.36)\qquad r_j(0,z) = r(z) \qquad \forall j = 1,\dots,m,$$

$$(2.3.37)\qquad r^\dagger(z)a(z)r(z) = a^0 = \begin{pmatrix} \lambda_1(z)I & \dots & 0 & 0 \\ \vdots & \ddots & \vdots & \vdots \\ 0 & \dots & \lambda_m(z)I & 0 \\ 0 & \dots & 0 & a'(z) \end{pmatrix},$$

$$(2.3.38)\qquad \operatorname{Spec} a'(z) \cap [\lambda_j(z) - \frac{\rho}{2}, \lambda_j(z) + \frac{\rho}{2}] = \emptyset \qquad \forall j = 1,\dots,m$$

and moreover for every j, *in* $\Omega_j(t)$,

$$(2.3.39)_j\qquad r_j^\dagger(t,z)a(z)r_j(t,z) = a^0_{j,t} = \begin{pmatrix} b'_j(t,z) & 0 & b''_j(t,z) \\ 0 & \lambda_j(z) & 0 \\ b'''_j(t,z) & 0 & b^{IV}_j(t,z) \end{pmatrix},$$

$$(2.3.40)_j\qquad \operatorname{Spec} a'_j(t,z) \cap [\lambda_j(z) - \frac{\rho}{2}, \lambda_j(z) + \frac{\rho}{2}] = \emptyset$$

with

$$a'_j(t,z) = \begin{pmatrix} b'_j(t,z) & b''_j(t,z) \\ b'''_j(t,z) & b^{IV}_j(t,z) \end{pmatrix}$$

and

$$(2.3.41)\qquad |\partial_t^k r_{j(\beta)}^{(\alpha)}| \le C\omega^{2(|\alpha|+|\beta|)+3k} \qquad \forall \alpha,\beta,k : |\alpha| + |\beta| + k \le 2N + K - 1.$$

Proof. Let us define

$$\Pi_j(z) = \frac{1}{2\pi i}\int_{\Gamma_j} (\zeta - a(z))^{-1} d\zeta$$

for $z \in \Omega_j(t)$ where Γ_j is a closed curve lying in the ring $\{\frac{1}{8}\rho \le |\zeta - \lambda_j(z)| \le \frac{1}{4}\rho\} \subset \mathbb{C}$ and going one time around $\lambda_j(z)$ in the counter-clockwise direction; then $\Pi_j(z)$ is the orthogonal projector to the eigenspace of $a(z)$ corresponding to the eigenvalue $\lambda_j(z)$; here we have used condition (2.3.34). Then in $\Omega_j(t)$

$$|\Pi^{(\alpha}_{j(\beta)})(z)| \le C\omega^{2(|\alpha|+|\beta|)} \qquad \forall \alpha,\beta : |\alpha| + |\beta| + \le 2N + K - 1;$$

here we have used condition (2.3.34) again and condition (2.3.27). These inequalities combined with the inequalities $|\partial_t \Phi_{j,t}(z)| \le C\omega$ (we again appeal to (2.3.27)) imply the statement of the lemma. □

It should be noted that the periodicity or approximate periodicity (i.e., dist($\Phi_{j,-t}(z), z) \le \epsilon\rho$) of some trajectories of the Hamiltonian flows possibly prevents the existence of r_j not depending on t.

Lemma 2.3.10. *Suppose the conditions of theorem 2.3.8 are fulfilled. Then there exist symbols* $r, r_{1t}, \dots, r_{mt} \in \mathrm{S}_{h,\gamma,\gamma,N}(T^*\mathbb{R}^d, \mathbb{H}, \mathbb{H})$ *satisfying* (2.3.36) *and such that* $\gamma^{2k}\partial_t^k r_{j,t} \in \mathrm{S}_{h,\gamma,\bar\gamma,N-[(k+1)/2]}(T^*\mathbb{R}^d, \mathbb{H}, \mathbb{H})$, $\forall k = 1, \dots, 2N-1$, *and the corresponding norms of these operators do not exceed* C, *and modulo operators with symbols vanishing in* Ω

$$R^*R \equiv RR^* \equiv I, \tag{2.3.42}$$

$$R^*AR \equiv \begin{pmatrix} A_1 I & \dots & 0 & 0 \\ \vdots & \ddots & \vdots & \vdots \\ 0 & \dots & A_m I & 0 \\ 0 & \dots & 0 & A' \end{pmatrix}, \tag{2.3.43}$$

and modulo operators with symbols vanishing in $\Omega_j(t)$

$$R^*_{j,t}R_{j,t} \equiv R_{j,t}R^*_{j,t} \equiv I, \tag*{(2.3.44)$_j$}$$

$$-ihR^*_{j,t}(\partial_t R_{j,t}) + R^*_{j,t}AR_{j,t} \equiv \begin{pmatrix} B'_{j,t} & 0 & B''_{j,t} \\ 0 & A_{j,t} & 0 \\ B'''_{j,t} & 0 & B^{IV}_{j,t} \end{pmatrix} \tag*{(2.3.45)$_j$}$$

where $R_{j,t} = \mathrm{Op}(r_{j,t})$, $R = \mathrm{Op}(r)$ *and the principal symbols of* $A_{j,t}, A', B'_{j,t}, \dots, B^{IV}_{j,t}$ *equal* $\lambda_j I, a', b'_{j,t}, \dots, b^{IV}_{j,t}$ *respectively.*

Proof. Let us use induction with respect to n:

$$V \in \gamma^{-2n}\omega^{-1}h^n \mathrm{S}_{h,\gamma,\gamma,N-n}(\Omega, \mathbb{H}, \mathbb{H}) \tag{2.3.46}$$

$$\partial_t^k V_{j,t} \in \gamma^{-2n-2k}\omega^{-1}h^n \mathrm{S}_{h,\gamma,\gamma,N-n-[\frac{k+1}{2}]}(\Omega_j(t), \mathbb{H}, \mathbb{H}) \tag*{(2.3.47)$_j$}$$

where V and $V_{j,t}$ are the differences between the symbols of the left-hand and right-hand expressions in (2.3.43) and (2.3.45)$_j$ respectively and here and below $k \le 2N - 2n - 1$.

Lemma 2.3.9 and theorem 1.1.12(ii) imply this assertion for $n = 1$. In order to make the induction step it is sufficient to construct Hermitian symbols l and $l_{j,t}$ such that

$$l \in \gamma^{-2n}\omega^{-1}h^n \mathrm{S}_{h,\gamma,\gamma,N-n}(\Omega, \mathbb{H}, \mathbb{H}), \tag{2.3.48}$$

$$\partial_t^k l_{j,t} \in \gamma^{-2n-2k}\omega^{-1}h^n \mathrm{S}_{h,\gamma,\gamma,N-n-[\frac{k+1}{2}]}(\Omega_j(t), \mathbb{H}, \mathbb{H}), \tag{2.3.49$_j$}$$

$l_{j,0} = l$ in Ω and

$$[l, a^0] = -iV \qquad \text{in } \Omega, \tag{2.3.50}$$

$$[l_{j,t}, a^0_{j,t}] = -iV_{j,t} \qquad \text{in } \Omega_j(t) \tag{2.3.51$_j$}$$

where a^0 and $a^0_{j,t}$ are defined in (2.3.37) and $(2.3.39)_j$ respectively. In fact, let us replace $R_{j,t}$ by $R_{j,t}(I + iL_{j,t})(I + iL'_{j,t})$ with $L_{j,t} = \mathrm{Op}(l_{j,t})$, $L'_{j,t} = \mathrm{Op}(l'_{j,t})$ with Hermitian symbols $l'_{j,t}$ satisfying $(2.3.48)_j$ with n replaced by $n+1$ and providing $(2.3.44)_j$ after the induction step; moreover, let us replace R by $R(I + iL)(I + L')$; then the induction step will obviously be made. On the other hand, the construction of l and $l_{j,t}$ satisfying (2.3.50) and $(2.3.51)_j$ is obvious and, moreover, Hermitian symbols l with diagonal blocks equal to 0 and $l_{j,t}$ of the form

$$l_{j,t} = \begin{pmatrix} 0 & * & 0 \\ * & 0 & * \\ 0 & * & 0 \end{pmatrix}$$

are uniquely defined from (2.3.50) and $(2.3.51)_j$. □

Let us introduce the operators $\tilde{R}_{j,t} = R_{j,t}\Pi_j$, $\tilde{R}_j = R\Pi_j$ where Π_j are projectors corresponding to the block decomposition in (2.3.35) and $j = 1, \dots, m+1$. Then (2.3.42)–$(2.3.45)_j$ imply that

$$\tilde{R}_j^* \tilde{R}_k \equiv \delta_{jk} I \qquad \text{in } \Omega \qquad (j,k = 1, \dots, m+1), \tag{2.3.52}$$

$$\tilde{R}_{j,t}^* \tilde{R}_{j,t} \equiv I \qquad \text{in } \Omega_j(t) \qquad (j = 1, \dots, m+1), \tag{2.3.52$'$}$$

$$A\tilde{R}_j \equiv \tilde{R}_j A_j \qquad \text{in } \Omega \qquad (j = 1, \dots, m+1), \tag{2.3.53}$$

$$A\tilde{R}_{j,t} - ih^{-1}[A_{j,t}, Q_{j,t}] \equiv \tilde{R}_{j,t} A_{j,t} \qquad \text{in } \Omega_j(t) \qquad (j = 1, \dots, m) \tag{2.3.53$'$}$$

with $A_{m+1} = A'$. Let us consider the equation

$$\partial_t Q_{j,t} \equiv -ih^{-1}[A_{j,t}, Q_{j,t}]. \tag{2.3.54}$$

The arguments used in proofs of lemmas 2.3.2 and 2.3.4 and proposition 2.3.5 easily imply that the proposition remains true for this equation also (not only for the approximate version of (2.3.5)); so the Cauchy problem for (2.3.54) is solvable. Moreover, (2.3.52)–(2.3.54) imply that the operator $Q^t = \tilde{R}_{j,t} Q_{j,t} \tilde{R}_{j,t}^*$ satisfies equation (2.3.5) approximately. Then the arguments of theorem 2.3.6 easily imply

Proposition 2.3.11. *Let*

$$Q = \tilde{R}_j Q_j \tilde{R}_j^* \qquad (j = 1, \dots, m) \tag{2.3.55}$$

and let $Q_{j,t}$ *be a solution of* (2.3.54) *with* $Q_{j,0} = Q_j$. *Then*

$$\|Q^t - e^{-ih^{-1}tA} Q e^{ih^{-1}tA}\| \le Ch^s \|\!|q|\!\|_\gamma$$

for $Q^t = \tilde{R}_{j,t} Q_{j,t} \tilde{R}_{j,t}^*$.

Proof of theorem 2.3.8. (i) First of all proposition 2.3.11 and the arguments of the proof of theorem 2.3.6 yield theorem 2.3.8 for Q of the form (2.3.55).

(ii) Let us note that

$$\sum_{1\le j\le m+1} \tilde{R}_j\tilde{R}_j^* \equiv I \qquad \text{in } \Omega$$

and hence the operator Q with symbol supported in the $\frac{\gamma}{2}$-neighborhood of $\bar{z}$ could be rewritten in the form

$$Q \equiv \sum_{1\le j\le m+1} \tilde{R}_j\tilde{R}_j^* Q \equiv \sum_{1\le j\le m+1} T'\tilde{R}_j Q = Q' + Q''$$

where T' is an operator with symbol supported in the $\frac{3}{4}\gamma$-neighborhood of $\bar{z}$, equal to 1 in the $\frac{2}{3}\gamma$-neighborhood of $\bar{z}$, $Q'' = \tilde{R}_{m+1}T'\tilde{R}_{m+1}Q$. Then (i) yields the theorem for Q replaced by Q'.

(iii) So, we need to prove theorem 2.3.8 only for the operator $Q = \tilde{R}_{m+1}Q'''$. However equation $(hD_t - A_x)U(t,x,y) = 0$ and (2.3.53) immediately yield that $(hD_t - A'_x)U'(t,x,y) \equiv 0$ where the prime corresponds to the $(m+1)$-th component in the block decomposition (2.3.43); moreover, for $z \in \Omega$, $\operatorname{Spec} a'(z) \cap [\tau_1, \tau_2] = \emptyset$, so we obtain (2.3.35) immediately by means of standard elliptic arguments (e.g., by a parametrix construction for $hD_t - A'$ in $\mathbb{R} \times [\tau_1 + \frac{\epsilon}{2}, \tau_2 - \frac{\epsilon}{2}] \times \Omega$). □

Comments

Theorem 2.1.2 was first proven for wave front sets [Ivrii 1]. A weaker version of theorem 2.1.14 (which is very important for applications in spectral asymptotics) was published only in Russian but there is a deep connection between this theorem and results on the normality of a singularity at the origin [Ivrii 1], [Melrose 1], [Hörmander 1]. The description of propagation of singularities in terms of generalized bicharacteristics is due to S. Wakabayashi [1]. There are many descriptions of this propagation in special cases (see [Ivrii 9] which attempts to provide a complete survey). The proofs of the results of section 2.3 have not been published before; they were obtained by me and my graduate student A. Kachalkina. Our starting equation (2.3.5) is due to W. Heisenberg. For wave front sets there is a paper due to A. Volovoi [1] where a parametrix construction by means of Fourier integral operators was examined very carefully; his approach is essentially more complicated. Moreover in standard microlocal analysis (when there is no exterior parameter such as h or a frequency) these results are senseless. I would like to suggest

Problem 2.1.** *Examine the construction in section 2.3 very carefully and obtain less restrictive conditions on $\bar{t}$, norms of the symbol and its derivatives, elements of the Jacobi matrix and γ.* ◇

3. Propagation of Singularities near the Boundary

In this chapter we perform the same task as in the previous chapter but near the smooth boundary. In section 3.1 we prove the basic theorem which is formulated in terms of auxiliary functions as was theorem 2.1.2. In section 3.2 we prove some statements similar to those of the second part of section 2.1 and section 2.2. Section 3.3 is similar to section 2.3 and section 2.4. Finally, in Appendix A we prove some auxiliary assertions.

3.1 General Theory

3.1.1 Statement of the basic theorem. I

In this section our geometric domain is $X = \bar{\mathbb{R}}^+ \times X' = \bar{\mathbb{R}}^+ \times \mathbb{R}^{d-1} \ni (x_1; x') = (x_1; x_2, \dots, x_d)$ with boundary $Y = 0 \times X'$ and interior $\overset{\circ}{X} = X \setminus Y$. We consider the operator

$$P = \sum_{0 \le k \le m} A_k (hD_1)^{m-k} \tag{3.1.1}$$

where

$$A_k \in \boldsymbol{\Psi}'_{h,N}(X, \mathbb{H}, \mathbb{H}) = \bigcap_{0 \le j \le N} C_b^j(\bar{\mathbb{R}}^+, \boldsymbol{\Psi}_{h,N-j}(X', \mathbb{H}, \mathbb{H})).$$

Let $a_k(x, \xi')$ and $p(x, \xi) = \sum_{0 \le k \le m} a_k(x, \xi') \xi_1^{m-k}$ be the principal symbols of A_k and P respectively. We assume that

(3.1.2) $p(\mathrm{z})$ is a Hermitian matrix for every $\mathrm{z} \in T^*X$,

i.e., that the system in question is *symmetric in the principal*. Let $\Omega \Subset T^*Y$ and let us further assume that

$$|a_0(z)w| \ge \epsilon_0 |w| \qquad \forall w \in \mathbb{H} \quad \forall z \in \Omega, \tag{3.1.3}$$

i.e., that *the boundary is not characteristic* in Ω.

Then (3.1.2) yields the equality

$$\langle p(z, D_1)u, v \rangle_+ - \langle u, p(z, D_1)v \rangle_+ = i \langle T(z) \eth u, \eth v \rangle_m \qquad \forall z \in T^*Y \qquad \forall u, v \in \mathcal{S}(\bar{\mathbb{R}}^+, \mathbb{H}) \tag{3.1.4}$$

where $\langle .,. \rangle_+$ and $\langle .,. \rangle_m$ are inner products in $L^2(\mathbb{R}^+, \mathbb{H})$ and $\mathbb{H}^m$ respectively, $\eth u = \eth_{(m)} u = (u, D_1 u, \dots, D_1^{m-1} u)|_{x_1=0}$ is the m-trace of u and $T(z) \in \mathcal{L}(\mathbb{H}^m, \mathbb{H}^m)$ is a matrix with blocks

(3.1.5) $$T_{jk}(z) = \begin{cases} a_{m-j-k-1} & \text{for } j+k \le m+1 \\ 0 & \text{for } j+k \ge m+2 \end{cases} \qquad j,k = 1,\dots,m$$

i.e.,

$$T(z) = \begin{pmatrix} a_{m-1} & a_{m-2} & \dots & a_1 & a_0 \\ a_{m-2} & a_{m-3} & \dots & a_0 & 0 \\ \vdots & \vdots & \ddots & \vdots & \vdots \\ a_1 & a_0 & \dots & 0 & 0 \\ a_0 & 0 & \dots & 0 & 0 \end{pmatrix}.$$

Then (3.1.3) implies the estimate

(3.1.6) $$|T(z)\underline{w}| \ge \epsilon|\underline{w}| \qquad \forall \underline{w} \in \mathbb{H}^m \quad \forall z \in \Omega$$

with a constant $\epsilon = \epsilon(m, c, \epsilon_0, c_2) > 0$, $c_2 = \max_{1\le k\le m} \sup_{z\in\Omega} |a_k(z)|$. Without loss of generality we may replace ϵ by ϵ_0.

Let $B = (B_1, \dots, B_m)$ be a boundary operator, $B_k \in \Psi_{h,N-k}(Y, \mathbb{H}_b, \mathbb{H})$ where $\mathbb{H}_b$ is another auxiliary Hilbert space with inner product $\langle .,. \rangle_b$ and norm $|.|_b$. Let us assume that there exist symbols $\beta_k \in C_b^N(Y, \mathcal{L}(\mathbb{H}_b, \mathbb{H}))$ and $\beta = (\beta_1, \dots, \beta_m)$ such that

(3.1.7) $$\langle T(z)\underline{w}, \underline{w}\rangle_m + 2\,\mathrm{Re}\,\langle b(z)\underline{w}, \beta(z)\underline{w}\rangle_b \le 0 \qquad \forall \underline{w} \in \mathbb{H}^m \qquad \forall z \in \Omega$$

where b_k and b are the principal symbols of B_k and B respectively; in this case the boundary value problem (P, B) is called *dissipative* in Ω; moreover, if

(3.1.7)* $$\langle T(z)\underline{w}, \underline{w}\rangle_m + 2\,\mathrm{Re}\,\langle b(z)\underline{w}, \beta(z)\underline{w}\rangle_b \le -\epsilon_0|\underline{w}|^2 \qquad \forall \underline{w} \in \mathbb{H}^m \qquad \forall z \in \Omega$$

then the boundary value problem (P, B) is called *strictly dissipative* Ω.

In what follows the most important case is $\mathbb{H} = \mathbb{C}^{\mathrm{D}}$; one can easily prove then that if $\mathrm{D}_\pm$ and $\bar{\mathrm{D}}_\pm$ are the numbers of positive (negative) eigenvalues of the matrices a_0 and T respectively then (3.1.3) yields that

(3.1.8) $$\bar{\mathrm{D}}_\pm = \begin{cases} \dfrac{1}{2}mD & \text{for even } m \\ \dfrac{1}{2}(m-1)\mathrm{D} + \mathrm{D}_\pm & \text{for odd } m. \end{cases}$$

Moreover, the most important case is $\mathbb{H}_b = \mathbb{C}^{\bar{\mathrm{D}}_+}$ when condition (3.1.7) is equivalent to the pair of conditions

(3.1.9) $$\mathrm{rank}\, b(z) = \bar{\mathrm{D}}_+ \qquad \forall z \in \bar{\Omega},$$

(3.1.10) $$\langle T(z)\underline{w}, \underline{w}\rangle_m \le 0 \qquad \forall \underline{w} \in \mathrm{Ker}\, b(z) \quad \forall z \in \bar{\Omega};$$

moreover, condition (3.1.7)* is equivalent to (3.1.9) and

(3.1.10)* $$\langle T(z)\underline{w}, \underline{w}\rangle_m \le -\epsilon_0|\underline{w}|^2 \qquad \forall \underline{w} \in \mathrm{Ker}\, b(z) \quad \forall z \in \bar{\Omega}.$$

Let us note that (3.1.3) implies that

(3.1.11) $$|\eta|^m|w| \le C(|p(z,\eta)w| + |w|) \qquad \forall w \in \mathbb{H} \quad \forall z \in \Omega \quad \forall \eta \in \mathbb{R}.$$

Let $\Sigma_\epsilon = \Sigma_\epsilon(P)$ be the set of all the points $z \in T^*X$ at which the ellipticity condition

(3.1.12) $$|p(z)w| \geq \epsilon|w| \qquad \forall w \in \mathbb{H}$$

is violated; here and below $\epsilon > 0$ is an arbitrarily small constant. Let $\Sigma'_\epsilon = \iota\Sigma_\epsilon|_Y$ be the set of all the points $z \in \Omega$ at which the ellipticity condition

(3.1.12)′ $$|p(z,\eta)w| \geq \epsilon|w| \qquad \forall w \in \mathbb{H} \quad \forall \eta \in \mathbb{R}$$

is violated; here and in what follows $\iota : T^*X|_Y \to T^*Y$ is a natural map.

Let us introduce the linear space $\mathbb{H}'(z) = \{v \in L^2(\mathbb{R}^+, \mathbb{H}) : p(z, D_1)v = 0\}$; then for $z \notin \Sigma'_\epsilon$ the following estimates hold:

(3.1.13) $$|e^{\sigma x_1} D_1^j v|_+ \leq C_j |v|_+ \qquad \forall j \quad \forall v \in \mathbb{H}'(z)$$

with an appropriate exponent $\sigma = \sigma(m, c, c_1, \epsilon_0, \epsilon) > 0$ and constants $C_j = C_j(m, c, c_1, \epsilon_0, \epsilon)$.

Let us introduce the sets $\Sigma_{b,\epsilon} = \Sigma_{b,\epsilon}(P, B)$: $z \in \Sigma_{b,\epsilon}$ if **either** $z \in \Sigma'_\epsilon$ **or** the following condition

(3.1.14) $$|b(z)\eth_m v|_{\mathbb{H}'} \geq \epsilon|v|_+^2 \qquad \forall v \in \mathbb{H}'(z)$$

is violated.

Let $\Sigma = \Sigma(P) = \cap_{\epsilon>0}\Sigma_\epsilon$, $\Sigma_b = \Sigma_b(P, B) = \cap_{\epsilon>0}\Sigma_{b,\epsilon}$; it is obvious that Σ and Σ_b are closed sets and that the space $\mathbb{H}'(z)$ depends continuously (in the natural sense) on $z \in \Omega \setminus \Sigma'$ where $\Sigma' = \iota\Sigma|_Y$.

Definition 3.1.1. A point $z \in \Sigma \subset T^*X$ ($z \in T^*X \setminus \Sigma$) is called a *characteristic* (*elliptic* respectively) point of the operator P. A point $z \in \Sigma_b \subset \Omega$ ($z \in \Omega \setminus \Sigma_b$) is called a *characteristic* (*elliptic* respectively) point of the problem (P, B). ◊

By means of standard elliptic arguments one can easily prove

Theorem 3.1.2. *Let condition* (3.1.3) *be fulfilled,* $N = N(d, s, l, q)$ *and*

(3.1.15) $$\|D_1^j u\|_X \leq h^{-l} \qquad \forall j = 1, \ldots, m + q.$$

Then

(3.1.16) $$\mathrm{OF}_b^{s,q+m}(u) \cap \Omega \subset \mathrm{OF}_b^{s,q}(Pu) \cup \mathrm{OF}^{s-\frac{1}{2}}(B\eth_h u) \cup \Sigma_b$$

where here and in what follows

$$\mathrm{OF}_b^{s,q}(u) = \bigcup_{0 \leq j \leq q} \mathrm{OF}^{s-j'}(D_1^j u)|_Y$$

is called the boundary oscillation front set of order (s, q) of u *and* $\eth_h u = \eth_{h(m)} u = (u, hD_1 u, \ldots, (hD_1)^{m-1} u)|_Y$.

We will prove this theorem later.

Let us now assume that

(3.1.17) $$\begin{aligned} \mathbb{H} &= \mathbb{C}^{D_1} \otimes \mathbb{H}_1^0 \oplus \ldots \oplus \mathbb{C}^{D_M} \otimes \mathbb{H}_M^0 \oplus \mathbb{H}_0 \\ p &= p_1 \otimes I_{\mathbb{H}_1^0} \oplus \ldots \oplus p_M \otimes I_{\mathbb{H}_M^0} \oplus p_0 \end{aligned}$$

p_0 on $\mathbb{H}_0$ satisfies condition (2.1.12)′ with $\epsilon = \epsilon_0$

where $\mathbb{H}_1^0, \ldots, \mathbb{H}_M^0, \mathbb{H}_0$ are auxiliary Hilbert spaces and $p_1, \ldots, p_M, p_0$ are symbols on $\mathbb{C}^{D_1}, \ldots, \mathbb{C}^{D_M}, \mathbb{H}_0$ respectively.

Then at every point $z \in \Omega$ the estimate (3.1.13) holds with an appropriate exponent $\sigma > 0$ and constants C_j depending on $p(z,.)$. Moreover, the space $\mathbb{H}''(z)$ depends semicontinuously on z:

$$\underline{\lim}_{\, z \to z_0} \mathbb{H}'(z) \supset \mathbb{H}'(z_0) \tag{3.1.18}$$

in the sense that

$$\lim_{z\to z_0} (I - \Pi'(z))\Pi'(z_0) = \Pi'(z_0) \lim_{z\to z_0} (I - \Pi'(z)) = 0 \tag{3.1.18$'$}$$

where $\Pi'(z)$ is the orthogonal projector to $\mathbb{H}'(z)$.

Let us introduce the *characteristic polynomial*

$$g(z,\eta) = \det p_1(z,\eta) \dots \det p_M(z,\eta). \tag{3.1.19}$$

Let $z \in \Omega$ and let $\eta_1, \dots, \eta_n$ be all the real distinct roots of $g(z,\eta)$ where n and $\eta_1, \dots, \eta_n$ depend on z.

Definition 3.1.3. Let conditions (3.1.1)–(3.1.3), (3.1.17) and (3.1.7) be fulfilled. Then the problem (P, B) is *microhyperbolic at the point $z \in \Omega$ in the multidirection $\mathcal{T} = (\mathcal{T}', \nu_1, \dots, \nu_n) \in T_zT^*Y \times \mathbb{R}^n$* if the following two conditions are fulfilled:

(i) For every $k = 1, \dots, n$ the operator P is microhyperbolic at the point $\mathrm{z}_k = (z, \eta_k)$ in the direction $\mathcal{T}_k = (\mathcal{T}', \nu_k, 0) \in T_{\mathrm{z}_k}(T^*X|_Y) \subset (T_{\mathrm{z}_k}T^*X)|_Y$;

(ii) The following estimate holds

$$\begin{aligned} - \operatorname{Re} \langle (\mathcal{T}'p)(z, D_1)v, v\rangle_+ + \operatorname{Re} i \langle (\mathcal{T}'b)(z)\eth v, \beta(z)\eth v\rangle_b \geq \\ \epsilon |v|_+^2 - c_1(c_1|b(z)\eth v|'^2 - \langle T(z)\eth v, \eth v\rangle_m) \qquad \forall v \in \mathbb{H}'(z) \end{aligned} \tag{3.1.20}$$

(compare with estimate (2.1.1) in definition 2.1.1). ◊

Below we prove the following

Proposition 3.1.4. *Let the problem (P, B) be microhyperbolic at the point z in the multidirection $\mathcal{T}$. Let the problem $(\tilde{P}, \tilde{B})$ also satisfy conditions* (3.1.1)–(3.1.3), (3.1.17) *and* (3.1.7) *and suppose $\|\tilde{p} - p\|_{C^1_b} \leq \varepsilon$, $\|\tilde{b} - b\|_{C^1_b} \leq \varepsilon$, $|\tilde{z} - z| \leq \varepsilon$, $|\tilde{\mathcal{T}} - \mathcal{T}| \leq \varepsilon$ with a small enough constant $\varepsilon > 0$. Then the problem $(\tilde{P}, \tilde{B})$ is also microhyperbolic at the point $\tilde{z}$ in the multidirection $\tilde{\mathcal{T}}$.*

Remark 3.1.5. *Surely the structure of $\iota^{-1}\tilde{z} \cap \Sigma(\tilde{P})$ does not necessarily coincide with that of $\iota^{-1}z \cap \Sigma(P)$. However one can avoid this difficulty by either of two methods:*

(i) If z_{ki} with $i = 1, \dots, t_k$ $(t_k \geq 0)$ are points of $\iota^{-1}\tilde{z} \cap \Sigma(\tilde{P})$ close to z_k then the condition $|\tilde{\mathcal{T}} - \mathcal{T}| \leq \varepsilon$ is understood in the sense that $|\tilde{\mathcal{T}}_{ki} - \mathcal{T}_k| \leq \varepsilon$, $\forall i = 1, \dots, q_k$, $\forall k = 1, \dots, n$;

(ii) The condition that η_k $(k = 1, \dots, n)$ are all the real roots of $g(z,\eta)$ is replaced by the condition that the set $\{\eta : (z,\eta) \in \Sigma_\epsilon\}$ is contained in the union of the intervals $(\eta_k - \mu, \eta_k + \mu)$ $(k = 1, \dots, n)$ where $|\eta_k - \eta_j| \geq 3\mu$ $\forall j \neq k$ and $0 < \mu \leq \bar{\mu} = \bar{\mu}(c, \epsilon_0, c_1, c_3)$ where ϵ_0 and c_1 are constants from the microhyperbolicity conditions (2.1.1) with $z, \mathcal{T}$ replaced by $\mathrm{z}_k, \mathcal{T}_k$ $(k = 1, \dots, n)$, (3.1.3) and (3.1.20), and $c_3 = \max_j \|a_j\|_{C^2_b} + \max_j \|b_j\|_{C^2_b}$. In this case $\varepsilon = \varepsilon(d, m, M, \mathrm{D}^*, c, \epsilon_0, c_1, c_2, \epsilon)$, $D^* = m \max_j D_j$. ◊

Now we can formulate the main result of this section (and even of this chapter):

Theorem 3.1.6. *Let conditions* (3.1.1)–(3.1.3), (3.1.17), (3.1.7) *be fulfilled and let $N = N(d, m, s, l, q, \mathrm{D}^*)$. Let $z^* \in \Omega$ and $\phi_j \in C^N(T^*Y)$, $\phi_{kj} \in C^N(T^*X' \times \bar{\mathbb{R}}^+)$ with $j = 1, \dots, J$, $k = 1, \dots, n$ be real-valued functions such that*

$$\phi_{1j} = \cdots = \phi_{nj} = \phi_j \qquad \text{at } T^*Y \quad \forall j \tag{3.1.21}$$

and for every $j = 1, \dots, J$ *the problem* (P, B) *is microhyperbolic at the point* z^* *in the multidirection* $\mathcal{T}_j = (H_{\phi_j}, -\partial_{x_1}\phi_{1j}, \dots, -\partial_{x_1}\phi_{nj})(z^*)$. *Let* u *satisfy* (3.1.15) *and*

$$(\mathrm{OF}_b^{s+1,q}(Pu) \cup \mathrm{OF}^{s+\frac{1}{2}}(B\eth_h u)) \cap \Omega' \cap \{\phi_1 \le t_1\} \cap \cdots \cap \{\phi_J \le t_J\} = \emptyset, \tag{3.1.22}$$

$$\mathrm{OF}^{s+1}(Pu) \cap \Omega'_k \cap \{\phi_{k1} \le t_1\} \cap \cdots \cap \{\phi_{kJ} \le t_J\} = \emptyset \quad \forall k = 1, \dots, n \tag{3.1.23}$$

$$\mathrm{OF}_b^{s,0}(u) \cap \partial\Omega' \cap \{\phi_1 \le t_1\} \cap \cdots \cap \{\phi_J \le t_J\} = \emptyset, \tag{3.1.24}$$

$$\mathrm{OF}^{s}(u) \cap \partial\Omega'_k \cap \{\phi_{k1} \le t_1\} \cap \cdots \cap \{\phi_{kJ} \le t_J\} = \emptyset \quad \forall k = 1, \dots, n \tag{3.1.25}$$

where Ω' *and* Ω'_k *are small enough neighborhoods of* z *and* z_k *in* T^*Y *and* T^*X *respectively, independend of* $u, t_1, \dots, t_J$, *such that* $\iota\Omega'_k|_Y = \Omega'$, $\forall k = 1, \dots, n$.

(i) *Then*

$$\mathrm{OF}_b^{s,q+m}(u) \cap \Omega' \cap \{\phi_1 \le t_1\} \cap \cdots \cap \{\phi_J \le t_J\} = \emptyset, \tag{3.1.26}$$

(ii) *Moreover, if condition* (3.1.7)* *is fulfilled then one can weaken condition* (3.1.22), *replacing* $\mathrm{OF}^{s+\frac{1}{2}}(B\eth_h u)$ *by* $\mathrm{OF}^{s}(B\eth_h u)$ *and in addition obtain at the same time that*

$$\mathrm{OF}^{s-i}(D_1^i u|_Y) \cap \Omega' \cap \{\phi_1 \le t_1\} \cap \cdots \cap \{\phi_J \le t_J\} = \emptyset \qquad \forall i = 0, \dots, m+q-1. \tag{3.1.27}$$

Remark 3.1.7. *(i) Conditions (3.1.3), (3.1.15), (3.1.22) and (3.1.24) again imply (3.1.24) with* $\mathrm{OF}^{s,0}(u)$ *replaced by* $\mathrm{OF}^{s,m+q}(u)$.

(ii) The uniform version of theorem 3.1.6 (similar to the result which was given in the proof of theorem 2.1.2) is obvious but very long in formulation. ◊

3.1.2 Reduction to a special case

In order to prove theorems 3.1.2 and 3.1.6 and proposition 3.1.4 we first of all reduce the general case to a special case.

(i) First of all, *one can assume that* $m = 1$. In fact, let us set $\underline{u} = (u_j)_{j=1,\dots,m}$ with $u_j = (hD_1)^{j-1}u$. Then in the intersection with the set in question

$$\mathrm{OF}^s(u) = \mathrm{OF}^s(\underline{u}), \qquad \mathrm{OF}^{s,q+m'}(u) = \mathrm{OF}^{s,q+1'}(\underline{u}),$$
$$\mathrm{OF}^{s'}(Pu) = \mathrm{OF}^{s'}(\underline{P}\,\underline{u}), \qquad \mathrm{OF}^{s,q'}(Pu) = \mathrm{OF}^{s,q'}(\underline{P}\,\underline{u})$$

provided $|s - s'| \le 1$ where $\underline{P} = \underline{A}_0 hD_1 + \underline{A}_1$,

$$\underline{A}_0 = (A_{m-j-k+1})_{j,k=1,\dots,m}$$

is an operator with principal symbol $T(z)$, $\underline{A}_1 = -\underline{A}_0 L$,

$$L = (L_{jk})_{j,k=1,\dots,m},$$

$L_{jk} = \delta_{j+1,k} - A'_0 A_{m-k+1}\delta_{j,m}$ where A'_0 is a parametrix of A_0 here (condition (3.1.3) guarantees its existence) and $\delta_{j,k}$ is the Kronecker symbol. Then modulo operators with symbols vanishing in the domain in question,

$$\underline{A}_1 = (-A_{m-j-k+2} + A_{m-j+1}\delta_{k,1} + A_{m-k+1}\delta_{j,1})_{j,k=1,\dots,m}.$$

Then condition (3.1.3) for the operator $\underline{P}$ is fulfilled (this condition for P was assumed). Moreover, condition (3.1.2) for P is obviously equivalent to the same condition for $\underline{P}$; as well condition (3.1.17) for P is equivalent to the same condition for $\underline{P}$ with D_j replaced by mD_j. It is easy to prove that $\Sigma_{\epsilon_1\epsilon}(P) \subset \Sigma_\epsilon(\underline{P}) \subset \Sigma_{C_1\epsilon}(P)$ for appropriate $\epsilon_1 = \epsilon_1(c, \epsilon_0, c_1, m) > 0$ and $C_1 = C_1(c, \epsilon_0, c_1, m)$ and arbitrary $\epsilon > 0$.

Moreover, $B\mathfrak{d}_{h(m)}u = B\mathfrak{d}_{h(1)}\underline{u}$ and hence condition (3.1.7) or (3.1.7)* for (P, B) is equivalent to to the same condition for $(\underline{P}, B)$; then it is easy to prove that $\Sigma_{\epsilon_1\epsilon}(P, B) \subset \Sigma_\epsilon(\underline{P}, B) \subset \Sigma_{C_1\epsilon}(P, B)$ for appropriate $\epsilon_1 = \epsilon_1(c, \epsilon_0, c_1, c_2, m) > 0$ and $C_1 = C_1(c, \epsilon_0, c_1, c_2, m)$ and arbitrary $\epsilon > 0$.

Finally, it is easy to prove the inequalities

$$|\langle(\mathcal{T}p)(z)w, w\rangle - \langle(\mathcal{T}\underline{p})(z)\underline{w}, \underline{w}\rangle_m| \le C|\underline{p}(z)\underline{w}| \cdot |\underline{w}| \cdot |\mathcal{T}| \qquad \forall \underline{w} \in \mathbb{H}^m : w = w_1$$

and

$$|\langle(\mathcal{T}'p)(z, D_1)v, v\rangle_+ - \langle(\mathcal{T}'\underline{p})(z, D_1)\underline{v}, \underline{v}\rangle_+| \le C\|\underline{p}(z, D_1)\underline{v}\|_+ \cdot \|\underline{v}\|_+ \cdot |\mathcal{T}| \qquad \forall \underline{v} \in \mathcal{S}(\bar{\mathbb{R}}^+, \mathbb{H}^m) : v = v_1.$$

Hence $\underline{P}$ is microhyperbolic at z in the direction $\mathcal{T}$ if and only if P has the same property; moreover, $(\underline{P}, B)$ is microhyperbolic at z in the multidirection $\mathcal{T}$ if and only if (P, B) has the same property. The reduction step is complete.

(ii) Let us now show that in the proofs of theorem 3.1.6 and proposition 3.1.4 *one can assume that*

$$|\det p_j(z^*, \eta)| \ge \epsilon \qquad \text{for } z \in \mathbb{C}, \quad |\eta - \eta_j| \ge \mu \tag{3.1.28}$$

where $0 < \mu$ is any arbitrarily chosen number, $\epsilon > 0$ depends on μ, η_j are not necessarily distinct,

$$p_0 = \begin{pmatrix} 0 & p_0' \\ p_0'' & 0 \end{pmatrix}, \qquad p_0' = \rho(\eta - \alpha), \quad p_0'' = (\eta - \alpha^\dagger)\rho^\dagger \tag{3.1.29}$$

and the matrix symbols α and ρ are such that

$$|(\eta - \alpha)w|| \ge \epsilon|w| \quad |\rho w| \ge \epsilon|w| \qquad \forall w \in \mathbb{H} \quad \forall \eta \in \mathbb{C} : \operatorname{Im}\eta \le \mu. \tag{3.1.30}$$

Similarly, in the proof of theorem 3.1.2 *one can assume that*

$$p_0 = \begin{pmatrix} 0 & p_0' \\ p_0'' & 0 \end{pmatrix}, \quad p_0' = \eta - \alpha', \quad p_0'' = \eta - \alpha \tag{3.1.29$'$}$$

and (3.1.30) holds for $\alpha = \alpha'$ and $\alpha = -\alpha''$.

Also in this reduction D^* does not increase.

In order to make the reduction step let us first consider the symbols $p_j(z, \eta)$ with $j = 1, \dots, M$ separately. Let η_{jk} be all the real distinct roots of $\det p_j(z^*, \eta)$, $k = 1, \dots, n_j$. Then η_{jk} are all the real distinct eigenvalues of $L_j(z^*)$ where $L_j(z) = -a_{0j}(z)^{-1}a_{1j}(z)$ (let us recall that $p_j(z, \eta) = a_{0j}(z)\eta + a_{1j}(z))$ and $z \in \Omega''$ where Ω'' is a small enough neighborhood of z^* in $T^*X' \times \bar{\mathbb{R}}^+$ which we shrink if necessary. Then there exist symbols $r_j(z)$ defined, smooth and invertible such that in Ω''

$$r_j^{-1}L_jr_j = \tilde{L}_j = \begin{pmatrix} \tilde{L}_{j1} & \dots & 0 & 0 \\ \vdots & \ddots & \vdots & \vdots \\ 0 & \dots & \tilde{L}_{jn_j} & 0 \\ 0 & \dots & 0 & \tilde{L}_{j0} \end{pmatrix},$$

where for $k \geq 1$ all the eigenvalues of $\tilde{L}_{jk}$ are equal to η_{jk} and all the eigenvalues of $\tilde{L}_{j0}$ are non-real. Then $r_j^\dagger a_{1j}r_j = r_j^\dagger a_{0j}r_j\tilde{L}_j$ and taking the Hermitian conjugate we obtain that $(r_j^\dagger a_{0j}r_j)\tilde{L}_j = \tilde{L}_j^\dagger(r_j^\dagger a_{0j}r_j)$. Since $\tilde{L}_j$ is of block-diagonal type and since $\eta \neq \bar{\zeta}$ if η and ζ belong to spectra of different blocks (here the bar means complex conjugation) we obtain that $r_j^\dagger a_{0j}r_j$ is also of block-diagonal type and hence $r_j^\dagger a_{1j}r_j$ is also of that type. Therefore multiplying $p(z,\eta)$ on the left and on the right by $r^\dagger(z)$ and $r(z)$ respectively we obtain

$$p = \oplus \sum_{1\leq j\leq M, 1\leq k\leq n_j} p_{jk} \otimes I_{\mathbb{H}_j^0} \oplus p_0 \tag{3.1.31}$$

provided

$$r = \begin{pmatrix} r_1 \otimes I_{\mathbb{H}_1^0} & \dots & 0 & 0 \\ \vdots & \ddots & r_M \otimes I_{\mathbb{H}_M^0} & \\ 0 & \dots & 0 & I \end{pmatrix};$$

here $\det p_{jk}(z^*,\eta)$ has no complex root except $\eta_{jk} \in \mathbb{R}$ and p_0 satisfies (3.1.12)$'$; after this reduction step the space $\mathbb{H}_0$ can be larger than before.

Similarly, let us consider $L_0(z) = -a_{00}(z)^{-1}a_{10}(z)$ (let us recall that $p_0(z,\eta) = a_{00}(z)\eta + a_{10}(z)$); then the spectrum of $L_0(z)$ lies in $\{\eta \in \mathbb{C}, |\operatorname{Im}\eta| \geq \epsilon\}$ and hence there exists a smooth and invertible symbol $r_0(z)$ such that in Ω''

$$r_0^{-1}L_0r_0 = \tilde{L}_0 = \begin{pmatrix} L_0' & 0 \\ 0 & L_0'' \end{pmatrix},$$

where the spectrum of L_0' lies in $\{\eta \in \mathbb{C}, \operatorname{Im}\eta \geq \epsilon\}$ and the spectrum of L_0'' lies in $\{\eta \in \mathbb{C}, \operatorname{Im}\eta \leq -\epsilon\}$. Then using the same arguments as before (but taking into account that $\eta \neq \bar{\zeta}$ provided η and ζ belong to the spectrum of the same block) we obtain that both symbols $r_0^\dagger a_{00}r_0$ and $r_0^\dagger a_{10}r_0$ are of the form $\begin{pmatrix} 0 & * \\ * & 0 \end{pmatrix}$. Then multiplying (3.1.31) on the left and on the right by $r^\dagger$ and r respectively with

$$r = \begin{pmatrix} I & \dots & 0 & 0 \\ \vdots & \ddots & \vdots & \vdots \\ 0 & \dots & I & 0 \\ 0 & \dots & 0 & r_0 \end{pmatrix},$$

we obtain p_0 of the form (3.1.29). So, the second step of the reduction is provided by multiplying P on the left and on the right by $\operatorname{Op}(r)^*$ and $\operatorname{Op}(r)$ respectively where r is now the product of the two matrices constructed in the two sub-steps of this step.

(iii) In the final reduction step we obtain

$$P = \begin{pmatrix} P_1 & \dots & 0 & 0 \\ \vdots & \ddots & \vdots & \vdots \\ 0 & \dots & P_n & 0 \\ 0 & \dots & 0 & P_0 \end{pmatrix}, \qquad P_0 = \begin{pmatrix} 0 & P_0' \\ P_0'' & 0 \end{pmatrix} \tag{3.1.32}$$

with $P_0'' = P_0'^*$ where for $k = 1, \ldots, n$ the principal symbols of P_k are of the form

$$p_k = \oplus \sum_{\leq j \leq M_k} p_{jk} \otimes I_{\mathbf{H}_j^0}, \tag{3.1.33}$$

p_{jk} are the same as in (3.1.31), and p_0' and p_0'' (the principal symbols of P_0' and P_0'' respectively) are of the same type as in (3.1.29).

In order to make this reduction step we assume that (3.1.32) is fulfilled modulo operators with symbol $h^n(F_0\eta + F_1)$; we then apply induction on n. For $n = 1$ these approximate equalities are fulfilled. In order to make the induction step we multiply P on the right and on the left by $(I + h^n Q^*)$ and $(I + h^n Q)$ respectively; then the step in the first equality (3.1.32) will be made provided for $j \neq k$

$$q_{kj}^\dagger a_{0k} + a_{0j} q_{jk} = F_{0jk},$$
$$q_{kj}^\dagger a_{1k} + a_{1j} q_{jk} = F_{1jk}$$

where a_{0k}, a_{1k} are diagonal blocks and q_{jk}, F_{0jk}, F_{1jk} are blocks of the corresponding symbols, $Q = \mathrm{Op}(q)$. Since F_0 and F_1 are Hermitian we need to fulfill these equalities only for $j < k$; for $j > k$ they will then be fulfilled automatically. Solving for $q_{kj}^\dagger$ in the first equality and substituting into the second we obtain after multiplication by a_{0j}^{-1} that we need only solve the equation

$$-q_{jk} L_k + L_j q_{jk} = F'_{jk}$$

for $j < k$ where $L_k = -a_{0k}^{-1} a_{1k}$ and F'_{jk} is obtained from F_{0jk} and F_{1jk}. This equation is solvable since the spectra of L_j and L_k are disjoint (see Appendix A). The choice of q_{jk} for $j = k$ is arbitrary.

In order to make the induction step in the second equality in (3.1.32) we need to solve the equations

$$q_{(21)}^\dagger \rho^\dagger + \rho q_{(21)} = F_{0(11)},$$
$$q_{(21)}^\dagger \alpha^\dagger \rho^\dagger + \rho \alpha q_{(21)} = F_{1(11)}$$

and similar equations for q_{12} where $\rho = p_0'$, $q_{(jk)}$ are blocks of

$$q_{00} = \begin{pmatrix} q_{(11)} & q_{(12)} \\ q_{(21)} & q_{(22)} \end{pmatrix},$$

$F_{0(jk)}$ and $F_{1(jk)}$ are similar blocks of the matrices F_{000} and F_{100}, and these blocks are Hermitian for $j = k$. The solution of the first equation is of the form

$$q_{(21)} = \rho^{-1}\left(\frac{1}{2} F_{0(11)} + i\Lambda\right)$$

where Λ is an arbitrary Hermitian symbol. Substituting this equality to the second equation we obtain that

$$\Lambda \alpha'^\dagger - \alpha' \Lambda = i F'_{(11)}$$

where $\alpha = \rho\alpha\rho^{-1}$ and $F'_{(11)}$ is some given Hermitian symbol. We can solve this equation because the spectra of α' and $\alpha'^\dagger$ are disjoint (again see Appendix A). Moreover, one can choose $q_{(11)}$ and $q_{(22)}$ arbitrarily. The induction step is done.

The reduction is complete.

3.1.3 Elliptic theory

The following series of assertions is crucial in the proof of theorem 3.1.2 and is important in the proof of theorem 3.1.6:

Proposition 3.1.8. *Let* $\alpha \in C_b^q(\mathbb{R}, \boldsymbol{\Psi}_{h,N}(\mathbb{R}^d, \mathbb{H}, \mathbb{H}))$ *be an operator with principal symbol* α_0 *such that*

$$|(\eta - \alpha_0(z))w| \geq \epsilon |w| \quad \forall w \in \mathbb{H} \quad \forall \eta \in \mathbb{C} : \operatorname{Im} \eta \leq \epsilon \quad \forall z \in \mathbb{R} \times T^*\mathbb{R}^d \tag{3.1.34}$$

and let $q \geq 1$ *and* $N = N(d, s, q, l)$. *Then for* $h \in (0, h_0]$ *with* $h_0 = h_0(d, c, \epsilon, c_2) > 0$, $c_2 = \|\alpha\|$ *there exists an operator family*

$$G \in C^{q+1}(\Delta, \mathcal{L}(L^2(\mathbb{R}^d, \mathbb{H}), L^2(\mathbb{R}^d, \mathbb{H})))$$

with $\Delta = \{(x_1, y_1) \in \mathbb{R}^2, x_1 \geq y_1\}$ *such that*

(i) The following relations hold:

$$G(x_1, x_1) = I, \tag{3.1.35}$$

$$\|D_{x_1}^j D_{y_1}^k G(x_1, y_1)\| \leq C h^{-j-k} e^{-\epsilon h^{-1}(x_1 - y_1)} \qquad \forall j, k : j + k \leq q + 1 \tag{3.1.36}$$

where here and below $C = C(d, q, s, l, \epsilon, c_2)$;

(ii) If the operators $G_\pm : L^2(\mathbb{R}^{d+1}, \mathbb{H}) \to L^2(\mathbb{R}^{d+1}, \mathbb{H})$ *are given by the formulas*

$$(G_+ f)(x_1) = ih^{-1} \int_{-\infty}^{x_1} G(x_1, y_1) f(y_1) dy_1, \tag{3.1.37$_+$}$$

$$(G_- f)(x_1) = -ih^{-1} \int_{x_1}^{\infty} G(y_1, x_1) f(y_1) dy_1 \tag{3.1.37$_-$}$$

then $G_+^* = G_-$ *and*

$$\|G_\pm\| \leq C, \tag{3.1.38}$$

$$PG_+ = G_+ P = I, \tag{3.1.39$_+$}$$

$$P^* G_- = G_- P^* = I \tag{3.1.39$_-$}$$

where $P = hD_1 - \alpha$ *(and then* $P^* = hD_1 - \alpha^*$*).*

Moreover, if f *is tempered (i.e.,* $\|D_1^j f\| \leq h^{-l}\ \forall j = 0, \dots, q$*) then*

$$\mathrm{OF}^{s,q+1\prime}(G_\pm f) = \mathrm{OF}^{s,q\prime}(f); \tag{3.1.40}$$

(iii) If the operator family $G_t : L^2(\mathbb{R}^d, \mathbb{H}) \to L^2(\bar{\mathbb{R}}_t^+ \times \mathbb{R}^d, \mathbb{H})$ *with* $\bar{\mathbb{R}}_t^+ = \{x_1, x_1 \geq t\}$ *is given by the formula*

$$(G_t f)(x_1) = G(x_1, t) f \tag{3.1.41}$$

then

$$\|G_t\| \leq C h^{\frac{1}{2}}, \tag{3.1.42}$$

$$PG_t = 0, \qquad \eth_t G_t = I \tag{3.1.43}$$

where $\eth_t u = u|_{x_1 = t}$ *and, moreover, if* f' *is tempered then*

$$\mathrm{OF}^{s,q+1\prime}(G_t f') = \{t\} \times \mathrm{OF}^{s-\frac{1}{2}}(f'). \tag{3.1.44}$$

Proof. Let us note that for $h \in (0, h_0]$ with an appropriate constant $h_0 > 0$ the following estimate holds:

$$\|(\eta - \alpha)v\| \geq \frac{1}{2}\epsilon\|v\| \qquad \forall v \in L^2(\mathbb{R}^d, \mathbb{H}) \quad \forall \eta \in \mathbb{C} : \operatorname{Im}\eta \leq \frac{1}{2}\epsilon \quad \forall x_1 \in \mathbb{R}.$$

Then (3.1.35) and (3.1.36) with $j = k = 0$ are fulfilled for $G_{(1)}(x_1, y_1) = e^{ih^{-1}(x_1-y_1)\alpha_{y_1}}$ with $\alpha_{y_1} = \alpha|_{x_1=y_1}$ and for $G_{(2)}(x_1, y_1) = e^{ih^{-1}(x_1-y_1)\alpha}$ and hence if we define $G'_\pm$ by formulas $(3.1.37)_\pm$ with $G_{(1)}$ and $G_{(2)}$ respectively instead of G then (3.1.38) will be true and $(3.1.38)_\pm$ will be true only partially and approximately:

$$PG'_+ = I + R_+, \quad P^*G'_- = I + R_-, \quad \|R_\pm\| \leq Ch.$$

Then for $h \in (0, h_0]$ with an appropriate constant $h_0 > 0$ the operators $G_\pm = G'_\pm(I + R_\pm)^{-1}$ satisfy $(3.1.39)_\pm$ partially

$$(3.1.39)' \qquad PG_+ = P^*G_- = I$$

and are of the form $(3.1.36)_\pm$ with appropriate functions G satisfying (3.1.35); (3.1.36) may be different for different indices $\pm$. However $(3.1.39)'$ yields that $G_+^*P^* = G_-^*P = I$ and these equalities and $(3.1.39)'$ again imply that $G_+ = G_-^*$ and $(3.1.39)_\pm$ hold; then the functions G in $(3.1.37)_\pm$ coincide. This implies (3.1.43) and (3.1.36).

It remains to prove (3.1.40) and (3.1.44); in fact, it is sufficient to prove that $\mathrm{OF}^{s'}(G_\pm f)$ and $\mathrm{OF}^{s'}(G_t f')$ are contained in the right-hand sets of (3.1.40) and (3.1.44) respectively; these inclusions and (3.1.39) and (3.1.43) imply (3.1.40) and (3.1.44). It is easy to see that it is sufficient to prove these inclusions for $G'_\pm f$ and $G'_t f'$ where G'_t is given by (3.1.41) with G replaced by $G_{(1)}$.

Note that

$$e^{ih^{-1}\tau\alpha}Q = \sum_{0\leq k\leq n} \frac{1}{k!}\partial_\lambda^k (e^{ih^{-1}\tau\lambda\alpha}Qe^{ih^{-1}\tau(1-\lambda)\alpha})|_{\lambda=0} +$$
$$\frac{1}{n!}\int_0^1 (1-\lambda)^n \partial_\lambda^{n+1}(e^{ih^{-1}\tau\lambda\alpha}Qe^{ih^{-1}\tau(1-\lambda)\alpha})d\lambda =$$
$$\sum_{0\leq k\leq n} \frac{1}{k!}(\mathrm{Ad}^k_{ih^{-1}\alpha} Q)\tau^k e^{ih^{-1}\tau\alpha} +$$
$$\frac{1}{n!}\int_0^1 (1-\lambda)^n (\mathrm{Ad}^{n+1}_{ih^{-1}\alpha} Q)\tau^{n+1}e^{ih^{-1}\tau(1-\lambda)\alpha}d\lambda$$

where $(\mathrm{Ad}^0_M Q) = Q$, $(\mathrm{Ad}^1_M Q) = [M, Q]$ and $(\mathrm{Ad}^k_M Q) = [M, (\mathrm{Ad}^{k-1}_M Q)]$ for $k \geq 1$.

It is easy to deduce from (3.1.36) that if Q is a scalar pseudo-differential operator then the norm of the last term does not exceed $C_n h^n e^{-\frac{1}{2}\epsilon\tau h^{-1}}$ and hence modulo negligible operators

$$e^{ih^{-1}\tau\alpha}Q \equiv \sum_{0\leq k\leq n} \frac{1}{k!}(\mathrm{Ad}^k_{ih^{-1}\alpha} Q)\tau^k e^{ih^{-1}\tau\alpha}$$

with large enough $n = n(s)$. This equality obviously implies the inclusions in question. □

Theorem 3.1.9. *Let $\tilde{\Omega}$ be a domain in $\mathbb{R} \times T^*\mathbb{R}^d$ and let $\tilde{\Omega}'$ be its ϵ-neighborhood. Let $\Omega = \tilde{\Omega} \cap \{x_1 = 0\}$ and $\Omega_+ = \tilde{\Omega} \cap \{x_1 \geq 0\}$, and let Ω' and Ω'_+ be defined like $\tilde{\Omega}'$. Let $\alpha \in C^q(\mathbb{R}, \boldsymbol{\Psi}_{h,N}(\mathbb{R}^d, \mathbb{H}, \mathbb{H}))$ satisfy (3.1.34) for all $z \in \tilde{\Omega}$; let $q \geq 1$ and $N = N(d, l, q, s)$. Then there exist operators $G_0 \in \mathcal{L}(L^2(Y, \mathbb{H}), L^2(X, \mathbb{H}))$ with $\|G_0\| \leq Ch^{1/2}$ and $G \in$*

$\mathcal{L}(L^2(X,\mathbb{H}), L^2(X,\mathbb{H}))$ with $\|G_0\| \le C$ with the same type of constant C as before such that for tempered u_0, f (i.e., with $\|u_0\|_Y \le h^{-l}, \|D_1^j f\|_X \le h^{-l}\ \forall j \le q$) the following relations hold:

$$\text{(3.1.45)} \qquad \begin{aligned} \mathrm{OF}^{s,q+1'}(G_0 u_0) &\subset \mathrm{OF}^{s-\frac{1}{2}}(u_0) \cap \Omega', \\ \mathrm{OF}^{s,q+1'}(Gf) &\subset \mathrm{OF}^{s,q'}(f) \cap \Omega'_+, \end{aligned}$$

$$\text{(3.1.46)} \qquad \begin{aligned} \mathrm{OF}^{s,q'}(PG_0 u_0) \cap \Omega_+ &= \emptyset, \\ \mathrm{OF}^{s-\frac{1}{2}}(\eth G_0 u_0 - u_0) \cap \Omega &= \emptyset, \\ \mathrm{OF}^{s,q'}(PGf - f) \cap \Omega_+ &= \emptyset, \\ \mathrm{OF}^{s-\frac{1}{2}}(\eth Gf) &= \emptyset, \end{aligned}$$

$$\text{(3.1.45)}' \qquad \mathrm{OF}^{s,q+1'}(G^* f) \subset \mathrm{OF}^{s,q'}(f) \cap \Omega'_+,$$

$$\text{(3.1.46)}' \qquad \mathrm{OF}^{s,q'}(P^* G^* f - f) \cap \Omega_+ = \emptyset.$$

Moreover, if $\|D_1^j u\|_X \le h^{-l}$ for all $j \le q+1$ then the following relations hold:

$$\text{(3.1.47)} \qquad \mathrm{OF}^{s,q+1'}(u - GPu - G_0 \eth u) \cap \Omega_+ = \emptyset,$$

$$\text{(3.1.47)}' \qquad \mathrm{OF}^{s,q+1'}(u - G^* P^* u) \cap \Omega_+ = \emptyset.$$

Proof. Without loss of generality one can assume that (3.1.34) is satisfied in $\mathbb{R} \times T^*\mathbb{R}^d$. In fact, (3.1.34) is satisfied in $\tilde{\Omega}'$ for a small enough constant $\epsilon > 0$ and hence $\bar{\alpha}(z) = \alpha(z)\varphi(z) + i(1-\varphi(z))I$ satisfies (3.1.34) in $\mathbb{R}\times T^*\mathbb{R}^d$ provided φ is a smooth scalar function equal to 1 in the $\frac{1}{3}\epsilon$-neighborhood of $\tilde{\Omega}$ and vanishing outside of the $\frac{2}{3}\epsilon$-neighborhood of $\tilde{\Omega}$ and such that $0 \le \varphi \le 1$.

Let us define $\bar{G}_\pm$ and $\bar{G}_0$ according to the previous proposition[1)] and let us set $\bar{G} = R\bar{G}_+ T$ where here R and T are the operators of restriction from $\mathbb{R}^{d+1}$ to X and extension from X to $\mathbb{R}^{d+1}$ by 0 in $\mathbb{R}^{d+1} \setminus X$ respectively. Then $\bar{G}^* = R\bar{G}_- T$ and operators $\bar{G}$ and $\bar{G}_0$ obviously satisfy (3.1.45)–(3.1.46)$'$ with $\Omega = \Omega' = T^*\mathbb{R}^d$ and with $\Omega_+ = \Omega'_+ = \bar{\mathbb{R}}^+ \times T^*\mathbb{R}^d$. Moreover, the obvious equalities

$$\begin{aligned} PTu &= TPu - ih(\eth u)\delta(x_1), \\ P^*Tu &= TP^*u - ih(\eth u)\delta(x_1) \end{aligned}$$

combined with $(3.1.28)_\pm$ and $(3.1.37)_\pm$ imply the equalities

$$u = \bar{G}Pu + \bar{G}_0 \eth u, \qquad u = \bar{G}^* P^* u;$$

these equalities combined with (3.1.40) and (3.1.43) imply (3.1.47) and (3.1.47)$'$ with $\Omega_+ = \bar{\mathbb{R}}^+ \times T^*\mathbb{R}^d$. Then $G = \mathrm{Op}(\varphi)\bar{G}\,\mathrm{Op}(\varphi)$ and $G_0 = \mathrm{Op}(\varphi)\bar{G}_0\,\mathrm{Op}(\varphi_0)$ are the operators in question provided $\varphi_0 = \varphi|_{x_1=0}$. □

Proof of theorem 3.1.2. The simplifying reduction of subsection 3.1.2 implies that without loss of generality one can assume that $\Sigma(P) = \emptyset$ and

$$P = \begin{pmatrix} P_1 & 0 \\ 0 & P_2 \end{pmatrix}, \qquad \mathbb{H} = \mathbb{H}_1 \oplus \mathbb{H}_2$$

1) We now use a bar for the operators constructed in that proposition, with $t = 0$ now in (3.1.43) and (3.1.44).

where $P_j = hD_1 - \alpha_j$ and α_1 and $\alpha_2^\dagger$ satisfy (3.1.34). Then if $\Omega \supset \omega$ and ω does not intersect $\mathrm{OF}_b^{s,q}(Pu)$ then theorem 3.1.9 yields that $\omega \cap \mathrm{OF}_b^{s,q+1}(u_2) = \emptyset$, $\omega \cap \mathrm{OF}_b^{s,q+1}(u_1 - G_0 v) = \emptyset$ and $\omega \cap \mathrm{OF}_b^{s,q+1}(u_1) \subset \mathrm{OF}^{s-\frac{1}{2}}(v)$ where G_0 is constructed for P_1, $v = \eth u_1$. Then

$$\omega \cap \mathrm{OF}^{s-\frac{1}{2}}(B\eth u) = \omega \cap \mathrm{OF}^{s-\frac{1}{2}}(B'v) \supset \omega \cap \mathrm{OF}^{s-\frac{1}{2}}(B'^* B'v)$$

where B' is the restriction of B to $\mathbb{H}_1$. It is easy to see that $\mathbb{H}'(z) = \mathbb{H}_1$ and hence the set of characteristic points of $B'^* B'$ is exactly $\Sigma_b(P, B)$. Hence if $\omega \cap \mathrm{OF}^{s-1/2}(B\eth u) = \emptyset$ then $\omega \cap \mathrm{OF}^{s-1/2}(v) \subset \Sigma_b(P, B)$, and hence $\omega \cap \mathrm{OF}_b^{s,q+1}(u) \subset \Sigma_b(P, B)$. □

3.1.4 Statement of the basic theorem. II

The rest of this section is almost entirely devoted to the proof of the following assertion which is slightly more precise than theorem 3.1.6:

Theorem 3.1.10. *Let P be a reduced operator satisfying the conditions of theorem 3.1.6, i.e., suppose conditions* (3.1.1) *with $m = 1$,* (3.1.2), (3.1.3), (3.1.29), (3.1.30), (3.1.32) *and* (3.1.33) *are satisfied and let us assume that $\det p_{jk}(z^*, \eta)$ with $k = 1, \dots, n$ and $j = 1, \dots, M_k$ has only one complex root (namely $\eta_k \in \mathbb{R}$) where $z^* \in \Omega \Subset T^*Y$. Let B satisfy* (3.1.7). *Let $\phi_j \in C^N(T^*Y)$, $\phi_{jk} \in C^N(\bar{\mathbb{R}}^+ \times T^*X')$ be real-valued functions satisfying* (3.1.21) *($j = 1, \dots, J$, $k = 1, \dots, n$). Moreover, let us assume that for every $j = 1, \dots, J$ the problem (P, B) is microhyperbolic at the point z^* in the multidirection $T_j = (H_{\phi_j}, -\partial_{x_1}\phi_{1j}, \dots, -\partial_{x_1}\phi_{nj})(z^*)$. Let $N = N(d, q, s, l)$ and $q \geq 2\mathrm{D}^* - 2$, $\mathrm{D}^* = \max_{jk} \mathrm{D}_{jk}$ where D_{jk} are the dimensions of $\mathbb{H}_{jk}$. Then*

(i) Let u satisfy (3.1.15) *and*

$$(\mathrm{OF}_b^{s+1,q}(P_0 u_0) \cup \mathrm{OF}^{s+\frac{1}{2}}(B\eth u)) \cap \Omega' \cap \{\phi_1 \leq t_1\} \cap \dots \cap \{\phi_J \leq t_J\} = \emptyset, \tag{3.1.48}$$

$$\mathrm{OF}^{s+1,q'}(P_k u_k) \cap \Omega_k'' \cap \{\phi_{k1} \leq t_1\} \cap \dots \cap \{\phi_{kJ} \leq t_J\} = \emptyset \qquad \forall k = 1, \dots, n, \tag{3.1.49}$$

$$\mathrm{OF}^{s-\frac{1}{2}}(\eth u_0) \cap \partial\Omega' \cap \{\phi_1 \leq t_1\} \cap \dots \cap \{\phi_J \leq t_J\} = \emptyset, \tag{3.1.50}$$

$$\mathrm{OF}^{s'}(u_k) \cap \partial\Omega_k'' \cap \{\phi_{k1} \leq t_1\} \cap \dots \cap \{\phi_{kJ} \leq t_J\} = \emptyset \qquad \forall k = 1, \dots, n, \tag{3.1.51}$$

where $u = (u_1, \dots, u_n, u_0)$, Ω' and Ω_k'' are small enough neighborhoods of z^* in T^*Y and $\bar{\mathbb{R}}^+ \times T^*X'$ respectively such that $\Omega' = \Omega_k''|_Y$ $\forall k = 1, \dots n$. Then

$$\mathrm{OF}^{s-\frac{1}{2}}(\eth u_0) \cap \Omega' \cap \{\phi_1 \leq t_1\} \cap \dots \cap \{\phi_J \leq t_J\} = \emptyset \tag{3.1.52}$$

and

$$\mathrm{OF}^{s,q+1'}(u_k) \cap \Omega_k'' \cap \{\phi_{k1} \leq t_1\} \cap \dots \cap \{\phi_{kJ} \leq t_J\} = \emptyset \quad \forall k = 1, \dots, n. \tag{3.1.53}$$

(ii) Moreover, if condition (3.1.7)* *is satisfied then one can weaken condition* (3.1.48) *by replacing $\mathrm{OF}_b^{s+1,q}(P_0 u_0)$ with $\mathrm{OF}_b^{s+1/2,q}(P_0 u_0)$ and $\mathrm{OF}^{s+1/2}(B\eth u)$ with $\mathrm{OF}^s(B\eth u)$, and omit condition* (3.1.50) *completely, and get statement* (3.1.27) *in addition to* (3.1.52) *and* (3.1.53).

Proof of theorem 3.1.6. Let conditions (3.1.15) and (3.1.22)–(3.1.25) be fulfilled. Then there obviously exists $\varepsilon > 0$ depending on $u, t_1, \dots, t_J$ such that if $\psi = \psi(x_1)$ is a smooth function equal to 1 for $x_1 \le \frac{1}{2}\varepsilon$ and vanishing for $x_1 \ge \varepsilon$ and if Q is a scalar h-pseudo-differential operator with symbol equal to 1 in the $\frac{1}{2}\varepsilon$-neighborhood of $\Sigma(P)$ and vanishing outside of the ε-neighborhood of $\Sigma(P)$ then $U = \psi u + Q(1-\psi)u$ satisfies (3.1.48)–(3.1.51) with $\Omega''_k = \tilde{\iota}\Omega'_k$ where $\tilde{\iota} : T^*X \to \bar{\mathbb{R}}^+ \times T^*X'$ is a natural map. Then theorem 3.1.10 applied to U yields (3.1.26) and even (3.1.27) (under condition (3.1.7)*) for u. □

3.1.5 Proof of theorem 3.1.10

In this proof we can assume without loss of generality that (3.1.52) and (3.1.53) hold with s replaced by $s - \frac{1}{2}$, i.e., that

$$\text{(3.1.54)} \qquad \mathrm{OF}^{s-1}(\eth u_0) \cap \Omega' \cap \{\phi_1 \le t_1\} \cap \dots \cap \{\phi_J \le t_J\} = \emptyset,$$

$$\text{(3.1.55)} \quad \mathrm{OF}^{s-\frac{1}{2}, q+1'}(u_k) \cap \Omega''_k \cap \{\phi_{k1} \le t_1\} \cap \dots \cap \{\phi_{kJ} \le t_J\} = \emptyset \qquad \forall k = 1, \dots, n.$$

Let us construct operators $Q_k \in \boldsymbol{\Psi}_{h,N}(X, \mathbb{C})$ in the same manner as Q_3 in the proof of theorem 2.1.2, i.e., let us set

$$Q_k = \mathrm{Op}^{\mathrm{w}}\bigl(\zeta_k(x,\xi') \prod_{1\le j\le J} \chi(\phi_{kj}(x,\xi'))\bigr)$$

where the real-valued function χ satisfies (2.1.8); for the sake of simplicity we assume that $t_1 = \dots = t_J = 0$, ϕ_{kj} are replaced everywhere by $\phi_{kj} - \frac{1}{4}\varepsilon$, $\varepsilon > 0$ is small enough, $\zeta_k \in C_0^N(\bar{\mathbb{R}}^+ \times T^*X')$, $0 \le \zeta_k \le 1$, ζ_k are supported in the $\frac{1}{2}\varepsilon$-neighborhood of Ω''_k and $\zeta_k = 1$ in the $\frac{1}{3}\varepsilon$-neighborhood of Ω''_k, and $\zeta_1 = \dots = \zeta_n = \zeta$ on T^*Y. Let us consider the following identities similar (2.1.9):

$$\text{(3.1.56)} \quad -\operatorname{Re} i([P_k, Q'_k]u_k, u_k) = \operatorname{Re} i((Q'_k + {Q'_k}^*)P_k u_k, u_k) + \operatorname{Re} i((P_k - P_k^*)Q'_k u_k, u_k) + \operatorname{Re} h({Q'}^* T_k u_k, u_k)_Y$$

where $T_k = A_{0k}|_Y$ (now T is an operator with principal symbol T_0 given by (3.1.5), so we have changed notation slightly), $Q'_k = Q_k^2$ and the index k in the last term is omitted because all the operators Q_k coincide at Y by condition (2.1.21).

Repeating the arguments of the proof of theorem 2.1.2 with obvious modifications we obtain the following inequalities similar to (2.1.10):

$$\text{(3.1.57)} \qquad |-\operatorname{Re} i([P_k, Q'_k]u_k, u_k) - \operatorname{Re}({Q'}^* T_k u_k, u_k)_Y| \le C_2 \|Q_k u_k\|^2 + Ch^{2s}.$$

Let us note that the principal symbol of the operator $-ih^{-1}[P_k, Q'_k]$ equals

$$\sum_{1\le j\le J} 2\{p_k, \phi_{kj}\} r_{kj}^2 - \{p_k, \zeta_k\} \prod_{1\le j\le J} \chi(\phi_{kj})^2$$

where $r_{kj} = \zeta \chi_1(\phi_{kj}) \chi_2(\phi_{kj}) \prod_{i\ne j} \chi(\phi_{ki})^2$ and the functions χ_1, χ_2 are defined in (2.1.8). Then similar to (2.1.11) we obtain the inequalities

$$\text{(3.1.58)} \quad \Bigl|\sum_{1\le j\le J} 2(S_{kj} R_{kj} u_k, R_{kj} u_k) - \operatorname{Re}({Q'}^* T_k u_k, u_k)_Y\Bigr| \le C_2 \|Q_k u_k\|^2 + C_2 h \|Q_k u_k\|_Y^2 + Ch^{2s}$$

where $R_{kj} = \mathrm{Op}(r_{kj})$, $S_{kj} = \mathrm{Op}^{\mathrm{w}}(\{p_k, \phi_{kj}\})$ and a boundary term (the second term) appears in the right-hand expression because the symbols $\{p_k, \phi_{kj}\}$ depend on ξ_1.

The microhyperbolicity of the symbols p_k in the directions $H_{\phi_{kj}}$ yields the existence of symbols q''_k vanishing in the ε-neighborhoods of the supports of r_{kj} for all k, j such that

$$\langle(\{p_k, \phi_{kj}\} + cp_k^\dagger p_k + (q''_k - \epsilon_0)I)w, w\rangle \geq 0$$
$$\forall w \in \mathbb{H}_k \quad \forall (x, \xi') \in \bar{\mathbb{R}}^+ \times T^*X' \qquad \forall j = 1, \dots, J \quad \forall k = 1, \dots, n.$$

Then passing from symbols to operators and applying the Gårding inequality we obtain an inequality similar to (2.1.12):

$$\mathrm{Re}(S_{kj}v, v) + c\|P_k v\|^2 + \mathrm{Re}(Q''_k v, v) \geq \epsilon_0\|v\|^2 - C_2 h\|v\|^2_Y - Ch\|v\|^2. \tag{3.1.59}$$

More precisely, we first obtain this inequality only for v with $\mathfrak{d}v = 0$; in order to get rid of this restriction we substitute $v - e^{-x_1 h^{-1}}\mathfrak{d}v$ for v in (3.1.59) to obtain (3.1.59) in the general case.

Substituting $v = R_{kj}u_k$ into (3.1.59) and repeating the arguments of the proof of theorem 2.1.2 we obtain the inequalities

$$\mathrm{Re}(S_{kj}R_{kj}u_k, R_{kj}u_k) \geq \epsilon_0\|R_{kj}u_k\|^2 - C_2 h\|R_j u_k\|^2_Y - Ch^{2s} \tag{3.1.60}$$

where the index k is omitted in the second term in the right-hand expression since R_{kj} do not depend on k in Y.

I have been repeating the arguments of the proof of theorem 2.1.2. However I now need some new ideas:

Proposition 3.1.11. *(i) Let conditions* (3.1.1)–(3.1.3) *with* $m = 1$ *and* (3.1.33) *be fulfilled and let us assume that* $\det p_{kj}(z^*, \eta)$ *has only one complex root (namely,* $\eta_k \in \mathbb{R}$ *). Then for every* $\epsilon > 0$ *there exists a neighborhood* Ω''' *of* z^* *such that if* $\mathcal{Q}$ *is an operator with principal symbol* $\mathcal{Q}^0$ *supported in* Ω''' *and satisfying the inequality* $|\mathcal{Q}^0| \leq 1$ *then the following estimate holds:*

$$h\|\mathcal{Q}v\|^2_Y \leq \epsilon\|w\|^2 + C_\epsilon \sum_{0\leq l\leq \bar{q}} \|D_1^l P_k v\|^2 + Ch \sum_{0\leq l\leq \bar{q}+1} \|D_1^l v\|^2 \tag{3.1.61}$$

where we omit the subscript "X" as before and $\bar{q} = 2\mathrm{D}^* - 1$ *where* D^* *is the maximal dimension of Jordan cells in the Jordan normal form of* $L_{kj}(z^*)$, $L_{kj} = -a_{0kj}^{-1}a_{1kj}$; *the maximum is taken for all* $j = 1, \dots, M_j$ *and all the Jordan cells.*

(ii) Moreover,

$$|v(0)|^2 \leq 2\epsilon|v|^2_+ + C_\epsilon \sum_{0\leq l\leq \bar{q}} |D_1^l p_k(z, D_1)v|^2_+ \tag{3.1.62}$$
$$\forall v \in \mathcal{S}(\bar{\mathbb{R}}^+, \mathbb{H}_k) \qquad \forall z \in \Omega'''.$$

This proposition will be proven later; let us now finish the proof of the theorem. Taking $\mathcal{Q}$ with symbol equal to 1 in a neighborhood of the support of the symbol of Q_k and using (3.1.60), (3.1.61), (3.1.49), (3.1.54) and (3.1.55) we obtain the estimate

$$\mathrm{Re}(S_{kj}R_{kj}u_k, R_{kj}u_k) \geq \frac{1}{2}\epsilon_0\|R_{kj}u_k\|^2 + c_4 h\|R_j u_k\|^2_Y - Ch^{2s} \tag{3.1.63}$$

with an arbitrary constant c_4 which will be chosen later. So the final estimate at this step in the proof is

$$(3.1.64) \qquad -\operatorname{Re} ih^{-1}([P_k, Q'_k]u_k, u_k) \geq$$
$$\frac{1}{2}\epsilon_0 \sum_{1\leq j\leq J} \|R_{kj}u_k\|^2 + c_4 h \sum_{1\leq j\leq J} \|R_j u_k\|_Y^2 - Ch^{2s} \quad \forall k = 1, \ldots, n.$$

Let us now consider

$$-\sum_{\leq k\leq n} \operatorname{Re}(Q'^* T_k u_k, u_k)_Y.$$

Modulo terms with absolute values not exceeding $C_2 h\|Qu\|_Y^2 + Ch^{2s}$ (where C_2 does not depend on the choice of c_3 and c_4) this expression equals

$$(3.1.65) \qquad -\sum_{0\leq k\leq n} \operatorname{Re}(T_k Q u_k, Q u_k)_Y = -\operatorname{Re}(TQu, Qu)_Y + \operatorname{Re}(T_0 Q u_0, Q u_0)_Y$$

(we have used the definitions of Q and Q' and (3.1.55)). Let us transform (3.1.65) modulo terms with absolute values not exceeding $C_2 h\|Qu\|_Y^2 + Ch^{2s}$ (where C_2 does not depend on the choice of c_3 and c_4); in these transformations we will use (3.1.55) repeatedly. Let us recall that in a neighborhood of z^*,

$$S_0 = -T_0 - \beta^\dagger b - b^\dagger \beta$$

is a non-negatively defined symbol by (3.1.7). Hence

$$-\operatorname{Re}(TQu, Qu)_Y \equiv \operatorname{Re}(SQu, Qu)_Y + 2\operatorname{Re}(BQu, \mathcal{B}Qu)_Y \equiv$$
$$-\operatorname{Re}(SQu, Qu)_Y + 2\operatorname{Re}(Q[B, Q]u, \mathcal{B}u)_Y$$

because (3.1.48) implies $\operatorname{Re}(QBu, \mathcal{B}Qu) \equiv 0$ where $S = \operatorname{Op}^{\mathrm{w}}(S_0)$ and $\mathcal{B} = \operatorname{Op}^{\mathrm{w}}(\beta)$.

On the other hand, the structure of P_0 yields that

$$|(T_0 Q u_0, Q u_0)_Y| \leq C_2 \|Qu'_0\|_Y \cdot \|Qu''_0\|_Y;$$

moreover, (3.1.48) and theorem 3.1.8 imply that $\|Qu''_0\|_Y \leq Ch^{s+\frac{1}{2}}$ and therefore this expression is equivalent to 0. Thus

$$(3.1.66) \qquad -\sum_{0\leq k\leq n} \operatorname{Re}(T_k Q u_k, Q u_k)_Y \equiv \operatorname{Re}(SQu, Qu)_Y + 2\operatorname{Re}(Q[B, Q]u, \mathcal{B}u)_Y.$$

Let S' be an h-pseudo-differential operator on Y with a non-negatively defined Weyl symbol coinciding with that of S in a neighborhood of z^*; then the Gårding inequality yields that

$$(3.1.67) \qquad \operatorname{Re}(S'v, v)_Y \geq -C_2\|v\|_Y^2$$

and hence

$$(3.1.68) \qquad \operatorname{Re}(SQu, Qu)_Y \geq -C_2 h\|Qu\|_Y^2 - Ch^{2s}$$

where we have used the fact that the symbol of Q is supported in a small neighborhood of z^*.

Let us now consider the second term in the right-hand expression in (3.1.66). Calculating the principal symbol of the operator $Q[B, Q]$ we see that this term is equivalent to

$$\sum_{1\le j\le J} h(L_jR_ju, R_ju)_Y$$

where the operators R_j are defined above and L_j are operators with principal symbols

$$L_{j0} = i\beta^\dagger\{b, \phi_j\} - i\{b^\dagger, \phi_j\}\beta.$$

Then (3.1.20) implies that in a neighborhood of z^* the symbols

$$L_{j0} + c(S_0 + b^\dagger b + \Pi) - 2\epsilon_0 I$$

are non-negative definite where Π is the projector to the orthogonal complement of $\mathbb{H}''(z^*)$. Therefore the Gårding inequality again implies that

$$\operatorname{Re}(L_jv, v)_Y = c\operatorname{Re}(Sv, v)_Y + c\|Bv\|_Y^2 + c\|\Pi v\|_Y^2 \ge$$
$$(2\epsilon_0 - Ch)\|v\|_Y^2 - \|Q'v\|_Y^2$$

where the symbol of Q' vanishes in some neighborhood of z^*. Setting $v = R_ju$ and multiplying by h we obtain that

$$h\operatorname{Re}(L_jR_ju, R_ju)_Y + ch\operatorname{Re}(SR_ju, R_ju)_Y + ch\|BR_ju\|_Y^2 + ch\|\Pi R_ju\|_Y^2 \ge \epsilon_0h\|R_ju\|_Y^2 - Ch^{2s}. \tag{3.1.69}$$

Moreover, condition (3.1.48) implies that the third term in the left-hand expression is equivalent to 0 and the second term is equivalent to $ch\operatorname{Re}(S''Qu, R'_ju)_Y$ where $S'' = S' + C_2h$ and R'_j is an operator with principal symbol $\zeta\chi_4(\phi_j)\prod_{k\ne j}\chi(\phi_k)$; $\chi_4 = (\chi')^2/\chi = (2\chi'_1)^2$ is a smooth function because of (2.1.8). Then (3.1.67) and the Cauchy inequality yield that

$$h|(S''Qu, R'_ju)_Y| \le h(S''Qu, Qu)_Y^{1/2}\cdot(S''R'_ju, R'_ju)_Y^{1/2} \le \frac{1}{2}\epsilon_0\operatorname{Re}h(SQu, Qu)_Y + C_2h\|Qu\|_Y^2 + Ch^{2s}. \tag{3.1.70}$$

On the other hand, (3.1.48) and theorem 3.1.8 yield that $\|R_ju''_0\|^2 \le Ch^{2s}$ and hence

$$\|\Pi R_ju\|_Y^2 \le C_2\sum_{1\le k\le n}\|R_ju_k\|_Y^2 + \varepsilon\|R_ju'_0\|_Y^2 + C_\varepsilon h^{2s}$$

with arbitrary small $\varepsilon > 0$ provided $\Omega' = \Omega'_\varepsilon$ is narrow enough.

Then (3.1.66), (3.1.68) and (3.1.69) imply that

$$-\sum_{1\le k\le n}\operatorname{Re}(Q'^*T_ku_k, u_k) \ge \epsilon_0h\sum_{\le j\le J}\|R_ju'_0\|_Y^2 - C_2h\sum_{1\le j\le J, 1\le k\le n}\|R_ju_k\|_Y^2 - C_2h\|Qu\|_Y^2 - Ch^{2s}. \tag{3.1.71}$$

This inequality combined with (3.1.57), (3.1.64) and (3.1.70) yields that, after an appropriate choice of constant c_4,

$$\sum_{1\le j\le J}\bigl(h\|R_ju\|_Y^2 + \sum_{1\le k\le n}\|R_{jk}u_k\|^2\bigr) \le C_2\bigl(h\|Qu\|_Y^2 + \sum_{1\le k\le n}\|Q_ku_k\|^2\bigr) + Ch^{2s}.$$

This estimate is similar to estimate (2.1.14) in the proof of theorem 2.1.2. Repeating the remaining arguments of the proof of theorem 2.1.2 associated with the choice of constant c_3 in (2.1.8) we finally obtain that (3.1.52) and (3.1.53) with $q = -1$ hold. Then (3.1.49) implies (3.1.53) in general form.

Let us now assume that condition (3.1.7)* is fulfilled. Then one can replace (3.1.68) by

$$\text{(3.1.68)}^* \qquad \operatorname{Re}(SQu, Qu) \geq (\epsilon_0 - Ch)\|Qu\|_Y^2 - Ch^{2s}$$

(we assume that (3.1.27) holds with $s - \frac{1}{2}$ instead of s). Moreover, we know that $\|Qu_0''\|_Y \leq Ch^s$ by theorem 3.1.8 and we can weaken condition (3.1.48) as is done in theorem 3.1.10(ii). But in this case, with no analysis of the principal symbols of the commutators, instead of (3.1.70) we obtain the following stronger inequality:

$$\text{(3.1.71)}^* \qquad -\sum_{1\leq k\leq n} \operatorname{Re}(Q'^* T_k u_k, u_k) \geq \frac{1}{2}\epsilon_0\|Qu\|_Y^2 - Ch^{2s}.$$

This inequality combined with (3.1.57) and (3.1.64) implies that

$$\|Qu\|_Y^2 + \sum_{1\leq j\leq J, 1\leq k\leq n} \|R_{jk}u_k\|^2 \leq C\sum_{1\leq k\leq n} \|Q_k u_k\|^2 + Ch^{2s}$$

and repeating the remaining arguments of the proof of theorem 2.1.2 we obtain (3.1.52), (3.1.53) and (3.1.27). □

3.1.6 Auxiliary estimates

The aim of this subsection is:

Proof of proposition 3.1.11. Obviously, it is sufficient to prove this proposition for $P = hD_1 - L$ with $L \in \Psi'_{h,N}(X, \mathbb{H}, \mathbb{H})$ where $\mathbb{H} = \mathbb{C}^r$ and $L_0(z^*)$ has exactly one complex eigenvalue (namely, $\eta \in \mathbb{R}$); now L_0 (the principal symbol of L) is not necessarily Hermitian. Obviously, estimates (3.1.61) and (3.1.62) do not depend on lower order terms; moreover, these estimates are stable with respect to small perturbations of L_0 for every fixed $\epsilon > 0$. Therefore it is sufficient to prove them only in the case of constant $L = L_0(z^*)$. Then (3.1.61) is reduced to (3.1.62) by the change of variables $x_{1\text{new}} = h^{-1}x_1$. So we need only prove (3.1.62) for $z = z^*$. Moreover, one can assume that $\eta = 0$; the general case is reduced to this case by the substitution $v = v_{\text{new}}e^{i\eta x_1}$. We will apply the following

Lemma 3.1.12. *Let L be an $r \times r$ matrix such that $L^r = 0$. Then for every $\mu > 0$ there exist $r \times r$ matrices ρ_j with $j = 0, \dots, 2r-2$ such that*

$$\text{(3.1.72)} \qquad \rho_j^\dagger = \rho_j \qquad \forall j,$$

$$\text{(3.1.73)} \qquad -\sum_{0\leq j\leq 2r-2} \operatorname{Re}\langle \rho_j L^{j+1}w, w\rangle \geq |Lw|^2 - \mu|w|^2$$

and

$$\text{(3.1.74)} \qquad -\sum_{0\leq k\leq j\leq 2r-2} \operatorname{Re}\langle i\rho_j L^{j-k}w, L^k w\rangle \leq \mu|w|^2 \qquad \forall w \in \mathbb{H}.$$

This lemma will be proven later; now I prove proposition 3.1.11 in reduced form. Let

$$P = D_1 - L, \qquad Q = \sum_{0\le k\le j\le 2r-2} \rho_j D_1^k L^{j-k};$$

then

$$PQ = \sum_{0\le j\le 2r-2} \rho_j (D_1^{j+1} - L^{j+1})$$

and hence

$$\begin{aligned}\operatorname{Re}\langle QPv, v\rangle_+ &= -\sum_{0\le j\le 2r-2} \langle \rho_j L^{j+1} v, v\rangle_+ + \sum_{0\le j\le 2r-2} \langle \rho_j D_1^{j+1} v, v\rangle_+ = \\ &- \sum_{0\le j\le 2r-2} \langle \rho_j L^{j+1} v, v\rangle_+ + \frac{1}{2}\sum_{0\le k\le j\le 2r-2} \operatorname{Re}\langle i\rho_j (D_1^{j-k} v)(0), (D_1^k v)(0)\rangle\end{aligned}$$

where we have applied (3.1.72). In the last term let us replace D_1^k and D_1^{j-k} by L^k and L^{j-k} respectively; then the absolute value of the error will not exceed

$$C\Big(\sum_{1\le j\le 2r-2} |(D_1 Pv)(0)|^2 + |v(0)|\Big) \le \mu|v|_+^2 + C\sum_{0\le j\le 2r-2} |D_1^j Pv|_+^2$$

because of the obvious estimates

$$|v(0)|^2 \le |v|_+ \cdot |D_1 v|_+, \quad |D_1 v|_+ \le |Pv|_+ + C|v|_+.$$

Hence for every $\mu > 0$ the following estimate holds:

$$\sum_{0\le j\le 2r-2} \operatorname{Re}\langle -\rho_j L^{j+1} v, v\rangle_+ + \frac{1}{2}\sum_{0\le k\le j\le 2r-2} \operatorname{Re}\langle i\rho_j L^{j-k} v(0), L^k v(0)\rangle \le$$

$$\mu|v|_+^2 + C\sum_{0\le j\le r-2} |D_1^j Pv|_+^2.$$

These estimates combined with (3.1.75) imply (3.1.62). □

Proof of lemma 3.1.12. By induction on $q = 1, \dots, r$ we will prove that for every q there exist Hermitian matrices ρ_j with $j = 0, \dots, 2q-2$ such that

$$(3.1.76)_q \qquad -\sum_{0\le j\le 2q-2} \operatorname{Re}\langle \rho_j L^{j+1} w, w\rangle \ge \epsilon_q |Lw|^2 - \mu|w|^2 - C_{q\mu}|L^q w|^2,$$

$$(3.1.77)_q \qquad -\sum_{0\le k\le j\le 2q-2} \operatorname{Re}\langle i\rho_j L^{j-k} w, L^k w\rangle \le \mu|w|^2 + C_{q\mu}|L^q w|^2 \qquad \forall w \in \mathbb{H}$$

where $\epsilon_q > 0$ do not depend on arbitrarily small $\mu > 0$.

Then for $q = r$ we obtain the assertion of the lemma. On the other hand, $(3.1.76)_1$ with $\mu = 0$ and $(3.1.77)_1$ with arbitrarily small $\mu > 0$ hold for $\rho_0 = L - L^\dagger$.

Let $q = 1, \dots, r-1$; let us assume that $(3.1.76)_q$ and $(3.1.77)_q$ are fulfilled with some $\epsilon_q > 0$ and matrices ρ_j ($j = 0, \dots, 2q-2$). Then $(3.1.76)_{q+1}$ and $(3.1.77)_{q+1}$ will be fulfilled with 2μ instead of μ and with $\epsilon_{q+1} = \frac{1}{2}\epsilon$ provided

$$(3.1.78)_{q+1} \quad \operatorname{Re}\langle -\rho_{2q-1}L^{2q} - \rho_{2q}L^{2q+1})w, w\rangle \geq -\frac{1}{2}\epsilon_q|Lw|^2 - \mu|w|^2 + C_{q\mu}|L^{2q}w|^2 - C'_{q+1\mu}|L^{2q+2}w|^2,$$

$$(3.1.79)_q \quad \operatorname{Re}\langle(-\rho_{2q}L^q w, L^q w\rangle \leq \mu|w|^2 - C_{q\mu}|L^q w|^2 + C'_{q+1\mu}|L^{q+1}w|^2 \qquad \forall w \in \mathbb{H}.$$

In fact, let us consider all the terms in the left-hand expression in $(3.1.77)_{q+1}$ which are new with respect to $(3.1.77)_q$ and different from $(3.1.79)_{q+1}$; obviously the absolute values of these terms do not exceed $\delta|w|^2 + C''_{q+1\mu\delta}|L^{q+1}w|^2$ where $\delta > 0$ is arbitrary, and the constant $C_{q\mu}$ is fixed but the constant $C'_{q+1,\mu}$ is not yet fixed.

Therefore we need only prove that for every $\delta > 0$ and $q < r$ there exist matrices $\alpha = -\alpha^\dagger$ and $\beta = -\beta^\dagger$ such that

$$-\operatorname{Re}\langle\alpha L^{2q}w, w\rangle - \operatorname{Re}\langle\beta L^{2q+1}w, w\rangle \geq -\delta|w|^2 + |L^{2q}w|^2 - C_\delta|L^{2q+2}w|^2$$

and

$$\operatorname{Re}\langle i\beta L^q w, L^q w\rangle \geq -\delta|w|^2 + |L^q w|^2 - C_\delta|L^{q+1}w|^2 \qquad \forall w \in \mathbb{H}.$$

Finally, let us note that $\alpha = L^{2q} - L^{\dagger 2q}$ and

$$\beta = 4\delta^{-1}(L^{2q+1} - L^{\dagger 2q+1}) - i(I - L^\dagger L)$$

are admissible matrices. □

3.1.7 Proof of proposition 3.1.4

Obviously it is sufficient to prove this proposition for the reduced operator. Moreover, it is sufficient to check only condition (ii) in definition 3.1.3.

The arguments of the proof of theorem 3.1.6 imply that if the operator P_k is microhyperbolic at (every) point $z_k \in \iota^{-1}z^*$ in the direction $(\mathcal{T}', \nu_k)$ then for every $t > 0$ the following estimate holds

$$-\operatorname{Re}\langle(\mathcal{T}'p_k)(z^*, D_1)v, v\rangle_+ \geq \epsilon_0|v|_+^2 + t|v(0)|^2 - C_t\sum_{0\leq j\leq q}|D_1^j P_k(z^*, D_1)v|_+^2 \qquad \forall v \in \mathcal{S}(\mathbb{R}^+, \mathbb{H}).$$

Hence the same estimate holds for all $z : |z - z^*| \leq \delta(t)$. Hence the estimate

$$-\operatorname{Re}\langle(\mathcal{T}'p_k)(z^*, D_1)v, v\rangle_+ \geq \epsilon_0|v|_+^2 + t|v(0)|^2 \qquad \forall v \in \mathbb{H}'_k(z)$$

holds where $\mathbb{H}'_k(z)$ is the space of all solutions of the equation $p_k(z, D_1)v = 0$ decreasing at $+\infty$. Therefore for large enough $t > 0$ and small enough $\delta > 0$ estimate (3.1.20) at the point $z = z^*$ implies the same estimate at points $z : |z - z^*| \leq \delta$. Moreover, $\mathbb{H}'(z^*) = \mathbb{H}'_0$ in our decomposition and $\mathbb{H}'(z) = \mathbb{H}'_0 \oplus \mathbb{H}'_1(z) \oplus \cdots \oplus \mathbb{H}'_n(z)$. Hence condition (ii) is stable with respect to perturbations of z and hence is stable with respect to perturbations of P, B also. Moreover, this condition is also obviously stable with respect to $\mathcal{T}'$. □

3.2 Related Results

In this section we consider the following issues: the finite speed of propagation of singularities and the decomposition of a solution into two components (the free space solution and the reflected wave); the structure in a neighborhood of $t = 0$ of the restriction of the fundamental solution to the diagonal and the structure of some related functions; and propagation of singularities along transversal bicharacteristics and its geometric interpretation (description in terms of generalized bicharacteristic billiards).

3.2.1 Finite speed of propagation

In this section we consider the operators

$$\begin{aligned} P &= hD_t - A, \qquad & B &= (B_j)_{j=1,\dots,m}, \\ A &\in \Psi'^{m}_{h,N}(X,\mathbb{H},\mathbb{H}), \qquad & B_j &\in \Psi_{h,N}(Y,\mathbb{H}',\mathbb{H}) \end{aligned} \tag{3.2.1}$$

where $\Psi'^{m}_{h,N}(X,\mathbb{H},\mathbb{H})$ is the class of operators of the form (3.1.1) with $A_k \in \Psi'_{h,N}(X,\mathbb{H},\mathbb{H})$.

Surely theorem 3.1.6 yields an assertion about the finite speed of propagation of singularities, but condition (3.1.17) of that theorem is not necessary for this special consequence, so I want to get rid of it and to prove the assertion directly.

Theorem 3.2.1. *Let P, B be operators of type* (3.2.1) *and let conditions* (3.1.2), (3.1.3) *and* (3.1.7) *be fulfilled in the domains $\Omega_+ \subset T^*(X' \times \mathbb{R}_t) \times \bar{\mathbb{R}}^+_{x_1}$ and $\Omega \subset T^*(Y \times \mathbb{R}_t)$ respectively where $\Omega = \Omega_+ \cap \{x_1 = 0\}$; the matrix T is now defined by the operator A but is taken with the opposite sign. Let us assume that*

$$|a_k(x,\xi')| + |\nabla_{x,\xi'} a_k(x,\xi')| \le c \qquad \forall (t,x,\tau,\xi') \in \Omega_+ \quad \forall k = 0,\dots,m, \tag{3.2.2$_1$}$$

$$|\beta_k(x',\xi')| + |b_k(x,\xi')| + |\nabla_{x',\xi'} b_k(x,\xi')| \le c \qquad \forall (t,x',\tau,\xi') \in \Omega \quad \forall k = 1,\dots,m, \tag{3.2.2$_2$}$$

$$|\tau| \le c \qquad \text{in } \Omega_+ \tag{3.2.2$_3$}$$

where $a_k(x,\xi')$ are the principal symbols of A_k, $A = \Sigma_{0\le k\le m} A_k (hD_1)^k$. Let $N = N(d,m,s,q,l)$ and let u satisfy condition (3.1.15). *Then*

(i) If $\bar{z} = (\bar{t},\bar{x},\bar{\tau},\bar{\xi}') \in \Omega_+$, $\bar{t} > 0$ and

$$\mathrm{OF}^{s+1,q'}(Pu) \cap K_-(\bar{z}) \cap \Omega_+ = \emptyset, \tag{3.2.3}$$

$$\mathrm{OF}^{s+\frac{1}{2}}(B\eth_h u) \cap K_-(\bar{z}) \cap \Omega = \emptyset, \tag{3.2.4}$$

$$\mathrm{OF}^{s,q+m'}(u) \cap K_-(\bar{z}) \cap \partial\Omega_+ = \emptyset \tag{3.2.5}$$

and if $K_-(\bar{z}) \cap \Omega_+$ is a bounded domain where

$$K_\pm(\bar{z}) = \{(t,x,\tau,\xi') \in T^*(X' \times \mathbb{R}) \times \bar{\mathbb{R}}^+ : \tau = \bar{\tau}, |x-\bar{x}| + |\xi' - \bar{\xi}'| \le \pm C_0(t-\bar{t})\}, \tag{3.2.6}$$

$C_0 = C_0(m,c)$, then

$$\mathrm{OF}^{s,q'} \cap K_-(\bar{z}) \cap \Omega_+ = \emptyset. \tag{3.2.7}$$

(ii) Moreover, if condition (3.1.7)* *is fulfilled then one can replace the index $s + \frac{1}{2}$ in* (3.2.4) *by s and obtain at the same time that*

$$\mathrm{OF}^{s-j}(D_1^j u|_Y) \cap K_-(\bar{z}) \cap \Omega = \emptyset \qquad \forall j = 0,\dots,q+m-1. \tag{3.2.8}$$

Proof. It is sufficient to prove this theorem with $K_-(\bar{z})$ replaced by $\{\phi \leq 0\}$ for $\phi = t + \psi(x,\tau,\xi')$ for every smooth real-valued function ψ with

$$|\nabla_{x,\xi'}\psi| \leq \varepsilon = \varepsilon(m,c) > 0; \tag{3.2.9}$$

the assertion modified in this way will be indicated by a prime. Moreover, it is sufficient to consider the case $q = 0$ and one can assume that

$$\mathrm{OF}^{s-\frac{1}{2},m'} \cap \{\phi \leq 0\} \cap \Omega_+ = \emptyset \tag{3.2.10}$$

and in the proof of (3.2.8) one can assume in addition that

$$\mathrm{OF}^{s-j-\frac{1}{2}}(D_1^j u|_Y) \cap \{\phi \leq 0\} \cap \Omega = \emptyset \qquad \forall j = 0,\dots,m-1. \tag{3.2.11}$$

Let $q(t,x,\tau,\xi') = \zeta\chi(\phi - \delta)$ where $\zeta \in C_0^\infty(\Omega_+)$ equals 1 in Ω_+ outside of the δ-neighborhood of $\partial\Omega_+$, χ is a function satisfying (2.1.8) and $\delta > 0$ is small enough. Let $Q = \mathrm{Op}^{\mathrm{w}}(q)$ and $Q' = Q^2$. Let us apply the identity

$$\begin{aligned} -\operatorname{Re} i([P,Q']u,u) = \operatorname{Re} i((Q'+Q'^*)Pu,u) + \\ \operatorname{Re} i((P^*-P)Q'u,u) + \operatorname{Re} h(Q'^*T\eth_h u, \eth_h u)_{Y\times\mathbb{R}} \end{aligned} \tag{3.2.12}$$

(compare with (3.1.49)) where the inner product $(.,.)$ is now taken on $X \times \mathbb{R}$ and the matrix T is defined by (3.1.5) with $-A_k$ instead of a_k; then after routine estimates we obtain the following inequality:

$$\begin{aligned} \|Q''u\|^2 \leq \operatorname{Re}([A,\Psi]Q''u, Q''u) - \\ \operatorname{Re} ih(\mathcal{B}[B,\Psi]Q''\eth_h u, Q''\eth_h u)_{Y\times\mathbb{R}} + Ch^{2s} \end{aligned} \tag{3.2.13}$$

where Q'' is a pseudo-differential operator with principal symbol $\zeta\chi_1(\phi-\delta)\chi_2(\phi-\delta)$, χ_1 and χ_2 are defined in (2.1.8) and $\Psi = \mathrm{Op}^{\mathrm{w}}(\psi)$; moreover, under condition (3.1.7)* one can add the term $\epsilon_1\|Q''\gamma_h u\|^2_{Y\times\mathbb{R}}$ to the left-hand expression in (3.2.13) where here and below $\epsilon_1 = \epsilon_1(m,c)$ and $C_1 = C_1(m,c)$. On the other hand, (3.2.9) yields that the right-hand expression in (3.2.13) does not exceed

$$C\varepsilon \sum_{0\leq j\leq m} \|(hD_1)^j Q''u\|^2 + Ch^{2s}$$

hence the theorem in question will be proven provided we prove the inequality

$$\sum_{0\leq j\leq m} \|(hD_1)^j Q''u\| \leq C_1\|Q''u\| + Ch^s.$$

This inequality follows from $(3.2.3)_3$ and the estimate

$$\sum_{0\leq j\leq m} \|(hD_1)^j v\| \leq C_1\bigl(\|v\| + \|Av\| + \|\mathcal{Q}v\|\bigr)$$

where $\mathcal{Q} = \mathcal{Q}(x,D_x) \in \boldsymbol{\Psi}'^{m}_{h,N}(X,\mathbb{H},\mathbb{H})$ is some operator with symbol vanishing in $\Pi\Omega_+ \times \mathbb{R}_{\xi_1}$ where Π is the map $(t,x,\tau,\xi) \to (x,\xi)$ and the norms are taken on X instead of $X \times \mathbb{R}$. This estimate is stable with respect to small variations of the principal symbol of A and arbitrary variations of the lower order terms and hence it is sufficient to prove the estimate

(3.2.14) $$\sum_{0\le j\le m} |(hD_1)^j v|_+ \le C_1\big(|v|_+ + |a(z,hD_1)v|_+\big)$$

$$\forall z \in \Pi\Omega_+ \quad \forall v \in \mathcal{S}(\bar{\mathbb{R}}^+, \mathbb{H});$$

the previous estimate follows from this one by a partition of unity in $\Pi\Omega_+$. In order to prove (3.2.14) we use the well known inequality

(3.2.15) $$|D_1^j v|_+ \le C_{mj} |D_1^m v|_+^{\frac{j}{m}} \cdot |v|_+^{1-\frac{j}{m}} \qquad \forall j = 0,\dots,m$$

(see, e.g., [Hörmander 1]); this inequality yields that for every $\delta > 0$,

$$\big|\big(a(z,hD_1) - a_0(z)(hD_1)^m\big)v\big|_+ \le \delta|(hD_1)^m v|_+ + C_\delta|v|_+$$

and hence condition (3.1.3) implies that

$$|(hD_1)^m v|_+ \le C_1\delta|(hD_1)^m v|_+ + C_\delta|v|_+ + C_\delta|a(z,hD_1)v|_+.$$

This inequality and (3.2.15) obviously imply (3.2.14). □

Let us replace condition (3.1.7) by the condition

(3.2.16) $$\langle T(z)\underline{w}, \underline{w}\rangle_m + 2\,\mathrm{Re}\,\langle \beta(z)b(z)\underline{w}, \underline{w}\rangle = 0 \qquad \forall \underline{w} \in \mathbb{H}^m.$$

Then one can replace t by $-t$, P by $-P$ and $K_-(\bar{z})$ by $K_+(\bar{z})$. Let us apply theorem 3.2.1 to $u^\pm = \theta(\pm t)u$ instead of u and let us use the obvious analogue of lemma 2.1.8; we then obtain

Theorem 3.2.2. *Let conditions* (3.2.1), (3.1.2), (3.1.3), (3.2.2) *and* (3.2.16) *be fulfilled and let u satisfy* (3.1.15). *Then if $\bar{z} = (\bar{t}, \bar{x}, \bar{\tau}, \bar{\xi}') \in \Omega_+$ and $\bar{t} > 0$,*

(3.2.17) $$\mathrm{OF}^{s+1,q'}(Pu) \cap K_-(\bar{z}) \cap \{t \ge 0\} \cap \Omega_+ = \emptyset,$$

(3.2.18) $$\mathrm{OF}^{s+\frac{1}{2}}(B\eth_h u) \cap K_-(\bar{z}) \cap \{t \ge 0\} \cap \Omega = \emptyset,$$

(3.2.19) $$\mathrm{OF}^{s,q+m'}(u) \cap K_-(\bar{z}) \cap \{t \ge 0\} \cap \partial\Omega_+ = \emptyset$$

(3.2.20) $$\mathrm{OF}^{s+1,q,0'}_{x_1,t}(Pu) \cap \Pi'(K_-(\bar{z}) \cap \{t = 0\} \cap \Omega_+) = \emptyset,$$

(3.2.21) $$\mathrm{OF}^{s+\frac{1}{2},'}(B\eth_h u) \cap (\Pi' K_-(\bar{z}) \cap \{t = 0\} \cap \Omega) = \emptyset,$$

(3.2.22) $$\mathrm{OF}^{s,q+m'}(u|_{t=0}) \cap \Pi'(K_-(\bar{z}) \cap \{t = 0\} \cap \partial\Omega_+) = \emptyset$$

where, in (3.2.20) *and* (3.2.21), *the oscillation front sets are calculated as oscillation front sets of functions on $X' \times (\mathbb{R}_t \times \bar{\mathbb{R}}^+)$ and $Y \times \mathbb{R}_t$ respectively and Π' is the map $(x,t,\xi,\tau) \to (x,\xi)$; then*

(3.2.23) $$\mathrm{OF}^{s,q+m'}(u) \cap K_-(\bar{z}) \cap \{t \ge 0\} \cap \Omega_+ = \emptyset.$$

This assertion remains true if we take $\bar{t} < 0$ and replace $K_-(\bar{z})$ by $K_+(\bar{z})$ and $\{t \ge 0\}$ by $\{t \le 0\}$.

3.2.2 Decomposition of the fundamental solution

At this point we assume that conditions (3.1.3) and $(3.2.2)_{1,2}$ are fulfilled for all $z \in T^*\mathbb{R}^d$. Let us assume that

(3.2.24) The operator A with domain $\mathrm{D}(A) = \{u : D_1^j u \in L^2(X,\mathbb{H})\,\forall j = 0,\dots,m$ and $B\eth_h u = 0\}$ is a self-adjoint operator in $L^2(X,\mathbb{H})$.

Let $U(t,x,y)$ be the Schwartz kernel of the operator $e^{ith^{-1}A}$. Then

$$P_x U = 0, \tag{3.2.25}$$

$$B_x \eth_{hx} U = 0, \tag{3.2.26}$$

$$U^t P_y = 0, \tag{3.2.25†}$$

$$U^t (B_y \eth_{hy}) = 0, \tag{3.2.26†}$$

$$U|_{t=0} = \delta(x-y) I, \tag{3.2.27}$$

$$U(t,x,y) = U^\dagger(-t,y,x). \tag{3.2.28}$$

Then theorem 3.2.2 immediately implies

Theorem 3.2.3. *Let conditions* (3.2.1), (3.1.2), (3.1.3), (3.2.16) *and* (3.2.24) *be fulfilled and* $N = N(d,m,s,q)$. *Then*

$$\mathrm{OF}^{s,q'}(U) \cap \{|\tau| \le c\} \subset \{|x-y| + |\xi' + \eta'| \le C_0|t|\} \tag{3.2.29}$$

where, in the calculation of the oscillation front set, U *is treated as a distribution on* $(\mathbb{R}_t \times X' \times X') \times \bar{\mathbb{R}}^{+2}_{x_1,y_1}$.

Let us now assume that

$(3.2.24)_0$ The operator A^0 with domain $\mathrm{D}(A^0) = \{u : D_1^j u \in L^2(X^0,\mathbb{H})\,\forall j = 0,\dots,m\}$ is a self-adjoint operator in $L^2(X^0,\mathbb{H})$, $X^0 = \mathbb{R}^d$ and coincides with A on X.

(If condition (3.1.3) is fulfilled in the whole space X^0 then the operator is self-adjoint provided it is symmetric, i.e., provided its Weyl symbol is Hermitian.) Let $U^0(t,x,y)$ be the Schwartz kernel of the operator $e^{ith^{-1}A^0}$; then U^0 satisfies (3.2.25), $(3.2.25)^\dagger$, (3.2.27) and (3.2.28) with X replaced by X^0. Then $U^1 = U - U^0$ satisfies equations (3.2.25), $(3.2.25)^\dagger$, (3.2.28) and

$$B_x \eth_{hx} U^1 = -B_x \eth_{hx} U^0, \tag{3.2.30}$$

$$U^{1t}(B_y \eth_{hy}) = -U^{0t}(B_y \eth_{hy}), \tag{3.2.30†}$$

$$U^1|_{t=0} = 0. \tag{3.2.31}$$

Hence $U^{1\pm} = \theta(\pm t)U^1$ satisfies (3.2.25), (3.2.25)† and (3.2.30), (3.2.30)† with U^0 replaced by $U^{0\pm}$. In particular, (3.2.25) and (3.2.25)† imply that $U^{1\pm}$ *is many times differentiable with respect to* x_1, y_1.

Theorem 2.1.2 implies that for $y_1 > 0$,

$$\mathrm{OF}^{s,q'}(\eth_{hx}U^{0\pm}) \cap \{|\tau| \le c\} \subset \{|x'-y'| + |\xi'+\eta'| + y_1 \le \pm C_0 t\}$$

and hence theorem 3.2.2 implies that

$$\mathrm{OF}^{s,q'}(U^{1\pm}) \cap \{|\tau| \le c\} \subset \{|x'-y'| + |\xi'+\eta'| + x_1 + y_1 \le \pm C_0 t\}; \tag{3.2.32}$$

moreover one can obviously drop the condition $y_1 > 0$ here.

3.2.3 Normal singularity property

In this subsection we investigate the structure near $t=0$ of the functions $\Gamma_x Q_{1x} U^t Q_{2y}$ and $\Gamma Q_{1x} U^t Q_{2y}$ [2)] and some related functions where $Q_1, Q_2 \in \Psi'^{m-1}_{h,N}(X \times \mathbb{R}, \mathbb{H}, \mathbb{H})$ are operators with symbols supported in a small neighborhood of $z^* \in T^*(Y \times 0)$. We now assume that condition (3.1.17) is fulfilled.

Let $\mathcal{R} \in \Psi'^{m-1}_{h,N}(X \times \mathbb{R}, \mathbb{H}^m, \mathbb{H})$ be an operator realizing all the reductions of point 3.1.2. Let us introduce $\underline{U} = \mathcal{R}_x U^t \mathcal{R}_y$ and let us consider its blocks $\underline{U}_{jk}$ with $j,k = 0,\dots,n$. Moreover, let us subdivide $\underline{U}_{00}$ into four subblocks

$$\underline{U}_{00} = \begin{pmatrix} \underline{U}_{00}^{(11)} & \underline{U}_{00}^{(12)} \\ \underline{U}_{00}^{(21)} & \underline{U}_{00}^{(22)} \end{pmatrix},$$

and subdivide $\underline{U}_{0k}$ and $\underline{U}_{j0}$ with $j,k = 1,\dots,n$ into two subblocks

$$\underline{U}_{0k} = \begin{pmatrix} \underline{U}_{0k}^{(1)} \\ \underline{U}_{0k}^{(2)} \end{pmatrix}, \qquad \underline{U}_{j0} = \begin{pmatrix} \underline{U}_{j0}^{(1)} & \underline{U}_{j0}^{(2)} \end{pmatrix}.$$

Then

$$\Gamma_x Q_{1x} U^t Q_{2y} \equiv \sum_{0 \le j,k \le n} \Gamma_x Q_{1,jx} \underline{U}_{jk} {}^t Q_{2,ky} \tag{3.2.33}$$

where $Q_{1,j}, Q_{2,j} \in \Psi'_{h,N}(X \times \mathbb{R}, \mathbb{H}, \mathbb{H}_j)$ are operators with symbols supported in a small neighborhood of z^*.

Theorem 3.1.2, (3.2.25) and (3.2.25)† imply that

$$\mathrm{OF}^{s,q'}(\underline{U}_{0k}^{(2)})^T \cap z^* \times X'_y \times \bar{\mathbb{R}}^{+2}_{x_1,y_1} = \emptyset, \tag{3.2.34}$$

$$^T\mathrm{OF}^{s,q'}(\underline{U}_{j0}^{(1)}) \cap X'_x \times z^* \times \bar{\mathbb{R}}^{+2}_{x_1,y_1} = \emptyset \tag{3.2.34†}$$

where ^{T} means that η is replaced by $-\eta$. Therefore in (3.2.32) one can replace $Q_{1,0x}\underline{U}_{0k}$, $\underline{U}_{j0}{}^t Q_{2,0y}$ and $Q_{1,0x}\underline{U}_{00}{}^t Q_{2,0y}$ by $Q^{(1)}_{1,0x}\underline{U}_{0k}$, $\underline{U}^{(2)}_{j0}{}^t Q^{(2)}_{2,0y}$ and $Q^{(1)}_{1,0x}\underline{U}^{(12)}_{00}{}^t Q^{(2)}_{2,0y}$ respectively where we use the obvious notation.

We need

---

2) Recall that $(\Gamma_x V)(x,t) = V(x,t)$ and $(\Gamma V)(t) = \int V(x,x,t)dx$.

Proposition 3.2.4. *The following formula holds:*

$$\Gamma'' Q_{1,x} U t^Q_{2,y} \equiv \sum_{1\le j\le n} \Gamma'' Q_{1,jx} \underline{U}_{jj} {}^t Q_{2,jy} + \sum_{0\le j,k\le n} h\Gamma_{x'} \mathcal{O}_{jk}(V_{jk}) \tag{3.2.35}$$

where $V = \underline{U}|_{x_1=y_1=0}$, $\mathcal{O}_{jk} \in \boldsymbol{\Psi}_{h,N}(Y\times Y\times \bar{\mathbb{R}}^+, \mathcal{L}(\mathbb{H},\mathbb{H}), \mathcal{L}(\mathbb{H}_j, bH_k))$ *are operators with symbols supported in a small neighborhood of* $z^{**} = (x^*, y^*, t^*, \xi^*, -\eta^*, \tau^*)$ *and*

$$(\Gamma'_{x'} V)(x',t) = V(x',x',t), \qquad (\Gamma'' W)(x',t) = \int_{\mathbb{R}^+} V(x,x,t)dx_1.$$

Proof. It is sufficient to prove that for $j\ne k$

$$\Gamma'' Q_{1,jx} \underline{U}_{jk} {}^t Q_{2,ky} \equiv h\mathcal{O}_{jk}(V_{jk}); \tag{3.2.36}$$

the equality

$$\Gamma'' Q^{(1)}_{1,jx} \underline{U}^{(12)t}_{00} {}^t Q_{2,ky} \equiv h\mathcal{O}_{00}(V^{(12)}_{00}) \tag{3.2.37}$$

is proven in the same way.

Let us note that modulo functions negligible in the vicinity of z^{**} the following equalities are due to (3.2.25) and (3.2.25)†:

$$(hD_{x_1} - L_{jx})\underline{U}_{jk} \equiv 0,$$

$$\underline{U}_{jk}(-hD_{y_1} -{}^t L_{ky}) \equiv 0$$

where $L_{jx} = L_j(x, hD'_x, hD_t) = -\underline{A}_{0j}^{-1}\underline{A}_{1j}$, $\underline{P}_j = \underline{A}_{0j} hD_1 + \underline{A}_{1j}$ are the diagonal blocks of the reduced operator and here $^{-1}$ means a parametrix in the neighborhood of z^* or z^{**}. Hence

$$h(D_{x_1} + D_{y_1})\underline{U}_{jk} \equiv L_{jx}\underline{U}_{jk} - \underline{U}_{jk} {}^t L_{ky}$$

and setting $W = \underline{U}_{jk}|_{x_1=y_1}$ we obtain that

$$hD_1 W \equiv \mathcal{N}_{jk}(W) \stackrel{\text{def}}{=} L_{jx} W - W^t L_{ky}. \tag{3.2.38}$$

Let us consider $\mathcal{N}_{jk}$ as an element of $\boldsymbol{\Psi}'_{h,N}(X'\times X'\times\bar{\mathbb{R}}^+, \mathcal{L}(\mathbb{H}_j,\mathbb{H}_k), \mathcal{L}(\mathbb{H}_j,\mathbb{H}_k))$. Let us note that if λ_j and λ_k belong to the spectra of the principal symbols of L_j and L_k respectively with $j\ne k$ then $|\lambda_j - \lambda_k| \ge \epsilon > 0$ at z^{**}; hence the principal symbol of the operator $\mathcal{N}_{jk}$ is invertible at z^{**} (here we use Appendix A); hence the operator is elliptic in a neighborhood of that point.[3)] Then (3.2.38) yields that

$$W \equiv \mathcal{N}_{jk}{}^{-1}(hD_1 W). \tag{3.2.39}$$

Hence

$$Q_{1,jx} W^t Q_{2,ky} \equiv -i\mathcal{O}'_{jk}(hD_1 W).$$

Let us permute $\mathcal{O}'_{jk} \in \boldsymbol{\Psi}'_{h,N}(X'\times X'\times\bar{\mathbb{R}}^+, \mathcal{L}(\mathbb{H},\mathbb{H}), \mathcal{L}(\mathbb{H}_j,\mathbb{H}_k))$ with hD_1 and substitute the right-hand expression of (3.2.29) instead of W into the term with the commutator. Moreover, let us repeat this procedure many times; we then obtain

$$Q_{1,jx} W^t Q_{2,ky} \equiv -ihD_1 \mathcal{O}''_{jk}(W).$$

Integration on $x_1\in\mathbb{R}^+$ and substitution $y'=x'$ gives us (3.2.36) with $\mathcal{O}_{jk} = \mathcal{O}''_{jk}|_{x_1=0}$. □

[3)] The same remains true if we replace L_j and L_k by $L^{(1)}_0$ and $L^{(2)}_0$ respectively and hence all the following analysis also remains true.

Theorem 3.2.5. *Let P and B satisfy conditions (3.2.1), (3.1.2), (3.1.3), (3.1.17), (3.2.16), (3.2.24) and $(3.2.24)_0$. Let us assume that the problem (P, B) is microhyperbolic at the point $z^* \in T^*(Y \times \mathbb{R})$, $t(z^*) = 0$ in the multidirection $\mathcal{T} = (T'_x, T'_\xi, \nu_1, \dots, \nu_n)$. Let $Q_1, Q_2 \in \Psi'_{h,N}(X \times \mathbb{R}, \mathbb{H}, \mathbb{H})$ have symbols supported in a small enough neighborhood of z^*. Finally, let $\chi \in C_0^N([\frac{1}{2}, 1])$, let $T > 0$ be a small enough constant, $N = N(d, m, q, s, \mathrm{D}^*)$ and let $|\alpha| \leq q$, $|\beta| \leq q$. Then*

(i) If $T'_x = 0, \nu_1 = \dots = \nu_n = \nu$ then

(3.2.40)
$$|\Gamma_x \chi_T(t) D_x^\alpha D_y^\beta Q_{1,x} U^t Q_{2,y}| \leq C' h^s$$

where C' depends on T here and below in this theorem;

(ii) If $T'_x = 0$ then

(3.2.41)
$$|\Gamma'' \chi_T(t) D_x^\alpha D_y^\beta Q_{1,x} U^t Q_{2,y}| \leq C' h^s,$$

(3.2.42)
$$|\Gamma'_{x'} \eth_x \eth_y \chi_T(t) D_x^\alpha D_y^\beta Q_{1,x} U^t Q_{2,y}| \leq C' h^s;$$

(iii) If $\nu_1 = \dots = \nu_n = \nu$ then

(3.2.43)
$$|\Gamma''' \chi_T(t) D_x^\alpha D_y^\beta Q_{1,x} U^t Q_{2,y}| \leq C' h^s$$

where $(\Gamma''' V)(x_1, t) = \int_{X'} (\Gamma_x V)(x_1, x', t) dx'$;

(iv) If $T'_x = T'_\xi = 0$ then

(3.2.44)
$$|\eth_x \eth_y \chi_T(t) D_x^\alpha D_y^\beta Q_{1,x} U^t Q_{2,y}| \leq C' h^s;$$

(v) In the general case

(3.2.45)
$$|\Gamma \chi_T(t) D_x^\alpha D_y^\beta Q_{1,x} U^t Q_{2,y}| \leq C' h^s,$$

(3.2.46)
$$|\Gamma' \eth_x \eth_y \chi_T(t) D_x^\alpha D_y^\beta Q_{1,x} U^t Q_{2,y}| \leq C' h^s.$$

Proof. Let us note that (iii) and (v) follow from (i) and (ii) respectively by means of a Fourier integral operator on x' with linear canonical transformation $\Psi : (x', \xi') \to \Psi(x', \xi')$ such that $(\Psi T')_\xi = 0$. Let us prove (i), (ii) and (v). Let $\underline{P}$ be a reduced operator, $\underline{U} = \mathcal{R}_x U^t \mathcal{R}_y$, $\underline{U}^0 = \mathcal{R}_x U^{0t} \mathcal{R}_y$, $\underline{U}^1 = \mathcal{R}_x U^{1t} \mathcal{R}_y$; then (3.2.25), $(3.2.25)^\dagger$ and the ellipticity of P_0 yield that

(3.2.47)
$$\mathrm{OF}^{s,q'}(\underline{U}^0_{0k}) \cap \pi(\Omega_+ \times \Omega_+^T) = \emptyset \qquad \forall k = 0, \dots, n,$$

$(3.2.47)^\dagger$
$$\mathrm{OF}^{s,q'}(\underline{U}^0_{j0}) \cap \pi(\Omega_+ \times \Omega_+^T) = \emptyset \qquad \forall j = 0, \dots, n,$$

(3.2.48)
$$\mathrm{OF}^{s,q'}(\underline{U}^1_{0k}) \cap \pi(\Omega_+ \times \Omega_+^T) \subset \{x_1 = 0\} \qquad \forall k = 0, \dots, n,$$

$(3.2.48)^\dagger$
$$\mathrm{OF}^{s,q'}(\underline{U}^1_{j0}) \cap \pi(\Omega_+ \times \Omega_+^T) \subset \{y_1 = 0\} \qquad \forall j = 0, \dots, n$$

where Ω and Ω_+ are small enough neighborhoods of z^* in $T^*(Y \times \mathbb{R})$ and $T^*(X' \times \mathbb{R}) \times \bar{\mathbb{R}}^+$ respectively and π is the map deleting the second copy of (t, τ) in $\Omega_+ \times \Omega_+^T$, etc. Moreover, theorem 3.1.2 yields that

(3.2.49)
$$\mathrm{OF}^{s,q'}(\underline{U}^{1(2)}_{0k}) \cap \pi(\Omega_+ \times \Omega_+^T) = \emptyset \qquad \forall k = 0, \dots, n,$$

$(3.2.49)^\dagger$
$$\mathrm{OF}^{s,q'}(\underline{U}^{1(1)}_{j0}) \cap \pi(\Omega_+ \times \Omega_+^T) = \emptyset \qquad \forall j = 0, \dots, n.$$

Let us note that corollary 2.1.11 implies that the set

$$\mathrm{OF}^s(\underline{U}^0_{jk}) \cap \pi(\Omega_+ \times \Omega^T_+) \times \mathbb{R}^2_{\xi_1,\eta_1} \cap \{|t| \le T\}$$

is contained in the C_0T-neighborhood of $\xi_1 = -\eta_1$ and equations (3.2.25) and (3.2.25)† imply that this set is contained in a small neighborhood of $\{\xi_1 = \zeta_j, \eta_1 = -\zeta_k\}$ (recall that ζ_j are characteristic roots) and hence this set is empty for small $T > 0$:

(3.2.50) $$\mathrm{OF}^s(\underline{U}^0_{jk}) \cap \pi(\Omega_+ \times \Omega^T_+) \times \mathbb{R}^2_{\xi_1,\eta_1} \cap \{|t| \le T\} = \emptyset \qquad \forall j \ne k.$$

Moreover, theorem 2.1.2 with $J = 2$, $\phi_1 = t$, $\phi_2 = \langle T'_\xi, x' - y'\rangle + \nu_k(x_1 - y_1) - \varepsilon t$ with a small enough constant $\varepsilon > 0$ and corollary 2.1.11 imply that

(3.2.51) $$\mathrm{OF}^s(\underline{U}^0_{kk}) \cap \pi(\Omega_+ \times \Omega^T_+) \times \mathbb{R}^2_{\xi_1,\eta_1} \cap \{0 \le t \le T, \langle T'_\xi, x' - y'\rangle + \nu_k(x_1 - y_1) < \varepsilon t\} = \emptyset \quad \forall k = 1,\dots,n.$$

Moreover, equations (3.2.25), (3.2.25)† and the ellipticity of $\underline{P}_{jx}$ and ${}^t\underline{P}_{ky}$ for $|\xi_1| \ge c, |\eta_1| \ge c$ (which is due to (3.1.3)) and (3.2.50), (3.2.51) imply that

(3.2.50)$'$ $$\mathrm{OF}^{s,q'}(\underline{U}^0_{jk}) \cap \pi(\Omega_+ \times \Omega^T_+) \cap \{|t| \le T\} = \emptyset \qquad \forall j \ne k,$$

(3.2.51)$'$ $$\mathrm{OF}^{s,q'}(\underline{U}^0_{kk}) \cap \pi(\Omega_+ \times \Omega^T_+) \cap \{0 \le t \le T, \langle T'_\xi, x' - y'\rangle + \nu_k(x_1 - y_1) < \varepsilon t\} = \emptyset \qquad \forall k = 1,\dots,n.$$

Therefore

(3.2.50)$''$ $$\mathrm{OF}^{s,q'}(\eth_x\underline{U}^0_{jk}) \cap \pi(\Omega \times \Omega^T_+) \cap \{|t| \le T\} = \emptyset \qquad \forall j \ne k,$$

(3.2.51)$''$ $$\mathrm{OF}^{s,q'}(\eth_x\underline{U}^0_{kk}) \cap \Omega \times \Omega^T_+ \cap \{0 \le t \le T, \langle T'_\xi, x' - y'\rangle - \nu_k y_1 < \varepsilon t\} = \emptyset \qquad \forall k = 1,\dots,n.$$

Let us apply theorem 3.1.10 with $J = 2$, $\phi_{j1} = \phi_1 = t$, $\phi_{j2} = \langle T'_\xi, x' - y'\rangle + \nu_j x_1 - \varepsilon t$, $\phi_2 = \langle T'_\xi, x' - y'\rangle - \varepsilon t$. Then (3.2.50)$''$, (3.2.51)$''$, (3.2.25), (3.2.26) and (3.2.32) yield that

(3.2.52) $$\mathrm{OF}^{s,q'}(\underline{U}^1_{jk}) \cap \pi(\Omega_+ \times \Omega^T_+) \cap \{0 \le t \le T, \langle T'_\xi, x' - y'\rangle - \nu_j y_1 + \nu_j x_1 < \varepsilon t\} = \emptyset \qquad \forall j,k = 1,\dots,n,$$

(3.2.52)$'$ $$\mathrm{OF}^{s,q'}(\underline{U}^1_{0k}) \cap \pi(\Omega_+ \times \Omega^T_+) \cap \{0 \le t \le T, \langle T'_\xi, x' - y'\rangle - \nu_k y_1 < \varepsilon t\} = \emptyset \qquad \forall k = 1,\dots,n;$$

similarly, (3.2.47)†, (3.2.25), (3.2.26) and (3.2.32) imply that

(3.2.52)$''$ $$\mathrm{OF}^{s,q'}(\underline{U}^1_{j0}) \cap \pi(\Omega_+ \times \Omega^T_+) \cap \{0 \le t \le T, \langle T'_\xi, x' - y'\rangle + \nu_j x_1 < \varepsilon t\} = \emptyset \qquad \forall j = 1,\dots,n;$$

(3.2.52)$'''$ $$\mathrm{OF}^{s,q'}(\underline{U}^1_{00}) \cap \pi(\Omega_+ \times \Omega^T_+) \cap \{0 \le t \le T, \langle T'_\xi, x' - y'\rangle < \varepsilon t\} = \emptyset.$$

These four relations combined with relations (3.2.50), (3.2.51) and (3.2.47) yield (i), (iii) and (iv); moreover, in the proof of (3.2.41) we also use proposition 3.2.4. □

The arguments used in the proof of theorem 2.1.15 yield

Theorem 3.2.6. *Let the conditions of assertions (i), (ii), (iii), (iv) or (v) of theorem 3.2.5 be fulfilled. Then the left-hand expressions in* (3.2.40), (3.2.41) *and* (3.2.42), (3.2.43), (3.2.44), (3.2.45) *and* (3.2.46) *respectively do not exceed*

$$Ch^{1-d-|\alpha|-|\beta|}\big(\frac{h}{T}\big)^s$$

where C does not depend on $T \in [h, T_0]$, where $T_0 > 0$ is a small constant.

(ii) Moreover, let $M = 0$ (i.e., let P be elliptic at $\iota^{-1}z^*$) and let the conditions of assertions (ii) or (v) of theorem 3.2.5 be fulfilled. Then the left-hand expressions in (3.2.41), (3.2.45) *respectively do not exceed*

$$Ch^{2-d-|\alpha|-|\beta|}\big(\frac{h}{T}\big)^s.$$

For the forthcoming applications inequality (3.2.45) is the most important.

3.2.4 Propagation of singularities along transversal bicharacteristics

Let P be an operator of the form (3.2.1) satisfying (3.1.17) at the point z^*. Let us assume that $g(z)$ (the characteristic symbol of P, see (3.1.19)) has characteristics of constant multiplicity r_k in a neighborhood of $\mathrm{z}_k \in \iota^{-1} \cap \Sigma(P)$; then in this neighborhood $g(z) = e_k(z)g_k(z)^{r_k}$ with $e_k(\mathrm{z}_k) \neq 0$ and $\partial_\tau g_k(\mathrm{z}_k) \neq 0$. Suppose that the bicharacteristics of g_k passing through z_k are transversal to Y, i.e., $\frac{dx_1}{dt} \neq 0$ along them; since $\frac{dx_1}{dt} = \partial_{\xi_1} g_k : \partial_\tau g_k$ we conclude that the multiplicity of the root $\zeta = \zeta_k$ of the polynomial $g(z,\zeta)$ is equal to r_k and hence in the reduction of subsection 3.1.2 $\mathrm{D}_k = (\sum_j \mathrm{D}_{kj}) = r_k$. On the other hand, for characteristics tangent to the boundary, $\mathrm{D}_k \geq 2r_k$. Bicharacteristics (and the point z_k) are called incoming (outgoing) if at this point $\frac{dx_1}{dt} < 0$ ($\frac{dx_1}{dt} > 0$ respectively); obviously at such a point P is microhyperbolic in the direction $-H_{\xi_1}$ (H_{ξ_1} respectively). This property can be used as a definition of the incoming (outgoing) points in the case when the multiplicities of characteristics are not constant. Let us make the reduction of subsection 3.1.2; then z_k is incoming (outgoing) if and only if $\underline{A}_{0k}$ is positive (negative) definite.[4)]

Let us assume that every point of $\iota^{-1} \cap \Sigma(P)$ is either incoming or outgoing. Let $\underline{u}^I$ and $\underline{u}^{II}$ be the corresponding components of $\underline{u} = \mathcal{R}\underline{u}$, with similar notation for all other objects. One can assume without loss of generality that $\underline{A}_0^I(z^*) = -I$ and $\underline{A}_0^{II} = I$. Let us assume for the moment that $\mathbb{H}''(z^*) = 0$. Then $\underline{u} = \begin{pmatrix} \underline{u}^I \\ \underline{u}^{II} \end{pmatrix}$ and $B = (B^I \ B^{II})$. Condition (3.1.7) means that

$$|\underline{w}^{II}|^2 \leq |\underline{w}^I|^2 + C|b^I\underline{w}^I + b^{II}\underline{w}^{II}|^2 \qquad \forall \underline{w}^I \in \mathbb{H}^I, \underline{w}^{II} \in \mathbb{H}^{II} \tag{3.2.53}$$

and yields that

$$|b^{II}\underline{w}^{II}| \geq \epsilon|\underline{w}^{II}| \qquad \forall \underline{w}^{II} \in \mathbb{H}^{II}$$

and hence $b^{II\dagger}b^{II}$ is an invertible matrix with bounded (and smooth) inverse; therefore one can multiply B on the left by an admissible operator and obtain $(-\mathcal{K} \ \ I)$. Since

4) Another equivalent definition: $\Gamma^\sigma(\mathrm{z}_k) \subset \{\mp x_1 > 0\} \cup 0$ where $\Gamma^\sigma(\mathrm{z})$ is defined in section 2.2. Based on $\Gamma^\sigma(\mathrm{z}_k)$ one can introduce a classification of all the remaining points (i.e., neither incoming nor outgoing) even without the constant multiplicities condition.

it seems to be rather foolish to place restrictions on incoming components we simply assume that

$$B = (-\mathcal{K} \quad I). \tag{3.2.54}$$

Then (3.2.53) means that $|\kappa| \leq 1$ where κ is the principal symbol of the operator $\mathcal{K}$ and $|\kappa| < 1$ under condition (3.1.7)*.

Now let $\mathbb{H}''(z^*) \neq 0$. Then $\underline{u}$ is a column vector with components $\underline{u}^I, \underline{u}^{II}, \underline{u}', \underline{u}''$ where $\underline{u}'$ and $\underline{u}''$ correspond to decomposition of P_0 and $B = (B^I \; B^{II} \; B' \; B'')$ and (3.1.7) means exactly that

$$|\underline{w}^{II}|^2 \leq |\underline{w}^I|^2 + 2\operatorname{Re}\langle \alpha^\dagger \underline{w}', \underline{w}''\rangle + C|b^I \underline{w}^I + b^{II}\underline{w}^{II} + b'\underline{w}' + b''\underline{w}''|^2 \qquad \forall \underline{w}^I, \underline{w}^{II}, \underline{w}', \underline{w}'' \tag{3.2.53$'$}$$

and this implies that the matrix b^{II} has a left inverse but does not imply that $(b^{II} \; b')$ has a left inverse. However the latter condition is reasonable if we do not want to consider propagation of singularities along the boundary connected with the degeneration of $(b^{II} \; b')$ and hence it is reasonable to replace (3.2.54) by the condition

$$B = \begin{pmatrix} -\mathcal{K} & I & 0 & -\mathcal{K}' \\ -\mathcal{K}'' & 0 & I & -\mathcal{K}''' \end{pmatrix}, \tag{3.2.54$'$}$$

and (3.2.53)′ implies that $|\mathcal{K}| \leq 1$. It should be noted that condition (3.1.7)* means exactly that (3.2.53)′ holds with strong inequality provided $(\underline{w}^I, \underline{w}^{II}, \underline{w}', \underline{w}'') \neq 0$ and therefore matrix $(b^{II} \; b')$ has a left inverse. Moreover, in this case one can assume without loss of generality that (3.2.54)′ holds with $|\kappa| < 1$ where κ is the principal symbol of $\mathcal{K}$.

On the other hand, (3.2.16) implies that $\kappa^\dagger\kappa = I$; a reasonable condition which is stronger than (3.2.16) is

$$\kappa^\dagger \kappa = \mathcal{K}\mathcal{K}^\dagger = I. \tag{3.2.55}$$

3.2.5 Geometric interpretation

This subsection is similar to section 2.2. First of all, we proved that the set $\mathrm{L} = \{(z, \mathcal{T} : z' \in T^*\partial X$, and the problem (P, B) is microhyperbolic at the point z in the multidirection $\mathcal{T}\}$ is open in the sense of proposition 3.1.4 and remark 3.1.5. Moreover, obviously

(3.2.56) The set $\mathrm{L}_z = \{\mathcal{T} : (z, \mathcal{T}) \in \mathrm{L}\}$ is convex for any $z \in T^*\partial X$.

Let us introduce the set $\Gamma_b(p, b, z) = \{\mathcal{T}', \exists(\nu_1, \dots, \nu_M)$ such that $(\mathcal{T}', \nu_1, \dots, \nu_M) \in \mathrm{L}_z \subset T_z T^*\partial X\}$. One can easily see that

$$\Gamma_b(p, b, z) = \bigcap_{\mathrm{z}_k \in {}^{-1}\iota z \cap \Sigma} \iota\Gamma(p, \mathrm{z}_k) \cap \Gamma'(p, b, z) \tag{3.2.57}$$

where

(3.2.58) $\Gamma'(p, b, z)$ is the set of all directions satisfying (3.1.20) ($\Gamma'(p, b, z)$ is an open convex cone).

Let us assume that

$$\Gamma_b(p, b, z) \neq \emptyset \quad \forall z \in \Omega, \qquad \Gamma(p, \mathrm{z}) \neq \emptyset \quad \forall \mathrm{z} \in \Omega_+, \qquad \Omega_+|_{\partial X} = \iota^{-1}\Omega. \tag{3.2.59}$$

Then

$$\Gamma_b^\sigma(p,b,z) = \oplus \sum_{z_k \in^{-1}\iota z \cap \Sigma} (\Gamma^\sigma(p,z) \cap T_z T^*\partial X) \oplus \Gamma'^\sigma(p,b,z) \tag{3.2.60}$$

and these sets have the properties described in lemma 2.2.1. Moreover, one can introduce sets $\Gamma_{b\rho}(p,b,z)$ and $\Gamma_{b\rho}^\sigma(p,b,z)$ in the obvious manner.

Without loss of generality one can assume that $X = \{x_1 > 0\}$ locally.

Definition 3.2.7. A *generalized bicharacteristic billiard* is a curve $\{\mathrm{z} = \mathrm{z}(t)\}$ in T^*X such that

(i) If $\mathrm{z}(t) \notin T^*X|_{\partial X}$ for some t then $|z(t') - z(t'')| \le C_0|t' - t''|$ for all t', t'' : $|t'-t| + |t''-t| \le \epsilon_0 \mathrm{dist}(\mathrm{z}(t), T^*X|_{\partial X})$ where the constants $\epsilon_0 > 0$ and C_0 do not depend on $t, \mathrm{z}(t)$.

(ii) In the general case $|\mathrm{z}'(t') - \mathrm{z}'(t'')| \le C_0|t' - t''|$ for all t', t'' where $\mathrm{z}' = (x, \xi')$, $\xi' = (\xi_2, \dots, \xi_d)$.

(iii) For a.e. t such that $z(t) \notin T^*X|_{\partial X}$ the function $\mathrm{z}(t)$ is differentiable and $\frac{dz}{dt} \in \Gamma^\sigma(p, \mathrm{z})$.

(iv) For a.e. t such that $\mathrm{z}(t) \in T^*X|_{\partial X}$ the function $\mathrm{z}'(t)$ is differentiable and $\frac{dz}{dt} \in \Gamma_b^\sigma(p, b, \mathrm{z}')$. ◊

Let $\mathrm{K}^\pm(p, b, z, \Omega_+)$ be the union of all generalized bicharacteristic billiards lying in $\bar\Omega_+$ and going out from the point $\mathrm{z} \in \Omega_+$ in the direction of increasing $\pm t$. Then

Theorem 3.2.8. *Let $\Omega_+ \Subset T^*X$ and $\Omega \Subset T^*\partial X$ and $\Omega_+|_{\partial X} = \iota^{-1}\Omega$. Let $H_{x_2} \in \Gamma^\sigma(p, \mathrm{z})$ at every point $\mathrm{z} \in \bar\Omega_+$ and $H_{x_2} \in \Gamma_b^\sigma(p, b, z)$ at every point $z \in \bar\Omega$. Let (P, B) satisfy the conditions of theorem 3.1.6 in Ω and let p be symmetric in Ω_+.*

(i) Let u be admissible (i.e., satisfy (3.1.15)*) and*

$$\begin{gathered}
\mathrm{OF}^{s+1}(Pu) \cap \mathrm{K}^-(p, b, \mathrm{z}, \Omega_+) = \emptyset \\
(\mathrm{OF}_b^{s+1,q}(Pu) \cup \mathrm{OF}^{s+\frac12}(B\eth_h u)) \cap \mathrm{K}^-(p, b, \mathrm{z}, \Omega_+) = \emptyset \\
\mathrm{OF}^s(u) \cap \partial\Omega_+ \cap \mathrm{K}^-(p, b, \mathrm{z}, \Omega_+) = \emptyset \\
\mathrm{OF}_b^{s,0}(u) \cap \partial\Omega \cap \mathrm{K}^-(p, b, \mathrm{z}, \Omega_+) = \emptyset.
\end{gathered}$$

Then

$$\begin{gathered}
\mathrm{OF}^s(u) \cap \mathrm{K}^-(p, b, \mathrm{z}, \Omega_+) = \emptyset, \\
\mathrm{OF}_b^{s,q+m}(u) \cap \mathrm{K}^-(p, b, \mathrm{z}, \Omega_+) = \emptyset.
\end{gathered}$$

(ii) Moreover, let condition (3.1.7)* *be fulfilled in Ω. Then one can weaken the conditions of (i) replacing $\mathrm{OF}^{s+\frac12}(B\eth_h u)$ by $\mathrm{OF}^s(B\eth_h u)$ and at the same time obtain in addition that*

$$\mathrm{OF}^{s-i}(D_1^i u|_{\partial X}) \cap \mathrm{K}^-(p, b, z, \Omega_+) = \emptyset \qquad \forall i = 0, \dots, m+q-1.$$

Proof. The proof repeats the proof of theorem 2.2.7 (including the proof of statements similar to lemma 2.2.6 and proposition 2.2.4) with no essential modifications. □

Remark 3.2.9. *(i) Under condition* (3.1.7)*, $\Gamma'(p, b, z) = \mathbb{R}^{2d-2}$ *and hence* $\Gamma_b'^\sigma(p, b, z) = 0$ *and there is no propagation linked with the boundary conditions.*

(ii) This theorem is not very precise. For example, let $a = \xi_1^2 + r(x, \xi')$, $\partial_{\xi_2} r > 0$. Then for appropriate boundary conditions (e.g., Dirichlet or Neumann) $\Gamma_b'^\sigma(p, b, z) = 0$ and $\Gamma_b^\sigma(p, z) = 0$ if $r(z) \ne 0$, $\Gamma_b^\sigma(p, z) = -H_{r_0}$ where $r_0 = r|_{\partial X}$. So generalized bicharacteristic billiards include boundary bicharacteristics *(i.e., bicharacteristics of r_0).*

However, singularities propagate along these trajectories only if the domain is bicharacteristically convex, i.e., $a = 0$, $x_1 = 0$, $\{a, x_1\} = 0$ *yield* $\{a, \{a, x_1\}\} \le 0$ *(or* $x_1 = 0$, $r = 0$ *yield* $\partial_{x_1} r \ge 0$*).*[5] *In order to exclude boundary bicharacteristics for a bicharacteristically concave domain one should refer to a more precise propagation theorem (see [Melrose 1], [Eskin 2,3], [Taylor 1,2] where the theorem was proven by an explicit parametrix construction, or [Ivrii 5] where more sophisticated energy estimates were applied; see also [Hörmander 1]) where the case of a strictly bicharacteristically concave domain was treated and then combine the results of these two theorems. In this way one can treat some difficult cases [Ivrii 5]. However, this additional theorem is proven under much more restrictive conditions and in this book we never apply such arguments.* ◊

3.3 Propagation of Singularities along Long Transversally Reflecting Bicharacteristics

3.3.1 Statement of the basic theorem

In this section we obtain results similar to those of section 2.3. We request certain conditions which are probably more restrictive than necessary. We assume that

(3.3.1) A is a second-degree h-differential operator, B is either a first-degree or 0-degree h-differential operator and under the boundary condition $Bu|_Y = 0$ the operator A is self-adjoint in $L^2(X, \mathbb{H})$ where X is the domain in $\mathbb{R}^d$ given by the inequality $\phi(x) > 0$, $Y = \partial X = \{\phi(x) = 0\}$;

(3.3.2) $\mathbb{H} = \mathbb{C}^D$, and the principal symbol of A is scalar and equals

$$a(x,\xi) = \sum_{1\le j,k\le d} g^{jk}(x)(\xi_j - V_j(x))(\xi_k - V_k(x)) + V(x)$$

where all the functions are real-valued;

(3.3.3) Condition (3.2.16) holds with β which is either a 0-degree or first-degree polynomial on ξ and $\mathbb{H}_b = \mathbb{C}^D$;

(3.3.4) All the coefficients of A, B, β satisfy the inequalities

$$|D^\alpha A \dots| \le c\rho^{-|\alpha|-1} \qquad \forall \alpha : |\alpha| \le K$$

etc., where here and in what follows in this section $\rho \in [h^{\delta'}, 1]$ with a small enough exponent $\delta' = \delta'(d, s, \delta) > 0$,

$$(3.3.5)_1 \qquad |D^\alpha \phi| \le c\rho^{-|\alpha|-1} \qquad \forall \alpha : |\alpha| \le K,$$

$$(3.3.5)_2 \qquad |\phi(x)| \le \epsilon_0 \rho \implies |\nabla_x \phi(x)| \ge \epsilon_0 \rho,$$

$$(3.3.6) \qquad c\rho^{-1} \ge \sum_{1\le j,k\le d} g^{jk}(x)\xi_j\xi_k|\xi|^{-2} \ge \epsilon_0 \rho \qquad \forall x \in X, \quad \forall \xi \in \mathbb{R}^d \setminus 0.$$

---

[5] Otherwise only analytic or Gevrey (with Gevrey index ≤ 3) singularities propagate along boundary bicharacteristics ([Sjöstrand 3–6], [Lebeau 1]). In this case these bicharacteristics are also called *creeping rays*.

Definition 3.3.1. A *Hamiltonian billiard* $\{z = z(t)\}$ with $t \in [t', t'']$ is a piecewise continuous curve consisting of $(n+1)$ continuous curves $z = z(t)$ with $t \in [t_j, t_{j+1}]$ $(j = 0, \dots, n)$, $t' = t_0 \le t_1 < \dots < t_n \le t_{n+1} = t''$ such that each of these sub-curves is a Hamiltonian trajectory of the symbol $-a(x,\xi)$, $z(t_j \pm 0) \in T^*X|_Y$, $\iota z(t_j - 0) = \iota z(t_j + 0)$, $a(z(t_j - 0)) = a(z(t_j + 0))$ $\forall j = 1, \dots, n$ (so $a(x,\xi) = \text{const}$ along this billiard), and $z(t) \in T^*X|_Y$ iff $t = t_j$ for some $j = 0, \dots, n$ in which case $\frac{d\phi}{dt}|_{z(t_j \pm 0)} > 0$. The points $z(t_j \pm 0)$ are *points of reflection.* ◊

One can introduce the *Hamiltonian flow with reflections* at Y, $\Phi_t(z)$, on T^*X; however it is not necessarily defined for all $z \in T^*X$ and $t \in \mathbb{R}$; in particular, it is not defined at the tangent points with $\{a, \phi\} = 0$.

It is well known [Cornfeld, Sinai, Fomin 1] that $\Phi_t(z)$ is N times differentiable with respect to (t, z) in a neighborhood of the point $(\bar{t}, \bar{z})$ provided $\bar{z}$ and $\Phi_{\bar{t}}(\bar{z})$ belong to $\overset{0}{T^*X}$, and it is piecewise N times differentiable provided either $\bar{z}$ or $\Phi_{\bar{t}}(\bar{z})$ belong to $T^*X|_Y$. It is also well known that Φ_t preserves the symplectic form and hence it preserves the volume form $dx \wedge d\xi$ on T^*X and the form $\mu_\tau = dx \wedge d\xi : da$ on surfaces $\Sigma_\tau = \{(x,\xi) : a(x,\xi) = \tau\}$. Moreover, it is known that in "reasonable" cases (for example, if the coefficients and the boundary are smooth; see Chapters 10 and 11 for more details) this Hamiltonian flow is defined for all t on the complement of the set Λ with $\mu_\tau(\Lambda) = 0$.

Our goal is to prove

Theorem 3.3.2. *Let A be a self-adjoint operator satisfying conditions* (3.3.1)–(3.3.6). *Let $\{z = z(t)\}$ with $t \in [0, \bar{t}]$ be a Hamiltonian billiard such that*

$$\operatorname{dist}(\Pi_x z(t), Y) \ge \epsilon_0 \rho^{-r} \gamma \qquad \text{for } t = 0, \bar{t}, \tag{3.3.7}$$

$$\operatorname{dist}(\Pi_x z(t), Y) \ge \epsilon_0 \rho^{-r} \gamma \min(1, \min_{j=1,\dots,n} |t - t_j|) \quad \forall t \in [0, \bar{t}], \tag{3.3.8}$$

$$\pm \frac{d\phi}{dt}(z(t_j \pm 0) \ge \epsilon_0 \rho \qquad (j = 1, \dots, n) \tag{3.3.9}$$

$$|D\Phi(z)| \le c\rho^{-1} \qquad \forall t \in [0, \bar{t}] \quad \forall z : |z - z(0)| \le \gamma, \tag{3.3.10$_1$}$$

$$\epsilon_0 \le \bar{t} \le c\rho^{-1} \tag{3.3.10$_2$}$$

where $D\Phi(z)$ is the Jacobi matrix of the map Φ_t, $\gamma \in [h^{1/2-\delta}, h^\delta]$, $\rho \in [h^{\delta'}, 1]$, $\delta > 0$ is arbitrary and $\delta' = \delta'(\delta) > 0$ is small enough and then $\gamma \le \rho^r$ with large enough $r = r(d)$. Then for every operator $Q^0 \in \boldsymbol{\Psi}_{h,\gamma,\gamma,N}(\mathbb{R}^d, \mathbb{H}, \mathbb{H})$ with symbol supported in the $\epsilon\gamma$-neighborhood $\mathcal{U}$ of $z(0)$ with $\epsilon \in (0, \epsilon_1)$ where ϵ_1 is a small constant there exists an operator $\bar{Q} \in \boldsymbol{\Psi}_{h,\gamma',\gamma',s}(\mathbb{R}^d, \mathbb{H}, \mathbb{H})$ with $\gamma' = \epsilon_1 \gamma \rho^{r'}$ and with symbol supported in $\Phi_t(\mathcal{U})$ (which is contained in the γ''-neighborhood of $z(\bar{t})$ with $\gamma'' = C_0 \epsilon \rho^{-r} \gamma$) such that for $t = \bar{t}$,

$$\|\bar{Q} - e^{-ith^{-1}A} Q^0 e^{-ith^{-1}A}\| \le C h^s \|Q^0\|_{\gamma}, \tag{3.3.11}$$

$$\|\bar{Q}\|_{\gamma'} \le C \|Q^0\|_{\gamma} \tag{3.3.12}$$

where $r' = r'(d, s)$ and $N = N(d, s, \delta)$ are large enough.

Moreover, it is sufficient to assume that conditions (3.3.1)–(3.3.6) *are fulfilled only in the γ''-neighborhood of the billiard $z(t)$.*

3.3.2 Proof of the basic theorem

If $n = 0$ then this theorem follows from the proof of theorem 2.3.6.

Let us first consider the case $n = 1$. According to theorem 2.3.6 there exists an operator Q^t with $t \in [t, t_1 - \rho^9]$ such that (3.3.11) and (3.3.12) are fulfilled. Let us consider the ρ^3-neighborhood of the point $z^1 = (z(t_1), \bar{\tau})$ with $\bar{\tau} = a(z(0))$ and let us realize here the reduction of subsection 3.1.2; let $\mathcal{R} = \begin{pmatrix} \mathcal{R}_1 \\ \mathcal{R}_2 \end{pmatrix}$ be the operator realizing this reduction where $\mathcal{R}_1$ and $\mathcal{R}_2$ correspond to incoming and outgoing bicharacteristics respectively. Then modulo negligible operators, the following decomposition is possible for $t \in [t_1 - 2\rho^9, t_1 - \rho^9]$:

$$\mathcal{R}_1 Q^t \equiv Q_1^t \mathcal{R}_1 + Q_2^t \mathcal{R}_2$$

with admissible operators $Q_j^t = Q_j(t, x, hD_t, hD_x, h)$ with supports not wider than $\operatorname{supp} Q^t \times \mathbb{R}_\tau$. Since the operator $\underline{P}_2$ is elliptic in a neighborhood of $(\operatorname{supp} Q^t) \times \mathbb{R}_\tau$, the equality $\underline{P}_2 \mathcal{R}_2 G \equiv 0$ implies that

$$\mathcal{R}_1 G Q^0 \equiv \mathcal{R}_1 Q^t G \equiv Q_1^t \mathcal{R}_1 G$$

for $t \in [t_1 - 2\rho^9, t_1 - \rho^9]$ where $G : f \to \{e^{ih^{-1}tA} f\}$ and we narrow the support of the symbol of $\mathcal{R}_1$ slightly. On the other hand, the following decomposition is also possible

$$Q_1^t \equiv Q_1'^{\,t} + S\underline{P}_1$$

where $Q_1'^{\,t} = Q_1'(t, x, hD_t, hD_x', h)$ and $S = S(t, x, hD_t, hD_x, h)$ and hence the equality $\underline{P}_1 \mathcal{R}_1 G = 0$ implies that for $t \in [t_1 - 2\rho^9, t_1 - \rho^9]$,

$$\mathcal{R}_1 G Q^0 \equiv Q_1'^{\,t} \mathcal{R}_1 G.$$

Let us consider $\underline{P}_1$ as an evolutionary operator with respect to x_1; surely, it depends on x_1 but the analysis of section 2.3 is applicable; moreover because of the short distance one can replace the condition of self-adjointness of the operator generating the evolution by the condition that its principal symbol is Hermitian; then theorem 2.3.6 yields that for $t \in [t_1 - \rho^9, t_1 + 2\rho^9]$,

$$\eth \mathcal{R}_1 G Q^0 \equiv Q_{(1)} \eth \mathcal{R}_1 G$$

where $Q_{(1)} \in \boldsymbol{\Psi}_{h,\gamma',\gamma',N}(X' \times \mathbb{R}_t, \mathbb{H}, \mathbb{H})$ is supported in the $\rho^{-r}\gamma$-neighborhood of (t_1, z^1). Moreover, this equality is true for $t \in [0, t_1 - \frac{1}{2}\rho^9]$ because the expressions on both sides are negligible.

Moreover, conditions (3.3.1)–(3.3.3) and the transversality condition yield that the boundary condition $Bv|_Y \equiv 0$ in a neighborhood of z^1 is equivalent to the condition $\eth \mathcal{R}_2 v \equiv \mathcal{K} \eth \mathcal{R}_1 v$ where $\mathcal{K}$ is an operator with unitary principal symbol and hence

$$\eth \mathcal{R}_2 G Q^0 \equiv Q'_{(1)} \eth \mathcal{R}_2 G$$

for $t \in [0, t_1 + 2\rho^9]$ where $Q'_{(1)} = \mathcal{K} Q_{(1)} \mathcal{K}^{-1}$ and $\mathcal{K}^{-1}$ means a parametrix of $\mathcal{K}$.

Let us consider $\underline{P}_2$ as an evolutionary operator with respect to x_1; again applying theorem 2.3.6 we obtain that

$$\mathcal{R}_2 G Q^0 \equiv Q''_{(1)} \mathcal{R}_2 G \qquad \text{for } t \in [0, t_1 + 2\rho^9]$$

with an admissible $Q''_{(1)} = Q''_{(1)}(t, x, hD_t, hD_x', h)$. On the other hand, theorem 2.3.6 yields that

$$\mathcal{R}_1 GQ^0 \equiv 0 \qquad \text{for } t \in [t_1+\rho^9, t_1+2\rho^9]$$

and that

$$GQ^0 \equiv 0 \qquad \text{for } t \in [t_1+\rho^9, t_1+2\rho^9],\ x_1 \notin [\epsilon_1\rho^{10}, C_1\rho^8].$$

These three equalities combined with the obvious equality $GQ^0 \equiv 0$ for $|\tau - \bar\tau| \geq C_0\epsilon\gamma$ yield that

$$GQ^0 \equiv Q'''_{(1)}G \qquad \text{for } t \in [t_1+\rho^9, t_1+2\rho^9]$$

with an admissible operator $Q'''_{(1)} = Q'''_{(1)}(t, x, hD_t, hD_x, h)$; representing this operator in the form $Q'''_{(1)} = Q^1 + S'P$ with $Q^1 = Q^1(t, x, hD_x, h)$ and $S' = S'(t, x, hD_t, hD_x, h)$ we obtain from the equality $PG = 0$ that (3.2.11) and (3.2.12) hold for $t \in [t_1+\rho^9, t_1+2\rho^9]$. Again applying theorem 2.3.6 we get rid of this condition. So *our theorem is proven for* $n = 1$ *and hence it remains true for arbitrary uniformly bounded* n.

In order to get rid of this restriction let us again analyze the construction of Q^1 and of Q^l obtained from Q^0 after l reflections. It is obvious that from the point of view of the *construction* of the symbol of Q^1 it is not necessary to consider $\underline{P}_k$ as evolutionary operators with respect to x_1 and hence one can apply theorem 2.3.6 for Q^1 only two times and make a "division with remainder" by $\underline{P}_1$ and by P at the point of reflection (the more sophisticated analysis which was done was necessary for the *justification* of this construction). Then we obtain the following simplified construction:

(i) For z in a neighborhood $\mathcal{U}$ of the point $z(0)$ we define a function $t_1(z)$ which is the time of the first destination of Y (and we also define $t_l(z)$ which is the time of the L-th destination of Y, $l = 1, \dots, n$);

(ii) For $z' = \phi_t(z)$ with $z \in \mathcal{U}$, $t \leq t_1(z)$ we define $Q^0_k(t, z')$ (which are the terms in the decomposition of the symbol of Q^0) as solutions to the recurrence equations $(2.3.7)_k$;

(iii) For $z'' = \psi^+(\Phi^0_{t_1(z)}(z))$ with $z \in \mathcal{U}$ and with $\psi^\pm(z^\mp) = z^\pm$, where $\iota z^+ = \iota z^-, a(z^+) = a(z^-)$ and z^+ and z^- correspond to outgoing and incoming bicharacteristics respectively, we define $Q^1_k(t, z'')$ with $t = t_1(z)$ by the formula

$$Q^1_k(t, z'') = \sum_{0\leq j\leq k} (\ell_{k-j}Q^0_j)(\psi^-(z'')) \tag*{$(3.3.13)^1_k$}$$

where ℓ_k are linear differential operators of degree not higher than k with respect to z and

$$\ell_0(q) = \kappa q \kappa^\dagger \tag{3.3.14}$$

where $\kappa(z) = \kappa(\iota z, \tau)$ $(\tau = a(z))$ is a unitary matrix;

(iv) For $z''' = \Phi_t(z)$ with $t_1(z) < t \leq t_2(z)$ we define $Q^1_k(t, z''')$ again as a solution to $(2.3.7)_k$ with the given initial data at $t = t_1(z)$;

(v) Repeating this procedure l times we obtain $Q^l_k(t, \tilde z)$ with $\tilde z = \Phi_t(z), z \in \mathcal{U}$, $t_l(z) \leq t \leq t_{l+1}(z)$.

Our conditions yield that the map ψ^+ and the coefficients of ℓ_k satisfy the estimates

$$\begin{aligned} &|D^\alpha \ell_{k,\beta}| \leq C\rho^{-4(k+|\alpha|)},\\ &|D^\alpha \psi^+| \leq C\rho^{-4|\alpha|} \qquad \forall k, \alpha : k + 2|\alpha| \leq K. \end{aligned} \tag{3.3.15}$$

Moreover, our conditions yield that $|\frac{d^2\phi}{dt^2}| \leq C\rho^{-4}$ and hence (3.3.9) and (3.3.7) imply that

(3.3.16) $$t_{l+1} - t_l \geq \epsilon' \rho^6 \qquad \forall l = 0, \dots, n \implies n \leq C\rho^{-7}.$$

Let us consider the Hamiltonian flow Φ_t and related mappings. Let $\mathcal{U}$ be a small neighborhood of $z(0)$. Then for $z' \in \mathcal{U}$ and $z = \Phi_{t'}(z')$ with $t' \in [0, \bar{t}]$, $\Phi_t(z)$ is defined for $t \in [-t', \bar{t} - t']$, and if $t_l(z') < t' < t_{l+1}(z')$ with $l = 0, \dots, n$ then there are functions $0 < t_1(z) < \dots < t_{n-l}(z)$ and $0 > t_{-1}(z) > \dots > t_{-l}(z)$ which are the times of reflections of the flow directed along either increasing or decreasing t.

Let us consider the maps

$$\Phi^+ : z \to (t_{\pm 1}(z), \Phi_{t_{\pm 1}(z)}) \in \bar{\mathbb{R}}_t^+ \times T^*X|_Y$$

defined for $z = \Phi_{t'}(z')$ with $z' \in \mathcal{U}$ and $t < t_n(z'), t > t_1(z')$ respectively. Since the Hamiltonian flow is symplectic, $\det(D\Phi_t(z)) = 1$ and (3.3.10) yields that

(3.3.10)* $$|D\Phi_t(z)| \leq c\rho^{1-2d}$$
$$\forall z = \Phi_{t'}(z'), z' \in \mathcal{U} \quad \forall t \in [-t', \bar{t} - t'] : z, \Phi_t(z) \in T^*\overset{0}{X}.$$

Lemma 2.3.2 implies that

(3.3.17) $$|D^\alpha \Phi_t(z)| \leq C\rho^{-r|\alpha|} \qquad \forall \alpha : |\alpha| \leq K - 1$$
$$\forall z = \Phi_{t'}(z') \in T^*\overset{0}{X} : z' \in \mathcal{U} \quad \forall t : \max(t_{-1}(z), -t') < t < \min(t_1(z), \bar{t} - t');$$

then (3.3.9) implies that

(3.3.18) $$|D^\alpha t_{\pm 1}(z)| \leq C\rho^{-r|\alpha|} \qquad \forall \alpha : |\alpha| \leq K - 1$$
$$\forall z = \Phi_{t'}(z') \in T^*\overset{0}{X} : z' \in \mathcal{U}, t' \leq t_n(z') \ (t > t_1(z'))$$

and hence

(3.3.19) $$|D^\alpha \Phi^\pm(z)| \leq C\rho^{-r|\alpha|}$$

under the same conditions.

Therefore the same estimates hold for $\Psi^\pm : z \to \Phi_{t_{\pm 1}(z)}(z)$ where now $z \in T^*X|_Y$: for $z = \Phi_{t'}(z')$ with $z' \in \mathcal{U}$ and $t' = t_l(z')$, $l < n$ ($l > 1$ respectively) we take $t_{\pm 1}(z) = t_{l\pm 1}(z') - t_l(z')$. Moreover, (3.3.9) implies that the same estimates hold for $\psi^\pm : z^\mp \to z^\pm$ where $\iota z^+ = \iota z^-$, $a(z^+) = a(z^-)$ and hence we do not distinguish these points.

For $z = \Phi_{t'}(z')$, $z' \in \mathcal{U}$, $t' = t_l'(z')$ let us consider the map $\Psi_j^\pm = (\Psi^\pm)^j$ with $1 \leq j \leq n - l$ ($1 \leq j \leq l - 1$ respectively). Then (3.3.10)* and the previous estimates imply that

(3.3.20) $$|D\Psi_j^\pm| \leq C\rho^{-r}$$

where r and C do not depend on j, n. Let us note that an m-th order derivative of $\Psi_j^\pm$ is the sum of fewer than $(m-1)!j^{m-1} + 1$ terms such that every term contains at most $(m-1)$ factors of the type $(D^\beta \Psi_q^\pm) \circ \Psi_p^\pm$ with $p \leq j - 1$ and $|\beta| \geq 2$ (if we sum the exponents $|\beta_i|$ for all these factors then $\sum(|\beta_i| - 1) = m - 1$) and at most $(2m - 1)$ factors of the type $(D\Psi_q^\pm) \circ \Psi_p^\pm$ with $p + q \leq j$. Then estimates (3.3.20) and (3.3.19) for $\Psi^\pm$ and (3.3.16) imply that estimate (3.3.19) for $\Psi_j^\pm$ holds with exponent r and constant C independent of j. This estimate combined with the estimates for $\Phi^\pm$ and for Φ_t with $t_{-1}(z) \leq t \leq t_1(z)$ yield

Lemma 3.3.3. *Let the conditions of theorem 3.3.2 be fulfilled and let $\mathcal{U}$ be the γ-neighborhood of $z(0)$. Then*

(i) If $z = \Phi_{t'}(z')$, $z' \in \mathcal{U}$, $t' = t_l(z')$, then for $1 \le j \le n-l$ ($1 \le j \le l-1$) the map $z \to \Psi_j^{\pm}(z)$ and the function $t_{\pm j}(z) = t_{l\pm j}(z') - t'$ satisfy estimate (3.3.19);

(ii) Moreover, for $t_j(z) \le t \le \min(t_{j+1}(z), \bar{t} - t')$ the map $z \to \Phi_t(z) = \Phi_{t-t_j(z)}(\Psi_j^+(z))$ satisfies estimate (3.3.19) and for $\max(-t', t_{-j-1}(z)) \le t \le t_{-j}(z)$ the map $z \to \Phi_t(z) = \Phi_{t-t_j(z)}(\Psi_j^-(z))$ satisfies the same estimate;

(iii) If $z = \Phi_{t'}(z')$, $z' \in \mathcal{U}$, $t_l(z') < t' < t_{l+1}(z')$ then for $1 \le j \le n-l$ ($1 \le j \le l-1$) the map $z \to \Phi_j^{\pm} = \Psi_{j-1}^{\pm}(\Phi^{\pm}(z))$ and the function $t_j(z) = t_{j+l}(z') - t'$ ($t_{-j}(z) = t_{l+1-j}(z') - t'$ respectively) satisfy (3.3.19);

(iv) Moreover, for $t_j(z) < t < \min(t_{j+1}(z), \bar{t} - t')$ the map $z \to \Phi_t(z) = \Phi_{t-t_j(z)}(\Phi_j^+(z))$ and for $\max(t_{-j-1}(z), -t') < t < t_{-j}(z)$ the map $z \to \Phi_t(z) = \Phi_{t-t_{-j}(z)}(\Phi_j^-(z))$ satisfy (3.3.19).

Let us recall that in (i) and (ii) all maps and functions are defined on a subdomain of $T^*X|_Y$ while in (iii) and (iv) they are defined on a subdomain of $T^*\overset{0}{X}$, and in (i) and (iii) the ranges of the maps are contained in $T^*X|_Y$ while in (ii) and (iv) they are contained in $T^*\overset{0}{X}$. Moreover, let us recall that r can increase at every step.

Let us now consider the principal symbol $Q_0(t,z)$ of the operator Q^t; according to the described procedure for D $= 1$ it is given by the equality $Q_0(t,z) = Q_0^0(\Phi_{-t}(z))$ provided $z' = \Phi_{-t}z \in \mathcal{U}$, $t \in [0, \bar{t}]$. This equality and lemma 3.3.3 and the formula for derivatives of composed functions (see section 2.3) yield the following estimates:

$$|D^\alpha Q_0^l| \le C\gamma'^{-|\alpha|} \qquad \forall \alpha : |\alpha| \le N-1 \tag{3.3.21}$$

where γ' is introduced in the statement of theorem. Moreover, according to the same procedure for D ≥ 2,

$$Q_0(t,z) = T(t,z) Q_0^0(\Phi_{-t}(z)) T^\dagger(t,z) \tag{3.3.22}$$

where the transforming matrix $T(t,z)$ is defined in the following way:

(i) According to section 2.3 we introduce $T(t,z) = T_{t,t}(z)$ for $z = \Phi_{t'}(z') \in T^*\overset{0}{X}$, $z' \in \mathcal{U}$ and $\max(t_{-1}(z), -t') \le -t \le \min(t_1(z), \bar{t} - t')$; the same formula is also applied for $z = \Phi_{t'}(z')$, $z' \in \mathcal{U}$, $t' = t_k(z')$ and $\max(t_{-1}(z), -t') \le -t \le 0$. Then according to section 2.3, $T(t,z)$ is a unitary matrix satisfying (3.3.19).

(ii) We set $T_{\pm}(z) = T(-t_{\mp 1}(z), z)$ and for $z = \Phi_{t'}(z')$, $z' \in \mathcal{U}$, $t_l(z') < t' < t_{l+1}(z')$, $t_j(z') < t' - t < t_{j+1}(z')$ with $j < l$, $t > 0$ we set

$$\begin{aligned} T(t,z) = T_+(z)\kappa(\Phi_1^-(z)) T_+(\Phi_1^-(z))\kappa(\Phi_2^-(z)) \dots T_+(\Phi_{l-j-1}^-(z)) \cdot \\ \kappa(\Phi_{l-j}^-(z)) T(t - t_{j-l}(z), \Phi_{l-j}^-(z)); \end{aligned}$$

then (i), the fact that the matrix κ is unitary and satisfies (3.3.19), lemma 3.3.3 and estimate (3.3.16) yield that $T(t,z)$ satisfies (3.3.19) and hence Q_0 also satisfies this estimate.

Let us prove by induction with respect to k that the terms Q_k of the symbol Q^t satisfy the estimate

$$|D^\alpha Q_k| \le C\gamma^{-|\alpha| - 2k} \rho^{r - r(k+1)(|\alpha|+1)} \qquad \forall k, \alpha : |\alpha| + 2k \le N-1. \tag{3.3.23}$$

Our construction implies that

$$
\begin{aligned}
Q_k(t,z) = {} & T(t,z)Q_k^0(\Phi_{-t}(z))T^\dagger(t,z) + \\
& \int_0^t T(t',z)F_m(t-t',\Phi_{-t'}(z))T^\dagger(t',z)dt' + \\
& \sum_{01\le j\le l} T_j(z)G_{k,j}(t+t_{-j}(z),\Phi_j^-(z))T_j^\dagger(z)
\end{aligned}
\tag{3.3.24}
$$

where

$$
T_j(z) = T_+(z)\kappa(\Phi_1^-(z))T_+(\Phi_1^-(z))\kappa(\Phi_2^-(z))\dots T_+(\Phi_{j-1}^-(z)),
\tag{3.3.25}
$$

F_k and $G_{k,j}$ are expressed in terms of derivatives of degrees not exceeding $k-i+1$ and $k-i$ respectively of Q_i with $i\le k-1$, and hence for admissible r these matrices satisfy (3.3.23) (our analysis is similar to that of section 2.3). Then we obtain that Q_k also satisfy (3.3.23).

So, *theorem 3.3.2 has been proven.* □

Remark 3.3.3. *(i) Theorem 3.3.2 obviously remains true if we replace $\bar t$ and $[0,\bar t]$ by $-\bar t$ and $[-\bar t,0]$ respectively in its conditions and statement.*

(ii) One can prove the following useful formula in this case and in the case treated in section 2.3:

$$
T(t+t',\Phi_{t+t'}(z)) = T(t,\Phi_{t+t'}(z))T(t',\Phi_{t'}(z)).
\tag{3.3.26}
$$

◊

3.3.3 Propagation of singularities

Theorem 3.3.2 immediately yields the following

Theorem 3.3.5. *Let $Q',Q \in \Psi_{h,\gamma,\gamma,N}(\mathbb{R}^d,\mathbb{H},\mathbb{H})$ be operators such that $\operatorname{dist}(\Pi_x \operatorname{supp} Q, Y)\ge \epsilon_0\gamma$, $\operatorname{dist}(\Pi_x \operatorname{supp} Q', Y)\ge \epsilon_0\gamma$ (where we do not distinguish the supports of an operator and its symbol). Let us assume that* ***either***

(i) For all $z\in\operatorname{supp} Q$ there passes a billiard $\Phi_t(z)$ through z with $t\in[0,\bar t]$, satisfying all the conditions of theorem 3.3.2, and for some $t=t^*\in[0,\bar t]$

$$
\operatorname{dist}(z',\Phi_{-t}(z))\ge\epsilon_0\gamma \qquad \forall z'\in\operatorname{supp} Q'
\tag{3.3.27}
$$

or

(ii) The conditions of (i) are fulfilled with Q, Q' and Φ_{-t} replaced by Q', Q, and Φ_t respectively.

Let all the remaining conditions of theorem 3.2.2 be fulfilled. Then estimates (2.3.30) *and* (2.3.31) *hold (the last estimate holds provided* $\operatorname{supp} Q\cup\operatorname{supp} Q'\subset\{|\xi|\le c\}$*).*

Proof. Without loss of generality one can consider case (i) and assume that $\operatorname{diam}\operatorname{supp} Q'\le\gamma$. Then, if

$$
\operatorname{dist}(\Pi_x\Phi_{-t}(\operatorname{supp} Q),Y)\ge\epsilon_0\rho^9
\tag{3.3.28}
$$

for $t=t^*$ then our theorem follows immediately from theorem 3.3.2. On the other hand, if condition (3.3.28) is violated for $t=t^*$ then it follows from the transversality condition that (3.3.28) is fulfilled for $t=t^*\pm\epsilon\rho^9$ with a small enough constant $\epsilon>0$; on the other hand, for $t'\in[-\epsilon\rho^9,\epsilon\rho^9]$, the flow $\Phi_{t'}(z')$ is defined for $z'\in\operatorname{supp} Q'$. Then our theorem follows from theorem 3.3.2, remark 3.3.4(i) and the equality

$$Qe^{ith^{-1}A}Q' = e^{i(t-t')h^{-1}A}Q^{t-t'} \cdot Q'^{t'}e^{it'h^{-1}A}$$

with

$$Q^{t-t'} = e^{-i(t-t')h^{-1}A}Qe^{i(t-t')h^{-1}A}, \qquad Q'^{t'} = e^{it'h^{-1}A}Q'e^{-it'h^{-1}A},$$

$t = t^*$, $t' = \epsilon\rho^9$. □

Comments

There are many papers devoted to the propagation of singularities near the boundary; for a survey I again send the reader to [Ivrii 9]. Theorems 3.1.6 and 3.1.9 were first proven for wave front sets in [Ivrii 2] in less general form. I have explained that these theorems, in contrast to theorem 2.1.2, are not very precise.

A certain analog of theorem 3.2.6 for the wave equation was first proven in [Ivrii 6], which was the first proof of this type of theorem without an explicit parametrix construction. The proof of theorem 3.2.6 in its present form has never been published before. Moreover, the geometric description is completely new.

One can formulate

Problem 3.1.** *Repeat Problem 2.1** in the case of propagation near the boundary. This problem seems to be very difficult even in the case in which only the transversality condition is fulfilled in the weakened form used here (and time, the Jacobi matrix and the coefficients with their derivatives are assumed to be uniformly bounded).* ◊

Appendix A. On Solutions to Linear Matrix Equations

Let $\mathbb{H}$ and $\mathbb{H}'$ be Hilbert spaces and let us consider the equation

(A.1) $$AX - XB = C$$

with respect to $X \in \mathcal{L}(\mathbb{H}, \mathbb{H}')$ with given $A \in \mathcal{L}(\mathbb{H}, \mathbb{H})$, $B \in \mathcal{L}(\mathbb{H}', \mathbb{H}')$, $C \in \mathcal{L}(\mathbb{H}, \mathbb{H}')$.

Let us assume that the *spectra of A and B are disjoint* and let $\mathbb{C} \supset \gamma$ be a closed counter-clockwise oriented curve, $\gamma = \partial G$ where G is a bounded domain in $\mathbb{C}$, $\mathrm{Spec}(A) \subset G$, $\mathrm{Spec}(B) \subset \complement\bar{G}$. Then

(A.2) $$X = \frac{1}{2\pi i} \int_\gamma (z - A)^{-1} C (z - B)^{-1} dz$$

is the (unique) solution of equation (A.1).

Moreover, if $\mathbb{C} \supset \gamma'$ is a closed counter-clockwise oriented curve, $\gamma' = \partial G'$, G' is a bounded domain in $\mathbb{C}$, $\mathrm{Spec}(B) \subset G'$, $\mathrm{Spec}(A) \subset \complement\bar{G}'$ then

(A.2)′ $$X = -\frac{1}{2\pi i} \int_{\gamma'} (z - A)^{-1} C (z - B)^{-1} dz.$$

Moreover, formula (A.2) ((A.2)′ respectively) remains true also in the case when B (A respectively) is an unbounded operator.

If $A \geq \epsilon I, B \leq -\epsilon I$ with $\epsilon > 0$ then the (unique) solution of (A.1) is also given by the formula

(A.3) $$X = \int_0^{+\infty} e^{-At} C e^{Bt} dt$$

and if $B \geq \epsilon I, A \leq -\epsilon I$ then

(A.3)′ $$X = \int_{-\infty}^{0} e^{-At} C e^{Bt} dt;$$

formula (A.3) ((A.3)′ respectively) remains true if $-A$ and B (A and $-B$ respectively) are unbounded operators which generate semigroups tending to 0 quickly enough as t tends to $+\infty$.

Let us note that if X is a solution of (A.1) then $X^\dagger$ is a solution of

(A.1)† $$X^\dagger A^\dagger - B^\dagger X^\dagger = C^\dagger$$

and hence if $\mathbb{H} = \mathbb{H}'$ and $A^\dagger = B$, $C^\dagger = -C$ then the uniqueness of solutions implies that $X^\dagger = X$.

Similarly, if $\mathbb{H}$ and $\mathbb{H}'$ are complexifications of real spaces then the operation of complex conjugation is defined and if A, B, C are real then X is also real; moreover, if $\mathbb{H} = \mathbb{H}'$ and $A^T = B$, $C^T = -C$ then $X^T = X$.

Let us emphasize that the fundamental condition is that the spectra of A and B are disjoint.

Part II

Local and Microlocal Semiclassical Asymptotics

4. Standard Local Semiclassical Spectral Asymptotics in the Interior of a Domain

In this chapter we begin the treatment of *local and microlocal semiclassical spectral asymptotics*. Now we are interested in *standard* spectral asymptotics, i.e., in asymptotics with respect to the Planck constant $h \to +0$ only; more precisely, in this chapter we are interested in asymptotics of the *spectral function* (restriction to the diagonal of the Schwartz kernel of the spectral projector) and its spatial means in the ball $B(0, \frac{1}{2})$ under the assumption that the ball $B(0,1)$ is contained in the domain. Moreover, averaging with respect to the spectral parameter is also possible. The only global condition we assume is that the operator A is self-adjoint; all other conditions are related either to the ball $B(0,1)$ or to the points which we can touch by some Hamiltonian trajectory of the necessary length starting in this ball (in the case when we derive spectral asymptotics with more accurate remainder estimates). One can consider this chapter as a preliminary to Chapter 5 in which the boundary will be taken into account.

In section 4.1 we introduce the basic notions and obtain the first rough estimates. In section 4.2 by means of successive approximations we construct $U(t,x,y)$ (the fundamental solution of the Cauchy problem for the corresponding evolution equation) in a bounded domain in phase space for $|t| \le h^{\frac{1}{2}+\delta}$ with an arbitrarily small exponent $\delta > 0$ and then we construct $(\Gamma_x U)(t,x)$ and related functions.

In section 4.3 we prove the basic general theorems. Namely, the construction of section 4.2 immediately yields the complete asymptotic decomposition of the *spectral means*[1]) of the spectral function. On the other hand, certain conditions associated with microhyperbolicity (see theorem 2.1.15) provide that the function $(\Gamma_x U)(t,x)$ (or some related function) is negligible for $h^{1-\delta} \le |t| \le T_0$ and for admissible τ with a small enough constant $T_0 > 0$; moreover, theorems 2.3.7 and 2.3.8 imply that under certain hypotheses one can replace this T_0 by a larger number T (under appropriate restrictions T can tend to infinity as $h \to +0$). Hence the construction of $(\Gamma_x U)(t,x)$ (or related functions) gives the correct answer for $|t| \le T$ and for admissible τ (surely, this is not true for $U(x,y,t)$ itself). The Tauberian technique then gives asymptotics for the spectral functions or their spatial means. Moreover, we obtain nice asymptotics for spectral means with a small mollification parameter as well as for spectral Riesz means.

In section 4.4 we consider operators with scalar principal symbols as well as more general operators for which we can improve our general basic theorems. Our main tools are dilatations (rescaling) on x and ξ and linked rescalings on t and a spectral parameter. After these procedures h is replaced by $h_{\text{eff}} = \frac{h}{\rho\gamma}$ where γ and ρ are some natural scales with respect to x and ξ respectively. After an appropriate dilatation we obtain that either the microhyperbolicity condition is fulfilled in the vicinity or $h_{\text{eff}} = 1$. Then we apply the accurate estimates of section 4.3 or the rough estimates of section 4.1 respectively. By means of an admissible partition of unity we obtain local and microlocal semiclassical spectral asymptotics in the case when the microhyperbolicity condition is replaced by some weaker conditions. This work is done step by step and at every step the condition

[1]) I.e., means with respect to a spectral parameter; here convolution is made with a smooth function.

is weaker. In order to obtain more accurate remainder estimates we also apply refined propagation theorems.

In section 4.5 we apply these results to the Schrödinger and Dirac operators. For the Schrödinger operator the conditions of these theorems are fulfilled automatically for any dimension d (if we are interested in spatial means; otherwise we assume that $d \geq 3$) but for the Dirac operator some additional treatment is necessary.

In section 4.6 we extend some of the above results to conjoint spectral asymptotics for families of commuting operators.

Finally, in section 4.7 we consider scalar operators with periodic Hamiltonian flow of the principal symbol (or some perturbations of such operators) and under certain conditions we obtain spectral asymptotics with highly accurate remainder estimates; these asymptotics are not necessarily completely Weylian. In the case of the cluster character of the spectrum we get the asymptotic distribution of the eigenvalues inside the clusters. Our main tools here are very long term propagation of singularities (the main result is that singularities propagate along periodic trajectories which drift along the Hamiltonian field of some auxiliary symbol) and conjoint spectral asymptotics of two commuting operators (namely, the operator in question can be represented as the sum of an operator A_0 with spectrum $\{2\pi hn, n \in \mathbb{Z}\}$ and a perturbing operator μB commuting with A_0, where μ is a small parameter).

In Appendix B we consider the boundary values of almost-analytic functions (which are useful for the calculation of successive approximations via the Fourier transform) and in Appendix C we list some properties of Dirac matrices.

4.1 Preliminary Analysis

4.1.1 General estimates

In this chapter we consider the operator

$$A = \mathrm{Op}^{\mathrm{w}}(A(x,\xi,h)) = \sum_{|\alpha|+|\beta|\leq m} \frac{(\alpha+\beta)!}{\alpha!\beta!}(hD)^\alpha a_{\alpha+\beta}(x,h)(hD)^\beta \tag{4.1.1}$$

with

$$A(x,\xi,h) = \sum_{|\alpha|\leq m} a_\alpha \xi^\alpha \tag{4.1.2}$$

and $h \mathrel{\dot{\in}} (0,1]$, given on the manifold X some part of which is identified with the ball $B(0,1)$ where the coefficients $a_\alpha(x,h)$ are linear (not necessarily bounded) symmetric operators in some auxiliary Hilbert space $\mathbb{H}$.[2)] We assume that A with the given domain $\mathrm{D}(A)$ is a self-adjoint operator in $L^2(X,\mathbb{H})$ (this condition will be formulated more precisely later). Let $E(\tau)$ be the spectral projector of A corresponding to the interval $(-\infty,\tau)$; then $E(\tau_1,\tau_2) = E(\tau_2) - E(\tau_1)$ for $\tau_1 < \tau_2$ is the spectral projector corresponding to the interval $[\tau_1,\tau_2)$. Our goal is to derive asymptotics as $h \to +0$ for various functions associated with $e(x,y,\tau_1,\tau_2)$, the Schwartz kernel of $E(\tau_1,\tau_2)$. Our main tool is the analysis of $U(t,x,y)$, the Schwartz kernel of the propagator $\mathbf{U}(t) = e^{ih^{-1}tA}$. Obviously

$$U(t,x,y) = \int e^{ith^{-1}\tau} d_\tau e(x,y,\tau_1,\tau) \tag{4.1.3}$$

---

2) Moreover, we consider operators $A \in \Psi_h(\mathbb{R}^d,\mathbb{H},\mathbb{H})$ and $A \in \tilde{\Psi}_h^{(m)}(\mathbb{R}^d,\mathbb{H},\mathbb{H})$ and similar pseudo-differential operators.

with an arbitrary $\tau_1 \in [-\infty, +\infty]$; this equality is understood in the weak sense and in the general case we cannot set $x = y$ here; however, theorem 1.1.8 and corollary 1.1.9 yield that we can set $x = y$ if we first apply operators Q_{1x} and ${}^tQ_{2y}$ on the left and right respectively provided the symbols of Q_1 and Q_2 are compactly supported. Namely, since the operator norms of $\mathbf{U}(t)$ and $E(\tau_1, \tau_2)$ do not exceed 1 corollary 1.1.9 yields the following

Proposition 4.1.1. *Let A be an operator self-adjoint in $L^2(X, \mathbb{H})$ and let functions $\psi_1, \psi_2 \in L^\infty$ be supported in $B(0,1)$ and $|\psi_j| \le 1$. Moreover, let $q_1, q_2 \in S_{h,N}(T^*\mathbb{R}^d, \mathbb{H}, \mathbb{H})$ be supported in $\{|\xi| \le \omega\}$ and $Q_i = \operatorname{Op}^w(q_i)$ $(i = 1, 2)$. Then*

$$(4.1.4)\quad |Q_{1x}\psi_1(x)e(x, y, \tau_1, \tau_2)\psi_2(y)^tQ_{2y}| \le CM\bigl(\frac{\omega}{h}\bigr)^d \qquad \forall x, y \in \mathbb{R}^d \quad \forall \tau_1, \tau_2 \in \mathbb{R},$$

$$(4.1.5)\qquad |Q_{1x}\psi_1(x)U(t, x, y)\psi_2(y)^tQ_{2y}| \le CM\bigl(\frac{\omega}{h}\bigr)^d \qquad \forall x, y \in \mathbb{R}^d \quad \forall \tau_1, \tau_2 \in \mathbb{R}$$

where $M = \|q_1\| \cdot \|q_2\|$, $\|.\|$ means the $S_{h,K}$-norm and $K = K(d)$, $C = C(d)$ here.

In the case of $\dim H = \infty$ we need to obtain a more refined assertion. Let us assume that B_1 is a (self-adjoint) operator in $\mathbb{H}$ such that

$$(4.1.6)\qquad B_1^\dagger = B, \qquad \langle B_1 w, w\rangle \ge |w|^2 \quad \forall w \in \mathrm{D}(B_1),$$

$$(4.1.7)\qquad \|B_1^l u\|_{L^2(B(0,1),\mathbb{H})} \le c(\|A^l u\|_{L^2(X,\mathbb{H})} + \|u\|_{L^2(X,\mathbb{H})}) \qquad \forall u \in \mathrm{D}(A^l)$$

with some $l > 0$.

Proposition 4.1.2. *Let the assumptions of proposition 4.1.1 and conditions (4.1.6) and (4.1.7) be fulfilled. Then*

$$(4.1.8)\quad |Q_{1x}\psi_1(x)B_1^j e(x, y, \tau_1, \tau_2)B_1^k\psi_2(y)^tQ_{2y}| \le CM\bigl(\frac{\omega}{h}\bigr)^d(|\tau_1| + |\tau_2| + 1)^{j+k} \qquad \forall x, y \in \mathbb{R}^d \quad \forall \tau_1, \tau_2 \in \mathbb{R} \quad \forall j, k \in [0, l]$$

where here and below K and C also depend on l and C also depends on c.

Moreover, if $\chi \in C_0^{k+j}([-1,1])$ and $T \ge h$, $T' \in \mathbb{R}$ then

$$(4.1.9)\quad |F_{t\to h^{-1}\tau}\chi_{T,T'}(t)Q_{1x}\psi_1(x)B_1^jU(t, x, y)B_1^k\psi_2(y)^tQ_{2y}| \le CT\|\chi\| \cdot \|Q_1\| \cdot \|Q_2\| \cdot \bigl(\frac{\omega}{h}\bigr)^d(|\tau| + 1)^{j+k} \qquad \forall x, y \in \mathbb{R}^d \quad \forall \tau \in \mathbb{R} \quad \forall j, k \in [0, l]$$

where $\|\chi\|$ means the C^{k+j}-norm of χ and $\chi_{T,T'}(t) = \chi(\frac{t-T'}{T})$. In particular, in the context of of lemma 4.1.1 this estimate holds for $j = k = 0$.

Proof. Let us note that the interpolation theorems of [Lions, Magenes 1] imply that (4.1.7) holds for l replaced by arbitrary $j \in [0, l]$ and hence $\|B_1^j(A \pm i)^{-j}\| \le c$; then the same inequality holds for the adjoint operator: $\|(A \mp i)^{-j}B_1^j\| \le c$ and therefore

$$\|B_1^jE(\tau_1, \tau_2)B_1^k\| \le c^2\|(A + i)^jE(\tau_1, \tau_2)(A + i)^k\| \le C(|\tau_1| + |\tau_2| + 1)^{j+k};$$

the resulting inequality combined with corollary 1.1.9 implies (4.1.8).

Similarly

$$\|F_{t\to h^{-1}\tau}\chi_{T,T'}(t)B_1^j\mathbf{U}(t)B_1^k\| \le C\|F_{t\to h^{-1}\tau}(A+i)^j\chi_{T,T'}(t)\mathbf{U}(t)(A+i)^k\| = C\|F_{t\to h^{-1}\tau}((-hD_t+i+\tau)^{j+k}\chi_{T,T'})\mathbf{U}(t)\| \le CT(|\tau|+1)^{j+k}\|\chi\|$$

where the inequality $T\ge h$ was also used. Applying corollary 1.1.9 again we obtain (4.1.9). □

4.1.2 Semiclassical spectral gaps. I

Let us now investigate the equations

(4.1.10) $$(hD_t - A_x)U = 0,$$

(4.1.10)† $$U(hD_t - {}^tA_y) = 0.$$

from the elliptic point of view.

Let us assume that

(4.1.11)$_1$ $$|D_x^\beta a_\alpha(x,h)B^{-(1-\frac{|\alpha|}{m})}| \le c \qquad \forall x\in B(0,1)\quad \forall\beta:|\beta|\le K\quad \forall\alpha:|\alpha|\le m,$$

(4.1.12)$_1$ $$|a_0(x,h)w| \ge \epsilon_0|Bw| - c|w| \qquad \forall x\in B(0,1)\quad \forall w\in \mathrm{D}(B)$$

where

(4.1.13) B is an operator in $\mathbb{H}$ satisfying (4.1.6)

and the meaning of the subscript "1" will be clear later. Moreover, let us assume that

(4.1.14) A is a self-adjoint operator in $L^2(X,\mathbb{H})$ and $\mathrm{D}(A)\supset C_0^m(B(0,1),\mathrm{D}(B))$

and

(4.1.15) $a_0(x,h)$ with $\mathrm{D}(a_0)=\mathrm{D}(B)$ is a self-adjoint operator in $\mathbb{H}$.

Then

(4.1.16) $a_\alpha(x,h)$ with $\mathrm{D}(a_\alpha)=\mathrm{D}(B)$ are symmetric operators in $\mathbb{H}$ for all $x\in B(0,1),h$ and $\alpha:|\alpha|\le m$

and

(4.1.17) $A(x,\xi,h)$ with the domain $\mathrm{D}(B)$ is a self-adjoint operator in $\mathbb{H}$ for all $x\in B(0,1),\xi\in\mathbb{R}^d,h$

and, moreover, the conditions

(4.1.11)$_\ell$ $$|D_x^\beta B^{-(1-\frac{|\alpha|}{m})(1-\ell)}a_\alpha(x,h)B^{-(1-\frac{|\alpha|}{m})\ell}| \le c \qquad \forall x\in B(0,1)\quad \forall\beta:|\beta|\le K\quad \forall\alpha:|\alpha|\le m,$$

(4.1.12)$_\ell$ $$\|B^{\ell-1}a_0(x,h)w\| \ge \epsilon_0|B^\ell w| - c|B^{\ell-1}w| \qquad \forall x\in B(0,1)\quad \forall w\in \mathrm{D}(B)$$

are fulfilled for all $\ell \in [0,1]$.

In fact, (4.1.16) follows from (4.1.14), and $(4.1.11)_\ell$ and $(4.1.12)_\ell$ follow from $(4.1.11)_1$ and $(4.1.12)_1$ by means of transition to the adjoint operator, (4.1.16) and interpolation. Finally, (4.1.17) follows from $(4.1.11)_1$ with $K=0$, $(4.1.12)_1$ and (4.1.15).

Moreover, let us assume that the condition

$(4.1.18)_\ell$ Condition (4.1.7) is fulfilled with $B_1 = B$ and $l = \ell$

is fulfilled with $\ell = 1$.

Conditions $(4.1.11)_1$, $(4.1.12)_1$, (4.1.13), (4.1.14), (4.1.15) and $(4.1.18)_1$ form *the first standard set of assumptions*. However, we can also deal with *the second standard set of assumptions* obtained by replacing the previous conditions by $(4.1.11)_{\frac{1}{2}}$, (4.1.13), $(4.1.18)_{\frac{1}{2}}$ and

$$(4.1.12)_\pm \qquad \pm\langle a_0(x,h)w, w\rangle \ge \epsilon_0 |B^{\frac{1}{2}}w|^2 - c|w|^2 \qquad \forall w \in \mathrm{D}(B)$$

$(4.1.14)'$ $\pm A$ is a self-adjoint semibounded from below operator in $L^2(X,\mathbb{H})$ and $\mathrm{D}(A^{\frac{1}{2}}) \supset C_0^m(B(0,1), \mathrm{D}(B^{\frac{1}{2}}))$.

Obviously, then (4.1.15) and (4.1.16) hold and, moreover,

$(4.1.17)_\pm$ $\pm A(x,\xi,h)$ is a self-adjoint semibounded from below operator in $\mathbb{H}$ with $\mathrm{D}(A(x,\xi,h)^{\frac{1}{2}}) = \mathrm{D}(B^{\frac{1}{2}})$ for all $x \in B(0,1)$, $\xi \in \mathbb{R}^d$, h

(in particular $(4.1.16)_\pm$ holds where $(4.1.16)_\pm$ is $(4.1.17)_\pm$ restricted to $\xi = 0$ and a_0).

If the first standard set of assumptions is fulfilled then multiplying the operator $(hD_t - A)$ by B^{-1} on the left or right we see that the parametrices G_1 and G_0 of the operators $P_1 = B^{-1}(hD_t - A)$ and $P_0 = (hD_t - A)B^{-1}$ can be constructed on $\operatorname{supp} q \times \mathbb{R}_t \times [\tau_1, \tau_2]$ provided

$$(4.1.19) \qquad |(\tau - A(x,\xi,h))wB^{-1}| \ge \epsilon_0 \nu \|w\|,$$

$$(4.1.20) \quad |A^{(\alpha)}_{(\beta)}(x,\xi,h)(\tau - A(x,\xi,h))^{-1}| \le c\rho^{-|\alpha|}\gamma^{-|\beta|}$$
$$\forall (x,\xi) \in \operatorname{supp} q \quad \forall \tau \in [\tau_1,\tau_2] \quad \forall w \in \mathbb{H} \quad \forall \alpha,\beta : |\alpha| + |\beta| \le K$$

where

$$(4.1.21) \qquad \gamma \le 1, \quad \rho\gamma \ge h^{1-\delta}\omega^\delta, \quad \omega \ge 1, \quad \tau_2 - \tau_1 \ge \nu \ge \left(\frac{h}{\omega}\right)^n$$

with arbitrarily small exponent $\delta > 0$ and arbitrary n (surely, K depends on n).

Similarly, if the second standard set of assumptions is fulfilled then multiplying the operator $(hD_t - A)$ by $B^{-\frac{1}{2}}$ on the left or right we see that the parametrix of the operator $P_{\frac{1}{2}} = B^{-\frac{1}{2}}(hD_t - A)B^{-\frac{1}{2}}$ can be constructed on $\operatorname{supp} q \times \mathbb{R}_t \times [\tau_1,\tau_2]$ provided

$$(4.1.19)_\pm \qquad \pm\langle (A(x,\xi,h) - \tau)w, w\rangle \ge \epsilon_0 \nu |B^{\frac{1}{2}}w|^2 \qquad \forall w \in \mathrm{D}(B),$$

$$(4.1.20)' \quad |(A(x,\xi,h) - \tau)^{-\frac{1}{2}} A^{(\alpha)}_{(\beta)}(x,\xi,h)(A(x,\xi,h) - \tau)^{-\frac{1}{2}}| \le c\rho^{-|\alpha|}\gamma^{-|\beta|}$$
$$\forall (x,\xi) \in \operatorname{supp} q \quad \forall \tau \in [\tau_1,\tau_2] \quad \forall w \in \mathbb{H} \quad \forall \alpha,\beta : |\alpha| + |\beta| \le K$$

where condition (4.1.21) is assumed to be fulfilled in this case also.

Let $q \in \mathrm{S}_{h,\rho,\gamma,N}(\mathbb{R}^d)$ and $Q = \mathrm{Op}^{\mathrm{w}}(q)$. Then in all three cases the operator norm of $Q(G_\ell P_\ell - I)$ does not exceed $C'\|Q\|(\frac{h}{\omega})^s$ where here and in what follows s is arbitrarily large and K, C' depend on d, m, δ, n, s and C' also depends on c. Then corollary 1.1.9 immediately implies

Proposition 4.1.3. *(i) Let conditions* $(4.1.11)_1$, $(4.1.12)_1$, (4.1.13), (4.1.14), (4.1.15), $(4.1.18)_1$ *and* (4.1.19)–(4.1.21) *be fulfilled with either* $q = q_1$ *or* $q = q_2$ *and let us assume that*

$$|\xi| \le c\omega \qquad \forall (x,\xi) \in \operatorname{supp} q_1 \cup \operatorname{supp} q_2 \tag{4.1.22}$$

where $q_1, q_2 \in \mathrm{S}_{h,\rho,\gamma,N}(\mathbb{R}^d)$ *and* $Q = \mathrm{Op}^{\mathrm{w}}(q)$ *and* $Q_i = \mathrm{Op}^{\mathrm{w}}(q_i)$. *Let* $\psi_1, \psi_2 \in C_0^K(B(0, 1-\epsilon))$, $\epsilon > 0$ *be an arbitrarily small constant,* $T \ge h^{1-\delta}\omega^\delta$ *and* $\chi \in C_0^K([-1,1])$. *Then*

$$\begin{aligned} |F_{t\to h^{-1}\tau}\chi_{T,T'}(t)Q_{1x}B^j\psi_1(x)U(t,x,y)B^k\psi_2(y)^tQ_{2y}| \le \\ C'TM(|\tau|+1)^{j+k}\Big(\frac{h}{\omega}\Big)^s \\ \forall x,y \in \mathbb{R}^d \quad \forall \tau \in [\tau_1,\tau_2] \quad \forall j,k \in [0,\ell] \end{aligned} \tag{4.1.23}$$

where $\ell = 1$ *and* $M = \|q_1\| \cdot \|q_2\| \cdot \|\psi_1\| \cdot \|\psi_2\| \cdot \|\chi\|$, *all the norms are calculated in the spaces indicated above, and* C' *also depends on* ϵ.

(ii) Suppose conditions $(4.1.11)_{\frac{1}{2}}$, $(4.1.12)_\pm$, (4.1.13), $(4.1.14)'$, $(4.1.18)_{\frac{1}{2}}$ *and* $(4.1.19)_\pm$, $(4.1.20)'$, (4.1.21) *are fulfilled with either* $q = q_1$ *or* $q = q_2$ *as well as the conditions of (i) on* $Q_1, Q_2, \psi_1, \psi_2, \chi$. *Then estimate* (4.1.23) *holds with* $\ell = \frac{1}{2}$.

(iii) In particular, conditions (4.1.19)–(4.1.20) *are automatically fulfilled in the* γ-*neighborhood of* $(\bar{x}, \bar{\xi})$ *with* $\gamma = h^{\frac{1}{2}-\delta}$ *provided* (4.1.19) *is fulfilled at this point with* $\nu = h^{\frac{1}{2}-\delta}$.

On the other hand, one can use the precise Gårding inequality (theorem 1.1.10 (i), (ii)) (combined with the fact that $B^{-\frac{1}{2}}[A,\psi]B^{-\frac{1}{2}}$ is a skew-self-adjoint operator modulo $O(h^2)$) instead of the parametrix construction and easily obtain

Proposition 4.1.3′. *Let conditions* $(4.1.11)_{\frac{1}{2}}$, $(4.1.12)_\pm$, (4.1.13), $(4.1.14)'$, $(4.1.19)_{\frac{1}{2}}$ *be fulfilled for* $\tau = \bar{\tau}$ *and let condition* $(4.1.19)_\pm$ *be fulfilled in the* ϵ-*neighborhood of* $\operatorname{supp} q_2$ *where* $\epsilon > 0$ *is an arbitrarily small constant. Then*

(i) Estimate (4.1.23) *remains true provided* $\nu = Ch$ *with a large enough constant* C *and* $\tau \in [\tau_1, \tau_2]$ *with* $[\tau_1, \tau_2] = (-\infty, \bar{\tau}]$ *or* $[\tau_1, \tau_2] = [\bar{\tau}, \infty)$ *respectively.*

(ii) Moreover, if a *is a scalar-valued symbol this estimate remain true provided* $\nu = Ch^{2-\delta}$ *with a large enough constant* C.[3)]

Finally, if $B = I$ and if in the ϵ-neighborhood of $(\bar{x}, \bar{\xi})$ either $\operatorname{Spec} A(x,\xi,h) \cap [\bar{\tau} - \epsilon, \bar{\tau} + \nu] = \emptyset$ or $\operatorname{Spec} A(x,\xi,h) \cap [\bar{\tau} - \nu, \bar{\tau} + \epsilon] = \emptyset$ then reducing the operator to the block structure $\begin{pmatrix} A_+ & 0 \\ 0 & A_- \end{pmatrix}$ with positively defined $\pm(A_\pm - \bar{\tau})$ and applying the arguments of lemma 4.1.3′ we obtain

Proposition 4.1.3″. *Let conditions* $(4.1.11)_1$, $(4.1.12)_1$, (4.1.13), (4.1.14) *and* (4.1.15), $(4.1.18)_1$ *be fulfilled with* $B = I$, *and let either* $\operatorname{Spec} A(x,\xi,h) \cap [\bar{\tau} - \epsilon, \bar{\tau} + \nu] = \emptyset$ *or* $\operatorname{Spec} A(x,\xi,h) \cap [\bar{\tau} - \nu, \bar{\tau} + \epsilon] = \emptyset$ *with* $\nu = Ch$ *in the general case and with* $\nu = h^{2-\delta}$ *in the case* $\operatorname{Spec} a(x,\xi) \cap [\bar{\tau} - \epsilon, \bar{\tau} + \epsilon] = \{\lambda(x,\xi)\}$. *Then for* τ *lying in the indicated interval estimate* (4.1.23) *holds.*

---

3) Moreover, reducing A to the block matrix $\begin{pmatrix} A_I & 0 \\ 0 & A_{II} \end{pmatrix}$ where $\operatorname{Spec} A_{II}$ does not intersect $[\tau_1 - \epsilon, \tau_2 + \epsilon]$ we obtain the same assertion provided $\operatorname{Spec} a(x,\xi) \cap [\tau_1 - \epsilon, \tau_2 + \epsilon] = \{\lambda(x,\xi)\}$.

In order to pass from estimates of $U(\cdot,\cdot,\cdot)$ to estimates of $e(\cdot,\cdot,\cdot,\cdot)$ we apply a *simplified Tauberian technique:*[4]

Proposition 4.1.4. *Under the assumptions of proposition 4.1.3(i), (ii) the estimate*

$$(4.1.24)\quad |Q_{1x}\psi_1(x)B^je(x,y,\tau',\tau'')B^k\psi_2(y)^tQ_{2y}| \le C'M\big(\frac{h}{\omega}\big)^s(|\tau'|+|\tau''|+1)^{j+k}(|\tau''-\tau'|+1)$$
$$\forall x,y\in\mathbb{R}^d\quad \forall\tau',\tau''\in[\tau_1,\tau_2]\quad \forall j,k\in[0,\ell]$$

holds with $\ell=1,\frac{1}{2}$ respectively with the same K and C' as before and with $M=|||q_1|||\cdot|||q_2|||\cdot\|\psi_1\|\cdot\|\psi_2\|$.

Moreover, a similar estimate holds in the context of propositions 4.1.3′ and 4.1.3″.

Proof. We apply Tauberian arguments. Namely, equality (4.1.3) implies the equality

$$(4.1.25)\quad F_{t\to h^{-1}\tau}\chi_T(t)Q_{1x}\psi_1(x)B^jU(t,x,y)B^k\psi_2(y)^tQ_{2y} = T\int\hat\chi\big(\frac{(\tau-\lambda)T}{h}\big)d_\lambda\mathcal{E}(\lambda)$$

with $\mathcal{E}(\lambda)=Q_{1x}\psi_1(x)B^je(x,y,\lambda_1,\lambda)B^k\psi_2(y)^tQ_{2y}$.

Let us pick $j=k$, $Q_2=Q_1^*$, $\psi_2=\bar\psi_1$ (where the bar means complex conjugation) and $x=y$; then the $\mathcal{L}(\mathbb{H},\mathbb{H})$-valued function $\mathcal{E}(\lambda)$ is monotone[5] with respect to λ (because in this case $\mathcal{E}(\lambda)$ is a restriction to the diagonal of the Schwartz kernel of the monotone operator-valued function $RE(\lambda_1,\lambda)R^*$ with $R=Q_1\psi_1B^j$). Let us pick a *Hörmander function* χ.[6] Then the right-hand expression in (4.1.25) will not increase if we take the integral only on $\lambda\in[\tau-\frac{h}{T},\tau+\frac{h}{T}]$ and replace here $\hat\chi(\frac{(\tau-\lambda)T}{h})$ by $\varepsilon=\min_{\tau\in[-1,1]}\hat\chi(\tau)>0$; then we obtain in the right-hand expression $\varepsilon\mathcal{E}(\tau-\frac{h}{T},\tau+\frac{h}{T})$ where here and in what follows $\mathcal{E}(\tau',\tau'')=\mathcal{E}(\tau'')-\mathcal{E}(\tau')$. Then estimate (4.1.23) yields that

$$\mathcal{E}(\tau-\frac{h}{T},\tau+\frac{h}{T})\prec C'M\big(\frac{h}{\omega}\big)^s(|\tau|+1)^{2j}I_{\mathbb{H}}\qquad\forall\tau\in[\tau_1,\tau_2]$$

where $I_{\mathbb{H}}$ is the identity operator and $\prec$ is the order relation for self-adjoint operators; then

$$\mathcal{E}(\tau',\tau'')\prec C'M\big(\frac{h}{\omega}\big)^s(T\frac{|\tau''-\tau'|}{h}+1)(|\tau'|+|\tau''|+1)^{2j}I_{\mathbb{H}}$$

and taking an appropriate constant T we obtain (4.1.25) in the special case $Q_2=Q_1^*$, $\psi_2=\bar\psi_1$ and $k=j$ provided the conditions of the proposition are fulfilled with $q=q_1$. But in this case the left-hand expression in (4.1.26) equals $|\mathcal{M}\mathcal{M}^*|$ where for fixed x $\mathcal{M}=Q_1\psi_1B^jE(\tau',\tau'')$ is treated as an operator from $L^2(X,\mathbb{H})$ to $\mathbb{H}$ (and hence $\mathcal{M}^*$ acts in the opposite direction) and hence the already proven version of (4.1.25) implies that

$$\|\mathcal{M}\|=\|\mathcal{M}^*\|\le C'M^{\frac{1}{2}}\big(\frac{h}{\omega}\big)^s(|\tau'|+|\tau''|+1)^j(|\tau''-\tau'|+1)^{\frac{1}{2}}.$$

On the other hand, proposition 4.1.2 and the same arguments as before yield the estimate

---

[4] In what follows using these words we refer to the proof of proposition 4.1.4.

[5] Recall that Hermitian operators on Hilbert space are partially ordered.

[6] That means that $\chi\in C_0^\infty([-1,1])$ is even and satisfies $\chi(0)=1$, $\hat\chi(\tau)>0\ \forall\tau\in\mathbb{R}$ where the hat means Fourier transform. Such a function exists (see, e.g., [Shubin 1]).

$$\|\mathcal{N}\| = \|\mathcal{N}^*\| \le C'M\big(\frac{\omega}{h}\big)^{\frac{d}{2}}(|\tau'|+|\tau''|+1)^k(|\tau''-\tau'|+1)^{\frac{1}{2}}$$

where $\mathcal{N}$ is defined in the same way as $\mathcal{M}$ was defined but with Q_1, ψ_1, x replaced by Q_2, ψ_2, y. Since the left-hand expression in (4.1.25) equals the norm of $\mathcal{M}\mathcal{N}^*$ we obtain (4.1.25) completely; recall that s is chosen arbitrarily. □

Let us consider explicitly the case in which the second standard set of assumptions holds. Then $(4.1.12)_\pm$ implies that if

$$(4.1.26)_\pm \qquad \pm\tau \le -C_0\langle\xi\rangle^m$$

with $C_0 = C_0(d,m,c)$ [7] then conditions $(4.1.19)_\pm$ and $(4.1.20)'$ are fulfilled with $\nu = 1$, $\rho = |\tau|^{\frac{1}{m}}$, $\gamma = 1$ and moreover

$$(4.1.27) \qquad |(\tau - A(x,\xi,h))^{-1}| \le C|\tau|^{-1}.$$

In fact, $(4.1.11)_{\frac{1}{2}}$ yields that

$$|\langle (A(x,\xi,h) - a_0(x,h))w, w\rangle| \le C_0|\xi|^m|w|^2 + C_0|\xi|\cdot|B^{\frac{m-1}{2m}}w|^2$$

and then $(4.1.12)_\pm$ and $(4.1.26)_\pm$ yield that

$$\pm\langle (A(x,\xi,h) - \tau)w, w\rangle \ge C_0^{-1}(|\tau|\cdot|w|^2 + |B^{\frac{1}{2}}w|^2)$$

and this inequality yields $(4.1.19)_\pm$, (4.1.27) and $(4.1.20)'$.

Hence if $q_i \in \mathrm{S}_{h,\rho,\gamma,N}(T^*\mathbb{R}^d)$, $Q_i = \mathrm{Op}^{\mathrm{w}}(q_i)$ and (4.1.21), (4.1.22) hold and if $\pm\tau \le -C_0\omega^m$, then (4.1.24) holds for $\tau' = \tau$, $\tau'' = 2\tau$; hence the estimate

$$(4.1.23)' \quad |Q_{1x}\psi_1(x)B^j e(x,y,\tau,2\tau)B^k\psi_2(y)^tQ_{2y}| \le C'Mh^s|\tau|^{-s} \qquad \forall x,y \in \mathbb{R}^d \quad \forall j,k \in [0,\ell]$$

holds with $Q_1 = \zeta(\frac{h}{\bar\omega}D)$ where $\bar\omega = C_0^{-1}|\tau|^{\frac{1}{m}}$, $\zeta \in C_0^K(B(0,1))$. It easily follows that this estimate holds for $Q_1 = \mathrm{Op}^{\mathrm{w}}(q_1)$ provided $q_1 \in \mathrm{S}_{h,\rho,\gamma,N}$ is supported in $\{|\xi| \le C_0^{-1}|\tau|^{\frac{1}{m}}\}$, $\pm\tau \le -C_0$ and condition (4.1.21) is fulfilled; here $\ell = \frac{1}{2}$. Moreover, if conditions $(4.1.11)_1$, $(4.1.12)_2$ and $(4.1.18)_1$ are fulfilled then one can take $\ell = 1$. Replacing τ by $2^p\tau$ and summing with respect to $p = 0,1,2,\dots$ we obtain

Proposition 4.1.5. *Let conditions $(4.1.11)_{\frac{1}{2}}$, $(4.1.12)_\pm$, (4.1.13), $(4.1.14)'$ and $(4.1.18)_{\frac{1}{2}}$ be fulfilled and let $q_i \in \mathrm{S}_{h,\rho,1,N}(T^*\mathbb{R}^d, \mathbb{H}, \mathbb{H})$ satisfy (4.1.22) and ρ and $\gamma = 1$ satisfy (4.1.21). Let $\psi_{1,2}$ be as in proposition 4.1.3. Then*

$$(4.1.28)_\pm \quad |Q_{1x}\psi_1(x)B^j e(x,y,\mp\infty,\tau)B^k\psi_2(y)^tQ_{2y}| \le C'Mh^s|\tau|^{-s} \qquad \forall x,y \in \mathbb{R}^d \quad \forall \tau : \pm\tau \le -C_0\omega^m \quad \forall j,k \in [0,\ell]$$

with $\ell = \frac{1}{2}$; moreover, if $(4.1.11)_1$, $(4.1.12)_2$ and $(4.1.18)_1$ are fulfilled then one can take $\ell = 1$.

The following two assertions show how one can set $x = y$ without multiplication by cutting operators.

---

[7] We use the notation C_0 instead of C in the case when in its definition c can be calculated for K replaced by $K_0 = 1,2,3$ where the definition of K_0 in every case we leave to the reader.

Proposition 4.1.6. *Let conditions* $(4.1.11)_1$ *and* $(4.1.12)_1$ *be fulfilled and let*

$$(4.1.29)\quad |\xi|^m|w| + |Bw| \le c(|A(x,\xi,h)w| + |w|) \qquad \forall x\in B(0,1)\quad \forall \xi\in\mathbb{R}^d\quad \forall w\in \mathrm{D}(B).$$

Then $(4.1.18)_1$ *holds.*

Let us assume that $q_1, q_2 \in \tilde{\mathrm{S}}^0_{h,\rho,\gamma,N}(T^*\mathbb{R}^d,\mathbb{H},\mathbb{H})$, $Q_i = \mathrm{Op}^{\mathrm{w}}(q_i)$, (2.1.21) *holds and* $\psi_{1,2}$ *are as in proposition 4.1.3. Then*

(i) If either q_1 *or* q_2 *is supported in* $\{|\xi|\ge\omega\}$ *with* $\omega\ge C_0$ *then*

$$(4.1.30)\quad |D_x^\alpha D_y^\beta Q_{1x}\psi_1(x)B^j e(x,y,\tau',\tau'')B^k\psi_2(y)^t Q_{2y}| \le C'M\Big(\frac{h}{\omega}\Big)^s \qquad \forall x,y\in\mathbb{R}^d\ \forall \tau',\tau'' : |\tau'|\le C_0^{-1}\omega,\ |\tau''|\le C_0^{-1}\omega \qquad \forall j,k\in[0,\ell]\ \forall\alpha,\beta : |\alpha|+|\beta|\le r$$

where K *and* C' *now also depend on* r, r *arbitrarily large, and the* $\tilde{\mathrm{S}}$*-norms of* $q_{1,2}$ *are now used in the definition of* M, *and* $\ell = 1$ *here.*

(ii) In the general case

$$(4.1.31)\quad |D_x^\alpha D_y^\beta Q_{1x}\psi_1(x)B^j e(x,y,\tau',\tau'')B^k\psi_2(y)^t Q_{2y}| \le C'Mh^{-d-|\alpha|-|\beta|}(|\tau'|+|\tau''|+1)^{\frac{1}{m}(d+|\alpha|+|\beta|)+j+k} \qquad \forall x,y\in\mathbb{R}^d\quad \forall\tau',\tau''\quad \forall j,k\in[0,\ell]\quad \forall \alpha,\beta : |\alpha|+|\beta|\le r.$$

Proof. We will prove the assertion that $(4.1.18)_1$ is fulfilled later (see subsection 4.1.3). Proposition 4.1.4 implies estimate (4.1.30) under the additional assumption that the symbols of Q_1 and Q_2 are supported in $\{|\xi|\le 4\omega\}$ and either the symbol of Q_1 or the symbol of Q_2 is supported in $\{|\xi|\ge\omega\}$. Replacing ω by 2ω and summing over a partition of unity by means of pseudo-differential operators with symbols supported in $\{2^p\omega\le|\xi|\le 2^{p+2}\omega\}$, $p=0,1,2,\dots$ we obtain (i); more precisely, we can take two partitions (one to the left of Q_1 and another to the right of Q_2, and p means the largest index).

Assertion (ii) follows from (i) and proposition 4.1.3. □

Proposition 4.1.7. *Let conditions* $(4.1.11)_{\frac{1}{2}}$, $(4.1.12)_\pm$, (4.1.13), (4.1.14)′ *be fulfilled and let*

$$(4.1.32)_\pm\quad \pm\langle A(x,\xi,h)w,w\rangle \ge \epsilon_0(|\xi|^m|w|^2 + |B^{\frac{1}{2}}w|^2) - c|w|^2 \qquad \forall x\in B(0,1)\quad \forall\xi\in\mathbb{R}^d\quad \forall w\in\mathrm{D}(B).$$

Then $(4.1.18)_{\frac{1}{2}}$ *holds.*

Let us assume that $q_1, q_2 \in \tilde{\mathrm{S}}^0_{h,\rho,\gamma,N}(T^*\mathbb{R}^d,\mathbb{H},\mathbb{H})$, $Q_i = \mathrm{Op}^{\mathrm{w}}(q_i)$, (2.1.21) *holds and* $\psi_{1,2}$ *are as in proposition 4.1.3. Then*

(i) The estimate

$$(4.1.33)_\pm\quad |D_x^\alpha D_y^\beta Q_{1x}\psi_1(x)B^j e(x,y,\mp\infty,\tau)B^k\psi_2(y)^t Q_{2y}| \le C'Mh^s|\tau|^{-s} \qquad \forall x,y\in\mathbb{R}^d\quad \forall\pm\tau\le -C_0\quad \forall j,k\in[0,\ell]$$

holds with $\ell=\frac{1}{2}$; *moreover, if either the symbol of* Q_1 *or the symbol of* Q_2 *is supported in* $\{|\xi|\ge\omega\}$ *then this estimate remains true with the additional factor* ω^{-s} *in the right-hand expression.*

(ii) In the general case the following estimate holds:

$$(4.1.34)_\pm \quad |D_x^\alpha D_y^\beta Q_{1x}\psi_1(x)B^j e(x,y,\mp\infty,\tau)B^k\psi_2(y)^t Q_{2y}| \leq C' M h^{-d-|\alpha|-|\beta|}(|\tau|+1)^{\frac{1}{m}(d+|\alpha|+|\beta|)+j+k} \qquad \forall x,y\in\mathbb{R}^d \quad \forall j,k\in[0,\ell]$$

with $\ell = \frac{1}{2}$;

Moreover, if conditions $(4.1.11)_1$, $(4.1.12)_1$ *are fulfilled then all these estimates hold with* $\ell = 1$.

Proof. The proof is similar to that of proposition 4.1.6 but also uses proposition 4.1.5. □

4.1.3 Semiclassical spectral gaps. II

The following assertions follow from propositions 4.1.4–4.1.7 under smoothness conditions which are too restrictive; here we give an independent proof.

Proposition 4.1.8. *Let conditions* $(4.1.11)_1$, $(4.1.12)_1$, (4.1.13), (4.1.14) *and* (4.1.15) *be fulfilled with* $K = \max(1+\frac{d}{2}, m) + r$. *Then*

(i) If condition (4.1.29) *is fulfilled then*

$$(4.1.35) \quad |D_x^\alpha D_y^\beta B^j e(x,y,\tau',\tau'')B^k| \leq C' h^{-d-|\alpha|-|\beta|}(|\tau'|+|\tau''|+1)^{\frac{1}{m}(d+|\alpha|+|\beta|)+j+k} \qquad \forall x,y\in B(0,1-\epsilon) \quad \forall \tau',\tau'' \quad \forall \alpha,\beta: |\alpha|\leq r, |\beta|\leq r \quad \forall j,k\in[0,1]$$

where here and below $C' = C'(d,m,r,c,\epsilon)$.

(ii) If the following condition

$$(4.1.36) \quad |(A(x,\xi,h)-\tau)w| \geq \epsilon((\omega+|\xi|^m)|w| + |Bw|) \qquad \forall x\in B(0,1) \quad \forall \xi\in\mathbb{R}^d \quad \forall \tau\in[\tau_1,\tau_2] \quad \forall w\in \mathrm{D}(B)$$

is fulfilled with $\omega \geq 1$ *then*

$$(4.1.37) \quad |D_x^\alpha D_y^\beta B^j e(x,y,\tau',\tau'')B^k| \leq C''\bigl(\frac{h}{\omega}\bigr)^s(|\tau'|+|\tau''|+1)^{j+k} \qquad \forall x,y\in B(0,1-\epsilon) \quad \forall \tau',\tau''\in[\tau_1,\tau_2] \quad \forall \alpha,\beta: |\alpha|\leq r, |\beta|\leq r \quad \forall j,k\in[0,1]$$

where s *is arbitrary and* $C'' = C''(d,m,r,s,c,\epsilon)$.

Proposition 4.1.9. *Let conditions* $(4.1.11)_{\frac{1}{2}}$, $(4.1.12)_\pm$, (4.1.13), $(4.1.14)'$ *and* $(4.1.32)_\pm$ *be fulfilled with* $K = \max(1+\frac{d}{2}, m) + r$. *Then*

$$(4.1.38) \quad |D_x^\alpha D_y^\beta B^j e(x,y,\mp\infty,\tau)B^k| \leq \begin{cases} C'' h^s (|\tau|+1)^{-s} & \forall \tau \leq -C_0 \\ C' h^{-d-|\alpha|-|\beta|}(|\tau|+1)^{\frac{1}{m}(d+|\alpha|+|\beta|)+j+k} & \end{cases} \qquad \forall x,y\in B(0,1-\epsilon) \quad \forall \alpha,\beta: |\alpha|\leq r, |\beta|\leq r \quad \forall j,k\in[0,\ell]$$

with $\ell = \frac{1}{2}$.

Moreover, if conditions $(4.1.11)_1$ *and* $(4.1.12)_1$ *are fulfilled then these estimates hold with* $\ell = 1$.

Proof of propositions 4.1.8, 4.1.9. Let us note that if (4.1.29) is fulfilled, $u \in C_0^m(B(0,1), \mathrm{D}(B))$, $l = 0, \dots, m$ and

$$Au = \sum_{|\beta| \le m-l} (hD)^\beta v_\beta$$

then

$$\sum_{|\alpha| \le l} \|(hD)^\alpha B^{\frac{1}{m}(l-|\alpha|)} u\| \le \sum_{|\beta| \le m-l} C\|B^{\frac{1}{m}(-m+l+|\beta|)} v_\beta\| + C\|u\|. \tag{4.1.39}$$

In fact, one can easily prove this estimate in the constant coefficient case by means of the Fourier transform. In the general case let ψ_n be a partition of unity in a neighborhood of $B(0,1)$ with small enough supports and let A_n be obtained from A by freezing coefficients at the point $x^n \in B(0,1) \cap \operatorname{supp} \psi_n$. Then

$$A_n \psi_n u = \sum_{|\beta| \le m-l} \psi_n (hD)^\beta v_\beta + (A_n - A)\psi_n u + [A, \psi_n] u;$$

representing the last two terms in the form

$$\sum_{|\alpha| \le l, |\beta| \le m-l} (hD)^\beta b_{\beta\alpha} (hD)^\alpha u$$

with small coefficients $b_{\beta\alpha}$ for $|\alpha| = l$, $|\beta| = m-l$ and applying estimate (4.1.39) to A_n we obtain that $M_1 \le \nu M_1 + C_\nu M_2$ where M_1 and M_2 are the left-hand and right-hand expressions respectively in (4.1.39) and one can choose $\nu > 0$ arbitrarily small (before choosing the partition of unity). Hence (4.1.39) holds in the general case.

Similarly one can prove that for $l = m+1, \dots, K$ the estimate

$$\sum_{l-m \le |\alpha| \le l} \|(hD)^\alpha B^{\frac{1}{m}(l-|\alpha|)u}\| \le C \sum_{|\beta| \le m-l} C\|(hD)^\beta A u\| + C\|u\| \tag{4.1.40}$$

holds for $u \in C_0^m(B(0,1), \mathrm{D}(B))$; in this deduction estimate (4.1.39) has been used.

Repeatedly substituting ψu for u where $\psi \in C_0^\infty(B(0, 1-\epsilon))$ in (4.1.39) and (4.1.40) we obtain that these estimates remain true for $u \in C^p(B(0,1), \mathrm{D}(B))$ with $p = \max(m, l)$ if in the left-hand and right-hand expressions we take the norms in the balls $B(0, 1-\epsilon)$ and $B(0, 1-\frac{\epsilon}{2})$ respectively; here C also depends on $\epsilon > 0$.

Iterating if necessary we obtain that for $|\alpha| \le l, j \in [0,1]$ and $\frac{1}{m}|\alpha| + j \in \mathbb{Z}$

$$\|(hD)^\alpha B^j u\|_{B(0,1-\epsilon)} \le C(\|A^{\frac{1}{m}|\alpha|+j} u\| + \|u\|).$$

Substituting $E(\tau', \tau'')u$ for u we conclude that the operator norm of the operator $(hD)^\alpha B^j E(\tau', \tau'') : L^2(X, \mathbb{H}) \to L^2(B(0, 1-\epsilon), \mathbb{H})$ does not exceed $C(|\tau'| + |\tau''| + 1)^{\frac{1}{m}|\alpha|+j}$; interpolation and estimate (4.1.40) again yield that this estimate remains true without the condition $\frac{1}{m}|\alpha| + j \in \mathbb{Z}$.

Then the embedding inequalities imply that the operator norm of the operator $\mathcal{M} : L^2(X, \mathbb{H}) \ni u \to ((hD)^\alpha B^j E(\tau', \tau''))(x) \in \mathbb{H}$ with $x \in B(0, 1-\epsilon)$ does not exceed $Ch^{-\frac{d}{2}-|\alpha|}(|\tau'| + |\tau''| + 1)^{\frac{d}{2m}+\frac{1}{m}|\alpha|+j}$. Obviously, $D_x^\alpha(-D_y^\beta)B^j e(x, y, \tau', \tau'')B^k = \mathcal{M}\mathcal{N}^*$ where $\mathcal{N}$ is defined in the same way as $\mathcal{M}$ but with β, k, y instead of α, j, x. Therefore we obtain (4.1.35).

In a similar way one can prove the two estimates

$$
\begin{aligned}
(4.1.41)\quad & \|D_x^\alpha D_y^\beta F_{t\to h^{-1}\tau}\chi_T(t)B^j\mathbf{U}(t,)B^k\| \le \\
& \qquad C''Th^{-|\alpha|-|\beta|}(|\tau|+1)^{\frac{1}{m}(|\alpha|+|\beta|)+j+k} \\
& \qquad\qquad \forall\tau\quad \forall\alpha,\beta:|\alpha|\le l,|\beta|\le l\quad \forall j,k\in[0,1]
\end{aligned}
$$

where $T\ge h^{-1}$, $\chi\in C_0^\infty([-1,1])$ is a fixed function, $\|\cdot\|$ means the operator $\mathcal{L}(L^2(B(0,1-\epsilon),\mathbb{H}),L^2(B(0,1-\epsilon),\mathbb{H}))$-norm and the operator is applied only to functions supported in $B(0,1-\epsilon)$; and

$$
\begin{aligned}
(4.1.42)\quad & |D_x^\alpha D_y^\beta F_{t\to h^{-1}\tau}\chi_T(t)B^jU(t,x,y)B^k| \le \\
& \qquad C''Th^{-d-|\alpha|-|\beta|}(|\tau|+1)^{\frac{1}{m}(d+|\alpha|+|\beta|)+j+k} \\
& \qquad \forall x,y\in B(0,1-\epsilon)\quad \forall\tau\quad \forall\alpha,\beta:|\alpha|\le r,|\beta|\le r\quad \forall j,k\in[0,1]
\end{aligned}
$$

where $|\cdot|$ means the $\mathcal{L}(\mathbb{H},\mathbb{H})$-norm.

In this deduction we use (4.1.40), (4.1.10), (4.1.10)† and (4.1.41) with $\alpha=\beta=0$, $j=k=0$ which follows from the fact that $\mathbf{U}(t)$ is a unitary operator.

On the other hand, if condition (4.1.36) is fulfilled then the arguments used to deduce (4.1.39) and (4.1.40) yield that these estimates remains true if in them as well as in their conditions A is replaced by $A-\tau$ with $\tau\in[\tau_1,\tau_2]$ and the last term in their right-hand expressions (namely $C\|u\|$) is replaced by $C''(\frac{h}{\omega})^s\|u\|$.

These modified estimates, (4.1.10), (4.1.10)† and (4.1.41), imply that for $T\ge h^{1-\delta}\omega^{-\delta}$, $\tau\in[\tau_1,\tau_2]$ the left-hand expressions in (4.1.41) and (4.1.42) do not exceed

$$
C''\bigl(\frac{h}{\omega}\bigr)^s(|\tau|+1)^{\frac{1}{m}(|\alpha|+|\beta|)+j+k-1},
$$
$$
C''\bigl(\frac{h}{\omega}\bigr)^s(|\tau|+1)^{\frac{1}{m}(d+|\alpha|+|\beta|)+j+k-1}
$$

respectively. Then the Tauberian arguments used in the proof of proposition 4.1.4 yield estimate (4.1.37). So proposition 4.1.8 is proven.

Let us assume that (4.1.32)$_\pm$ is fulfilled. Then the arguments used to deduce estimates (4.1.39) and (4.1.40) imply that

(i) If

$$
Au=\sum_{|\beta|\le m-l}(hD)^\beta v_\beta,
$$

$l=0,1,\dots,\frac{m}{2}$ (obviously m is even in this case) then

$$
\begin{aligned}
(4.1.43)\quad & \sum_{|\alpha|\le l}\|(hD)^\alpha B^{\frac{1}{m}(l-|\alpha|)}u\| \le \\
& \qquad\qquad \sum_{|\beta|\le m-l} C\|B^{-\frac{1}{2}+\frac{1}{m}(-\frac{m}{2}+l+|\beta|)+}v_\beta\| + C\|B^{\frac{1}{2}}u\|
\end{aligned}
$$

for $u\in C_0^m(B(0,1),\mathrm{D}(B^{1/2}))$.

(ii) For $l\ge\frac{m}{2}$

$$
(4.1.44)\qquad \sum_{\frac{m}{2}\le|\alpha|\le l}\|(hD)^\alpha B^{\frac{1}{m}(l-|\alpha|)}u\| \le \sum_{|\beta|\le l-\frac{m}{2}-k} C\|B^{\frac{k}{m}-\frac{1}{2}}Au\| + C\|B^{\frac{1}{2}}u\|
$$

where $u\in C_0^p(B(0,1),\mathrm{D}(B^{\frac{1}{2}}))$, $k=0,1,\dots,\min(\frac{m}{2},l-\frac{m}{2})$, $p=\max(m,l)$ is arbitrary;

(iii) Moreover, these estimates remain true if we replace A by $A-\tau$ with $\pm\tau \le -C_0$, add

$$|\tau| \sum_{|\alpha|\le l-\frac{m}{2}-\frac{k}{2}} \|(hD)^\alpha B^{-\frac{1}{2}+\frac{k}{m}} u\|$$

to the left-hand expression and in the right-hand expression replace $C\|B^{\frac{1}{2}}u\|$ by $C'h^s|\tau|^{-s}\|B^{\frac{1}{2}}u\|$.

(iv) Finally, all these estimates remain true for $u \in C^p(B(0,1), \mathrm{D}(B^{\frac{1}{2}}))$ if in their left-hand and right-hand expressions we take norms on $B(0,1-\epsilon)$ and $B(0,1-\frac{\epsilon}{2})$ respectively.

Then the arguments used in the proof of proposition 4.1.8 yield proposition 4.1.9. □

4.1.4 Estimates in the Hilbert scale

In conclusion we return to the scale of Hilbert spaces $\mathbb{H}_j = \mathrm{D}(B_1^j)$ with norms $|u|_j = |B_1^j u|$, $j \in [0,l]$ where B_1 satisfies (4.1.7). Moreover, we introduce the spaces $\mathbb{H}_j = \mathbb{H}_{-j}^\dagger$ for $j \in [-l,0]$. We assume that

$$\text{(4.1.45)} \qquad |(a_0(x,h)+i)^{-1}w|_j \le c|w|_{j-1} \qquad \forall w \in \mathbb{H}_{j-1},$$

$$\text{(4.1.46)} \quad |(a_0(x,h)+i)^{-1}D^\beta a_\alpha(x,h)w|_j \le c|w|_{j-\frac{|\alpha|}{m}}$$
$$\forall w \in \mathbb{H}_{j-\frac{|\alpha|}{m}} \cap \mathrm{D}(B^\ell) \quad \forall x \in B(0,1) \quad \forall \beta : |\beta| \le K \quad \forall j \in [0,l]$$

and we assume that

$$\text{(4.1.47)} \qquad |B_1^\ell w| \le c|B^\ell w| \qquad \forall w \in \mathrm{D}(B^\ell)$$

where $\ell = 1, \frac{1}{2}$.

Now, if $K = K(d,m,s)$ is large enough and $u \in H^s_{\mathrm{loc}}(B(0,1), \mathbb{H}_k \cap \mathrm{D}(B^{1-\ell}))$, $Au \in H^s_{\mathrm{loc}}(B(0,1), \mathbb{H}_j)$ then for $k = j$ we conclude that $u \in H^{s'}_{\mathrm{loc}}(B(0,1), \mathbb{H}_{j+1})$ with $s' = s-m, s \in \mathbb{R}$. This deduction remains true for $k = 0$, $s' = s-(j+1)m$. Therefore if $u \in H^s_{\mathrm{loc}}(B(0,1), \mathrm{D}(B^{1-\ell}))$ and $A^l u \in H^s_{\mathrm{loc}}(B(0,1), \mathbb{H})$, $l \in \mathbb{Z}^+$ then $u \in H^{s'}_{\mathrm{loc}}(B(0,1), \mathbb{H}_l)$ with an admissible s'. Then these hypotheses imply the inequality

$$\text{(4.1.48)} \qquad \|\psi B_1^l u\|_{(s)} \le C'M(\|A^l u\| + \|u\|)$$

where $\psi \in C_0^\infty(B(0,1-\epsilon))$; here $\|\cdot\|_{(s)}$ means the Sobolev $H^s(\mathbb{R}^d, \mathbb{H})$-norm, and $s = s(m,l) < 0$, $M = \|\psi\|_{C^K}$, $C' = C'(d,m,l,c,\epsilon)$.

On the other hand, let us assume that condition (4.1.48) is fulfilled for all $\psi \in C_0^K(B(0,1-\frac{\epsilon}{2}))$ and that under certain conditions we have proven either estimate (4.1.43) or estimate (4.1.44) for some q_1 and q_2 and for all $\psi_1, \psi_2 \in C_0^K(B(0,1-\frac{\epsilon}{2}))$ with $q_1 = I$, $q_2 = I$ in neighborhoods of $\operatorname{supp} q_1'$, $\operatorname{supp} q_2'$ respectively. Then one can easily see that this estimate ((4.1.43) or (4.1.44)) will also be fulfilled for q_1', q_2', B_1^j, B_1^k with $j,k \in [0,l]$ instead of q_1, q_2, B^j, B^k with $j,k \in [0,\ell]$ respectively; however, one should replace T, ϵ, s by $T(1-\epsilon), 2\epsilon, s - K_0(m,l)$ respectively.

In particular, *under conditions* (4.1.45)–(4.1.47) *we obtain propositions* 4.1.4*–4.1.6* *where * means the following modification: we replace* B^j, B^k *with* $j,k \in [0,\ell]$ *by* B_1^j, B_1^k *with* $j,k \in [0,l]$.

Moreover, if the conditions

(4.1.49) $$|(A(x,\xi,h)+i)^{-1}w|_j + |\xi|^m|(A(x,\xi,h)+i)^{-1}w|_{j-1} \le c|w|_{j-1} \qquad \forall w\in \mathbb{H}_{j-1},$$

(4.1.50) $$|(A(x,\xi,h)+i)^{-1}D^\beta a_\alpha(x)w|_j \le c|w|_{j-\frac{|\alpha|}{m}} \qquad \forall w\in \mathbb{H}_{j-\frac{|\alpha|}{m}}\cap \mathrm{D}(B^\ell) \quad \forall x\in B(0,1) \quad \forall \xi\in\mathbb{R}^d$$

are fulfilled then one can easily prove by standard elliptic methods that for $B(0,1-\epsilon)$ (4.1.7) holds instead of (4.1.48) and then it is easy to prove propositions 4.1.6*–4.1.10* by the same methods. We leave all the details to the reader.

4.2 Construction of Fundamental Solutions by the Method of Successive Approximations

4.2.1 The method of successive approximations in the standard case

Let us temporarily assume that $X=\mathbb{R}^d$ and that $A\in\Psi_{h,K}(\mathbb{R}^d,\mathbb{H},\mathbb{H})$. Moreover, let us assume that A is a self-adjoint operator in $L^2(\mathbb{R}^d,\mathbb{H})$ and let $U(t,x,y)$ be the Schwartz kernel of the operator $\mathbf{U}(t)=e^{ih^{-1}tA}$. Then (4.1.10) and the initial condition $U|_{t=0}=\delta(x-y)I$ imply the equations

(4.2.1) $$(hD_t-A)U^\pm = \mp ih\delta(x-y)\delta(t)I$$

where $U^\pm=\theta(\pm t)U$, $\theta(.)$ is the Heaviside function, $I=I_\mathbb{H}$ is the identity operator in $\mathbb{H}$ and here and in what follows *all the operators without index y act on x and t*. Let $\bar{A}=a(y,hD_x)$ so the operator $\bar{A}$ is the *principal part of operator A frozen at the point $x=y$*; then

(4.2.2) $$A=\bar{A}+R$$

where

(4.2.3) $$R=\sum_{1\le|\alpha|+p\le N}(x-y)^\alpha h^j A_{\alpha,p},$$

$A_{\alpha,p}=A_{\alpha,p}(y,hD_x)$ for $|\alpha|+p<N$ and $A_{\alpha,p}=A_{\alpha,p}(x,y,hD_x,h)$ for $|\alpha|+p=N$ belong to $\Psi_{h,K-N}(\mathbb{R}^d,\mathbb{H},\mathbb{H})$.

Then (4.2.1) yields the equality

(4.2.4) $$(hD_t-\bar{A})U^\pm = \mp ih\delta(x-y)\delta(t)I + RU^\pm$$

and hence

(4.2.5) $$U^\pm = \mp ih\bar{G}^\pm\delta(x-y)\delta(t)I+\bar{G}^\pm RU^\pm$$

where $\bar{G}^\pm$ and $G^\pm$ are parametrices of the problems

(4.2.6)$_\pm$ $$(hD_t-\bar{A})u=f, \qquad \operatorname{supp} u\subset\{\pm(t-t_0)\ge 0\},$$

(4.2.6)$'_\pm$ $$(hD_t-A)u=f, \qquad \operatorname{supp} u\subset\{\pm(t-t_0)\ge 0\}$$

respectively with $\operatorname{supp} f\subset\{\pm(t-t_0)\ge 0\}$ for some $t_0\in\mathbb{R}$.

Moreover, equation (4.2.1) yields that

$$U^{\pm} = \mp ih G^{\pm}\delta(x-y)\delta(t)I. \tag{4.2.7}$$

Iterating (4.2.5) N times and then substituting (4.2.7) we obtain the equality

$$U^{\pm} = \mp ih \sum_{0\le n\le N-1} (\bar{G}^{\pm}R)^n \bar{G}^{\pm}\delta(x-y)\delta(t)I \mp ih(\bar{G}^{\pm}R)^n G^{\pm}\delta(x-y)\delta(t)I. \tag{4.2.8}$$

So, the method of successive approximations works provided the operator norm of $\bar{G}^{\pm}R$ does not exceed $h^{\delta'}$ with $\delta'>0$. On the time interval $[0,T]$ the operator norms of the parametrices $\bar{G}^{\pm}$ do not exceed $C\frac{T}{h}$ (due to the Duhamel principle; see below) and $R\sim |x-y|+h\sim T$ for $T\ge h$ due to the finite speed of propagation property. So, the restriction to T is $T^2\le h^{1+\delta'}$ or $T\le h^{\frac12+\delta}$ with $\delta>0$. This is the heuristic idea but the actual implementation is slightly different.

Namely, let us consider the factors (x_j-y_j) entering into all the copies of R and let us transfer them from the left to the right (to $\delta(x-y)$). If some factor reaches $\delta(x-y)$ then the corresponding term vanishes and so only those terms survive in which all factors (x_j-y_j) are killed. There are two possibilities for killing (x_j-y_j):

(i) Commuting it with a pseudo-differential operator; then there arises a pseudo-differential operator again and an additional factor h;

(ii) Commuting it with either $\bar{G}^{\pm}$ or $G^{\pm}$. Let us note that for $\bar{P}=hD_t-\bar{A}$

$$\bar{P}(x_j-y_j)u=(x_j-y_j)\bar{P}u-[\bar{A},(x_j-y_j)]u$$

and $\operatorname{supp}(x_j-y_j)u$ is contained in $\{\pm(t-t_0)\ge 0\}$ provided $\operatorname{supp}u$ is contained here; therefore

$$(x_j-y_j)u=\bar{G}^{\pm}(x_j-y_j)\bar{P}u-\bar{G}^{\pm}[\bar{A},(x_j-y_j)]u$$

and we finally obtain the equality

$$(x_j-y_j)\bar{G}^{\pm}=\bar{G}^{\pm}(x_j-y_j)-\bar{G}^{\pm}[\bar{A},(x_j-y_j)]\bar{G}^{\pm}; \tag{4.2.9}$$

similarly

$$(x_j-y_j)G^{\pm}=G^{\pm}(x_j-y_j)-G^{\pm}[A,(x_j-y_j)]\bar{G}^{\pm}. \tag{4.2.9$'$}$$

So as a result of commuting of (x_j-y_j) with $\bar{G}^{\pm}$ there appear a pseudo-differential operator, a factor h and an additional factor $\bar{G}^{\pm}$; the same is also true for $G^{\pm}$.

Now our goal is to prove that *all the terms in* (4.2.8) *containing more than* $N-1$ *factors* h *and* (x_j-y_j) *are negligible in some appropriate sense;* in particular, one can then drop the last term in (4.2.8) and in all the remaining terms in R (defined by (4.2.3)) we can drop all terms with $|\alpha|+p=N$; all the remaining terms we call *regular*.

Recall that we consider the case $X=\mathbb{R}^d$, $A\in\boldsymbol{\Psi}_{h,N}(X,\mathbb{H},\mathbb{H})$; later we will justify the same procedure in other situations. Then the Duhamel equality

$$(\bar{G}^{\pm}u)(t,.)=ih^{-1}\int_{\pm t'\le\pm t}\bar{\mathbf{U}}(t-t')u(t',.)\,dt' \tag{4.2.10}$$

with the unitary operator $\bar{\mathbf{U}}(t)=e^{ih^{-1}t\bar{A}}$ and a similar formula for $G^{\pm}$ imply that the $\mathcal{L}(L^2([-T,T]\times X,\mathbb{H}),L^2([-T,T]\times X,\mathbb{H}))$-norms of $\bar{G}^{\pm}$ and $G^{\pm}$ do not exceed $\frac{2T}{h}$. Moreover, the equality

$$\bar{G}^{\pm}(\delta(t)v)=\pm ih^{-1}\theta(\pm t)\bar{\mathbf{U}}(t)v$$

and a similar equality for $G^{\pm}$ imply that the $\mathcal{L}(L^2([-T,T]\times X,\mathbb{H}),L^2(X,\mathbb{H}))$-norms of the operators $v\to \bar{G}^{\pm}(\delta(t)v)$, $v\to G^{\pm}(\delta(t)v)$ do not exceed $\frac{\sqrt{T}}{h}$. Hence we obtain that the $\mathcal{L}(L^2([-T,T]\times X,\mathbb{H}),L^2([-T,T]\times X,\mathbb{H})$-norm[8] of every term of decomposition (4.2.8) does not exceed $C'h^{r-k}T^{p+k+\frac{1}{2}}$ where r, p and $k+1$ are the original numbers of factors h, (x_j-y_j), $\bar{G}^{\pm}$ (or $G^{\pm}$) respectively; obviously $k\le p+r$ and hence the norm in question does not exceed $C'h^{\sigma(p+r)}$ with $\sigma=\sigma(\delta)>0$ provided $T\le h^{\frac{1}{2}+\delta}$ with an arbitrarily small exponent $\delta>0$. Applying corollary 1.1.9 we obtain

Proposition 4.2.1. *Let $A\in\boldsymbol{\Psi}_{h,K}(\mathbb{R}^d,\mathbb{H},\mathbb{H})$, $q_1,q_2\in \mathrm{S}_{h,K}(T^*\mathbb{R}^d,\mathbb{H},\mathbb{H})$ be supported in $\{|\xi|\le c\}$, $Q_i=\mathrm{Op}^{\mathrm{w}}(q_i)$, $i=1,2$, $\chi\in C_0^K([-1,1])$ and $T\in[h^{1-\delta},h^{\frac{1}{2}+\delta}]$ with $\delta>0$. Then*

$$(4.2.11)\quad |F_{t\to h^{-1}\tau}\chi_T(t)Q_{1x}\big(U^{\pm}\pm ih\sum_{0\le n\le N-1}(\bar{G}^{\pm}R')^n\delta(x-y)\delta(t)\big)^tQ_{2y}|\le C'Mh^s \qquad \forall x,y\in\mathbb{R}^d\quad \forall\tau\in\mathbb{R}$$

where

$$(4.2.3)'\qquad R'=\sum_{1\le|\alpha|+p\le N-1}(x-y)^{\alpha}h^jA_{\alpha,p}(y,hD_x),$$

$\delta>0$ and s are arbitrary, $N=N(d,s,\delta)$ and $K=K(d,s,\delta,N)$ are large enough, $C'=C'(d,s,\delta,N,c)$, $M=|\!|\!|q_1|\!|\!|\cdot|\!|\!|q_2|\!|\!|\cdot\|\chi\|$ and all the norms are calculated in the corresponding spaces.

Let us make minor modifications to the standard case. Namely, let us replace condition $A\in\boldsymbol{\Psi}_{h,K}(\mathbb{R}^d,\mathbb{H},\mathbb{H})$ by one one of the following conditions:

(4.2.12)$_1$ $\zeta A\in\boldsymbol{\Psi}_{h,\rho,\gamma,K}(\mathbb{R}^d,\mathbb{H},\mathbb{H})$ with $\rho,\gamma\in[0,1]$, $\zeta\ge h^s$, $\rho\gamma\ge h^{1-\delta}$;

(4.2.12)$_2$ $A\in\tilde{\boldsymbol{\Psi}}^m_{h,K}(\mathbb{R}^d,\mathbb{H},\mathbb{H})$

(4.2.12)$_3$ A is a differential operator with coefficients satisfying (4.1.11) with $B=I$;

we consider the domain $\Omega\subset T^*X$ (or $\Omega\subset T^*B(0,1-\epsilon_0)$ in the third case) and we assume that

(4.2.13) In Ω the symbol of A coincides with that of $A'\in\boldsymbol{\Psi}_{h,K}(\mathbb{R}^d,\mathbb{H},\mathbb{H})$.

Moreover, let us assume that

$$(4.2.14)\qquad |\nabla_{x,\xi}A(x,\xi,h)|\le c_0\qquad \forall(x,\xi)\in\Omega$$

Then corollary 2.1.6 yields that if $q_1,q_2\in\boldsymbol{\Psi}_{h,\rho,\gamma,K}(\mathbb{R}^d,\mathbb{H},\mathbb{H})$ with $\rho,\gamma\in[0,1]$, $\rho\gamma\ge h^{1-\delta}$ are supported in $\Omega_{(\epsilon_0)}$ and $Q_1=\mathrm{Op}^{\mathrm{w}}(q_1)$, $Q_2=\mathrm{Op}^{\mathrm{w}}(q_2)$, then for $t\in[-T_0,T_0]$ with $T_0=T_0(c_0,\epsilon_0)>0$ $U(t,x,y)^tQ_{2y}$ [9] is negligible outside of $\Omega_{\frac{1}{2}\epsilon_0}$ where ϵ_0 is arbitrarily small and $\Omega_\varepsilon=\{(x,\xi):\mathrm{dist}((x,\xi),\complement\Omega)\ge\varepsilon\}$. Hence (4.2.1) yields that

[8] One can consider this term as the Schwartz kernel of an operator and calculate its operator norm.

[9] In the case (4.2.12)$_3$ one should replace Q_i by $Q_i\psi_i$ with ψ_i supported in $B(0,1-\epsilon_0)$; usually we do it without special mention.

$$(4.2.1)' \qquad (hD_t - A')U^{\pm t}Q_{2y} \equiv \mp ih\delta(t)\mathcal{K}(x,y) \qquad \text{in } [-T_0,T_0]\times\mathbb{R}^d\times\mathbb{R}^d$$

modulo negligible functions where $\mathcal{K}(x,y) = \delta(x-y)^t Q_{2y}$ is the Schwartz kernel of Q_2; hence one can replace (4.2.8) by

$$(4.2.8)' \quad U^{\pm t}Q_{2y} \equiv \mp ih \sum_{0\le n\le N-1} (\bar{G}^{\pm}R)^n \bar{G}^{\pm}\mathcal{K}(x,y)\delta(t)I \mp ih(\bar{G}^{\pm}R)^n G^{\pm}\mathcal{K}(x,y)\delta(t)I$$

where $\bar{A}$, R, $\bar{G}^{\pm}$ and $G^{\pm}$ are constructed for A' instead of A.[10)] Let us apply the previous procedure of the transfer of factors $(x_j - y_j)$ to the right; now these factors can be also killed in the commutator with ${}^tQ_{2y}$ but it is still in consistent with (i). So we obtain the estimate

$$|F_{t\to h^{-1}\tau}\chi_T(t)Q_{1x}\bigl(U^{\pm\, t}Q_{2y}\pm ih^{-1}\sum_{0\le n\le N-1}(\bar{G}^{\pm}R')^n\bar{G}^{\pm}\delta(t)(\delta(x-y)^tQ_{2y})| \le C'Mh^s.$$

Moreover, let us replace here ϵ_0 by $\frac{1}{3}\epsilon_0$ and q_2 by q_2' equal to 1 in $\Omega_{\frac{2}{3}\epsilon_0}$; let us multiply the function inside the norm by ${}^tQ_{2y}$; then corollary 2.1.6 yields that the obtained function will not change modulo negligible terms if we replace the operator ${}^tQ'_{2y}$ by I.

So we have proven

Proposition 4.2.2. *(i) The assertion of proposition 4.2.1 remains true for operators A of the type $(4.2.12)_j$–(4.2.14), $j = 1,2$, with K, C' depending on m and C' also depending on c_0, ϵ_0, $c = \|A'\| + \|A\|$ where all the norms are calculated in the corresponding spaces.*

(ii) Moreover, for operators A of the type $(4.2.12)_3$–(4.2.14) this assertion remains true if we replace Q_1, Q_2 by $Q_1\psi_1, Q_2\psi_2$ respectively with $\psi_i \in C_0^K(B(0,1-\epsilon_0))$; i.e., the estimate

$$(4.2.11)' \quad |F_{t\to h^{-1}\tau}\chi_T(t)Q_{1x}\psi_1(x)\bigl(U^{\pm}\pm ih\sum_{0\le n\le N-1}(\bar{G}^{\pm}R')^n\delta(x-y)\delta(t)\bigr)\psi_2(y)^tQ_{2y}\| \le C'Mh^s \qquad \forall x,y\in\mathbb{R}^d \quad \forall\tau\in\mathbb{R}$$

holds, where now $M = \|q_1\| \cdot \|q_2\| \cdot \|\psi_1\| \cdot \|\psi_2\|$.

4.2.2 The framing procedure

Let us consider the general case of a differential operator with coefficients which are unbounded operators; we use the successive approximations method with the operator A replaced by the operator

$$(4.2.15) \qquad A' = a_0(x,h) + \psi_3(x)A''\psi_3(x)$$

where here and in what follows $\psi_3, \psi_4, \psi_5 \in C_0^K$ are real-valued functions supported in $B(0,1-\frac{7}{9}\epsilon_0)$, $B(0,1-\frac{5}{9}\epsilon_0)$, $B(0,1-\frac{3}{9}\epsilon_0)$ and are equal to 1 in $B(0,1-\frac{8}{9}\epsilon_0)$, $B(0,1-\frac{2}{9}\epsilon_0)$, $B(0,1-\frac{4}{9}\epsilon_0)$ respectively (recall that ψ_1 and ψ_2 are supported in $B(0,1-\epsilon_0)$) and **either**

(i) $a_0(x,h)$ satisfies $(4.1.1)_1$, (4.1.15) and $(4.1.12)_1$ in $\mathbb{R}^d$ and for every $\nu\in[0,1]$

10) Later we will see that in the final answer this replacement is not necessary.

$$(4.2.16)_\nu \qquad B^{\nu-1}A''B^{-\nu}\in \Psi_{h,K}(\mathbb{R}^d,\mathbb{H},\mathbb{H})$$

or

(ii) $a_0(x,h)$ satisfies $(4.1.11)_{\frac{1}{2}}$, (4.1.15) and $(4.1.12)_\pm$ in $\mathbb{R}^d$ and, moreover, $(4.2.16)_{\frac{1}{2}}$ holds; we assume that

(4.2.17) A'' is a symmetric operator

and hence in (i) one need check (4.2.16) only for $\nu=1$.

Such an extension for $a_0(x,h)$ can be easily chosen provided in case (i) we replace $B(0,1)$ by any ball from its ϵ-covering (if necessary); the choice of A'' is trivial (as before, the symbols of A and A' should coincide in Ω).

The *principal difficulty now is that $A_{\alpha,j}$ are non-bounded operators and parametrices $G^\pm$ and $\bar{G}^\pm$ do not compensate this unboundedness in an explicit way.* However we apply the *framing procedure.* Namely, let us note that

$$(4.2.18) \qquad \bar{G}^\pm=-(\bar{A}'+i)^{-1}+\bar{G}^\pm(hD_t+i)(\bar{A}'+i)^{-1}$$

and that the operators $\bar{G}^\pm,(\bar{A}'+i)^{-1},(hD_t+i)$ commute. Let us transform in this way all the terms in the right-hand expression in (4.2.8)′ and let us move all the copies of the factor (hD_t+i) to the left; in this way we can replace $\bar{G}^\pm$ in the original version of the right-hand expression in (4.2.8)′ by $-(\bar{A}'+i)^{-1}$ or $(\bar{A}'+i)^{-2}$ or $(\bar{A}'+i)^{-1}\bar{G}^\pm(\bar{A}'+i)^{-1}$ (with the additional factor $(hD_t+i)^k$ with $k=0,1,2$ respectively). One can easily see that for $k\ge 1$ the $\mathcal{L}(L^2(\mathbb{R}^d,\mathbb{H}),L^2(\mathbb{R}^d,\mathbb{H}))$-norms of the operators $B^{1-\nu}(\bar{A}'+i)^{-k}B^\nu$ do not exceed C_0 where here and in what follows $\nu\in[0,1]$ is arbitrary in case (i) and $\nu=\frac{1}{2}$ in case (ii). Therefore the $\mathcal{L}(L^2([-T,T]\times\mathbb{R}^d,\mathbb{H}),L^2([-T,T]\times\mathbb{R}^d,\mathbb{H}))$-norm of the operator $B^{1-\nu}(\bar{A}'+i)^{-1}\bar{G}^\pm(\bar{A}'+i)^{-1}B^\nu$ does not exceed $C_0\frac{T}{h}$. Similarly we can work with $G^\pm$. Hence if we consider a term (in the right-hand expression in (4.2.8)′) which was negligible in the context of subsection 4.2.1 then after multiplication by $F_{t\to h^{-1}\tau}\chi_T(t)Q_{1x}\psi_1(x)$ with $Q_1=\mathrm{Op}^\mathrm{w}(q_1)$, q_1 supported in $\{|\xi|\le c\}$ we obtain an expression with $\mathcal{L}(\mathbb{H},\mathbb{H})$-norm not exceeding $C'Mh^s(|\tau|+1)^\mu$ with $\mu=\mu(d,m,s,\delta)$ for all $x,y\in\mathbb{R}^d$. Moreover, this assertion remains true if we replace $Q_{1x},{}^tQ_{2y}$ by $Q_{1x}B^j,B^{kt}Q_{2y}$ with $j,k\in[0,1]$, $j+k\le 1$ in case (i), $j,k\in[0,\frac{1}{2}]$ in case (ii).

However, we should remember that instead of equation (4.2.1)′ we now have

$$(4.2.1)'' \qquad (hD_t-A')U^\pm B^k\psi_2(y)tQ_{2y}=\mp ih\delta(t)\delta(x-y)B^k\psi_2(y)tQ_{2y}+f^\pm$$

where

$$f^\pm=(A-A')U^\pm B^k\psi_2(y)^tQ_{2y}$$

and hence we need to examine corresponding terms which were automatically negligible in the context of subsection 4.2.1. Let us note that if we replace $U^\pm$ by $U_1^\pm=\psi_4(x)U^\pm$ then (4.2.1)″ remains true with $f^\pm$ given by formula

$$f^\pm=\psi_4(x)(A-A')\psi_5U^\pm B^k\psi_2(y)^tQ_{2y}$$

where

$$U^\pm=\mp ih\hat{G}^\pm\delta(t)\delta(x-y)I_\mathbb{H}$$

and $\hat{G}^\pm$ is a parametrix of the problem $(4.2.6)'_\pm$ (for the operator hD_t-A; $G^\pm$ are reserved for parametrices of hD_t-A'). One can apply the framing procedure to the parametrices $\hat{G}^\pm$ too; then we obtain terms in which $\hat{G}^\pm$ is framed by $\psi_5(A+i)^{-1}$ from the left and by $(A+i)^{-1}\psi_4$ from the right and by the terms containing no factor $\hat{G}^\pm$.

Condition $(4.1.18)_\ell$ yields that for $\psi \in C_0^K(B(0,1-\varepsilon))$ the operator $\psi(A\pm i)^{-1}$ acts from $H^r(X,\mathbb{H}_{\nu-1})$ to $H^{r-m}(X,\mathbb{H}_\nu)$ where H^r is a Sobolev space and r is arbitrary (but K depends on r). Then the same is true for the adjoint operator $(A\pm i)^{-1}\psi$ (condition $(4.2.16)_\nu$ is assumed to be fulfilled for given ν and for $1-\nu$ simultaneously). Let us consider the commutator of $(A\pm i)^{-1}$ with an operator of the form $\psi Q\psi$ where Q is a quantization of a compactly supported symbol. Then $[(A\pm i)^{-1},\psi Q\psi] = -(A\pm i)^{-1}[A,\psi Q\psi](A\pm i)^{-1}$ and hence the previous property yields that its operator norm from $L^2(X,\mathbb{H}_{\nu-1})$ to $L^2(X,\mathbb{H}_\nu)$ does not exceed C_0h. This assertion immediately yields that if $q_1,q_2\in \mathrm{S}_{h,K}(T^*\mathbb{R}^d)$, q_2 is compactly supported with respect to (x,ξ) and its support does not intersect that of q_1 then the operator norms from $L^2(X,\mathbb{H}_{\nu-1})$ to $L^2(X,\mathbb{H}_\nu)$ of the operator $\psi Q_1\psi(A\pm i)^{-1}\psi Q_2\psi$ do not exceed $C'Mh^s$ and the same is true for $(1-\psi')(A\pm i)^{-1}\psi Q_2\psi$ provided $\psi'=1$ in the neighborhood of $\operatorname{supp}\psi$; here $Q_i=\mathrm{Op}^{\mathrm{w}}(q_i)$. This assertion and corollary 2.1.6 imply that

$$(4.2.19)\qquad \|F_{t\to h^{-1}\tau}B^{-\nu}\chi_{2T}(t)f^{\pm}w\|\le C'Mh^s|w| \qquad \forall y\in\mathbb{R}^d\quad \forall\tau:|\tau|\le 2c\quad \forall w\in\mathbb{H}$$

where $\chi\in C_0^K([-1,1])$ is arbitrary and we estimate the $L^2(\mathbb{R}^d,\mathbb{H})$-norm with respect to x. Therefore the extra terms arising in $(4.2.8)'$ in connection with $f^\pm$ after multiplication by $F_{t\to h^{-1}\tau}\chi_T(t)Q_{1x}\psi_1(x)B^j$ with $|\tau|\le c$ have $\mathcal{L}(\mathbb{H},\mathbb{H})$-norms not exceeding $C'Mh^s$. So we have proven

Proposition 4.2.3. *(i) Let conditions* $(4.1.11)_1$, $(4.1.12)_1$, (4.1.13), (4.1.14), (4.1.15), $(4.1.18)_1$ *be fulfilled and let* $T,\chi,\psi_1,\psi_2,Q_1,Q_2,N$ *be as in proposition 4.2.1. Then the estimates*

$$(4.2.20)_\pm\qquad |F_{t\to h^{-1}\tau}\chi_T(t)Q_{1x}\psi_1(x)B^j\big(U^\pm \pm ih\sum_{0\le n\le N-1}(\bar G^\pm R')^n\bar G^\pm\delta(t)\delta(x-y)\big)B^k\psi_2(y)^tQ_{2y}|\le C'Mh^s \qquad \forall x,y\in\mathbb{R}^d\quad\forall\tau:|\tau|\le c\quad \forall j,k\in[0,\ell],j+k\le 1$$

hold with $\ell=1$.

(ii) Let conditions $(4.1.11)_{\frac12}$, $(4.1.12)_\pm$, (4.1.13), $(4.1.14)'$, (4.1.15) *and* $(4.1.18)_{\frac12}$ *be fulfilled and let* $T,\chi,\psi_1,\psi_2,Q_1,Q_2,N$ *be as in proposition 4.2.1. Then estimates* $(4.2.20)_\pm$ *hold with* $\ell=\frac12$.

Let us assume that conditions (4.1.45)–(4.1.47) are fulfilled. Then (4.2.18) yields that every copy of $\bar G^\pm$ can either be framed by $(\bar A+i)^{-p}$ with arbitrarily large p or replaced by $(\bar A+i)^{-q}$ with $q=1,\dots,2p$, and the same is true for $G^\pm$ and $\hat G^\pm$. Let us perform this procedure for the leftmost and rightmost copies of $\bar G^\pm$ and $G^\pm$ (for now we analyze terms without $\hat G^\pm$). If a factor $\bar G^\pm$ or $G^\pm$ disappears we go to the next factor of this type. Then two final answers are possible:

(a) The term contains a factor $(\bar A+i)^{-p}$ or a factor $(A'+i)^{-p}$ with no factor $\bar G^\pm$, $G^\pm$ to the left of it, and the term contains a factor $(\bar A+i)^{-p}$ or a factor $(A'+i)^{-p}$ with no factor $\bar G^\pm$, $G^\pm$ to the right of it;

(b) The term contains no factor $\bar G^\pm$, $G^\pm$.

In case (a) (with large enough p) after multiplication on the left by an operator $F_{t\to h^{-1}\tau}\chi_T(t)Q_{1x}\psi_1(x)B_1^j$ and replacing B^k with $k\in[0,\ell]$ by B_1^k with $k\in[0,l]$ we conclude that the $\mathcal{L}(\mathbb{H},\mathbb{H})$-norm of the term obtained does not exceed $C'Mh^s$ provided this term was negligible under the assumptions of subsection 4.2.1. In case (b) it is true only if $j+k\le 1$ but *if we consider* $U=U^++U^-$ *then two copies of this term will be annihilated.*

Let us consider terms containing $\hat{G}^{\pm}$ (and hence originating from $f^{\pm}$). Let us note that

$$(A+i)^{-1} = (a_0+i)^{-1} \sum_{0\le n\le N-1} ((a_0-A)(a_0+i)^{-1})^n + ((a_0-A)(a_0+i)^{-1})^N (A+i)^{-1};$$

then the arguments used in the proof of proposition 4.2.3 and conditions (4.1.45)–(4.1.47) imply that if q_2 is compactly supported with respect to (x,ξ) then the operator norm from $L^2(X,\mathbb{H}_{\nu-1})$ to $L^2(X,\mathbb{H}_\nu)$ of the operator $(A\pm i)^{-1}\psi Q_2\psi$ does not exceed C_0M; moreover, if the support of q_2 does not intersect the support of q_1 then the operator norms from $L^2(X,\mathbb{H}_{\nu-1})$ to $L^2(X,\mathbb{H}_\nu)$ of the operators $\psi Q_1\psi(A\pm i)^{-1}\psi Q_2\psi$ do not exceed $C'Mh^s$, and the same is true for $(1-\psi')(A\pm i)^{-1}\psi Q_2\psi$ provided $\psi'=1$ in a neighborhood of $\operatorname{supp}\psi$; here $Q_i=\operatorname{Op}(q_i)$ and $\nu\in[0,l]$ is arbitrary.

Then applying the framing procedure in the recently modified form and referring in case (a) to corollary 2.1.6 we obtain that terms of type (a) satisfy the same estimate as before and terms of type (b) annihilate when we add U^+ and U^-. So we have proven

Proposition 4.2.4. *Suppose that conditions* (4.1.11)$_1$, (4.1.12)$_1$, (4.1.13), (4.1.14), (4.1.15), (4.1.18)$_1$ *or that conditions* (4.1.11)$_{\frac12}$, (4.1.12)$_\pm$, (4.1.13), (4.1.14)$'$, (4.1.15), (4.1.18)$_{\frac12}$ *are fulfilled and let* $T,\chi,\psi_1,\psi_2,Q_1,Q_2,N$ *be as in proposition 4.2.1. Moreover, let* (4.1.45)–(4.1.47) *be fulfilled (surely* N,K *and* C *now also depend on* l*). Then*

$$|F_{t\to h^{-1}\tau}\chi_T(t)Q_{1x}\psi_1(x)B^j\big(U+ ih\sum_{\varepsilon=\pm}\sum_{0\le n\le N-1}\varepsilon(\bar{G}^\varepsilon R')^n\bar{G}^\varepsilon\delta(t)\delta(x-y)\big)B^k\psi_2(y)^tQ_{2y}| \le C'Mh^s \qquad \forall x,y\in\mathbb{R}^d \quad \forall\tau:\tau\le c \quad \forall j,k\in[0,l]. \tag{4.2.21}$$

4.2.3 Calculations

Let us go from estimates to calculations. Let us apply the Fourier transform $x\to h^{-1}\xi$ with $\xi\in\mathbb{R}^d$ and the Fourier-Laplace transform $t\to h^{-1}\tau$ with $\tau\in\mathbb{C}_\mp\setminus\mathbb{R}$. Then $\delta(x-y)\psi_2(y)^tQ_{2y}$ will be transformed into $e^{ih^{-1}\langle x-y,\xi\rangle}q_2(y,\xi,h)$ where $q_2(y,\xi,h)$ is the pq-symbol of operator Q_2 (replaced by ψ_2Q_2 in the formulas and in this definition if necessary),[11] parametrices $\bar{G}^\pm$ will be transformed into operators of multiplication by $(\tau-a(y,\xi))^{-1}$ (where $a(x,xi)=A(x,\xi,0)$ is a principal symbol of A) and operators $b(y,hD_x,h)$ and (x_j-y_j) will be transformed into $b(y,\xi,h)$ and $ih\partial_{\xi_j}$ respectively. So for $\tau\in\mathbb{C}_\mp\setminus\mathbb{R}$

$$F_{t\to h^{-1}\tau}\sum_{0\le n\le N-1}(\bar{G}^\pm R')^n\bar{G}^\pm\delta(t)\delta(x-y))^tQ_{2y} = \mp i\sum_{0\le n\le N'}(2\pi)^{-d-1}h^{-d+n}\int e^{ih^{-1}\langle x-y,\xi\rangle}F_n(y,\tau,\xi)[q_2(y,\xi,h)]d\xi \tag{4.2.22}$$

where F_n are differential operators[12] with respect to ξ of degree not exceeding n with $\mathcal{L}(\mathbb{H},\mathbb{H})$-valued coefficients which are holomorphic with respect to $\tau\in\mathbb{C}\setminus\mathbb{R}$ and C^s with respect (x,ξ) where s is arbitrarily large. More precisely, under the assumptions

[11] In the previous analysis the method of quantization was not essential.

[12] Applied to q_2.

of propositions 4.2.1, 4.2.2 the $\mathcal{L}(\mathbb{H},\mathbb{H})$-norms of these coefficients[13] do not exceed $C|\operatorname{Im}\tau|^{-2n-1-|\beta|}$. The framing procedure yields that under the assumptions of proposition 4.1.3 the $\mathcal{L}(\mathbb{H}_\nu,\mathbb{H}_{\nu-1})$-norms (with the same ν as in this proposition) of these coefficients[13] do not exceed $C(|\tau|+1)^{2r}|\operatorname{Im}\tau|^{-r}$ with large enough r depending on n. Finally, the modified framing procedure yields that under the assumptions of proposition 4.1.4 these coefficients are sums of functions for which the same estimates remain true with $\mathcal{L}(\mathbb{H}_l,\mathbb{H}_{-l})$-norms and polynomials with respect to τ with $\mathcal{L}(\mathbb{H}_\nu,\mathbb{H}_{\nu-1})$-norms of the coefficients[13] not exceeding C.

Moreover, $F_n(y,\tau,\xi)$ depend only on the values of $A'(.,.,h)$ in a neighborhood of (y,ξ) and hence the replacement of A by A' was not essential from the formal point of view.

Remark 4.2.5. *The way in which we obtained formula (4.2.22) yields that its right-hand expression is approximately equal*[14] *to*

$$\mp i(2\pi)^{-d-1}h^{-d}\int e^{ih^{-1}\langle x-y,\xi\rangle}\operatorname{symb}_{pq(N')}\big((\tau-A)^{-1}\psi_2Q_2\big)d\xi \tag{4.2.23}$$

where $(\tau-A)^{-1}$ means a parametrix formally constructed in terms of the symbol calculus (or in terms of the formal calculus of pseudo-differential operators), "symb" means symbol, the subscript pq means the type of the symbol and the subscript (N') means that we cut off the decomposition at $h^{N'}$; this subscript is not written in what follows. In fact, for every fixed $\tau\in\mathbb{C}\setminus\mathbb{R}$ the construction of this parametrix by the method of successive approximations gives exactly the same formulas (without the factor $\pm\frac{1}{2\pi i}$). Therefore (4.2.23) is the Schwartz kernel of the operator $\mp\frac{1}{2\pi i}(\tau-A)^{-1}\psi_2Q_2$. In particular,

$$F_0(y,\tau,\xi)=(\tau-a(y,\xi))^{-1}, \tag{4.2.24}$$

$$\begin{aligned}F_1(y,\tau,\xi)[q_2]=(\tau-a(y,\xi))^{-1}\cdot\\ \Big(a^s(y,\xi)+\frac{i}{2}\sum_j a^{(j)}_{(j)}(y,\xi)+\sum_j a_{(j)}(y,\xi)(i\partial_{\xi_j})\Big)\big[(\tau-a(y,\xi))^{-1}q_2\big]\end{aligned} \tag{4.2.25}$$

where a^s is the subprincipal symbol of A. ◊

Formula (4.2.22) yields that

$$\begin{aligned}F_{t\to h^{-1}\tau}\sum_{0\le n\le N-1}Q_{1x}(\bar G^{\pm}R')^n\bar G^{\pm}\delta(t)\delta(x-y))^tQ_{2y}=\\ \mp i\sum_{0\le n\le N'}(2\pi)^{-d-1}h^{-d+n}\int e^{ih^{-1}\langle x-y,\xi\rangle}q_1(x,\xi,h)F_n(y,\tau,\xi)[q_2(y,\xi,h)]d\xi\end{aligned} \tag{4.2.26}$$

where q_1 is the qp-symbol of Q_1 (replaced by $Q_1\psi_1$ in the formulas and in this definition if necessary).

On the other hand, Appendix B yields that one can set $\tau\in\mathbb{R}$ in formulas (4.2.22) and (4.2.26) and then one should replace the τ-analytic functions $F_n(y,\tau,\xi)$ by the distributions $F_n^{\mp}(y,\tau,\xi)=F_n(y,\tau\mp i0,\xi)\in\mathcal{S}'(\mathbb{R}_\tau,C^s(\Omega,\mathcal{L}(\mathbb{H}_\nu,\mathbb{H}_{\nu-1})))$.[15] Moreover, under the assumptions of proposition 4.2.4 $\mathcal{F}_n=\operatorname{Res}_{\mathbb{R}}(F_n)=\frac{1}{2\pi i}(F_n^--F_n^+)\in\mathcal{S}'(\mathbb{R}_\tau,C^s(\Omega,\mathcal{L}(\mathbb{H}_l,\mathbb{H}_{-l})))$.

Therefore propositions 4.2.1–4.2.4 yield

[13] And their derivatives of degree $\beta,|\beta|\le s$ with respect to $(x,\xi)\in\Omega$.

[14] Errors will only be in terms with $n\ge s+d$.

[15] More precisely, the coefficients of $F_n^{\mp}$ belong to this space; recall that F_n and hence $F_n^{\mp}$ and $\mathcal{F}_n$ are differential operators.

Proposition 4.2.6. *(i) Under the assumptions of proposition 4.2.1 the following estimate holds:*

$$(4.2.27)\quad |F_{t\to h^{-1}\tau}\chi_T(t)Q_{1x}U^{\pm t}Q_{2y} \pm i \sum_{0\le n\le N'} (2\pi)^{-d-1}h^{-d+n+1} \times$$

$$\iint T\hat{\chi}(\frac{(\tau-\tau')T}{h})e^{ih^{-1}\langle x-y,\xi\rangle}q_1(x,\xi,h)F_n^{\mp}(y,\tau',\xi)[q_2(y,\xi,h]d\tau'd\xi| \le$$

$$C'Mh^s \qquad \forall x,y\in\mathbb{R}^d \quad \forall\tau\in\mathbb{R};$$

(ii) Under the assumptions of proposition 4.2.2 this estimate also remains true; in terms of condition (4.2.12)$_3$ *one should replace* Q_1 *and* Q_2 *by* $Q_1\psi_1$ *and* ψ_2Q_2 *respectively;*

(iii) Under the assumptions of proposition 4.2.3 this estimate also holds with Q_1, Q_2 *and* q_1, q_2 *replaced by* $Q_1\psi_1B^j$, $\psi_2B^kQ_2$ *and* q_1B^j, B^kq_2 *respectively (with* $j,k\in[0,\ell]$*) and with* $\tau\in[-2c,2c]$*;*

(iv) Under the assumptions of proposition 4.2.4 the following estimate holds:

$$(4.2.28)\quad |F_{t\to h^{-1}\tau}\chi_T(t)Q_{1x}\psi_1(x)B_1^jUB_1^k\psi_2(y)^tQ_{2y} -$$

$$\sum_{0\le n\le N'} (2\pi)^{-d}h^{-d+n+1}\iint T\hat{\chi}(\frac{(\tau-\tau')T}{h})e^{ih^{-1}\langle x-y,\xi\rangle}q_1(x,\xi,h)\times$$

$$B_1^j\mathcal{F}_n(y,\tau',\xi)[B_1^kq_2(y,\xi,h)]d\tau'd\xi| \le C'Mh^s$$

$$\forall x,y\in\mathbb{R}^d \quad \forall\tau\in[-2c,2c] \quad \forall j,k\in[0,l].$$

Surely, remark 4.2.5 remains true with obvious modifications for the further formulas. In particular, the sum of the left-hand expressions in (4.2.22) (with both signs) is approximately equal[14)] to

$$(4.2.29)\qquad (2\pi)^{-d}h^{-d}e^{ih^{-1}\langle x-y,\xi\rangle}\,\mathrm{symb}_{pq(N')}\,\mathrm{Res}_{\mathbb{R}}((\tau-A)^{-1}\psi_2Q_2)d\xi$$

which is the Schwartz kernel of the operator $\mathrm{Res}_{\mathbb{R}}((\tau-A)^{-1}\psi_2Q_2)$. We also have the nice formula

$$(4.2.30)\qquad \mathcal{F}_0(y,\tau,\xi)=\delta(\tau-a(y,\xi))=\partial_\tau\theta(\tau-a(y,\xi)),$$

and a formula for $\mathcal{F}_1$ will be discussed in the next section.

4.2.4 Increasing T

Let us return to estimates (4.2.27) and (4.2.28) and discuss the possibility of increasing T in their corollaries; it will be very important for the Tauberian method.

Proposition 4.2.7. *Let the operator* $P=B^{-\frac12}(hD_t-A)B^{-\frac12}$ *be microhyperbolic at the point* $(\bar{x},\bar{\tau},\bar{\xi})\in B(0,\frac12)\times\mathbb{R}\times\mathbb{R}^d$ *in the direction* $\mathcal{T}=(\mathcal{T}_x,\mathcal{T}_\xi)\in T^*\mathbb{R}^d$ *with* $|\mathcal{T}|\le c$. *Let* $q_1,q_2\in \mathrm{S}_{h,K}(T^*\mathbb{R}^d,\mathbb{H},\mathbb{H})$ *and let* q_2 *be supported in the* ϵ*-neighborhood of* $(\bar{x},\bar{\xi})$ *with a small constant* $\epsilon>0$. *Let all the conditions of proposition 4.2.4 be fulfilled. Then*

(i) If $\mathcal{T}_x=0$ *then for* $T\in[h^{1-\delta},T_0]$ *with a small constant* $T_0>0$ *the following estimate holds:*

$$(4.2.31)\quad |\Gamma_xF_{t\to h^{-1}\tau}\chi_T(t)Q_{1x}\psi_1(x)B_1^jUB^k\psi_2(y)^tQ_{2y} -$$

$$\sum_{0\le n\le N'} h^{-d+n}\int T\hat{\chi}(\frac{(\tau-\tau')T}{h})\kappa'_n(x,\tau')d\tau'| \le C'Mh^s$$

$$\forall x\in\mathbb{R}^d \quad \forall\tau\in[\bar{\tau}-\epsilon,\bar{\tau}+\epsilon]$$

where $j,k\in[0,l]$ are arbitrary and

$$\kappa'_n(x,\tau)=(2\pi)^{-d}\int q_1(x,\xi,h)B_1^j\mathcal{F}_n(x,\tau,\xi)[B_1^k q_2(x,\xi,h)]d\xi \tag{4.2.32}$$

belong to $C_0^s(\mathbb{R}^d\times[\bar\tau-\epsilon,\bar\tau+\epsilon],\mathcal{L}(\mathbb{H},\mathbb{H}))$;

(ii) *In the general case for the same T the following estimate holds:*

$$\begin{aligned}(4.2.33)\quad &|\Gamma F_{t\to h^{-1}\tau}\chi_T(t)Q_{1x}\psi_1(x)B_1^jUB^k\psi_2(y)^tQ_{2y}-\\ &\sum_{0\le n\le N'}h^{-d+n}\int T\hat\chi(\frac{(\tau-\tau')T}{h})\varkappa'_n(\tau')d\tau'|\le C'Mh^s\\ &\qquad\qquad\qquad\forall\tau\in[\bar\tau-\epsilon,\bar\tau+\epsilon]\end{aligned}$$

where $j,k\in[0,l]$ are arbitrary and

$$\varkappa'_n(\tau)=(2\pi)^{-d}\iint q_1(x,\xi,h)B_1^j\mathcal{F}_n(x,\tau,\xi)[B_1^kq_2(x,\xi,h)]d\xi dx \tag{4.2.34}$$

belong to $C_0^s([\bar\tau-\epsilon,\bar\tau+\epsilon],\mathcal{L}(\mathbb{H},\mathbb{H}))$;

(iii) *Under the assumptions of proposition 4.2.3 this estimate holds for U, $\mathcal{F}_n$, B_1^j, B_1^k with $j,k\in[0,l]$ replaced by $U^\pm$, $\mp\frac{1}{2\pi i}F_n^\mp$, B^j, B^k respectively with $j,k\in[0,\vartheta]$, $j+k\le 1$.*

Proof. Let us note that under the microhyperbolicity condition the assertions on the regularity properties of κ'_n and $\varkappa_n$ follow from Appendix B.

Obviously, for $T\in[h^{1-\delta},h^{\frac12+\delta}]$ the estimates (4.2.31) and (4.2.33) are due to (4.2.28). Let us take $\chi=\bar\chi$ where here we assume that $\bar\chi\in C_0^K([-2,2])$ equals 1 on $[-\frac32,\frac32]$. Then theorem 2.1.14 yields that in estimates (4.2.31) and (4.2.33) one can replace $\bar\chi_T$ by $\chi_{T'}$ with arbitrary $T'\in[h^{1-\delta},T_0]$; moreover, in terms of proposition 4.2.4 we also use the arguments of subsection 4.1.4.

Let us prove that these estimates remain true if we replace all the remaining copies of T by T'. Since $\hat\chi\in\mathcal{S}$ then in (4.2.31) and (4.2.33) for $\tau\in[\bar\tau-\epsilon,\bar\tau+\epsilon]$ we can insert an additional factor $\phi(\tau')$ with $\phi\in C_0^K([\bar\tau-2\epsilon,\bar\tau+2\epsilon]$ equal to 1 on $[\bar\tau-\frac32\epsilon,\bar\tau+\frac32\epsilon]$. Let us note that if $\chi(t)$ is replaced by $t\chi(t)$ then $\hat\chi(\frac{(\tau-\tau')T}{h})$ should be replaced by $-\frac{h}{T}\partial_{\tau'}\hat\chi(\frac{(\tau-\tau')T}{h})$ and integration by parts moves $\partial_{\tau'}$ from $\hat\chi(\frac{(\tau-\tau')T}{h})$ to $\phi(\tau')\kappa_n(x,\tau')$ or $\phi(\tau')\varkappa_n(\tau')$. Repeating this procedure we conclude from the regularity properties of κ_n or $\varkappa_n$ that if we replace $\chi_T(t)$ by $t^r\chi_T(t)$ then we obtain in (4.2.31) and (4.2.33) negligible terms (for now we discuss only terms in which we have not replaced T by T'). Taking $\chi(t)=t^{-r}(\bar\chi(\frac{t}{\zeta})-\bar\chi(t))$ we conclude that we can replace T by ζT in (4.2.31) and (4.2.33) provided $\chi=\bar\chi,|\zeta|\le 2$. Repeating this procedure $[\log_2(\frac{T'}{T})]+1$ times we conclude that we can replace T by T' provided $\chi=\bar\chi$; obviously, we can drop $\phi(\tau')$ now. So for $\chi=\bar\chi$ the proposition is proved.

Since replacing a factor $\bar\chi$ by a factor $\chi=\chi\bar\chi$ (with χ supported in $[0,1]$) is equivalent to convolving the Fourier transform with $2\pi iT\hat\chi(\frac{\tau T}{h})$ we conclude that the proposition has also been proven for χ. □

Proposition 4.2.7 and theorem 2.3.8 immediately yield

Proposition 4.2.8. *Let the operator A satisfy the conditions of proposition 4.2.2 and let the operator $P=B^{-\frac12}(hD_t-A)B^{-\frac12}$ be microhyperbolic at the point $(\bar x,\bar\tau,\bar\xi)\in B(0,\frac12)\times\mathbb{R}\times\mathbb{R}^d$ in the direction $\mathcal{T}=(\mathcal{T}_x,\mathcal{T}_\xi)\in T^*\mathbb{R}^d$ with $|\mathcal{T}|\le c$ where $\epsilon_0=c^{-1}$ and c also enter into the microhyperbolicity condition. Let $\mathcal{U}$ be the γ-neighborhood of $(\bar x,\bar\xi)$ with $\gamma\in[h^{\frac12-\delta},\gamma_0]$ where $\gamma_0=\gamma_0(c,d)$ is a small constant. Moreover, let $\tau_1=\bar\tau-\rho$*

, $\tau_2 = \bar{\tau} + \rho$, $\rho = \omega^{-1}$ *and let condition* (2.3.33) *be fulfilled for all* $z \in \mathcal{U}$ *with* r *instead of* m. *Let* $\Phi_{j,t}$ *be the Hamiltonian flows generated by* $\lambda_j(x,\xi)$ *with* $t \in \mathcal{J}$, $\mathcal{J} = [-T,0]$ *or* $\mathcal{J} = [0,T]$. *Let conditions* (2.3.34) *and* (2.3.15)′ *be fulfilled for all* $z \in \mathcal{U}, t \in \mathcal{J}$ *(for* $|\alpha| + |\beta| = 1$, $k = 0$ *also) as well as condition* (2.3.16)″ *and assume*

$$\text{dist}(\Phi_{j,t}(z), \partial X) \ge \epsilon_0 \gamma, \tag{4.2.35}$$

$$\text{dist}(\Phi_{j,t}(z), z) \ge \epsilon_0 |t| \gamma \qquad \forall j = 1, \dots, r \quad \forall z \in \mathcal{U} \quad \forall t \in \mathcal{J}. \tag{4.2.36}$$

Moreover, let us assume that $T \le \omega$, $1 \le \omega \le h^{-\delta'}$ *where* $\delta' = \delta'(d,\delta,s) > 0$ *is small enough and* $K = K(d,m,s,\delta)$ *is large enough,* $q_1, q_2 \in \mathrm{S}_{h,\gamma,\gamma,N}(T^*\mathbb{R}^d, \mathbb{H}, \mathbb{H})$ *are supported in* $\mathcal{U}$. *Then for all* $\tau \in [\tau_1, \tau_2]$ *estimate* (4.2.33) *holds with the* T *given here.*

Remark 4.2.9. *(i) One can easily see from the proofs of propositions 4.2.1–4.2.8 that these propositions remain true for* $q_1, q_2 \in \mathrm{S}_{h,\rho',\gamma',K}(\mathbb{R}^d, \mathbb{H}, \mathbb{H})$ *with* $\rho' \ge h^{1-\delta}$, $\gamma' \ge h^{1-\delta}$, $\rho'\gamma' \ge h$ *in propositions 4.2.1–4.2.8 and with* $\rho'\gamma \ge h^{1-\delta}$, $\gamma'\gamma \ge h^{1-\delta}$, $\rho'\gamma' \ge h$ *in proposition 4.2.8.*

(ii) One can replace condition (4.2.35) by some condition in the spirit of sections 3.2 and 3.3 but in this case we should consider Hamiltonian flows with reflections and assume certain conditions in the spirit of theorem 3.3.2, as is done in the next chapter.

(iii) Under the condition $\mathcal{T}_x = 0$ *one can consider* Γ_x *instead of* Γ *but in this case Hamiltonian trajectories with* $t \in [-T,T]$ *should be considered and condition (4.2.36) should be replaced by*

$$\text{dist}(\pi_x \Phi_{j,t}(z), \pi_x z) \ge \epsilon_0 |t| \gamma \qquad \forall j = 1, \dots, r \quad \forall z \in \mathcal{U} \quad \forall t \in \mathcal{J} \tag{4.2.36$'$}$$

where π_x *is the projection to the* x*-coordinate. Surely one can say that condition (4.2.36) is the non-periodicity condition and condition (4.2.36)′ is the condition of transient movement in* x*-space or no loop in* x*-space condition.* ◊

Appendix B. On the Traces of Almost Analytic Functions

Let $\Omega \subset \mathbb{R}^d$ be an open domain, Γ an open convex domain with a vertex at 0 and let $\tilde{\Omega} = \Omega + i\Gamma z \subset \mathbb{C}^d$. Let $\mathbb{B}$ be a Banach space. A function $F(.) \in C^1(\tilde{\Omega}, B)$ is called r-*almost analytic* if

$$|F(z)| \le M|\operatorname{Im} z|^{-l+\delta} \qquad \forall z \in \tilde{\Omega}, \tag{B.1}$$

$$|\bar{\partial}_{z_j} F(z)| \le M|\operatorname{Im} z|^{r-1+\delta} \qquad \forall z \in \tilde{\Omega} \quad \forall j = 1, \dots, d \tag{B.2}$$

where $\bar{\partial}_{z_j} = \frac{1}{2}(\partial_{x_j} + i\partial_{y_j})$ and $\delta > 0$, $l \in \mathbb{N}$.

Proposition B.1. *Let conditions* (B.1), (B.2) *with* $r = 0$ *be fulfilled. Then*

(i) For every $y \in \Gamma$ *with* $|y| = 1$ *there exists a limit*

$$f(x) = F(x + i0y) = \lim_{t \to +0} F(x + ity)$$

in the space of $\mathbb{B}$*-valued distributions* $\mathcal{D}'(\Omega, \mathbb{B})$;

(ii) Moreover $f = \sum_{0 \le k \le l} \mathcal{M}^k f_k$ *where* $f_k \in C(\Omega, \mathbb{B})$ *and* $|f_k| \le CM$ *with* $C = C(l,\delta)$, $\mathcal{M} = \langle y, \partial_x \rangle$.

(iii) $f(x)$ *does not depend on the choice of* y.

Proof. For $l = 0$ this assertion is obvious. Let us note that

$$\partial_t F(x + ity) = -\mathcal{M}F(x + ity) + 2\langle y, \bar{\partial}_z\rangle F(x + ity)$$

and hence

$$F(x + ity) = F(x + iy) + 2\int_1^t \langle y, \bar{\partial}_z\rangle F(x + it'y)dt' - \int_1^t \mathcal{M}F(x + it'y)dt'. \tag{B.3}$$

Then (B.1) and (B.2) yield that the first and the second terms in the right hand expression belong to $C(\Omega_x \times J_t \times \Gamma_y, \mathbb{B})$, $J = [0, 1]$; the same is true for the "shortened" third term $\int_1^t F(x + it'y)dt'$ also provided $l = 1$. On the other hand, for $l \geq 2$ we replace $F(x + it'y)dt'$ in (B.3) by the right-hand expression of (B.3) with t, t' replaced by t', t'' respectively. Repeating this procedure $l - 1$ times we obtain that

$$F(x + ity) = \tag{B.4}$$

$$\sum_{0 \leq k \leq l-1} \mathcal{M}^k F_k(x) + \frac{1}{(l-1)!}(-\mathcal{M})^l \int_1^t (t - t')^{l-1} F(x + it'y)dt'$$

where F_k with $k = 0, \dots, l - 1$ as well as

$$F_l(x) = \frac{1}{(l-1)!}(-\mathcal{M})^l \int_1^t (t - t')^{l-1} F(x + it'y)dt'$$

obviously belong to $C(\Omega_x \times J_t \times \Gamma_y, B)$. Therefore the limit $F(x + i0y)$ exists in the distribution sense and is the sum of $\mathcal{M}^k f_k$ with $f_k(x) = F_k(x + i0y)$. So (i),(ii) are proven.

In order to prove (iii) let us note that

$$\partial_s F(x + iy^0 - is\omega) = \mathcal{M}' F(x + iy^0 - is\omega) - 2\langle \omega, \bar{\partial}_z\rangle F(x + iy^0 - is\omega)$$

with $\mathcal{M}' = -\langle \omega, \partial_x\rangle$; setting $y, y^0 \in \Gamma$, $s, t \in [0, 1]$ and $\omega = y^0 - ty$ we obtain a formula similar to (B.3) with $F(x + ity)$, $\mathcal{M}$, $\langle y, \bar{\partial}_z\rangle$, $t, t', 1$ replaced by $F(x + iy^0 - is\omega)$, $-\mathcal{M}'$, $-\langle \omega, \bar{\partial}_z\rangle$, $s, s', 0$ respectively. Then formula (B.4) also holds modified in the same way:

$$F(x + ity) = \sum_{0 \leq k \leq l-1} (-\mathcal{M}')^k F_k(x, y^0, \omega, t) +$$

$$\frac{1}{(l-1)!}\mathcal{M}'^l \int_0^1 (1 - s')^{l-1} F(x + iy^0 - is'\omega)ds';$$

moreover, in this sum we can take the limit as $t \to +0$ (and $\omega \to y^0$, $\mathcal{M}' \to -\mathcal{M}$), and this limit obviously does not depend on y. □

Let us consider a symbol $a \in C^k(\Omega, \mathbb{H}, \mathbb{H})$ (where $\mathbb{H}$ is an auxiliary Hilbert space) such that

$$a^\dagger(x) = a(x) \qquad \text{in } \Omega, \tag{B.5}$$

$$\langle(\ell a)(x)w, w\rangle \geq \epsilon_0 |w|^2 - c|(a(x) - \tau)w|^2 \qquad \forall x \in \Omega \quad \forall w \in \mathbb{H} \tag{B.6}$$

where we write now x instead of (x, ξ) and $\ell \in T\Omega$. Let $\tilde{a}(x + iy)$ be the q-analytic extension of $a(x)$:

$$\tilde{a}(x+iy) = \sum_{|\alpha|\le q} \frac{1}{\alpha!}(\partial_x^\alpha a)(x)(iy)^\alpha. \tag{B.7}$$

We assert that for small enough $\epsilon \ge 0$, $\nu \ge 0$

$$|\tilde{P}w| \ge \epsilon_1(\varepsilon+\nu)|w| \tag{B.8}$$

where here and below $\tilde{P} = \tilde{a}(x+i\varepsilon\ell) - \tau + i\nu I$ and $P = a(x) - \tau$. In fact,

$$\tilde{P}(x)w = P(x)w + i(\varepsilon(\ell a)(x) + \nu)w + O(\varepsilon^2|w|)$$

and hence (B.5) and (B.6) yield that

$$\begin{aligned}|\tilde{P}(x)w|\cdot|w| \ge &\operatorname{Im}\langle\tilde{P}(x)w, w\rangle \ge \\ &\epsilon_0(\varepsilon+\nu)|w|^2 - C_0\varepsilon^2|w|^2 - c|P(x)w|^2 \ge \\ &(\frac{1}{2}(\varepsilon+\nu) - C\varepsilon^2 - C\nu^2)|w|^2 - c|\tilde{P}(x)w|^2,\end{aligned}$$

and (B.8) trivially follows from this inequality.

Therefore $F(z) = (\tilde{a}(z') - z_0)^{-1}$ satisfies (B.1) and (B.2) with $l=2$, $\delta=1$, $r=q-2$, $z=(z_0,z')$ and $\Gamma = \{y_0 \le 0, y' = t\ell, t \ge 0, y_0 - t < 0\}$. Moreover,

$$(a(x) - \tau \pm i0)^{-1} = (\tilde{a}(x \pm i0\ell) - \tau)^{-1} \tag{B.9}$$

(we have analyzed the "+" case; the "−" case can be analyzed in the same way) and the same is true for all products of the factors $(\tilde{a}(z') - z_0)^{-1}$ (the number of which should not exceed j) and regular[16)] q-almost analytic factors, $q = q(j)$.

We will use these properties in the forthcoming sections and chapters.

4.3 General Basic Theorems

In this section we obtain and improve local and microlocal semiclassical spectral asymptotics. We consider here the most important and sophisticated case of partial differential operators with unbounded coefficients. Moreover, the results obtained here remain true for both classes of pseudo-differential operators introduced in section 4.2[17)] in which case a cutting function ψ_i is necessary. We assume that

$$\begin{aligned}(4.3.1)\quad Q_i = \operatorname{Op}(q_i)\psi_i,\ &\operatorname{supp} q_i \subset \Omega \subset \{|\xi| \le c\},\\ &\operatorname{dist}(\operatorname{supp} q_i, \complement\Omega) \ge \epsilon_0, \psi_i \in C_0^K(B(0,\frac{1}{2})),\ \phi, \phi_i \in C_0^K([-c,c]).\end{aligned}$$

We start from the general case in which no microhyperbolicity condition is assumed, then treat the microhyperbolic case and then the most special case when conditions on Hamiltonian trajectories are assumed. In each case we derive asymptotics of smooth spectral means and in the second and third cases we also derive asymptotics without averaging on the spectral parameter or on the Riesz means. Recall that the coefficients $\mathcal{K}_n$ depend on Q_1, Q_2, j, k and the same is true for the coefficients κ_n and $\varkappa_n$ defined below.

---

16) I.e., of the class C^q up to $\Omega \subset \mathbb{R}^d$.

17) See conditions $(4.2.12)_1$ and $(4.2.12)_2$.

4.3.1 General theory: $T \asymp h^{\frac{1}{2}-\delta}$

Let us start with the most general case. In this case we only derive asymptotics for (smooth) spectral means:

Theorem 4.3.1. *Suppose conditions* $(4.1.11)_1$, $(4.1.12)_1$, (4.1.13), (4.1.14), (4.1.15), $(4.1.18)_1$ *or conditions* $(4.1.11)_{\frac{1}{2}}$, $(4.1.12)_\pm$, (4.1.13), $(4.1.14)'$, (4.1.15), $(4.1.18)_{\frac{1}{2}}$ *are fulfilled. Moreover, suppose conditions* (4.3.1) *and* (4.1.45)–(4.1.47) *are fulfilled. Let* $q_i \in \mathrm{S}_{h,\rho,\gamma,K}(T^*\mathbb{R}^d)$, $\bar\tau \in [-c,c]$,

$$L \in [h^{\frac{1}{2}-\delta}, c], \tag{4.3.2}$$

$\rho \ge h^{1-\delta}$, $\gamma \ge h^{1-\delta}$, $\rho\gamma \ge h$. *Then*

(i) The following estimates hold:

$$\begin{aligned} \mathcal{R}'_1(\bar\tau) \overset{\mathrm{def}}{=} |\int \phi(\frac{\tau-\bar\tau}{L}) d_\tau \big(Q_{1x} B_1^j e(x,y,\tau_1,\tau) B_1^{k\ t}Q_{2y}\big) - \\ \sum_{0\le n\le N'-1} h^{-d+n} \int \phi(\frac{\tau-\bar\tau}{L}) \mathcal{K}'_n(x,y,\tau)\,d\tau| \le CMh^s, \end{aligned} \tag{4.3.3}$$

$$\begin{aligned} \mathcal{R}'_2(\bar\tau) \overset{\mathrm{def}}{=} \frac{1}{L^2} |\iint (Q_{1x} B_1^j e(x,y,\tau',\tau'') B_1^{k\ t}Q_{2y} - \\ \sum_{0\le n\le N'-1} h^{-d+n} \mathcal{K}_n(x,y,\tau)) \phi_1(\frac{\tau'-\bar\tau}{L}) \phi_2(\frac{\tau''-\bar\tau}{L})\,d\tau' d\tau''| \le CMh^s \\ \forall x,y \in \mathbb{R}^d \quad \forall j,k \in [0,l] \quad \forall L \ge h^{\frac{1}{2}-\delta} \end{aligned} \tag{4.3.4}$$

where s *and* $\delta > 0$ *are arbitrary,* $N' = N'(d,m,s,\delta)$, $K = K(d,m,s,N',\delta)$, $M = \|\!|q_1|\!\| \cdot \|\!|q_2|\!\| \cdot \|\psi_1\| \cdot \|\psi_2\| \cdot \|\phi\|$ *here,*

$$\mathcal{K}'_n(x,y,\tau) = \int e^{ih^{-1}\langle x-y,\xi\rangle} q_1(x,\xi,h) B_1^j \mathcal{F}_n(y,\tau,\xi) B_1^k [q_2(y,\xi,h)]\,d\xi, \tag{4.3.5}$$

$$\mathcal{K}_n(.,.,\tau',\tau) = \int_{\tau'}^{\tau} \mathcal{K}'_n(.,.,\tau'')\,d\tau''; \tag{4.3.6}$$

$\mathcal{R}$ *with various sub- and superscripts means the remainder estimate.*[18)]

(ii) If we reject conditions (4.1.45)–(4.1.47) *then the corresponding estimates remain true with* B_1^j, B_1^k *replaced by* B^j, B^k, $j,k \in [0,\ell]$;

(iii) Moreover, the corresponding estimates remain true for $|\bar\tau| \le cL$, $L \ge 1$, *and in the case* $B = I$ *one can take arbitrary* $\bar\tau \in \mathbb{R}$.

Proof. Let us pick $T \in [h^{1-\delta}, h^{\frac{1}{2}+\delta}]$ such that

$$TL \ge h^{1-\delta} \tag{4.3.7}$$

(which can be done by condition (4.3.2); we decrease $\delta > 0$ if necessary). Applying (4.1.25) and (4.2.28) we obtain that estimate (4.3.3) holds with $T\hat\chi(\frac{(\tau'-\tau)T}{h})$ instead of $\phi(\frac{\tau-\bar\tau}{L})$ and

18) Let us recall that $\mathcal{K}'_n$ are distributions with respect to τ; so $\mathcal{K}_n$ are distributions with respect to (τ',τ).

with $\tau' \in [-c, c]$. Let us multiply this estimate by $h^{-1}\phi(\frac{\tau-\bar\tau}{L})$ and integrate over $\tau \in \mathbb{R}$.[19] Then (4.3.3) holds with the function $\phi(\lambda)$ replaced by $\phi_\varepsilon(\lambda) = \frac{1}{\varepsilon}\int \phi(\lambda')\hat\chi(\frac{\lambda-\lambda'}{\varepsilon})d\lambda'$, $\varepsilon = \frac{h}{LT}$. Pick $\chi = \bar\chi$. Recall that $\bar\chi = 1$ on $[-\frac{1}{2}, \frac{1}{2}]$. Then one can easily prove that

$$\phi_\varepsilon(\lambda) = \phi(\lambda) + O\big(\varepsilon^s(1+|\lambda|)^{-s}\big).$$

Therefore it is sufficient to estimate the norm of

$$\varepsilon^s \int \Big(1 + \frac{|\tau-\bar\tau|T}{h}\Big)^{-s} d_\tau \mathcal{E}(\tau) \tag{4.3.8}$$

with $\mathcal{E} = Q_{1x}B_1^j e(x,y,\tau_1,\tau)B_1^{k\,t}Q_{2y}$. Dividing this integral into a sum of integrals over intervals $[n, n+1]$ with $n \in \mathbb{Z}$ and applying proposition 4.1.1 we obtain that the norm of the n-th integral does not exceed $Ch^s(|n|+1)^{-s}$ (recall that $\varepsilon \le h^\delta$) and hence estimate (4.3.3) is proven.

Then integrating by parts and replacing $\partial_\tau\phi(\tau)$ by $\phi(\tau)$ we obtain that under the additional restriction

$$\int_{-\infty}^{\infty} \phi(\tau)d\tau = 0 \tag{4.3.9}$$

the following estimate holds:

$$\begin{aligned} &\mathcal{R}'_3(\tau_1,\bar\tau) \overset{\text{def}}{=} \\ &\qquad L^{-1}\Big|\int \phi(\frac{\tau-\bar\tau}{L})Q_{1x}B_1^j e(x,y,\tau_1,\tau)B_1^{k\,t}Q_{2y}d\tau - \\ &\qquad \sum_{0\le n\le N'-1} h^{-d+n}\int \phi(\frac{\tau-\bar\tau}{L})\mathcal{K}_n(x,y,\tau_1,\tau)d\tau\Big| \le CMh^s \\ &\qquad\qquad \forall x,y\in\mathbb{R}^d \quad \forall j,k\in[0,l] \quad \forall L\in[h^{\frac{1}{2}-\delta},1] \end{aligned} \tag{4.3.10}$$

where under condition (4.3.9) $\mathcal{R}'_3$ does not depend on τ_1.

Let us note that the function

$$\phi(\tau) = \phi_2(\tau)\int_{-\infty}^{\infty}\phi_1(\tau''')d\tau''' - \phi_1(\tau)\int_{-\infty}^{\infty}\phi_2(\tau''')d\tau'''$$

satisfies condition (4.3.6); substituting it into (4.3.19) we obtain the difference of two double integrals; let us replace (τ,τ''') in the first integral by (τ'',τ') and in the second one by (τ',τ''); then using the equality $e(.,.,\tau_1,\tau'') - e(.,.,\tau_1,\tau') = e(.,.,\tau',\tau'')$ and similar equalities for $\mathcal{K}_n$ we obtain (4.3.4).

So, statements (i) and (ii) have been proven. In order to prove (iii) let us first note that in the case $B = I$, $L \in [h^{\frac{1}{2}-\delta}, 1]$ one can choose $\bar\tau$ arbitrarily (because for $|\bar\tau| > C_0$ we are in the semiclassical gap; recall that q_i are compactly supported). Finally, for arbitrary B one can reduce the case $L \ge 1$ to the case $L = 1$ by dividing A by L and replacing h by $h' = L^{-\frac{1}{m}}h$ and B by $B' = L^{-\frac{1}{m}}B + I$ and changing B_1 in the same way. Surely, one then obtains the required estimates with h'-pdo $Q'_j = q'_i(h'D)\psi_i(x)$ with compactly supported symbols; multiplying by the h-pdo $Q_{1,2}$ on the left and right we obtain the required estimates with these operators. □

This theorem combined with propositions 4.1.3–4.1.3″ immediately yields

[19] This means exactly that we apply operator $\phi(\frac{1}{L}(hD_t - \bar\tau))$ to $Q_{1x}B_1^jUB_1^{kt}Q_{2y}$ and to its approximation obtained in the previous section and set $t = 0$ afterwards.

Theorem 4.3.2. *Let the conditions of theorem 4.3.1 be fulfilled.*

(i) Moreover, let either conditions of proposition 4.1.3 be fulfilled in an interval of length $\geq h^{\frac{1}{2}-\delta}$ *containing* τ_1 *or let the conditions of proposition 4.1.3′ or 4.1.3″ be fulfilled with* $\bar{\tau}$ *replaced by* $\tau_1 \in \mathbb{R}$. *Finally, let* $|\tau_1| \leq cL + c$. *Then*

$$\mathcal{R}'_3(\tau_1, \bar{\tau}) \leq CMh^s; \tag{4.3.11}$$

here $\mathcal{R}'_3$ *does not depend (modulo* $O(h^s)$*) on* τ_1 *lying in the indicated interval.*

(ii) Moreover, let the assumptions of (i) be fulfilled for τ_1 *as well as for* τ_2. *Then*

$$\mathcal{R}'_4(\tau_1, \bar{\tau}) \stackrel{\mathrm{def}}{=} |Q_{1x} B_1^j e(x, y, \tau_1, \tau_2) B_1^{k\,t} Q_{2y} - \sum_{0 \leq n \leq N'-1} h^{-d+n} \mathcal{K}_n(x, y, \tau_1, \tau_2)| \leq CMh^s; \tag{4.3.12}$$

here $\mathcal{R}'_3$ *does not depend (modulo* $O(h^s)$*) on* τ_1, τ_2 *lying in the indicated intervals.*

(iii) Moreover, if conditions $(4.1.12)_\pm$, $(4.1.14)'$ *are fulfilled then in statements (i) and (ii) one can take* $\tau_1 = \mp\infty$.

(iv) If $B = I$ *then in (i) and (ii) one can take* $\tau_1 \leq -C_0$ *(or* $\tau_1 = -\infty$ *and* $\tau_2 \geq C_0$ *(or* $\tau_2 = \infty$*) with sufficiently large* C_0.

(v) Assertions (ii) and (iii) of theorem 4.3.1 remain true for the estimates of this theorem.

Proof. The corresponding proposition of subsection 4.1.2 yields that in the function to be estimated one can replace τ_1 and (or) τ_2 by any other value from the neighborhood indicated above and hence by the mean value on this interval. On the other hand, for the spectral mean the corresponding estimate is obvious. □

4.3.2 General microhyperbolic theory: $T \asymp 1$

Let us now assume that the operator A is microhyperbolic on $\operatorname{supp} q_1$ at the energy level $\bar{\tau}$ in the direction $\mathcal{T}$. First of all, let us apply the arguments of the previous subsection to $\Gamma Q_{1x} B_1^j e(x, y, ., .) B_1^{k\,t} Q_{2y}$ in the general case and to $\Gamma_x Q_{1x} B_1^j e(x, y, ., .) B_1^{k\,t} Q_{2y}$ provided $\mathcal{T}_x = 0$. Let us then note that in these arguments one can pick a small constant T by proposition 4.2.7 and since the only condition on L is (4.3.7) we obtain

Remark 4.3.3. *Let the operator* A *be microhyperbolic on* $\operatorname{supp} q_1$ *at the energy level* $\bar{\tau}$ *in direction* $\mathcal{T}$. *Then theorems 4.3.1 and 4.3.2 remain true for* $L \geq h^{1-\delta}$ *provided* $Q_{1x} B_1^j e(x, y, ., .) B_1^{k\,t} Q_{2y}$ *is replaced by* $\Gamma Q_{1x} B_1^j e(x, y, ., .) B_1^k$ *in the general case and by* $\Gamma_x Q_{1x} B_1^j e(x, y, ., .) B_1^{k\,t} Q_{2y}$ *in the case* $\mathcal{T}_x = 0$. *Surely, one should replace the coefficients* $\mathcal{K}_n$ *(and their derivatives on* τ*) by the coefficients* $\kappa_n(x, \tau', \tau) = \Gamma_x \mathcal{K}_n(x, ., \tau', \tau)$ *and* $\varkappa_n(\tau', \tau) = \Gamma \mathcal{K}_n(., ., \tau', \tau)$ *(and their derivatives on* τ*) respectively. Later we will improve this statement.* ◊

Now let us apply the Tauberian technique (see the proof of proposition 4.1.4). In the analysis of this subsection $T > 0$ is a small constant but we do not remove it from the estimates because in other situations when these arguments apply T is not necessarily a constant.

Proposition 4.3.4. *Suppose that conditions* $(4.1.11)_1$, $(4.1.12)_1$, $(4.1.13)$, $(4.1.14)$, $(4.1.15)$, $(4.1.18)_1$ *or conditions* $(4.1.11)_{\frac{1}{2}}$, $(4.1.12)_\pm$, $(4.1.13)$, $(4.1.14)'$, $(4.1.15)$, $(4.1.18)_{\frac{1}{2}}$ *are fulfilled. Moreover, suppose conditions* (4.3.1) *and* (4.1.45)–(4.1.47) *are fulfilled. Let* A *be microhyperbolic at the point* $\bar{z} = (\bar{x}, \bar{\xi})$ *at the energy level* $\bar{\tau}$ *in the direction* $\mathcal{T}$, $\bar{\tau} \in [-c, c]$. *Let* $\gamma_0 \leq \epsilon$, $\rho_0 \leq \epsilon$.[20)]

[20)] This assumption is made in this and the following subsections.

(i) Let $q_i \in \mathrm{S}_{h,\gamma,\gamma,K}(T^*\mathbb{R}^d,\mathbb{H},\mathbb{H})$ be supported in the γ-neighborhood of $\bar{z}$, $\gamma \ge h^{\frac{1}{2}}$. Then the estimate

$$(4.3.13)\quad \Gamma Q_{1x} B_1^j e(x,y,\tau',\tau) B_1^k\, {}^t Q_{2y} \prec CMh^{-d}\gamma^{2d-1}\bigl(|\tau-\tau'| + \frac{h}{T}\bigr) \qquad \forall \tau,\tau' \in [\bar\tau-\epsilon,\bar\tau+\epsilon]$$

holds where for $\mathcal{Q}_1, \mathcal{Q}_2 \in \mathcal{L}(\mathbb{H},\mathbb{H})$ the relation[21] *$\mathcal{Q}_1 \prec \mathcal{Q}_2$ means that*

$$|\langle \mathcal{Q}_1 w, w\rangle| \le \langle \mathcal{Q}_2 w, w\rangle \qquad \forall w \in \mathbb{H}.$$

(ii) Let $\mathcal{T}_x = 0$ and let $q_i \in \mathrm{S}_{h,\rho,\gamma,K}(T^*\mathbb{R}^d,\mathbb{H},\mathbb{H})$ be supported in the (γ,ρ)-neighborhood of $(\bar{x},\bar{\xi})$, $h^{1-\delta} \le \rho \le \gamma$, $\rho\gamma \ge h$. Then the estimate

$$(4.3.14)\quad \Gamma_x Q_{1x} B_1^j e(x,y,\tau',\tau) B_1^k\, {}^t Q_{2y} \prec CMh^{-d}\rho^{d-1}\bigl(|\tau-\tau'| + \frac{h}{T}\bigr) \qquad \forall \tau,\tau' \in [\bar\tau-\epsilon,\bar\tau+\epsilon]$$

holds.

Proof. Let us prove statement (i); the proof of the statement (ii) is the same. Let us note that due to Appendix B

$$(4.3.15)\quad |\partial_\tau^p B_1^j \varkappa_n(\tau) B_1^k| \le CM\gamma^{2d-1-p-2n} \qquad \forall \tau \in [\bar\tau - 3\epsilon_0, \bar\tau + 3\epsilon_0] \quad \forall n = 0,\dots,N' \quad \forall p = 1,\dots,N'.$$

Therefore proposition 4.2.7 yields that $|F_{t\to h^{-1}\tau}\chi_T(t)\Gamma Q_{1x} B_1^j U B_1^k\, {}^t Q_{2y}|$ does not exceed $CMh^{1-d}\gamma^{2d-1}$ for $\tau \in [\bar\tau - 2\epsilon, \bar\tau + 2\epsilon]$.

Then picking $Q_2 = Q_2^*$ and a Hörmander function χ we obtain the required estimate for $|\tau-\tau'| \le \frac{h}{T}$ by means of Tauberian arguments; in this case the right-hand expression of (4.3.13) is $Ch^{1-d}T^{-1}\gamma^{2d-1}$. The transition to Q_2 which is not adjoint to Q_1 also repeats the proof of proposition 4.1.4. In the case $\frac{h}{T} \le \tau - \tau' \le 2\epsilon$ we divide the interval $[\tau',\tau]$ into $[\frac{(\tau-\tau')T}{h}]$ subintervals with the required properties. □

Proposition 4.3.5. *(i) Under the conditions of proposition 4.3.4(i) the estimate*

$$(4.3.16)\quad |F_{t\to h^{-1}\tau}\chi_{T',T''}\Gamma Q_{1x} B_j^l U B_k^l\, {}^t Q_{2y}| \prec CMh^{1-d}\gamma^{2d-1}\bigl(1 + \frac{T'}{T}\bigr) \qquad \forall \tau \in [\bar\tau-\epsilon,\bar\tau+\epsilon],$$

holds for arbitrary $T' \ge h^{1-\delta}$ and T''.

(ii) Under the assumptions of proposition 4.3.4(ii) similar estimates (with ρ^{d-1} instead of γ^{2d-1} and with Γ_x instead of Γ) hold; here $\chi_{T',T''}(t) = \chi(\frac{t-T''}{T})$.

Proof. Applying (4.1.3) one can write the left-hand expression of (4.3.16) as

$$T'\int_0^\infty \hat\chi\bigl(\frac{(\tau-\tau')T'}{h}\bigr) e^{-ih^{-1}(\tau-\tau')T''} d_{\tau'}\bigl(\Gamma Q_{1x} B_1^j e(x,y,\tau,\tau') B_1^k\, {}^t Q_{2y}\bigr).$$

Let us divide this integral into a sum of integrals over segments of the length $\frac{h}{T}$. Then proposition 4.3.4 yields that the n-th integral does not exceed

[21] The advantage of this form of the estimate will be seen later.

$$CM(1+\frac{|n|T'}{T})^{-s}T^{-1}h^{1-d}\gamma^{2d-1}$$

for $|n|\le \epsilon\frac{T}{h}$ and proposition 4.1.2 yields that this expression does not exceed $CMn^{-s}h^s$ otherwise. After summation we obtain (4.3.16). The other estimates are proven in the same way. □

Now we can prove the following statement which improves remark 4.3.3 and is the most important statement of this subsection:

Theorem 4.3.6. *Suppose conditions* $(4.1.11)_1$, $(4.1.12)_1$, (4.1.13), (4.1.14), (4.1.15), $(4.1.18)_1$ *or conditions* $(4.1.11)_{\frac{1}{2}}$, $(4.1.12)_\pm$, (4.1.13), (4.1.14)′, (4.1.15), $(4.1.18)_{\frac{1}{2}}$ *are fulfilled. Moreover, suppose conditions* (4.3.1) *and* (4.1.45)–(4.1.47) *are fulfilled. Let A be microhyperbolic at the point $\bar z=(\bar x,\bar\xi)$ at the energy level $\bar\tau$ in the direction T, $\bar\tau\in[-c,c]$.*

(i) Let $q_i\in \mathrm{S}_{h,\gamma,\gamma,K}(T^*\mathbb{R}^d,\mathbb{H},\mathbb{H})$ be supported in the γ-neighborhood of $\bar z$, $\gamma\ge h^{\frac12}$. Then the estimate

$$\mathcal{R}_1(\bar\tau)\overset{\text{def}}{=}\int\phi(\frac{\tau-\bar\tau}{L})d_\tau(\Gamma Q_{1x}B_1^je(x,y,\tau',\tau)B_1^{k\,t}Q_{2y})-\sum_{0\le n\le N'-1}h^{-d+n}\int\phi(\frac{\tau-\bar\tau}{L})\varkappa_n'(\tau)d\tau\prec CM\frac{h^{1-d}}{T}\gamma^{2d-1}(\frac{h}{TL})^\vartheta \tag{4.3.17}$$

holds for $L\in[\frac{h}{T},1]$ where here and in (ii) $\vartheta=s$ is arbitrarily large. In particular, for constant γ and T the estimate

$$\mathcal{R}_1(\bar\tau)\prec CMH^{1-d}(\frac{h}{L})^\vartheta \tag{4.3.18}$$

holds for $L\in[h,c]$.

(ii) Let $T_x=0$, and let $q_i\in \mathrm{S}_{h,\rho,\gamma,K}(T^*\mathbb{R}^d,\mathbb{H},\mathbb{H})$ be supported in the (γ,ρ)-neighborhood of $(\bar x,\bar\xi)$, $h^{1-\delta}\le\rho\le\gamma$, $\rho\gamma\ge h$. Then the estimate

$$\mathcal{R}_{1x}(\bar\tau)\overset{\text{def}}{=}\int\phi(\frac{\tau-\bar\tau}{L})d_\tau(\Gamma_x Q_{1x}B_1^je(x,y,\tau',\tau)B_1^{k\,t}Q_{2y})-\sum_{0\le n\le N'-1}h^{-d+n}\int\phi(\frac{\tau-\bar\tau}{L})\kappa_n'(x,\tau)d\tau\prec CM\frac{h^{1-d}}{T}\rho^{d-1}(\frac{h}{TL})^\vartheta \tag{4.3.19}$$

holds for $L\in[\frac{h}{T},1]$. In particular, for constant γ,ρ and T the estimate

$$\mathcal{R}_{1x}(\bar\tau)\prec CMH^{1-d}(\frac{h}{L})^\vartheta \tag{4.3.20}$$

holds for $L\in[h,c]$.

Proof. I will prove statement (i). The proof of (ii) is identical. Applying the arguments of the proof of theorem 4.3.1(i) with $T=\text{const}$ we obtain that it is sufficient to prove that the norm of (4.3.8) does not exceed $CMh^{1-d}\gamma^{2d-1}\varepsilon^s$; now $\mathcal{E}(\tau)=\Gamma Q_{1x}B_1^je(.,.,\tau_1,\tau)B_1^{k\,t}Q_{2y}$. Dividing the integral into a sum of integrals over the intervals $[\bar\tau+n\frac{h}{T},\bar\tau+(n+1)\frac{h}{T}]$ and applying proposition 4.3.4 with $|n|\le\epsilon\frac{T}{h}$ we obtain that the norm of the n-th integral does not exceed $CM\frac{h^{1-d}}{T}\varepsilon^s(|n|+1)^{-s}$ and the total sum is estimated properly. On the other hand, for $|n|\ge\epsilon\frac{T}{h}$ one can apply proposition 4.1.1 and easily obtain that the total contribution of these integrals does not exceed CMh^s. □

Let us consider averaging functions with singularities:[22]

Theorem 4.3.7. *Let φ be a function such that*

$$|D^i\varphi(\lambda)| \le c|\lambda|^{\vartheta-i} \qquad \forall i \le K \tag{4.3.21}$$

where $\vartheta \ge 0$ and K and all the constants depend on ϑ also. Then statements (i) and (ii) of theorem 4.3.6 remain true with φ instead of ϕ and with ϑ given above.

Proof. I will prove estimate (4.3.17). The other proofs are identical. Let us take an admissible partition of unity. Then we should treat

$$\int \varphi(\frac{\tau-\bar\tau}{L})\zeta(\frac{\tau-\bar\tau}{L'})d_\tau\mathcal{E}$$

and

$$\int \varphi(\frac{(\tau-\bar\tau)T}{h})\bar\zeta(\frac{\tau-\bar\tau}{L'})d_\tau\mathcal{E}$$

with $L' \in [\frac{h}{T}, L]$, $L' \asymp 2^p\frac{h}{T}$, $\mathcal{E}$ defined above and smooth functions ζ and $\bar\zeta$ supported in $[-2,-1]\cup[1,2]$ and $[-2,2]$ respectively. Theorem 4.3.6 yields that the contribution of the first integral to the remainder estimate does not exceed $CM\frac{h^{1-d}}{T}(\frac{h}{L'T})^s(\frac{L'}{L})^\vartheta$ and after summation over $p=0,1,\dots,N'$ we obtain the proper estimate. Proposition 4.3.4 provides the same estimate for the norm of the second integral. □

Now we can get rid of the averaging:

Theorem 4.3.8. *(i) Under the assumptions of theorem 4.3.6(i) the estimate*

$$\mathcal{R}_2(\tau',\tau) \overset{\text{def}}{=} \Gamma Q_{1x}B_1^j e(.,.,\tau',\tau)B_1^k{}^tQ_{2y} - \sum_{0\le n\le N'-1} h^{-d+n}\varkappa_n(\tau',\tau) \prec CM\frac{h^{1-d}}{T}\gamma^{2d-1} \tag{4.3.22}$$

holds for $\tau',\tau \in [\bar\tau-\epsilon,\bar\tau+\epsilon]$. In particular, for constant γ and T the estimate

$$\mathcal{R}_2(\tau',\tau) \prec CMh^{1-d} \tag{4.3.23}$$

holds.

(ii) Under the assumptions of theorem 4.3.6(ii) the estimate

$$\mathcal{R}_{2x}(\bar\tau) \overset{\text{def}}{=} \Gamma_x Q_{1x}B_1^j e(.,.,\tau',\tau)B_1^k{}^tQ_{2y} - \sum_{0\le n\le N'-1} h^{-d+n}\kappa_n(x,\tau',\tau) \prec CM\frac{h^{1-d}}{T}\rho^{d-1} \tag{4.3.24}$$

holds for $\tau',\tau \in [\bar\tau-\epsilon,\bar\tau+\epsilon]$. In particular, for constant γ,ρ and T the estimate

$$\mathcal{R}_{2x}(\tau',\tau) \prec CMH^{1-d} \tag{4.3.25}$$

holds.

(iii) Moreover, all these estimates remain true with mollification on τ' with an arbitrary parameter L.

---

[22] The term "mollifying" is reserved for smooth functions.

Proof. Setting $\varphi = \theta\bar{\zeta}$ where θ is the Heaviside function and $\bar{\zeta}$ is a smooth function equal to 1 on $[-1,1]$, applying theorem 4.3.7 and integrating by parts we obtain the proper estimate for the contribution to the remainder from

$$-\mathcal{E}(\tau',\tau'') - \int \zeta(\tau-\tau'')\mathcal{E}(\tau',\tau)d\tau \tag{4.3.26}$$

with $\mathcal{E}(\tau',\tau)$ as introduced above and some smooth function ζ supported in $[1,c]$. Here $\bar{\tau}$ was replaced by τ''. This does not violate the microhyperbolicity condition since τ',τ'' are assumed to lie in $[\bar{\tau}-\epsilon,\bar{\tau}+\epsilon]$. In particular, picking $\tau'=\tau''$ we obtain the proper estimate for $\int\zeta(\tau-\tau'')\mathcal{E}(\tau',\tau)d\tau$. Shifting the function $\bar{\zeta}$ a small amount one can replace $\zeta(\tau-\tau'')$ by $\zeta(\tau-\tau')$ here and hence get the proper estimate for the contribution to the remainder from $\mathcal{E}(\tau',\tau)$.

Finally, in order to prove (iii) let us note that for $L\le\epsilon$ the estimates with averaging are due to estimates without it. Let $L\ge\epsilon$ and let us consider mollification with the same function ϕ and with L replaced by $2L$. Then the difference is the mollification with function satisfying (4.3.9) and is negligible modulo $O(h^sL^{-s})$. Then, starting from $L=\epsilon$ we can reach arbitrarily large L. □

Remark 4.3.9. *Usually statements similar to theorem 4.3.8 and theorem 4.3.7 with $L=1$ are considered a main goal. In this case our arguments in this subsection are a bit more complicated than the routine Tauberian technique (see [Hörmander 1], [Shubin 1], [Ivrii 6], etc.). However, we have proven a few assertions which seem to be interesting and which will be useful in what follows. From the technical point of view the extra parameter L is useful.* ◊

4.3.3 Asymptotics with enhanced remainder estimates: $T \gg 1$

Recall that all the statements of the previous subsection were proven with a small enough constant $T>0$. However, applying proposition 4.2.8 and remark 4.2.11 we can take larger T (and therefore smaller L) in all these statements:

Theorem 4.3.10. *(i) Let the assumptions of theorem 4.3.6(i) be fulfilled as well as the assumptions of proposition 4.2.8. Then propositions 4.3.4(i), 4.3.5(i) and theorems 4.3.6(i),[23] 4.3.8(i), (iii) remain true with the indicated T and with $L\in[\frac{h}{T},1]$.*

(ii) Let $\rho=\gamma$. Let the assumptions of theorem 4.3.6(ii) be fulfilled as well as the assumptions of proposition 4.2.8 with Hamiltonian trajectories with $t\in[-T,T]$ and condition (4.2.36)′ instead of (4.2.36). Then propositions 4.3.4(ii), 4.3.5(ii) and theorems 4.3.6(ii)[23], 4.3.8(ii), (iii) remain true with the indicated T and with $L\in[\frac{h}{T},1]$. T and with $L\in[\frac{h}{T},1]$.

Now let us consider the case $q_i\in \mathrm{S}_{h,K}$. Surely one can simply pick $\rho=\gamma=1$ but it would be better to apply a partition of unity.

Remark 4.3.11. *(i) Let us consider the set*

$$\Sigma_\tau = \{(x,\xi) : \tau \textit{ is an eigenvalue of } a(x,\xi)\}. \tag{4.3.27}$$

Let us note that the $\mathcal{L}(\mathbb{H},\mathbb{H})$-valued distribution $\partial_\tau\theta(\tau-a(x,\xi))$ is supported in Σ_τ; moreover, under the microhyperbolicity condition this distribution is a $\mathcal{L}(\mathbb{H},\mathbb{H})$-valued measure supported at Σ_τ. Let $\wp_\tau$ be this measure. One can easily see that if in the vicinity of $(\bar{x},\bar{\xi})$ the only eigenvalue of $a(x,\xi)$ close to τ is $\lambda(x,\xi)$ then in this vicinity[24]

[23] Moreover, one can replace ϕ by φ satisfying (4.3.21).

[24] We call a point of this type regular; the multiplicity of λ is constant in the vicinity of regular points.

$d\wp_\tau = \Pi(x,\xi)dxd\xi : d\lambda$ where here $\Pi(x,\xi)$ is a projector on the corresponding eigenspace (surely, those who prefer exterior form notation should write $dx\wedge d\xi$ instead of $dxd\xi$).

(ii) Moreover, one can fix x and introduce $\Sigma_{\tau,x}$ and the corresponding measure $\wp_{\tau,x}$, etc., provided $\mathcal{T}_x = 0$ in the microhyperbolicity condition. ◊

Remark 4.3.12. *Let us consider all the proofs of this and the previous subsections again. We consider only (i); the same arguments work for (ii) also.*

(i) First of all, in proposition 4.3.4 for $|\tau-\tau'|\le \frac{h}{T}$ and $Q_2 = Q_1^*$, $j=k$ one can replace the right-hand expression in estimate (4.3.13) by

$$C\varkappa_0'(\tau'')\frac{h^{1-d}}{T} + CM\frac{h^{2-d}}{T}\gamma^{2d-3} \tag{4.3.28}$$

where $0 \prec \varkappa_0'$ in this case and τ'' is an arbitrary value such that $|\tau''-\tau|\le\frac{h}{T}$ (the choice of τ'' is not essential). Therefore, under general restrictions on $\tau>\tau'$ we obtain that (4.3.13) holds with the right-hand expression

$$Ch^{-d}(\varkappa_0(\tau)-\varkappa_0(\tau')) + C\frac{h^{1-d}}{T}\varkappa_0'(\tau) + CMh^{1-d}\bigl(|\tau-\tau'|+\frac{h}{T}\bigr)\gamma^{2d-3}. \tag{4.3.29}$$

Let Q_1 be a scalar operator and $|q_1|\le c$. Then taking $j=k=l$ and multiplying by B_1^{-l} on the left and right we obtain that for $j=k=0$ estimate (4.3.13) holds with the right-hand expression

$$h^{-d}\int_{\mathcal{U}} \mathbf{e}(x,\xi,\tau',\tau)dxd\xi + \frac{h^{1-d}}{T}\int_{\mathcal{U}\cap\Sigma_\tau} d\wp_\tau + CMh^{1-d}\bigl(|\tau-\tau'|+\frac{h}{T}\bigr)\gamma^{2d-3}B_1^{-2l} \tag{4.3.30}$$

where $\mathbf{e}(x,\xi,\tau',\tau)$ is the spectral projector of the symbol $a(x,\xi)$ and $\mathcal{U} = \operatorname{supp} q_1 \cup \operatorname{supp} q_2$.

Applying the arguments used in the proof of proposition 4.1.4 we can get rid of the assumption $Q_2 = Q_1^*$. However, the assumptions that Q_1 and Q_2 are scalar operators, $|q_1|\le c$, $|q_2|\le c$ and $j=k=0$ remain here and below.

(ii) Then, in proposition 4.3.5 we obtain that estimate (4.3.16) holds with the right-hand expression

$$Ch^{1-d}\bigl(1+\frac{T'}{T}\bigr)\int_{\mathcal{U}\cap\Sigma_\tau} d\wp_\tau + CMh^{2-d}\gamma^{2d-3}\bigl(1+\frac{T'}{T}\bigr)B_1^{-2l}. \tag{4.3.31}$$

(iii) Moreover, in theorems 4.3.6–4.3.7 estimate (4.3.17) remains true with the right-hand expression

$$C\frac{h^{1-d}}{T}\bigl(\frac{h}{TL}\bigr)^{\vartheta}\int_{\mathcal{U}\cap\Sigma_\tau} d\wp_\tau + CM\frac{h^{2-d}}{T}\gamma^{2d-3}\bigl(\frac{h}{TL}\bigr)^{\vartheta}B_1^{-2l}. \tag{4.3.32}$$

(iv) Hence in theorem 4.3.8 estimate (4.3.23) remains true with the same right-hand expression with $\vartheta = 0$, i.e., with

$$C\frac{h^{1-d}}{T}\int_{\mathcal{U}\cap\Sigma_\tau} d\wp_\tau + CM\frac{h^{2-d}}{T}\gamma^{2d-3}B_1^{-2l}. \tag{4.3.33}$$

(v) Finally, theorem 4.3.10 obviously remains true. ◊

Then, after summation on a partition of unity theorems 4.3.6–4.3.8 and remarks 4.3.11 and 4.3.12 yield the following

Theorem 4.3.13. *(i) Let the assumptions of theorem 4.3.6(i) be fulfilled in the vicinity of every point of* $\operatorname{supp} q_1 \cup \operatorname{supp} q_2$. *Then the estimates*

$$\mathcal{R}_1(\tau) \prec Ch^{1-d}\big(\frac{h}{L}\big)^{\vartheta} \int_{\Sigma_\tau \cap \mathcal{U}} d\wp_\tau + CMh^{1-d+\delta''}\big(\frac{h}{L}\big)^{\vartheta} B_1^{-2l} \tag{4.3.34}$$

$$\mathcal{R}_2(\tau',\tau) \prec CMh^{1-d} \int_{(\Sigma_\tau \cup \Sigma_{\tau'})\cap \mathcal{U}} d\wp_* + CMh^{1-d+\delta''} B_1^{-2l} \tag{4.3.35}$$

hold where $\mathcal{U} = \operatorname{supp} q_1 \cap \operatorname{supp} q_2$ *and* $* = \tau, \tau'$ *on* Σ_τ *and* $\Sigma_{\tau'}$ *respectively and* $\delta'' > 0$ *is small enough.*

(ii) Let the assumptions of theorem 4.3.6(ii) be fulfilled in the vicinity of every point of $\mathcal{U} = \operatorname{supp} q_1 \cup \operatorname{supp} q_2$. *Then these estimates hold for every* x *and for* $\mathcal{R}_*$, Σ_τ *and* $d\wp_\tau$ *replaced by* $\mathcal{R}_{*x}$, $\Sigma_{\tau x}$ *and* $d\wp_{\tau x}$ *respectively.*

Moreover, after summation on a partition of unity theorem 4.3.10 yields the following

Theorem 4.3.14. *(i) Let the assumptions of theorem 4.3.6(i) be fulfilled in the vicinity of every point of* $\mathcal{U} = \operatorname{supp} q_1 \cup \operatorname{supp} q_2$. *Moreover, let* $\Lambda_{\tau,j}$ *be closed subsets of* Σ_τ *and let the conditions of theorem 4.3.10(i) be fulfilled at every point of* $\Lambda_{\tau,j}$ *with* $T_j \ge 1$ *instead of* T $(j = 1, \dots, p)$. *Let* $\Lambda_{\tau,0} = \Sigma_\tau \setminus (\Lambda_{\tau,1} \cap \dots \cap \Lambda_{\tau,p})$ *and* $T_0 = 1$. *Then the estimates*

$$\mathcal{R}_1(\tau) \prec \sum_{0\le j\le p} C\frac{h^{1-d}}{T_j} \min\big(\big(\frac{h}{LT_j}\big)^{\vartheta}, 1\big) \int_{\Lambda_{j,\tau}\cap \mathcal{U}} d\wp_\tau + CMh^{1-d+\delta''}\big(\frac{h}{L}\big)^{\vartheta} B_1^{-2l} \tag{4.3.36}$$

$$\mathcal{R}_2(\tau,\tau') \prec \sum_{0\le j\le p} C\frac{h^{1-d}}{T_j} \int_{(\Lambda_{j,\tau}\cup\Lambda_{j,\tau'})\cap \mathcal{U}} d\wp_* + CMh^{1-d+\delta''} B_1^{-2l} \tag{4.3.37}$$

hold where $\delta'' > 0$ *is small enough.*

(ii) Let the assumptions of theorem 4.3.6(ii) be fulfilled in the vicinity of every point of $\mathcal{U} = \operatorname{supp} q_1 \cup \operatorname{supp} q_2$. *Moreover, let* $\Lambda_{\tau,j}$ *be closed subsets of* Σ_τ *and let the conditions of theorem 4.3.10(ii) be fulfilled at every point of* $\Lambda_{\tau,j}$ *with* $T_j \ge 1$ *instead of* T $(j = 1, \dots, p)$. *Let* $\Lambda_{\tau,0} = \Sigma_\tau \setminus (\Lambda_{\tau,1} \cap \dots \cap \Lambda_{\tau,p})$ *and* $T_0 = 1$.

Then these estimates hold for every x *and for* $\mathcal{R}_*$, Σ_τ *and* $d\wp_\tau$ *replaced by* $\mathcal{R}_{*x}$, $\Sigma_{\tau x}$ *and* $d\wp_{\tau x}$ *respectively.*

Remark 4.3.15. *(i) Now let* $\gamma = \rho = 1$ *in the definition of* M *(recall that* M *is the product of norms* q_1, q_2, ψ_1, ψ_2 *and perhaps* ϕ *or* φ*). Then in theorem 4.3.13 one can drop all terms with* $n \ge \vartheta + 1$ *in* $\mathcal{R}_*$ *and in theorem 4.3.14 one can drop all terms with with* $n > \vartheta + 1$ *in* $\mathcal{R}_*$ *where* $\vartheta \overset{\text{def}}{=} 0$ *if we consider* $\mathcal{R}_{2*}$. *Recall that* $\max T_j \le h^{-\delta'}$ *with sufficiently small exponent* δ'.

(ii) Applying the arguments used in the proof of (4.3.4) one can easily prove that all the estimates with mollification also hold for

$$\iint \varphi\big(\frac{\lambda-\tau}{L}\big)\varphi_1\big(\frac{\lambda'-\tau'}{L'}\big)\mathcal{E}'(\tau',\tau)\,d\tau' d\tau - \dots$$

where $\frac{h}{T} \le L \le L' \le c$ *(*T *is specified in theorems 4.3.6 and 4.3.7; otherwise* $T = 1$*), the microhyperbolicity condition should also be fulfilled at the level* τ' *and in the right-hand expressions one should replace* $\Lambda_{\tau,j}$ *by* $\Lambda_{\tau,j} \cup \Lambda_{\tau',j}$*; both functions* φ *and* φ_1 *should fulfill (4.3.21) with* ϑ *replaced by* $\vartheta - 1$.

(v) Moreover, for a lacunary value τ' these estimates remain true without mollification on τ', i.e., for

$$\int \varphi(\frac{\lambda-\tau}{L})\mathcal{E}(\tau',\tau)d\tau - \dots$$

(vi) In particular, under condition (4.1.12)$_\pm$ *one can pick* $\tau' = \mp\infty$. ◊

4.3.4 Formulas for coefficients

Starting from (4.2.24) one can easily prove the formula

$$\varkappa_0(x,\tau',\tau) = (2\pi)^{-d}\int q_1^0 q_2^0 \mathbf{e}(x,\xi,\tau',\tau)dxd\xi \tag{4.3.38}$$

and a similar formula for $\kappa_0(\tau',\tau)$ (without integration with respect to dx) where a^s is the subprincipal symbol of A, $\mathbf{e}(x,\xi,\tau',\tau)$ is the spectral projector for $a(x,\xi)$ and q_j^0 are the principal symbols of Q_j. Obtaining formulas for the other coefficients is more difficult. For example, (4.2.25) yields that $\kappa_1 = \kappa_{11} + \kappa_{10}$ where

$$\operatorname{tr}\varkappa_{11} = -(2\pi)^{-d}\operatorname{tr}\int q_1^0 q_2^0 a^s d\wp_{\tau x} \tag{4.3.39}$$

(and a similar formula holds for κ_{11} with $\wp_{\tau x}$ instead of $d\wp_\tau$) and κ_{10} depends only on a. Moreover, $\operatorname{tr}\varkappa_{10} = 0$ for a real-valued symbol a and $q_1^s = q_2^s = 0$; in particular, this is true for scalar symbol a.

4.4 Spectral Asymptotics for Scalar Operators

In this section we consider operators with scalar principal symbols for which we weaken the restrictions on L of subsection 4.3.1 and the microhyperbolicity condition we had in subsection 4.3.2. Our main tool is rescaling applied to the results of the previous section. This theory is applied to more general operators as well. Namely, all the results remain true for operators such that

$$\operatorname{Spec} A^0(x,\xi)\cap[-\epsilon_0,\epsilon_0] = \{a(x,\xi)\} \qquad \text{in } \Omega \tag{4.4.1}$$

where here $A^0(x,\xi)$ is the principal symbol of A and $a(x,\xi)$ is the scalar function. Then in a fixed neighborhood of $\operatorname{supp} q_1 \cup \operatorname{supp} q_2$ the operator A can be reduced[25] to the form $A \equiv \begin{pmatrix} A_I & 0 \\ 0 & A_{II}\end{pmatrix}$ where the principal symbol of A_I equals $a(x,\xi)$ and $\operatorname{Spec} a_{II}\cap[-\epsilon_0,\epsilon_0] = \emptyset$; let Ω be this neighborhood. If $U = \begin{pmatrix} U_{I,I} & U_{I,II} \\ U_{II,I} & U_{II,II}\end{pmatrix}$ corresponds to this reduction then standard ellipticity arguments yield that

(4.4.2) All the components of U apart from $U_{I,I}$ are negligible in $\Omega\times\Omega^T\times[-\epsilon_0,\epsilon_0]_\tau$;

in what follows we will consider only $\tau\in[-\epsilon_0,\epsilon_0]$ and so we can assume without loss of generality that

(4.4.3) $B = I$ and A is an operator with scalar principal symbol aI.

---

[25] By means of multiplication on the right and left by G^* and G respectively where G is a pseudo-differential operator with $G^*G \equiv GG^* \equiv I$.

In the final subsections we consider scalar operators. For operators satisfying (4.4.3) the microhyperbolicity condition at the energy level τ is

$$|a-\tau| + |\nabla_{x,\xi} a| \ge \epsilon_0 \qquad \text{in } \Omega \tag{4.4.4}$$

i.e., that τ is not the critical value of a.

One can easily see that

$$d\wp_\tau = I_{\mathbb{H}} dx d\xi : da\big|_{\Sigma_\tau} \tag{4.4.5}$$

in the sense of exterior forms.

4.4.1 From the condition "no critical point" to "rank Hess $a \ge 2$ at critical points." I

Let us introduce the function

$$\gamma(x,\xi) = \epsilon_1 \bigl(|a(x,\xi)-\tau| + |\nabla_{(x,\xi)} a|^2\bigr)^{\frac{1}{2}} + \bar\gamma \tag{4.4.6}$$

with $\bar\gamma = h^{\frac{1}{2}}$; then one can easily check that

$$|\nabla_{(x,\xi)} \gamma(x,\xi)| \le 1 \tag{4.4.7}$$

provided $\epsilon_1 > 0$ is a small enough constant. Let us consider a γ-admissible partition of unity by pseudo-differential operators Q'_l.[26] Let $Q''_l = \mathrm{Op}(q''_l)$ where q''_l is the symbol of the same type as q'_l. Then one can easily see that

(4.4.8) $\gamma_l{}^{-2} Q'_l A_I \in \Psi_{h,\gamma_l,\gamma_l,K}(T^*\mathbb{R}^d, \mathbb{H}, \mathbb{H})$ uniformly with respect to l.

Then one needs to consider the separate elements of the partition and to treat two cases:

(i) $\gamma \overset{\text{def}}{=} \gamma_l \ge c_0 \bar\gamma$; then for this element all the conditions including the microhyperbolicity condition will be fulfilled after the change $x \to x_{\text{new}} = \frac{x-x^l}{\gamma}$, $\xi \to \xi_{\text{new}} = \frac{\xi-\xi^l}{\gamma}$, $h \to h_{\text{new}} = \frac{h}{\gamma^2}$ and $A \to A_{\text{new}}\gamma^{-2} = e^{ih^{-1}\langle x,\xi_l\rangle} A_I e^{-ih^{-1}\langle x,\xi_l\rangle}$. So, there are (apart from inessential shifts in (x,ξ)) rescalings in (x,ξ) and in the spectral parameter τ as well. Therefore one can apply all the results of subsection 4.3.2 with L replaced by $L_{\text{new}} = \frac{L}{\gamma^2}$ after rescaling. Moreover, one can apply the results of subsection 4.3.3 as well noticing that now no rescaling on t is necessary because $h^{-1}_{\text{new}} A_{\text{new}} = h^{-1}A$ modulo an inessential shift. So, let us apply all the results of subsections 4.3.2 and 4.3.3 and then return back to the original variables. Recall that $\frac{h_{\text{new}}}{L_{\text{new}}} = \frac{h}{L}$ and $\frac{h_{\text{new}}}{L_{\text{new}}T} = \frac{h}{LT}$ and therefore

$$\frac{h_{\text{new}}^{1-d}}{T}\Bigl(\frac{h_{\text{new}}}{L_{\text{new}}T}\Bigr)^{\vartheta} \int_{\mathcal{U}\cap\Sigma_\tau} dx_{\text{new}} d\xi_{\text{new}} : da_{\text{new}} = \frac{h^{1-d}}{T}\Bigl(\frac{h}{LT}\Bigr)^{\vartheta} \int_{\mathcal{U}\cap\Sigma_\tau} dx d\xi : da \tag{4.4.9}$$

(surely in the left-hand expression not $(x_{\text{new}},\xi_{\text{new}})$ but (x,ξ) belongs to $\mathcal{U}\cap\Sigma_\tau$) and

$$h_{\text{new}}^{1-d+\delta}\Bigl(\frac{h_{\text{new}}}{L_{\text{new}}T}\Bigr)^{\vartheta} = \Bigl(\frac{h}{\gamma^2}\Bigr)^{1-d+\delta}\Bigl(\frac{h}{L}\Bigr)^{\vartheta}. \tag{4.4.10}$$

We obtain that

---

[26] That means that $Q'_l = \mathrm{Op}(q'_l)$ where $q_l \in \mathrm{S}_{h,\gamma(z^l),\gamma(z^l),K}(T^*\mathbb{R}^d)$ are supported in the balls $B(z^l, \frac{1}{2}\gamma_l)$ with $\gamma_l = \gamma(z^l)$ and the multiplicity of the covering by balls $B(z^l,\gamma_l)$ does not exceed $K_0 = K_0(d)$. Such a partition exists (see [Hörmander 1], for example).

(4.4.11) For this element of the partition all the estimates of subsections 4.3.2 and 4.3.3 remain true with an additional term

$$CMh^{1-d+\delta}\left(\frac{h}{L}\right)^{\vartheta}\int_{\mathcal{U}}\gamma^{-2-2\delta}\,dxd\xi \tag{4.4.12}$$

in the right-hand expression where $\mathcal{U}=B(z^l,\gamma_l)$ now.

(ii) $\gamma\overset{\text{def}}{=}\gamma_l\le c_0\bar{\gamma}$; then for this element all the conditions apart from the microhyperbolicity condition will be fulfilled after the same change of variables as before. However, the microhyperbolicity condition provides no advantage because $h_{\text{new}}=1$ now. So one ought to apply proposition 4.1.1 and theorem 4.3.1. Returning to the original variables and operator we see that the contribution of this element to the remainder does not exceed

$$CM\left(\frac{h}{L}\right)^{\vartheta}\asymp CMh^{-d}\left(\frac{h}{L}\right)^{\vartheta}\int_{\mathcal{U}'}dxd\xi \tag{4.4.13}$$

where $\mathcal{U}'=B(z^l,\gamma_l)$ now.[27)]

Now let us sum the contributions of all the elements of this partition. We obtain that all the estimates of subsections 4.3.2 and 4.3.3 remain true with Σ_τ integrals restricted to $\Sigma_\tau\cap\{\gamma(x,\xi)\ge c_0\bar{\gamma}\}$ and with additional terms (4.4.12) with $\mathcal{U}=\Omega\cap\{\gamma(x,\xi)\ge c_0\bar{\gamma}\}$ and (4.4.13) with $\mathcal{U}'=\Omega\cap\{\gamma(x,\xi)\ge c_0\bar{\gamma}\}$.

First of all we obtain immediately

Theorem 4.4.1. *Let condition* (4.4.3) *be fulfilled. Then the statements of theorems 4.3.1 and 4.3.2 remain true for* $L\ge h^{1-\delta}$ [28)] *Moreover, the estimates*

$$\mathcal{R}'_1\le Ch^{-d}\left(\frac{h}{L}\right)^{\vartheta}+CMh^{s} \tag{4.4.14}$$

$$\mathcal{R}'_2\le Ch^{-d}\left(\frac{h}{L}\right)^{\vartheta+1}+CMh^{s} \tag{4.4.15}$$

hold for an averaging function φ *satisfying* (4.3.21) *and for* $L\ge h$.

Let us assume that

(4.4.16)$_r$ $|a(x,\xi)-\tau|+|\nabla_{x,\xi}a(x,\xi)|\le\epsilon_0\implies$ Hess $a(x,\xi)$ has r eigenvalues (counting multiplicities) $f_1,\dots,f_r$ such that $|f_1|\ge\epsilon_0,\dots,|f_r|\ge\epsilon_0$.[29)]

Then in the vicinity of every point with $|a(x,\xi)-\tau|+|\nabla_{a(x,\xi}a(x,\xi)|\le\epsilon_0$ one can choose a coordinate system $z=(z';z'')=(z_1,\dots,z_r;z_{r+1},\dots,z_{2d})$ such that

$$a(z)=\sum_{1\le j\le r}\zeta_j z_1^2+V(z'')$$

with $\zeta_j=\pm1$ (this system is not assumed to be symplectic). Then in a smaller vicinity

$$\gamma\asymp|z'|+\left(|V(z')-\tau|+|\nabla V(z')|^2\right)^{\frac{1}{2}}+\bar{\gamma} \tag{4.4.17}$$

---

27) Surely, we obtained these results only for a smooth averaging function ϕ and $\vartheta=s$. However, applying the arguments of the proof of theorem 4.3.7 we extend them to an arbitrary averaging function φ satisfying (4.3.21).

28) Recall that the averaging function is assumed to be smooth in these statements.

29) The other eigenvalues are arbitrary.

and for $r \geq 3$ all the additional terms are $O(h^{1-d+\vartheta+\delta'}L^{-\vartheta})$ with a small enough exponent $\delta' > 0$ and

$$\int_{\Sigma_\tau \cap \Omega \cap \{\gamma(x,\xi) \leq c_0 \bar{\gamma}\}} dx d\xi : da = O(h^{\frac{r-2}{2}} \log h). \tag{4.4.18}$$

On the other hand, under the condition

$(4.4.16)_2^+$ $|a(x,\xi) - \tau| + |\nabla_{a(x,\xi}a(x,\xi)| \leq \epsilon_0 \implies \operatorname{Hess} a(x,\xi)$ has 2 eigenvalues (counting multiplicities) f_1, f_2 of the same sign such that $|f_1| \geq \epsilon_0, |f_2| \geq \epsilon_0$.[29)]

these additional terms are $O(h^{1-d+\vartheta}L^{-\vartheta})$ and (4.4.17) holds without a logarithmic factor.[30)] Therefore we obtain

Theorem 4.4.2. *Let condition* (4.4.3) *be fulfilled. Then*
(i) Under condition $(4.4.16)_2^+$ *statement (i) of theorem 4.3.10 remains true with* $\mathcal{R}_1, \mathcal{R}_2$ *defined by* (4.3.17) *and* (4.3.22) *with* $T \asymp 1$ *and* $N' = 2 + [\vartheta]$.[31)]
(ii) Under condition $(4.4.16)_3$ *the statements (i) of theorems 4.3.10, 4.3.13 remain true with* $\mathcal{R}_1$ *defined by* (4.3.17) *with* $N' = 2 + [\vartheta]$.

Remark 4.4.3. *(i) Surely, condition* $(4.4.16)_2^+$ *is weaker than* $(4.4.16)_3$.
(ii) Under condition $(4.4.16)_2$ *we obtain the estimate*

$$|\mathcal{R}_1| \leq Ch^{1-d}\Big(\frac{h}{L}\Big)^\vartheta (|\log h| + 1)$$

and for $a = z_1^2 - z_2^2$, $\tau = 0$ *integral (4.4.18) diverges logarithmically.*
(iii) Under condition $(4.4.16)_1$ *we obtain the estimate*

$$|\mathcal{R}_1| \leq Ch^{\frac{1}{2}-d}\Big(\frac{h}{L}\Big)^\vartheta$$

and for $a = z_1^2$, $\tau = 0$ *integral (4.4.18) strongly diverges.* ◇

However, below we recover good estimates even in this case.

Let us recover the statements of theorem 4.3.13 (with reasonable modifications) under condition $(4.4.16)_2^+$. Let us introduce the set

$$\Sigma'_\tau = \{(x,\xi) : \tau \text{ is an eigenvalue of } a(x,\xi) + ha^s(x,\xi)\} \tag{4.4.19}$$

and the $\mathcal{L}(\mathbb{H},\mathbb{H})$-valued[32)] measure

$$d\wp'_\tau = \partial_\tau \theta(\tau - a(x,\xi) - a^s(x,\xi)h) \tag{4.4.20}$$

supported on Σ'_τ.

Theorem 4.4.4. *Let conditions* (4.4.3) *and* $(4.4.16)_2^+$ *be fulfilled. Let us assume that* $\mathrm{D} = \dim H < \infty$. *Then*
(i) The estimates

30) In this case $\int_{\Sigma_\tau} dx d\xi : da$ is uniformly bounded. However, for $a = z_1^2 + z_2^2$ this integral is not uniformly convergent.

31) For $\vartheta \in \mathbb{N}$ one can take $N' = 1 + \vartheta$ as well. However for large N' the integral defining the coefficients is not necessarily convergent.

32) We do not assume that the subprincipal symbol a^s is scalar.

$$(4.4.21)\quad \mathcal{R}_1(\tau) \prec Ch^{1-d}\big(\frac{h}{L}\big)^{\vartheta}\int_{\Sigma'_\tau\cap\mathcal{U}\cap\{\gamma\ge h^{\frac12-\delta}\}} d\wp'_\tau + Ch^{1-d}\big(\frac{h}{L}\big)^{\vartheta}\int_{\Sigma'_\tau\cap\mathcal{U}\cap\{\gamma\le h^{\frac12-\delta}\}} |\{z_1,z_2\}|^{1+\vartheta}d\wp'_\tau + CMh^{1-d+\delta''}\big(\frac{h}{L}\big)^{\vartheta}$$

$$(4.4.22)\quad \mathcal{R}_2(\tau',\tau) \prec Ch^{1-d}\int_{(\Sigma'_\tau\cup\Sigma'_{\tau'})\cap\mathcal{U}\cap\{\gamma\ge h^{\frac12-\delta}\})} d\wp'_* + Ch^{1-d}\int_{(\Sigma'_\tau\cup\Sigma'_{\tau'})\cap\mathcal{U}\cap\{\gamma\le h^{\frac12-\delta}\})} |\{z_1,z_2\}|d\wp'_* + CMh^{1-d+\delta''}$$

hold where $\mathcal{U} = \operatorname{supp} q_1 \cap \operatorname{supp} q_2$ *and* $* = \tau,\tau'$ *on* Σ'_τ *and* $\Sigma'_{\tau'}$ *respectively,* $\delta''>0$ *is small enough and the function* ϕ *(or* φ*) is assumed to be supported in* $\bar{\mathbb{R}}^{\mp}$ *for the positive (negative) indicated eigenvalues of* Hess V*,* M *depends on* D*,* $N' = 2+[\vartheta]$.

(ii) In the general case (4.4.22) *holds with* a^s *replaced by* $\mp C_0 I$ *in the definition of* Σ_τ *and* $\wp_\tau$.

Proof. Let the indicated eigenvalues of Hess V be positive. Introducing appropriate local coordinates $z=(z';z'')=(z_1,z_2;z'')$ in $\mathbb{R}^{2d}$ we obtain that $a-\tau=(z_1^2+z_2^2+V(z''))$ and then (4.4.17) holds for γ defined by (4.4.6); let us take $\bar\gamma = h^{\frac12-\delta}$ and consider a γ-admissible partition of unity. Then the arguments of the proof of theorem 4.4.1 provide us with an appropriate estimate in the subdomain $\{\gamma\ge c_0\bar\gamma\}$ and we need to consider only the subdomain $\{|z'|\le\bar\gamma\}\times\Omega''$ with $\Omega''=\{z'',|d_{z''}V|+|V-\tau|^{\frac12}\le\bar\gamma\}$. Let us cover Ω'' with balls of radius $\gamma_1=h^{\frac12-\delta_1}$ with $\delta_1>\delta$. Let $|\operatorname{Hess} V|\ge c_0\gamma_1$ at some point $z''\in\Omega''$. Then this remains true in $B(z'',\gamma_1)$ (with c_0 replaced by $\frac12 c_0$) and $\operatorname{mes}(\Omega''\cap B(z'',\gamma_1))\le \operatorname{mes} B(z'',\gamma_1)h^{\delta_1-\delta}$ and hence we need to consider only balls with $|\operatorname{Hess} V|\le\gamma_1$.[33] Let us consider balls with $|\{z_1,z_2\}|\le c\omega$, where $\omega\ge h^{\delta_0}$ and exponents $\delta_i>0$ are small enough. This inequality holds at all points of this ball simultaneously. Then in the appropriate *symplectic* coordinates

$$(a-\tau)=x_1^2+(\xi_2+b_1(x,\xi))^2+b_2(x,\xi)$$

where $|\nabla b_1|\le c\omega$ and $|\operatorname{Hess} b|\le c\omega^2$ in this ball. Let us redefine the scaling functions γ and ρ in the ball $B(z'',\gamma_1)$. Namely, pick

$$(4.4.23)\qquad \rho=|\xi_1|+|\xi_2+b_1|+\big(|b_2-\tau|+\omega^{-2}|\nabla b_2|^2\big)^{\frac12}+\bar\rho', \qquad \gamma=\rho\omega^{-1},\qquad \bar\rho'=h^{\frac12}\omega^{\frac12}.$$

Then $|\nabla\rho|\le 1$ (due to the condition on the Hessian) and $\rho\gamma\ge h$. Moreover, one can check that if $\operatorname{dist}_x(z',z'')\le\gamma(z'')$ and $\operatorname{dist}_\xi(z,z'')\le\rho(z'')$ then $\rho(z')\asymp\rho(z'')$ and $\gamma(z')\asymp\rho(z'')$. Let us rescale $x_{\mathrm{new}}=\frac{x}{\gamma}$, $\xi_{\mathrm{new}}=\frac{\xi}{\rho}$. Then we should replace h by $h_{\mathrm{new}}=\frac{h}{\rho\gamma}$ and L by $L_{\mathrm{new}}=\frac{L}{\rho^2}$ because we divide A by ρ^2 now. Obviously, $\frac{h_{\mathrm{new}}}{L_{\mathrm{new}}}=\frac{h}{L}\omega$. One can easily check that after this rescaling the derivatives (up to order K) of the symbol of the operator are bounded.[34] Moreover, the microhyperbolicity condition is fulfilled for

---

[33] Moreover, the same arguments yield that one need only treat balls with $\sum_{2\le|\alpha|\le n}|D^\alpha V|\le c_0\gamma_1$.

[34] Surely, the symbol itself is not necessary bounded because of the term $a^s\frac{h}{\rho^2}$ but in this case one can introduce the operator $B=a^s(\bar z)\frac{h}{\rho^2}$ where $\bar z$ is a fixed point in the original ball of radius γ_1. Then the standard theory holds. For a scalar symbol a^s one can avoid this difficulty simply by replacing a with $a+ha^s$ in the definitions of γ and ρ.

$\rho \ge c_0\bar{\rho}'$.[35] Then theorem 4.3.10 provides all the necessary estimates in this domain and its total contribution to $\mathcal{R}_1$ surely does not exceed the right-hand expression of estimate (4.4.22) with ω instead of $|\{z_1, z_2\}|$ and with $-C_0 h$ instead of a^s. Theorem 4.4.1 applies in the domain $\rho \le \bar{\rho}'$, the total contribution to $\mathcal{R}_1$ of which does not exceed

$$CM\bigl(\frac{h}{L}\bigr)^{\vartheta}\bar{\rho}'^{2-d}\bar{\gamma}'^{-d} \le CM\bigl(\frac{h}{L}\bigr)^{\vartheta}h^{1-d}\omega.$$

Therefore (ii) is proven and in the proof of (i) the contribution of balls with $\omega \le h^{\delta_0}$ is estimated properly.

Finally, let us consider balls of radius $\gamma_2 = h^{\frac{1}{2}-\delta_2}$ where $|\operatorname{Hess} V| \le c_0\gamma_1$ and $|\{z_1, z_2\}| \asymp \omega \ge h^{\delta_0}$. In these balls the variation of b_2 is less than $h^{1+\delta_3}$ and we simply include $h^{-1}b_2$ in a^s. Without loss of generality one can assume that in the ball, $h^{-1}b_2 + a^s - \tau = \begin{pmatrix} B_+ & 0 \\ 0 & B_- \end{pmatrix}$ where $B_+ \ge \nu I$, $B_- < (1-\epsilon)\nu I$ and $\epsilon''\omega \le \nu \le \epsilon'\omega$ with small enough positive constants $\epsilon, \epsilon', \epsilon''$. Then one can easily see that

$$\int_{\Sigma'_\tau \cap B(z,\gamma_2)} d\wp'_\tau \succ \Pi_-\gamma_2^{2-2d}$$

where $\Pi_- = \begin{pmatrix} 0 & 0 \\ 0 & I \end{pmatrix}$. Let us consider $A_+ = Z_1^2 + Z_2^2 + B_+$. The Gårding inequality yields that

$$\begin{aligned}(A_+u, u) \ge \|Z_1u\|^2 + \|Z_2u\|^2 - C_0h^{1+2\delta_0}\|u\|^2 - C\|Qu\|^2 \ge \\ \frac{1}{2}(\|Z_1u\|^2 + \|Z_2u\|^2 + |([Z_1, Z_2]u, u)|) - C_0h^{1+2\delta_0}\|u\|^2 - C\|Qu\|^2 \ge \\ \epsilon(\|Z_1u\|^2 + \|Z_2u\|^2 + \omega h\|u\|^2) - C\|Qu\|^2\end{aligned}$$

where Q is negligible in the ball in question. Therefore for $\tau' \le \tau$ the operator $A_+ - \tau'$ is invertible in this ball and the norm of the inverse does not exceed $Ch^{-1}\omega^{-1}$. Therefore $F_{t\to h^{-1}\tau}\chi_T Q_1 U^{++}$ is negligible for $\tau \le \epsilon\omega h$, $T \ge h^{-1-\delta'''}\omega^{-1}$ with an arbitrarily small exponent $\delta''' > 0$ and Tauberian arguments yield that $Q_1E(\tau', \tau)$ is negligible for $\tau', \tau'' \le \tau + \epsilon\omega h$. This and estimates from below of the right-hand expressions of (4.4.21) and (4.4.22) yield these estimates (for ϕ or φ supported in the proper half-line). □

Remark 4.4.5. *One can check that for the harmonic oscillator $h^2D_1^2 + x_1^2$ estimate (4.4.2) fails for $L = h$ unless ϕ is supported in $\{\tau \le \frac{1}{2}\}$.* ◊

4.4.2 From the condition "no critical point" to "rank Hess $a \ge 2$ at critical points." II

Now we begin treating the more complicated case in which condition $(4.4.16)_2$ is fulfilled but condition $(4.4.16)_2^+$ fails. I want to obtain remainder estimate $O(h^{1-d+\vartheta})$ in this case. Without loss of generality one can assume that

$(4.4.16)_2^-$ $|a(x,\xi) - \tau| + |\nabla_{x,\xi}a(x,\xi)| \le \epsilon_0 \implies \operatorname{Hess} a(x,\xi)$ has two eigenvalues (counting multiplicities) f_1, f_2 such that $f_1 \ge \epsilon_0, -f_2 \ge \epsilon_0$.[29]

Then in appropriate coordinates locally

[35] According to our procedures, one should check this condition for the symbol $a + a^s h$ after rescaling, but the derivatives of $a^s h$ are negligible, so one may check the condition for the scalar symbol a instead.

$$a - \tau = z_1 z_2 + V(z') \tag{4.4.24}$$

with $z' = (z_3, \dots, z_{2d})$.

So, the goal of this subsection is

Theorem 4.4.6. *Suppose conditions (4.4.3) and $(4.4.16)_2^-$ are fulfilled. Then asymptotics (4.4.21) with $N' = 2 + [\vartheta]$ and (4.4.22) hold.*

Remark 4.4.7. *Let conditions (4.4.3), $(4.4.16)_2^-$ be fulfilled. Then for $\vartheta \in \mathbb{N}$, in particular, for $\mathcal{R}_2$ the integral defining $\varkappa_n$ is not necessarily convergent even after mollification so one should regularize it by replacing $\mathcal{F}_n$ with $\mathcal{F}_n(1-\zeta)$ where the function ζ is supported in $\{\gamma \le ch^{\frac{1}{2}}\}$ and equal to 1 for $\gamma \le \epsilon h^{\frac{1}{2}}$. However, if $a^s = 0$ (which can always be arranged by replacing the principal symbol by the complete Weyl symbol) then for even n this integral converges and one can take $N' = 1 + n$. This is true for $\mathcal{R}_2$ in particular. On the other hand, for odd n the divergent (after mollification) part of this integral is given by the formula*

$$\varkappa_{n,h} = c_{d,n} \partial_\tau^n h^{-1} \int Q(x,\xi) \zeta^n \bigl(\theta(a + \zeta h) - \theta(a)\bigr) dx d\xi \tag{4.4.25}$$

where

$$\zeta(x,\xi) = \bigl(\sum f_j^2\bigr)_+^{\frac{1}{2}} \tag{4.4.26}$$

and f_j are all the eigenvalues of the skew-Hessian of $a(x,\xi)$. I leave the calculations to the reader. ◊

Proof of theorem 4.4.6. Let us introduce the functions

$$\zeta_n(z') = \epsilon \Bigl(\sum_{|\alpha| \le n-1} |D^\alpha V|^{\frac{n}{n-|\alpha|}} \Bigr)^{\frac{1}{n}} + h^{\frac{1}{n}} \tag{4.4.27}$$

with a small enough constant $\epsilon > 0$. Then

$$\gamma(z) \asymp |z_1| + |z_2| + \zeta_2(z')$$

and it follows from the arguments of the previous subsection that

(4.4.28) The contribution of the zone $\{\gamma(z) \le C_0 h^{\frac{1}{2}}\}$ to the remainder estimate does not exceed $Ch^{1-d+\vartheta} \operatorname{mes}\{z' : \zeta_2(z') \le C_1 h^{\frac{1}{2}}\}$

and

(4.4.29) The contribution of the zone $\{\gamma(z) \le C_0 \zeta_2(z')\}$ to the remainder estimate does not exceed $Ch^{1-d+\vartheta}$.

Moreover, the same arguments yield that

(4.4.30) The contribution of the zone $\{z' \in \Omega, \gamma(z) \le \mu(z')\}$ to the remainder estimate does not exceed

$$Ch^{1-d+\vartheta} \int \Bigl(\log \frac{\mu(z')}{\zeta_2(z')} + 1\Bigr) dz'$$

with the $(2d-2)$-dimensional domain Ω provided $\zeta_2(z') \le \mu(z')$ and $|\nabla \mu| \le 1$.

Let R_n be the integral taken over the domain in the right-hand expression with $\mu_n = \sum_{2\le k\le n} \zeta_k^{p_k}$, $p_k = 2^{k-2}$. I want to prove by induction on $n \ge 2$ that R_n does not exceed C_n provided $\operatorname{diam} \Omega \le c$. For $n = 2$ this is obvious. Let us assume that it is true for $n = m$; we should prove that

$$\int \log\Bigl(\frac{\mu_{m+1}(z')}{\mu_m(z')}\Bigr) dz'$$

is bounded. Let us consider a ζ-admissible covering of Ω (with $\zeta = \zeta_{m+1}$) and let us consider the contribution of each element. Surely, one should consider only elements $B(\bar z', \zeta)$ with $|D^\alpha V| \le c\zeta^{m-|\alpha|+1}$ $\forall \alpha : |\alpha| \le m$ under the assumption that for some β with $|\beta| = m$ the inequality $|D^\beta V| \ge \epsilon_0 \zeta$ holds. Making the dilatation $y = \frac{z'}{\zeta}$ and introducing the function $W(y) = V(\zeta y)\zeta^{-m-1}$ we see that the contribution of each element does not exceed $C\zeta^{2d-2} \int |\log \phi(y)| dy$ with $\phi(y) = \sum_{\alpha:|\alpha|=m} |D_y^\alpha W|$ and

(4.4.31) For some β with $|\beta| = m+1$, $|D_y^\beta W| \ge \epsilon_0$.

Therefore it is sufficient to prove that $\int |\log \phi(y)| dy \le C$ but this is obvious because in the appropriate coordinate system $\phi(y) \ge |y_1|$ (due to (4.4.29)). So, the induction step is complete and R_n are estimated. As a corollary we obtain that

(4.4.32) The contribution of the zone $\{z' \in \Omega, \gamma(z) \le \mu_n(z') = \sum_{2\le k\le n} \zeta_k^{p_k}, p_k = 2^{k-2}\}$ to the remainder estimate does not exceed $Ch^{1-d+\vartheta}$.

Thus, we have estimated the contribution of the zone $\{\sum_{|\alpha|=2} |D^\alpha V| \ge \gamma^\delta\}$ with arbitrarily small $\delta > 0$ properly. Now we want to estimate the contribution of the zone satisfying the opposite inequality

$$|D^\alpha V| \le c\gamma^\delta \quad \forall \alpha : |\alpha| = 2. \tag{4.4.33}$$

First of all we consider the part of this zone with

$$|\nabla V| \ge \gamma^{1+\delta'} \tag{4.4.34}$$

with $\delta' < \delta$. Let us consider a point $\bar z'$ satisfying both these inequalities. Without loss of generality one can assume that all the components of $\nabla V(\bar z')$ vanish excluding $\sigma = \partial_{z_3} V$. Then in the ρ-vicinity $\mathcal{U}$ of $\bar z'$

$$|\nabla V| \le C\sigma, \quad V(z) = V(\bar z') + \sigma(z_3 - \bar z_3) + O(\rho^2\gamma^\delta)$$

provided $\rho \le \min(\sigma\bar\gamma^{-\delta}, \bar\gamma^\delta)$. Let $\mathcal{V} = \mathcal{U} \cap \{|V| \le \sigma^{1-\delta''}\}$. One can see easily that $\operatorname{mes}_{2d-2} \mathcal{V} \le \gamma^{\delta''} \operatorname{mes}_{2d-2} \mathcal{U}$ with $\delta'' > 0$ provided $\frac{\rho}{\sigma} \le \gamma^{\delta''-\delta}$, $\frac{\sigma}{\rho} \le \gamma^{\delta''}$. These conditions are not contradictory for small enough $\delta' > 0$. Therefore, for any $\bar\gamma$ the contribution of the zone $\{z : \frac{1}{2}\bar\gamma \le \gamma(z) \le 2\bar\gamma, |\nabla^2 V| \le \gamma^\delta, |\nabla V| \ge \gamma^{1+\delta'}\}$ to the remainder estimate does not exceed $Ch^{1-d+\vartheta}\bar\gamma^{\delta''}$ and therefore the total contribution of the zone in question does not exceed $Ch^{1-d+\vartheta}$.

So, we should consider only the zone in which (4.4.33) and

$$|\nabla V| \le \gamma^{1+\delta} \tag{4.4.35}$$

hold with a small enough exponent $\delta > 0$. In this zone we apply theorem 4.3.13 after rescaling. One can easily see that the time rescaling coefficient is equal to $\frac{\rho}{\gamma} = 1$ in this case. If these inequalities are fulfilled at the point $\bar z$ they hold in its $\gamma^{1-\delta'}$-vicinity $\mathcal{V}$

with sufficiently small $\delta' > 0$ (we replace δ by $\frac{1}{2}\delta$ if it is necessary). Let us consider the Hamiltonian trajectory starting at the point $\bar{z}$ while it remains in $\mathcal{V}$. Inequalities (4.4.32) yield that $\frac{d}{dt}D^\alpha V = O(\gamma^{\delta+1})\ \forall\alpha : |\alpha| = 1$ and hence along the trajectory $\{V, z_j\} = O(\gamma^{1+\delta}) = O(\bar{\gamma}^{1+\delta'})$ and therefore

$$\frac{d}{dt}z_j = \pm\omega z_j + O(\gamma^{1+\delta'}) = O(\omega\gamma + \bar{\gamma}^{1+\delta'}) \tag{4.4.36}$$

where $\bar{\gamma} = \gamma(\bar{z})$, $\omega = |\{z_1, z_2\}(\bar{z})|$, $j = 1, 2$ and without any loss of generality one can assume that $\pm$ corresponds to $j = 1, 2$. Therefore $\gamma \le Ce^{\omega t}\bar{\gamma} \le \bar{\gamma}^{1-\delta''}$ with arbitrarily small $\delta'' > 0$ provided $|t| \le T = \epsilon\min(\omega^{-1}, \bar{\gamma}^{-\delta'''})$ with sufficiently small $\delta''' > 0$. Therefore $|\frac{d}{dt}z| \le C\gamma$ and hence $|z - \bar{z}| \le C\bar{\gamma}^{1-2\delta''}$. Therefore, for $t \le T$ the trajectory remains in an even smaller vicinity of $\bar{z}$ than $\mathcal{V}$ and hence surely does not leave $\mathcal{V}$.

Moreover, for infinitesimal variations we obtain the inequality $|\frac{d}{dt}dz| \le C(\gamma^\delta + \omega)|dz|$ and therefore $|dz(t)| \le C|dz(0)|$. Hence, $|D\Phi_t| \le C$.

Let us check the non-periodicity condition for $|t| \le T$. First of all, (4.4.33) yield that

$$z_j(t) = z_j(0)e^{\pm\omega t} + O(\bar{\gamma}^{1+\delta})$$

and therefore this condition is surely fulfilled unless

$$\omega|z_j(0)| \le C\bar{\gamma}^{1+\delta'} \qquad \forall j = 1, 2. \tag{4.4.37}$$

Let (4.4.37) be fulfilled. In this case along the trajectory $\gamma \asymp \bar{\gamma}$ and $z_j = \bar{z}_j + O(\gamma^{1+\delta})$. One should consider only points with $|\bar{z}_1| \ge \epsilon_0\bar{\gamma}$ (or with the same inequality for $\bar{z}_2$). Obviously there exists a linear function ℓ such that $\{\ell, z_1\} = \{\ell, V\} = 0$ and $\{\ell, z_2\} = 1$ at $\bar{z}$ in which case $\frac{d}{dt}\ell = z_1 + O(\gamma^{1+\delta})$ and therefore $\ell(z(t)) = \ell(z(0)) + z_1(0)t + O(\gamma^{1+\delta'})$ and the non-periodicity condition is fulfilled as well.

Therefore applying theorem 4.3.13 to elements of the partition one can take $T = \min(\gamma^{-\delta'''}, \omega^{-1}, (\frac{h}{\gamma^2})^{-\delta'''})$ (recall that the condition $T \le h_{\text{new}}^{-\delta'''}$ should be fulfilled). The corresponding remainder estimate for an element of the partition is $O(\frac{h}{\gamma^2})^{1-d+\vartheta}\gamma^{2\vartheta}T^{-1-\vartheta}$ and after summation over elements with

$$\omega^{1+\vartheta} \le |\log\gamma|^{-1} \tag{4.4.38}$$

we obtain the total remainder estimate $O(h^{1-d+\vartheta})$ and hence

(4.4.39) The total contribution of the points satisfying (4.4.38) to the remainder estimate is $O(h^{1-d+\vartheta})$.

So, we should now consider only points satisfying (4.4.32), (4.4.33) and

$$\omega = \{z_1, z_2\} \ge \gamma^{\delta'} \tag{4.4.40}$$

where one can choose $\delta' > 0$ arbitrarily.

Let us again consider the Hamiltonian flow. Recall that one can choose the time direction arbitrarily. Let us first consider the positive time direction assuming that

$$|z_1| \ge C_0\zeta \tag{4.4.41}$$

with $\zeta = \zeta_2$; otherwise we consider the opposite time direction and replace t by $-t$. We consider a trajectory while it does not leave the $\bar{\gamma}^\delta$-vicinity of $\bar{z}$ and while condition (4.4.41) remains true along it. Along such a trajectory $\omega = \bar{\omega} + O(\bar{\gamma}^\delta)$ and hence $|\frac{\omega}{\bar{\omega}} - 1| \le \varepsilon$ with arbitrarily chosen $\varepsilon > 0$ and C_0 depending on ε and $|z_2| \le c_0\zeta^2/|z_1| \le \epsilon|z_1|$. Therefore, along this trajectory

$$-\bar{\omega}(1+\varepsilon) \le \frac{d}{dt}\log|z_1| \le -\bar{\omega}(1-\varepsilon)$$

with an arbitrarily small constant $\varepsilon > 0$ and therefore

$$|\bar{z}_1|e^{-\bar{\omega}(1+\varepsilon)t} \le |z_1| \le |\bar{z}_1|e^{-\bar{\omega}(1-\varepsilon)t}. \tag{4.4.42}$$

Therefore condition (4.4.41) is the first condition to be violated. Moreover, at the moment $t = T_+(z)$ of the violation $\bar{\omega}t \asymp \log\frac{\bar{\gamma}}{\rho}$ and $|\frac{d}{dt}\frac{\zeta}{|z_1|}| \ge \frac{1}{2}\bar{\omega}$ where $\rho = \zeta(T_+(z'))$.

On the other hand, let us consider the negative time direction while $\gamma \le \epsilon_0$. One can see easily that this inequality holds for $0 \ge t \ge -T_-$ with $T_- = \epsilon_0\frac{1}{\bar{\omega}}|\log|\bar{z}_1||$.

Moreover, one can easily see that along trajectories $|D\Phi_t| \le Ce^{c\bar{\omega}|t|}$. One can easily check that the non-periodicity condition (4.2.38) is also fulfilled. Moreover, along a trajectory $h_{\text{eff}} = \frac{h}{|z_1|^2}$.

Therefore, in order to fulfill all the conditions of theorem 4.3.13 one should take positive or negative directions with $T_\pm$ replaced by $\min(T_\pm, T')$ with $T' = \epsilon'\frac{1}{\bar{\omega}}|\log\frac{h}{\bar{\gamma}^2}|$.

Then the contribution to the remainder estimate of the zone in question is equal to

$$h^{1-d+\vartheta}\int_{\Sigma_0\cap\{|z_1|\ge C_0\zeta^2\}} T^{-1-\vartheta}dxd\xi : da + \int\Bigl(\frac{h}{\gamma^2}\Bigr)^{1-d+\vartheta+\delta'''}\gamma^{2\vartheta-2d}dxd\xi$$

where $T = \min(T_- + T_+, T')$ with $\delta''' > 0$. The second term is obviously $O(h^{1-d+\vartheta})$. Replacing T by T' we obtain that the first integral is $O(1)$ provided either $\vartheta > 0$ or only the zone $\gamma \ge h^{\frac{1}{2}-\delta''}$ with arbitrarily small $\delta'' > 0$ is under consideration. Replacing T by $T_- + T_+$ we notice that this integral is asymptotically equivalent to

$$\int_{\Sigma_0\cap\{C_0\zeta^2\le|z_1|\le 2C_0\zeta^2\}} \omega^{1+\vartheta}|\log\rho|^{\vartheta}dxd\xi : da$$

because the Hamiltonian flow preserves Σ_0 and the measure $dxd\xi : da$ and the time necessary to intersect the zone $C_0\zeta \le \gamma \le \epsilon$ is $T_- + T_+ \asymp \frac{1}{\bar{\omega}}\log\rho$ while the time necessary to intersect the zone $C_0\zeta \le \gamma \le 2C_0\zeta$ is $\asymp \frac{1}{\bar{\omega}}$. Therefore *the contribution of the zone described by conditions* (4.4.32), (4.4.33), (4.4.40) *and* $\gamma \ge h^{\frac{1}{2}-\delta''}$ *to the remainder estimate does not exceed* $Ch^{1-d+\vartheta}$ *and we need only treat zones with* $\gamma \le h^{\frac{1}{2}-\delta''}$ *and only for* $\vartheta = 0$.

In order to treat such zones we first consider the same type of zone with δ_0 instead of δ''. Recall that δ_0, δ'' and δ' are arbitrarily small and should be chosen after δ. Let us apply the routine dilatation $x_{\text{new}} = \frac{x-\bar{x}}{\bar{\gamma}}$ and $h_{\text{new}} = \frac{h}{\bar{\gamma}^2}$ with $\bar{\gamma} = h^{\frac{1}{2}-\delta_0}$. We get an operator of the same type but with complete symbol $a = z_1z_2 + \bar{V} + \dots$ where $\{z_1, z_2\} = \bar{\omega} + \dots$ where the dots denote symbols all the derivatives of which do not exceed h^M with arbitrarily large M and are therefore negligible, $\bar{V}$ and $\bar{\omega}$ are constant and $\bar{V}$ is not necessarily a scalar; here h means h_{new} and $\bar{\omega} \ge h^{\delta'}$ and only the contribution of the zone $h^{\frac{1}{2}}\gamma \le h^{\frac{1}{2}-\delta''}$ need be treated (surely $\gamma = \gamma_{\text{new}} = \frac{\gamma_{\text{old}}}{\bar{\gamma}}$ as well and δ' and δ'' are new but still arbitrarily small). Now $\delta_0 = 0$.

Without loss of generality one can assume that $z_1 = hD_1$ and either $z_2 = \bar{\omega}x_1$ or $z_2 = \bar{\omega}x_1 + x_2$ (which can be arranged otherwise by an appropriate symplectic change of variables). Moreover, changing coordinates $x_{1\text{new}} = x_1 + \bar{\omega}^{-1}x_2$, $x'_{\text{new}} = x'$ if necessary and multiplying the operator by $\frac{1}{\bar{\omega}}$ we get exactly the operator $x_1hD_1 + \frac{ih}{2} + \bar{V}$ (modulo negligible) with $\bar{V}_{\text{new}} = \frac{1}{\bar{\omega}}\bar{V}$.

All that remains to do is to consider points $(\bar{x}, \bar{\xi})$ with $\bar{\gamma} = |\bar{x}_1| \ge |\bar{\xi}_1|$ (points with the opposite inequality are treated similarly with negative time direction) and to prove that one can take $T = \epsilon|\log h|$ instead of $T = \epsilon|\log(\frac{h}{\bar{\gamma}^2})|$. But for this very specific operator the proof is direct: one can consider operators of multiplication; then $e^{-ith^{-1}A}\phi e^{ith^{-1}A} \equiv \phi_t$ with $\phi_t(x) = \phi(e^{-t}x)$ and for the indicated $t \ge \epsilon$ we obtain that $\phi_t \phi = 0$ provided $\operatorname{supp}\phi$ is small enough. This proves that the contribution of the element of the partition in question to the remainder estimate is $O(h^{1-d}\gamma^{2d-2}\log^{-1}h)$ (recall that $\vartheta = 0$) and the total contribution of such elements is $O(h^{1-d})$ as well. □

Remark 4.4.8. *In the previous version of the proof given under the assumption that $V = 0$ identically, the canonical form of such operators was used. Namely, it was proven (for similar proofs see, e.g., section 6.2) that*

$$\mathcal{F}^* A \mathcal{F} \equiv \sum_{m,n:m+n\ge 1} a_{mn}(x', hD')(x_1 hD_1 - \frac{1}{2}ih)^m h^n \tag{4.4.43}$$

for an appropriate Fourier integral operator $\mathcal{F}$ such that in the vicinity of the point in question $\mathcal{F}^*\mathcal{F} \equiv I$. ◊

4.4.3 From the condition "rank Hess $a \ge 2$" to "rank Hess $a \ge 1$" at critical points

Now we will consider operators with stronger degenerations and we consider only scalar operators. I will indicate later what order of matrix symbol can be added. So, let us assume that

(4.4.44) A is a scalar operator.[36)]

In this subsection we assume that condition $(4.3.16)_1$ is fulfilled. Then locally in the appropriate coordinate system

$$a = z_1^2 + V(z') \tag{4.4.45}$$

where here and below a means the complete symbol of A. However we need to pass to symplectic coordinates with $z_1 = \xi_1$. By the Malgrange preparation theorem [Malgrange 1] $a = b(x,\xi)((\xi_1 - \alpha(x,\xi'))^2 + W(x,\xi'))$ and after the symplectic transformation $\xi_{1\text{new}} = \xi_1 - \alpha(x,\xi')$, $(x,\xi')_{\text{new}} = \varphi(x,\xi')$ we obtain the same formula with $\alpha = 0$. Thus, without loss of generality one can assume that

$$a(x,\xi) = b(x,\xi)(\xi_1^2 + V(x,\xi')), \qquad b(x,\xi) \ge \epsilon_0. \tag{4.4.46}$$

Theorem 4.4.9. *Let conditions* (4.4.40) *and* $(4.4.16)_1$ *be fulfilled.*

(i) Then for $\vartheta > 0$ estimate (4.3.18) *holds.*

(ii) If

$$\sum_{|\alpha|\le m} |D^\alpha V| \ge \epsilon \tag*{(4.4.47)$_m$}$$

with some $m \ge 2$ then the estimate

$$\mathcal{R}_2 \le Ch^{1-d}|\log h|^{m-2} \tag*{(4.4.48)$_m$}$$

holds for $\vartheta = 0$. Moreover, estimate (4.3.23) *remains true.*

(iii) In the general case the estimate

$$\mathcal{R}_2 \le Ch^{1-d-\delta} \tag*{(4.4.49)$_\infty$}$$

holds for $\vartheta = 0$ and arbitrarily small $\delta > 0$.

In the smoothness conditions K depends on m and δ.

36) In fact, one need assume that this condition is fulfilled only modulo $O(h^2)$.

Proof. First of all, this theorem is proven under condition $(4.4.47)_m$ for $m = 2$ (see above). Let us assume that it is proven for given $m = \bar{m} - 1 \geq 2$ and let us prove it for $m = \bar{m}$. Let us introduce the function $\gamma = \zeta_m$ by formula (4.4.27); then $|\nabla \gamma| \leq \frac{1}{2}$ and

(4.4.49) $$|D^\alpha V| \leq C\rho^2 \gamma^{-|\alpha|} \qquad \forall \alpha : |\alpha| \leq K$$

with $\rho = \gamma^{\frac{m}{2}}$. Let us consider the point $(\bar{x}, \bar{\xi}')$. Let $\mathcal{U}'$ be its $\bar{\gamma}$-vicinity (with respect to (x, ξ')) with $\bar{\gamma} = \gamma(\bar{x}, \bar{\xi}') \geq h^{\frac{1}{m}}$. First let $|\bar{\xi}_1| \geq C_0 \bar{\rho}$ with $\bar{\rho} = \bar{\gamma}^{\frac{m}{2}}$ and let $\mathcal{U} = \{(x, \xi) : (x, \xi') \in \mathcal{U}', |\xi_1 - \bar{\xi}_1| \leq \frac{1}{2}|\bar{\xi}_1|\}$. Let us rescale $x_{\text{new}} = \frac{x - \bar{x}}{\gamma}$, divide the operator by ρ^{-2} and introduce $h_{\text{new}} = \frac{h}{\rho\gamma}$ where we skip the bar over γ and ρ for the sake of simplicity. Let us note that the ellipticity condition is fulfilled and $\frac{h}{\gamma^2} \leq h^{1 - \frac{2}{m}}$. Therefore the contribution of $\mathcal{U}$ to the remainder estimate does not exceed $C\rho^{2\vartheta}(\frac{h}{\rho\gamma})^s (\frac{\gamma}{\rho})^{d-1}$ (where the factor $(\frac{\gamma}{\rho})^{d-1}$ appears since $\mathcal{U}'$ is covered by this number of balls of radius ρ on ξ') with arbitrarily large s and after summation on a ρ-admissible partition of unity in $\{(x, \xi) : (x, \xi') \in \mathcal{U}', |\xi_1| \geq c\rho\}$ we obtain that the contribution of this zone to the remainder estimate does not exceed

$$R_2 = C\bar{\rho}^{2\vartheta} \bigl(\frac{h}{\rho\gamma}\bigr)^{s-1} \bigl(\frac{\gamma}{\rho}\bigr)^{d-1}.$$

Let us consider $\mathcal{U} = \{(x, \xi) : (x, \xi') \in \mathcal{U}', |\xi_1| \leq c\bar{\rho}|\}$. Let us rescale and multiply the operator as above with $\rho = \bar{\rho}$. Let us note that if $\bar{\gamma} \geq ch^{\frac{1}{m}}$ then condition $(4.4.47)_{m-1}$ is fulfilled[37] and the contribution of $\mathcal{U}$ to the remainder estimate does not exceed

$$R_1 = C\rho^{2\vartheta} \bigl(\frac{h}{\gamma\rho}\bigr)^{1-d+\vartheta} \bigl(\frac{\gamma}{\rho}\bigr)^{d-1}$$

for $\vartheta > 0$; for $\vartheta = 0$ the additional factor $\ell_{m-1} = |\log \frac{h}{\rho\gamma}|^{m-3}$ should be added.

If $\bar{\gamma} \leq ch^{\frac{1}{m}}$ then $h_{\text{new}} \asymp 1$ and one need not check condition $(4.4.47)_{m-1}$ and therefore the contribution of this element of the partition to the remainder estimate does not exceed R_1 as well.

So, the contribution of $\mathcal{V} = \{(x, \xi) \in \Omega, (x, \xi') \in \mathcal{U}'\}$ to the remainder estimate does not exceed $R_1 + R_2$ and after summation on a γ-admissible partition we get the remainder estimate

(4.4.50) $$C \int (R_1 + R_2)\gamma^{1-2d} dx d\xi' \asymp h^{1-d+\vartheta} \int \gamma^{-1-\vartheta} \rho^\vartheta dx d\xi';$$

for $\vartheta = 0$ one should add factor ℓ_{m-1} as well.

In this deduction we have not yet applied condition $(4.4.47)_m$. Now, let us note that this condition yields that $\gamma \geq |z_2|$ in the appropriate coordinates in (x, ξ')-space and then for $\vartheta > 0$ the integral in the right-hand expression of (4.4.50) is bounded (since $\rho = \gamma^{\frac{m}{2}}$ with $m \geq 3$) and for $\vartheta = 0$ this integral does not exceed $C|\log h|$ (multiplied by ℓ_{m-1} it gives ℓ_m). Therefore the induction step is done. So, *the theorem is proven under condition* $(4.4.47)_m$ *with arbitrary* n.

In order to treat the general case let us apply remainder estimate (4.4.50) with large enough n. Since the integrand in the right-hand expression of this estimate does not exceed 1 and $h^{-\delta}$ for $\vartheta > 0, m\vartheta \geq 1$ and for $\vartheta = 0, m\delta \geq 1$ respectively we obtain the corresponding remainder estimate.

---

[37] To be exact, this condition is fulfilled if $\gamma' \geq \epsilon\bar{\gamma}$ where γ' is introduced by the same formula (4.4.27) with differentiation only on x. However before rescaling one can apply the symplectic transformation $(x_j, \xi_j) \to (\xi_j, -x_j)$ with some $j = 2, \dots, d$ if necessary and fulfill this condition.

Now, our goal is to get rid of logarithmic factor for $\vartheta = 0$ under condition $(4.4.47)_m$. First of all, without loss of generality one can assume that

$$(4.4.51)_m \qquad |\partial_{z_2}^m V| \ge \epsilon_0$$

(condition $(4.4.47)_m$ yields this condition after ϵ-partition and appropriate choice of coordinate system). The Malgrange preparation theorem yields that

$$V = \nu(z')(z_2^m + \mu_1(z'')z_2^{n-1} + \dots + \mu_m(z''))$$

with $|\nu(z'')| \ge \epsilon$, $z'' = (z_3, \dots, z_{2d})$. Without loss of generality one can assume that $\mu_1 = 0$. Then on the induction step $\gamma_m = \epsilon_1|z_2| + \gamma'_m(z'')$ with

$$(4.4.52) \qquad \gamma'_{ml} = \epsilon\Big(\sum_{2\le k\le m} \sum_{|\alpha|\le l(k-1)} |D^\alpha \mu_k|^{\frac{ml}{kl-|\alpha|}}\Big)^{\frac{1}{ml}}$$

with $l = 1$. Then $|\nabla \gamma'_m| \le \frac{1}{2}$. These arguments coincide with the previous proof. However, let us apply the induction hypothesis only to the estimate of the contribution of zone $\{|z_2| \le C_0\gamma'_m(z'')\}$. Then no additional logarithmic factor appears. But in the remaining zone $|V| \ge \epsilon_2\rho_m^2$ and we should refer to theorem 4.3.13 rather than the induction hypothesis. This yields immediately that *estimate* $(4.4.48)_1$ *holds for any* $m \ge 3$.

In order to get rid of the logarithmic factor here we need to apply propagation of singularities arguments. It is important that after rescaling on x and ξ, τ (the latter two mean multiplication of the operator by ρ_m^{-2}) the time is also rescaled: $t_{\text{new}} = t_{\text{old}}\rho_m\gamma_m^{-1} = t_{\text{old}}\gamma_m^{\frac{m}{2}-1}$. However, first let us note that exactly the same arguments as in the proof of theorem 4.4.6 yield that *the contribution of the zone* $\{|z_2| \le C\gamma'^M_{ml}\}$ *to the remainder estimate does not exceed* Ch^{1-d} *for any* $l > 1$ *and for sufficiently large* $M = M_l$; moreover, *the contribution to the remainder estimate of the complement of the zone, given by the inequalities*

$$(4.4.53) \qquad |D^\alpha \mu_k| \le C|z_2|^{k-|\alpha|(1-\delta)} \qquad \forall k\ \forall \alpha : |\alpha| \le k$$

with sufficiently small $\delta > 0$, *does not exceed* Ch^{1-d} as well.

So, we should consider only starting points satisfying (4.4.53) and

$$(4.4.54) \qquad |z_1| \ge C_0\zeta$$

with arbitrarily large C_0 and $\zeta = \gamma'_2$. Then

$$(4.4.55) \qquad |V(z')| \ge \epsilon_0\rho_m^2.$$

Let us consider the Hamiltonian trajectory while $|t_{\text{new}}| \le \bar\gamma^{-\delta''} = t_1$ with sufficiently small $\delta'' > 0$, (4.4.54) holds and $\gamma \le c\bar\gamma$ (which will be violated first). Then along this trajectory $\{\xi_1, z_1\} = \bar\omega + O(\bar\gamma^\delta)$ where $\bar\omega = \{\xi_1, z_1\}(\bar z)$. One can easily see that for $|\bar\omega| \le \bar\gamma^{\delta'''}$ with arbitrarily small $\delta''' > 0$ one can take $t_{\text{new}} = t_1$ with sufficiently small $\delta'' > 0$. Furthermore, one can easily see that $\int_\Sigma \frac{1}{t_1} dxd\xi : da$ is bounded.

On the other hand, for $|\bar\omega| \ge \bar\gamma^{\delta'''}$ a trajectory launched in the appropriate direction meets $\{|z_1| = C_0\zeta\}$ (transversally provided C_0 is large enough). So, let us take $t = \max(t_+, t_-)$. We obtain that

$$\int_\Sigma \frac{1}{t_{\text{new}}(z)} dxd\xi : da \asymp$$

$$\int_{\Sigma\cap\{C_0\zeta\le|z_1|\le 2C_0\zeta\}} \Big(\int_\zeta^1 z_1^{-\frac{m}{2}}\zeta^{\frac{m}{2}-1}dz_1\Big)\frac{1}{t_{\text{new}}(z)} dxd\xi : da$$

because the $dxd\xi : da$ measure along the trajectory is invariant and because along the trajectory $|\frac{d}{dt}z_1| \asymp \bar{\omega}|z_1|^{\frac{m}{2}}$; here $t_\pm$ means the time to leave the appropriate zone in the appropriate direction and $t = t_+ + t_-$. The integral in parentheses is bounded and hence the right-hand expression does not exceed $C\int_{\{C_0\zeta \le |z_1| \le 2C_0\zeta\}} dxd\xi : da$ which is bounded as well.

We neglected three other conditions; the first is the condition $|t_{\text{new}}| \le t_2 = h_{\text{new}}^{-\delta}$ with $h_{\text{new}} = \frac{h}{\bar{\gamma}^{\frac{m}{2}+1}}$ and with sufficiently small $\delta' > 0$. One can see easily that $\int_\Sigma \frac{1}{t_2} dxd\xi : da$ is bounded. The second is the condition on the Jacobi matrix. I leave to the reader the easy proof that for $|\bar{\omega}| \le \bar{\gamma}^{\delta'''}$ and for $|t_{\text{new}}| \le \bar{\gamma}^{-\delta''}$ with sufficiently small $\delta'' > 0$ the Jacobi matrix is bounded. For $|\bar{\omega}| \ge \bar{\gamma}^{\delta'''}$ one can introduce $t' = \bar{\omega}t$ and it is easy to prove that the norm of the Jacobi matrix does not exceed $\langle t' \rangle^M$ and hence this condition is fulfilled for $|t_{\text{new}}| \le t_2$. So this condition does not cause any trouble. The third condition is the non-periodicity condition. Since $\frac{d}{dt_{\text{old}}} z_1 = (\omega + O(z_1^\delta))z_1^{\frac{m}{2}}$ we obtain that for $|\bar{\omega}| \ge \bar{z}_1^{\delta_0} + h_{\text{new}}^{\delta_0}$ with sufficiently small $\delta_0 > 0$ this condition is surely fulfilled. On the other hand, if $|\bar{\omega}| \le \bar{z}_1^{\delta_0} + h_{\text{new}}^{\delta_0}$ and $\bar{\gamma}$ and h_{new} are less than ε' then along the trajectory introduced above $|z(t) - \bar{z} - H_a(\bar{z})t_{\text{old}}| \le \varepsilon |t_{\text{old}}|$ where $\varepsilon > 0$ is an arbitrarily small constant and $\varepsilon' > 0$ is a constant depending on ε. Hence the non-periodicity condition is fulfilled again. So, the non-periodicity condition also does not cause any trouble. □

Remark 4.4.10. *In the case* $V = \text{const}$ *(treated as a parameter) one can divide the operator by* $\rho^2 = \max(|V|, h^2)$ *and introduce* $h_{\text{new}} = \frac{h}{\rho}$ *(no rescaling on* x *is necessary). Then for* $\vartheta = 0$ *the remainder estimate* $O(h^{1-d})$ *holds as well. Therefore for operators with analytic symbols the remainder estimate* $O(h^2)$ *holds under condition* $(4.4.16)_1$. ◇

4.4.4 From the condition "rank Hess $a \ge 1$ at critical points" to a weak non-degeneracy condition

Now even the condition "rank Hess $a \ge 1$" seems to be too restrictive for us and I am going to get rid of it and to obtain the estimate

$$\mathcal{R}_\vartheta \le Ch^{1-d+\vartheta-\sigma} \tag{4.4.56}$$

with $\sigma = 0$ for $\vartheta > 0$ and with arbitrarily small $\sigma > 0$ for $\vartheta = 0$ under the condition

$$\sum_{|\alpha| \le n} |D^\alpha a| \ge \epsilon_0 \tag*{$(4.4.57)_n$}$$

with $n \ge 3$. We apply induction on n. So suppose the case with condition $(4.4.53)_{n-1}$ has been treated and let us consider some point $\bar{z}$ at which condition $(4.4.56)_n$ holds but $(4.4.56)_{n-1}$ fails. Then in the vicinity in the appropriate symplectic coordinates

$$a(x,\xi) = b(x,\xi) \sum_{0 \le j \le n} \xi_1^{n-j} a_j(x,\xi'), \qquad |b(x,\xi)| \ge \epsilon_0 \tag{4.4.58}$$

with

$$a_0 = 1, \qquad a_1 = 0 \tag{4.4.59}$$

where we have applied the Malgrange preparation theorem again. Now instead of V we have a few functions a_j and we want to prove remainder estimate (4.4.56) under the condition

$$\sum_{2 \le j \le n, |\alpha| \le \nu j} |D^\alpha a_j| \ge \epsilon_0 \tag*{$(4.4.60)_{n,\nu}$}$$

with $\nu \geq 1$ (later we avoid it for large ν). First let us treat the case $\nu = 1$. Let us introduce the functions

$$\gamma = \epsilon\Big(\sum_{2\leq j\leq n, |\alpha|\leq j-1} |D^\alpha a_j|^{\frac{n}{j-|\alpha|}}\Big)^{\frac{1}{n}} + h^{\frac{1}{2}} \tag{4.4.61}$$

and $\rho = \gamma$. Then $|\nabla\gamma| \leq \frac{1}{2}$ for small enough $\epsilon > 0$ and

$$|D^\beta a_j| \leq c\rho^j \gamma^{-|\beta|} \qquad \forall j, \beta. \tag{4.4.62}$$

Let us rescale $x_{\text{new}} = \frac{x-\bar{x}}{\gamma}$, multiply the operator by ρ^{-n} and introduce $h_{\text{new}} = \frac{h}{\rho\gamma}$. Then condition $(4.4.60)_{n,1}$ yields that for $\gamma \geq \bar{\gamma} = ch^{\frac{1}{2}}$ after this procedure the smoothness conditions and condition $(4.4.57)_{n-1}$ will be fulfilled and hence the contribution of the zone $\{z, z' \in B(\bar{z}', \gamma(\bar{z}')), |z_1| \leq c\rho\}$ to the remainder estimate will be

$$\asymp \Big(\frac{h}{\rho\gamma}\Big)^{-d+1+\vartheta-\sigma} \gamma^{d-1}\rho^{n\vartheta-d+1} = h^{-d+1+\vartheta-\sigma}\rho^{(n-1)\vartheta+\sigma}\gamma^{2d-1-\vartheta+\sigma} \tag{4.4.63}$$

where later ρ will be different from γ. Therefore the total contribution of the zone $\{z : |z_1| \leq c\rho, \rho\gamma \geq h\}$ to the remainder estimate is

$$Ch^{1-d+\vartheta-\sigma} \int_{\rho\gamma\geq h} \gamma^{-1-\vartheta+\sigma}\rho^{(n-1)\vartheta+\sigma}\, dz'. \tag{4.4.64}$$

For $\rho\gamma \asymp h, |z_1| \leq c\rho$ we obtain $h_{\text{new}} = 1$ and the contribution of the corresponding element of the partition to the remainder estimate does not exceed $C\rho^{m\vartheta}$ and the total contribution of this zone does not exceed

$$C\bar{\rho}^{n\vartheta}\bar{\gamma}^{1-2d}\,\text{mes}_{2d-1}\{z : \rho\gamma \leq ch\} \tag{4.4.65}$$

with $\bar{\rho} = \bar{\gamma} = h^{\frac{1}{2}}$ here. Finally, in the zone $\{|z_1| \geq c\rho\}$ we apply an $\epsilon|z_1|$-admissible partition and standard elliptic arguments and then the contribution of each element does not exceed $C|z_1|^{n\vartheta}\big(\frac{h}{z_1^2}\big)^s$ with arbitrarily large s and therefore the total contribution does not exceed the sum of (4.4.64) and (4.4.65).

Remark 4.4.11. *(i) Let us note that under condition* $(4.4.60)_{nm,1}$ *in the appropriate coordinate system* $\gamma \geq \epsilon'|z_2|$ *and then (4.4.64) and (4.4.65) with* $\rho = \gamma$ *do not exceed* $Ch^{1-d+\vartheta-\sigma}$. *Therefore, assuming that under condition* $(4.4.57)_{n-1}$ *remainder estimate (4.4.56) was proven, we have proven this estimate under conditions* $(4.4.57)_n$ *and* $(4.4.60)_{n,1}$. *In the next step we apply induction on* ν *as well.*

(ii) Moreover, let us note that for $\vartheta > 0$ *we obtain remainder estimate (4.4.56) even without condition* $(4.4.60)_{n,1}$ *for* $n > n_\vartheta$. *So, for* $\vartheta > 0$ *the estimate will be proven for any* n *if it was proven for* $n \leq n_\vartheta$. *Later for* $\vartheta > 0$ *we will get rid of condition* $(4.4.57)_n$ *completely!* ◊

Now, *let us assume that estimate (4.4.56) was proven under* $(4.4.57)_{n-1}$ *and under* $(4.4.57)_n$ *and* $(4.4.60)_{n,\nu}$ *and now our goal is to prove it under conditions* $(4.4.57)_n$ *and* $(4.4.60)_{n,\nu+\mu}$ *and to get rid of the latter condition for* $\nu \geq \nu_{n,\vartheta}$. Here $\mu = \frac{1}{n!}$. Let us introduce the function

$$\gamma(z') = \epsilon\Big(\sum_{2\leq j\leq n, |\alpha|\leq j\nu} |D^\alpha a_j|^{\frac{L}{j\nu-|\alpha|}}\Big)^{\frac{1}{L}} + h^{\frac{1}{\nu+1}} \tag{4.4.66}$$

with $L = n\nu$ and small enough $\epsilon > 0$. Let $\rho = \gamma^\nu$. One can easily see that $|\nabla\gamma| \leq \frac{1}{2}$ and that

$$|D^\alpha a_j| \le c\rho^j \gamma^{-|\alpha|} \qquad \forall j,k : 2 \le j \le n, |\alpha| \le K. \tag{4.4.67}$$

Moreover, for some j and $\alpha : |\alpha| < \nu j$ the opposite inequality holds (with $\epsilon_1 > 0$ instead of c) provided $\gamma \ge ch^{\frac{1}{1+\nu}}$. Therefore, after the routine rescaling procedure with $x_{\text{new}} = \frac{x-\bar{x}}{\gamma}$, $h_{\text{new}} = \frac{h}{\rho\gamma}$ and with multiplication operator ρ^{-n} we obtain the routine smoothness conditions and either condition $(4.4.60)_{n,\nu}$ or $h_{\text{new}} \asymp 1$; we consider now the zone $\{|\xi_1| \le c\rho\}$ and if necessary apply a linear symplectic transformation for (x', ξ'). In the both cases the contributions of the corresponding balls was described above and the contributions of the corresponding zones are given by (4.4.64) and (4.4.65). In the zone $\{|\xi_1| \ge c\rho\}$ elliptic arguments are applicable and taking into account that $\gamma^2 \ge h^{1-\sigma}$ for $\nu > 1$ we easily obtain that the contribution of this zone does not exceed the sum of (4.4.64) and (4.4.65) as well. So, the total remainder estimate is given by the sum of (4.4.64) and (4.4.65). Recall that this deduction holds under the assumption that for given $\nu \ge 1$ condition $(4.4.60)_{n,\nu}$ yields remainder estimate (4.4.56). Let us assume now that condition $(4.4.60)_{n,\bar{\nu}}$ holds. Then there exist α and j with $|\alpha| \le \bar{\nu} j$ such that

$$|D^\alpha a_j| \ge \epsilon.$$

Then in appropriate coordinates

$$|D^\beta a_j| \ge \epsilon |z_2|$$

for some $\beta : |\beta| = |\alpha| - 1$ and picking $\nu = \frac{|\alpha|-1}{j}$ we obtain that $\gamma \ge \epsilon|z_2|$ and the remainder estimate (4.4.56) for $\nu = \bar{\nu}$ provided it holds for ν. We can restrict ourselves to numbers $\nu = \frac{m}{n!}$ with integers $m \ge n$ and apply induction on m. Moreover, for $\vartheta > 0$ and $m \ge m_{\vartheta,n}$ none of the final estimates linked with z_2 and condition $(4.4.60)_{n,\nu}$ are necessary. These estimates can also be avoided in the case $\vartheta = 0$ as well (for $m \ge m_{\sigma,n}$) by doubling σ. Therefore the induction step from $n-1$ to n is complete and the following statement is proven:

Theorem 4.4.11. *Suppose condition* $(4.4.57)_n$ *is fulfilled for some* n. *Then estimate* (4.4.56) *holds with* $\sigma = 0$ *for* $\vartheta > 0$ *and arbitrarily small* $\sigma > 0$ *for* $\vartheta = 0$.

Now let us perform the case $\vartheta > 0$ completely:

Theorem 4.4.12. *For* $\vartheta > 0$ *estimate* (4.4.56) *holds with no additional assumption.*

Proof. Let us introduce the functions

$$\gamma = \epsilon\Bigl(\sum_{\alpha:|\alpha|\le n-1} |D^\alpha a|^{\frac{n}{n-|\alpha|}}\Bigr)^{\frac{1}{n}} + h^{\frac{1}{2}}.$$

Then $|\nabla\gamma| \le \frac{1}{2}$ and

$$|D^\alpha a| \le c\gamma^{n-|\alpha|} \qquad \forall \alpha$$

and either $\gamma \le ch^{\frac{1}{2}}$ or the opposite estimate holds for some α with $|\alpha| < n$. Then applying the rescaling procedure $x_{\text{new}} = \frac{x-\bar{x}}{\gamma}$, $h_{\text{new}} = \frac{h}{\gamma^2}$ and multiplying the operator by γ^{-n} we get into the situation of $(4.4.57)_n$ or $h \asymp 1$ respectively and applying estimate (4.4.56) we obtain after summation on the partition the remainder estimate (4.4.56) again for $\vartheta > 0$ and $n \ge n_\vartheta$. However, for $\vartheta = 0$ we gain nothing here. □

Remark 4.4.13. *(i) We obtain the unpleasant factor* $h^{-\sigma}$ *for* $\vartheta = 0$ *only when we get rid of condition* $(4.4.60)_{n,\nu}$. *All other steps of our analysis give only additional factors*

$|\log h|^M$. *However, we cannot go from condition* $(4.4.60)_{n-1,\nu}$ *to* $(4.4.60)_{n,1}$ *unless we assume that the condition*

$$(4.4.68)_s \qquad \sum_{2\le j\le n-s, |\alpha|\le m} |D^\alpha a_j| \ge \epsilon$$

holds with $s=1$. *Therefore in order to provide the remainder estimate* $O(h^{1-d}|\log h|^M)$ *one should assume that condition* $(4.4.68)_s$ *is fulfilled for* $s=n-2$*; under this condition one can easily prove the estimate.*

(ii) On the other hand, if we assume that $a_2,\dots,a_{n-s-1}$ *vanish identically while condition* $(4.4.68)_s$ *holds, we notice that the problem is reduced (after a few steps of the induction described above) to the problem with* $a=b\xi_1^r$ *with* $r=n-s-1$ *and non-vanishing symbol* b. *Then the same partition arguments yield the remainder estimate* $O(h^{1-d})$ *because the zone* $\{|\xi_1|\ge h\}$ *is elliptic. Therefore estimate (4.3.23) remains true but I do not know whether it is uniform with respect to the spectral parameter.*

(iii) In certain cases (under some additional conditions) propagation of singularities arguments provide the remainder estimate $O(h^{1-d})$ *but I leave this analysis to the reader.*
◊

Remark 4.4.14. *(i) One can replace condition* $(4.4.56)_m$ *by the conditions*

$$(4.4.69) \qquad |D^\alpha a| \le c\nu_{|\alpha|} \qquad \forall\alpha : |\alpha|\le K,$$

$$(4.4.70) \qquad \sum_{\alpha:|\alpha|\le m} \frac{1}{\nu_{|\alpha|}}|D^\alpha a| \ge \epsilon_0$$

with $\nu_1\ge\nu_2\ge\dots\ge h^l$ *and* $K=K(d,l,m,\vartheta)$. *In fact, under these conditions either* $\nu_1\le C\nu_m$ *or the conditions are fulfilled with* $m-1$ *instead of* m *and one can launch induction on* m.

(ii) Similarly, one can replace condition $(4.4.47)_m$ *by (4.4.69)–(4.4.70) for* V *with* $\nu_1\ge\nu_2\ge\dots$. ◊

4.4.5 From microlocal to local spectral asymptotics

Let us go from microlocal to local spectral asymptotics. Then the results of the previous subsections combined with the results of section 4.1 imply

Theorem 4.4.15. *Let condition (4.1.29) be fulfilled. Then the assertions of all the above statements remain true for* $Q_1=Q_2=I$ *provided their conditions are fulfilled for* $\Omega=T^*B(0,1)$.[38] *Moreover if condition* $(4.1.12)_\pm$ *holds then in every estimate without mollification with respect to* τ_1 *one can take* $\tau_1=\mp\infty$ *and in every estimate with mollification one can drop the mollification and take* $\tau_1=\mp\infty$.

4.5 Local Spectral Asymptotics for the Schrödinger and Dirac Operators

In this section we treat the Schrödinger operator

$$(4.5.1) \qquad A = P_j g^{jk} P_k + V, \qquad P_j = hD_j - A_j$$

[38] In fact, it is necessary to check them only for $\Omega=B(0,1)\times B(0,C_0)$ with a large enough C_0.

and Dirac operator

$$A = \frac{1}{2}\sigma_l(\omega^{jl}P_j + P_j\omega^{jl}) + \sigma_0 m + V \cdot I \tag{4.5.2}$$

where we use the Einstein summation convention and *the components* g^{jk} *of the inverse metric tensor,* ω^{jl}*, the components of the vector-potential* V_j and *the scalar potential* V are real-valued functions, σ_l are *Dirac matrices* (see subsection 4.5.2 and Appendix C), $m = \text{const} \ge 0$ is a parameter, and $h \in [0,1]$.

For technical reasons we also consider the operator

$$\mathbf{A} = J^{-\frac{1}{2}}(A + W\sigma_0)J^{-\frac{1}{2}} \tag{4.5.3}$$

with $J = \frac{1}{2}J_+(I+\sigma_0) + \frac{1}{2}J_-(I-\sigma_0)$ where $J_\pm$ are positive-valued functions. We assume that

(4.5.4) A is self-adjoint operator in $L^2(X,\mathbb{H})$, $\mathbb{H} = \mathbb{C}^D$, $D = 1$ for the Schrödinger operator.

4.5.1 The Schrödinger operator

Let us start with the Schrödinger operator assuming that

$$B(0,1) \subset X, \tag{4.5.5}$$

$$|D^\alpha g^{jk}| \le c, \quad |D^\alpha V_j| \le c, \quad |D^\alpha V| \le c \quad \forall \alpha : |\alpha| \le K, \tag{4.5.6$_{1-3}$}$$

$$\epsilon_0 \le g^{jk}(x)\xi_j\xi_k|\xi|^{-2} \le c \quad \forall x \in B(0,1) \quad \forall \xi \in \mathbb{R}^d \setminus 0. \tag{4.5.7}$$

Remark 4.5.1. *A (unitary) gradient transformation* $u \to e^{ih^{-1}\varphi(x)}u$ *with a real-valued function* φ *can be applied which for the Schrödinger and Dirac operators is equivalent to the transformation* $V_j \mapsto V_j + \partial_j\varphi$*. Hence one can weaken condition* (4.5.6)$_2$*, replacing it with*

$$|D^\alpha F_{jk}| \le c \quad \forall \alpha : |\alpha| \le K \tag{4.5.6$'_2$}$$

where $F_{jk} = \partial_j V_k - \partial_k V_j$ *are components of the tensor magnetic intensity.* ◊

Here the principal symbol of A equals

$$a(x,\xi) = g^{jk}(\xi_j - V_j)(\xi_k - V_k) + V \tag{4.5.8}$$

and the subprincipal symbol vanishes and hence the coefficients in the spectral asymptotics are

$$\kappa_0(x,\tau) = (2\pi)^{-d}\omega_d(\tau - V)_+^{d/2}\sqrt{g}, \tag{4.5.9}$$

$$\kappa_1(x,\tau) = 0 \tag{4.5.10}$$

where $\tau_1 = -\infty$, $\tau_2 = \tau$, ω_d is the volume of the unit ball in $\mathbb{R}^d$ and $g = \det(g^{jk})^{-1}$.

Then the results of the previous section immediately yield the following assertions.

Theorem 4.5.2. *Let conditions (4.5.1), (4.5.4), (4.5.5), (4.5.6) and (4.5.7) be fulfilled and $d \geq 2$. Then*

(i) The following estimates hold:

$$\mathcal{R} = |\int \psi(x)(e(x,x,\tau) - h^{-d}\kappa_0(x,\tau))dx| \leq C_0 M^0 h^{1-d} + CMh^{1-d+\delta'}, \tag{4.5.11}$$

$$\mathcal{R}_\vartheta = |\tau_+^{\vartheta-1} * \int \psi(x)\bigl(e(x,x,\tau) - \sum_{0\leq n\leq 1+\vartheta} h^{-d+n}\kappa_n(x,\tau)\bigr)dx| \leq C_0 M^0 h^{1-d+\vartheta} + CMh^{1-d+\vartheta+\delta'}, \tag{4.5.12}$$

where here and below $C = C(d,c)$, $K = K(d)$, $\delta' = \delta'(d) > 0$ is small enough and $M^0 = \sup_x |\psi(x)|$, $M = \|\psi\|_{C^K}$;

(ii) Let $\Lambda_1,\dots,\Lambda_n$ be closed subsets of $\Sigma_\tau = \{(x,\xi) \in T^*B(0,1), a(x,\xi) = \tau\}$ and $\Lambda_j \subset \{\gamma^2 = |\tau - V| + |\nabla_x V|^2 \geq h^{1-\delta}\}$ with $\delta > 0$ and let there pass through each point of Λ_j a Hamiltonian trajectory of $a(x,\xi)$ with $t \in \mathcal{J}_j, \mathcal{J}_j = [0,T_j]$ or $\mathcal{J}_j = [-T_j,0]$, $T_j \in [1, h^{-\delta'}]$ along which

$$|D^\alpha g^{jk}| \leq ch^{-\delta'}, \quad |D^\alpha V_j| \leq ch^{-\delta'}, \quad |D^\alpha V| \leq ch^{-\delta'} \qquad \forall \alpha : |\alpha| \leq K', \tag{4.5.13$_{1-3}$}$$

$$\operatorname{dist}(x(t), \partial X) \geq \epsilon h^{\delta'}, \tag{4.5.14}$$

$$\operatorname{dist}((x(t),\xi(t)),\ (x,\xi)) \geq \epsilon |t| h^{\frac{1}{2}-\delta} \tag{4.5.15}$$

and let condition (2.3.16)″ hold with $\omega = h^{-\delta'}$ with small enough $\delta' = \delta'(d,\delta) > 0$. Then

$$\mathcal{R}_\vartheta \leq C_0 M^0 \sum_{0\leq j\leq n} \frac{h^{1-d+\vartheta}}{T_j^{1+\vartheta}} \int_{\Lambda_j} dxd\xi : da + CMh^{1-d+\vartheta+\delta'} + C'M'h^s \tag{4.5.16}$$

where $C' = C'(d,s,\delta,c,c_1), K' = K'(d,s,\delta), M' = \|\psi\|_{C^{K'}}$.

Theorem 4.5.3. *Let conditions (4.5.1), (4.5.4), (4.54.5), (4.5.6) and (4.5.7) be fulfilled and $d = 1$. Then*

(i) The following estimate holds for $\vartheta = 0$

$$\mathcal{R} \leq C_0 M h^{-\delta}; \tag{4.5.17}$$

(ii) Moreover, if either

$$\sum_{|\alpha|\leq n} |D^\alpha V| \geq \epsilon_0 \tag{4.5.18}$$

or $\vartheta > 0$ then assertions (i) and (ii) of theorem 4.5.2 remain true. However, for $\vartheta \in \mathbb{N}$ one should regularize the coefficient $\kappa_{1+\vartheta}$ and take it depending on h (and it is $O(|\log h|^M)$ by the previous section).

Let us consider spectral asymptotics without spatial mollification.

Theorem 4.5.4. *Let conditions* (4.5.1), (4.5.4), (4.5.5), (4.5.6) *and* (4.5.7) *be fulfilled. Then*

(i) If condition

$$|V-\tau| \geq \epsilon_0 \qquad \forall x \in B(0,1) \tag{4.5.19}$$

is fulfilled then

$$\mathcal{R}_x = |e(x,x,\tau) - h^{-d}\kappa_0(x,\tau)| \leq C_0 h^{1-d} \qquad \forall x \in B(0,\frac{1}{2}) \tag{4.5.20}$$

and

$$\mathcal{R}_{x\vartheta} = |\tau_+^{\vartheta-1} * (e(x,x,\tau) - \sum_{0\leq n<1+\vartheta} h^{-d+n}\kappa_n(x,\tau))| \leq C_0 h^{1-d+\vartheta} \qquad \forall x \in B(0,\frac{1}{2}). \tag{4.5.21}$$

(ii) For $d \geq 3$ estimate (4.5.20) *remains true even if condition* (4.5.19) *is violated and for $d \geq 3+\vartheta$ estimate* (4.5.21) *remains true as well;*

(iii) For $d < 3+\vartheta$ estimate

$$\mathcal{R}_x \leq C_0 h^{\frac{2}{3}(-d+\vartheta)} \qquad \forall x \in B(0,\frac{1}{2}); \tag{4.5.22}$$

holds.

Proof. Statement (i) follows from theorem 4.3.3. In order to prove statements (ii) and (iii) let us introduce

$$\gamma(x) = \epsilon^{-1}|V-\tau| + \bar{\gamma}, \qquad \rho(x) = \gamma(x)^{\frac{1}{2}} \tag{4.5.23}$$

with $\bar{\gamma} = h^{\frac{2}{3}}$ here; then $|\nabla_x \gamma| \leq \frac{1}{2}$ and $\rho\gamma \geq h$. Let us consider a γ-admissible partition of unity; then repeating the arguments used in the proof of theorem 4.4.10 but without mollification with respect to x one can easily prove (ii) and (iii); here $h_{\text{eff}} = h\gamma^{-\frac{3}{2}}$. In this proof it is necessary to remember that $e(x,x,\tau)$ is a density but not a function. □

The reader can reformulate theorem 4.4.9 for the Schrödinger operator. Using the same arguments but with the *scaling functions* $\gamma(x)$ and $\rho(x)$ given by (4.5.22) with $\bar{\gamma} = h^{\frac{2}{3}-\delta}$ one can easily prove

Theorem 4.5.5. *Let conditions* (4.5.1), (4.5.4), (4.5.5), (4.5.6) *and* (4.5.7) *be fulfilled. Then the estimate*

$$\frac{1}{L}|\int \phi(\frac{\tau-\bar{\tau}}{L})(e(x,x,\tau) - \sum_{0\leq n\leq N'-1} h^{-d+n}\kappa_n(x,\tau))d\tau| \leq C'h^s \tag{4.5.24}$$

holds for $L \geq h^{\frac{2}{3}-\delta}$.

I leave it to the reader to treat the case $L \geq h^{\frac{2}{3}}$ (c.f. theorem 4.4.4). Finally the results of section 4.1 yield

Theorem 4.5.6. *Let conditions* (4.5.1), (4.5.4), (4.5.5), (4.5.6) *and* (4.5.7) *be fulfilled. Then for $\tau \leq V_* = \inf_{B(0,1)} V$ the following estimate holds*

$$|e(x,y,\tau)| \leq C'h^{-d}(1+\frac{|V_*-\tau|}{h})^{-s} \quad \forall x,y \in B(0,1-\epsilon) \tag{4.5.25}$$

where $C' = C'(d,c,s,\epsilon)$ and $\epsilon > 0$ is arbitrary.

Remark 4.5.7. *By a partition of unity one can easily prove that the exponent "$-d$" of h can be replaced by "$-\frac{d}{2}$;" moreover, if in the neighborhood of points with $V - V_* \le \epsilon$ the Hessian of V has an r-minor with absolute value greater than ϵ then after mollification with respect to x one can take the exponent to be "$\frac{r-d}{2}$."* ◊

4.5.2 The Dirac operator: calculations

Let us go now to the Dirac operator. Let us assume that

$$|D^\alpha \omega^{jl}| \le c, \quad |D^\alpha V_j| \le c \qquad \forall \alpha : |\alpha| \le K, \tag{4.5.6$'_{1-2}$}$$

$$|D^\alpha V| \le c \min(1, \frac{1}{m}) \qquad \forall \alpha : 1 \le |\alpha| \le K \tag{4.5.6$'_3$}$$

$$\begin{aligned}(\tau_1 - m - V)_- &\le c \min(1, \frac{1}{m}),\\ (\tau_2 + m - V)_+ &\le c \min(1, \frac{1}{m}),\end{aligned} \tag{4.5.6$'_4$}$$

$$|D^\alpha W| \le c \min(1, \frac{1}{m}) \qquad \forall \alpha : |\alpha| \le K \tag{4.5.6$'_5$}$$

$$|D^\alpha J_\varepsilon| \le c \quad \forall \alpha : |\alpha| \le K, \varepsilon = \pm, \tag{4.5.6$'_6$}$$

$$\epsilon_0 \le J_\varepsilon \le c \tag{4.5.6$'_7$}$$

where $\tau_1 < \tau_2$ are the energy levels in question. Moreover let us assume that the metric tensor

$$g^{jk} = \omega^{jl} \omega^{kr} \delta_{lr} \tag{4.5.25}$$

satisfies (4.5.7).

Let us recall that

$$\sigma_l^\dagger = \sigma_l, \tag{4.5.26}$$

$$\sigma_l \sigma_k + \sigma_k \sigma_l = 2\delta_{kl} I \qquad (k, l = 0, \dots, d), \tag{4.5.27}$$

i.e., that σ_l are Hermitian, unitary and anti-commuting. The properties of Dirac matrices are presented more explicitly in Appendix C.

Lemma 4.5.8. *For $d \ge 1$ the principal symbol of the Dirac operator*

$$a(x, \xi) = \sigma_l \omega^{jl} (\xi_j - V_j) + \sigma_0 m + V \cdot I \tag{4.5.28}$$

has two eigenvalues

$$\lambda_\mp = V \mp (g^{jk} (\xi_j - V_j)(\xi_k - V_k) + m^2)^{\frac{1}{2}} \tag{4.5.29}$$

of multiplicity $\frac{D}{2}$.

Proof. Obviously it is sufficient to treat the case $V = V_1 = \cdots = V_d = 0$. Then (4.5.27) implies that $a(x,\xi)^2 = g^{jk}\xi_j\xi_k + m^2$ and hence there is no eigenvalue of a different from $\lambda_\pm$. Since $\lambda_\pm$ coincide only if $\xi = 0, m = 0$ then in the calculation of multiplicities one can assume that $\xi = 0, m > 0$. But in this case the multiplicities of $\lambda_\pm$ are equal to $\frac{D}{2}$ (see Appendix C). □

As a corollary we obtain that for the Dirac operator

(4.5.30) $$\operatorname{tr}\kappa_0(x,\tau_1,\tau_2) = \frac{D}{2}(2\pi)^{-d}\sum_{\varepsilon=\pm}\int_{\tau_1<\lambda_\varepsilon<\tau_2} d\xi$$

and for the operator (4.5.3)

(4.5.31) $$\operatorname{tr}\kappa_0(x,\tau_1,\tau_2) = (2\pi)^{-d}\int \mathbf{n}(x,\xi,\tau_1,\tau_2)d\xi$$

where $\mathbf{n}(x,\xi,\tau_1,\tau_2)$ is the eigenvalue counting function for the symbol $\mathbf{a}(x,\xi) = J(x)^{-\frac{1}{2}}(a(x,\xi)+\sigma_0 W(x))J(x)^{-\frac{1}{2}}$ and a is defined by (4.5.28).

Let us calculate $\kappa_1'(y,\tau)$. The arguments of section 4.2 imply that $\kappa_1'(y,\tau)$ depends only on the values of $\omega^{jl}, V_j, V, m+W, J_1, J_2$ and their first-degree derivatives at the point y and the dependence on the derivatives is linear. Since the gradient transformation is equivalent to multiplying the spectral function $e(x,y,\tau_1,\tau_2)$ by $\exp ih^{-1}(\varphi(x)-\varphi(y))$ and hence does not change $e(y,y,\tau_1,\tau_2)$ then $\kappa_1'(y,\tau)$ depends on V_j only through $F_{jk}(y)$. Hence it is sufficient to be able to calculate $\kappa_1'(y,\tau)$ in two cases:

(a) $V_j = 0$ and ω^{jl}, V, W, J_1 and J_2 are linear functions;

(b) V_j are linear functions and ω^{jl}, V, W, J_1 and J_2 are constants.

But in case (a) the replacement $A \to \sigma_0 A\sigma_0$, $x_j \to 2y_j - x_j$ (which obviously does not change $\operatorname{tr} e(y,y,\tau_1,\tau_2)$) is equivalent to replacing the coefficients of $x_j - y_j$ in ω^{kl}, V, W, J_1 and J_2 by their negatives and hence $\operatorname{tr} e(y,y,\tau_1,\tau_2)$ does not depend on the derivatives of these functions. In case (b) we reach $g^{jk} = \delta_{jk}$ after an appropriate linear change of variables; then one can assume that $\omega^{jl} = \delta_{jl}$ (otherwise we obtain these equalities by replacing σ_l with $\sigma_l' = \omega^{lk}\sigma_k, l = 1,\dots,d, \sigma_0' = \sigma_0$; one can check that these new matrices also satisfy (4.5.26) and (4.5.27). Let us now apply an explicit calculation based on (4.2.23), including for the sake of simplicity $-\frac{1}{2}(J_+ + J_-)$ in V and $\frac{1}{2}(J_- - J_+) + W$ in m; then taking into account that $a^s = 0$ we obtain that

$$\operatorname{tr}\kappa_1'(y,\tau) = \frac{1}{2}(2\pi)^{-d-1}\int \operatorname{Res}_{\mathbf{R}}(\tau - a)d\xi$$

where

$$\mathcal{F} = \operatorname{tr} J(\tau-a)^{-1}a_{(j)}(\tau-a)^{-1}a^{(j)}(\tau-a)^{-1}$$

and $a_{(j)}$ and $a^{(j)}$ are the derivatives of a with respect to x_j and ξ_j respectively.

Let us note that $(\alpha I + \alpha^j\sigma_j)^{-1} = (\alpha^2 - \sum_k \alpha^{k2})^{-1}(\alpha I - \alpha^j\sigma_j)$ for $\alpha, \alpha^1, \dots, \alpha^d \in \mathbb{C}$; therefore

$$\mathcal{F} = -\Delta^{-3}\operatorname{tr} J(\tau + a' - V)\sigma_k V_{k(j)}(\tau + a' - V)\sigma_j(\tau + a' - V)$$

where $\Delta = (\tau - V)^2 - |\xi|^2 - m^2$, $a' = \sigma_j\xi_j + \sigma_0 m$, $V_{k(j)} = \partial_j V_k$ and we have taken into account that $a^{(j)} = \sigma_j, a_{(j)} = -\sigma_k V_{k(j)}$ because of simplifying assumptions. Since $(\tau + a' - V)\sigma_k = \sigma_k(\tau + a' - V) + 2\xi_k I$ and $(\tau + a' - V)(\tau - a' - V) = \Delta I$ and these two factors commute we obtain that

$$\mathcal{F} = -\Delta^{-2}\operatorname{tr} J\sigma_k\sigma_j(\tau + a' - V)V_{k(j)} - 2\Delta^{-2}\operatorname{tr} J\sigma_j\xi_k V_{k(j)} - 2\Delta^{-2}\operatorname{tr} J(\tau + a' - V)V_{k(j)}\xi_k\xi_j.$$

The properties of Dirac matrices (see Appendix C) yield that the second term vanishes. Without loss of generality one can assume that $V_{k(j)} = -V_{j(k)} = -\frac{1}{2}F_{kj}$; we can always arrange this by a gradient transformation; then the third term also vanishes and

$$\mathcal{F} = \frac{1}{2}\Delta^{-2}\operatorname{tr}\sigma_k\sigma_j F_{kj} J(\tau + a' - V).$$

Then the properties of Dirac matrices yield that F vanishes provided $d \neq 2$, while for $d = 2$ only terms of type $f\sigma_0$ with scalar f in the product $J(\tau + a' - V)$ are essential and hence

$$\mathcal{F} = \frac{1}{2}\Delta^{-2}\operatorname{tr}\sigma_1\sigma_2\sigma_0 F_{12}\bigl((J_+ + J_-)m + (J_+ - J_-)(\tau - V)\bigr).$$

Explicit calculation in polar coordinates in $\mathbb{R}^2_\xi$ yields that

$$\int \Delta^{-2} d\xi = \pi((\tau - V)^2 - m^2)^{-1}$$

and the real residue of this function is equal to

$$\frac{\pi^2 i}{m}(\delta(V + m - \tau) - \delta(V - m - \tau))$$

and hence recalling that some terms were "hidden" in V and m we conclude that

$$\operatorname{tr}\kappa_1'(x,\tau) = \frac{1}{8\pi}F_{12}\operatorname{tr} i\sigma_1\sigma_2\sigma_0 \sum_{\varepsilon=\pm} \varepsilon J_\varepsilon \delta(J_\varepsilon\tau - \varepsilon(m + W) - V)$$

and hence finally

$$\operatorname{tr}\kappa_1(x,\tau_1,\tau_2) = \frac{1}{8\pi}F_{12}\operatorname{tr} i\sigma_1\sigma_2\sigma_0 \sum_{\varepsilon=\pm} \varepsilon\theta(J_\varepsilon\tau - \varepsilon(m + W) - V)\Big|_{\tau=\tau_1}^{\tau=\tau_2}. \tag{4.5.32}$$

So we have proven

Lemma 4.5.9. *For the generalized Dirac operator* (4.5.3) $\operatorname{tr}\kappa_1(x,\tau_1,\tau_2)$ *vanishes for* $d \neq 2$ *and* $\operatorname{tr}\kappa_1(x,\tau_1,\tau_2)$ *is given by* (4.5.32) *for* $d = 2$.

4.5.3 The Dirac operator: asymptotics for $m \geq c_0$

Let us pass from calculations to asymptotics. We assume that $\tau_2 = 0$, which is not a restriction for operator (4.5.2) but is a restriction for the more general operator (4.5.3). Let us first consider the case

$$m \geq c_0. \tag{4.5.33}$$

Then at least one of the two conditions

$$W + m \mp V \geq \epsilon_0 m \tag{4.5.34$_\pm$}$$

is fulfilled. Moreover let us assume that

(4.5.35) $$m + W \ge |V - \tau_1| + \epsilon_0 m;$$

we can always satisfy this condition by appropriate choice of τ_1 but we cannot guarantee that $\tau_1 < \tau_2$.

Without loss of generality one can assume that $\sigma_0 = \begin{pmatrix} I & 0 \\ 0 & -I \end{pmatrix}$ (see Appendix C). Separating the $\mathbb{C}^D$-valued functions into two components and the $\mathbb{C}^D \times \mathbb{C}^D$-valued functions into four blocks we obtain that the non-stationary generalized Dirac equation

$$(JhD_t - A)J^{-\frac{1}{2}}U = 0$$

can be rewritten in the form of two equations

(4.5.36)$_+$ $$(J_+hD_t - m - W - V)U^+ - \mathcal{M}U^- = 0,$$

(4.5.36)$_-$ $$-\mathcal{M}^\dagger U^+ + (J_-hD_t + m + W - V)U^- = 0$$

where $\mathcal{M} = \frac{1}{2}\sigma_l'(\omega^{jl}P_j + P_j\omega^{jl})$, $\mathcal{M}^\dagger = \frac{1}{2}\sigma_l'^\dagger(\omega^{jl}P_j + P_j\omega^{jl})$, and we assume that $\sigma_l = \begin{pmatrix} 0 & \sigma_l' \\ \sigma_l'^\dagger & 0 \end{pmatrix}$.

In this case if condition (4.5.34)$_\pm$ is fulfilled then the second (the first respectively) of these equations is elliptic for $|\tau| \le \epsilon_0 m$ and hence we can exclude the corresponding component. Let us assume that condition (4.5.34)$_+$ is fulfilled. Then substituting

(4.5.37) $$U^- \equiv (J_-hD_t + m + W - V)^{-1}\mathcal{M}^\dagger U^+$$

into (4.5.36)$_+$ we obtain that

(4.5.38) $$m((J_+hD_t - m - W - V) - \mathcal{M}(J_-hD_t + m + W - V)^{-1}\mathcal{M}^\dagger)U^+ \equiv 0$$

where $(J_-hD_t + m + W - V)^{-1}$ is a parametrix of $(J_-hD_t + m + W - V)$ in the zone $\{|\tau| \le \epsilon_0 m\}$ and the norm of the error does not exceed $C'\frac{h^s}{m}$.

Let us consider the operator

(4.5.39) $$\mathcal{A} = m\mathcal{M}(J_-hD_t + m + W - V)^{-1}\mathcal{M}^\dagger;$$

the properties of Dirac matrices yield that its principal symbol is equal to

$$m(J_-\tau + m + W - V)^{-1}g^{jk}(\xi_j - V_j)(\xi_k - V_k).$$

Rescaling the time $t_{\text{new}} = \frac{1}{m}t$ we go to the operator with Schrödinger principal part and conditions (4.5.6)$'_*$ for the original operator $\mathbf{A}$ yield conditions (4.5.6)$_*$ for $\mathcal{A}$. All the arguments for operators scalar in principal and all the arguments for the Schrödinger operator remain true excluding arguments linked with the subprincipal symbol. First of all we have the standard elliptic

Lemma 4.5.10. *Let conditions* (4.5.3), (4.5.4), (4.5.6)$'$, (4.5.7), (4.5.33) *and* (4.5.34)$_\pm$ *be fulfilled and let*

(4.5.40)$_\pm$ $$m + W \pm V \ge \epsilon\frac{r}{m} \qquad \forall x \in B(0,1),$$

with $\epsilon > 0$, $m^2 \ge r \ge 1$. *Then*

(4.5.41) $$|F_{t\to h^{-1}\tau}\chi_T(t)U| \le C'\bigl(\frac{h}{r}\bigr)^s T \qquad \forall x,y \in B(0,1-\epsilon) \quad \forall \tau \in [-\epsilon' r, \epsilon' r]$$

where $\epsilon' = \epsilon'(d,c,\epsilon_0,\epsilon) > 0$ *is a small constant and* $C' = C'(d,s,c,\epsilon_0,\epsilon)$. *Moreover, the same estimate with an additional factor* m^{-1}, m^{-1}, m^{-2} *in the right-hand expression holds for the blocks* $U^{\mp,\cdot}$, $U^{\cdot,\mp}$, $U^{\mp,\mp}$ *respectively.*

Then standard Tauberian arguments yield

Corollary 4.5.11. *In the context of lemma 4.5.10*

$$|e(x,y,\tau',\tau'')| \leq C'\big(\frac{h}{r}\big)^s \qquad \forall x,y \in B(0,1-\epsilon) \quad \forall \tau',\tau'' \in [-\epsilon' r, \epsilon' r]. \tag{4.5.42}$$

Moreover, the same estimate with a factor m^{-1}, m^{-1}, m^{-2} in the right-hand expression holds for the blocks $e^{\mp,\cdot}$, $e^{\cdot,\mp}$, $e^{\mp,mp}$ respectively.

Remark 4.5.12. *(i) The arguments of subsection 4.1.3 yield that in this lemma and corollary one can take $K = K(d) = [\frac{d}{2}] + 1$.*

(ii) Corollary 4.5.11 yields that we can assume in what follows that

$$m + W \pm V \leq \frac{c}{m}. \tag{4.5.43$_\pm$}$$

(iii) Without loss of generality one can assume that $D = \bar{D} \overset{\mathrm{def}}{=} 2^{[\frac{d+1}{2}]}$ because $\sigma_k = q^\dagger(\bar{\sigma}_k \otimes I_{D'})q$ with the unitary matrix q, $\bar{D}\times\bar{D}$ matrices $\bar{\sigma}_k$ and $D' = D\bar{D}^{-1}$. In particular, for $d = 1,2$ we get $\bar{D} = 2$ and the eigenvalues of $a(x,\xi)$ are simple. ◊

Then the following assertions similar to those for the Schrödinger operator are proven easily:

Theorem 4.5.13. *Let conditions* (4.5.3), (4.5.4), (4.5.6)′, (4.5.7), (4.5.33) *and* (4.5.34)$_\pm$, (4.5.43)$_\pm$ *be fulfilled. Let $\psi \in C_0^K(B(0,\frac{1}{2}))$, $|\psi| \leq 1, \phi \in C_0^K([-c,c])$ and $\int_{-\infty}^{\infty} \phi(\tau)d\tau = 1$. Let $L \in (0,c]$. Then*

(i) For $d \geq 2$ or $\vartheta > 0$ estimates (4.5.11), (4.5.12) *hold and for $d = 1, \vartheta = 0$ estimate* (4.5.17) *holds in the general case and estimate* (4.5.11) *holds under condition* (4.5.18) *for*

$$\mathcal{R} = \frac{1}{L}\Big|\int \psi(x)\phi(\frac{\tau}{L})\,\mathrm{tr}\big(e(x,x,\tau,0) - h^{-d}\kappa_0(x,\tau,0)\big)\,dxd\tau\Big|; \tag{4.5.44}$$

moreover, a similar estimate holds for $L \geq 1$ but in this case one should also include in $\mathcal{R}$ terms with larger numbers with appropriate regularization;

(ii) Moreover, let $d \geq 2$, $\Lambda_1^\pm, \dots, \Lambda_n^\pm$ be closed subsets of $\Sigma^\pm = \{(x,\xi) : \lambda_\pm = 0\}$ where $\lambda_\pm$ are given by (4.5.29), *$\Lambda_j^\pm \subset \{\gamma_\pm^2 = m| \pm V + m + W| + m^2|\nabla(V \pm W)|^2 \geq h^{1-\delta}\}$ and suppose that through each point of the $h^{\frac{1}{2}-\delta}$-neighborhood of $\Lambda_j^\pm$ there passes a Hamiltonian trajectory generated by $mJ_\pm(x)^{-1}\lambda_\pm(x,\xi)$ with $t \in \mathcal{J}_j = [0,T_j]$ or $\mathcal{J}_j = [-T_j, 0]$, $T_j \in [1, h^{-\delta'}]$, along which*

$$|D^\alpha \omega^{jl}| \leq c_1 h^{-\delta'}, \quad |D^\alpha V_j| \leq c_1 h^{-\delta'}, \tag{4.5.12$'_{1,2}$}$$

$$\begin{aligned} &|D^\alpha V| \leq c_1 h^{-\delta'} \min(1,\frac{1}{m}) \qquad (|\alpha| \geq 1)\\ &|D^\alpha W| \leq c_1 h^{-\delta'} \min(1,\frac{1}{m}), \qquad |D^\alpha J_\varepsilon| \leq c_1 h^{-\delta'} \quad \forall \alpha : |\alpha| \leq K', \end{aligned} \tag{4.5.12$'_{3,5,6}$}$$

$$\epsilon_1 h^{\delta'} \leq J_\varepsilon \leq c_1 h^{-\delta'} \tag{4.5.12$'_7$}$$

and conditions (2.3.16)″, (4.5.13), (4.5.14), (4.5.15) *are fulfilled. Then for $d \neq 2$ estimate* (4.5.16) *with $\Lambda_j = \Lambda_j^+ \cup \Lambda_j^-$, $da = d\lambda_\pm$ at $\Sigma^\pm$ remains true and for $d = 2$ the estimate*

$$(4.5.16)' \quad \mathcal{R}_1 \le C_0 \sum_{0\le j\le n, \varepsilon=\pm} \frac{1}{T_j h}\int_{\Lambda_j^\pm} dxd\xi : d\lambda_\pm + C_0\frac{1}{h}\int_{\{\gamma_\pm \le h^{1/2-\delta}\}} |F_{12}|dx + CMh^{\delta'-1} + C'M'h^s$$

holds where the notation of theorem 4.5.2 is used;

(iii) For $\mathcal{R}_0$ and for $\mathcal{R}_x$ introduced in a similar way the estimates of theorems 4.5.3, 4.5.4 and 4.5.5 remain true but in conditions (4.5.18) *and* (4.5.19) *one must replace V by mV and in theorem 4.5.5 $L \ge \frac{1}{m}h^{\frac{2}{3}-\delta}$ now.*

(iv) Moreover, the same assertions remain true if we replace τ by τ_1 satisfying (4.5.35) *and omit mollification with respect to τ.*

Remark (Remarks on the proof). *In this modification only the proof of (ii) with $d=2$ must be changed, only the part at which we consider a ball $B(z',\gamma_1)$ with $|\operatorname{Hess} V^*| \le \gamma_1$, $|\{z_1,z_2\}| \asymp \gamma_3$ where $V^* = (m+W\pm V)m$, $\gamma_3 \ge h^{\delta_3}$ with arbitrarily small $\delta_3>0$ and we refer to the proof of theorem 4.3.12.*

Let us consider the Dirac operator in the ball of radius $\gamma_4 = h^{\frac{1}{2}-\delta_4}$ with respect to x with a small enough exponent $\delta_4>0$. Let us introduce new coordinates $x_{\text{new}} = \frac{x}{\gamma_4}$, divide the Dirac operator by $\gamma_4\gamma_3$ and replace h,m by $h_{\text{new}} = \frac{h}{\gamma_4^2\gamma_3} < h^{\delta_5}$, $m_{\text{new}} = \frac{m}{\gamma_3\gamma_4}$. One can easily check that we remain consistent with all the assumptions of this subsection provided $V^* \ge -c_0\gamma_3^2\gamma_4^2$ in the ball in question and in this case the contribution of this ball to the remainder estimate does not exceed

$$C_0 h_{\text{new}}^{-1} + CMh_{\text{new}}^{\delta'-1} \le C_0 h^{-1}\gamma_4^2|F_{12}| + CMh_{\text{new}}^{\delta'-1}$$

where s and the constants are changed if necessary; after summation with respect to all the balls of this type we obtain the second and the fourth terms of the right-hand expression in (4.5.15)$'$.

On the other hand, balls with $V^* \le -c_0\gamma_3^2\gamma_4^2$ are consistent with the proof of theorem 4.4.4 : after dilatation we divide by $(-V^*)^{\frac{1}{2}}$ and obtain the case in which the microhyperbolicity condition is fulfilled; so the total contribution of the balls of this type does not exceed the sum of the first and the fourth terms of the right-hand expression in (4.5.15)$'$.
□

4.5.4 The Dirac operator: asymptotics for $m \le c_0$

Let us now consider the case

$$(4.5.45) \qquad m \le c_0.$$

In this case the eigenvalues $\lambda_\pm$ of $a(x,\xi)$ are not necessarily disjoint. Let us note that the microhyperbolicity condition at the level $\tau=0$ for operator (4.5.3) means exactly that

$$(4.5.46) \qquad |m+W\pm V| + |\nabla(W\pm V)| \ge \epsilon_0$$

for both signs. Let us now assume that $W=0$. Then theorems 4.3.6 and 4.3.8 yield

Theorem 4.5.14. *Let conditions* (4.5.3), (4.5.4), (4.5.5), (4.5.6)$'$, (4.5.7) *and* (4.5.45) *be fulfilled. Let $W=0$ and*

$$(4.5.47) \qquad |m\pm V| + |\nabla V| \ge \epsilon_0$$

and let ϕ and ψ be the same as before. Then

(i) Estimate (4.5.11) *holds with* $\mathcal{R}_0$ *given by* (4.5.44).

(ii) Let $\Lambda_j^{\pm}$ *be closed subsets of* $\Sigma^{\pm}$ *and suppose that through every point of* $\Lambda_j^{\pm}$ *there passes a Hamiltonian trajectory generated by* $J_{\pm}^{-1}\lambda_{\pm}$ *with* $t \in \mathcal{J}_j$, $\mathcal{J}=[0,T_j]$ *or* $\mathcal{J}=[-T_j,0]$, $T_j \in [1, h^{-\delta'}]$ *in the* $h^{\frac{1}{2}-\delta}$*-neighborhood for which conditions* (4.5.12)′, (4.5.13), (4.5.14), (2.3.16)″ *are fulfilled and*

$$m^2 + g^{jk}(\xi_j - V_j)(\xi_k - V_k) \geq \epsilon_1 h^{\delta'}. \tag{4.5.48}$$

Then estimate (4.5.15) *with* $\Lambda_j = \Lambda_j^+ \cup \Lambda_j^-$, $d\lambda = d\lambda_{\pm}$ *on* $\Sigma^{\pm}$ *holds.*

(iii) Moreover, if

$$|m \pm V| \geq \epsilon_0 \tag{4.5.49}$$

for both signs then estimate (4.5.19) *holds.*

Proposition 4.5.15. *Let* $d \geq 2$ *and suppose the condition*

$$g^{jk}\eta_j\eta_k \geq \epsilon_0|\eta|^2 \quad \forall \eta \in \mathbb{R}^d \tag{4.5.7$'$}$$

holds along trajectories. Then in assertion (ii) of theorem 4.5.14 one can drop condition (4.5.48) *by replacing* $CMh^{1-d+\delta'}$ *by* $C'M'h^{1-d+\delta'}$ *in* (4.5.15).

Proof. It is sufficient to prove that the measure[39)] of the set $\mathcal{Z}$ of the points $(X,\xi) \in \Lambda$ (we omit index j for the sake of simplicity) such that (4.5.48) is violated somewhere along trajectory does not exceed $C'h^{\delta''}$ with appropriate $\delta'' > 0$. Let us first note that because of (4.5.47) this estimate holds for the measure of the set of all points of $\Sigma^{\pm} \subset T^*B(0,1)$ at which condition (4.5.48) is violated. Hence we should consider points in which condition (4.5.47) holds but is violated somewhere along the trajectory. Let $\mathcal{Z}$ be the set of these points. Then $|V| \geq \epsilon_1 h^{\delta'}$ on $\mathcal{Z}$. Let X' be the union of all the Hamiltonian trajectories of length not exceeding T passing through $\mathcal{Z}$. Then every trajectory intersects $\mathcal{Y}_1$ where

$$\mathcal{Y}_\nu = \{(x,\xi) \in \Sigma, 2^{-\nu}\rho \leq |V| \leq 2^\nu \rho\}, \quad \rho = h^{\delta'}$$

and one can assume that $m \leq \frac{1}{8}\rho$. Along these trajectories

$$|\frac{dz}{dt}| \leq c_1 h^{-\delta'}; \tag{4.5.50}$$

hence $X' \subset B(0, h^{-l\delta'})$ where $l = l(d)$ changes if necessary and the trajectory intersecting $\mathcal{Y}_1$ belongs to $\mathcal{Y}_2$ in a time interval of length $t_0 = C_1^{-1}\rho$. Hence

$$\operatorname{mes}\mathcal{Z} \leq \frac{T}{t_0}\operatorname{mes}\mathcal{Y}_2 \leq C'T\rho^{d-2}\operatorname{mes}\{x \in X', |V| \leq 4\rho\}$$

and taking $T \leq h^{-\delta''}$ with small enough $\delta'' > 0$ we obtain the estimate in question provided $d \geq 3$.

For $d = 2$ a more sophisticated analysis is necessary. Let $\mathcal{Y}'_\nu = \{z \in \mathcal{Y}_\nu, |\nabla V| \leq 2^\nu h^{\delta''}\}$, $\mathcal{Y}''_\nu = \{z \in \mathcal{Y}_\nu, |\nabla V| \geq 2^{-\nu}h^{\delta''}\}$, with a small enough $\delta'' = \delta''(\delta') > 0$. Then the previous arguments are applicable to $\mathcal{Y}'_1$, $\mathcal{Y}'_2$ and t'_0 instead of $\mathcal{Y}_1$, $\mathcal{Y}_2$ and t_0 respectively and hence

$$\operatorname{mes}\mathcal{Z} \leq C'\rho^{d-1}\frac{T}{t'_0}\operatorname{mes}\{x \in X', |V| \leq 4\rho\} + \\ C'\frac{T}{t_0}\operatorname{mes}\{x \in X', |V| \leq 4\rho, |\nabla V| \geq h^{\frac{\delta''}{4}}\} \leq C'h^{\delta''}$$

provided δ'' is small enough. □

39) We always speak of the measure with density $dxd\xi : J_{\pm}d\lambda_{\pm}$.

Theorem 4.5.16. *Let conditions* (4.5.3), (4.5.4), (4.5.5), (4.5.6)′, (4.5.7) *and* (4.5.45) *be fulfilled. Let* $W = 0$ *and* ψ, ϕ *and* ϕ_j *be the same as before. Then*

(i) For $d \geq 2$ *estimate* (4.5.11) *for* $\mathcal{R}_0$ *holds and for* $d = 1$

$$|\mathcal{R}| \leq Ch^{-\delta} + C + C'Mh^s \tag{4.5.51}$$

in the general case and

$$|\mathcal{R}| \leq C|\log(h+m)|^M + C + C'Mh^s \tag{4.5.52}$$

under condition (4.5.18).

(ii) Let $\Lambda_j^\pm$ *be closed subsets of* $\Sigma^\pm$ *such that* $\Lambda_j^\pm \subset \{\gamma = |\nabla V| + \min(|V|, \frac{|V|^2}{m}) \geq h^{\frac{1}{2}-\delta}\}$ *and through every point of* $\Lambda_j^\pm$ *there passes a Hamiltonian trajectory generated by* $J_\pm^{-1}\lambda_\pm$ *with* $t \in \mathcal{J}_j$, $\mathcal{J}=[0,T_j]$ *or* $\mathcal{J}=[-T_j,0]$, $T_j \in [1, h^{-\delta'}]$ *in the* $h^{\frac{1}{2}-\delta}$*-neighborhood for which conditions* (4.5.12)′, (4.5.13), (4.5.14), (2.3.16)″ *and* (4.5.7)′ *are fulfilled. Then for* $d \geq 3$ *estimate* (4.5.16) *holds with* $\Lambda_j = \Lambda_j^+ \cup \Lambda_j^-$ *and* $da = d\lambda_\pm$ *on* $\Sigma^\pm$, *while for* $d = 2$ *the estimate*

$$\mathcal{R}_1 \leq C_0 \sum_{0\leq j\leq n} \frac{1}{T_j h}\int_{\Lambda_j} dx d\xi : d\lambda + C_0 \int_{\{\gamma \leq h^{\frac{1}{2}-\delta}\}} |F_{12}| + CMh^{\delta'-1} + C'M'h^s \tag{4.5.16''}$$

holds where $\Lambda_j = \Lambda_j^+ \cup \Lambda_j^-$, $d\lambda = d\lambda_\pm$ *at* $\Sigma^\pm$;

(iii) For $d \geq 3$ *estimate* (4.5.19) *for* $\mathcal{R}_x$ *holds; for* $d = 2$ *the estimate*

$$|\mathcal{R}_x| \leq Ch^{-\frac{4}{3}}m^{\frac{8}{3}} + Ch^{-1} \tag{4.5.53}$$

holds and for $d = 1$ *the estimate*

$$|\mathcal{R}_x| \leq Ch^{-\frac{2}{3}}m^{\frac{4}{3}} + C \tag{4.5.54}$$

holds.

(iv) The complete asymptotics hold for

$$\iiint \phi_2(\frac{\tau'}{L})\phi_1(\frac{\tau}{L})\psi(x)e(x,x,\tau',\tau)dxd\tau'd\tau \tag{4.5.55}$$

and

$$\iint \phi_2(\frac{\tau'}{L})\phi_1(\frac{\tau}{L})e(x,x,\tau',\tau)d\tau'd\tau \tag{4.5.56}$$

provided $L \geq h^{1-\delta}$ *and* $L \geq \min(h^{\frac{2}{3}-\delta}m^{-\frac{1}{3}}h^{\frac{1}{2}-\delta})$ *respectively.*

Proof. On $B(0,1)$ let us consider the function

$$\gamma(x) = \epsilon_0(|V(x)| + |\nabla V(x)|) + \bar{\gamma} \tag{4.5.57}$$

with $\bar{\gamma} = h^{\frac{1}{2}} + \epsilon_0 m$; then $|\nabla\gamma(x)| \leq 1$ for $\epsilon_0 = \epsilon_0(d,c)$; let us introduce a γ-admissible partition of unity. Let $x^n \in B(0,\frac{1}{2})$ and $\gamma_n = \gamma(x^n) \geq m$; let us make the dilatation $x_{\text{new}} = \frac{x-x^n}{\gamma_n}$ in $B(x^n,\gamma_n)$ and replace A by $\frac{1}{\gamma_n}A$, h by $h_{\text{new}} = \frac{h}{\gamma_n^2}$ and m by $m_{\text{new}} = \frac{m}{\gamma_n}$; then the microhyperbolicity condition will surely be fulfilled and the contribution of this ball in the remainder estimate does not exceed Ch_{new}^{1-d}. On the other hand, if $\gamma_n \leq m$

then after the same procedure we are in the situation of the case $m_{\text{new}} \geq 1$ and one of conditions $(4.5.36)_\pm$ will be fulfilled provided C_0 is large enough. Hence the contribution of this ball in the remainder estimate can be estimated in the same way for $d \geq 2$; for $d = 1$ the contribution to the remainder estimate does not exceed $Ch_{\text{new}}^{-\delta}$. Finally, if $\gamma_n + m \leq c_0 h^{\frac{1}{2}}$ then we can apply the rough estimate of lemma 4.1.1 improved by lemmas 4.1.4, etc.; then the remainder estimate will be less than C. After summation on this partition of unity we obtain assertion (i) for $d \geq 2$ and for $d = 1$ under some additional conditions. Assertion (iii) is proven in the same way with

$$\gamma(x) = \epsilon |V(x)| + \bar{\gamma} \tag{4.5.58}$$

with the same $\bar{\gamma}$ and assertion (iv) is also proven in the same way with $\bar{\gamma} = h^{\frac{1}{2}-\delta} + \epsilon_0 m$ and with γ defined by (4.5.57) and (4.5.58) in the first and the second cases respectively.

In order to complete (i) for $d = 1$ let us assume that under condition $(4.4.43)_{n-1}$ estimate (4.5.52) is proven and we must prove it under condition $(4.4.43)_n$. Let us introduce functions γ_l by $(4.4.48)_n$ with $\bar{\gamma} = m^{\frac{1}{n}} + h^{\frac{1}{n+1}}$ and $\rho_n = \gamma_n^l$. Then after dilatation and division by ρ_n^{-1} we are in the situation of the induction assumption provided $\gamma_n \geq C_0 \bar{\gamma} l$ and the contribution to the remainder estimate does not exceed $C|\log(m+h)|^M$. After summation over the partition in the zone $\{\gamma_n \geq \bar{\gamma}\}$ we obtain an additional logarithmic factor while under condition $(4.4.43)_n$ the contribution of the zone $\{\gamma_n \leq \bar{\gamma}\}$ surely does not exceed C. We refer to theorem 4.5.13 for $m \geq h^{\frac{n}{n+1}}$ (with condition (4.4.43) fulfilled) and to the rough estimate otherwise and notice that under condition $(4.4.43)_n$ the measure of this zone does not exceed $C\bar{\gamma}$. So, we obtain estimate (4.5.52). We get rid of condition $(4.4.43)_n$ and obtain the remainder estimate (4.5.51) exactly as was done in the proof of theorem 4.4.9.

Let us prove (ii); we need consider only balls of radius $\gamma_1 = h^{\frac{1}{2}-\delta}$ in which

$$|V| \leq c_0 h^{\frac{1}{2}-\delta'}, \quad |\nabla V| \leq c_0 h^{\frac{1}{2}-\delta'}, \quad m \leq c_0 h^{\frac{1}{2}-\frac{1}{4}\delta'}; \tag{4.5.59$'_{1,2,3}$}$$

the last condition follows from the fact that $\gamma \geq \epsilon_0 m$ in this proof. Let us consider various cases:

(i) If $|F_{12}(x^l)| = \mu \geq h^{\frac{1}{4}\delta'}$ and $v = |V(x^l)| \leq \mu\gamma_1$ then $|F_{12}| \asymp \mu$ and $|V| \leq c\mu\gamma_1$ in the ball $B(x^l, \gamma_1)$; let us change variables $x_{\text{new}} = \frac{x - x^n}{\gamma_1}$, divide A by $\mu\gamma_1$ and set $h_{\text{new}} = \frac{h}{\mu\gamma_1^2}$, $m_{\text{new}} = \frac{m}{\mu\gamma_1}$; then we remain in the situation of our theorem and assertion (i) of this theorem yields that the contribution of this ball to the remainder estimate does not exceed

$$C\gamma_1^2 \mu h^{-1} + CM\gamma_1^2 h^{\delta''-1} + C'M'h^s.$$

(ii) Let $|F_{12}(x^l)| = \mu \geq h^{\frac{1}{4}\delta'}$ and $v = |V(x^l)| \geq \mu\gamma_1$; let us change variables $x_{\text{new}} = \frac{x - x^n}{\gamma_1}$, divide A by v and set $h_{\text{new}} = \frac{h}{v\gamma_1}$, $m_{\text{new}} = \frac{m}{v}$; then we remain in the situation of this theorem and the microhyperbolicity condition is fulfilled; applying theorem 4.5.13 we obtain that the contribution of this ball to the remainder estimate does not exceed

$$C\gamma_1 v h^{-1} + CM\gamma_1^2 h^{\delta''-1} + C'M'h^s.$$

(iii) If $|F_{12}(x^l)| \leq \mu = h^{\frac{1}{4}\delta'}$ then the analysis of (i) and (ii) is applicable for $v = |V(x^l)| \leq \mu\gamma_1$ and $v = |V(x^l)| \geq \mu\gamma_1$ respectively and we obtain that the estimate of (ii) remains true.

Calculating the total contribution of all balls of this type we obtain (4.5.16)$''$. □

Remark 4.5.17. *(i) One can improve these remainder estimates by replacing C with $C_0 + Ch^{\delta''}$ in them.*

(ii) In all the assertions proven before we assumed that $\tau_2 = 0$ and either τ_1 is a classically forbidden level or there is mollification with respect to τ_1. However all these results yield that for operator (4.5.2) and for general τ_1, τ_2 in the case with no mollification the following estimate holds

$$\mathcal{R}(\tau_1, \tau_2) \le R(\tau_1) + R(\tau_2) \tag{4.5.63}$$

where the meaning of $\mathcal{R}(\tau_1, \tau_2)$ is obvious and $R(\bar\tau)$ is the remainder estimate (obtained above) for $A' = A - \bar\tau$ at the level $\tau_2' = 0$ with mollification with respect to τ_1'. I hope that the reader will write out these estimates. ◊

In order to finish the case $d = 1$ let us prove

Theorem 4.5.18. *Let $d = 1$ and suppose condition* (4.5.18) *is fulfilled for $V \pm m$. Then estimates* (4.5.11) *and* (4.5.12) *hold.*

Proof. Let us repeat the proof of the previous theorem and treat the behavior of the Hamiltonian trajectories in the zone $\{\rho \ge C_0 m\}$ (the only zone which yields to the logarithmic factor).[40] Condition (4.4.43) yields that $V = \omega(x)(x_1 - a_1)\dots(x_1 - a_l)$ with ω disjoint from 0 and $a_1 \le a_2 \le \dots \le a_l$. We should get the estimate uniformly with respect to $a_1, \dots, a_l \in [-c, c]$. One can assume without loss of generality that $a_1 = 0$ in which case on each step of the induction a logarithmic factor appears only due to the zone $\{|x_1| \ge C_0\nu\}$ with $\nu = a_l + m^{\frac{1}{n}} + h^{\frac{1}{n+1}}$ and an arbitrarily large constant C_0. But in this zone one should not apply the induction assumption but only the microhyperbolic theorem; so $M = 1$. Moreover, let us note that after rescaling $t_{\mathsf{new}} = \frac{t}{\gamma}$ and therefore the only term to be treated is

$$\int_{\{|x_1| \ge C_0\nu\}} \frac{1}{T(x_1)} dx_1 \tag{4.5.63}$$

(and a logarithmic factor appears if we take $T = \epsilon_0\gamma$); here $\xi_1 = \pm\bigl((V - \tau)^2 - m^2\bigr)^{\frac{1}{2}}$ and T depends on τ as well but we can include τ in V (so a_j and ω depend on τ). Let us consider the trajectory of the Hamiltonian system (after ξ_1 is excluded)

$$\frac{dx_1}{dt} = \pm \frac{\bigl(V^2 - m^2\bigr)^{\frac{1}{2}}}{V}$$

starting from the point $\bar x_1$ in the time direction in which $|x_1|$ increases. Then the time until x_1 remains in $\{|x_1| \le \epsilon_0\}$ is disjoint from 0 and one can see easily that in this time interval $|x_1(t)| \asymp |\bar x_1| + |t|$ and

$$|dx_1(t)| \le C dx_1(0)| + C(|\bar x_1| + |t|)^L \bar x_1^{-L-l+1} |d\tau|$$

and so in order to fulfill condition on the Jacobi matrix we should take $T = \epsilon_1 \min(1, |\bar x_1|^{1+\delta(L+1)} h^{-\delta})$ with sufficiently large L. But then integral (4.5.63) is bounded. The non-periodicity condition is fulfilled trivially. □

Appendix C. On Dirac Matrices

The results of this appendix are important for section 4.5 and for Chapter 7.

[40] Surely, these trajectories coincide geometrically with the trajectories of the Schrödinger operator (depending on the energy level) but dynamically there is an essential difference.

Definition C.1. *Dirac matrices are matrices $\sigma_0, \dots, \sigma_d$ of $D \times D$ matrices such that*

$$\sigma_j^\dagger = \sigma_j, \tag{C.1}$$

$$\sigma_j \sigma_k + \sigma_k \sigma_j = 2\delta_{jk} I \qquad j,k = 0,\dots,d. \Diamond \tag{C.2}$$

We have the key statement

Theorem C.2. *(i) A set $\sigma_0, \dots, \sigma_d$ of $D \times D$ Dirac matrices exists if and only if $D \in \bar{D}\mathbb{Z}$ where $\bar{D} = 2^{[\frac{d+1}{2}]}$;*

(ii) Moreover, if $D = \bar{D}$ and d is odd then all sets of Dirac matrices are unitarily equivalent.[41] *On the other hand, for $D = \bar{D}$ and even d there exist precisely two sets of Dirac matrices which are not unitarily equivalent;*

(iii) If $D = \bar{D}$ and d is even then

$$\sigma_0 \cdots \sigma_d = \varsigma i^{\frac{d}{2}} I \quad \textit{with } \varsigma = \pm 1 \tag{C.3}$$

and two sets are equivalent if and only if $\varsigma = \varsigma'$. If $D = \bar{D}$ and d is odd then $\sigma_0, \dots, \sigma_d, \sigma_{d+1}$ with

$$\sigma_{d+1} = \varsigma (-i)^{\frac{d+1}{2}} \tag{C.4}$$

are also Dirac matrices;

(iv) If $p = \frac{D}{\bar{D}} \in \mathbb{Z}$ and d is odd then every set of Dirac matrices is unitarily equivalent to the set $\sigma_j = \bar{\sigma}_j \otimes I_p$ $j = 0,\dots,d$ where $\{\bar{\sigma}_j \quad j = 0,\dots,d\}$ is a standard set in dimension $\bar{D}$. If $p = \frac{D}{\bar{D}} \in \mathbb{Z}$ and d is even then every set of Dirac matrices is unitarily equivalent to the set $\sigma_j = \bar{\sigma}_j \otimes J_{p,q}$ $j = 0,\dots,d$ where $J_{p,q} = \operatorname{diag}(1,\dots,1,-1,\dots,-1)$ is a diagonal $p \times p$ matrix, $q = 0,\dots,p$. Here sets with different q are not unitarily equivalent. So a set of Dirac matrices is irreducible if and only if $D = \bar{D}$;

(v) The property "Every matrix commuting with all the Dirac matrices is scalar" is fulfilled if and only if $D = \bar{D}$.

Proof. Property (C.2) yields that $\sigma_j^2 = I$ and hence $\operatorname{Spec} \sigma_j \subset \{-1, 1\}$. Let us reduce σ_0 to diagonal form; then for $d = 0$ a set of Dirac matrices is unitarily equivalent to the set $\sigma_0 = \begin{pmatrix} I & 0 \\ 0 & -I \end{pmatrix}$ and hence for $d = 0$ we obtain all the statements of the theorem and $\bar{D} = 1$ and for $D = 1$ $\sigma_0 = \varepsilon = \pm 1$.

Without loss of generality one can assume that $\sigma_0 = \begin{pmatrix} I & 0 \\ 0 & -I \end{pmatrix}$ and for $d \geq 1$ (C.1) and (C.2) yield that $-\sigma_0 = \sigma_1^{-1} \sigma_0 \sigma_1$ and hence the multiplicities of the eigenvalues ± 1 are equal to $\frac{D}{2}$. Moreover, (C.1) and (C.2) yield that $\sigma_j = \begin{pmatrix} 0 & \alpha_j \\ \alpha_j^\dagger & 0 \end{pmatrix}$ for $j = 1, \dots, d$ where α_j are unitary. Let $q = \begin{pmatrix} I & 0 \\ 0 & q' \end{pmatrix}$ where q' is a unitary $\frac{D}{2} \times \frac{D}{2}$ matrix; then $q^\dagger \sigma_0 q = \sigma_0$ and $\sigma_j' = q^\dagger \sigma_j q$ $(j \geq 1)$ have the same form as σ_j but with α_j replaced by $\alpha_j' = \alpha_j q$. Hence without loss of generality one can assume that $\alpha_1 = iI$, i.e., that $\sigma_1 = \begin{pmatrix} 0 & iI \\ -iI & 0 \end{pmatrix}$. So we obtain (i) and (ii) for $d = 1$ with $\bar{D} = 2$.

Moreover, (C.1) and (C.2) yield that if σ_0 and σ_1 have the indicated form then the matrices α_j with $j = 2, \dots, d$ also satisfy (C.1) and (C.2). The proof is complete. □

[41] The sets $\sigma_0, \dots, \sigma_d$ and $\sigma_0', \dots, \sigma_d'$ are unitarily equivalent if $\sigma_j' = q^\dagger \sigma_j' q \quad \forall j = 0, \dots, d$ with appropriate unitary matrix q. Obviously the transition to a unitarily equivalent set preserves properties (C.1) and (C.2).

Lemma C.3. *If $d \geq 2$ and conditions (C.1) and (C.2) are fulfilled then by a unitary transformation the set $\sigma_0, \dots, \sigma_d$ can be reduced to the form*

$$\sigma_0 = \begin{pmatrix} I & 0 \\ 0 & -I \end{pmatrix}, \quad \sigma_1 = \begin{pmatrix} 0 & iI \\ -iI & 0 \end{pmatrix}, \quad \sigma_j = \begin{pmatrix} 0 & \alpha_j \\ \alpha_j & 0 \end{pmatrix}$$

($j = 2, \dots, d$) where $\alpha_2, \dots, \alpha_d$ are $\frac{D}{2} \times \frac{D}{2}$ matrices satisfying (C.1) and (C.2) also.

Conversely, if $\alpha_2, \dots, \alpha_d$ are $\frac{D}{2} \times \frac{D}{2}$ matrices satisfying (C.1) and (C.2) then the $D \times D$ matrices $\sigma_0, \dots, \sigma_d$ introduced in this way also satisfy (C.1) and (C.2).

Moreover if $\sigma_0, \dots, \sigma_d$ satisfy (C.3) if and only if $\alpha_2, \dots, \alpha_d$ do with the same ς and with d replaced by $d-2$.

Theorem C.2 can be easily proven by induction with respect to d with step 2.

We have obtained in particular that for $d = 2$ $\bar{D} = 2$ and

$$\sigma_0 = \begin{pmatrix} 1 & 0 \\ 0 & -1 \end{pmatrix}, \quad \sigma_1 = \begin{pmatrix} 0 & i \\ -i & 0 \end{pmatrix}, \quad \sigma_j = \varsigma \begin{pmatrix} 0 & 1 \\ 1 & 0 \end{pmatrix}$$

modulo unitary equivalence; we have obtained the *Pauli matrices;* moreover,

(C.6) $$[\sigma_j, \sigma_k] = 2i\varsigma\bar{\varepsilon}^{jkl}\sigma_l$$

where $\bar{\varepsilon}^{jkl}$ is an absolutely skew-symmetric tensor with $\bar{\varepsilon}^{123} = 1$ (we do not discuss the change of coordinates) and we write σ_3 instead of σ_0 here. We have also obtained that for $d = 3, 4$ $\bar{D} = 4$ and we have a method to construct the corresponding sets. In particular, for $d = 3$ one can assume that

(C.7) $$\sigma_0 = \begin{pmatrix} I & 0 \\ 0 & -I \end{pmatrix}, \quad \sigma_j = \begin{pmatrix} 0 & \alpha_j \\ \alpha_j & 0 \end{pmatrix} \quad j = 1, 2, 3$$

where $\alpha_3 = \alpha_0, \alpha_1, \alpha_2$ are Pauli matrices.

We do not prove the two following simple assertions here:

Proposition C.4. *Let $\sigma_0, \dots, \sigma_d$ be Dirac matrices with $d \geq 2$. Let $\mathbb{K}^{\pm} = \operatorname{Ker}(\sigma_1 \pm i\sigma_2)$. Then*

(i) $\dim \mathbb{K}^{\pm} = \frac{D}{2}$, $\mathbb{K}^+$ and $\mathbb{K}^-$ are orthogonal, and $\mathbb{K}^+ \oplus \mathbb{K}^- = \mathbb{H}$;
(ii) $\sigma_j \mathbb{K}^{\pm} = \mathbb{K}^{\pm}$ for $j \neq 1, 2$ and $\sigma_j \mathbb{K}^{\pm} = \mathbb{K}^{\mp}$ for $j = 1, 2$;
(iii) If (C.3) is fulfilled then

$$\sigma_0 \cdots \sigma_d|_{\mathbb{K}^{\pm}} = \pm\varsigma i^{\frac{d-2}{2}} I.$$

Proposition C.5. *Let $\sigma_0, \dots, \sigma_d$ be Dirac matrices and $(\omega^{jk})_{j,k=0,\dots,d}$ be a real orthogonal matrix. Then*

$$\sigma'_j = \sum_{0 \leq k \leq d} \omega^{jk}\sigma_k \quad j = 0, \dots, d$$

are also Dirac matrices. Moreover, (C.3) for $\sigma_0, \dots, \sigma_d$ yields the same equality for $\sigma'_0, \dots, \sigma'_d$ with $\varsigma' = \varsigma \operatorname{sign} \det(\omega^{jk})$.

Moreover the following assertion is useful:

Proposition C.6. *Let $\sigma_0, \dots, \sigma_d$ be Dirac matrices. Let us introduce Hermitian matrices $\beta_j = \frac{1}{2i}[\sigma_{2j-2}, \sigma_{2j-1}]$ with $j = 1, \dots, r \leq q = [\frac{d+1}{2}]$. Then*

(i) $\beta_j^2 = I$, $\beta_1, \dots, \beta_r$ commute;
(ii) For every set $(\varepsilon_1, \dots, \varepsilon_r) \in \{-1, 1\}^r$

$$\dim\Big(\bigcap_{1 \leq j \leq r} \operatorname{Ker}(\beta_j - \varepsilon_j)\Big) = 2^{-r}D.$$

Proof. (i) is obvious; (ii) is easily proven by induction with respect to r with a constant $d-2r$ where the inductive construction of the proof of theorem C.2 is used. □

Propositions C.5 and C.6 yield

Corollary C.7. *Let $\sigma_0, \dots, \sigma_d$ be Dirac matrices, $(F_{jk})_{j,k=0,\dots,d}$ be a skew-symmetric real matrix and $\pm i f_j$, $j = 1, \dots, q$ be its non-zero eigenvalues (we take multiplicities into account), $f_j > 0$. Then*

$$\mathrm{Spec}(\sum_{j,k} F_{jk}[\sigma_j, \sigma_k]) = \{\nu_\varepsilon = \sum_{1\le j\le r} \varepsilon_j f_j, \quad \varepsilon \in \{-1,1\}^r\}.$$

Further, the multiplicity of each eigenvalue on the left coincides with the multiplicity of each number on the right multiplied by $2^{-r}D$, assuming that if ν_ε and $\nu_{\varepsilon'}$ coincide then their multiplicities are combined.

4.6 Spectral Asymptotics for Families of Commuting Operators

In this section we generalize the results of the previous sections to the case of families of commuting operators.

4.6.1 Introduction

Let $A_j = A_j(x, hD, h)$ $(j = 1, \dots, l)$ be h-pseudo-differential operators on a manifold X. We assume that

(4.6.1) Either $X = \mathbb{R}^d$ or X is a compact closed C^K manifold or X is a compact C^K manifold with boundary $\partial X \in C^K$ with large enough $K = K(d)$, $d = \dim X$. In the third case A_j are differential operators and their original domains are given by some boundary value conditions.

Moreover, let us assume that

(4.6.2) The operators A_j are self-adjoint operators in $L^2(X, \mathbb{H})$ where $\mathbb{H}$ is a Hermitian space[42] and commute (i.e., their spectral projectors commute).

Then

$$A_j = \int_{\mathbb{R}^l} \tau_j d_\tau E(\tau) \tag{4.6.3}$$

where $\tau = (\tau_1, \dots, \tau_l) \in \mathbb{R}^l$, $E(\tau) = E_1(\tau_1)\dots E_l(\tau_l)$ is the spectral projector of the family $\{A_1, \dots, A_l\}$ and $E_j(\tau_j)$ is the spectral projector of A_j. The standard question is how to get the semiclassical asymptotics of the number of conjoint eigenvalues τ lying in the (compact) domain $\Xi \subset \mathbb{R}^l$. Usually, however, the case of a domain with smooth boundary is treated when $\Xi = \{\tau, \phi(\tau) \le 0\}$ with $\phi \in C^K$ and $|\phi| + |\nabla\phi|$ disjoint from 0. But results of this type are due to the previous theory applied to the operator $A = \phi(A_1, \dots, A_l)$. The real problem arises when $\partial\Xi$ is not smooth, e.g., if Ξ is a diffeomorphic image of the box $[\tau_1', \tau_1] \times [\tau_2', \tau_2] \times \dots \times [\tau_l', \tau_l]$ with $\tau_j' < \tau_j$. The above arguments yield that without loss of generality one can assume that Ξ is the box. In this case one can consider the sizes of this box as parameters as well. In this section I am going to treat the semiclassical

[42] Surely, one can consider operators acting in Hermitian bundles.

asymptotics of $\operatorname{Tr} QE(\tau';\tau)$ where $E(\tau';\tau) = (E_1(\tau_1) - E_1(\tau_1'))\dots(E_l(\tau_l) - E_l(\tau_l'))$ and Q is an appropriate h-pseudo-differential operator.

To give a more precise description of the operators in question we assume that

(4.6.4) Either A_j are h-differential operators of orders m_j or $A_j \in \tilde{\boldsymbol{\Psi}}^{(m_j)}_{\rho,\gamma,h}$ where the classes of h-pseudo-differential operators $\boldsymbol{\Psi}_{\rho,\gamma,h}$ and $\tilde{\boldsymbol{\Psi}}^{(m)}_{\rho,\gamma,h}$ were introduced in Chapter 1 for $\rho,\gamma \in (0,1]$, $\rho\gamma \ge h^{1-\delta}$, $\delta > 0$ is arbitrarily small.

Let us introduce the propagator $U(t) = \exp ih^{-1}(t_1A_1 + \dots + t_lA_l) = \exp(ih^{-1}t_1A_1)\cdots\exp(ih^{-1}t_lA_l)$ with $t = (t_1,\dots,t_l) \in \mathbb{R}^l$ to attack this problem. Then

$$U(t) = \int_{\mathbb{R}^l} e^{ih^{-1}\langle t,\tau\rangle} d_\tau E(\tau). \tag{4.6.5}$$

U satisfies the "Cauchy" problem

$$(hD_{t_j} - A_j)U = 0 \quad (j = 1,\dots,l), \qquad U|_{t=0} = \delta(x-y)I \tag{4.6.6}$$

where we use the same notation for the operator U and its Schwartz kernel.

Applying routine elliptic arguments one can easily prove the following

Proposition 4.6.1. *Let the ellipticity condition*

$$\sum_{1\le j\le l} |a_j(z)v| \ge 2\rho|v| \qquad \forall v \in \mathbb{H} \tag{4.6.7}$$

be fulfilled in the γ-neighborhood $\mathcal{U}$ of $\bar{z} \in T^*\mathbb{R}^d$ with $\gamma \ge h^{\frac12-\delta}$, $\rho \ge \gamma^2$ where a_j are the principal symbols of A_j (surely, $a_1,\dots,a_l$ commute). Let $Q \in \boldsymbol{\Psi}_{h,\gamma,\gamma}$ and its symbol be supported in $\mathcal{U}$. Then

$$|F_{t\to h^{-1}\tau}\chi_T(t)QU| \le Ch^sT_1\cdots T_l \qquad \forall \tau : |\tau| \le \frac{1}{2}\rho \;\forall x,y \in \mathbb{R}^d \tag{4.6.8}$$

provided $T_j\rho \ge h^{1-\delta}$ $\forall j$ where $\chi \in C_0^K(B_l(0,1))$ is any fixed function, subscript l means dimension here, $T = \operatorname{diag}(T_1,\dots,T_l)$, $\chi_T(t) = \chi(T^{-1}t)$.

4.6.2 Successive approximations

As in section 4.2 let us first construct

$$F_{t\to h^{-1}\tau}\chi_T(t)\operatorname{Tr} QU(t)$$

with $\chi_T(t) = \chi_{1T_1}(t_1)\dots\chi_{lT_l}(t_l)$ and $h^{1-\delta} \le T \le h^{\frac12+\delta}$ by the successive approximation method. To do it we do not need to repeat the construction of that section. An easy induction on l with proposition 4.2.6 at each step gives us the following

Proposition 4.6.2. *Let the above conditions be fulfilled and $\chi_j \in C_0^K[-1,1]$ $T_j \in [h^{1-\delta}, h^{\frac12+\delta}]$ $\forall j$. Let $Q \in \boldsymbol{\Psi}(\mathbb{R}^d)$ be an operator with compactly supported symbol (if $X \ne \mathbb{R}^d$ we replace Q by $Q\psi$ with $\psi \in C_0^K(B(\bar{x},\frac12))$ where $B(\bar{x},1) \subset X$). Then*

$$\begin{aligned}(4.6.9)\quad &|F_{t\to h^{-1}\tau}\chi_T(t)QU(t,x,y) - \\ &h^{-d}T_1\dots T_l \iint e^{ih^{-1}\langle x-y,\xi\rangle}\hat{\chi}(h^{-1}T(\tau-\tau'))\mathcal{F}(y,\xi,\tau',h)d\xi d\tau'| \le Ch^s\end{aligned}$$

where $\mathcal{F}(y,\xi,\tau',h) = \operatorname{Res}_{\mathbb{R}} F(y,\xi,\tau',h)$ *with* l*-dimensional operation* $\operatorname{Res}_{\mathbb{R}}$ *and* F *is the symbol of the operator* $(2\pi)^{-d}(\tau_1 - A_1)^{-1}\dots(\tau_l - A_l)^{-1}Q$ *constructed in the context of the formal symbol calculus, and* s *is arbitrarily large. Then* $F = \sum_{0\le n\le M} F_n(y,\xi,\tau)$ *with* $M = M(s)$ *and* F_n *are sums of terms of the type*

$$\prod_{1\le i\le L} (\tau_{j_i} - a_{j_i}(y,\xi))^{-1} b_{j_i}(y,\xi),$$

b_* *are pseudo-differential symbols supported in* $\operatorname{supp} Q$, $l+1\le L\le l+2n$ *for* $n\ge 1$,

$$F_0 = (\tau_1 - a_1)^{-1}\dots(\tau_l - a_l)^{-1} q(y,\xi),$$

and q *is the complete symbol of* Q.

4.6.3 The microhyperbolicity condition

Let us replace the routine microhyperbolicity condition by the condition

$(4.6.10)_r$ For each $\zeta\in\mathbb{S}^l$ with $\zeta''=0$ (where $\zeta = (\zeta';\zeta'') = (\zeta_1,\dots,\zeta_r;\zeta_{r+1},\dots,\zeta_l)$) and each $z\in\Omega$ there exists $\mathcal{T}\in T_zT^*X$ with $|\mathcal{T}|\le 1$ such that

$$\langle \mathcal{T} a_\zeta(z)v, v\rangle \ge \epsilon_0|v|^2 - c\sum_{1\le j\le l} |a_j(x,\xi)v|^2 \qquad \forall v\in\mathbb{H} \tag{4.6.11}$$

where $a_\zeta = \sum_{1\le j\le l}\zeta_j a_j(x,\xi)$.

Let us first assume that condition (4.6.11) is fulfilled at the point $\bar z$ for $\zeta=\bar\zeta$ only. Let $\mathcal{T}$ be a corresponding direction. Let us assume that $\mathcal{T}_x = 0$. Then the routine propagation of singularities arguments (combined with elliptic arguments) yield that if $\chi\in C_0^K$ is supported in the ϵ-vicinity of $\bar\zeta$ with sufficiently small $\epsilon>0$ and if $\mathcal{T}_x=0$ then

$$|F_{t\to h^{-1}\tau}\chi_T(t)\Gamma_x QU| \le Ch^s \qquad \forall x\forall\tau : |\tau|\le\epsilon$$

with any $T\in[\epsilon_1, T_0]$ and with $Q\in\Psi$ with symbol supported in the ϵ-neighborhood of $\bar z$ where $\epsilon_1>0$ is arbitrarily small, and T_0 is a sufficiently small constant which does not depend on ϵ_1 or the choice of χ. Repeating the arguments of theorem 2.1.15 we then obtain the inequality

$$|F_{t\to h^{-1}\tau}\chi_T(t)\Gamma_x QU| \le Ch^{l-d}\Bigl(\frac{h}{T}\Bigr)^s \qquad \forall x\ \forall\tau : |\tau|\le\epsilon\ \forall T\in[h,T_0].$$

For general $\mathcal{T}$ we then obtain the same inequality with Γ_x replaced by Γ. Taking ϵ-admissible partitions of unity in $C_0^K(B_l(0,1)\setminus B_l(0,\frac12))$ and in Ω we obtain the following

Proposition 4.6.3. *Let condition* $(4.6.10)_l$ *be fulfilled in* Ω. *Let* $Q\in\Psi$ *and the symbol of* Q *be supported in* Ω *and let* $\chi\in C_0^K(B_l(0,1)\setminus B_l(0,\frac12))$. *Then*

$$|F_{t\to h^{-1}\tau}\chi_T(t)\Gamma QU| \le Ch^{l-d}\Bigl(\frac{h}{|T|}\Bigr)^s \qquad \forall x\ \forall\tau : |\tau|\le\epsilon\ \forall T : |T|\in[h,T_0] \tag{4.6.12}$$

where ϵ *and* T_0 *are sufficiently small constants,* $T = \operatorname{diag}(T_1,\dots,T_l)$ *and* $|T| = \max_j T_j$.

Then we immediately obtain

Proposition 4.6.4. *Estimate* (4.6.9) *holds under the conditions of proposition 4.6.3 for* $T_j\in[h^{1-\delta},T_0]$ *with a small constant* T_0 *and with* $\chi\in C_0^K(B_l(0,1))$ *(now* $\operatorname{supp}\chi$ *is not necessarily disjoint from* 0*).*

Remark 4.6.5. *(i) The same arguments yield that if only a partial microhyperbolicity condition (with $r < l$) is fulfilled then estimate (4.6.12) holds for χ with $\chi \in C_0^K((B_r(0,1) \setminus B_r(0,\frac{1}{2})) \times [-\frac{1}{2},\frac{1}{2}])$, $T_j \le \epsilon|T'|$.*

(ii) In this case estimate (4.6.9) holds provided $T_j \in [h^{1-\delta}, T_0]$ $\forall j = 1,\dots,r$ and $T_j \in [h^{1-\delta}, h^{\frac{1}{2}+\delta}]$ $\forall j = r+1,\dots,l$.

(iii) One can extend these estimates to larger T_j $(j = 1,\dots,r' \le r)$ as well. In this case the multiple Hamiltonian flow $\Phi_{t_1,\dots,t_{r'}}$ should be treated and in all conditions one should take $\gamma \ge c_0 \max_{j \ge r'+1} T_j$ in order to compensate for the perturbations caused by the propagators on $t_{r'+1},\dots,t_l$. I leave the details to the reader. ◊

4.6.4 Principal results (general case)

First of all, there are rough estimates due to theorem 1.1.8 and proposition 4.6.1 (in this case we pick a Hörmander function χ (i.e., a product of one-dimensional Hörmander functions)):

Theorem 4.6.6. *Let conditions* (4.6.2) *and* (4.6.4) *be fulfilled. Then the estimate*

$$|(Qe)(x,y,\tau)| \le Ch^{-d} \tag{4.6.13}$$

holds and under the assumptions of proposition 4.6.1 the estimate

$$|(Qe)(x,y,\tau',\tau)| \le Ch^{s} \qquad \forall \tau',\tau : |\tau| \le \rho, |\tau'| \le \rho \tag{4.6.14}$$

holds.

Now let us use the successive approximations output. First of all, picking $\chi = 1$ in $B_l(0,\frac{1}{2})$, applying operator $\phi(hL^{-1}D_t)$ with $L = \operatorname{diag}(L_1,\dots,L_l)$ and setting $t = 0$ we obtain as in section 4.3 the following

Theorem 4.6.7. *Suppose the operators $A_1,\dots,A_l$ satisfy conditions* (4.6.2) *and* (4.6.4)*. Then*

(i) The estimate

$$\mathcal{R}_1 \stackrel{\mathrm{def}}{=} \Big| \int \phi(\frac{h}{L}(\bar\tau - \tau))\big(d_\tau \int (Qe)(x,x,\tau)dx - h^{-d}\iint \mathcal{F}(x,\xi,\tau)dxd\xi d\tau\big)\Big| \le Ch^{s} \tag{4.6.15}$$

holds provided $h^{\frac{1}{2}-\delta} \le L_j \le h^{-n}$ $\forall j$.

(ii) Moreover, under microhyperbolicity condition $(4.6.10)_r$ *at energy level $\bar\tau$ this estimate remains true provided $h^{1-\delta} \le L_j \le \epsilon$ $\forall j = 1,\dots,r$, $h^{\frac{1}{2}-\delta} \le L_j \le \epsilon$ $\forall j = r+1,\dots,l$ with a small enough constant $\epsilon > 0$.*

On the other hand, under microhyperbolicity condition $(4.4.10)_r$ the function

$$\int \phi(L^{-1}\tau'')\mathcal{F}(x,\xi,\tau',\tau'')d\tau''dxd\xi$$

is smooth with respect to $\tau' = (\tau_1,\dots,\tau_r)$; this follows from the theory of almost analytic functions as in the case $l = 1$. Then picking a Hörmander function (i.e., a product of one-dimensional Hörmander functions) for χ we obtain the following

Proposition 4.6.8. *Suppose that the operators $A_1,\dots,A_l$ satisfy conditions* (4.6.2) *and* (4.6.4) *and let microhyperbolicity condition* $(4.6.10)_r$ *be fulfilled. Then the estimate*

$$(4.6.16)\qquad |\int (Qe)(x,x,\lambda,\tau)| \le Ch^{-d+r}\prod_{1\le j\le r}\bigl(\frac{|\tau_j-\lambda_j|}{h}+1\bigr)$$
$$\forall \tau,\lambda : |\lambda|\le\epsilon, |\tau|\le\epsilon$$

holds with a small enough constant $\epsilon > 0$.

Estimate (4.6.16) immediately yields that

$$(4.6.17)\qquad |F_{t\to h^{-1}\tau}\chi_T(t)\Gamma QU| \le Ch^{r-d}T_1\dots T_l \qquad \forall\tau : |\tau|\le\epsilon$$

provided $T_j \ge h^{1-\delta}\ \forall j$. Applying $\phi(hL^{-1}D_t)$ to $\chi_T(t)\Gamma(QU)$ calculated by the successive approximations method (with $T_j = T_0$ for $j\le r$, $T_j = h^{\frac12+\delta}$ for $j\ge r+1$) and setting $t=0$ we then obtain

Theorem 4.6.7. *Suppose operators* $A_1,\dots,A_l$ *satisfy conditions* (4.6.2) *and* (4.6.4) *and let microhyperbolicity condition* $(4.6.10)_r$ *be fulfilled. Then the estimate*

$$(4.6.18)\qquad \mathcal{R}_1 \le Ch^{-d}L_1\dots L_r\sum_{1\le j\le r}\bigl(\frac{h}{L_j}\bigr)^s$$

holds provided $h\le L_j\le\epsilon\ \forall j=1,\dots,r$, $h^{\frac12-\delta}\le L_j\le\epsilon\ \forall j=r+1,\dots,l$.

Now we can to prove the main general statement of this section:

Theorem 4.6.8. *Suppose operators* $A_1,\dots,A_l$ *satisfy conditions* (4.6.2) *and* (4.6.4) *and let microhyperbolicity condition* $(4.6.10)_r$ *be fulfilled. Let the function* φ, *supported in* $\{\tau, |\tau|\le\epsilon\}$, *satisfy the condition*

$$(4.6.19)\qquad |D^\alpha_\tau\varphi| \le C\prod_{1\le j\le r}|\tau_j|^{\vartheta_j-(1-\sigma_j)\alpha_j} \qquad \forall\alpha : |\alpha|\le K$$

with $\vartheta_j\ge 0\ \forall j=1,\dots,r$ *and arbitrarily small* $\sigma_j\ge 0$ *such that* $\sigma_j+\vartheta_j>0\ \forall j$. *Then with* ϕ *replaced by* φ *the estimate*

$$(4.6.20)\qquad \mathcal{R}_1 \le Ch^{-d}L_1\dots L_r\sum_{1\le j\le r}\bigl(\frac{h}{L_j}\bigr)^{\vartheta_j}$$

holds provided $h\le L_j\le\epsilon\ \forall j=1,\dots,r$, $h^{\frac12-\delta}\le L_j\le\epsilon\ \forall j=r+1,\dots,l$.

Proof. The proof follows the proof of theorem 4.3.10 with obvious modifications. □

Remark 4.6.9. *Let* $a_1,\dots a_p$ *be scalar symbols. Then*

(i) Condition $(4.6.10)_p$ *means exactly that*

$$(4.6.21)_p\qquad |\sum_j\zeta_j\nabla a_j| + \sum_j|\zeta_j|\cdot|a_j| \ge \epsilon_0 \qquad \forall\zeta\in\mathbb{S}^p.$$

In this case in the microlocal study one can ignore symbols a_j *with* $|a_j|\ge\epsilon$ *(we ignore them in what follows); then we obtain the condition*

$$(4.6.22)_p\qquad |a_j|\le\epsilon,\qquad |\sum_j\zeta_j\nabla a_j|\ge\epsilon_0 \qquad \forall\zeta\in\mathbb{S}^p.$$

(ii) Moreover, without loss of generality one can assume that $a_j=\xi_j$ *for* $j=1,\dots,p$. *Moreover, without any loss of generality one can assume that* $A_j=hD_j$ *for* $j=1,\dots,p$.

(iii) Condition $(4.6.10)_r$ *with* $r>p$ *means exactly that microhyperbolicity condition* $(4.6.10)_{r-p}$ *is fulfilled for symbols* $a_{p+1},\dots,a_l$ *restricted to the manifold* $\Sigma=\{a_1=\dots=a_r=0\}$ *which is smooth under condition (4.6.21).* ◊

4.6.5 Principal results (scalar case)

Let us assume that

(4.6.23) all the symbols $a_1, \dots, a_l$ are scalar.

The first problem is to replace the condition $L_j \ge h^{\frac{1}{2}-\delta}$ by the condition $L_j \ge h^{1-\delta}$ in estimate (4.6.15):

Theorem 4.6.10. *Let conditions* (4.6.2), (4.6.4) *and* (4.6.23) *be fulfilled. Then*
(i) Estimate (4.6.20) *holds provided* $L_j \ge h^{1-\delta}$ $\forall j$.
(ii) Moreover, under condition $(4.6.21)_r$ *estimate* (4.6.15) *holds provided* $L_j \ge h^{1-\delta}$ $\forall j \ge r+1$.

Proof. Let us apply induction with respect to (q,r) where q is the number of $j = 1, \dots, r$ such that $L_j \le h^{1-\delta}$. Under condition $(4.6.22)_r$ one can assume that $A_j = hD_j$ for $j = 1, \dots, r$ and $A_j = A_j(x'', hD', hD'')$ for $j = r+1, \dots, l$ with $x = (x'; x'') = (x_1, \dots, x_r; x_{r+1}, \dots, x_d)$. Let us introduce the function

$$\gamma = \epsilon\Big(\sum_{r+1\le j\le l} (|b_j| + |\nabla b_j|^2) + |\xi'| + C_0|L|\Big)^{\frac{1}{2}} + \bar{\gamma} \tag{4.6.24}$$

with $b_j = a_j(x'', 0, \xi'')$, $|L| = \sum_{1\le j\le r} L_j$, $\bar\gamma = h^{\frac{1}{2}-\delta'}$, a small constant $\epsilon > 0$ and an arbitrarily large constant $C_0 = C_0(\epsilon)$; then $|\nabla\gamma| \le \frac{1}{2}$. Let us make a $(\epsilon, \gamma, \gamma, \gamma^2)$-admissible partition with respect to (x', x'', ξ', ξ'') and consider the contribution of one element of this partition. Obviously, we need to rescale $x''_{\text{new}} = \frac{x'' - \bar{x}''}{\gamma}$ and introduce $h_{\text{new}} = \frac{h}{\gamma^2}$. This yields the rescaling $\xi'_{\text{new}} = \frac{\xi'}{\gamma^2}$, $\xi''_{\text{new}} = \frac{\xi''}{\gamma}$. Moreover, let us divide the operators A_j by γ^2 for all $j = 1, \dots, l$ (so, $L_{j,\text{new}} = \frac{L_j}{\gamma^2}$).

Let us first consider elements of the partition with $|\xi'| \ge \epsilon_0\gamma^2$, $\gamma \ge C_0 h^{\frac{1}{2})-\delta'}$. Then theorem 4.6.8 yields that the contribution of such elements to the principal part is 0 and the contribution to the remainder estimate is $O(h^s)$ provided C_0 is large enough.

Now, if $|L| \ge \epsilon_0\gamma^2$ then after the rescaling procedure we are in the case $(q-1, l-r)$ and we obtain that the contribution of such an element does not exceed the right-hand expression of estimate (4.6.20) multiplied by $\gamma^{2(\vartheta_1+\dots+\vartheta_r - r + d)}$ while the number of such elements does not exceed $C\gamma^{-2d+2r}$.

On the remaining elements with $\gamma \ge h^{\frac{1}{2}-\delta'}$ condition $(4.6.22)_{r+1}$ is fulfilled after rescaling and permutation of A_j with $j = r+1, \dots, l$, and therefore we can also apply the induction hypothesis. We obtain that the contribution of such elements to the remainder estimate does not exceed the right-hand expression of estimate (4.6.20) multiplied by $\gamma^{2(\vartheta_1+\dots+\vartheta_r - r + d - 1)}|L|$ while the number of such elements with $\gamma \in [\varepsilon, 2\varepsilon]$ does not exceed $C\varepsilon^{-2d+2r}$.

Finally, if $\gamma \le C\bar\gamma$ then after rescaling we get $L_j \ge \epsilon_1$ for all $j \ge r+1$ and we can then apply theorem 4.6.8. We immediately obtain that the contribution of such elements also does not exceed the right-hand expression in (4.6.20) multiplied by $\bar\gamma^{2(\vartheta_1+\dots+\vartheta_r - r + d)}$ while the number of such elements does not exceed $C\bar\gamma^{-2d+2r}$. □

Now we can get rid of the mollification.

Theorem 4.6.11. *Let conditions* (4.6.2) *and* (4.6.4) *be fulfilled. Moreover, let condition* $(4.6.10)_{l-1}$ *be fulfilled (in which case* $\Sigma = \{a_1 = \dots = a_{l-1} = 0\}$ *is a smooth manifold). Let* $b = a_l|_\Sigma$ *satisfy the condition*

$(4.6.25)_r$ *If* $|b| + |\nabla b| \le \epsilon$ *then* $\operatorname{Hess} b$ *has* q *eigenvalues* $f_1, \dots, f_q$ *disjoint from* 0.

Let $L_j = 1$. *Then*

(i) Estimate (4.6.20) *holds provided* $2(\vartheta_1 + \cdots + \vartheta_{l-1}) + r - 2 > 0$.

(ii) The estimate

$$\mathcal{R}_1 \le Ch^{1-d+\vartheta_1+\cdots+\vartheta_l}|\log h| \tag{4.6.26}$$

holds provided either $r = 1$, $2(\vartheta_1 + \cdots + \vartheta_{l-1}) = 1$ *or* $r = 2$, $\vartheta_1 + \cdots + \vartheta_{l-1} = 0$.

(iii) The estimate

$$\mathcal{R}_1 \le Ch^{\frac{1}{2}-d+\vartheta_1+\cdots+\vartheta_l} \tag{4.6.27}$$

holds for $r = 1$, $2(\vartheta_1 + \cdots + \vartheta_{l-1}) < 1$.

Proof. The proof is the easy combination of the proofs of theorems 4.4.10 and 4.6.10: we introduce γ by (4.6.24) with $\bar\gamma = h^{\frac{1}{2}}$ and with $|L| = 0$ and apply a γ-admissible partition on (x'', ξ'') and a γ^2-admissible partition on ξ', and make the corresponding rescalings and divisions. I leave the details to the reader. □

Remark 4.6.12. *In the general case terms with* $|\alpha| = 1$ *in the decomposition* $b \sim \sum b_\alpha(x, \xi')\xi'^\alpha$ *prevent us from applying the other more advanced techniques of theorem 4.4.10.* ◊

4.7 Scalar Operators with Periodic Hamiltonian Flow: Very Long-Time Propagation of Singularities and Spectral Asymptotics

In this section we treat scalar operators with periodic Hamiltonian flow and weak perturbations of such operators. We consider very long-time propagation of singularities. Then under some restrictions we prove a microlocal version of a theorem concerning lacunas in the spectrum, and under other restrictions we obtain highly accurate semiclassical spectral asymptotics which may contain a non-Weylian term.

4.7.1 Introduction

Let $A \in \boldsymbol{\Psi}_{h,K}(\mathbb{R}^d)$ be a self-adjoint operator with principal symbol $a(x,\xi)$, let $\bar\tau$ be a fixed energy level (usually for a sake of simplicity $\bar\tau = 0$), and let Ω and Ω_0 be connected domains in $T^*\mathbb{R}^d$ with

$$\operatorname{dist}(\Omega, \complement\Omega_0) \ge \epsilon_0. \tag{4.7.1}$$

Let us assume that

$$|\nabla_{(x,\xi)}a(x,\xi)| \ge \epsilon_0 \quad \text{in } \Omega_0, \tag{4.7.2}$$

$$\Phi^a_{T_0}(x,\xi) = (x,\xi) \quad \forall (x,\xi) \in \Omega_0 \qquad \text{for } T_0 = T_0(x,\xi) \tag{4.7.3}$$

where Φ^a_t is a Hamiltonian flow generated by a. Then one can replace Ω_0 by $\cup_{t\in[0,T_0]}\Phi^a_{T_0}(\Omega_0)$ and hence without loss of generality we can assume that

$$\Phi^a_t(\Omega_0) = \Omega_0 \quad \forall t. \tag{4.7.4}$$

It is well known [Arnold 1] that (4.7.3) then holds with $T_0(x,\xi) = T_0(a(x,\xi))$ (i.e., the period depends on the energy level)[43] with a smooth function $T_0(\tau)$, $\tau \in [\tau_1, \tau_2]$ provided this interval is contained in the set of values of $a(x,\xi)$ at Ω_0. Moreover, it is well known that the quantity

(4.7.5) $$\frac{1}{T_0}\alpha_1 = \frac{1}{T_0}\int_0^{T_0} (-\xi dx + a dt)$$

(where the integral is taken along the Hamiltonian trajectory $(x(t), \xi(t)) = \Phi_t^a(x,\xi)$) is a constant. Moreover, we assume that

(4.7.6) $$T_0(\tau) = 1, \quad \text{i.e.} \quad \Phi_1^a = I \quad \text{in } \Omega_0$$

then the *action* α_1 will also be a constant; later we will get rid of condition (4.7.6) by means of the functional calculus.

Furthermore, it is also well known [Hörmander 1] that under conditions (4.7.4) and (4.7.6),

(4.7.7) $$e^{ih^{-1}A} \equiv \mathcal{M} \quad \text{in } \Omega$$

with operator $\mathcal{M} \in \Psi_{h,K}(\mathbb{R}^d)$ (where we decrease K if necessary). Recall that (4.7.7) means that $\|(e^{ih^{-1}A} - \mathcal{M})Q\| \le C'Mh^s$ for $Q = \mathrm{Op}(q)$, $q \in S_{h,K}(T^*\mathbb{R}^d)$ supported in Ω,[44] $M = \|\!|q|\!\|$. Moreover, the principal symbol of $\mathcal{M}$ is given in Ω by the formula

(4.7.8) $$\mathcal{M}^0 = \exp i\ell(x,\xi), \quad \ell(x,\xi) = h^{-1}\alpha + \int_0^1 a^s(\Phi_t^a(x,\xi))dt$$

where $\alpha = \alpha_1 - \frac{\pi}{2}\alpha_0 h$, α_0 is the *Maslov index* of the trajectory $\{\Phi_t^a(x,\xi),\ t \in [0,1]\}$ and α_0 is a constant; we call α the *Maslov constant*. Without loss of generality one can assume that

(4.7.9) $$|\mathcal{M}^0| \le 1 \qquad \forall (x,\xi) \in T^*\mathbb{R}^d.$$

Let us assume that

(4.7.10) $$\mathcal{M} = \eta\mathcal{M}' + \mathrm{const}\cdot I, \quad \mathcal{M}' \in \Psi_{h,K}(\mathbb{R}^d), \quad \eta \in [h^r, 1];$$

for $r = 1$ this condition can be checked effectively because we know $\mathcal{M}$ modulo $h\Psi$. Here K depends on d, s, r and s is decreasing if necessary but remains arbitrarily large.

Let us consider the self-adjoint operators $\mathcal{M}_1 = \frac{1}{2}(\mathcal{M}+\mathcal{M}^*)$ and $\mathcal{M}_2 = \frac{1}{2i}(\mathcal{M}-\mathcal{M}^*)$; then (4.7.8) remains true for $\exp ih^{-1}A$ and $\mathcal{M}$ replaced by either $\cos ih^{-1}A$ and $\mathcal{M}_1$ or $\sin ih^{-1}A$ and $\mathcal{M}_2$ respectively.

Then lemma 4.1.4 and footnote [44] yield that if

$$|\mathcal{M}_j(x,\xi,h) - \mu_j| \ge \eta \qquad \forall (x,\xi) \in \Omega_0$$

for $j = 1$ or $j = 2$, where $\zeta \in [C_0h^2\eta, 1]$ and $\mathcal{M}_j(x,\xi,h)$ and $\mathcal{M}(x,\xi,h)$ are the Weyl symbols of $\mathcal{M}_j$ and $\mathcal{M}$ respectively, then

$$E'_j(\mu_j - \frac{1}{2}\eta\zeta, \mu_j + \frac{1}{2}\eta\zeta) \equiv 0 \qquad \text{in } \Omega$$

[43] Surely there can be *subperiodic points* (x,ξ) with $T_0(x,\xi) = \frac{1}{n}T_0(a(x,\xi))$, $n \in \mathbb{Z}$ but condition (4.7.2) forbids stationary points of Φ_t^a.

[44] Surely without loss of generality one can replace Ω by its ϵ_0-neighborhood.

where E_1' and E_2' are the spectral projectors of $\cos h^{-1}A$, $\sin h^{-1}A$ respectively. Taking into account the link between these spectral projectors and the spectral projector E of A we obtain that

$$E(\tau - \frac{1}{2}h\eta\zeta, \tau + \frac{1}{2}h\eta\zeta) \equiv 0 \quad \text{in } \Omega$$

provided $\cos h^{-1}\tau = \mu_1$ or $\sin h^{-1}\tau = \mu_2$.

So we have proven

Theorem 4.7.1. *Let conditions* (4.7.7) *and* (4.7.10) *be fulfilled and*

$$|\mathcal{M}(x,\xi,h) - \exp ih^{-1}\tau| \geq \eta\zeta \quad \forall (x,\xi) \in \Omega_0 \quad \forall \tau \in [\tau_1, \tau_2] \tag{4.7.11}$$

where $\zeta \in [C_0 h^2 \eta, 1]$. *Then*

$$E(\tau_1, \tau_2) \equiv 0 \qquad \text{in } \Omega. \tag{4.7.12}$$

Remark 4.7.2. *(i) If* $[\tau_1, \tau_2]$ *is contained in the set of values of a then (4.7.11) yields that* $\eta \leq C_0$ *and* $|\tau_2 - \tau_1| \leq 2\pi h + O(h^2)$.

(ii) Condition (4.7.11) holds provided

$$|\ell(x,\xi)h - \tau - 2\pi m h| \geq C_0 h^2 \eta \quad \forall (x,\xi) \in \Omega_0 \quad \forall m \in \mathbb{Z}. \tag{4.7.13}$$

◇

We call intervals $[\tau_1, \tau_2]$ satisfying (4.7.11) *(semiclassical) spectral gaps.* Later we obtain complete asymptotics in the case when the spectral parameters belong to different spectral gaps.

4.7.2 Generalization

First of all, let us note that without loss of generality one can assume that

$$e^{ih^{-1}A} \equiv e^{i\eta B} \quad \text{in } \Omega, \qquad B = B^* \in \boldsymbol{\Psi}_{h,K}(\mathbb{R}), \quad \eta \in [h^r, 1]. \tag{4.7.14}$$

In fact, let $B_0 = \operatorname{Op}(i\ell)$ where Weyl quantization is used. Then $B^* = B$ and (4.7.7) and (4.7.8) yield that $e^{ih^{-1}A} \equiv e^{iB}(I + ih^p\mathcal{M}_p)$ in Ω where $B = B_0$, $p = 1$ and $\mathcal{M}_p \in \boldsymbol{\Psi}_{h,K}$; moreover, the self-adjointness of A and B yields that $\mathcal{M}_p^0$ is real-valued. Replacing B by $B + h^p \operatorname{Op}(\mathcal{M}_p^0)$ we obtain the same statement with p replaced by $p+1$. Induction on p yields (4.7.14) for $\eta = 1$. Comparing this equality with (4.7.10) we get (4.7.14) in the general case.

Let us treat the operator $A = A_0 + \varepsilon A'$ where A_0 is the previous operator, A' also belongs to $\boldsymbol{\Psi}_{h,K}$ and

$$\varepsilon \in [h, h^{\bar\delta}], \qquad \bar\delta > 0. \tag{4.7.15}$$

So we now treat *weakly perturbed operators with periodic Hamiltonian flow.* Let us start with

Proposition 4.7.3. *Let conditions* (4.7.2), (4.7.4), (4.7.6) *and* (4.7.15) *be fulfilled and suppose that for operator* A_0

$$a^s = 0, \quad \alpha = 0 \tag{4.7.16}$$

(Otherwise one can replace A_0 *by* $A_0 - \alpha - h\operatorname{Op}(a^s)$ *and* A' *by* $A' + h\varepsilon^{-1}\operatorname{Op}(a^s)$*). Then* (4.7.14) *is fulfilled with* $\eta = \varepsilon h^{-1} \in [1, h^{\bar\delta - 1}]$ *and* $B = \operatorname{Op}^{\mathrm{w}}(B(x,\xi,\varepsilon,h))$,

$$B(x,\xi,0,0) = \int_0^1 a'(\Phi_t^a(x,\xi))dt. \tag{4.7.17}$$

Proof. Let $U_0(t) = \exp ith^{-1}A_0$ and $U(t) = \exp ith^{-1}A$. Let us consider the operator $V(t) = U(t)U_0(-t)$; then

$$D_tV(t) = h^{-1}U(t)A'U_0(-t) = \varepsilon h^{-1}V(t)A''(t)$$

with $A''(t) = U_0(t)A'U_0(-t) \equiv A'(\operatorname{mod} \varepsilon\boldsymbol{\Psi}_{h,K})$ and $V(0) = I$. We want to construct a self-adjoint operator $A'''(t) \in \boldsymbol{\Psi}_{h,K}$ such that

$$V(t) \equiv \exp\bigl(i\varepsilon h^{-1}\int_0^t A'''(t')dt'\bigr) \qquad \text{in } \Omega. \tag{4.7.18}$$

To construct $A'''(t)$ let us consider

$$W(\sigma,t) = \exp\bigl(i\mu h^{-1}\sigma\int_0^t A'''(t')dt'\bigr).$$

Then

$$\partial_\sigma W = i\varepsilon h^{-1}\int_0^t A'''(t')dt' \times W$$

and differentiating with respect to t we obtain that

$$(\partial_\sigma - i\varepsilon h^{-1}\int_0^t A'''(t')dt')\partial_t W = i\varepsilon h^{-1}tA'''(t)W.$$

Since $\partial_t W = 0$ for $\sigma = 0$ we conclude that

$$\partial_t W = i\varepsilon h^{-1}\int_0^\sigma W(\sigma-\sigma',t)A'''(t)W(\sigma',t)d\sigma' = i\varepsilon h^{-1}A^{IV}W.$$

Setting $\sigma = 1$ we obtain that $\partial_t W_1 = i\varepsilon h^{-1}A^{IV}W_1$ for $W_1(t) = W(1,t)$ and

$$A^{IV}(t) = \int_0^1 W^*(\sigma,t)A'''(t)W(\sigma,t)d\zeta.$$

Since $V(0) = W_1(0) = I$ we conclude that (4.7.18) will be fulfilled provided

$$A^{IV}(t) \equiv A''(t) \quad \forall t \in [-1,1] \tag{4.7.19}$$

modulo negligible operators in a fixed neighborhood of Ω. Theorem 2.3.6 yields that A^{IV} and A'' belong to $\boldsymbol{\Psi}_{h,K}$ provided A''' and A' belong to this class; moreover, if $A''' \equiv \sum_{0\le n\le N} A'''_n\varepsilon^n$ with $A'''_n \in \boldsymbol{\Psi}_{h,K}$ then the same decomposition holds for A^{IV} also and for their symbols the following recurrent relation holds: $A^{IV}_n = A'''_n + \mathcal{N}(A'''_0,\dots,A'''_{n-1})$ where $\mathcal{N}$ is a non-linear partial differential operator. Then for given $A''(t)$ we always can construct $A'''(t)$ of the indicated type such that (4.7.19) is fulfilled; hence (4.7.18) holds; one can easily check that all the operators $A'',\dots,A^{IV}$ are self-adjoint. Setting $t = 0$ in (4.7.18) we obtain that $U(1) \equiv e^{ih^{-1}\varepsilon B_1}e^{ihB_0}$ where $B_1 = \int_0^1 A'''(t)dt$ and B_0 is an operator such that (4.7.14) holds with A, B, η replaced by A_0, B_0, h (condition (4.7.15) provides this possibility). Then applying the above arguments again in simplified form one can easily prove that $e^{ih^{-1}\varepsilon B_1}e^{ihB_0}Q \equiv e^{ih^{-1}\varepsilon B}Q$ for an admissible $B \equiv B_1 + \varepsilon^{-1}h^2B_0(\operatorname{mod} h^2\boldsymbol{\Psi}_{h,K})$ which yields (4.7.17). The self-adjointness of all the operators B_* is obvious. □

Remark 4.7.4. *(i) Surely if the right-hand expression in (4.7.17) vanishes then (4.7.14) holds with $\eta = \varepsilon^2h^{-1}$ etc. Therefore it is possible that condition (4.7.14) is fulfilled with $\eta \le 1$ but one should remember that B depends on ε.*

(ii) Proposition 4.7.1 remains true for this type of operator also. ◊

Proposition 4.7.5. *In the context of proposition 4.7.3*

$$[A, B] \equiv 0 \quad \text{in } \Omega_0. \tag{4.7.20}$$

Proof. Let us assume that $C = [A, B] = h^p C_1$ in Ω_0 with $C_j \in \Psi$. Let us apply the formula

$$e^{-it\eta B} A e^{it\eta B} = A - i\eta \int_0^t e^{-it'\eta B} [A, B] e^{it'\eta B} dt' \tag{4.7.21}$$

to C_1 instead of A. We then obtain that $e^{-it\eta B} C_1 e^{it\eta B} \equiv C_1 + h\eta C_2$ in Ω_0. Applying (4.7.21) to A we then obtain that $e^{-itB} A e^{itB} \equiv A - it\eta h^p C_1 + \eta^2 h^{p+1} C_3(t)$ in Ω_0; taking $t = 1$ and replacing $e^{\pm i\eta B}$ by $e^{\pm ih^{-1}A}$ we obtain that $C_1 \equiv -ih\eta C_3(1)$ and hence p can be replaced by $p + \bar{\delta}$; induction concludes this analysis. □

4.7.3 Propagation of singularities

Now we consider the long-range propagation of singularities:

Proposition 4.7.6. *Let conditions* (4.7.4), (4.7.6) *and* (4.7.14) *be fulfilled with*

$$\eta \in [h^r, h^{\bar{\delta}-1}]. \tag{4.7.22}$$

Then for all $n \in \mathbb{Z}$, $|n| \le \frac{\epsilon}{h\eta}$ with small enough constant $\epsilon > 0$ the uniform equality

$$e^{inh^{-1}A} \equiv e^{in\eta B} \tag{4.7.23}$$

holds in Ω. Moreover, for $|t| \le c$ the uniform equalities

$$e^{i(n+t)h^{-1}A} \equiv e^{in\eta B} e^{ith^{-1}A}, \tag{4.7.24}$$

$$[e^{ith^{-1}A}, e^{in\eta B}] \equiv 0 \tag{4.7.25}$$

hold in Ω.

Proof. Let us first note that $e^{in\eta B} = e^{ih^{-1}t'B}$ with $t' = n\eta h$ is a Fourier integral operator for finite t'. Then the finite speed of propagation of singularities theorem (see section 2.1) yields that $(I - Q')e^{in\eta B} Q \equiv 0$ provided $|t'| \le \epsilon$ and $Q = \operatorname{Op}(q)$, $Q' = \operatorname{Op}(q')$, $\operatorname{supp} q \subset \Omega$, $q' = 1$ in the ϵ_0-neighborhood of Ω. Then (4.7.14), (4.7.15) and footnote [44] imply that

$$e^{ih^{-1}A} e^{in\eta B} Q \equiv e^{i\eta B} e^{in\eta B} Q$$

and then (4.7.23) follows from the equality

$$(e^{ih^{-1}nA} - e^{in\eta B})Q = \sum_{0 \le j \le n-1} e^{ih^{-1}jA}(e^{ih^{-1}A} - e^{i\eta B}) e^{i(n-j-1)\eta B} Q.$$

Equality (4.7.24) follows from (4.7.23), (4.7.4) and the fact that $e^{ih^{-1}tA}$ propagate singularities along the trajectories of Φ_t^a. Similarly, from (4.7.22) we conclude that for $|t| \le c$ the equality $[e^{ih^{-1}tA}, e^{ih^{-1}t'B}]Q \equiv 0$ holds; this equality, the finite speed of propagation of singularities and footnote [44] yield that

$$[e^{ih^{-1}tA}, e^{ih^{-1}t'B}] e^{ih^{-1}t''A} Q \equiv 0 \tag{4.7.26}$$

for $|t'| \le \epsilon$, $|t''| \le \epsilon$, $|t| \le c$. This inequality yields (4.7.25). □

This corollary combined with results of section 2.3 yields

Corollary 4.7.7. *The singularities of $e^{ih^{-1}tA}$ with $|t| \le \frac{\epsilon}{h\eta}$ propagate along the (periodic) flow $\Phi^a_{t''}$ drifting along the flow $\Phi^b_{t'}$ where $t'' = t - [t]$ and $t' = [t]h\eta$ are quick and slow times respectively. These flows commute.*

The functional calculus of pseudo-differential operators of section 1.4 implies that for $n \in \mathbb{Z}$, $|n| \le h^{\delta-\frac{1}{2}}\eta^{-1}$

$$e^{in\eta B} \equiv \mathcal{G}_n \in \boldsymbol{\Psi}_{h,\gamma,\gamma,K}(\mathbb{R}^d) \quad \text{in } \Omega \tag{4.7.27}$$

and the symbol of this operator is

$$\mathcal{G}_n(x,\xi,h) = e^{in\eta b(x,\xi)} \sum_{0\le j\le N, 0\le k\le 2j} P_{j,k}(\eta n)^k h^j \tag{4.7.28}$$

where b is the principal symbol of B, $|D^\beta_{(x,\xi)} P_{j,k}| \le C' \ \forall \beta, j : |\beta| \le \frac{1}{2}K$, $j \le \frac{1}{2}K$ and $P_{0,0} = 1$, $\gamma = \frac{1}{|n|\eta}$. This statement is empty for $\eta \ge h^{\frac{1}{2}+\delta}$.

Moreover, (4.7.26), (4.7.25), (4.7.2) and the results of section 4.2 yield that if $q_i \in S_{h,\gamma,\gamma,K}$ are supported in $\{|x| + |\xi| \le c\}$ and $\rho \in C^K_0([-\epsilon, \epsilon))$ with a sufficiently small constant $\epsilon > 0$ then

$$\begin{aligned} &|F_{t\to h^{-1}\tau}\rho(t-n)\Gamma Q_{1x}U(t,x,y)^t Q_{2y} - \\ &\quad \sum_{0\le j\le N'} h^{j-d} e^{-inh^{-1}\tau} \int \hat{\rho}(\frac{\tau-\tau'}{h}) \varkappa'_{n,j}(\tau')d\tau'| \le C'Mh^s \\ &\qquad\qquad \forall n : |n| \le h^{\delta-\frac{1}{2}}\eta^{-1} \quad \forall \tau \in \mathbb{R} \end{aligned} \tag{4.7.29}$$

where $\varkappa'_{n,j}$ are the coefficients of section 4.2 constructed for the operators $Q_1\mathcal{M}_n$, Q_2 (or for Q_1, $\mathcal{M}^*_n Q_2$) instead of Q_1, Q_2.

Let Λ_1 be a closed subset in Ω_0 such that

$$\operatorname{dist}(\Phi^a_t(x,\xi),(x,\xi)) \ge \epsilon_0\gamma t(1-t) \quad \forall t \in [0,1] \quad \forall (x,\xi) \in \Lambda_1, \gamma \ge h^{\frac{1}{2}-\delta}. \tag{4.7.30}$$

This condition excludes subperiodic trajectories starting from $\operatorname{supp} q_1$. Later we will modify our arguments in order to permit them. The most important case below is when in this condition $\gamma = \epsilon'$. Then the same property holds for points of the $\epsilon_1\gamma$-neighborhood of Λ_1 intersected with Ω_0. Moreover, let us assume that q_2 is supported in the $\epsilon_1\gamma$-neighborhood of Λ_1. Then (4.7.29) remains true for $\rho \in C^K_0([-1+\epsilon, 1-\epsilon))$.

This yields that if $\chi \in C^K_0([-1,1])$ and $T \in [\frac{1}{2}, h^{\delta-\frac{1}{2}}\eta^{-1}]$ then

$$\begin{aligned} &|F_{t\to h^{-1}\tau}\chi_T(t)\Gamma Q_{1x}U(t,x,y)^t Q_{2y} - \\ &\quad \sum_{0\le j\le N', |n|\le N} h^{j-d} e^{-inh^{-1}\tau} \int T\hat{\chi}(\frac{(\tau-\tau')T}{h}) \varkappa'_{n,j}(\tau')d\tau'| \le C'Mh^s \\ &\qquad\qquad \forall \tau \in \mathbb{R} \end{aligned} \tag{4.7.31}$$

where $N = [T]$.

Let us assume that $\chi = 1$ at $[-\frac{1}{2}, \frac{1}{2}]$ and $\phi \in C^K_0([-c,c])$, $L \ge T^{-1}h^{1-\delta}$. In this case we know that

$$\begin{aligned} &|\int \phi(\frac{\tau}{L}) d_\tau \Gamma Q_{1x} e(.,.,\tau_1,\tau)^t Q_{2y} - \\ &\quad h^{-1} \int \phi(\frac{\tau}{L})(F_{t\to h^{-1}\tau}\chi_T \Gamma Q_{1x} U(.,.,t)^t Q_{2y} d\tau| \le C'Mh^s. \end{aligned} \tag{4.7.32}$$

These two inequalities yield that

$$(4.7.33)\quad |\int \phi(\frac{T}{L})d_\tau \Gamma Q_{1x}e(.,.,\tau_1,\tau)^t Q_{2y} - \sum_{0\le j\le N',|n|\le N} h^{j-d}\int \phi(\frac{T}{L})e^{-inh^{-1}}\varkappa'_{n,j}d\tau| \le C'Mh^s$$

where we have applied the estimate

$$|\phi(\frac{\tau}{L}) - \frac{T}{h}\int \hat{\chi}(\frac{(\tau-\tau')T}{h})\phi(\frac{\tau'}{L})d\tau'| \le C'h^s.$$

In fact, replacing $\phi(\frac{\tau'}{L})$ by

$$\sum_{0\le k\le N'-1} \phi^{(k)}(\frac{\tau}{L})\frac{1}{k!}(\tau'-\tau)^k L^{-k} + O((\tau'-\tau)^{N'}L^{-N'}),$$

estimating $\hat{\chi}(t)$ by $C(|t|+1)^{-2N'}$ and recalling that $\chi(0)=1$, $\chi'(0)=\chi''(0)=\cdots=0$ and that $LT\ge h^{1-\delta}$ we obtain this inequality.

Taking into account the construction of $\varkappa'_j(\tau)$ in section 4.2 and the construction of $\varkappa'_{n,j}(\tau)$ and $\mathcal{G}_n$ here we conclude that

$$(4.7.34)\quad \varkappa'_{n,j}(\tau) = \sum_{m+k\le 2j}\int P_{j,k,m}(n\eta)^m e^{in\eta b}\delta^{(k)}(\tau-a)dxd\xi$$

where

$$(4.7.35)\quad |D^\beta_{(x,\xi)}P_{j,k,m}| \le C'\gamma^{-2j+m+k-|\beta|}$$

and hence

$$\sum_{|n|\le T}\int \phi(\frac{T}{L})e^{-inh^{-1}\tau}\varkappa'_{n,j}(\tau)d\tau$$

is a sum of terms of the form

$$(4.7.36)\quad \sum_{|n|\le T}\iint \phi(\frac{T}{L})e^{in(\eta b-h^{-1}\tau)}P_{j,k,m}(n\eta)^m\delta^{(k)}(\tau-a)dxd\xi$$

with $m+k\le 2j$.

Let us replace in this sum "$|n|\le T$" by "$n\in\mathbb{Z}$;" we claim that the error will be negligible. In fact, let us consider a term with $|n|\ge T$. Then integrating by parts p times with respect to τ and changing τ we obtain a sum of terms of the form

$$c_{p,p'}L^{-p'}n^{-p}h^p(n\eta)^m\iint \phi^{(p')}(\frac{\tau+\eta hb}{L})P_{j,k,m}\delta^{(k+p-p')}(\tau-\tilde{a})dxd\xi d\tau$$

with $p'\le p$, $\tilde{a}=a-\eta hb$ and with absolute constants $c_{p,p'}$. Let us consider the integral

$$\int \phi^{(p')}(\frac{(\tau+\eta hb}{L})P_{j,k,m}\delta^{(k+p-p')}(\tau-\tilde{a})dxd\xi;$$

condition (4.7.2) yields that $|\nabla_{(x,\xi)}\tilde{a}|\ge\frac{1}{2}\epsilon_0$; this inequality and (4.7.35) yield that the absolute value of this integral does not exceed

$$C' \max_{p''=0,\dots,k+p-p'} (\frac{h\eta}{L})^{p''} \gamma^{-2j+m-p+p'+p''}$$

and hence taking p large enough we obtain that the absolute value of any term with $|n| \geq T$ does not exceed $C'h^s n^{-2}$ and hence this assertion is true.

Let us note that $\Upsilon(z) = \sum_{n\in\mathbb{Z}\setminus 0}(in)^{-1}e^{inz}$ is a 2π-periodic function equal to $\pi - z$ at $[0, 2\pi]$; hence after all these transformations we obtain from (4.7.33) that

$$(4.7.37)\quad |\int \phi(\frac{\tau}{L})d_\tau \Gamma Q_{1x}e(.,.,\tau_1,\tau)^t Q_{2y} - \sum_{0\leq j\leq N'} h^{j-d}\int \phi(\frac{\tau}{L})(\varkappa'_j(\tau) + h\frac{d}{d\tau}F_j(h^{-1}\tau,\tau))d\tau| \leq C'Mh^s$$

where $\varkappa'_j(\tau)$ are the coefficients of theorem 4.3.1 (and condition (4.7.2) yields that these coefficients are smooth) and

$$(4.7.38)\qquad F_j(\lambda,\tau) = \sum_{k+m\leq 2j}\int S_{j,k,m}\Upsilon^{(m)}(\lambda - \eta b)\eta^m \delta^{(k)}(\tau - a)dxd\xi$$

and $|D^\beta_{(x,\xi)}S_{j,k,m}| \leq C'\gamma^{-2j+k+m-|\beta|}$; in particular,

$$(4.7.39)\qquad F_0(\lambda,\tau) = (2\pi)^{-d}\int_{\Sigma_\tau} \Upsilon(\lambda - \eta b)q_1^0 q_2^0 dxd\xi : da.$$

Recall that $\Sigma_\tau = \{a = \tau\}$, $dxd\xi : da$ is a natural density on Σ_τ and q_1^0, q_2^0 are the principal symbols of Q_1, Q_2.

Since $T \leq h^{\delta-\frac{1}{2}}\eta^{-1}$ and $\delta > 0$ can be chosen arbitrarily we obtain

Theorem 4.7.8. *Let conditions* (4.7.2), (4.7.4), (4.7.6), (4.7.14), (4.7.22) *and* (4.7.30) *be fulfilled and let* $q_i \in S_{h,\gamma,\gamma,K}(T^*\mathbb{R}^d)$ *be supported in* $\Omega \cap \{|\xi| \leq c\}$ *intersected with the* $\epsilon_1\gamma$*-neighborhood of* Λ_1*; let* $Q_i = \mathrm{Op}(q_i)$. *Then for* $\phi \in C_0^K([-c,c]), L \geq \eta h^{\frac{3}{2}-\delta}$ *estimate* (4.7.36) *holds with* $M = |||q_1||| \cdot |||q_2||| \cdot \|\phi\|_{C^K}$.

Combining this theorem and theorem 4.7.1[45] and replacing condition (4.7.11) by the equivalent condition

$$(4.7.40)\qquad |B(x,\xi,h)\eta - ih^{-1}\tau - 2\pi m| \geq \eta\zeta \qquad \forall m\in\mathbb{Z}$$

we obtain

Corollary 4.7.9. *Let the conditions of theorem 4.7.8 be fulfilled and let condition* (4.7.40) *with* $\zeta \geq \epsilon_0\gamma$ *be fulfilled on* $\operatorname{supp} q_2$ *for* $\tau = \tau_1$ *and* $\tau = \tau_2$. *Then*

$$(4.7.41)\quad |\Gamma Q_{1x}e(.,.,\tau_1,\tau)^t Q_{2y} - \sum_{0\leq j\leq N'} h^{j-d}(\varkappa'_j(\tau) + hF_j(h^{-1}\tau,\tau))|_{\tau=\tau_1}^{\tau=\tau_2}| \leq C'Mh^s.$$

[45] Modified in the obvious way in order to cover the case when condition (4.7.11) is fulfilled only on a small set.

4.7.4 Spectral trace on a larger interval

In order to obtain asymptotics without mollification and with mollification with a smaller L we construct $\Gamma U(t,x,y)$ for $|t| \le \frac{\epsilon}{\eta h}$.

Let us note that

$$\|(e^{ih^{-1}tA}e^{ih^{-1}t'B} - e^{ih^{-1}(tA+t'B)})Q\| \le C'Mh^s \tag{4.7.42}$$

where here and in what follows $|t| \le c, |t'| \le \epsilon$, $\operatorname{supp} q \subset \Omega$.

In fact, (4.7.26) yields that for $|t''| \le 1$

$$\|\frac{d}{d\zeta}e^{-ih^{-1}\zeta(tA+t'B)}e^{ih^{-1}t\zeta A}e^{ih^{-1}t'\zeta B}Q\| \le C'Mh^s$$

and integrating with respect to ζ from 0 to 1 we obtain that

$$\|(e^{-ih^{-1}(tA+t'B)}e^{ih^{-1}tA}e^{ih^{-1}t'B} - I)Q\| \le C'Mh^s$$

and hence (4.7.42) is proven.

Let us now assume that

$$|\nabla_{\Sigma_\tau} b| \ge c\epsilon_0 \tag{4.7.43}$$

where ∇_{Σ_τ} means the gradient along Σ_τ for $\tau = a(x,\xi)$ (i.e., we have assumed that ∇a and ∇b are linearly independent; then $d \ge 2$ automatically); later we will weaken this condition.

Then theorem 2.1.15 implies that for $2\epsilon \ge |t| + |t'| \ge h^{1-\delta}$

$$|\Gamma Q_{1x} W(.,.,t,t')^t Q_{2y}| \le C'Mh^s$$

where $W(.,.,t,t')$ is the Schwartz kernel of the operator $e^{-ih^{-1}(tA+t'B)}$ and $Q_i = \operatorname{Op}(q_i)$, $q_i \in S_{h,\gamma,\gamma,K}(T^*\mathbb{R}^d)$ are supported in $\Omega \cap \{|x| + |\xi| \le c_1\}$, $\gamma \ge h^{\frac{1}{2}-\delta}$.

Moreover, theorem 2.3.8 implies that if condition (4.7.30) holds and q_2 is supported in the ϵ_1-neighborhood of Λ_1 then the same estimate holds for $\epsilon \le |t| \le 1-\epsilon$, $|t'| \le \epsilon\gamma$ with a small enough constant $\epsilon > 0$. This fact and (4.7.14), (4.7.26) and (4.7.43) imply

Theorem 4.7.10. *Let conditions* (4.7.2), (4.7.4), (4.7.6), (4.7.14), (4.7.30) *and* (4.7.43) *be fulfilled and let $q_i \in S_{h,\gamma,\gamma,K}(T^*\mathbb{R}^d)$ be supported in $\Omega \cap \{|x| + |\xi| \le c_1\}$ intersected with the $\epsilon_1\gamma$-neighborhood of Λ_1. Then*

(i) For $h^{-\delta}\eta^{-1} \le |t| \le T = \epsilon\gamma h^{-1}\eta^{-1}$ with $\epsilon = \epsilon(d,c) > 0$ the estimate

$$|\Gamma Q_{1x} U(.,.,t)^t Q_{2y}| \le C'Mh^s$$

holds.

(ii) Moreover, if

$$\operatorname{dist}(\Phi^a_t \circ \Phi^b_{t'}(x,\xi),(x,\xi)) \ge \epsilon_0\gamma t(1-t) \qquad \forall (x,\xi) \in \Lambda_1 \quad \forall t \in (0,1) \quad \forall t' \in (-\gamma',\gamma') \tag{4.7.44}$$

with $\gamma' \in [\gamma, 1]$ then in this assertion one can take $T = \epsilon\gamma' h^{-1}\eta^{-1}$.

Let us recall that the flows Φ^a_t and $\Phi^b_{t'}$ commute for $|t'| \le \epsilon_1$ at Ω.

So we need to construct $\Gamma Q_{1x} U(.,.,t)^t Q_{2y}$ only for $|t| \le h^{-\delta}\eta^{-1}$ but in this interval the construction of the previous subsection is applicable. So in the context of proposition

4.7.10 estimate (4.7.31) remains true for $T = \epsilon\gamma' h^{-1}\eta^{-1}$ but in this estimate one can take $N = [T']$ with arbitrary $T' \in (h^{-\delta}\eta^{-1}, h^{-\frac{1}{2}+\delta}\eta^{-1})$.

Let us consider the right-hand expression in (4.7.34) for arbitrary $n \in \mathbb{Z}$. Without loss of generality we can assume that q_1 and q_2 are supported in a small enough neighborhood $\mathcal{U}$ of the point $\bar{z}$ and then $P_{j,k,m}$ are also supported in $\mathcal{U}$. Condition (4.3.43) yields that we can introduce a coordinate system $(z_1, z') = (z_1, z_2, z'')$ in $\mathcal{U}$ with $z_1 = a$, $z_2 = b$; then

$$\varkappa'_{n,j}(\tau) = \sum_{m+k\le 2j} (-1)^k (n\eta)^m \int P^{(k)}_{j,k,m}(\tau, z') e^{in\eta z_2} dz' \tag{4.7.45}$$

where the superscript "(k)" means the k-th order derivative with respect to τ.

Substituting into (4.7.31) we obtain that, modulo terms not exceeding $C'Mh^s$,

$$\begin{aligned} F_{t\to h^{-1}\tau}\chi_T(t)\Gamma Q_{1x}U(t,x,y)^tQ_{2y} \equiv \\ \sum_{0\le j\le N', |n|\le N, m+k\le 2j} h^{j-d}(-1)^k(\eta n)^m \iint T\hat{\chi}\Bigl(\frac{(\tau-\tau')T}{h}\Bigr)\times \\ P^{(k)}_{j,k,m}(\tau, z')e^{-in(h^{-1}\tau-\eta z_2)}dz'd\tau. \end{aligned} \tag{4.7.46}$$

Let us assume that $q_i = \bar{q}_i + \tilde{q}_i$ where $\bar{q}_i \in S_{h,K}$, $\tilde{q}_i \in S_{h,\gamma,\gamma,K}$ and $\tilde{q}_i$ are supported in $\Lambda_0 = \Omega\setminus\Lambda_1$. Then $P_{j,k,m}$ and $\varkappa_{n,j}$ have similar representations. It is easy to see that if we replace $P_{j,k,m}$ in (4.7.45) by $\bar{P}_{j,k,m}$ then the error will not exceed $CTh^{1-d}M_0(\tau) + C'Mh^s$ with

$$M_0(\tau) = \int_{\Lambda_0\cap\Sigma_\tau} dxd\xi : da \tag{4.7.47}$$

(surely this estimate is very rough but improving it is useless in what follows). On the other hand, if $|n|\eta \ge 1$ then integrating by parts sufficiently many times with respect to z_2 in (4.7.44) with $P_{j,k,m}$ replaced $\bar{P}_{j,k,m}$ we obtain that

$$|\bar{\varkappa}^{(k)}_{n,j}| \le C(1+|n|\eta)^{-p} \tag{4.7.48}$$

with arbitrarily large p; therefore the sum of all terms in the right-hand expression of (4.7.46) with $P_{j,k,m}$ replaced by $\bar{P}_{j,k,m}$ does not exceed $C(1+\eta^{-1})h^{1-d}M_1(\tau) + C'Mh^s$ with

$$M_1(\tau) = \int_{\Sigma_\tau} dxd\xi : da. \tag{4.7.49}$$

So, for small M_0 the principal contribution to this estimate is due to the terms with $|n| \le C_1\eta^{-1}$ and for $\eta > C_1$ only $n = 0$ satisfies this condition. Finally, we get that

$$\begin{aligned} |F_{t\to h^{-1}\tau}\chi_T(t)\Gamma Q_{1x}U^tQ_{2y}| \le \\ Ch^{1-d}((1+\eta^{-1})M_1(\tau) + TM_0(\tau)) + C'Mh^s. \end{aligned} \tag{4.7.50}$$

A partition of unity permits us to drop the condition that $\operatorname{supp} q_i$ is small.

Applying the Tauberian arguments used in section 4.3 we obtain from (4.7.39) and (4.7.46) the same estimate (4.3.13) for $|\Gamma Q_{1x}e(.,.,\tau',\tau)^tQ_{2y}|$ with $|\tau'-\tau| \le c\frac{h}{T}$ first and then in the general case. Applying the arguments used in the proofs of theorems 4.3.6, 4.3.7 and 4.3.8 we obtain the estimates

(4.7.51) $$\mathcal{R}_0 = |\Gamma Q_{1x} e(.,.,-\infty,\tau)^t Q_{2y} - \sum_{0\le j\le N', |n|\le N} h^{j-d}\int_{-\infty}^{\tau} e^{-inh^{-1}\tau'}\varkappa'_{n,j}(\tau')d\tau'| \le Ch^{1-d}((1+\eta)\eta^{-1}T^{-1}M_1(\tau)+M_0(\tau)) + C'Mh^s \quad \forall\tau\in\mathbb{R}$$

and

(4.7.52) $$\mathcal{R}_\vartheta = |\int \varphi(\frac{\tau-\tau'}{L})d_\tau \Gamma Q_{1x} e(.,.,-\infty,\tau)^t Q_{2y} - \sum_{0\le j\le N', |n|\le N} h^{j-d}\int_{\mathbb{R}} e^{-inh^{-1}\tau'}\varphi(\frac{\tau-\tau'}{h})\varkappa'_{n,j}(\tau')d\tau'| \le Ch^{1-d}((1+\eta)\eta^{-1}T^{-1}M_1(\tau)+M_0(\tau))(\frac{h}{LT})^\vartheta + C'Mh^s \quad \forall\tau\in\mathbb{R}$$

with $N=[h^{-\delta}\eta^{-1}]$ and $\eta^{-1}T^{-1}\asymp h\gamma'^{-1}$; the coefficients $\varkappa_{n,j}$ here are associated with symbols q_i.

On the other hand, we have the routine estimates of section 4.3 for Q_i replaced by $\tilde{Q}_i$:

(4.7.53) $$\mathcal{R}_\vartheta \le Ch^{1-d}M_0(\tau)\min(1,(\frac{h}{L})^\vartheta) + Ch^s$$

(it is not now natural to assume that $L\ge h$). Surely, in this estimate there is only one term with $n=0$. However, one can calculate terms with $|n|\ge 1$ formally (with the above formula (4.7.45)). We will prove later that their total contribution does not exceed the right-hand expression in (4.7.53) and hence this estimate holds for these terms also included. The sum of the estimates for Q and $\tilde{Q}$ yields the estimate

(4.7.53) $$\mathcal{R}_\vartheta = |\int \varphi(\frac{\tau-\tau'}{L})d_\tau \Gamma Q_{1x} e(.,.,-\infty,\tau)^t Q_{2y} - \sum_{0\le j\le N', |n|\le N} h^{j-d}\int_{\mathbb{R}} e^{-inh^{-1}\tau'}\phi(\frac{\tau-\tau'}{h})\varkappa'_{n,j}(\tau')d\tau'| \le Ch^{1-d}(1+\eta^{-1})T^{-1}M_1(\tau)(\frac{h}{LT})^\vartheta + Ch^{1-d}M_0(\tau)\min(1,(\frac{h}{L})^\vartheta) + C'Mh^s \quad \forall\tau\in\mathbb{R}$$

and a similar estimate for $\mathcal{R}_0$ where now and to the end of this subsection $\mathcal{R}_0$, $\mathcal{R}_\vartheta$ are defined by $\bar{Q}$ instead of Q and the coefficients $\varkappa'_{n,j}$ correspond to $\bar{Q}$ as well.

Let us consider the non-Weylian terms (i.e., terms with $n\ne 0$) in these estimates more carefully. Their sum is given by (4.7.36) with $1\le|n|\le T$. However, due to condition (4.7.43) these terms are negligible for $|n|\ge T_1=h^{-\delta}\eta^{-1}$ and therefore we can replace T in this formula by infinity or by T_1. Considering the terms corresponding to symbols $\tilde{q}_i$ we integrate by parts on τ for $nL\ge h$ sufficiently many times and easily obtain that their total contribution does not exceed $CM_0(\tau)h^{1-d}\min(1,(\frac{h}{L})^s)$ for smooth φ with arbitrarily large s. That yields the same estimate with s replaced by ϑ for φ satisfying (4.3.19) with $\vartheta>0$. For $\vartheta=0$ the same is true (one need calculate the contribution of the "suspicious" terms only and note that the partial sums of the Fourier series for Υ are uniformly bounded).

Let us consider the non-Weylian terms defined by $\bar{Q}$. First of all, for $\eta\ge h^{-\delta}$ their contribution is negligible. Moreover, due to condition (4.7.43) the error will be negligible if we sum over all $n\in\mathbb{Z}$ instead of $|n|\le T$. Therefore we obtain the estimates

$$(4.7.55)\quad |\int \varphi(\frac{\tau}{L})d_\tau \Gamma Q_{1x}e(.,.,\tau_1,\tau)^t Q_{2y} - \sum_{0\le j\le N'} h^{j-d}\int \varphi(\frac{\tau}{L})\bigl(\varkappa'_j(\tau)+h\frac{d}{d\tau}F_j(h^{-1}\tau,\tau)\bigr)d\tau| \le R_\vartheta$$

and

$$|\Gamma Q_{1x}e(.,.,\tau_1,\tau)^t Q_{2y} - \sum_{0\le j\le N'} h^{j-d}\bigl(\varkappa'_j(\tau)+hF_j(h^{-1}\tau,\tau)\bigr)|_{\tau=\tau_1}^{\tau=\tau_2} \le R_0$$

where R_ϑ is introduced above and τ_1 is a lacunary value for symbol a.

Recall that F'_j are sums of terms $(n\eta)^p e^{in(\eta b - h^{-1}\tau)}$ with $p\le 2j$ and hence after summation on n and one integration on τ we see that F_j are sums of $\eta^p \Upsilon^{(p)}$ with mollification on $(x,\xi)\in\Sigma_\tau$. Moreover, due to condition (4.7.43) after integration by parts on (x,ξ) we can only obtain terms with $p=0$. Therefore, under condition (4.7.43) functions F_j are uniformly bounded.

For $\vartheta=0$ we immediately obtain that the contribution of the non-Weylian terms with $j\ge 1$ does not exceed CR_0 and hence we obtain the estimate

$$(4.7.56)\quad |\Gamma Q_{1x}e(.,.,\tau_1,\tau)^t Q_{2y} - \sum_{0\le j\le 1} h^{j-d}\varkappa_j(\tau_1,\tau + h^{1-d}F_0(h^{-1}\tau,\tau)|_{\tau=\tau_1}^{\tau=\tau_2} \le R_0.$$

For $\vartheta>0$ the situation is more difficult because R_ϑ can be small (until $h^{2-d+2\vartheta}\eta^\vartheta(1+\eta)$). However, we know that functions F_j are sums of functions $\Upsilon^{(k)}(h^{-1}\tau-\eta b)$ with some weights integrated over Σ_τ. On the other hand, the construction yields that

$$(4.7.57)\quad \sum_j \int h^{1-d+j}\varphi(\frac{\tau}{L})d_\tau F_j(h^{-1}\tau,\tau)\sim \int \varphi(\frac{\tau}{L})\,\mathrm{Res}_{\mathbb{R}}\,\mathrm{Tr}\,Q_1(\tau-A)^{-1}\sum_{n\ne 0}e^{in\eta B}Q_2 \sim \mathrm{Tr}\,Q_1(\tau-A)^{-1}\tilde\varphi(h^{-1}\tau-\eta B)Q_2$$

with $\tilde\varphi(\tau)=\int\phi(\frac{\tau}{L})*\Upsilon$. This equality yields that the contribution of all the Weylian and non-Weylian terms together is

$$\mathrm{Res}_{\mathbb{R}}\,\mathrm{Res}_{\mathbb{R}}\,\mathrm{Tr}\,Q_1(\tau-A)^{-1}\sin^{-1}(h^{-1}\tau'-\eta B)|_{\tau'=\tau}$$

in the context of the formal symbol calculus. In particular, we obtain a simple formula for $\eta=0$ and that for general η the deviation from the formula with $\eta=0$ is $O(h^{1-d+\vartheta}\eta^\vartheta)$.

Therefore the following theorem is proven:

Theorem 4.7.11. *Let conditions* (4.7.1), (4.7.2), (4.7.4), (4.7.6), (4.7.14) *and* (4.7.22) (4.7.43), (4.7.44) *be fulfilled and let* $Q_1=\mathrm{Op}(q_1)$, $Q_2=\mathrm{Op}(q_2), q_1,q_2\in S_{h,K}(T^*\mathbb{R}^d)$ *be supported in* $\Omega\cap\{|x|+|\xi|\le c_1\}$. *Then estimates* (4.7.51) *and* (4.7.52) *hold for*

$$(4.7.58)\quad \mathcal{R}_0^* = |\Gamma Q_{1x}e(.,.,-\infty,\tau)^t Q_{2y} - h^{-d}\varkappa_0(-\infty,\tau) - h^{1-d}\varkappa_1(-\infty,\tau) - h^{1-d}F_0(h^{-1}\tau,\tau)|$$

($\vartheta=0$, $L=1$) and for

$$(4.7.59)\quad \mathcal{R}_\vartheta^* = |\int \phi(\frac{\tau-\tau'}{L})d_{\tau'}(\Gamma Q_{1x}e(.,.,.,\tau')^t Q_{2y} - \sum_{0\le j<1+\vartheta} h^{j-d}\varkappa_j(.,\tau') - h^{1-d}F_0(h^{-1}\tau',\tau)|$$

respectively.

Remark 4.7.12. *(i) Recall that in the situation of theorem 4.7.10 Weylian asymptotics hold with remainder estimates* $O(h^{1-d})$, $O(h^{1-d}(\frac{h}{L})\vartheta)$ *for* $\mathcal{R}_0$ *and* $\mathcal{R}_\vartheta$ *respectively.*

(ii) If we apply the arguments of the proof of theorem 4.7.5 but in (4.7.31) and in the definition of L *we take* $T = \epsilon\frac{\gamma'}{h\eta}$ *then we obtain that in the context of theorem 4.7.10 for* $q_1, q_2 \in S_{h,\gamma,\gamma,K}$ *supported in* $\Omega \cap \{|x| + |\xi| \le c_1\}$ *and* q_2 *supported in* $\epsilon_0\gamma$*-neighborhood of* Λ_1 *estimate (4.7.37) remains true for* $L \ge h^{2-\delta}\eta\gamma'^{-1}$.

(ii) Applying an admissible partition of unity and using theorem 4.7.1 we conclude that one can replace condition (4.7.43) in proposition 4.7.10 by the condition

$$\frac{1}{\eta}|\frac{\tau}{h} - \eta b(x,\xi) - 2\pi m| + |\nabla_{\Sigma_\tau} b| \ge \epsilon_0. \tag{4.7.43$'$}$$

(iii) The following question arises naturally: can we decrease the coefficient of $M_1(\tau)$ *in the right-hand expression of estimate (4.7.50) surely including the next terms* $-h^{2-d}\varkappa_2(-\infty,\tau)$, $-h^{2-d}F_1(h^{-1}\tau,\tau)$ *in the left-hand expression? The answer will be affirmative if we manage to increase* T *in proposition 4.7.10. It is not a difficult task and the additional arguments are completely in terms of the Tauberian method provided the following conditions are fulfilled:*

$$\Phi^b_{t'}\Omega' \subset \Omega_0 \qquad \forall t' \in [-T', T'], \tag{4.7.58}$$

$$|D\Phi^b_{t'}/D(x,\xi)| \le h^{-\delta'} \qquad \forall (x,\xi) \in \Omega' \quad \forall t' \in [-T', T'], \tag{4.7.59}$$

$$\operatorname{dist}(\Phi^a_t \circ \Phi^b_{t'}(x,\xi), (x,\xi) \ge \epsilon_0\gamma|t'|t(1-t) \qquad \forall (x,\xi) \in \Lambda_1 \quad \forall t \in (0,1) \quad \forall t' \in [-T', T'] \tag{4.7.60}$$

where Ω' *contains the* ϵ_1*-neighborhood of* Ω, $\Phi^a_t\Omega_0 = \Omega_0$ *and* $1 \le T' \le h^{-\delta'}$, $\delta' > 0$ *is a small enough exponent. Then in the corresponding estimates one can replace* γ' *by* T'. *In this case* $d \ge 3$ *automatically.*

(iv) On the other hand, one can investigate the case when the flow $\Phi^b_{t'}$ *on* $\tilde{\Omega}_0 = \Omega_0/\sim$ *is* T_1*-periodic where* $\sim$ *is the equivalence relation given by* $z \sim z'$ *if* $\Phi^a_t(z) = z'$ *for some* t, *and if* $\tilde{\Omega}_0$ *contains the* ϵ*-neighborhood of* $\tilde{\Omega} = \Omega/\sim$; *in this case the arguments of this section work; moreover this chain could be prolonged but these conditions seem crazy.* ◇

4.7.5 The case of degenerate symbols *B*

Let us now assume that all the conditions of theorem 4.7.11 excluding (4.7.43) are fulfilled. In order to treat this case we apply our standard method of partition of unity. Let us first assume that in the 2ρ-neighborhood of the point $\bar{z} = (\bar{x}, \bar{\xi})$ the condition

$$c\rho \ge |\nabla_{\Sigma_\tau} b| \ge \epsilon_0\rho \tag{4.7.61}$$

is fulfilled with $\rho \ge h^{\frac{1}{2}-\delta}$. Let us repeat our arguments and explain how we should modify them; in this analysis q_i and $\bar{q}_i$ belong to $S_{h,\rho,\rho,K}$ and are supported in the ρ-neighborhood of $\bar{z}$, $\tilde{q}_i$ belong to $S_{h,\gamma,\gamma,K}$ and are supported in the γ-neighborhood of $\bar{z}$, the supports of $\bar{q}_i$ are contained in the $\frac{1}{4}\epsilon_1\gamma$-neighborhood of Λ_1, $\rho \ge \gamma \ge h^{\frac{1}{2}-\delta}$; everywhere, particularly in definitions of $M_k(\tau)$, the sets Σ_τ, Λ_1 and Λ_0 are replaced by their intersections with the 2ρ-neighborhood of $\bar{z}$, none of the constants depend on ρ and the norms of q_i, $\bar{q}_i$ and $\tilde{q}_i$ are calculated in the indicated classes, and $\delta' = \delta'(d, \delta, \vartheta) > 0$ is small enough.

First of all, proposition 4.7.10 remains true for $h^{-\delta}\eta^{-1}\rho^{-1} \le |t| \le T = \gamma' h^{-1}\eta^{-1}$ where now $1 \ge \gamma' \ge \epsilon\gamma\rho^{-1}$ (because for $\gamma' = \frac{\gamma}{\rho}$ condition (4.7.44) is fulfilled automatically). Therefore (4.7.46) holds with $N = [\frac{h^{-\delta'}}{\eta\rho^2}]$.

Therefore *estimate* (4.7.33) *holds for* $L \ge h^{2-\delta}\frac{\eta}{\gamma'}$. Moreover, the standard Weylian asymptotics with improved remainder estimate (see below) corresponding to the indicated T hold for $\eta\rho^2 \ge h^{-\delta}$. On the other hand, statement (4.7.27) is now replaced by

Proposition 4.7.13. *Let condition* (4.7.61) *be fulfilled and* $Q = \mathrm{Op}(q)$, $q \in S_{h,\rho,\rho,K}$ *be supported in the* ρ*-neighborhood of* $\bar z$. *Let* $T' = \eta^{-1}\rho^{-1}h^{\delta-\frac{1}{2}}$. *Then for* $t \in [-T', T']$ *there exists an operator* $\mathcal{Q}_t \in \boldsymbol{\Psi}_{h,\bar\gamma,\bar\gamma,K}$ *with* $\gamma' = h^{\delta'-\frac{1}{2}}$ *with a small enough exponent* $\delta' > 0$ *such that*

$$e^{i\eta t B}Q \equiv \mathcal{Q}_t Q$$

and the symbol of $\mathcal{Q}_t$ *is of the form*

$$\mathcal{Q}_t(x,\xi,.,.) = e^{it\eta b}\sum_{0\le j\le N, 0\le k\le 2j} h^j P_{j,k}(\eta t)^k \tag{4.7.62}$$

(compare with (4.7.28)*)* *and* $|D^\beta_{(x,\xi)}P_{j,k}| \le C'\rho^{-2j+|\beta|+k}\ \forall j,k,\beta$.

Proof. The proof follows obviously for $|t| \le T'' = \eta^{-1}h^{\delta-\frac{1}{2}}$ from the functional calculus. In the general case the proof follows from this special version of the assertion and the formula for the composition of the $m \le \rho^{-1}$ operators obtained with $|t| \le T''$. □

On the other hand, it is easy to see that (4.7.48) is now replaced by

$$|\bar{\varkappa}'^{(k)}_{n,j}| \le C'\rho^{2d-1-2j-k}(1+|n|\eta\rho^2)^{-p} \tag{4.7.63}$$

and the estimate for $|\tilde{\varkappa}'^{(k)}_{n,j}|$ remains true. Hence now the left-hand expression in (4.7.50) does not exceed

$$Ch^{1-d}((\rho^{-2}\eta^{-1}+1)M_1(\tau) + TM_0(\tau)) + C'Mh^s$$

and therefore the left-hand expressions in (4.7.51) and (4.7.52) do not exceed

$$Ch^{1-d}(\rho^{-2}\eta^{-1}+1)T^{-1}(\frac{h}{TL})^\vartheta M_1(\tau) + Ch^{1-d}M_0(\tau)\min((\frac{h}{L})^\vartheta, 1) + C'Mh^s$$

provided $TL \ge h$ (recall that $\vartheta = 0$ in (4.7.51)).

The arguments associated with the procedure which simplifies the left-hand expressions of (4.7.51) and (4.7.52) continue to hold with the obvious modifications and instead of (4.7.55) we obtain the estimate

$$\begin{aligned}\mathcal{R}_\vartheta \le Ch^{1-d}(\rho^{-2}\eta^{-1}+1)\frac{h\eta}{\gamma'}(\frac{h\eta}{L\gamma'})^\vartheta M_1(\tau) + \\ Ch^{1-d}\min((\frac{h}{L})^\vartheta, 1)M_0(\tau) + C'Mh^s\end{aligned} \tag{4.7.64}$$

provided $h\eta \le L\gamma'$ (and the same estimate for $\vartheta = 0$).

Moreover, if condition (4.7.61) is replaced by the condition

$$c\rho \ge |\nabla_{\Sigma_\tau} b| \tag{4.7.65}$$

in the 2ρ-neighborhood of $\bar z$ then $e^{in\eta B}Q_j \in \boldsymbol{\Psi}_{h,\bar\gamma,\bar\gamma,K}$ provided $|n| \le h^{\delta'-\frac{1}{2}}\rho^{-1}\eta^{-1}$, $\bar\gamma = h^{\frac{1}{2}-\delta'}$ and therefore (4.7.32) remains true for $L \ge h^{\frac{3}{2}-\delta}\rho\eta$ where $\delta > 0$ is arbitrary and in (4.7.32) q_j are supported in the ϵ_1-neighborhood of Λ_1.

So we have proven

Proposition 4.7.14. *Suppose conditions* (4.7.2), (4.7.4), (4.7.6), (4.7.14), (4.7.22) *and* (4.7.44) *are fulfilled and let* $Q_i = \mathrm{Op}(q_i)$, *etc.*, $\bar{q}_i \in S_{h,\rho,\rho,K}$, *let* q_i *be supported in the* ρ*-neighborhood of* $\bar{z}$, *let* $\bar{q}_i$ *be supported in the* $\frac{1}{4}\epsilon_1\gamma$*-neighborhood of* Λ_1, *and let* $\tilde{q}_i = \bar{q}_i - q_i \in S_{h,\gamma,\gamma,K}$ *be supported in the the* $\frac{1}{4}\epsilon_1\gamma$*-neighborhood of* Λ_0. *Then*

(i) If in the 2ρ*-neighborhood of* $\bar{z}$ *condition* (4.7.65) *is fulfilled then estimate* (4.7.32) *remains true for* $L \ge h^{\frac{3}{2}-\delta}\rho\eta$;

(ii) If in the 2ρ*-neighborhood of* $\bar{z}$ *condition* (4.7.61) *is fulfilled then estimate* (4.7.32) *remains true for* $L \ge h^{2-\delta}\frac{\eta}{\gamma'}$ *and estimate* (4.7.64) *holds;*

Using a ρ-admissible partition of unity with the function

$$\rho = \epsilon(|\nabla_{\Sigma_\tau} b|^2 + \min_{m\in\mathbf{Z}} \frac{1}{\eta}|h^{-1}\tau - b\eta - 2\pi m|)^{\frac{1}{2}} + \bar{\gamma} \qquad \bar{\gamma} = h^{\frac{1}{2}-\delta} \tag{4.7.66}$$

we immediately obtain

Theorem 4.7.15. *Let conditions* (4.7.2), (4.7.4), (4.7.6), (4.7.14), (4.7.22) *and* (4.7.44) *be fulfilled and let* $Q_i = \mathrm{Op}(q_i)$, $q_i \in S_{h,K}$ *be supported in* $\Omega \cap \{|x| + |\xi| \le c_1\}$. *Then*

$$\begin{aligned}\mathcal{R}_\vartheta \le Ch^{1-d}&\int_{\Sigma_\tau\cup\{\rho\ge\bar{\gamma},\gamma' L\ge h\eta\}} (\rho^{-2}\eta^{-1}+1)\frac{h\eta}{\gamma'}(\frac{h\eta}{L\gamma'})^\vartheta dxd\xi : da + \\ &Ch^{1-d}\min((\frac{h}{L})^\vartheta, 1)\int_{\Sigma_\tau\cap(\Lambda_0\cup\{\rho\le 2\bar{\gamma}\}\cup\{\gamma' L\le h\eta\})} dxd\xi : da + C'Mh^s\end{aligned} \tag{4.7.67}$$

with $\rho, \bar{\gamma}$ *given by* (4.7.66), $\gamma \in [\bar{\gamma}, \rho]$ *and* γ' *given by* (4.7.44).

Moreover, if $q_i \in S_{h,\gamma,\gamma,K}$ *and* $\operatorname{supp} q_i \cap \Sigma_\tau$ *are contained in the* $\epsilon_1\gamma$*-neighborhood of* Λ_1 *then estimate* (4.7.37) *holds for* $L \ge h^{2-\delta}\frac{\eta}{\gamma'}$.

4.7.6 Corollaries

Let us discuss various consequences. First of all, theorem 4.7.15 implies

Corollary 4.7.16. *Let the conditions of theorem 4.7.15 be fulfilled. Moreover, let us assume that*

$(4.7.68)_r$ If $\rho \le \epsilon_0$ then $\mathrm{Hess}_{\Sigma_\tau} b$ has eigenvalues ω_i $(i = 1, \dots, r)$ with $|\omega_i| \ge \epsilon_0$ where $\mathrm{Hess}_{\Sigma_\tau}$ denotes the Hessian along Σ_τ with $\tau = a(x,\xi)$.

Moreover let us assume that

$(4.7.69)_{r'}$ Condition (4.7.44) is fulfilled with $\gamma' = \epsilon$, $\gamma(z) = \epsilon_0 \min(\mathrm{dist}(z, \Xi), h^{\frac{1}{2}-\delta})$ where Ξ is a C^K submanifold of Σ_τ of codimension $2r'$.

Then $\mathcal{R}_\vartheta \le R_1 + R_2$ *with*

(i) $R_1 = Ch^{2-d}(\eta + 1)(\frac{h\eta}{L})^\vartheta$ *for* $r = 3$, $R_1 = Ch^{2-d}(\eta + |\log h|)(\frac{h\eta}{L})^\vartheta$ *for* $r = 2$ *and* $R_1 = Ch^{2-d}(\eta + h^{\delta-\frac{1}{2}})(\frac{h\eta}{L})^\vartheta$ *for* $r = 1$. *So, for* $r = 1, 2$ R_1 *coincides with the optimal remainder estimate for* $\eta \ge h^{\frac{1}{2}-\delta}$, $\eta \ge |\log h|$ *respectively.*

(ii) $R_2 = Ch^{1-d+r'-\delta'}\min(\frac{h}{L})^\vartheta$.

Remark 4.7.17. *This corollary is slightly less precise than the corresponding statements of section 4.4; the reason is that in theorem 4.7.15 we cannot take* $\bar{\gamma} = h^{\frac{1}{2}}$. *I think that only technical difficulties prevent this. Later we will use an alternative approach.* ◇

4.7.7 Some remarks

Let us made a few final remarks on the first part of this section.

Remark 4.7.18. *We can reduce the case when condition (4.7.6) is violated to the case treated here: if we replace A by $\tilde{A} = f(A)$ where f is a monotone increasing function given near the energy level $\bar{\tau}$ by the formula*

$$f(\tau) = \int_{\bar{\tau}}^{\tau} \frac{1}{T_0(\tau')} d\tau'; \tag{4.7.70}$$

in section 1.4 we proved that the operator A' then belongs to $\boldsymbol{\Psi}_{h,K}$ provided A does and it is easy to prove the following relations for symbols:

$$\tilde{a}(x,\xi) = f(a(x,\xi)), \tag{4.7.71}$$

$$\tilde{a}^s(x,\xi) = f'(a(x,\xi))a^s(x,\xi), \tag{4.7.72}$$

and hence $\tilde{T}_0(\tilde{\tau}) = T_0(\tau)f'(\tau)$ where $\tilde{\tau} = f(\tau)$ and the prime means derivative; then $\tilde{T}_0(\tilde{\tau}) = 1$ near 0 because of the choice of f.

Moreover, the constants α_1 (see (4.7.5)) and

$$\frac{1}{T_0(a(x,\xi))} \int a^s(\Phi_t^a(x,\xi))dt \tag{4.7.73}$$

are invariant.

Furthermore, if $A = A_0 + \varepsilon A'$ with $\varepsilon \in [h, h^\delta]$ then $f(A) = f(A_0) + \varepsilon A''$ where $A'' \in \boldsymbol{\Psi}_{h,K}$ uniformly and $A'' \equiv f'(A_0)A' \pmod{\varepsilon u \boldsymbol{\Psi}_{h,K}}$. ♢

Remark 4.7.19. *All the results obtained here can also be derived for operators of the class $\tilde{\boldsymbol{\Psi}}_{h,K}^{m}$, for operators in domains which satisfy*

$$\operatorname{dist}(\partial X, \Phi_t^a(x,\xi)) \ge \epsilon_0 \qquad \forall t \in (0,1) \quad \forall (x,\xi) \in \Omega_0,$$

for operators on compact manifolds, and for operators on manifolds of the form $X' \times \mathbb{R}^{d''}$ where X' is a compact manifold. ♢

Remark 4.7.20. *All the asymptotics with remainder estimate greater than $h^{2-\delta}(\eta+1)$ remain true if we assume that all the operators belong to $\boldsymbol{\Psi}_{h,\nu,\nu,K}$ with $\nu = h^{\delta'}$ with small enough $\delta' > 0$ depending on d, δ, r, r'; moreover we can replace condition (4.7.2) by*

$$|\nabla_{(x,\xi)} a(x,\xi)| \ge \epsilon_0 \nu \qquad \forall (x,\xi) \in \Omega_0; \tag{4.7.2$'$}$$

in all these cases it is necessary to assume that $|D\Phi_t^a(x,\xi)/D(x,\xi)| \le c\nu^{-1}$. The same is true for all the other estimates provided we add the factor $h^{-\delta}$ to their right-hand expression.

4.7.8 Operators with completely periodic evolution

Now I want to present a quite different approach. However, the first step of the new approach is based on the previous ideas. Let us assume in this subsection that

$$e^{ih^{-1}A} \equiv I \qquad \text{in } \Omega. \tag{4.7.74}$$

Theorem 4.7.1 immediately yields that the spectral zones in Ω lie in very narrow vicinities of the points $2\pi nh$ with $n \in \mathbb{Z}$. Let $\bar{\tau} = 2\pi nh$; let $\tau_1 \in (2\pi(n-1)\pi h, 2\pi nh)$, $\tau_2 \in (2\pi n\pi h, 2\pi(n+1)h)$ be lacunary values and let $\phi \in C_0^K([-1,1])$ equal 1 on $[-\frac{1}{3},\frac{1}{3}]$. Then

$$QE(\tau_1,\tau_2) \equiv \int \phi(\frac{\tau-\bar{\tau}}{L})Qd_\tau E(*,\tau) \equiv \phi(\frac{1}{L}(hD_t-\bar{\tau}))\bar{\chi}_T QU|_{t=0} \tag{4.7.75}$$

for $L=\frac{h}{2}$, $\bar{\chi}$ the same type of function as ϕ, and $LT \geq h^{1-\delta}$. One can replace $\bar{\chi}_T(t)$ here by $\sum_{|n|\leq T}\bar{\chi}(t-n)$ for $\bar{\chi} \in C_0^K([-\frac{2}{3},\frac{2}{3}])$ such that the last sum with n running over $\mathbb{Z}$ equals 1. Then applying (4.7.74) we obtain that the right-hand expression of (4.7.75) is equal to

$$\int_{\mathbb{R}} \phi(\frac{\tau-\bar{\tau}}{L}))\sum_{|n|\leq T} e^{ih^{-1}n\tau}F_{t\to h^{-1}\tau}\bar{\chi}(t)QU\,d\tau \tag{4.7.76}$$

provided $T \leq h^{-m}$ with arbitrarily large m. Recall that QU is a Fourier integral operator. Let us write its Schwartz kernel in the form of a Lagrangian distribution of the class $I_{h,M}$ [46] where temporarily $M = \{(x,\xi,y,\eta,t,\tau) : (x,\xi) = \Phi_t(y,-\eta), \tau = a(x,\xi)\}$. Then one can easily see that the Schwartz kernel of $F_{t\to h^{-1}\tau}\bar{\chi}(t)QU$ is a Lagrangian distribution of the class $I_{h,\Lambda}$ with Lagrangian manifold

$$\Lambda_\tau = \{(x,\xi;y,\eta),\ a(x,\xi)=\tau, \exists t : (y,-\eta) = \Phi_t(x,\xi)\} \tag{4.7.77}$$

depending smoothly on $h^{-1}\tau$ provided there are no subperiodic trajectories.[47] Then one can take the sum on n running over $\mathbb{Z}$; the error will be negligible. The sum of the exponents then equals $\sum_m \delta(t-2\pi mh)$. Moreover, one can take here only $m=0$ (other terms vanish because of the choice of ϕ, L). Then the right-hand expression of (4.7.76) is a Lagrangian distribution of the class $hI_{h,\Lambda_{\bar{\tau}}}$ depending smoothly on $\bar{\tau}$ with Lagrangian manifold $\Lambda_{\bar{\tau}}$.[48] So, the following statement is proven:

Theorem 4.7.21. *Let conditions* (4.7.1), (4.7.2), (4.7.4), (4.7.6) *and* (4.7.74) *be fulfilled. Then*

(i) Let there be no subperiodic trajectory. Then $\forall \tau = 2\pi nh$ *with* $|\tau| \leq \epsilon$ *the Schwartz kernel of* $E_A(\tau-\zeta;\tau+\zeta)$ *is a Lagrangian distribution with Lagrangian manifold* $\Lambda_{\bar{\tau}}$*;* $\zeta = C_0 h^s$ *here.*[49]

(ii) In the general case Λ *is (locally) the union of a finite number of Lagrangian manifolds and the distribution in question is microlocally the sum of the corresponding Lagrangian distributions.*

Now let us consider $\operatorname{Tr} QE(\tau_1,\tau_2)$. The same techniques are applicable (it seems to be more convenient to go from $\operatorname{Tr} F_{t\to h^{-1}\tau}\chi_T QU$ than from QE). First of all, let us assume that there are no subperiodic trajectories. Then one can take $\bar{\chi}$ supported in a small vicinity of 0. Then applying the calculations of subsection 4.2.3 we immediately obtain that modulo $O(h^s)$ the answer is

$$(2\pi h)^{1-d}\iint \operatorname{Res}_{\mathbb{R}}(Q(\tau-A)^{-1})(x,\xi)\,dxd\xi \tag{4.7.77}$$

46) Recall that this means that it is given by the expression $h^{-l}\int b(x,y,t,\theta)e^{ih^{-1}\varphi(x,y,t,\theta)}d\theta$ with l-dimensional variable θ and phase function φ satisfying standard conditions.

47) If there are subperiodic trajectories then Λ_τ is a locally finite union of Lagrangian manifolds and we obtain a sum of the corresponding Lagrangian distributions.

48) By definition $\bar{\tau}$ runs over a lattice; nevertheless we can obviously extend the phase functions and amplitudes to the interval.

49) However, $E(\tau-\zeta;\tau+\zeta)$ surely is not a Fourier integral operator because $\dim\operatorname{Ker} d\pi_X = \dim\operatorname{Ker} d\pi_Y = 1(\neq 0)$ for projectors $\pi_X : \Lambda \to T^*X$, $\pi_Y : \Lambda \to T^*Y$, $Y = X$ here.

for $t = \bar{\tau}$. On the other hand, let us assume now that there are subperiodic trajectories with the (minimal) subperiod $T = \frac{1}{p}$, $p \in \mathbb{N}$ (then one should also take into account periods $\frac{n}{p}$, $n \in \mathbb{Z}$). Let Ξ be the set of all the subperiodic points. Due to [Colin de Verdiere 2], proposition 1.3,[50] we have the following

Proposition 4.7.22. *The set of all subperiodic points Ξ is a union of a finite number of symplectic manifolds $\Xi_{p,r}$ of codimension $2r$ consisting of $\frac{1}{p}$-subperiodic points ((p,r) runs over a certain index set) with $T_z\Xi_{p,r} = \operatorname{Ker}(D\Phi_{\frac{1}{p}}(z) - I)$ for all $z \in \Xi_{p,r}$. Moreover, due to the equality $\Phi_T^p = I$ the same equality holds for $D\Phi_T(z)$ for $z \in \Xi_p$ and hence $\operatorname{Spec} D\Phi_{\frac{1}{p}}(z) \subset \{e^{2\pi i \frac{q}{p}}, q = 0, \dots, p-1\}$ and all the Jordan cells of $D\Phi$ are 1-dimensional. Moreover, $\Xi_{p,r}$ and $\Xi_{p,r'}$ are disjoint for $p \ne p'$, $\Xi_{p'} \subset \Xi_p$ if and only if p is a factor of p' and $\Xi_{p'} \cap \Xi_{p''} = \Xi_p$ where p is the largest common factor of p' and p'' and the intersection is clean.*[51] *Here $\Xi_p = \cup_r \Xi_{p,r}$ is the set of $\frac{1}{p}$-periodic points. Moreover, the intersection of $\Xi_{p,r}$ and Σ_τ is clean and $\operatorname{codim} \Xi_{p,r} \cap \Sigma_\tau = 2r + 1$.*

Therefore in order to take into account subperiodic trajectories one should in addition treat

$$\int \phi(\frac{\tau - \bar{\tau}}{L}) F_{t \to h^{-1}\tau} \chi_T(t - T') \operatorname{tr} QU(t) d\tau$$

with $T = k + \frac{q}{p}$ and irreducible integers p and q, $1 \le q \le p-1$. The easy calculations of oscillatory integrals give us an answer of the form

$$\int \phi(\frac{\tau - \bar{\tau}}{L}) h^{-d+r} \int_{\Xi_{p,r} \cap \Sigma_\tau} F_{p,q}(x, \xi, \tau, h) e^{ih^{-1}T'\tau} d\mu d\tau$$

with smooth functions F_* where $d\mu = d\mu_\tau$ is a smooth density on $\Xi_{p,r} \cap \Sigma_\tau$. One can see easily that the error will be negligible if we sum over $k \in \mathbb{Z}$ instead of $|k| \le N$ with $N = h^{-K}$ with large enough K. After summation on $k \in \mathbb{Z}$ we get a sum of $h\delta(\tau - 2\pi mh)$ and only $m = n$ gives a contribution to the final answer which at this step is equal to

$$h^{1-d+r} \int_{\Xi_{p,r} \cap \Sigma_{\bar{\tau}}} F_{p,q}(x, \xi, \bar{\tau}, h) e^{2\pi i n \frac{q}{p}} d\mu. \tag{4.7.79}$$

So, the following statement has been proven:

Theorem 4.7.23. *Let conditions* (4.7.1), (4.7.2), (4.7.4), (4.7.6) *and* (4.7.74) *be fulfilled. Then* $\operatorname{Tr} QE(2\pi nh - \zeta, 2\pi nh + \zeta)$ *is the sum of the terms* (4.7.77) *and* (4.7.79) *where (p,r) runs over the set of the admissible indices and q runs over* $(0), 1, \dots, p-1$.[52]

4.7.9 Operators with global decomposition

In this subsection we treat operators of the form

(4.7.80) $A = A^0 + \eta hB$ where $e^{ih^{-1}A^0} = I$ and $B \in \boldsymbol{\Psi}_h$ and A^0 commute.

Then (4.7.14) holds with the same operator B. Later I will show that the general case can be reduced to this one. Then

[50] V. Ginzburg explained this proof and the linearization property to me.

[51] See [Duistermaat, Guillemin 1] for the definition.

[52] One can easily see that the character of the answer does not change if we include $q = 0$ and (or) q with reducible ratio $\frac{q}{p}$.

(4.7.81)
$$E_A(\tau',\tau'') = \sum_m \pi_n E_B(\eta^{-1}(\tau' - 2\pi mh), \eta^{-1}(\tau'' - 2\pi mh))$$

where π_m is the spectral projector of A to the eigenspace corresponding to eigenvalue $2\pi mh$. We are interested in the eigenvalues of A close to $\bar\tau = 2\pi nh$. For $\eta \le \epsilon_2$ only the term with $m = n$ is essential in the right-hand expression of (4.7.81). Without loss of generality one can assume that microlocally $a = \xi_1$. Let us consider a $(\gamma,\gamma,\varepsilon,\gamma^2)$-admissible partition of unity with respect to (x',ξ',x_1,ξ_1) with $\gamma = \gamma(x',\xi')$ given by (4.7.44), $\bar\gamma = h^{\frac{1}{2}}$. For $\gamma^2\eta \le \epsilon_2$ only the term with $m = n$ is essential in the right-hand expression of (4.7.80) after application of the operator Q with the corresponding support. We treat only this case below. One can treat the opposite case by the same method but the results of the previous subsections are sufficient for our needs in this case. Let us consider the two-parameter group $\mathbf{U}(t,t') = e^{ih^{-1}(tA+t'B)}$. This group is 1-periodic with respect to t. The following statement is due to the results of section 4.6:

Proposition 4.7.24. *In the previous assumptions for small enough positive constants* T, T'

$$F_{(t,t')\to h^{-1}(\tau,\tau')}\bar\chi_T(t)\bar\chi_{T'}(t')\operatorname{Tr} QU$$

is given modulo $O(h'^s)$ *by the successive approximations method where* $h' = \frac{h}{\gamma^2}$ *and the absolute value of this quantity does not exceed* Ch'^{-d}.

Then the same answer with an additional factor $e^{imh^{-1}\tau}$ holds for this expression with $\bar\chi_T(t-m)$ instead of $\bar\chi$. Therefore in the absence of subperiodic trajectories

$$\int \phi(\frac{\tau-\bar\tau}{L_0})F_{(t,t')\to h^{-1}(\tau,\tau')}\bar\chi_T(t)\bar\chi_{T'}(t')\operatorname{Tr} QU\,d\tau$$

is given modulo $O(h'^s)$ by the successive approximations method and further calculations: it is equal to

$$\int \operatorname{Res}_{\mathbb{R}} \operatorname{Tr} Q(\tau - A)^{-1}(\tau' - B)^{-1}\phi(\frac{\tau-\bar\tau}{L_0})d\Upsilon(h^{-1}\tau)$$

provided $TL_0 \ge h^{-\delta}$; here $\operatorname{Res}_{\mathbb{R}}$ is taken with respect to (τ,τ'). For $L \le h$ the absolute value of the last expression does not exceed Ch'^{2-d}. Applying the Tauberian method we obtain then that for a function φ satisfying (4.3.19)

$$\int \varphi(\frac{\tau'}{L})d_{\tau'}QE_B(*,\tau')\pi_n = \int \varphi(\frac{\tau'}{L})d_{\tau'}\operatorname{Res}_{\mathbb{R}}\operatorname{Tr} Q(\bar\tau - A)^{-1}(\tau'-B)^{-1}d\tau'$$

modulo $O(h'^{2-d}h^\vartheta L^\vartheta)$.

This immediately yields (after summation with respect to the partition)

Proposition 4.7.25. *Let the previous assumptions be fulfilled and let there be no subperiodic trajectory. Let condition* $(4.7.68)_r$ *be fulfilled. Let* $L = 1$. *Then for* $r \ge 3$ *we obtain the estimate*

$(4.7.82)_3$
$$R_\vartheta \le h^{2-d+2\vartheta}\eta^\vartheta(1+\eta)$$

again and we obtain estimates

$(4.7.82)_2$
$$R_\vartheta \le h^{2-d+2\vartheta}\eta^\vartheta|\log h| + h^{2-d+2\vartheta}\eta^{\vartheta+1},$$

$$R_\vartheta \le h^{\frac{3}{2}-d+2\vartheta}\eta^\vartheta + h^{2-d+2\vartheta}\eta^{\vartheta+1} \tag{4.7.82$_1$}$$

and

$$R_\vartheta \le Ch^{1-d+\frac{3}{2}\vartheta}\eta^\vartheta \tag{4.7.82$_0$}$$

for $r = 2, 1, 0$ *respectively. The same estimates hold for* $\vartheta = 0$ *as well.*

Now let us consider the contribution of the subperiodic trajectories. Direct calculation based on proposition 4.7.22 yields that the contribution of such trajectories to $F_{t\to h^{-1}\tau}\chi_T(t)QU$ does not exceed $Ch^{2-d}T$ (which does not exceed the standard estimate of this Fourier transform) and therefore their contribution to the remainder estimate does not exceed $Ch^{2-d+\vartheta}T^{-\vartheta}$ which does not exceed the final remainder estimate. Furthermore, their contributions to the principal parts are given by the sum of the terms

$$\begin{aligned} &h^{1-d+r+\vartheta+j}\eta^\vartheta \times \\ &\iint_{\Sigma_\tau\cap\Xi_{p,r}} \varphi(\frac{\tau-\bar\tau}{L})d_\tau \sum_{0\le q\le p-1} e^{2\pi i\frac{p}{q}}\Upsilon(p^{-1}h^{-1}(\tau-\eta hb))d\rho_{p,q,r} \end{aligned} \tag{4.7.83}$$

where $\rho_{p,q,r}$ are appropriate densities on $\Xi_{p,r}$.

Therefore the following statement is proven:

Theorem 4.7.26. *Let operator* A *be given by* (4.7.80) *where conditions* (4.7.1), (4.7.2), (4.7.4), (4.7.6) *and* (4.7.74) *are fulfilled for* A_0. *Furthermore, let condition* (4.7.68)$_r$ *be fulfilled. Then estimates* (4.7.82)$_{3,2,1,0}$ *hold for* $r \ge 3$, $r = 2$, $r = 1$ *and* $r = 0$ *respectively where now in the principal part of the asymptotics terms of type* (4.7.83) *are also included.*

4.7.10 General operators

Let us consider the general operator A satisfying conditions (4.7.1), (4.7.2), (4.7.4) and (4.7.6). Then condition (4.7.14) is also fulfilled and without loss of generality one can assume that B is a self-adjoint operator. Let $A_0 = A - \eta B$. Then

$$e^{ih^{-1}A_0} \equiv I, \quad [A_0, B] \equiv 0 \qquad \text{in } \Omega_1 \tag{4.7.84}$$

where Ω_1 is the intermediate domain between Ω and Ω_0. Then for the operator A_0 the results of subsection 4.7.8 hold. Let us set

$$\bar A_0 = \sum_n \tau_n E(\tau_n - \zeta, \tau_n + \zeta) \tag{4.7.85}$$

with $\epsilon h \ge \zeta \ge h^K$ and $\tau_n = 2\pi nh$. Then

$$\bar A_0 \equiv A_0 \quad \text{in } \Omega_1, \qquad e^{ih^{-1}\bar A_0} = I.$$

Furthermore, let us set

$$\bar B = \frac{1}{2\pi}\int_0^{2\pi} e^{-ih^{-1}t\bar A_0} B e^{ih^{-1}t\bar A_0}\, dt.$$

Then

$$\bar B \equiv B \quad \text{in } \Omega_1, \qquad [\bar A_0, \bar B] = 0$$

and all the operators are self-adjoint.

Then the operator $\bar{A} = \bar{A}_0 + \bar{B}$ satisfies the assumptions of the previous section and we can apply those results to it. On the other hand, $A \equiv \bar{A}$ in Ω_1. The propagation of singularities results of subsection 4.7.3 yield that $e^{ith^{-1}tA}Q \equiv e^{ith^{-1}t\bar{A}}Q$ for $|t| \le T = \frac{\epsilon}{h\eta}$ and therefore

$$\int \hat{\chi}(\frac{(\tau-\tau')T}{h})d_{\tau'}\nu(\tau') \equiv \int \hat{\chi}(\frac{(\tau-\tau')T}{h})d_{\tau'}\bar{\nu}(\tau')$$

with $\nu(\tau) = \operatorname{Tr} QE_A(-\infty,\tau)$ and $\bar{\nu}(\tau) = \operatorname{Tr} QE_{\bar{A}}(-\infty,\tau)$. Picking a Hörmander function for χ we obtain by the Tauberian method that

$$|\nu(\tau+\frac{h}{T}) - \nu(\tau)| \le C\sup_\tau |\bar{\nu}(\tau+\frac{h}{T}) - \bar{\nu}(\tau)| + Ch^s$$

and one can see easily that the right-hand expression does not exceed the estimate for $\mathcal{R}_0$. Then applying the arguments of the proof of theorem 4.3.6 one can see easily that

$$|\int \varphi(\frac{\tau-\bar{\tau}}{L})d_\tau(\nu(\tau) - \bar{\tau}(\tau))|$$

does not exceed the remainder estimate for $\mathcal{R}_\vartheta$ and then the following statement is proven:

Theorem 4.7.27. *Theorem 4.7.26 remains true under the present assumptions.*

4.8 Mollification with Respect to Non-Spectral Parameters

4.8.1 Linear dependence

Let operator $A = A_\zeta$ depend on parameter $\zeta \in (-1,1)$ smoothly and microhyperbolically, i.e.,

$$\operatorname{Re}\langle \partial_\zeta av, v\rangle \ge \epsilon_0|v|^2 - c|av|^2 \qquad \forall\zeta \quad \forall(x,\xi)\in\Omega \quad \forall v\in\mathbb{H} \tag{4.8.1}$$

where a is the principal symbol of A.

Then the propagation of singularities theory yields that U is negligible in zones $\{|t| \le T_0, \pm t > 0, \pm z \le \pm\epsilon_1 t - T, |\tau| \le \epsilon_2\}$ with appropriate constants $\epsilon_j > 0$ and $T_0 > 0$ where z is dual to the ζ variable and $T > h^{1-\delta}$ is a parameter, $\delta > 0$.

In particular, U is negligible for $t = \pm T_0$, $\pm z \le \epsilon_3$. Moreover, let us assume that

(4.8.2) A_ζ is a linear function of ζ,

$$\operatorname{Re}\langle \partial_\zeta A_\zeta u, u\rangle \ge 0 \quad \forall u. \tag{4.8.3}$$

Then using the standard energy estimates technique with $\psi = \psi(z)$ one can prove easily that U is negligible in zones $\{T_0 \le \pm t \le h^{-l} = T_1, \pm z \le \frac{1}{2}\epsilon_3\}$ with an arbitrary exponent l.

In the z-representation, let us apply the operator $\phi_1(hL^{-1}D_z)$ where $\phi_1 \in C_0^K([-\frac{1}{2},\frac{1}{2}])$ and set $z = 0$; in the ζ-representation this procedure means multiplication by $\phi_1(\frac{\zeta}{L})$ and integration on $\zeta \in \mathbb{R}$. The resulting function is negligible for $T_1 \ge |t| \ge T, \tau \in [-\epsilon_2, \epsilon_2]$ and for $|t| \le h^{\frac{1}{2}+\delta}$ this function was constructed by a successive approximation method. Thus, for $\tau, \tau' \in [-\frac{1}{2}\epsilon_2, \frac{1}{2}\epsilon_2]$ we obtain the complete asymptotic expansion for

(4.8.4) $$\frac{1}{L}\int \phi_1(\frac{\zeta}{L})(\Gamma Q'_x{}^t Q''_y e)(\tau',\tau;\zeta)d\zeta$$

and therefore for $\tau \in [-\frac{1}{2}\epsilon_2, \frac{1}{2}\epsilon_2]$ we obtain the complete asymptotic expansion for

(4.8.5) $$\frac{1}{LL_1}\iint \phi(\frac{\tau'}{L_1})\phi_1(\frac{\zeta}{L})(\Gamma Q'_x{}^t Q''_y e)(\tau',\tau;\zeta)d\zeta d\tau'.$$

We assume in these expansions that L and L_1 are such that $1 \geq L \geq h^{\frac{1}{2}-\delta}$ (because the only condition on L is $LT \geq h^{1-\delta}$ and we can take $T = h^{\frac{1}{2}+\delta}$) and $1 \geq L_1 \geq 0$.

Moreover, if a has only one eigenvalue in the interval $[-h^{\delta'}, h^{\delta'}]$ with a small enough exponent $\delta' > 0$ (in which case this eigenvalue is of constant multiplicity) then the same arguments as for mollification on the spectral parameter in section 4.4 yield that we can weaken the restrictions on L and obtain the condition $1 \geq L \geq h^{1-\delta}$. Moreover, later for the two-dimensional Schrödinger and Dirac operators with strong magnetic field we will be able to weaken this condition also, replacing it by the same condition as for mollification with respect to spectral parameter. So the following statement has been proven:

Proposition 4.8.1. *Under conditions (4.8.1)–(4.8.3) all the theorems and complete asymptotics for mollification with respect to spectral parameter remain true if we replace mollification with respect to spectral parameter by mollification with respect to ζ with the same restrictions on L; here we assume that $\tau', \tau \in [-\epsilon, \epsilon]$ with a small constant $\epsilon > 0$. Moreover, these theorems and asymptotics remain true if we also include mollification with respect to the spectral parameter τ' with $L_1 \in [0,1]$ where we take $e(\tau',\tau,\zeta)d\tau' d\zeta$ instead of $d_\tau(.,\tau)$ in the original mollification.*

On the other hand, if we drop conditions (4.8.2) and (4.8.3) then we are forced to take $T_1 = T_0$.[53)] Then the Tauberian method provides the following

Proposition 4.8.2. *Under condition (4.8.1) in the theorems on asymptotics without mollification we can drop the microhyperbolicity condition provided we mollify with respect to ζ and take $\tau', \tau \in [-\epsilon, \epsilon]$ (we can mollify also with respect to ζ, τ' and take $\tau \in [-\epsilon, \epsilon]$). Here we have the same restrictions on L and L_1 as before (recall that in the scalar and similar cases these restrictions are weaker than in the general case).*

Let us notice that applying the same energy estimates technique as before in the context of (4.8.1) and (4.8.2) we can reach $T_1 = \epsilon h^{-2}$ and so we can improve the remainder estimate in proposition 4.8.2 under the additional condition (4.8.2).

4.8.2 Microhyperbolic case

Let us treat the general case and assume that

(4.8.6) At level 0 the operator A is microhyperbolic with respect to (x,ξ) in Ω for every $\zeta \in (-1,1)$.

Then taking into account that under this condition for $T = T_0$, $\tau_1, \tau_2 \in [-\epsilon, \epsilon]$ with $|\tau_2 - \tau_1| \leq \frac{h}{T}$ the trace norm of the operator $Q'E(\tau_1,\tau_2)Q''$ does not exceed Ch^{1-d} and that this norm does not exceed $Ch^{-m}(|\tau_1| + |\tau_2| + 1)^m$ for general $\tau_1, \tau_2 \in \mathbb{R}$ with appropriate m, we obtain from the formula

[53)] But we can increase T_1 under standard conditions on the Hamiltonian flow or under the conditions of section 4.7.

$$F_{t\to h^{-1}\tau}\chi_T(t)U = \int \hat{\chi}_T((\tau-\tau')h)d_{\tau'}e(\tau_1,\tau')$$

that (after multiplication by Q'_x and ${}^tQ''_y$) the trace norm of the left-hand expression does not exceed Ch^{1-d} for $T=T_0$ and $\tau\in[-\frac{1}{2}\epsilon,\frac{1}{2}\epsilon]$.

This and the previous arguments easily yield that for $|T|\ge 2T_0$, $|\tau|\le\frac{1}{2}\epsilon$ the trace norm of

$$F_{t\to h^{-1}\tau}\chi_{T_0,T}(t)MQ'_xU^tQ''_y$$

does not exceed $CTh^{1-d}(\frac{h}{L})^{3s}$ where $L\in[h,1]$ is arbitrary, $M=M_L: Mv = L^{-1}\int\phi(\frac{\zeta}{L})v_\zeta d\zeta$. Moreover, for $|T|\le 2T_0$ this norm does not exceed Ch^{1-d}. But then the trace norm of

$$F_{t\to h^{-1}\tau}\chi_T(t)MQ'_xU^tQ''_y$$

does not exceed $Ch^{1-d}+CT^2h^{1-d}(\frac{h}{L})^{3s}$ and therefore the standard Tauberian arguments with $T_1=(\frac{h}{L})^{-s}$ yield that

$$|M\Gamma Q'_x(e-\frac{1}{h}\int_{\tau_1}^{\tau_2}F_{t\to h^{-1}\tau}\bar{\chi}_{T_1}(t)U^tQ''_yd\tau)| \tag{4.8.7}$$

does not exceed $Ch^{1-d}(\frac{h}{L})^s$.

Let us now consider

$$|M\Gamma Q'_x\frac{1}{h}\int_{\tau_1}^{\tau_2}F_{t\to h^{-1}\tau}(\bar{\chi}_{T_1}(t)-\bar{\chi}_{T_0}(t))U^tQ''_yd\tau| =$$
$$|M\Gamma Q'_xF_{t\to h^{-1}\tau}\frac{1}{t}(\bar{\chi}_{T_1}(t)-\bar{\chi}_{T_0}(t))U^tQ''_y|_{\tau=\tau_1}^{\tau=\tau_2}|.$$

The previous analysis yields that the right-hand expression of this equality does not exceed $CT_1^2h^{1-d}(\frac{h}{L})^{3s}=Ch^{1-d}(\frac{h}{L})^s$ and therefore we can replace T_1 by T_0 in (4.8.7) without increasing the error. However, under condition (4.8.6) for $T=T_0$ we have complete asymptotics for $\Gamma Q'_xF_{t\to h^{-1}\tau}\bar{\chi}_T(t)U^tQ''_y$ and hence we have obtained *complete asymptotics for $M\Gamma Q'_xe^tQ''_y$ with remainder estimate $Ch^{1-d}(\frac{h}{L})^s$.*

Remark 4.8.3. *(i) Without condition (4.8.6) we would obtain remainder estimate $Ch^{-d}(\frac{h}{L})^s$.*

(ii) Let us assume that condition (4.4.1) is fulfilled for all $\zeta\in(-1,1)$. Using a partition of unity and the arguments of section 4.4 we obtain that under conditions $(4.4.16)_1$ the remainder estimate is $O(h^{\frac{1}{2}-d}(\frac{h}{L})^s)$; under condition $(4.4.16)_2^+$ the remainder estimate is $O(h^{1-d}(\frac{h}{L})^s)$ and under conditions $(4.4.16)_2$ the remainder estimate is $Ch^{1-d}(\frac{h}{L})^s\log h$. Moreover, the further arguments of section 4.4 yield that under condition $(4.4.16)_2$ the remainder estimate is $O(h^{1-d}(\frac{h}{L})^s)$ as well and, moreover, we get the same remainder estimate under conditions $(4.4.16)_1$ and $(4.4.43)_m$. Furthermore, under condition $(4.4.16)_1$ only the remainder estimate is $Ch^{1-d-\delta}(\frac{h}{L})^s$ with an arbitrarily small exponent $\delta>0$ and the same estimate holds under condition (4.4.53).

(iii) Similarly, under conditions of a global nature linked to the Hamiltonian trajectories the remainder estimate can be improved; now in the remainder estimates $T_1=(\frac{h}{L})^{-s}T_2$ where T_2 is the time controlled by the Hamiltonian trajectories.

(iv) For Schrödinger and Dirac operators with strong magnetic field we will be able to obtain similar results. ◊

4.8.3 Riesz means

Starting from the previous results and applying a partition of unity with respect to ζ we obtain that if $L = 1$ but the condition $\phi \in C_0^K([-\frac{1}{2}, \frac{1}{2}])$ is replaced by the condition

$$\operatorname{supp}\phi \subset [0, \frac{1}{2}], \qquad |D_\zeta^k \phi| \le C\zeta^{\vartheta - k} \quad \forall k \le K \tag{4.8.8}$$

then in general under condition (4.8.6) the remainder estimate is $O(h^{\vartheta+1-d})$, etc.: in all cases the remainder estimates are as good as those we obtained for Riesz means of order ϑ with respect to τ.

So under conditions (4.8.1)–(4.8.3) the estimates for Riesz means remain true if they are taken with respect to ζ (instead of the spectral parameter).

5. Standard Local Semiclassical Spectral Asymptotics near the Boundary

In this chapter we continue the treatment of local and microlocal semiclassical spectral asymptotics. Now we consider the case where our unit ball intersects the boundary of the domain in a part of the boundary which is smooth. This chapter consists of four sections corresponding to first four sections of Chapter 4. Namely, section 5.1 is devoted to the preliminary analysis; here we obtain rough estimates of the spectral function. More precisely, we find that there are three different cases; two of them (when the spectral function on a fixed interval has order h^{-d} and $O(h^{\infty})$ respectively) we have met before; in the third case the operator A is elliptic but the boundary value problem is not and the spectral function has order h^{1-d}. In section 5.2 we find successive approximations, which are now essentially more complicated; in section 5.3 we prove the basic theorems of this chapter. Furthermore, in section 5.4 we improve our results for the Schrödinger operator as we did in the second part of section 4.3 and in section 4.4. Finally, in Appendix D we prove that the set of "bad" energy levels where the microhyperbolicity property is violated has measure 0; this result is reasonable even if there is no boundary. It is assumed that the reader is familiar with the material of the previous chapters excluding sections 4.5–4.7.

5.1 Preliminary Analysis

5.1.1 The general case

In this chapter we treat the operator

(5.1.1) $$A = \sum_{|\alpha|\le m} a_\alpha(x,h)(hD)^\alpha$$

in the domain $X \subset \mathbb{R}^d$ assuming that

(5.1.2) $$X \cap B(0,1) = \mathbb{R}^d_+ \cap B(0,1)$$

where $\mathbb{R}^d_+ = \mathbb{R}_+ \times \mathbb{R}^{d-1}$; the slightly more general case

(5.1.3) $$X \cap B(0,1) = \{x_1 \ge \phi(x')\} \cap B(0,1), \quad |D^\beta \phi| \le c \quad \forall \beta, |\beta| \le K$$

is reduced to this one after a regular change of coordinates which preserves in particular the condition

$(5.1.4)_1$ $$|D^\beta a_\alpha| \le c \qquad \forall \alpha : |\alpha| \le m, \quad \forall \beta : |\beta| \le K;$$

we assume that

(5.1.5) $\mathbb{H} = \mathbb{C}^D$ and so a_α are $D \times D$ matrices.

We also assume that the boundary operator B on $Y = \partial X$ is given; more precisely we assume that

(5.1.6) $u \in C_0^m(B(0,1), \mathbb{H})$ belongs to $\mathrm{D}(A)$ if and only if

$$B_k u|_Y = 0 \qquad k = 1, \dots, l$$

where

$$B_k = \sum_{|\alpha| \le m_k} b_{k\alpha}(x', h)(hD)^\alpha \tag{5.1.7}$$

where $0 \le m_1 < m_2 < \cdots < m_l \le m-1$, $b_{k\alpha}$ are $D_k \times D$ matrices and

$$|D^\beta b_{k\alpha}| \le c \qquad \forall \alpha, |\alpha| \le m_k, \quad \forall \beta, |\beta| \le K \quad \forall k = 1, \dots l. \tag{5.1.4$_2$}$$

Moreover, we assume that the boundary is not characteristic, i.e., we assume that

$$|a_{(m,0,\dots,0)}(x)^{-1}| \le c \qquad \forall x \in B(0,1) \tag{5.1.8}$$

and that

(5.1.9) A is a self-adjoint operator in $L^2(X, \mathbb{H})$ where $\mathrm{D}(A)$ satisfies condition (5.1.6).

One can then prove that

$$D_1 + \cdots + D_l = \frac{mD}{2}. \tag{5.1.10}$$

Let us introduce a condition similar to (3.1.7): there exist symbols $\beta_j(z)$ with $j = 1, \dots, m$ and $\beta = (\beta_1, \dots, \beta_m)$ such that

$$T(z) + (\beta^\dagger b)(z) + (b^\dagger \beta)(z) = 0 \qquad \forall z \in T^*(Y \cap B(0,1)) \tag{5.1.11}$$

where $T(z)$ is written in the form (3.1.5) after A has been written in the form (3.1.1) and operators A_k have been defined, with principal symbols a_k, and b is the principal symbol of the boundary operator $B = (B_{(1)}, \dots, B_{(m)})$ where the boundary conditions have been rewritten in the form

$$Bu = \sum_{1 \le j \le m} B_{(j)}(hD_1)^{j-1} u|_Y = 0.$$

Condition (5.1.11) follows from the equality

$$(Au, v)_X + \sum_{1 \le k \le l} (B_k u, B'_k v)_Y = (u, Av)_X + \sum_{1 \le k \le l} (B'_k u, B_k v)_Y \tag{5.1.12}$$
$$\forall u, v \in C_0^m(B(0,1), \mathbb{H})$$

where B'_k are h-differential operators of degrees $m - m_k - 1$ with coefficients satisfying condition (5.1.4)$_2$. Let us note that conditions (5.1.12), (5.1.8) and (5.1.10) yield that

$$\operatorname{rank} b(z) = \frac{mD}{2} \qquad \forall z \in T^*(Y \cap B(0,1)), \tag{5.1.13}$$

$$\operatorname{rank} b_{k(m_k,0,\dots,0)}(x) = D_k \qquad \forall x \in Y \cap B(0,1) \tag{5.1.14}$$

where these conditions are satisfied uniformly (i.e., there exits a minor of the corresponding dimension the absolute value of which is not less than ϵ_0).

Let $e(x, y, \tau)$ be the Schwartz kernel of the spectral projector of A and let $e(x, y, \tau_1, \tau_2) = e(x, y, \tau)|_{\tau=\tau_1}^{\tau=\tau_2}$.

Let us start from an assertion similar to lemma 4.1.1:

Theorem 5.1.1. *Let conditions* (5.1.1), (5.1.2), $(5.1.4)_1$, (5.1.8) *and* (5.1.9) *be fulfilled. Let* $\psi_1, \psi_2 \in C_0^K(B(0,1-\epsilon))$, $\epsilon > 0$ *and let* $q_1, q_2 \in S'_{h,K}$ *be supported in* $\{|\xi'| \le c_1\}$, $Q_i = \mathrm{Op}(q_i)$. *Then*

$$(5.1.15)\quad |Q_{1x}\psi_1(x)e(.,.,\tau_1,\tau_2)\psi_2(y)^t Q_{2y}| \le C'Mh^{-d}(1+|\tau_1|+|\tau_2|)^{\frac{1}{m}} \qquad \forall x,y \in \mathbb{R}^d \quad \forall \tau_1,\tau_2 \in \mathbb{R}$$

with $K = K(d,m)$, $C' = C'(d,m,c,c_1,\epsilon)$ *and with* M *defined in the same way as in lemma 4.1.1 (with the corresponding norms of* q_i*).*

Proof. Let us note first of all that the conditions of this theorem imply the very rough estimate

$$(5.1.16)\qquad \|D_1^j\psi v\|_{(-n)} \le C'h^{-n}(\|Av\| + \|v\|) \qquad \forall j = 0,\dots,m-1$$

where in the left-hand expression the L^2-norm with respect to x_1 and the $H^{(-n)}$-norm with respect to x' is calculated, $\psi \in C_0^K(B(0,1-\epsilon'))$, C' also depends on $\epsilon' > 0$ and ψ, and $n = n(d,m)$.

In fact, (5.1.8) yields that

$$(hD_1)^m\psi u = \psi A u + \sum_{|\alpha|\le m, \alpha_1 \le m-1} (hD)^\alpha a'_\alpha u$$

where $a'_\alpha \in C_0^K(B(0,1-\epsilon'))$ and hence

$$(5.1.17)_j\quad (hD_1)^j\psi u = \mathcal{K}^{m-j}\psi A u + \sum_{|\alpha|\le m, m-j\le\alpha_1\le m-1} (hD')^{\alpha'}(hD_1)^{\alpha_1-m+j}a'_\alpha u + \sum_{|\alpha|\le m,\alpha_1\le m-j} \mathcal{K}^{m-j-\alpha_1}(hD')^{\alpha'}a'_\alpha u$$

where $(\mathcal{K}u)(x) = ih^{-1}\int_\infty^{x_1} u(t,x')dt$. Substituting $(5.1.17)_1$ into $(5.1.17)_2$ and continuing this process we delete from the right-hand expression of $(5.1.17)_m$ every degree of hD_1; in this expression there will only be multiple compositions of operators of multiplication by smooth compactly supported functions, hD' and $\mathcal{K}$; this representation yields (5.1.16) with $n = n(m)$.

Let $v \in C_0^m(B(0,1),\mathbb{H})$. Then (5.1.8) yields that

$$\|(hD_1)^m v\| \le C\|Av\| + C\sum_{|\alpha|\le m,\alpha_1\le m-1}\|(hD)^\alpha v\| \le C\|Av\| + C\sum_{|\alpha|\le m,\alpha_1=0}\|(hD)^\alpha v\|$$

where we have applied the inequality

$$\sum_{|\alpha|\le m,\alpha_1\le m-1}\|(hD)^\alpha v\| \le \epsilon\|(hD_1)^m v\| + C_\epsilon \sum_{|\alpha|\le m,\alpha_1=0}\|(hD)^\alpha v\|$$

with an arbitrarily small $\epsilon > 0$. Then substituting $v = Q_1\psi_1 u$ we again obtain by means of (5.1.16) that

$$\sum_{j\le m}\|(hD_1)^jQ_1\psi_1u\|\le C(\|Au\|+\|u\|)+\sum_{j\le m-1}Ch\|(hD_1)^jQ_1'\psi_1'u\|$$

where Q_1' and ψ_1' are an operator and function of the same type as Q_1 and ψ_1 but with c_1 and ϵ replaced by $2c_1$ and $\frac{\epsilon}{2}$. Applying the same arguments to Q_1' and ψ_1' repeatedly we finally obtain the estimate

$$\sum_{j\le m}\|(hD_1)^jQ_1\psi_1u\|\le C(\|Au\|+\|u\|).$$

Applying the inequality

$$\|v(x_1,.)\|'\le Ch^{-\frac12}\|v\|^{\frac{2m-1}{2m}}\sum_{j\le m}\|(hD_1)^jv\|^{\frac{1}{2m}}$$

with the L^2-norm with respect to x' in the left-hand expression we conclude that

$$\|(Q_1\psi_1u)(x_1,.)\|'\le Ch^{-\frac12}(\|u\|+\|Au\|^{\frac{1}{2m}}\|u\|^{\frac{2m-1}{2m}}).$$

Noting that instead of Q_1 we could take in this analysis $(hD)^\alpha Q_1$ with any fixed α we obtain from the Sobolev embedding theorems that

$$\|(Q_1\psi_1u)(x)\|\le Ch^{-\frac{d}{2}}(\|u\|+\|Au\|^{\frac{1}{2m}}\|u\|^{1-\frac{1}{2m}}).$$

Substituting $u=Q_1\psi_1E(\tau_1,\tau_2)v$ we conclude that the operator norm of the operator $\mathcal{M}=Q_1\psi_1E(\tau_1,\tau_2):L^2(X,\mathbb{H})\to\mathbb{H}$ does not exceed $Ch^{-\frac{d}{2}}(1+|\tau_1|+|\tau_2|)^{\frac{1}{2m}}$, and since $Q_{1x}\psi_1(x)e(.,.,\tau_1,\tau_2)\psi_2^tQ_2=\mathcal{M}\mathcal{N}^*$ (where $\mathcal{N}=Q_2^*\bar\psi_2E(\tau_1,\tau_2)$ is an operator of the same type as $\mathcal{M}$) we obtain (5.1.15). □

Remark 5.1.2. *Our proof implies the estimate*

$$(5.1.15)'\quad \|D^j_{x_1}D^k_{y_1}Q_{1x}\psi_1(x)e(.,.,\tau_1,\tau_2)\psi_2(y)^tQ_{2y}\|\le C'h^{-d-j-k}(1+|\tau_1|+|\tau_2|)^{\frac{j+k+1}{m}}$$

for $j\le m-1, k\le m-1$. If we apply our analysis to the degrees of A instead of A itself we obtain this estimate for arbitrary j,k but now K and C' also depend on j,k. ◊

5.1.2 The elliptic case

Our next step is to treat the zone where A is elliptic:

Theorem 5.1.3. *Let conditions* (5.1.1), (5.1.2), (5.1.4)$_1$, (5.1.8) *and* (5.1.9) *be fulfilled and let Q_i and ψ_i be the same as in theorem 5.1.1.*[1)] *Let us assume that*

$$(5.1.18)\quad |(\tau-a(x,\xi))^{-1}|\le c\qquad \forall(x,\xi')\in\operatorname{supp}q_1\cup\operatorname{supp}q_2\quad\forall\xi_1\in\mathbb{R}\quad\forall\tau\in[\tau_1,\tau_2].$$

Then for $\tau_1',\tau_2'\in[\tau_1,\tau_2]$ the following estimates hold:

$$(5.1.19)\quad \|D^j_{x_1}D^k_{y_1}Q_{1x}\psi_1(x)e(.,.,\tau_1',\tau_2')\psi_2(y)^tQ_{2y}\|\le C'h^{-d-j-k}(1+|\tau_1'|+|\tau_2'|)^{\frac{j+k+1}{m}}\bigl(1+(x_1+y_1)/h\bigr)^{-s},$$

$$(5.1.20)\quad \|\Gamma''Q_{1x}\psi_1(x)e(.,.,\tau_1',\tau_2')\psi_2(y)^tQ_{2y}\|\le C'h^{1-d}(1+|\tau_1'|+|\tau_2'|)^{\frac{j+k+1}{m}}$$

where Γ'' is introduced in section 3.2.

1) Including the condition $\operatorname{supp}q_i\subset\{|\xi'|\le c_1\}$.

Proof. Let $\chi \in C_0^K([-1,1]$ be fixed. Then partially repeating the arguments of the proof of theorem 5.1.1 (see also the proof of lemma 4.1.2) we obtain that

$$\|F_{t\to h^{-1}\tau}\chi_{T,T'}(t)_x D_{x_1}^j Q_1'\psi_1' u\| \le C'Th^{-\frac{1}{2}-j}(1+|\tau|)^{\frac{1}{m}(j+\frac{1}{2})}\|v\| \tag{5.1.21}$$

where $v \in L^2(X,\mathbb{H})$ is arbitrary, $u = \exp(ih^{-1}tA)\cdot v$, $T \ge h$, $T' \in \mathbb{R}$ and $\tau \in \mathbb{R}$ are arbitrary, Q_i' and ψ_i' are operators and functions of the same type as Q_i and ψ_i, equal to 1 in the neighborhoods of the supports of Q_i and $\psi_i$respectively, and $\mathfrak{d}$ is the operator of restriction to Y.

Now let C be the operator realizing the reduction of subsection 3.1.2 in a neighborhood of $\operatorname{supp} Q_1' \times [\tau_1,\tau_2]$; then estimate (5.1.21) also holds for u replaced by $\tilde u = Cu$. Moreover, $\tilde u$ is divided into two blocks and theorem 3.1.1 and proposition 3.1.7 yield that one of them is negligible while the second satisfies the estimate

$$\begin{gathered}\|F_{t\to h^{-1}\tau}\chi_{T,T'}(t)D_{x_1}^j Q_1\psi_1 u'\|' \le \\ C'Th^{-\frac{1}{2}-j}(1+|\tau|)^{\frac{1}{m}(j+\frac{1}{2})}\|v\|(1+\frac{x_1}{h})^{-s} \quad \forall \tau \in [\tau_1,\tau_2]\end{gathered} \tag{5.1.22}$$

where in the left-hand expression the L^2-norm with respect to x' is calculated; then the same estimate holds also for u' replaced either by Cu or by u itself. Further, here we can replace Q_1 by $(hD')^\alpha Q_1$ and hence the embedding theorems yield that

$$\begin{gathered}|F_{t\to h^{-1}\tau}\chi_{T,T'}(t)Q_1\psi_1 u| \le C'Th^{-\frac{d}{2}}(1+|\tau|)^{\frac{1}{2m}}(1+\frac{x_1}{h})^{-s}\|v\| \\ \forall x \in \bar{\mathbb{R}}^+\times\mathbb{R}^{d-1} \quad \forall \tau \in [\tau_1,\tau_2].\end{gathered}$$

Taking into account the connection between u, v and U we conclude that

$$\begin{gathered}|F_{t\to h^{-1}\tau}\chi_{T,T'}(t)Q_{1x}\psi_1(x)f^\dagger U| \le C'T^{-\frac{d}{2}}(1+|\tau|)^{\frac{1}{2m}}(1+\frac{x_1}{h})^{-s} \\ \forall x \in \bar{\mathbb{R}}^+\times\mathbb{R}^{d-1} \quad \forall \tau \in [\tau_1,\tau_2] \quad \forall f \in \mathbb{H}\end{gathered}$$

where in the left-hand expression the L^2-norm with respect to y is calculated. Repeating the same arguments with x replaced by y we obtain from this estimate and from the rough estimate with $s=0$ (which holds for all $\tau \in \mathbb{R}$) the estimate

$$\begin{gathered}|F_{t\to h^{-1}\tau}\chi_{T,T'}(t)Q_{1x}\psi_1(x)U\psi_2(y)^tQ_{2y}| \le C'T^{-d}(1+|\tau|)^{\frac{1}{2}}\Bigl(1+\frac{(x_1+y_1)}{h}\Bigr)^{-s} \\ \forall x,y \in \bar{\mathbb{R}}^+\times\mathbb{R}^{d-1} \quad \forall \tau \in [\tau_1,\tau_2].\end{gathered}$$

Obviously, one can assume that $|\tau_2-\tau_1|\ge 1$. Applying the standard simplified Tauberian arguments we obtain estimate (5.1.19) from which estimate (5.1.20) follows. □

Let us now treat the completely elliptic zone:

Theorem 5.1.4. *Let conditions* (5.1.1), (5.1.2), $(5.1.4)_{1,2}$, (5.1.8), (5.1.9) *and* (5.1.11) *be fulfilled. Let Q_i and ψ_i be the same as in theorem 5.1.1. Let us assume that condition* (5.1.18) *is fulfilled and that*

$$\begin{gathered}\|\mathfrak{d}b(x',D_{x_1},\xi')v\| \ge \epsilon_0\|v\|_+ \\ \forall v \in \mathcal{S}(\bar{\mathbb{R}}^+): \quad (a(0,x',D_{x_1},\xi')v-\tau)=0 \\ \forall (x',\xi') \in (\operatorname{supp} q_1 \cup \operatorname{supp} q_2)\cap Y \quad \forall \tau \in [\tau_1,\tau_2].\end{gathered} \tag{5.1.23}$$

Then the following estimate holds for $\tau_1',\tau_2' \in [\tau_1,\tau_2]$:

$$\|D_{x_1}^jD_{y_1}^kQ_{1x}\psi_1(x)e(.,.,\tau_1',\tau_2')\psi_2(y)^tQ_{2y}\| \le C'h^s(1+|\tau_1'|+|\tau'2|)^{\frac{j+k+1}{m}}. \tag{5.1.24}$$

Proof. The proof is similar to the proof of lemma 4.1.4 but also uses theorem 3.1.3 (see also the end of the proof of the next theorem). □

Theorem 5.1.5. *Let conditions* (5.1.1), (5.1.2), $(5.1.4)_{1,2}$, (5.1.8), (5.1.9) *and* (5.1.11) *be fulfilled. Let us assume that*

$$\langle a(x,\xi_1,\xi')v,v\rangle \geq (\epsilon_0|\xi|^m - c)|v|^2 \qquad \forall x\in B(0,1)\cap X \quad \forall \xi_1\in\mathbb{R}, \xi'=0 \quad \forall v\in\mathbb{H} \tag{5.1.25}$$

and that

$$\sum_{1\leq j\leq m} |\tau|^{-\frac{1}{m}(m_j+\frac{1}{2})}|\eth b_j(x',D_1,\xi')v| \geq \epsilon_0|v|_+^2 \qquad \forall x'\in B(0,1)\cap Y,\ \xi'=0 \quad \forall\tau\leq -c \quad \forall v\in\mathcal{S}(\bar{\mathbb{R}}^+,H):\ (a(0,x',D_1,\xi')-\tau)v=0. \tag{5.1.26}$$

Then if Q_i and ψ_i are the same as in theorem 5.1.1 then for $\tau\leq -C_1 = C_1(d,c,c_1)$ the following estimate holds:

$$\|D_{x_1}^j D_{y_1}^k Q_{1x}\psi_1(x)e(.,.,-\infty,\tau)\psi_2(y)^t Q_{2y}\| \leq C'h^s|\tau|^{-s}. \tag{5.1.27}$$

Proof. Obviously (5.1.25) and (5.1.26) imply the same conditions for symbols a and b_j from which all the terms not containing ξ_1^m and $\xi_1^{m_j}$ respectively are excluded. Then for these symbols the following estimate holds:

$$\sum_j |\tau|^{\frac{1}{m}(2m-2m_j-1)}|\eth b_j(x',D_1,\xi')v|^2 + |(a(0,x',D_1,\xi')-\tau)v|_+^2 \geq \epsilon_0\sum_{0\leq k\leq m}|\tau|^{\frac{2(m-k)}{m}}|D_1^k v|_+^2 - c|\xi'|^m|v|_+^2 \tag{5.1.28}$$

and hence the same estimate holds also for the original symbols a and b_j where now $\xi'\in\mathbb{R}^d$, $\tau\leq -C_1$ and $v\in\mathcal{S}(\bar{\mathbb{R}}^+,\mathbb{H})$ are arbitrary and c, C_1 are increased if necessary. Then the h-Fourier transform with respect to x' and dilatations with respect to x_1 yield that if a and b_j do not depend on x and $A=a(hD)$, $B=b(hD)$ then

$$\sum_j |\tau|^{\frac{2m-2m_j-1}{m}}\|\eth B_j w\|_Y^2 + \|(A-\tau)w\|^2 \geq \epsilon_0\sum_{0\leq k\leq m}|\tau|^{\frac{2(m-k)}{m}}\|(hD_1)^k w\|_+^2 - c\sum |\alpha|\leq m\|(hD')^\alpha w\|^2 \tag{5.1.29}$$

for $w\in C_0^K(B(0,1),\mathbb{H})$.

This estimate is obviously preserved when we perturb the operators A and B_j by arbitrary bounded lower order terms; moreover, it is preserved when we perturb these operators by operators of the same orders with small coefficients (i.e., with small maxima of the $\mathcal{L}(\mathbb{H},\mathbb{H})$-norms of the coefficients). Hence for general operators A and B_j estimate (5.1.29) remains true provided we replace w by $\psi_\nu w$ where $\psi_\nu\in C_0^K(B(y_\nu,\epsilon))$, $y_\nu\in B(0,1)$ and $\epsilon>0$ is a small constant. Taking real-valued functions ψ_ν such that $\sum_\nu\psi_\nu^2 = 1$ in $B(0,1+\frac{\epsilon}{2})$ and summing estimates (5.1.29) for $\psi_\nu w$ with respect to ν we obtain inequality (5.1.29) for u in the general case.

Let us substitute

$$w = \phi(\frac{T}{L}) F_{t\to h^{-1}\tau} \chi_T(t) \bar{\psi}_1 Q_1 \psi_1 u$$

into (5.1.29) where $u = \exp(ih^{-1}tA)\cdot v$, $\chi \in C_0^K([-1,1])$, $\phi \in C_0^K([-c,-\epsilon_1,])$, $Q_1 = \mathrm{Op}(q_1)$, $q_1 \in S_{h,\rho,1,K}(\mathbb{R}^{d-1},\mathbb{C})$ is supported in $\{|\xi'| \le c_2^{-1}\rho\}$, and $\rho = L^{\frac{1}{M}}$ with an appropriate constant c_2; then if $\bar{\psi}_1$, $\bar{q}_1$, $\bar{\phi}$ and $\bar{\chi}$ are functions of the same type as ψ_1, q_1, ϕ and χ, equal to 1 in the neighborhoods of their respective supports, (5.1.29) yields that for $L \le -C_1$, $T = L$

$$\sum_{0\le k\le m} \|(hD_1)^k \phi(\frac{h}{L}D_t)\chi_T(t) Q_1\psi_1 u\| \le$$
$$C'hL^{-1} \sum_{0\le k\le m} \|(hD_1)^k \bar{\phi}(\frac{h}{L}D_t)\bar{\chi}_T(t)\bar{Q}_1\bar{\psi}_1 u\| +$$
$$C'h^sL^{-s} \sum_{0\le k\le m-1} \|(hD_1)^k \phi(\frac{h}{L}D_t)\chi_T(t) u\|_{(-m)}$$

where we calculate the L^2-norm with respect to (x,t) everywhere but the last term in which we calculate the L^2-norm with respect to (x_1,t) and the $H^{(-m)}$-norm with respect to x', which obviously does not exceed

$$C'h^sL^{-s}(\|\phi(\frac{h}{L}D_t)\chi_T(t)u\| + \|A\phi(\frac{h}{L}D_t)\chi_T(t)u\|);$$

replacing Au by hD_tu and permuting hD_t with $\chi_T(t)$ we obtain that none of these expressions exceed $C'h^sL^{-s}\|v\|$. Thus,

$$\sum_{0\le k\le m} \|(hD_1)^k \phi(\frac{h}{L}D_t)\chi_T(t) Q_1\psi_1 u\| \le$$
$$C'hL^{-1} \sum_{0\le k\le m} \|(hD_1)^k \bar{\phi}(\frac{h}{L}D_t)\bar{\chi}_T(t)\bar{Q}_1\bar{\psi}_1 u\| + C'h^sL^{-s}\|v\|.$$

Then a repeated argument yields that

$$\|(hD_1)^j \phi(\frac{h}{L}D_t)\chi_T(t) Q_1\psi_1 u\| \le C'h^sL^{-s} + C'h^sL^{-s}\|v\|$$

for $j = 0,\dots,m$ and therefore the non-characteristic property of Y yields that this estimate holds for all j. Then the embedding theorems yield that

$$\|(hD_1)^j \phi(\frac{h}{L}D_t)\chi_T(t) Q_1\psi_1 u\| \le C'h^sL^{-s}\|v\| \qquad \forall x \in B(0,1)$$

and this inequality and the connection between u, v and U imply that

$$\|(hD_{x_1})^j \phi(\frac{h}{L}D_t)\chi_T(t) Q_{1x}\psi_1(x) f^\dagger U\| \le C'h^sL^{-s}\|f\|$$
$$\forall x \in B(0,1) \quad \forall f \in \mathbb{H}$$

where in the left-hand expression the L^2-norm with respect to y is calculated.

Repeating all these arguments for y instead of x we easily obtain the inequality

$$\|(hD_{x_1})^j (hD_{y_2})^k \phi(\frac{h}{L}D_t)\chi_T(t) Q_{1x}\psi_1(x) f^\dagger U \psi_2(y)^t Q_{2y}\| \le C'h^sL^{-s}\|f\|$$
$$\forall x,y \in B(0,1) \quad \forall f \in \mathbb{H}$$

and therefore we obtain the estimate

$$|F_{t\to h^{-1}\tau}(hD_{x_1})^j(hD_{y_2})^k\chi_T(t)Q_{1x}\psi_1(x)U\psi_2(y)^tQ_{2y}| \le C'h^s|\tau|^{-s}$$
$$\forall x,y\in B(0,1)\quad \forall \tau\le -C_1.$$

The rest of the proof coincides with the end of the proof of lemma 4.1.5. □

Remark. *In fact we have proven this theorem under the restriction* $\operatorname{supp} q_i\subset\{|\xi'|\le C_0^{-1}|\tau|^{\frac{1}{m}}\}$ *which is weaker than* $\operatorname{supp} q_i\subset\{|\xi'|\le c_1\}$. ◊

The following two statements are similar to lemmas 4.1.6 and 4.1.7 and are proven in the same way (taking into account the proof of theorem 5.1.5).

Theorem 5.1.6. *Let conditions* (5.1.1), (5.1.2), $(5.1.4)_{1,2}$, (5.1.8), (5.1.9) *and* (5.1.11) *be fulfilled and let*

(5.1.30)
$$|\xi|^m|v|\le c(|a(x,\xi)v|+|v|)\quad \forall x\in B(0,1)\cap X\quad \forall\xi\in\mathbb{R}^d\quad \forall v\in\mathbb{H},$$

(5.1.31)
$$|\xi'|^m|v|_+\le c(|a(0,x',D_1,\xi')v|_+ +\sum_j|\xi'|^{m-m_j-\frac12}|\eth b_j(x',D_1,\xi')v|+|v|_+)$$
$$\forall x'\in B(0,1)\cap Y\quad \forall\xi'\in\mathbb{R}^{d-1}\quad \forall v\in\mathcal{S}(\bar{\mathbb{R}}^+,\mathbb{H}).$$

(Obviously, it is sufficient to check this condition just for $v\in\mathcal{S}(\bar{\mathbb{R}}^+,\mathbb{H})$ *such that* $a(0,x'D_1,\xi')v=0$.*) Let us assume that* $Q_i=\operatorname{Op}(q_i)$, $q_i\in\tilde{S}^{0\prime}_{h,\rho,\gamma,K}$,

(5.1.32)
$$\rho\gamma\ge h^{1-\delta}\omega^\delta,\qquad \omega\ge 1,\gamma\le 1$$

where $\delta>0$ *is arbitrary; let* ψ_1 *and* ψ_2 *be the same as above. Then*

(i) If the symbol q_i *is supported in* $\{|\xi'|\ge\omega\}$ *for some* $i=1,2$ *then for* $\omega\ge C_0$, $|\tau_1'|+|\tau_2'|\le\epsilon_2\omega^m$, $x,y\in\mathbb{R}^+\times\mathbb{R}^{d-1}$,

(5.1.33)
$$|D^j_{x_1}D^k_{y_1}Q_{1x}\psi_1(x)e(.,.,\tau_1',\tau_2')\psi_2(y)^tQ_{2y}|\le C'M'h^s\omega^{-s}.$$

(ii) In the general case

(5.1.34)
$$|D^j_{x_1}D^k_{y_1}Q_{1x}\psi_1(x)e(.,.,\tau_1',\tau_2')\psi_2(y)^tQ_{2y}|\le CMh^{-d-j-k}(|\tau_1'|+|\tau_2'|+1)^{\frac{d+j+k}{m}}$$

where M *and* M' *are products of the norms of* ψ_i, q_i $(i=1,2)$ *in the corresponding spaces.*

(iii) If condition (5.1.25) *is fulfilled then*

(5.1.35)
$$\|D^j_{x_1}D^k_{y_1}Q_{1x}\psi_1(x)e(.,.,\tau_1',\tau_2')\psi_2(y)^tQ_{2y}\|\le CMh^{1-d-j-k}(|\tau_1'|+|\tau_2'|+1)^{\frac{d+j+k-1}{m}}$$
$$\forall\tau_1'\le -C_0,\tau_2'\le -C_0.$$

Theorem 5.1.7. *Suppose that conditions* (5.1.1), (5.1.2), $(5.1.4)_{1,2}$, (5.1.8), (5.1.9), (5.1.11), (5.1.30) *and* (5.1.31) *are fulfilled and let conditions* (5.1.25) *and* (5.1.26) *be* *fulfilled for all* $\xi \in \mathbb{R}^d$ *and* $\xi' \in \mathbb{R}^{d-1}$ *respectively.*[2] *Let* Q_i *and* ψ_i *be the same as in theorem 5.1.6. Then*

(i) For $\tau \le -C_0$ *estimate* (5.1.21) *holds. Moreover, if* q_i *is supported in* $\{|\xi'| \ge \omega\}$ *for some* $i = 1, 2$ *then estimate* (5.1.21) *remains true with an additional factor* ω^{-s} *in the right-hand expression.*

(ii) In the general case

$$|D^j_{x_1} D^k_{y_1} Q_{1x}\psi_1(x) e(.,.,-\infty,\tau)\psi_2(y) {}^tQ_{2y}| \le CMh^{d-j-k}(|\tau|+1)^{\frac{d+j+k}{m}}. \tag{5.1.36}$$

Remark. *Condition (5.1.30) means exactly that the senior symbol*[3] *of a is elliptic and condition (5.1.31) means exactly that the senior symbols of* a, b_1, ..., b_l *satisfy the Shapiro-Lopatinskii condition. If (5.1.31) holds then condition (5.1.26) for all* ξ' *means exactly that the bottom of the spectrum of the operator* $a(0,x',D_1,\xi')$ *on* $L^2(\mathbb{R}^+,\mathbb{H})$ *with boundary condition* $b(x',D_1,\xi')v = 0$ *is not less than* $C_0^{-1}|\xi'|^m - C_0$. ◇

5.1.3 Operators with minimal smoothness conditions

The following step of our analysis will prove assertions similar to lemmas 4.1.8 and 4.1.9, i.e., obtain estimates under minimal restrictions on smoothness:

Theorem 5.1.8. *Suppose that conditions* (5.1.1), (5.1.2), $(5.1.4)_{1,2}$ *with* $K = \max(\frac{d}{2}+1, m) + n$, (5.1.8), (5.1.9), (5.1.12), (5.1.30) *and* (5.1.31) *are fulfilled. Then*

(i) The following estimate holds:

$$\|D^\alpha_x D^\beta_y e(x,y,\tau_1',\tau_2')\| \le C'' h^{-d-|\alpha|-|\beta|}(1+|\tau_1'|+|\tau_2'|)^{\frac{1}{m}(d+|\alpha|+|\beta|)} \tag{5.1.37}$$
$$\forall x,y \in B(0,1-\epsilon) \quad \forall \alpha,\beta : |\alpha| \le n, |\beta| \le n.$$

(ii) If the condition

$$|(a(x,\xi)-\tau)v| \ge \epsilon_0(\omega + |\xi|^m)|v| \tag{5.1.38}$$
$$\forall x \in B(0,1)\cap X \quad \forall \xi \in \mathbb{R}^d \quad \forall v \in \mathbb{H} \quad \forall \tau \in [\tau_1,\tau_2]$$

is fulfilled with $\omega = |\tau| + 1$ *then the estimate*

$$\|D^\alpha_x D^\beta_y e(x,y,\tau_1',\tau_2')\| \le \tag{5.1.39}$$
$$C'' h^{-d-|\alpha|-|\beta|}(1+|\tau_1'|+|\tau_2'|)^{\frac{1}{m}(d+|\alpha|+|\beta|-1)}(1+\frac{(x_1+y_1)}{h})^{-s}$$

holds with the same x, y, α, β *as before and with* $\tau_1', \tau_2' \in [\tau_1,\tau_2]$.

(iii) Moreover, if condition (5.1.38) *is fulfilled and*

$$|(a(0,x',D_1,\xi')-\tau)v|_+ + \tag{5.1.40}$$
$$\sum_j (|\xi'| + \omega^{\frac{1}{m}})^{m-m_j-\frac{1}{2}} |\mathfrak{d}b_j(x',D_1,\xi')v| \ge \epsilon_0(|\xi'|^m + \omega)|v|_+$$
$$\forall x' \in B(0,1)\cap Y \quad \forall \xi' \in \mathbb{R}^{d-1} \quad \forall v \in \mathcal{S}(\bar{\mathbb{R}}^+,\mathbb{H}) \quad \forall \tau \in [\tau_1,\tau_2]$$

then the estimate

$$\|D^\alpha_x D^\beta_y e(x,y,\tau_1',\tau_2')\| \le C' h^s \omega^{-s}$$

holds with the same x, y, α, β *as before and with* $\tau_1', \tau_2' \in [\tau_1,\tau_2]$.

---

[2] That means exactly that the operator $a(0,x',D_1,\xi')$ in $L^2(\bar{\mathbb{R}}^+,\mathbb{H})$ with boundary condition $\mathfrak{d}b(x',D_1,\xi')v = 0$ is semibounded from below uniformly with respect to (x',ξ').

[3] I.e., the higher degree homogeneous symbol.

Theorem 5.1.9. *Suppose that conditions* (5.1.1), (5.1.2), $(5.1.4)_{1,2}$ *with* $K = \max(\frac{d}{2}+1, m)+n$, (5.1.8), (5.1.9) *and* (5.1.12) *are fulfilled. Let us assume that*

(5.1.42)
$$\epsilon_0|\xi|^m|v|^2 \le (\langle a(x,\xi)v,v\rangle + c|v|^2) \qquad \forall x\in B(0,1)\cap X \quad \forall \xi\in\mathbb{R}^d \quad \forall v\in\mathbb{H},$$

(5.1.43)
$$\epsilon_0|\xi'|^m|v|_+^2 \le (\langle a(0,x',D_1\xi')v,v\rangle_+ + c|v|_+^2)$$
$$\forall x'\in B(0,1)\cap Y \quad \forall \xi'\in\mathbb{R}^{d-1} \quad \forall v\in\mathcal{S}(\bar{\mathbb{R}}^+,\mathbb{H}):\ \mathfrak{d}b(x'D_1,\xi')v = 0.$$

Then the estimate

(5.1.44)
$$\|D_x^\alpha D_y^\beta e(x,y,-\infty,\tau)\| \le \begin{cases} C''h^s(1+|\tau|)^{-s} & \forall\tau\le -C_0\\ C''h^{-d-|\alpha|-|\beta|}(1+|\tau|)^{\frac{1}{m}(d+|\alpha|+|\beta|)} \end{cases}$$

holds with the same x, y, α, β *as before.*

Proof. Let us note first of all that in the context of theorem 5.1.8(i)

(5.1.45)
$$\sum_{|\alpha|\le n}\|(hD)^\alpha u\| \le C\Bigl(\sum_{|\beta|\le n-m}\|(hD)^\beta Au\| + \|u\| + \sum_j\sum_{|\beta|\le n-m_j-1} h^{\frac12}\|(hD')^\beta B_ju\|_{Y,\frac12}\Bigr) \qquad \forall u\in C_0^K(B(0,1),\mathbb{H})$$

where

$$\|v\|_{Y,\frac12} = \inf\{\sum_{1\le k\le d}\|hD_kw\|:\quad \mathfrak{d}w = v\}$$

and $m\le n\le K$; this estimate is similar to (4.1.38). It is proven in the constant coefficient case by means of the Fourier transform with respect to x' and then is proven in the general case by means of a partition of unity; then we easily obtain (see the proof of lemma 4.1.8) that estimate (5.1.45) remains true for $u\in C^n(B(0,1),\mathbb{H})$ provided that in the left and right-hand expressions the norms on $B(0,1-\epsilon)\cap X$ and $B(0,1-\frac{\epsilon}{2})\cap X$, $B(0,1-\frac{\epsilon}{2})\cap Y$ are calculated. Iterating if necessary we obtain for $n\le K$, $k=\frac{n}{m}\in\mathbb{Z}$ the estimate

(5.1.46)
$$\sum_{|\alpha|\le n}\|(hD)^\alpha u\|_{B(0,1-\epsilon)} \le C(\|A^{\frac{n}{m}}u\|_{B(0,1-\epsilon')} + \|u\|_{B(0,1-\epsilon')}) \qquad \forall u\in \mathrm{D}(A^{\frac{n}{m}});$$

on the other hand, for $0\le n\le m$, (5.1.45) yields the inequality

(5.1.46)′
$$\sum_{|\alpha|\le n}\|(hD)^\alpha u\|_{B(0,1-\epsilon)} \le C\|u\|^{1-\frac{n}{m}}_{B(0,1-\frac{\epsilon}{2})}\Bigl(\sum_{|\alpha|\le m}\|(hD)^\alpha u\|_{B(0,1-\frac{\epsilon}{2})}\Bigr)^{\frac{n}{m}} + C\|u\|_{B(0,1-\frac{\epsilon}{2})}) \qquad \forall u\in\mathrm{D}(A).$$

Substituting $E(\tau_1',\tau_2')u$ for u we obtain from (5.1.46) and (5.1.46)′ that the norm of the operator

$$(hD)^\alpha E(\tau_1', \tau_2') : L^2(X, \mathbb{H}) \to L^2(B(0, 1-\epsilon) \cap X, \mathbb{H})$$

does not exceed $C(|\tau_1'| + |\tau_2'| + 1)^{\frac{|\alpha|}{m}}$ for all $\alpha : |\alpha| \le K$. The rest of proof of theorem 5.1.8(i) repeats the end of the proof of lemma 4.1.8.

In a similar way the estimate

$$\|(hD)^\alpha F_{t\to h^{-1}\tau} \chi_T(t) u\|_{B(0,1-\epsilon)} \le C'' T(|\tau| + 1)^{\frac{|\alpha|}{m}} \|v\| \tag{5.1.47}$$

can be proven for $u = e^{ih^{-1}tA}v$, $v \in \mathrm{D}(A^{\frac{n}{m}})$, $|\alpha| \le n$ and $\frac{n}{m} \in \mathbb{Z}$ where $T \ge h^{-1}$ and $\chi \in C_0^\infty([-1, 1])$ is fixed. Starting from this estimate one can prove theorems 5.1.8(iii) and 5.1.9 in the same way as we proved lemma 4.1.8(ii); namely, in estimates (5.1.45) and (5.1.46) for $\tau \in [\tau_1, \tau_2]$, the term $\|u\|$ in the right hand expression is replaced by $(1 + |\tau|)^{-s} h^s \|u\|$ and A is replaced by $A - \tau$; all the other arguments repeat those of the proof of lemma 4.1.8(ii).

Let us now prove theorem 5.1.8(ii). Let us first note that without loss of generality we can assume that $|\tau_1| + |\tau_2| \le c$. In fact, for $d \ge 2$ it is sufficient to prove estimate (5.1.38) under this restriction or the restriction $c\epsilon_0 \le \tau_1/\tau_2 \le c$, $|\tau_j| \ge 1$; the general case can be reduced to these special cases by division of the interval $[\tau_1, \tau_2]$ into intervals $[-2^l, -2^{l-1}]$ and two other intervals, the first of which satisfies the first restriction and the second of which satisfies the second restriction. But if $\epsilon_0 \le \pm\tau_j/\omega \le c \quad \forall j = 1, 2$ with $\omega \ge 1$ then dividing A by ω and replacing τ_j and h by τ_j/ω and $h\omega^{-\frac{1}{m}}$ respectively we reduce the second case to the first one. On the other hand, one can easily see that if $d = 1$, $|\tau| \ge C_0$ and (5.1.38) holds then (5.1.40) also holds and hence it is sufficient for us to treat the case $|\tau_1| + |\tau_2| \le c$.

But if condition (5.1.38) is fulfilled then

$$\sum_{|\alpha|\le n} \|(hD)^\alpha w\| \le C\Big(\sum_{|\beta|\le n-m} \|(hD)^\beta (A - \tau) w\| + h^s \|w\|\Big) \tag{5.1.48}$$
$$\forall \tau \in [\tau_1, \tau_2] \quad \forall w \in C_0^n(B(0,1) \cap \{x_1 \ge -\epsilon\}, \mathbb{H})$$

where here (and only here) we calculate the L^2-norms on $B(0,1)$ instead of its intersection with X. Let us substitute $\phi(\frac{x_1}{\nu})w$ in place of w in (5.1.48) where $\phi \in C^\infty(\mathbb{R})$, $\phi(t) = 1$ for $t \ge 1$, $\phi(t) = 0$ for $t \le 1 - \epsilon$ and $\nu \in [h, 1]$ is arbitrary; we then obtain the inequality

$$\sum_{|\alpha|\le n} \|(hD)^\alpha w\|_{X_\nu} \le C\Big(\sum_{|\beta|\le n-m} \|(hD)^\beta (A - \tau) w\|_{X_{\nu(1-\epsilon)}} + \sum_{|\beta|\le n-1} \frac{h}{\nu} \|(hD)^\beta w\|_{X_{\nu(1-\epsilon)}}\Big)$$

where $X_\eta = \{x, x_1 \ge \eta\}$. Iterating this inequality we obtain (since $\epsilon > 0$ is arbitrary) that

$$\sum_{|\alpha|\le n} \|(hD)^\alpha w\|_{X_\nu} \le C\Big(\sum_{|\beta|\le n-m} \|(hD)^\beta (A - \tau) w\|_{X_{\nu(1-\epsilon)}} + \sum_{|\beta|\le n-1} h^s \nu^{-s} \|(hD)^\beta w\|_{X_{\nu(1-\epsilon)}}\Big); \tag{5.1.49}$$

moreover, standard arguments yield that we can take $w \in C^n((B(0,1), \mathbb{H})$ but in this case the norms in the left and right-hand expressions should be calculated on $B(0, 1-\epsilon) \cap X_\nu$ and $B(0, 1 - \frac{\epsilon}{2}) \cap X_{\nu(1-\epsilon)}$ respectively.

Let us substitute into this inequality $w = F_{t\to h^{-1}\tau}\chi_T(t)u$ with $u = e^{ih^{-1}t}Av$ and let us note that $(A-\tau)w = ihT^{-1}w_1$ with $w_1 = F_{t\to h^{-1}\tau}\chi'_T(t)u$ where χ' is the derivative of χ. Applying this inequality to w_1, etc., we finally obtain that

$$\sum_{|\alpha|\le n} \|(hD)^\alpha F_{t\to h^{-1}\tau}\chi_T(t)u\|_{B(0,1-\epsilon)\cap X_\nu} \le$$
$$Ch^s\nu^{-s}\sum_{|\beta|\le n-1} h^s\nu^{-s}\|(hD)^\beta F_{t\to h^{-1}\tau}\bar\chi_T(t)u\|_{B(0,1-\frac{\epsilon}{2})\cap X_{\frac{\nu}{2}}}$$
$$\le Ch^s\nu^{-s}T\|v\|$$

where $\bar\chi \in C_0^\infty([-2,2])$ is equal to 1 on $[-\frac32,\frac32]$ and we used (5.1.48). The embedding theorems imply that

$$|(hD)^\alpha F_{t\to h^{-1}\tau}\chi_T(t)u| \le Ch^{-\frac{d}{2}+s}T\nu^{-s}\|v\|$$
$$\forall x\in B(0,1-\epsilon): x_1\ge\nu \quad \forall\tau\in[\tau_1,\tau_2] \quad \forall\alpha: |\alpha|\le n \quad \forall T\ge\nu$$

where in the left-hand expression the norms in $\mathbb{H}$ are calculated. Since v is arbitrary we obtain the estimate

$$\|(hD_x)^\alpha F_{t\to h^{-1}\tau}\chi_T(t)f^\dagger U(t,x,y)\| \le \tag{5.1.50}$$
$$Ch^{-\frac{d}{2}+s}T\nu_1^{-s}\|f\|$$
$$\forall x\in B(0,1-\epsilon): x_1\ge\nu \quad \forall\tau\in[\tau_1,\tau_2] \quad \forall\alpha: |\alpha|\le n \quad \forall T\ge\nu \quad \forall f\in\mathbb{H}$$

where in the left-hand expression the L^2-norm with respect to y is calculated. Starting with estimates (5.1.50) and

$$\sum_{|\alpha|\le n, |\beta|\le k} \|(hD_x)^\alpha (hD_y)^\beta F_{t\to h^{-1}\tau}\chi_T(t)f^\dagger U(t,x,y)\| \le \tag{5.1.51}$$
$$Ch^{-\frac{d}{2}}(|\tau|+1)^{\frac{1}{m}(d+|\alpha|+|\beta|)}\|f\|$$
$$\forall x\in B(0,1-\epsilon): x_1\ge\nu \quad \forall\tau\in\mathbb{R} \quad \forall T\ge\nu \quad \forall f\in\mathbb{H}$$

and repeating the same arguments with y instead of x we obtain that

$$|(hD_x)^\alpha (hD_y)^\beta F_{t\to h^{-1}\tau}\chi_T(t)U(t,x,y)| \le Ch^{-d+s}T\nu^{-s}$$
$$\forall x,y\in B(0,1-\epsilon): x_1\ge\nu, y_1\ge\nu \quad \forall\tau\in[\tau_1,\tau_2] \quad \forall T\ge\nu$$
$$\forall\alpha,\beta: |\alpha|\le n, |\beta|\le n.$$

Restricting to $x=y$, $\alpha=\beta$ and applying standard Tauberian arguments we obtain estimate (5.1.39) in this case; then the estimate also holds without the restriction due to the arguments used in the proof of (4.1.37). □

5.1.4 Estimates for the Schrödinger operator

One of the most important assertions of subsection 5.1.2 from the point of view of applications is theorem 5.1.6 which permits cutting with respect to ξ' (for $|\tau|\le c$) taking into account only the zone $\{|\xi'|\le C_0\}$. For this we were forced to assume that the operator and boundary value problem are elliptic in the classical sense. Let us demonstrate that in a certain special case one can weaken this restriction. Let us treat the Schrödinger operator (4.4.1) satisfying (4.4.6) and (4.4.7). Let us choose a coordinate system in a neighborhood of Y in which x_1 is the Riemannian distance in X from x to Y corresponding to the Riemannian metric g^{jk}; in this system $g^{1k} = \delta_{1k}$. By means of a gradient transform we can reduce the general case to $V_1 = 0$. So we treat the operator

(5.1.52) $$A = (hD_1)^2 + A'(x', hD', h) = (hD_1)^2 + \sum_{j,k}(hD_j - V_j)g^{jk}(hD_k - V_k) + V;$$

here and below we sum with respect to $j, k = 2, \dots, d$ and

(5.1.53) $$\sum_{j,k} g^{jk}\xi'_j\xi'_k \geq \epsilon_0|\xi'|^2.$$

Let us consider the boundary operator

(5.1.54) $$B = -i\alpha(x')hD_1 + \mathcal{M}(x', hD') + \beta(x') + h\beta'(x'), \qquad \mathcal{M}(x', \xi') = \sum_j m^j(x')(\xi_j - V_j).$$

It is easy to see that the self-adjointness of A (with the boundary condition $Bu = 0$) yields that functions $\alpha^{-1}m^j$ and $\alpha^{-1}\beta$ are real-valued (at points where $\alpha \neq 0$). Let us assume that

(5.1.55) α, β, m^j are real-valued functions.

Proposition 3.1.7 yields that if $(A - \tau)u = 0$ and $Q = \mathrm{Op}(q)$, $q \in S'_{h,\rho,1,K}$ with $\rho \geq C_0$ and if q is supported in $\{\epsilon_0\rho \leq |\xi'| + |\tau|^{\frac{1}{2}} \leq c\rho\}$ and if either $|\xi'| \geq C_0|\tau|^{\frac{1}{2}}$ on $\operatorname{supp} q$ or $\tau \leq 0$ then

$$\|Q\eth(ihD_1 + L)u\| \leq C'\rho^{-s}h^s\|u\|$$

where $L = \mathrm{Op}(\ell)$, $\ell \in \tilde{S}^{1\prime}_{h,K}$ does not depend on Q and its principal symbol equals

$$\ell_0 = (a'(x', \xi') - \tau)^{\frac{1}{2}}, \quad a'(x', \xi') = \sum_{jk} g^{jk}(\xi_j - V_j)(\xi_k - V_k).$$

Therefore in the zone $\Omega = \{|\xi'| \geq C_0(|\tau| + 1)\} \cup \{\tau \leq -C_0\}$ the boundary condition $\eth Bu = 0$ is equivalent to the condition $\tilde{B}\eth u = 0$ where

(5.1.56) $$\tilde{B} = \alpha L + \mathcal{M} + \beta + h\beta'.$$

It is easy to see that for $d \geq 3$ the Shapiro-Lopatinskii condition means exactly that

(5.1.57) $$|\alpha| \geq (g_{jk}m^j m^k)^{\frac{1}{2}} + \epsilon_0$$

where (g_{jk}) is the tensor inverse to (g^{jk}); in the case $d = 2$ this condition is stronger than the Shapiro-Lopatinskii condition which in this case means exactly that

(5.1.57)′ $$|\alpha \pm \sqrt{g_{22}}m| \geq \epsilon_0$$

for both signs $\pm$; moreover, speaking about this condition we assume that $|\alpha| \geq \epsilon_0$. We replace condition (5.1.57) by the conditions

(5.1.58)$_1$ $$\alpha \geq (1 + \epsilon_0)(g_{jk}m^j m^k)^{\frac{1}{2}},$$

$$(5.1.58)_2 \qquad \beta \geq \epsilon_0.$$

Let us note that $(5.1.58)_2$ and (5.1.55) provide (5.1.10). Moreover, $(5.1.58)_{1,2}$ and (5.1.55) imply that in Ω

$$\tilde{B}_0(x', \xi', \tau) \geq \epsilon_1(\alpha(x')(|\xi'| + |\tau|^{\frac{1}{2}}) + 1)$$

with $\epsilon_1 = \epsilon_0(d, c, \epsilon_0) > 0$.

We need the following assertion:

Lemma 5.1.10. *Let conditions* (5.1.52)–(5.1.55), (5.1.58) *and* (5.1.4) *be fulfilled. Let* $Q = q(hD')$ *with* $q \in S'_{h,\rho,1,K}$ *supported in* $\{\epsilon_0\rho \leq |\xi'| + |\tau|^{\frac{1}{2}} \leq c\rho\} \cap \Omega$; *moreover, let* Q' *and* q' *be an operator and symbol of the same nature, with* $q' = 1$ *in the* $\epsilon\rho$*-neighborhood of* $\operatorname{supp} q$; *let* $\psi, \psi' \in C_0^K(B(0,1))$ *and assume* ψ' *equals* 1 *in the* ϵ*-neighborhood of* $\operatorname{supp}\psi$. *Then*

$$\|Q\psi u\|_Y \leq C(\|Q'\tilde{B}\psi' u\|_Y + \rho^{-s}h^s\|\psi' u\|). \tag{5.1.59}$$

Proof. Obviously it is sufficient to prove the estimate

$$\|Q\psi u\|_Y \leq C(\|Q'\tilde{B}\psi' u\|_Y + \rho^{-\frac{1}{2}}h^{\frac{1}{2}}\|Q'\psi' u\|_Y + \rho^{-s}h^s\|\psi' u\|). \tag{5.1.60}$$

Applying this estimate to Q', ψ' instead of Q, ψ and repeating this process we obtain (5.1.59). We can assume without loss of generality that ψ, ψ', Q, Q' are real-valued.

Then (5.1.55), (5.1.58) and the precise Gårding inequality yield that

$$\operatorname{Re}(\tilde{B}Q\psi u, Q\psi u)_Y \geq \epsilon_0\|Q\psi u\|_Y^2 - C\rho^{-s}h^s\|\psi' u\|_Y^2$$

and we need to consider

$$\begin{aligned}\operatorname{Re}(\tilde{B}Q\psi u, Q\psi u)_Y = {} & \operatorname{Re}(Q\psi\tilde{B}\psi' u, Q\psi u)_Y + {} \\ & \operatorname{Re}([\tilde{B}, Q]\psi u, Q\psi u)_Y + \operatorname{Re}(Q[\tilde{B}, \psi]\psi' u, Q\psi u)_Y.\end{aligned} \tag{5.1.61}$$

The absolute value of the first term in the right-hand expression does not exceed

$$\|Q\psi u\|_Y \cdot \|Q'\tilde{B}\psi' u\|_Y + C\rho^{-s}h^s\|\psi' u\|_Y^2.$$

The second term in the right-hand expression of (5.1.61) can be rewritten in the form $\frac{1}{2}\operatorname{Re}(\mathcal{K}_1\psi u, \psi u)_Y$ where $\mathcal{K}_1 = [Q, [\tilde{B}, Q]] - [\tilde{B}^* - B, Q]Q \in \rho^{-1}h\boldsymbol{\Psi}'_{h,\rho,1,K}$ uniformly with respect to ρ and hence the absolute value of this term does not exceed

$$C\rho^{-1}h\|Q'\psi' u\|_Y^2 + \rho^{-s}h^s\|\psi' u\|_Y^2. \tag{5.1.62}$$

The third term in the right-hand expression of (5.1.61) is equal to $\operatorname{Re}(\mathcal{K}_2\psi' u, \psi' u)_Y$ where

$$\mathcal{K}_2 = \psi Q^2[B, \psi] = Q\psi[\tilde{B}, \psi]Q + [\psi, Q][\tilde{B}, \psi]Q + Q\psi[Q, [\tilde{B}, \psi]];$$

here the second and the third terms obviously belong to $\rho^{-1}h\boldsymbol{\Psi}'_{h,\rho,1,K}$ and hence modulo an expression with absolute value not exceeding (5.1.62) the third term in the right-hand expression of (5.1.61) is

$$\operatorname{Re}(\psi[\tilde{B}, \psi]Q\psi' u, Q\psi' u)_Y = \frac{1}{2}\operatorname{Re}(\mathcal{K}_3\psi' u, \psi' u)_Y$$

where

$$\mathcal{K}_3 = [\psi, [\tilde{B}, \psi]] - [\tilde{B}^* - \tilde{B}, \psi] \in \rho^{-1}h\boldsymbol{\Psi}'_{h,\rho,1,K}.$$

Therefore the absolute value of the third term in the right-hand expression of (5.1.61) also does not exceed (5.1.62). These estimates clearly yield (5.1.60). □

This lemma, proposition 3.1.7 and standard Tauberian techniques obviously yield

Theorem 5.1.11. *Suppose conditions* (5.1.52)–(5.1.55), $(5.1.58)_{1,2}$, (5.1.4) *and* (5.1.5) *are fulfilled. Then statements (i), (ii) of theorem 5.1.7 remain true with* $\omega = \rho$.

5.2 The Method of Successive Approximations

The goal of this section is the same as the goal of section 4.2. There is an additional difficulty arising from the fact that there is no mollification with respect to x_1 and y_1 because there is no cutting with respect to ξ_1 and η_1 (so smoothness with respect to x_1 and y_1 should be treated carefully).

5.2.1 Approximating the free space term

It will be convenient for us to represent U as a sum $U = U^0+U^1$ as was done in section 3.2; to construct U^0 we continue A to the domain $\{x_1 \geq -\epsilon\}\cap B(0,1-\epsilon)$ with a small enough constant $\epsilon > 0$; then for U^0 the construction of subsection 4.2.1 is applicable with the modifications indicated in subsection 4.2.2 (but the coefficients are bounded operators now) and with the calculations of subsection 4.2.3. Namely, for U^0 a slightly modified form of proposition 4.2.6 holds:

Proposition 5.2.1. *Suppose conditions* (5.1.1), (5.1.2), $(5.1.4)_1$, (5.1.8) *and* (5.1.9) *are fulfilled, let* $Q_i = \operatorname{Op}(q_i)$, $q_i \in S'_{h,K}$ *be supported in* $\{|\xi'| \leq c\}$ *for* $i = 1, 2$, *and let* $\psi_i \in C_0^K(B(0,1-\epsilon))$. *Then*

$$(5.2.1)\quad \mathcal{R} = \|D_x^\alpha D_y^\beta\Bigl(F_{t\to h^{-1}\tau}\chi_T(t)Q'_{1x}\psi_1(x)U^0\psi_2(y)^tQ'_{2y} - \sum_{0\leq n\leq N'} h^{-d+n}\iint T\hat{\chi}(\frac{(\tau-\tau')T}{h})e^{ih^{-1}\langle x-y,\xi\rangle}q'_1(x,\xi')\mathcal{F}_n^0(y,\tau',\xi)\times [q'_2(y,\xi')]d\tau'd\xi'\Bigr)\| \leq C'M'h^s$$

$$\forall x,y\in\mathbb{R}^d \quad \forall\tau : |\tau|\leq c \quad \forall\alpha,\beta : |\alpha|\leq r, |\beta|\leq r$$

where

$$\mathcal{F}_n^0(y,\tau',\xi) = \operatorname{Res}_{\mathbb{R}} F_n^0(y,\tau',\xi),$$

$T \in [h^{1-\delta}, h^{\frac{1}{2}+\delta}]$, $\delta > 0$ *is an arbitrarily small exponent,* r *is arbitrary, and* K *and the constants depend on* r *also.*

Proof. Proposition 4.2.6(ii) yields this estimate for Q'_{1x} and Q'_{2y} replaced by $Q_{1x} = Q'_{1x}Q''_{1x}$ and $Q_{2y} = Q'_{2y}Q''_{2y}$ respectively where $Q''_{1x} = q''_1(hD_{x_1})$, $Q''_{2y} = q''_2(hD_{y_1})$ with q''_1 supported in $\{|\xi_1| \leq c_1\}$ and with similar q''_2; the constant c_1 will be chosen later. Surely $q'_1(x,\xi')$, $q'_2(x,\xi')$ should be replaced by $q_1(x,\xi)$, $q_2(y,\xi)$ respectively.

On the other hand, standard elliptic arguments yield that

$$\|D_x^\alpha D_y^\beta(F_{t\to h^{-1}\tau}\chi_T(t)Q_{1x}\psi_1(x)U^0\psi_2(y)^tQ_{2y})\| \leq C'M'h^s\rho^{-s}$$

provided Q_1 and Q_2 have the same structure as before and $q''_i \in S_{h,K}(T^*\mathbb{R})$ are supported in $\{|\xi_1| \leq c\rho\}$ for $i = 1, 2$ and $q''_i \in S_{h,\rho,1,K}(T^*\mathbb{R})$ is supported in $\{|\xi_1| \geq \epsilon_0\rho\}$ for some $i = 1, 2$ where $\rho \geq c_1$ with an appropriate constant c_1. Moreover, the regularity property of F_n^0 for $|\tau| \leq c$, $|\xi_1| \geq c_1\epsilon_0$ and $|\xi'| \leq c$ yields that all the other terms in $\mathcal{R}$ vanish.

Summing with respect to an admissible partition of unity on $\mathbb{R}_{\xi_1}$, we obtain estimate (5.2.1). □

Now our goal is to obtain a decomposition for U^1. The first step is to obtain a decomposition for $\eth_x D^j_{x_1} U^\pm_0$ for $j = 0, \dots, m-1$. For this purpose let us use proposition 4.2.6(ii) again. However, now we modify the construction of subsection 4.2.1. First of all, we now freeze the symbol of A at the point $(0, y')$ instead of y; then the factor $x_1 - y_1$ will be replaced by the factor x_1 which we now move to the right (to $\eth_x$) instead of to the left (to $\delta(x-y)$) as was done before with the factor $x_1 - y_1$, but the final result is the same: if the factor x_1 reaches $\eth_x$ then the corresponding term vanishes. Moreover, we now cut only with respect to ξ', we do not cut with respect to ξ_1. In order to control the smoothness with respect to x_1 and y_1 we apply the construction of subsection 4.2.2 with $B^{\frac{1}{m}}$ replaced by hD_1 (therefore we assume that the prolonged operator A' satisfies condition (5.1.8) at every point). All other details of construction are preserved without change and then the calculations of subsection 4.2.3 yield the estimate

$$
\begin{aligned}
(5.2.2)\quad \mathcal{R}'_\pm = \|\eth_x D^\alpha_x D^\beta_y \Big(& F_{t\to h^{-1}\tau} \chi_T(t) Q'_{1x} \psi_1(x) U^0 \psi_2(y) {}^t Q'_{2y} \pm \\
& \sum_{0\le n\le N'} h^{-d+n} \iint T\hat{\chi}(\frac{\tau' T}{h}) e^{ih^{-1}\langle x-y,\xi\rangle} q'_1(x,\xi') F^1_n(y, \tau - \tau', \xi) \times \\
& [q'_2(y,\xi')] d\tau' d\xi' \Big) \| \le C' M' h^s (1 + |\tau|)^N \\
& \forall x', y' \in \mathbb{R}^{d-1} \quad \forall \tau : \pm \operatorname{Im} \tau < 0 \quad \forall \alpha, \beta : |\alpha| \le r, |\beta| \le r, \alpha_1 + \beta_1 \le m-1
\end{aligned}
$$

where F^1_n are $2n$-th order differential operators with respect to ξ' with coefficients analytic in $\mathbb{C}_\mp \setminus \mathbb{R}$ and $F^1_0 = F^0_0$; the norm in the left-hand expression is the operator norm from $L^2(\mathbb{R}_{y_1}, \mathbb{H})$ to $\mathbb{H}$ and we assume that $Q'_i = \operatorname{Op}(q'_i)$ with $q'_i \in S'_{h,K}$ supported in $\{|\xi| \le c\}$ for $i = 1, 2$.

Proposition 5.2.2. *Suppose conditions* (5.1.1), (5.1.2), (5.1.4)$_1$, (5.1.8) *and* (5.1.9) *are fulfilled and let* Q_i, ψ_i *be the same as in proposition 5.2.1. Then estimate* (5.2.2) *remains true under the restriction "*$|\alpha| \le r, |\beta| \le r, \alpha_1 \le m-1$*" instead of the previous restriction provided* $y_1 > 0$ *and in the left-hand expression the operator norm from* $\mathbb{H}$ *to* $\mathbb{H}$ *is calculated.*

Proof. Obviously it is sufficient to prove that this estimate holds for the $\mathcal{L}(\mathbb{H}, L^2(\mathbb{R}^+, \mathbb{H}))$-norm with the condition "$\alpha_1 + \beta_1 \le m-1$" replaced by "$\alpha_1 \le m-1$." If we perform the framing procedure of subsection 4.2.2 then we see that the only terms preventing us from obtaining this estimate are terms which contain no copy of $\bar{G}^\pm$, $G^\pm$, $\hat{G}^\pm$, each of which contains copies of $(\bar{A}+i)^{-1}$, $(A'+i)^{-1}$, $(A+i)^{-1}$ and contains at least one copy of the first operator and at most one copy of the third operator; every copy of $(\bar{A}+i)^{-1}$ is to the left of every copy of $(A'+i)^{-1}$ and $(A+i)^{-1}$ is to the right of every copy of $(A'+i)^{-1}$; moreover, these terms contain h-pseudo-differential operators with compact supports with respect to (x, ξ'), functions with compact supports with respect to x and operators hD_t and hD_{x_1}, hD_{y_1}. All these terms vanish if we sum U^{0+} and U^{0-}. Let us replace all copies of the operator D_{y_1} applied to $\delta(x-y)$ by $-D_{x_1}$. Let us first treat all terms which do not contain $(A+i)^{-1}$. Let us move all the copies of hD_{x_1} from the left to the right by applying the commutation formula

$$
[(A'+i)^{-1}, T] = -(A'+i)^{-1}[A', T](A'+i)^{-1};
$$

we know that $(\bar{A}+i)^{-1}$ and hD_1 commute. Moreover, if there is a copy of $(hD_1)^k$ with $k \ge m$ near either $(A'+i)^{-1}$ or $(\bar{A}+i)^{-1}$ then (5.1.8) yields that this term equals a sum of terms one of which does not contain a copy of $(A'+i)^{-1}$ or $(\bar{A}+i)^{-1}$ while the other contains hD_1 to a degree not exceeding $m-1$. Therefore we obtain terms of two types as the result of these transformations:

(a) Terms which are products of factors $(hD_1)^j(\bar{A}+i)^{-1}$, $(hD_1)^j(A'+i)^{-1}$ with $j \le m-1$, operators belonging to $\Psi'_{h,K}$, smooth functions and hD_t;

(b) Terms which contain neither $(\bar{A}+i)^{-1}$ nor $(A'+i)^{-1}$.

Furthermore, terms originating from terms which we threw away before contain a factor h^L with arbitrary large L. Then the norms of *these* terms of type (a) do not exceed $C'M'h^s(1+|\tau|)^N$ because of (5.1.8) and terms of type (b) will vanish for $x_1 \ne y_1$ because they are differential operators with respect to x_1, y_1.

Let us now treat terms containing $(A+i)^{-p}$; these terms necessarily contain a factor $A-A'$ as well. Applying the commutation formula for $(A+i)^{-1}$ and $\psi Q\psi$ and repeating the arguments of subsection 4.2.2, we can reduce our analysis to the case in which there is no factor $A-A'$ but there is a high degree of h and the factor $(A+i)^{-p}$ either preserves or is replaced by a product of factors $(A+i)^{-1}\psi_4$ and $\psi_4(A+i)^{-1}$, where there are at least p factors of these forms and ψ_4 is defined in subsection 4.2.2; there are also factors hD_1, hD_t, operators belonging to $\Psi'_{h,K}$ with compactly supported symbols and compactly supported functions. If we consider the commutators of $(A+i)^{-1}$ with ψD_1^q and apply division of ψD_1^q by $A+i$ with remainder (we have $\psi \in C_0^K(B(0,1-\epsilon))$ decreasing $\epsilon > 0$ if necessary) we obtain as before three types of terms:

(a) Terms which are products of factors $(hD_1)^j(\bar{A}+i)^{-1}$, $(hD_1)^j(A'+i)^{-1}$, $\psi(hD_1)^j(A+i)^{-1}$ with $j \le m-1$, operators belonging to $\Psi'_{h,K}$, smooth functions and hD_t;

(a′) Terms which are products of factors $(hD_1)^j(\bar{A}+i)^{-1}$, $(hD_1)^j(A'+i)^{-1}$ with $j \le m-1$, $(A+i)^{-p}$ (only one copy), operators belonging to $\Psi'_{h,K}$, smooth functions and hD_t;

(b) Terms which contain neither $(\bar{A}+i)^{-1}$, nor $(A'+i)^{-1}$, nor $(A+i)^{-1}$.

All these terms contain a factor h to a high degree. Let us note that for $0 \le j \le m$ the operators $(hD_1)^j(\bar{A}+i)^{-1}$, $(hD_1)^j(A'+i)^{-1}$ and $\psi(hD_1)^j(A+i)^{-1}$ act from $L^2(\mathbb{R}^d,\mathbb{H})$ ($L^2(B(0,1),\mathbb{H})$ in the third case) into $L^2(\mathbb{R}^d,\mathbb{H})$. Then we can estimate terms of types (a) and (a′); terms of type (b) vanish. $\square$

Remark. *Surely one could refer to the theory of h-pseudo-differential operators with the transmission property and prove that $\psi(A+i)^{-p}\psi$ is an operator of this type, but one should first develop this theory!* $\diamond$

Corollary 5.2.3. *Let conditions* (5.1.1), (5.1.2), (5.1.4)$_1$, (5.1.8) *and* (5.1.9) *be fulfilled; let Q'_i, ψ'_i be the same as in proposition 5.2.1. Then*

$$(5.2.3)\quad \mathcal{R}''_\pm = \|D_x^\alpha D_y^\beta\Big(F_{t\to h^{-1}\tau}\chi_T(t)Q'_{1x}\psi_1(x)V^\pm\psi_2(y)^tQ'_{2y} - \sum_{0\le n\le N'} h^{-d+n}\iint T\hat{\chi}(\frac{\tau\tau' T}{h})e^{ih^{-1}\langle x-y,\xi\rangle}q'_1(x,\xi')F_n^2(y,\tau-\tau',\xi)\times [q'_2(y,\xi')]d\tau' d\xi'\Big)\| \le C'M'h^s(|\tau|+1)^{N''}$$

$$\forall x',y' \in \mathbb{R}^{d-1}\quad \forall y_1 > 0\quad \forall \tau : \pm\tau < 0\quad \forall \alpha,\beta : |\alpha| \le r, |\beta| \le r$$

where $T \in [h^{1-\delta}, h^{\frac{1}{2}+\delta}]$ as before, $V^\pm = -\eth_x B(x,hD_x)U^{0\pm}$, and F_n^k are functions of the same type as before; here they are sums of terms

$$\Big(\prod_{1\le j\le l_n}(\tau - a(0,y',\xi))^{-1}p_j(y',\xi)\Big)\cdot(\tau - a(0,y',\xi))^{-1}$$

where p_j are polynomials with respect to ξ the degrees of which with respect to ξ_1 do not exceed $m-1$.

5.2.2 Approximations for the reflected wave

Let us again calculate $U^{1\pm}$ by the method of successive approximations. For this purpose we write the problem (3.2.25), (3.2.30), (3.2.31) in the previous notation:

$$(hD_t - A)U^{1\pm} = 0, \tag{5.2.4}$$

$$\eth BU^{1\pm} = -\eth BU^{0\pm}, \tag{5.2.5}$$

$$U^{1\pm}|_{\pm t<0} = 0; \tag{5.2.6}$$

therefore

$$(hD_t - A')U^{1\pm} = F^{\pm}, \tag{5.2.7}$$

$$\eth B'U^{1\pm} = -\eth B'U^{0\pm} + f^{\pm} \tag{5.2.8}$$

and then

$$(hD_t - \bar{A})U^{1\pm} = F^{\pm} + RU^{1\pm}, \tag{5.2.9}$$

$$\eth \bar{B}U^{1\pm} = -V^{\pm} + \eth R_1 U^{1\pm} + f^{\pm} \tag{5.2.10}$$

where $F^{\pm} = (A - A')U^{1\pm}$, $f^{\pm} = -\eth(B - B')(U^{0\pm} + U^{1\pm})$, $R = A' - \bar{A}$, $R_1 = B' - \bar{B}$, $\bar{A}$ and $\bar{B}$ are operators A' and B' with symbols frozen at the point $(0, y')$, A' and B' are obtained from A and B by cutting outside of a neighborhood of $\operatorname{supp} q_2'$, and all the operators act with respect to x.

Then (5.2.9), (5.2.10) and (5.2.6) yield that

$$U^{1\pm} = (\bar{G}^{\pm}R + \bar{G}'^{\pm}\eth R_1)U^{1\pm} + \bar{G}^{\pm}F^{\pm} + \bar{G}'^{\pm}(-V^{\pm} + f^{\pm}). \tag{5.2.11}$$

Furthermore, (5.2.7), (5.2.8), (5.2.6) yield that

$$U^{1\pm} = G^{\pm}F^{\pm} + G'^{\pm}(-V^{\pm} + f^{\pm}) \tag{5.2.12}$$

where $\bar{G}^{\pm}$ and $\bar{G}'^{\pm}$ are parametrices of the problems

$$(hD_t - \bar{A})u = f, \quad \operatorname{supp} f \subset \{\pm(t - t_0) \geq 0\}, \tag{5.2.13}$$

$$\eth \bar{B}u = 0, \tag{5.2.14}$$

$$u|_{\pm(t-t_0)<0} = 0 \tag{5.2.15}$$

and

$$(hD_t - \bar{A})u = 0, \tag{5.2.13$'$}$$

(5.2.14)′ $$\eth\bar{B}u = f, \quad \operatorname{supp} f \subset \{\pm(t-t_0)\ge 0\},$$

and (5.2.15) respectively; $G^\pm$ and $G'^\pm$ are defined in the same way for A' and B' instead of $\bar{A}$ and $\bar{B}$ and in what follows we will use $\hat{G}^\pm$ and $\hat{G}'^\pm$ defined in the same way for A and B.

Iterating (5.2.11) and applying (5.2.12) we obtain that

(5.2.16) $$\begin{aligned} U^{1\pm} = -&\sum_{0\le n\le N'-1} (\bar{G}^\pm R + \bar{G}'^\pm \eth R_1)^n \bar{G}'^\pm \eth B' U^{0\pm} - \\ &(\bar{G}^\pm R + \bar{G}'^\pm \eth R_1)^{N'} G'^\pm \eth U^{0\pm} + \\ &\sum_{0\le n\le N'-1} (\bar{G}^\pm R + \bar{G}'^\pm \eth R_1)^n (\bar{G}^\pm F^\pm + \bar{G}'^\pm f^\pm) + \\ &(\bar{G}^\pm R + \bar{G}'^\pm \eth R_1)^{N'} (G^\pm + G'^\pm f^\pm). \end{aligned}$$

Let us first treat the first and second terms (the third and fourth terms will be called "extra terms").

For this task we need the commutation relations for $\bar{G}^\pm$ and $\bar{G}'^\pm$ with $x_j - y_j$ ($j = 2,\dots,d$) and x_1 and similar relations for $G^\pm$, $G'^\pm$. Let us take u supported in $\{t : \pm(t-t_0)\ge 0\}$ and let us consider the equalities

$$\bar{P}x_j u = x_j \bar{P}u - [\bar{A}, x_j]u, \quad \eth\bar{B}x_j u = \eth[\bar{B}, x_j]u + x_j\eth\bar{B}u$$

where $\bar{P} = hD_t - \bar{A}$ and for $j = 1$ the last term should be omitted; then

$$x_j u = \bar{G}^\pm x_j \bar{P}u - \bar{G}[\bar{A}, x_j]u + \bar{P}'^\pm[\bar{B}, x_j]u + \bar{G}'^\pm x_j \eth\bar{B}u$$

where for $j = 1$ the last term should be omitted. Substituting $u = \bar{G}^\pm v$ or $u = \bar{G}'^\pm w$ with arbitrary v, w supported in $\{t : \pm(t-t_0)\ge 0\}$ we obtain that

(5.2.17) $$x_j\bar{G}^\pm = \bar{G}^\pm x_j - \bar{G}^\pm[\bar{A}, x_j]\bar{G}^\pm + \bar{G}'^\pm\eth[\bar{B}, x_j]\bar{G}^\pm,$$

(5.2.18) $$x_j\bar{G}'^\pm = -\bar{G}^\pm[\bar{A}, x_j]\bar{G}'^\pm + \bar{G}'^\pm\eth[\bar{B}, x_j]\bar{G}'^\pm + \bar{G}'^\pm x_j$$

where for $j = 2$ one can replace x_j by $x_j - y_j$ and for for $j = 1$ the last term in (5.2.18) should be omitted. The same relations hold for $G^\pm$ and $G'^\pm$.

Let us recall that

(5.2.19) $$\begin{aligned} \eth B' U^{0\pm} = \mp i &\sum_{0\le n\le N'-1} \eth B' (\bar{G}_0^\mp R)^n \bar{G}_0^\mp \delta(t)\delta(x-y) \mp \\ &ih\eth B'(\bar{G}_0^\mp R)^{N'} G_0^\mp \delta(t)\delta(x-y) + \\ &\sum_{0\le n\le N'-1} \eth B'(\bar{G}_0^\mp R)^n \bar{G}_0^\mp F_0^\mp + \eth B'(\bar{G}_0^\mp R)^{N'} \bar{G}_0^\mp F_0^\mp \end{aligned}$$

where $\bar{G}_0^\mp$ and $G_0^\mp$ are parametrices of the Cauchy problems (i.e., of problems $(4.2.6)_\pm$, $(4.2.6)'_\pm$), $F_0^\pm = (A - A')U^{0\pm}$ and for $\bar{G}_0^\mp$ and $G_0^\mp$ the commutation relations (4.2.9), (4.2.9)′ hold. Let us apply commutation relations (4.2.9), (4.2.9)′, (5.2.17), (5.2.18) to the first and second terms in the right-hand expression of (5.2.16); we transfer $(x_j - y_j)$ with $j \ge 2$ and x_1 to the right, i.e., to $B'U^{0\pm}$. When x_1 reaches it the corresponding term vanishes (which is equivalent to the absence of the last term in in (5.2.18) for $j = 1$); in (5.2.19) we transfer x_1 to the left (i.e., to $\eth$); when x_1 reaches it the corresponding term vanishes; in (5.2.16), (5.2.19) we transfer $(x_j - y_j)$ with $j \ge 2$ to the right (i.e., to

$\delta(x-y)$) with the same result. Therefore each such factor should be killed this way, and then a factor h arises. Let us substitute (5.2.19) into (5.2.16) and let us omit all terms containing $F^\pm$, $f^\pm$, $F_0^\pm$ which will be analyzed later. Let us transfer all factors x_j-y_j with $j\ge 2$ and x_1 (or large number of them). In addition, let us frame all the operators $G^\pm$, $\bar G^\pm$, $G_0^\pm$, $\bar G_0^\pm$ (for $G^\pm$ and $\bar G^\pm$; the framing procedure is the same as for $G_0^\pm$ and $\bar G_0^\pm$). Moreover we can "frame" the operators $\bar G'^\pm$ and $G'^\pm$ on the left. In fact, the equalities

$$(\bar A+i)\bar G'^\pm=(-hD_t+\bar A)\bar G'^\pm+(hD_t+i)\bar G'^\pm=(hD_t+i)\bar G'^\pm$$

and $\eth\bar B\bar G'^\pm=I$ yield that

$$\bar G'^\pm=(hD_t+i)(\bar A+i)^{-1}\bar G'^\pm+\bar E' \tag{5.2.20}$$

where $\bar E'$ is a parametrix of the problem

$$(\bar A+i)u=0,\quad \eth\bar Bu=f$$

and $(\bar A+i)^{-1}$ means a parametrix of the problem

$$(\bar A+i)u=f,\quad \eth\bar Bu=0$$

as it was in the framing procedure for $\bar G^\pm$. The same method is also applicable to $G'^\pm$. Moreover, we also have the procedure of division by $\bar A$ (or A) with remainder which we apply to $(hD_1)^k$ with $k\ge m$. Namely, if we have a factor $(hD_1)^k\bar G_0^\pm$ then we can transform this factor to a sum of factors of the same type with $k\le m-1$ and an operator which is h-differential with respect to x_1 and h-pseudo-differential with respect to x'. The same is true for $\bar G^\pm$, $\bar G_0^\pm$, $G^\pm$ and for operators $\bar G_0^\pm$, $G_0^\pm$ staying to the left of $(hD_1)^k$. Moreover, we can transform the operator $(hD_1)^k\bar G'^\pm$ to a sum of operators of the same type with $k\le m-1$ multiplied on the left by operators belonging to $\Psi'_{h,K}$. Let us note that if between the rightmost copies of $\eth$ and $\delta(x-y)$ there is no parametrix[4] then the corresponding term vanishes for $y_1>0$. Therefore treating the commutators of $(hD_1)^k$ with $(A'+i)_0^{-1}$ we conclude that to the right of the rightmost copy of $\eth$ we can obtain a product of operators $(hD_1)^k(A'+i)_0^{-1}$, $(hD_1)^k(\bar A+i)_0^{-1}$ with $k\le m-1$, $G_0^\pm$, $\bar G_0^\pm$, h-pseudo-differential operators with respect to x' with symbols compactly supported with respect to (x,ξ'), and compactly supported functions (even if we have applied hD_{y_1} to an arbitrarily high degree before this procedure).

Similarly, if we apply hD_{x_1} to an arbitrarily high degree we can provide that to the left of the rightmost copy of $\eth$ is a product of operators $(hD_1)^k(A'+i)^{-1}$, $(hD_1)^k(\bar A+i)^{-1}$ with $k\le m-1$, $\eth h^{\frac12}(hD_1)^k(A'+i)^{-1}$, $\eth h^{\frac12}(hD_1)^k(\bar A+i)^{-1}$, $h^{-\frac12}(hD_1)^kE'$, $h^{-\frac12}(hD_1)^k\bar E'$, $\eth(hD_1)^k\bar E'$, $\eth(hD_1)^kE'$ with $k\le m-1$, $\bar G^\pm$, $G^\pm$, hD_t, h-pseudo-differential operators with respect to x' with symbols compactly supported with respect to (x,ξ') and compactly supported functions; moreover, the total quantity of copies of $\bar G^\pm$, $G^\pm$, $\bar G_0^\pm$ and $G_0^\pm$ in every product does not exceed $2\nu+1$ where ν is the degree of h gained in the commutation process. All these operators, excluding hD_t, $\bar G^\pm$, $G^\pm$, $\bar G_0^\pm$ and $G_0^\pm$, have operator norms from $L^2(Z,.)$ to $L^2(Z',.)$ not exceeding C' where Z and Z' are corresponding domains from the following list: $\mathbb{R}^d$, $\mathbb{R}^{d-1}$, $\bar{\mathbb{R}}^+\times\mathbb{R}^{d-1}$. Moreover, the operator norms of $\bar G^\pm$, $G^\pm$, $\bar G_0^\pm$ and $G_0^\pm$ in $L^2([-T,T]\times Z)$ do not exceed $C'h^{-1}T$. Therefore *all terms which contain neither $F_0^\pm$ nor $F^\pm$ nor $f^\pm$ and contain either $G^\pm$ or $G'^\pm$ or terms of the decomposition of $\bar A-A'$ or $\bar B-B'$ with $x_1^{\alpha_1}(x'-y')^{\alpha'}h^\nu$ with $|\alpha|+\nu\ge N$ are negligible;* we say that a term is negligible if after applying $Q'_{1x}\psi_1(x)(hD_{x_1})^j$ and $Q'_{2y}\psi_2(y)(hD_{y_1})^j$

[4] I.e., no operator $G_0^\pm$, $\bar G^\pm$, $(A'+i)_0^{-1}$, $(\bar A+i)_0^{-1}$; there can be no other parametrix at this place and the index "0" means that the parametrix is constructed in $\mathbb{R}^d$.

to it with $j,k \le s$ the corresponding norms will be less than $C'h^s$ where s is arbitrarily large and $N' = N'(d,m,s)$.

Let us consider terms originating from $F_0^\pm$, $F^\pm$ and $f^\pm$. However, we change the first step of our construction slightly. Namely, we write the problems for $U^{0\pm}$ and $U^{1\pm}$ in the form

$$(hD_t - A')\psi_4 U^{0'\pm} = \mp ih\delta(t)\delta(x-y)\psi_4(x) + F_0^\pm,$$

$$(hD_t - A')\psi_4 U^{1'\pm} = F^\pm,$$

$$\eth B'\psi_4 U^{1'\pm} = -\eth B'\psi_4 U^{0'\pm} + f^\pm$$

where now

$$F_0^\pm = \psi_4(A - A')\psi_5 U^{0\pm} - [A,\psi_4]\psi_5 U^{0\pm},$$

$$F^\pm = \psi_4(A - A')\psi_5 U^{1\pm} - [A,\psi_4]\psi_5 U^{1\pm},$$

$$f^\pm = \eth\psi_4(B - B')\psi_5(U^{0\pm} + U^{1\pm}) + \eth[B,\psi_4](U^{0\pm} + U^{1\pm}).$$

Then when we commute h-pseudo-differential operators with $\bar{G}^\pm$ and $G^\pm$, frame and divide $(hD_1)^k$ with $k \ge m$ by A' (or $\bar{A}$) with remainder we obtain[5] from $F_0^\pm$, $F^\pm$ and $f^\pm$ terms which after multiplication by $Q'_{1x}\psi_1(x)$ [6] are of one of the following forms:

(a) $(hD_t)^{N'''}\mathcal{K}^\pm(hD_1)^j\psi_5 U^{1\pm}$ with $j \le m-1$;

(c) $(hD_t)^{N'''}\mathcal{K}'^\pm\eth(hD_1)^j\psi_5 U^{i\pm}$ with $i = 0,1$ and $j \le m-1$ where the operator norms of the operators $\mathcal{K}^\pm$ and $\mathcal{K}'^\pm$ acting from $L^2(\mathbb{R}^+ \times [-T,T], H_{-\mu}(\mathbb{R}^{d-1},.))$ and $L^2([-T,T], H_{-\mu}(\mathbb{R}^{d-1},.))$ respectively to $L^2(\mathbb{R}^+ \times [-T,T], H_\mu(\mathbb{R}^{d-1},.))$ will be less than $C'h^\mu$ where μ is arbitrary and does not depend on N' (but now K and N''' depend on N' and μ).

On the other hand, the following lemma is a consequence of the proof of theorem 5.1.1.

Lemma 5.2.4. *Let conditions* (5.1.1), (5.1.2), $(5.1.4)_1$, (5.1.8) *and* (5.1.9) *be fulfilled and let ψ_i be the same as before. Then*

$$\|\psi_1(x)(hD_{x_1})^j(hD_{y_1})^k\psi_2(y)U^{1\pm}\| \le C'h^{-L},$$

$$\|\eth_x\psi_1(x)(hD_{x_1})^j(hD_{y_1})^k\psi_2(y)U^{0\pm}\| \le C'h^{-L},$$

where j and k are arbitrary (less than L), in the left-hand expression the operator norms from $L^2(\mathbb{R}^+ \times [-T,T], H_L(\mathbb{R}^{d-1},.))$ to $L^2(\mathbb{R}^+ \times [-T,T], H_{-L}(\mathbb{R}^{d-1},.))$ and to $L^2([-T,T], H_\mu(\mathbb{R}^{d-1},.))$ respectively are calculated, and $L = L(d,m,j,k)$.

The proof is very simple and is based on the equalities

$$(hD_t - A_x)U^{i\pm} = 0, \quad U^{i\pm t}(hD_t - A_y) = 0$$

(provided $x_1 \ne y_1$ for $i = 0$).

Therefore *all terms originating from $F_0^\pm$, $F^\pm$ and $f^\pm$ are negligible.* Therefore we have proven

---

[5] Even if we apply hD_{x_1} to a high degree before this procedure.

[6] Let us recall that $Q'_1 = \mathrm{Op}(q'_1)$ where $\operatorname{supp} q'_1$ does not intersect the support of $A' - A$.

Proposition 5.2.5. *Let conditions (5.1.1), (5.1.2), (5.1.4)–(5.1.9) be fulfilled and let the operators Q_i and functions ψ_i be the same as before. Let $\chi \in C_0^K([-1,1])$. Then*

$$(5.2.21)\quad \|(hD_x)^\alpha (hD_y)^\beta Q'_{1x}\psi_1(x) F_{t\to h^{-1}\tau}\chi_T(t) \times \\ (U^{1\pm} - \sum_{n\le N'-1} (\bar{G}^\pm R' + \bar{G}'^\pm \eth R'_1)^n \bar{G}'^\pm V^\pm)\psi_2(y)^t Q'_{2y}\| \le \\ C' h^s (1+|\tau|)^{N'''}$$

with

$$(5.2.22)\quad V^\pm = \mp ih\eth(\bar{B}+R'_1) \sum_{k\le N'-1} (\bar{G}_0^\pm R')^k \delta(t)\delta(x-y)$$
$$\forall x,y \in \bar{\mathbb{R}}^+ \times \mathbb{R}^{d-1} \quad \forall \tau : \pm \operatorname{Im}\tau < 0 \quad \forall \alpha,\beta : |\alpha|\le s, |\beta|\le s$$

where $N' \ge N'(d,m,s,\delta)$, $K' = K = K(d,m,s,\delta,N')$, and R', R'_1 are obtained from R, R_1 by dropping terms containing $x_1^{\alpha_1}(x'-y')^{\alpha'} h^\nu$ with $|\alpha|+\nu \ge N'$; all operators without the index "y" act with respect to x.

5.2.3 Calculations

Let us go from estimates to calculations. Let us recall that

$$(5.2.23)\qquad F_{t\to h^{-1}\tau} F_{x'\to h^{-1}\xi'} V^\pm = \pm \sum h^{-d+k} \int_{-\infty}^{\infty} F_k(y',\eta_1,\xi',\tau) e^{-iy_1\eta_1} d\eta_1$$

where F_k are functions smooth with respect to (y',ξ') and holomorphic with respect to (η_1,τ) in the domain containing $\mathbb{R}\times\mathbb{C}_\mp$; these functions tend to 0 in this domain as $|\eta_1|\to\infty$.

Let us construct parametrices $\bar{G}'^\pm$ and $\bar{G}^\pm$. Let us note that

$$(5.2.24)\qquad F_{(t,x')\to h^{-1}(\tau,\xi')} \bar{G}'^\pm = \mathcal{G}'(z, h^{-1}x_1, \tau) F_{(t,x')\to h^{-1}(\tau,\xi')} \qquad (\pm \operatorname{Im}\tau < 0)$$

where $z = (x',\xi')$ and $\mathcal{G}'$ is a solution of the problem

$$(5.2.25)\qquad (\tau - \bar{A}(z,D_1))\mathcal{G}' = 0,$$

$$(5.2.26)\qquad \eth\bar{B}(z,D_1)\mathcal{G}' = 0,$$

$$(5.2.27)\qquad \mathcal{G}' \in \mathcal{S}(\mathbb{R}^+,.)$$

and for $z \in \operatorname{supp} q_2$ (as will be assumed without being specially mentioned) we can replace $\bar{A}$ and $\bar{B}$ by a and b respectively.

Using $\bar{G}'^\pm$ and parametrices in the whole space $\mathbb{R}^d$ for the same operator $\tau - \bar{A}$ we can construct the parametrices $\bar{G}^\pm$. Namely, let $\mathcal{G}(z,\tau)$ be the one-dimensional (on $\mathbb{R}^+$) integral operator with Schwartz kernel

$$(5.2.28)\quad \mathcal{K}_{\mathcal{G}}(x_1,y_1;z,\tau) = \\ (2\pi)^{-1} \operatorname{vrai} \int_{-\infty}^{\infty} (e^{i(x_1-y_1)\xi_1} - \mathcal{G}'(z,x_1,\tau)\bar{B}(z,\xi_1)e^{-iy_1\xi_1}) \times \\ (\tau - \bar{A}(z,x_1,\xi_1))^{-1} d\xi_1;$$

then

$$F_{(t,x')\to h^{-1}(\tau,\xi')} L\bar{G}^{\pm} = \mathcal{G}(z,\tau) L F_{(t,x')\to h^{-1}(\tau,\xi')} \qquad (\pm \operatorname{Im}\tau < 0) \tag{5.2.29}$$

where $(Lv)(x_1) = v(x_1 h)$.

It is easy to see that for $\eta_1 \in \mathbb{R}$

$$\mathcal{G}(e^{ix_1\eta_1} v) = (e^{ix_1\eta_1} - \mathcal{G}(z,x_1,\tau)\bar{B}(z,\eta_1))(\tau - \bar{A}(z,\eta_1))^{-1} v. \tag{5.2.30}$$

We need to establish certain properties of $\mathcal{G}'$ and $\mathcal{G}$ for $z \in \operatorname{supp} q_2$.

Lemma 5.2.6. *(i) The functions $W_j(z,\tau) = \eth D_1^j \mathcal{G}'(z,x_1,\tau)$ are regular on $\mathbb{R}^{2d-2} \times (\mathbb{C} \setminus \mathbb{R}) \ni (z,\tau)$ and satisfy the inequalities*

$$|D_z^\alpha D_\tau^\beta W_j| \le C |\operatorname{Im}\tau|^{-N} (1+|\tau|)^N \qquad \forall \alpha,\beta,j : |\alpha|+\beta+j \le s \tag{5.2.31}$$

where $N = N(d,m,s)$ and s is arbitrary.

(ii) $\eth_y (D_{y_1})^j \mathcal{K}_\mathcal{G} = \mathcal{G}' W_j'$ where the functions $W_j'(z,\tau)$ have the same properties as W_j.

Proof. Since $\mathcal{G}$ is a parametrix for $\tau - a(z,D_1)$ with boundary condition $\eth b(z,D_1)u = 0$ and since the operator $a(z,D_1)$ with this condition is self-adjoint we conclude that the operator norm in $L^2(\mathbb{R}^+,\mathbb{H})$ of the operator $\mathcal{G}$ does not exceed $|\operatorname{Im}\tau|^{-1}$. On the other hand, (5.1.8) and the definition of $\mathcal{G}$ as a parametrix yield that

$$\|D_1^m \mathcal{G}\| \le C\Big(\sum_{0\le j\le m-1} \|D_1^j \mathcal{G}\| + |\tau| \cdot \|\mathcal{G}\| + 1\Big);$$

this inequality, the inequality $\|\phi\| \le C|\operatorname{Im}\tau|^{-1}$ and interpolation yield that

$$\|D_1^j \mathcal{G}\| \le C |\operatorname{Im}\tau|^{-1} (|\tau|+1)^{\frac{j}{m}} \qquad (j = 0,\dots,m). \tag{5.2.32}$$

But then

$$\|\eth_y D_{y_1}^j \mathcal{K}_\mathcal{G}\| \le C |\operatorname{Im}\tau|^{-1} (|\tau|+1)^{\frac{j+1}{m}} \qquad (j = 0,\dots,m-1) \tag{5.2.33}$$

where in the left-hand expression the operator norm from $\mathbb{H}$ to $L^2(\mathbb{R}^+,\mathbb{H})$ is calculated; then the equality $(\tau - a(z,D_{y_1}))\mathcal{K}_\mathcal{G}^\dagger = 0$ for $y_1 \ne x_1$ yields that (5.2.33) holds for all j.

Starting from (5.2.33) and treating the equation

$$(\tau - a(z,D_{x_1}))\eth_y D_{y_1}^j \mathcal{K}_\mathcal{G} = 0 \tag{5.2.34}$$

for $x_1 > 0$ we obtain that

$$\|D_{x_1}^k \eth_y D_{y_1}^j \mathcal{K}_\mathcal{G}\| \le C |\operatorname{Im}\tau|^{-1} (|\tau|+1)^{\frac{j+k+1}{m}} \tag{5.2.35$_1$}$$

and

$$\|\eth_x D_{x_1}^k \eth_y D_{y_1}^j \mathcal{K}_\mathcal{G}\| \le C |\operatorname{Im}\tau|^{-1} (|\tau|+1)^{\frac{j+k+2}{m}} \tag{5.2.35$_2$}$$

for $j+k \le s$.

Moreover, equation (5.2.34) yields the equality $\eth_y D_{y_1}^j \mathcal{K}_\mathcal{G} = \mathcal{G}' W_j'$ with $W_j' = \eth_x b(z,D_{x_1})\eth_y D_{y_1}^j \mathcal{K}_\mathcal{G}$; therefore assertion (ii) holds with (5.2.31) proven only for $\alpha = \beta = 0$ so far. Furthermore, the equality $\mathcal{G}' = \mathcal{R} + \mathcal{G}(\tau - a(z,D_1)\mathcal{R}$ holds for every operator $\mathcal{R}$ from $\mathbb{H}'$ to $\mathcal{S}(\bar{\mathbb{R}}^+,\mathbb{H})$ such that $\eth b(z,D_1)\mathcal{R} = I$ (e.g., for $\mathcal{R} = \mathcal{G}(z,i)$). Therefore assertion (i) holds (but with (5.2.31) proven only for $\alpha = \beta = 0$ so far). For arbitrary α and β (5.2.31) follows from the following formulas similar to (5.2.17) and (5.2.18):

$$\partial \mathcal{G} = -\mathcal{G}(\partial(\tau - a))\mathcal{G} - \mathcal{G}'\eth(\partial b)\mathcal{G},$$

$$\partial \mathcal{G}' = -\mathcal{G}(\partial(\tau - a))\mathcal{G}' - \mathcal{G}'\eth(\partial b)\mathcal{G}'$$

where $\partial = \partial_{x_j}, \partial_{\xi_j}$ with $j = 2, \dots, d$ or $\partial = \partial_\tau$; in the last case the last terms in these equalities vanish. □

Remark 5.2.7. *Operator $\mathcal{G}$ is bounded in a neighborhood of (z^*, τ^*) if and only if the problem $(\tau - a(z, D_1), \eth b(z, D_1))$ is elliptic, i.e., if $z^* \notin \Sigma_{\tau^*,b}$. In this case the functions W_j and W'_j are regular in a neighborhood of (z^*, τ^*).* ◊

Let us assume that the problem $(\tau - a(z, D_1), \eth b(z, D_1))$ is microhyperbolic at (z^*, τ^*) in the multidirection $\mathcal{T} = (\mathcal{T}', \nu_1, \dots, \nu_n)$; in particular, in this case (3.1.19) is fulfilled. Without loss of generality one can assume that either

(a) $\mathcal{T}' = \partial_{z_1}$ or

(b) $\mathcal{T}' = 0$ (otherwise we can arrange this by a uniformly non-degenerate change of coordinates in $\mathbb{R}^{2d-2}$).

In the first case let us treat the almost analytic (with respect to z_1) extensions of $a(z, D_1)$ and $b(z, D_1)$ to the neighborhood $\tilde{\Omega} \subset \mathbb{C} \times \mathbb{R}^{2d-3} \times \mathbb{C}$ of the point (z^*, τ^*). Let us make the reduction of subsection 3.2.1 but now the coefficients of powers of hD_1 are functions of (z, τ) but not h-pseudo-differential operators. Let $\mathrm{C} = (\mathrm{C}_j) = (\mathrm{C}_j(z, D_1, \tau))_{j=0,\dots,n}$ be the operator realizing this reduction.

Proposition 5.2.8. *Let the problem $(\tau - a, \eth b)$ be microhyperbolic at (z^*, τ^*) in the multidirection $\mathcal{T} = (\mathcal{T}', \nu_1, \dots, \nu_n)$. Then*

(i) If $\mathcal{T}' = 0$ then W_j and W'_j are regular in a neighborhood of (z^*, τ^*), i.e., (5.2.31) holds with $N = 0$;

(ii) If $\mathcal{T}' = \partial_{z_1}$ then there are almost analytic (with respect to z_1) extensions of W_j and W'_j in $\tilde{\Omega}_\pm = \tilde{\Omega} \cap \{\pm \operatorname{Im} \tau < 0, \mp \operatorname{Im} z_1 < 0\}$ and here

(5.2.36) $$\|D_z^\alpha D_\tau^\beta W_j\| \le C(|\operatorname{Im} \tau| + |\operatorname{Im} z_1|)^{-N} \qquad \forall \alpha, \beta, j : |\alpha| + \beta + j \le s.$$

Proof. Let us note that case (b) is reduced to (a) by introducing an additional variable z_1 (on which a and b do not depend); therefore let us treat case (a). Let us define the almost analytic extensions of $\mathcal{G}'_j = \mathbf{R}_j \mathcal{G}'$; these functions should satisfy (5.2.25) and (5.2.26) but (5.2.27) does not necessarily remain true. Let us note that the microhyperbolicity condition yields that

(5.2.37) $g_j(z, \xi_1, \tau) = \det \underline{P}_j(z, \xi_1, \tau)$ has no root ξ_1 in the strip $\{|\operatorname{Im} \xi_1 - \nu_j \operatorname{Im} z_1| \le \epsilon(|\operatorname{Im} \tau| + |\operatorname{Im} z_1|)\}$ provided $(z, \tau) \in \tilde{\Omega}_\pm$.

Therefore we can replace (5.2.27) by the condition

(5.2.38) $$e^{\nu_j \sigma x_1} \mathbf{R}_j \mathcal{G}' \in \mathcal{S}(\mathbb{R}^+, .) \qquad \forall j = 0, \dots, n$$

where $\sigma = \operatorname{Im} z_1$, $\nu_0 = 0$ (one can choose σ and ν_0 arbitrarily). Introducing $u = (u_0, \dots, u_n)$ with $u_j = e^{-\nu_j z_1 x_1} \mathbf{R}_j \mathcal{G}'$ we obtain the problem

(5.2.39) $$\underline{P}_j(z, D_1 - \nu_j z_1, \tau) u_j = 0,$$

(5.2.40) $$\underline{B}(z,\tau)u = I,$$

(5.2.41) $$u \in \mathcal{S}(\mathbb{R}^+,.);$$

let us recall that for $z_1 \in \mathbb{R}$, $\operatorname{Im}\tau \neq 0$ this problem is solvable. □

Let us prove

Lemma 5.2.9. *Problem* (5.2.39)–(5.2.41) *is solvable for* $(z,\tau) \in \tilde{\Omega}_\pm$; *its solution is almost analytic with respect to* z_1 *and satisfies the estimates*

(5.2.42) $$|D_1^k u|_+ \leq C_k(|\operatorname{Im}\tau| + |\operatorname{Im} z_1|)^{-1} \qquad \forall k$$

Proof. Since this problem is solvable for $\operatorname{Im} z_1 = 0$, $\operatorname{Im}\tau \neq 0$, its solubility in the general case follows from (5.2.42). Let us include τ in z. The deduction of the inequality in the proof of theorem 3.1.9 implies that for real z and τ

(5.2.43) $$\begin{aligned} M = &-\operatorname{Re}(\langle \mathcal{T}'\underline{P}(z,D_1)\underline{v},\underline{v}\rangle_+ \\ &+ \sum_{1\leq j\leq n} \nu_j \langle \mathcal{T}'\underline{P}'_j(z,D_1)\underline{v}_j,\underline{v}_j\rangle_+ - i\langle \mathcal{T}'\underline{B}(z)\eth\underline{v}, \underline{B}'(z)\eth\underline{v}\rangle_b) \geq \\ &C^{-1}|\underline{v}|_+^2 - C|\underline{P}(z,D_1)\underline{v}|_{(r)+} - C|\underline{B}(z)\eth\underline{v}|^2 \\ &\qquad \forall \underline{v} \in \mathcal{S}(\bar{\mathbb{R}}^+, \mathbb{H}^m) \end{aligned}$$

with some q and C where $\underline{P}'_j(z) = \partial_{\xi_1}\underline{P}_j$, $\underline{v}_j$ are the components of $\underline{v}$ after the reduction and

$$|w|_{(r)+} = \sum_{0\leq k\leq r} |D_1^k w|_+.$$

Let us take into account that for real z the expression

$$\langle \underline{P}(z,D_1)\underline{v},\underline{v}\rangle_+ - i\langle \underline{B}(z)\eth\underline{v}, \underline{B}'(z)\eth\underline{v}\rangle_b$$

is real-valued by (5.1.11) and let us take into account the inequality

$$\begin{aligned} |\operatorname{Im}(\langle \underline{P}_\sigma(z,D_1)\underline{v},\underline{v}\rangle_+ - i\langle \underline{B}_\sigma(z)\eth\underline{v}, \underline{B}'_\sigma(z)\eth\underline{v}\rangle_b) - M\sigma - M_0\operatorname{Im}\tau| \leq \\ C|\sigma||\underline{B}_\sigma(z)\eth\underline{v}|_b \cdot |\eth\underline{v}|_b + C(|\sigma| + |\operatorname{Im}\tau|)^2|\underline{v}|_{(1)+}^2 \end{aligned}$$

where all the component of z excluding τ are real, $\underline{P}_\sigma(z,D_1) = (\underline{P}_j(z+i\sigma\mathcal{T}', D_1 - i\sigma\nu_j))_{j=0,\dots,n}$, $\underline{B}'_\sigma(z) = \underline{B}'(z+i\sigma\mathcal{T}')$, M_0 is given as M by (5.2.43) with $\mathcal{T}'$ and ν_j replaced by ∂_τ and 0 respectively; these inequalities yield the estimate

$$\begin{aligned} (|\operatorname{Im} z_1| + |\operatorname{Im}\tau|)|\underline{v}|_+ \leq C(\epsilon|\operatorname{Im} z_1|)^{-1}(|\underline{P}_{z_1}(z,D_1)\underline{v}|_{(q)+}^2 + |\underline{B}_{z_1}(z)\eth\underline{v}|_b^2) + \\ C((|\operatorname{Im} z_1| + |\operatorname{Im}\tau|)^2 + \epsilon(|\operatorname{Im} z_1| + |\operatorname{Im}\tau|))|\underline{v}|_{(q+1)+} \end{aligned}$$

with an arbitrarily small $\epsilon > 0$. In the left-hand expression of this estimate one can replace $|\underline{v}|_+$ by $|\underline{v}|_{(q+1)+}$ because the coefficient of hD_1 in $\underline{P}_{z_1}(z,D_1)$ is non-degenerate and so for small $|\operatorname{Im} z_1| + |\operatorname{Im}\tau|$ one can drop the last term in the right-hand expression, and hence

(5.2.44) $$|\underline{v}|_{(q+1)+} \leq (|\operatorname{Im} z_1| + |\operatorname{Im}\tau|)^{-1}(|\underline{P}_{z_1}(z,D_1)\underline{v}|_{(q)+} + |\underline{B}_{z_1}(z)\eth\underline{v}|_b).$$

This inequality yields (5.2.42). Moreover, (5.2.44) and the equalities

$$(\partial \underline{P}_{z_1})u + \underline{P}_{z_1}(\partial u) = 0,$$

$$\eth(\partial \underline{B}_{z_1})u + \eth\underline{B}_{z_1}(\partial u) = 0$$

for every vector field ∂ yield that u is almost analytic with respect to τ and z_1 and that the estimate

$$|D_z^\alpha D_\tau^\beta D_1^j u| \le (|\operatorname{Im} z_1| + |\operatorname{Im}\tau|)^{-N} \tag{5.2.45}$$

holds for $|\alpha| + \beta = 1$. If we make a few differentiations we can easily obtain the same estimate for $|\alpha| + \beta \le s$ with $N = N(s)$. □

Proof of proposition 5.2.8 (end). Lemma 5.2.9, estimate (5.2.45) and the definition of W_j immediately imply that for W_j the assertion of proposition 5.2.8 holds. Let us prove proposition 5.2.8 for W_j'. Formula (5.2.28) yields that for real z, $\operatorname{Im}\tau \ne 0$ and $x_1 + y_1 > 0$ the following formula holds:

$$\mathcal{K}_{\mathcal{G}}(x_1, y_1; z, \tau) = (2\pi)^{-1}\int_{\gamma_+} e^{i(x_1 - y_1)\xi_1}(\tau - a(z,\xi_1))^{-1} d\xi_1 +$$
$$\mathcal{G}'(z, x_1, \tau)(2\pi)^{-1}\int_{\gamma_-} b(z,\xi_1)e^{-iy_1\xi_1}(\tau - a(z,\xi_1))^{-1} d\xi_1$$

where

(5.2.46) $\gamma_\pm$ are closed contours in $\mathbb{C}$ passing once around the poles of $(\tau - a(z,\xi_1))^{-1}$ lying in $\mathbb{C}_\pm$ and not passing around the poles lying in $\mathbb{C}_\mp$; for real z and $\operatorname{Im}\tau \ne 0$ there is no real pole.

The coefficient near $\mathcal{G}'$ in the second term in the right-hand expression is a sum of terms of the form

$$\int_{\gamma_-} \xi_1^k e^{-iy_1\xi_1}(\tau - a(z,\xi_1))^{-1} d\xi_1$$

with regular coefficients and with $k \le m - 1$. Differentiating j times with respect to y_1 and setting $y_1 = 0$ we obtain terms of the form

$$W_k''(z,\tau) = \int_{\gamma_-} \xi_1^k(\tau - a(z,\xi_1))^{-1} d\xi_1$$

with coefficients regular with respect to z, τ. In order to treat $\operatorname{Im} z_1 \ne 0$ let us replace the contours $\gamma_\pm$ by homotopic contours $\gamma_\pm^\sigma$ such that for $0 \le \pm\sigma' \le \pm\sigma$ there is no pole of $(\tau - a_{z_1})^{-1}$ in $\gamma_\pm^\sigma$ where $\zeta = z_1$ and $\sigma = \operatorname{Im}\zeta$. Property (5.2.37) yields that one can assume without loss of generality that in the intersection with the ϵ-neighborhood of ζ_j the contour $\gamma_\pm^\sigma$ is $\operatorname{Im}\xi_1 = \nu_j \operatorname{Im} z_1$; then the microhyperbolicity condition yields that on $\gamma_\pm^\sigma$

$$|(\tau - a_{z_1}(z,\xi_1))^{-1}| \le C(|\operatorname{Im}\tau| + |\operatorname{Im} z_1|)^{-1}$$

and hence W_k'' also satisfies the assertion of proposition 5.2.7. Hence we need to treat only the first term in $\mathcal{K}_{\mathcal{G}}$. Applying $\eth D_{y_1}^j$ we obtain a sum of terms of the form

$$(2\pi)^{-1}\int_{\gamma_+} e^{ix_1\xi_1}\xi_1^j(\tau - a(z,\xi_1))^{-1}d\xi_1 = v;$$

these terms satisfy the equation $(\tau - a(z, D_1))v = 0$ for $x_1 > 0$ and hence they are of the form $v = \mathcal{G}'(z, x_1, \tau)\eth b(z, D_1)v$; repeating the same arguments as before we see that $W_k''' = \eth b(z, D_1)v$ satisfy assertion of proposition 5.2.8; hence this assertion is valid for W_k' also. □

Let us represent $\mathcal{G}'$ as an oscillatory integral:

Lemma 5.2.10. *The following formula holds:*

$$\mathcal{G}'(z, x_1, \tau) = \text{vrai}\int_{-\infty}^{\infty} \sum_{0\le j\le m-1} \xi_1^{m-j-1}(\tau - a(z,\xi_1)^{-1})d\xi_1 \cdot W_j''(z,\tau) \tag{5.2.46}$$

where W_j'' *satisfy* (5.2.32);

(ii) In this formula one can replace integration along $\mathbb{R}$ *by integration along* γ_+;

(iii) Hence the same representation holds for $D_1^k\mathcal{G}'(z, x_1, \tau)$ *for all* $k \ge 0$;

(iv) Under the microhyperbolicity condition for W_j'' *the assertion of proposition 5.2.8 holds.*

Proof. Starting from the equality

$$(\tau - a(z, D_1))(\mathcal{G}'\theta(x_1)) = \sum_{j+k\le m-1} \delta^{(k)}(x_1)a_{kj}(z)\eth D_1^j\mathcal{G}'$$

with regular coefficients a_{kj} and applying the Fourier transform with respect to x_1, multiplication by $(\tau - a(z,\xi_1))^{-1}$, and the inverse Fourier transform we obtain (5.2.46); estimate (5.2.32) follows from lemma 5.2.6 and assertion (iv) follows from proposition 5.2.8. Assertion (iii) is obvious. Applying D_1^k to (5.2.46) with γ_+ instead of $\mathbb{R}$ and dividing by $(\tau - a(z,\xi_1))$ with remainder we obtain the same representation for $D_1^k\mathcal{G}'$, and at the last moment we can replace γ_+ by $\mathbb{R}$ again. □

Now we can prove

Proposition 5.2.11. *(i) For* $\operatorname{Im}\tau \ne 0$ *the operator*

$$\mathcal{A} = c_1(z)D_1^{j_1}\mathcal{G}c_2(z)D_1^{j_2}\dots\mathcal{G}c_\mu(z)D_1^{j_\mu}\mathcal{G}'$$

containing $m-1$ *factors* $\mathcal{G}$ *is a sum of terms of the form*

$$\int_{\gamma_+} e^{ix_1\xi_1}\xi_1^k F_{kl}(z,\xi_1,\tau)d\xi_1 \cdot W_{lk}(z,\tau) \tag{5.2.47}$$

where $k = 0,\dots,m-1$, F_{lk} *is an* l*-multiple product of* $c'_-(z,\tau)(\tau - a(z,\xi_1))^{-1}$ $(l \le \mu)$, *and* W_{lk} *satisfy* (5.2.31).

(ii) Moreover, $\eth\mathcal{A} = W(z,\tau)$ *satisfies* (5.2.31).

(iii) Under the microhyperbolicity condition W_{lk} *and* W *satisfy the assertion of proposition 5.2.8.*

Proof. Let us argue by induction on μ. For $\mu = 1$ this proposition follows from lemma 5.2.10 and proposition 5.2.8. Let us write (5.2.47) as the principal value of an integral along $\mathbb{R}$ and apply (5.2.30); we then obtain terms of the same form with F_{lk} replaced by $(\tau - a(z,\xi_1))^{-1}F_{lk}$ and terms of the form $\mathcal{G}(z,\xi_1,\tau)W_{lk}'$ with

$$W'_{lk} = \text{vrai} \int_{-\infty}^{\infty} b(z,\xi_1)\xi_1^k(\tau - a(z,\xi_1))^{-1} F_{lk}(z,\xi_1,\tau).$$

In this integral we can take integration over the contour γ_+ coinciding with the real axis in the ϵ-neighborhood of ζ_j and therefore W'_{lk} satisfy (5.2.31); moreover, under the microhyperbolicity condition taking $\operatorname{Im} z_1 \neq 0$ we can deform this contour into γ_+^σ (see the proof of proposition 5.2.8) and obtain that W'_{lk} satisfy the assertion of proposition 5.2.8. We rewrite $\mathcal{G}'$ in the form (5.2.46), pass to integration along γ_+, apply the operator $c_1(z)D_1^{j_1}$ and divide ξ_1^k with k exceeding $m-1$ by $(\tau - a(z,\xi_1))$ with remainder; since the quotient contains no pole its integral is 0, so only the remainder matters and we finish the induction step. Then (i) and the first part of (iii) is proven. (In fact (iii) consists of two parts corresponding to the two parts of proposition 5.2.8.)

Applying the operator $\bar{\partial}$ to (5.2.47) and estimating $\int_{\gamma_+} \xi_1^k F_{kl}(z,\xi_1,\tau)d\xi_1$ (and its derivatives) in a manner similar to the previous arguments (and with $\operatorname{Im} z_1 \neq 0$ under the microhyperbolicity condition) we obtain (ii) and the second part of (iii). □

5.2.4 Calculations (end)

Now we can finish our calculations. Applying proposition 5.2.5, the possibility of killing all the degrees of x_1, $x' - y'$ by commutation, proposition 5.2.11, formulas (5.2.24) and (5.2.29) and the formulas for $\bar{G}_0^\pm$ we obtain

Theorem 5.2.12. *Let conditions* (5.1.1), (5.1.2), (5.1.4), (5.1.8) *and* (5.1.9) *be fulfilled, let $Q'_i = \operatorname{Op}(q'_i)$ with symbols $q'_i \in S'_{h,K}$ supported in $\{|\xi'| \le c\}$, and let $\psi_i \in C_0^K(B(0,1-\epsilon))$. Then*

$$(5.2.48)\quad \Big|D_x^\alpha D_y^\beta \Big(F_{t\to h^{-1}\tau}\chi_T(t) Q'_{1x}\psi_1(x) U^{1\pm} \psi_2(y)\, {}^t Q'_{2y} \mp \sum_{0\le n\le N'-1} h^{-d+n} \int T\hat{\chi}(\frac{\tau' T}{h})d\tau' \int d\xi' \int_{\gamma_+} d\xi_1 \int_{\gamma_-} d\eta_1 \times e^{ih^{-1}(x_1\xi_1 - y_1\eta_1 + \langle x'-y',\xi'\rangle)} q'_1(x,\xi') F_n(y',\xi',\xi_1,\eta_1,\tau-\tau')[q'_2(y,\xi')]\Big)\Big| \le C' h^s (|\tau|+1)^{N''}$$

$$\forall \alpha,\beta : |\alpha| \le s, |\beta| \le s \quad \forall \tau \in \mathbb{C}_\mp \quad \forall x,y \in \bar{\mathbb{R}}^+ \times \mathbb{R}^{d-1}$$

provided $N' = N'(d,m,s,\delta)$, $K = K(d,m,s,\delta,N')$ and similarly for N'', $T \in [h^{1-\delta}, h^{\frac{1}{2}+\delta}]$, and F_n are sums of terms of the form

$$(5.2.49)\quad \xi_1^j \eta_1^k \prod_{1\le\mu\le L} (\tau - a(y',\xi_1,\xi'))^{-1} c_\mu(y',\xi') \times W(y',\xi',\tau) \prod_{1\le\nu\le L'} c'_\nu(y',\xi')(\tau - a(y',\eta_1,\xi'))^{-1}$$

with $j,k \le m-1$, $L + L' \le L_0(n) = 2n+2$, and functions W satisfying (5.2.31); *here q'_1 and q'_2 are symbols of $Q'_1\psi_1$ and $Q'_2\psi_2$ respectively and there are no more than $L_0(n)$ differentiations with respect to (ξ', x_1).*

Moreover, under the microhyperbolicity condition the functions W satisfy the assertion of proposition 5.2.8.

Remark 5.2.13. *(i) In formula (5.2.48) we have replaced integration (on ξ_1) along $\mathbb{R}$ by integration along γ_-.*

(ii) Multiplying all the terms of decomposition (5.2.48) by x_1 or y_1 is equivalent (after integration by parts) to the application to F_n of the operator $ih\partial_{\xi_1}$ or $-ih\partial_{\eta_1}$ respectively. Application of ∂_{ξ_1} or ∂_{η_1} increases the singularity of F_n for $\xi_1, \eta_1 \in \mathbb{R}$ and $\tau \to \mathbb{R} \pm i0$ by no more than an extra factor $(\operatorname{Im}\tau)^{-1}$ and hence the singularity of the distribution $F_n(y', \xi', \xi_1, \eta_1, \tau \pm i0)$ increases by no more than by 1; hence when we estimate the corresponding integral we should replace the C^N-norm of $\hat{\chi}(\frac{\tau' T}{h})$ by its C^{N+1}-norm, i.e., there arises an extra factor $\frac{T}{h}$; also multiplication by x_1 or y_1 essentially multiplies the corresponding integral by T; therefore for $x_1 + y_1 \geq th^{-\delta}$ with $\delta > 0$ all terms in the decomposition are negligible.

Let us set $x_1 = y_1$ and integrate over x_1 along $\mathbb{R}^+$; let us take the Taylor decomposition of q'_i with respect to x_1 and y_1 and apply remark 5.2.13(ii); then we can drop all the remainder terms and delete factors x_1 and y_1 by means of integration by parts. As a result there arises inside the exterior integrals with respect to τ' and ξ' a term

$$W'_n(z, \tau - \tau') = ih \int_{\gamma_+} \int_{\gamma_-} F_n(y', \xi', \xi_1, \eta_1, \tau - \tau')(\xi_1 - \eta_1)^{-1} d\xi_1 d\eta_1$$

where we have taken non-intersecting contours γ_+ and γ_- and we do not indicate q'_1 and q'_2. Moreover, we can choose γ_+ and γ_- (depending on τ) such that $\operatorname{dist}(\gamma_+, \gamma_-) \geq \epsilon |\operatorname{Im}\tau|$ and at these contours $|(\tau - a(y', \xi', \xi_1)^{-1}| \leq C|\operatorname{Im}\tau|^{-1}$. Hence W'_n satisfy (5.2.31).

Moreover, under the microhyperbolicity condition we can pass to non-real z_1 and choose γ_+ and γ_- such that in the preceding inequalities $|\operatorname{Im}\tau|$ will be replaced by $|\operatorname{Im}\tau| + |\operatorname{Im} z_1|$ (let us recall the that signs of $\operatorname{Im}\tau$ and $\operatorname{Im} z_1$ should be different). But then for W'_n the assertion of proposition 5.2.7 holds. So we have proven

Theorem 5.2.14. *Let conditions (5.1.1), (5.1.2), (5.1.4), (5.1.8) and (5.1.9) be fulfilled, let $Q'_i = \operatorname{Op}(q'_i)$ with symbols $q'_i \in S'_{h,K}$ supported in $\{|\xi'| \leq c\}$, and let $\psi_i \in C_0^K(B(0, 1-\epsilon))$. Then*

$$\begin{aligned}
(5.2.50)\quad &\Big| D_{x'}^{\alpha} D_{y'}^{\beta} \Big(F_{t \to h^{-1}\tau} \chi_T(t) \int_0^{\infty} (Q'_{1x} \psi_1(x) U^{1\pm} \psi_2(y) {}^t Q'_{2y})|_{x_1 = y_1} dy_1 \mp \\
&\sum_{0 \leq n \leq N'-1} h^{-d+n+1} \int T \hat{\chi}(\frac{\tau' T}{h}) d\tau' \int d\xi' \times \\
&e^{ih^{-1}\langle x' - y', \xi' \rangle} q'_1(x, \xi') W'_n(y', \xi', \tau - \tau') [q'_2(y, \xi')] \Big) \Big| \leq C' h^s (|\tau| + 1)^{N''} \\
&\forall \alpha, \beta : |\alpha| \leq s, |\beta| \leq s \quad \forall \tau \in \mathbb{C}_{\mp} \quad \forall x', y' \in \mathbb{R}^{d-1}
\end{aligned}$$

with the same N', N'' and K as before where $T \in [h^{1-\delta}, h^{\frac{1}{2}+\delta}]$ with $\delta > 0$ and W'_n satisfying (5.2.31). Moreover, under the microhyperbolicity condition for these functions the assertion of proposition 5.2.8 holds. Furthermore, the same assertion remains true if we replace ψ_1 and (or) ψ_2 by $\psi_1 \psi'_1$ and (or) $\psi_2 \psi'_2$ with $\psi'_i(x_1) = \psi''_i(x_1 h^{-\delta} T^{-1})$, $\psi''_i \in C_0^K([-1, 1])$ equal to 1 on $[-\frac{1}{2}, 1\frac{1}{2}]$; moreover none of the terms of the decomposition will change by remark 5.2.13.

Applying the operator $\phi(hD_t)$ and taking $t = 0$ we obtain

Theorem 5.2.15. *Let conditions (5.1.1), (5.1.2), (5.1.4), (5.1.8) and (5.1.9) be fulfilled and let $Q'_i - \operatorname{Op}(q'_i)$ with symbols $q'_i \in S'_{h,K}$ supported in $\{|\xi'| \leq c\}$, $\psi_i \in C_0^K(B(0, 1-\epsilon))$, $\phi \in C_0^K(\mathbb{R})$. Let $\chi = 1$ at $[-\frac{1}{2}, \frac{1}{2}]$. Then for $L \geq h^{1-\delta'} T^{-1}$, $T \in [h^{1-\delta}, h^{\frac{1}{2}+\delta}]$ with $\delta > 0$, $\delta' > 0$*

$$(5.2.51)\quad |D_x^\alpha D_y^\beta \Big(\phi(L^{-1}hD_t)\chi_T(t)(Q'_{1x}\psi_1(x)U^{1\pm}\psi_2(y)^t Q'_{2y})|_{t=0} \mp \sum_{0\le n\le N'-1} h^{-d+n+1}\int \phi(\frac{\tau'}{L})d\tau' \int d\xi' \int_{\gamma_+} d\xi_1 \int_{\gamma_-} d\eta_1 \times e^{ih^{-1}(x_1\xi_1 - y_1\eta_1 + \langle x'-y',\xi'\rangle)} q'_1(x,\xi')F_n(y',\xi',\xi_1,\eta_1,\tau'\mp i0)[q'_2(y,\xi')]\Big)| \le C'h^s \quad \forall \alpha,\beta: |\alpha|\le s, |\beta|\le s \quad \forall x,y\in \bar{R}^+\times\mathbb{R}^{d-1},$$

$$(5.2.52)\quad |D_{x'}^\alpha D_{y'}^\beta \Big(\phi(\frac{h}{L}D_t)\chi_T(t) \int_0^\infty (Q'_{1x}U^{1\pm}(y)^t Q'_{2y})(x_1 = y_1)dy_1|_{t=0} \mp \sum_{0\le n\le N'-1} h^{-d+n+2}\int \phi(\frac{\tau'}{L}))d\tau' \int d\xi' \times e^{ih^{-1}\langle x'-y',\xi'\rangle} q'_1(x,\xi')W'_n(y',\xi',\tau'\mp i0)[q'_2(y,\xi')]\Big)| \le C'h^s \quad \forall \alpha,\beta: |\alpha|\le s, |\beta|\le s \quad \forall x',y'\in\mathbb{R}^{d-1}$$

with the same N', N'', K, δ, δ' and T as before.[7)]

5.2.5 The microhyperbolicity condition

Now let us apply the microhyperbolicity condition and try to prove that inequalities (5.2.48) and (5.2.50) remain true (under some assumptions) for $T \in [h^{1-\delta}, T_0]$ with a small enough constant $T_0 > 0$; then the same will be true for (5.2.51) and (5.2.52) as well.

Theorem 5.2.16. *Let conditions* (5.1.1), (5.1.2), (5.1.4), (5.1.8) *and* (5.1.9) *be fulfilled and let $Q'_i = \mathrm{Op}(q'_i)$ with symbols $q'_i \in S'_{h,K}$ supported in $\{|\xi'|\le c\}$, $\psi_i \in C_0^K(B(0,1-\epsilon))$. Let us assume that the problem $(\tau - A, B)$ is microhyperbolic on $\operatorname{supp} q'_2$ in the multidirection $\mathcal{T} = (T',\nu_1,\dots,\nu_n)$ for every $\tau\in[\tau_1,\tau_2]$. Then for $T\in[h^{1-\delta},T_0]$, $T_0 = T_0(d,m,D,c,\epsilon)$, $|\alpha|\le s$, $|\beta|\le s$, $\tau\in[\tau_1,\tau_2]$ the following estimates hold:*

(i) If $T'_x = 0$, $\nu_1 = \dots = \nu_n = \nu$ then

$$(5.2.53)\quad |F_{t\to h^{-1}\tau}\chi_T(t)\Gamma_x(D_x^\alpha D_y^\beta Q'_{1x}\psi_1(x)U^{1\pm}\psi_2(y)^t Q'_{2y})|_{t=0} \mp \sum_{0\le n\le N'-1} h^{-d+n-|\alpha|-|\beta|}\int T\hat{\chi}(\frac{\tau' T}{h})d\tau' \int_{\gamma_+} d\xi_1 \int_{\gamma_-} d\eta_1 \times e^{ih^{-1}y_1(\xi_1-\eta_1)} F^1_{n,\alpha,\beta}(y',\xi_1,\eta_1,\tau-\tau'\mp i0)| \le C'h^s \quad \forall y\in\bar{\mathbb{R}}^+\times\mathbb{R}^{d-1},$$

where $F^1_{n,\alpha,\beta}(y',\xi_1,\eta_1,\tau\mp i0)$ are for $\tau\in[\tau_1,\tau_2]$ regular functions with respect to all variables and analytic with respect to ξ_1 and η_1 such that

$$F^1_{n,\alpha,\beta}(y',\xi_1,\eta_1,\tau\mp i0) \sim \sum_{0\le k} c_{nk\alpha\beta}(y',\eta_1,\tau\mp i0)\xi_1^{-k-1}$$

as $|\xi_1|\to\infty$ (and a similar decomposition holds as $|\eta_1|\to\infty$), coefficients $c_{nk\alpha\beta}$ are regular and for them similar decompositions hold as $|\eta_1|\to\infty$ or $|\xi_1|\to\infty$; moreover,

---

7) We assume for the sake of simplicity that Q'_i and ψ'_i do not depend on x_1 for $x_1 \le Th^{-\delta'}$.

functions $F^1_{n,\alpha,\beta}(y',\xi_1,\eta_1,\tau+i0)$ *and* $F^1_{n,\alpha,\beta}(y',\xi_1,\eta_1,\tau-i0)$ *coincide provided* $(y',\xi')\notin \Sigma_{\tau,b}\cap\operatorname{supp}q'_1\cap\operatorname{supp}q'_2$ *for every* $\xi'\in\mathbb{R}^{d-1}$.

(ii) *If* $T'_x=0$ *then*

$$(5.2.54)\quad |F_{t\to h^{-1}\tau}\chi_T(t)\Gamma''(D^\alpha_x D^\beta_y Q'_{1x}\psi_1(x)U^{1\pm}\psi_2(y)^tQ'_{2y})|_{t=0}\mp \sum_{0\le n\le N'-1} h^{-d+n-|\alpha|-|\beta|+1}\int T\hat\chi(\frac{\tau' T}{h})d\tau' W_{n,\alpha,\beta}(y',\tau-\tau'\mp i0)|\le C'h^s,$$

$$(5.2.55)\quad |F_{t\to h^{-1}\tau}\chi_T(t)\Gamma'_{x'}(\eth_x\eth_y D^\alpha_x D^\beta_y Q'_{1x}\psi_1(x)U^{1\pm}\psi_2(y)^tQ'_{2y})|_{t=0}\mp \sum_{0\le n\le N'-1} h^{-d+n-|\alpha|-|\beta|}\int T\hat\chi(\frac{\tau' T}{h})d\tau' W'_{n,\alpha,\beta}(y',\tau-\tau'\mp i0)\|\le C'h^s \qquad \forall y'\in\mathbb{R}^{d-1}$$

where $W_{n,\alpha,\beta}(y',\tau\mp i0)$ *and* $W'_{n,\alpha,\beta}(y',\tau\mp i0)$ *are for* $\tau\in[\tau_1,\tau_2]$ *regular functions which (with the signs "+" and "−") coincide provided* $(y',\xi')\notin\Sigma_{\tau,b}\cap\operatorname{supp}q'_1\cap\operatorname{supp}q'_2$ *for every* $\xi'\in\mathbb{R}^{d-1}$.

(iii) *If* $\nu_1=\dots=\nu_n=\nu$, *then*

$$(5.2.56)\quad |F_{t\to h^{-1}\tau}\chi_T(t)\Gamma'''(D^\alpha_x D^\beta_y Q'_{1x}\psi_1(x)U^{1\pm}\psi_2(y)^tQ'_{2y})|_{t=0}\mp \sum_{0\le n\le N'-1} h^{-d+n-|\alpha|-|\beta|}\int T\hat\chi(\frac{\tau' T}{h})d\tau'\int_{\gamma_+}d\xi_1\int_{\gamma_-}d\eta_1\times e^{ih^{-1}y_1(\xi_1-\eta_1)}F^2_{n,\alpha,\beta}(\xi_1,\eta_1,\tau-\tau'\mp i0)|\le C'h^s \qquad \forall y_1\in\bar{\mathbb{R}}^+$$

where $F^2_{n,\alpha,\beta}(\xi_1,\eta_1,\tau-\tau'\mp i0)$ *are functions of the same type as in (i) which (with the signs "+" and "−") coincide provided* $(y',\xi')\notin\Sigma_{\tau,b}\cap\operatorname{supp}q'_1\cap\operatorname{supp}q'_2$ *for every* $(y',\xi')\in T^*Y$.

(iv) *If* $T'=0$ *then*

$$(5.2.57)\quad |F_{t\to h^{-1}\tau}\chi_T(t)(\eth_x\eth_y D^\alpha_x D^\beta_y Q'_{1x}\psi_1(x)U^{1\pm}\psi_2(y)^tQ'_{2y})|_{t=0}\mp \sum_{0\le n\le N'-1} h^{-d+n-|\alpha|-|\beta|}\int T\hat\chi(\frac{\tau' T}{h})d\tau'\times e^{ih^{-1}\langle x'-y',\xi'\rangle}W''_{n,\alpha,\beta}(y',\tau-\tau'\mp i0)d\xi'|\le C'h^s \qquad \forall x',y'\in\mathbb{R}^{d-1}$$

where $W''_{n,\alpha,\beta}(y',\tau\mp i0)$ *are functions of the same type as in (ii) which (with the signs "+" and "−") coincide provided* $(y',\xi')\notin\Sigma_{\tau,b}\cap\operatorname{supp}q'_1\cap\operatorname{supp}q'_2$.

(v) *In the general case*

$$(5.2.58)\quad |F_{t\to h^{-1}\tau}\chi_T(t)\Gamma(D^\alpha_x D^\beta_y Q'_{1x}\psi_1(x)U^{1\pm}\psi_2(y)^tQ'_{2y})|_{t=0}\mp \sum_{0\le n\le N'-1} h^{-d+n-|\alpha|-|\beta|+1}\int T\hat\chi(\frac{\tau' T}{h})d\tau'\times e^{ih^{-1}\langle x'-y',\xi'\rangle}W'''_{n,\alpha,\beta}(\tau-\tau'\mp i0)d\xi'|\le C'h^s,$$

(5.2.59) $$\|F_{t\to h^{-1}\tau}\chi_T(t)\Gamma'\eth_x\eth_y D_x^\alpha D_y^\beta Q'_{1x}\psi_1(x)U^{1\pm}\psi_2(y)^tQ'_{2y})|_{t=0} \mp \sum_{0\le n\le N'-1} h^{-d+n-|\alpha|-|\beta|}\int T\hat{\chi}(\frac{\tau' T}{h})d\tau' \times \\ e^{ih^{-1}\langle x'-y',\xi'\rangle}W^{IV}_{n,\alpha,\beta}(y',\tau-\tau'\mp i0)d\xi'\| \le C'h^s$$

where $W'''_{n,\alpha,\beta}(\tau\mp i0)$ *and* $W^{IV}_{n,\alpha,\beta}(\tau\mp i0)$ *are functions of the same type as in (ii) which (with the signs "+" and "−") coincide provided* $(y',\xi')\notin\Sigma_{\tau,b}\cap\operatorname{supp}q'_1\cap\operatorname{supp}q'_2$ *for every* $(y',\xi')\in T^*Y$.

Moreover, for $n=0$ *all the functions* $F^{\cdot}_{\cdot}$, $W^{\cdot}_{\cdot}$ *contain* q^0_1, q^0_2, ψ_1 *and* ψ_2 *as factors. Moreover in (ii), (iv) and (v) one can replace* ψ_1 *and* ψ_2 *by* $\psi_1\psi'_1(x_1h^{\delta'}T^{-1})$ *and* $\psi_2\psi'_2(y_1h^{\delta'}T^{-1})$ *respectively.*

Proof. The proof evidently follows from theorems 5.2.12, 5.2.14 and 3.2.5 if we prove that for χ supported in $[-1,-\epsilon']\cup[\epsilon',1]$ with arbitrary $\epsilon'>0$ all the terms of the decomposition are $O(h^s)$ and prove that the functions $F^{\cdot}_{\cdot}$, $W^{\cdot}_{\cdot}$ are as regular as is announced in the corresponding assertions. Repeating the arguments of the proof of proposition 4.2.9 we obtain that it is sufficient to prove only the assertions concerning regularity.

All the assertions concerning regularity of the functions $W^{\cdots}_{\cdot}$ follow from the corresponding assertions of theorems 5.2.12 and 5.2.14 and from Appendix B; therefore it is sufficient to prove regularity properties of $F^{\cdot}_{\cdot}$ in the cases (i) and (iii); we treat (i), and case (iii) can be treated in the same way. Let us replace the contour γ_+ by the union of contours $\gamma_{+,j}$ lying near $\xi_1=\zeta_j$ for $j\ge1$ and $\gamma_{+,0}$ lying in $\operatorname{Im}\xi_1\ge\epsilon''$; similarly we replace γ_- where $\gamma_{-,0}$ lies in $\operatorname{Im}\xi_1\le-\epsilon$; as a result we obtain

$$\int_{\gamma_{+,j}} d\xi_1\int_{\gamma_{-,k}} d\eta_1\xi_1^p\eta_1^r e^{ix_1(\xi_1-\eta_1)}f_1(\xi_1,\xi',\tau)W(\xi',\tau)f_2(\eta_1,\xi',\tau)=\mathcal{K}(\xi',\tau)$$

where f_1 is a product of $(\tau-a(\xi_1,\xi'))^{-1}$ separated by regular functions of ξ' and f_2 has a similar form, $p\le m-1$, $r\le m-1$ and for the sake of simplicity we do not indicate here dependence on y' and do not write the exterior integral with respect to ξ'.

Without loss of generality one can assume that $\mathcal{T}'=\partial_{\xi_2}$ (if $\mathcal{T}'=0$ we can introduce an additional variable). Then in W the replacement of $\tau\in\mathbb{R}$ by $\tau\mp i0$ is equivalent to the replacement of $\xi_2\in\mathbb{R}$ by $\xi_2\pm i0$ and in f_1 and f_2 the replacement of $\tau\in\mathbb{R}$ by $\tau\mp i0$ is equivalent to the replacement

$$\xi_2\to\xi_2\pm i0,\quad \xi_1\to\xi_1+i\nu\operatorname{Im}\xi_2,\quad \eta_1\to\eta_1+i\nu\operatorname{Im}\xi_2$$

(let us recall that $\nu_1=\dots=\nu_n=\nu$ and ν_0 can be chosen arbitrarily). Therefore the replacement of $\tau\in\mathbb{R}$ by $\tau\mp i0$ is equivalent to the replacement of $\xi_2\in\mathbb{R}$ by $\xi_2\pm i0$ with the deformation of contours $\gamma_\pm\to\gamma_\pm+i\nu\operatorname{Im}\xi_2$ and without any loss of generality one can assume that $\gamma_\pm\subset\mathbb{C}_\pm\cup\mathbb{R}$; then the absolute value of the exponent does not exceed 1 and on the deformed contours $|(\tau-a(\xi',\xi_1))^{-1}|\le C|\operatorname{Im}\xi_2|^{-1}$. Therefore we obtain that there exists an almost analytic (with respect to ξ_2) extension of $\mathcal{K}_n(y',\xi',\tau\mp i0)$ in the domain $\{\pm\operatorname{Im}\xi_2>0\}$ such that

$$|D^\alpha_{y'}D^\beta_\tau\mathcal{K}_n(y',\xi',\tau\mp i0)|\le C'|\operatorname{Im}\xi_2|^{-N}\qquad\forall\alpha,\beta:|\alpha|+\beta\le s.$$

Therefore after multiplication by functions compactly supported with respect to ξ' and integration with respect to ξ_2 we obtain functions $F^{\cdot}_{\cdot}(y',\tau\mp i0)$ which are smooth by Appendix B. □

5.3 Basic Theorems

In this section we obtain the principal microlocal and local asymptotics. Proofs which are completely similar to those of section 4.3 and section 4.6 are usually left to the reader.

5.3.1 Asymptotics with mollification

We begin with microlocal asymptotics with mollification with respect to the spectral parameter.

Theorem 5.3.1. [8] *Let conditions* (5.1.1), (5.1.2), (5.1.4), (5.1.8) *and* (5.1.9) *be fulfilled and let* $Q_i = \mathrm{Op}(q_i)$, $q_i \in S_{h,K}$ *be supported in* $\{|\xi'| \le c\}$, $\psi_i \in C_0^K(B(0,1-\epsilon))$; *let* $\phi \in C_0^K([-c,c])$ *and* $\bar\tau \in [-c,c]$. *Then*

(i) *For* $L \in [h^{\frac12-\delta},1]$ *the following estimates hold:*

$$\text{(5.3.1)}\quad \frac{1}{L}\Big\| \int \phi(\frac{\tau-\bar\tau}{L})\Big(d_\tau Q'_{1x}\psi_1(x)e(x,y,\tau_1,\tau)\psi_2(y)^tQ'_{2y} - \sum_{0\le n\le N'-1} h^{-d+n}\mathcal{K}'_n(x,y,\tau)d\tau\Big)\Big\| \le C'M'h^s \qquad \forall x,y \in \bar{\mathbb{R}}^+\times\mathbb{R}^{d-1},$$

$$\text{(5.3.2)}\quad \frac{1}{L}\Big\| \int \phi(\frac{\tau-\bar\tau}{L})\Big(d_\tau \Gamma_x Q'_{1x}\psi_1(x)e(x,y,\tau_1,\tau)\psi_2(y)^tQ'_{2y} - \sum_{0\le n\le N'-1} h^{-d+n}\kappa'_n(x,\tau)d\tau\Big)\Big\| \le C'M'h^s \qquad \forall x \in \bar{\mathbb{R}}^+\times\mathbb{R}^{d-1},$$

$$\text{(5.3.3)}\quad \frac{1}{L}\Big\| \int \phi(\frac{\tau-\bar\tau}{L})\Big(d_\tau \Gamma Q'_{1x}\psi_1(x)e(x,y,\tau_1,\tau)\psi_2(y)^tQ'_{2y} - \sum_{0\le n\le N'-1} h^{-d+n}\varkappa'_n(x,y,\tau)d\tau\Big)\Big\| \le C'M'h^s$$

where $\mathcal{K}'_n = \mathcal{K}'^0_n + \mathcal{K}'^1_n$, $\kappa'_n = \kappa'^0_n + \kappa'^1_n$, $\varkappa'_n = \varkappa'^0_n + \varkappa'^1_n$, *the functions with superscript "0" are defined by* (4.3.4) *and* (4.3.12) *for* U^0; *in particular* $\mathcal{K}'^0_0$, κ'^0_0, $\varkappa'_0$ *are are derivatives with respect to* τ *of monotone functions,* M *and* M' *depend on the corresponding norms of* q_1, q_2, ψ_1, ψ_2, ϕ, M', C' *and* K *all depend on* s,

$$\text{(5.3.4)}\quad \mathcal{K}'^1_n(x,y,\tau) = -2\pi i \int d\xi' \int_{\gamma_+} d\xi_1 \int_{\gamma_-} d\eta_1 \times e^{ih^{-1}(x_1\xi_1 - y_1\eta_1 + \langle x'-y',\xi'\rangle)}\, \mathrm{Res}_{\mathbb{R}}\, F^1_{n,0,0}(x',\xi_1,\eta_1,\xi',\tau),$$

$$\text{(5.3.5)}\quad \kappa'^1_n(x,\tau) = \Gamma_x\mathcal{K}'^1_n(x,y,\tau) = -2\pi i \int d\xi' \int_{\gamma_+} d\xi_1 \int_{\gamma_-} d\eta_1 \times e^{ih^{-1}x_1(\xi_1-\eta_1)}\, \mathrm{Res}_{\mathbb{R}}\, F^1_{n,0,0}(x',\xi_1,\eta_1,\xi',\tau),$$

[8] Compare with theorem 4.3.1.

$$(5.3.6)\qquad \varkappa'^{1}_{n}(\tau) = \Gamma K'^{1}_{n}(x,y,\tau) = -2\pi i \operatorname{Res}_{\mathbb{R}} W_{n-1,0,0}(\tau), \quad \varkappa'^{1}_{0}(\tau) = 0,$$

and functions $F^{1}_{n,0,0}(x',\xi_1,\eta_1,\xi',\tau)$ and $W_{n,0,0}$ are defined in theorem 5.2.16;

(ii) If the operator $\tau - A$ is microhyperbolic for $\tau = \bar\tau$ at $\operatorname{supp} q'_1 \cap \operatorname{supp} q'_2 \cap T^*(X \cap B(0,1))$ in some direction ℓ and if the problem $(\tau - A, B)$ is microhyperbolic for $\tau = \bar\tau$ at $\operatorname{supp} q'_1 \cap \operatorname{supp} q'_2 \cap T^*(Y \cap B(0,1))$ in some multidirection $\mathcal{T}$ then (5.3.3) remains true for $L \in [h^{1-\delta},1]$; moreover, if $\ell_x = \mathcal{T}'_x = 0$ then the same is true for (5.3.2).

5.3.2 Asymptotics under the microhyperbolicity condition

Let us now consider non-mollified asymptotics and asymptotics with smaller L.

Theorem 5.3.2. [9] *Let conditions (5.1.1), (5.1.2), (5.1.4), (5.1.8) and (5.1.9) be fulfilled and let $Q_i = \operatorname{Op}(q_i)$, $q_i \in S_{h,K}$ be supported in $\{|\xi'| \le c\}$, $\psi_i \in C_0^K(B(0,\frac12))$. Let us assume that the operator $\tau - A$ is microhyperbolic for $\tau = \bar\tau$ at $\operatorname{supp} q'_1 \cap \operatorname{supp} q'_2 \cap T^*(X \cap B(0,1))$ in some direction ℓ and that the problem $(\tau - A, B)$ is microhyperbolic for $\tau = \bar\tau$ at $\operatorname{supp} q'_1 \cap \operatorname{supp} q'_2 \cap T^*(Y \cap B(0,1))$ in some multidirection $\mathcal{T}$. Then the estimates*

$$(5.3.7)\qquad \mathcal{R} = |\Gamma Q'_{1x}\psi_1(x) e(x,y,\tau_1,\tau_2)\psi_2(y)^t Q'_{2y} - h^{-d}\varkappa_0(\tau_1,\tau_2) - h^{1-d}\varkappa_1(\tau_1,\tau_2)| \le C(M_0(\tau_1)+M_0(\tau_2))h^{1-d} + CMh^{2-d} \qquad \forall \tau_1,\tau_2 \in [\bar\tau-\epsilon_0, \bar\tau+\epsilon_0],$$

$$(5.3.8)\qquad \mathcal{R}_1 = \frac{1}{L}\Bigl|\int \phi\Bigl(\frac{\tau'-\bar\tau}{L}\Bigr)\Bigl(\Gamma Q'_{1x}\psi_1(x)e(x,y,\tau',\tau)\psi_2(y)^t Q'_{2y} - h^{-d}\varkappa_0(\tau',\tau) - h^{1-d}\varkappa_1(\tau',\tau)\Bigr)d\tau'\Bigr| \le M_0(\tau)h^{1-d} + CMh^{2-d} \qquad \forall \tau \in [\bar\tau-\epsilon_0, \bar\tau+\epsilon_0]$$

$$(5.3.9)\qquad \mathcal{R}_2 = \Bigl|\int \phi\Bigl(\frac{\tau'-\bar\tau}{L}\Bigr)d_\tau\Bigl(\Gamma Q'_{1x}\psi_1(x)e(x,y,\tau_1,\tau)\psi_2(y)^t Q'_{2y} - h^{-d}\varkappa_0(\tau_1,\tau) - h^{1-d}\varkappa_1(\tau_1,\tau)\Bigr)\Bigr| \le C\Bigl(M_0(\bar\tau)h^{1-d} + CMh^{2-d}\Bigr)\Bigl(\frac{h}{L}\Bigr)^s$$

hold where $\epsilon_0 = \epsilon_0(d,m,D,c) > 0$ is small enough, $\phi \in C_0^K([-c,c])$ is such that $\phi \ge 0$, $\int_{-\infty}^{\infty}\phi(\tau)d\tau = 1$, $L \in [h,1]$ is arbitrary,

$$M_0(\tau) = |\varkappa'_{0,q_1}(\tau)|^{\frac12}\cdot|\varkappa'_{0,q_2}(\tau)|^{\frac12} \qquad \varkappa_n(\tau_1,\tau_2) = \int_{\tau_1}^{\tau_2}\varkappa'_n(\tau)d\tau$$

and the notation of section 4.3 is used.

Remark 5.3.3. *(i) Surely, the list of assertions could be extended in correspondence with the assertions of theorem 5.2.16.*

(ii) Obviously, the remainder estimates with extra factor $\left(\frac{h}{L}\right)^{\vartheta}$ hold for ϑ-th order Riesz means with respect to the spectral parameter (see section 4.3) and $L \ge h$. ◊

We leave these assertions to the reader.

---

[9] Compare with theorem 4.3.2.

Theorem 5.3.4. [10] *Let the conditions of theorem 5.3.2 be fulfilled and let us assume that* $m = 2$ *and*

$$a = \Bigl(\sum_{j,k} g^{jk}(\xi_j - V_j)(\xi_k - V_k) + V\Bigr)\cdot I \tag{5.3.10}$$

where $g^{jk} = g^{kj}$, V_j, V *are real-valued functions satisfying* (4.4.6) *and* (4.4.7) *in* $B(0,1)$. *Let* $Q_i = \mathrm{Op}(q_i)$, $q_i \in S_{h,\rho,\rho,K}$ *be supported in* $\{|\xi'| \le c\}$, $\rho \ge h^{\frac12-\delta}$ *with* $\delta > 0$, *and* $\mathrm{dist}(\operatorname{supp} q_i, Y) \ge \rho' = h^{\delta'}$ *with a small small enough exponent* $\delta > 0$. *Let* $\mathcal{U}$ *be the* ρ-*neighborhood of* $\operatorname{supp} q_2$, $T \in [1, \rho'^{-1}]$. *Let us assume that through every point* $(\bar{x},\bar{\xi}) \in \mathcal{U}$ *there passes a Hamiltonian billiard* $(x(t),\xi(t))$ *with* $(x(0),\xi(0)) = (\bar{x},\bar{\xi})$ *and* $t \in [0,T]$ *or* $t \in [-T,0]$ *in the* ρ'-*neighborhood for which conditions* (3.2.56), (3.2.58)–(3.2.60), (3.2.62)–(3.2.64) *are fulfilled and*

$$\mathrm{dist}\bigl((x(t),(\xi(t)),(\bar{x},\bar{\xi})\bigr) \ge \epsilon_0\rho|t|. \tag{5.3.11}$$

Let $\psi_i \in C_0^K(B(0,\frac12))$. *Then*

$$\mathcal{R} \le C\frac{1}{T}\Bigl(M_0(\tau_1) + M_0(\tau_2)\Bigr)h^{1-d} + CMh^{2-d} \qquad \forall \tau_1,\tau_2 \in [\bar{\tau} - h^{\frac12-\delta''}, \bar{\tau} + h^{\frac12-\delta''}], \tag{5.3.12}$$

$$\mathcal{R}_1 \le C\frac{1}{T}M_0(\tau_1)h^{1-d} + CMh^{2-d} \qquad \forall \tau \in [\bar{\tau} - h^{\frac12-\delta''}, \bar{\tau} + h^{\frac12-\delta''}], \tag{5.3.13}$$

$$\mathcal{R}_2 \le C\Bigl(M_0(\bar{\tau})h^{1-d} + CMh^{2-d}\Bigr)\Bigl(\frac{h}{LT}\Bigr)^s \tag{5.3.14}$$

provided $\delta'' = \delta''(d,\delta) > 0$ *is small enough.*

Theorem 5.3.5. *Let the conditions of theorem 5.3.2 be fulfilled. Let us assume that* $\Lambda_{\tau 1},\dots,\Lambda_{\tau n}$ *are closed subsets of* Σ_τ *contained in* $\{(x,\xi) : \mathrm{dist}(x,Y) \ge \epsilon\}$. *Let us assume that in* $\mathcal{X} \subset X$

$$\Sigma_\tau = \cup_{1\le j\le r}\Sigma_{\tau,j}, \qquad \Sigma_{\tau,j} = \{a_j(x,\xi) = \tau\}$$

and $|a_j - a_k| \ge \epsilon \quad \forall j \ne k$ *and in* $\mathcal{X}$ *all the conditions* (5.1.1)–(5.1.4) *are fulfilled with* c *replaced by* c_1. *Let us assume that through every point* $(\bar{x},\bar{\xi}) \in \mathcal{U}$ *there pass Hamiltonian billiards* $(x(t),\xi(t)) \subset \mathcal{X}$ *with* $(x(0),\xi(0)) = (\bar{x},\bar{\xi})$ *and* $t \in [0,T_j]$ *or* $t \in [-T_j,0]$ *in the* ρ'-*neighborhood for which the conditions*

$$|\frac{dx_1}{dt}| + |x_1| \ge \epsilon, \tag{5.3.15}$$

$$\mathrm{dist}((x(t),\xi(t)),(\bar{x},\bar{\xi})) \ge \epsilon\min(|t|,1) \tag{5.3.16}$$

are fulfilled (where $x_1 = \mathrm{dist}(x,Y)$*); and if* $x(t) \in Y$ *then the characteristic polynomial* $\det(\tau - a(x,\xi',\zeta))$ *has no real roots different from* $\frac{dx_1}{dt}|_{t\pm 0}$ *at the point* $(x(t),\xi(t))$ *for this or for any other billiard from the indicated family and there is no root with the absolute value of its imaginary part less than* ϵ, *and finally at this point* $x(t) \in Y$

[10] Compare with theorem 4.3.3; in its proof theorem 3.2.10 is used.

(5.3.17) $|\eth B(x',D_1,\xi')v| \geq \epsilon |v|_+$

$$\forall v \in \mathcal{S}(\bar{\mathbb{R}}^+, \mathbb{H}) : (\tau - a(0,x',D_1,\xi'))V = 0.$$

Then

(5.3.18) $\mathcal{R} \leq$

$$C \sum_{0\leq j\leq n} \sum_{\tau=\tau_i} D\frac{h^{1-d}}{T_j}\int_{\Lambda_{\tau,j}\cap\mathcal{U}} dxd\xi : da' + \epsilon h^{1-d} + C'h^{2-d}$$

$$\forall \tau_1,\tau_2 \in [\bar{\tau}-\epsilon', \bar{\tau}+\epsilon']$$

(5.3.19) $$\mathcal{R}_1 \leq C \sum_{0\leq j\leq n} D\frac{h^{1-d}}{T_j}\int_{\Lambda_{\tau,j}\cap\mathcal{U}} dxd\xi : da' + \epsilon h^{1-d} + C'h^2 - d$$

$$\forall \tau \in [\bar{\tau}-\epsilon', \bar{\tau}+\epsilon']$$

(5.3.20) $$\mathcal{R}_2 \leq C \sum_{0\leq j\leq n} D\frac{h^{1-d}}{T_j}\Bigl(\bigl(\frac{h}{LT_j}\bigr)^s\bigl(\int_{\Lambda_{\tau,j}\cap\mathcal{U}} dxd\xi : da' + \epsilon + C'h\Bigr)$$

where $\mathcal{U}$ is the ϵ-neighborhood of $\operatorname{supp} q_1 \cap \operatorname{supp} q_2$, *$\epsilon'$ is small enough, $a' = a_j$ at $\Sigma_{\tau,j}$, C' depends also on ϵ and $T_1,\dots,T_n$,*

$$\Lambda_{\tau,0} = \Sigma_\tau \setminus \bigcup_{1\leq j\leq n} \Lambda_{\tau,j}$$

$T_0 = 1$.

Remark 5.3.6. *Under the assumptions of the two previous theorems we can estimate Riesz means; in this case we need to take into account more terms than in the statement of the theorems, and in each term of the right-hand expression we can take the additional factor $(\frac{h}{LT_j})^\vartheta$.* ◊

5.3.3 Asymptotics in the elliptic zone

Let us now treat the case in which A is elliptic for $\tau \in [\tau_1,\tau_2]$ on $\operatorname{supp} q_2'$. In this case $\varkappa_0(\tau',\tau'') = 0$ for $\tau',\tau'' \in [\tau_1,\tau_2]$. Let us calculate $\varkappa_1'(\tau)$, $\varkappa(\tau',\tau'')$ and $\mathcal{K}_0'(x,y,\tau)$, $\mathcal{K}_0(x,y,\tau',\tau'')$ for these τ, τ', τ''. Construction of these functions yields that in our case it is sufficient to do the special case $A = a(hD_1)$, $B = b(hD_1)$; this easily yields

Lemma 5.3.7. *Let the conditions of theorem 5.3.1 be fulfilled and let A be elliptic for $\tau \in [\tau_1,\tau_2]$ on* $\operatorname{supp} q_2'$. *Then for $\tau',\tau'' \in [\tau_1,\tau_2]$*

(5.3.21) $$\mathcal{K}_0(x,y,\tau',\tau'') = (2\pi)^{1-d}\int e^{ih^{-1}\langle x'-y',\xi'\rangle} q_1^0(x',\xi') \times \Upsilon(h^{-1}x_1, h^{-1}y_1; y',\xi',\tau',\tau'')q_2^0(x',\xi')d\xi',$$

(5.3.22) $$\kappa_0(x,y,\tau',\tau'') = (2\pi)^{1-d}\int \upsilon(y',\xi',\tau',\tau'')q_2^0(x',\xi')d\xi'$$

where τ', τ'' are spectral parameters, $\Upsilon(x_1,y_1;y',\xi',.,.)$ is the Schwartz kernel of the spectral projector of $a(0,y',D_1,\xi')$ under the boundary condition $\eth b(y',D_1,\xi')u = 0$ on $L^2(\mathbb{R}^+,\mathbb{H})$, and

$$\upsilon = \int_0^\infty \Upsilon(x_1,x_1)dx_1.$$

Then repeating the proof of theorem 4.3.2 we obtain

Theorem 5.3.8. *Let conditions* (5.1.1), (5.1.2), (5.1.4), (5.1.8) *and* (5.1.9) *be fulfilled and let* $Q_i = \mathrm{Op}(q_i)$, $q_i \in S_{h,K}$ *be supported in* $\{|\xi'| \le c\}$, $\psi_i \in C_0^K(B(0,\frac{1}{2}))$. *Let us assume that the operator* $\tau - A$ *is elliptic for* $\tau = \bar\tau$ *at* $\operatorname{supp} q_1' \cap \operatorname{supp} q_2' \cup T^*(X \cap B(0,1))$ *and that the problem* $(\tau - A, B)$ *is microhyperbolic for* $\tau = \bar\tau$ *at* $\operatorname{supp} q_1' \cap \operatorname{supp} q_2' \cap T^*(Y \cap B(0,1))$ *in some multidirection* $\mathcal{T}$.[11] *Then*

$$(5.3.23)\quad \mathcal{R}_3 = |\Gamma Q'_{1x}\psi_1(x)e(x,y,\tau_1,\tau_2)\psi_2(y)^t Q'_{2y} - h^{1-d}\varkappa_1(\tau_1,\tau_2) - h^{2-d}\varkappa_2(\tau_1,\tau_2)| \le C\big(M_1(\tau_1)+M_1(\tau_2)\big)h^{2-d} + CMh^{3-d} \quad \forall \tau_1,\tau_2 \in [\bar\tau-\epsilon_0,\bar\tau+\epsilon_0],$$

$$(5.3.24)\quad \mathcal{R}_4 = \frac{1}{L}\Big|\int \phi\big(\frac{\tau'-\bar\tau}{L}\big)\Big(\Gamma Q'_{1x}\psi_1(x)e(x,y,\tau',\tau)\psi_2(y)^t Q'_{2y} - h^{1-d}\varkappa_1(\tau',\tau) - h^{2-d}\varkappa_2(\tau',\tau)\Big)d\tau'\Big| \le C(M_1(\tau)+M_1(\bar\tau))h^{2-d} + CMh^{3-d} \quad \forall,\tau \in [\tau-\epsilon_0,\bar\tau+\epsilon_0]$$

$$(5.3.25)\qquad \mathcal{R}_2 \le C(M_1(\bar\tau)+Mh)h^{1-d}\big(\frac{h}{L}\big)^s$$

where ϵ_0, ϕ, M, L *are the same as in theorem 5.3.2,*

$$M_1(\tau) = |\varkappa'_{1q_1}(\tau)|^{\frac{1}{2}} \cdot |\varkappa'_{1q_2}(\tau)|^{\frac{1}{2}};$$

(*ii*) *If* $\mathcal{T}'_x = 0$ *then*

$$(5.3.26)\quad \mathcal{R}_5 = |\Gamma_{x'} Q'_{1x}\psi_1(x)e(x,y,\tau_1,\tau_2)\psi_2(y)^t Q'_{2y} - h^{-d}\mathcal{K}_0(\tau_1,\tau_2)| \le C'M'h^{1-d}\big(1+\frac{(x_1+y_1)}{h}\big)^{-s} \quad \forall x_1,y_1 \in [0,1] \quad \forall x' \in \mathbb{R}^{d-1} \quad \forall \tau_1,\tau_2 \in [\bar\tau-\epsilon_0,\bar\tau+\epsilon_0].$$

Remark 5.3.9. (*i*) *In the proof of* (*ii*) *we also use the estimate*

$$|\Gamma_{x'}\mathcal{K}_n(x,y,\tau)| \le C'\big(1+\frac{(x_1+y_1)}{h}\big)^{-s} \quad \forall n \le s$$

for $\tau \in [\bar\tau-\epsilon_0,\bar\tau+\epsilon_0]$; *this estimate follows from the fact that contours* $\gamma_\pm$ *can be chosen (by the ellipticity condition) lying in the zone* $\{\pm \operatorname{Im}\xi_1 \ge \epsilon > 0\}$.

(*ii*) *Remark* (*5.3.6*) *concerning Riesz means remains true.*

(*iii*) *In estimates* (*5.3.7*) *and* (*5.3.11*) *one can take* $\tau_1 = \tau' = -\infty$ *provided the following conditions are fulfilled:*

$$(5.3.27)_1\quad \langle a(x,\xi)v,v\rangle \ge -c|v|^2 \quad \forall v \in \mathbb{H} \quad \forall (x,\xi') \in (\operatorname{supp} q_1' \cup \operatorname{supp} q_2') \cap (X \cap B(0,1)) \times \mathbb{R}^{d-1} \quad \forall \xi_1 \in \mathbb{R},$$

$$(5.3.27)_2\quad \langle a(0,x',D_1,\xi')v,v\rangle_+ \ge -c|v|_+^2 \quad \forall v \in \mathcal{S}(\bar{\mathbb{R}}^+,\mathbb{H}) : \mathfrak{d}b(x',D_1,\xi')v = 0 \quad \forall (x',\xi') \in (\operatorname{supp} q_1' \cup \operatorname{supp} q_2') \cap (Y \cap B(0,1)) \times \mathbb{R}^{d-1};$$

[11] In this case $\mathcal{T} = \mathcal{T}' \in T^*Y$.

moreover, these conditions imply the same inequalities with the additional terms $\epsilon_0|\xi_1|^m|v|^2$ *and* $\epsilon_0|D_1^{\frac{m}{2}}v|_+^2$ *respectively in their right-hand expressions (because* $a(0,x',D_1,\xi')$ *is elliptic as a one-dimensional differential operator; we assume that either (5.1.11) or the conditions of lemma 5.1.10 are fulfilled). Furthermore, in all the other estimates mollification with respect to* τ' *can be replaced by setting* $\tau' = -\infty$ *in this situation.* ◊

Examples associated with the Schrödinger operator are treated in the next section. There is also a reasonable example from elasticity theory (with free boundary) associated with Rayleigh waves.

5.3.4 Local asymptotics

Let us proceed from microlocal asymptotics to local asymptotics based on the results of section 5.1.

Theorem 5.3.10. *Let conditions* (5.1.1), (5.1.2), (5.1.4), (5.1.8), (5.1.9) *and* (5.1.30), (5.1.31) *be fulfilled and let* $\phi \in C_0^K([-c,c])$, $\bar\tau \in [-c,c]$ *and* $L \in [h^{\frac{1}{2}-\delta},1]$. *Then*

(i) The following estimate holds:

$$(5.3.28)\quad \frac{1}{L}\Bigl|\int (\phi(\frac{\tau-\bar\tau}{L})d_\tau e(x,y,\tau_1,\tau) - \sum_{0\le n\le N'-1} h^{-d+n}\mathcal{K}'_n(x,y,\tau)d\tau)\Bigr| \le C'M'h^s \qquad \forall x,y\in B(0,1-\epsilon)$$

where $M = \|\phi\|$ *and* $\mathcal{K}'_n$, $\mathcal{K}_n$ *are defined above for* $q_1=q_2=\psi_1=\psi_2=1$;

(ii) Moreover, if $a(x,\xi)$ *is semibounded from below uniformly with respect to* $(x,\xi)\in T^*(X\cap B(0,1))$ *and if the operator* $a(0,x',D_1,\xi')$ *is semibounded from below uniformly with respect to* $(x',\xi')\in T^*(Y\cap B(0,1))$ *then*

$$(5.3.29)\quad \frac{1}{L}\Bigl|\int (\phi(\frac{\tau-\bar\tau}{L})(e(x,y,-\infty,\tau) - \sum_{0\le n\le N'-1} h^{-d+n}\mathcal{K}_n(x,y,-\infty,\tau))d\tau\Bigr| \le C'M'h^s \qquad \forall x,y\in B(0,1-\epsilon).$$

Theorem 5.3.11. *Let the conditions of the previous theorem be fulfilled and let us assume that the operator* $\tau - A$ *is microhyperbolic for* $\tau=\bar\tau$ *at* $T^*(X\cap B(0,1))$ *in some direction* ℓ *and that the problem* $(\tau - A, B)$ *is microhyperbolic for* $\tau=\bar\tau$ *at* $T^*(Y\cap B(0,1))$ *in some multidirection* $\mathcal{T}$. *Then*

$$(5.3.30)\quad \mathcal{R}_6 = |\Gamma\psi_1(x)e(x,x,\tau_1,\tau_2) - h^{-d}\varkappa_0(\tau_1,\tau_2) - h^{1-d}\varkappa_1(\tau_1,\tau_2)| \le C\Bigl(M_0(\tau_1)+M_0(\tau_2)\Bigr)h^{1-d} + CMh^{2-d} \qquad \forall \tau_1,\tau_2\in[\bar\tau-\epsilon_0,\bar\tau+\epsilon_0],$$

$$(5.3.31)\quad \mathcal{R}_7 = \frac{1}{L}\Bigl|\int \phi(\frac{\tau-\bar\tau}{L})\Bigl(\Gamma\psi_1(x)e(x,x,\tau',\tau) - h^{-d}\varkappa_0(\tau',\tau) - h^{1-d}\varkappa_1(\tau',\tau)\Bigr)d\tau'\Bigr| \le M_0(\tau)h^{1-d} + CMh^{2-d} \qquad \forall \tau\in[\bar\tau-\epsilon_0,\bar\tau+\epsilon_0]$$

where $\epsilon_0 = \epsilon_0(d,m,D,c) > 0$ *is small enough,* $\phi \in C_0^K([-c,c])$ *is such that* $\phi \geq 0$, $\int_{-\infty}^{\infty} \phi(\tau)d\tau = 1$, $L \in [h^{1-\delta},1]$ *is arbitrary,* $M_0(\tau)$ *and* $\varkappa_n$ *were defined above for* $q_1 = q_2 = 1$*; moreover, the estimates for* $\mathcal{R}_2$ *with proper modifications remain true as well.*

(ii) If the operator $\tau - A$ *is elliptic for* $\tau = \bar\tau$ *at* $T^*(X \cap B(0,1))$ *then*

$$\text{(5.3.32)}\quad \mathcal{R}_8 = |\Gamma\psi_1(x)e(x,x,\tau_1,\tau_2) - h^{1-d}\varkappa_1(\tau_1,\tau_2) - h^{2-d}\varkappa_2(\tau_1,\tau_2)| \leq C(M_1(\tau_1)+M_1(\tau_2))h^{2-d} + CMh^{3-d} \qquad \forall \tau_1,\tau_2 \in [\bar\tau-\epsilon_0,\bar\tau+\epsilon_0],$$

$$\text{(5.3.33)}\quad \mathcal{R}_9 = \frac{1}{L}\Big|\int \phi(\frac{\tau-\bar\tau}{L})\Big(\Gamma\psi_1(x)e(x,x,\tau',\tau) - h^{1-d}\varkappa_1(\tau',\tau) - h^{1-d}\varkappa_1(\tau',\tau)\Big)d\tau'\Big| \leq M_1(\tau)h^{2-d} + CMh^{3-d} \qquad \forall \tau \in [\bar\tau-\epsilon_0,\bar\tau+\epsilon_0]$$

where $M_1(\tau)$ *is defined above for* $q_1 = q_2 = 1$*; moreover, the estimates for* $\mathcal{R}_2$ *with proper modifications remain true as well.*

(iii) In the context of assertion (ii) of the previous theorem estimates (5.3.30), (5.3.32) *remain true for* $\tau_1 = -\infty$*;*

(iv) Assertions (i) and (ii) of the previous theorem remain true for $L \in [h^{1-\delta},1]$.

Theorem 5.3.12. *(i) Suppose the conditions of theorem 5.3.4 and conditions* (5.1.34), (5.1.35) *are fulfilled for* $q_1 = q_2 = 1$. *Then all the estimates of theorem 5.3.4 remain true for for* $q_1 = q_2 = 1$*;*

(ii) Suppose the conditions of theorem 5.3.5 and conditions (5.1.34), (5.1.35) *are fulfilled for* $q_1 = q_2 = 1$. *Then all the estimates of theorem 5.3.5 remain true for for* $q_1 = q_2 = 1$*;*

(iii) In the context of assertion (iii) of remark 5.3.9 all the estimates with no mollification with respect to τ_1 *remain true for* $\tau_1 = -\infty$.

Remark 5.3.13. *If we assume that the conditions of some theorem (or some part of a theorem) are fulfilled for* $\tau = \bar\tau_i$, $i = 1,2$, *then the corresponding estimates (without mollification) remain for* τ_i *belonging to the* ϵ_0*-neighborhoods of* $\bar\tau_i$. ◇

5.4 Spectral Asymptotics for the Schrödinger Operator

5.4.1 Application of general results

In this section we treat operator (4.5.1) assuming that conditions $(4.5.6)_{1-3}$, (4.5.7) and (5.1.3) are fulfilled and that the boundary operator is either $B = I$ or B given by (5.1.54)

$$B = -i\alpha(x')hD_1 + \mathcal{M}(x',hD',h) + \beta(x',hD',h)$$

in the coordinate system in which $X = \{x_1 \geq 0\}$ locally and the operator A has form (5.1.52); we assume that conditions $(5.1.4)_2$, (5.1.55) and either (5.1.57) ((5.1.57)′ for $d = 2$) or $(5.1.58)_{1,2}$ are fulfilled. Let us recall that (5.1.57) ((5.1.57)′ for $d = 2$). is the Shapiro-Lopatinskii condition which we assumed in theorems 5.3.10–5.3.12. However, theorem 5.1.11 yields

Remark 5.4.1. *(i) For the Schrödinger operator in theorems 5.3.10–5.3.12 one can replace condition (5.1.31) by conditions (5.1.55) and* $(5.1.58)_{1,2}$.

(ii) Under conditions (5.1.55) and either (5.1.57) or $(5.1.58)_{1,2}$, *one can take* $\tau_1 = -\infty$ *in all the asymptotics of section 5.3.*

(iii) If $d = 2$ *and condition* (5.1.57)′ *is fulfilled (but condition (5.1.57) is violated) the operator* A *is not semibounded from below in the "semiclassical sense."* ◇

The microhyperbolicity condition at the interior point (x,ξ) is

$$|\tau - V(x)| - \sum_{1\le j\le d} |\xi_j - V_j(x)| + |\nabla_x V(x)| \ge \epsilon_0; \tag{5.4.1}$$

the microhyperbolicity condition for $\tau - A$ at the boundary point (x,ξ') is

$$|\tau - V(x)| - \sum_{2\le j\le d} |\xi_j - V_j(x)| + |\nabla_{x'} V(x)| \ge \epsilon_0 \quad (x_1 = 0); \tag{5.4.2}$$

moreover, one can take $\mathcal{T}_x = 0$ and $\mathcal{T}'_x = 0$ in these conditions respectively if and only if

$$|\tau - V(x)| - \sum_{1\le j\le d} |\xi_j - V_j(x)| \ge \epsilon_0 \tag{5.4.1$'$}$$

and

$$|\tau - V(x)| - \sum_{2\le j\le d} |\xi_j - V_j(x)| \ge \epsilon_0 \quad (x_1 = 0); \tag{5.4.2$'$}$$

respectively.

Let us treat the microhyperbolicity condition for the problem $(\tau - A, B)$ in the same multidirection; it is easy to see that this condition is fulfilled with $\mathcal{T}'_x = 0$ if and only if one of the following four conditions is fulfilled:

(a) The point (x',ξ') is either hyperbolic or "almost hyperbolic," i.e.,

$$\tau \ge \sum_{2\le j,k\le d} g^{jk}(x)(\xi_j - V_j)(\xi_k - V_k) + V - \epsilon' \tag{5.4.3$_+$}$$

with $\epsilon' = \epsilon'(d,c,\epsilon_0) > 0$;

(b_1)
$$\alpha\beta \ge -\epsilon',$$

(b_2)
$$|\tau - W(x)| \ge \epsilon_0$$

where

$$W = V - \beta^2(\alpha^2 - g'(\mathbf{m}))^{-1}, \quad g'(\mathbf{m}) = \sum_{2\le j,k\le d} g'_{jk} m^j m^k \tag{5.4.4}$$

where $\mathbf{m} = (m^2,\dots,m^d)$ is the vector field corresponding to $\mathcal{M}$ and (g'_{jk}) is the tensor inverse to $(g^{jk})_{j,k=2,\dots d}$;

(b_3)
$$\sum_{2\le j\le d} |\xi_j - V_j + (\tau - V)\frac{m_j}{\beta}| \ge \epsilon_0, \quad m_j = \sum_{2\le k\le d} g'_{jk} m^j;$$

in cases (b_1), (b_2), and (b_3) the operator $\tilde{B}$ given by (5.1.56) and the problem $(\tau - A, B)$ are both either elliptic at this point or microhyperbolic in the multidirection $(\mathcal{T}'_x, \mathcal{T}'_\xi) = (0, \xi_j - V_j)_{j=2,\dots d})$; we assume that condition (5.4.3)$_+$ is violated and hence operator $\tau - A$ is elliptic at this point.

Finally, if all four conditions (a)–(b_3) are violated then the microhyperbolicity condition is fulfilled if and only if

$$|\nabla_{x'} W| \ge \epsilon_0; \tag{5.4.5}$$

however $\mathcal{T}'_x \ne 0$ in this case.

5.4.2 General estimates without the microhyperbolicity condition

Let us first obtain estimates (5.3.23)–(5.3.25). First of all let us prove

Theorem 5.4.2. *Suppose that conditions* (4.5.1), $(4.5.6)_{1_3}$, (4.5.7), (5.1.3), $(5.1.4)_2$, (5.1.9), (5.1.55) *and either* (5.1.57) *((5.1.57)′ for $d = 2$) or* $(5.1.58)_{1,2}$ *are fulfilled. Moreover, let us assume that in $X \cap B(0,1)$ the condition*

$$|\tau - V(x)| + |\nabla_x V(x)| \ge \epsilon_0 \tag{5.4.6}$$

and in $Y \cap B(0,1)$ the condition

$$|\tau - V(x)| + |\nabla_{x'} V(x)| \ge \epsilon_0 \tag{5.4.7}$$

are fulfilled. Then

(i) Estimates (5.3.23)–(5.3.25) *hold;*

(ii) Moreover, if $d \ge 2$ and the additional conditions of theorem 5.3.12 are fulfilled then the estimates of this theorem (i.e., the estimates of theorem 5.3.4 for $Q_1 = Q_2 = I$) hold;

(iii) If either condition (5.1.57) *or condition* $(5.1.58)_{1,2}$ *is fulfilled then one can take $\tau_1 = -\infty$ in all the above estimates.*

Proof. Theorems 5.1.6 and 5.1.11 yield that it is sufficient to prove estimates (5.3.7) and (5.3.8) in case (i), and (5.3.18) and (5.3.19) in case (ii) for symbols q_1' and q_2' supported in a small enough neighborhood of the point $(\bar{x}', \bar{\xi}')$ running over $T^*(Y \cap B(0,1)) \cap \{|\xi'| \le C_0\}$ with a large enough C_0.

If at this point (i.e., $(\bar{x}', \bar{\xi}')$) condition $(5.4.3)_+$ is fulfilled then we can apply theorems 5.3.4 and 5.3.5 in cases (i) and (ii) respectively. If condition $(5.4.3)_+$ is violated then (i) follows from theorem 5.1.3 but in the proof of (ii) we need to apply more refined arguments: (ii) follows from theorem 5.4.6 which will be proven later independently of the results of this subsection. □

Let us now get rid of conditions (5.4.6) and (5.4.7); our first step is to get rid of condition (5.4.7):

Proposition 5.4.3. *All the assertions of theorem 5.4.2 remain true for $d = 2$ even if condition* (5.4.7) *is violated and $CMh^{1-d+\sigma'}$ is added to the right-hand expressions of all the estimates where $\sigma' > 0$ is a small enough exponent. In other words, the new term is added only if condition (5.4.7) is violated.*

Proof. Let all the conditions of theorem 5.4.2 including (5.4.6) but excluding (5.4.7) be fulfilled. Applying an ϵ-admissible partition of unity with $\epsilon = \epsilon(d, c, \epsilon_0) > 0$ and a gradient transformation if necessary we see that we can assume $\tau = 0$,

$$|\partial_1 V| \ge \frac{1}{2}\epsilon_0, \quad |V| + |\nabla_{x'} V| \le \frac{1}{4}\epsilon_0, \quad V_1 = 0 \quad \text{in } B(0,1), \quad V_k(0) = 0. \tag{5.4.8}$$

In $X \cap B(0,1)$ let us introduce the function

$$\gamma(x) = \epsilon_0(x_1 + |V(x)|) + h^{\frac{2}{3}} \tag{5.4.9}$$

with a small constant $\epsilon_0 > 0$; then $|\nabla_x \gamma| \le 1$ and after the dilatation $x \to x_{\text{new}} = \frac{x - \bar{x}}{\bar{\gamma}}$ with $\bar{\gamma} = \gamma(\bar{x})$, $h_{\text{eff}} = h\bar{\gamma}^{-\frac{3}{2}} \le 1$ and division of A by $\bar{\gamma}$ the following cases are possible:

(a) If $\bar{x}_1 \ge 2\bar{\gamma} \ge 2h^{\frac{2}{3}}$ then we are in the situation of section 4.4. In this case the contribution of the ball $B(\bar{x}, \bar{\gamma})$ to the remainder estimate is $O(h_{\text{eff}}^{1-d})$ and the total contribution of all balls of this type does not exceed

$$Ch^{1-d} \int x_1^{\frac{3d}{2}-\frac{3}{2}} x_1^{-d} dx_1$$

and does not exceed Ch^{1-d} provided $d \geq 2$.

(b) If $2\bar{\gamma} \geq \max(\bar{x}_1, 2h^{\frac{2}{3}})$ then operator A satisfies all the conditions of theorem 5.4.2. Let us treat B. It is easy to see that α and $\mathbf{m}$ are multiplied by $\bar{\gamma}^{\frac{1}{2}}$ and hence condition $(5.1.58)_{1,2}$ is preserved. Similarly, if condition (5.1.57) (or (5.1.57)′ for $d = 2$) was fulfilled and if $|\beta(\bar{x}')| \leq C_0\bar{\gamma}^{\frac{1}{2}}$ then dividing B by $\bar{\gamma}^{\frac{1}{2}}$ we obtain that condition (5.1.57) (or (5.1.57)′ for $d = 2$) is preserved. In both cases the contribution of the ball $B(\bar{x}, \bar{\gamma})$ to the remainder estimate is $O(h_{\mathrm{eff}}^{1-d})$. Let us note that the number of such balls with $\frac{1}{2}\rho \leq \gamma \leq 2\rho$ in the covering does not exceed $C\rho^{1-d}$ (in fact, the volume of every ball intersected with X is not less than ρ^d and the volume of the covering zone does not exceed $C\rho$ because $\gamma \geq \epsilon x_1$ in these balls; moreover, the covering is of finite multiplicity); hence the total contribution of all balls of this type also does not exceed Ch^{1-d}.

(c) If $\bar{\gamma} \leq 2h^{\frac{2}{3}}$ and either condition $(5.1.58)_{1,2}$ or (5.1.57) (or (5.1.57)′ for $d = 2$) was fulfilled and $|\beta(\bar{x}')| \leq C_0 h^{\frac{1}{3}}$ in the last case then we are in the situation of theorem 5.1.6 and the contribution of such a ball does not exceed Ch^{-d} and the number of such balls does not exceed $Ch^{\frac{3}{2}(1-d)}$; hence the total contribution of all balls of this type does not exceed $Ch^{1-d+\sigma'}$ provided $d \geq 2$;

(d) Let the conditions of either (b) or (c) be fulfilled excluding the restriction on β which is now replaced by $|\beta(\bar{x}')| \geq C_0\bar{\gamma}^{\frac{1}{2}}$. Then theorems 5.3.4 (in case (b)) and 5.1.1 (in case (c)) yield that the previous arguments apply to estimate the remainder in the Weylian asymptotics for expression

$$\frac{1}{L}\int \Gamma\Big(Q'_{1x}\psi_1(x)e(.,.,\tau',0)\psi_2(y)\Big)\phi(\frac{\tau'}{L})d\tau' \tag{5.4.9}$$

where $Q'_1 = \mathrm{Op}(q'_1)$, $q'_1 \in S'_{h,\bar{\rho},\bar{\rho},K}$ is supported in $\{|\xi'| \leq \epsilon_1^{-1}\bar{\rho}\}$ with $\bar{\rho} = |\beta(\bar{x}')|$ and with small constants $\epsilon_j = \epsilon_j(d, c, \epsilon_0) > 0$. After summation with respect to a $\bar{\gamma}$-admissible partition of unity in $\mathbb{R}^{d-1}_{x'}$ we obtain that for this expression with ψ_j supported in $B(\bar{x}', \tilde{\gamma})$, $\tilde{\gamma} = \epsilon_2\bar{\rho}^2$ the Weylian approximation holds with the remainder estimate not exceeding $Ch^{1-d}\bar{\rho}^{d-1+\sigma'}$.

Let us consider $q''_1 \in S'_{h,\bar{\rho},\bar{\rho},K}$ supported in $\{|x'| \geq \frac{1}{2}\epsilon_1\bar{\rho}\}$. Then applying the previous procedure with $\tilde{\gamma}$ instead of $\bar{\gamma}$ we see that for a large enough constant C_0 operators A and B will satisfy the conditions of theorem 5.4.6(ii) after this procedure (see below; here only derivatives with respect to ξ' are calculated) and hence the remainder estimate will not exceed $C(h\tilde{\gamma}^{\frac{-3}{2}})^{2-d}$. So the contribution of the ball $B(\bar{x}, \bar{\gamma})$ does not exceed

$$Ch^{1-d}\bar{\gamma}^{d-1+\delta'} + Ch^{2-d}\bar{\gamma}^{\frac{3d}{2}-3}$$

where $Q'_1 = I$ now.

Let us define

$$\gamma(x') = \epsilon_3\beta^2(x') + h^{\frac{2}{3}}$$

and let us sum with respect to a γ-admissible partition of unity in $Y \cap B(0,1)$; taking into account that the number of the balls of this type with $\frac{1}{2}\bar{\gamma} \leq \gamma \leq 2\bar{\gamma}$ does not exceed $C\bar{\gamma}^{1-d}\,\mathrm{mes}\{x', \frac{1}{2}\bar{\gamma} \leq \gamma \leq 2\bar{\gamma}\}$ we obtain that the total contribution to the remainder estimates of balls of type (d) does not exceed Ch^{1-d}.

So *we have proven the remainder estimate* $\mathcal{R}_3 \leq Ch^{1-d}$. Surely it is not the assertion of proposition 5.4.3 yet; moreover, it is not even assertion (i) of this proposition but it also follows from our analysis that *the total contribution of the balls with* $\gamma(x) \leq h^{\sigma''}$

with $\sigma''>0$ does not exceed $Ch^{1-d+\sigma'}$ with $\sigma'>0$ even if no restriction on $\beta(x)$ [12] *is made and hence we need to treat balls with $\gamma\ge h^{\delta''}$ with arbitrarily small $\sigma''>0$ more carefully.*

Moreover, *in order to prove proposition 5.4.3 we need to treat these balls with additional cutting operators $Q'_{1x}=\mathrm{Op}(q'_1)$ with $q'_1\in S_{h,\gamma,\gamma,K}$ supported in $\{|\xi'|\le C_1\gamma^{\frac12}\}$* because for cutting operators Q''_{1x} with symbols supported in $\{|\xi'|\ge \frac12 C_1\gamma^{\frac12}\}$ our procedure gives the same remainder estimate as was indicated in the proposition.

Let us apply theorem 5.3.4 to every ball of this type. In order to do this let us analyze the bicharacteristics generated by the operator A after the dilatation $x\to\tilde x=\frac{x-\bar x}{\bar\gamma}$, $\tilde\xi=\xi\bar\gamma$, $\tilde h=h\bar\gamma^{-\frac32}$ and division of A by $\bar\gamma$. The principal symbol of the new operator is $\tilde a(\tilde x,\tilde\xi)=\bar\gamma^{-1}a(\tilde x\bar\gamma+\bar x,\tilde\xi\bar\gamma^{\frac12})$ so writing down the equations of the bicharacteristics and returning back to (x,ξ) we obtain the equations

$$\frac{dx}{dt}=\bar\gamma^{\frac12}\nabla_\xi a,\qquad \frac{d\xi}{dt}=-\bar\gamma^{\frac12}\nabla_x a,$$

i.e., if we assumed that the conditions of theorem 5.3.4 were fulfilled for a with T_j then now these conditions are fulfilled with $\tilde T_j=T_j\bar\gamma^{-\frac12}$ for $\tilde a$. Therefore one can take $\tilde T_j=T_j\bar\gamma^{-\frac12}$ in the estimate of this theorem. Moreover

$$\tilde h^{1-d}d\tilde x d\tilde\xi : d\tilde a = h^{1-d}dxd\xi : da\bar\gamma^{-\frac12}$$

because of the connection between $\tilde x,\tilde\xi,\tilde a,\tilde h$ and x,ξ,a,h.

Therefore the contribution of Λ_j with $j=1,\dots,n$ to the remainder estimate does not exceed (for the ball $B(\bar x,\bar\gamma)$)

$$C\frac{h^{1-d}}{T_j}\int_{\Lambda_j\cap B(\bar x,\bar\gamma)\times\mathbb{R}^d} dxd\xi : da + Ch^{1-d+\sigma'}\bar\gamma^{d-1}$$

and after summation with respect to all these balls we obtain

$$C\frac{h^{1-d}}{T_j}\int_{\Lambda_j} dxd\xi : da + Ch^{1-d+\sigma'}; \tag{5.4.10}$$

this answer is indicated in proposition 5.4.3(ii).

So we need to treat trajectories starting from Λ_0. If we assume that

$$\partial_1 V\le -\epsilon_0 \quad \text{in } B(0,1) \tag*{(5.4.11)$_-$}$$

then for $\gamma(x)\le\epsilon_3=\epsilon_3(d,c,\epsilon_0)>0$ along the trajectory $\frac{d\xi_1}{dt}\ge\frac14\epsilon_0$ and therefore every trajectory generated by $a(x,\xi)$ going from any point $(\bar x,\bar\xi)\in\Sigma\cap\{x\in B(0,\frac23),\gamma(x)\le\epsilon_3\}\times\mathbb{R}^d$ with $x_1>0$ in the direction $t>0$ ($t<0$) for $\bar\xi_1>0$ ($\bar\xi_1<0$ respectively) of length $T_0=T_0(d,c)>0$ does not intersect Y and satisfies all the conditions of theorem 5.3.4. Hence the previous arguments are applicable to Λ_0 with $T=T_0$, $\tilde T=\tilde T_0=T_0\bar\gamma^{-\frac12}$ and the hence the contribution of Λ_0 to the remainder estimate does not exceed expression (5.4.10) with $j=0$; this answer is indicated in proposition 5.4.3(i), (ii). One can say that the boundary is *repulsive* in this case.

Let us consider an *attractive* boundary, i.e., let us assume that

$$\partial_1 V\ge\epsilon_0 \quad \text{in } B(0,1). \tag*{(5.4.11)$_+$}$$

---

[12] It follows from our analysis that the contribution of the zone $\{\gamma\le h^{\sigma'},\ |\xi'|\ge C_0\gamma(x)^{\frac12}\}$ also does not exceed $Ch^{1-d+\sigma'}$.

Let us recall that we need to treat only the Hamiltonian trajectories starting at points $(\bar{x}, \bar{\xi})$ with $\bar{x}_1 \geq h^{\sigma''}$. Moreover, there are also restrictions $|\bar{x}_1| \leq \epsilon_4$, $|\bar{\xi}_1| \leq \epsilon_4$ with a constant $\epsilon_4 > 0$ as small as we wish. In fact, if neither of these inequalities are fulfilled then through the point $(\bar{x}, \bar{\xi})$ there passes a Hamiltonian trajectory of $a(x,\xi)$ with either $t \in [-T_0, 1]$ or $t \in [0, T_0]$ and with a constant $T_0 = T_0(d, c, \epsilon_0, \epsilon_4) > 0$; along this trajectory all the conditions of theorem 5.3.4 are fulfilled and therefore all the arguments used in the case $(5.4.11)_-$ are applicable. Finally we can assume that $\gamma(\bar{x}) \leq \epsilon_4$; otherwise theorem 5.3.4 is applicable with $\tilde{T}_0 = 1$.

We assert that *under this restriction the Hamiltonian flow Φ_t is defined everywhere for $|t| \leq T_0 = T_0(d, c, \epsilon_0)$ and satisfies all the conditions of theorem 5.3.5, perhaps excluding the non-periodicity condition.* It is sufficient to check this assertion for the symbol $\tau - a(x,\xi)$ with $\tau = a(\bar{x}, \bar{\xi})$ on $T^*(\xi \times \mathbb{R})$. It is well known [Melrose 1,2] that if condition $(5.4.11)_+$ holds and if $(\bar{x}, \bar{\xi})$ is a point such that $|a(\bar{x}, \bar{\xi})| \leq \epsilon_4$, $\gamma(\bar{x}) \leq \epsilon_4$, $\bar{x}_1 \leq \epsilon_4$, $|\bar{\xi}_1| \leq \epsilon_4$ then there exists a symplectomorphism Ψ of the ϵ_4-neighborhood of this point to a neighborhood of 0 in $\mathbb{R}^{2d+2}$ such that

$$|D^\alpha \Psi| \leq C_\alpha \qquad \forall \alpha : |\alpha| \leq K,$$

$$x_1 \circ \Psi^{-1} = f_1 x_1, \quad (\tau - a) \circ \Psi^{-1} = f_2(\tau - x_1 - \xi_1^2)$$

with f_1, f_2 satisfying the same estimates as Ψ and, moreover,

$$|\det D\Psi| \geq \epsilon_6, \quad f_1(0) \geq \epsilon_6, \quad f_2(0) \geq \epsilon_6$$

with a constant $\epsilon_6 = \epsilon_6(d, c, \epsilon_0) > 0$. Therefore it is sufficient to check all the conditions of theorem 5.3.4 excluding the non-periodicity condition only in the special case $d = 1$, $a = x_1 + \xi_1^2$, $\tau = a(x,\xi) \in [h^{\sigma''}, \epsilon_0]$. In this case $x_1 + |\frac{dx_1}{dt}| \geq \epsilon_6 h^{\delta'}$ (i.e., the almost transversality condition is fulfilled) and the time between two consequent reflections equals $2\sqrt{\tau}$. Therefore the time interval $[-T_0, T_0]$ contains a number of reflections $\asymp \frac{T_0}{\sqrt{\tau}}$ and, moreover, if $(\bar{x}, \bar{\xi}) \notin T^*X|_Y$ and $\Phi_{\bar{t}}(\bar{x}, \bar{\xi}) \notin T^*X|_Y$ then in a neighborhood of $(\bar{t}, \bar{x}, \bar{\xi})$

$$\Phi_t(x,\xi) = \Phi_{t - 2n(x_1 + \xi_1^2)^{\frac{1}{2}}}(x,\xi)$$

with a number $n \in \mathbb{Z}$ which is constant in a neighborhood of this point and satisfies the inequality $|t - 2n(x_1 + \xi_1^2)^{\frac{1}{2}}| < 2\sqrt{\tau}$; then the flow in the right-hand expression contains no reflection. Calculating the Jacobi matrix of the right-hand expression we obtain that $|D\Phi_t| \leq C(\frac{|n|}{\sqrt{\tau}} + 1) \leq Ch^{-2\sigma'} T_0$ and our assertion is proven.

Let us treat now the non-periodicity condition. If it is fulfilled then we can apply theorem 5.3.4 with $T = T_0$ and $\tilde{T} = T_0 \bar{\gamma}^{-\frac{1}{2}}$. Conditions (5.4.8) and $(5.4.11)_+$ yield that

$$a(x,\xi) = \xi_1^2 - R(x', \xi') + x_1 S(x_1, x', \xi') \tag{5.4.12}$$

where in the set under consideration $S \geq \frac{1}{4}\epsilon_0$. Then for $a(x,\xi) = 0$

$$\xi_1^2 + x_1 \leq C_0 R \qquad (R \geq 0) \tag{5.4.13}$$

and, moreover, along trajectories

$$\frac{dR}{dt} = |\{a, R\}| \leq C_0 R. \tag{5.4.14}$$

Therefore

$$(5.4.15) \qquad (1 - C_0T_0) \le \frac{R(z'(t))}{R(z(0))} \le 1 + C_0T_0$$

where $C_0 = C_0(d, c, \epsilon_0)$ is increased if necessary, $C_0T_0 \le \frac{1}{2}$ and $z' = (x', \xi')$. Furthermore, $R_1 = R + |\nabla_{x,\xi}R|^2$ also satisfies (5.4.14) and therefore R_1 satisfies (5.4.15):

$$(5.4.15)' \qquad (1 - C_0T_0) \le \frac{R_1(z'(t))}{R_1(z(0))} \le 1 + C_0T_0.$$

Then treating the Poisson brackets of a and $\partial_{z_k} R_1$ we see that if

$$\eta = |\nabla_{x',\xi'} R(z(0))| \ge \epsilon_7 (R(z(0)))^{\frac{1}{2}}$$

then

$$|H_R(z(t)) - H_R(z(0))| \le C_0|t|\eta, \quad R(z(t)) \le \epsilon_7\eta^2$$

and therefore

$$|\frac{dz'}{dt} + H_R(z'(0))| \le C_1 t\eta + C_1\eta^2$$

and then for $\eta \le \epsilon_8$, $|t| \le T_0$ with appropriate constants ϵ_8, T_0 the inequality

$$|z'(t) - z(0)| \ge \epsilon_8 \eta |t|$$

holds and the non-periodicity condition is fulfilled because $\eta \ge \sqrt{R(z(0))}$ and we assumed that this quantity is not less than $h^{\sigma'}$.

Therefore we need to consider only points with

$$(5.4.16) \qquad |\nabla_{x',\xi'} R| \le \epsilon_7 (R(z(0)))^{\frac{1}{2}};$$

then this inequality remains true along the trajectory (maybe with the extra factor $(1 + C_0T_0)$ in the right-hand expression).

Let us note that the time between two consequent reflections is $\asymp \sqrt{R(z')}$. Therefore, it has the same order all along the trajectory and therefore the number of reflections is $\asymp \frac{T_0}{\sqrt{R(z')}}$ where we assume that $\sqrt{R(z')} \le C_0^{-1}T_0$.

Let us now consider the set $\Xi_0 \subset \Sigma$ of points z such that the inequality $\operatorname{dist}(\Phi_t(z), z) \ge |t|h^{\frac{1}{2}-\sigma'}$ fails for some $t \in [-T_0, T_0]$; for the set $\Lambda_0 \setminus \Xi_0$ we can apply theorem 5.3.4 with $\tilde{T}_0 = T_0\bar{\gamma}^{-\frac{1}{2}}$. It is easy to see that the $dxd\xi : da$-measure of Ξ_0 does not exceed the Lebesgue measure of the set

$Z_0 = \{z' \in T^*B(0, \frac{3}{4}) : R(z') \in [h^{\sigma''}, \epsilon_8]$, condition (5.4.16) is fulfilled and there exists $n \in \mathbb{Z}$: $|n| \le C_0 \frac{T_0}{\sqrt{R(z')}}$, $\operatorname{dist}(\Phi^n(z'), z') \le C_0^{-1} h^{\frac{1}{2}-\sigma'}\}$

where $\sigma' = 3\sigma''$ and Φ is *the boundary mapping* defined in the following manner: we lift the point $z' \in T^*Y$ to the point $(z', 0, \sqrt{R(z')}) = (x', \xi', x_1, \xi_1) \in \Sigma|_Y$, we consider the Hamiltonian curve passing from this point in the positive direction[13)] until it meets $\Sigma|_Y$ again and then we throw away x_1 and $\xi_1 < 0$ and obtain $(x', \xi') = \Phi(z)$.

Let us treat Φ and $D\Phi$ carefully. It is easy to see that the time t_1 until the first meeting with Σ_Y can be calculated approximately by the formula

13) We have taken $\xi_1 > 0$.

(5.1.17)
$$|t_1 - 2\frac{\sqrt{R}}{S_0}| \le C_0 R$$

where $S_0 = S|_{x_1=0}$. Furthermore, it is also easy to see that if we take a variation $\mathrm{d}z'$ of point z' then

(5.4.18)
$$|\mathrm{d}t_1 - (S_0\sqrt{R})^{-1}\mathrm{d}R| \le C_0 t_1 |\mathrm{d}z'|.$$

In fact, if we vary the initial point z' then the equality $\frac{dx_1}{dt} = 2\xi_1$ along the trajectory yields that $\frac{d(\mathrm{d}x_1)}{dt} = 2\mathrm{d}\xi_1 = 2\mathrm{d}\xi_1(0) + O(t|\mathrm{d}z'|)$ and hence $\mathrm{d}x_1(t) = 2t\mathrm{d}\xi_1(0) + O(t^2|\mathrm{d}z'|)$ because $\mathrm{d}x_1(0) = 0$; therefore $\mathrm{d}t_1 = -(\mathrm{d}x_1)(t_1)/\xi_1(t_1)$; moreover, it is easy to see that here $\xi_1(t_1) = -2\xi_1(0) + O(t_1^2)$, $t \asymp \xi_1(0)$; therefore

$$\mathrm{d}t_1 = 2t_1\frac{(\mathrm{d}\xi_1(0))}{\xi_1(0)} + O(t_1|\mathrm{d}z'| + t_1|\mathrm{d}\xi_1(0)|);$$

but $\xi_1(0) = \sqrt{R}$, $\mathrm{d}\xi_1(0) = \frac{1}{2\sqrt{R}}(\mathrm{d}R)$ and (4.5.16) yields that $|\xi_1(0)| \le |\mathrm{d}z'|$; taking (5.4.17) into account we obtain (5.4.18).

Inequalities (5.4.17) and (5.4.18) yield that

(5.4.19)
$$\mathrm{d}\Phi(z') = \mathrm{d}z' + t_1\mathrm{d}H'_R + (\mathrm{d}t_1)H'_R + (\mathrm{d}t_1)t_1\frac{d^2z}{dt^2} + O(t_1^2|\mathrm{d}z'|)$$

where H'_R is the Hamiltonian field generated by R and the second derivatives are calculated along Hamiltonian trajectories. This estimate and (5.4.16)–(5.4.18) yield the rougher estimate

(5.4.20)
$$|\mathrm{d}\Phi(z') - \mathrm{d}z' - t_1(z')F(z')\mathrm{d}z'| \le \epsilon t_1(z')|\mathrm{d}z'|$$

where F is the fundamental matrix (skew-Hessian) of R at the point z': $\mathrm{d}H'_R = F\mathrm{d}z'$ and $\epsilon > 0$ can be chosen arbitrarily small by the choice constants in (5.4.16) and ϵ_j.

Moreover, equation $\frac{dz'}{dt} = H'_R + x_1 H'_S$ along the Hamiltonian trajectories and (5.4.16) yield the rough estimate

(5.4.21)
$$\mathrm{dist}(\Phi^n(z'), z') \le \epsilon\sqrt{R(z')}$$

with the same type of ϵ as before; then (5.4.20) and (5.4.18) yield that for $z'' = \Phi^n(z')$

$$|\mathrm{d}\Phi(z'') - \mathrm{d}z'' - t_1(z')F(z')\mathrm{d}z''| \le \epsilon\sqrt{R(z')}|\mathrm{d}z''|,$$

i.e., $D\Phi(z'') \equiv I + t_1(z')F(z')$ modulo a matrix with norm not exceeding $\epsilon\sqrt{R(z')}$. Taking the product $\prod_{k=n}^{1} D\Phi(z'_k)$ with $z'_k = \Phi^{k-1}(z')$ we obtain that

(5.4.22)
$$|D\Phi^n(z') - I - nt_1(z')F(z')| \le \epsilon|n|t_1(z').$$

Let us represent the set Z_0 as union of sets Z_{0n} with $|n| \le C_0 h^{-\sigma'}$. Let us recall that on the set in question $|n|t_1(z') \le 1$ and $t_1(z') \ge C_0^{-1}h^{\sigma'}$. It is easy to prove by explicit calculation that $|F(z')| \ge C_0^{-1}$. Then (5.4.22) yield that $\mathrm{mes}\, Z_{0n} \le C_0 h^{\frac{1}{2}-3\sigma'}$. Therefore $\mathrm{mes}\, Z_0 \le C_0 h^{\frac{1}{2}-4\sigma'}$ and hence the $dxd\xi : da$-measure of Ξ_0 does not exceed $C_0 h^{\frac{1}{2}-4\sigma'}$. Then taking $\bar{T} = 1$ for Ξ_0 in theorem 5.3.4 we obtain that for appropriate σ' the contribution of Ξ_0 to the remainder estimate does not exceed $Ch^{1-d+\sigma'}$. Proposition 5.4.3 is proven. □

Remark. *Surely we were forced to apply the results of section 5.3 with $L \geq 1$ at every step; in order to avoid this under conditions (5.1.57) or* (5.1.58)$_{1,2}$ *we could consider $\mathcal{R}$ with $\tau_1 = -\infty$ instead of $\mathcal{R}_2$; the same is true under condition* (5.1.57)$'$ *but only if we apply operator Q'_{1x} to $\psi_1 e \psi_2$ before we apply Γ where $Q'_1 = \mathrm{Op}(q'_1)$, $q'_1 \in S_{h,K}$ is supported in $\{|\xi'| \leq c_0\}$. Therefore we can go from $\mathcal{R}$ to $\mathcal{R}_2$ with $L \leq \epsilon = \epsilon(d, c, \epsilon_0)$ and then with arbitrary $L \leq C_0$ by theorem 5.3.1. In order to omit the cutting operator Q'_1 in the case of condition* (5.1.57)$'$ *(and $d = 2$) it is sufficient to note that if we replace Q'_1 by $Q'_2 = \mathrm{Op}(q'_2)$ with $q'_2 \in S'_{h,K}$ supported in $\{|\xi'| \geq \frac{1}{2}c_0\}$ then for a large enough c_0 the microhyperbolicity condition is fulfilled and we have no need to make dilatations and treat $L \geq 1$.* ◊

The proof of proposition 5.4.3 implies

Proposition 5.4.4. *In the context of proposition 5.4.3 estimates* (5.3.28) *and* (5.3.29) *hold for $L \in [h^{\frac{2}{3}-\sigma}, 1]$.*

Proof. In fact, for every element of the covering with $\bar{\gamma} \geq h^{\frac{2}{3}-\sigma}$ one can take $L \in [\bar{\gamma} h^{-\sigma}(h\bar{\gamma}^{-\frac{3}{2}}), 1]$, i.e., one can take $L \in [h^{1-\sigma}\bar{\gamma}^{-\frac{1}{2}}, 1] \supset [h^{\frac{2}{3}-2\sigma}, 1]$; on the other hand, for $\gamma \geq h^{\frac{2}{3}-\sigma}$ one can take $L \in [\bar{\gamma} h^{-}\sigma, 1] \supset [h^{\frac{2}{3}-2\sigma}, 1]$ where we have applied theorems 5.3.11(iv) and 5.3.10 respectively. □

At the following step we throw away condition (5.4.5).

Theorem 5.4.5. *Suppose that conditions* (4.5.1), (4.5.6)$_{1_3}$, (4.5.7), (5.1.3), (5.1.4)$_2$, (5.1.9), (5.1.55) *and either* (5.1.57) *((5.1.57)$'$ for $d = 2$) or* (5.1.58)$_{1,2}$ *are fulfilled. Let us assume that $d \geq 2$. Then*

(i) Estimates (5.3.30) *and* (5.3.31) *hold.*

(ii) Under the additional conditions of theorem 5.3.4 the estimates of this theorem remain true for $q_1 = q_2 = 1$ with no microhyperbolicity condition;

(iii) Moreover under conditions (5.1.57) *or* (5.1.58)$_{1,2}$ *one can take $\tau_1 = -\infty$ in all the estimates;*

(iv) Estimates (5.3.24) *and* (5.3.25) *hold for $L \in [h^{\frac{2}{3}-\sigma}, 1]$.*

Proof. Let us introduce the function

$$\gamma(x) = \epsilon_0(|\nabla_x V|^2 + |V|)^{\frac{1}{2}} + h^{\frac{1}{2}-\sigma'} \tag{5.4.23}$$

with $\sigma' > 0$ and let us consider the corresponding partition of unity; we assume that $\tau = 0$. Then after the dilatation $x_{\text{new}} = \frac{x-\bar{x}}{\bar{\gamma}}$, $h_{\text{eff}} = h\bar{\gamma}^{-2} \leq h^{\sigma}$ with $\bar{\gamma} = \gamma(\bar{x})$ and division of the operator A by $\bar{\gamma}^2$ we obtain the following cases:

(a) If $\bar{x}_1 \geq 2\bar{\gamma}$ then we are in the situation of section 4.4. Then the contribution of every ball to the remainder estimate $\mathcal{R}_2$ does not exceed Ch_{eff}^{1-d} and after summation with respect to all balls of this type we see that their total contribution does not exceed Ch^{1-d}. Moreover, let us note that under the additional conditions of theorem 5.3.4 the Hamiltonian trajectories of $\tilde{a}(x_{\text{new}}, \xi_{\text{new}}) = \bar{\gamma}^{-2} a(x_{\text{new}}\bar{\gamma} + \bar{x}, \xi_{\text{new}}\bar{\gamma})$ have the same equations as Hamiltonian trajectories of $a(x, \xi)$; surely we need to return to coordinates (x, ξ) in the first case), i.e., that now $\tilde{T}_j = T_j$; also noting that

$$h_{\text{eff}}^{1-d} dx_{\text{new}} d\xi_{\text{new}} : d\tilde{a} = h^{1-d} dx d\xi : da$$

we obtain that the total contribution of balls of this type to the remainder estimate does not exceed the right-hand expression of (5.3.17).

(b) If $\bar{\gamma} \geq 2h^{\frac{1}{2}-\sigma'}$ and $B(\bar{x}, \bar{\gamma}) \cap Y \neq \emptyset$ then we are in the situation of proposition 5.4.3 provided $|\beta(\bar{x})| \leq C_0\bar{\gamma}$; in this case all the arguments of (a) are applicable.

On the other hand, if condition (5.1.57) ((5.1.57)$'$ for $d=2$) is fulfilled and if $|\beta(\bar{x})| \geq C_0\bar{\gamma}$ then all the previous arguments are applicable in order to obtain the remainder estimate for the Weylian semiclassical approximation to expression (5.4.9). Applying the additional arguments of part (d) of the proof of proposition 5.4.3 we obtain that the total contribution to the remainder of the balls of this type can be properly estimated also.

(c) We need to consider only balls with $\bar{\gamma} \leq 2h^{\frac{1}{2}-\sigma'}$ intersecting Y. In this case if either $|\beta(\bar{x})| \leq C_0\bar{\gamma}$ or condition $(5.1.58)_{1,2}$ is fulfilled then we can apply theorem 5.1.6; then the contribution of this ball to the remainder estimate does not exceed $Ch_{\text{eff}}^{-d} = Ch^{-d}\bar{\gamma}^{2d}$ and the total number of these balls does not exceed $C\bar{\gamma}^{1-d}$ because the volume of each ball is greater than $C_0^{-1}\bar{\gamma}^d$ and the volume of the covered domain is less than $C_0\bar{\gamma}$ (and the covering is of finite multiplicity); then the total contribution of balls of this type does not exceed $Ch^{-d}\bar{\gamma}^{d+1} \leq Ch^{1-d+\sigma'}$ for appropriate σ'.

On the other hand, if condition (5.1.57) ((5.1.57)$'$ for $d=2$) is fulfilled and $|\beta(\bar{x})| \geq C_0\bar{\gamma}$ then the additional arguments of part (d) of the proof of proposition 5.4.3 give the same result.

So assertions (i)–(iii) have been proven; moreover the remark on the proof of proposition 5.4.3 remains true. Let us note that all the special difficulties in the proof of (ii) in the case $d=2$ were overcome when we proved theorem 4.3.13.

The proof of assertion (iv) is standard and it is based on proposition 5.4.4: if $\bar{\gamma} \geq 2h^{\frac{1}{2}-\sigma}$ we can take $L \in [\bar{\gamma}^2(h\bar{\gamma}^{-2})^{\frac{2}{3}-\sigma}, 1] \supset [h^{\frac{2}{3}-\sigma}, 1]$ and if $\bar{\gamma} \leq 2h^{\frac{1}{2}-\sigma}$ we can take $L \in [\bar{\gamma}^2, 1]$. □

5.4.3 Estimates in the elliptic zone

Let us now consider the elliptic zone of the operator $\tau - A$. In this case the ellipticity or microhyperbolicity of problem $(\tau - A, B)$ is equivalent to the same property of the symbol

$$\tilde{b} = \alpha(x')(a' - \tau)^{\frac{1}{2}} + \mu + \beta \tag{5.4.24}$$

where $a'(x', \xi') = \sum_{j,k} g^{jk}(\xi_j - V_j)(\xi_k - V_k) + V$, $\mu = \sum_j m^j\xi_j$, we sum with respect to $j, k = 2, \dots, d$ and we assume that $g^{1k} = \delta_{1k}$; let us recall that the elliptic zone is given by the inequality

$$a'(x', \xi') \geq \tau + \epsilon \tag{5.4.25}$$

with a constant $\epsilon > 0$.

Let us note that condition $(5.1.58)_{1,2}$ yields that $\tilde{b}$ is elliptic everywhere and hence we are not interested in this case.

If condition (5.1.57) is fulfilled then $\tilde{b}$ is elliptic for $\tau \leq -C_0$ but this assertion is wrong if $d=2$ and (5.1.57)$'$ holds but (5.1.57) is violated. In all these cases $\tilde{b}$ is elliptic for $|\tau| \leq c$ and $|\xi'| \geq C_0$ with arbitrary c and appropriate C_0 depending on c.

I would like to give an analysis in the elliptic zone (of $\tau - A$) which is independent (to a certain extent) of the previous arguments. We assume that $|\xi'| \leq C_0$ in the zone in question. Let us consider $U^{0\pm}$ first; since

$$(hD_t - A)U^{0\pm} = \mp ih\delta(t)\delta(x-y),$$

in this zone

$$U^{0\pm} \equiv \mp ihG\delta(t)\delta(x-y)$$

where G is the parametrix of the elliptic operator $hD_t - A$ and hence G is an h-pseudo-differential operator; therefore

(5.4.26)
$$U^{0\pm} \equiv \mp ih(2\pi h)^{-d-1}\iint G(x,\xi,\tau,h)e^{ih^{-1}(\langle x-y,\xi\rangle+it\tau)}d\xi d\tau$$

where $G(x,\xi,\tau,0) = (\tau - a(x,\xi))^{-1}$ and $G(x,\xi,\tau,h) = -\xi_1^{-2} + O(\xi_1^{-4} + |\xi_1|^{-3}h)$ as $|\xi_1| \to \infty$ and $|x| + |\xi'| + |\tau|$ is bounded.

Therefore taking $x_1 = y_1 = 0$ we obtain that

(5.4.27)
$$\eth_x\eth_y U^{0\pm} \equiv \mp i(2\pi h)^{-d}\iint G_0(x',\xi',\tau,h)e^{ih^{-1}(\langle x'-y',\xi'\rangle+it\tau)}d\xi' d\tau$$

where $G_0 \in S'_{h,K}$ and, moreover,

(5.4.28)
$$G_0(x',\xi',\tau,0) = -\frac{1}{2}(a'(x',\xi')-\tau)^{-\frac{1}{2}}.$$

On the other hand, taking $y_1 = 0$ in (5.4.26), applying $\eth_x hD_{x_1}$ and taking the limit as $y_1 \to +0$ we obtain that

(5.4.29)
$$\eth_y\eth_x hD_{x_1}U^{0\pm} \equiv \mp i(2\pi h)^{-d}\iint G_1(x',\xi',\tau,h)e^{ih^{-1}(\langle x'-y',\xi'\rangle+it\tau)}d\xi' d\tau$$

where $G_1 \in S'_{h,K}$ and, moreover,

(5.4.30)
$$G_1(x',\xi',\tau,0) = \frac{i}{2}.$$

Finally, (5.4.27) and (5.4.29) yield that

(5.4.31)
$$\eth_y\eth_x B_x U^{0\pm} \equiv \mp i\delta(x'-y')\delta(t)^t S_y$$

where $S \in \boldsymbol{\Psi}'_{h,K}$ and its principal symbol equals

(5.4.32)
$$S^0 = \frac{1}{2}\tilde{b}'(x',\xi')(a'(x',\xi')-\tau)^{-\frac{1}{2}}$$
$$\text{with}\quad \tilde{b}'(x',\xi') = -\alpha(a(x',\xi')-\tau)^{\frac{1}{2}} + \mu + \beta.$$

Let us consider $U^{1\pm}$; we know that

$$(hD_t - A)U^{1\pm} = 0.$$

Therefore reducing to the first-order 2×2-system and applying proposition 3.1.7 we obtain that

(5.4.33)
$$\eth_x hD_{x_1}U^{1\pm} \equiv i\eth_x G_2 U^{1\pm}$$

where $G_2 \in \boldsymbol{\Psi}'_{h,K}$ and its principal symbol equals

$$G_{20} = (a'(x',\xi')-\tau)^{-\frac{1}{2}}.$$

Now (5.4.29), (5.4.32) and the equality $\eth_x B_x(U^{0\pm}+U^{1\pm}) = 0$ imply that in the elliptic zone intersected with $\{|\xi'| \le C_0\} \cap B(0,1-\epsilon)\times\mathbb{R}^d$

(5.4.34)
$$\mathcal{B}_1 V^{\pm} \equiv h \mp i\delta(t)\delta(x'-y')^t S_y$$

where $V^{\pm} = h\eth_x\eth_y\psi(y)U^{1\pm}$, $\psi \in C_0^K(B(0,1-\frac{1}{3}\epsilon))$ equals 1 in $B(0,1-\frac{1}{2}\epsilon)$ and $\mathcal{B}_1 \in \boldsymbol{\Psi}'_{h,K}$ has principal symbol

$$\mathcal{B}_{10} = (a'(x',\xi') - \tau)^{1/2} + \mu(x',\xi') + \beta(x')$$

where we assume that $\alpha = 1$.[14] Applying an h-pseudo-differential operator with principal symbol

$$-(a'(x',\xi') - \tau)^{\frac{1}{2}} + \mu(x',\xi') + \beta(x')$$

to (5.4.34) we obtain the equation

$$\Big(hD_t - \mathcal{A}^0(x', hD_{x'}, h) - h\mathcal{A}^1(x', hD_t, hD_{x'}, h)\Big)V^{\pm} \equiv \mp ih\delta(t)\delta(x'-y')^t S_{1y} \tag{5.4.35}$$

where $\mathcal{A}^0, \mathcal{A}^1, S_1 \in \Psi'_{h,K}$, the principal symbol of $\mathcal{A}^0$ is equal to

$$\mathcal{A}^0_0 = a(x',\xi') - (\mu(x',\xi') + \beta(x'))^2 \tag{5.4.36}$$

and the principal symbol of S_1 is equal to

$$S_{10} = \frac{1}{2}(a'(x',\xi') - \tau)^{-\frac{1}{2}}(-(a'(x',\xi') - \tau)^{\frac{1}{2}} + \mu(x',\xi') + \beta(x'))^2. \tag{5.4.37}$$

Rewriting $\mathcal{A}^1(x', hD_t, hD_{x'}, h)$ in the form

$$\mathcal{A}^1 = \mathcal{A}^{1\prime}(x', hD_{x'}, h) + \mathcal{A}^{1\prime}(x', hD_t, hD_{x'}, h)(hD_t - \mathcal{A}^0(x', hD_{x'}, h)) + \mathcal{A}^{1\prime\prime\prime}(x', hD_t, hD_{x'}, h),$$

applying the operator $(I + h\mathcal{A}^{1\prime\prime})$ to (5.4.35) and including $\mathcal{A}^{1\prime}$ in $\mathcal{A}^0$ we obtain an equation of the same type but with the term $h^2\mathcal{A}^2$ instead of $\mathcal{A}^1$; here the operator S_1 changes but its principal symbol stays the same. Repeating the same arguments we obtain $h^s\mathcal{A}^s$ instead of $h\mathcal{A}^1$, i.e., we obtain that

$$(hD_t - \mathcal{A}^0(x', hD_{x'}, h))V^{\pm} \equiv \mp ih\delta(t)\delta(x'-y')^t S_{1y} \tag{5.4.38}$$

where the principal symbols of $\mathcal{A}^0$ and S_1 are given by (5.4.36) and (5.4.37).

On the other hand, proposition 3.2.5 implies that for $\psi_1 \in C_0^K(B(0, 1-\frac{1}{3}\epsilon))$ equal to 1 in $B(0, 1-\frac{1}{2}\epsilon)$

$$\int \psi_1(x) U^{1\pm}(x_1, x_1, x', y', t)\,dx_1 \equiv G_3(x', y', hD_{x'}, hD_{y'}, hD_t, h)V^{\pm} \tag{5.4.39}$$

where $G_3 \in \Psi'_{h,K}$ has principal symbol

$$G_{30} = ((a'(x',\xi') - \tau)^{\frac{1}{2}} + (a'(y',\eta') - \tau)^{\frac{1}{2}})^{-1}; \tag{5.4.40}$$

if we replace $U^{1\pm}$ by U^0 in (5.4.39) we obtain $\equiv 0$ because $U^0 \equiv 0$ in the elliptic zone. Let us recall that $U^{1\pm}$ is the Schwartz kernel of the operator with operator norm not exceeding C for every t. Moreover, equation $(hD_t - A)U^{1\pm} = 0$ and the corresponding equation with respect to y yield that $(hD_{x_1})^j(hD_y)^k U^{1\pm}$ are also the Schwartz kernels of the operators with operator norms not exceeding C for every t multiplied by an h-differential operator with respect to t, x', y' provided $j \le s, k \le s$. Therefore the embedding theorem and the presence of the factor h in the definition of $V^{\pm}$ yield that $V^{\pm}$ are the Schwartz kernels of operators with operator norms not exceeding C multiplied by an h-differential operator $\mathcal{E}$ acting with respect to t, x', y'.

Therefore we are in the situation of Chapter 4; it is easy to see that the presence of the additional operator factors S_1 (acting with respect to t, y'), G_3 and $\mathcal{E}$ (acting with respect to t, x', y') is not essential and therefore the following assertions hold:

---

[14] (5.1.57) or (5.1.57)′ imply that we can arrange this without loss of generality.

Theorem 5.4.6. *Suppose that conditions* (4.5.1), $(4.5.6)_{1-3}$, (4.5.7), (5.1.3) ,(5.1.9), (5.1.55) *and* (5.1.57) *((5.1.57)′ for d = 2) are fulfilled. Let* $\alpha = 1$, $Q_j = \mathrm{Op}(q_j)$, $q_j \in S'_{h,K}$ *be supported in* $\{a'(x',\xi') \ge \bar\tau + \epsilon\}$ *with* $\epsilon > 0$,[15] $\psi_j \in C_0^K(B(0,\frac{1}{2}))$, $0 \le \psi_j \le 1$. *Then*

(i) For d = 2 the following estimates hold:

$$\mathcal{R}_3 \le Ch^{-\frac{1}{2}} \qquad \forall \tau_1,\tau_2 \le \bar\tau : |\tau_j| \le c_1 (j=1,2), \tag{5.4.40}$$

$$\mathcal{R}_4 \le Ch^{-1/2} \qquad \forall \tau \le \bar\tau : |\tau| \le c_1; \tag{5.4.41}$$

(ii) For d ≥ 3 with no additional restriction the following estimates hold:

$$\mathcal{R}_3 \le Ch^{2-d} \qquad \forall \tau_1,\tau_2 \le \bar\tau; |\tau_j| \le c_1 (j=1,2), \tag{5.4.45}$$

$$\mathcal{R}_4 \le Ch^{d-2} \qquad \forall \tau \le \bar\tau : |\tau| \le c_1; \tag{5.4.46}$$

moreover, the same estimates hold for d = 2 provided at every point of $\operatorname{supp} q_1 \cup \operatorname{supp} q_2$ *either condition* (b_1), *or* (b_2), *or* (b_3) *or*

$$|\partial_{x_2} W| + |\partial^2_{x_2} W| \ge \epsilon_0 \tag{b_4}$$

is fulfilled and $\tau_j, \tau \in [\bar\tau - \epsilon', \bar\tau + \epsilon']$ *where* $\epsilon' = \epsilon'(d,c,\epsilon) > 0$ *is small enough;*

(iii) Moreover, if condition (5.1.57) *is fulfilled then all these estimates remain true if we take* $\tau_1 = -\infty$ *in* $\mathcal{R}_3$; *if condition* (5.1.57)′ *is fulfilled then one can do the same provided* q_j *are supported in* $\{|\xi'| \le c_1\}$ *for* $j = 1,2$;

(iv) Estimate (5.3.24) *remains true for* $L \in [h^{1-\delta}, 1]$; *the same is true for* (5.3.25) *provided* (5.1.57) *is fulfilled.*

Remark 5.4.7. *(i) This theorem provides an analysis of all the previous subsections.*

(ii) For $q_1 = q_2 = 1$ *the leading term in the asymptotics is* $\varkappa_1(\tau_1,\tau_2)h^{1-d}$ *with*

$$\varkappa_1 = -(2\pi)^{-d} \iint_{\tau_1 \le a' - (\mu+\beta)^2 \le \tau_2, \beta < 0} (\mu+\beta)\psi_1\psi_2 \, dx' d\xi' \tag{5.4.47}$$

for $\tau_1 < \tau_2 \le \bar\tau$.

(iii) Under some conditions on the Hamiltonian trajectories of $a' - (\mu+\beta)^2$ *one can obtain two-term asymptotics; however one should keep in mind that these trajectories can leave the elliptic zone, so it is necessary either to forbid this or to use the complete analysis of propagation of singularities in the hyperbolic zone. In the latter case one can only obtain results similar to theorem 5.3.5; it is not very likely that results similar to theorem 5.3.4 can be obtained.*

(iv) The arguments of this section yield that all the results of section 5.3 for Riesz means remains true for the Schrödinger operator with no microhyperbolicity condition (see remarks 5.3.3, 5.3.6 and 5.3.9). ◊

Appendix D. On the Structure of Certain Sets Associated with the Microhyperbolicity Condition

In this section we study the structure of certain sets associated with the microhyperbolicity condition; namely let $\sigma(x,\xi)$ be the set of all the eigenvalues of $J^{-1}(x)a(x,\xi)$, $\Sigma_\tau = \{(x,\xi) \in T^*\bar X, \tau \in \sigma(x,\xi)\}$, $\Sigma'_{\tau b} = \iota\Sigma_\tau$ where $\iota : T^*\bar X|_Y \to T^*Y$ is a natural map,

[15] For appropriate $\bar\tau$ one can take $Q_j = I$ for $\tau \cdots \le \bar\tau$.

$$\Xi_0 = \overline{\bigcup_{T^*\bar{X}} \sigma(x,\xi)}.$$

Let $(x',\xi') \in T^*Y$ and let $\sigma_d(x',\xi')$ be the discrete spectrum of the problem

$$(a(x',D_1,\xi') - \tau J(x'))v = 0, \tag{D.1}$$

$$\eth b(x',D_1,\xi')v = 0, \tag{D.2}$$

$$v = o(1) \qquad \text{as } x_1 \to +\infty; \tag{D.3}$$

let

$$\Sigma''_{\tau b} = \{(x',\xi') \in T^*Y, \tau \in \sigma_d(x',\xi')\}, \qquad \Xi_1 = \overline{\bigcup_{T^*Y} \sigma_d(x,\xi)}$$

where we omit $x_1 = 0$ in the notation for a, J, $\Sigma_{\tau b} = \Sigma'_{\tau b} \cup \Sigma''_{\tau b}$.

Moreover, let us introduce the sets $\mathcal{Y}_0 = \{\tau \in \mathbb{R}, \Sigma_\tau$ contains at least one point at which the symbol $\tau J(x) - a(x,\xi)$ is microhyperbolic in no direction $\mathcal{T} \in T(T^*\bar{X})\}$, and $\mathcal{Y}_1 = \{\tau \in \mathbb{R}$ and $\Sigma_{\tau b}$ contains at least one point at which the problem $(\tau J(x') - a(x',D_1,\xi'), \eth b(x',D_1,\xi'))$ is microhyperbolic in no multidirection$\}$.

Finally, let $\mathcal{Z}_0$ be the set of the points $(x,\tau) \in \mathring{X} \times \mathbb{R}$ such that for some $\xi \in T^*_x\mathring{X}$ the symbol $\tau J(x) - a(x,\xi)$ is microhyperbolic at (x,ξ) in no direction $\mathcal{T}$ with $\mathcal{T}_x = 0$; let $\mathcal{Z}_1$ be the set of points $(x',\tau) \in Y \times \mathbb{R}$ such that for some $\xi' \in T^*_{x'}Y$ the problem $(\tau J(x') - a(x',D_1,\xi'), \eth b(x',D_1,\xi'))$ is microhyperbolic at (x',ξ') in no multidirection $\mathcal{T}$ with $\mathcal{T}'_x = 0$.

We assume that

(D.4) $J(x)$ and $a(x,\xi)$ are Hermitian (matrix-valued) symbols and $J(x)$ is non-degenerate,

(D.5) Either $J(x)$ or $a(x,\xi)$ is positive definite,

(D.6) The operator $a(x',D_1,\xi')$ with boundary condition $\eth b(x',D_1,\xi')v = 0$ is self-adjoint on $L^2(\mathbb{R}^+,\mathbb{H})$,

(D.7) Either $J(x')$ or operator $a(x',D_1,\xi')$ (with the indicated boundary condition) is positive definite,

(D.8) The coefficient of D_1^m in $a(x',D_1,\xi')$ is non-degenerate.

Here and in what follows $\mathring{X}$, Y and $\bar{X}$ mean the the previous domain, its boundary and its closure respectively intersected with $\overline{B(0,1)}$.

Obviously,

(D.9) Ξ_0 and Ξ_1 are closed sets.

Let us recall that if a symbol is microhyperbolic at some point in some direction then it is microhyperbolic in a neighborhood of this point in all the directions close to original one. Similarly, if the problem is microhyperbolic at some point in some multidirection then it is microhyperbolic in a neighborhood of this point in all multidirections close to original one (see proposition 3.1.4). Hence

(D.10) $\mathcal{Y}_0$, $\mathcal{Y}_1$ are closed subsets of $\mathbb{R}$, $\mathcal{Z}_0$ is a closed subset of $\mathring{X} \times \mathbb{R}$, and both sides of $\mathcal{Z}_0 \cap (Y \times \mathbb{R}) \subset \mathcal{Z}_1$ are closed subsets of $Y \times \mathbb{R}$.

Our main goal is

Proposition D.1. *Suppose conditions* (D.4)–(D.8) *are fulfilled and let* $K = K(d, D)$ *in the smoothness conditions. Then*

(i) The Lebesgue measures of the sets $\mathcal{Y}_0$ *and* $\mathcal{Y}_1$ *equal* 0;

(ii) For every $x \in \mathring{X}$, $y \in Y$ *the Lebesgue measure of the set* $\{\tau, (x, \tau) \in \mathcal{Z}_0\}$ *equals* 0;

(iii) The Lebesgue measures of the sets $\mathcal{Z}_0$ *and* $\mathcal{Z}_1$ [16] *equal* 0;

(iv) The Lebesgue measure of the set $\Sigma''_{\tau b}$ *(in* T^*Y*) equals* 0 *provided* $\tau \notin \mathcal{Y} = \mathcal{Y}_0 \cup \mathcal{Y}_1$.

Proof. We will prove (i) here; assertion (ii) can be proven similarly. Obviously $\mathcal{Z}_0$ and $\mathcal{Z}_1$ are measurable sets and hence (iii) follows from (ii). In the proof of (iv) one can assume that $(\partial_{x_2}, \nu_1, \dots, \nu_M)$ is the multidirection of microhyperbolicity; then it follows from the proof below that for every $x'' = (x_3, \dots, x_d)$ and ξ' the Lebesgue measure of the set $\{x_2, (x_2, x'', \xi') \in S''_{\tau b}\}$ equals 0; this fact and the obvious fact that $\Sigma''_{\tau b}$ is measurable imply (iv).

So let us prove that $\operatorname{mes} \mathcal{Y}_1 = 0$ (the equality $\operatorname{mes} \mathcal{Y}_0 = 0$ can be proven in the same way but with no preliminary analysis; this equality follows from lemma D.2). Let $\varepsilon > 0$ and Ω be an open subset in $T^*Y \times \mathbb{R}$; let us introduce the set $\mathcal{Y}_1(\Omega, \varepsilon)$ as we defined $\mathcal{Y}_1$ but with $(z, \tau) \in \Omega$, $z = (x', \xi')$ and with condition (D.3) in the problem (D.1)–(D.3) replaced by

$$\text{(D.3)}' \qquad u = O(e^{-\varepsilon x_1}) \quad \text{as } x_1 \to +\infty;$$

then it is sufficient to prove that every point (z^*, τ^*) has a neighborhood $\Omega = \Omega(\varepsilon)$ such that $\operatorname{mes} \mathcal{Y}_1(\Omega, \varepsilon) = 0$. Without loss of generality one can assume that the polynomial $g(z, \zeta, \tau)$ has no root with imaginary part equal to ε for $(z, \tau) \in \Omega$.

Let us reduce problem (D.1),(D.2) to a first-order problem and let us replace $\underline{u}$ by $\underline{u}_{\text{new}} = Q\underline{u}$ as we did in subsection 3.1.2; then we obtain the problem $(\underline{p}, \mathrm{r}\underline{b})$ where

$$\underline{p} = \begin{pmatrix} \underline{p}_I & 0 & 0 \\ 0 & 0 & \underline{p}_{II} \\ O & \underline{p}^*_{II} & 0 \end{pmatrix}, \qquad \underline{b} = (\underline{b}_I, \underline{b}_{II}, \underline{p}_{III})$$

and all the roots of the polynomials $\det \underline{p}_I(z, \zeta, \tau)$ and $\det \underline{p}_{II}(z, \zeta, \tau)$ lie in the domains $\{|\operatorname{Im} \zeta| \le \varepsilon\}$, $\{|\operatorname{Im} \zeta| \ge \varepsilon\}$ of $\mathbb{C}$ respectively. Moreover let $\underline{b}' = (\underline{b}'_I, \underline{b}'_{II}, \underline{b}'_{III})$ be defined by the equality

$$\begin{aligned} &\langle \underline{p}(z, D_1, \tau)U, V\rangle_+ - \langle U, \underline{p}(z, D_1, \tau)V\rangle_+ = \\ &\qquad \langle \underline{b}'(z, \tau)U(0), \underline{b}(z, \tau)V(0)\rangle - \langle \underline{b}(z, \tau)U(0), \underline{b}'(z, \tau)V(0)\rangle; \end{aligned}$$

then $S = \underline{b}'^*_{II}\underline{b}_{II}$ is a Hermitian matrix.

The microhyperbolicity of $(\tau J(x') - a(x', D_1, \xi'), \mathrm{r}b(x', D_1, \xi'))$ means exactly that

$$\text{(D.11)} \quad \langle (T' + \nu_k \partial_\zeta)\underline{p}_I(z, \zeta, \tau)V, V\rangle < 0 \qquad \forall V \in \operatorname{Ker} \underline{p}_I(z, \zeta_k, \tau) \setminus O \quad \forall k = 1, \dots, M$$

---

16) In $\mathring{X} \times \mathbb{R}$ and $Y \times \mathbb{R}$ respectively.

and

(D.12) $\langle \mathcal{T}' S(z,\tau)V, V\rangle \; < 0$

$$\forall V \in \operatorname{Ker} \underline{b}(z,\tau) \setminus 0.$$

Moreaver, (D.11) and (D.12) are fulfilled for $\mathcal{T}' = \partial_\tau$ and $\nu_1 = \cdots = \nu_M = 0$.

Let us introduce

$$\mathcal{L}(z,\zeta_1,\ldots,\zeta_M,\tau) = \begin{pmatrix} \underline{p}_I(z,\zeta_1,\tau) & \ldots & 0 & 0 \\ \vdots & \ddots & \vdots & \vdots \\ 0 & \ldots & \underline{p}_I(z,\zeta_M,\tau) & 0 \\ 0 & \ldots & 0 & \underline{b}_{II}(z,\tau) \end{pmatrix},$$

$$\mathcal{M}(z,\tau) = \begin{pmatrix} I & \ldots & 0 & 0 \\ \vdots & \ddots & \vdots & \vdots \\ 0 & \ldots & I & 0 \\ 0 & \ldots & 0 & \underline{b}'^{*}_{II}(z,\tau) \end{pmatrix};$$

then the equality $\operatorname{mes} \mathcal{Y}_1(\Omega(\varepsilon),\varepsilon) = 0$ follows from □

Lemma D.2. *Let $\mathcal{L}(z,\tau)$ and $\mathcal{M}(z,\tau)$ be C^K matrices of dimensions $N \times D$ and $D \times N$ respectively for $(z,\tau) \in \Omega$ where $\Omega \subset \Omega_x \times \mathbb{R}$ and Ω_x are open domains. Let us assume that*

(D.13) *$\mathcal{ML}$ is a Hermitian matrix*

and for $\mathcal{T} = \partial_\tau$ the condition

$$\text{(D.14)} \qquad \langle \mathcal{T}(\mathcal{ML}(z,\tau)v, v\rangle < 0 \qquad \forall v \in \operatorname{Ker} \mathcal{L}(z,\tau) \setminus 0 \quad \forall (z,\tau) \in \Omega.$$

Let $\mathcal{N} = \{(z,\tau) :$ there exists no vector field $\mathcal{T} = \mathcal{T}_z \in T_x\Omega_x$ satisfying condition (D.14)$\}$. *Then*

$$\text{(D.15)} \qquad \operatorname{mes}\{\tau, \exists z : (z,\tau) \in \mathcal{N}\} = 0.$$

Proof. Let $\mathcal{X}_s = \{(z,\tau) : \operatorname{rank} \mathcal{L}(z,\tau) \le s\}$. Then it is sufficient to prove that

$$\text{(D.16)} \qquad \operatorname{mes}\{\tau, \exists z : (z,\tau) \in \mathcal{N} \cap (\mathcal{X}_s \setminus \mathcal{X}_{s-1}\} = 0.$$

Moreover, it is sufficient to treat the case $s = 0$. In fact, in the proof of (D.16) we can assume without loss of generality that $\mathcal{L} = \begin{pmatrix} \mathcal{L}' & 0 \\ 0 & \mathcal{L}'' \end{pmatrix}$ where $\mathcal{L}''$ is an invertible $s \times s$ matrix. This case can obviously be reduced to the case $s = 0$.

Let $t(z,\tau) = \operatorname{tr}(\mathcal{ML})(z,\tau)$; then condition (D.14) with $\mathcal{T} = \partial_\tau$ yields that $t(z,\tau) = 0$, $\partial_\tau t(z,\tau) > 0$ for $(z,\tau) \in \mathcal{X}_0$. Then without loss of generality one can assume that $\tau = \lambda(z)$ for $(z,\tau) \in \mathcal{X}_0$ where λ is a real-valued C^K function. Let $\alpha_1(z), \ldots, \alpha_p(z)$ be elements of matrix $S(z) = (\mathcal{ML})(z,\lambda(z))$ arbitrarily enumerated. Then

$$\mathcal{X}_0 = \{(z,\tau) : \tau = \lambda(z), \alpha_1(z) = \cdots = \alpha_p(z) = 0\}.$$

Let us note that for $(z,\tau) \in \mathcal{N} \cap \mathcal{X}_0$ and for arbitrary dz, $d\lambda$ is a linear combination of $d\alpha_1, \ldots, d\alpha_p$.

In fact, if that is not the case then there exists a direction $\mathcal{T} \in T_x\Omega_x$ such that $\mathcal{T}\lambda < 0$ and $\mathcal{T}\alpha_1 = \cdots = \mathcal{T}\alpha_p = 0$. Then

$$\mathcal{T}(\mathcal{ML})(z,\tau) = -\partial_\tau(\mathcal{ML})(z,\tau)\cdot\mathcal{T}\lambda(z)$$

and hence (D.14) holds for this direction because it holds for ∂_τ.

Therefore $\mathcal{N}\cap\mathcal{X}_0$ is contained in the union of sets $\mathcal{N}_\mathcal{K}$ for $\mathcal{K}\subset\{1,\dots,p\}$ where $\mathcal{N}_\mathcal{K} = \{\tau=\lambda(z), \alpha_j(z)=0\quad \forall j\in\mathcal{K},\ d\alpha_j\ (j\in\mathcal{K})$ are linearly independent and $d\lambda$ is their linear combination$\}$.

Then Sard's theorem yields that

$$\operatorname{mes}\{\tau, \exists z, (z,\tau)\in\mathcal{N}_\mathcal{K}\} = 0$$

and this equality yields (D.16) with $s=0$. □

The following assertion also seems to be of interest:

Proposition D.3. *Let conditions* (D.4)–(D.8) *be fulfilled. Then*

(i) If $\tau^*\in\partial\Xi_0$ and $z^*\in\Sigma_{\tau^*}$ then the symbol (τ^*J-a) is microhyperbolic in no direction at (z^*,τ^*);

(ii) If $\tau^*\in(\partial\Xi_1)\setminus\Xi_0$ and $z^*\in\Sigma''_{\tau^*b}$ then the problem $(\tau^*J-a(x',D_1,\xi'),\mathfrak{d}b)$ is microhyperbolic in no multidirection at (z^*,τ^*).

Proof. In the proof of (i) or (ii) let $\tau_1,\dots,\tau_M$ be continuous eigenvalues of the matrix $J^{-1}(x)a(x,\xi)$ (or those of the problem (D.1)–(D.3)) depending on $z\in T^*X$ or $z\in T^*Y$ equal to τ^* for $z=z^*$. Let $\Pi=\Pi(z)$ be the corresponding spectral projector in $\mathbb{H}$ or in the weighted space $L^2(\mathbb{R}^+,\mathbb{H},Jdx)$ respectively. Then both assertions will be proven if we prove (i) under the additional restriction $\tau^*J(z^*)-a(z^*)=0$.

Conditions (D.1)–(D.4) yield that τ_j are one time differentiable at z^* in every direction ([Friedrichs 1], e.g.) and hence it is sufficient to prove that if $\mathcal{T}$ is a microhyperbolicity direction then $\mathcal{T}\tau_j(z^*)\neq 0$ and therefore Ξ_0 contains a neighborhood of τ^*. But the equalities

$$\mathcal{T}(\tau_1+\cdots+\tau_M)(z^*) = \operatorname{tr}\mathcal{T}(J^{-1}a)(z^*) = -\operatorname{tr}J^{-1}(\mathcal{T}(\tau^*J-a))(z^*)$$

yield that if the matrix $(\mathcal{T}(\tau^*J-a))(z^*)$ is either positive or negative definite then $\mathcal{T}(\tau_1+\cdots+\tau_M)\neq 0$ provided J is positive definite; if a is positive definite we can permute a and J and use the same arguments. □

Comments

The results of section 5.1 are rather trivial, but the proofs given here were developed specially for this book. The construction of section 4.2 (the method of successive approximations) in its present form was suggested first in [Ivrii 6,7] for (pseudo-)differential operators (in fact in [6] the more complicated case of a domain with boundary was treated and namely the presence of the boundary was the reason that the traditional method did not work properly); the more sophisticated and general form which was used here was suggested in [Ivrii 8]; for h-(pseudo)-differential operators this construction and its justification is more transparent. The results of section 5.3 are now well known in certain special cases: for scalar operators for which $h=1$ and the large spectral parameter τ is considered (then this problem can be reduced to a semiclassical one with $h=|\tau|^{1/m}$ where m is the degree of the operator), asymptotics with remainder estimate $O(|\tau|^{(d-1)/m})$ are due to Seeley [1]; later by other methods they were obtained in

[Ivrii 6–8], [Melrose 1] in the case $m = 2$; for scalar operators with $m > 2$ and for matrix operators with constant multiplicities of the eigenvalues of the principal symbol these asymptotics were first obtained by [Vassiliev 1] and in the general case by [Ivrii 8]; we do not discuss asymptotics with less precise remainder estimate because they are outside of the domain of the method of hyperbolic operators and we only mention that in the paper of Babich and Levitan this estimate was obtained under the very restrictive condition that the domain is strictly concave in the corresponding metrics (which is fulfilled for no domain in $\mathbb{R}^d$ for the standard Laplacian) before it was obtained by Seeley in the general case. More precise asymptotics with remainder estimate $o(|\tau|^{(d-1)/m})$ were first obtained (under conditions on the Hamiltonian flow) by Ivrii [1] and in all the papers mentioned above excluding those of Seeley such asymptotics were also obtained; this estimate was obtained by Melrose simultaneously with Ivrii [6] but under the condition that the domain is strictly concave in the corresponding metric.

The method of dilatation which permitted the treatment of domains with some singularities of the boundary was first suggested in [Ivrii 11] and [Ivrii, Fedorova 1] and for h-differential operators it was first applied in [Ivrii 12] (we now discuss section 5.4 and we do not mention papers in which less precise remainder estimates were obtained). Another approach which works in slightly different situations was suggested by Vassiliev.

I am aware that these comments are very poor and inadequate. I send the reader to the excellent survey of Rozenblioum, Solomyak and Shubin [1] for more information.

There are many open problems. I list only those which seem to be the most important for me:

Problem 5.1.** *Asymptotics with remainder estimate $O(h^{d-1+\delta'})$ with positive δ' were obtained in section 5.3; the problem is to calculate this δ'.* $\Diamond$

Problem 5.2.** *Consider the operator $h^2\Delta + x_1$ and demonstrate that the estimates for $\mathcal{R}_x$ and $\mathcal{R}_{\vartheta,x}$ which are sometimes worse than $O(h^{1-d})$ and $O(h^{1-d+\vartheta})$ are in these cases precise and cannot be improved (unless additional terms are included in the asymptotics).* $\Diamond$

Problem 5.3.** *Develop the method of dilatation for operators different from the Schrödinger operator.* $\Diamond$

Some less difficult problems were suggested earlier in the chapter.

6. Local Spectral Asymptotics for the Schrödinger Operator with Strong Magnetic Field

In this chapter we analyze two- and three-dimensional Schrödinger operators with strong magnetic field; now we have not only a small parameter h but a large parameter μ (a coupling constant with the magnetic field or simply a magnetic parameter); moreover, a natural condition $\mu \le h^{-1}$ arises. In section 6.1, which has a preliminary character, we obtain a formula for $e(x, x, \tau)$ for operators in $\mathbb{R}^d$ with constant g^{jk}, V and F_{jk} (which is the tensor intensity of the magnetic field) and we also prove some preliminary assertions. In section 6.2 we obtain microlocal canonical forms for $d = 2, 3$ which are the basis for the advanced analysis. In section 6.3 we derive (quasi-)Weylian and related formulas for $d = 2, 3$ in the case of a weak magnetic field ($\mu \le h^{\delta-1}$) and in section 6.4 we derive non-Weylian formulas in the case of a strong magnetic field ($\mu \ge h^{-\delta}$); these formulas permit us to justify the (quasi-)Weylian formulas under weaker restrictions than those of section 6.3.

All these results are essentially different in the cases $d = 2$ and $d = 3$. For example, in $\mathbb{R}^d$ for constant g^{jk}, V and $(F_{jk}) \ne 0$ the spectrum of the Schrödinger operator for $d = 2$ is pure point of infinite multiplicity while it is absolutely continuous for $d = 3$. The remainder estimate in sections 6.3 and 6.4 in the case $d = 2$ under appropriate restrictions is $O(\mu^{-1}h^{-1})$ or even better (the best possible is $O(1)$); so it is better than in the standard case (without magnetic field); however, without such restrictions the remainder estimate is very bad ($O(\mu h^{-1})$) which can be derived from the results in the standard case. On the other hand, in the case $d = 3$ the remainder estimate is $O(h^{-2})$ provided $\mu \le h^{-1+\delta}$ (i.e., we obtain the same estimate as in the standard case); under appropriate restrictions the same estimate holds for $h^{-1+\delta} \le \mu \le h^{-1}$ but in the general case we obtain the remainder estimate $O(\mu h^{-1-\delta})$ which is not so good but in any case it is less than the principal part $\asymp h^{-3}$. So in the case $d = 2$ the remainder estimate is essentially more sensitive to additional restrictions.

In section 6.5 under additional restrictions of a global nature we treat the long-term propagation of singularities and justify the classical mechanics answer: in the three-dimensional case singularities propagate along magnetic lines while in the two-dimension case they propagate along drift lines (with average velocities $\asymp 1$ and $\asymp \nu\mu^{-1}$). This provides us with slightly improved remainder estimates under conditions on these lines including conditions on the growth of the Jacobi matrix elements, avoidance of the boundary and irrevocability of the movement.

In section 6.6 we treat the case of a strong magnetic field for $d = 2$ but we treat spectral gaps not containing the semiclassical approximation of the spectrum and we obtain complete spectral asymptotics (i.e., $O(h^s)$ with arbitrary s). Moreover for $d = 2, 3$ and $1 \le \mu \le h^{-1}$ we treat the lower spectral gap (from $-\infty$ to the bottom of the semiclassical approximation of the spectrum). Finally, in section 6.7 we apply the method of partition of unity and dilatation and we weaken the conditions on the assertions proven earlier.

6.1 Preliminary Analysis

6.1.1 The model operator

In this chapter we treat the Schrödinger operator

$$A = P_j g^{jk} P_k + V, \quad P_j = hD_j - \mu V_j \tag{6.1.1}$$

where we use the Einstein summation rule and $g^{jk} = g^{kj}$, V_j and V are real-valued and in $B(0,1) \subset X$ satisfy the following conditions:

$$|D^\alpha g^{jk}| \le c, \quad |D^\alpha V_j| \le c, \quad |D^\alpha V| \le c \qquad \forall \alpha : |\alpha| \le K, \tag*{(6.1.2)$_{1-3}$}$$

$$\epsilon_0 \le g^{jk}\eta_j\eta_k \cdot |\eta|^{-2} \le c \quad \forall \eta \in \mathbb{R}^d \setminus 0 \quad \forall x \in B(0,1); \tag{6.1.3}$$

we always assume that

$$X \supset B(0,1), \qquad h \in (0,1], \quad \mu \in [1,\infty]. \tag{6.1.4}$$

Let us note that $e(x,x,\tau)$ does not change if we replace A by the unitarily equivalent operator $e^{-i\mu h^{-1}\phi} A e^{i\mu h^{-1}\phi}$ with an arbitrary real-valued function $\phi(x)$ which means precisely that we replace V_j by $V_j + \partial_{x_j}\phi$ (*gradient transformation*) and $e(x,y,\tau)$ by $e(x,y,\tau)e^{-i\mu h^{-1}(\phi(y)-\phi(x))}$; therefore one can replace condition $(6.1.2)_2$ by

$$|D^\alpha F_{jk}| \le c \quad \forall \alpha : |\alpha| \le K-1 \tag*{(6.1.2)$'_2$}$$

where

$$F_{jk} = \partial_k V_j - \partial_j V_k \tag{6.1.5}$$

is the *tensor intensity* of the magnetic field.

Let us start with the model case in which $X = \mathbb{R}^d$, g^{jk}, V_j, V and F_{jk} are constant:

Theorem 6.1.1. *Let $X = \mathbb{R}^d$, g^{jk}, V_j, V and F_{jk} be real constants and let condition (6.1.3) be fulfilled. Then*

$$e(x,x,\tau) = h^{-d} S(\tau, \mu h) \tag{6.1.6}$$

where

$$S(\tau,\mu h) = (2\pi)^{r-d}\omega_{d-2r}\sqrt{g} \times \sum_{\alpha\in\mathbb{Z}^{+r}} \Bigl(\tau - \sum_{1\le j\le r}(2\alpha_j+1)\mu h f_j - V\Bigr)_+^{\frac{d-2r}{2}} f_1 \cdots f_r (\mu h)^r, \tag{6.1.7}$$

$2r = \operatorname{rank}(F_{jk})$, $f_j > 0$ and $\pm i f_j$ $(j = 1,\dots,r)$ are all the eigenvalues (counting multiplicities) of the matrix $F^l_j = g^{lj}F_{jk}$, $g = \det\bigl(g^{jk}\bigr)^{-1}$.

In particular, for $2r = d$ the spectrum of the operator A is pure point of infinite multiplicity and is given by the formula

$$\operatorname{Spec} A = \{ \sum_{1\le j\le r} (2\alpha_j+1)\mu h f_j, \quad \alpha \in \mathbb{Z}^{+r}\} \tag{6.1.8}$$

and in order to have equality (6.1.6) at every point $\tau \in \mathbb{R}$ we take the function τ_+^0 semicontinuous from the left in (6.1.7). On the other hand, if $2r < d$ then the spectrum of A is absolutely continuous and is given by the formula

$$\operatorname{Spec} A = [\mu h(f_1 + \cdots + f_r), \infty). \tag*{(6.1.7)$'$}$$

The numbers $(2\alpha_j+1)\mu h f_j$ are called *Landau levels.*

Corollary 6.1.2. *(i) Let $d=2$ and $F_{12}\neq 0$. Then the spectrum of A is pure point of infinite multiplicity, is equal to $\{(2j+1)\mu hF,\ j\in\mathbb{Z}^+\}$ and*

$$S(\tau,\mu h)=\#\{j\in\mathbb{Z}^+ : (2j+1)\mu hF+V\le\tau\}\cdot(2\pi)^{-1}\sqrt{g}(\mu h)F, \tag{6.1.7$_{(2)}$}$$

where $F=\frac{1}{\sqrt{g}}|F_{21}|$ is the scalar intensity *of the magnetic field;*

(ii) Let $d=3$ and $(F_{jk})\neq 0$. Then the spectrum of A is absolutely continuous, is equal to $[\mu hF,\infty]$ and

$$S(\tau,\mu h)=\sum_{0\le j\le\infty}\big(\tau-(2j+1)\mu hF\big)_+^{\frac12}\,\mu hF\cdot(2\pi)^{-2}\sqrt{g} \tag{6.1.7$_{(3)}$}$$

where $F=(g_{jk}F^jF^k)^{\frac12}$ is the scalar intensity of the magnetic field, *$F^j=\frac12\varepsilon^{jkl}F_{kl}$ is the* vector intensity of the magnetic field; *(ε^{jkl}) is the absolutely skew-symmetric tensor with $\varepsilon^{123}=1\sqrt{g}$, $(g_{jk})=(g^{jk})^{-1}$.*

Proof of theorem 6.1.1. It is easy to see that under a coordinate change S changes as a density; we know that $e(x,x,\tau)$ also changes as a density and therefore we can make a linear coordinate change by which we can arrange $g^{jk}=\delta_{jk}$; and then by an orthogonal change of coordinates we can reduce (F_{jk}) to the canonical form for skew-symmetric matrices: $F_{2l-1,2l}=f_l$, $F_{2l,2l-1}=-f_l$ for $l=1,\dots,r$ and all the other elements of F_{jk} vanish. Moreover, one can assume that V_j are linear functions, $V_{2l}=-f_lx_{2l-1}$, $l=1,\dots,r$ and all the other components of the vector potential vanish; we can obtain this by a gradient transform. Without loss of generality we can assume that $V=0$ (otherwise we replace V by 0 and τ by $\tau-V$), that $h=1$ (otherwise we divide A and τ by h^2, i.e., we replace h,μ,τ,V by 1, $\frac{\mu}{h}$, $\frac{\tau}{h^2}$ and $\frac{V}{h^2}$ respectively) and that $\mu=1$ (otherwise we include μ in V_j); our theorem is stable with respect to all these transformations. So we need to consider the operator

$$A=\sum_{1\le j\le r}\Big(D_{2j-1}^2+(D_{2j}+x_{2j-1}f_j)^2\Big)+\sum_{2r+1\le j\le d}D_j^2. \tag{6.1.9}$$

Let $x'=(x_1,x_3,\dots,x_{2r-1})$, $x''=(x_2,x_4,\dots,x_{2r})$ and $x'''=(x_{2r+1},\dots,x_d)$. Let us make the Fourier transformation $x''\to\xi'',x'''\to\xi'''$; then A will be transformed into the operator

$$A_I=\sum_{1\le j\le r}\Big(D_{2j-1}^2+(x_{2j-1}f_j+\xi_{2j})^2\Big)+|\xi'''|^2.$$

Let us introduce coordinates $\tilde{x}_{2j-1}=\frac{1}{\sqrt{f_j}}(x_{2j-1}f_j+\xi_{2j})$; then we obtain the operator

$$A_{II}=\sum_{1\le j\le r}f_j(D_{2j-1}^2+x_{2j-1}^2)+|\xi'''|^2.$$

The operators $(D_{2j-1}^2+x_{2j-1}^2)$ are harmonic oscillators with eigenvalues $2k+1$, $k\in\mathbb{Z}^+$ and with the corresponding eigenfunctions Υ_k which are Hermite functions.[1)] Then the Schwartz kernel of the spectral projector of A_{II} is

$$e_{II}(x',\xi'',\xi''';y',\eta'',\eta''';\tau)=\sum_\alpha\Upsilon_\alpha(x')\Upsilon_\alpha(y')\delta(\xi''-\eta'')\delta(\xi'''-\eta''')$$

[1)] We take orthonormal and real-valued Hermite functions.

where

$$\Upsilon_\alpha(x') = \prod_{1\le j\le r} \Upsilon_{\alpha_j}(x_{2j-1})$$

and we sum with respect to $\alpha \in \mathbb{Z}^{+r}$ such that

$$\sum_{1\le j\le r} (2\alpha_j + 1) f_j + |\xi'''|^2 < \tau. \tag{6.1.10}$$

Then the Schwartz kernel of the spectral projector of A_I is

$$e_I(x', \xi'', \xi'''; y', \eta'', \eta'''; \tau) = \sum_\alpha \Upsilon_\alpha(\tilde{x}')\Upsilon_\alpha(\tilde{y}')(f_1 \cdots f_r)^{\frac{1}{2}}\delta(\xi'' - \eta'')\delta(\xi''' - \eta''')$$

with $\tilde{x}_{2j-1} = \frac{1}{\sqrt{f_j}}(x_{2j-1}f_j + \xi_{2j})$ and therefore

$$e(x, y, \tau) = (2\pi)^{r-d}\sum_\alpha \iint \Upsilon_\alpha(\tilde{x}')\Upsilon_\alpha(\tilde{y}')(f_1 \cdots f_r)^{\frac{1}{2}} \times e^{i\langle x''-y'',\xi''\rangle + \langle x'''-y''',\xi'''\rangle} dx''d\xi'''$$

where we sum with respect to $\alpha \in \mathbb{Z}^{+r}$ and integrate with respect to $\xi'' \in \mathbb{R}^r$ and $\xi''' \in \mathbb{R}^{d-2r}$ such that (6.1.10) holds. Let us set $x = y$; going from integration with respect to ξ'' to integration with respect to $\tilde{x}'$ we obtain that

$$\int \Upsilon_\alpha(\tilde{x}')^2 d\xi'' = (f_1 \cdots f_r)^{\frac{1}{2}}$$

and therefore

$$e(x, x, \tau) = (2\pi)^{r-d} \sum_{\alpha\in(\mathbb{Z}^+)^r} \int d\xi''' \cdot f_1 \cdots f_r$$

where we sum with respect to $\alpha \in (\mathbb{Z}^+)^r$ and integrate with respect to $\xi'' \in \mathbb{R}^r$ and $\xi''' \in \mathbb{R}^{d-2r}$ such that (6.1.10) holds; this formula implies (6.1.6)–(6.1.7). □

Let us discuss a connection between (6.1.6)–(6.1.7) and the Weylian formula. Without loss of generality one can assume that $V = 0$. Let us note that the right-hand expression in (6.1.7) can be treated as a Riemann integral sum for the r-dimensional integral of the function $\varphi(z) = (2\pi)^{r-d}2^{-r}\omega_{d-2r}\sqrt{g}(\tau - f_1z_1 - \cdots - f_rz_r)_+^{\frac{d-2r}{2}} f_1\cdots f_r$ with respect to $z \in \mathbb{R}^{+r}$ with step $\varepsilon = 2\mu h$ on each variable z_j. Therefore as $\varepsilon \to 0$, $S(\tau, \varepsilon)$ tends to the corresponding integral which is equal to $S_0(\tau) = (2\pi)^{-d}\omega_d\tau_+^{\frac{d}{2}}$. The corresponding integral formula is accurate to second order but the deduction that $S(\tau, \varepsilon) = S_0(\tau) + O(\varepsilon^2)$ is not necessarily true because the function φ is not necessarily smooth enough. So we must treat this formula more carefully. Let us note that if $\rho(z) = \tau - (f_1z_1 + \cdots + f_rz_r) > \varepsilon$ for $z_j = (2\alpha_j + 1)\varepsilon$ for all j then the inequalities $\epsilon_0\rho(z) \le \rho(z') \le c\rho(z)$ hold for all points z, z' belonging to the box $\Omega_\alpha = \{(2\alpha_j - 1)\varepsilon \le z_j \le (2\alpha_j)\varepsilon \ \forall j\}$ and hence the error of our integral formula in this box is $O(\varepsilon^2 \int_{\Omega_\alpha} \rho(z)^{\frac{d-2r}{2}-2}dz)$; summing we obtain that the contribution of all boxes of this type to the error is $O(\varepsilon^2)$ for $d \ge 2r + 2$, $O(\varepsilon^{\frac{3}{2}})$ for $d = 2r + 1$ and $O(\varepsilon)$ for $d = 2r$. On all other boxes $|\varphi(z)| \le C\varepsilon^{\frac{d-2r}{2}}$ and the total volume of these boxes is $O(\varepsilon)$; hence we conclude that

(6.1.11) $$S(\tau,\varepsilon)=S_0(\tau)+O(\varepsilon^\nu)$$

with $\nu=1$ for $d=2r$, $\nu=\frac{3}{2}$ for $d=2r+1$, $\nu=2$ for $d=2r+2$.

Moreover, the error in the box with $\rho(z)\ge\varepsilon$ at each point is

$$c_0\int_{\Omega_\alpha}\sum_j \partial^2_{z_j}\varphi(z)dz+O(\varepsilon^4\int_{\Omega_\alpha}\rho(z)^{\frac{d-2r}{2}-4}dz)$$

with the absolute constant c_0 and therefore repeating the same arguments we obtain that

(6.1.12) $$S(\tau,\varepsilon)=S_0(\tau)+c_{d,r}\tau_+^{\frac{d}{2}-2}(f_1^2+\dots+f_r^2)\varepsilon^2+O(\varepsilon^\nu)$$

$\nu=4$ for $d\ge 2r+6$, $\nu=2+\frac{j}{2}$ for $d=2r+2+j$ with $j=1,2,3$

and with $c_{d,r}\ne 0$; hence for $d\ge 2r+3$ one can improve estimate (6.1.12) at no fixed point $\tau>0$. Moreover, we assert that for $d=2r+2$ one can improve estimate (6.1.12) on no fixed interval (τ_1,τ_2) with $0<\tau_1<\tau_2$. In fact, if one could improve this estimate on some interval then the improved error estimate would be true for $\tau=\tau^*=\frac{1}{2}(\tau_1+\tau_2)$ for function $\varphi_1=\varphi*\phi$ where $\phi\in C_0^K((-\delta,\delta))$ is a fixed function with $\delta=\frac{1}{2}(\tau_2-\tau_1)2$ and convolution is made with respect to τ. But φ_1 is a smooth function and hence for this function the error is

$$c_0\varepsilon^2\int_{\mathbb{R}^{+r}}\sum_j \partial_{z_j}\varphi_1(z)dz+O(\varepsilon^4);$$

here the coefficient of ε^2 is equal to

$$c_0(\sum_j f_j^2)\int_{\mathbb{R}^{+r}}\theta(\lambda-z_1-\dots-z_r)\phi'(\tau-\lambda)d\lambda=$$

$$c_{d,r}(\sum_j f_j^2)\int_0^\infty \lambda^r\phi'(\tau-\lambda)d\lambda$$

with $c_{d,r}\ne 0$ and hence it does not vanish for appropriate ϕ.

Let us now treat the case $d=2r+j$, $j=0,1$. Here the main source of error is the non-smoothness of φ. We treat the most interesting case $r=1$; for $r\ge 2$ the analysis and the answer are essentially more complicated and depend on the commensurability of the numbers $f_1,\dots,f_r$. Without loss of generality one can assume that $f=1$.

I assert that one can improve the estimates $O(\varepsilon)$ and $O(\varepsilon^{\frac{3}{2}})$ for $d=2$ and $d=3$ respectively on no fixed interval $[\tau_1,\tau_2]$ with $0<\tau_1<\tau_2$. For $d=2$ this obviously follows from formula $(6.1.7)_{(2)}$. For $d=3$ formula $(6.1.7)_{(3)}$ yields that for $\tau=(2n+1)\varepsilon\asymp 1$, $\delta>0$

$$S(\tau+\delta\varepsilon)\ge S(\tau,\varepsilon)+c_1\sqrt{\tau}\delta\varepsilon+c_2\varepsilon^{\frac{3}{2}}\sqrt{\delta}+O(\varepsilon^{\frac{3}{2}}\delta)$$

with positive constants c_1 and c_2; on the other hand, $S_0(\tau+\delta\varepsilon)=S_0(\tau)+c_1\sqrt{\tau}\delta\varepsilon+O(\varepsilon^2\delta^2)$ and therefore the error estimate $O(\varepsilon^{\frac{3}{2}})$ is the best possible in either τ or $\tau+\delta\varepsilon$.

I leave it to the reader to consider Riesz means.

6.1.2 Rough estimates

Let us now turn to functional analytic arguments.

Theorem 6.1.3. *Let (6.1.1)–(6.1.3) be fulfilled with $K>\frac{d}{2}$. Let $h\in(0,1]$, $\mu\ge 1$, let the operator A be self-adjoint and suppose $u\in C_0^2(B(0,1))$ belongs to $\mathrm{D}(A)$ if and only if $u|_{\partial X}=0$.*[2)] *Then*

2) In this theorem we do not assume that (6.1.4) is fulfilled.

$$(6.1.13)\quad |F_{t\to h^{-1}\tau}\chi_{T,T'}(t)U(x,y,t)| \le C'Th^{s}\mu^{-s}(|\tau|+1)^{-s}$$
$$\forall x,y\in X\cap B(0,1-\epsilon)\quad \forall\tau\le \inf_{X\cap B(0,1)}(V+\mu h\operatorname{Tr}^{+}(gF))-\epsilon(1+\mu h)$$

where $\operatorname{Tr}^{+}(gF)=\sum_l f_l$, $\pm i f_l$ *are all the eigenvalues of* $((gF)_j^k)$, $f_l>0$ *and* s, $\epsilon>0$, $T>\epsilon_0 h^{1-\delta}$, $\delta>0$, T' *are arbitrary and* $C'=C'(d,c,\delta,\epsilon,s)$.

Proof. Let us note that the proof of lemma 4.1.1 yields the estimate

$$(6.1.14)\quad \|F_{t\to h^{-1}\tau}\chi_{T,T'}(t)U\| \le C'Th^{-L}\mu^{L}(|\tau|+1)^{L}$$
$$\forall x\in X\cap B(0,1-\epsilon)\quad \forall\tau\in\mathbb{R}$$

where $L=\frac{d}{2}$ and in the left-hand expression the L^2-norm with respect to y is calculated; a similar estimate with permuted x and y; and the estimate

$$(6.1.15)\quad |F_{t\to h^{-1}\tau}\chi_{T,T'}(t)U(x,y,t)| \le C'Th^{-L}\mu^{L}(|\tau|+1)^{L}$$
$$\forall x,y\in X\cap B(0,1-\epsilon)\quad \forall\tau\in\mathbb{R}$$

with $L=d$; moreover, this proof yields similar estimates with $F_{t\to h^{-1}\tau}\chi_{T,T'}(t)U$ replaced by $e(x,y,-\infty,\tau)$. □

The key in the forthcoming arguments is

Lemma 6.1.4. *If the conditions of theorem 6.1.3 are fulfilled with* $K=2$ *then*

$$(6.1.16)\quad \epsilon\sum_j \|P_jw\|^2+\epsilon(\mu h+|\tau|+1)\|w\|^2\le$$
$$C'\|(\tau-A)w\|\cdot\|w\|+C'h\|w\|^2$$
$$\forall\tau\le \inf_{X\cap B(0,1)}(V+\mu h\operatorname{Tr}^{+}(gF))-\epsilon(\mu h+1)$$
$$\forall w\in C^2(X\cap B(0,1)):\ \operatorname{supp} w\subset B(0,1-\epsilon),\ w|_{\partial X}=0.$$

This lemma will be proven later; now let us finish the proof of theorem 6.1.3. Let $v\in \mathrm{D}(A)$ and $u=U(t)v$. Then the equation $(hD_t-A)u=0$, the boundary condition $u|_{\partial X\cap B(0,1)}=0$, lemma 6.1.4 and standard elliptic arguments with induction with respect to s yield the estimate

$$\sum_j \|P_jW\|+\|W\|\le C'Th^{s}\mu^{-s}(|\tau|+1)^{-s}\|v\|$$

for $W=F_{t\to h^{-1}\tau}\chi_{T,T'}(t)\psi u$ with an arbitrary fixed function $\psi\in C_0^\infty(B(0,1-\epsilon))$ where C' also depends on ψ; at every step the support of ψ is decreased but remains arbitrary. Then treating our equation and boundary condition again we conclude by induction with respect to $m=1,\dots,K$ that

$$\sum_{|\alpha|\le K}\|D^\alpha\psi W\|\le C'Th^{s}\mu^{-s}(|\tau|+1)^{-s}\|v\|$$

where s decreases at every step but remains arbitrary. Then the embedding inequalities yield that the same estimate holds for $|W(x,\tau)|$ with arbitrary $x\in X\cap B(0,1-\epsilon)$; since v is arbitrary we obtain that

$$\|F_{t\to h^{-1}\tau}\chi_{T,T'}(t)U(x,.,t)\|\le C'Th^{s}\mu^{-s}(|\tau|+1)^{-s}$$

for every $x \in X \cap B(0, 1-\epsilon)$ and for the indicated τ where in the left-hand expression the L^2-norm with respect to y is calculated. The previous arguments with y instead of x and with $u(y,t) = U(x,y,t)$ with fixed x, the dual equation and estimate (6.1.15) prove theorem. □

Proof of lemma 6.1.4. It is sufficient to prove that if all the assumptions of the lemma excluding the conditions on V are fulfilled and if

$$\inf(V + \mu h \operatorname{Tr}^+(gF)) \geq \epsilon(\mu h + 1) \tag{6.1.17}$$

then

$$(Aw, w) \geq C^{-1}\epsilon\bigl(\sum_j \|P_j w\|^2 + (\mu h + 1)\|w\|^2\bigr) - Ch\|w\|^2. \tag{6.1.18}$$

It is sufficient to prove this inequality under the additional assumption that w is supported in $B(\bar{x}, \varepsilon)$ with an arbitrarily small constant $\varepsilon > 0$. In fact, let ψ_ν^2 be an ε-admissible partition of unity; then

$$|(Aw, w) - \sum_\nu (A\psi_\nu w, \psi_\nu w)|$$

and

$$|\|P_j w\|^2 - \sum_\nu \|P_j \psi_\nu w\|^2|$$

do not exceed $Ch\varepsilon^{-1}\|w\|^2$ and therefore estimate (6.1.18) for w follows from this estimate for $\psi_\nu w$.

Without loss of generality one can assume that $g^{jk} = \delta_{jk}$ at $\bar{x}$. Let $G = (G^{jk})$ be a real skew-symmetric matrix with operator norm (in $\mathcal{L}(\mathbb{R}^d, \mathbb{R}^d)$) equal to 1. Then

$$\sum_j \|P_j w\|^2 \geq |(G^{jk} P_j w, P_k w)| = |(G^{jk}[P_k, P_j]w, w)| = \frac{1}{2}\mu h |(G^{jk} F_{kj} w, w)|.$$

It is easy to see that one can choose G of the indicated type such $G^{jk}F_{jk} = 2\operatorname{Tr}^+(gF)$; this is easy to do in the coordinate system in which (F_{kj}) is in the canonical form for skew-symmetric matrices, in which case $G^{jk} = -\operatorname{sign}(F_{jk})$ is an appropriate matrix. Then

$$\sum_j \|P_j w\|^2 \geq \mu h(\operatorname{Tr}^+(gF)w, w) - C\varepsilon\mu h\|w\|^2;$$

since $g^{jk} = \delta_{jk}$ we obtain that

$$(Aw, w) \geq \bigl((\mu h \operatorname{Tr}^+(gF) + V)w, w\bigr) - C\varepsilon\mu h\|w\|^2.$$

This inequality and (6.1.17) yield (6.1.18). □

Theorem 6.1.3 and standard Tauberian arguments (see the proof of proposition 4.1.4) yield

Theorem 6.1.5. *Let conditions* (6.1.1)–(6.1.3) *with $K > \frac{d}{2}$ be fulfilled, let $h \in [0,1)$, $\mu \geq 1$, let operator A be self-adjoint and suppose $u \in C_0^2(B(0,1))$ belongs to* $\mathrm{D}(A)$ *if and only if $u|_{\partial X} = 0$. Then*

$$|e(x,y,\tau)| \leq C' h^s \mu^{-s}(|\tau|+1)^{-s} \tag{6.1.19}$$
$$\forall x, y \in X \cap B(0, 1-\epsilon) \quad \forall \tau \leq \inf_{(X\cap B(0,1))} (V + \mu h F) - \epsilon(\mu h + 1)$$

with the same s, ϵ and C' as in theorem 6.1.3.

Remark 6.1.6. *(i) The condition on τ in (6.1.19) can be weakened for $d = 2, 3$ (see the following sections).*

(ii) Now the condition

$$\operatorname{Tr}^+(gF) \geq \epsilon_0 \tag{6.1.20}$$

and the condition $\tau \leq c$ on the energy level yield that one can assume that $\mu h \leq C_0$. Hence without loss of generality one can assume that

$$\mu h \leq 1. \tag{6.1.21}$$

◊

Let us now prove that singularities propagate with finite speed with respect to x at energy levels $\tau \leq c$.

Theorem 6.1.7. *Let the conditions of theorem 6.1.3 be fulfilled. Then*

$$|F_{t\to h^{-1}\tau}\chi_T(t)U(x,y,t)| \leq C'h^s \qquad \forall T \geq h^{1-\delta} \quad \forall x,y \in X \cap B(0, 1-C_0T) : |x-y| \geq C_0T \quad \forall \tau \leq c \tag{6.1.22}$$

where $C_0 = C_0(c, \epsilon_0)$, $C' = C'(d, c, \epsilon_0\delta, s)$, s and $\delta > 0$ are arbitrary.

Proof. Let us note that it is sufficient to prove this estimate only for $T \geq \epsilon$ with an arbitrarily small constant $\epsilon > 0$ and with C' also depending on ϵ. The generalization is made by the dilatation method (see the proof of theorem 2.1.15).

Moreover, elliptic arguments (see the proof of theorem 6.1.3) yield that it is sufficient to prove that if $\bar{x}, \bar{y} \in B(0, 1-C_0T)$, $|\bar{x} - \bar{y}| \geq C_0T$, $\zeta \in L^2(X \cap B(\bar{y}, \epsilon_1))$ with a small enough constant $\epsilon_1 > 0$ and $u = U(t)\zeta$ then

$$\|F_{t\to h^{-1}\tau}\chi_T(t)u\|_{X\cap B(\bar{x},\epsilon_1)} \leq C'h^s\|\zeta\| \qquad \forall T \in [\epsilon, T_0] \quad \forall \tau \leq c. \tag{6.1.23}$$

In order to prove this estimate we apply the method of the proof of theorem 2.1.2. Let $u^\pm = \theta(\pm t)u$, $v^\pm = \Phi(hD_t)\bar{\chi}(t)u^\pm$ where $\Phi \in C^\infty(\mathbb{R})$ is fixed and supported in $[-\infty, 2c]$, $\Phi(\tau) = 1$ for $\tau \leq \frac{3}{2}c$, $\bar{\chi} \in C_0^\infty([-2T, 2T])$ and $\bar{\chi} = 1$ at $[-\frac{3}{2}T, \frac{3}{2}T]$. Then

$$(hD_t - A)v^\pm = \mp ih\Phi(hD_t)\delta(t)\zeta(x) \quad \text{for } |t| \leq T \tag{6.1.24}$$

modulo terms with L^2-norm not exceeding $C'h^s$ where here and in what follows C' depends on all the auxiliary functions and constants.

Let $\phi \in C^\infty(B(0,1))$ be real-valued and satisfy

$$|\nabla_x\phi| \leq \Delta, \tag{6.1.25}$$

$$\phi < 0 \quad \text{provided either } |x| \geq 1 - 2\epsilon_1 \text{ or } |x-y| \leq 2\epsilon_1 \tag{6.1.26}$$

where $\Delta > 0$ will be chosen later.

Let $\chi_0 \in C^\infty(\mathbb{R})$ be supported in $[-\infty, 1]$, $\chi(t)_0 = 1$ for $t \leq 0$ and $\chi_0 = \chi_1^2$, $\chi' = \frac{d\chi_0}{dt} = -\chi_2^2$ with real-valued functions $\chi_1, \chi_2 \in C^\infty(\mathbb{R})$. Then for $\alpha = \alpha(c, \epsilon_0, \epsilon_1)$ the function $\psi(x,t) = \chi_0(\alpha(t - \phi(x))$ belongs to $C^\infty(B(0,1) \times \mathbb{R})$ and $\operatorname{supp}\psi \cap \{t \geq 0\} \subset \{|x| \leq 1 - \epsilon_1, |x-y| \geq \epsilon_1\}$.

Finally, let $\chi(t) = \chi_0(\gamma t)$ and $\chi_3(t) = \chi_0(-\gamma_1 t - 1)$. Then modulo terms with absolute value not exceeding $C'h^{2s}\|\zeta\|^2$

(6.1.27)
$$
\begin{aligned}
0 \equiv 2\operatorname{Re}\bigl(i\Phi(hD_t)\delta(t)\zeta(x), e^{-2t}\chi\chi_3\psi v^+\bigr) \equiv \\
-h^{-1}\operatorname{Re} i([hD_t - A, e^{-2t}\chi\chi_3\psi]v^+, v^+) \equiv \\
\operatorname{Re}(2\chi\chi_3\psi e^{-2t}v^+, v^+) - \operatorname{Re}(\chi'\chi_3\psi e^{-2t}v^+, v^+) - \\
\operatorname{Re} i(h^{-1}[hD_t - A, \psi]e^{-2t}\chi\chi_3 v^+, v^+) \geq \\
2\|(\chi\chi_3\psi)^{\frac{1}{2}}e^{-t}v^+\|^2 - \operatorname{Re} i(h^{-1}[hD_t - A, \psi]e^{-2t}\chi\chi_3 v^+, v^+)
\end{aligned}
$$

where $(.,.)$ and $\|.\|$ denote the inner product and the norm in $L^2(X \times \mathbb{R})$.

Let us consider the second term in the right-hand expression; it is equal to

$$
\alpha \operatorname{Re}\bigl((I + B)w^+, w^+\bigr) + h(\Psi v^+, v^+)
$$

where $B = -2g^{jk}(D_j\phi)P_k$, $w^+ = e^{-t}\psi_2(\chi\chi_3)^{\frac{1}{2}}v^+$, $\psi_2 = \chi_2(\alpha(t-\phi))$ and Ψ... here and in what follows denotes a C^∞ function supported in $\operatorname{supp}\chi\chi_3\psi$. Then (6.1.25) yields that

(6.1.28)
$$
\begin{aligned}
|(Bw^+, w^+)| \leq C_0\Delta\bigl\{\|w^+\|^2 + (g^{jk}P_j w^+, P_k w^+)\bigr\} \leq \\
C_0\Delta\bigl\{\|w^+\|^2 + (Aw^+, w^+)\bigr\} \leq \\
C_0\Delta\bigl\{\|w^+\|^2 + (Av^+, e^{-2t}\chi\chi_3\psi_2^2 v^+)\bigr\} + h(\Psi' v^+, v^+)
\end{aligned}
$$

where we change C_0 if necessary.

On the other hand, (6.1.24), the definition of ψ_2 and (6.1.26) yield that

(6.1.29)
$$
\begin{aligned}
(Av^+, e^{-2t}\chi\chi_3\psi_2^2 v^+) \equiv (hD_t v^+, e^{-2t}\chi\chi_3\psi_2^2 v^+) \equiv \\
\Bigl(hD_t\Phi(hD_t)(\chi\chi_3)^{\frac{1}{2}}\psi_2 u^+, \Phi(hD_t)(\chi\chi_3)^{\frac{1}{2}}\psi_2 u^+\Bigr) + \\
\sum_{n\geq 1} h^n(\Phi_n(hD_t)\Psi_n u^+, u^+)
\end{aligned}
$$

where $\Phi_n \in C_0^\infty([c, 2c])$ and $\Psi_n \in C^\infty$ are supported in $\operatorname{supp}\chi\chi_3\psi$. The first term in the right-hand expression of (6.1.29) does not exceed

$$
\|\Phi(hD_t)(\chi\chi_3)^{1/2}\psi_2 u^+\| \equiv 2c\|w^+\|^2 + \sum_{n\geq 1} h^n(\Phi_n(hD_t)\Psi_n u^+, u^+)
$$

where we change Φ_n and Ψ_n if necessary. Then (6.1.28) and (6.1.29) yield that

$$
|(Bw^+, w^+)| \leq \frac{1}{4}C_0\Delta\|w^+\|^2 + \sum_{n\geq 1} h^n(\Phi_n(hD_t)\Psi_n u^+, u^+)
$$

and then (6.1.27) yields that for $\Delta \leq \frac{1}{3C_0}$

$$
\|(\chi\chi_3\psi)^{\frac{1}{2}}e{-t}v^+\|^2 \leq \sum_{n\geq 1} h^n(\Phi_n(hD_t)\Psi_n u^+, u^+) + C'h^2 s\|\zeta\|^2;
$$

moreover, one can allow Φ and Ψ to be smooth functions here (nevertheless these functions were not smooth when they appeared the first time).

Using this inequality and applying induction with respect to s (where the supports of χ_0 and Φ are decreased at each step) we can easily prove that if ϕ satisfies (6.1.25) and (6.1.26) with $\Delta = \frac{1}{3C_0}$ then for $T \in [\epsilon, T_0]$

$$
\|\Phi(hD_t)\chi_T(t)\chi_1(\alpha(t-\phi))u^+\| \leq C'h^s\|\zeta\|;
$$

since we can arbitrarily choose ϕ satisfying (6.1.25) and (6.1.26) with $\Delta = \frac{1}{3C_0}$ we obtain (6.1.23). □

In the cases $d = 2,3$ using microlocal canonical form we will essentially improve theorems 6.1.3 and 6.1.7 (assuming that the coefficients are smooth enough). Let us conclude this section with the following assertion on the support of singularities:

Theorem 6.1.8. *Let conditions* (6.1.1)–(6.1.3) *and* (6.1.4) *be fulfilled with* $K = K(d,s)$. *Let* $q \in \tilde{S}^0_{h,\rho,1,K}$ *and* $\psi \in C_0^K(B(0,1-\epsilon))$. *Let* $\mu h \le 1$, $\rho \ge 1$, $\mu\rho^{-2}h^{1-\delta} \le 1$ *with* $\delta > 0$ *and let* q *be supported in* $\{|\xi| \ge C_0\rho\}$ *and* $Q = q(x, hD_1 - \mu V_1, \dots, hD_d - \mu V_d)$ *be an operator with symbol* $Q(x,\xi) = q(x, h\xi_1 - \mu V_1(x), \dots, h\xi_d - \mu V_d(x))$. *Then*

$$|F_{t\to h^{-1}\tau}\chi_{T,T'}(t)Q\psi U| \le C'MTh^s(|\tau|+1)^{-s} \qquad \forall x,y \in X\cap B(0,1-\epsilon) \quad \forall \tau \le c \tag{6.1.30}$$

where $T \ge h^{1-\delta}$, T', $\epsilon > 0$, $\delta > 0$ *and* s *are arbitrary,* $C_0 = C_0(c,\epsilon_0)$, $K = K(d,\delta,s)$, $C' = C'(d,c,\epsilon_0,\epsilon,\delta,s)$ *and* $M = \|\psi\|\cdot|\!|\!|q|\!|\!|$.

Proof. The proof is standard elliptic, based on the equation $(hD_t - A)U = 0$, estimates (6.1.13), (6.1.14) and the ellipticity of the operator $hD_t - A$ for $\tau \le c$, $\sum_j |\xi_j - \mu V_j|^2 \ge C_0^2$; it should be mentioned that the condition $\rho \ge \mu^{\frac12}h^{\frac12-\delta}$ is necessary to obtain $Q(x,\xi) \in S^0_{h,\rho_1,\rho_2,K}$ with $\rho_1 \ge 1$, $\rho_2 \in (0,1]$, $\rho_1\rho_2 \ge h^{-\delta'}$ with $\delta' > 0$; this condition is essential only for $\mu \in [h^{\delta-1}, h^{-1}]$. □

6.2 Microlocal Canonical Form

In this section for $d = 2,3$ we construct the microlocal canonical form for operator (6.1.1) satisfying conditions (6.1.2), (6.1.3) and

$$F \ge \epsilon_0 \tag{6.2.1}$$

in the ball $B(0,1) \subset X$ where F is the scalar intensity of the magnetic field.

6.2.1 Symplectic canonical form. I

Let us multiply operator (6.1.1) by μ^{-2} and then note that $\mu^{-2}A$ is a $\mu^{-1}h$-differential operator

$$\mu^{-2}A = (\mu^{-1}hD_j - V_j)g^{jk}(\mu^{-1}hD_k - V_k) + \mu^{-2}V$$

with principal symbol $a_\varepsilon(x,\xi) = a(x,\xi) + \varepsilon V$ where

$$a(x,\xi) = g^{jk}(\xi_j - V_j)(\xi_k - V_k) \tag{6.2.2}$$

and $\varepsilon = \mu^{-2}$.

Let us fire treat the symbol $a(x,\xi)$; conditions (6.1.1)–(6.1.3) yield that $a \ge 0$ and the stationary set of a is

$$\Lambda = \{da = 0\} = \{a = 0\} = \{(x,\xi), \xi_j = V_j(x) \quad \forall j = 1,\dots,d\} \tag{6.2.3}$$

which is a d-dimensional C^K manifold. Moreover, (6.2.1) yields that for $d = 2$, Λ is a symplectic manifold and for $d = 3$ the restriction to Λ of the symplectic form has constant rank 2 and therefore there exists a unique (modulo a scalar factor) vector field on Λ which is tangent and symplectically orthogonal to Λ at the same time. Let us note that this field is the Hamiltonian field generated by $F^j(\xi_j - V_j)$ where F^j are the components of the vector intensity of the magnetic field.

For $d \ge 4$ conditions (6.1.1)–(6.1.3) and (6.2.1) do not completely fix the structure of Λ.

Proposition 6.2.1. *Let conditions (6.1.1)–(6.1.3) and (6.2.1) be fulfilled. Then*

(i) For $d = 2$ there exists a symplectomorphism Ψ^{-1} of a neighborhood $\mathcal{W}$ of every point $z \in \Lambda$ to a neighborhood of

$$\Lambda^0 = \{(x,\xi), x_2 = \xi_2 = 0, (x_1,\xi_1) \in \bar{\Omega}\}$$

where Ω is a domain in $\mathbb{R}^2$ such that the inverse symplectomorphism Ψ has the following properties:

$$\Psi(\Lambda^0) = \Lambda \cap \mathcal{U}, \tag{6.2.4}$$

$$a \circ \Psi \equiv \sum_{1 \le n \le N-1} b_n(x_1,\xi_1)(x_2^2 + \xi_2)^n \qquad (\operatorname{mod} O_{2N-1}(\Lambda^0)) \tag{6.2.5}$$

where $O_M(\Lambda)$ is the set of functions vanishing together with all their derivatives up to order $M-1$ on Λ; moreover,

$$b_1 = F \circ \bar{\Psi}, \tag{6.2.6}$$

$$\bar{\Psi}^{-1*}(dx'd\xi_1) = F\sqrt{g}dx \tag{6.2.7}$$

where $\bar{\Psi} : \Lambda^0 \to \Lambda$ is the restriction of Ψ to Λ^0 and $\bar{\Psi}^*$ is the induced map of densities, and $x' = x_1$, $dx' = dx_1$ for $d = 2$.

(ii) For $d = 3$ there exists a symplectomorphism Ψ^{-1} of a neighborhood $\mathcal{W}$ of every non-closed Hamiltonian trajectory Γ of finite length generated by the Hamiltonian $F^j(\xi_j - V_j)$ and lying on Λ to a neighborhood of

$$\Lambda^0 = \{(x,\xi), x_3 = \xi_3 = 0, (x_1,x_2,\xi_1) \in \bar{\Omega}\}$$

where Ω is a domain in $\mathbb{R}^3$ such that the inverse symplectomorphism Ψ satisfies (6.2.4),

$$a \circ \Psi \equiv \xi_2^2 + \sum_{1 \le n \le N-1} b_n(x',\xi_1)(x_3^2 + \xi_3)^n \qquad (\operatorname{mod} O_{2N-1}(\Lambda^0)), \tag{6.2.8}$$

(6.2.6) and (6.2.7) where $x' = (x_1,x_2)$, $dx' = dx_1 dx_2$ for $d = 3$.

(iii) In (i) for $F =$ const

$$b_2 = \frac{1}{8} F\kappa \circ \bar{\Psi} \tag{6.2.9}$$

where κ is the scalar curvature corresponding to the metric $g^{jk}F^{-1}$.

Remark 6.2.2. *(i) In this proposition N is arbitrary and $K = K(N)$ in condition (6.1.2); moreover, $b_n \in C^N$. Therefore (6.2.5) and (6.2.8) hold with O_{2N} instead of O_{2N-1}.*

(ii) It follows from the proof that for $d = 3$ one can also take another canonical form

$$a \circ \Psi \equiv (\xi_2^2 + x_3^2 + \xi_3) \sum_{1 \le n \le N-1} b_n(x',\xi_1)(x_3^2 + \xi_3)^{n-1} \qquad (\operatorname{mod} O_{2N-1}(\Lambda^0)). \tag{6.2.8$'$}$$

This remark will be essential in the next chapter. ◊

Proof. (i) Let us first treat the simpler case $d=2$. Since Λ is a 2-dimensional symplectic manifold, $a \geq 0$ and a has a zero on Λ of precisely the second order then there exist functions $w_j(x,\xi)$ $(j=1,2)$ such that

$$w_j = 0 \quad \text{at } \Lambda\cap\mathcal{W} \qquad (\forall j=1,\dots,d), \tag{6.2.10}$$

$$\{w_1,w_2\}=1 \qquad \text{at } \Lambda\cap\mathcal{W}, \tag{6.2.11}$$

$$a \equiv b_1(w_1^2+w_2^2) \qquad (\operatorname{mod} O_3(\Lambda\cap\mathcal{W})); \tag{6.2.12}$$

moreover, it is easy to see that here one can take $\mathcal{W}=B(0,1)\times\mathbb{R}^2$. Moreover, one can easily replace (6.2.11) by

$$\{w_1,w_2\}\equiv 1 \qquad (\operatorname{mod} O_M(\Lambda\cap\mathcal{W})) \tag{6.2.11$'_M$}$$

with arbitrary M (in which case $K=K(M)$). In fact, if (6.2.11)$'_M$ holds then replacing w_2 by $w_2'=w_2+\mathcal{P}(w;x,\xi)$ where $\mathcal{P}$ is a homogeneous polynomial of order $M+1$ with respect to $w=(w_1,w_2)$ and noting that

$$\{w_1,\mathcal{P}\}\equiv\partial_{w_2}\mathcal{P} \qquad (\operatorname{mod} O_{M+1}(\Lambda\cap\mathcal{W}))$$

we obtain that for appropriate $\mathcal{P}$ condition (6.2.11)$'_{M+1}$ will be fulfilled. Moreover, one can replace (6.2.11)$'$ by

$$\{w_1,w_2\}=1 \qquad \text{in } \mathcal{W}. \tag{6.2.11$''$}$$

In fact, one can reach this equality by replacing w_2 with $w_2'\equiv w_2\,(\operatorname{mod} O_M(\Lambda))$ because $H_{w_1}=\{w_1,.\}$ is a non-vanishing vector field tangent to Λ in $\mathcal{W}$.

But if (6.2.10) and (6.2.11)$''$ are fulfilled then there exists a symplectomorphism $\Psi_{\frac{3}{2}}$ of a neighborhood of Λ^0 onto $\mathcal{W}$ (where $\mathcal{W}$ is decreased if necessary) such that $w_1\circ\Psi_{\frac{3}{2}}=\xi_2$, $w_2\circ\Psi_{\frac{3}{2}}=x_2$. This yields (6.2.5) with $\Psi=\Psi_{\frac{3}{2}}$ and $N=2$.

Let us calculate the non-zero eigenvalues of the fundamental matrices (skew-Hessians) of $a(x,\xi)$ and $x_2^2+\xi_2^2$ at points of Λ and Λ^0 respectively; we obtain $\pm 2iF$ and $\pm 2i$ respectively; this yields (6.2.6) and (6.2.7).[3)]

Let us now prove that if (6.2.5) is fulfilled with some $N\geq 2$ then there exists a symplectomorphism Ψ_N of a neighborhood of Λ^0 onto a neighborhood of Λ^0 such that $\Psi_N\equiv I\,(\operatorname{mod} O_{2N-2}(\Lambda^0))$ and (6.2.5) holds for Ψ replaced by $\Psi\circ\Psi_N$ with N replaced by $N+\frac{1}{2}$. Then for $\Psi=\Psi_{\frac{3}{2}}\circ\Psi_2\circ\cdots\circ\Psi_{N-\frac{1}{2}}$ we have (6.2.5) with arbitrarily chosen N.

Let us construct this symplectomorphism Ψ_N as an exponential map $\Psi_N=\exp H_S$ where H_S is the Hamiltonian field generated by $S\in O_{2N-1}(\Lambda^0)$. Then for any smooth enough function $f(x,\xi)$

$$f\circ\Psi_N\sim f+\{S,f\}+\frac{1}{2}\{S,\{S,f\}\}+\dots \tag{6.2.13}$$

where the order of the zero at Λ^0 increases when the number of the term increases. In particular,

$$x_2\circ\Psi_N\equiv x_2+\{S,x_2\},\quad \xi_2\circ\Psi_N\equiv\xi_2+\{S,\xi_2\},\quad x_1\circ\Psi_N\equiv x_1,\quad \xi_1\equiv\xi_1 \qquad (\operatorname{mod} O_{2N-1}(\Lambda^0))$$

3) We recall that symplectomorphisms preserve the spectra of fundamental matrices at stationary points and the canonical density $dxd\xi$.

and

$$a \circ \Psi \circ \Psi_N \equiv a \circ \Psi_N + b_1\{S, x_2^2 + \xi_2^2\} \pmod{O_{2N}(\Lambda^0)}.$$

Therefore the existence of the appropriate generating function S follows from

Lemma 6.2.3. *Let $d = 2$. Then for every $M \in \mathbb{Z}^+ \setminus 0$ and for every function $Q \in O_M(\Lambda^0)$ there exist functions $S \in O_M(\Lambda^0)$ and ω such that*

$$Q \equiv \{S, x_2^2 + \xi_2^2\} + \omega(x_2^2 + \xi_2^2)^{\frac{M}{2}} \pmod{O_{M+1}(\Lambda^0)}.$$

Moreover, for odd M, $\omega = 0$ and $S \pmod{O_{M+1}(\Lambda^0)}$ is defined uniquely, while for even M, $S \pmod{O_{M+1}(\Lambda^0)}$ is defined uniquely modulo the term $\omega'(x_2^2+\xi_2^2)^{\frac{M}{2}}$ and $\omega \pmod{O_1(\Lambda^0)}$ is defined uniquely.

Proof. The proof of this lemma is obvious. □

(ii) Let us now treat the case $d = 3$. Then the symplectic structure of Λ and the properties of a yield that there exist functions $w_j(x,\xi)$ $(j = 1,2,3)$ such that (6.2.1) holds and

(6.2.14) dw_j $(j = 1,2,3)$ are linearly independent on $\Lambda \cap \mathcal{W}$,

(6.2.15) $$\{w_1, w_2\} = 1, \quad \{w_1, w_3\} = \{w_2, w_3\} = 0 \quad \text{at } \Lambda \cap \mathcal{W},$$

(6.2.16) $$a \equiv w_3^2 + b_1(w_1^2 + w_2^2) \pmod{O_3(\Lambda^0 \cap \mathcal{W})};$$

then on $\Lambda \cap \mathcal{W}$, $F^j d(\xi_j - V_j)$ and dw_3 differ by a scalar factor. Here one can take $\mathcal{W} = B(0,1) \times \mathbb{R}^3$.

We assert that one can replace (6.2.15) by

(6.2.15)′ $$\{w_1, w_2\} \equiv 1, \quad \{w_1, w_3\} \equiv \{w_2, w_3\} \equiv 0 \qquad \pmod{O_2(\Lambda \cap \mathcal{W})}.$$

In fact, (6.2.15) means precisely that

$$\{w_j, w_k\} \equiv \epsilon_{jk} + \sum_l \alpha_{jkl} w_l \qquad \pmod{O_2(\Lambda \cap \mathcal{W})}$$

with $\epsilon_{12} = 1$, $\epsilon_{13} = \epsilon_{23} = 0$; the Jacobi identity yields that $\alpha_{132} = -\alpha_{231}$ on $\Lambda \cap \mathcal{W}$. It is easy to reach (6.2.15)′ by replacing w_j with $w_j + \mathcal{P}_j(w; x, \xi)$ with appropriate homogeneous second degree polynomials $\mathcal{P}$ of $w = (w_1, w_2, w_3)$; here we use the equalities $\{w_1, \mathcal{P}\} \equiv \partial_{w_2}\mathcal{P}$, $\{w_2, \mathcal{P}\} \equiv -\partial_{w_1}\mathcal{P}$, $\{w_3, \mathcal{P}\} \equiv 0 \pmod{O_2(\Lambda \cap \mathcal{W})}$ and the equality $\alpha_{132} = -\alpha_{231}$ on $\Lambda \cap \mathcal{W}$.

Moreover, one can replace (6.2.15)′ by

(6.2.15)″ $$\{w_1, w_2\} = 1, \quad \{w_1, w_3\} = \{w_2, w_3\} = 0 \qquad \text{in } \mathcal{W}.$$

In fact, since H_{w_3} is a non-vanishing vector field in $\mathcal{W}$ tangent to Λ, Γ is the non-closed trajectory of H_{w_3} and $\mathcal{W}$ is a narrow neighborhood of Γ then the second of these equalities will be fulfilled if we replace w_1 by an appropriate function $w_1' \equiv w_1 \pmod{O_2(\Lambda)}$. Moreover, the first equality in (6.2.15)′ remains true and the third equality remains true automatically. Then the vector fields H_{w_1} and H_{w_3} commute and are linearly independent and it is easy to reach the first and third equalities in (6.2.15)″ by replacing w_2 with an appropriate function $w_2' \equiv w_2 \pmod{O_2(\Lambda)}$.

Moreover, the same arguments with vector fields yield that there exist functions w_4, w_5, w_6 in $\mathcal{W}$ such that

$\{w_j, w_k\} = 1$ for $j = k-1$ and even k; $\{w_j, w_k\} = 0$ for $j < k-1$ and for $j = k-1$ and odd k

and therefore there exists a symplectomorphism $\Psi_{\frac{3}{2}}$ such that

$$w_1 \circ \Psi_{\frac{3}{2}} = \xi_3, \quad w_2 \circ \Psi_{\frac{3}{2}} = x_3, \quad w_3 \circ \Psi_{\frac{3}{2}} = \xi_2;$$

these equalities yield (6.2.8) with $\Psi = \Psi_{\frac{3}{2}}$ and $N = 2$. Calculating the non-vanishing eigenvalues of the fundamental matrices of a and $x_3^2 + \xi_3^2$ on Λ and Λ^0 respectively we obtain $\pm 2iF$ and $\pm 2i$ respectively and this yields (6.2.6) and (6.2.7).

Let us prove that if (6.2.8) is fulfilled with $N \geq 2$ then there exists a symplectomorphism Ψ_N of a neighborhood of Λ^0 onto a neighborhood of Λ^0 such that $\Psi_N \equiv I$ $(\operatorname{mod} O_{2N-2(\Lambda^0)})$ and (6.2.8) holds with Ψ replaced by $\Psi \circ \Psi_N$ and N replaced by $N + \frac{1}{2}$ in the sum in the right-hand expression and in $R'_n \in O_{2N-2n-1}(\Lambda^0)$.

Let us first treat the most delicate case $N = 2$. Let $\Psi' = \exp H_{S'}$, $S' = \alpha(x', \xi_1)\xi_2^2 + \beta(x', \xi_1)(x_3^2 + \xi_3^2)$. Then formula (6.2.13) is wrong but it is easy to prove that

$$(\xi_3^2 + \xi_3^2) \circ \Psi' = (\xi_3^2 + \xi_3^2), \quad x_1 \circ \Psi' \equiv x_1, \quad \xi_1 \circ \Psi' \equiv \xi_1, \quad x_2 \circ \Psi' \equiv x_2 + \partial_{\xi_2} S' \qquad (\operatorname{mod} O_2(\Lambda^0))$$

and $\xi_2 \equiv \xi_2 - \partial_{x_2} S'$ $(\operatorname{mod} O_3(\Lambda^0))$; then

$$\xi_3 \circ \Psi' \equiv \xi_2 \cos 2\beta - x_3 \sin 2\beta, \quad x_3 \circ \Psi' \equiv x_3 \cos 2\beta + \xi_3 \sin 2\beta \qquad (\operatorname{mod} O_2(\Lambda^0)).$$

Therefore

$$a \circ \Psi_{\frac{3}{2}} \circ \Psi' \equiv a \circ \Psi_{\frac{3}{2}} + \{S', a^0\}_2 + \sum_{1 \leq k \leq 3} T_k(x', \xi_1; x_3, \xi_3)\xi_2^{3-k} \qquad (\operatorname{mod} O_4(\Lambda^0))$$

where $a^0 = \xi_2^2 + b_1(x_3^2 + \xi_3^2)$, $\{f, g\}_2 = \partial_{\xi_2} f \cdot \partial_{x_2} g - \partial_{x_2} f \cdot \partial_{\xi_2} g$ and T_k are homogeneous k-th order polynomials of (x_3, ξ_3) with coefficients depending on (x', ξ_1); moreover, the trace of the matrix of the quadratic form T_2 vanishes. This yields that for appropriate α and β

$$a \circ \Psi_{\frac{3}{2}} \circ \Psi' \equiv a^0 + \sum_{1 \leq k \leq 3} T'_k(x', \xi_1; x_3, \xi_3)\xi_2^{3-k} \qquad (\operatorname{mod} O_4(\Lambda^0))$$

where T'_k are homogeneous k-th order polynomials of (x_3, ξ_3) with coefficients depending on (x', ξ_1); moreover, the trace of the matrix of the quadratic form T'_2 vanishes.

Let $S'' \in O_3(\Lambda^0)$ and $\Psi'' = H_{S''}$. Then (6.2.13) yields that

$$a \circ \Psi_{\frac{3}{2}} \circ \Psi' \circ \Psi'' \equiv a^0 + b_1\{S'', x_3^2 + \xi_3^2\} + \sum_{1 \leq k \leq 3} T'_k(x', \xi_1; x_3, \xi_3)\xi_2^{3-k} \qquad (\operatorname{mod} O_4(\Lambda^0))$$

and it is easy to see that for appropriate S''

$$a \circ \Psi_{3/2} \circ \Psi' \circ \Psi'' \equiv a^0 (\operatorname{mod} O_4(\Lambda^0)).$$

Let us consider the case $N \geq \frac{5}{2}$. Let us look for Ψ_N in the form $\Psi_N = \exp H_S$, $S = S' + S''$ where

$$S' = \sum_{0 \leq k \leq [N]} \alpha_k(x', \xi_1)(x_3^2 + \xi_3^2)^k \xi_2^{2N-2k-2} \tag{6.2.17$_N$}$$

belongs to $O_{2N-2}(\Lambda^0)$ and $S'' \in O_{2N-1}(\Lambda^0)$. Then (6.2.12) yields that

$$a \circ \Psi \circ \Psi_N \equiv a \circ \Psi + \{S', a^0\}_2 + b_1\{S'', x_3^2 + \xi_3^2\} \qquad (\operatorname{mod} O_{2N}(\Lambda^0))$$

and the existence of the appropriate generating function S follows from

Lemma 6.2.4. *Let $d = 3$, $a^0 = \xi_2^2 + b_1(x_3^2 + \xi_3^2)$, $b_1 = b_1(x', \xi_1) > 0$. Then for every $M \in \mathbb{Z}^+$ and for every function $Q \in O_M(\Lambda^0)$ there exists S' of the form (6.2.16)$_N$ with $N = \frac{M+1}{2}$, $S'' \in O_M(\Lambda^0)$ and b ($b = 0$ for odd M) such that*

$$Q \equiv \{S', a^0\}_2 + b_1\{S'', x_3^2 + \xi_3^2\} + b(x_3^2 + \xi_3^2)^{\frac{M}{2}} \qquad (\operatorname{mod} O_{M+1}(\Lambda^0)).$$

(iii) Without loss of generality one can assume that $F = 1$; then $b_1 = 1$. Let us introduce a local conformal coordinate system [Dubrovin, Fomenko, Novikov 1]. Then $g^{11} = g^{22} = \rho^2$, $g^{12} = 0$ and $a(x, \xi) = s^2 + t^2$, $\{s, t\} = \pm 1$ at Λ where $s = \rho(\xi_1 - V_1)$, $t = \rho(\xi_2 - V_2)$. Permuting x_1 and x_2, s and t we obtain that $\{s, t\} = 1$ at Λ. Then

$$\{s, t\} = 1 + \beta s - \alpha t, \quad \alpha = -\partial_1 \rho, \quad \beta = -\partial_2 \rho. \tag{6.2.18}$$

So $w_1 = s$, $w_2 = t$ fulfill the starting relations (6.2.10)–(6.2.12). Let us take $w' - 1 = s + ut$, $w_2' = t - us$ with $u \in O_1(\Lambda)$; then $a \equiv w_1'^2 + w_2'^2$ $(\operatorname{mod} O_4(\Lambda))$ and

$$\{w_1', w_2'\} \equiv \{s, t\} - s\{s, u\} - t\{t, u\} \qquad (\operatorname{mod} O_2(\Lambda)).$$

Then (6.2.6) yields that $\{w_1', w_2'\} \equiv 1$ $(\operatorname{mod} O_2(\Lambda))$ for $u = \alpha s + \beta t$. More detailed calculation then gives that

$$\begin{aligned}\{w_1', w_2'\} \equiv 1 + s^2(2\alpha^2 - \beta^2 - \{s, \alpha\}) + st(6\alpha\beta - \{t, \alpha\} - \{s, \beta\}) +\\ t^2(2\beta^2 - \alpha^2 - \{t, \beta\}) \qquad (\operatorname{mod} O_3(\Lambda)).\end{aligned} \tag{6.2.19}$$

Let us construct functions w_1'' and w_2'' coinciding modulo $O_3(\Lambda)$ with w_1' and w_2' such that

$$\{w_1'', w_2''\} \equiv 1 \quad (\operatorname{mod} O_3(\Lambda)), \tag{6.2.20}$$

$$a \equiv w_1''^2 + w_2''^2 + \omega(w_1''^2 + w_2''^2)^2 \qquad (\operatorname{mod} O_5(\Lambda)), \tag{6.2.21}$$

and let us calculate the coefficient ω. In order to fulfill (6.2.21) let us note that

$$a = (w_1'^2 + w_2'^2)(1 + u^2)^{-1} \equiv (w_1'^2 + w_2'^2)(1 - u^2) \qquad (\operatorname{mod} O_5(\Lambda))$$

and let us set

$$w_1'' = s + ut - \frac{1}{2}(\omega(s^2 + t^2) + u^2) + vt,$$

$$w_2'' = t - us - \frac{1}{2}(\omega(s^2+t^2)+u^2) - vs$$

with $v \in O_2(\Lambda)$. Then (6.2.7) yields that

(6.2.22) $$\{w_1'', w_2''\} \equiv 1 + s^2(-\beta^2 - \{s,\alpha\} - 2\omega) + st(2\alpha\beta - \{t,\alpha\} - \{s,\beta\}) + t^2(-\alpha^2 - \{t,\beta\} - 2\omega) - s\{s,v\} - t\{t,v\} \qquad (\operatorname{mod} O_3(\Lambda)).$$

Then lemma 6.2.3 yields that (6.2.20) holds for appropriate $v \in O_2(\Lambda)$ if and only if the sum of coefficients at s^2 and t^2 in (6.2.21) vanishes, i.e.,

$$\omega = \frac{1}{4}(-\{s,\alpha\} - \{t,\beta\} - \alpha^2 - \beta^2);$$

equalities (6.2.18) and the definitions of s and t then yield that

$$\omega = \frac{1}{4}(\rho\Delta\rho - (\nabla\rho)^2) = \frac{1}{4}\rho^2\Delta\log\rho$$

where Δ is the two-dimensional Laplace operator and therefore the formula for the scalar curvature in a conformal coordinate system [Dubrovin, Fomenko, Novikov 1] yields that $\omega = \frac{1}{8}\kappa$. This equality yields (6.2.9) if we take a symplectomorphism Ψ such that $w_1''\circ\Psi = \xi_2$, $w_2''\circ\Psi \equiv x_2 \pmod{O_4(\Lambda^0)}$; moreover, (6.2.5) is fulfilled with $N=3$. Taking this symplectomorphism as the starting point of the construction used in (i) we obtain (6.2.5) for arbitrary N; moreover (6.2.9) is also fulfilled.

On the other hand, if Φ is an arbitrary symplectomorphism of a neighborhood of Λ_1^0 onto a neighborhood of Λ_2^0 where in the definition of Λ_j^0 domains Ω_j are taken instead of Ω ($j=1,2$) and if Φ preserves the decomposition (6.2.5) then $\Phi = \Phi'\circ\Phi''$ where $\Phi'(x_1,\xi_1,x_2,\xi_2) = (\bar\Phi'(x_1,\xi_1),x_2,\xi_2)$, $\bar\Phi' : \Lambda_1^0 \to \Lambda_2^0$ is a symplectomorphism and Φ'' is also a symplectomorphism and $\Phi''|_{\Lambda_1^0} = I$. Then $\Phi'' = \exp H_S$ with $S \in O_2(\Lambda_1^0)$. Since Φ' obviously preserves the decomposition (6.2.5) the same also is true for Φ''; then the arguments of (i) yield that

$$S \equiv \sum_{1\le n\le N-1} \phi_n(x_1,\xi_1)(x_2^2+\xi_2^2)^n \qquad (\operatorname{mod} O_{2N-1}(\Lambda_1^0)).$$

It is easy to see that all the coefficients b_n are invariant with respect to Φ' (i.e., $b_n \to b_n\circ\bar\Phi'$) and that Φ'' transforms the coefficient b_2 to $b_2 + \{\phi_1, b_1\}$ (so it is also invariant provided b_1 is constant, i.e., provided F is constant). Therefore (6.2.9) holds for every symplectomorphism Ψ reducing $a(x,\xi)$ to canonical form. Proposition 6.2.1 is proven. □

6.2.2 Symplectic canonical form. II

Let us now treat the case $\varepsilon \ne 0$.

Theorem 6.2.5. *Let conditions* (6.1.1)–(6.1.3) *and* (6.2.1) *be fulfilled. Then*

(i) Let $d=2$. *Then for every* $\varepsilon \in [-\epsilon_0, \epsilon_0]$ *(where* $\epsilon_0 > 0$ *is a small enough constant) there exists a symplectomorphism* Ψ_ε^{-1} *of a neighborhood* $\mathcal{W}$ *of every point* $z \in \Lambda$ *to a neighborhood of* Λ^0 *depending smoothly on* ε, *coinciding with* Ψ^{-1} *for* $\varepsilon = 0$ *such that the inverse symplectomorphism* Ψ *has the following properties:*

(6.2.23) $$a_\varepsilon\circ\Psi_\varepsilon \equiv \sum_{m+n\le N-1} b_{mn}(x_1,\xi_1)\varepsilon^m(x_2^2+\xi_2)^n + \sum_{1\le n\le N-1}\varepsilon^n R_n(x,\xi,\varepsilon)$$

where the coefficients b_{mn} *and* R_n *are smooth functions; moreover,*

$$(6.2.24)_{1-3} \qquad b_{00}=0, \quad b_{01}=F\circ\bar{\Psi}_0, \quad b_{10}=V\circ\bar{\Psi}_0$$

and $R_n \in O_{2N-2n}(\Lambda^0)$.

(ii) Let $d=3$. *Then for every* $\varepsilon \in [-\epsilon_0, \epsilon_0]$ *there exists a symplectomorphism* Ψ_ε^{-1} *of a neighborhood* $\mathcal{W}$ *of every non-closed Hamiltonian trajectory* Γ *of finite length generated by the Hamiltonian* $F^j(\xi_j - V_j)$ *and lying on* Λ *to a neighborhood of*

$$\Lambda^0 = \{(x,\xi), x_3=\xi_3=0, (x_1,x_2,\xi_1)\in\bar{\Omega}\},$$

depending smoothly on ε, *coinciding with* Ψ^{-1} *for* $\varepsilon=0$ *such that the inverse symplectomorphism* Ψ_ε *has the following properties:*

$$(6.2.25) \quad a_\varepsilon\circ\Psi_\varepsilon \equiv \xi_2^2 + \sum_{1\le m+n\le N-1} b_{mn}(x',\xi_1)\varepsilon^m(x_3^2+\xi_3)^n + \sum_{0\le n\le N}\varepsilon^n R_n(x,\xi,\varepsilon)$$

where the coefficients b_{mn} *and* R_n *are smooth functions; moreover,* $(6.2.24)_{1-3}$ *hold and* $R_n \in O_{2N-2n}(\Lambda^0)$;

(iii) If $d=2$ *and* $F=\mathrm{const}$ *then*

$$(6.2.26)_1 \qquad b_{02} = \frac{1}{8}F\kappa\circ\bar{\Psi}_0$$

$$(6.2.26)_2 \qquad b_{11} = (\frac{1}{4}g^{jk}F^{-1}\partial_j\partial_k V + k^j\partial_j V)\circ\bar{\Psi}_0,$$

$$(6.2.26)_3 \qquad b_{20} = l^j\partial_j V\circ\bar{\Psi}_0$$

where the coefficients k^j *and* l^j *depend on the method of reduction and can be chosen arbitrarily at every point separately.*

Proof. If we take the symplectomorphism Ψ constructed in proposition 6.2.1 instead of Ψ_ε then (6.2.23) or (6.2.25) will be fulfilled modulo $\varepsilon O_1(\Lambda^0)$ and the coefficients in these decompositions will fulfill $(6.2.24)_{1-3}$. Therefore it is sufficient to prove that if (6.2.23) or (6.2.25) are fulfilled modulo $\varepsilon^L O_1(\Lambda^0)$ with $L\ge 1$ then there also exists a symplectomorphism $\Psi_{\varepsilon,L}$ of a neighborhood of Λ^0 onto a neighborhood of Λ^0, depending smoothly on ε and such that $\Psi_\varepsilon \equiv I \pmod{O(\varepsilon^L)}$ and (6.2.23) or (6.2.25) is fulfilled modulo $\varepsilon^{L+1}O_1(\Lambda^0)$ for $\Psi\varepsilon$ replaced by $\Psi_\varepsilon\circ\Psi_{\varepsilon,L}$; equalities $(6.2.24)_{1-3}$ automatically remain true at each step.

We assert that there exists a symplectomorphism $\Psi_{\varepsilon,L,0}\equiv I \pmod{O(\varepsilon^L)}$ such that $a_\varepsilon\circ\Psi_\varepsilon\circ\Psi_{\varepsilon,L,0}$ is equal to the right-hand expression of (6.2.23) or (6.2.25) modulo $\varepsilon^L O_2(\Lambda^0)+O(\varepsilon^{L+1})$.

In fact, for $d=2$ the term preventing us from concluding the last equality is $\varepsilon^L(\alpha\xi_2+\beta x_2)$ with $\alpha=\alpha(x_1,\xi_1)$ and $\beta=\beta(x_1,\xi_1)$. Let us set $w_1=\xi_2+\alpha_1\varepsilon^L$, $w_2=x_2+\beta_1\varepsilon^L$, $w_3=\xi_1$, $w_4=x_1$ with $\alpha_1=\alpha/2b_1$, $\beta_1=\beta/2b_1$. Then the necessary commutation relations are fulfilled modulo $O(\varepsilon^L)$. Furthermore, perturbing w_j with $j=1,2$ by functions of the class $\varepsilon^L O_1(\Lambda^0)$ we obtain a commutation relation between w_1 and w_2 fulfilled modulo $\varepsilon^L O_1(\Lambda^0)+O(\varepsilon^{L+1})$. Then there exist functions w'_j fulfilling the same commutation relations exactly such that $w'_j\equiv w_j \pmod{(\varepsilon^L O_1(\Lambda^0)+O(\varepsilon^{L+1}))}$ for $=1,2$ and $w'_j\equiv w_j \pmod{O(\varepsilon^L)}$ for $j=3,4$. Then the symplectomorphism $\Psi_{\varepsilon,L,0}$

transforming (w_1', w_2', w_3', w_4') to (ξ_2, x_2, ξ_1, x_1) is obviously the required symplectomorphism.

Similarly, for $d = 3$ the term preventing us from concluding the last equality is $\varepsilon^L(\alpha\xi_3 + \beta x_3 + \gamma\xi_2)$ with $\alpha = \alpha(x', \xi_1)$, $\beta = \beta(x', \xi_1)$ and $\gamma = \gamma(x', \xi_1)$. Let us set $w_1 = \xi_3 + \alpha_1\varepsilon^L$, $w_2 = x_3 + \beta_1\varepsilon^L$, $w_3 = \xi_2 + \gamma_1\varepsilon^L$ with $\alpha_1 = \alpha/2b_1$, $\beta_1 = \beta/2b_1$, $\gamma_1 = \gamma/2b_1$ and, moreover, let us set $w_4 = x_2$, $w_5 = \xi_1$, $w_6 = x_1$. Then all the remaining arguments used for $d = 2$ remain true with obvious modifications.

Therefore it is sufficient to prove that if (6.2.23) or (6.2.25) is fulfilled modulo $\varepsilon^L O_M(\Lambda^0) + O(\varepsilon^{L+1})$ with $2 \le M \le 2(N-L)-1$ then there exists a symplectomorphism $\Psi_{\varepsilon,L,M} \equiv I \ (\mathrm{mod}(\varepsilon^L O_{M-1}(\Lambda^0) + O(\varepsilon^{L+1})))$ such that for Ψ_ε replaced by $\Psi_\varepsilon \circ \Psi_{\varepsilon,L,M}$ the same equality is fulfilled modulo $\varepsilon^L O_{M+1}(\Lambda^0) + O(\varepsilon^{L+1})$.

Repeating the arguments used in the proof of proposition 6.2.1(i) or (ii) it is easy to construct this symplectomorphism in the form $\Psi_{\varepsilon,L,M} = \exp \varepsilon^L H_S$ where $S \in O_M(\Lambda^0)$ for $d = 2$ and $S = S' + S''$ where S' is of the form (6.2.17) with $\frac{M+1}{2}$ instead of N and $S'' \in O_M(\Lambda^0)$ for $d = 3$. We only note that (6.2.13) is replaced by formula

$$f \circ \Psi_{\varepsilon,L,M} \sim f + \varepsilon^L\{S, f\} + \frac{1}{2}\varepsilon^{2L}\{S, \{S, f\}\} + \dots$$

which is always applicable. So assertions (i), (ii) are proven. Let us prove (iii). Let us note that $(6.2.26)_1$ follows from (6.2.9). Without loss of generality one can assume in the proof of $(6.2.26)_{2,3}$ that $F = 1$ and let us again consider a conformal coordinate system. Let Ψ be the symplectomorphism constructed in proposition 6.2.1(iii); let us consider the decomposition of $W = V \circ \Psi$ in powers of x_2, ξ_2:

$$W \sim \sum_\gamma v_\gamma(x_1, \xi_1) x_2^{\gamma_1} \xi_2^{\gamma_2}.$$

Let us return to the origin of the proof of (i). We took $w_1 = \xi_2 + \alpha_1\varepsilon$, $w_2 = x_2 + \beta_1\varepsilon$, $w_3 = \xi_1$ and $w_4 = x_1$ and noted that the commutation relations were fulfilled modulo $O(\varepsilon)$. It is easy to see that one can change $w_1, \dots, w_4$ by functions of the class $\varepsilon O_1(\Lambda^0)$ such that for new $w_1, \dots, w_4$ all the commutation relations (not only between w_1 and w_2) will be fulfilled modulo $\varepsilon O_1(\Lambda^0) + O(\varepsilon^2)$ and therefore $w_j' \equiv w_j \ (\mathrm{mod}\, \varepsilon O_1(\Lambda^0) + O(\varepsilon^2))$ for the appropriate method of reduction. Therefore for the symplectomorphism constructed at this step

$$a_\varepsilon \circ \Psi_\varepsilon \equiv a \circ \Psi + \varepsilon Q(x_1, \xi_1; x_2, \xi_2) - \varepsilon^2(\alpha_1^2 + \alpha_2^2) \qquad (\mathrm{mod}\, \varepsilon O_2(\Lambda^0) + \varepsilon^2 O_1(\Lambda^0) + O(\varepsilon^3))$$

with $Q(x_1, \xi_1; x_2, \xi_2) = v_{(20)} x_2^2 + 2v_{(21)} x_2\xi_2 + v_{(02)}\xi_2^2$; symplectomorphisms of the remaining part of the proof of (i) kill the traceless part of the quadratic form Q and preserve the remaining terms indicated here. Since $\alpha_1 = \frac{1}{2}v_{(01)}^2$, $\beta_1 = \frac{1}{2}v_{(10)}$ then

$$b_{11} = \frac{1}{2}(v_{(20)} + v_{(02)}), \quad b_{20} = -\frac{1}{4}(v_{(10)}^2 + v_{(01)}^2) \tag{6.2.27}$$

and we only need to calculate $v_{(10)}, v_{(01)}, v_{(20)}, v_{(02)}$. But

$$v_{(10)} = -\{W, \xi_2\} \circ \Pi = -\{V, w_1\} \circ \Psi \circ \Pi$$

where here and below $w_1 = \xi_2 \circ \Psi^{-1}$, $w_2 = x_2 \circ \Psi^{-1}$, $\Pi(x, \xi) = (x_1, \xi_1, 0, 0)$; since $w_1 \equiv s$, $w_2 \equiv t \ (\mathrm{mod}\, O_2(\Lambda^0))$ (see the proof of proposition 6.2.1 (iii)) then

$$v_{(10)} = \{s, V\} \circ \bar{\Psi} = \rho\partial_1 V \circ \bar{\Psi}$$

and similarly $v_{(01)} = -\rho\partial_2 V \circ \bar{\Psi}$; these equalities yield $(6.2.25)_3$ with $l^j = -\frac{1}{4}\rho^2\partial_j V$.

Moreover,

$$v_{(20)} = \frac{1}{2}\{\{W,\xi_2\},\xi_2\} \circ \Pi = \frac{1}{2}\{\{V,w_1\},w_1\} \circ \Psi \circ \Pi$$

and a similar equality is true for $v_{(02)}$. Let us recall that $w_1 \equiv s + ut$, $w_2 \equiv t - us$ $(\operatorname{mod} O_3(\Lambda))$ with $u = \alpha s + \beta t$; then (6.2.18) and (6.2.27) yield that

$$b_{11} = \frac{1}{4}(\rho^2 \Delta V) \circ \bar{\Psi};$$

this equality yields $(6.2.26)_2$ with $k^j = 0$.

Finally, if Φ_ε is an arbitrary symplectomorphism of a neighborhood of Λ_1^0 onto a neighborhood of Λ_2^0, smoothly depending on ε, mapping Λ_1^0 onto Λ_2^0 and preserving decomposition (6.2.23) then repeating the arguments of the proof of proposition 6.2.1(iii) with obvious modifications we obtain that $\Phi_\varepsilon \equiv \Phi' \circ \Phi''_\varepsilon \ (\operatorname{mod} \sum_n \varepsilon^n O_{2N-2n-2}(\Lambda_1^0))$ where Φ' is a symplectomorphism preserving (x_2, ξ_2) (as in the end of the proof of proposition 6.2.1(iii)) and $\Phi''_\varepsilon = \exp H_{S_\varepsilon}$,

$$S_\varepsilon = \sum_{1 \le m+n \le N-1} \phi_{mn}(x_1,\xi_1)\varepsilon^m (x_2^2 + \xi_2^2)^n$$

with arbitrary functions ϕ_{mn}; all the coefficients b_{mn} are invariant with respect to Φ'; on the other hand, Φ''_ε transforms b_{20} and b_{11} to $b_{20} + \{\phi_{10}, b_{10}\}$ and $b_{11} + \{\phi_{01}, b_{10}\}$ respectively. This and $(6.2.24)_3$ yield $(6.2.26)_{2,3}$ and the possibility of choosing k^j and l^j arbitrarily at every fixed point. □

At the end of this subsection let us get rid of the condition that F is constant:

Proposition 6.2.6. *Let Ψ_ε be the symplectomorphism constructed in theorem 6.2.5(i). Then*

$$a_\varepsilon \circ \Psi_\varepsilon = \sum_{m+n \le N-1} b'_{mn}(x_1,\xi_1)\varepsilon^m \bar{a}_\varepsilon^n + \sum_{n \le N} \varepsilon^n R'_n(x,\xi,\varepsilon) \tag{6.2.28}$$

where b'_{mn}, R'_n are smooth functions and $R'_n \in O_{2N-2n-1}(\Lambda^0)$,

$$\bar{a}_\varepsilon = x_2^2 + \xi_2^2 + \varepsilon b(x_1,\xi_1), \tag{6.2.29}$$

$$b = v \circ \bar{\Psi}_0, \quad v = \frac{V}{F}, \tag{6.2.30}$$

$$b'_{00} = b'_{10} = 0, \quad b'_{01} = F \circ \bar{\Psi}_0, \tag{6.2.31}$$

$$b'_{20} = \Bigl(\frac{1}{8}F\kappa v^2 - \frac{1}{4}g^{jk}v\partial_j\partial_k v + m^j\partial_j v\Bigr) \circ \bar{\Psi}_0, \tag{6.2.32}$$

and m^j depend on the method of reduction and can be chosen arbitrarily at every fixed point.

Proof. Formulas (6.2.28)–(6.2.31) obviously follow from theorem 6.2.4(i); formula (6.2.32) for constant F follows from assertion (iii) of the same theorem and $m^j = l^j - k^j v$ here. Let us prove (6.2.31) in the general case; the proof of assertion (i) of theorem 6.2.4(i) yields that it is sufficient to do the reduction with $N = 3$. Let Ψ'_ε be a symplectomorphism, reducing $F^{-1}a_\varepsilon$; then modulo terms negligible for $N = 3$

$$F^{-1}a_\varepsilon \circ \Psi'_\varepsilon \equiv \bar{a}_\varepsilon + \gamma(x_1,\xi_1)\varepsilon^2 + f(x_1,\xi_1)\bar{a}_\varepsilon$$

with $\gamma = b_{20} - b_{11}b + b_{02}b^2$ and $f \in O_2(\Lambda^0) + O(\varepsilon)$. Theorem 6.2.5(iii) yields that γ coincides with the right-hand expression of (6.2.32) multiplied by $F^{-1} \circ \bar{\Psi}_0$. Then

$$a_\varepsilon \circ \Psi'_\varepsilon = (F \circ \Psi'_\varepsilon)(\bar{a}_\varepsilon + \gamma\varepsilon^2 + f\bar{a}_\varepsilon) \equiv (b'_{01} + f')\bar{a}_\varepsilon + b'_{20}\varepsilon^2$$

where b'_{10} and b'_{20} are given by (6.2.30) and (6.2.31) and $f' \in O_1(\Lambda^0) + O(\varepsilon)$. It is easy to construct a symplectomorphism $\Psi''_\varepsilon = \exp H_{S_\varepsilon}$ with $S_\varepsilon = (S' + S'')\bar{a}_\varepsilon$, where $S' \in O_1(\Lambda^0)$ and $S'' \in O_2(\Lambda^0)$ are chosen consequently, such that

$$a_\varepsilon \circ \Psi'_\varepsilon \equiv (b'_{01} + f'')\bar{a}_\varepsilon + b'_{20}\varepsilon^2$$

with $f'' = \phi_1(x_1,\xi_1)\varepsilon + \phi_2(x_1,\xi_1)\bar{a}_\varepsilon$; the proposition is proven. □

6.2.3 Final reduction

Let us proceed from symplectomorphisms to Fourier integral operators. Let us consider the η-differential operator

$$A'_{\eta,\varepsilon} = (\eta D_j - V_j)g^{jk}(\eta D_k - V_k) + \varepsilon V,$$

depending on a small parameter ε. In what follows $\eta = \mu^{-1}h$ and $\varepsilon = \mu^{-2}$.

Theorem 6.2.7. *Let conditions (6.1.1)–(6.1.4) and (6.2.1) be fulfilled. Then for $d = 2, 3$ and small enough ε there exists an (η-)Fourier integral operator $\mathcal{F}_{\eta,\varepsilon}$, depending smoothly on ε and corresponding to the symplectomorphism Ψ_ε constructed in theorem 6.2.6, such that modulo η-pseudo-differential operators depending smoothly on ε with symbols vanishing in a neighborhood of Λ^0 (which does not depend on ε)*

$$\mathcal{F}^*_{\eta,\varepsilon}\mathcal{F}_{\eta,\varepsilon} \equiv I, \tag{6.2.33}$$

$$\mathcal{F}^*_{\eta,\varepsilon}A'_{\eta,\varepsilon}\mathcal{F}_{\eta,\varepsilon} \equiv \mathcal{A}'_{\eta,\varepsilon} \tag{6.2.34}$$

and modulo η-pseudo-differential operators depending smoothly on ε with symbols vanishing in a neighborhood $\mathcal{W}$ of (for $d = 2$) any point $z \in \Lambda$ or (for $d = 3$) any non-closed Hamiltonian trajectory Γ of finite length generated by $F^j(\xi_j - V_j)$

$$\mathcal{F}_{\eta,\varepsilon}\mathcal{F}^*_{\eta,\varepsilon} \equiv I, \tag{6.2.35}$$

$$\mathcal{F}_{\eta,\varepsilon}\mathcal{A}'_{\eta,\varepsilon}\mathcal{F}^*_{\eta,\varepsilon} \equiv A'_{\eta,\varepsilon} \tag{6.2.36}$$

where for $d = 2$

$$\mathcal{A}'_{\eta,\varepsilon} = \sum_{0\le n\le N-1} B^w_n(x_1,\eta D_1,\varepsilon,\eta)\mathcal{L}^n_0 + R \tag{6.2.37}$$

and for $d = 3$

$$\mathcal{A}'_{\eta,\varepsilon} = \eta^2 D_2^2 + \sum_{0\le n\le N-1} B^w_n(x',\eta D_1,\varepsilon,\eta)\mathcal{L}^n_0 + R \tag{6.2.38}$$

and

(6.2.39)
$$\mathcal{L}_0 = x_d^2 + \eta^2 D_d^2,$$

(6.2.40)
$$R = \sum_{m+n\le N} R_{mn}(x,\eta D,\varepsilon,\eta)\varepsilon^m\eta^n\mathcal{L}_0^{N-m-n},$$

B_n^w and R_{mn} are η-pseudo-differential operators depending smoothly on ε, the Weyl symbols of B_n are real-valued and

(6.2.41)$_1$
$$B_0(.,.,\varepsilon,\eta) = \varepsilon V\circ\bar{\Psi}_0 + O(\varepsilon^2+\eta^2),$$

(6.2.41)$_2$
$$B_1(.,.,0,0) = F\circ\bar{\Psi}_0.$$

So, we have the *microlocal canonical form* for two- and three-dimensional Schrödinger operators with magnetic field.

Proof. Let $\mathcal{F}'_{\eta,\varepsilon}$ be a compactly supported (η-)Fourier integral operator corresponding to the symplectomorphism Ψ_ε, depending smoothly on ε and satisfying (6.2.33); then (6.2.35) is fulfilled automatically. Theorem 6.2.5 yields that the operator $\mathcal{B}'_{\eta,\varepsilon} = \mathcal{F}'^{\,*}_{\eta,\varepsilon}A'_{\eta,\varepsilon}\mathcal{F}'_{\eta,\varepsilon}$ differs from the operator $\mathcal{A}'_{\eta,\varepsilon}$ of the form (6.2.37) or (6.2.38) by a term $\eta r(x,\eta D,\varepsilon,\eta)$ where r is an η-pseudo-differential operator; all operators here and in what follows depend smoothly on ε.

Since $A'_{\eta,\varepsilon}$ is formally self-adjoint the same is true for $\mathcal{B}'_{\eta,\varepsilon}$, i.e., its Weyl symbol is real-valued. It is easy to see that the subprincipal symbol of $A'_{\eta,\varepsilon}$ is equal to 0; therefore the subprincipal symbol of $B'_{\eta,\varepsilon}$ is a linear combination of the first-order derivatives of its principal symbol. Hence r is an η-pseudo-differential operator and $r(.,.,0,0)=0$ at Λ^0. This yields (6.2.41)$_2$ automatically in what follows.

It is sufficient to prove that there exists an η-pseudo-differential operator $Q_{\eta,\varepsilon}$ such that

(6.2.42)
$$Q^*_{\eta,\varepsilon}Q_{\eta,\varepsilon}\equiv I, \qquad Q_{\eta,\varepsilon}Q^*_{\eta,\varepsilon}\equiv I,$$

(6.2.43)
$$Q^*_{\eta,\varepsilon}\mathcal{B}'_{\eta,\varepsilon}Q_{\eta,\varepsilon}\equiv I,$$

modulo η-pseudo-differential operators with symbols vanishing in a neighborhood of Λ^0 (which does not depend on ε). Then $\mathcal{F}_{\eta,\varepsilon} = \mathcal{F}'_{\eta,\varepsilon}Q_{\eta,\varepsilon}$ satisfies (6.2.33)–(6.2.36). Obviously, we need to fulfill only first equalities in (6.2.42) and (6.2.43); the second equalities follow from them automatically.

In order to construct this operator $Q_{\eta,\varepsilon}$ it is sufficient to prove that if the operator $\mathcal{B}'_{\eta,\varepsilon}$ differs from an operator of the form (6.2.37) or (6.2.38) by a term $\eta^J\varepsilon^L r^w(x,\eta D,\varepsilon,\eta)$ where r is an η-pseudo-differential operator with real-valued symbol such that $r(.,.,0,0)\in O_M(\Lambda^0)$ and $J\ge 1$, $M\ge 1$, $L\ge 0$ then there exists an η-pseudo-differential operator $q_{\eta,\varepsilon}$ satisfying (6.2.42) such that $q^*_{\eta,\varepsilon}\mathcal{B}'_{\eta,\varepsilon}q_{\eta,\varepsilon}$ differs from an operator of the form (6.2.37) or (6.2.38) by a term $\eta^J\varepsilon^L r'^w(x,\eta D,\varepsilon,\eta)$ where r' is an η-pseudo-differential operator (automatically with real-valued symbol) such that $r'(.,.,0,0)\in O_{M+1}(\Lambda^0)$. We start with $M=1$ because the term $\eta^J\varepsilon^L r^w(x',\eta D_1,\varepsilon,\eta)$ can be included in decomposition (6.2.37) or (6.2.38).

It is easy to construct $q_{\eta,\varepsilon}$ by means of simplifications of the arguments of the proof of proposition 6.2.1; we search for $q_{\eta,\varepsilon}$ of the form $\exp(i\eta^{J-1}\varepsilon^L S^w(x,\eta D))$ with a real-valued function S; here $S\in O_M(\Lambda^0)$ for $d=2$ and $S=S'+S''$ where $S'\in O_{M-1}(\Lambda^0)$ is of the form (6.2.17) with $N=(M+1)/2$ and $S''\in O_M(\Lambda^0)$ for $d=3$; then for $J=1$, $q_{\eta,\varepsilon}$ is an operator with principal symbol $\exp(i\varepsilon^L S(x,\xi))$ and for $J\ge 2$ $q_{\eta,\varepsilon}$ differs from I by an operator with principal symbol $i\eta^{J-1}\varepsilon^L S(x,\xi)$. □

For $d = 2$ one can improve this theorem:

Proposition 6.2.8. *Let conditions (6.1.1)–(6.1.4) and (6.2.1) be fulfilled and let $\mathcal{A}'_{\eta,\varepsilon}$ be a reduced operator of the form (6.2.37). Then one can rewrite it in the form*

$$\mathcal{A}'_{\eta,\varepsilon} \equiv \sum_{1\le n\le N-1} b_n(x_1, \eta D_1, \varepsilon, \eta)\mathcal{L}^n + R' \tag{6.2.44}$$

where R' is an operator of the form (6.2.40), b_n are now not necessarily real-valued, $b_1(.,.,0,0) = F \circ \bar{\Psi}_0$,

$$\mathcal{L} = x_2^2 + \eta^2 D_2^2 + \varepsilon B^w(x_1, \eta D_1, \varepsilon, \eta), \tag{6.2.45}$$

$$B(.,.,\varepsilon,\eta) = v \circ \bar{\Psi}_0 + \varepsilon\omega_1 + \eta\omega_2 + O(\varepsilon^2) + O(\eta^2\varepsilon^{-1}) + O(\varepsilon\eta), \tag{6.2.46}$$

$$\omega_1 = (\frac{1}{8}\kappa v^2 - \frac{1}{4} g^{jk} F^{-1} v \partial_j \partial_k v + m^j \partial_j v) \circ \bar{\Psi}_0, \tag{$6.2.47)_1$}$$

$$\omega_2 = r^j \partial_j v \circ \bar{\Psi}_0; \tag{$6.2.47)_2$}$$

let us recall that $v = \frac{V}{F}$ and κ is the scalar curvature corresponding to the metric $\frac{1}{F} g^{jk}$; here r^j and m^j can be chosen arbitrarily at every fixed point.

Proof. All of the proposition excluding formulas $(6.2.47)_{1,2}$ follows immediately from theorem 6.2.7; $(6.2.47)_1$ follows from proposition 6.2.6 and $(6.2.47)_2$ follows from the fact that the subprincipal symbol of $\mathcal{A}'_{\eta,\varepsilon}$ is a linear combination of the first-order derivatives of its principal symbol. The possibility of choosing m^j arbitrarily was proven and the possibility of choosing r^j arbitrarily then follows if we multiply $\mathcal{A}'_{\eta,\varepsilon}$ by q^* and $q = \exp(iS(x_1, \eta D_1))$ on the left and right respectively. □

Remark 6.2.9. *(i) No part of the reduction process (from proposition 6.2.1 to proposition 6.2.8) is uniquely determined; however, it can be taken stable in the natural sense; surely there are natural limitations on $k^j, \dots, r^j$: $|k^j| \le c$, etc.*

(ii) It is easy to see that (as in, e.g., the construction of theorem 6.1.1) if g^{jk}, F_{jk} and V are constant then (6.2.37) and (6.2.38) are replaced by

$$\mathcal{A}'_{\eta,\varepsilon} \equiv F(x_2^2 + \eta^2 D_2^2) + \varepsilon V, \tag{6.2.37)'}$$

$$\mathcal{A}'_{\eta,\varepsilon} \equiv \eta^2 D_2^2 + F(x_3^2 + \eta^2 D_3^2) + \varepsilon V. \tag{6.2.38)'}$$

This fact and stability yields that if

$$(6.2.48)_{1-3}\quad |D^\alpha(g^{jk} - \bar{g}^{jk})| \le c\bar{\nu}, \quad |D^\alpha(F_{jk} - \bar{F}_{jk})| \le c\bar{\nu},$$
$$|D^\alpha(V - \bar{V})| \le c\bar{\nu} \qquad \forall \alpha : |\alpha| \le K \quad \forall x \in B(0,1)$$

with constant $\bar{g}^{jk}$, $\bar{F}_{jk}$, $\bar{V}$ then in (6.2.37), (6.2.38) and (6.2.39) the operators $\bar{\nu}^{-1}(B_0 - \bar{V})$, $\bar{\nu}^{-1}(B_1 - \bar{F})$ and $\bar{\nu}^{-1}B_n$ with $n \ge 2$ are η-pseudo-differential operators uniformly with respect to $\bar{\nu} \in (0,1]$; in (6.2.44), (6.2.45) the same is true for the operators $\bar{\nu}^{-1}(b_1 - \bar{F})$, $\bar{\nu}^{-1}(B - \frac{\bar{V}}{\bar{F}})$ and $\bar{\nu}^{-1}b_n$ with $n \ge 2$; moreover, one can replace $O(\varepsilon^2)$, etc., by $O(\varepsilon^2\bar{\nu})$, etc., in (6.2.46). ◊

Remark 6.2.10. *In what follows we assume without loss of generality and without special mention that the symbol of the η-pseudo-differential operator $\mathcal{F}^*_{\eta,\varepsilon}\mathcal{F}_{\eta,\varepsilon}$ is supported in a small enough neighborhood of Λ^0.* ◊

6.2.4 Reduced problems

Let us consider what we have obtained in the investigation of U, etc. Let $\mu \geq 1$, $h \in (0,1]$, $\mu h \leq 1$ and $\eta = \mu^{-1}h$, $\varepsilon = \mu^{-2}$. Let the operator $\mathcal{F} = \mathcal{F}_{\eta,\varepsilon}$ be the operator constructed in theorem 6.2.7 and let $\mathcal{U} = \mathcal{F}_x^* U^t \mathcal{F}_y^*$. Then

$$(hD_t - 2\mathcal{A})\mathcal{U} \equiv 0, \tag{6.2.49}$$

$$\mathcal{U}|_{t=0} \equiv \delta(x-y) \tag{6.2.50}$$

modulo functions f and f' respectively such that if Q is a $\mu^{-1}h$-pseudo-differential operator with symbol q supported in a small enough neighborhood of Λ^0 then

$$\|F_{t\to h^{-1}\tau}\chi_{T,T'}(t)Q_x f\| \leq C'Th^s, \tag{6.2.51}$$

$$\|Q_x f'\| \leq C'h^s \tag{6.2.52}$$

for arbitrary s (provided $K = K(N,s,\delta)$ in condition (6.1.2)), $T \geq h^{1-\delta}$, T' and $\delta > 0$ are arbitrary, C' depends on $c, \epsilon_0, N, s, \delta$ and contains factors $\|\!|q|\!\|_{S_{\mu^{-1}h}}$ and $\|\chi\|_{C^K}$ (only in (6.2.51)); moreover, $\mathcal{A} \equiv \mu^2 \mathcal{A}'_{\mu,h}$ where for $d=2$, $\mathcal{A}'_{\mu,h}$ is given by (6.2.37) or (6.2.44) and for $d=3$, $\mathcal{A}'_{\mu,h}$ is given by (6.2.38), $\eta = \mu^{-1}h$, $\varepsilon = \mu^{-2}$.

Moreover, $\mathcal{U}$ satisfies the dual equation and the properties of U yield similar properties of $\mathcal{U}$:

$$\mathcal{U}(x,y,t) = \mathcal{U}^\dagger(y,x,-t), \tag{6.2.53}$$

$$|F_{t\to h^{-1}\tau}\chi_{T,T'}(t)\mathcal{U}| \leq C'\mu T h^{-d} \tag{6.2.54}$$

provided $T \geq h^{1-\delta}$ where C' is the same constant as above and one can assume without loss of generality that

$$\operatorname{supp}\mathcal{U} \subset B(0,c_1) \times B(0,c_1) \times \mathbb{R}. \tag{6.2.55}$$

Finally, the definition of $\mathcal{U}$, (6.2.33), (6.2.34) and theorem 6.1.8 yield that

$$|F_{t\to h^{-1}\tau}\chi_{T,T'}(t)(U - \mathcal{F}_x\mathcal{U}^t\mathcal{F}_y)| \leq C'Th^s \qquad \forall x,y \in \mathcal{W}_x \quad \forall \tau \leq 4c \tag{6.2.56}$$

where $\mathcal{W}_x$ is the x-projection of $\mathcal{W}$.

6.3 Spectral Asymptotics in the Case of a Weak Magnetic Field

In this section we obtain basic asymptotics under the assumption that the magnetic field is not very strong, i.e.,

$$\mu \in [\mu_0, h^{\delta-1}] \tag{6.3.1}$$

where here and in what follows μ_0 is a large enough constant and the exponent $\delta > 0$ is arbitrarily small. Other assumptions will appear later. The results obtained here make it possible to assume in the next sections that $\mu \geq h^{-\delta}$; in this case the remainder R in (6.2.37), (6.2.38) as well as all the terms containing large negative degrees of μ will be negligible. In *this* section the microlocal canonical form is applied only to obtain a more refined results on the propagation of singularities; the calculations are taken from section 4.2 and are made for the original (non-reduced to canonical form) operator; a more accurate analysis of the propagation of singularities permits us to take larger values of T in the Tauberian arguments. In the following sections the microlocal canonical form will also be used for calculations. We always assume that all the assumptions used in section 6.2 are fulfilled.

6.3.1 Analysis of singularities

Let us start with an analysis of singularities. First of all let us analyze their location:

Proposition 6.3.1. *Let $d = 2,3$ and let condition (6.3.1) be fulfilled. Let $q \in S_{\mu^{-1}h,\mu^{-1},\mu^{-1},K}(T^*\mathbb{R}^d)$ and $Q = \mathrm{Op}(q) = q(x,\mu^{-1}hD,\mu^{-1}h)$; let us assume that*

$$(6.3.2)_{(2)} \qquad \operatorname{supp} q \cap \{\tau_1 \le \bar{B}(x_1,\xi_1) + \bar{B}_1(x_1,\xi_1)(x_2^2+\xi_2^2)\mu^2 \le \tau_2\} = \emptyset,$$

$$(6.3.2)_{(3)} \qquad \operatorname{supp} q \cap \{\tau_1 \le \bar{B}_0(x',\xi_1) + \bar{B}_1(x',\xi_1)(x_3^2+\xi_3^2)\mu^2 + \mu^2\xi_2^2 \le \tau_2\} = \emptyset$$

for $d=2,3$ respectively where the bar means that the corresponding symbol is calculated for $\eta=\varepsilon=0$ and where $-\infty \le \tau_1 < \tau_2 \le \infty$. Then

$$(6.3.3) \quad |F_{t\to h^{-1}\tau}\chi_{T,T'}(t)Q_x\mathcal{U}| \le C'Th^s\bigl(1+\min(\tau-\tau_1,\tau_2-\tau)\bigr)^{-s} \qquad \forall x,y\in\mathbb{R}^d \quad \forall\tau\in[\tau_1+\epsilon,\tau_2-\epsilon]$$

where $s,\epsilon>0, \delta>0, T\ge h^{1-\delta}$ and T' are arbitrary, $K=K(\delta,s)$, $N=N(\delta,s)$ in (6.2.37), (6.2.38), $\mu_0=\mu_0(c,\epsilon)$ in (6.3.1) and $C'=C'(c,\epsilon,\delta,s)$ also includes factors $\|q\|$ and $\|\chi\|$. This estimate remains true for Q_x replaced by ${}^tQ_y$.

Proof. Let us consider equation (6.2.49). The principal symbol of the η-pseudo-differential operator $\mathcal{A}$ is

$$\bar{B}_0 + \bar{B}_1(x_2^2+\xi_2^2)\mu^2 + O\bigl((x_2^2+\xi_2^2)^2\mu^2+\mu^{-2}\bigr)$$

for $d=2$ and

$$\bar{B}_0 + \bar{B}_1(x_3^2+\xi_3^2)\mu^2 + \mu^2\xi_2^2 O\bigl((x_3^2+\xi_3^2)^2\mu^2+\mu^{-2}\bigr)$$

for $d=3$. Therefore $hD_t-\mathcal{A}$ is elliptic under the above conditions in the zone $\{\lambda(x,\xi)\le\epsilon',\tau_1+\epsilon\le\tau\le\tau_2-\epsilon\}$ where $\epsilon'>0$ is a small enough constant and $x'=x_1$, $\lambda(x,\xi)=x_2^2+\xi_2^2$ for $d=2$ and $x'=(x_1,x_2)$, $\lambda(x,\xi)=x_3^2+\xi_3^2+\xi_2^2$ for $d=3$. On the other hand, for any $\epsilon'>0$ and $\mu\ge\mu_0(c,\delta,\epsilon,\epsilon')$ this operator is automatically elliptic in the zone $\{\lambda(x,\xi)\le\epsilon',\tau_1+\epsilon\le\tau\le\tau_2-\epsilon\}$ (recall that $\bar{B}_1\ge\epsilon_0$). Applying standard elliptic arguments, estimate (6.2.54) and remark 6.2.10, one easily proves this proposition. □

Remark 6.3.2. *Let us note that for $\tau_1=-\infty$, $\tau_2\le c$ and q supported in $\{\lambda(x,\xi)\ge C_0\mu^{-2}\}$ with $C_0=C_0(c)$ condition (6.3.2) is fulfilled automatically; therefore proposition 6.3.1 asserts in particular that $\mathcal{U}$ is negligible outside of the zone $\{\lambda(x,\xi)\le C_0\mu^{-2}\}$ and hence in decompositions (6.2.37), (6.2.38) and (6.2.44) $R=O(\mu^{2-2N})$ in the heuristic sense. Replacing $\mathcal{U}$ by $Q_x\mathcal{U}\equiv\mathcal{U}$ with $q=1$ for $\lambda(x,\xi)\le C_0\mu^{-2}$ and supported in $\{\lambda(x,\xi)\le 2C_0\mu^{-2}\}$ we reach $R=O(\mu^{2-2N})$ in the sense that $R\mathcal{U}\equiv R'\mathcal{U}$ with $\mu^{2N-2}R'\in\boldsymbol{\Psi}_{\mu^{-1}h,\mu^{-1},\mu^{-1},K}$. Moreover, remark 6.2.9(ii) yields that $\bar{\nu}^{-1}\mu^{2N-2}R'\in\boldsymbol{\Psi}_{\mu^{-1}h,\mu^{-1},\mu^{-1},K}$ under condition $(6.2.48)_{1-3}$.* ◊

Let us proceed to an analysis of propagation of singularities; here we obtain the principal difference between $d=2$ and $d=3$:

Theorem 6.3.3. *Let $d=2,3$ and condition (6.3.1) be fulfilled with $\mu_0=\mu_0(c,\delta)$. Let $Q'=\mathrm{Op}(q')$, $Q''=\mathrm{Op}(q'')$, $q',q''\in S_{\mathbf{h},\boldsymbol{\rho},\boldsymbol{\gamma},K}(T^*\mathbb{R}^d)$ where*

$$(6.3.4) \quad \mathbf{h}=(h_1,\dots,h_d),\ \boldsymbol{\gamma}=(\gamma_1,\dots,\gamma_d),\ \boldsymbol{\rho}=(\rho_1,\dots,\rho_d), \qquad h_1=\mu^{-1}h,\ h_2=h_3=h,\ \rho_j,\gamma_j\in(0,1],\ \rho_j\gamma_j\ge h_jh^{-\delta}\quad\forall j,\ \gamma_d\le\mu^{-1}$$

and quantization is taken with the vector semiclassical parameter $\mathbf{h}$. *Then*

(i)$_{(3)}$ Let $d=3$ *and one of the following conditions*

$$|x_1-y_1|\ge C_0\mu^{-1}T+\gamma_1, \tag{6.3.5$_1$}$$

$$|\xi_1-\eta_1|\ge C_0\mu^{-1}T+\rho_1, \tag{6.3.5$_2$}$$

$$|x_2-y_2|\ge C_0T+\gamma_2, \tag{6.3.5$_3$}$$

$$|\xi_2-\eta_2|\ge C_0T+\rho_2 \tag{6.3.5$_4$}$$

be fulfilled for all $(x,\xi)\in\operatorname{supp}q'$ *and for all* $(y,\eta)\in\operatorname{supp}q''$ *where*

$$T\in[h^{1-\delta},T_0], \tag{6.3.6$_{(3)}$}$$

$C_0=C_0(c)$, $T_0=T_0(c,\delta)>0$. *Then*

$$|F_{t\to h^{-1}\tau}\chi_T(t)Q'_x\mathcal{U}^tQ''_y|\le C'h^s \qquad \forall x,y\in\mathbb{R}^d \quad \forall\tau\le c. \tag{6.3.7}$$

(i)$_{(2)}$ Let $d=2$ *and either* (6.3.5)$_1$ *or* (6.3.5)$_2$ *be fulfilled with*

$$T\in[h^{1-\delta},\mu T_0]. \tag{6.3.6$_{(2)}$}$$

Then with C_0, T_0 *the same as before, estimate* (6.3.7) *remains true.*

(ii) Moreover, if $\rho_d=\mu^{\frac12}h^{\frac12(1-\delta)}$, $\gamma_d=\mu^{-\frac12}h^{\frac12(1-\delta)}$ *and if the condition*

$$|\mu^2x_d^2+\xi_d^2-\mu^2y_d^2-\eta_d^2|\ge C\mu^{-l}+\rho_d \tag{6.3.8}$$

is fulfilled for all $(x,\xi)\in\operatorname{supp}q'$ *and for all* $(y,\eta)\in\operatorname{supp}q''$ *with* $C=C(c,\delta,l)$ *then for* T *satisfying* (6.3.6)$_{(d)}$ *estimate* (6.3.7) *holds.*

Assertions (i) and (ii) mean respectively that the velocity of propagation of singularities with respect to (x_1,ξ_1) does not exceed $C_0\mu^{-1}$ and for $d=3$ the velocity with respect to (x_2,ξ_2) does not exceed C_0 and that quantity $\mu^2x_d^2+\xi_d^2$ is almost constant for large μ (all these assertions hold for energy levels not exceeding c and for quantization corresponding to the vector $\mathbf{h}$) indicated above. Let us emphasize that for $d=2$ and $d=3$ the restrictions on T from above are different, which will be essential for the Tauberian arguments.

For $d=3$ a more accurate theorem will be proven in section 6.5. For $d=2$ we have the more accurate

Theorem 6.3.4. *Let* $d=2$ *and* $q',\dots,Q''$ *be as in theorem 6.3.4. Let us assume that conditions* (6.2.48)$_{(1-3)}$ *are fulfilled and*

$$\nu^{-1}(B-\tilde{B})\in\boldsymbol{\Psi}_{\mu^{-1}h,K}(T^*\mathbb{R}^d) \tag{6.3.9}$$

where B *is the operator introduced in* (6.2.45) *and* $\tilde{B}=\mathrm{const}$. *Furthermore, let us assume that*

$$1\ge\bar\nu\ge h^{l'}, \quad \bar\nu\ge\nu\ge\mu^{-l}\bar\nu \tag{6.3.10$_{1,2}$}$$

and in what follows N,K,μ_0,C_0 *and* C *also depend on* l *and* K' *and* C' *also depend on* l,l'. *Then*

(i) Let one of the following conditions

(6.3.5)$'_1$
$$|x_1 - y_1| \geq C_0 \nu\mu^{-1} T + \gamma_1,$$

(6.3.5)$'_2$
$$|\xi_1 - \eta_1| \geq C_0 \nu\mu^{-1} T + \rho_1$$

be fulfilled for all $(x,\xi) \in \operatorname{supp} q'$ *and for all* $(y,\eta) \in \operatorname{supp} q''$ *with*

(6.3.6)$'_{(2)}$
$$T \in [\bar{\nu} h^{1-\delta}\nu^{-1}, \mu T_0 \nu^{-1}],$$

and with $T_0 = T_0(c,\delta,l) > 0$*. Then estimate* (6.3.7) *remains true for* $|\tau| \leq c\nu\bar{\nu}^{-1}$*;*

(ii) Moreover, if $\rho_2 = \mu^{\frac{1}{2}} h^{\frac{1}{2}(1-\delta)}$, $\gamma_2 = \mu^{-\frac{1}{2}} h^{\frac{1}{2}(1-\delta)}$ *and if the condition*

(6.3.8)$'$
$$|\mu^2 x_2^2 + \xi_2^2 - \mu^2 y_2^2 - \eta_2^2| \geq C\mu^{-l}\bar{\nu} + \rho_2$$

is fulfilled for all $(x,\xi) \in \operatorname{supp} q'$ *and for all* $(y,\eta) \in \operatorname{supp} q''$ *then for the* T *and* τ *indicated in (i) estimate* (6.3.7) *holds.*

Let us note that for $\bar{\nu} = \nu$ condition (6.3.9) is fulfilled automatically by remark 6.2.9(ii).

Proof of theorems 6.3.3, 6.3.4. The proof is similar to that of theorem 2.1.1. For $d = 2$ theorem 6.3.3 obviously follows from theorem 6.3.4; let us prove the latter theorem. Dividing equation (6.2.49) by ν and introducing $t' = \frac{\mu}{\nu} t$ in place of t we obtain the equation

(6.3.11)
$$(h_1 D_{t'} - M(x_1, h_1 D_1, h_1, \mu^{-2})\mathcal{B} - R')\mathcal{U} \equiv 0$$

with $\mathcal{B} = \nu^{-1}\mu^2\mathcal{L} = \nu^{-1}(\mu^2 x_2^2 + h^2 D_2^2) + \nu^{-1} B^w$ and with

$$M = \sum_{0 \leq n \leq N-2} b_{n+1}(x_1, h_1 D_1, h_1, \mu^{-2})\mathcal{L}^n.$$

Moreover, remark 6.3.2 yields that we can assume without loss of generality that $\mu^N R' \in \boldsymbol{\Psi}_{\mathbf{h},\boldsymbol{\rho},\boldsymbol{\gamma},K}$; condition (6.3.10) justifies all these operations.

Let us note that the operator M is elliptic in the indicated zone and therefore (6.3.11) yields that

(6.3.12)
$$\mathcal{B}\mathcal{U} \equiv M'(h_1 D_{t'} - R')\mathcal{U},$$

where M' is a parametrix of M. Let us note that if $\omega(x',\xi',\lambda)$ satisfies the inequalities

(6.3.13)
$$|D^\alpha_{x'} D^\beta_{\xi'} D^\sigma_\lambda \omega| \leq c_1 \rho'^{-\beta}\gamma' - \alpha\bar{\rho}^{-\sigma} \qquad \forall \alpha,\beta,\sigma : |\alpha| + |\beta| + \sigma \leq K_0 + K$$

with $\boldsymbol{\rho}' = (\rho_1)$, $\boldsymbol{\gamma}' = (\gamma_1)$ for $d = 2$ [4] and with $\bar{\rho} = \mu^{\frac{1}{2}} h^{\frac{1}{2}(1-\delta)}$ and if condition (6.3.4) is fulfilled then the symbol $\psi(x,\xi) = \omega(x_1,\xi_1,\mu^2 x_2^2 + \xi_2^2)$ belongs to $S_{\mathbf{h},\boldsymbol{\rho},\boldsymbol{\gamma},K}$ in the zone $\{\mu^2 x_2^2 + \xi_2^2 \leq C_0\}$ where $\boldsymbol{\rho} = (\rho_1,\rho_2)$, $\boldsymbol{\gamma} = (\gamma_1,\gamma_2)$, ρ_1 and γ_1 were introduced earlier and $\rho_2 = \bar{\rho}$, $\gamma_2 = \mu^{-1}\bar{\rho}$ for $d = 2$. Moreover, the absolute values of the Poisson brackets $\{\psi,\mathcal{B}\}$ and $\{\psi, M\}$ do not exceed $C_0|\nabla_{x_1,\xi_1}\omega|$ and the absolute value of the Poisson bracket $\{\psi, R'\}$ does not exceed $C_0\mu^{-l}|\nabla_{x_1,\xi_1,\lambda}\omega|$ with arbitrarily large l.[5] Taking into account these observations we can apply the arguments used in the proof of theorem

[4] And with $\boldsymbol{\rho}' = (\rho_1,\rho_2)$, $\boldsymbol{\gamma}' = (\gamma_1,\rho_2)$ for $d = 3$ below.

[5] After the calculation of the gradients one should substitute $\lambda = \mu^2 x_2^2 + \xi_2^2$.

2.1.1 to $\phi(x,\xi,t') = \psi(x,\xi) + t' = \omega(x_1,\xi_1,\mu^2x_2^2+\xi_2^2) + t'$ provided conditions (6.3.13) and (6.3.4) with $j=1$ are fulfilled and

$$(6.3.14)_{(2)} \qquad |\nabla_{x_1,\xi_1}\omega| \le C_0^{-1}, \quad |\partial_\lambda \omega| \le C^{-1}\mu^l.$$

Similarly for $d=3$ theorem 6.3.3 is proven but now we do not divide by ν, we introduce the variable t' and apply a parametrix M' of M; we need only apply the arguments used in the proof of theorem 2.1.1 to $\phi(x,\xi,t) = \omega(x',\xi',\mu^2x_3^2+\xi_3^2)+t$ where ω satisfies (6.3.13) with $\boldsymbol{\rho}' = (\rho_1,\rho_2)$, $\gamma' = (\gamma_1,\gamma_2)$ and

$$(6.3.14)_{(3)} \qquad |\nabla_{x_1,\xi_1}\omega| \le C_0^{-1}\mu, \quad |\nabla_{x_2,\xi_2}\omega| \le C_0^{-1}, \quad |\partial_\lambda\omega| \le C^{-1}\mu^l,$$

and (6.3.4) with $j=1,2$ are fulfilled. □

Theorems 6.3.3, 6.3.4 and 6.1.8 and formula (6.2.56) immediately imply

Theorem 6.3.5. *Let conditions* (6.1.1)–(6.1.4), (6.2.1) *and* (6.3.1) *be fulfilled. Then*

(i) If $d=2$ and condition $(6.3.6)_{(2)}$ is fulfilled then

$$(6.3.15) \quad |F_{t\to h^{-1}\tau}\chi_T(t)U(x,y,t)| \le C'h^s \qquad \forall x,y \in B(0,\frac{3}{4}) : |x-y| \ge C_0\mu^{-1}(T+1) \quad \forall \tau \le c;$$

(ii) Moreover, if $d=2$ and conditions $(6.2.48)_{1-3}$, (6.3.9), (6.3.10), $(6.3.6)'_{(2)}$ are fulfilled then this estimate holds for all $x,y \in B(0,\frac{3}{4}),\tau$ such that $|x-y| \ge C_0\mu^{-1}(\nu T+1)$ and $|\tau| \le c\nu\bar{\nu}^{-1}$;

(iii) If $d=3$ and condition $(6.3.6)_{(3)}$ is fulfilled then estimate (6.3.15) *holds for all $x,y \in B(0,\frac{3}{4}),\tau \le c$ such that $\operatorname{dist}(x,\Gamma(y)) \ge C_0\mu^{-1}$ where $\Gamma(y)$ passes through y* magnetic line, *i.e., an integral curve of the vector field $\mathbf{F} = (F^1,F^2,F^3)$ (more precisely, its segment lying in $B(0,1)$).*

Thus for $d=2$ the average (in time intervals greater than ϵ_0) spatial velocity of propagation of singularities does not exceed $C_0\mu^{-1}$ which permits us to control U on a time interval of length $T_0\mu$ (under the assumptions of assertion (ii) this statement can be improved); on the other hand, for $d=3$ singularities propagate almost along magnetic lines. Moreover, let us recall that the average (on time intervals greater than $h^{1-\delta}$) spatial velocity of propagation of singularities does not exceed C_0 (see theorem 6.1.7); this is important for $d=3$ and for $d=2$ for short time intervals.

6.3.2 Refined estimates

Let us proceed with a more refined analysis. It will be convenient for us to use symbols of the form

(6.3.16) $q = q_1q_2 \in S_{\mathbf{h},\boldsymbol{\rho},\gamma,K}(T^*\mathbb{R}^d)$ with $q_1 \in S_{\mathbf{h}',\gamma,\gamma,K}(T^*\mathbb{R}^{d-1})$ and $q_2 = \omega(\mu^2x_d^2+\xi_d^2) \in S_{h,\bar\rho,\mu^{-1}\bar\rho,K}(T^*\mathbb{R})$ where $\boldsymbol{\rho} = (\gamma,\gamma,\mu\bar\rho)$, $\boldsymbol{\gamma} = (\gamma,\gamma,\bar\rho)$, $\bar\rho \ge \mu^{\frac12}h^{\frac12(1-\delta)}$, ω satisfies (6.3.13), i.e., $|D_\lambda^\sigma\omega(\lambda)| \le C\bar\rho^{-\sigma}$ for $\sigma : |\sigma| \le K$ and $\operatorname{diam}\operatorname{supp}\omega \le c\bar\rho$, $\operatorname{diam}\operatorname{supp} q_1 \le c\gamma$;

restrictions on γ depending on d will be given later; let us recall that $\mathbf{h}$ and $\mathbf{h}'$ are defined in (6.3.4).

Standard elliptic methods combined with the rough estimate (6.2.54) easily imply

Proposition 6.3.6. *Let conditions* (6.1.1)–(6.1.4), (6.2.1) *and* (6.3.1) *be fulfilled.*

(i) Let us assume that $d = 2$ *and conditions* $(6.2.48)_{1-3}$, (6.3.9) *and* (6.3.10) *are fulfilled and* $Q = \mathrm{Op}(q)$, q *is of the form* (6.3.16) *and*

$$\gamma \ge \mu^{-\frac{1}{2}} h^{\frac{1}{2}(1-\delta)} + C\mu^{-l}. \tag{6.3.17$_{(2)}$}$$

Moreover, let us assume that

$$|\lambda + B(x_1,\xi_1,\mu^{-2},0)| \ge \zeta \ge \epsilon_0(\bar\rho + \nu\gamma^2), \tag{6.3.18$_{(2)}$}$$

$$|\nabla B(x_1,\xi_1,\mu^{-2},0)| \le c\nu\gamma \qquad \forall (x_1,\xi_1) \in \operatorname{supp} q_1 \quad \forall \lambda \in \operatorname{supp} \omega. \tag{6.3.19$_{(2)}$}$$

Then

$$|F_{t\to h^{-1}\tau} \chi_{T,T'}(t) Q_x \mathcal{U}| \le C' T h^s \qquad \forall \tau : |\tau| \le \epsilon\zeta \quad \forall x,y \in \mathbb{R}^d \tag{6.3.20}$$

where $T \ge h^{1-\delta}\zeta^{-1}$, T' *and* s *are arbitrary, and here and in what follows* $\epsilon = \epsilon(c,l,\epsilon_0,\delta) > 0$ *is a small enough constant;*

(ii) Let us assume that $d = 3$, $Q = \mathrm{Op}(q)$ *where* q *is of the form* (6.3.16) *and*

$$\gamma \ge h^{\frac{1}{2}(1-\delta)} + C\mu^{-l}. \tag{6.3.17$_{(3)}$}$$

Moreover, let us assume that

$$|\xi_2^2 + \mathcal{B}(x',\xi_1,\lambda,\mu^{-2},0)| \ge \zeta \ge \epsilon_0(\bar\rho + \gamma^2), \tag{6.3.18$_{(3)}$}$$

$$|\xi_2| + |\nabla \mathcal{B}(x',\xi_1,\lambda,\mu^{-2},0)| \le c\gamma \quad \forall (x',\xi') \in \operatorname{supp} q_1 \quad \forall \lambda \in \operatorname{supp} \omega \tag{6.3.19$_{(3)}$}$$

where here and below

$$\mathcal{B} = \sum_{0 \le n \le N-1} B_n(x',\xi_1,.,.)\lambda^n \tag{6.3.21}$$

and B_n *are taken from decomposition* (6.2.38). *Then estimate* (6.3.20) *holds.*

Let us now treat propagation of singularities.

Proposition 6.3.7. *Let conditions* (6.1.1)–(6.1.4), (6.2.1) *and* (6.3.1) *be fulfilled.*

(i) Let us assume that $d = 2$, *conditions* $(6.2.48)_{1-3}$, (6.3.9) *and* (6.3.10) *are fulfilled,* $Q' = \mathrm{Op}(q')$, $Q'' = \mathrm{Op}(q'')$, q', q'' *are of the form* (6.3.16) *and* γ *satisfies* $(6.3.17)_{(2)}$. *Moreover, let us assume that* $(6.3.19)_{(2)}$ *is fulfilled on* $\operatorname{supp} q'$ *and that*

$$|x_1 - y_1| + |\xi_1 - \eta_1| \ge C_0\nu\mu^{-1}\gamma T + \gamma \qquad \forall (x_1,\xi_1) \in \operatorname{supp} q' \quad \forall (y_1,\eta_1) \in \operatorname{supp} q''. \tag{6.3.22$_{(2)}$}$$

Then estimate (6.3.7) *holds for all* $x,y \in \mathbb{R}^2$ *and all* $\tau : |\tau| \le c\nu\gamma/\bar\nu$ *provided*

$$T \in [h^{1-\delta}\frac{\bar\nu}{\gamma\nu}, \mu\frac{T_0}{\nu}]. \tag{6.3.23$_{(2)}$}$$

(ii) Let us assume that $d = 3$, $Q' = \mathrm{Op}(q')$, $Q'' = \mathrm{Op}(q'')$, q', q'' *are of the form* (6.3.16) *and* γ *satisfies* $(6.3.17)_{(3)}$. *Moreover, let us assume that* $(6.3.19)_{(3)}$ *is fulfilled on* $\operatorname{supp} q'$ *and that*

$$\mu(|x_1 - y_1| + |\xi_1 - \eta_1|) + |x_2 - y_2| + |\xi_2 - \eta_2| \ge C_0\gamma T + \gamma \qquad \forall (x',\xi') \in \operatorname{supp} q' \quad \forall (y',\eta') \in \operatorname{supp} q''. \tag{6.3.22$_{(3)}$}$$

Then estimate (6.3.7) *holds for all* $x,y \in \mathbb{R}^3$ *and all* $\tau : |\tau| \le c\gamma$ *provided*

$$T \in [h^{1-\delta}\gamma^{-1}, T_0]. \tag{6.3.23$_{(3)}$}$$

Proof. The proof is similar to that of propositions 2.1.9 and 2.1.10; moreover, we take into account the remarks on the proofs of theorems 6.3.3 and 6.3.4. □

Moreover, the arguments used in the proofs of proposition 2.1.12 and theorem 2.1.14 and the same remarks yield the following assertion similar to theorem 2.1.14:

Proposition 6.3.8. *Let conditions (6.1.1)–(6.1.4), (6.2.1) and (6.3.1) be fulfilled.*

(i) Let us assume that $d = 2$, conditions $(6.2.48)_{1-3}$, (6.3.9) and (6.3.10) are fulfilled, $Q' = \mathrm{Op}(q')$, $Q'' = \mathrm{Op}(q'')$, q', q'' are of the form (6.3.16) and γ satisfies $(6.3.17)_{(2)}$. Moreover, let us assume that on $\operatorname{supp} q'$ condition $(6.3.19)_{(2)}$ is fulfilled and

$$(6.3.24)_{(2)} \qquad |\nabla B| \ge \epsilon_0 \nu\gamma \quad \text{on } \operatorname{supp} q'.$$

Let $\chi, \bar\chi \in C_0^K([-1,1])$, $\bar\chi = 1$ on $[-\frac{1}{2}, \frac{1}{2}]$ and let T', T satisfy $(6.3.23)_{(2)}$ and

$$(6.3.25) \qquad \frac{h^{1-\delta}}{\nu\gamma^2} \le T' \le T.$$

Then

$$(6.3.26) \qquad |F_{t\to h^{-1}\tau}\chi_T(t)(1-\bar\chi_{T'})\Gamma Q'_x \mathcal{U}^t Q''_y| \le C' h^s \qquad \forall \tau : |\tau| \le \epsilon \frac{\nu\gamma}{\bar\nu}.$$

(ii) Let us assume that $d = 3$, $Q' = \mathrm{Op}(q')$, $Q'' = \mathrm{Op}(q'')$, q', q'' are of the form (6.3.16) and γ satisfies $(6.3.17)_{(3)}$. Moreover, let us assume that on $\operatorname{supp} q'$ condition $(6.3.19)_{(3)}$ is fulfilled and

$$(6.3.24)_{(3)} \qquad |\xi_2| + |\nabla\mathcal{B}| \ge \epsilon_0\gamma \quad \text{on } \operatorname{supp} q'.$$

Let $\chi, \bar\chi$ be the same as before and let T', T satisfy conditions $(6.3.23)_{(3)}$ and (6.3.25) where $\bar\nu = \nu = 1$ automatically for $d = 3$ in this section. Then estimate (6.3.26) holds.

Remark (Remark on the proof). *For $d = 3$ we treat four cases, namely when one of the following conditions holds on $\operatorname{supp} q'$:*

$$(6.3.27)_1 \qquad |\partial_{x_1}\mathcal{B}| \ge \epsilon_0\gamma,$$

$$(6.3.27)_2 \qquad |\partial_{\xi_1}\mathcal{B}| \ge \epsilon_0\gamma,$$

$$(6.3.27)_3 \qquad |\partial_{x_2}\mathcal{B}| \ge \epsilon_0\gamma,$$

$$(6.3.27)_4 \qquad |\xi_2| \ge \epsilon_0\gamma.$$

Moreover, our arguments repeat those of the proof of proposition 2.1.12 with no change if either $(6.3.27)_3$ or $(6.3.27)_4$ is fulfilled; the same is true under conditions $(6.3.27)_{1,2}$ but in the latter case we also need to change coordinates: $x_{1\text{new}} = \mu x_1$. Moreover, in the auxiliary partition of unity (see proof of proposition 2.1.12) we use standard h- (but not **h**-*) pseudo-differential operators. For $d = 2$ we treat two cases and our arguments repeat those of the proof of proposition 2.1.12.* ◊

Remark 6.3.9. *Proposition 6.3.6 yields that we can replace Q', Q'' by Q'_1, Q''_1 respectively in estimates (6.3.7) and (6.3.26) of propositions 6.3.7 and 6.3.8, where $Q'_1 = \mathrm{Op}(q'_1)$, $Q''_1 = \mathrm{Op}(q''_1)$, and q'_1 and q''_1 are taken from decomposition (6.3.16). In this case we need to check conditions $(6.3.19)_{(3)}$ and $(6.3.24)_{(3)}$ with λ violating condition $(6.3.18)_{(3)}$ with $\zeta = \epsilon(\bar\rho + \gamma)$; the particular choice of λ is not essential. For $d = 2$ the corresponding conditions do not include λ at all.* ◊

6.3.3 Calculations

Let us proceed from estimates to calculations. In this subsection $d \geq 2$ is arbitrary and the crucial statement is the following:

Proposition 6.3.10. *Let conditions* (6.1.1)–(6.1.4) *and* (6.3.1) *be fulfilled and let* $Q' = \mathrm{Op}(q')$, $Q'' = \mathrm{Op}(q'')$, $q', q'' \in S_{h,\gamma,\gamma,K}$,

$$\gamma \geq (\mu h)^{\frac{1}{2}-\delta} \tag{6.3.28}$$

and $\psi', \psi'' \in C_0^K(B(0, \frac{3}{4}))$. *Then*

(i) For $T \in [h^{1-\delta}, (\mu^{-1}h)^{\frac{1}{2}+\delta}]$ *the following estimate holds:*

$$\begin{aligned} &|F_{t\to h^{-1}\tau}\chi_T(t)\bigl(Q'_x\psi'(x)U\psi''(y)^tQ''_y\bigr)(x,x,t) - \\ &\quad \sum_{m+n\leq M} Th^{-1}\int \hat{\chi}(\frac{(\tau-\tau')T}{h})\varkappa'_{mn}(x,\tau')\mu^m h^{1-d+m+n}d\tau'| \leq \\ &\qquad C'h^{1-d}(\mu h)^s \qquad \forall x\in\mathbb{R}^d \quad \forall\tau: |\tau|\leq c. \end{aligned} \tag{6.3.29}$$

Moreover, for $L \in [(\mu h)^{1-\delta}, 1]$, $T' \geq h^{1-\delta}L^{-1}$ *the following estimate holds:*

$$\begin{aligned} &|\int \phi(\frac{\tau}{L})\bigl(F_{t\to h^{-1}\tau}\bar{\chi}_T(t)(Q'_x\psi'(x)U)(x,x,t) - \\ &\quad \sum_{m+n\leq M}\varkappa'_{mn}(x,\tau)\mu^m h^{1-d+m+n}\bigr)d\tau| \leq C'h^{1-d}(\mu h)^s \qquad \forall x\in\mathbb{R}^d \end{aligned} \tag{6.3.30}$$

where $\varkappa_{mn} \in C^s(B(0,1), \mathcal{S}'(\mathbb{R}))$ *are supported in* $\{\tau \geq V(x)\}$, $\varkappa'_{mn} = \partial_\tau\varkappa_{mn}$ *and we assume that* $\chi, \bar{\chi} \in C_0^K([-1,1])$ *are fixed,* $\bar{\chi} = 1$ *in* $[-\frac{1}{2}, \frac{1}{2}]$, $\phi \in C_0^K([-c,c])$ *is also fixed,* $\delta > 0$ *and* s *are arbitrary and* $K = K(d,\delta,s)$, $M = M(d,\delta,s)$. *Moreover, terms in both decompositions become smaller as* $m+n$ *increases; in particular, for* $\gamma = L = 1$ *in* (6.3.30) *one can take* $s = M+1 \in \mathbb{Z}^+$.

(ii) Moreover, if

$$V \leq -\epsilon_0 \qquad \forall x \in B(0,1) \tag{6.3.31}$$

then for $|\tau| \leq \frac{1}{2}\epsilon_0$ *estimate* (6.3.29) *holds with every* $T \in [h^{1-\delta}, \mu^{-1}T_0]$ *where* $T_0 = T_0(d,c,\epsilon_0,\delta) > 0$ *is a small enough constant; moreover for* $\gamma = 1$ *one can take* $s = M+1 \in \mathbb{Z}^+$.

In particular, if $\operatorname{diam}\operatorname{supp} q' \leq c\gamma$ *then under condition* (6.3.31) *the following estimate holds:*

$$\begin{aligned} &|F_{t\to h^{-1}\tau}\chi_T(t)\bigl(Q'_x\psi'(x)U\psi''(y)^tQ''_y\bigr)(x,x,t)| \leq C'h^{1-d}\gamma^{d-2} \\ &\qquad \forall x\in\mathbb{R}^d \quad \forall\tau: |\tau|\leq \frac{1}{2}\epsilon_0 \end{aligned} \tag{6.3.32}$$

where T *is as before.*

(iii) On the other hand, if

$$|V| + |\nabla V| \geq \epsilon_0 \qquad \forall x \in B(0,1) \tag{6.3.33}$$

then for $\gamma = 1$ *the estimate*

$$(6.3.34)\quad \Bigl|\int F_{t\to h^{-1}\tau}\chi_T(t)\bigl((Q'_x\psi'(x)U\psi''(y)^tQ''_y)(x,x,t) - \sum_{m+n\le M} Th^{-1}\int\hat\chi(\frac{(\tau-\tau')T}{h})\varkappa'_{mn}(x,\tau')\mu^m h^{1-d+m+n}d\tau'\bigr)dx\Bigr| \le C'h^{1-d}(\mu h)^s \qquad \forall\tau:|\tau|\le\epsilon$$

(which is due to (6.3.29)) holds for $\epsilon=\epsilon(d,c,\epsilon_0)>0$ and $s=M+1\in\mathbb{Z}^+$ with the same T as in (i). In particular, (6.3.34) holds for $M=-1$, $s=0$.

(iv) Moreover,

$$(6.3.35)\qquad \varkappa_{00}=(2\pi)^{-d}\int_{a(x,\xi)<\tau} Q'_0(x,\xi)q''_0(x,\xi)\psi'(x)\psi''(x)d\xi$$

where

$$(6.3.36)\qquad a(x,\xi)=g^{jk}(\xi_j-\mu V_j)(\xi_k-\mu V_k)+V$$

and Q'_0, Q''_0 denote the principal symbols of Q', Q'' in (6.3.35).

(v) Moreover, for $Q'=Q''=I$

$$(6.3.37)\qquad \varkappa_{00}=(2\pi)^{-d}\omega_d(\tau-V)_+^{d/2}\sqrt{g}\psi'\psi''$$

and $\varkappa_{mn}$ vanishes provided either m or n is odd.

Proof. Let us make the dilatation $x_{\text{new}}=\mu(x-\bar x)$; we then obtain the case in which h and μ are replaced by $h_{\text{new}}=\mu h$ and $\mu_{\text{new}}=1$ respectively. In this case assertions (i) and (ii) are due to proposition 4.2.6 and proposition 4.2.9 respectively; we obtain series with respect to powers of μh and h also (because the basic operator, cutting operators and functions depend on μ^{-1} (which is a coefficient of some copies of x). One can obtain the same results simply repeating constructions of section 4.2 (the method of successive approximations) in which case no dilatation is necessary. Moreover, an analysis based on Appendix B (see the proof of theorem 4.3.3) yields that under condition (6.3.33) and $\gamma=1$ the coefficients $\varkappa'_{mn}$ integrated with respect to $x\in B(0,1)$ are bounded functions, even regular with respect to $\tau\in[-\epsilon,\epsilon]$; this yields (iii). Formulas (6.3.35) and (6.3.37) for $\varkappa_{00}$ are well known (see sections 4.3 and 4.4).

Let us note that replacing h and μ by $-h$ and $-\mu$ respectively we preserve the basic operator and therefore we also preserve $e(x,x,\tau)$; on the other hand, the functions

$$(6.3.38)\qquad \int\phi(\frac{\tau}{L})(F_{t\to h^{-1}\tau}\bar\chi_T(t)U(x,x,t))d\tau$$

and

$$(6.3.39)\qquad h\int\phi(\frac{\tau}{L})d_\tau e(x,x,\tau)d\tau$$

coincide modulo $O(h^s)$ and hence for $Q'=Q''=I$ all the terms in our decompositions with odd $m-n$ vanish (really, the term with $m=n=0$ does not vanish; obviously the factor h^{1-d} arises as $|h|^{1-d}$ through the Jacobian). Moreover, let us note that replacing μ by $-\mu$ we replace the basic operator by its complex conjugate and therefore we also preserve $e(x,x,\tau)$ in this case; since $e(x,x,\tau)$ is real-valued, all terms with odd n in our decompositions vanish for $Q'=Q''=I$. Assertion (v) is also proven. □

Let us treat the terms with $n = 0$ for $Q' = Q'' = I$, $\psi' = \psi'' = 1$. The standard calculations of the method of successive approximations yield that for $x = \bar{x}$ these coefficients depend only on $g^{jk}(\bar{x})$, $\partial_k V_j(\bar{x})$ and $V(\bar{x})$ and therefore calculating these coefficients one can assume without loss of generality that g^{jk} and V are constants and V_j are linear functions. Moreover, the same calculations yield that if g^{jk} and V are constants and V_j are linear functions then only terms with $n = 0$ will appear in our decompositions and since (6.3.38) and (6.3.39) coincide modulo $O(h^s)$ we obtain from theorem 6.1.1 that

$$S(\bar{x}, \tau, \mu h) \sim \sum_{m \in \mathbb{Z}^+} \varkappa_{m0}(\bar{x})(\mu h)^m \tag{6.3.40}$$

in the weak sense (i.e., this asymptotic decomposition will be true if we multiply it by $\phi \in C_0^\infty(\mathbb{R})$ and integrate over $\mathbb{R}$; surely we proved this only for equality (6.3.40) differentiated with respect to τ but (6.3.40) is also true without differentiation because the supports of both sides are semibounded from below with respect to τ).

We noted earlier that $S(\tau, \mu h)$ can be treated as a Riemann sum for the function

$$\mathcal{K}(t, z) = (2\pi)^{r-d} 2^{-r} \omega_{d-2r} (\tau - f_1 z_1 - \cdots - f_r z_r)_+^{\frac{d-2r}{2}} f_1 \cdots f_r \sqrt{g}$$

but this function was not smooth; however after multiplication by $\phi(\tau)$ and integration on $\mathbb{R}_\tau$ the resulting function will be smooth and hence we can obtain a complete decomposition with respect to powers of μh. Therefore we obtain

Proposition 6.3.11. *For $Q' = Q'' = I$ and $\psi' = \psi'' = 1$*

$$\varkappa_{m0} = \partial_\tau (\tau - V)_+^{\frac{d}{2}} \mathcal{P}_m(f_1^2, \ldots, f_r^2)\sqrt{g} \tag{6.3.41}$$

where $\mathcal{P}_m$ are homogeneous symmetric polynomials of degrees $\frac{m}{2}$ also depending on d and r. In particular,

$$\varkappa_{20} = -c\partial_\tau^2 (\tau - V)_+^{\frac{d}{2}} F^2 \sqrt{g} \tag{6.3.42}$$

where $c = \frac{1}{3}(2\pi)^{r-d} 2^{-r-1} \omega_{d-2r}$ and

$$F = (f_1^2 + \cdots + f_r^2)^{\frac{1}{2}} = \frac{1}{2}(g^{jk} g^{lm} F_{jl} F_{km})^{\frac{1}{2}} \tag{6.3.43}$$

is the scalar intensity of the magnetic field.

Remark 6.3.12. *Let us note that if Q' is an operator with Weyl symbol $q'(x, \xi)$ even with respect to ξ and h then Q is preserved if we replace h by $-h$ and for Q complex and Hermitian conjugation coincide. Therefore for this operator Q' and for $Q'' = I$, $\psi' = \psi'' = 1$ in a neighborhood of* $\operatorname{supp} q'$, *$\varkappa_{mn}$ vanish unless m and n are even.* ◊

6.3.4 Improved remainder estimates

Let us improve the results of the previous subsection by means of proposition 6.3.8. Namely, let the conditions of that proposition be fulfilled and let $T_1 = h^{1-\delta} \nu^{-1} \gamma^{-2}$ (the minimal possible value of T' in the proposition; let us recall that $\nu = 1$ for $d = 3$) be less than T_2 (the maximal possible value of T in proposition 6.3.10; let us recall that $T_2 = \mu^{-\frac{1}{2}} h^{\frac{1}{2}+\delta}$ in (iii) and $T_2 = T_0 \mu^{-1}$ in (ii)); then in estimate (6.3.34) one can take $T = T_3$ where T_3 is the maximal possible value of T in proposition 6.3.8 (namely, $T_3 = \mu T_0 / \nu$ for $d = 2$ and $T_3 = T_0$ for $d = 3$). Therefore we obtain

Proposition 6.3.13. *Let conditions* (6.1.1)–(6.1.4), (6.2.1) *and* (6.3.1) *be fulfilled. Then*

(i) Under the condition

$$|\nabla \frac{V}{F}| \geq \epsilon_0 \qquad \text{in } B(0,1) \tag{6.3.44}$$

estimate (6.3.34) *holds with* $T = \mu T_0$ *for* $d = 2$ *and with* $T = T_0$ *for* $d = 3$.[6)]

(ii) Furthermore, let $d = 2$, *conditions* $(6.2.48)_{1-3}$, (6.3.9), (6.3.10) *and* (6.3.31) *be fulfilled, let* $\tilde{Q}' = \mathrm{Op}(\tilde{q}')$, $\tilde{Q}'' = \mathrm{Op}(\tilde{q}'')$, $\tilde{q}', \tilde{q}'' \in S_{h_1,\gamma,\gamma,K}(T^*\mathbb{R})$,

$$1 \geq \gamma \geq (\frac{\mu h^{1-\delta}}{\nu})^{\frac{1}{2}}, \tag{6.3.45}$$

$$c\nu\gamma \geq |\nabla B| \geq \epsilon_0 \nu\gamma \qquad \text{on } \operatorname{supp} \tilde{q}' \tag{6.3.46}$$

and $Q' = \mathcal{F}\tilde{Q}'\mathcal{F}^*$, $Q'' = \mathcal{F}\tilde{Q}''\mathcal{F}^*$. *Then estimate* (6.3.34) *holds for* $\tau : |\tau| \leq \epsilon \frac{\nu\gamma}{\bar{\nu}}$ *with* $T = \mu T_0/\nu$.[7)]

(iii) On the other hand, let $d = 3$, $\tilde{Q}' = \mathrm{Op}(\tilde{q}')$, $\tilde{Q}'' = \mathrm{Op}(\tilde{q}'')$, $\tilde{q}', \tilde{q}'' \in S_{\mathbf{h}',\gamma,\gamma,K}(T^*\mathbb{R}^2)$, *let conditions* (6.3.31) *and* (6.3.45) *be fulfilled with* $\nu = 1$ *and let*

$$\begin{aligned} &|\xi_2^2 + \mathcal{B}(x', \xi_1, \lambda, \mu^{-2}, 0)| \leq \epsilon_0\gamma \implies \\ &c\gamma \geq |\xi_2| + |d\mathcal{B}(x', \xi_1, \lambda, \mu^{-2}, 0)| \geq \epsilon_0\gamma \qquad \forall (x', \xi') \in \operatorname{supp}\tilde{q}' \quad \forall \lambda. \end{aligned} \tag{6.3.47}$$

Then estimate (6.3.34) *holds with* $T = T_0$.

Combining this proposition with standard Tauberian arguments we immediately obtain the following two statements:

Theorem 6.3.14. *Let conditions* (6.1.1)–(6.1.4), (6.2.1), (6.3.1) *and* (6.3.44) *be fulfilled. Then*

(i) For $d = 2$ *the estimates*

$$\begin{aligned} \mathcal{R}_1 = |\int \psi(x)\Big(e(x,x,-\infty,\tau) - \\ h^{-d}\sum_{m\leq M} \varkappa_{m0}(x,\tau)(\mu h)^m\Big)dx| \leq C\mu^{-1}h^{-1} + C'h, \end{aligned} \tag{6.3.48$_{(2)}$}$$

$$\begin{aligned} \mathcal{R}_{1\vartheta} = |\int \tau_+^{\vartheta-1} * \Big(\psi(x)e(x,x,-\infty,.) - \\ h^{-d}\sum_{m,n\leq M} \varkappa_{\psi,mn}(x,.)(\mu h)^m h^n\Big)dx| \leq C(\mu h)^{-1+\vartheta} + C'h^s, \end{aligned} \tag{6.3.49$_{(2)}$}$$

hold for $|\tau| \leq \epsilon$, $\vartheta > 0$ *where* M *depends on* ϑ *here and below.*

(ii) for $d = 3$ *the estimates*

$$\mathcal{R}_1 \leq Ch^{-2} + C', \tag{6.3.48$_{(3)}$}$$

$$\mathcal{R}_{1\vartheta} \leq Ch^{-2+\vartheta} + C'h^s, \tag{6.3.49$_{(3)}$}$$

hold for $|\tau| \leq \epsilon$, $\vartheta > 0$.

---

6) Recall formulas $(6.2.41)_{1,2}$ linking B with $\frac{V}{F}$.

7) Thus T does not depend on γ satisfying (6.3.45), (6.3.46).

We dropped terms with $n \geq 1$ for $\vartheta = 0$ because they do not exceed the remainder estimate due to proposition 6.3.10(v). On the other hand, because the coefficients $\varkappa_{\psi,m0}$ contain ψ only as a factor we separated this factor.

Proposition 6.3.15. *Suppose conditions* (6.1.1)–(6.1.4), (6.2.1), (6.3.1) *and* (6.3.31) *are fulfilled. Let* $\tilde{Q}' = \operatorname{Op}(\tilde{q}')$, $\tilde{q}' \in S_{\mathbf{h}',\gamma,\gamma,K}(T^*\mathbb{R}^{d-1})$ *where* γ *satisfies* (6.3.45), $\operatorname{diam}\operatorname{supp}\tilde{q}' \leq c\gamma$ *and* $Q' = \mathcal{F}\tilde{Q}'\mathcal{F}^*$. *Then*

(i) Let $d = 2$ *and conditions* $(6.2.48)_{1-3}$, (6.3.9) *and* (6.3.10) *be fulfilled. Moreover, let condition* (6.3.46) *be fulfilled on* $\operatorname{supp}\tilde{q}'$. *Then the estimates*

$$(6.3.50)_{(2)} \quad \mathcal{R}_2 = \Big| \int \Big((Q'_x\psi(x)e)(x,x,-\infty,0) - h^{-d} \sum_{m+n\leq M} \varkappa_{mn}(x,0)(\mu h)^m h^n \Big) dx \Big| \leq C\nu\gamma^2\mu^{-1}h^{-1} + C'h^s,$$

$$(6.3.51)_{(2)} \quad \mathcal{R}_{2\vartheta} = \Big| \int \tau_+^{\vartheta-1} * \Big((Q'_x\psi(x)e)(x,x,-\infty,\tau) - h^{-d} \sum_{m+n\leq M} \varkappa_{Q'\psi,mn}(x,\tau)(\mu h)^m h^n \Big) dx \Big| \leq C\nu^{1+\vartheta}\gamma^2(\mu h)^{-1+\vartheta} + C'h^s$$

hold for $\tau = 0$.

(ii) Let $d = 3$, $\nu = 1$ *and condition* (6.3.47) *be fulfilled on* $\operatorname{supp}\tilde{\omega}'$. *Then the estimates*

$$(6.3.50)_{(3)} \qquad \mathcal{R}_2 \leq C\gamma^4 h^{-2} + C'h^s,$$

$$(6.3.51)_{(3)} \qquad \mathcal{R}_{2\vartheta} \leq C\gamma^4 h^{-2+\vartheta} + C'h^s,$$

hold for $\tau = 0$.

On the other hand, if we drop condition (6.3.46) or (6.3.47) in the context of this proposition then the results of section 4.4 (or proposition 6.3.10 and standard Tauberian arguments) yield the estimate

$$(6.3.52) \qquad \mathcal{R}_{2\vartheta} \leq C\mu^{1+\vartheta}\gamma^{2d-2}h^{1-d+\vartheta} + C'h^s$$

for $\tau = 0$, $\vartheta \geq 0$ [8] (here we have no need of condition (6.2.1) and the additional conditions of assertions (i) and (ii)). Summing with respect to an appropriate partition of unity and applying estimate $(6.3.50)_{(d)}$, $(6.3.51)_{(d)}$ under condition (6.3.46) or (6.3.47) and estimate (6.3.52) otherwise we immediately obtain the following

Proposition 6.3.16. *Suppose conditions* (6.1.1)–(6.1.4), (6.2.1), (6.3.1) *and* (6.3.31) *be fulfilled. Then*

(i) Let $d = 2$ *and let conditions* $(6.2.48)_{1-3}$, (6.3.9) *and* (6.3.10) *be fulfilled. Then the estimate*

$$(6.3.53)_{(2)} \qquad \mathcal{R}_{1\vartheta} \leq C\nu^{1+\vartheta}(\mu h)^{-1+\vartheta} + C\mu^{1+\vartheta}h^{-1+\vartheta}M_2 + C'h^s$$

holds for $\tau = 0$, $\vartheta \geq 0$ *where* M_2 *is the two-dimensional measure of the set* $\mathcal{M}_2 = \{(x_1,\xi_1) : |\nabla B| \leq \nu\bar{\gamma}\}$ *with* $\bar{\gamma} = (\frac{\mu h}{\nu})^{\frac{1}{2}-\delta}$.

(ii) For $d = 3$ *the estimate*

[8] Recall that $\vartheta = 0$ means that there is no mollification on τ.

$$\mathcal{R}_{1\vartheta} \le Ch^{-2+\vartheta} + C\mu^{1+\vartheta}h^{-2}M_3 + C'h^s \tag{6.3.53$_{(3)}$}$$

holds for $\tau = 0$, $\vartheta \ge 0$ *where* M_3 *is the four-dimensional measure of the set* $\mathcal{M}_3 = \{(x', \xi') : \xi_2^2 + \mathcal{B}(x', \xi_1, \lambda, \mu^{-2}, 0) = 0 \implies |\xi_2| + |\nabla \mathcal{B}(x', \xi_1, \lambda, \mu^{-2}, 0)| \le \bar\gamma\}$ *with* $\bar\gamma = (\mu h)^{\frac{1}{2}-\delta}$.

I leave it to the reader to treat the remainder estimates after convolution with the function $\phi(\frac{\tau}{L})$ where ϕ is either a smooth function or satisfies (4.3.21). In this case an additional factor $(\frac{h}{LT})^\vartheta$ arises in the right-hand expression in (6.3.48), (6.3.50) with corresponding T while the right-hand expression in estimates $(6.3.53)_{(2,3)}$ will be $h^{1-d}\gamma^{2d-2}\min\bigl((\frac{h\mu}{L})^\vartheta, 1\bigr)$ and we assume that $L \ge \frac{h}{T}$. This yields the obvious modifications of the further estimates.

It is easy to see that $M_3 \le \bar\gamma M_3'$ where M_3' is the three-dimensional measure of set $\mathcal{M}_3' = \{(x', \xi_1) : \inf_\lambda(|\mathcal{B}(x', \xi_1, \lambda, \mu^{-2}, 0)|^{\frac{1}{2}} + |\nabla \mathcal{B}(x', \xi_1, \lambda, \mu^{-2}, 0)|) \le \bar\gamma\}$. This observation immediately yields that for $d = 3$ and $\mu \le h^{-\frac{1}{3+2\vartheta}+\delta}$ estimate $(6.3.49)_{(3)}$ holds with no additional condition. In particular, for $\mu \le h^{-\frac{1}{3}+\delta}$ estimate $(6.3.48)_{(3)}$ holds with no additional condition. On the other hand, let us recall that $\mathcal{B} = (V + F\lambda) \circ \bar\Psi + O(\mu^{-2})$ and therefore for $\mu \ge h^{-\frac{1}{5}-\delta}$ the estimate $M_3' \le C \operatorname{mes}\{x : |\nabla \frac{V}{F}| \le C\bar\gamma\}$ holds. Therefore we have proven

Theorem 6.3.17. *Let* $d = 3$ *and let conditions* (6.1.1)–(6.1.4), (6.2.1), (6.3.1) *and* (6.3.31) *be fulfilled. Then*

(i) For $\mu \le h^{-\frac{1}{3}+\delta}$ *estimate* $(6.3.48)_{(3)}$ *holds and for* $\mu \le h^{-\frac{1}{3+2\vartheta}+\delta}$ *estimate* $(6.3.49)_{(3)}$ *holds.*

(ii) For $\vartheta \ge 0$, $\mu \le h^{-\frac{1}{3+2\vartheta}+\delta}$ *the estimate*

$$\mathcal{R}_{1\vartheta} \le Ch^{-2+\vartheta} + C'\mu^{\frac{3}{2+\vartheta}}h^{-\frac{3}{2}+\vartheta-\delta} \operatorname{mes}\{x : |\nabla \frac{V}{F}| \le \bar\gamma_1\} + C'h \tag{6.3.54}$$

holds with $\bar\gamma_1 = \bar\gamma + C_0\mu^{-2}$; *for* $\mu \ge h^{-\frac{1}{5}-\delta}$ *one can replace* $\bar\gamma_1$ *by* $\bar\gamma$. *In particular, for* $\vartheta < 1$ *one always can replace* $\bar\gamma_1$ *by* $\bar\gamma$.

The treatment of the case $d = 3$ is almost complete (later we will modify $\mathcal{R}_1$). The case $d = 2$ is more complicated. Taking into account that $B = \frac{V}{F} \circ \bar\Psi + O(\mu^{-2}\bar\nu)$ we obtain from proposition 6.3.16(i) the following

Theorem 6.3.18. *Let* $d = 2$ *and let conditions* (6.1.1)–(6.1.4), (6.2.1), (6.3.1), (6.3.31) *and* $(6.2.48)_{1-3}$, (6.3.9), (6.3.10) *be fulfilled (for* $\nu = \bar\nu = 1$ *these three conditions are fulfilled automatically). Then*

(i) The estimate

$$\mathcal{R}_{1\vartheta} \le C\nu^{1+\vartheta}\mu^{-1-\vartheta}h^{-1+\vartheta} + C\mu^{1+\vartheta}h^{-1+\vartheta} \operatorname{mes}\{x : |\nabla \frac{V}{F}| \le \bar\gamma\} + C'h \tag{6.3.55}$$

holds for $\vartheta \ge 0$, $\tau = 0$ *with* $\bar\gamma = (\mu h\nu)^{\frac{1}{2}-\delta} + C_0\bar\nu\mu^{-2} + C'h^l$ *with a sufficiently large constant* C_0. *In particular, under the conditions*

$$|\nabla \frac{V}{F}| \ge \epsilon_0\nu \ge C_0\bar\nu\mu^{-2} + h^{1-\delta} \tag{6.3.56}$$

the estimate

$$\mathcal{R}_{1\vartheta} \le C\nu^{1+\vartheta}(\mu h)^{-1+\vartheta} + Ch^s \tag{6.3.57}$$

holds for $\vartheta \ge 0$, $|\tau| \le \epsilon\nu\bar\nu^{-1}$.

(ii) Let $\nu \ge h^{1-\delta}$ and

$$\sum_{1\le|\alpha|\le m} |D^\alpha \frac{V}{F}| \ge \epsilon_0 \nu \tag{6.3.58}$$

and $\bar\nu \le \epsilon\nu\mu^2$, $\mu \le h^{-\delta'}$ with $\delta' = \delta'(\delta,\vartheta,m) > 0$, $\epsilon = \epsilon(c,m,\epsilon_0,\delta,\vartheta) > 0$. Then the estimate

$$\mathcal{R}_{1\vartheta} \le C\nu^{1+\vartheta}\mu^{-1-\vartheta}h^{-1+\vartheta} + C'h^s \tag{6.3.59}$$

holds. Moreover, this statement remains true even without assumption (6.3.9).

The proof of (ii) is based on the fact that under condition (6.3.58) the same is true for B (provided $\bar\nu \le \epsilon\nu\mu^2$) and then $\operatorname{mes}\{|\nabla B| \le \nu\varepsilon\}$ does not exceed $C\varepsilon^\sigma$ with $\sigma = \sigma(m,\delta) > 0$. The reader can consider the case in which the stronger non-degeneracy conditions

$(6.3.60)_r$ If $|\nabla \frac{V}{F}| \le \epsilon_0\nu$ then $\nu^{-1}\operatorname{Hess}\frac{V}{F}$ has r eigenvalues disjoint from 0

are fulfilled with $r = 1, 2$. I will do that in the next section; (ii) is what I really need here. Moreover, in these and all other conditions one can replace $\frac{V}{F}$ by $\frac{V}{F} + \omega_1\mu^{-2}$ and replace the assumption $\bar\nu \le \epsilon\nu\mu^2$ by the assumption $\bar\nu \le \epsilon\nu\mu^4$. Here we take into account that $B = (\frac{V}{F} + \omega_1\mu^{-2}) \circ \bar\Psi + O(\mu^{-4}\bar\nu)$ where ω_1 is given in section 6.2.

6.3.5 Miscellaneous modifications of $\mathcal{R}_1$

At the end of this section let us consider various modifications of $\mathcal{R}_1$. So $\vartheta = 0$ here. The case $\vartheta > 0$ is left to the reader, except that I want to note that for $\mu \le h^{-\sigma}$ with $\sigma = \sigma(\vartheta) > 0$ all the arguments below remain true.

First of all, for $d = 3$, $\mu \le h^{-\frac12}$ and for $d = 2$, $\mu \le h^{-\frac13}$ in estimate $(6.3.48)_{(d)}$ one can obviously take $M = 0$ and therefore replace $\mathcal{R}_1$ by the Weyl expression

$$\mathcal{R}^w = |\int \psi(x)\Big(e(x,x,-\infty,0) - (2\pi)^{-d}\omega_d h^{-d} V_-^{\frac d2}\sqrt{g}\Big)dx|.$$

Moreover, for $d = 2$ and condition (6.3.31) formula (6.3.41) yields that $\varkappa_{m0} = 0$ for $m \ge 1$ in which case $\mathcal{R}_1$ and $\mathcal{R}^w$ coincide.

Let us discuss the possibility of replacing $\mathcal{R}_1$ by

$$\mathcal{R} = |\int \psi(x)\Big(e(x,x,-\infty,0) - h^{-d}S(x,0,\mu h)\Big)dx|. \tag{6.3.61}$$

Let us recall that equality (6.3.40) holds only in the weak sense. However, if condition (6.3.44) is fulfilled then after multiplication by $\psi'(x)$ and integration over $\mathbb{R}^d_x$ the function $\mathcal{K}$ introduced in the proof of proposition 6.3.11 will be smooth with respect to (τ, z) and therefore the equality

$$\int \psi'(x)S(x,\tau,\mu h)dx \equiv \sum_{0\le m\le M}\int \psi'(x)\varkappa_{m0}(x,\tau)(\mu h)^m dx \qquad (\operatorname{mod} O(h^s)) \tag{6.3.62}$$

holds in the strong sense for $|\tau| \le \epsilon$ and $M = M(d,s)$. Therefore theorem 6.3.14 remains true for $\mathcal{R}_1$ replaced by $\mathcal{R}$.

Moreover, one can see easily that if $\psi' \in C_0^K(B(0,1))$, diam supp $\psi' \le c\gamma$,

$$|D^\alpha \psi'| \le c\gamma^{-|\alpha|} \qquad \forall \alpha : |\alpha| \le K, \tag{6.3.63}$$

and if conditions (6.3.31),

$$|D^\alpha \frac{V}{F}| \le c\nu \qquad \forall \alpha : 1 \le |\alpha| \le K, \tag{6.3.64}$$

$$c\nu\gamma \ge |\nabla \frac{V}{F}| \ge \epsilon_0 \nu\gamma \qquad \text{on supp}\,\psi' \tag{6.3.65}$$

are fulfilled then

$$|D_z^\alpha \int \psi' \mathcal{K}(x,z,\tau)dx| \le C'\gamma^{d-2|\alpha|}\nu^{-|\alpha|} \qquad \forall \alpha : |\alpha| \le K' \quad \forall \tau : |\tau| \le \epsilon\nu\gamma$$

for arbitrary K' (K depends on K') and therefore (6.3.63) holds.

On the other hand, if condition (6.3.65) is violated then the difference between the left-hand and right-hand expressions in (6.3.63) does not exceed $C\gamma^d\mu h$. Summing with respect to a partition of unity we then obtain that the estimate

$$|\mathcal{R}_1 - \mathcal{R}| \le C\mu h^{1-d}\,\text{mes}\{x : |\nabla \frac{V}{F}| \le \bar{\gamma}\} + C'h \tag{6.3.66$_{(d)}$}$$

holds with $\bar{\gamma} = (\mu h\nu)^{\frac{1}{2}-\delta}$.

Moreover, for $d=3$, rewriting the function $\mathcal{K}(x,z,\tau)$ in the form

$$\mathcal{K}(x,z,\tau) = \text{const}\int \theta(\tau - V - Fz - \eta^2)d\eta\sqrt{g}$$

with $\eta \in \mathbb{R}$, $z \in \mathbb{R}$ and making an appropriate partition of unity with respect to (x,η) with condition (6.3.65) replaced by the condition

$$c\gamma \ge |\eta| + |\nabla \frac{V}{F}| \ge c^{-1}\gamma \qquad \text{at supp}\,\psi' \tag{6.3.65$'$}$$

we obtain the estimates

$$|\mathcal{R}_1 - \mathcal{R}| \le C\mu h^{-2}\,\text{mes}\{(x,\eta) : |\eta| + |\nabla \frac{V}{F}| \le \bar{\gamma}\} + C'h \le \\ C'\mu^{\frac{3}{2}}h^{-\frac{3}{2}-\delta}\,\text{mes}\{x, |\nabla \frac{V}{F}| \le \bar{\gamma}\} + C'h \tag{6.3.66$'_{(3)}$}$$

with $\bar{\gamma} = (\mu h)^{\frac{1}{2}-\delta}$.

Therefore theorems 6.3.14, 6.3.17 and 6.3.18, proposition 6.3.15 and formulas $(6.3.66)_{(2)}$ and $(6.3.66)'_{(3)}$ yield the following

Theorem 6.3.19. *(i) Under the assumptions of theorem 6.3.14,* $(6.3.48)_{(2,3)}$ *hold for* $\mathcal{R}$ *as well.*

(ii) Under the assumptions of proposition 6.3.16(i) with the additional assumption (6.3.64) *the estimate*

$$\mathcal{R}_\vartheta \le C\nu^{1+\vartheta}\mu^{-1-\vartheta}h^{-1+\vartheta} + \\ C\mu^{1+\vartheta}h^{-1+\vartheta}(\text{mes}\{x : |\nabla \frac{V}{F}| \le \bar{\gamma}\} + M_2) + C'h^s \tag{6.3.67$_{(2)}$}$$

holds with $\bar{\gamma} = (\mu h\nu)^{\frac{1}{2}-\delta}$.

(iii) Under the assumptions of theorem 6.3.17(i),(ii) estimates (6.3.48), (6.3.54) *respectively hold for* $\mathcal{R}$ *also; under the assumptions of theorem 6.3.18(i),(ii) estimates* (6.3.55), (6.3.57) *respectively hold for* $\mathcal{R}$ *also.*

Remark 6.3.20. *(i) One can easily see that the corresponding estimates remain true for* $\mathcal{R}_\vartheta$ *with* $\vartheta = 1$ *for* $d = 3$ *and any* $\vartheta < 1$, $\mu \le h^{-\delta'}$ *with sufficiently small* $\delta' = \delta'(\vartheta) > 0$ *for* $d = 2$.

(ii) Under the restriction

$$1 \ge \nu \ge C_0 \bar{\nu} \mu^{-2}, \quad \bar{\nu} \ge C_0 \mu^{-l'} \tag{6.3.68}$$

condition (6.3.9) is equivalent to (6.3.64). ◊

6.4 Spectral Asymptotics in the Case of a Strong Magnetic Field

6.4.1 Decomposition of the fundamental solution and propagation of singularities

In this section we will always assume that $d = 2, 3$ and

$$h^{-\delta} \le \mu \le h^{-1} \qquad (\delta > 0). \tag{6.4.1}$$

Then the remainder term R in decomposition (6.2.37), (6.2.38) or (6.2.40) is negligible by theorem 6.1.8 and remark 6.3.2. But then x_d and D_d are included in the operator $\mathcal{A}$ only through the harmonic oscillator $\mathcal{L}_0 = x_d^2 + \mu^{-2}h^2D_d^2$. Decomposing all functions of x_d into series with respect to eigenfunctions of $\mathcal{L}_0$ we decompose $\mathcal{A}$ into a direct sum of $\mathbf{h}'$-pseudo-differential operators with respect to x' which we treat by the methods of Chapter 4. In particular, all of the analysis of section 6.3 can be repeated with weakened restrictions on γ and ζ.

Let us recall that the operator $\mathcal{L}_0$ has simple eigenvalues $(2j+1)\mu^{-1}h$ with eigenfunctions $\Upsilon_j(x_d) = (\frac{\mu}{h})^{\frac14} v_j(\sqrt{\frac{\mu}{h}}x_d)$ which are orthonormal in $L^2(\mathbb{R})$ where v_j are real-valued orthonormal Hermite functions and here and in what follows $j \in \mathbb{Z}^+$.

Theorem 6.4.1. *Let condition* (6.4.1) *be fulfilled.*

(i) For some $j \in \mathbb{Z}^+$ *let*

$$(2j+1)\mu h \bar{B}_1 + \bar{B}_0' \ge \bar{\tau} + \epsilon_0 \qquad \forall (x', \xi_1) \tag{6.4.2}$$

with $\epsilon_0 > 0$, $\bar{\tau} \le c$ *where the bar means that* $\varepsilon = \eta = 0$ *and in* $\bar{B}_0'$ *we divided* B_0 *by* ε *before taking* $\varepsilon = \eta = 0$. *Then the estimate*

$$\begin{aligned} &|F_{t\to h^{-1}\tau}\chi_{T,T'}(t)\int Q_x'\mathcal{U}(x,y,t)\Upsilon_j(x_d)dx_d| \le \\ &\qquad C'Th^s(1+j\mu h)^{-s}(1+\bar{\tau}-\tau)^{-s} \\ &\qquad\qquad \forall x' \in \mathbb{R}^{d-1} \quad \forall y \in \mathbb{R}^d \quad \forall \tau \le \bar{\tau} \quad \forall T \ge h^{1-\delta} \end{aligned} \tag{6.4.3}$$

holds with $Q' = I$ *where here and in what follows* $C' = C'(c, \epsilon_0, \delta, s)$, $N = N(\delta, s)$, $K = K(\delta, s)$ *and* $\delta > 0$ *and* s *are arbitrary.*

(ii) On the other hand, let $d = 3$, $Q' = \operatorname{Op}(q')$, $q' \in S_{\mathbf{h}',K}(T^*\mathbb{R}^2)$, $\operatorname{supp}(1-q') \subset \{|\xi_2| \le c\}$ *and for some* $j \in \mathbb{Z}^+$ *let*

$$\xi_2^2 + (2j+1)\mu h \bar{B}_1 + \bar{B}_0' \ge \bar{\tau} + \epsilon_0 \qquad \forall (x', \xi') \in \operatorname{supp} q' \tag{6.4.4}$$

with the same ϵ_0 *and* $\bar{\tau}$; *then estimate* (6.4.3) *remains true;*

(iii) The following decomposition holds

$$\mathcal{U} \equiv \sum_{j,k\le J} \mathcal{U}_{j,k}(x',y',t)\Upsilon_j(x_d)\Upsilon_k(y_d) \tag{6.4.5}$$

modulo functions f satisfying (6.2.51) *with arbitrary $q \in S_{\mathbf{h}',K}(T^*\mathbb{R}^{d-1})$ (and $Q = \mathrm{Op}(q)$); here $\bar\tau \le c$ is arbitrary and*

$$J = J(\bar\tau) = \min\{j \in \mathbb{Z}^+ : (2j+1)\mu h\bar B_1 + \bar B_0' \ge \bar\tau + \epsilon_0 \quad \forall (x',\xi_1)\}; \tag{6.4.6}$$

(iv) Furthermore,

$$\mathcal{U}_{j,k}(x',y',t) = \mathcal{U}^{\dagger}_{k,j}(y',x',-t), \tag{6.4.7}$$

the estimate

$$|F_{t\to h^{-1}\tau}\chi_{T,T'}(t)\mathcal{U}_{j,k}| \le CT\mu h^{1-d}(1+|\tau|)^{d-1} \qquad \forall x',y' \in \mathbb{R}^{d-1} \quad \forall j,k \quad \forall\tau \quad \forall T \ge h^{1-\delta} \tag{6.4.8}$$

holds and $\mathcal{U}_{j,k}$ satisfy the Cauchy problems

$$(hD_t - \mathcal{A}_{jx})\mathcal{U}_{j,k} \equiv 0, \tag{6.4.9}$$

$$\mathcal{U}_{j,k}|_{t=0} \equiv \delta_{jk}\delta(x'-y') \tag{6.4.10}$$

modulo functions f and f' respectively satisfying (6.2.51) *and* (6.2.52) *with arbitrary $q \in S_{\mathbf{h}',K}$ supported in Ω ($d=2$) and in $\Omega \times \{|\xi_2| \le c\}$ ($d=3$) where δ_{jk} is the Kronecker symbol and*

$$\mathcal{A}_j = \sum_{0\le n\le N-1} B_n(x_1,h_1D_1,\mu^{-2},h_1)\mu^{2-2n}\left((2j+1)\mu h\right)^n \qquad \text{for } d=2 \tag{6.4.11}$$

and

$$\mathcal{A}_j = h^2D_2^2 + \sum_{0\le n\le N-1} B_n(x',h_1D_1,\mu^{-2},h_1)\mu^{2-2n}\left((2j+1)\mu h\right)^n \qquad \text{for } d=3; \tag{6.4.12}$$

let us recall that $h_1 = \mu^{-1}h$.

Proof. First of all (6.4.7) and (6.4.8) follow from (6.2.53) and (6.2.54) respectively if we set

$$\mathcal{U}_{j,k} = \iint \mathcal{U}(x,y,t)\Upsilon_j(x_d)\Upsilon_k(y_d)dx_ddy_d.$$

If $q' \in S_{\mathbf{h},\rho,\gamma,K}(T^*\mathbb{R}^d)$ with $\rho_j,\gamma_j \in (0,1]$, $\rho_j\gamma_j \ge h_jh^{-\delta}$ $\forall j \le d-1$, $\rho_d = \mu^{\frac12}h^{\frac12-\delta'}$, $\gamma_d = \mu^{-\frac12}h^{\frac12-\delta'}$, $\delta' > 0$ and if

$$\operatorname{supp} q' \cap \{\mu^2x_2^2 + \xi_2^2 \le h^{-\delta''}\} = \emptyset \quad \text{for } d=2,$$

$$\operatorname{supp} q' \cap \{\mu^2x_3^2 + \xi_3^2 \le h^{-\delta''}, \xi_2^2 \le C_0\} = \emptyset \quad \text{for } d=3$$

with $\delta'' > 0$ then it is easy to prove by standard elliptic arguments that estimate (6.2.51) holds for $f = \mathcal{U}$, $q = Q'$; in this deduction we also cut $\mathcal{U}$ (see remark 6.2.10).

This yields that for $j \ge J' = \mu^{-1}h^{-1-\delta''}$ estimate (6.4.3) holds and therefore decomposition (6.4.5) with J' instead of J holds. Then formulas (6.2.49), (6.2.50) and (6.2.37), (6.2.38) yield (6.4.9), (6.4.10) and (6.4.11), (6.4.12) respectively. Furthermore, (6.4.9), (6.4.8) and standard elliptic arguments yield the implications (6.4.2) $\implies$ (6.4.3) and (6.4.4) $\implies$ (6.4.3); therefore in decomposition (6.4.5) one can replace J' by J given by formula (6.4.6). □

We can now apply all the methods of section 2.1, Chapter 4 and section 6.3 to the analysis of $\mathcal{U}_{j,k}$; moreover, when we apply the methods of section 6.3 we can weaken restrictions on γ and ζ associated with the uncertainty principle for **h**-pseudo-differential operators because we now have only **h**′-pseudo-differential operators and therefore condition (6.3.4) should be replaced by

$$\begin{aligned}(6.4.13)\quad & \mathbf{h}' = (h_1, \dots, h_{d-1}),\ \gamma' = (\gamma_1, \dots, \gamma_{d-1}),\ \boldsymbol{\rho}' = (\rho_1, \dots, \rho_{d-1}), \\ & h_1 = \mu^{-1}h,\ h_2 = h,\ \rho_j, \gamma_j \in (0,1],\ \rho_j\gamma_j \ge h_j h^{-\delta} \quad \forall j = 1, \dots, d-1.\end{aligned}$$

We leave to the reader the proof of the following two statements:

Proposition 6.4.2. *Let $d = 2$, $Q' = \operatorname{Op}(q')$, $Q'' = \operatorname{Op}(q'')$ with $q', q'' \in S_{h_1,\rho_1,\gamma_1,K}(T^*\mathbb{R})$ and let condition (6.4.13) be fulfilled. Then*

(i) If

$$(6.4.14)\qquad \operatorname{supp} q' \cap \{\tau_1 \le (2j+1)\mu h + \bar{B}(x_1, \xi_1) \le \tau_2\} = \emptyset$$

with $-\infty \le \tau_1 < \tau_2 \le \infty$ (where here and in what follows B is taken from (6.2.45)) then the estimate

$$\begin{aligned}(6.4.15)\quad & |F_{t\to h^{-1}\tau}\chi_{T,T'}(t)Q'_x\mathcal{U}_{j,k}| \le C'Th^s\Big(1 + \min(\tau - \tau_1, \tau_2 - \tau)\Big)^{-s} \\ & \qquad \forall x', y' \in \mathbb{R}^{d-1} \quad \forall \tau \in [\tau_1 + \epsilon, \tau_2 - \epsilon] \quad \forall T \ge h^{1-\delta}\end{aligned}$$

holds.

(ii) For $h^{1-\delta} \le T \le \mu T_0$ the estimate

$$(6.4.16)\qquad |F_{t\to h^{-1}\tau}\chi_T(t)\mathcal{U}_{j,k}| \le C'h^s \qquad \forall x', y' \in \mathbb{R}^{d-1} \quad \forall \tau \le c \quad \forall j \ne k$$

holds.

(iii) If $h^{1-\delta} \le T \le \mu T_0$ and one of the conditions $(6.3.5)_{1,2}$ is fulfilled then the estimate

$$(6.4.17)\qquad |F_{t\to h^{-1}\tau}\chi_T(t)Q'_x\mathcal{U}_{j,k}{}^tQ''_y| \le C'h^s \qquad \forall x', y' \in \mathbb{R}^{d-1} \quad \forall \tau \le c$$

holds.

(iv) Moreover, if conditions $(6.2.48)_{1-3}$, (6.3.9) and (6.3.10) are fulfilled then in assertion (i) we can replace ϵ by $\epsilon\nu$ and in assertions (ii), (iii) we can take $T \in [h^{1-\delta}\bar{\nu}\nu^{-1}, \mu T_0\nu^{-1}]$, $|\tau| \le c\nu\bar{\nu}^{-1}$ and replace conditions $(6.3.5)_{1,2}$ by $(6.3.5)'_{1,2}$.

Proposition 6.4.3. *Let $d = 3$, $Q' = \operatorname{Op}(q')$, $Q'' = \operatorname{Op}(q'')$ with $q', q'' \in S_{\mathbf{h}',\rho',\gamma',K}(T^*\mathbb{R}^2)$ and let condition (6.4.13) be fulfilled. Then*

(i) If

$$(6.4.18)\qquad \operatorname{supp} q' \cap \{\tau_1 \le \xi_2^2 + (2j+1)\mu h\bar{B})_1(x', \xi_1) + \bar{B}'_0(x', \xi_1 \le \tau_2\} = \emptyset$$

with $-\infty \le \tau_1 < \tau_2 \le \infty$ *then estimate* (6.4.15) *holds;*

(ii) *For* $h^{1-\delta} \le T \le T_0$ *estimate* (6.4.16) *holds;*

(iii) *If* $h^{1-\delta} \le T \le T_0$ *and one of conditions* $(6.3.5)_{1-4}$ *is fulfilled then estimate* (6.4.17) *holds.*

Proof. The proof of assertion (i) is based on standard elliptic arguments, and assertions (ii) and (iii) follow from the theory of propagation of singularities (see section 2.1 and proof of proposition 6.3.3). □

6.4.2 Calculations

Let us proceed from estimates to calculations;[9] we need to prove the following statements:

Proposition 6.4.4. *Let* $d = 2$ *and* $Q' = \mathrm{Op}(q')$, $Q'' = \mathrm{Op}(q'')$, $q', q'' \in S_{h_1,\rho_1,\gamma_1,K}(T^*\mathbb{R})$; *let condition* (6.4.13) *be fulfilled and let* $\tilde{Q}' = \mathcal{F}Q'\mathcal{F}^*$, $\tilde{Q}'' = \mathcal{F}Q''\mathcal{F}^*$ *and* $\psi', \psi'' \in C_0^K(B(0,\frac{1}{2}))$ *be fixed functions. Then*

(i) *The following equality holds*

$$(6.4.19)_{(2)}\quad F_{t\to h^{-1}\tau}\chi_T(t)\Gamma\bigl(\tilde{Q}'_x\psi'(x)U\psi''(y)^t\tilde{Q}''_y\bigr) \equiv \\ \sum_{j\le J}\sum_{r+m+p\le N} F_{t\to h^{-1}\tau}\chi_T(t)\Gamma'\bigl(Q'_x\mathcal{M}_{rmp}(x_1,y_1,h_1D_{x_1},h_1D_{y_1})\mathcal{U}_{j,j}{}^tQ''_y\bigr)\times \\ \bigl((2j+1)\mu h\bigr)^r\mu^{-2(m+r)-p}h^p \qquad (\mathrm{mod}\, O(h^s))$$

for $\tau \le \epsilon_1$, $T \in [h^{1-\delta}, \mu T_0]$ *where* J *is defined by* (6.4.6) *with* $\bar{\tau} = 0$ *and* $\epsilon = 2\epsilon_1$, Γ' *is the operator of restriction to the diagonal and integration acting with respect to the variables* x' *and* y',

$$(6.4.20)_{(2)}\quad \mathcal{M}_{rmp} = \sum_{|\alpha|+|\beta|\le 2p+2r+m} \varsigma_{rmp\alpha\beta}(x_1,y_1,\xi_1,\eta_1)\times \\ (D^\alpha\psi')\circ\bar{\Psi}_0(x_1,\xi_1)\times(D^\beta\psi'')\circ\bar{\Psi}_0(y_1,\eta_1),$$

$\varsigma_{\dots}$ *are smooth symbols independent of* ψ', ψ'', $\varsigma_{\dots} = 1$ *for* $m = p = r = \alpha = \beta = 0$ *and* $\varsigma_{\dots} = 0$ *for* $m + p + r > 0$, $\alpha = \beta = 0$.

(ii) *Moreover, if conditions* $(6.2.48)_{1-3}$, (6.3.9) *and* (6.3.10) *are fulfilled then equality* $(6.4.19)_{(2)}$ *holds for* $|\tau| \le \epsilon_1\nu\bar{\nu}^{-1}$, $T \in [h^{1-\delta}\bar{\nu}\nu^{-1}, \mu T_0\nu^{-1}]$.

(iii) *If* $\psi'' = 1$ *in a neighborhood of* $\operatorname{supp}\psi'$ *then one can take* $\mathcal{M}_{rmp}$ *depending only on* (x_1,ξ_1) *and* $\varsigma_{\dots} = 0$ *for* $\beta \ne 0$.

(iv) *In particular, if* $F = \mathrm{const}$ *and we apply the reduction method described in proposition 6.2.1(iii) then in the conformal coordinate system*

$$(6.4.21)_1\qquad \mathcal{M}_{010} = -\frac{1}{2}(\rho^2\nabla\psi'\cdot\nabla v)\circ\bar{\Psi}_0,$$

$$(6.4.21)_2\qquad \mathcal{M}_{100} = \frac{1}{4}(\rho^2\Delta\psi')\circ\bar{\Psi}_0;$$

(v) *If conditions* $(6.2.48)_{1-3}$ *are fulfilled then* $|D^\sigma\varsigma_{\dots}| \le C\bar{\nu}$ *for all* r, m, p, α, β *indicated above and for all* $\sigma : 1 \le \sigma \le K_1$ *with arbitrary* K_1 *(*K *depends on* K_1*).*

---

[9] More refined calculations will be made in the next subsection.

Proposition 6.4.5. *Let $d = 3$ and $Q' = \operatorname{Op}(q')$, $Q'' = \operatorname{Op}(q'')$ with $q', q'' \in S_{\mathbf{h}',\rho',\gamma',K}(T^*\mathbb{R}^2)$; let condition (6.4.13) be fulfilled and let $\tilde{Q}' = \mathcal{F}Q'\mathcal{F}^*$, $\tilde{Q}'' = \mathcal{F}Q''\mathcal{F}^*$ and $\psi', \psi'' \in C_0^K(B(0, \frac{1}{2}))$ be fixed functions. Then*

(i) The following equality holds

$$(6.4.19)_{(3)}\quad F_{t\to h^{-1}\tau}\chi_T(t)\Gamma\bigl(\tilde{Q}'_x\psi'(x)U\psi''(y)^t\tilde{Q}''_y\bigr) \equiv$$
$$\sum_{j\le J}\ \sum_{r+m+p+k'+k''\le N} F_{t\to h^{-1}\tau}\chi_T(t)\Gamma'\bigl(Q'_x\mathcal{M}_{rmpk'k''}(x',y',\mathbf{h}'D_{x'},\mathbf{h}'D_{y'})\times$$
$$(hD_{x_2})^{k'}(hD_{y_2})^{k''}\mathcal{U}_{j,j}{}^tQ''_y\bigr)\cdot\bigl((2j+1)\mu h\bigr)^r\mu^{-2(m+r)-p-k'-k''}h^p$$
$$(\operatorname{mod} O(h^s))$$

for $\tau \le \epsilon_1$, $T \in [h^{1-\delta}, T_0]$ with the same J and Γ' as before,

$$(6.4.20)_{(3)}\quad \mathcal{M}_{rmpk'k''} = \sum_{|\alpha|+|\beta|\le 2p+2r+m+k'+k''} \varsigma_{rmpk'k''\alpha\beta}(x',y',\xi_1,\eta_1)\times$$
$$(D^\alpha\psi')\circ\bar{\Psi}_0(x',\xi_1)\times(D^\beta\psi'')\circ\bar{\Psi}_0(y',\eta_1),$$

where $\varsigma_{\dots}$ are smooth symbols independent of ψ', ψ'' and $\varsigma_{\dots} = 1$ for $m = p = r = k' = k'' = \alpha = \beta = 0$ and $\varsigma_{\dots} = 0$ for $m + p + r + k' + k'' > 0$, $\alpha = \beta = 0$.

(ii) If $\psi'' = 1$ in a neighborhood of $\operatorname{supp}\psi'$ then one can take $\mathcal{M}_{rmp}$ depending only on (x_1, ξ_1) and $\varsigma_{\dots} = 0$ for $(\beta, k') \ne 0$.

(iii) Assertion (v) of proposition 6.4.4 remains true; moreover, it is true for $k'+k'' > 0$ and $\sigma = 0$.

Proof of propositions 6.4.4 and 6.4.5. Let us first consider the operators $\mathcal{Q}'(x, \mu^{-1}hD, \mu^{-2}, \mu^{-1}h) = \mathcal{F}^*\psi'\mathcal{F}$ and $\mathcal{Q}''$ (which is defined in a similar way and has similar properties). We know that

$$\mathcal{Q}' \sim \sum_p \mathcal{Q}'_p(x, \mu^{-1}hD, \mu^{-2})(\mu^{-1}h)^p,$$

where $\mathcal{Q}'_0 = \psi' \circ \Psi_{\mu^{-2}}$ and $\mathcal{Q}'_p$ is a linear combination of $D^\alpha\psi' \circ \Psi_{\mu^{-2}}$ with $1 \le |\alpha| \le 2p$ for $p \ge 1$. Let us decompose $\mathcal{Q}'_p$ with respect to powers of μ^{-2}. Moreover, let us decompose $\mathcal{Q}'_p$ with respect to powers of x_d and $\mu^{-1}hD_d$; for $d = 3$ let us also decompose $\mathcal{Q}'_p$ with respect to powers of $\mu^{-1}hD_2$; remark 6.3.2 and (6.4.1) yield that we can take only a finite number of terms provided we apply the operator $\mathcal{Q}'_x$ or ${}^t\mathcal{Q}''_y$ to $\mathcal{U}$ and then we apply the operator $F_{t\to h^{-1}\tau}\chi_T(t)$ with $\tau \le c$. So for $d = 3$

$$\mathcal{Q}' \sim \sum_{m,\sigma,k,p} \mathcal{Q}'_{mp\sigma k}(x', \mu^{-1}hD_1)\mu^{-2m-p-k}h^p x_3^{\sigma_1}(\mu^{-1}hD_3)^{\sigma_2}(hD_2)^k$$

where $\mathcal{Q}'_{0000} = \psi \circ \bar{\Psi}_0$ and $\mathcal{Q}'_{mp\sigma k}$ is a linear combination of $D^\alpha\psi' \circ \bar{\Psi}_0$ with $1 \le |\alpha| \le m + 2p + |\sigma| + k$ for $m + 2p + |\sigma| + k \ge 1$. It will be convenient for us to pass from this formula to the formula

$$\mathcal{Q}' \sim \sum_{m,\sigma,k,p} \mathcal{Q}'_{mp\sigma k}(x', \mu^{-1}hD_1)\mu^{-2m-p-k}h^p(\mu^{-1}h)^{\frac{1}{2}|\sigma|}Z_+^{\sigma_1}Z_-^{\sigma_2}(hD_2)^k$$

with $Z_\pm = \sqrt{\frac{\mu}{h}}(x_3 \pm i\mu^{-1}hD_3)$ and with coefficients $\mathcal{Q}'_{mp\sigma k}$ of the same nature as before; then $[Z_+, Z_-] = 2$. A similar but simpler formula holds for $d = 2$.

Let us apply $F_{t\to h^{-1}\tau}\mathcal{Q}'_x$ and ${}^t\mathcal{Q}''_y$ to $\mathcal{U}$ and let us take decomposition (6.4.5) for $\mathcal{U}$; let us take into account that proposition 6.4.2 yields that only terms with $j = k \le J$ are not necessarily negligible in this decomposition with the T and τ indicated in proposition 6.4.4(i),(ii) for $d=2$ and in proposition 6.4.5(i) for $d=3$. Furthermore, let us apply the equalities

$$(6.4.22)_{1,2} \qquad (z+\partial_z)\upsilon_j=\sqrt{2j}\upsilon_{j-1}, \qquad (z-\partial_z)\upsilon_j=\sqrt{2(j+1)}\upsilon_{j+1},$$

for our Hermite functions. We obtain that in the resulting formula for the left-hand expression in $(6.4.19)_{(d)}$ terms with multi-indices σ' and σ'' from the decompositions of $\mathcal{Q}'$ and $\mathcal{Q}''$ are negligible unless $\sigma'_1-\sigma'_2=\sigma''_1-\sigma''_2$; therefore $|\sigma'|+|\sigma''|$ is even. Moreover, the inner product in $L^2(\mathbb{R})$ of functions $Z_+^{\sigma'_1}Z_-^{\sigma'_2}\Upsilon_j$ and $Z_+^{\sigma''_1}Z_-^{\sigma''_2}\Upsilon_j$ is a $\frac{1}{2}(|\sigma'|+|\sigma''|)$-degree polynomial of $(2j+1)$; this yields $(6.4.19)_{(d)}$ for the indicated T and τ. The same calculations also yield $(6.4.20)_{(d)}$. So proposition 6.4.4(i),(ii) and proposition 6.4.5(i) are proven. Moreover, the same arguments yield propositions 6.4.4(iii) and 6.4.5(ii).

Furthermore, the same arguments also yield propositions 6.4.4(v) and 6.4.5(iii). In fact, if $d=3$, $g^{jk}=\delta_{jk}$, $F^1=F^3=0$ and $F^2=\text{const}$, $V=\text{const}$ then in the context of the proof of theorem 6.1.1

$$\mathcal{Q}'=\psi'(x_1-\mu^{-1}h\ell D_3, x_2, x_3-\mu^{-1}h\ell D_1)$$

with constant ℓ; a similar equality holds for $d=2$, $g^{jk}=\delta_{jk}$ and $F=\text{const}$, $V=\text{const}$. Therefore in these cases $\varsigma_{\dots}=\text{const}$; moreover, $\varsigma_{\dots}=0$ for $d=3$ and $k'+k''\geq 1$. Then the stability of all the procedures yields propositions 6.4.4(v) and 6.4.5(iii).

Finally, in order to prove proposition 6.4.4(iv) it is sufficient to note that under the indicated method of reduction

$$\mathcal{M}_{100}=\frac{1}{2}\Big(\{x_2,\{x_2,\mathcal{Q}'\}\}+\{\xi_2,\{\xi_2,\mathcal{Q}'\}\}\Big)|_{\Lambda^0,\varepsilon=0}=$$

$$\frac{1}{2}\Big(\{w''_2,\{w''_2,\psi'\}\}+\{w''_1,\{w''_1,\psi'\}\}\Big)\Big|_{\Lambda,\varepsilon=0}\circ\bar{\Psi}_0,$$

$$\mathcal{M}_{010}=\partial_\varepsilon\psi'|_{\Lambda,\varepsilon=0}\circ\bar{\Psi}_0$$

where w''_1, w''_2 are defined in the proof of proposition 6.2.1(iii); the remaining arguments are the same as in the proof of theorem 6.2.4(iii). □

6.4.3 Estimates in the zone $\{\gamma\geq\bar{\gamma}\}$

Now we are able to apply the results of Chapter 4 in complete measure. Our next step is

Proposition 6.4.6. *Let $d=2$ or $d=3$ and $q',q'',Q',Q'',\tilde{Q}',\tilde{Q}''$ be the same as in proposition 6.4.4 or proposition 6.4.5 respectively with $\rho_j=\gamma_j=\gamma\geq\bar{\gamma}=\sqrt{\frac{\mu h}{\nu}}$. Let* $\operatorname{diam}\operatorname{supp} q'\leq c\gamma$, $\operatorname{diam}\operatorname{supp} q''\leq c\gamma$ *and suppose that on* $\operatorname{supp} q'$ *conditions* $(6.3.19)_{(d)}$, $(6.3.24)_{(d)}$ [10] *be fulfilled where $\nu=1$ for $d=3$. Furthermore, let conditions* $(6.2.48)_{1-3}$, (6.3.9) *and* (6.3.10) *be fulfilled if $d=2$ and $\nu<1$. Then the estimate*

$$(6.4.23)\qquad |F_{t\to h^{-1}\tau}\chi_T(t)\Gamma\tilde{Q}'_x\psi'(x)U\psi''(y)^t\tilde{Q}''_y|\leq Ch^{1-d}\gamma^{2d-2}+C'h^s$$

$$\forall\tau:|\tau|\leq\epsilon\nu\gamma$$

holds with arbitrary $T\in[h^{1-\delta}\nu^{-1}\gamma^{-2},\mu T_0\nu^{-1}]$ for $d=2$, $T\in[h^{1-\delta}\gamma^{-2},T_0]$ for $d=3$.

10) Let us recall that in these conditions for $d=3$ we take λ violating condition $(6.3.18)_{(3)}$ with ζ replaced by $\epsilon\gamma$.

Proof. The arguments of subsection 4.3.2 yield that

$$|F_{t\to h^{-1}\tau}\chi_T(t)\Gamma' Q'_x \mathcal{U}_{j,j}{}^t Q''_y| \le C\mu h^{2-d}\gamma^{2d-4}\nu^{-1} + C'h^s \tag{6.4.24}$$

for the T and τ indicated above and for all $j\in\mathbb{Z}^+$; in fact, we should replace d by $d-1$ in all the corresponding estimates and add a factor $\mu\nu^{-1}$ to the right-hand expressions because of the other scale with respect to t; moreover, for $d=3$ we should add a factor μ in the right-hand expressions because of the additional $\mu^{-1}\gamma$-admissible partition of unity with respect to ξ_1 or x_1 (which is necessary in order to be in the context of this subsection when its arguments are applied).

Moreover, if the condition

$$|-\bar\tau + (2j+1)\mu h + B(x_1,\xi_1,\mu^{-2})| \ge \epsilon_0\nu\gamma^2, \tag{6.4.25$_{(2)}$}$$

$$|-\bar\tau + \xi_2^2 + \mathcal{B}(x',\xi_1,(2j+1)\mu h,\mu^{-2},0)| \ge \epsilon_0\gamma^2, \tag{6.4.25$_{(3)}$}$$

is fulfilled on $\operatorname{supp} q'$ for $d=2,3$ respectively (compare these conditions with $(6.3.18)_{(2,3)}$) then for $|\bar\tau-\tau|\le\epsilon\nu\gamma^2$ and for indicated T the left-hand expression in (6.4.24) does not exceed $C'h^s$. Finally, for every τ from the indicated interval the quantity of numbers j violating condition (6.4.25) does not exceed $C_0\nu\gamma^2(\mu h)^{-1}$.[11)] This fact, estimate (6.4.24) and propositions 6.4.4 and 6.4.5 yield estimate (6.4.23). □

Now applying the standard Tauberian arguments of section 4.3 we immediately obtain that under the assumptions of proposition 6.4.6 the estimates

$$\begin{aligned}\mathcal{R}_3 = |\Gamma(\tilde Q'_x\psi'(x)e\psi''(y){}^t\tilde Q''_y)(-\infty,\tau) - \\ h^{-1}\int_{-\infty}^{\tau}(F_{t\to h^{-1}\tau}\bar\chi_T(t)\Gamma(\tilde Q'_x\psi'(x)U\psi''(y){}^t\tilde Q''_y))\,d\tau| \le R_{30}\end{aligned} \tag{6.4.26}$$

and

$$\mathcal{R}_{3\vartheta} \le R_{3\vartheta} \tag{6.4.26$_\vartheta$}$$

hold for $\vartheta>0$, $|\tau|\le\epsilon\nu\gamma$ where for $\vartheta>0$, $\mathcal{R}_{3\vartheta}$ is the absolute value of the convolution of the corresponding expression in $\mathcal{R}_3$ and $\tau_+^{\vartheta-1}$ and

$$R_{3\vartheta} = \begin{cases} Ch^{\vartheta-1}\gamma^2(\dfrac{\nu}{\mu})^{\vartheta+1} + C'h^s & \text{for } d=2,\\ Ch^{-2+\vartheta}\gamma^4 + C'h^s & \text{for } d=3,\end{cases} \tag{6.4.27}$$

for $\vartheta\ge0$, $T=\mu\nu^{-1}T_0$ for $d=2$, $T=T_0$ for $d=3$ and $\bar\chi=1$ on $[-\frac12,\frac12]$.

Applying proposition 6.4.4 or 6.4.5 to reduce the calculation of the second term in the middle expression in (6.4.26) to the calculation of $F_{t\to h^{-1}\tau}\chi_T(t)\Gamma' Q'_x\mathcal{U}_{j,j}{}^tQ''_y$ (with other operators Q' and Q'') and calculating these expressions by means of the results of section 4.2, we obtain the following

Proposition 6.4.7. *Under the assumptions of proposition 6.4.6 the following estimates hold:*

$$\begin{aligned}\mathcal{R}_4 = |\Gamma(\tilde Q'_x\psi'(x)e)(-\infty,\tau) - \\ \sum_{j\le J}\sum_{m,p,l\le N}\mu^{1-2m-2p-l}h^{1+l-d}((2j+1)\mu h)^m\Theta_{j,mpl}(\tau)| \le R_{30},\end{aligned} \tag{6.4.28}$$

11) The condition $C_0\nu\gamma^2(\mu h)^{-1}\ge1$ defines $\bar\gamma$.

$$(6.4.28)_\vartheta \qquad \mathcal{R}_{4\vartheta} \le R_{3\vartheta}$$

where

$$(6.4.29) \qquad \Theta_{j,mpl}(\tau) = \int \theta(\tau - \mathcal{A}_j(x', \xi', \mu^{-2}, \mu^{-1}h, h^2)) f_{mpl}(x', \xi') dx' d\xi',$$

$$(6.4.30) \qquad \operatorname{supp} F_{mpl} \subset \operatorname{supp} q',$$

$$(6.4.31) \qquad |D^\alpha_{x',\xi'} f_{mpl}| \le C\gamma^{-|\alpha|-2l} \qquad \forall \alpha : |\alpha| \le K',$$

K' is arbitrary and

$$(6.4.32) \qquad f_{000} = (2\pi)^{1-d} q'(x', \xi')(\psi' \circ \bar{\Psi}_0)(x', \xi_1);$$

the definition of $\mathcal{R}_{4\vartheta}$ is obvious and left to the reader. In particular, for $d = 2$ and $\tau = 0$ we can replace the symbol $\mathcal{A}_j$ by $(2j+1)\mu h + B(x_1, \xi_1, \mu^{-2}, \mu^{-1}h, h^2)$ in (6.4.29), and for $d = 3$ only terms with $l = 0$ are present in these and the forthcoming arguments (if ϑ is less than some critical value).

Our goal is to estimate $\mathcal{R}_{4\vartheta}$ with $Q' = I$. In this case for $d = 3$, $\mu \ge h^{-1+\delta}$, $\vartheta = 0$ the desired estimate will be more precise under conditions $(6.2.48)_{1-3}$, (6.3.68) and (6.3.64).

In order to use this condition let us prove

Proposition 6.4.8. *Let $d = 3$ and conditions (6.3.64), (6.3.68) be fulfilled with $\bar{\nu} = 1$, $\nu \ge \mu h$. Let $Q' = \operatorname{Op}(q')$, $q' \in S_{\mathbf{h}',\rho',\gamma',K}(T^*\mathbb{R}^2)$ with $\rho_1 = \gamma_1 = \gamma_2 = \sqrt{\frac{\mu h}{\nu}}$, $\rho_2 = \sqrt{\mu h}$. Let $\operatorname{diam} \operatorname{supp} q' \le c\gamma_1$ and let the following conditions be fulfilled on $\operatorname{supp} q'$:*

$$(6.4.33)_1 \qquad |\xi_2| \le c\rho_2,$$

$$(6.4.33)_2 \qquad |\mathcal{A}_j| \le \epsilon_0 \mu h \implies \epsilon_0 \sqrt{\mu h \nu} \le |\nabla_{x',\xi_1} \mathcal{A}_j| \le c\sqrt{\mu h \nu}.$$

Then

$$(6.4.34) \qquad \mathcal{R}_4 \le C\mu h^{-1}\gamma_1^4 + C'h^s \qquad \forall \tau : |\tau| \le \epsilon \mu h.$$

Proof. First of all, applying an $\epsilon(\rho', \gamma')$-admissible partition of unity on $\operatorname{supp} q'$ we obtain that without loss of generality we can assume that condition $|\mathcal{A}_j| \ge \epsilon \mu h$ can be violated on $\operatorname{supp} q'$ for only one number $j = \bar{\jmath}$. But then standard elliptic arguments yield the estimates

$$|F_{t \to h^{-1}\tau} \chi_T(t) Q'_x \mathcal{U}_{j,k}| \le C'h^s \qquad \forall j \ne \bar{\jmath},$$

$$|F_{t \to h^{-1}\tau} \chi_T(t) \mathcal{U}_{j,k} {}^t Q'_y| \le C'h^s \qquad \forall k \ne \bar{\jmath}$$

with arbitrary $T \in [\mu^{-1}h^{-\delta}, h^{-l}]$, $\tau : |\tau| \le \epsilon \mu h$ and therefore for these T and τ estimate $(6.4.19)_{(3)}$ holds. Let us multiply the operator $\mathcal{A}_j$ with $j = \bar{\jmath}$ by a parametrix of $\bar{B}_1$ and by $\frac{1}{\nu}$; we then obtain a $(\mu^{-1}h, \frac{h}{\sqrt{\nu}})$-pseudo-differential operator with respect to x' and we can apply the arguments of section 4.3; in the estimates we obtain an additional factor $\frac{\mu}{\sqrt{\nu}}$ in the right-hand expressions because we need an additional $\sqrt{\nu}\gamma_1\mu^{-1}$-partition of unity with respect to ξ_1 or x_1. □

Let us note that under the assumptions of proposition 6.4.8 (excluding condition $(6.4.33)_2$) we can replace ν by an arbitrary $\nu' \in [\nu, 1]$ and therefore we can make an appropriate partition of unity in the zone $\{(x', \xi_1) : \max(\mu h, \sqrt{\mu h \nu}) \le |\nabla_{x',\xi_1} \mathcal{A}_j| \le \sqrt{\mu h}\}$ where $j \in \mathbb{Z}^+$ is defined in order to satisfy condition $|\mathcal{A}_j| \le \epsilon\mu h$ (in which case the choice of j is no longer unique).

Let us note that for $d = 2$ proposition 6.4.7 leaves only the zone $\{(x_1, \xi_1) : |\nabla B| \le \epsilon_0\sqrt{\mu h \nu}\}$ uncovered, and for $d = 3$ the proposition leaves only the zone $\{(x', \xi') : |\xi_2| \le \epsilon_0\sqrt{\mu h}, |\nabla_{x',\xi_1} \mathcal{A}_j| \le \epsilon_0\sqrt{\mu h \nu}\}$ uncovered where $j \in \mathbb{Z}^+$ is always taken by the indicated method. Moreover, for $d = 3$, $\nu < 1$, $\vartheta = 0$ both these propositions together leave only the zone $\{(x', \xi') : |\xi_2| \le \epsilon_0\sqrt{\mu h}, |\nabla_{x',\xi_1} \mathcal{A}_j| \le \epsilon_0\sqrt{\mu h \nu}\}$ uncovered. After summation with respect to a partition of unity we therefore obtain from propositions 6.4.7 and 6.4.8 the following statement:

Proposition 6.4.9. *(i) Let $d = 2$, suppose conditions $(6.2.48)_{1-3}$, (6.4.9), (6.4.10)[12] and (6.3.64) are fulfilled and let $Q'' = \mathrm{Op}(q'')$ where $q'' \in S_{h_1,\bar\gamma,\bar\gamma,K}(T^*\mathbb{R})$ with $\bar\gamma = \sqrt{\frac{\mu h}{\nu}}$; moreover, let us assume that*

$$\operatorname{supp} q'' \subset \{|\nabla B| \le c\sqrt{\mu h \nu}\}, \qquad q'' = 1 \quad \text{at } \{|\nabla B| \le \epsilon_0\sqrt{\mu h \nu}\}. \tag{6.4.35}$$

Then for $Q' = I - Q''$, $\vartheta \ge 0$ the estimate

$$\mathcal{R}_{4\vartheta} \le C\nu^{1+\vartheta}\mu^{-\vartheta-1}h^{-1+\vartheta} + C'h^s \tag{6.4.36}$$

holds.

(ii) Let $d = 3$ and suppose conditions (6.3.64), (6.4.32) with $\bar\nu = 1$ and $\nu \ge \mu h$ are fulfilled and let $Q'' = \mathrm{Op}(q'')$ where $q'' \in S_{\mathbf{h}',\rho',\bar\gamma,K}(T^*\mathbb{R}^2)$ with $\rho_1 = \bar\gamma = \sqrt{\frac{\mu h}{\nu}}$, $\rho_2 = \sqrt{\mu h}$ and

$$\begin{aligned} &\operatorname{supp} q'' \subset \{|\xi_2| \le c\sqrt{\mu h}, |\nabla_{x',\xi_1} \mathcal{A}_j| \le c\sqrt{\mu h}\}, \\ &q'' = 1 \quad \text{at } \{|\xi_2| \le \epsilon_0\sqrt{\mu h}, |\nabla_{x',\xi_1} \mathcal{A}_j| \le \epsilon_0\sqrt{\mu h \nu}\} \end{aligned} \tag{6.4.37}$$

where j is defined for given (x', ξ') by the method indicated above. Then for $Q' = I - Q''$, $\vartheta = 0$ the estimate

$$\mathcal{R}_4 \le Ch^{-2} + C'h^s \tag*{$(6.4.38)_0$}$$

holds.

(iii) Furthermore, for $\nu = \mu h$ we obtain $\bar\gamma = 1$ and condition (6.4.37) means exactly that

$$\operatorname{supp} q'' \subset \{|\xi_2| \le c\sqrt{\mu h}\}, \quad q'' = 1 \quad \text{at } \{|\xi_2| \le \epsilon_0\sqrt{\mu h}\}. \tag*{$(6.4.37)'$}$$

(iv) Finally, let $d = 3$ and let $Q'' = \mathrm{Op}(q'')$ where $q'' \in S_{\mathbf{h}',\rho',\bar\gamma,K}(T^*\mathbb{R}^2)$ with $\rho_1 = \rho_2 = \bar\gamma = \sqrt{\mu h}$. Let condition (6.4.37) with $\nu = 1$ be fulfilled, $Q' = I - Q''$. Then the estimate

$$\mathcal{R}_{4\vartheta} \le Ch^{-2+\vartheta} + C'h^s \tag*{$(6.4.38)_\vartheta$}$$

holds for $\vartheta > 0$.

---

[12] For $\nu \le \mu h$ this statement is empty.

6.4.4 Estimates in the zone $\{\gamma \le \tilde\gamma\}$

Let us note that if Q'' satisfies the conditions of the last proposition and if $\operatorname{diam}\operatorname{supp} q'' \le \epsilon\tilde\gamma$ and, moreover, if for $d = 3$ the symbol q'' is supported in the zone $\{|\xi_2| \le \epsilon\sqrt{\mu h}\}$ then for $\epsilon = \epsilon(c, \epsilon_0) > 0$ the number j is constant on $\operatorname{supp} q''$ and hence the results of section 4.4 are applicable without change. Namely, theorem 4.4.11 yields that for $Q' = Q'' = \operatorname{Op}(q'')$, $\tau = 0$ under the assumptions of proposition 6.4.9(i) (i.e., for $d = 2$) the following estimates hold without the first term in their right-hand expressions:

$$\mathcal{R}_4 \le C\frac{\nu}{\mu h} + Ch^{-\delta'} + C'h^s \tag{6.4.39$_0$}$$

provided

$$\sum_{|\alpha|\le m} |D^\alpha \mathcal{B}_j| \ge \epsilon_0\nu \qquad \forall (x', \xi_1) \tag{6.4.40}$$

where $\mathcal{B}_j = B + (2j+1)\mu h$ and $\delta' > 0$ and m are arbitrary and theorem 4.4.6 yields that estimate

$$\mathcal{R}_4 \le C\frac{\nu}{\mu h} + C + C'h^s \tag{6.4.41$_0$}$$

holds provided

$$|\mathcal{B}_j| + |\nabla B| \le \epsilon_0\nu \implies |\det \operatorname{Hess} B| \ge \epsilon_0\nu^2. \tag{6.4.42}$$

Remark. *Moreover, the last estimate holds under more general but more sophisticated conditions corresponding to conditions (4.4.45) and (4.4.47) for* $\frac{V}{F} - (2j+1)\mu h$ *instead of* a. *On the other hand, due to remark 4.4.14 estimate* (6.4.39)$_0$ *holds under conditions* (4.4.69)–(4.4.70) *for the same function provided* $\nu = \nu_m$ *satisfies (6.4.10) and in the right-hand expression of* (6.4.39)$_0$ $\nu = \nu_1$. ◇

Furthermore, for $\vartheta > 0$ theorem 4.4.12 yields the estimate

$$\mathcal{R}_{4\vartheta} \le C\nu^{1+\vartheta}\mu^{-1-\vartheta}h^{-1+\vartheta} + C\nu^{\vartheta}\mu^{-\vartheta}h^{\vartheta} + C'h^s \tag{6.4.41$_\vartheta$}$$

without the first term in the right-hand expression.

Moreover, theorem 4.4.9(iii) yields that for $Q' = Q'' = \operatorname{Op}(q'')$, $\tau = 0$, $\vartheta = 0$ in frames of proposition 6.4.9(ii) or (iii) (i.e., for $d = 3$) estimate

$$\mathcal{R}_4 \le Ch^{-2} + C\mu\nu^{\frac{1}{2}\delta'}h^{-1-\delta'} + C'h^s \tag{6.4.43}$$

holds without the first term in the right-hand expressions in the general case; furthermore, theorem 4.4.9(ii) yields that under condition (6.4.40) the estimate

$$\mathcal{R}_4 \le Ch^{-2} + C'h^s \tag{6.4.44$_0$}$$

holds where now $\mathcal{B}_j = \mathcal{A}_j - \xi_2^2$, and j is defined no more than uniquely by condition $|\mathcal{B}_j| \le \epsilon\mu h$. Moreover, the same theorem yields estimate for $\vartheta > 0$.

$$\mathcal{R}_{4\vartheta} \le Ch^{-2+\vartheta} + C'h^s. \tag{6.4.44$_\vartheta$}$$

Applying propositions 6.4.8 and 6.4.9 we obtain the following key statement:

Theorem 6.4.10. *Let $Q' = I$, $\tau = 0$.*

(i) Let $d = 2$ and let conditions $(6.2.48)_{1-3}$, (6.3.9) and (6.3.10) be fulfilled. Then for $\vartheta = 0$ estimate $(6.4.39)_0$ holds; under the additional condition (6.4.40) estimate $(6.4.41)_0$ holds. Finally, for $\vartheta > 0$, $(6.4.41)_\vartheta$ holds.

(ii) Let $d = 3$ and let either $\vartheta > 0$ or $\mu \le h^{-1+\delta}$. Then estimate $(6.4.44)_\vartheta$ holds.

(iii) Moreover, for $d = 3$, $\vartheta = 0$ estimate (6.4.43) with $\nu = 1$ holds. Furthermore, let conditions (6.3.68) and (6.3.64) be fulfilled with $\bar\nu = 1$. Then estimate $(6.4.44)_0$ holds provided either $\nu \le h(\mu h)^{-s}$ or condition (6.3.58) is fulfilled. In the general case estimate (6.4.43) holds.

Finally, propositions 6.4.4 and 6.4.5 and the same results of section 4.4 yield the following

Proposition 6.4.11. *(i) Let $d = 2$ and conditions $(6.2.48)_{1-3}$, (6.3.9) and (6.3.10) be fulfilled. Let ϕ satisfy condition (4.3.21) with $\vartheta > 0$. Then the estimate*

$$
\begin{aligned}
(6.4.45)\quad \mathcal{R}_{5\vartheta} = | \int \phi(\frac{\tau-\tau'}{L}) d_{\tau'} \big((\Gamma\psi(x)e)(-\infty,\tau') - \\
\sum_{j\le J} \sum_{m,p,l,n\le N, n\le 2l} \mu^{1-2(m+p)-l} h^{l-1} ((2j+1)\mu h)^m \Theta^{(n)}_{j,mpln}(\tau') \big) | \le \\
C R_4 (\frac{h\nu}{\mu L})^\vartheta \qquad \forall \tau : |\tau| \le c\nu \quad \forall L \in [\nu h \mu^{-1}, c]
\end{aligned}
$$

holds where $\Theta_{j,mpln}(\tau)$ are given by (6.4.28) with f_{mpln} instead of f_{mpl} with $f_{mpln} \in C^{K'}$ and $f_{mpl0} = f_{mpl}$, the superscript (n) means the n-th order derivative with respect to the spectral parameter, and here and below R_4 is the right-hand expression in the corresponding estimate for $\mathcal{R}_{40}$ (in the general case or under some additional condition).

(ii) Let $d = 3$ and ϕ satisfy condition (4.3.21) with $\vartheta > 0$. Then the estimate

$$
(6.4.46)\qquad \mathcal{R}_{5\vartheta} \le C' R_4 (\frac{h}{L})^\vartheta \qquad \forall \tau \le c \quad \forall L \in [h, c]
$$

holds.

(iii) In particular, for $\phi \in C_0^K(\mathbb{R})$ and $L \ge h^{1-\delta}\nu\mu^{-1}$, $L \ge h^{1-\delta}$ the estimate $\mathcal{R}_5 = O(h^s)$ holds without any additional conditions.

Remark 6.4.12. *By remark 4.4.7, for odd integers ϑ one should also include the extra term described in that remark in $\mathcal{R}_{4\vartheta}$. For $d = 3$ the extra term appears only if the condition*

$$
(6.4.47)_+ \quad |\tfrac{V}{F} + (2j+1)\mu h| + |\nabla \tfrac{V}{F}| \le \epsilon_0 \implies \quad \operatorname{Hess} \tfrac{V}{F} \text{ has an eigenvalue } f \ge \epsilon_0
$$

is violated in which case it does not exceed $C\mu h^{-1+\vartheta} |\log h|$ (so it is negligible for $\mu \le h|\log h|^{-1}$ anyway). For $d = 2$ the extra term appears only if the condition

$$
(6.4.48)_+ \qquad |\frac{V}{F} + (2j+1)\mu h| + |\nabla \frac{V}{F}| \le \epsilon_0 \nu \implies \det \operatorname{Hess} \frac{V}{F} \ge \epsilon_0 \nu^2
$$

is violated in which case it does not exceed $C\nu\vartheta\mu^{-\vartheta} h^\vartheta |\log \mu^{-1} h|$ (so it is negligible for $\nu \ge \mu h |\log \mu^{-1} h|$ anyway). The reader can easily prove that one can replace the eigenvalues of the skew-Hessian of the reduced symbol by $\pm\sqrt{-\zeta_k}$ where ζ_k are the eigenvalues of the matrix $g^{jk}\partial_{kl} V$. ◊

6.4.5 Effective asymptotics. I

The asymptotics obtained in theorem 6.4.10 are optimal in a reasonable sense, but we cannot effectively calculate all the terms of the decompositions of $\mathcal{A}_j$ and amplitudes $f_{\dots}$. Moreover, the right-hand expression is not always effectively calculable. Therefore our purpose through the end of this section is to obtain simpler (though more special and maybe not so optimal) asymptotics and estimates. We deduce them from theorem 6.4.10, proposition 6.4.11 and the results of section 6.3 by means of a certain universal approach which will be described just now. First of all we want to estimate

$$\mathcal{R} = |\int \psi(x)(e(x,x,-\infty,\tau) - h^{-d}S(x,\tau,\mu h))dx| \tag{6.4.49}$$

and

$$\mathcal{R}_\vartheta = |\iint (\tau-\tau')_+^\vartheta \psi(x) d_{\tau'}(e(x,x,-\infty,\tau') - h^{-d}S(x,\tau',\mu h))dx| \tag{6.4.50}$$

where $S(x,\tau,\mu h)$ is given by $(6.1.7)_{(d)}$ for $d=2,3$.

Let us skip all the terms with $m+p+l \geq 1$ in formula (6.4.28) and replace $\mathcal{A}_j(x',\xi',\mu^{-2},\mu^{-1}h,h^2)$ by $\bar{\mathcal{A}}_j(x',\xi') = \mathcal{A}_j(x',\xi',0,0)$ in (6.4.29). Furthermore, let us integrate with respect to ξ_2 for $d=3$. So we obtain an integral over Λ^0 in both cases $d=2,3$. Then going from an integral with respect to $(x',\xi_1)\in\Lambda^0$ to an integral with respect to $x\in\Lambda$ (by means of the symplectomorphism $\bar{\Psi}_0$) and taking into account formulas (6.4.32) with $\psi'=\psi$, $Q'=I$ and

$$\bar{\mathcal{A}}_j = \bigl((2j+1)\mu h F + V\bigr)\circ\bar{\Psi}_0 \tag{6.4.51}$$

(which follows from proposition 6.2.1 and the definition of $\mathcal{A}_j$) we obtain precisely

$$h^{-d}\int \psi(x)S(x,\tau,\mu h)dx.$$

Therefore we need to consider

$$\mathcal{J} = \sum_{j\leq J} \mu h^{1-d}\Bigl(-\bar{\Theta}_{j,000}(\tau) + \sum_{r+m+p+l\leq N} h^{l+r}\mu^{-2(m+p)-l}\bigl((2j+1)\mu h\bigr)^m \Theta_{j,mplr}(\tau)\Bigr) \tag{6.4.52}$$

where here and in what follows the functions $\bar{\Theta}_{j,\dots}(\tau)$ are given by (6.4.29) with $\bar{\mathcal{A}}_j$ instead of $\mathcal{A}_j$. Let us note that if we introduce $z=(2j+1)\mu h$ and formally apply formula $(\mathrm{E}.3)_1$ (replacing the sum with respect to j by an integral with respect to z divided by $2\mu h$) then under condition (6.3.33) we obtain for $|\tau|\leq\epsilon$ an expression of the form

$$h^{-d}G_0(\tau,\mu^{-1}) + h^{1-d}G_1(\tau,\mu^{-1}) + O(h^{2-d})$$

with functions $G_k(\tau,\zeta)$ smooth in the neighborhood of $(0,0)$; here we have integrated with respect to $z\in(0,C_0)$ and for $z\in(\frac{1}{3}C_0,3C_0)$ the integrand vanishes. Here condition (6.3.33) provides the smoothness in a neighborhood of $(0,0)$ of the functions $\Theta_{\dots}(z,\tau)$ and $\bar{\Theta}_{\dots}(z,\tau)$ obtained from $\Theta_{j,\dots}(\tau)$ and $\bar{\Theta}_{j,\dots}(\tau)$ respectively by the substitution $z=(2j+1)\mu h$. Therefore the following estimate holds:

$$|\int \psi(x)(e(x,x,-\infty,\tau) - h^{-d}S(x,\tau,\mu h)dx - h^{-d}G_0(\tau,\mu^{-1}) - h^{1-d}G_1(\tau,\mu^{-1})| \leq \mathcal{R}_4 + \mathcal{R}_6 + C'h^{2-d} \tag{6.4.53}$$

where $\mathcal{R}_4$ is the same as above and $\mathcal{R}_6$ is the absolute value of the error obtained when we replaced the sum with respect to j by the integral with respect to z (divided by $2\mu h$) in the definition of $\mathcal{J}$. A similar estimate holds for $\vartheta > 0$.

Moreover, if we perform the same procedure but with preliminary mollification with respect to τ with $L = 1$ then proposition 6.4.11 yields the estimate

$$(6.4.54)\quad |\int \phi(\tau)(\int \psi(x)(e(x,x,-\infty,\tau) - h^{-d}S(x,\tau,\mu h))dx - h^{-d}G_0(\tau,\mu^{-1}) - h^{1-d}G_1(\tau,\mu^{-1}))d\tau| \le C'h^{2-d}$$

with the same functions G_0 and G_1; here application of formula $(E.3)_1$ is justified rather than merely formal because

$$|D_z^k \int \phi(\tau)\Theta_{\dots}(z,\tau)d\tau| \le C,$$

$$|D_z^k \int \phi(\tau)(\Theta_{\dots}(z,\tau) - \bar{\Theta}_{\dots}(z,\tau))d\tau| \le C\mu^{-2}$$

for all $k \le K'$ and $z \in [0, C_0]$ provided ϕ is smooth enough.

On the other hand, proposition 6.3.10 and formula (6.3.40) (in the weak sense) yield that the estimate

$$(6.4.55)\qquad \mathcal{R}_7 = |\iint \phi(\tau)(e(x,x,-\infty,\tau) - h^{-d}S(x,\tau,\mu h))dxd\tau| \le C'h^{2-d}$$

holds for $1 \le \mu \le h^{\delta-1}$. Comparing (6.4.53) and (6.4.54) for $h^{-\delta} \le \mu \le h^{\delta-1}$ we obtain that the functions

$$\int \phi(\tau)G_j(\tau,\zeta)d\tau$$

with $j = 1,2$ have a zero of arbitrarily high order at $\zeta = 0$; since the function ϕ is arbitrary and $G_j(\tau,\zeta)$ are smooth near $(0,0)$ we obtain that the functions $G_j(z,\tau)$ have a zero of arbitrarily high order at $\zeta = 0$ provided $|\tau| \le \epsilon$. Then (6.4.52) and (6.4.53) yield that for $h^{-\delta} \le \mu \le h^{-1}$ estimates (6.4.54) and

$$(6.4.56)\qquad \mathcal{R} \le \mathcal{R}_4 + \mathcal{R}_6 + C'h^{2-d} \qquad \forall \tau : |\tau| \le \epsilon$$

hold; let us recall that in order to obtain estimate (6.4.55) we assumed that condition (6.3.33) is fulfilled. A similar estimate holds for $\vartheta > 0$ but for $\vartheta > 1$ one should add the correction terms $-\kappa_2(\tau,\mu h)h^{2-d} - \kappa_4(\tau,\mu h)h^{4-d} - \dots$, and for odd integers ϑ the extra term described in remark 6.4.12 should be included as well.

So in order to estimate $\mathcal{R}$ we should estimate $\mathcal{R}_6$.

First of all let us note that under condition (6.3.44) the following inequalities hold:

$$|D_z^k\Theta(z,\tau)| \le C', \qquad |D_z^k(\Theta(z,\tau) - \bar{\Theta}(z,\tau))| \le C'\mu^{-2} \qquad \forall \tau : |\tau| \le \epsilon \quad \forall z \in [0, C_0] \quad \forall k \le K';$$

therefore $\mathcal{R}_6 \le C'h^{2-d}$ in this case; so we obtain

Theorem 6.4.13. *Suppose conditions* (6.1.1)–(6.1.4), (6.2.1), (6.3.33) *and* (6.3.44) *are fulfilled. Then for* $d = 2,3$ *respectively the following estimates hold:*

$$(6.4.57)_{(2)}\qquad \mathcal{R}_\vartheta \le C\mu^{-1-\vartheta}h^{\vartheta-1} + C + C'h^s,$$

$$(6.4.57)_{(3)}\qquad \mathcal{R}_\vartheta \le Ch^{-2+\vartheta} + Ch^{-1} + C'$$

where $\vartheta \ge 0$ *and theorems 6.3.19(i) and 6.3.14 permitted us to get rid of condition* (6.4.1).

In what follows we assume that condition (6.3.31) holds; we will get rid of this restriction in section 6.7. Let us first treat the case $d = 3$.

Theorem 6.4.14. *Let $d = 3$, $\vartheta = 0$, $\tau = 0$ and let conditions (6.1.1)–(6.1.4), (6.3.31) and (6.2.1) be fulfilled with $\bar{\nu} = 1$. Then for $\mathcal{R}$ the estimates of theorem 6.4.10 remain true under the indicated restrictions.*

Proof. Recall the equality

$$\mathcal{B}_j = \Big(V + (2j+1)\mu h F\Big) \circ \bar{\Psi}_0 + O(\mu^{-2}).$$

We want to estimate $\mathcal{R}_6$; for this purpose let us introduce

$$\gamma_1 = \gamma_1(x', \xi') = \min_j (|\bar{\mathcal{A}}_j| + |\nabla \bar{\mathcal{A}}_j|^2)^{\frac{1}{2}}$$

and let us introduce γ_2 in the same way with $\mathcal{A}_j$ instead of $\bar{\mathcal{A}}_j$. Let us treat various zones and various γ_1- and γ_2-admissible partitions of unity in those zones:

(a) The zone $\{(x', \xi') : \gamma_1 \geq \max(C_0\mu^{-1}, \sqrt{\mu h})\}$; in this zone $\gamma_1 \asymp \gamma_2$. One can easily see that if $\Theta(z)$ and $\bar{\Theta}(z)$ are functions associated with the same element of the partition of unity ($\tau = 0$ and m, p, l are omitted here for the sake of simplicity) then

$$|D_z^k \bar{\Theta}(z)| \leq C_k \gamma_1^{2d-2-2k} + C' h^s \qquad \forall k \leq K' \tag{6.4.59$_1$}$$

and the same estimate with γ_1 replaced by γ_2 (this replacement is not essential in this zone) holds for $\Theta(z)$. Moreover, let us apply the formula

$$\Theta(z) - \bar{\Theta}(z) = \int_0^1 \partial_t \Theta(z, t) dt$$

with $\Theta(t, z)$ defined by (6.4.29) with $\tau = 0$ and with $\mathcal{A}_j t + \bar{\mathcal{A}}_j(1 - t)$ instead of $\mathcal{A}_j$; we then easily obtain the estimate

$$|D_z^k (\Theta(z) - \bar{\Theta}(z))| \leq C_k \mu^{-2} \gamma_1^{2d-4-2k} + C' h^s \qquad \forall k \leq K'. \tag{6.4.59$_2$}$$

Furthermore, let us note that outside of an interval of length $C_0\gamma_1^2$ the functions $\Theta(z)$ and $\bar{\Theta}(z)$ are constant and that for $\vartheta = 0$ the integrands are polynomial with respect to z; since the integral formula (E.3)$_1$ is exact on polynomials of arbitrary fixed degree we obtain in the error estimate an additional factor $C_0\gamma_1^2$. So the contribution to $\mathcal{R}_6$ of this element of the partition of unity does not exceed

$$C_k h^{-d} \mu^{-2} \gamma_1^{2d-4-2k} (\mu h)^k \gamma_1^2 + C' h^s.$$

On the other hand, in the zone $\{(x', \xi') : 2C_0\mu^{-1} \geq \gamma_1\}$ a similar inequality[13] holds for γ_2; in this zone we treat $\bar{\Theta}$ and Θ separately; we do not apply (6.4.59)$_2$ in this case and so we take γ_1- and γ_2-admissible partitions of unity for $\bar{\Theta}$ and Θ respectively. We consider the contribution of $\bar{\Theta}$; the contribution of Θ is treated similarly:

(b) Let us treat the subzone $\{(x_1, \xi_1) : 2C_0\mu^{-1} \geq \gamma_1 \geq \sqrt{\mu h}\}$. In this case we apply (6.4.59)$_1$ and obtain that the contribution to $\mathcal{R}_6$ of such an element of the partition does not exceed

$$C_k h^{-d} \gamma_1^{2d-2-2k} (\mu h)^k \gamma_1^2 + C' h^s$$

[13] I.e., $C_1\mu^{-1} \geq \gamma$.

if we treat $\bar{\Theta}$ only; a similar estimate (for the corresponding element) holds if we treat Θ only.

(c) Let us treat the subzone $\{(x_1, \xi_1) : 2C_0\mu^{-1} \le \gamma_1 \le \sqrt{\mu h}\}$. In this case we treat the integrals and Riemann sums separately; however since the integral formulas are exact on polynomials we must integrate over an interval of length not exceeding $C_0\mu h$ and we must take less than C_0 terms in the Riemann sum. This remark and estimates $(6.4.59)_{1,2}$ yield that the contribution of this element of the partition does not exceed

$$Ch^{-d}\mu^{-2}\gamma_1^{2d-4}\mu h + C'h^s;$$

let us recall that in this case $\gamma_1 \asymp \gamma_2$ and we take the same partition for $\bar{\Theta}$ and Θ.

(d) Finally, let us consider the subzone $\{(x_1, \xi_1) : 2\min(\sqrt{\mu h}, C_0\mu^{-1}) \ge \gamma_1\}$; then a similar inequality holds for γ_2. In this zone we apply the same arguments as in (c) but we do not use $(6.4.59)_2$ and consider $\bar{\Theta}$ and Θ separately. Then the contribution to $\mathcal{R}_4$ of this element does not exceed

$$Ch^{-d}\gamma_1^{2d-2}\mu h + C'h^s$$

if we treat $\bar{\Theta}$ only; a similar estimate (for the corresponding element) holds if we treat Θ only. Summing with respect to partitions of unity and observing that the estimates of the contribution of the common element of two (sub)zones calculated by different recipes are equivalent we obtain that

$$\text{(6.4.60)}\quad \mathcal{R}_6 \le \sum_{\iota=1,2} Ch^{-d}\int \min\bigl(\mu^{-2}(\frac{\mu h}{\gamma_\iota^2})^k, \mu h(\frac{\mu h}{\gamma_\iota^2})^k, \mu^{-1}h\gamma_\iota^{-2}, \mu h\bigr)dx'd\xi' + C'h^s.$$

The right-hand expression in this estimate can only increase if we replace γ_ι by functions which are smaller. Then for $d = 3$ we can replace γ_1 and γ_2 by $|\xi_2|$ and then separately treating the cases $\mu \le h^{-\frac{1}{3}}$ and $\mu \ge h^{-\frac{1}{3}}$ we obtain that in both cases $\mathcal{R}_4 \le Ch^{-2} + C'h^s$. So our theorem is proven for $\vartheta = 0$ under condition (6.4.1) (but theorems 6.3.19(iii) and 6.3.17 yield that we can skip this condition). □

Obviously, the same arguments hold for $\vartheta > 0$ but we obtain estimate $(6.4.40)_\vartheta$ with the additional term Ch^{-2} in the right-hand expression which is obviously bad. In order to get rid of this term let us consider mollification with the function $\phi(\frac{\tau}{L})$ with a smooth function ϕ. Obviously in this case all the arguments of the proof remain true with γ_ι replaced by $\gamma_\iota + L^{\frac{1}{2}}$. Then, summing over a partition of non-smooth φ via $\phi(\frac{\tau}{L})$, with smooth ϕ for $L \ge \mu h$ and with non-smooth ϕ for $L = \mu h$ (as we did in section 4.3), we obtain the estimates

$$\text{(6.4.61)}\quad \mathcal{R}_6 \le C\mu h^{1-d}\int_{\{\gamma \ge \sqrt{\mu h}\}} (\mu h)^\vartheta \bigl(\frac{\mu h}{\gamma^2}\bigr)^s dx d\xi' + C\mu h^{1-d}\int_{\{\gamma \le \sqrt{\mu h}\}} (\mu h)^\vartheta dx d\xi'$$

provided $\mu^3 h \le 1$, $\vartheta \le 1$ and

$$\text{(6.4.62)}\quad \mathcal{R}_6 \le C\mu^{-2}h^{-d}\int_{\{\gamma \ge \sqrt{\mu h}\}} (\mu h)^\vartheta \bigl(\frac{\mu h}{\gamma^2}\bigr)^s dx d\xi' + C\mu^{-2}h^{-d}\int_{\{\gamma \le \mu^{-1}\}} (\mu h)^\vartheta dx d\xi'$$

provided $\mu^3 h \ge 1$, $\vartheta \le 1$. For $d = 3$ these estimates yield the estimate

$$\text{(6.4.63)}\quad \mathcal{R}_6 \le R_6 = \begin{cases} C\mu^{\vartheta+\frac{3}{2}}h^{\vartheta-\frac{3}{2}} & \text{for } \mu \le h^{-\frac{1}{3}}, \\ C\mu^{-\frac{3}{2}+\vartheta}h^{-\frac{5}{2}+\vartheta} & \text{for } \mu \ge h^{-\frac{1}{3}} \end{cases}$$

in the general case and the estimate $\mathcal{R}_6 \le Ch^{-2+\vartheta}$ under the condition

$$|\frac{V}{F} + (2j+1)\mu h| + |\nabla \frac{V}{F}| + |\operatorname{Hess} \frac{V}{F}| \ge \epsilon_0. \tag{6.4.47}$$

Recall that $\vartheta \le 1$. Under more restrictive conditions one can improve this estimate (taking into account the term $\kappa_2 h^{-1}$ with $\kappa_2 = \kappa_2(\tau, \mu h)$). I leave this task to the reader as well as the analysis in the case $\vartheta > 1$ when this term should always be taken into account. Moreover, I leave to the reader the case $\bar{\nu} < 1$ when the above estimates can be improved.

Therefore, the following statement is proven:

Theorem 6.4.14′. *Let $d = 3$, $0 < \vartheta \le 1$, $\tau = 0$. Then*

(i) For $\mu \ge h^{-\frac{1}{3-2\vartheta}}$ and for $\mu \le h^{-\frac{1}{3+2\vartheta}}$ and under condition (6.4.47) estimate $\mathcal{R}_\vartheta \le Ch^{-2+\vartheta}$ holds.

(ii) For $h^{-\frac{1}{3+2\vartheta}} \le \mu \le h^{-\frac{1}{3-2\vartheta}}$ estimate $\mathcal{R}_\vartheta \le Ch^{-2+\vartheta} + R_6$ holds where R_6 is the right-hand expression of (6.4.61), (6.4.62) for $\mu \le h^{-\frac{1}{3}}$, $\mu \ge h^{-\frac{1}{3}}$ respectively. Moreover, one can define R_6 by (6.4.63).

I remind the reader that for $\vartheta = 1$, $\mathcal{R}_\vartheta$ contains an additional term by remark 6.4.12.

Finally, one can obtain the estimate $\mathcal{R}_6 = o(h^{-2+\vartheta})$ calculating $\mathcal{A}_j$ modulo $O(\mu^{-4}) + O(\mu^{-2}\gamma^2)$ instead of $O(\mu^{-2})$ as we did. In this case in $\mathcal{R}$ the term $S(x, \tau', \mu h)$ is replaced by $S(x, \tau' - \rho(x)\mu^{-2}, \mu h)$ and additional terms of the type $\varsigma_j(x)\mu^{-2j}h^{-3}$ appear. However, I have failed to calculate them.

Obviously, our arguments apply to the case $d = 2$ as well. In particular, estimate (6.4.60) remains true for $d = 2$, $\nu = 1$, $\vartheta = 0$ as well but we want to improve it for $\mu \ge C_0 h^{-\frac{1}{3}}$. Namely, we want to estimate the contribution of the (sub)zone $\{\epsilon_0\sqrt{\mu h} \ge \gamma_1 \ge C_1\mu^{-1}\}$ more accurately where C_1 depends on $\epsilon > 0$. More precisely, if $|v + (2j+1)\mu h| \ge \epsilon\gamma_1^2$ on the element of the partition of unity (with $j \in \mathbb{Z}^+$ for which this absolute value is minimal) then Θ_j and $\bar{\Theta}_j$ coincide for all $j \le J$ and therefore the contributions to the error of Riemann sums vanish for $m = p = l = 0$ and since $\Theta(z)$ and $\bar{\Theta}(z)$ coincide outside of intervals of length $C_0\mu^{-2}$, the contribution of $\Theta(z) - \bar{\Theta}(z)$ to the error does not exceed

$$Ch^{1-d}\gamma_1^{2d-4}\mu^{-1} + C'h^s.$$

The contributions of all the other terms do not exceed the previous estimate in zone (c) with the additional factor μ^{-2} and so they are less than this estimate.

Now applying estimate (6.4.60) with this improvement and theorem 6.4.10 we immediately obtain that under condition (6.4.1) (which could be skipped by theorem 6.3.19) the following theorem holds:

Theorem 6.4.15. *Let $d = 2$, $\vartheta = 0$, $\tau = 0$ and let conditions (6.1.1)–(6.1.4), (6.2.1), (6.3.31) and (6.4.1) be fulfilled. Then*

(i) For $\mu^3 h \le C_0$ the following estimate holds:

$$\mathcal{R} \le \frac{C}{\mu h} + \frac{C'}{\mu h}\int_{\{\gamma \ge C_0\mu^{-1}\}} \left(\frac{\mu h}{\gamma}\right)^s \frac{dx}{\gamma^2} + C\frac{\mu}{h}\int_{\{\gamma \le C_0\mu^{-1}\}} dx + C'; \tag{6.4.64}$$

moreover, under condition (6.4.47) the estimate

$$\mathcal{R} \le \frac{C}{\mu h} + C\mu^{\frac{3}{2}}h^{-\frac{1}{2}} \tag{6.4.65}$$

holds and under the condition

(6.4.66) $$|\frac{V}{F}+(2j+1)\mu h|+|\nabla\frac{V}{F}|+|\det\operatorname{Hess}\frac{V}{F}|\geq\epsilon_0$$

estimate $(6.4.57)_{(2)}$ *holds.*

(ii) For $\mu^3h\geq C_0$ *the following estimate holds:*

(6.4.67) $$\mathcal{R}\leq\frac{C}{\mu h}+\frac{C'}{\mu h}\int_{\{\gamma\geq\sqrt{\mu h}\}}\bigl(\frac{\mu h}{\gamma^2}\bigr)^s\frac{dx}{\gamma^2}+\frac{C}{\mu h}\int_{\{\sqrt{\mu h}\geq\gamma\geq C_1\mu^{-1},|v+(2j+1)\mu h|\leq\epsilon\mu h\}}\frac{dx}{\gamma^2}+C\frac{\mu}{h}\int_{\{\gamma\leq C_0\mu^{-1}\}}dx+C';$$

moreover, under condition (6.4.47) *estimate*

(6.4.68) $$\mathcal{R}\leq Ch^{-1}+C'$$

holds and under the condition

$(6.4.66)_+$ $$|\frac{V}{F}+(2j+1)\mu h|+|\nabla\frac{V}{F}|\leq\epsilon_0\implies\det\operatorname{Hess}\frac{V}{F}\geq\epsilon_0$$

estimate $(6.4.57)_{(2)}$ *holds; furthermore, under condition* (6.4.66) *the estimates*

(6.4.69) $$\mathcal{R}\leq\frac{C}{\mu h}(1+|\log\frac{\hat\tau}{\mu h}|)+C'\leq\frac{C}{\mu h}(1+\log\mu^3h)+C'$$

hold where $\hat\tau=\min_x\gamma^2(x)+\mu^{-2}$.

Remark. *(i) Theorem 6.3.19 yields that in the general case and under conditions (6.4.47) and (6.4.66) the corresponding estimates have a place even if condition (6.4.1) is violated.*

(ii) I leave the case $\vartheta>0$ *to the reader: in this case even condition* $(6.4.66)_+$ *does not provide estimate* $(6.4.57)_{(2)}$ *for* $\mathcal{R}_6$ *for* μ *close to* $h^{\frac{1}{3}}$. ◇

6.4.6 Effective asymptotics. II

In order to improve estimate (6.4.69) we need more sophisticated arguments which are applicable in other situations as well. So let us assume that $\mu^3h\geq C_0$ and condition (6.4.66) holds but $(6.4.66)_+$ does not. Without loss of generality one can assume that $v=\frac{V}{F}$ has only one stationary point $\bar x$ in $B(0,1)$ (in which case this point is saddle point) and that $\tau=0$; let us replace $\mathcal{A}_j(x_1,\xi_1,\dots)$ in the argument of the Heaviside function by $(2j+1)\mu h+B(x,\xi,\dots)$. In our previous arguments we dropped $O(\mu^{-2})$ here; let us now drop $O(\mu^{-4})$, $O(h^2)$ and $O(\mu^{-2}\gamma)$; equalities $(6.2.47)_{1,2}$ yield that this is equivalent to replacing B by $v\circ\bar\Psi_0+\bar\omega_1\mu^{-2}$ where $\bar\omega_1$ is the value of the coefficient ω_1 in $\bar x$,

(6.4.70) $$\bar\omega_1=(\frac{1}{8}\kappa v^2-\frac{1}{4}g^{jk}F^{-1}\partial_j\partial_k v)(\bar x).$$

Moreover, we skip all terms with $m+p+l\geq1$ as before. If in the difference (between original and final expressions) we replace the sum with respect to j by the corresponding integral then the previous arguments yield that the error is $O(\mu^{-1}h^{-1})+O(\mu h|\log\hat\tau\mu^{-1}h^{-1}|)$ and therefore theorem 6.4.10(ii) yields the estimate

(6.4.71) $$|\int\psi(x)\bigl(e(x,x,-\infty,0)-h^{-2}S(x,-\bar\rho\mu^{-2},\mu h)\bigr)dx-h^{-2}G(\mu^{-2})|\leq\frac{C}{\mu h}+C\mu h|\log\frac{\hat\tau}{\mu h}|+C'$$

with $\bar{\rho} = \bar{\omega}_1 F(\bar{x})$; this estimate holds for $h^{-\delta} \le \mu \le h^{-1}$ and moreover, for $\mu \le h^{-\frac{1}{3}}$ this estimate holds without the second term in the right-hand expression. However, we know that under condition (6.3.31) and for $\mu \le h^{\delta-\frac{1}{3}}$, $|\tau| \le \epsilon$

$$\int \psi(x)\bigl(S(x,\tau,\mu h) - \frac{1}{4}(\tau - V)\sqrt{g}\bigr)dx = O(h^s)$$

and, moreover, we know that in this case Weyl asymptotics hold with remainder estimate $O(\mu^{-1}h^{-1})$ (see theorems 6.3.19(iii) and 6.3.18(ii)). Then comparing (6.4.71) and the Weyl asymptotics we deduce that the function $G(\zeta) - \varsigma\zeta$ has at $\zeta = 0$ a zero of arbitrarily high order with coefficient $\varsigma = \frac{1}{4}\int \psi\sqrt{g}dx$. Therefore the following estimate holds:

$$\begin{aligned}(6.4.72)\quad \mathcal{R}_7 = \\ |\int \psi(x)(e(x,x,-\infty,0) - h^{-2}S(x,-\bar{\rho}\mu^{-2},\mu h) - \bar{\rho}\sqrt{g}\mu^{-2}h^{-2})dx| \le \\ \frac{C}{\mu h} + C\mu h|\log\frac{\hat{\tau}}{\mu h}| + C'\end{aligned}$$

where $\bar{\rho} = \bar{\omega}_1 F(\bar{x})$ and $\bar{\omega}_1$ is given by (6.4.70). This estimate is effective (and is worse than corresponding estimate for $\mathcal{R}_4$ only if $(\mu h)^2|\log\frac{\hat{\tau}}{\mu h}| \ge 1 \implies \mu \ge h^{-1}|\log h|^{-\frac{1}{2}}$); however, we will transform this estimate slightly.

The same arguments with the Riemann sum replaced by an integral and with the same error estimate as in the proof of theorem 6.4.13 yield that if $\psi_1 \in C^K$ coincides with ψ for $|x-\bar{x}| \ge \frac{1}{2}\gamma$, vanishes in $B(\bar{x},\frac{1}{2}\gamma)$ and satisfies the inequalities $|D^\alpha\psi_1| \le c\gamma^{-|\alpha|}$ and if $\gamma = \sqrt{\mu h}$, $\mu^3 h \ge 1$ then

$$|\int \psi_1(x)\bigl(S(x,-\bar{\rho}\mu^{-2},\mu h) - S(x,0,\mu h) + \frac{1}{4\pi}\bar{\rho}\mu^{-2}\sqrt{g}\bigr)dx| \le C\mu^{-1}h + C'h^2.$$

Let us consider a similar expression for ψ_1 supported in $B(\bar{x},\gamma)$, $\psi_1 = \psi$ in $B(\bar{x},\frac{1}{2}\gamma)$; it is easy to prove that in this case

$$\begin{aligned}|\int \psi_1(x)\bigl(S(x,-\bar{\rho}\mu^{-2},\mu h) - S(x,0,\mu h) + \frac{1}{4\pi}\bar{\rho}\mu^{-2}\sqrt{g}\bigr)dx - \\ \Delta(\bar{x})\mu^{-1}h| \le C\mu^{-1}h + C'h^2\end{aligned}$$

where

$$(6.4.73)\qquad \Delta(\bar{x}) = \frac{\mu^2}{\pi}\bigl(f(\bar{\tau}') - f(\bar{\tau})\bigr)F\sqrt{g}|\det\operatorname{Hess} v|^{-\frac{1}{2}}\psi\bigr|_{x=\bar{x}},$$

$$(6.4.74)\qquad f(\tau) = \tau\log(\frac{|\tau|}{\mu h}),$$

$$(6.4.75)\qquad \bar{\tau} = v(\bar{x}) + (2\bar{\jmath}+1)\mu h,\quad \bar{\tau}' = \bar{\tau} + \bar{\rho}\mu^{-2},\quad \bar{\rho} = \bar{\omega}_1 F(\bar{x})$$

and for $j = \bar{\jmath}$ the quantity $|v(\bar{x} + (2j+1)\mu h|$ is minimal on $\mathbb{Z}^+$. We leave this proof to the reader; let us note only that it is convenient to use the coordinate system in which $\bar{x} = 0$, $v = x_1x_2|\det\operatorname{Hess} v(\bar{x})|^{-\frac{1}{2}}$ in this proof. Passing from the case with one critical point to general case we obtain the following

Theorem 6.4.16. *Let $d = 2$ and let conditions (6.1.1)–(6.1.4), (6.2.1), (6.3.31) and (6.4.47) be fulfilled. Then for $\mu^3 h \geq 1$ the following estimate holds:*

$$(6.4.76)\quad \mathcal{R}_I = \Big| \int \psi(x)\big(e(x,x,-\infty,0) - h^{-2}S(x,0,\mu h)\big)\,dx - \sum_{x_k\in \mathrm{M}} \frac{1}{\mu h}\Delta(x_k)\Big| \leq \frac{C}{\mu h} + C\mu h\Big|\log\frac{\hat{\tau}}{\mu h}\Big| + C'$$

where M *is the set of saddle points of $v = \frac{V}{F}$ and at every point $x_k \in \mathrm{M}$, $\Delta(x_k)$ is given by formulas (6.4.73)–(6.4.75), (6.4.70) with x_k instead of $\bar{x}$.*

Let us note that here and in estimates (6.4.69) and (6.4.72) we can redefine $\hat{\tau} = \min_{x_k\in\mathrm{M}} |\tau(x_k)| + \mu^{-2}$.

Similarly one can treat the case of $\vartheta \in [0,1]$. In this case we replace $\mathcal{A}_j$ by $\bar{\mathcal{A}}'_j = \mathcal{A}_j + O(\mu^{-4}) + O(\mu^{-2}\gamma^2)$ and note that by a shift one can get rid of the term which is linear with respect to x in $\bar{\mathcal{A}}'_j$ (which is $O(\mu^{-2}\gamma)$). One can easily prove the required estimate for $\mathcal{R}_{7\vartheta}$. However, I have failed to derive an estimate for something similar to $\mathcal{R}_I$ (with $\vartheta > 0$). Moreover, the same approach works when condition (6.4.47) is fulfilled; the Morse theorem yields that without loss of generality one can assume that

$$(6.4.77)_\pm\qquad v(x) = \bar{v}(x_1) \pm x_2^2.$$

We leave the detailed proof of the following two statements to the reader:

Theorem 6.4.17. *Under the assumptions of theorem 6.4.16, estimate $(6.4.36)_\vartheta$ holds for $\mathcal{R}_{7\vartheta}$ with $\vartheta \in [0,1]$.*

Theorem 6.4.18. *Let $d = 2$, $\vartheta = 0$ and let conditions (6.1.1)–(6.1.4), (6.2.1) and $(6.4.77)_\pm$ be fulfilled. Then*

(i) For $h^{-\frac{1}{5}} \leq \mu \leq h^{-\frac{1}{3}}$ the following estimate holds:

$$(6.4.78)\qquad \mathcal{R}_7 \leq C\mu^{-1}h^{-1} + C'$$

where $\mathcal{R}_7$ is given by (6.4.71), ω_1 is given by (6.4.70) and in this theorem the bar means that the corresponding function is calculated at $x_2 = 0$.

(ii) For $\mu \geq h^{-\frac{1}{3}}$ the following estimate holds:

$$(6.4.79)\qquad \mathcal{R}_7 \leq C\mu^{-1}h^{-1} + C\mu^{\frac{1}{2}}h^{-\frac{1}{2}} + C'$$

and moreover the following estimate holds as well:

$$(6.4.80)\qquad \mathcal{R}_7 \leq \frac{C}{\mu h} + C\int_{\{\sqrt{\mu^{-1}h}\leq\gamma_1\leq\sqrt{\mu h}\}} \frac{dx_1}{\gamma_1} + C\mu^{\frac{1}{2}}h^{-\frac{1}{2}}\int_{\{\gamma_1\leq\sqrt{\mu^{-1}h}\}} dx_1 + C'$$

where $\gamma_1 = \gamma_1(x_1) = \min_{j\in\mathbb{Z}^+}\big(|\partial_1\bar{w}|^2 + |\bar{w} + (2j+1)\mu h|\big)^{\frac{1}{2}}$ and $\bar{w} = \bar{v} + \mu^{-2}\bar{\omega}_1$.

Let us note that for $\vartheta = 0$, $\mu \leq h^{-\frac{1}{5}}$ estimate $(6.4.57)_{(2)}$ holds. I want to mention that the extra term in $\mathcal{R}_1$ has the same nature as the extra term in $\mathcal{R}_\vartheta$ for $\vartheta = 1$.

6.4.7 Effective asymptotics. III

Let us now treat the case $d = 2$ and $\nu < 1$. In this case the optimal estimate (6.4.36) for $\mathcal{R}_3$ is better than in the case $\nu = 1$ but the transition to estimates for $\mathcal{R}$ and similar effective estimates can make it essentially worse. In this case it is important whether

$\nu \ge \mu h$ or $\nu \le \mu h$; in the second case the transition from summation to integration is completely useless in the analysis of the term with $m = p = l = 0$ but in this case the replacement of $\mathcal{A}_j$ by $\bar{\mathcal{A}}_J$ changes only terms with less than C_0 numbers j. Taking into account this observation we prove the following two theorems (the second theorem improves the first):

Theorem 6.4.19. *Let $d = 2$, $\vartheta = 0$ and let conditions (6.1.1)–(6.1.4), (6.2.1), $(6.2.48)_{1-3}$, (6.3.68), (6.2.32), (6.4.1), (6.3.31) and (6.3.56) be fulfilled. Then*

(i) For $\nu \ge \mu h$ the following estimate holds:

$$\mathcal{R} \le C\frac{\nu}{\mu h} + C'\bar{\nu}\nu^{-2}\Big(\frac{\mu h}{\nu}\Big)^s + C' \qquad \forall \tau : |\tau| \le \epsilon\nu \tag{6.4.81}$$

with arbitrary s and with $\epsilon = \epsilon(c, \epsilon_0) > 0$.

(ii) For $\nu \le \mu h$ the following estimate holds:

$$\mathcal{R} \le C\frac{\bar{\nu}}{\mu h \nu} + C' \qquad \forall \tau : |\tau| \le \epsilon\nu \tag{6.4.82}$$

with $\epsilon = \epsilon(c, \epsilon_0) > 0$.

Theorem 6.4.20. *Let $d = 2$, $\vartheta = 0$ and let conditions (6.1.1)–(6.1.4), (6.2.1), $(6.2.48)_{1-3}$, (6.4.1) and (6.3.31) be fulfilled. Let us assume that*

$$|D^\alpha(\frac{V}{F} + \omega_1\mu^{-2})| \le c\nu \qquad \text{in } B(0,1) \quad \forall \alpha : 1 \le |\alpha| \le K, \tag{6.4.83}$$

$$\nu \ge C_0\bar{\nu}(\mu^{-4} + h^2), \tag{6.4.84}$$

$$\min_j |\frac{V}{F} + (2j+1)\mu h + \omega_1\mu^{-2}| + |\nabla(\frac{V}{F}v + \omega_1\mu^{-2})| \ge \epsilon_0\nu \quad \text{in } B(0,1) \tag{6.4.85}$$

where

$$\omega_1 = \frac{1}{8}\kappa v^2 - \frac{1}{4}v\mathcal{L}v - \frac{1}{4}g^{jk}\partial_j v \cdot \partial_k v, \qquad v = \frac{V}{F}, \tag{6.4.86}$$

$$\mathcal{L}v = \frac{1}{F\sqrt{g}}\partial_j(g^{jk}\sqrt{g}\partial_k v), \tag{6.4.87}$$

(i.e., $\mathcal{L}$ is the Laplace-Beltrami operator in the metric $\frac{1}{F}g^{jk}$). Then

(i) For $\mu h \le \nu$ and $\tau = 0$ the following estimate holds

$$\mathcal{R}_{II} = |\int (\psi(x)e(x,x,-\infty,\tau) - h^{-2}S'_\psi(x,\mu h,\mu^{-2}))dx| \le \\ C\frac{\nu}{\mu h} + C'\bar{\nu}\mu^{-2}\nu^{-2}(\frac{\mu h}{\nu})^s + C' \tag{6.4.88}$$

where

$$S'_\psi = \frac{1}{2\pi}\Big(\sum_{j\in\mathbb{Z}^+} \theta(-v - (2j+1)\mu h - \omega_1\mu^{-2})(\psi + \psi'\mu^{-2})\mu h + \\ \frac{1}{4}\mu^{-2}(\omega_1\psi + v\psi') + \frac{1}{4}\mu^{-4}\omega_1\psi'\Big)F\sqrt{g}, \tag{6.4.89}$$

$$\psi' = -\frac{1}{4}v\mathcal{L}\psi - \frac{1}{2}vg^{jk}F^{-2}\partial_j F\cdot\partial_k\psi. \tag{6.4.90}$$

(ii) For $\mu h \ge \nu$ and $\tau = 0$ the following estimate holds:

$$\mathcal{R}_{II} \le C'(1+\bar{\nu}\mu^{-3}h^{-1}\nu^{-1}+\bar{\nu}\mu h\nu^{-1}). \tag{6.4.91}$$

Proof. We will prove the second theorem; the proof of the first theorem is similar but simpler. In the argument of the Heaviside function in the terms with $m+p\le 1$, $l=0$ let us drop items $O(\mu^{-4}\bar{\nu})$, $O(h^2\bar{\nu})$, $O(\mu^{-1}h\nu)$; then repeating the previous arguments we obtain that modulo a term with absolute value not exceeding the right-hand expression of estimate (6.4.88) for $\mu h\le\nu$ (or the right-hand expression of estimate (6.4.91) for $\mu h\ge\nu$) the correction is of the form $h^{-2}G_0(\mu^{-2})+\mu^{-1}h^{-1}G_1(\mu^{-2})$. The same arguments give the same answer if we drop all terms with $m+p+2l\ge 2$. Let us treat terms with $m+p=1$, $l=0$; they are of the form

$$\frac{1}{2\pi h^2}\int\sum_j \theta(-v-(2j+1)\mu h-\omega_1\mu^{-2})\times$$
$$(\psi'+(v+(2j+1)\mu h+\omega_1\mu^{-2})\psi'')\mu^{-2}F\sqrt{g}dx;$$

here we have replaced integration with respect to (x_1,ξ_1) with integration with respect to x by means of the symplectomorphism $\bar{\Psi}_0$. Then similar arguments show that if we drop items which are $O(\nu)$ in ψ' and if we skip ψ'' completely then the correction has the same form with the same error estimate; in the second case we used the fact that the function $f(z)$ (which is integrated with respect to z) satisfies the estimate $|D_z^k|\le C\nu^{1-k}\quad\forall k\le K'$ on the "bad" interval of length $C_0\nu$ (if $\nu\ge\mu h$); moreover on the "bad" interval of length $C_0\mu h$ the estimate $|f|\le C\nu$ holds for $\nu\le\mu h$. Thus the quantity

$$\mathcal{R}_8 = \Bigl|\int\psi(x)\Bigl(e(x,x,-\infty,0)-$$
$$\frac{1}{2\pi h^2}\sum_{j\in\mathbb{Z}^+}\theta(-v-(2j+1)\mu h-\omega_1\mu^{-2})(\psi+\psi'\mu^{-2})\mu h\Bigr)dx-$$
$$h^{-2}G_0(\mu^{-2})-\mu^{-1}h^{-1}G_1(\mu^{-2})\Bigr|$$

does not exceed the right-hand expression of (6.4.88) or (6.4.91) for $\mu h\ge\nu$ and $\mu h\le\nu$ respectively. Furthermore, the same estimate remains true if we replace $\tau=0$ by $\tau\in[-\epsilon\nu,\epsilon\nu]$ (with a small constant $\epsilon>0$) and if we simultaneously replace V by $V-\tau$ in the argument of the Heaviside function and in the definition of ω_1 and ψ'; moreover, in this case the functions G_0 and G_1 depend smoothly on τ. Finally, if we have mollified (inside the absolute value) with respect to τ with a small constant $L>0$ and with ϕ supported in $[-1,1]$ we would obtain the remainder estimate C'. But then proposition 6.3.10 and formula (6.3.40) yield that for $1\le\mu\le h^{\delta-1}$ the same estimate (with mollification) remains true with

$$\bar{G}_0(\zeta)=\frac{1}{4}\zeta(\omega_1+v\psi')+\frac{1}{4}\zeta^2\omega_1\psi',\quad \bar{G}_1(\zeta)=0$$

instead of G_0 and G_1 respectively; here $\tau\in[-\epsilon,\epsilon]$ enters into the definition of these functions through v, ω_1 and ψ'. Comparing these two estimates for $h^{-\delta}\le\mu\le h^{\delta-1}$ we conclude that the functions $G_0-\bar{G}_0$ and G_1 have zeros of arbitrarily high order at $\zeta=0$, $\tau\in[-\epsilon,\epsilon]$; we first make this conclusion for functions mollified with respect to τ and then we use the arbitrary choice of mollifying function and the smoothness of the functions G_0, $\bar{G}_0$, G_1 (see proof of theorem 6.4.13). But then we can replace G_0 and G_1 by $\bar{G}_0$ and $\bar{G}_1 = 0$ in the definition of $\mathcal{R}_8$ (without mollification); so the theorem is almost proven: we only need to calculate ω_1 and ψ'.

In order to calculate ω_1 and ψ' let us note that in the conformal coordinate system for $F = 1$ and for the method of reduction used in proposition 6.2.1(iii)

$$\omega_1 = \frac{1}{8}\kappa v^2 - \frac{1}{4}v\rho^2\Delta v - \frac{1}{4}\rho^2(\nabla v)^2,$$

i.e., ω_1 is given by (6.4.86); moreover, in this case

$$\psi' \equiv -\frac{1}{4}v\rho^2\Delta\psi \pmod{|\nabla v|},$$

i.e., ψ' is given by (6.4.90). Furthermore, let us note that in the transition from $F = 1$ to arbitrary F (see the procedure used in proposition 6.2.5) the arguments in the Heaviside functions do not change and the trace of the matrix of quadratic form

$$(S_1w_1 + S_2w_2)\{(\{\psi, w_2\}w_1 - \{\psi, w_1\}w_2), w_1^2 + w_2^2\} \pmod{O(\nu)}$$

is added to ψ'; here S_1 and S_2 are chosen from the equation

$$F\{S_1w_1 + S_2w_2, w_1^2 + w_2^2\} + \{F, w_2\}w_1 - \{F, w_1\}w_2 = 0 \qquad \text{at } \Lambda;$$

then explicit calculation gives that this quantity is equal to $-\frac{1}{2}v\rho^2\partial_jF\cdot\partial_k\psi$ and therefore ψ' is again given by (6.4.90). Since formulas (6.4.86) and (6.4.90) are invariant with respect to coordinate transformations we conclude that ω_1 and ψ' are given by these formulas in an arbitrary coordinate system. □

Remark 6.4.21. *(i) In a method of reduction different from the method used in propositions 6.2.1(iii) and 6.2.5 the term $O(\nu)$ is added to ω_1 and the term $O(1)$ is added to ψ'; I have not managed to work with these additional terms and to preserve estimate (6.4.88) (or (6.4.91)); namely in this estimate the term $C'\nu^{-1}(\mu h\nu^{-1})^s$ (or $C'\mu^{-1}h^{-1}$ respectively) should be added to the right-hand expression.*

(ii) On the other hand item $O(\nu^2)$ in ω_1 could be skipped; so item $-\frac{1}{4}g^{jk}\partial_jv\cdot\partial_kv$ could also be skipped. ◊

I suggest to the reader the investigation of the case when

$$|\nabla(v + \omega_1\mu^{-2})| + \min_j|(2j+1)\mu h + v + \omega_1\mu^{-2}| \le \epsilon_0\nu \implies$$
$$\det\operatorname{Hess}(v + \omega_1\mu^{-2}) \ge \epsilon_0\nu^2 \qquad \text{in } B(0,1)$$

or when this condition is fulfilled for the absolute value of the Hessian determinant. I formulate without proof only a simpler statement:

Theorem 6.4.22. *Let $d = 2$, $\vartheta = 0$ and let conditions (6.1.1)–(6.1.4), (6.2.1), $(6.2.48)_{1-3}$, (6.4.1), (6.3.68), (6.4.32) and (6.3.31) be fulfilled. Let us assume that*

$$|\frac{V}{F} + (2j+1)\mu h| + |\nabla\frac{V}{F}| \le \epsilon_0\nu \implies |\det\operatorname{Hess}\frac{V}{F}| \ge \epsilon_0\nu^2 \tag{6.4.48}$$

Then

(i) For $\nu \ge \mu h$, $\bar{\nu} \ge \mu^3 h$ the following estimate holds:

$$\mathcal{R} \le C(\frac{\nu}{\mu h} + \frac{\bar{\nu}\mu^2}{\nu}) + C'. \tag{6.4.92}$$

(ii) For $\nu \ge \mu h$, $\bar\nu \le \mu^3 h$ the following estimate holds:

$$\mathcal{R} \le C' \frac{\bar\nu}{\mu h \nu}(1 + \log \frac{\mu^3 h}{\bar\nu}) + C'. \tag{6.4.93}$$

Moreover, under condition (6.4.91)$_+$ *estimate* (6.4.82) *remains true in this case.*

(iii) For $\nu \le \mu h$ (and then $\bar\nu \le \mu^3 h$ automatically) the following estimate holds:

$$\mathcal{R} \le C \frac{\bar\nu}{\mu h \nu}(1 + \log \frac{\nu \mu^2}{\bar\nu} + 1) + C'; \tag{6.4.94}$$

moreover, under condition (6.4.48)$_+$ *estimate* (6.4.82) *holds.*

6.5 Propagation of Singularities and Improved Asymptotics

In this section we consider the two- and three-dimensional cases and under some assumptions of a global nature we obtain slightly more precise estimates than in the previous sections; this improvement will be by no more than an additional factor $\varsigma \in [\mu^{-\delta'}, 1]$ with a small enough exponent $\delta' > 0$ in the right-hand expression of the estimates. These results are not very important from the point of view of applications but they seem to be interesting enough by themselves. Here we need to revise all the arguments of sections 6.2–6.4. We start with the three-dimensional case when singularities propagate along magnetic lines and then proceed to the two-dimensional case when singularities propagate along drift lines.

6.5.1 Improved reduction ($d = 3$)

It is well known from physics that the classical three-dimensional particles in a strong magnetic field move almost along magnetic lines. So it is not surprising that in our analysis magnetic lines are also important. Let us recall that a magnetic line of the magnetic field $\mathbf{F} = (F^1, F^2, F^3)$ is an integral curve $x = x(t)$ of the field $\frac{1}{F}\mathbf{F}$. Thus, $x(t)$ is given by the system

$$\frac{dx_j}{dt} = \frac{2}{F} F^j \qquad (j = 1, 2, 3) \tag{6.5.1}$$

where $F = (g_{jk} F^j F^k)^{\frac{1}{2}}$ is the scalar intensity; let us introduce the corresponding flow ϕ_t on X. It follows from the definition of $\mathbf{F}$ that this field is solenoidal, i.e.,

$$\partial_j (F^j \sqrt{g}) = 0 \tag{6.5.2}$$

and therefore this flow preserves the density $F\sqrt{g}dx$ on X. Moreover, the flow ϕ_t can be lifted to the Hamiltonian flow $\tilde\phi_t$ on Λ which is given by the Hamiltonian $\frac{1}{F}F^j(\xi_j - \mu V_j)$.[14)]

The proof of proposition 6.2.1 and the analysis which follows in section 6.2 easily yield the following statement (the detailed proof of which we leave to the reader):

---

14) We could define the flow ϕ_t by the system

$$\frac{dx_j}{dt} = \frac{2}{F}\omega F^j \quad (j = 1, 2, 3) \tag{6.5.1$'$}$$

with functions ω and ω^{-1} regular along the trajectory; then $\tilde\phi_t$ would be given by the Hamiltonian $\frac{\omega}{F}F^j(\xi_j - \mu V_j)$ and the invariant density would be $\omega^{-1}F\sqrt{g}dx$.

Theorem 6.5.1. *Let conditions (6.1.1)–(6.1.4) and (6.2.1) be fulfilled. Let $\hat{x} \in B(0, \frac{1}{2})$ and for all $x \in B(0, \zeta^l)$ let the flow ϕ_t be defined by system (6.5.1) with $t \in [-T_1, T_2]$ where $\zeta \in [\mu^{-\delta'}, 1]$, $\delta' > 0$ is small enough, $0 \le T_i \le \zeta^{-1}$ and $T_1 T_2 = 0$. Moreover, let us assume that the following conditions are fulfilled along this flow:*

$$|D^\alpha g^{jk}| \le c\zeta^{-l}, \quad |D^\alpha F^j| \le c\zeta^{-l}, \quad |D^\alpha V| \le c\zeta^{-l} \qquad \forall \alpha : |\alpha| \le K, \tag{6.5.3$_{1-3}$}$$

$$F \ge \epsilon_0 \zeta^l, \tag{6.5.4}$$

$$g^{jk}\eta_j\eta_k \ge \epsilon_0 \zeta^l |\eta|^2 \qquad \forall \eta \in \mathbb{R}^3, \tag{6.5.5}$$

$$\mathrm{dist}(\phi_t(x), \partial X) \ge \epsilon_0 \zeta^l, \tag{6.5.6}$$

$$\mathrm{dist}(\phi_t(x), x) \ge \epsilon_0 \zeta^l \min(|t|, 1), \tag{6.5.7}$$

$$|D_x \phi_t(x)| \le c\zeta^{-l} \tag{6.5.8}$$

where $D_x\phi_t(x)$ denotes the Jacobi matrix of $\phi_t(x)$. Then

(i) All the reductions of section 6.2 can be done in the subdomain

$$\bigcup_{t \in [-T_1, T_2]} \phi_t(B(\hat{x}, \epsilon\zeta^l)) \tag{6.5.9}$$

of X where $\epsilon = \epsilon(l, c, \epsilon_0) > 0$ is a small enough constant, and then all the symplectomorphisms and symbols satisfy the estimates

$$|D^\alpha \Psi_{\dots}(x, \xi)| \le C' \zeta^{-L} \qquad \forall \alpha : |\alpha| \le K' \tag{6.5.10$_1$}$$

$$|D^\alpha b_{\dots}(x, \xi)| \le C' \zeta^{-L} \qquad \forall \alpha : |\alpha| \le K' \tag{6.5.10$_2$}$$

with arbitrary N and K' (and then K, L, δ' depend on N, K', l); moreover, for $x \in \phi_t(B(\hat{x}, \epsilon\zeta^l))$ with $|t| \le \epsilon$ both these inequalities hold with $L = 0$.

(ii) Moreover, the same statement remains true for the flow ϕ_t defined by (6.5.1)′ provided that on $\cup_{t \in [-T_1, T_2]} \phi_t(B(\hat{x}, \zeta^l))$

$$|D^\alpha \omega| \le c_1 \zeta^{-l} \qquad \forall \alpha : |\alpha| \le K, \tag{6.5.11$_1$}$$

$$\omega \ge \epsilon_1^{-1} \zeta^l. \tag{6.5.11$_2$}$$

(iii) Finally, this statement remains true without condition (6.5.7) if we introduce coordinates (x_2, x') in the pipe (6.5.9) transforming it into a cylinder (thus removing self-intersections).

6.5.2 The case of a weak magnetic field ($d = 3$)

Let us assume that (6.3.1) holds. In this subsection we improve certain statements of section 6.3. For this goal let us first assume that the reduction of the previous subsection is done and let us take $(\bar{x}', \bar{\xi}_1) \in \Omega$.[15] Let us introduce the sets

$$\Lambda^0_{(\bar{x}_1,\bar{\xi}_1)} = \{(x',\xi') \in \Omega \times \mathbb{R}, x_1 = \bar{x}_1, \xi_1 = \bar{\xi}_1\}, \tag{6.5.12}$$

the one-dimensional Hamiltonians

$$H_{(\bar{x}_1,\bar{\xi}_1)r}(x_2,\xi_2) = \xi_2^2 + \bar{B}_0'(\bar{x}_1, x_2, \bar{\xi}_1) + \bar{B}_1(\bar{x}_1, x_2, \bar{\xi}_1) r \tag{6.5.13}$$

and the corresponding Hamiltonian flows $\Phi_{t,r}$ on $\Lambda^0_{(\bar{x}_1,\bar{\xi}_1)}$. Here $(\bar{x}_1, \bar{\xi}_1, r)$ is considered constant when we define the Hamiltonian field.

Theorem 6.5.2. *Let $q', q'' \in S_{\mathbf{h},\rho,\gamma,K}(T^*\mathbb{R}^3)$ and $\mathbf{h}, \rho, \gamma$ satisfy (6.3.4), $\rho_1 = \rho_2 = \gamma_1 = \gamma_2 = \gamma$ and $\rho_3 = \mu^{\frac12} h^{\frac12(1-\delta)}$, $\gamma_3 = \mu^{-\frac12} h^{\frac12(1-\delta)}$. Let $\epsilon_0 \le T^* \le \zeta^{-1}$ and let $\delta' = \delta'(\delta, s, l)$ in the condition $\zeta \ge \mu^{-\delta'}$ to ζ. Furthermore, let us assume that*

(6.5.14)$_\pm$ *If the point (x,ξ) belongs to the (ρ,γ)-neighborhood of* $\operatorname{supp} q'$ *and if $|r - (\mu^2 x_3^2 + \xi_3^2)| \le C\mu^{-l} + \rho_3$ then for all t: $\pm t \in [0, 2T^*]$*

$$\Phi_{t,r}(x',\xi') \in \Omega \times \mathbb{R}$$

and along $\Phi_{t,r}(x',\xi')$ conditions (6.5.3)–(6.5.8) *and the following conditions are fulfilled:*

$$|D^\alpha B_n| \le c\zeta^{-l}, \quad |D^\alpha R| \le c\mu^{-N}\zeta^{-l} \qquad \forall \alpha : |\alpha| \le K \quad \forall n \le N \tag{6.5.15$_{1,2}$}$$

where B_n and R are taken from (6.2.38), *$N = N(\delta, s, l)$, $K = K(\delta, s, l)$ and*

$$|D_{(x',\xi')}\Phi_{t,r}(x',\xi')| \le c\zeta^{-l} \tag{6.5.16}$$

where $D_Z f$ means the Jacobi matrix of F at the point Z. Then

(i) Let one of the following conditions be fulfilled:

$$|x_1 - y_1| \ge C_0 T^* \mu^{-1}\zeta^{-L} + \gamma, \tag{6.5.17$_1$}$$

$$|\xi_1 - \eta_1| \ge C_0 T^* \mu^{-1}\zeta^{-L} + \gamma, \tag{6.5.17$_2$}$$

$$|x_2 - y_2| \ge C_0 T^* \zeta^{-L}, \tag{6.5.17$_3$}$$

$$|\xi_2 - \eta_2| \ge C_0 T^* \zeta^{-L}, \tag{6.5.17$_4$}$$

for all $(x,\xi) \in$ $\operatorname{supp} q'$ *and for all $(y,\eta) \in$* $\operatorname{supp} q''$ *or let condition* (6.3.8) *be fulfilled where $L = L(\delta, s, l)$ is large enough here. Then for all $x, y \in \mathbb{R}^3$, $\tau \le c$ estimate* (6.3.7) *with $\chi_T(t)$ replaced by $\chi_{T^*,\pm T^*}(t)$ holds; recall that χ is supported in $[-\frac12, \frac12]$, $Q' = \operatorname{Op}(q')$, $Q'' = \operatorname{Op}(q'')$.*

(ii) Let $\gamma \ge \mu^{\frac12} h^{\frac12(1-\delta)}$. Moreover, let us assume that if $(x,\xi) \in$ $\operatorname{supp} q'$ *and $|r - (\mu^2 x_3^2 + \xi_3^2)| \le C_0\mu^{-l} + \rho_3$ and*

$$|-\bar{\tau} + \xi_2^2 + \bar{B}_0'(x',\xi_1) + \bar{B}_1(x',\xi_1) r| \le C\rho_3 + C\zeta^{-L}\mu^{-2} = C\bar{\gamma} \tag{6.5.18}$$

---

[15] Let us recall that $\Lambda^0 = \{(x,\xi), (x',\xi_1) \in \Omega, \xi_2 = \xi_3 = x_3 = 0\}$.

then

(6.5.19) $$\mathrm{dist}(\Phi_{t,r}(x',\xi'),(y',\eta'))\ge C\bar\gamma$$
$$\forall(y,\eta)\in\operatorname{supp}q'':|\mu^2y_3^2+\eta_3^2-r|\le C_0\mu^{-l}+\rho_3\quad\forall t:\pm t\in[T',T'']$$

where $0\le T'<T''\le 2T^*$. *Then estimate (6.3.7) with* $\chi_T(t)$ *replaced by* $\chi_{T,\pm T'''}(t)$ *with* $T=\frac12(T''-T')$, $T'''=\frac12(T''+T')$ *holds for all* $x,y\in\mathbb{R}^3$, $\tau\in[\bar\tau-\epsilon\bar\gamma,\bar\tau+\epsilon\bar\gamma]$ *where* $\epsilon>0$ *is a small enough constant.*

Proof. Let us apply theorem 2.3.7. In order to do this we need to compare the Hamiltonian flow generated by the three-dimensional Hamiltonian $\mathcal{A}(x,\xi)$ with the flow $\Phi_{t,r}$. Let us introduce the variables

$$Z=(x_1,\mu^{-1}\xi_1,x_2,\xi_2,r,\mu^{-2}x_3,\mu^{-2}\xi_3)$$

with $r=\mu^2x_3^2+\xi_3^2$, which will be treated as an "independent" variable in what follows. In these variables the flow $\Phi_{t,r}$ is described by the system $\frac{dZ}{dt}=f(Z)$ and the Hamiltonian flow generated by $\mathcal{A}(x,\xi)$ is described by the system $\frac{dZ}{dt}=\tilde f(Z)$, and the first, second, sixth and seventh components of f vanish, and for $(x',\xi_1)\in\Omega$ with $\xi_2^2+\xi_3^2+\mu^2x_3^2\le C_0\zeta^{-l}$ the estimate

(6.5.20) $$|D^\alpha(f(Z)-\tilde f(Z))|\le C\mu^{-1}\zeta^{-L}\qquad\forall\alpha:|\alpha|\le K$$

holds with appropriate L.

Let $Z(t)$ and $\tilde Z(t)$ be trajectories corresponding to these two systems originating from the same point. We assert that

(6.5.21) $$|Z(t)-\tilde Z(t)|<C\mu^{-1}\zeta^{-L'}\qquad\forall t:\pm t\in[0,2T^*]$$

with appropriate $L'=L'(\delta,s,l)$. In fact, if this statement is wrong then (6.5.21) holds for $\pm t\in[0,T)$ with $T<2T^*$ and for $\pm t=T$ there is equality in (6.5.21). But in this case (6.5.20) yields that for $\pm t\in[0,T]$

$$|\frac{d}{dt}(Z-\tilde Z)(t)-(D_{Z(t)}f)(Z-\tilde Z)(t)|\le C\mu^{-1}\zeta^{-L}$$

where $(D_Zf)(Z')$ means the differential of f at the point Z applied to Z'. Therefore (6.5.16) yields that

$$|\frac{d}{dt}\mathcal{J}^{-1}(Z-\tilde Z)(t)|\le C\mu^{-1}\zeta^{-2L}$$

where $\mathcal{J}$ is the 7×7 Jacobi matrix corresponding to the system $\frac{dZ}{dt}=f(Z)$; its elements are equal to δ_{jk} for $j\ne3,4$, to 0 for $j=3,4$ and $k=6,7$ and the upper left 4×5-submatrix of $\mathcal{J}$ is equal to $D_{(x',\xi',r)}\Phi_{t,r}(x',\xi')$; the determinant of $\mathcal{J}$ is equal to 1. Therefore (6.5.16) yields that the norms of $\mathcal{J}$ and $\mathcal{J}^{-1}$ do not exceed $C\zeta^{-2L}$. Since $\mathcal{J}(0)=I$, $Z(0)=\tilde Z(0)$ and $T\le\zeta^{-1}$ it follows that

$$|\mathcal{J}^{-1}(Z-\tilde Z)(t)|\le C\mu^{-1}\zeta^{-3L}$$

and therefore $|(Z-\tilde Z)(t)|\le C\mu^{-1}\zeta^{-5L}$. But then for $\zeta\le\zeta_0=\mathrm{const}$ condition (6.5.21) with $L'=9L$ cannot be violated and for $\zeta\in[\zeta_0,1]$ this condition is provided by the appropriate choice of C_0. So condition (6.5.21) is fulfilled and for the sake of simplicity one can assume that $L'=L$.

Let us now check the conditions of theorem 2.3.7 on the Jacobi matrix. Let us vary the initial data in which case we obtain the system

$$\frac{d}{dt}\mathrm{d}\tilde{Z} = (D_{\tilde{Z}}\tilde{f})\mathrm{d}\tilde{Z}$$

where d denotes variation; this equation and (6.5.21) yield that

$$\frac{d}{dt}\mathrm{d}\tilde{Z} = (D_Z\tilde{f})\mathrm{d}\tilde{Z} + O(\mu^{-1}\zeta^{-2L}|\mathrm{d}\tilde{Z}|)$$

and therefore

$$|\frac{d}{dt}\mathcal{J}(t)^{-1}\mathrm{d}\tilde{Z}| \le C\mu^{-1}\zeta^{-3L}|\mathcal{J}(t)^{-1}\mathrm{d}\tilde{Z}|.$$

This inequality yields that $|\mathcal{J}(t)^{-1}\mathrm{d}\tilde{Z}| \le C|\mathrm{d}\tilde{Z}(0)|$ provided $\delta' > 0$ is small enough in the condition on ζ. Then

$$|\frac{d}{dt}\mathcal{J}(t)^{-1}\mathrm{d}\tilde{Z}| \le C\mu^{-1}\zeta^{-3L}|\mathrm{d}\tilde{Z}(0)|$$

and therefore

$$|\mathrm{d}\tilde{Z}(t) - \mathcal{J}(t)\mathrm{d}\tilde{Z}(0)| \le C\mu^{-1}\zeta^{-5L}|\mathrm{d}\tilde{Z}(0)|.$$

This inequality obviously yields that

$$|D_{(x(0),\xi(0))}(x'(t),\xi(t))| \le C\zeta^{-5L}$$

and

$$|D_{(x(0),\xi(0))}((x_3(t),\xi_3(t))| \le C\mu\zeta^{-5L}.$$

Therefore for $\mu \in [\mu_0, h^{-\delta'}]$ with a small enough exponent $\delta' > 0$ our theorem follows from theorem 2.3.7. On the other hand, for $\mu \in [h^{-\delta}, h^{\delta-1}]$ this theorem is due to proposition 6.5.7 which will be proven independently below and which is also due to theorem 2.3.7; excluding (x_3,ξ_3) from the analysis we will be able to check conditions only for part of the Jacobi matrix. □

Remark 6.5.3. *One can replace condition (6.5.16) in theorem 6.5.2 by the condition*

(6.5.22) $$|\xi_2| \ge \epsilon\zeta^l \qquad \text{along } \Phi_{t,r}(x',\xi').$$

In order to prove this let us first of all note that the definition of $\Phi_{t,r}$ yields that in $\Phi_{t,r}(x',\xi')$ only the (x_2,ξ_2)-components need to be checked. Moreover, the fact that $H_{(\bar{x}_1,\bar{\xi}_1)r}(x_2,\xi_2)$ is constant along trajectories, $(6.5.15)_{1,2}$ and (6.5.22) yield that one need check only the x_2-component (we change l if necessary). Let $z = (x_1,\xi_1,r)$ and $W(z,x_2) = \bar{B}_0'(z,x_2) + \bar{B}_1(z,x_2)r$. Then

(6.5.23) $$\pm 2t = \int_{x_2(0)}^{x_2(t)} (E - W(z,s))^{-\frac{1}{2}}ds$$

with $E = \xi_2^2(0) + W(z,x_2(0))$. Moreover, (6.5.22) yields that $E - W(z,s) = \xi_2^2 \ge \epsilon_0\zeta^{2l}$. Let us treat $x_2(t)$ as an implicit function of $E, z, x_2(0)$ and t (where E is a function of $\xi_2(0)$, $x_2(0)$ and r but the derivatives of E with respect to these variables are easily estimated). Then (6.5.23) and (6.5.22) yield that the absolute values of the first order partial derivatives of t with respect to E and z do not exceed $C\zeta^{-2l}|t|$ and that the absolute values of the first order partial derivatives of t with respect to $x_2(t)$ and $x_2(0)$ do not exceed $C\zeta^{-l}$ and are greater than $\epsilon\zeta^l$ (we have again used (6.5.15)). This yields that the absolute values of the first order partial derivatives of $x_2(t)$ with respect to $x_2(0), z, \xi_2(0)$ do not exceed $C\zeta^{-5l}$. ◊

Remark 6.5.4. *(i) Condition (6.5.22) provides that the movement is irrevocable:*

$$\operatorname{dist}(\Phi_{t,r}(x',\xi'),(x',\xi'))\ge \epsilon\zeta^l \min(|t|,1)\qquad \forall t:\pm t\in[0,2T^*]. \tag{6.5.24}$$

(ii) One can rewrite condition (6.5.22) in the form

$$|F^j(\xi_j-\mu V_j)|\ge\epsilon\zeta^l \tag{6.5.25}$$

or

$$|\tau-V-rF|\ge\epsilon\zeta^l \tag{6.5.26}$$

along the trajectory in the original coordinates.

(iii) Moreover, returning to the original coordinates one can write the trajectory equation in the form

$$\frac{dx_j}{dt}=\frac{2}{F}\sqrt{\tau-V-rF}F^j\qquad (j=1,2,3) \tag{6.5.1$''$}$$

where we reverse the time if necessary.

(iv) Finally, returning to $U(t)$ one can pick $Q'=\psi'$, $Q''=Q'''\psi''$ where $Q'''=\mathcal{F}\tilde{Q}'''\mathcal{F}^*$ and r runs through all the possible values of $x_3^2+\mu^2\xi_3^2$ on $\operatorname{supp}\tilde{q}'''\psi''$ where small and capital letters denote symbols and corresponding operators. Alternatively, r runs through all the possible values of $F^{-1}\bigl(\tau-V-F^{-2}\bigl(F^j(\xi_j-\mu V_j)\bigr)^2\bigr)$ on $\operatorname{supp} q'''$.◊

Therefore theorem 6.5.2 yields that if its conditions and condition (6.5.22) are fulfilled then estimate (6.3.26) holds with $T=T^*$ and with $T'\in[h^{1-\delta},T^*]$ for all $\tau\in[-\epsilon'\zeta^l,\epsilon'\zeta^l]$. Therefore we can apply Tauberian arguments with $T=T^*$ instead of $T=T_0$.

Let us now assume that condition (6.3.33) is fulfilled. Then the analysis of section 6.3 yields that for $\mu\le h^{\delta-\frac13}$ the main contribution to $\mathcal{R}_1$ is given by the zone $\{(x,\xi):|\xi_2|\ge c\zeta\}$. More precisely, if we calculate $\mathcal{R}_2$ for the operator $Q'=\mathcal{F}\tilde{Q}'\mathcal{F}^*$ with $\tilde{Q}'=\operatorname{Op}(\tilde{q}')$, $\tilde{q}'\in S_{\mathbf{h}',\zeta,\zeta,K}(T^*\mathbb{R}^2)$ supported in $\{|\xi_2|\le 2c\zeta\}$ and for the function ψ_1 supported in $B(0,\frac12)$ and satisfying (6.3.63) then the estimate

$$\mathcal{R}_2\le C\zeta\gamma^3h^{-2}+C'h^s \tag{6.5.27}$$

holds provided $\operatorname{diam}\operatorname{supp}\psi_1\le c\gamma$.

Let us estimate the contribution of the zone $\{(x,\xi),|\xi_2|\ge c\zeta\}$ to $\mathcal{R}_1$ by means of Tauberian arguments. Namely, let us estimate $\mathcal{R}_2$ for the operator $Q'=\mathcal{F}\tilde{Q}'\mathcal{F}^*$ with $\tilde{Q}'=\operatorname{Op}(\tilde{q}')$, $\tilde{q}'\in S_{\mathbf{h}',\zeta,\zeta,K}(T^*\mathbb{R}^2)$ supported in $\{|\xi_2|\ge c\zeta\}$ and for the function ψ_1 supported in $B(0,\frac12)$ and satisfying (6.3.63) with $\gamma\ge\mu^{-1}$ and $\operatorname{diam}\operatorname{supp}\psi_1\le c\gamma$. Then under conditions on the flows $\phi_{\tilde t}$ and $\Phi_{t,r}$ estimate (6.5.27) holds; here $\phi_{\tilde t}$ is given by (6.5.1)$''$ with $(\pm)\tilde t\in[0,\tilde T]$ and $\tilde T\le\zeta^{-l}$, conditions $(6.5.11)_{1,2}$, $(6.5.14)_\pm$ are assumed to be fulfilled and the signs $(\pm)$ of $\tilde t$ and $\pm$ of t are not connected; moreover, one should take $r\in[0,r^*]$ where r^* is the minimal value of $r\in\mathbb{R}$ for which the inequality

$$V+Fr+c^2\zeta^2\ge0\quad \text{at } B(\hat x,\zeta^l) \tag{6.5.28}$$

is fulfilled (we now pick $\tau=0$).

One can easily see that all the conditions excluding irrevocability should be checked for $r=0$ only. Thus, let us consider the flow defined by (6.5.1)$''$ with $r=0$, $\tau=0$:

$$\frac{dx_j}{dt}=\frac{2}{F}\sqrt{-V}F^j\qquad (j=1,2,3). \tag{6.5.1$'''$}$$

Let ϕ_t be the corresponding flow.

Then the irrevocability for $r \in [r^*, r]$ is due to the condition

$$v(\phi_t(x)) \le v(x) + c^2\zeta^2 \qquad \forall t : \pm t \in [0,T] \tag{6.5.29}$$

with $v = \frac{V}{F}$.

Then our arguments yield estimate (6.5.27) in this case also; thus it is true for $Q = I$ as well. Summing with respect to a partition of unity in $B(0, \frac{2}{3})$ we obtain the principal statement of this step:

Theorem 6.5.5. *Let conditions (6.1.1)–(6.1.4), (6.2.1), (6.3.1) and (6.3.33) be fulfilled. Let $\Xi_1, \dots, \Xi_n$ be closed subsets of $B(0,1)$ and let $1 \le T_1 \le \dots \le T_n \le \mu^{\delta'}$ with a small enough exponent $\delta' = \delta'(\delta, l) > 0$. Let us assume that through every point x belonging to the γ_i-neighborhood of Ξ_i (with $\gamma_i = (\mu^{\frac{1}{2}} h^{\frac{1}{2}(1-\delta)} + C_0\mu^{-1})\zeta_i^{-L}$, $\zeta_i = T_i^{-1}$ where $L = L(\delta, l)$ is large enough) there passes a trajectory of the flow $\phi_t(x)$ given by the system (6.5.1)‴ with $\phi_0(x) = x$, $\pm t \in [0, T_i]$ and along this trajectory conditions (6.5.3)–(6.5.5), (6.5.8), (6.5.29),*

$$V \le -\epsilon\zeta^2, \tag{6.5.30}$$

(6.5.3)$_{1-3}$, (6.5.8) and

$$\operatorname{dist}(\phi_t(x), \partial X) \ge \gamma_i, \tag{6.5.6$'$}$$

$$\operatorname{dist}(\phi_t(x), x) \ge \gamma_i \min(|t|, 1), \tag{6.5.7$'$}$$

are fulfilled with $\zeta = \zeta_i$. Then

(i) For $\mu \le h^{\delta - \frac{1}{3}}$ the estimate

$$\mathcal{R}_1 \le \sum_{0 \le i \le n} C T_i^{-1} h^{-2} \operatorname{mes} \Xi_i + C' \tag{6.5.31}$$

holds with $T_0 = 1$, $\Xi_0 = B(0,1) \cap \{V \le 0\} \setminus (\Xi_1 \cup \dots \cup \Xi_n)$;

(ii) Under condition (6.3.44) this estimate holds for $\mu \le h^{\delta - 1}$.

Remark 6.5.6. *(i) In the theorem we have replaced $\gamma_i = \zeta_i^l$ which follows from the previous analysis by $\gamma_i = \mu^{-1}\zeta_i^{-L}$. One can do this for the following reasons. Let us note that despite the absence of condition (6.5.7) which prevents the global (i.e., in $\cup_t \phi_t(B(\hat{x}, \zeta^{\frac{1}{2}}))$) reduction of our operator to canonical form, we may still reduce the operator to canonical form locally (i.e., in $\cup_t \phi_{t \in [T_1, T_2]}(B(\hat{x}, \zeta^{\frac{1}{2}L'})$ provided $|T_2 - T_1| \le \zeta^{L'}$ with appropriate exponent L'). Moreover, the absence of condition (6.5.7) does not prevent us from making the global reduction in the canonical coordinates described in theorem 6.5.1(iii). But then the singularities of $U\psi''(y)^t Q''_y$ (intersected with $\{|\tau| \le \epsilon\gamma\}$) lie in the pipe consisting of the lines described below starting in the γ-vicinity of $\operatorname{supp}\psi''$. Moreover, for $t \in [T_1, T_2]$ with $|T_2 - T_1| \le \zeta^{L'}$ these singularities lie in the part of the pipe where reduction is possible. This permits us to exclude condition (6.5.7) and prove the estimate*

$$|F_{t \to h^{-1}\tau} \chi_{T, \pm T'}(t) Q'_x \psi'(x) U \psi''^t_y Q''_y| \le C' h^s \qquad \forall \tau \in [-\epsilon\gamma\zeta^L, \epsilon\gamma\zeta^L]$$

provided $Q' = \operatorname{Op}(q')$, $Q'' = \operatorname{Op}(q'')$, $q', q'' \in S_{h,\gamma,\gamma,K}(T^*\mathbb{R}^3)$, ψ', ψ'' satisfy (6.3.63), $\operatorname{diam} \operatorname{supp} \psi' \le c\gamma$, $\operatorname{dist}(\operatorname{supp}\psi', \partial X) \ge \epsilon_0\gamma$ (and the same is true for ψ''),

$$\operatorname{supp} q' \subset \iota\{|\omega| \ge c\zeta^l\} \tag{6.5.32}$$

where $\iota : T^*X \to X \times \mathbb{R}$ is the natural projection $\iota(x,\xi) = (x, \omega)$ with $\omega = \frac{1}{F} F^j(\xi_j - \mu V_j)$, and the following condition is fulfilled:

(6.5.33) *If we consider trajectories of the system*

$$\frac{dx_j}{dt} = 2\omega\frac{1}{F}F^j \quad (j=1,2,3), \tag{6.5.34$_1$}$$

$$F\frac{d\omega}{dt} = -(\omega^2 F^j\partial_j F + F^j\partial_j\frac{V}{F}) \tag{6.5.34$_2$}$$

passing through $\iota(\operatorname{supp} q')$, *then along them* $\pm\omega \ge \zeta^l$ *and for* $t \in [T'-T, T'+T]$

$$\operatorname{dist}((x(t),\omega(t),\iota(y,\eta)) \ge \gamma \qquad \forall (y,\eta)\in \operatorname{supp} q'',$$

or if similar conditions with x, y, q', q'', t *replaced by* $y, x, q'', q', -t$ *respectively are fulfilled. System* $(6.5.34)_{1,2}$ *is an alternative form of* (6.5.1)″.

Results of this type permit us to apply Tauberian arguments.

(ii) Let us replace condition (6.5.6) by (6.5.6)′. *In order to do this let us apply the previous arguments but in each "small" time interval let us add the following idea: let us "move" the boundary* ∂X *by the distance* ζ^l *and let us treat the new fundamental solution* U^0; *our previous arguments yield that* $Q'_x\psi'(x)U^0$ *is negligible provided* $y \in X$ *and* $\operatorname{dist}(y,\partial X) \ge \mu^{-1}\zeta^{-L'}$ *where* X *means the "old"* X; *it is then easy to obtain that* $Q'_x\psi'(x)U \equiv Q'_x\psi'(x)U^0$, *i.e., the analysis of propagation of singularities along trajectories satisfying* (6.5.6)′ *can be reduced to the analysis of propagation of singularities of* U^0 *(along trajectories satisfying (6.5.6) in* X_{new}*).*

(iii) The connections between $\mathcal{R}_1$ *and* $\mathcal{R}^w$ *(see section 6.3) and between* $\mathcal{R}_1$ *and* $\mathcal{R}$ *(see section 6.4) yield that in theorem 6.5.3 estimate (6.5.31) remains true for* $\mathcal{R}^w$ *and* $\mathcal{R}$ *if* $\mu \le h^{\delta-\frac{1}{3}}$; *moreover, under condition (6.3.44) this estimate holds for* $\mathcal{R}$ *if* $\mu \le h^{\delta-1}$.

(iv) I leave the case $\vartheta > 0$ *to the reader. Recall that in this case the contribution of the zone* $\{(x,\xi) : |\xi_2| \le c\zeta\}$ *is negligible without a condition on* $\frac{V}{F}$ *only if* $\mu \le h^{-\delta''}$ *with* $\delta' = \delta''(\vartheta)$. ◊

6.5.3 The case of a strong magnetic field ($d = 3$)

Let us assume that condition (6.4.1) is fulfilled. Now I want to improve the arguments of section 6.4. In order to do this let us introduce the one-dimensional Hamiltonian (6.5.13) with $r = (2j+1)\mu h$; let us denote it by $H_{(x_1,\xi_1)j}(x_2,\xi_2)$ and let us denote the corresponding Hamiltonian flow by $\Phi_{t,j}(x_2,\xi_2)$. Let us improve proposition 6.4.3 (in what follows we will improve it again):

Proposition 6.5.7. *Let* $j \in \mathbb{Z}^+$, $Q' = \operatorname{Op}(q')$, $Q'' = \operatorname{Op}(q'')$ *where* $q', q'' \in S_{\mathbf{h}',\gamma,\gamma,K}(T^*\mathbb{R}^2)$ *with* $\gamma \ge h^{\frac{1}{2}(1-\delta)}$ *and* $\mathbf{h}'$ *is defined in* (6.4.13). *Let us assume*

$(6.5.35)_\pm$[16] *If* (x',ξ') *belongs to the* γ*-neighborhood of* $\operatorname{supp} q'$ *then for all* $t : \pm t \in [0, 2T^*]$, $\Phi_{t,j}(x',\xi') \in \Omega\times\mathbb{R}$

and along $\Phi_{t,j}(x',\xi')$ *conditions* (6.5.3)–(6.5.5), (6.5.8), (6.5.15)–(6.5.16) *are fulfilled.*

Let $\zeta \in [\mu^{-\delta'}, 1]$ *with* $\delta' = \delta'(\delta, l, s) > 0$. *Then*

(i) Let one of conditions $(6.5.17)_{1-4}$ *be fulfilled or let* $j \ne k$. *Then estimate* (6.4.17) *holds with* $\chi_T(t)$ *replaced by* $\chi_{T^*,\pm T^*}$.

(ii) For all $(x',\xi') \in \operatorname{supp} q'$, $(y',\eta') \in \operatorname{supp} q''$ *let*

$$\operatorname{dist}(\Phi_{t,j}(x',\xi'),(y',\eta')) \ge C\bar\gamma \qquad \forall t : \pm t \in [T', T''], \tag{6.5.36$_\pm$}$$

[16] Compare with $(6.5.14)_\pm$.

$\bar{\gamma} = h^{\frac{1}{2}(1-\delta)} + C_0\zeta^{-L}\mu^{-2}$, $0 \le T' < T'' \le 2T^*$. *Then estimate* (6.4.17) *holds with* $\chi_T(t)$ *replaced by* $\chi_{T,\pm T'''}(t)$ *where* $x, y \in \mathbb{R}^3$, $\tau \in [\bar{\tau} - \epsilon\bar{\gamma}, \bar{\tau} + \epsilon\bar{\gamma}]$, $T = \frac{1}{2}(T'' - T')$, $T''' = \frac{1}{2}(T'' + T')$.

Proof. This proof repeats the proof of theorem 6.5.2 but now the variables (x_3, ξ_3, r) are absent and the proof is simplified. Moreover, for $h^{-\delta} \le \mu \le h^{\delta-1}$ theorem 6.5.2 obviously follows from this proposition. □

Remark 6.5.8. *(i) Obviously, one need only check conditions* $(6.5.35)_\pm$ *and* $(6.5.36)_\pm$ *for* (x', ξ') *such that*

$$|H_{(x_1,\xi_1),j}(x_2, \xi_2) - \bar{\tau}| \le C\bar{\gamma} \tag{6.5.37}$$

and for (y', η') *satisfying this restriction.*

(ii) One should remember the relationship $r = (2j+1)\mu h$ *between* $r \in \mathbb{R}^+$ *from the previous subsection and* $j \in \mathbb{Z}^+$ *used here.* ◊

Proposition 6.5.9. *Let the conditions of proposition 6.4.5 be fulfilled and let* $\rho_1 = \rho_2 = \gamma_1 = \gamma_2 \ge h^{\frac{1}{2}(1-\delta)}$. *Let us assume that for every* $j \le J$ *conditions* $(6.5.35)_\pm$ *are fulfilled and along the trajectory conditions* (6.5.3)–(6.5.5), (6.5.8), (6.5.15), (6.5.16) *are fulfilled (we take into account remark 6.5.8). Then formula* $(6.4.19)_{(3)}$ *remains true for* $\tau \in [\bar{\tau} - \epsilon\bar{\gamma}, \bar{\tau} + \epsilon\bar{\gamma}]$ *and* $T \in [1, 2T^*]$.

Proof. Let us recall that in proposition 6.4.5 the restriction $T \le T_0$ appeared only in order to get rid of terms with $j \ne k$. But now proposition 6.5.5(i) permits us to do this for $T \le 2T^*$. In fact, applying an admissible partition of unity we see that we should prove $(6.4.19)_{(3)}$ under the additional condition $\operatorname{diam}(\operatorname{supp} q' \cup \operatorname{supp} q'') \le c\gamma$. □

Let us assume that the magnetic field is not very strong, i.e., that condition (6.3.1) is fulfilled. Then by section 6.4 the contribution to $\mathcal{R}_4$ of the zone $\{|\xi_2| \le \zeta^L\}$ with $\zeta \le \mu^{-\delta'}$ does not exceed $Ch^{-2}\zeta$ where $\delta > 0$ is arbitrarily small, L is large enough and $\delta' = \delta'(\delta, L) > 0$ is small enough. So, only the zone $\{|\xi_2| \ge \zeta^L\}$ should be treated and in this zone all the arguments of the previous subsection remain true. Then we obtain the following

Proposition 6.5.10. *Let the conditions of theorem 6.5.5 be fulfilled as well as condition* (6.4.1). *Then estimate* (6.5.31) *holds for* $\mathcal{R}_4$ *without any additional assumptions.*

However, the transition from $\mathcal{R}_4$ to $\mathcal{R}$ is not very smooth: one can easily prove that $\mathcal{R}_6$ does not exceed

$$R_6 = \frac{C}{h^2}\int_{\{\gamma \ge \mu^{-1}\}} \Bigl(\frac{\mu h}{\gamma^2}\Bigr)^s dx + \frac{C}{h^2}(\mu^3 h)^s \operatorname{mes}\{\gamma \le \mu^{-1}\} \tag{$(6.5.38)_1$}$$

for $\mu^3 h \le 1$ and does not exceed

$$R_6 = \frac{C}{h^2}\int_{\{\gamma \ge \sqrt{\mu h}\}} \Bigl(\frac{\mu h}{\gamma^2}\Bigr)^s dx + \frac{C}{h^2\sqrt{\mu^3 h}} \operatorname{mes}\{\gamma \le \sqrt{\mu h}\} \tag{$(6.5.38)_2$}$$

for $\mu^3 h \ge 1$; recall that

$$\gamma = |\nabla \frac{V}{F}| + \min_j(|\frac{V}{F} + (2j+1)\mu h|^{\frac{1}{2}}. \tag{6.5.39}$$

Therefore we obtain the following

Proposition 6.5.11. *Let the conditions of theorem 6.5.3 be fulfilled as well as condition* (6.4.1). *Then the estimate*

$$\mathcal{R} \leq \sum_{0 \leq i \leq n} CT_i^{-1} h^{-2} \operatorname{mes} \Xi_i + R_6 + C' \tag{6.5.40}$$

holds with R_6 *given by* $(6.5.38)_{1,2}$ *for* $\mu^3 h \leq 1$, $\mu^3 h \geq 1$ *respectively. In particular, one can skip the second term for* $h^{-1+\delta} \geq \mu \geq h^{\frac{1}{3}-\delta}$. *Moreover, one can skip that term under condition* (6.3.58) *with* $\nu = 1$.

Remark. *I again leave the case* $\vartheta > 0$ *to the reader.* ◊

Let us assume that the magnetic field is very strong, i.e.,

$$\mu \geq h^{1-\delta} \tag{6.5.41}$$

with an arbitrarily small exponent $\delta > 0$. Then $\mathcal{R}_6$ definitely does not exceed $Ch^{-2+\delta'}$ but in order to get an appropriate estimate for $\mathcal{R}_4$ one should assume that either condition (6.3.64) with $\nu \leq ch^2(\frac{\zeta}{\mu h})^L$ with arbitrarily large L or conditions (6.3.64) an (6.3.58) are fulfilled. Then the results of section 6.4 yield that the contribution of the zone $\{(x', \xi') : |\xi_2| \leq 2c\zeta\}$ to $\mathcal{R}_4$ does not exceed $C(\mu h^{-1} + \zeta h^{-2})$. Let us first assume that

$$\zeta \geq c\mu h. \tag{6.5.42}$$

Then the contribution of this zone to $\mathcal{R}_6$ does not exceed $C\zeta h^{-2} + C'$. Then all the above arguments remain true and we easily get

Proposition 6.5.12. *Let all the conditions of proposition 6.5.11 excluding condition* (6.3.1) *be fulfilled. Moreover, let either condition (6.3.64) with* $\nu \leq ch^2(\frac{\zeta}{\mu h})^L$ *with arbitrarily large* L *or conditions (6.3.64) and (6.3.58) be fulfilled. Finally, let condition* (6.5.42) *be fulfilled.*[17] *Then estimate* (6.5.40) *remains true.*

In these constructions we have taken an estimate for the contribution to $\mathcal{R}_4$ of the zone $\{(x', \xi') : |\xi_2| \leq 2c\zeta\}$ from section 6.4. Now I am going to improve it for

$$\mu h = 1.$$

Then in our decompositions $j \leq \bar{\jmath}$. Without loss of generality one can assume that

$$V + (2\bar{\jmath} + 3)\mu h \geq \epsilon_0 \tag{6.5.43$_1$}$$

$$V + (2\bar{\jmath} - 1)\mu h \leq -\epsilon_0 \tag{6.5.43$_2$}$$

in $B(0,1)$ (otherwise we can arrange this by an ϵ-partition of unity. If $\bar{\jmath} \geq 2$ then the contribution of terms with $j = 0, \dots, \bar{\jmath} - 1$ can be estimated according to the above results. One can easily see that only trajectories with $j = \bar{\jmath} - 1$ need to be treated. Let us consider the contribution of $j = \bar{\jmath}$ and for the sake of simplicity in the following statements let us assume that $\bar{\jmath} = 1$. Moreover, let us assume that

$$|D^\alpha V_{\bar{\jmath}}| \leq c\nu \qquad \forall \alpha : |\alpha| \leq K \tag{6.5.44}$$

and

$$|V_{\bar{\jmath}}| + |\nabla V_{\bar{\jmath}}| \geq \epsilon_0 \nu \tag{6.5.45}$$

---

[17] All these conditions should be fulfilled for all $\zeta = T_i^{-1}$, $i = 1, \dots, n$.

with $\nu \ge h^2$, $V_j = V + (2j+1)\mu h F$. Then in order to improve the results of section 6.4 one should treat the trajectories of the system

$$(6.5.1)^{IV} \qquad \frac{dx_i}{dt} = \frac{1}{F\sqrt{\nu}}(-V_{\bar{j}})^{\frac{1}{2}} F^i \qquad (i = 1, 2, 3).$$

One should assume that $\zeta_i^{-1} = T_i \le h'^{-\delta'}$ with $h' = (\frac{h}{\sqrt{\nu}})$ and that along these trajectories all the conditions (6.5.3), (6.5.8), (6.5.6)$'$, (6.5.7)$'$ and

$$(6.5.3)'_3 \qquad |D^\alpha V_{\bar{j}}| \le c\nu\zeta^{-l},$$

$$(6.5.29)' \qquad V_{\bar{j}} \ge \epsilon_0 \zeta^l$$

are fulfilled. Under these assumptions the estimate

$$(6.5.40)' \qquad \mathcal{R} \le \sum_{0\le i\le n} C T_i^{-1} \operatorname{mes} \Xi_i + C' h^{-2} h'^s$$

holds.

I leave the possible generalizations to the reader.

6.5.4 Propagation of singularities ($d = 2$)

Let us now consider the two-dimensional case. Then there are no magnetic lines and in the reduced operator there are no "quick" variables like ξ_2 (we do not care about the "very quick" variables (x_d, ξ_d)). Therefore we can take a much larger characteristic time (under condition (6.3.64) it is $\mu\nu^{-1}$) as we did in sections 6.3 and 6.4; then the slow variables (x_1, ξ_1) are no longer constant. One can easily see that for the reduced symbol one should consider Hamiltonian trajectories of the symbol $(a \circ \Psi - (F \circ \Psi)(x_2^2 + \xi_2^2))$ at the energy level 0, so the Hamiltonian field is $(F \circ \Psi) H_{(F^{-1}a)\circ\Psi}$. Returning to the original operator we obtain trajectories in x-space along which $\frac{du}{dt} = F H_a(u + \alpha_1 s_1 + \alpha_2 s_2)|_{s_1=s_2=0}$ with $s_j = \xi_j - V_j$ and α_j defined by the equations $\{s_k, u + \alpha_1 s_1 + \alpha_2 s_2\}|_{s_1=s_2=0} = 0$; this should be true for any smooth function u. Therefore this field is $\frac{1}{\sqrt{g}}(\partial_{x_2}\frac{V}{f}, -\partial_{x_1}\frac{V}{f})$ where $f = \frac{1}{\sqrt{g}} F_{12}$ is a pseudo-scalar, $F = |f|$. So, we have the dynamical system

$$(6.5.46) \qquad \frac{dx_1}{dt} = \frac{1}{\sqrt{g}}\partial_{x_2}\frac{V}{f}, \quad \frac{dx_2}{dt} = -\frac{1}{\sqrt{g}}\partial_{x_1}\frac{V}{f}.$$

The integral lines of this system we will call *drift lines;* in the two-dimensional case we will use them instead of magnetic lines.

Remark 6.5.13. *Even from the heuristic point of view there is an essential difference:*

(i) Magnetic lines depend only on $\mathbf{F}$ *while drift lines depend on* V *as well. In order to the get drift lines at the energy level* τ *one should replace* V *by* $V - \tau$. *Then, for* $F \ne \text{const}$ *the drift lines depend on the energy level as well.*

(ii) As soon as magnetic lines are defined, the movement along them is defined by the quantum number j *and the energy level* τ. *Moreover, the direction of the movement depends on* $(\xi_k - V_k)F^k$. *In contrast, in the two-dimensional case the choice of* j *is automatic and the movement is defined only by the energy level which should be fixed. System (6.5.46) is a dynamical system while (6.5.1) only describes geometrical shape.*

(iii) The movement along drift lines is given in scaled time; the real time is μt. *According to the theory of sections 2.2 and 2.3,* t *should not be too large. Moreover,*

under condition (6.3.64) it is sufficient to assume that νt is not too large (because we can divide operator by ν and then the scaled time will be νt).

(iv) $\frac{V}{f}$ is constant along drift lines and for big enough μ (e.g., $\mu \ge h^{-\frac{1}{2}}$) one need only consider lines with $-\frac{V}{\mu h F} \in 2\mathbb{Z}^+ + 1$. ◊

Theorem 6.5.14. *Let ψ_1, ψ_2 satisfy (6.3.63) with $\gamma = \mu^{-1}$. Let us assume that along trajectories of (6.5.46) starting in the intersection of the $C_0\gamma$-vicinity of* supp ψ_1 *and $\{-\frac{V}{\mu h F} \in \mathbb{Z}^+\}$ (where here and below C_0 is a large enough constant, $L = L(l)$ is large enough and $\delta = \delta(l) > 0$ is small enough) conditions $(6.5.3)_{1-3}$ (with f instead of F^j), (6.5.4), (6.5.5), (6.5.6)′, (6.5.8) are fulfilled where ϕ_t is the corresponding flow and $\zeta \ge h^\delta$, $0 \le \pm t \le \pm T'$, $|T'| \le T = \zeta^{-1}$, $\delta > 0$ is sufficiently small. Moreover, let us assume that for $t = T$ the distance between these trajectories and* supp ψ_2 *is larger than $C_0\mu^{-1}\zeta^{-L}$. Then*

(i) The estimate

$$|D_x^\alpha D_y^\beta F_{t\to h^{-1}\tau}\chi_{\mu T,\mu T'}(t)\psi_1(x)U\psi_2(y)| \le Ch^s \qquad \forall \tau : \mu T|\tau| \le h^{1-\delta} \quad \forall \alpha,\beta : |\alpha| \le s, |\beta| \le s. \tag{6.5.47}$$

(ii) Moreover, if the conditions

$$|D^\alpha g^{jk}| \le c\nu\mu\zeta^{-l}, \quad |D^\alpha F| \le c\nu\mu\zeta^{-l}, \quad |D^\alpha \frac{V}{F}| \le c\nu\zeta^{-l} \qquad \forall \alpha : 1 \le |\alpha| \le K, \tag{$(6.5.3)'_{1-3}$}$$

hold along this trajectory then one can replace $T = \zeta^{-1}$ by $T = \nu^{-1}\zeta^{-1}$

I leave the proof to the reader. The main obstacle is the absence of the global reduction. However, reduction can be done locally and then one can use "pipe coordinates" along drift lines and apply the same arguments as in proof of theorem 6.5.2 and proposition 6.5.7.

6.5.5 Improved spectral asymptotics ($d = 2$)

Applying standard Tauberian arguments one can easily prove the following

Theorem 6.5.15. *Let $d = 2$ and let conditions (6.1.1)–(6.1.4), (6.2.1) and (6.3.3) be fulfilled. Moreover, let conditions (6.3.64), (6.2.48) and (6.3.68) be fulfilled and*

$$\min_j |V + (2j+1)\mu h F| + |\nabla(V + (2j+1)\mu h F)| \ge \epsilon_0 \nu. \tag{6.5.48}$$

Furthermore, let **either** $\mu \ge h^{-\delta}$ **or** $\nu \ge \mu h^{1-\delta}$ *and condition (6.3.33) be fulfilled. Finally, let $\Xi_1, \dots, \Xi_n$ be closed subsets in $B(0,1)$ and let $1 \le T_1 \le \dots \le T_n \le \mu^{\delta'}$ with a small enough exponent $\delta' = \delta'(\delta, l) > 0$. Let us assume that through every point x belonging to the γ_i-neighborhood of Ξ_i (with $\gamma_i = \mu^{-1}\zeta_i^{-L}$, $\zeta_i = T_i^{-1}$ where $L = L(\delta, l)$ is large enough) there passes a trajectory of the flow $\phi_t(x)$ given by the system (6.5.46) with $\phi_0(x) = x$, $\pm t \in [0, \nu^{-1}T_i]$ along which conditions (6.5.3)–(6.5.5), (6.5.7), (6.5.8) and (6.5.47) are fulfilled. Then the estimate*

$$\mathcal{R}_{4\vartheta} \le C \sum_{0 \le i \le n} \sum_j \frac{h^\vartheta \mu^{-\vartheta}}{T_i^{1+\vartheta}} \int_{\Xi_i \cap \{V_j = 0\}} dx : dV_j + C'h^s \tag{6.5.49}$$

holds where $V_j = V + (2j+1)\mu h$ and $\Xi_0 = B(0,1) \setminus (\Xi_1 \cup \dots \cup \Xi_n)$.

Moreover, for $\nu = 1$, $\vartheta = 0$, the same estimate holds for $\mathcal{R}$ if one includes the term $-\kappa_2$ in $\mathcal{R}$.

6.6 Lacunas in the Semiclassical Approximation to the Spectrum

6.6.1 Existence of spectral gaps

In this section we assume that $d = 2$ and the case $d = 3$ is specially indicated sometimes. The reason is that for $d = 3$ there exists only one spectral gap in the semiclassical approximation to the spectrum from $-\infty$ to $\inf(V + \mu h F)$, while for $d = 2$ many spectral gaps can exist as follows from theorem 6.1.1. First of all by the standard elliptic methods (for the reduced operator) the following two propositions can be easily proven:

Proposition 6.6.1. *Let $d = 2$, suppose conditions* (6.1.1)–(6.1.4), (6.2.1), $(6.2.48)_{1-3}$, (6.4.1) *and* $(6.3.10)_1$ *are fulfilled and assume that for some $j \in \mathbb{Z}^+$ and $\varsigma = \pm 1$*

$$\text{(6.6.1)}\quad \varsigma\big((2j+1)\mu h + B(x_1, \xi_1, \mu^{-2}, \mu^{-1}h, h^2)\big) \ge \zeta^2 \ge \bar{\nu}\mu^{-2}h^{2-\delta} \qquad \forall (x_1, \xi_1) \in \Omega.$$

Then

$$\text{(6.6.2)}\quad |F_{t\to h^{-1}\tau}\chi_T(t)Q_x\mathcal{U}_{jk}| \le C'Th^s(1+|\tau|)^{-s} \qquad \forall x, y, \in \mathbb{R}^d \quad \forall T \ge h^{1-\delta}\zeta^{-2} \quad \forall \tau : \varsigma\tau \le \epsilon\zeta^2.$$

In particular, if condition (6.6.1) *holds for $j = 0$ and $\varsigma = 1$ (in which case it holds for $\varsigma = 1$ for all $j \in \mathbb{Z}^+$) then*

$$\text{(6.6.3)}\quad |F_{t\to h^{-1}\tau}\chi_T(t)U| \le C'Th^s(1+|\tau|)^{-s} \qquad \forall x, y, \in B(0, \frac{1}{2}) \quad \forall T \ge h^{1-\delta}\zeta^{-2} \quad \forall \tau : \varsigma\tau \le \epsilon\zeta^2.$$

Proposition 6.6.2. *Let conditions* (6.1.1)–(6.1.4) *and* (6.2.1) *be fulfilled. Then*

(i) Let $d = 2$ and let conditions $(6.2.48)_{1-3}$, $(6.3.10)_1$ *and* (6.6.1) *with $j = 0$ and $\varsigma = 1$ be fulfilled. Then estimate* (6.6.3) *holds.*

(ii) Let $d = 3$ and

$$\text{(6.6.4)}\qquad \mathcal{B}_0(x', \xi_1, \mu^{-2}, \mu^{-1}h, h^2) \ge \zeta^2 \ge h^2 \qquad \forall (x', \xi_1) \in \Omega.$$

Then the estimate

$$\text{(6.6.5)}\quad |F_{t\to h^{-1}\tau}\chi_T(t)U| \le C'T\mu h^{-1}\Big(\frac{h}{\zeta}\Big)^s \qquad \forall x, y, \in B(0, \frac{1}{2}) \quad \forall T \ge h^{1-\delta}\zeta^{-2} \quad \forall \tau : \tau \le \epsilon\zeta^2.$$

holds. In particular, for $\zeta \ge h^{\frac{1}{2}-\delta}$ estimate (6.6.3) *holds.*

By standard Tauberian methods, propositions 6.6.1 and 6.6.2 imply the following two statements:

Proposition 6.6.3. *Let $d = 2$ and let conditions* (6.1.1)–(6.1.4), (6.2.1), $(6.2.48)_{1-3}$, (6.4.1), $(6.3.10)_1$ *and* (6.6.1) *be fulfilled. Then*

(i) For $j = k$ the following estimate holds:

$$\text{(6.6.6)}\quad |\varepsilon_{Q,jk}(\tau_2) - \varepsilon_{Q,jk}(\tau_1)| \le C'h^s(1 + \min(|\tau_1|, |\tau_2|))^{-s} \qquad \forall \tau_1, \tau_2 : \varsigma\tau_1 \le \epsilon\zeta^2, \varsigma\tau_2 \le \epsilon\zeta^2 \quad \forall x_1, y_1 \in \mathbb{R}$$

where

$$\varepsilon_{Q,jk}(x_1,y_1,\tau)=\iint Q_x\mathcal{F}_x^*e(x,y,-\infty,\tau)^t\mathcal{F}_y^*\Upsilon_j(x_2)\Upsilon_k(y_2)dx_2dy_2 \tag{6.6.7}$$

and the functions Υ_j are introduced in the beginning of section 6.4.

(ii) Therefore estimate (6.6.6) holds for $j\neq k$ as well.

(iii) Therefore making a $\bar\gamma$-admissible partition of unity in Ω with $\gamma=\mu^{-\frac12}h^{\frac12-\delta}$ and applying a similar theorem with x and y permuted we obtain that estimate (6.6.6) holds for $j\neq k$ even if condition (6.6.1) is violated but $\mu^{\frac32}h^{\frac12+\delta}\ge\bar\nu$. In particular, this estimate holds for $\mu\ge h^{-\frac13-\delta}$.

(iv) Therefore if condition (6.6.1) with $\varsigma=1$ is fulfilled then the estimate

$$|\varepsilon_{Q,jk}(x_1,y_1,\tau)|\le C''h^s(1+|\tau|)^{-s}\qquad\forall\tau\le\epsilon\zeta^2\quad\forall x_1,y_1\in\mathbb{R}\quad\forall k \tag{6.6.8}$$

holds.

(v) On the other hand, if condition (6.6.1) with $\varsigma=-1$ is fulfilled then estimate

$$|\varepsilon_{Q,jk}(x_1,y_1,0)-\int\phi(\tau)d_\tau\varepsilon_{Q,jk}(\tau)|\le C'h^s\qquad\forall x_1,y_1\in\mathbb{R}\quad\forall k \tag{6.6.9}$$

holds where ϕ is an arbitrary fixed function, equal to 0 for $\tau\le0$ and equal to 1 for $\tau\ge2c$.

(vi) Therefore if

$$|(2j+1)\mu h+B(x_1,\xi_1,\mu^{-2},\mu^{-1}h,h^2)|\ge\zeta^2\ge\bar\nu\mu^{-2}h^{2-\delta}\qquad\forall(x_1,\xi_1)\in\Omega\quad\forall j\in\mathbb{Z}^+ \tag{6.6.10}$$

then the estimate

$$|\int\psi(x)\Big(e(x,x,-\infty,0)-\int_{-\infty}^{\infty}h^{-1}\phi(\tau)\big(F_{t\to h^{-1}\tau}\chi_T(t)U^J(x,y,t)\big)d\tau\Big)dx|\le C'h^s \tag{6.6.11}$$

holds for $T\in[h^{1-\delta},1]$ where

$$U^J=\mathcal{F}_x\sum_{j,k\le J-1}\mathcal{U}_{jk}(x_1,y_1,t)\Upsilon_j(x_2)\Upsilon_k(y_2)^t\mathcal{F}_y, \tag{6.6.12}$$

$$J=\max\{j\in\mathbb{Z}^+:(2j-1)\mu h+B<0\qquad\forall(x_1,\xi_1)\}; \tag{6.6.13}$$

in particular, $J=0$ if and only if condition (6.6.1) with $\varsigma=1$ and $j=0$ is fulfilled.

Proposition 6.6.4. *(i) Under the assumptions of proposition 6.6.2(i) the following estimate holds:*

$$|e(x,y,-\infty,\tau)|\le C'h^s(1+|\tau|)^{-s}\qquad\forall x,y\in B(0,\frac12)\quad\forall\tau\le\epsilon\zeta^2. \tag{6.6.14}$$

(ii) Under the assumptions of proposition 6.6.2(iii) the following estimate holds:

$$|e(x,y,-\infty,\tau)|\le C'\mu h^{-1}(\frac{h}{\zeta})^s(1+|\tau|)^{-s}\qquad\forall x,y\in B(0,\frac12)\quad\forall\tau\le\epsilon\zeta^2. \tag{6.6.15}$$

In particular, for $\zeta\ge h^{1-\delta}$ estimate (6.6.14) holds.

Remark 6.6.5. *(i) One can prove that assertion (iii) of proposition 6.6.3 remains true for* $d = 3$ *as well.*

(ii) The construction of section 6.2 yields that the conditions of proposition 6.6.2(iii) are fulfilled provided $d = 3$ *and*

$$V + \mu h F \geq \zeta^2 \geq C_0(\mu^{-1}h + h^2) \qquad \text{in } B(0,1).$$

In this way one can prove that the assertion of proposition 6.6.2(ii) remains true provided

$$V + \mu h F \geq \zeta^2 \geq C_0 h^2; \tag{6.6.16}$$

moreover, one can weaken this condition. However, we will use another method which is simpler (see remark 6.6.10(ii)).

(iii) Moreover, the same construction yields that if $d = 2$ *and conditions* $(6.2.48)_{1-3}$, $(6.3.10)_1$ *are fulfilled then the conditions of proposition 6.6.2(i) are fulfilled provided*

$$V + \mu h F + \frac{1}{4} g^{jk}\partial_j\partial_k\bigl(\frac{V}{F}\bigr)\mu^{-1}h \geq \zeta^2 \geq C_0\bar{\nu}h^2 \qquad \forall x \in B(0,1). \tag{6.6.17}$$

◊

6.6.2 Calculations in spectral gaps

Let us pass from the analysis of the lower ("underbottom") spectral gap to other gaps. So, let $d = 2$ and let us assume that conditions (6.4.1) and (6.6.10) are fulfilled and $J \geq 1$. In this case proposition 6.4.2(ii) yields that in decomposition (6.6.12) all the terms with $j \neq k$ are negligible. Thus, repeating the decomposition of section 6.4 in very rough form one can replace estimate (6.4.45) by

$$|\int \bigl(e(x,x,-\infty,0)\psi(x) - \\ \int \phi(\tau)\sum_{j<J}\sum_{m,p,l,n\leq N} \mu^{1-2(m+p)-l}h^{l-1}\bigl((2j+1)\mu h\bigr)^m \Theta^{(n)}_{j,mpln}(\tau)d\tau| \leq Ch^s$$

provided $\phi \in C_0^\infty(\mathbb{R}^+)$ and $\int_0^\infty \phi d\tau = 1$. Recall that the functions $\Theta_{\dots}$ are given by (6.4.29). Let us note that in this decomposition one can replace $\Theta_{\dots}$ by 1 in which case all terms with $n \geq 1$ disappear and we obtain

$$|\int \bigl(e(x,x,-\infty,0)\psi(x) - \\ \sum_{j<J}\sum_{m,p,l,\leq N} \mu^{1-2(m+p)-l}h^{l-1}\bigl((2j+1)\mu h\bigr)^m \int f_{mpl}\,dx_1 d\xi_1| \leq Ch^s.$$

Let us note that in this case the only source of terms with $l + m + p \geq 1$ is the transformation of ψ: we introduce the $\mu^{-1}h$-pseudo-differential operator $\mathcal{F}^*\psi\mathcal{F} = q(x, \mu^{-1}hD, \mu^{-2}, \mu^{-1}h)$ which we then decompose in powers of x_2, $\mu^{-1}hD_2$ and μ^{-2}, $\mu^{-1}h$. Therefore, f_{mpl} with $m + p + l \geq 1$ are linear combinations of $D^\alpha\psi \circ \bar{\Psi}$ with $1 \leq |\alpha| \leq \frac{1}{2}(p+l) + m$ and only terms with $p + l \in 2\mathbb{Z}^+$ arise. Therefore we obtain the estimate

$$|\int \bigl(e(x,x,-\infty,0)\psi(x) - \\ (2\pi)^{-1}\sum_{k+m+p\leq N} J^{k+1}\mu^{1-2m-k-p}h^{k+p-1}(\mathcal{L}_{kmp}\psi)F\sqrt{g}\bigr)dx| \leq C'h^s \tag{6.6.18}$$

where $N = N(s,\delta)$, J is defined by (6.6.13), $\mathcal{L}_{kmp}$ are $2(k+m+p)$-th order linear partial differential operators with coefficients depending on $D^\alpha g^{\cdot\cdot}$, $D^\alpha F$, $D^\alpha V$ with $|\alpha| \le 2(k+m+p)$ and $\mathcal{L}_{000} = I$, $\mathcal{L}_{kmp}1 = 0$ for $k+m+p>0$. In this deduction we have also used proposition 6.4.4 and formula (6.2.7). Moreover, proposition 6.4.4(v) yields that

$$|\mathcal{L}_{kmp}(F\sqrt{g})| \le C'\bar\nu \qquad \forall k,m,p: 1 \le k+m+p \le K. \tag{6.6.19}$$

Therefore we have proven

Theorem 6.6.6. *Let $d = 2$ and let conditions (6.1.1)–(6.1.4), $(6.2.48)_{1-3}$, $(6.3.10)_1$, (6.4.1) and (6.6.10) be fulfilled. Then estimates (6.6.18) and*

$$|e(x,y,\tau_1,\tau_2)| \le C'h^s \qquad \forall x,y \in B(0,\frac{1}{2}) \quad \forall \tau_1,\tau_2 \in [-\epsilon\zeta^2, \epsilon\zeta^2] \tag{6.6.20}$$

hold.

Theorem 6.6.7. *Let $d = 2$ and let conditions (6.1.1)–(6.1.4), $(6.2.48)_{1-3}$, $(6.3.10)_1$, (6.4.1) be fulfilled. Furthermore, let*

$$|\frac{V}{F} + (2j+1)\mu h| \ge \zeta^2 \ge C_0\mu^{-2}\bar\nu \qquad \forall x \in B(0,1). \tag{6.6.21}$$

Then estimates (6.6.18) and (6.6.20) hold with

$$J = \max\{j \in \mathbb{Z}^+ : V + (2j-1)\mu h F \le 0 \quad \forall x \in B(0,1)\} \tag{6.6.22}$$

and $J = 0$ if the right-hand set is empty.

Remark 6.6.8. *Under condition (6.6.21) one can apply the method of successive approximations to the operator which is reduced to canonical form modulo $O(\nu(|x_2|+|\xi_2|)+\mu^{-1}h)$ provided*

$$\zeta^2 \ge \nu\mu^{-1}h^{-\delta'}$$

with arbitrarily small δ'. Then one can take $\mathcal{F}$ and q which do not depend on μ^{-2} explicitly. Then calculations easily show that all the terms with $m \ge 1$ vanish in (6.6.18). I leave the details to the reader (see also the proof of proposition 7.7.8). However, one can easily see that condition (6.6.21) with $j \ge 1$ yields condition (6.3.64) with $\nu = (\mu h)^{1-\delta'}$ with arbitrarily small $\delta' > 0$, and this ν fulfills the above inequality. ◊

Let us note that if $|D^\alpha f| \le c\bar\nu \quad \forall\alpha : |\alpha| \le 2$ and $f \ge \zeta^2$ in $B(0,1)$ then $|\nabla f| \le c_1(\bar\nu f)^{\frac{1}{2}}$ and

$$f + tm^k\partial_k f \ge \zeta^2(1-\epsilon) - C_2t^2\bar\nu$$

where $C_2 = C_2(c,\epsilon)$, $\epsilon > 0$ is arbitrary and we assume that $|\mathbf{m}| \le c$; therefore the term $m^k\partial k(\frac{V}{F}) \circ \bar\Psi_0$ in ω_1 is not essential and therefore theorem 6.6.6 yields the following

Theorem 6.6.9. *Let $d = 2$ and let conditions (6.1.1)–(6.1.4), $(6.2.48)_{1-3}$, $(6.3.10)_1$ and (6.4.1) be fulfilled. Furthermore, let*

$$|\frac{V}{F} + (2j+1)\mu h + \mu^{-2}\omega_1| \ge \zeta^2 \ge C_0\bar\nu(\mu^{-4}+h^2) \qquad \forall x \in B(0,1) \quad \forall j \in \mathbb{Z}^+ \tag{6.6.23}$$

where now

(6.6.24) $$\omega_1 = \frac{1}{8}\kappa v^2 - \frac{1}{4} v\mathcal{L}v, \qquad v = \frac{V}{F};$$

let us recall that κ *and* $\mathcal{L}$ *are the scalar curvature and the Laplace-Beltrami operator in the metric* $\frac{1}{F}g^{jk}$.

Then estimates (6.6.18) *and* (6.6.20) *hold with*

(6.6.25) $$J = \max\{j \in \mathbb{Z}^+ : v + (2j-1)\mu h + \mu^{-2}\omega_1 \le 0 \quad \forall x \in B(0,1)\}$$

and $J = 0$ *if the right-hand set is empty.*

6.6.3 Remarks

Let us make a few final remarks.

Remark 6.6.10. *(i) If there is a spectral gap described by (6.6.23) with* $J \ge 1$ *and* $\bar\nu \ge \mu h$ *then the following inequality holds:*

(6.6.26) $$|D^\alpha v| \le C\bar\nu\mu^{-2} + C'\bar\nu(\frac{\mu h}{\bar\nu})^{1-\delta} \qquad \forall\alpha : |\alpha| \le K_0 \quad \forall x \in B(0, \frac{9}{10})$$

where $C' = C'(c,\delta)$ *and* $\delta > 0$ *is arbitrary and* C *depends only on* K_0 *and on a constant* c *for which the basic inequalities hold with* K *replaced by* $K_0 + 2$. *Therefore for* $\zeta^2 \ge C_0\bar\nu\mu^{-4} + C'\mu^{-1}h(\frac{\bar\nu}{\mu h})^\delta$ *in the definition of the spectral gap (6.6.23) one can set* $\omega_1 = \frac{1}{8}\kappa v^2$ *(conversely if condition (6.6.23) holds with* $\omega_1 = \frac{1}{8}\kappa v^2$ *then this condition also holds with* ω_1 *given by (6.6.24)).*

(ii) On the other hand, for $\zeta^2 \ge C_0\bar\nu(\mu^{-4} + \mu^{-1}h)$ *the last term in the left-hand expression of (6.6.17) can be dropped. Therefore for the restrictions on* ζ *indicated in (i) one can always set* $\omega_1 = \frac{1}{8}\kappa v^2$. ◊

Remark 6.6.11. *Direct calculation shows that* $\int(\mathcal{L}_{kmp}\psi)F\sqrt{g}dx$ *with* $p = 0$ *and* $k+m = 1$ *does not necessarily vanish.* ◊

Remark 6.6.12. *(i) If*

(6.6.27) $$\zeta \ge (\mu^{-1}h)^{\frac{4}{5}}\bar\nu^{-\frac{1}{2}}$$

then in the proof of lemma 6.6.1 we can replace the Gårding inequality with the Gårding-Melin inequality (see theorem 6.2 of [Hörmander 1]) by means of energy inequalities (see section 1.1); this leads us to the condition

(6.6.1)′ $$\varsigma\big((2j+1)\mu h + B(x_1,\xi_1,\mu^{-2},\mu^{-1}h,h^2)\big) + \frac{1}{2}\operatorname{Tr}^+(\operatorname{Hess}^\sigma B(x_1,\xi_1,\mu^{-2},0,0)) \ge \zeta^2 \ge C_0\bar\nu(\mu^{-2}h^{2-\delta} + \mu^{-\frac{8}{5}}h^{\frac{8}{5}})$$

instead of (6.6.1) where $\operatorname{Hess}^\sigma$ *is the fundamental matrix of* B *and* $\operatorname{Tr}^+ M$ *is the semi-sum of absolute values of the purely imaginary eigenvalues of the matrix* M. *Since for* $d = 2$, $\operatorname{Hess}^\sigma B$ *has only two eigenvalues* η *and* $-\eta$ *then*

$$\operatorname{Tr}^+ \operatorname{Hess}^\sigma B = (-\frac{1}{2}\operatorname{Tr}(\operatorname{Hess}^\sigma B)^2)_+^{\frac{1}{2}} = \frac{1}{\sqrt2}(\det \operatorname{Hess} B)_+^{\frac{1}{2}}.$$

Formula (6.2.7) yields that the latter expression is equal to $2\omega_3 \circ \bar\Psi_0 + O(\mu^2\bar\nu)$ *where*

(6.6.28) $$\omega_3 = \frac{1}{F\sqrt{8g}}(\det \operatorname{Hess} v)_+^{\frac{1}{2}}$$

and therefore (6.6.23) should here be replaced by

(6.6.29) For every $j \in \mathbb{Z}^+$ one of the following two conditions holds:

$$\pm(v + (2j+1)\mu h + \mu^{-2}\omega_1) + \mu^{-1}h\omega_3 \geq \zeta^2 \geq C_0(\mu^{-4} + h^2)$$

(in which case $\zeta^2 \geq C(\mu^{-2}h^{2-\delta} + \mu^{-\frac{8}{5}}h^{\frac{8}{5}})$ automatically) and condition (6.6.17) should be replaced here by

$$v + \mu h + \mu^{-1}h\omega_4 \geq \zeta^2 \geq C_0\bar{\nu}(h^2 + \mu^{-\frac{8}{5}}h^{\frac{8}{5}}) \tag{6.6.30}$$

where

$$\omega_4 = \frac{1}{4}g^{jk}F^{-1}\partial_j\partial_k v + \frac{1}{F\sqrt{8g}}(\det \operatorname{Hess} v)_+^{\frac{1}{2}}. \tag{6.6.31}$$

(ii) It is easy to see that for $d = 2,3$ and conditions (6.1.1)–(6.1.3) and (6.2.1) operator A can be written in the form

$$A = \mathcal{M}_1^*\mathcal{M}_1 + V + \mu hF + h^2V',$$

$$A = \mathcal{M}_2^*\mathcal{M}_2 + \mathcal{M}_1^*\mathcal{M}_1 + V + \mu hF + h^2V'$$

respectively with $\mathcal{M}_k = \sum_j \rho_{jk}P_j$ with complex-valued (real-valued for $k=2$) coefficients ρ_{kj} and with $|D^\alpha V'| \leq C'$ $\forall \alpha : |\alpha| \leq K' - 2$. This easily yields that the assertion of proposition 6.6.2(ii) holds for $d = 2,3$ provided conditions (6.1.1)–(6.1.3), (6.2.1) and (6.6.16) are fulfilled in $B(0,1) \cap \bar{X}$ even if $B(0,1) \cap \partial X \neq \emptyset$ but all the conditions of theorem 6.1.5 (excluding the restriction on τ in (6.1.19)) are fulfilled and the Dirichlet boundary condition is fulfilled on $B(0,1)\cap\partial X$. However, this is not true for the Neumann boundary condition. ◊

6.7 Certain Generalizations

In this section we discuss what happens to the results of the previous sections (sections 6.3–6.5) in the case in which one of the conditions "$v \leq -\epsilon$," "$F \geq \epsilon_0$" and "$B(0,1) \subset X$" is violated where $v = \frac{V}{F}$. We also discuss the Weylian estimates following from the non-Weylian estimates of sections 6.4 and 6.5.

6.7.1 Getting rid of the condition "$v \leq -\epsilon$"

Let us treat the case in which the condition $v \leq -\epsilon$ is violated. Recall that for $\vartheta = 0$ this condition was assumed to be fulfilled only for $\mu \leq h^{-1+\delta}$. Let us note that one can assume that $\nu = 1$.[18] In fact, if $\nu < 1$ then one of the following conditions is fulfilled:

(a) $v \geq \epsilon\nu$ in $B(0,1)$;
(b) $v \leq -\epsilon\nu$ in $B(0,1)$;
(c) $|v| \leq c\nu$ in $B(0,1)$.

In cases (a) and (b) dividing the operator A by $r = |v(0)|$ and introducing $h' = \frac{h}{\sqrt{r}}$, $\mu' = \frac{\mu}{\sqrt{r}}$, $\nu' = \frac{\nu}{r}$ we get into the situation of conditions

(a)′ $v' \geq \epsilon$,
(b)′ $v' \leq -\epsilon$

[18] This is important only for $d = 2$.

respectively. Moreover, theorem 6.1.5 yields estimate (6.1.19) for $\tau \le 0$ in case (a) and in cases (b), (c) for $\mu h \ge C_0 r$. Thus we can assume that $\mu h \le C_0 r$ and we should discuss the results of sections 6.4 and 6.5. It is easy to see that $\frac{\nu}{\mu h} = \frac{\nu'}{\mu' h'}$, $\frac{\nu'}{\mu'} = \frac{\nu\sqrt{r}}{\mu}$ and that the conditions $\mu \ge h^{-\delta}$, $\nu \ge C_0\mu^{-2}\bar{\nu}$ are also fulfilled for $h', \mu', \nu', \bar{\nu}' = \bar{\nu}$. Moreover, it is easy to see that the conditions of all our statements remain true after this reduction and that the estimates with h', μ', ν' instead of h, μ, ν are stronger than the original. Therefore in cases (a) and (b) the condition $v \le -\epsilon$ in the statements indicated above is not necessary.

Let us treat case (c). Dividing the operator A by $r = \nu$ and introducing $h' = \frac{h}{\sqrt{r}}$, $\mu' = \frac{\mu}{\sqrt{r}}$ and $\nu' = 1$ we get into the situation of the case $\nu' = 1$; recall that $\nu \ge C_0^{-1}\mu h$ and therefore $\mu' h' \le C_0$. It is easy to check that after this reduction all the conditions and estimates are also preserved.

So we can assume in what follows that $\nu = 1$. Let us note that a similar reduction is applicable to theorems 6.3.18, 6.3.18′ and 6.3.19 provided either $\nu \ge \mu h^{1-\delta'}$ or $|v(0)| \ge \mu h^{1-\delta'}$ with arbitrarily small $\delta' > 0$.

Proposition 6.7.1. *Theorem 6.4.14 and estimate* (6.4.61) *remain true without condition* (6.3.31). *For $\vartheta > 0$ estimate* (6.4.61) *holds.*

Proof. Recall that this condition was assumed to be fulfilled for $\mu \le h^{\delta-1}$ only and in this case we assumed that $\nu = 1$. So, without loss of generality one can assume that these conditions are fulfilled. Let us introduce the function

$$\rho = \max((|v| + |\nabla v|^2)^{\frac{1}{2}}, \bar{\rho}), \tag{6.7.1}$$

on $B(0,1)$ and let us consider a ρ-admissible partition of unity. Let us set $\bar{\rho} = \epsilon_0 \mu^{\frac{1}{2}} h^{\frac{1}{2}-\sigma}$ with arbitrarily small $\sigma > 0$ for $\vartheta = 0$ and $\sigma = 0$ for $\vartheta > 0$. Let us consider the contribution to $\mathcal{R}, \mathcal{R}_4, \mathcal{R}_6$ of every element of this partition.

(i) $\rho \ge \bar{\rho}$ and $|\nabla v| \ge \epsilon\rho$. Then after division by ρ^2 and dilatation we get into the situation of theorem 6.4.12 with $h' = h\rho^{-2}$, $\mu' = \mu$. Then the contribution to $\mathcal{R}_\vartheta$, $\mathcal{R}_{4\vartheta}$ of this element does not exceed $C\rho^{4-2\vartheta}h^{-2+\vartheta} \times \rho^{2\vartheta}$ and the total contribution of all elements of this type does not exceed $Ch^{-2+\vartheta}$.

(ii) $\rho \ge \bar{\rho}$ and $-v \ge \epsilon\rho^2$. Then after division by ρ^2 and dilatation we get into the situation of theorem 6.4.14 and estimate (6.4.61) with $h' = h\rho^{-2}$, $\mu' = \mu$. Moreover, $\mu' \le h'^{-1+\delta}$ for $\vartheta = 0$. Then the contribution of this element to the remainder does not exceed $Ch'^{-2+\vartheta}\rho^{2\vartheta}$ plus the contribution to $\mathcal{R}_{6\vartheta}$. One can easily see that the total contribution to $\mathcal{R}_{4\vartheta}$ does not exceed $Ch^{-2+\vartheta}$. Applying estimate (6.4.61) for the contribution of the element to $\mathcal{R}_{6\vartheta}$ we obtain $\min(\mu^{\frac{3}{2}+\vartheta}h^{-\frac{3}{2}+\vartheta}, \mu^{-\frac{3}{2}+\vartheta}h^{-\frac{5}{2}+\vartheta}r) \times r^3$ and therefore the total contribution of these elements does not exceed the right-hand expression in (6.4.61).

(iii) $\rho \le \bar{\rho}$. Then for $\vartheta > 0$ after division by $\bar{\rho}^2$ and dilatation we get into the situation of theorem 6.1.5. The contribution of the indicated element does not exceed $C'\mu^{-s}$ and the total contribution of all such elements does not exceed $C'\mu^{-s}(\mu h)^{-\frac{3}{2}}$. For $\vartheta = 0$ we get $\mu' \asymp h'^{-1+\delta}$ and then the contribution of this element to $\mathcal{R}$ does not exceed Ch'^{-2} and the total contribution does not exceed Ch^{-2}.

Summing all these estimates we obtain the estimate (6.4.61). □

The same arguments prove the following

Proposition 6.7.2. *Theorem 6.5.3 and propositions 6.5.8, 6.5.9 and 6.5.10 remain true without condition* (6.3.33).

Let us now treat the case $d = 2$.

Proposition 6.7.3. *Theorem 6.4.15 remains true without condition* (6.3.31).

Proof. Let us introduce a function ρ by means of (6.7.1) with $\bar{\rho} = \epsilon_0\sqrt{\mu h}$ and let us make a ρ-admissible partition of unity, division and dilatation. Let us treat various elements of this partition. Then

(i) Let $\rho \geq \bar{\rho}$ and $|\nabla v| \geq \epsilon\rho$. Then we get into the situation of theorem 6.4.12 and the contribution to $\mathcal{R}$ of this element does not exceed $C\rho^2\mu^{-1}h^{-1}$ and the total contribution to $\mathcal{R}$ of such elements does not exceed $C\mu^{-1}h^{-1}$.

(ii) Let $\rho \geq \bar{\rho}$ and $-v \geq \epsilon\rho^2$. Then we get into the situation of theorem 6.4.15. Let us treat various subvariants:

(ii)$_1$ Let $\mu^{\frac{3}{2}}h^{\frac{1}{2}} \leq C_0\rho$. Then we get into the situation of assertion (i) of theorem 6.4.15. Substituting $\mu' = \mu$, $h' = h\rho^{-2}$ and $\gamma' = \gamma\rho^{-1}$ in place of μ, h, γ in (6.4.64) we obtain that the contribution to $\mathcal{R}$ of this element does not exceed

$$C\rho^2\mu^{-1}h^{-1} + C'\mu^{-1}h^{-1}\int_{\{\gamma\geq C_0\mu^{-1}\rho\}} \bigl(\frac{\mu h}{\gamma^2}\bigr)^s\gamma^{-2}dx + C\mu h^{-1}\int_{\{\gamma\leq C_0\mu^{-1}\rho\}} dx$$

where we have taken into account that $dx' = \rho^{-2}dx$ and where here and in what follows we integrate only over one element of the covering if we discuss a separate element of the partition (we do not indicate this in what follows). Obviously the expression obtained does not decrease if the domains in the second and third terms are replaced by $\{\gamma \geq C_0\mu^{-1}\}$ and $\{\gamma \leq C_0\mu^{-1}\}$ respectively. Then the total contribution of such elements of the partition does not exceed the right-hand expression of (6.4.64).

(ii)$_2$ Let $\mu^{\frac{3}{2}}h^{\frac{1}{2}} \geq C_0\rho$. Then we get into the situation of assertion (ii) of theorem 6.4.14. According to (6.4.67) the contribution to $\mathcal{R}$ of such an element does not exceed

$$(6.7.2)\quad C\rho^2\mu^{-1}h^{-1} + C'\mu^{-1}h^{-1}\int_{\{\gamma\geq\sqrt{\mu h}\}} \bigl(\frac{\mu h}{\gamma^2}\bigr)^s\gamma^{-2}dx + C\mu^{-1}h^{-1}\int_{\{\sqrt{\mu h}\geq\gamma\geq C_1\mu^{-1}\rho,|v+(2j+1)\mu h|\leq\epsilon_2\mu h\}} \gamma^{-2}dx + C\mu h^{-1}\int_{\{\gamma\leq C_1\mu^{-1}\rho\}} dx.$$

If $\mu^{\frac{3}{2}}h^{\frac{1}{2}} \geq C_0$ here then expression (6.7.2) does not decrease if in the restrictions on γ from below in the third term and in the restrictions on γ from above in the fourth term we replace ρ by 1; then summing with respect to the partition of unity we obtain that the total contribution of all such terms does not exceed the right-hand expression of (6.4.67). On the other hand, if $\mu^{\frac{3}{2}}h^{\frac{1}{2}} \leq C_0$ then expression (6.7.2) does not decrease if we replace the domain of integration in the second term by $\{\gamma \geq C_0\mu^{-1}\}$ and we replace the integral over the remaining part of the original domain and two other integrals (in the third and fourth terms) by

$$C\mu h^{-1}\int_{\{\gamma\leq C_0\mu^{-1}\}} dx;$$

then the total contribution of all such terms does not exceed the right-hand expression of (6.4.64).

(iii) Finally, theorem 6.1.5 yields that the contribution of the zone $\{\rho \leq 2\bar{\rho}\}$ does not exceed $C'h'^s\bar{\rho}^{-2}$. Therefore assertion (i) of theorem 6.4.15 is proven (without condition (6.3.31)).

Let us now assume that condition (6.4.47) is fulfilled. Then without loss of generality we can assume that $v = \pm x_1^2 + w(x_2)$. Let us introduce the function $\rho'(x_2)$ by formula

(6.7.1) with v replaced by w. Then applying a $\frac{1}{2}|x_1|$-admissible partition of unity and referring to theorem 6.4.12 we obtain that the contribution to $\mathcal{R}$ of the zone $\{|x_1| \ge \frac{1}{2}C_0\rho'\}$ does not exceed $C\mu^{-1}h^{-1} + C'$. Let us consider a ρ'-admissible partition of unity in the zone $\{|x_1| \le C_0\rho'\}$. Then repeating the previous arguments and referring to theorem 6.4.15 we obtain that the contribution to $\mathcal{R}$ of the separate element does not exceed

$$C\rho'^2\mu^{-1}h^{-1} + C\mu^{\frac{3}{2}}h^{-\frac{1}{2}}\rho' \qquad \text{or} \qquad C'\rho'^2h^{-1}$$

respectively for $\rho' \ge \mu^{\frac{3}{2}}h^{\frac{1}{2}}$ or $\rho' \le \mu^{\frac{3}{2}}h^{\frac{1}{2}}$.

If $\mu^{\frac{3}{2}}h^{\frac{1}{2}} \le 1$ then elements of both types can arise and one can easily see that the total contribution does not exceed the right-hand expression of (6.4.65). If $\mu^{\frac{3}{2}}h^{\frac{1}{2}} \ge 1$ then only elements of the second type can arise and one can easily see that the total contribution does not exceed the right-hand expression of (6.4.68). So assertion (ii) of theorem 6.4.15 is proven (without condition (6.3.31)).

Let us now assume that condition (6.4.66) is fulfilled. Without loss of generality one can assume that $x = 0$ is the unique critical point of v. Let us set $r = \sqrt{-v(0)}$ if $v(0) \le -\epsilon_0\mu h$ and $r = \sqrt{\epsilon_0\mu h}$ otherwise. Then the previous arguments which refer to theorem 6.4.12 give that the contribution to $\mathcal{R}$ of the zone $\{|x| \ge \epsilon_0 r\}$ does not exceed $C\mu^{-1}h^{-1} + C'$. Moreover, if $v(0) \le -\epsilon_0\mu h$ then referring to theorem 6.4.12 yields that the contribution to $\mathcal{R}$ of the zone $\{|x| \le 2\epsilon_0 r\}$ does not exceed $C\mu^{-1}h^{-1}$ provided either condition $(6.4.66)_+$ is fulfilled or $\mu^3h \le r^2$. On the other hand, if condition $(6.4.66)_+$ is violated and $\mu^3h \ge r^2$ then this contribution does not exceed $Cr^2\mu^{-1}h^{-1}(|\log\mu^3hr^{-2}| + 1)$ again due to theorem 6.4.15; this expression only increases if we replace r by $C_0\min(\mu^{\frac{3}{2}}h^{\frac{1}{2}}, 1)$. We then obtain that the indicated contribution does not exceed $C\mu^{-1}h^{-1}$ for $\mu^3h \le 1$ and $C\mu^{-1}h^{-1}(|\log\mu^3h| + 1)$ for $\mu^3h \ge 1$. Moreover, if condition $(6.4.66)_+$ is violated, $\mu^3h \ge 1$ and $\hat\lambda = \min_j(|v(0) + (2j+1)\mu h| \ge \mu^{-2}$ then estimate (6.4.69) yields that the indicated contribution does not exceed $Cr^2\mu^{-1}h^{-1}(|\log\hat\lambda\mu^{-1}h^{-1}| + 1)$ and the last expression is maximal for $r = 1$.

Finally for $v(0) \ge -\epsilon_0\mu h$ we can apply theorem 6.1.5. In this case we obtain estimate $(6.4.57)_{(2)}$ provided either $\mu^3h \le 1$ or condition $(6.4.66)_+$ is fulfilled and estimate (6.4.69) otherwise. □

Proposition 6.7.4. *Theorem 6.4.16 remains true without condition* (6.3.31) *also.*

Proof. The arguments used in the proof of the previous proposition provide the estimate

$$\mathcal{R}_I \le C\mu^{-1}h^{-1} + C|\log\frac{\hat\lambda}{\mu h}|$$

which does not coincide with estimate (6.4.76) only if $\mu h \ge |\log\frac{\hat\lambda}{\mu h}|^{-1}$ $(\ge |\log h|^{-1})$. So for $\mu \le h^{\delta-1}$ this proposition is proven.

In order to prove it for $\mu \in [h^{\delta-1}, h^{-1}]$ let us recall that we had the estimate

$$\mathcal{R}_4 \le C\mu^{-1}h^{-1}$$

without condition (6.3.31) which appeared only when we replaced Riemann sums with integrals because we wanted to obtain a regular function G. However, if we apply the integral formula $(E.3)_1$ of first order of accuracy we have no need of this condition. On the other hand, for $\mu \ge h^{-\frac{1}{2}}$ we need only an integral formula of first order of accuracy and therefore for $\mu \le h^{-\frac{1}{2}}$ we obtained (6.4.76) without condition (6.3.31) from the beginning. □

We leave to the reader the proof of the following

Proposition 6.7.5. *Theorem 6.4.17 remains true without condition* (6.3.31).

Summing all the achieved results and taking into account remark 4.4.14 we obtain

Theorem 6.7.6. *Theorems 6.4.13–6.4.21, 6.5.5 and 6.5.15 and propositions 6.5.10–6.5.12 remain true without condition* (6.3.31) *(or* (6.3.33) *also).*

I leave it to the reader to start from theorem 6.3.19 and use (6.7.1) with $h^{\frac{1}{2}-\delta''}$ instead of $\sqrt{\mu h}$ and to prove

Theorem 6.7.7. *Remark 6.3.20 remains true without condition* (6.3.31) *(or* (6.3.33) *also).*

Moreover if either $d = 3$ or $d = 2$ and one of the conditions (6.4.63), (6.4.66), $(6.4.66)_+$ *is fulfilled then one can skip condition* (6.4.1).

6.7.2 Getting rid of the condition "$F \geq \epsilon$"

Let us now treat the case when condition (6.2.1) is violated. Then the restriction $\mu h \leq 1$ is no longer natural and assumed; however, cases $\mu h \leq 1$ and $\mu h \geq 1$ differ in the order of the principal part of the asymptotics. Our methods work properly, however, only under the condition

$$\mu \leq h^{-2}. \tag{6.7.3}$$

Obviously, the two-dimensional case should be more sensitive to this violation. Let us start with the case $d = 3$.

Theorem 6.7.8. *Let $d = 3$ and let conditions* (6.1.2)–(6.1.4) *be fulfilled. Then*

(i) Let $\vartheta = 0$, $\mu \leq h^{-1+\delta}$ and let

$$\sum_{1\leq j\leq 3}\ \sum_{\alpha:|\alpha|\leq m} |D^\alpha F^j| \geq \epsilon_0 \tag*{$(6.7.4)_m$}$$

with a constant $\epsilon_0 > 0$. Then the estimate

$$\mathcal{R} \leq Ch^{-2}\log^m \mu + C' \tag{6.7.5}$$

holds. The same estimate holds for $\mathcal{R}_4$.

(ii) For $0 < \vartheta \leq 1$, $\mu \leq h^{-2}$ the estimates

$$\mathcal{R}_{4\vartheta} \leq Ch^{-2+\vartheta}\mu^{\frac{\vartheta}{2}}, \tag{6.7.6}$$

$$\mathcal{R}_\vartheta \leq Ch^{-2+\vartheta}\mu^{\frac{\vartheta}{2}} + C\min\bigl(\mu^{\frac{3}{2}+\vartheta}h^{-\frac{3}{2}+\vartheta}, \mu^{\frac{2}{5}\vartheta}h^{-2+\frac{4}{5}\vartheta}\bigr) \tag{6.7.7}$$

hold.

(iii) In the general case for $\vartheta = 0$, $\mu \leq h^{-2}$ the estimate

$$\mathcal{R} \leq Ch^{-2}\mu^\delta \tag{6.7.8}$$

holds with arbitrarily small $\delta > 0$. The same estimate holds for $\mathcal{R}_4$.

Proof. Let us assume that under condition $(6.7.4)_{m-1}$ this theorem is proven. Let us introduce the function

$$\gamma(x)=\epsilon\sum_j\Bigl(\sum_{|\alpha|\le m-1}|D^\alpha F_j|^{\frac{m}{m-|\alpha|}}\Bigr)^{\frac{1}{m}}+\bar\gamma,\qquad \bar\gamma=\mu^{-\frac{1}{m+1}}+h. \tag{6.7.9}$$

Then $|\nabla\gamma|\le\frac12$ for sufficiently small constant ϵ. Let us introduce a γ-admissible partition and consider its separate element. After the standard dilatation we obtain that conditions (6.1.2)–(6.1.4) are fulfilled with $h'=h\gamma^{-1}$ and $\mu'=\mu\gamma^{m+1}$ instead of h,γ. Let us first consider the case $\mu\le h^{-2}$.

(a) Moreover, for $\gamma\ge 2\bar\gamma$ condition $(6.7.4)_{m-1}$ is fulfilled as well and so we can apply an induction assumption. Thus, if $0<\vartheta\le 1$ the contribution of this element to $\mathcal{R}_{4\vartheta}$ does not exceed $Ch^{-2+\vartheta}\gamma^{2-\vartheta}$, $Ch^{-2+\vartheta}\mu^{\frac{\vartheta}{2}}\gamma^{2-\vartheta+\frac{\vartheta}{2}(m+1)}$ for $m=1$, $m\ge2$ respectively. Moreover, the contribution of this element to $\mathcal{R}_{6\vartheta}$ does not exceed

$$C\min\bigl(\mu^{\frac32+\vartheta}h^{-\frac32+\vartheta}\gamma^{(\frac32+\vartheta)(m+1)+(-\frac32+\vartheta)},\\ \mu^{-\frac32+\vartheta}h^{-\frac52+\vartheta}\gamma^{(-\frac32+\vartheta)(m+1)+(-\frac52+\vartheta)}\bigr),$$

$$C\min\bigl(\mu^{\frac32+\vartheta}h^{-\frac32+\vartheta}\gamma^{(\frac32+\vartheta)(m+1)+(-\frac32+\vartheta)},\mu^{\frac25\vartheta}h^{-2+\frac45\vartheta}\gamma^{\frac25\vartheta(m+1)+(2-\frac45\vartheta)}\bigr)$$

for $m=1$, $m\ge2$ respectively. Therefore, the total contribution of such elements of the partition does not exceed these expressions multiplied by γ^{-3} and integrated over x. By condition $(6.7.4)_m$ in an appropriate coordinate system $\gamma\ge\epsilon|x_1|$ and therefore the total contribution of such elements to $\mathcal{R}_{4\vartheta}$ does not exceed $Ch^{-2+\vartheta}\mu^{\frac{\vartheta}{2}}$ while the total contribution to $\mathcal{R}_{6\vartheta}$ does not exceed

$$C\min(\mu^{\frac32+\vartheta}h^{-\frac32+\vartheta},\mu^{\frac25\vartheta}h^{-2+\frac45\vartheta}).$$

The same arguments apply to $\vartheta=0$ as well provided $\mu\le h^{-1+\delta}$. However, at each step of the induction there appears an additional factor $\log\mu$ in the total contribution to $\mathcal{R}_4$. Moreover, the same arguments are applicable for $h^{-1+\delta}\le\mu\le h^{-1}$ but in this case the remainder estimate is $O(h^{-2}\mu^{\delta})$ and this estimate does not change under the induction transition.

(b) Let us consider elements with $\gamma\le2\bar\gamma$. On these elements $\mu'=1$ and their contributions do not exceed $Ch^{-2+\vartheta}\bar\gamma^{2-\vartheta}$. Due to condition $(6.7.4)_m$ their number does not exceed $C\bar\gamma^{-2}$ and their total contribution to the remainder estimate does not exceed $Ch^{-2+\vartheta}\mu^{\frac{\vartheta}{2}}$ for $0\le\vartheta\le1$. Thus statement (i) is proven and (ii) is proven under the additional condition (6.7.4). In order to get rid of this condition one can apply the same arguments with large m without reference to condition $(6.7.4)_m$. Then keeping in mind that $\gamma\ge\mu^{\delta}$ with $\delta=\frac{1}{m+1}$ we obtain statements (ii) and (iii). □

Remark 6.7.9. *(i) One can easily see that for $\mu h\gg1$ the principal part of the asymptotics does not exceed $h^{-d}\operatorname{mes}\{x,\mu hF+V\le t\}$ with $t=\epsilon$ and is larger than the same expression with $t=-\epsilon$.*

(ii) Surely, one can improve the remainder estimate for (6.7.5) for $\vartheta=0$, $\mu\ge h^{-1+\delta}$ under an appropriate condition associated to (6.3.58) (as will be done for $d=2$; see condition (6.7.10)). However, for $d=3$ and degenerate $\mathbf{F}$ the function F is not necessarily smooth.

(iii) One can improve remainder estimate (6.7.5) and obtain the standard remainder estimate $O(h^{-2})$ for $\vartheta=0$, $\mu\le h^{-1+\delta}$ and fulfilled condition (6.7.3). First of all,

applying the same arguments as in theorem 4.4.6 one can see that the total contribution of the complement of the set

$$\mathcal{X} = \{x, \exists \mathbf{e}, \mathbf{f}, \bar{x} : \mu\gamma^{m+1} \geq 1, |D^\alpha(\mathbf{F} - \mathbf{f}\ell(x)^m)| \leq \gamma^{m-|\alpha|+\delta'} \quad \forall \alpha : |\alpha| \leq m, -V \geq \gamma^{\delta_1}\}$$

to $\mathcal{R}$ does not exceed Ch^{-2} where $\ell(x) = \langle \mathbf{e}, x - \bar{x}\rangle$, $\gamma(x) = |\ell(x)|$, $\delta' > 0$ is sufficiently small and $\delta_1 > 0$ is arbitrarily small. Let us consider the $\gamma^{1-\delta''}$-vicinity of the point $\bar{x}$. Without loss of generality one can assume that $g^{jk} = \delta_{jk}$ in $\bar{x}$ and therefore $|g^{jk} - \delta_{jk}| \leq \gamma^\delta$ in this vicinity. Then the condition $\langle \nabla, \mathbf{F}\sqrt{g}\rangle = 0$ yields that without loss of generality one can assume that $\langle \mathbf{e}, \mathbf{f}\rangle = 0$ and therefore in an appropriate coordinate system this "approximate" field is $(x_2^m, 0, 0)$. Considering magnetic lines of length $\gamma^{1-\delta'''}$ with sufficiently small $\delta''' > 0$ one can easily see that they are non-closed and that all the assumptions of theorem 6.5.5 are fulfilled with $T = \min(\gamma^{-\delta'''}, (\mu\gamma^{m+1})^{\delta'''})$. Then the contribution of the γ-vicinity of $x \in \mathcal{X}$ to the remainder estimate does not exceed $Ch^{-2}\gamma^2(\gamma^{\delta'''} + (\mu\gamma^{m+1})^{-\delta'''})$ and therefore the total contribution of $\mathcal{X}$ is $O(h^{-2})$ as well. I leave all the details to the reader. ◊

Let consider the two-dimensional case. Now the function F is smooth. Let us assume that condition (6.3.31) is fulfilled and let us replace condition (6.3.58) by

$$\sum_{1\leq|\alpha|\leq m} |D^\alpha \frac{F}{V}| \geq \epsilon_0. \tag{6.7.10$_m$}$$

Applying the same arguments as above one can easily obtain that under conditions (6.7.3), (6.3.31) and $(6.7.10)_1$ the estimate

$$\mathcal{R} \leq Ch^{-1} \tag{6.7.11}$$

holds. The same estimate holds for $\mathcal{R}_4$.

Theorem 6.7.10. *Let $d = 2$ and let conditions (6.1.2)–(6.1.4), (6.7.3) and $(6.7.10)_m$ be fulfilled. Then the estimate*

$$\mathcal{R}_4 \leq Ch^{-1} \tag{6.7.12}$$

holds.

Proof. Let us apply induction on m. First of all, without loss of generality one can assume that

$$\frac{F}{V} = \rho(x)P(x_1, x_2), \qquad P(x) = x_1^m + \sum_{2\leq k\leq m} \nu_k(x_2)x_1^{m-k}$$

where

$$|D^\alpha(\rho - \bar{\rho})| \leq \epsilon \quad \forall \alpha : |\alpha| \leq m, \qquad \bar{\rho} = \text{const}, \quad |\bar{\rho}| \geq \epsilon_0$$

with arbitrarily small $\epsilon > 0$. Let us introduce the function $\zeta(x_2) = \gamma'$ in the same manner as in the proof of theorem 4.4.6, and let $\gamma = \epsilon_1(|x_1| + \zeta(x_2)) + \bar{\gamma}$, $\bar{\gamma} = \mu^{-\frac{1}{m+1}}$. Let us consider points with $\gamma \geq C_0(\zeta(x_2) + \bar{\gamma})$ with a sufficiently large constant C_0. One can easily see that (after dilatation) in the γ-vicinity of such a point we can refer to theorem 6.4.12 and the contribution of this vicinity to $\mathcal{R}_4$ does not exceed $C(\frac{1}{\mu h\gamma^m} + 1)$ and the total contribution of such points does not exceed Ch^{-1}. On the other hand, let us consider points with $\gamma \ge c\bar\gamma$, $\gamma \le C_0(\zeta(x_2)+\bar\gamma)$; then either condition $(6.7.10)_{m-1}$ is fulfilled and one can refer to the induction assumption or $|P(x)| \ge \epsilon_2\gamma^m$ in the γ-vicinity and we can refer to theorem 6.4.10 with condition $(6.3.58)_m$. In any case, the contribution of this vicinity to $\mathcal{R}_4$ does not exceed $C\frac{\gamma}{h}$ and the total contribution of such points to $\mathcal{R}_4$ does not exceed Ch^{-1} as well. At points $\gamma \le c\bar\gamma$ after dilatation we can again refer to the results of section 4.5 and obtain that the total contribution of such points does not exceed $Ch^{-1}\bar\gamma^{-1}\operatorname{mes}\{x, \gamma(x) \le c\bar\gamma\} \le Ch^{-1}$ under condition $(6.7.10)_m$. □

Remark. *Due to the necessary reference to condition* $(6.3.58)_m$ *we cannot get rid of condition* $(6.7.10)_m$. ◊

6.7.3 Getting rid of the condition "$B(0,1) \subset X$"

Finally, let us treat the case in which $B(0,1)$ is not contained in X but $\partial X \cap B(0,1)$ is regular (i.e., condition (5.1.3) is fulfilled). Without loss of generality one can assume that X coincides with $\{x_1 > 0\}$ in the intersection with $B(0,1)$. Let us consider a γ-admissible partition of unity with $\gamma = \max(\frac{1}{2}x_1, \bar\gamma)$, $\bar\gamma = C\mu^{-1}$. Then if $\gamma \ge \bar\gamma$ then $B(x,\rho) \subset X$ and after dilatation conditions $(6.2.48)_{1-3}$ are fulfilled with $\bar\nu = x_1$.

Let us first consider the case $d = 3$ and let us assume that condition (6.3.64) is fulfilled. Then after dilatation the same condition is fulfilled with $\mu' = \mu x_1$, $h' = x_1^{-1}h$, $\nu' = \max(\nu x_1, \mu'^{-2})$. Moreover, if we check condition (6.3.64) only for α with $|\alpha| \ge 2$ (see remark 6.7.2) then one can take $\nu' = \max(\nu x_1^2, \mu'^{-2})$. Then estimates (6.4.43), $(6.4.44)_\vartheta$ with h', μ' and ν' instead of h, μ and ν are applicable to the contribution to $\mathcal{R}$ of the separate element with $x_1 \ge \bar\gamma$. Thus, the contribution to $\mathcal{R}$ and $\mathcal{R}_{4\vartheta}$ of this element does not exceed

$$C\bigl(h^{-2} + \mu h^{-1}\bigl(\frac{h}{\sqrt{\nu}}\bigr)^{-\delta}\bigr)x_1^2$$

and $Ch^{-2+\vartheta}x_1^{2-\vartheta}$ respectively. Then the total contributions to $\mathcal{R}$, $\mathcal{R}_{4\vartheta}$ of the zone $\{x_1 \ge \bar\gamma\}$ do not exceed

$$C\bigl(h^{-2} + \mu h^{-1}\bigl(\frac{h}{\sqrt{\nu}}\bigr)^{-\delta}\bigr)\log\mu$$

and $Ch^{-2+\vartheta}\mu^{\vartheta}$ respectively.

On the other hand, after dilatation we can apply the standard asymptotics of section 4.5 in the zone $\{x_1 \le \bar\gamma\}$ (assuming that the regular boundary condition is given) and therefore the contribution of the separate element to $\mathcal{R}_\vartheta$ with $0 \le \vartheta \le 1$ does not exceed $Ch^{-2+\vartheta}\mu^{\vartheta}$. The transition from $\mathcal{R}_{4\vartheta}$ to $\mathcal{R}_\vartheta$ can be made by the arguments of the end of section 6.4. So, we have proven

Theorem 6.7.12. *Let* $d = 3$, $\tau = 0$ *and let conditions* (6.1.1)–(6.1.3), (6.2.1), (6.3.64) *and* (6.4.32) *with* $\bar\nu = 1$ *and* (5.1.3) *be fulfilled. Suppose that a regular boundary condition*[19] *is given on* $\partial X \cap B(0,1)$. *Then the estimates*

$$\mathcal{R} \le C(h^{-2} + \mu h^{-1}(\frac{h}{\sqrt{\nu}})^{-\delta})\log\mu \tag{6.7.13}$$

and

$$\mathcal{R}_\vartheta \le Ch^{-2+\vartheta}\mu^{\vartheta} \tag{6.7.14}$$

[19] E.g., Dirichlet or Neumann boundary condition.

hold for $\vartheta = 0$ *and* $\vartheta > 0$ *respectively. In particular, for* $\mu \le h^{-1+\delta}$ *or* $\nu \le h(h\mu)^{-n}$ *the estimate*

$$\mathcal{R} \le Ch^{-2}\log\mu \tag{6.7.15}$$

holds.

In order to get rid of the logarithm for $\vartheta = 0$ we need to apply the results of section 6.5. Let us assume that condition (6.3.31) is fulfilled and let us consider the magnetic line passing from the point $\bar{x} \in B(0, \frac{9}{10})$ of length T' less then $\epsilon = \epsilon(c) > 0$. It lies in $B(0,1)$ and it is easy to see that at least one of its parts $\{\pm t \in [0, T']\}$ with $T' = 2\epsilon\bar{x}_1^{\frac{1}{2}}$ lies in the zone $\{x_1 > \frac{1}{2}\bar{x}_1\}$. After dilatation we obtain a time interval of length $T = \epsilon\bar{x}_1^{-\frac{1}{2}}$. In order to apply the results of section 6.5 it is necessary to fulfill conditions (6.5.24) and (6.5.26). The first of is fulfilled automatically and the second is fulfilled provided $T \le \epsilon\bar{x}_1^{-\frac{1}{3}}$. Finally, we need to fulfill the condition $T \le \mu'^{\delta'}$ with a small enough exponent $\delta' > 0$. All the other conditions (excluding the conditions of subsection 6.5.4) are fulfilled automatically because the trajectory does not leave $B(0,1)$. Then the contribution to $\mathcal{R}$ of the separate element of partition with $x_1 \ge \bar{\gamma}$ does not exceed

$$C\Big(h^{-2} + \mu h^{-1}(\frac{h}{\sqrt{\nu}})^{-\delta}\Big)(x_1^{\delta'} + (\mu x_1)^{-\delta'})x_1^2 + C\mu h^{-1}x_1^2.$$

Then the total contribution of this zone does not exceed

$$C\Big(h^{-2} + \mu h^{-1}(\frac{h}{\sqrt{\nu}})^{-\delta}\Big) + C\mu h^{-1}\log\mu.$$

Finally, the arguments of subsection 6.7.2 permit us to get rid of condition (6.3.31). So we have proven

Theorem 6.7.12. *Under the assumptions of theorem 6.7.11, estimate* (6.4.43) *holds with the additional term* $C\mu h^{-1}\log\mu$ *in the right-hand expression. In particular, for* $\mu \le h^{-1+\delta}$ *estimate* (6.4.44) *holds.*

So we see that the presence of the boundary in the three-dimensional case does not yield an essential deterioration of our estimates (in contrast to the two-dimensional case; see below). We can even get rid of the term $C\mu h^{-1}(\frac{h}{\sqrt{\nu}})^{-\delta}$ and improve the term $C\mu h^{-1}\log\mu$. First of all, let us note that condition (6.3.58) does not survive under our dilatation but transfers to a condition of (4.4.69)–(4.4.70)-type with $\nu_k = \nu\gamma^k$. According to remark 4.4.14 one can apply the better remainder estimate (6.4.44) instead of (6.4.43) after dilatation (which means the above estimates with $\delta = 0$). Thus we obtain

Theorem 6.7.13. *Suppose the conditions of theorem 6.7.11 and condition* (6.3.58) *are fulfilled. Then estimate* (6.4.44) *holds with the additional term* $C\mu h^{-1}\log\mu$ *in the right-hand expression.*

In order to improve the term $C\mu h^{-1}\log\mu$ let us check the condition of non-reversal of the movement with $j = \bar{\jmath}(x)$. This condition is fulfilled in a time interval of length $T = \varepsilon|v + (2j+1)\mu h|(\bar{\nu}\bar{x}_1)^{-1}$ where v is calculated in the corresponding element of the covering. Moreover, in what follows one can replace $T < 1$ by $T = 1$ because the estimate with this T is guaranteed in any case. We assume that $|\nabla v| \le c\bar{\nu}$ in $B(0,1)$ because we need to check condition (6.3.64) only for $|\alpha| \ge 2$. Therefore the extra term is equal to $C\mu' h'^{-1} \cdot \min(T^{-1}, 1)$ (all the other terms which appear because of other restriction on T from above are covered by other terms in the right-hand expression). Moreover, in the formula for T one can calculate v at the point $(0, x')$ (it is easy to see that the ratio of T_{new} and T_{old} is disjoint from 0 and infinity; recall that $T_. \geq 1$). As a result, after integration with respect to the measure $x_1^{-3}dx$ over the domain $\{x_1 \geq C_0\mu^{-1}\}$, we obtain the additional term

$$R_8 = C\mu h^{-1} \int_{\partial X \cap B(0,1)} |\log \zeta| dx' \tag{6.7.16}$$

with $\zeta = \min_j \bigl(|v(0,x')+(2j+1)\mu h|+\mu^{-2}\bigr)$. So one can replace the extra term $C\mu h^{-1}\log\mu$ by $\mathcal{R}_8$. We used condition (6.3.31) but we can get rid of it by applying standard arguments. So we have proven

Theorem 6.7.14. *All the assertions of theorems 6.7.11 and 6.7.13 with $\mu \geq h^{-1}|\log h|^{-1}$ remain true with the term $C\mu h^{-1}\log\mu$ replaced by R_8 in the right-hand expressions of all the estimates.*

Remark 6.7.15. *(i) Assuming that*

$$\sum_{\alpha:1\leq|\alpha|\leq m} |D_Y^\alpha v_Y| \geq \epsilon_0$$

where the subscript Y means that v is restricted to Y and all the derivatives are taken along Y, one can easily see that $R_8 \leq C\mu h^{-1}$.

(ii) If we assume that $\partial_1 v \geq 0$ on $\{x_1 = 0\}$ then in the definition of T one can replace $\bar{\nu}x_1$ by $\bar{x}_1^{1+\delta'}$ with small enough $\delta' > 0$ and then the corresponding extra term does not exceed Ch^{-2}. So if $\mathbf{n}$ is an interior normal to $Y = \partial X$ then under the condition $\langle \mathbf{n}, \nabla v\rangle \geq 0$ on Y estimate (6.4.44) remains true with no extra term (we get rid of condition (6.3.31) using the standard arguments). Other modifications are also possible; in particular, one can replace the condition $\langle \mathbf{n}, \nabla v\rangle \geq 0$ by the condition $\langle \mathbf{n}, \nabla v\rangle \geq -c\zeta^\delta$ with $\delta > 0$, etc. ◊

Let $d = 2$. Then if the condition (6.3.44) is fulfilled then theorem 6.4.14 with $\nu' = \bar{\nu}' = x_1$, $\mu' = \mu x_1 \geq 1$, $h' = hx_1^{-1} \leq 1$ yields that the contribution to $\mathcal{R}_{4\vartheta}$ (with $0 \leq \vartheta \leq 1$) of the separate element is $O(\mu'^{-1-\vartheta}h'^{-1+\vartheta}\nu'^{1+\vartheta}) = O(\mu^{-1-\vartheta}h^{-1+\vartheta}x_1^{1-\vartheta})$ for $\mu'h' \geq \nu'$ (i.e., for $x_1 \geq \mu h$) and is $O(\mu'^{-\vartheta}h'^{\vartheta}\nu'^{\vartheta}) = O(\mu^{-\vartheta}h^{\vartheta}x_1^{-\vartheta})$ otherwise.

After multiplication by x_1^{-2} and integration we obtain that the total contribution of these zones does not exceed $\mu^{-1-2\vartheta}h^{-1}$ (with an additional factor $|\log\mu h|$ for $\vartheta = 0$) and μh^ϑ respectively.

On the other hand, theorem 4.4.12 yields that the contributions to $\mathcal{R}^w$ and $\mathcal{R}$ of the zone $\{x_1 \leq 2\mu^{-1}\}$ does not exceed $(\mu h)^{-1+\vartheta}\cdot\mu = Ch^{-1+\vartheta}\mu^\vartheta$. One can easily prove that $\mathcal{R}_6$ does not exceed this quantity as well.

Therefore we have proven the following

Theorem 6.7.16. *Let $d = 2$ and let conditions* (6.1.1)–(6.1.3), (6.2.1), (6.3.44) *and* (5.1.3) *be fulfilled. Then for $0 \leq \vartheta \leq 1$ the estimate*

$$\mathcal{R}_\vartheta \leq Ch^{-1+\vartheta}\mu^\vartheta \tag{6.7.17}$$

holds.

Remark 6.7.17. *(i) Thus, for $d = 2$ the most essential contribution to the remainder estimate is given by the near-the-boundary strip.*

(ii) One can generalize this theorem, replacing condition (6.3.44) by a weaker one; for example, by (6.3.66).

(iii) One can combine the results of this subsection and the previous subsections. I leave all this work to the reader. ◊

Appendix E. On Two Integral Formulas

Let us define numbers A_k and B_k with $k \in \mathbb{Z}^+$ such that

$$\text{(E.1)}_1 \qquad \sum_{0\le k\le m-1} A_k \frac{(2m+1)!}{(2m-2k-1)!} = 1,$$

$$\text{(E.1)}_2 \qquad \sum_{0\le k\le m-1} B_k \frac{(2m+1)!}{(2m-2k-1)!} = -2m;$$

in particular, $A_0 = \frac{1}{6}$, $B_0 = -\frac{1}{3}$.

Then for $m \in \mathbb{Z}^+$ the formulas

$$\text{(E.2)}_1 \qquad I(c-h, c+h) \sim 2hf(c) + \sum_{0\le k\le m-1} A_k h^{2k+2} f^{(2k+1)}(x)|_{x=c-h}^{x=c+h},$$

$$\text{(E.2)}_2 \qquad I(c-h, c+h) \sim h(f(c-h)+f(c+h)) + \sum_{0\le k\le m-1} B_k h^{2k+2} f^{(2k+1)}(x)|_{x=c-h}^{x=c+h}$$

with $I(a,b) = \int_a^b f(x)dx$ are exact provided $f(x)$ is a polynomial of degree not exceeding $2m+1$. This is easy to prove for $c = 0$ by first observing that these formulas are exact for all odd functions; then we translate. Applying the Taylor formula we therefore obtain that the error $\mathcal{R}'_m$ of formula $(E.2)_j$ (i.e., the absolute value of the difference between its left- and right-hand expressions) does not exceed

$$C_m h^{2m+1+i} \max_{c-h\le x\le c+h} |f^{(2m+i)}(x)|$$

with $i = 0, 1, 2$.

Let us treat the segment $[a,b]$ of length $L = b-a$; let us divide it into N subsegments of length $2h = \frac{L}{N}$. Let us consider the integral formulas

$$\text{(E.3)}_1 \qquad I(c-h, c+h) \sim \sum_{0\le j\le N-1} f(a+(2j+1)h)\cdot 2h + \sum_{0\le k\le m-1} A_k h^{2k+2} f^{(2k+1)}(x)|_{x=a}^{x=b},$$

$$\text{(E.3)}_2 \qquad I(c-h, c+h) \sim (f(a)+f(b))h + \sum_{1\le j\le N-1} f(a+2jh)\cdot 2h + \sum_{0\le k\le m-1} B_k h^{2k+2} f^{(2k+1)}(x)|_{x=a}^{x=b}$$

which follow from $(\text{E.2})_{1,2}$ by the substitution $c = a+(2j+1)h$ and summation with respect to $j = 0, \ldots, N-1$.

Then the previous statements imply that these formulas are exact on all polynomials of degree not exceeding $2m+1$ and their errors satisfy the estimates

$$\text{(E.4)} \qquad \mathcal{R}_m \le C_m L h^{2m+i+1} \sum_{0\le j\le N-1} \max_{a+2jh\le x\le a+2(j+1)h} |f^{(2m+i)}(x)| \le C_m L h^{2m+i} \max_{a\le x\le b} |f^{(2m+i)}(x)|$$

with $i = 0, 1, 2$.

In particular, for $m = 0$ we have formulas of second order accuracy.

Appendix F. Trajectories of the Classical Particles in a Magnetic Field

Let us treat the movement of a classical particle of mass $m = \frac{1}{2}$ and energy τ in various situations. We assume that $g^{jk} = \delta_{jk}$.

F.1. $V = 0, \mathbf{F} = \text{const}$

Let there be no electric field (i.e., $V = 0$) and let the magnetic field be constant: $F_{12} = \mu$ (and $F_{13} = F_{23} = 0$ in the three-dimensional case). Then the equations

$$\frac{dp_1}{dt} = -2\mu p_2 - \partial_{x_1} V, \quad \frac{dp_2}{dt} = 2\mu p_1 - \partial_{x_2} V, \quad \frac{dp_3}{dt} = -\partial_{x_3} V,$$
$$\frac{dx_j}{dt} = 2p_j \qquad (j = 1, 2, 3)$$

yield that a two-dimensional particle moves along the circle

$$x_1 = \bar{x}_1 + \frac{a}{\mu}\cos 2\mu t, \quad x_2 = \bar{x}_2 + \frac{a}{\mu}\sin 2\mu t$$

with $a^2 = \tau$ where here and in what follows we replace t by $t - t_0$ if necessary. The corresponding drift line degenerates into a point.

A three-dimensional particle moves along the spiral

$$x_1 = \bar{x}_1 + \frac{a}{\mu}\cos 2\mu t, \quad x_2 = \bar{x}_2 + \frac{a}{\mu}\sin 2\mu t, \quad x_3 = \bar{x}_3 + 2bt$$

with $a^2 + b^2 = \tau$. The magnetic lines are parallel to the x_3-axis.

F.2. $\nabla V = 0, \mathbf{F} = \text{const}$

Let us assume that electric field is constant (and the magnetic field as the same as before until otherwise indicated). Then for $d = 2$ one can assume without loss of generality that $V = Ex_2$ where E is the scalar intensity of the electric field; then the particle moves along the (epi-, hypo-) cycloid

$$x_1 = \bar{x}_1 + \frac{a}{\mu}\cos 2\mu t + \frac{E}{\mu}t, \quad x_2 = \bar{x}_2 + \frac{a}{\mu}\sin 2\mu t$$

with $a^2 + \frac{1}{4\mu^2}E^2 + E\bar{x}_2 = \tau$. So the average speed of the particle is equal to $\mu^{-1}E$ and is orthogonal to electric field. The corresponding drift lines are parallel to the x_1-axis.

For $d = 3$ one can assume without loss of generality that $V = E_1x_1 + E_3x_3$ in which case

$$x_1 = \bar{x}_1 + \frac{a}{\mu}\cos 2\mu t, \quad x_2 = \bar{x}_2 + \frac{a}{\mu}\sin 2\mu t - \frac{E_1}{\mu}t, \quad x_3 = \bar{x}_3 + 2bt - E_3t^2$$

with $a^2 + \frac{1}{4\mu^2}E_1^2 + E_1\bar{x}_1 + E_3\bar{x}_3 + \frac{1}{4}b^2 = \tau$. The corresponding magnetic lines are parabolas (for $E_3 \neq 0$) or straight lines (for $E_3 = 0$).

F.3. $V = \frac{1}{4}k^2|x|^2$, F = const

Let $V = \frac{1}{4}k^2|x|^2$ with $k > 0$. Then the two-dimensional particle moves along the cycloidal curve

$$\begin{cases} x_1 = \dfrac{A_1}{\omega_1}\cos\omega_1 t + \dfrac{A_2}{\omega_2}\cos(\omega_2 t + \varphi), \\ x_2 = \dfrac{A_1}{\omega_1}\sin\omega_1 t + \dfrac{A_2}{\omega_2}\sin(\omega_2 t + \varphi) \end{cases}$$

with $\omega_{1,2} = (k^2 + \mu^2)^{\frac{1}{2}} \mp \mu$ and

$$A_1^2(1 + \frac{k^2}{\omega_1^2}) + A_2^2(1 + \frac{k^2}{\omega_2^2}) = 4\tau.$$

It is obvious that for $k \in (0, 1]$ and $\mu \gg 1$ we obtain $\omega_1 \asymp \frac{k^2}{\mu}$ and $\omega_2 \asymp \mu$. The corresponding drift lines are circles.

In the three-dimensional case one should to add the third equation

$$x_3 = A_3\cos(kt + \psi)$$

and the energy equation is now

$$A_1^2(1 + \frac{k^2}{\omega_1^2}) + A_2^2(1 + \frac{k^2}{\omega_2^2}) + k^2 A_3^2 = 4\tau.$$

The corresponding magnetic lines are parallel to the x_3-axis.

Let us note that in case F.2 the trajectories are non-periodic and in F.3 the generic trajectories are non-periodic; less generic trajectories (with commensurable ω_1 and ω_2 for $d = 2$ and with commensurable ω_1, ω_2 and k for $d = 3$) are long-periodic (with period $\geq \epsilon\frac{\mu}{k^2}$). On the other hand, in F.1 the trajectories are short-periodic (with period $2\frac{\tau}{\mu}$).

F.4. $V = \frac{1}{4}k^2x_3^2$, F = const

One can easily prove that in this case generic trajectories are non-periodic and less generic trajectories (with commensurable k and μ) are periodic (with period $\geq \epsilon\max(\frac{1}{k}, \frac{1}{\mu})$).

Let us now treat only the two-dimensional case.

F.5. Linear F

Let $V = 0$ and suppose F is linear: $F = \mu x_1$. Then the movement equations can be reduced to

$$\frac{d^2}{dt^2}x_1 = 2\mu x_1(\xi_2 - \frac{1}{2}x_1^2\mu), \quad \frac{d^2}{dt^2}x_2 = 2\xi_2 - x_1^2\mu, \quad \xi_2 = \text{const}$$

with

$$4\tau = (\frac{dx_1}{dt})^2 + (\xi_2 - \frac{1}{2}\mu x_1^2)^2.$$

The first equation describes the movement of the one-dimensional particle with the potential

$$v(x_1) = (\xi_2 - \frac{1}{2}\mu x_1^2)^2$$

and with energy τ.

For $\xi_2 > 0$ there are two potential wells $x_1 = \pm\sqrt{\frac{2\xi_2}{\mu}}$ and for $\xi_2 < 0$ there is one potential well $x_1 = 0$. So a generic particle oscillates along the x_1-axis and moves along the x_2-axis. It is easy to see that a generic particle starting in the zone $\{|x_1| \le \frac{\epsilon_0}{\sqrt{\mu}}\}$ has period of oscillation (with respect to x_1) $\asymp \frac{1}{\sqrt{\mu}}$ during which its shift along x_2 is $\asymp \frac{1}{\sqrt{\mu}}$, i.e., its average velocity is $\asymp 1$.

F.6. Movement in the Half-Plane. I

Let us treat a particle with constant magnetic field and with no electric field in the half-plane $X = \{x_2 \ge 1\}$; we assume that the particle reflects from ∂X according to the law of geometrical optics. Then trajectories intersecting the boundary consist of segments of circles with radius $R = \sqrt{\frac{\tau}{\mu}}$ and with centers at points $(., R\sin\varphi)$ with constant R and $\varphi \in (-\frac{\pi}{2}, \frac{\pi}{2})$. The time between two consequent reflections is equal to $\frac{1}{\mu}(\frac{\pi}{2} + \varphi)$ and the shift along x_1 is equal to $\frac{2}{\mu}\sqrt{\tau}\cos\varphi$. Thus the average velocity is equal to $2\sqrt{\tau}\cos\varphi(\frac{\pi}{2} + \varphi)^{-1} \asymp 1$ for $\tau \asymp 1$, $\varphi \le \frac{\pi}{2} - \varepsilon$ with $\varepsilon > 0$. This velocity tends to $2\sqrt{\tau}$ as $\varphi \to -\frac{\pi}{2}$.

F.7. Movement in the Half-Plane. II

Trajectories close to the boundary have approximately the same appearance if we add the potential $V = Ex_2$ but segments of circles will be replaced by segments of (hypo-, hyper-)cycloids. Moreover, for $E < 0$ (i.e., field is directed from the boundary) there is no periodic trajectory at all. So if we consider only trajectories passing from $B(0, \frac{2}{3})$ at $t = 0$ then during a time interval of length $T = \epsilon_0\mu$ with $\epsilon_0 > 0$ trajectories lying in the zone $\{x_2 > \frac{C_0}{\mu}\}$ do not leave $B(0,1)$ and are not periodic. The same is true with $T = \epsilon_0$ for trajectories lying in $\{x_2 < \frac{C_0}{\mu}\}$. But then the arguments and results of section 4.4 yield the remainder estimate $\mathcal{R}^w \le C\mu^{-1}h^{-1}$ for $\mu \in [1, h^{-\delta}]$ with small enough $\delta > 0$ which improves the estimate $\mathcal{R}^w \le Ch^{-1}$ obtained in subsection 6.7.3.

Similarly, in case F.5 the same arguments yield the remainder estimate $\mathcal{R}^w \le C\mu^{-\delta}h^{-1}$ for $\mu \in [1, h^{-\delta}]$ which improves the estimate $\mathcal{R}^w \le Ch^{-1}$ obtained in subsection 6.7.3.

Comments

The results obtained in this chapter were announced in a few notes and in my lecture at the International Congress of Mathematicians (Berkeley, 1986) but in less precise and general form. I know no other paper devoted to LSSA for the Schrödinger (or Dirac) operator with strong magnetic field. The papers of Y. Colin de Verdiere [2,3] and H. Tamura [1,2] devoted to eigenvalue asymptotics will be discussed in Chapter 11; these papers, implicitly containing approximate formulas for $e(x, x, 0)$, were very stimulating for me. Moreover, there are papers of A. Sobolev and G. Raikov also devoted to eigenvalue asymptotics of another type but they will also be discussed in Chapter 11 and even in Chapter 12. In section 12.5 I will return to the analysis of this chapter. Surely I know many papers devoted to the self-adjointness of the Schrödinger operator but they will also be discussed in the comments to Chapter 11. Finally, I would like to mention the paper [Helffer, Sjöstrand 1] in which a reduction similar to those made in section 6.2 in the analytical situation was made independently.

There are many open problems. Some of them are

Problem 6.1****.** *Obtain non-trivial asymptotics without spatial mollification.* ◊

Problem 6.2****.** *Obtain some results in dimensions* $d \geq 4$*. In this situation a complete reduction similar to those of section 6.2 seems to be difficult unless some unpleasant restrictions excluding resonances are assumed. However, an incomplete reduction may be sufficient?* ◊

Problem 6.3****.** *Improve the results of section 6.7 for the cases described in Appendix F and generalize those results.* ◊

7. Local Spectral Asymptotics for the Dirac Operator with Strong Magnetic Field

In this chapter we analyze two- and three-dimensional Dirac operators with strong magnetic field. The results and structure of this chapter are similar to those of the previous chapter: sections 7.1–7.4 and sections 7.6–7.8 correspond to sections 6.1–6.4 and sections 6.5–6.7 respectively. However, there appears a new section 7.5 devoted to the case of a very strong magnetic field. Namely, the condition $\mu h \le \text{const}$ is not natural for the Dirac operator and there appears a new case $\mu h \ge C_0$ which is treated in section 7.5. In this case the only contribution to the principal part of the asymptotics and the main contribution to the remainder are given by $j = 0$. Moreover, the principal part is $\asymp \mu h^{-1}$ and $\asymp \mu h^{-2}$ for $d = 2$ and $d = 3$ respectively.[1)] Furthermore, under a fairly strong non-degeneracy condition the remainder estimate is $O(1)$ for $d = 2$ while under a weak non-degeneracy condition it is $O(\mu^\delta h^{-\delta})$ or $O(\mu h^{-1})$ for $d = 2$ and $d = 3$ respectively. In the general case the remainder estimate is $O(\mu h^{-1})$ and $O(\mu h^{-\delta} \nu^{\frac{1}{2}\delta})$ respectively. Moreover, in the case $d = 2$ we now do not assume that $h \ll 1$; this condition is replaced by the condition $h \ll \mu$. This is reasonable from the technical point of view: after reduction for $d = 2$ our semiclassical parameter is $\mu^{-1}h$. This generalization will be useful in Chapter 11.

7.1 Preliminary Analysis

7.1.1 The model operator

In this chapter we consider the operator

$$A = \frac{1}{2}\sigma_l(\omega^{jl}P_j + P_j\omega^{jl}) + \sigma_0 m + I \cdot V, \quad P_j = hD_j - \mu V_j \tag{7.1.1}$$

where we use the Einstein summation convention, ω^{jl}, V_j, V are real-valued, $m = \text{const} \ge 0$ and the following conditions are fulfilled in $B(0,1) \subset X$:

$$|D^\alpha \omega^{jk}| \le c, \quad |D^\alpha V_j| \le c, \forall \alpha : |\alpha| \le K$$
$$|D^\alpha V| \le c\min(1, \frac{1}{m}) \quad \forall \alpha : 1 \le |\alpha| \le K, \tag{7.1.2$_{1-3}$}$$

$$(\tau_1 - m - V)_- \le c\min(1, \frac{1}{m}),$$
$$(\tau_2 + m - V)_- \le c\min(1, \frac{1}{m}), \tag{7.1.2$_4$}$$

1) For $d = 3$ under condition (6.3.64), fulfilled for all $\alpha : |\alpha| \le K$ (in fact, here we need it only for $\alpha = 0$) and modified for the Dirac operator (that means that V is replaced by V^* defined below), the principal part is $\asymp \mu h^{-2}\nu^{\frac{1}{2}}$.

(7.1.3) $$g^{jk}\xi_j\xi_k \ge \epsilon_0|\xi|^2 \qquad \forall \xi \in \mathbb{R}^d, \qquad \det(\omega^{jk}) > 0$$

where $g^{jk} = \omega^{jl}\omega^{kr}\delta_{lr}$ (we have rewritten conditions $(4.4.6)'_{1-4}$ and (4.4.7)). Furthermore, we assume that the matrices $\sigma_0, \dots, \sigma_d$ satisfy conditions (C.1) and (C.2). Let us note that $e(x,x,\tau_1,\tau_2)$ does not change if we replace A by the operator $e^{-i\mu h^{-1}\phi}Ae^{i\mu h^{-1}\phi}$, i.e., if we replace V_j by $V_j + \partial_j\phi$. Therefore condition $(7.1.2)_2$ can be replaced by condition $(7.1.2)'_2 = (6.1.2)'_2$ where the tensor intensity of the magnetic field (F_{jk}) is given by (6.1.5).

Let us first treat the case when $X = \mathbb{R}^d$ and ω^{jl}, V and F_{jk} are constant.

Theorem 7.1.1. *Let $X = \mathbb{R}^d$, let ω^{jk}, V, F_{jk} be constant and let condition (7.1.3) be fulfilled. Then for $\tau_1 < \tau_2$*

(7.1.4) $$\operatorname{tr} e(x,x,\tau_1,\tau_2) = h^{-d}S(\tau_1,\tau_2,\mu h)$$

where

(i) If $r = \frac{1}{2}\operatorname{rank}(F_{jk}) < \frac{d}{2}$ then

$(7.1.5)_1$ $$S(\tau_1,\tau_2,\mu h) = \operatorname{sign}(\tau - V) \times \sum_{\alpha\in(\mathbb{Z}^+)^r} \Bigl((\tau-V)^2 - \sum_{1\le j\le r} 2\alpha_j f_j\mu h - m^2\Bigr)_+^{\frac{1}{2}(d-r)} \times \frac{D}{2}(2\pi)^{r-d}\omega_{d-2r}2^{-\#\{j,\alpha_j=0\}} f_1\dots f_r(\mu h)^r\sqrt{g}\Big|_{\tau=\tau_1}^{\tau=\tau_2}$$

where $f_j > 0$, $\pm if_j$ with $j = 1,\dots,r$ are all the eigenvalues of the tensor $(F_{kl}g^{ln})_{k,n}$ disjoint from 0 and multiplicities are taken into account.

(ii) If $r = \frac{d}{2}$ then for $\tau_1 < \tau_2$

$(7.1.5)_2$ $$S(\tau_1,\tau_2,\mu h) = \sum_{(\varkappa,\alpha)\in\mathbf{Z}:\tau_1\le H_{\varkappa\alpha}<\tau_2} c_{\varkappa\alpha}\frac{D}{2}(2\pi)^{-r}f_1\dots f_r(\mu h)^r\sqrt{g}$$

where

(7.1.6) $$H_{\varkappa\alpha} = V + \varkappa\bigl(m^2 + \sum_k 2\alpha_k f_k\mu h\bigr)^{\frac{1}{2}},$$

$c_{\varkappa\alpha} = 2^{-\#\{j,\alpha_j=0\}}$ for $\alpha \ne 0$ and if

(7.1.7) $$\sigma_0\sigma_1\dots\sigma_d = \varsigma i^{\frac{d}{2}}$$

with $\varsigma = \pm 1$ then

(7.1.8) $$\mathbf{Z} = \{-1,1\}\times(\mathbb{Z}^+)^r \setminus (\bar\varkappa, 0)$$

and $c_{-\bar\varkappa,0} = 2^{1-r}$ where $\bar\varkappa = \pm\varsigma$ if by a coordinate transformation with positive (negative respectively) determinant one can reduce the matrix $(F_{kl}g^{ln})$ to the canonical form

(7.1.9) $$\begin{aligned} &F_{kl}g^{ln} = +f_j \quad \text{for } k = 2j-1, n = 2j,\\ &F_{kl}g^{ln} = -f_j \quad \text{for } k = 2j, n = 2j-1,\\ &(j = 1,\dots,r) \quad \text{and} \quad F_{kl}g^{ln} = 0 \quad \text{otherwise.}\end{aligned}$$

In particular, for $r < \frac{d}{2}$ the spectrum of A is absolutely continuous and equals $(-\infty, V-m] \cup [V+m,\infty)$ and for $r = \frac{d}{2}$ the spectrum of A is pure point of infinite multiplicity and equals $\{H_{\varkappa\alpha}, (\varkappa,\alpha)\in\mathbf{Z}\}$ where $\mathbf{Z}$ has been introduced under condition (7.1.7) and $\mathbf{Z} = \{-1,1\}\times(\mathbb{Z}^+)^r$ if (7.1.7) holds with neither $\varsigma = 1$ nor $\varsigma = -1$.

Remark. *Appendix C yields that without loss of generality one can assume condition (7.1.7) provided d is even.* ◊

Proof. Proposition C.5 yields that without loss of generality one can assume that $g^{kn} = \delta_{kn}$, that F_{kn} has structure (7.1.9), and that $\mu = h = 1$, and therefore without loss of generality one can assume that $P_{2j} = D_{2j} + f_j x_{2j-1}$ for $j = 1, \dots, r$ and $P_k = D_k$ for $k \neq 2, 4, \dots, 2r$. Then the arguments of the proof of theorem 6.1.1 yield that for $\tau_1 < \tau_2$

$$\operatorname{tr} e(x, x, \tau_1, \tau_2) = (2\pi)^{r-d} f_1 \dots f_r \int \mathbf{n}(\tau_1, \tau_2, \xi'') d\xi''$$

where $\xi'' = (\xi_{2r+1}, \dots, \xi_d) \in \mathbb{R}^{d-2r}$, $\mathbf{n}(\tau_1, \tau_2, \xi'')$ is the number of eigenvalues of the operator

$$\mathcal{L}(\xi'') = \sum_{1 \leq k \leq r} \sqrt{f_j} (\sigma_{2k-1} D_k + \sigma_{2k} x_k) + \sum_{2r+1 \leq k \leq d} \sigma_k \xi_k + \sigma_0 m + V \cdot I$$

in $L^2(\mathbb{R}^r, \mathbb{C}^D)$ lying in $[\tau_1, \tau_2)$. Let us rewrite the parenthesis in the definition of $\mathcal{L}(\xi'')$ in the form

$$\frac{1}{2}(\sigma_{2k} - i\sigma_{2k-1}) Z_k + \frac{1}{2}(\sigma_{2k} + i\sigma_{2k-1}) Z_k^*$$

with $Z_k = x_k + \partial_k$, $Z_k^* = x_k - \partial_k$.

Let us construct a unitary operator $\mathcal{T} : L^2(\mathbb{R}^r, \mathbb{C}^D) \to \oplus \sum_\alpha \mathcal{H}_\alpha$ with $\alpha \in (\mathbb{Z}^+)^r$ in the following way: let us decompose $u(x') \in L^2(\mathbb{R}^r, \mathbb{C}^D)$ into the series $u(x') = \sum_\alpha \Upsilon_\alpha(x') w_\alpha$ where $\Upsilon_\alpha(x') = \upsilon_{\alpha_1}(x_1) \dots \upsilon_{\alpha_r}(x_r)$, $\upsilon.(.)$ are real-valued orthonormal Hermite functions, $w_\alpha \in \mathbb{C}^D$; then using proposition C.4 let us set $w_\alpha = \oplus \sum_\varepsilon v_{\alpha,\varepsilon}$ with $\varepsilon \in \{-1, 1\}^r$, $w_{\alpha,\varepsilon} \in \bigcap_{1 \leq j \leq r} \operatorname{Ker}(\sigma_{2j} - i\varepsilon_j \sigma_{2j-1})$; finally let us set $u_\alpha = \oplus \sum_\varepsilon w_{\beta(\alpha,\varepsilon),\varepsilon}$ where $\beta_j(\alpha, \varepsilon) = \alpha_j + \frac{1}{2}(1 - \varepsilon_j)$. Then $\mathcal{T} u = (u_\alpha)_{\alpha \in (\mathbb{Z}^+)^r}$ defines the operator in question.

Then proposition C.4 yields that

$$\mathcal{H}_0 = \bigcap_{1 \leq j \leq r} \operatorname{Ker}(\sigma_{2j} - i\sigma_{2j-1}), \tag{7.1.10$_0$}$$

$$\dim \mathcal{H}_\alpha = 2^{-\#\{j, \alpha_j = 0\}} D \tag{7.1.10$_\alpha$}$$

for $\alpha \neq 0$; moreover, for $d > 2r$ the same is true for $\alpha = 0$. Furthermore, (6.4.22) yields that

$$\mathcal{T}\mathcal{L}\mathcal{T}^*|_{\mathcal{H}_\alpha} = \sum_{1 \leq k \leq r} \sqrt{2\alpha_k f_k} \sigma'_{k\alpha} + \sum_{2r+1 \leq k \leq d} \sigma_{k\alpha} \xi_k + \sigma_{0\alpha} + V \cdot I$$

where $\sigma_{0\alpha}, \sigma'_{1\alpha}, \dots, \sigma'_{r\alpha}, \sigma_{(2r+1)\alpha}, \dots, \sigma_{d\alpha}$ transform $\mathcal{H}_\alpha$ into itself and form a system of Dirac matrices. Then the eigenvalue counting function of the restriction of the operator $\mathcal{T} L(\xi'') \mathcal{T}^*$ to $\mathcal{H}_\alpha$ is equal to

$$\frac{1}{2} \dim \mathcal{H}_\alpha \operatorname{sign}(\tau - V) \theta\Big((\tau - V)^2 - \sum_{1 \leq j \leq r} 2\alpha_j f_j - m^2 - |\xi''|^2\Big)\Big|_{\tau=\tau_1}^{\tau=\tau_2}$$

provided either $d > 2r$ or $d = 2r$, $\alpha \neq 0$; moreover, if $d = 2r$, $\alpha = 0$ and condition (7.1.7) is fulfilled then this function is equal to $\dim \mathcal{H}_0$ for $\tau_1 \leq H_{-\zeta,0} < \tau_2$ and vanishes otherwise.

This immediately yields $(7.1.5)_{1,2}$. □

Remark 7.1.2. *Let us note that under the assumptions of theorem 7.1.1 for* $V=0$

$$A^2 = g^{jk}P_jP_k + m^2 + \frac{1}{4i}\mu h\omega^{jl}\omega^{ks}F_{jk}[\sigma_k,\sigma_s]. \tag{7.1.11}$$

Then corollary C.7 yields that multiplying by $q^\dagger$ *and* q *on the left and right respectively with an admissible unitary matrix* q *we transform* A^2 *into a direct sum of the Schrödinger operators* B_ε *(where* ε *runs over* $\{-1,1\}^r$*) with potentials* $V_\varepsilon = m^2 + \sum_j \varepsilon_j f_j \mu h$ *and with multiplicity* $2^{-r}D$ *and therefore theorem 6.1.1 is applicable to the operator* A^2.

On the other hand, let $2r < d$. *Then one can assume without loss of generality that* $g^{jk} = \delta_{jk}$, $F_{jd} = 0$. *Then multiplying* A *by* σ_d *on the left and on the right and replacing* x_d *by* $x'_d = -x_d$ *we obtain that for* $V=0$ *the operator* $-A$ *and therefore* $\operatorname{tr} e(x,x,0,\tau) = \operatorname{tr} e(x,x,-\tau,0) = \frac{1}{2}\operatorname{tr} e_S(x,x,-\infty,\tau^2)$ *for* $\tau > 0$ *where* $e_S(.,.,.,.)$ *is the Schwartz kernel for the Schrödinger operator. These arguments easily yield* $(7.1.5)_1$. *Similar arguments yield* $(7.1.5)_1$ *for* τ_1 *and* τ_2 *not in the spectrum of* A *provided* $d = 2r$ *and there exists a unitary matrix* q *such that* $q^\dagger\sigma_j q = -\sigma_j$ *for* $j = 0,\dots,d$. ◊

Remark 7.1.3. *For odd d one can treat the "principally massless" Dirac operator*

$$A = \frac{1}{2}\sigma_l(\omega^{jl}P_j + P_j\omega^{jl}) + I\cdot V \tag{7.1.1$'$}$$

where $\sigma_1,\dots,\sigma_d$ *are Dirac matrices and there exists no admissible matrix* σ_0. *Then the proof of theorem 7.1.1 yields that formula* $(7.1.5)_1$ *also remains true in this case.*

Let us note that the connection between formulas obtained here and Weylian formulas is the same as for the Schrödinger operator (see section 6.1) but the approximate formula $(E.2)_1$ *is replaced by* $(E.2)_2$ *and for* $d = 2r$ *the term*

$$\frac{1}{2}\varsigma D\Bigl(\theta(\tau - V + m) - \theta(\tau - V - m)\Bigr)\Big|_{\tau=\tau_1}^{\tau=\tau_2} \times (4\pi)^{-r}(\mu h)^r f_1\dots f_r\sqrt{g} \tag{7.1.12}$$

should be subtracted from $S(\tau_1,\tau_2,\mu h)$.

Moreover, one can replace ςD *by* $i^{-r}\operatorname{tr}(\sigma_0\dots\sigma_d)$ *and then one can get rid of condition (7.1.7). Then for* $d \ge 4$ *term (7.1.12) is* $O(h^2)$ *for* $\mu = 1$ *while for* $d = 2$ *that term is connected with formula (4.4.34) for the coefficient* κ_1 *in the spectral asymptotic decomposition.*

7.1.2 Rough estimates

Let us pass from calculations to analytical arguments as we did in section 6.1. Let us first treat the main spectral gap $(-m+\max V, m+\min V)$ which is empty if $\max V - \min V \ge 2m$; theorem 7.1.1 yields that it can be extended in only one direction and only in the case $d = 2r$.

Theorem 7.1.4. *Let conditions* (7.1.1)–(7.1.3) *and* (6.1.4) *be fulfilled and let* $\tau_1 = -m + \max_{x\in B(0,1)} V$ *and* $\tau_2 = m + \min_{x\in B(0,1)} V$. *Let us assume that* $\tau_2 - \tau_1 \ge 2\epsilon(1+\mu h)(m+1)^{-1}$ *with* $\epsilon > 0$. *Then the following estimates hold:*

$$|F_{t\to h^{-1}\tau}\chi_{T,T'}(t)U(x,y,t)| \le C'Th^s\mu^{-s}\bigl(\min(\tau_2-\tau,\tau-\tau_1)(m+1)+1\bigr)^{-s}, \tag{7.1.13}$$

$$|e(x,y,\tau_1',\tau_2')| \le C'h^s\mu^{-s}\bigl(\min(\tau_2-\tau_2',\tau_1'-\tau_1)(m+1)+1\bigr)^{-s} \tag{7.1.14}$$
$$\forall x,y\in B(0,1-\epsilon)\ \forall \tau,\tau_1',\tau_2' \in [\tau_1+\epsilon(1+\mu h)(m+1)^{-1}, \tau_2-\epsilon(1+\mu h)(m+1)^{-1}]$$

where $T \ge h^{1-\delta}$.

Proof. The proof of the first estimate is similar to the proof of lemma 4.5.10, and the second follows from the first and standard Tauberian arguments. □

Theorem 6.1.7 should be replaced by the exact equality

$$U(x,y,t)=0 \qquad \forall x,y\in B(0,1-C_0T):|x-y|\geq C_0T \tag{7.1.15}$$

which is due to the standard uniqueness theorem for symmetric hyperbolic systems.

Finally, theorem 6.1.8 should be replaced by

Theorem 7.1.5. *Let conditions* (7.1.1)–(7.1.3) *and* (6.1.4) *be fulfilled. Let* $q\in\tilde{S}^0_{h,\rho,1,K}$ *with* $\rho\geq\max(1,\mu h,\mu^{\frac12}h^{\frac12-\delta})$ *be supported in* $\{|\xi|\geq C_0\rho\}$, *let* $Q=\operatorname{Op}\bigl(q(x,h\xi_1-\mu V_1(x),\dots,h\xi_d-\mu V_d(x))\bigr)$ *and let* $\psi\in C_0^K(B(0,1-\epsilon))$ *with* $\epsilon>0$. *Then the estimate*

$$|F_{t\to h^{-1}\tau}\chi_{T,T'}(t)Q\psi U|\leq C'Th^s\mu^{-s} \qquad \forall x,y\in B(0,1-\epsilon)\quad \forall\tau:|\tau|\leq m+c(m+1)^{-1} \tag{7.1.16}$$

holds with the same $C_0,C',K,\delta,s,T\geq h^{1-\delta}$ *as in theorem 6.1.8.*

Proof. This proof repeats the proof of theorem 6.1.8, and for $m\geq1$ we apply the reduction made in the proof of lemma 4.4.10. □

7.2 Microlocal Canonical Form

In this section we reduce operators (7.1.1) and (7.1.1)′ to microlocal canonical form assuming that $d=2,3$ and conditions (7.1.1)–(7.1.3) and

$$F\geq\epsilon_0 \qquad \text{in } B(0,1) \tag{7.2.1}$$

are fulfilled. Let us recall that $B(0,1)\subset X$. Then Appendix C yields that we need to treat the following three cases:

(i) $d=2$, $D=2$;
(ii) $d=3$, $D=4$;
(iii) $d=3$, $D=2$ and the operator is of the form (7.7.1)′.

7.2.1 Principal symbol reduction

Then one can assume that in cases (i) and (iii) $\sigma_1,\sigma_2,\sigma_0$ and $\sigma_1,\sigma_2,\sigma_3=\sigma_0$ respectively are Pauli matrices (see Appendix C) while in case (ii) the matrices $\sigma_0,\dots,\sigma_3$ are given by (C.7) with α_j denoted as σ'_j now. We will construct a reducing operator commuting with σ_0 in cases (i) and (ii). Moreover, in case (ii) this operator will commute with σ_4 where in the context of (ii), (C.7) the following auxiliary matrices are used:

$$\sigma_4=\begin{pmatrix}0&I\\I&0\end{pmatrix},\qquad \sigma_5=\begin{pmatrix}0&iI\\-iI&0\end{pmatrix},\qquad \hat\sigma_j=\sigma_4\sigma_j=\begin{pmatrix}\sigma'_j&0\\0&\sigma'_j\end{pmatrix}\quad(j=1,2,3). \tag{7.2.2}$$

But then case (ii) is reduced to (iii) at *this* step of the analysis. So, $D=2$ and we treat operators (7.1.1) with $d=2$ and (7.1.1)′ with $d=3$. Let us start with the case $m=V=0$, i.e., let us treat the operator

(7.2.3) $$A^0 = \frac{1}{2}\sigma_l(\omega^{jl}P_j + P_j\omega^{jl}).$$

Then the $\mu^{-1}h$-pseudo-differential operator $\mu^{-1}A^0$ has principal symbol

(7.2.4) $$a^0(x,\xi) = \sigma_l\omega^{jl}(\xi_j - V_j)$$

in which case

(7.2.5) $$\bigl(a^0(x,\xi)\bigr)^2 = g(x,\xi)I, \qquad g(x,\xi) = g^{jk}(\xi_j - V_j)(\xi_k - V_k).$$

We use the notation of section 6.2 in this section.

Proposition 7.2.1. *Let conditions* (6.1.4), (7.1.1)–(7.1.3) *and* (7.2.1) *be fulfilled with* $K = K(N)$. *Let* Ψ *be the symplectomorphism constructed in proposition 6.2.1(i) (or (iii)) for* $d = 2$ *and in proposition 6.2.1(ii) and remark 6.2.2(ii) for* $d = 3$ *with* N *replaced by* $N+2$. *Then there exists a smooth unitary matrix* $T = T(x,\xi)$ *in the neighborhood of* Λ^0 *such that*

(i) For $d = 2$

(7.2.6) $$T = \exp(iw\sigma_0),$$

(7.2.7) $$T^\dagger(a^0 \circ \Psi)T = (\sigma_1\xi_2 + \sigma_2 x_2)\Bigl(\sum_{0\le n\le N} B_n(x_1,\xi_1)(x_2^2 + \xi_2^2)^n\Bigr) + \sum_{1\le j\le 2}\sigma_j R_j, \qquad R_j \in \mathcal{O}_{2N+3}(\Lambda^0) \quad (j = 1,2)$$

where here and below w, w_k, B_n, R_j *are real-valued scalar symbols.*

(ii) For $d = 3$

(7.2.8) $$T = \exp(iw^j\sigma_j),$$

(7.2.9) $$T^\dagger(a^0 \circ \Psi)T = (\sigma_1\xi_3 + \sigma_2 x_3 + \sigma_3\xi_2)\Bigl(\sum_{0\le n\le N} B_n(x',\xi_1)(x_2^2 + \xi_2^2)^n\Bigr) + \sum_{1\le j\le 3}\sigma_j R_j, \qquad R_j \in \mathcal{O}_{2N+3}(\Lambda^0)\ (j = 1,2,3).$$

Moreover, in the both cases

(7.2.10) $$B_0 = F^{\frac{1}{2}} \circ \bar{\Psi}$$

and under the assumptions of proposition 6.2.1(iii)

(7.2.11) $$B_1 = \frac{1}{8}F^{\frac{1}{2}}\kappa \circ \bar{\Psi}.$$

Proof. Let Ψ be the indicated symplectomorphism. Then

(7.2.12) $$a^0 \circ \Psi = \sum_j \sigma_j U_j \qquad (j = 1,\dots,d)$$

where $U_j \in \mathcal{O}_1(\Lambda^0)$, the linear hulls of the sets $\{dU_j,\ j = 1,\dots,d\}$ and $\{du_j,\ j = 1,\dots,d\}$ at points of Λ^0 coincide,

(7.2.13)
$$\sum_j U_j^2 \equiv \sum_j u_j^2\varphi^2 \pmod{O_{2N+3}(\Lambda^0)},$$
$$\varphi = \Big(\sum_{1\le n\le N+1} b_n(x_d^2+\xi_d^2)^{n-1}\Big)^{\frac12},$$

$u_1=\xi_2$, $u_2=x_2$ for $d=2$ and $u_1=\xi_3$, $u_2=x_3$, $u_3=\pm\xi_2$ for $d=3$ and the sign of ξ_2 here will be fixed later.

Let us prove that in a neighborhood of Λ^0 there exists a smooth matrix $\rho=(\rho^{jk})\in \mathrm{SO}(d)$ [2] such that

(7.2.14)$_n$
$$u_j \equiv \rho^{jk}U_k\varphi^{-1} \qquad \pmod{O_n(\Lambda^0)}$$

with $n=2N+2$. Let us apply induction on n. Equality (7.2.13) yields that for $n=2$, (7.2.14)$_n$ obviously holds with appropriate $\rho\in \mathrm{O}(d)$. Moreover, if $d=2$ then $\rho\in\mathrm{SO}(2)$ necessarily; this follows from the positiveness of $F\cdot\det(\omega^{jk})$ and analysis of the Poisson brackets $\{U_1,U_2\}$ and $\{u_1,u_2\}$. For $d=3$ one can obtain $\rho\in\mathrm{SO}(3)$ choosing the sign at ξ_2.

Let us assume that (7.2.14)$_n$ holds, $2\le n\le 2N+1$. Then
$$u_j\equiv \rho^{jk}U_k\varphi^{-1}+v_j \qquad \pmod{O_{n+1}(\Lambda^0)}$$
where $v_j\in O_n(\Lambda^0)$ and (7.2.13) yields that

(7.2.15)
$$\sum_j u_jv_j\equiv 0 \qquad \pmod{O_{n+2}(\Lambda^0)}.$$

Let $\phi\in O_{n-1}(\Lambda^0)$ be a real-valued skew-symmetric matrix. Then $\mathrm{SO}(d)\ni \exp\phi\equiv I+\phi \pmod{O_{n+1}(\Lambda^0)}$ and (7.2.14)$_{n+1}$ holds with ρ replaced by $(\exp\phi)\rho$ provided

(7.2.16)
$$\phi^{jk}u_k\equiv -v_j \qquad \pmod{O_{n+1}(\Lambda^0)}.$$

Therefore the possibility of making the induction step is due to the following

Lemma 7.2.2. *Let $d=2,3$ and $v_j\in O_n(\Lambda^0)$ satisfy* (7.2.15) *with $n\ge 2$. Then there exists a real-valued skew-symmetric matrix $(\phi^{jk})\in O_{n-1}(\Lambda^0)$ such that* (7.2.16) *holds.*

This lemma will be proven later; let us first complete the proof of proposition 7.2.1. Thus, $U_k\equiv\varphi\rho^{jk}u_j \pmod{O_{2N+2}(\Lambda^0)}$. Substituting into (7.2.12) we obtain that

(7.2.17)
$$a^0\Psi\equiv\sum_{jk}\varphi\rho^{jk}\sigma_k u_j.$$

Let us set $\bar\sigma_j=\rho^{jk}\sigma_k$ $(j=1,\dots,d)$. Moreover, for $d=2$ let us introduce $\bar\sigma_3=\sigma_3$ $(=\sigma_0)$. Then $\bar\sigma_1,\bar\sigma_2,\bar\sigma_3$ are Dirac matrices and $\bar\sigma_1\bar\sigma_2\bar\sigma_3=\sigma_1\sigma_2\sigma_3=\varsigma iI$ and hence there exists a unitary unimodular matrix $T=T(\rho)$ such that $T^\dagger\bar\sigma_jT=\sigma_j$ $(j=1,2,3)$. This matrix depends analytically on ρ, is defined uniquely modulo a factor ±1 [3] and it can be chosen as a one-valued smooth function of (x,ξ) in a neighborhood of Λ^0 because Ω is simply connected. Every unitary unimodular 2×2 matrix is of the form (7.2.8), and one can take smooth real-valued functions w^j. For $d=2$, the matrix T commutes with σ_3 (because $\bar\sigma_3=\sigma_3$) and hence it is of the form (7.2.6). Equalities (7.2.7) and (7.2.9) follow from (7.2.17) and the definition of T if B_n are chosen such that

[2] Here and in what follows $\mathrm{O}(d)$ is the group of real orthogonal $d\times d$ matrices and $\mathrm{SO}(d)$ is its subgroup of matrices with determinant equal 1.

[3] The map $\mathrm{SO}(3)\ni\rho\to T(\rho)$ is called a spinor representation.

$$\sum_{0\le n\le N} B_n(x_d^2+\xi_d^2)^n \equiv \varphi^{\frac12} \qquad (\operatorname{mod} O_{2N+1}(\Lambda^0)).$$

Then $B_0=\sqrt{b_1}$, $B_1=\frac{b_2}{2\sqrt{b_1}}$ and (7.2.10), (7.2.11) follow from (6.2.6), (6.2.9). □

Proof of lemma 7.2.2. Instead of $O_n(\Lambda^0)$ one can use $\mathcal{P}_n$, spaces of n-th degree homogeneous polynomials of $u=(u_1,\dots,u_d)$ with coefficients depending smoothly on the remaining coordinates in which case one should replace modulo equalities by exact equalities.

(i) If $d=2$ then $\phi=\begin{pmatrix}0 & \psi\\ -\psi & 0\end{pmatrix}$ and (7.2.16) means exactly that $v_1=\psi u_2$, $v_2=-\psi u_1$; one can fulfill these equalities with an appropriate $\psi\in\mathcal{P}_{n-1}$ because (7.2.15) yields that $u_1v_1=-u_2v_2$.

(ii) Let $d=3$; then every real skew-symmetric matrix is given by a vector product: $\phi v=\psi\times v\ \forall v$ with $\psi\in\mathcal{P}^3_{n-1}=(\mathcal{P}_{n-1})^3$. In this case (7.2.15) and (7.2.16) mean that $\langle u,v\rangle=0$ and $\psi\times u=-v$ respectively.

Now let $\mathcal{P}_n$ be the spaces of n-th degree homogeneous polynomials of $u=(u_1,\dots,u_d)$ with constant coefficients. Let us consider the maps $\Phi_s:\mathcal{P}^3_s\ni v\to u\times v\in\mathcal{P}^3_{s+1}$, $\mathcal{L}_s:\mathcal{P}^3_s\ni v\to\langle u,v\rangle\in\mathcal{P}_{s+1}$ and $\mathcal{L}'_s:\mathcal{P}_s\ni w\to wu\in\mathcal{P}^3_{s+1}$. Obviously, $\operatorname{Ran}\Phi_s\subset\operatorname{Ker}\mathcal{L}_{s+1}$ and the lemma will be proven if we prove that $\operatorname{Ran}\Phi_s=\operatorname{Ker}\mathcal{L}_{s+1}$. In order to prove this equality it is sufficient to prove that the dimensions of these spaces are equal. It is an easy exercise if one notes that $\dim\mathcal{P}_s=\frac12(s+1)(s+2)$, $\operatorname{Ker}\Phi_s=\operatorname{Ran}\mathcal{L}'_{s-1}$ and hence $\dim\operatorname{Ker}\Phi_s=\dim\mathcal{P}_{s-1}$ (because $\operatorname{Ker}\mathcal{L}'_{s-1}=0$) and $\operatorname{Ran}\mathcal{L}_{s+1}=\mathcal{P}_{s+1}$. The calculation of the dimensions in question we leave to the reader; they are equal to $(s+1)(s+3)$. □

Remark 7.2.3. *For $d=2$, the function w is uniquely defined modulo $O_{2N}(\Lambda^0)$; for $d=3$ the functions w^j are defined modulo the term λu_j (with scalar real-valued function λ) and modulo $O_{2N}(\Lambda^0)$.* ◊

7.2.2 Final reduction

Let $\mathcal{F}$ be a scalar $\mu^{-1}h$-Fourier integral operator corresponding to the symplectomorphism Ψ and satisfying (6.2.33) and (6.2.35). Let T be a $\mu^{-1}h$-pseudo-differential operator with principal symbol $T(x,\xi)$ such that in a neighborhood of Λ^0

$$T^*T\equiv TT^*\equiv I. \tag{7.2.18}$$

Since $T(x,\xi)$ commutes with σ_0 for $d=2$ one can also take T commuting with σ_0 (for $d=2$). Then in a neighborhood of Λ^0

$$T^*\mathcal{F}^*\mathcal{A}^0\mathcal{F}T\equiv\mathcal{A}^0 \tag{7.2.19}$$

and in a neighborhood $\mathcal{W}$ of the point $\rho\in\Lambda$ for $d=2$ (and in a neighborhood $\mathcal{W}$ of the magnetic line Γ for $d=3$)

$$\mathcal{F}T\mathcal{A}^0T^*\mathcal{F}^*\equiv A^0 \tag{7.2.20}$$

where for $d=2$

$$\mathcal{A}^0=\sum_{0\le n\le N}B_n(x_1,h_1D_1)\times \\ \bigl((\sigma_1hD_2+\sigma_2\mu x_2)(\mu^2x_2^2+h^2D_2^2)^n\mu^{-2n}\bigr)^w+\sum_{1\le j\le 2}\sigma_jR_j \tag{7.2.21}$$

and for $d = 3$

$$\text{(7.2.22)} \quad \mathcal{A}^0 = \sum_{0\le n\le N} B_n(x', h_1D_1) \times$$
$$\Big((\sigma_1 hD_3 + \sigma_2 \mu x_3 + \sigma_3 hD_2)(\mu^2 x_3^2 + h^2 D_3^2)^n \mu^{-2n}\Big)^w + \sum_{1\le j\le 3} \sigma_j R_j + R_4 \cdot I$$

where here and in what follows $h_1 = \mu^{-1}h$, $R_j = R_{0j} + h_1 R_{1j}$, $R_{\dots}$ are scalar Hermitian h_1-pseudo-differential operators uniformly with respect to all the parameters. Moreover, the symbols of R_{0j} belong to $O_{2N+3}(\Lambda^0)$, $R_{04} = 0$. Here we quantize the total expression in the parentheses with the exterior superscript w. Moreover, for $d = 3$ we have written lower order terms of the general form and for $d = 2$ we noted that these terms are combinations of σ_1 and σ_2 because σ_0 skew-commutes with $\mathcal{A}^0$. In fact, σ_0 skew-commutes with A^0 and commutes with $\mathcal{F}$ and T. Calculation shows that for $d = 3$ the term R_{14} does not necessarily vanish.

For $d = 3$ one can apply the additional construction motivated by remark 7.2.3. Let T' be an h_1-pseudo-differential operator satisfying (7.2.18) with Weyl symbol $\exp(i\sum_j \sigma_j u_j w)$ (modulo $O(h_1^2)$); this expression commutes with the principal symbol of $\mu^{-1}\mathcal{A}^0$; this fact and (C.5) yield that

$$T'^{*}\mu^{-1}\mathcal{A}^0 T' = \mu^{-1}\mathcal{A}^0 + \mu^{-1}h\Big(\sum_j \sigma_j R'_j + R'_4\Big)$$

where the principal symbol of R'_4 is equal to $\frac{1}{2}\{b_0(u_1^2 + u_2^2 + u_3^2), w\}$. Therefore replacing T by TT', choosing an appropriate function w and repeating the proof of proposition 6.2.1 we can obtain

$$R_4 = h_1 \sum_{0\le n\le N} B'_n(x', h_1D_1)(x_3^2 + h_1^2 D_3^2)^n + R''_4 + h_1^2 R_{24}$$

in (7.2.22), where R''_4 is an h_1-pseudo-differential operator with principal symbol belonging to $O_{2N+1}(\Lambda^0)$; N is decreased if necessary but remains arbitrarily large.

Now by induction with respect to L one can construct operators $T = T_L$ satisfying (7.2.18) such that for $d = 2$ they commute with σ_0 and (7.2.21) holds with $B_n = B_n(x_1, h_1D_1, h_1)$ and

$$\text{(7.2.23)} \qquad R_j = \sum_{0\le l\le L} R_{jl} h_1^l.$$

For $d = 3$ we obtain that

$$\text{(7.2.24)} \quad \mathcal{A}^0 =$$
$$\sum_{0\le n\le N} \Big(B_n(x', h_1D_1, h_1)(\sigma_1 hD_3 + \sigma_2 \mu x_3 + \sigma_3 hD_2)(x_3^2 + h_1^2 D_3^2)^n\Big)^w +$$
$$\sum_j \sigma_j R_j + R_4 \cdot I.$$

Here R_j ($j = 1, \dots, 4$) satisfy (7.2.23) and in both cases R_{jl} are operators with symbols belonging to $O_{N_l}(\Lambda^0)$ and $N_l \to \infty$ as $N \to \infty$ for every fixed l.

Namely, in this construction $T_L = T_{L-1}T'_L T''_L T'''_L$ where T'_L, T''_L and T'''_L satisfy (7.2.18), T'_L is a scalar operator, T''_L and T'''_L are matrix operators and $(T'_L - I)h_1^{1-L}$

and $(T''_L - I)h_1^{-L}$ are h_1-pseudo-differential operators uniformly with respect to all the parameters, for $d = 2$ the matrix T''_L commutes with σ_0 and $T'''_L \equiv I$ and for $d = 3$

$$(T'''_L - I - ih_1^L\alpha^j\sigma_j)h_1^{-L-1}$$

is an h_1-pseudo-differential operator uniformly with respect to all parameters and α^j are scalar h_1-pseudo-differential operators with real-valued symbols. Checking this construction is left to the reader. So we have proven

Proposition 7.2.4. *Let conditions (7.1.1)–(7.1.3) and (7.2.1) be fulfilled with $K = K(N)$ and let Ψ be the symplectomorphism constructed in proposition 6.2.1(i) or (iii) for $d = 2$ and in proposition 6.2.1(ii) and remark 6.2.2(ii) for $d = 3$. Then there exists a $\mu^{-1}h$-pseudo-differential operator T satisfying (7.2.18) in a neighborhood of Λ^0 such that*

(i) For $d = 2$, T commutes with σ_0 and (7.2.19), (7.2.20), (7.2.21) hold with $B_n = B_n(x_1, h_1D_1, h_1)$ and with R_j with symbols belonging to $O_{2N+1}(\Lambda^0)$.

(ii) For $d = 3$ and $D = 2$, equalities (7.2.19), (7.2.20), (7.2.24) hold with $B_n = B_n(x', h_1D_1, h_1)$ and with R_j with symbols belonging to $O_{2N+1}(\Lambda^0)$; moreover the principal symbol of R_4 vanishes.

Now we can also treat the case $d = 3$, $D = 4$:

Corollary 7.2.5. *Let $d = 3$, $D = 4$ and $\sigma_0, \dots, \sigma_3$ be defined by (C.6). Let us construct $T'(= T)$ as in proposition 7.2.4(ii) for the operator $A^{0\prime}$ defined by (7.2.3) with σ'_j instead of σ_j and let us now set $T = \begin{pmatrix} T' & 0 \\ 0 & T' \end{pmatrix}$. Then for A^0 formulas (7.2.18)–(7.2.20) and (7.2.24)$^\#$ hold where the superscript $\#$ means that for R_4 the matrix I should be replaced by σ_4 with σ_4 defined by (7.2.2).*

Remark 7.2.6. *For constant F_{jk} and ω^{jk} one can choose a constant matrix $T(x, \xi)$. Therefore the stability of the reduction procedure yields that if*

$$(7.2.25)_{1-3} \quad |D^\alpha \omega^{jk}| \le c\bar{\nu}, \quad |D^\alpha F_{jk}| \le c\bar{\nu} \quad |D^\alpha V| \le c\bar{\nu}(m+1)^{-1} \qquad \forall \alpha : 1 \le |\alpha| \le K$$

then

$$B_0 = \bar{F} + \bar{\nu}\tilde{B}_0,\ B_n = \bar{\nu}\tilde{B}_n\ (n \ge 1),\ B'_n = \bar{\nu}\tilde{B}'_n,\ R_j = \bar{\nu}\tilde{R}_j \tag{7.2.26}$$

where $\bar{F} =$ const, $\tilde{B}_n$, $\tilde{B}'_n$ and $\tilde{R}_j$ are $\mu^{-1}h$-pseudo-differential operators uniformly with respect to all parameters and the symbols of $\tilde{R}_j$ belong to $O_{2N+1}(\Lambda^0)$ uniformly with respect to all parameters. Surely condition $(7.2.25)_3$ is not used here. Condition $(7.2.25)_4$ will be introduced in the next section. ◊

Let us consider

$$\hat{V} = T^*\mathcal{F}^*(V \cdot I)\mathcal{F}T. \tag{7.2.27}$$

For $d = 2$ this operator commutes with σ_0 and for $D = 4$ it is of the form $\hat{V}' \otimes I_2$. Therefore we obtain

Proposition 7.2.7. *Under the assumptions of proposition 7.2.4*

$$(7.2.28)_1 \qquad \hat{V} = W \cdot I + \mu^{-1}hW_0\sigma_0 \qquad (d = 2),$$

$$(7.2.28)_2 \qquad \hat{V} = W \cdot I + \sum_j \mu^{-1}hW_j\sigma_j \qquad (d = 3,\ D = 2),$$

$$(7.2.28)_3 \qquad \hat{V} = W \cdot I + \sum_j \mu^{-1}hW_j\hat{\sigma}_j \qquad (d = 3,\ D = 4)$$

where $(m+1)W - \text{const}$ *and* $(m+1)W_j$ $(j=0,\dots,3)$ *are scalar* $\mu^{-1}h$*-pseudo-differential operators uniformly with respect to all parameters and the principal symbol of* W *is equal to* $V \circ \Psi$. *Moreover, if conditions* $(7.2.25)_{1-3}$ *are fulfilled then*

$$W_j = \bar{\nu}\tilde{W}_j \tag{7.2.29}$$

for all j *where* $\tilde{W}_j$ *are* $\mu^{-1}h$*-pseudo-differential operators uniformly with respect to all parameters.*

Let us note that we can make a more advanced (but incomplete, unfortunately) reduction of the operator

$$\mathcal{A} = \mathcal{A}^0 + \sigma_0 m + \hat{V} \tag{7.2.30}$$

but in the forthcoming sections we will use another method.

7.2.3 Reduced propagators

Let us consider what we obtained in the treatment of U, etc. Let $\mu \ge 1$, $h \in (0,1]$ and $\mu h \le 1$ (we will get rid of this condition in section 7.5) and let $\mathcal{F}$ and T be constructed above. Moreover, let us assume that remark 6.2.9 is fulfilled. Let us set $\mathcal{U} = T_x^* \mathcal{F}_x^* U^t \mathcal{F}_y^{*t} T_y^*$ where U is the Schwartz kernel of operator $\exp(ith^{-1}(m+1)A)$. Then

$$(hD_t - (m+1)\mathcal{A})\mathcal{U} \equiv 0, \tag{7.3.31}$$

$$\mathcal{U}|_{t=0} \equiv \delta(x-y)\cdot I \tag{7.2.32}$$

modulo functions f and f' satisfying (6.2.51) and (6.2.52) respectively where the operator $\mathcal{A}$ is given by (7.2.30) (for $d=3$, $D=2$ term $\sigma_0 m$ is absent). Moreover, $\mathcal{U}$ satisfies (6.2.54)–(6.2.56) and (6.2.57) is replaced by

$$|F_{t\to h^{-1}\tau}\chi_{T,T'}(t)(U - \mathcal{F}_x T_x \mathcal{U}^t T_y{}^t\mathcal{F}_y)| \le C'Th^s \quad \forall x,y \in \mathcal{W}_x \quad \forall \tau : |\tau| \le 4c \tag{7.2.33}$$

where $\mathcal{W}_x$ is the x-projection of $\mathcal{W}$.

7.3 The Case of a Weak Magnetic Field

In this section we consider the case when condition (6.3.1)[4)] is fulfilled and we obtain some results similar to basic results of section 6.3. Surely some of the auxiliary results of section 6.3 do not remain true here because the reduction of the operator $\mathcal{A}$ is now incomplete. For the sake of simplicity here and in sections 7.4–7.7 we assume that

$$m \le h^{-\bar{l}}\mu^{\bar{l}} \tag{7.3.1}$$

with some (arbitrarily large) exponent $\bar{l}$. Only in section 7.8 will we get rid of this condition, reducing the treatment of the Dirac operator to the treatment of the Schrödinger operator in the case when this condition is not fulfilled.

---

4) I.e., $\mu \in [\mu_0, h^{\delta-1}]$ with $\delta > 0$.

7.3.1 Analysis of singularities. I

We start with an analysis of singularities. Let us first prove

Proposition 7.3.1. [5] *Let $d = 2,3$ and suppose condition (6.3.1) is fulfilled. Let $Q = \operatorname{Op}(q)$, $q \in S_{\mu^{-1}h,\mu^{-1},\mu^{-1},K}(T^*\mathbb{R}^d)$ and let us assume that for both signs*

$$(7.3.2)_{(2)} \qquad \operatorname{supp} q \cap \{\tau_1 \le \bar{W} \pm (m^2 + B_0^2\mu^2(x_2^2+\xi_2^2))^{\frac12} \le \tau_2\} = \emptyset,$$

$$(7.3.2)_{(3)} \qquad \operatorname{supp} q \cap \{\tau_1 \le \bar{W} \pm (m^2 + B_0^2\mu^2(x_3^2+\xi_3^2+\xi_2^2))^{\frac12} \le \tau_2\} = \emptyset,$$

for $d = 2,3$ respectively where W is taken from $(7.2.28)_{1-3}$ and the bar means that in this symbol we set $h_1 = 0$ and $x_2 = \xi_2 = 0$, $x_3 = \xi_3 = x_2 = 0$ for $d = 2,3$ respectively. Then estimate (6.3.3) remains true. Moreover, this estimate remains true if we replace Q_x by ${}^tQ_y$.

Proof. The proof is by standard elliptic methods; it is direct for $m \le C_0$ while for $m \ge C_0$ the reduction of subsection 4.5.3 is used. □

It will be convenient for us to assume in what follows that for $m \ge 1$ one of the following conditions is fulfilled:

$$(7.3.3)_\pm \qquad |V \pm m| \le \frac{c}{(m+1)}.$$

In this case we obtain

Corollary 7.3.2. *Under condition $(7.3.3)_\pm$ remark 6.3.2 remains true for $\tau_{1,2} : |\tau_j| \le 4c$.*

Theorem 7.3.3. [6] *Let $d = 2,3$ and let condition (6.3.1) be fulfilled with $\mu_0 = \mu_0(c,\delta)$. Let $Q' = \operatorname{Op}(q')$, $Q'' = \operatorname{Op}(q'')$, $q', q'' \in S_{\mathbf{h},\rho,\gamma,K}(T^*\mathbb{R}^d)$ and condition (6.3.4) be fulfilled. Furthermore, let **either** $d = 3$ and condition $(6.3.6)_{(3)}$ and one of the conditions $(6.3.5)_{1-4}$ be fulfilled **or** $d = 2$ and condition $(6.3.6)_{(2)}$ and one of the conditions $(6.3.5)_{1,2}$ be fulfilled. Then estimate (6.3.7) holds for $\tau : |\tau| \le 2c$.*

Proof. The proof is the standard energy (and commutator) proof (see the proof of theorem 2.1.1) and for $m \ge C_0$ the reduction of subsection 4.5.3 is also used. □

Remark 7.3.4. *Theorem 7.3.3 and standard elliptic arguments yield a statement similar to assertion (ii) of theorem 6.3.3 but with an essentially stronger condition than (6.3.8).*[7]

Similarly, let us use the standard energy (and commutator) arguments with the reduction of subsection 4.5.3 for $m \ge C_0$ and with the equality

$$i[\sigma_1 hD_d + \sigma_2\mu x_d\,,\, h^2D_d^2 + \mu^2x_d^2] = 2\mu h\omega(\sigma_1 hD_d + \sigma_2\mu x_d)$$

for $m \le C_0$. In the latter case we write the terms in the parenthesis in the right-hand expression using $(m+1)hD_t - \hat{V}$, $m\sigma_0$ and other terms of $\mathcal{A}^0$ and we use the fact that the matrices $\omega = -\sigma_2\sigma_1$, $\omega\sigma_0$ and $\omega\sigma_3$ are skew-Hermitian. Then we can prove a similar assertion under a restriction stronger than (6.3.8), coinciding with (6.3.8) provided

---

[5] Compare with proposition 6.3.1.

[6] Compare with theorem 6.3.3.

[7] The left-hand expression of the inequality is the same as in (6.3.8) but the right-hand expression is essentially larger. Finding it is left to the reader.

$$|D^\alpha V| \le \frac{c}{(m+1)}\mu^{|\alpha|-l+d-3} \qquad \forall \alpha : 1 \le |\alpha| \le l+2-d.$$

Moreover, under some conditions one can prove a weakened version of theorem 6.3.4. However we have no need for it; we will treat the case $d=2$, $\nu<1$ later by another method. ◊

Theorem 7.3.3 immediately yields that assertions (ii) and (iii) of theorem 6.3.5 remain true with $\tau : |\tau| \le c$ in estimate (6.3.15).

One can prove some statements similar to propositions 6.3.6–6.3.8 with $\nu = 1$. However, we will now prove only a special statement; the general case will be treated in the following subsection by another method under an additional condition (which we will get rid of later).

Proposition 7.3.5. [8] *(i) Let one of the following three conditions be fulfilled:*

(7.3.4) $$|\nabla(\frac{V}{\sqrt{F}})| \ge \epsilon_0 \quad \text{in } B(0,1), \qquad m \le \epsilon = \epsilon(c,\epsilon_0),$$

(7.3.5) $$|\nabla(\frac{V^2-m^2}{F})| \ge \epsilon_0 \qquad \text{in } B(0,1),$$

(7.3.6) $$d=3, \quad |\xi_2| \ge \epsilon_0 \qquad \text{on } \operatorname{supp} q'$$

where $q', q'' \in S_{h,\rho,\rho,K}(T^*\mathbb{R}^d)$ with $\rho \ge \mu^{\frac12}h^{\frac12(1-\delta)}$. Let $\chi, \bar\chi \in C_0^K([-1,1])$, $\bar\chi = 1$ on $[-\frac12, \frac12]$ and

(7.3.7)(2) $$h^{1-\delta} \le T' \le T \le \mu T_0 \qquad \text{for } d=2,$$

(7.3.7)(3) $$h^{1-\delta} \le T' \le T \le T_0 \qquad \text{for } d=2,$$

with a sufficiently small constant $T_0 = T_0(c) > 0$. Then estimate (6.3.26) holds for $\tau \in [-\epsilon, \epsilon]$.

Therefore this estimate remains true if we replace $\mathcal{U}$, Q'_x and ${}^tQ''_y$ by U, $Q'_x\psi'(x)$ and $\psi''(y){}^tQ''_y$ respectively with $\psi', \psi'' \in C_0^K(B(0,\frac12))$. Here conditions (7.3.4), (7.3.5) do not change and condition (7.3.6) should be replaced by

(7.3.6)′ $$|F^j(\xi_j - V_j)| \ge \epsilon_0 \qquad \text{on } \operatorname{supp} q'\psi'.$$

Proof. The proof is similar to the proof of theorem 2.1.14 but one should take into account the remarks made in the proof of theorem 6.3.3. Here under condition (7.3.4) the reduced Dirac operator is treated directly while under conditions (7.3.5) and (7.3.6) the reduction of subsection 4.5.3 is used. Let us note that (7.3.5) yields that $m+|V| \ge \epsilon_1$ in $B(0,1)$ and that (7.3.6) yields that the operator $hD_t - (m+1)\mathcal{A}$ is elliptic provided $|\tau| \le \epsilon$, $m + |V| \le \epsilon_0$. □

7.3.2 Analysis of singularities. II

Let us now assume that

(7.3.8) $$m + |V| \ge \epsilon_0 \qquad \text{in } B(0,1).$$

[8] Compare with proposition 6.3.8.

This condition provides a reduction and we can assume that one of the conditions $(7.3.3)_\pm$ is fulfilled. Let us assume for definiteness that condition $(7.3.3)_+$ is fulfilled; then we can assume that the condition

$$(m-V)\geq \epsilon_0(m+1) \tag{7.3.9}$$

is fulfilled as well. For $m\geq C_0$ this condition is fulfilled automatically while for $m\leq C_0$ one can reach it by an ϵ-admissible partition of unity in $B(0,1)$ with a small enough constant $\epsilon>0$; in the latter case we replace A by $-A$ if necessary. Under this assumption one can apply the reduction to the equation

$$\bigl(-hD_t-(m+1)A\bigr)U^\pm=\pm ih\delta(x-y)\delta(t)I \tag{7.3.10}$$

directly. Recall that $U^\pm=\theta(\pm t)U$. Dividing U into four blocks U_{jk} and assuming that for $d=2,3$ the matrices σ_j are given by formulas (C.5), (C.6) respectively we obtain the following system in the zone $\{|\tau|\leq\epsilon\}$:

$$\Bigl(-hD_t+(m+1)(V+m)-\mathcal{L}^*E\mathcal{L}\Bigr)U^\pm_{1r}\equiv\pm ihf_{1r}, \tag{7.3.11}$$

$$U^\pm_{2r}\equiv -E\mathcal{L}U^\pm_{1r}\pm ihf_{2r}. \tag{7.3.12}$$

Here E is a parametrix of the operator $-h(m+1)^{-1}D_t+(V-m)$ which is elliptic in the indicated zone, $\mathcal{L}=\frac{1}{2}\sigma'_l(\omega^{jl}P_j+P_j\omega^{jl})$, $\sigma_l=\begin{pmatrix}0&\sigma_l'^{\dagger}\\ \sigma'_l&0\end{pmatrix}$ and $f_{21}=0$, $f_{22}=E\delta(x-y)\delta(t)I$, $f_{11}=\delta(x-y)\delta(t)I$, $f_{12}=-\mathcal{L}^*E\delta(x-y)\delta(t)I$.

In the case $d=3$, $D=2$ we can formally "double" our system setting $U'=U''=U$:

$$(-hD_t+V)\begin{pmatrix}U'\\ U''\end{pmatrix}=\begin{pmatrix}0&-A_0\\ -A_0&0\end{pmatrix}\begin{pmatrix}U'\\ U''\end{pmatrix}\pm ih\delta(x-y)\delta(t)\begin{pmatrix}I\\ I\end{pmatrix};$$

then the above reduction can be applied to this system as well. Here and below $A_0=A-V\cdot I$.

Let us note that without loss of generality one can assume that

$$|V|\geq\frac{\epsilon_0}{(m+1)}\qquad\text{in } B(0,1). \tag{7.3.13}$$

In fact, for $m\geq C_0$ this condition follows from $(7.3.3)_\pm$ and for $m\leq C_0$ one can easily see that under condition (7.3.8) the operator in the left-hand expression of (7.3.11) is elliptic for $|\tau|\leq\epsilon$, $|V|\leq\epsilon$ in which case all the statements below can be easily proven by means of simple elliptic arguments (instead of standard hyperbolic arguments).

Let E' be an operator with symbol equal to

$$F^{-\frac{1}{2}}\bigl((m-V)+(m+1)^{-1}\tau\bigr)^{\frac{1}{2}}$$

in the zone in question. Let us substitute $U_{1r}=E'u$ and note that

$$\mathcal{L}^*F^{-1}\mathcal{L}=P_jF^{-1}g^{jk}P_k\cdot I+\mu h\mathcal{M}$$

where here and below $\mathcal{M}$ is a $(\mu^{-1}h,h)$-pseudo-differential operator with respect to (x,t) which we change if necessary. Then multiplying (7.3.11) by E' we obtain the equation

$$\Bigl(P_jF^{-1}g^{jk}P_k+\mu h\mathcal{M}-F^{-1}\Bigl(\bigl(h(m+1)^{-1}d_t-V\bigr)2-m^2\Bigr)\Bigr)u\equiv f$$

with $f = -E' f_{1r}$. One can rewrite this equation in the form

$$\Big(M + 2F^{-1}(m+1)^{-1}VhD_t - F^{-1}(m+1)^{-2}(hD_t)^2 + \mu h\mathcal{M}\Big)u \equiv f \tag{7.3.14}$$

with

$$M = P_j(F^{-1}g^{jk})P_k + v_*, \qquad v_* = \frac{1}{F}(m^2 - V^2). \tag{7.3.15}$$

Let us note that the operator in the left-hand expression in (7.3.14) is microhyperbolic in one of the directions $\pm\partial_\tau$ by condition (7.3.13); this statement remains true in what follows. Let us consider the operator M as the Schrödinger operator of the previous chapter; then the scalar magnetic intensity equals 1. Let $\mathcal{F}$ be the corresponding reducing Fourier integral operator constructed in section 6.2. Then (7.3.14) yields the equation

$$\Big(M' + M''hD_t + M'''(hD_t)^2 + \mu h\mathcal{M}'\Big)u' \equiv f' \tag{7.3.16}$$

where $u' = \mathcal{F}_x^* u\, {}^t\mathcal{F}_y^*$, $f' = \mathcal{F}_x^* {}^t\mathcal{F}_y^* f$ and $M' = \mathcal{F}^* M\mathcal{F}$ is the reduced operator, $M'' = 2(m+1)^{-1}\mathcal{F}^* F^{-1}V\mathcal{F}$, $M''' = -(m+1)^{-2}\mathcal{F}^* F^{-1}\mathcal{F}$ are scalar $\mu^{-1}h$-pseudo-differential operators and $\mathcal{M}'$ is an operator of the same type as $\mathcal{M}$.

We consider this equation in the zone $\{\epsilon\mu^{-1} \le |t| \le T^*,\ |\tau| \le \mu h^{1-\delta'}\}$ with $T^* = T_0$ for $d = 3$, $T^* = \mu T_0$ for $d = 2$ and with $T^* = \nu^{-1}\mu T_0$ for $d = 2$ and some additional restriction where $T_0 = T_0(c) > 0$ is a small enough constant. The restriction $\epsilon\mu^{-1} \le T$ is associated with the same circumstances as in section 6.3.

First of all the arguments used in the proof of propositions 2.1.9 and 2.1.10 with the additional arguments used in the proof of theorem 6.3.3 imply the following assertion:

Proposition 7.3.6. [9] *(i) Let $Q' = \operatorname{Op}(q'), Q'' = \operatorname{Op}(q'')$ where q', q'' are symbols of type (6.3.16) and let*

$$|\nabla_{x',\xi'} M'(x,\xi)| \le c\gamma \tag{7.3.17}$$

on $\operatorname{supp} q'$ *where*

$$\gamma \ge \mu^{-\frac12}h^{\frac12(1-\delta')} + C_0\mu h^{1-\delta'} + C_0\mu^{-l} \qquad \text{for } d = 2, \tag{7.3.18$_{(2)}$}$$

$$\gamma \ge h^{\frac12(1-\delta')} + C_0\mu h^{1-\delta'} + C_0\mu^{-l} \qquad \text{for } d = 3. \tag{7.3.18$_{(3)}$}$$

Then estimate (6.3.7) remains true for u' (instead of $\mathcal{U}$) for all $x, y \in \mathbb{R}^d$ and for all $\tau : |\tau| \le \mu h^{1-\delta}$ with arbitrary T such that

$$T \in [\epsilon\mu^{-1}, \mu T_0] \qquad \text{for } d = 2, \tag{7.3.19$_{(2)}$}$$

$$T \in [\epsilon\mu^{-1}, T_0] \qquad \text{for } d = 3. \tag{7.3.19$_{(3)}$}$$

(ii) Moreover, let $d = 2$, let conditions $(7.2.25)_{1-3}$ be fulfilled and suppose

$$\nu \ge \bar\nu\mu h^{1-\delta'}, \tag{7.3.20}$$

$$\nu^{-1}(M'' - \tilde M) \in \boldsymbol{\Psi}_{\mu^{-1}h,K}(T^*\mathbb{R}) \tag{7.3.21}$$

where M'' is obtained from M' by the substitution $\mu^2 x_2^2 + h^2 D_2^2 \to \lambda \in [0, C_0]$ and $\tilde M = \text{const}$. *Then under the conditions*

[9] Compare with proposition 6.3.7.

(7.3.17)′ $$|\nabla_{x',\xi'} M'(x,\xi)| \le c\nu\gamma$$

on supp q',

(7.3.18)′$_{(2)}$ $$\gamma \ge \mu^{-\frac{1}{2}} h^{\frac{1}{2}(1-\delta')} + C_0\bar{\nu}\mu h^{1-\delta'} + C_0\mu^{-l} \qquad \text{for } d=2$$

estimate (6.3.7) *for* u' *remains true for* $|\tau| \le \mu h^{1-\delta'}$ *and*

(7.3.19)′$_{(2)}$ $$T \in [\epsilon\mu^{-1}, \nu^{-1}\mu T_0] \qquad \text{for } d=2.$$

Remark 7.3.7. *(i) Condition (7.3.21) means exactly that multiplying M' by an appropriate operator E'' we can obtain*

$$E''M' = B(x_1, h_1 D_1) + \mu^2 x_2^2 + h^2 D_2^2$$

where B satisfies condition (6.3.9). One can use this reformulation even for $\nu = 1$ in the proof.

(ii) If $\nu \ge \epsilon_0\mu^{-2}\bar{\nu}$ and (7.3.20) holds then condition (7.3.21) is equivalent to

(7.2.25)$_4$ $$|D^\alpha v_*| \le c\nu \qquad \forall \alpha : 1 \le |\alpha| \le K;$$

recall that $v_* = \frac{1}{F}(m^2 - V^2)$.

Proposition 7.3.8. [10] *(i) Let $Q' = \mathrm{Op}(q'), Q'' = \mathrm{Op}(q'')$ where q', q'' are symbols of type* (6.3.16) *and let*

(7.3.22) $$\epsilon_0\gamma \le |\nabla_{x',\xi'} M'(x,\xi)| \le c\gamma$$

on supp q' *where*

(7.3.23) $$1 \ge \gamma \ge \mu^{\frac{1}{2}} h^{\frac{1}{2}(1-\delta')} + C_0\mu^{-l}.$$

Then estimate (6.3.26) *remains true for u' (instead of $\mathcal{U}$) for all $x, y \in \mathbb{R}^d$ and for all $\tau : |\tau| \le \mu h^{1-\delta}$ with arbitrary T satisfying* (7.3.19)$_{(d)}$.

(ii) Moreover, let $d = 2$, let conditions (7.2.25)$_{1-3}$, (7.3.20) *and* (7.3.21) *be fulfilled and let*

(7.3.22)′ $$\epsilon_0\nu\gamma \le |\nabla_{x',\xi'} M'(x,\xi)| \le c\nu\gamma,$$

(7.3.23)′ $$1 \ge \gamma \ge \mu^{\frac{1}{2}} h^{\frac{1}{2}(1-\delta')}\nu^{-\frac{1}{2}} + C_0\mu^{-l}.$$

Then estimate (6.3.26) *remains true for $|\tau| \le \mu h^{1-\delta'}$ and T satisfying* (7.3.19)′$_{(2)}$.

Remark 7.3.9. *(i) Both propositions remain true if we multiply Q' and Q'' by operators* $\mathrm{Op}(q_1')$, $\mathrm{Op}(q_1'')$ *with* $q_1', q_1'' \in S_{(\mathbf{h},h),K}(T^*(\mathbb{R}^d_x \times \mathbb{R}_t))$;

(ii) In what follows in conditions (7.3.22), (7.3.22)′ *the parameter $\lambda = \mu^2 x_d^2 + h^2\xi_d^2$ is taken such that $|M'(x,\xi)| \le \mu h^{1-\delta'}$ in which case the concrete choice of λ is not essential; compare with remark 7.3.7(i).* ◊

[10] Compare with proposition 6.3.8.

7.3.3 Calculations

Let us pass from estimates to calculations. Here the basic result is the following statement (with arbitrary $d \geq 2$):

Proposition 7.3.10. [11] *Suppose that conditions* (7.1.1)–(7.1.3), (6.3.1), (6.3.28), $(7.3.3)_\pm$ *are fulfilled. Then*

(i) Assertion (i) of proposition 6.3.10 remains true but now

$$\operatorname{supp} \varkappa'_{kn} \subset \{\tau \leq V(x) - m\} \cup \{\tau \geq V(x) + m\}$$

and we introduce only $\varkappa'_{kn}$ *and* $\varkappa_{kn}(x, \tau', \tau'')$ *with finite* τ', τ''.

(ii) Moreover, if the condition

$$V^2 - m^2 \geq \epsilon_0 \qquad \forall x \in B(0,1) \tag{7.3.24}$$

is fulfilled (instead of condition (6.3.31)*) then assertion (ii) of proposition 6.3.10 also remains true.*

(iii) On the other hand, if the condition

$$|V^2 - m^2| + (m+1)|\nabla V| \geq \epsilon_0 \qquad \forall x \in B(0,1) \tag{7.3.25}$$

is fulfilled (instead of condition (6.3.33)*) then assertion (iii) of proposition 6.3.10 also remains true.*

(iv) Furthermore, for $\tau_1 < \tau_2$

$$\begin{aligned} \operatorname{tr} \varkappa_{00}(x, \tau_1, \tau_2) = \int_{\tau_1}^{\tau_2} \operatorname{tr} \varkappa'_{00}(x, \tau) d\tau = \\ (2\pi)^{-d} \frac{D}{2} \sum_{\eta = \pm 1} \int_{\tau_1 \leq H_\eta(x,\xi) \leq \tau_2} q'_0(x,\xi) q''_0(x,\xi) \psi'(x) \psi''(x) d\xi \end{aligned} \tag{7.3.26}$$

where

$$H_\eta(x,\xi) = V + \eta\big(g^{jk}(\xi_j - V_j)(\xi_k - V_k) + m^2\big)^{\frac{1}{2}} \tag{7.3.27}$$

and q'_0, q''_0 *are the principal symbols of the operators* Q', Q''.

(v) Furthermore, for $q'_0 = q''_0 = 1$

$$\begin{aligned} \operatorname{tr} \varkappa_{00}(x, \tau_1, \tau_2) = (2\pi)^{-d} \omega_d \frac{D}{2} \sqrt{g} \times \\ \operatorname{sign}(\tau - V)\big((\tau - V)^2 - m^2\big)_+^{\frac{d}{2}} \Big|_{\tau=\tau_1}^{\tau=\tau_2} \psi'\psi''. \end{aligned} \tag{7.3.28}$$

(vi) Moreover, let $Q' = Q'' = I$ *and let the set of Dirac matrices contain* σ_0. *Then* $\operatorname{tr} \varkappa'_{kn} = 0$ *for odd* n. *Moreover, if* $d \not\equiv 2 \pmod 4$ *then* $\operatorname{tr} \varkappa'_{kn} = 0$ *for even* k. *On the other hand, if* $d \equiv 2 \pmod 4$ *then* $\operatorname{tr} \varkappa'_{kn}$ *does not depend on the choice of Dirac matrices of the given dimension for even* k, *and* $\operatorname{tr} \varkappa'_{kn}$ *depends only on the class of unitary equivalence of the set* $\{\sigma_0, \dots, \sigma_d\}$ *for odd* k: *namely, the replacement* $\sigma_0 \to -\sigma_0$ *yields the replacement* $\operatorname{tr} \varkappa'_{kn} \to -\operatorname{tr} \varkappa'_{kn}$.

(vii) Finally, let $Q' = Q'' = I$ *and suppose that the set of Dirac matrices does not contain* σ_0 *principally (principally massless particle; this is possible only for odd* d*). Then for* $d \equiv 3 \pmod 4$, $\operatorname{tr} \varkappa'_{kn} = 0$ *for odd* n*; while for even* n $\operatorname{tr} \varkappa'_{kn}$ *does not depend on the choice of the Dirac matrices of the given dimension for odd* k, *and* $\operatorname{tr} \varkappa'_{kn}$ *depends only on*

[11] Compare with proposition 6.3.10.

the unitary equivalence class of the set $\{\sigma_1,\dots,\sigma_d\}$ *for even* k*: namely, the replacement* $\sigma_1 \to -\sigma_1$ *yields the replacement* $\operatorname{tr}\varkappa'_{kn} \to -\operatorname{tr}\varkappa'_{kn}$.

On the other hand, if $d \equiv 1 \pmod 4$ *then* $\operatorname{tr}\varkappa'_{kn} = 0$ *for odd* n *and* $\operatorname{tr}\varkappa'_{kn}$ *does not depend on choice of Dirac matrices of the given dimension for even* k*, and* $\operatorname{tr}\varkappa'_{kn}$ *depends only on the class of the unitary equivalence of the set* $\{\sigma_1,\dots,\sigma_d\}$ *for odd* k*: namely, the replacement* $\sigma_1 \to -\sigma_1$ *yields the replacement* $\operatorname{tr}\varkappa'_{kn} \to -\operatorname{tr}\varkappa'_{kn}$.

Proof. The proof of assertions (i)–(v) repeats the proof of proposition 6.3.10 with the following modification: for $m \le C_0h^{-l}$ we use direct calculation and for $m \ge C_0h^{-l}$ we use reduction; we can choose $l \ge 0$ arbitrarily.

In order to prove (vi), (vii) let us note that the replacement $\sigma_j \to q^\dagger\sigma_j q$ for all j with the unitary matrix q yields the replacement $U \to q^\dagger U q$ and does not change $\operatorname{tr}\varkappa'_{kn}$. Moreover, let us note that the complex conjugation of operator A also does not change $\operatorname{tr}\varkappa'_{kn}$ because A is a self-adjoint operator.

Furthermore, in the case (vi) let us note that the replacement $(\mu,h) \to (-\mu,-h)$ is equivalent to the replacement $\sigma_j \to \sigma_0\sigma_j\sigma_0$ for all j and therefore does not change $\operatorname{tr}\varkappa'_{kn}$; then $\operatorname{tr}\varkappa'_{kn} = 0$ for odd n (see also the arguments of the proof of proposition 6.3.10 associated with the Jacobian and mollification). Let us note that the replacement $h \to -h$ is equivalent to complex conjugation and the replacement $\{\sigma_0,\dots,\sigma_d\} \to \{\bar\sigma_0,\dots,\bar\sigma_d\}$ where the bar means complex conjugation. For $d \not\equiv 2 \pmod 4$ the latter set of Dirac matrices is unitarily equivalent to $\{\sigma_0,\dots,\sigma_d\}$ and for $d \not\equiv 0 \pmod 4$ it is unitarily equivalent to the set $\{-\sigma_0,\dots,-\sigma_d\}$ (see formula (C.4)). This observation yields all the remaining statements in (vi).

In case (vii) let us note that the replacement $\mu \to -\mu$ is equivalent to the complex conjugation of operator and the replacement $\{\sigma_1,\dots,\sigma_d\} \to \{\bar\sigma_1,\dots,\bar\sigma_d\}$; for $d \equiv 3 \pmod 4$ these sets are equivalent and therefore $\operatorname{tr}\varkappa'_{kn} = 0$ for odd k. Moreover, the replacement $(\mu,h) \to (-\mu,-h)$ is equivalent to the replacement $\sigma_j \to -\sigma_j$ for all j and this observation concludes the proof of (vii) for $d \equiv 3$.

Finally, let us note that the replacement $h \to -h$ is equivalent to the complex conjugation of the operator A and the replacement $\{\sigma_1,\dots,\sigma_d\} \to \{\bar\sigma_1,\dots,\bar\sigma_d\}$. For $d \equiv 1$ these sets are equivalent and therefore $\operatorname{tr}\varkappa'_{kn} = 0$ for odd n. Using the replacement $\mu \to -\mu$ (see above) one can easily conclude the proof of (vii) for $d \equiv 1$. □

Repeating the corresponding arguments used in the proof of (6.3.40) one can prove the similar formula

$$S'(\tau,\mu h) = \partial_\tau S(x,\tau_1,\tau,\mu h) \sim \sum_{k\in\mathbb{Z}^+} \operatorname{tr}\varkappa'_{k0}(x,\tau)(\mu h)^k \tag{7.3.29}$$

in the weak sense.

In the calculation below the dependence on x is not essential and without loss of generality one can assume that $V = 0$. In the case $d > 2r = \operatorname{rank}(F_{jk})$, $S'(\tau,\mu h)$ is an even function of τ and therefore $\operatorname{tr}\varkappa'_{k0}(\tau)$ are also even functions of τ. Then one can introduce their odd (on τ) primitive functions $S(\tau,\mu h)$ and $\operatorname{tr}\varkappa_{k0}(\tau)$. The same is true for $d = 2r$ if we treat the even (with respect to τ) parts of $S'(\tau,\mu h)$ and $\operatorname{tr}\varkappa'_{k0}(\tau)$ and the odd parts of $S(\tau,\mu h)$ and $\operatorname{tr}\varkappa_{k0}(\tau)$. In any case one can treat $S(\tau,\mu h)$ as Riemann sums given by (E.4) for the function

$$\mathcal{K}(\tau,z) = c_0\bigl(\tau^2 - z_1 - \dots - z_r\bigr)_+^p \operatorname{sign}\tau$$

with $p = \frac12(d-2r)$, $c_0 = (2\pi)^{r-d}2^{-r-1}\omega_{d-2r}D\sqrt{g}$. We assert that the function $\mathcal{K}(\tau,z)$ mollified with respect to τ is smooth with respect to z. In fact, any mollification with an even mollifying function $\phi(\tau)$ gives 0 while mollification with an odd mollifying function $\phi(\tau)$ (equal to $\phi_0(\tau^2)\tau$ with $\phi_0 \in C_0^K(\mathbb{R})$) is equal to

$$\text{const}\int_0^\infty \phi_0(\tau)(\tau-t)_+^p\,d\tau$$

with $t=z_1+\cdots+z_r+m^2$ where we change constants if necessary. Integrating by parts n times we obtain for $t\geq 0$

$$\text{const}\int_0^\infty \phi_0^{(n)}(\tau)(\tau-t)_+^{p+n}\,d\tau$$

and the smoothness with respect to t is obvious. Therefore we have proven the first assertion of the following

Proposition 7.3.11. [12] *Let $Q'=Q''=I$, $\psi'=\psi''=1$. Then*
(i) For $d>2r$ the following equalities are true

(7.3.30) $$\operatorname{tr}\varkappa_{k0}=D\partial_t^k\operatorname{sign}(\tau-V)\big((\tau-V)^2-m^2+t\big)_+^{\frac{d}{2}}\big|_{t=-0}\times\mathcal{P}_k(f_1^2,\ldots,f_r^2)\sqrt{g}$$

where $\mathcal{P}_k$ are homogeneous symmetric polynomials of degree $\frac{k}{2}$ also depending on d and r and vanishing for odd k. In particular,

(7.3.31) $$\operatorname{tr}\varkappa_{20}=2cD\partial_t^2\operatorname{sign}(\tau-V)\big((\tau-V)^2-m^2+t\big)_+^{\frac{d}{2}}\big|_{t=-0}F^2\sqrt{g}$$

where c and F are introduced in proposition 6.3.11.

(ii) For $d=2r$ these equalities hold for the odd with respect to $\tau-V$ parts of $\varkappa_{k0}(\tau)$ and the even part of $\varkappa_{k0}(\tau)$ vanishes for $k\neq r$, while for $k=r$ under condition (7.1.8) it is equal to

(7.3.32) $$\operatorname{tr}\varkappa_{r0\text{even}}=-\bar\varkappa\frac{D}{2}(4\pi)^{-r}f_1\ldots f_r\sqrt{g}\times\big(\theta(\tau-V+m)-\theta(\tau-V-m)\big)$$

with $\bar\varkappa$ introduced in theorem 7.1.1. In particular, for $d=2$ the coefficient in the right-hand expression is equal to $-\frac{D}{8\pi}\varsigma F_{12}$.

In order to prove (ii) it is sufficient to observe that outside of the discontinuity points the even part of $S(\tau,\mu h)$ is equal to the right-hand expression of (7.3.32) multiplied by $(\mu h)^r$; this is due to formula (7.1.6).

Remark 7.3.12. *In the manner of remark 6.3.12 we conclude that if Q' is an operator with Weyl symbol q' even with respect to ξ and h and if $q''=1$, $\psi'=\psi''=1$ in a neighborhood of* $\operatorname{supp}q'$ *then assertions (vi), (vii) of proposition 7.3.9 remain true.* ◊

7.3.4 Estimates

Now, repeating the corresponding arguments of section 6.3 we easily obtain the following statements:

Proposition 7.3.13. [13] *Suppose that conditions* (7.1.1)–(7.1.3), (7.2.1) *and* (6.3.1) *are fulfilled. Then*
(i) If one of the conditions (7.3.4), (7.3.5) *is fulfilled then estimate* (6.3.34) *holds for $|\tau|\leq\epsilon$ with $T=\mu T_0$ for $d=2$ and $T=T_0$ for $d=3$.*
(ii) Let $d=2$ and conditions $(7.2.25)_{1-3}$, (7.3.20), (7.3.21), (7.3.24) *be fulfilled and let $\tilde Q'=\operatorname{Op}(\tilde q')$, $\tilde Q''=\operatorname{Op}(\tilde q'')$ with $\tilde q',\tilde q''\in S_{h_1,\gamma,\gamma,K}(T^*\mathbb{R})$. Moreover, let conditions*

12) Compare with proposition 6.3.11.
13) Compare with proposition 6.3.13.

(7.3.22)′, (7.3.23)′ *be fulfilled on* $\operatorname{supp}\tilde{q}'$ *and* $Q' = \mathcal{F}\tilde{Q}'\mathcal{F}^*$, $Q'' = \mathcal{F}\tilde{Q}''\mathcal{F}^*$. *Then estimate* (6.3.34) *holds for* $|\tau| \le \mu h^{1-\delta'}$ *with* $T = \nu^{-1}\mu T_0$.

(iii) Let $d = 3$, $\tilde{q}', \tilde{q}' \in S_{\mathbf{h}',\gamma,\gamma,K}(T^*\mathbb{R}^2)$ *and let conditions* (7.3.24), (7.3.22), (7.3.23) *be fulfilled on* $\operatorname{supp}\tilde{q}'$. *Let* $Q' = \mathcal{F}\tilde{Q}'\mathcal{F}^*$, $Q'' = \mathcal{F}\tilde{Q}''\mathcal{F}^*$. *Then estimate* (6.3.34) *holds for* $|\tau| \le \mu h^{1-\delta'}$ *with* $T = T_0$.

Theorem 7.3.14. [14] *Suppose conditions* (7.1.1)–(7.1.3), (7.2.1), (6.3.1) *and one of conditions* (7.3.4), (7.3.5) *are fulfilled. Let us define*

$$\mathcal{R}_1 = \Bigl| \iint \phi(\tau')\psi(x)\operatorname{tr}\bigl(e(x,x,\tau',\tau) - h^{-d}\sum_{0\le k\le M} \varkappa_{k0}(x,\tau',\tau)(\mu h)^k\bigr)\,dx d\tau\Bigr| \tag{7.3.33}$$

with $\phi \in C_0^K([-c,c])$. *Then estimates* $(6.3.48)_{(2,3)}$ *hold for* $d = 2,3$ *respectively,* $|\tau| \le \epsilon(m+1)^{-1}$.

Proposition 7.3.15. [15] *Suppose conditions* (7.1.1)–(7.1.3), (7.2.1), (6.3.1) *and* (7.3.24) *are fulfilled. Let* $\tilde{Q}' = \operatorname{Op}(\tilde{q}')$ *where* $\tilde{q}' \in S_{\mathbf{h}',\gamma,\gamma,K}(T^*\mathbb{R}^{d-1})$, γ *satisfies* (7.3.23)′, $\operatorname{diam}\operatorname{supp}\tilde{q}' \le c\gamma$, *and let* $Q' = \mathcal{F}\tilde{Q}'\mathcal{F}^*$. *Then*

(i) Let $d = 2$, *let conditions* $(7.2.25)_{1-3}$, (7.3.20), (7.3.21) *be fulfilled and let condition* (7.3.22)′, (7.3.23) *be fulfilled on* $\operatorname{supp}\tilde{q}'$. *Let us define*

$$\mathcal{R}_2 = \Bigl| \iint \phi(\tau')\operatorname{tr}\bigl((Q'_x\psi(x)e)(x,x,\tau',\tau) - h^{-d}\sum_{k+n\le M} \varkappa_{Q'\psi kn}(x,\tau',\tau)(\mu h)^k h^n\bigr)\,dx d\tau'\Bigr|. \tag{7.3.34}$$

Then estimate $(6.3.50)_{(2)}, (6.3.50)_{(2)}$ *holds.*

(ii) Let $d = 3$, $\nu = 1$ *and let condition* (7.3.22) *be fulfilled on* $\operatorname{supp}\tilde{q}'$. *Then estimate* $(6.3.50)_{(3)}$ *holds.*

Proposition 7.3.16. [16] *Suppose conditions* (7.1.1)–(7.1.3), (7.2.1), (6.3.1) *and* (7.3.24) *are fulfilled. Let* $\vartheta = 0$. *Then*

(i) Let $d = 2$ *and let conditions* $(7.2.25)_{1-3}$, (7.3.20), (7.3.21) *be fulfilled. Then estimate* $(6.3.53)_{(2)}$ *holds where* M_d, $\mathcal{M}_d$, $\bar{\gamma}$ *are introduced in theorem 6.3.16(i) and* B *is introduced for the reduced Schrödinger operator.*

(ii) For $d = 3$ *estimate* $(6.3.53)_{(3)}$ *holds.*

Theorem 7.3.17. [17] *Let* $d = 3$ *and suppose conditions* (7.1.1)–(7.1.3), (7.2.1), (6.3.1) *and* (7.3.24) *are fulfilled. Then for* $\mu \le h^{-\frac{1}{3}+\delta}$ *estimate* $(6.3.48)_{(3)}$ *holds.*

Theorem 7.3.18. [18] *Let* $d = 2$ *and let conditions* (7.1.1)–(7.1.3), (7.2.1), (6.3.1) *and* (7.3.24) *be fulfilled. Moreover, let conditions* $(7.2.25)_{1-3}$, (7.3.20) *and* (7.3.21) *be fulfilled. Then for* $\vartheta = 0$ *assertions (i), (ii) of theorem 6.3.18 remain true with* V *replaced by*

$$V_* = m^2 - V^2 \tag{7.3.35}$$

in their right-hand expressions.

---

[14] Compare with theorem 6.3.14.
[15] Compare with proposition 6.3.15.
[16] Compare with proposition 6.3.16.
[17] Compare with theorem 6.3.17.
[18] Compare with theorem 6.3.18.

Theorem 7.3.19 (compare theorem 6.3.19). *(i) Under the conditions of theorem 7.3.14 estimates* (6.3.48)$_{(2,3)}$ *(with* $\vartheta = 0$*) remain true for*

$$\mathcal{R} = \Big| \iint \phi(\tau)\psi(x)\,\mathrm{tr}\big(e(x,x,\tau,0) - h^{-d}S(x,\tau,0,\mu h)\big)\,dx d\tau \Big|. \tag{7.3.36}$$

(ii) Under the assumptions of proposition 7.3.16(i) estimate (6.3.67)$_{(2)}$ *(with* $\vartheta = 0$*) remains true with* M_2 *and* $\mathcal{M}_2$ *defined in this section and with* V *replaced by* V_* *in the right-hand expression; moreover, condition* (6.3.59) *is now replaced by the corresponding condition on* V_*:

$$|D^\alpha \frac{V_*}{F}| \le c\nu \qquad \forall \alpha : 1 \le |\alpha| \le K. \tag{7.3.37}$$

(iii) Under the assumptions of theorem 7.3.17(i),(ii) estimates (6.3.55), (6.3.57) *(with* $\vartheta = 0$*) respectively hold for* $\mathcal{R}$ *also.*

7.4 The Case of a Strong Magnetic Field

7.4.1 Separation of variables and decomposition

In this section we assume that

$$h^{-\delta} \le \mu \le \epsilon_1 h^{-1} \tag{7.4.1}$$

where small enough ϵ_1 will be chosen later. The case $\mu \ge \epsilon_1 h^{-1}$ will be treated in the next section. Later conditions (7.3.24) and (6.3.1) will appear in some statements of this section. Let us make a decomposition similar to those of the proof of theorem 7.1.1. Namely, let us note that every $\mathbb{H}$-valued function v on $\mathbb{R}$ (with $\mathbb{H} = \mathbb{C}^D$) can be written in the form

$$w = \sum_{j\in\mathbb{Z}^+} (w_j^+ \Upsilon_{j-1} + w_j^- \Upsilon_j), \qquad w_j^\pm \in \mathbb{H}^\pm, \quad v_0^+ = 0 \tag{7.4.2}$$

where $\mathbb{H}^\pm = \mathrm{Ker}(\sigma_1 \pm i\sigma_2)$, $\Upsilon_j = \Upsilon_j(x_d) = \left(\frac{\mu}{h}\right)^{\frac14} v(\sqrt{\frac{\mu}{h}}x_d)$, where v_j are real-valued orthonormal Hermite functions satisfying (6.4.22). Let us set $w_j = (w_j^+, w_j^-) \in \mathbb{H}$; so the isomorphism

$$\mathrm{R} : L^2(\mathbb{R}) \to \mathbb{H}^- \oplus \sum_{j\ge1} \mathbb{H}$$

is introduced.

Formulas (6.4.22) and the definition of $\mathbb{H}^\pm$ yield that for $w_j^\pm \in \mathbb{H}^\pm$

$$(hD_d\sigma_1 + \mu x_d \sigma_2)(w_j^+\Upsilon_{j-1} + w_j^-\Upsilon_j) = \sigma_2\sqrt{2j\mu h}(w_j^+\Upsilon_{j-1} + w_j^-\Upsilon_j). \tag{7.4.3}$$

Similarly, every $\mathrm{Hom}(\mathbb{H},\mathbb{H})$-valued function $\mathcal{U}$ on $\mathbb{R}^2$ can be written in the form

$$\mathcal{U} = \sum_{j,k\in\mathbb{Z}^+} \big(\mathcal{U}_{jk}^{++}\Upsilon_{j-1}(x_d)\Upsilon_{k-1}(y_d) + \mathcal{U}_{jk}^{+-}\Upsilon_{j-1}(x_d)\Upsilon_k(y_d) + \mathcal{U}_{jk}^{-+}\Upsilon_j(x_d)\Upsilon_{k-1}(y_d) + \mathcal{U}_{jk}^{--}\Upsilon_j(x_d)\Upsilon_k(y_d)\big) \tag{7.4.4}$$

with $\mathcal{U}_{jk}^{\varsigma\varsigma'} \in \mathrm{Hom}(\mathbb{H}^{\varsigma}, \mathbb{H}^{\varsigma'})$ where for $j=0$, $\varsigma=+$ and for $k=0$, $\varsigma'=+$ the corresponding components vanish. Using these four blocks (with different ς and ς') one can introduce the blocks $\mathcal{U}_{jk}^{\varsigma;} \in \mathrm{Hom}(\mathbb{H}^{\varsigma}, \mathbb{H})$, $\mathcal{U}_{jk}^{;\varsigma'} \in \mathrm{Hom}(\mathbb{H}, \mathbb{H}^{\varsigma'})$ and $\mathcal{U}_{jk} \in \mathrm{Hom}(\mathbb{H}, \mathbb{H})$.

Now let $\mathcal{U} = \mathcal{U}(x,y,t)$ be defined in subsection 7.2.3. Then $\mathcal{U}_{jk} = \mathcal{U}_{jk}(x',y',t)$, etc., and the properties of $\mathcal{U}$ yield that

$$\mathcal{U}_{jk}(x',y',t) = \mathcal{U}_{kj}^{\dagger}(y',x',-t) \tag{7.4.5}$$

and

$$|F_{t\to h^{-1}\tau}\chi_{T,T'}(t)\mathcal{U}_{jk}| \le CT\mu h^{1-d}(1+|\tau|)^{d-1} \qquad \forall x',y' \in \mathbb{R}^{d-1} \quad \forall j,k \in \mathbb{Z}^{+} \quad \forall \tau \in \mathbb{R}, \tag{7.4.6}$$

and theorem 6.1.8 which obviously holds under condition (6.4.1) for the Dirac operator also yields that

$$|F_{t\to h^{-1}\tau}\chi_{T,T'}(t)\mathcal{U}_{jk}| \le CTh^{s}(1+j+k)^{-s} \qquad \forall x',y' \in \mathbb{R}^{d-1} \quad \forall \tau \in \mathbb{R}: |\tau| \le c \quad \forall j,k \in \mathbb{Z}^{+}: j+k \ge C_0\mu^{-1}h^{-1} + h^{-\delta} \tag{7.4.7}$$

where $T \ge h^{1-\delta}$ and T' are arbitrary.

Formula (7.4.3) yields that the operator $\mathcal{A} = T^{*}\mathcal{F}^{*}A\mathcal{F}T$ applied to $\mathcal{U}$ can be replaced[19] *in the indicated representation* by the operator

$$\mathcal{A} = \mathcal{A}_0 + \mathcal{A}_1 + \mathcal{A}_2 \tag{7.4.8}$$

where we use the same notation, $\mathcal{A}_0$ and $\mathcal{A}_1$ are of block-diagonal type[20] and

(i) For $d = D = 2$ the blocks of $\mathcal{A}_0$ and $\mathcal{A}_1$ are

$$\mathcal{A}_{0(j)} = \sum_{0\le n\le N} B_n(x_1, h_1D_1, h_1)\mu^{-2n}(2j\mu h)^{n+\frac{1}{2}}\sigma_2 + \sigma_0 m + \bar{W}(x_1,h_1D_1,h_1)I \qquad (j \ge 1), \tag*{$(7.4.9)_j$}$$

$$\mathcal{A}_{0(0)} = -\varsigma m + \bar{W}(x_1,h_1D_1,h_1)I, \tag*{$(7.4.9)_0$}$$

and

$$\mathcal{A}_{1(j)} = \sum_{0\le n\le N} B_n(x_1,h_1D_1,h_1)\mu^{1-2n}h \times \bigl(\mathcal{P}'_{n-1}(2j\mu h)\sigma_1 + \mathcal{P}''_{n-1}(2j\mu h)\sigma_2\bigr)\sqrt{2j\mu h} \qquad (j \ge 1), \tag*{$(7.4.10)_j$}$$

$$\mathcal{A}_{1(0)} = 0 \tag*{$(7.4.10)_0$}$$

respectively, $\mathcal{P}'_{n-1}$ and $\mathcal{P}''_{n-1}$ are absolute (i.e., independent of the operator) polynomials of degree $n-1$ and the operator

$$\mathcal{A}_2 = T^{*}\mathcal{F}^{*}(V\cdot I)\mathcal{F}T - \bar{W}\cdot I \tag{7.4.11}$$

has blocks

19) With error $O(h^{s})$ after this application.

20) In what follows the blocks of an operator $\mathcal{B}$ of block-diagonal type are denoted by $\mathcal{B}_{(j)}$.

(7.4.12) $$\mathcal{A}_{2,jk} = W_{jk} \cdot I + W^0_{jk}\sigma_0,$$

which for $j = 0$ and (or) for $k = 0$ have an additional factor Π_- and (or) $\Pi_+^\dagger$ from the left and (or) right respectively where $\Pi_\pm : \mathbb{H} \to \mathbb{H}^\pm$ are projectors and the adjoint operators $\Pi_\pm^\dagger : \mathbb{H}^\pm \to \mathbb{H}$ are embedding operators; all these blocks can contain the following small factors: h_1 arising from commutation and surely contained in W^0_{jk} and (or) μ^{-1} (there are at least $|j-k|$ copies of this factor); moreover, there is surely a factor $(m+1)^{-1}$ and under conditions $(7.2.25)_{1-3}$ there is surely a factor $\bar\nu$. Therefore

$(7.4.13)_1$ $$W_{jk} = \bar\nu(m+1)^{-1}\mu^{-|j-k|}\bigl((1+j+k)\mu h\bigr)^{\frac{1}{2}|j-k|}\tilde W_{jk},$$

$(7.4.13)_2$ $$W^l_{jk} = \bar\nu(m+1)^{-1}h\mu^{-|j-k|-1}\bigl((1+j+k)\mu h\bigr)^{\frac{1}{2}|j-k|}\tilde W^l_{jk}$$

where $W_{\dots}$ and $\tilde W_{\dots}$ here and below are **h**-pseudo-differential operators uniformly with respect to all parameters. Moreover, for $j = k$ the operator W_{kk} surely contains either a factor $h_1 = \mu^{-1}h$ or μ^{-2} and therefore

$(7.4.13)_3$ $$W_{kk} = \bar\nu(m+1)^{-1}\mu^{-2}\bigl((1+j+k)\mu h\bigr)\tilde W_{kk}.$$

(ii) For $d = 3$, $D = 4$ the blocks of $\mathcal{A}_0$ and $\mathcal{A}_1$ are

$(7.4.14)_j$ $$\mathcal{A}_{0(j)} = \sum_{0\le n\le N} \bigl(B_n(x', h_1D_1, h_1)\mu^{-2n}(2j\mu h)^n \times \bigl(\sqrt{2j\mu h}\sigma_2 + hD_2\sigma_3)\bigr)^w + \sigma_0 m + \bar W(x', h_1D_1, h_1)I \quad (j\ge 1),$$

$(7.4.14)_0$ $$\mathcal{A}_{0(0)} = \bigl(B_0(x', h_1D_1, h_1)\sigma'_3 hD_2\bigr)^w + \sigma'_0 m,$$

and

$(7.4.15)_j$ $$\mathcal{A}_{1(j)} = \sum_{0\le n\le N} B_n(x', h_1D_1, h_1)\mu^{1-2n}h \times \bigl(\mathcal{P}'_{n-1}(2j\mu h)\sigma_1 + \mathcal{P}''_{n-1}(2j\mu h)\sigma_2\bigr)\sqrt{2j\mu h} + \sum_{0\le n\le N'} B'_n(x', h_1D_1, h_1)\mu^{-1-2n}h(2j\mu h)^n\sigma_4 \quad (j\ge 1),$$

$(7.4.15)_0$ $$\mathcal{A}_{1(0)} = B'_0(x', h_1D_1, h_1)\mu^{-1}h\sigma'_4$$

respectively, where σ'_0, σ'_3 and σ'_4 are the restrictions of σ_0, σ_3 and σ_4 to $\mathbb{H}^-$, σ_4 is defined in (7.2.2), $\mathcal{A}_2$ is defined by (7.4.11) and its blocks are

(7.4.16) $$\mathcal{A}_{2,jk} = W_{jk} \cdot I + \sum_{l=1,2,3} W^l_{jk}\hat\sigma_l,$$

where W_{jk} and W^l_{jk} satisfy $(7.4.13)_{1-3}$ with $\bar\nu = 1$ and for $j = 0$ and (or) $k = 0$ the right-hand expression of (7.4.16) has an additional factor Π_- and (or) $\Pi_+^\dagger$ from the left and (or) right respectively.

(iii) Similar formulas are valid for $d = 3$, $D = 2$ but in $(7.4.14)_j$, $(7.4.14)_0$ the terms $\sigma_0 m$ and $\sigma'_0 m$ are absent, in $(7.4.15)_j$, $(7.4.15)_0$ the matrices σ_4, σ'_4 are replaced by I and in (7.4.16) the matrices $\hat\sigma_l$ are replaced by σ_l.

Now for $d = 3$, $W_{jk} = W_{jk}(x', h_1D', h_1)$ but later we will get rid of h_1D_2.

7.4.2 Further reduction

Thus we have the problem

$$\Bigl(hD_t - (m+1)(\mathcal{A}_{0(j)} + \mathcal{A}_{1(j)})\Bigr)\mathcal{U}_{jl} - (m+1)\sum_{|k-j|\le N} \mathcal{A}_{2,jk}\mathcal{U}_{kl} \equiv 0, \tag{7.4.17}$$

$$\mathcal{U}_{jl}|_{t=0} = \delta_{jl}\delta(x'-y')I \tag{7.4.18}$$

with $N = N(s,\delta)$.

Formulas (7.4.7), (7.4.6) and the uniform (with respect to all parameters) ellipticity of the operator $hD_t - (m+1)(\mathcal{A}_{0(j)} + \mathcal{A}_{1(j)})$ for the τ, j and ξ_2 (for $d=3$) indicated below and the fact that the operator $\mathcal{A}_{2,jk}$ surely contains the factor $(m+1)^{-1}\mu^{-1}$ yield the following

Proposition 7.4.1. *Suppose that condition (7.4.1) is fulfilled with $\epsilon_1 = \epsilon_1(c,\epsilon_0,s)$. Then*

(i) The estimate

$$|F_{t\to h^{-1}\tau}\chi_{T,T'}(t)\mathcal{U}_{jk}| \le C'Th^s(1+j\mu h)^{-s} \qquad \forall x',y' \in \mathbb{R}^{d-1} \quad \forall \tau : |\tau| \le \epsilon = \epsilon(c) \quad \forall j \ge C_0\mu^{-1}h^{-1} \tag{7.4.19}$$

holds with arbitrary $T \ge h^{1-\delta}$, T', s.

(ii) Moreover, if $d=2$ and condition (7.3.24) is fulfilled then this estimate also holds for all $j \le \epsilon\mu^{-1}h^{-1}$ with a small enough constant $\epsilon = \epsilon(c,\epsilon_0) > 0$.

(iii) If $d=3$, condition (7.3.24) is fulfilled and $Q = q(x',\mathbf{h}'D')$ with q is supported in the domain $\{|\xi_2| \le \epsilon\}$ with a small enough constant $\epsilon = \epsilon(c,\epsilon_0) > 0$ then the estimate

$$|F_{t\to h^{-1}\tau}\chi_{T,T'}(t)Q_x\mathcal{U}_{jk}| \le C'Th^s \qquad \forall x',y' \in \mathbb{R}^{d-1} \quad \forall \tau : |\tau| \le \epsilon \quad \forall j \le \epsilon\mu^{-1}h^{-1} \tag{7.4.20}$$

holds with arbitrary $T \ge h^{1-\delta}$, T', s.

(iv) On the other hand, if q is supported in the domain $\{|\xi_2| \ge \frac{1}{2}C_0\}$ then estimate (7.4.20) holds for all $j \in \mathbb{Z}^+$.

In what follows $\tilde{\Omega}$ denotes the domain Ω for $d=2$ and the domain $\Omega \times \{\xi_2, |\xi_2| \le C_0\}$ for $d=3$.

Proposition 7.4.2. *There exist operators $\mathcal{T}'_{(j)} = \mathcal{T}'_{(j)}(x',\mathbf{h}'D',\mu^{-1},\mathbf{h}')$ with $j \in \mathcal{J} = \{\epsilon\mu^{-1}h^{-1} \le j \le C_0\mu^{-1}h^{-1}\}$ such that in $\tilde{\Omega}$ the following equalities hold:*

$$\mathcal{T}'^{\,*}_{(j)}\mathcal{T}'_{(j)} \equiv \mathcal{T}'_{(j)}\mathcal{T}'^{\,*}_{(j)} \equiv I, \tag{7.4.21}$$

$$(m+1)\mathcal{T}'^{\,*}_{(j)}(\mathcal{A}_{0(j)} + \mathcal{A}_{1(j)})\mathcal{T}'_{(j)} \equiv \mathcal{B}_{(j)} = \begin{pmatrix} \mathcal{B}^+_{(j)} & 0 \\ 0 & \mathcal{B}^-_{(j)} \end{pmatrix} \tag{7.4.22}$$

where $\mathcal{B}^+_{(j)}$ and $\mathcal{B}^-_{(j)}$ are operators with principal symbols $(m+1)\bigl(\bar{W}(x',\xi') \pm L_{(j)}(x',\xi')\bigr)I$,

$$\bar{W} = V \circ \bar{\Psi}, \tag{7.4.23}$$

$$L_{(j)} = \bigl(m^2 + 2j\mu h\bar{B}^2\bigr)^{\frac{1}{2}} \qquad \text{for } d=2, \tag{7.4.24$_{(2)}$}$$

$$L_{(j)} = \left(m^2 + (2j\mu h + \xi_2^2)\bar{B}^2\right)^{\frac{1}{2}} \qquad \text{for } d = 3, \tag{7.4.24$_{(3)}$}$$

$$\bar{B} = \sum_{0\le n\le N} B_n(x', \xi', 0)(2j\mu h)^n. \tag{7.4.25}$$

(ii) Moreover, the inequalities

$$|\mathcal{D}^\alpha D_z^\beta(\mathcal{T}'_{(j)} - I)| \le c_{\alpha\beta}\bar{\nu}_\alpha(m+1)^{-1}, \tag{7.4.26$_1$}$$

$$|\mathcal{D}^\alpha D_z^\beta(\mathcal{B}_{(j)} - \sigma_0 m(m+1) - (m+1)\bar{W}\cdot I)| \le c_{\alpha\beta}\bar{\nu}_\alpha(m+1)^{-1} \quad \forall \alpha, \beta : |\alpha| \le K', \beta \le K' \tag{7.4.26$_2$}$$

hold where here and below $\mathcal{D} = \nabla_{(x,\xi,\mathbf{h}',\mu^{-1})}$, $z = 2j\mu h$, *j is now any real number in the interval* $(\epsilon\mu^{-1}h^{-1}, C_0\mu^{-1}h^{-1})$, $\bar{\nu}_\alpha = 1$ *for* $\alpha = 0$ *and* $\bar{\nu}_\alpha = \bar{\nu}$ *for* $|\alpha| \ge 1$ *and we assume that conditions* $(7.2.25)_{1-3}$ *are fulfilled for* $d = 2$ *and* $\bar{\nu} = 1$ *for* $d = 3$.

Remark. *In all the inequalities below* $j, k \in \mathbb{R}^+$ *(i.e., all the functions of* $j, k \in \mathbb{Z}^+$ *are extended to functions of* $j, k \in \mathbb{R}^+$ *satisfying the indicated inequalities); here* $C_{j,j}$ *and* $C_{j,k}$ *are extended separately.* ◊

Proof. The proof is obvious because for $j \ge 1$ the principal symbol of $\mathcal{A}_{0(j)}$ is unitarily equivalent to the matrix $\bar{W}\cdot I + L_{(j)}\sigma_0$ [21] and the operator realizing this unitary equivalence obviously also satisfies $(7.4.26)_1$. □

Let us define $\mathcal{T}'_{(j)}$ for all $j \in \mathbb{R}^+$ so that inequalities $(7.4.26)_{1,2}$ are fulfilled for all j and so that $\mathcal{T}'_{(j)} = I$ for $j \le \frac{1}{2}\epsilon(\mu h)^{-1}$ and for $j \ge 2C_0(\mu h)^{-1}$. Let $\mathcal{T}' = (\mathcal{T}'_{(j)})_{j\in\mathbb{Z}^+}$ be an operator of block-diagonal type. Let us set $\mathcal{A}^I = \mathcal{T}'^*\mathcal{A}\mathcal{T}'$.

Proposition 7.4.3. *There exists an operator* $\mathcal{K} = (\mathcal{K}_{jk})_{j,k\in\mathbb{Z}^+}$ *such that in* $\tilde{\Omega}$

$$\mathcal{K}^*\mathcal{K} \equiv \mathcal{K}\mathcal{K}^* \equiv I, \tag{7.4.27}$$

$$(m+1)\mathcal{K}^*\mathcal{A}^I \equiv \mathcal{B} + \mathcal{B}' + \mathcal{R}, \tag{7.4.28}$$

$$\mathcal{R}_{jk} \equiv 0 \qquad \forall (j,k) \in \mathcal{J}\times\mathbb{Z}^+ \cup \mathbb{Z}^+\times\mathcal{J} \tag{7.4.29}$$

where $\mathcal{B}$ *is an operator of block-diagonal type with blocks* $\mathcal{B}_{(j)}$,

$$\mathcal{B}'_{jk} = \begin{pmatrix} \mathcal{B}'^{+}_{jk} & 0 \\ 0 & \mathcal{B}'^{-}_{jk} \end{pmatrix} \tag{7.4.30}$$

and for $|\alpha| \le K'$, $\beta \le K'$

$$|\mathcal{D}^\alpha D^\beta_{(z)}\mathcal{K}_{jk}| \le c_{\alpha\beta}\mu^{-|j-k|}\bar{\nu}(m+1)^{-1} \qquad \text{for } j \ne k, \tag{7.4.31$_1$}$$

$$|\mathcal{D}^\alpha D^\beta_z(\mathcal{K}_{jj} - I)| \le c_{\alpha\beta}\mu^{-2}\bar{\nu}(m+1)^{-1} \qquad \text{for } j = k, \tag{7.4.31$_2$}$$

$$|\mathcal{D}^\alpha D^\beta_{(z)}\mathcal{B}'_{jk}| \le c_{\alpha\beta}\mu^{-|j-k|}\bar{\nu}, \tag{7.4.31$_3$}$$

$$|\mathcal{D}^\alpha D^\beta_{(z)}\mathcal{B}'_{jj}| \le c_{\alpha\beta}\mu^{-2}\bar{\nu} \tag{7.4.31$_4$}$$

where we set $z = 2j\mu h$, $z' = 2k\mu h$, $j, k \in \left(\epsilon(\mu h)^{-1}, C_0(\mu h)^{-1}\right)$ *are not necessarily integers and* $D_{(z)} = D_z + D_{z'}$.

[21] Where here and below for $d = 3, D = 2$ the notation σ_0 is used instead of σ_4.

Proof. By induction on M we construct an operator $\mathcal{K} = \mathcal{K}_M$ satisfying (7.4.27), (7.4.28), (7.4.30) and $(7.4.31)_{1-4}$ with $\mathcal{R}$ such that

$$(7.4.32)_M \quad |\mathcal{D}^\alpha D^\beta_{(z)} \mathcal{R}_{jk}| \le c_{\alpha\beta}\mu^{-\max(M,|j-k|)}\bar{\nu}$$
$$\forall (j,k) \in \mathcal{J}' \times \mathbb{R}^+ \cup \mathbb{R}^+ \times \mathcal{J}', \quad \mathcal{J}' = \bigl(\epsilon(\mu h)^{-1}, C_0(\mu h)^{-1}\bigr).$$

The latter inequality with large M provides (7.4.29). Obviously, one can set $\mathcal{K}_1 = I$. Let us assume that $(7.4.32)_M$ is fulfilled and let us set $\mathcal{K}_{M+1} = \mathcal{K}_M \exp i\mathcal{L}$ where the operator $\mathcal{L}$ is Hermitian, skew-commutes with σ_0 (i.e., $\mathcal{L} = \begin{pmatrix} 0 & * \\ * & 0 \end{pmatrix}$) and the symbols of its blocks satisfy the inequalities

$$(7.4.33)_1 \qquad |\mathcal{D}^\alpha D^\beta_{(z)} \mathcal{L}_{jk}| \le c_{\alpha\beta}\mu^{-|j-k|}\bar{\nu}(m+1)^{-1},$$

$$(7.4.33)_2 \qquad |\mathcal{D}^\alpha D^\beta_z \mathcal{L}_{jj}| \le c_{\alpha\beta}\mu^{-2}\bar{\nu}(m+1)^{-1}$$

and their principal symbols are defined by the equations

$$(7.4.34) \quad i\sigma_0(L_{(j)} + L_{(k)})(x',\xi')\mathcal{L}^0_{jk} = -(m+1)^{-1}\mathcal{R}^0_{jk}$$
$$\forall (j,k) \in \mathcal{J} \times \mathbb{Z}^+ \cup \mathbb{Z}^+ \times \mathcal{J}$$

where $\mathcal{R}^0_{jk}$ are the principal symbols of $\mathcal{R}_{jk}$. One can assume without loss of generality that they also skew-commute with σ_0 [22] and the operator $\mathcal{R}$ is Hermitian.

It is easy to prove that $\mathcal{K}_{M+1}$ also satisfies (7.4.27), (7.4.28), (7.4.30), (7.4.31) and $(7.4.32)_{M+1}$. □

Remark 7.4.4. *Here and in what follows the principal symbols of operators can contain the factors μ^{-1} and h.* ◇

Let us set $\mathcal{U}' = \mathcal{K}^*_x \mathcal{T}'^*_x \mathcal{U}^t \mathcal{T}'^{*t}_y \mathcal{K}^*_y$. Then in $\tilde{\Omega}$

$$(7.4.35) \qquad (hD_t - \mathcal{B}_{(j)})\mathcal{U}'_{jl} - \sum_{|j-k|\le N} \mathcal{B}'_{jk}\mathcal{U}'_{kl} \equiv 0,$$

$$(7.4.36) \qquad \mathcal{U}'_{jl}|_{t=0} \equiv \delta_{jl}\delta(x'-y')I$$

for $j \in \mathcal{J}$. Let us decompose $\mathcal{U}'_{jl}$ into four blocks $\mathcal{U}'^{\varsigma\varsigma'}_{kl}$ with $\varsigma, \varsigma' = \pm$. Then (7.4.22), (7.4.30), (7.4.29) yields that system (7.4.35) is decomposed into four separate subsystems. Moreover, if condition (7.3.24) is fulfilled then two of them are elliptic in the zone in question. Namely, subsystems with ς such that

$$(7.4.37)_\varsigma \qquad \varsigma V > \epsilon_0 \qquad \text{in } B(0,1)$$

are elliptic. Taking into account that $\mathcal{U}'_{jk}$ also satisfies (7.4.5) and (7.4.6) we then obtain the following

Proposition 7.4.5. *Under conditions (7.3.24), $(7.4.37)_\varsigma$ the distribution $\mathcal{U}'^{\varsigma\cdot}_{jk}$ satisfies (7.4.20) for all $j \in \mathcal{J}$ and $\mathcal{U}'^{\cdot\varsigma}_{jk}$ satisfies (7.4.20) for all $k \in \mathcal{J}$. Let us recall that the symbol of Q is supported in $\tilde{\Omega}$.*

[22] Otherwise the part of $\mathcal{R}$ commuting with σ_0 can be included in $\mathcal{B}'$.

So we need to treat the problem

$$(hD_t - \mathcal{B}^+_{(j)})\mathcal{U}''_{jl} - \sum_{|k-j|\le N} \mathcal{B}'^+_{jk}\mathcal{U}''_{kl} \equiv 0, \tag{7.4.38$_+$}$$

$$\mathcal{U}''_{jk}|_{t=0} \equiv \delta_{jk}\delta(x'-y')I \tag{7.4.39}$$

where for definiteness we have assumed that condition (7.4.37)$_-$ is fulfilled and we set $\mathcal{U}''_{jk} = \mathcal{U}'^{++}_{jk}$.

Proposition 7.4.6. *There exists an operator* $\mathcal{K}' = (\mathcal{K}'_{jk})_{j,k\in\mathbb{Z}^+}$ *with scalar blocks satisfying* (7.4.27) *such that*

$$\mathcal{K}'^*(\mathcal{B}^+ + \mathcal{B}'^+ + \mathcal{R})\mathcal{K}' \equiv \mathcal{B}^+ + \mathcal{B}'' + \mathcal{R}' \tag{7.4.40}$$

where $\mathcal{R}'$ *satisfies* (7.4.29),

$$\mathcal{K}_{jk} \equiv 0 \qquad \text{for } |j-k| \ge N(1+\mu^{-2}h^{-1-\delta'}), \tag{7.4.41}$$

$$|\mathcal{D}^\alpha D^\beta_{(z)}\mathcal{B}''_{jk}| \le c_{\alpha\beta}\mu^{-|j-k|}\bar{\nu}\mu h \qquad \text{for } j\ne k \tag{7.4.42}$$

and $\mathcal{B}''_{jj}$ *satisfies* (7.4.31)$_4$, $N = N(s,\delta,\delta')$ *and* $\delta' > 0$ *is arbitrary.*

Proof. We again apply induction on M fulfilling inequality (7.4.32)$_M$ at every step. For $D=2$ all the blocks are one-dimensional and for $D=4$ they are two-dimensional but every block $\mathcal{B}'^+_{jk}$ is the sum of a scalar operator $\mathcal{B}'^+_{jkI}$ and an operator $\mathcal{B}'^+_{jkII}$ satisfying (7.4.42). Let us again set $\mathcal{K}'_{M+1} = \mathcal{K}'_M \exp i\mathcal{L}'$ where $\mathcal{L}'$ is a Hermitian operator with scalar blocks $\mathcal{L}'_{jk}$ such that

$$|\mathcal{D}^\alpha D^\beta_{(z)}\mathcal{L}'_{jk}| \le c_{\alpha\beta}\mu^{-\max(M,|j-k|)}\bar{\nu}(\mu h)^{-1}, \tag{7.4.43}$$

$\mathcal{L}'_{jj} = 0$ and the principal symbols of $\mathcal{L}'_{jk}$ are defined by the equations

$$i(L_{(j)} - L_{(k)})\mathcal{L}'^0_{jk} + (m+1)^{-1}\mathcal{B}'^{+0}_{jkI} = 0 \qquad \text{for } j\ne k. \tag{7.4.44}$$

We see that $\mathcal{L}_{jk}$ satisfy (7.4.42).

On the other hand, even without (7.4.44) it is easy to prove that if $\mathcal{L}'$ is a Hermitian operator with scalar blocks and $\mathcal{Q}_{(l)} = (\operatorname{Ad}\mathcal{L}')^l\mathcal{B}^+$, $\mathcal{Q}'_{(l)} = (\operatorname{Ad}\mathcal{L}')^l\mathcal{B}'^+_I$, $\mathcal{Q}'_{(l)} = (\operatorname{Ad}\mathcal{L}')^l\mathcal{B}''^+_{II}$ then

$$|\mathcal{D}^\alpha D^\beta_{(z)}\mathcal{Q}_{(l),jk}| \le c_{\alpha\beta l}\mu^{-\max(M+l-1,|j-k|)}\bar{\nu}, \tag{7.4.45$_1$}$$

$$|\mathcal{D}^\alpha D^\beta_{(z)}\mathcal{Q}'_{(l),jk}| \le c_{\alpha\beta l}\mu^{-\max(M+l,|j-k|)}\bar{\nu} \tag{7.4.45$_2$}$$

and the second estimate with an additional factor μh in the right-hand expression holds for $\mathcal{Q}''_{(l),jk}$. This yields that if $\mathcal{K}'_{M+1} = \exp i\mathcal{L}'$ then

$$\mathcal{K}'^*_{(M+1)}(\mathcal{B}^+ + \mathcal{B}''^+ + \mathcal{R}_M)\mathcal{K}'_{(M+1)} = \mathcal{B}^+ + \mathcal{B}''^+ + i[\mathcal{B}^+,\mathcal{L}'] + \mathcal{R}'_{M+1} \tag{7.4.46}$$

where $\mathcal{R}'_{M+1}$ satisfies (7.4.32)$_{M+1}$.

Let us note that if we define $\mathcal{L}'$ by formula (7.4.44) then $(\exp(it\mathcal{L}'))_{jk} \equiv 0$ for $|j-k| > N$, $|t| \le \min(1,\mu^{1+M}h^{1+\delta'})$ and therefore $\mathcal{K}'_{(M+1)jk} \equiv 0$ for $|j-k| \ge N(1+\mu^{-1-M}h^{-1-\delta'})$. This fact and (7.4.46) yield that $\mathcal{K}'_{M+1} = \mathcal{K}'_M\mathcal{K}'_{(M+1)}$ provides the induction step. □

Proposition 7.4.7. *There exists an operator $\mathcal{K}'' = (\mathcal{K}''_{jk})_{j,k\in\mathbb{Z}^+}$ which satisfies* (7.4.27), *$\mathcal{K}''_{jk} = 0$ for $|j-k| \ge N$, and*

$$\mathcal{K}'^*(\mathcal{B}^+ + \mathcal{B}'' + \mathcal{R}')\mathcal{K}' \equiv \mathcal{B}^+ + \mathcal{B}''' + \mathcal{R}'' \tag{7.4.47}$$

where $\mathcal{R}''$ satisfies (7.4.29),

$$|\mathcal{D}^\alpha D^\beta_{(z)}\mathcal{K}''_{jk}| \le c_{\alpha\beta}\mu^{-|j-k|}\bar{\nu} \qquad \text{for } j \ne k \tag{7.4.48$_1$}$$

$$|\mathcal{D}^\alpha D^\beta_z \mathcal{K}''_{jj}| \le c_{\alpha\beta}\mu^{-2}\bar{\nu} \qquad \text{for } j = k; \tag{7.4.48$_2$}$$

$\mathcal{B}'''$ is operator of block-diagonal type and satisfies (7.4.31)$_4$.

Proof. The proof is similar to the proof of the previous proposition but now we have matrix operators $\mathcal{L}''_{jk}$ for $D = 4$ which satisfy (7.4.43) with an additional factor μh in the right-hand expression; in particular, all of them are $O(\mu^{-1})$. □

Let us now set $\mathcal{U}''' = \mathcal{K}'''^*_x \mathcal{U}''\, {}^t\mathcal{K}'''^*_y$ with $\mathcal{K}''' = \mathcal{K}'\mathcal{K}''$. Then in $\tilde{\Omega}$

$$(hD_t - \mathcal{C}_{(j)})\mathcal{U}'''_{jl} \equiv 0 \quad \forall j \in \mathcal{J}, \tag{7.4.49}$$

$$\mathcal{U}_{jl}|_{t=0} \equiv \delta_{jl}\delta(x'-y')I \tag{7.4.50}$$

where $\mathcal{C}_{(j)} = \mathcal{B}^+_{(j)} + \mathcal{B}'''_{(j)}$. The reduction is complete.

7.4.3 Estimates

Let $H_{(j)} = H_{(j)}(x',\xi',\mu h,\mu^{-1},\mathbf{h}')$ be the Weyl symbols of the operators $\mathcal{C}_{(j)}$. Our constructions yield that $H_{(j)}$ are smooth functions of $x',\xi', z = 2j\mu h, \mu^{-1}, \mathbf{h}'$ and

$$H_{(j)} = (m+1)(\bar{W} + L_{(j)}) + O(\mu^{-2}\bar{\nu}). \tag{7.4.51}$$

It is easy to see that for $d = 2$ the operators $\mathcal{F}$ and $\mathcal{T}$ depend on μ and h only through $\mu^{-1}h$ and μ^{-2} and the operators are smooth functions of those arguments. Therefore the same is true for the operators W_{jk} and for $H_{(j)}$ as well. In particular,

$$H_{(j)} = (m+1)(\bar{W} + L_{(j)}) + \omega'_1\mu^{-2} + \omega'_2\mu^{-1}h + O(\bar{\nu}\mu^{-4} + \bar{\nu}\mu^{-2}h^2) \qquad (d=2) \tag{7.4.52}$$

with as yet unknown coefficients ω'_1 and ω'_2.

Repeating the corresponding arguments of section 6.4 we see that the following statements similar to theorem 6.4.1 and propositions 6.4.2–6.4.3, 6.4.6 are true:

Theorem 7.4.8. *Let conditions* (7.3.24) *and* (7.4.1) *be fulfilled and let q' be supported in $\{|\xi_2| \le 2\epsilon\}$ for $d = 3$. Then*

(i) Let $|H_{(j)}| \ge \epsilon_0$ for all $(x,\xi') \in \tilde{\Omega}$. Then for $\mathcal{U}'''_{jk}$ estimate (7.4.20) *holds for $|\tau| \le \epsilon$.*

(ii) For $T \in [h^{1-\delta}, T^*]$ with $T^* = \mu T_0$ for $d = 2$ and $T^* = T_0$ for $= 3$, estimate (7.4.20) *holds for $\mathcal{U}'''_{jk}$ with $j \ne k$; hence this estimate holds for $\mathcal{U}_{jk}$ with $|j-k| \ge N(1+\mu^{-2}h^{-1-\delta'})$ as well; recall that $T_0 = T_0(c,_0\,\epsilon) > 0$ and $\epsilon = \epsilon(c,\epsilon_0) > 0$ are small enough constants.*

(iii) Let **either** $d = 2$ *and one of the conditions* $(6.3.5)_{1,2}$ *be fulfilled* **or** $d = 3$ *and one of the conditions* $(6.3.5)_{1-4}$ *be fulfilled. Then for* $T \in [h^{1-\delta}, T^*]$ *the following estimate holds:*

$$(7.4.53) \qquad |F_{t\to h^{-1}\tau}\chi_T(t)Q'_x\mathcal{U}'''^{t}_{y}Q''_{jk}| \le C'h^s \qquad \forall x', y' \in \mathbb{R}^{d-1} \quad \forall \tau : |\tau| \le \epsilon;$$

therefore this estimate holds for $\mathcal{U}_{jk}$ *also.*

(iv) Let $d = 2$ *and let conditions* $(7.2.25)_{1-3}$, $(6.3.10)$ *and* $(7.3.21)$ *be fulfilled. Then all the previous and the following statements remain true for* $T^* = \mu\nu^{-1}T_0$, $T \in [h^{1-\delta}\nu^{-1}, T^*]$, $|\tau| \le \nu$.

(v) Under the assumptions of proposition 6.4.6 (with conditions (6.3.19) *and* (6.3.24) *replaced by* (7.3.22)*) estimate* (6.4.23) *holds for* $\mathcal{U}'''$.

7.4.4 Calculations

Let us pass from estimates to calculations. Let us note first of all that the operator $q = q(x, \mathbf{h}'D', \mathbf{h}')$ in the representation R is transfered to the block-operator $Q = (Q_{jk})_{j,k\in\mathbb{Z}^+}$ where the symbols of the operators Q_{jk} satisfy the estimates

$$(7.4.54) \qquad |\mathcal{D}^\alpha D^\beta_{(z)} Q_{jk}| \le c_{\alpha\beta}\mu^{-|j-k|} \qquad \forall j,k \in \mathcal{J}'.$$

Moreover, one can decompose these symbols into asymptotic series in μ^{-2} and $\mathbf{h}'$. It is easy to prove that if $\mathcal{M}$ is one of the operators $\mathcal{T}'$, $\mathcal{K}$, $\mathcal{K}'$, $\mathcal{K}'''$ constructed in propositions 7.4.2, 7.4.3, 7.4.6 and 7.4.7 respectively (or if $\mathcal{M}$ is a product of these operators) then the operator $\mathcal{M}^*Q\mathcal{M}$ also has these properties. Therefore using the arguments used in the proofs of propositions 6.4.4, 6.4.5 and applying theorem 7.4.8 we obtain the following

Proposition 7.4.9. *Let conditions* (7.3.24) *and* (7.4.1) *be fulfilled. Then*

(i) Let $d = 2$ *and* $T \in [h^{1-\delta}\nu^{-1}, T^*]$, $Q' = I$ *and let conditions* $(7.2.25)_{1-3}$, (6.3.10) *and* (7.3.21) *be fulfilled. Then*

$$(7.4.55) \qquad \sigma(t) = F_{t\to h^{-1}\tau}\chi_T(t)\operatorname{tr}\Gamma\bigl(Q'_x\psi(x)U\bigr) \equiv F_{t\to h^{-1}\tau}\chi_T(t)\sum_j \Gamma'(f_j\mathcal{U}'''_{jj}) \qquad \forall \tau \in [-\epsilon\nu, \epsilon\nu]$$

modulo $O(h^s)$ *with the summation over* $j \in \mathcal{J}$,

$$(7.4.56) \qquad f_j = f(x', \mathbf{h}'D', 2j\mu h, \mu^{-2}, \mathbf{h}')$$

where f *is a smooth function of its arguments,*

$$(7.4.57)_{(2)} \qquad f = f_0 \cdot I + O(\mu^{-2}\bar{\nu}), \qquad f_0 = \psi \circ \bar{\Psi}.$$

(ii) Let $d = 3$, *let* $Q = Q(x', \mathbf{h}'D', \mathbf{h}')$ *be a scalar operator with symbol supported in* $\{|\xi_2| \le \epsilon\}$, $Q' = \mathcal{F}TQT^*\mathcal{F}^*$ *and* $T \in [h^{1-\delta}, T^*]$. *Then equality* (7.4.55) *holds for* $\tau \in [-\epsilon, \epsilon]$. *Moreover,* (7.4.56) *remains true and*

$$(7.4.57)_{(3)} \qquad f = f_0 \cdot I + O(\mu^{-2} + h), \qquad f_0 = q_0(\psi \circ \bar{\Psi})$$

where q *and* q_0 *are the symbol and the principal symbol of the operator* Q *respectively. Finally,*

$$(7.4.58) \qquad \operatorname{supp} f \subset \operatorname{supp} q.$$

7.4.5 Spectral asymptotics

Repeating the corresponding considerations of section 6.4 we can easily prove that under the assumptions of proposition 6.4.7 and the additional condition (7.3.24)[23] for

$$\text{(7.4.59)}\quad \mathcal{R}_4 = \Big| \int \phi(\tau')\Big(\operatorname{tr} \Gamma\big(\tilde{Q}'_x\psi'(x)e\big)(\tau',\tau) - \sum_{j\in\mathcal{J}}\sum_{p,r\le N} \Theta_{j,pr}(\tau',\tau)h^{1+r-d}\mu^{-1-2p-r}\Big)d\tau'\Big|$$

estimate (6.4.28) holds where

$$\Theta_{j,pr}(\tau',\tau) = \Theta_{j,pr}(\tau)\big|_{\tau=\tau'}^{\tau=\tau},$$

$\Theta_{j,pr}$ are given by formulas (6.4.29) with $H_{(j)}$ instead of $\mathcal{A}_j$ and with $f_{j,pr} = f_{pr}(x',\xi',2j\mu h)$ instead of $f_{j,pr}$, (6.4.30) are fulfilled and, moreover, the inequality

$$\text{(7.4.60)}\qquad |D^\alpha_{x',\xi'}D^\beta_z f_{pr}| \le c_{\alpha\beta}\gamma^{-|\alpha|-2r} \qquad \forall\alpha : |\alpha|\le K', \beta\le K'$$

holds instead of (6.4.31) and equality

$$\text{(7.4.61)}\qquad f_{00} = (2\pi)^{1-d}\frac{D}{2}q'(\psi'\circ\bar{\Psi}_0)$$

holds instead of (6.4.32) and we assume that $\phi\in C_0^K([-\epsilon,\epsilon])$; $R_3 = R_{30}$ is given by (6.3.27). Finally, under the indicated additional conditions in the assumptions of proposition 6.4.8 estimate (6.4.34) holds for $\mathcal{R}_3$.

This yields that in the context of proposition 6.4.9 with $H_{(j)}$ instead of $\mathcal{A}_j$ and with $Q' = Q(I-Q'')$ instead of $Q' = I-Q''$ for $d=3$ (provided the symbol of Q is supported in $\{|\xi_2|\le 2\epsilon\}$, $Q\sim I$ in $\{|\xi_2|\le\epsilon\}$) estimates $(6.4.36)_0$ and $(6.4.38)_0$ (with $\vartheta=0$) hold for $d=2$ and for $d=3$ respectively; we assume that condition (7.3.24) is fulfilled.

Moreover, applying the results of section 4.4 we see that for $Q'=Q''=I$ the following estimates hold:

(i) For $d=2$ estimate $(6.4.41)_0$ holds under condition (6.4.42) and estimate $(6.4.39)_0$ holds for $\vartheta=0$ under condition (6.4.40).

(ii) For $d=3$ estimate $(6.4.44)_0$ holds under condition (6.4.40) and estimate (6.4.43) holds in the general case.

Consequently we obtain the following

Theorem 7.4.10. [24] *Suppose conditions* (7.3.24) *and* (7.4.1) *are fulfilled and let* $\phi\in C_0^K([-\epsilon,\epsilon])$, $\tau=0$. *Then*

(i) Let $d=2$, $Q'=I$ *and let conditions* $(7.2.25)_{1-3}$, (7.4.20) *and* (6.4.10) *be fulfilled. Then estimate* $(6.4.41)_0$ *holds under condition (6.4.42) and estimate* $(6.4.39)_0$ *holds under condition (6.4.40).*

(ii) Let $d=3$. *Then estimate* $(6.4.44)_0$ *holds under condition (6.4.40) and estimate (6.4.43) holds in the general case.*

(iii) In particular, under the condition

$$|\nabla v_*|\ge\epsilon_0 \qquad \text{in } B(0,1)$$

estimates (6.4.36) *with* $\nu=1$ *for* $d=2$ *and* (6.4.44) *for* $d=3$ *hold.*

[23] And under the condition $\operatorname{supp} q\subset\{|\xi_2|\le 2\epsilon\}$ in the case $d=3$.

[24] Compare with theorem 6.4.10.

7.4.6 Effective asymptotics. I

Now our goal is to pass from estimates for $\mathcal{R}_4$ to estimates for

$$\mathcal{R} = \Big|\iint \phi(\tau')\psi(x)\big(\operatorname{tr} e(x,x,\tau',\tau) - h^{-d}S(x,\tau',\tau,\mu h)\big)dxd\tau'\Big|. \tag{7.4.62}$$

To this end we apply the corresponding arguments of section 6.4. Surely, now the functions $f_{..}$ are not necessarily polynomials of $z = 2j\mu h$; however, these functions depend smoothly on z. Taking into account equalities $(7.4.57)_{(2,3)}$ and the fact that f_{00} does not depend on z we easily obtain that the transformations of section 6.4 and the estimate for $\mathcal{R}_6$ obtained in section 6.4 also remain true with an additional term in the right-hand expression; this term is due to the non-polynomial character of $f_{..}$. The additional term is $O(1)$ for $d=2$ and $O(h^{-2})$ for $d=3$ and, moreover, it is $O(h^s)$ under condition (6.3.1).

Surely, for $d=3$ we cannot set $Q'=I$; we can only take Q' with symbol supported in $\{|\xi_2|\le 2\epsilon\}$. In order to avoid this difficulty we temporarily assume that condition (6.3.1) is fulfilled; the case $\mu\in[h^{1-\delta},\epsilon h^{-1}]$ will be treated later. Then for the operator $I-Q'$ with symbol supported in $\{|\xi_2|\ge\epsilon\}$ the results of section 7.3 apply and therefore we obtain the corresponding estimate for

$$\mathcal{R}_7 = \Big|\iint \phi(\tau)\phi_1(\tau')\Big(\int \psi(x)\big(\operatorname{tr} e(x,x,\tau',\tau) - h^{-d}S(x,\tau',\tau)\big)dx$$
$$h^{-d}G_0(\tau,\mu^{-1}) - h^{1-d}G_1(\tau,\mu^{-1}) - h^{-d}G_2(\tau,\mu h) - h^{1-d}G_3(\tau,\mu h)\Big)d\tau d\tau'\Big|$$

(i.e., for $Q'=I$) where G_0,G_1 are polynomials with respect to μ^{-1} and G_2,G_3 are polynomials with respect to μh. Then applying arguments similar to those of section 6.4 we easily prove that these polynomials vanish. In this proof we start with the results of section 7.3 and the equality

$$\iint \phi_0(\tau)\phi(\tau')e(.,.,\tau',\tau)d\tau'd\tau = \int \Phi(\tau)d_\tau e(.,.,\tau_1,\tau)$$

where the choice of τ_1 does not matter and $\Phi\in C_0^K([-\epsilon,\epsilon])$ is a primitive of the function $k_0\phi(\tau) - k\phi_0(\tau)$ with constants $k_. = \int_{\mathbb{R}}\phi_.(\tau)d\tau$.

As a result we obtain the following series of statements where *the following notational convention is used:* * *means that* V *and* $v=\frac{V}{F}$ *are replaced by* $V_* = m^2 - V^2$ *and* $v_* = \frac{V_*}{F}$ *respectively and* $(2j+1)\mu h$ *is replaced by* $2j\mu h$*:*

Theorem 7.4.11. [25] *Let* $d=3$, $\vartheta=0$, $\tau=0$ *and let conditions* (6.1.1)–(6.1.4), (6.2.1), (7.3.24) *be fulfilled. Then for* $\mathcal{R}$ *estimate* $(6.4.44)_0$ *holds under the additional assumption* (6.3.1).

Theorem 7.4.12. [26] *Let* $d=2$, $\tau=0$, $\vartheta=0$ *and conditions* (7.1.1)–(7.1.3), (7.2.1), (7.3.24) *and* (7.4.1) *be fulfilled. Then*

(i) For $\mu^3 h\le C_0$ *estimate* $(6.4.64)^*$ *holds; moreover, under condition* $(6.4.47)^*$ *estimate* (6.4.65) *holds and under condition* $(6.4.66)^*$ *estimate* $(6.4.55)_{(2)}$ *holds.*

(ii) For $\mu^3 h\ge C_0$ *estimate* $(6.4.67)^*$ *holds. Moreover, under condition* $(6.4.47)^*$ *estimate* (6.4.68) *holds, under condition* $(6.4.66)^*_+$ *estimate* $(6.4.55)_{(2)}$ *holds and under condition* $(6.4.66)^*$ *estimate* (6.4.68) *holds.*

(iii) Moreover, for $\mu^3 h\ge C_0$ *estimate* (6.4.76) *holds for*

---

[25] Compare with theorem 6.4.14.

[26] Compare with theorems 6.4.15, 6.4.16.

$$\text{(7.4.64)}\quad \mathcal{R}_I = \Big| \int \phi(\tau')\Big(\int \psi(x)\big(\operatorname{tr} e(x,x,\tau',0) - h^{-2}S(x,\tau',0,\mu h)\big)dxd - \sum_{x_l\in \mathrm{M}} \frac{1}{\mu h}\Delta(x_l)\Big)d\tau'\Big|$$

where M *is the set of saddle points of the function* v_* *and at every point* $\bar{x}\in$ M, $\Delta(\bar{x})$ *is defined by formulas* (6.4.73)* *with the additional factor* $\frac{D}{2}$ *in the right-hand expression,* (6.4.74) *and*

$$\text{(7.4.65)}\qquad \bar{\tau} = v_*(\bar{x}) + 2j\mu h, \quad \bar{\tau}' = \bar{\tau} + \bar{\rho}_1\mu^{-2} + \bar{\rho}_2\mu^{-1}h, \quad \bar{\rho} - j = -2\bar{\omega}_j V_*$$

where we do not yet know ω_1, ω_2.

Theorem 7.4.13. [27] *(i) Under the assumptions of theorem 7.4.12 estimate* $(6.4.36)_0$ *holds for* $\mathcal{R}_7$ *with* $\vartheta = 0$.

(ii) Under the additional condition (7.4.1) *both statements (i), (ii) of theorem 6.4.18) remain true.*

Theorem 7.4.14. *Under additional condition* (7.4.1) *the statements of theorems 6.4.19–6.4.22 remain true with obvious modifications. Without this condition they remain true only with an additional term* $O(\bar{\nu})$ *in the right-hand expression.*

Remark 7.4.15. *One can see easily that for* $\vartheta = 0$ *all the above results hold without condition (6.3.1); the only exceptions are the effective asymptotics for* $d = 3$ *where we applied the results of the previous case in the zone* $\{|\xi_2| \geq \epsilon_0\}$.

7.4.7 Effective asymptotics. II

In order to finish our analysis in the case $d = 2$ we need to calculate the coefficients ω_1, ω_2 at a critical point of the function v_*. To this end let us consider the estimate

$$\text{(7.4.66)}\qquad (m+1)\|Au\| \geq \eta\mu^{-1}h\|u\| \qquad \forall u\in C_0^K(B(\bar{x},\zeta\mu^{-1}),\mathbb{C}^D)$$

with $\eta > 0$ and with small enough $\zeta = \zeta(\eta) > 0$ independent of μ and h provided $\mu h \leq \epsilon = \epsilon(\eta)$. By reduction one can easily prove that this estimate is equivalent to the estimate

$$\text{(7.4.67)}\qquad \|(A_{\mathrm{S}}I + \sigma_0\mu h)u\| \geq \eta\mu^{-1}h\|u\| \qquad \forall u\in C_0^K(B(\bar{x},\zeta\mu^{-1}),\mathbb{C}^D)$$

for the Schrödinger operator

$$A_{\mathrm{S}} = P_j(F^{-1}g^{jk})P_k + v_*;$$

namely, estimate (7.4.66) yields estimate (7.4.67) and *vice versa* (in each implication we replace η by $C_0^{-1}\eta$).

On the other hand, the constructions of section 6.4 and this section yield that estimates (7.4.66) and (7.4.67) are equivalent in the sense indicated above to the inequalities

$$\text{(7.4.68)}\qquad (m+1)\big|(m^2+2j\mu hF)^{\frac{1}{2}} + V + \omega_1\mu^{-2} + \omega_2\mu^{-1}h\big| \geq \eta\mu^{-1}h$$

and

$$\text{(7.4.69)}\qquad \big|v_* + 2j\mu h + \omega_1'\mu^{-2} + \omega_2'\mu^{-1}h\big| \geq \eta\mu^{-1}h$$

[27] Compare with theorems 6.4.17, 6.4.18.

respectively; these inequalities should be checked for all $j \in \mathbb{Z}^+$ and for $x = \bar{x}$. Let us take $j \in \mathbb{Z}^+$ such that

$$(m+1)\big|(m^2 + 2j\mu hF)^{\frac{1}{2}} + V\big| \le C^{-1}\mu h \tag{7.4.70$_1$}$$

and

$$|v_* + 2j\mu h| \le C^{-1}\mu h; \tag{7.4.70$_2$}$$

one can easily see that for $\mu^3 h \ge C$ these inequalities define the same number j.

Let us replace v_* by $v_*(1+\lambda)$ and let us alter V in the corresponding manner. Then the variations of ω_k and ω'_k are $O(|\lambda|(m+1)^{-1})$ and $O(|\lambda|)$ respectively and for $|\lambda| \le \epsilon'''$ these variations are are not essential in the sense that they do not change estimates (7.4.68) and (7.4.69). Thus estimates (7.4.68) and (7.4.69) remain equivalent after replacing v_* by $v_*(1+\lambda)$ (preserving the critical points), and the corresponding replacement of V. Taking j satisfying (7.4.70) and taking λ such that $|\lambda| \le C\mu h$ and the left-hand expression of (7.4.68) vanishes we obtain (because $\eta > 0$ is arbitrary) that $\omega_k = -\omega'_k F(2V)^{-1}$ at the point $\bar{x}$. However, we have formula (6.4.70) for ω'_1 at the point $\bar{x}$ and we have proven that ω'_2 vanishes at this point. So we have proven

Proposition 7.4.16. *In formula* (7.4.65) *we can calculate* $\bar{\rho}_1 = \bar{\rho}$ *according to* (6.4.75), (6.4.70) *with* v_* *instead of* v *and take* $\bar{\rho}_2 = 0$.

7.4.8 Effective asymptotics. III

Now our goal is to obtain similar results in the case when condition (7.3.24) is replaced by one of the conditions (7.3.4)–(7.3.6) and, in particular, to get rid of condition (6.3.1) in theorem 7.4.11. Recall that we used condition (7.3.24) not only in the transition from $\mathcal{R}_4$ to $\mathcal{R}$ but also in the reduction done in subsection 7.4.2.

Let us first of all note that if at some point $(\bar{x}', \bar{\xi}')$ the inequality $|v_* + 2j\mu h| \ge \epsilon > 0$ holds, if condition (7.4.1) is fulfilled with $\epsilon_1 = \epsilon_0(c, \epsilon_0, \epsilon, s) > 0$ and if the symbol of Q is supported in the ϵ_1-neighborhood of $(\bar{x}', \bar{\xi}')$ then estimate (7.4.20) is true for all τ with $|\tau| \le \epsilon_1$. Therefore our standard arguments yield that if condition (7.4.1) and one of the conditions (7.3.4)–(7.3.6) are fulfilled then the following estimate holds:

$$|F_{t\to h^{-1}\tau}\chi_{T,\pm T}(t)\Gamma' Q'_x \mathcal{U}_{jk}| \le C'h^s \qquad \forall \tau : |\tau| \le \epsilon_1 \tag{7.4.71}$$

where $T \in [h^{1-\delta'}, T^*]$ is arbitrary, $T^* = \mu T_0$ for $d = 2$, $T^* = T_0$ for $d = 3$ and $T_0 = T_0(c, \epsilon_0, s) > 0$ as before.

On the other hand, let us apply the standard method of successive approximations to calculate

$$F_{t\to h^{-1}\tau}\chi_T(t)\,\mathrm{tr}\,\Gamma' Q'_x \mathcal{U}_{jk}{}^t Q''_y$$

and

$$F_{t\to h^{-1}\tau}\chi_T(t)\,\mathrm{tr}\,\Gamma \tilde{Q}'_x \psi'(x) U \psi''(y)^t \tilde{Q}''_y$$

for $T = h^{1-\delta'}$, $|\tau| \le \epsilon_0$ with the unperturbed operator

$$\bar{A} = \sigma_1\sqrt{2j\mu h B_0} + \sigma_0 m + \bar{W}\cdot I \qquad \text{for } d = 2,$$

$$\bar{A} = (\sigma_1\sqrt{2j\mu h B_0} + \sigma_3 h D_2 B_0)^w + \sigma_0 m + \bar{W}\cdot I \qquad \text{for } d = 3.$$

Moreover, for $m \ge C_0$ one can either do the reduction beforehand or skip this procedure. Then applying the arguments of the proofs of propositions 6.4.4, 6.4.5 we obtain the equality

$$F_{t\to h^{-1}\tau}\chi_T(t)\operatorname{tr}\varGamma Q'_x\mathcal{U}^tQ''_y \equiv \sum_{j,n} T\mu h^{1-d}\int \hat{\chi}\Big(\frac{(\tau-\tau')T}{h}\Big)\Theta^{(n+1)}_{j,n}(\tau')d\tau' \tag{7.4.72}$$

and a similar equality for $\tilde{Q}'_x\psi'(x)U\psi''(y)^t\tilde{Q}''_y$ modulo $O(h^s)$ where $\Theta_{j,n}(\tau)$ are given by (6.4.29) with $f_{j,n}(x',\xi',\mu h,\mathbf{h}',\mu^{-1})$ instead of f and

$$|\mathcal{D}^\alpha f_{j,n}| \le c'\mu^{-n} \qquad \forall\alpha : |\alpha|\le K.$$

Therefore, in particular, the absolute values of

$$F_{t\to h^{-1}\tau}\chi_T(t)\operatorname{tr}\varGamma Q'_x\mathcal{U}^tQ''_y$$

and

$$F_{t\to h^{-1}\tau}\chi_T(t)\operatorname{tr}\varGamma\tilde{Q}'_x\psi'(x)U\psi''(y)^t\tilde{Q}''_y$$

do not exceed Ch^{1-d}.

Let us calculate the second expression for $\psi''=1$, $Q''\equiv I$; this calculation will be done modulo $O(\mu^{-1})$ for $d=2$ and modulo $O(h^{-1+\delta'})$ with $\delta'>0$ for $d=3$; in this section calculations with $\delta'=0$ also apply. We assume that

$$\mu \ge h^{\delta-1} \tag{7.4.73}$$

with a small enough exponent $\delta>0$. Then we should treat terms of the following five types:

(i) Principal terms of type

$$\sum_j \mu h^{2-d}\Theta^{(1)}_{j,0}(\tau)$$

with $\Theta_{j,0}(\tau)$ given by (6.4.29) with

$$f_{j,0}=(2\pi)^{1-d}q'(x',\xi')(\psi'\circ\bar{\Psi})(x',\xi_1);$$

these terms originate from the terms

$$\pm ih\bar{G}^\pm\delta(x-y)\delta(t)I$$

of the successive approximation series for $\mathcal{U}$ where $\bar{G}^\pm$ are parametrices of the operator $\bar{P}=hD_t-(m+1)\bar{A}$ with coefficients frozen for $x=y$ and with $\mu^{-1},\mathbf{h}'$ replaced by 0. The total contribution of these terms is of order h^{1-d}.

(ii) Terms with total contribution of order $\mu^{-2}h^{1-d}$. These terms are associated with $\mathcal{U}_{jj}$ and originate from the terms

$$\pm ih\bar{G}^\pm(W-\bar{W}\cdot I)\bar{G}^\pm(W-\bar{W}\cdot I)\bar{G}^\pm\delta(x-y)\delta(t)I$$

of the successive approximation series for $\mathcal{U}$. The total contribution of the other terms of the successive approximation series is $O(\mu^{-3}h^{1-d})$ and is less than required remainder estimate (because of (7.4.73)). Terms of the type

$$\pm ih\bar{G}^\pm(W-\bar{W}\cdot I)\bar{G}^\pm\delta(x-y)\delta(t)I$$

have zero diagonal blocks because $W - \bar{W}\cdot I$ has the same property; so these terms make no contribution to the final answer.

One should consider terms of this type as well as terms of the following type only for $d = 2$ and one can skip all blocks of $W - \bar{W}I$ with $j \neq k \pm 1$ (we use (7.4.73) again; recall that W_{jk} are of order $O((m+1)^{-1}\mu^{-|j-k|})$). Moreover, one can consider only terms with $j \geq 2$ because the total contribution of the terms with $j = 0, 1$ is $O(\mu^{-1})$. Let us take into account that

$$W_{j,j+1} \equiv \mu^{-1}\bigl(\sqrt{2j\mu h}\,\Pi^{+} + \sqrt{2(j-1)\mu h}\,\Pi^{-}\bigr)w$$

modulo $O(h(m+1)^{-1})$ where w is a scalar operator and $\Pi^{\pm}$ are orthogonal projectors to $\mathbb{H}^{\pm}$, and let us take into account that $W_{j,j-1} = W^{*}_{j-1,j}$ and that for $d = 2$

$$\bigl(\tau - (m+1)\bar{A}\bigr)^{-1} = \bigl(\tau_1^2 - (m+1)^2(2j\mu h B_0^2 + m^2)\bigr)^{-1} \times \bigl(\tau_1 + (m+1)(\sigma_1\sqrt{2j\mu h}B_0 + \sigma_0 m)\bigr)$$

with $\tau_1 = \tau - (m+1)\bar{W}$. Then we obtain terms of type

$$D_\tau^3 \int \operatorname{sign}\tau_1 \cdot \theta(\tau_1^2 - (m+1)^2(2j\mu h b_0 + m^2)) \times \mathcal{P}(j\mu h, x_1, \xi_1, \mu h, \mu^{-1}, h)\mu^{-1}dx_1 d\xi_1$$

where $\mathcal{P}$ is a second order polynomial of $j\mu h$ with coefficients depending smoothly on all the other variables and $b_0 = B_0^2$.

Obviously, for $m + |\tau_1| \geq 3\epsilon_1$ the integrand is a smooth function of τ and $z = j\mu h$ and is a distribution with respect to (x_1, ξ_1); this follows from condition (7.3.5) and the inequality $|\tau| \leq \epsilon_1$. Using the arguments of the proof of proposition 7.3.11 one can easily show that for fixed (x_1, ξ_1) and for $|\tau_1| \leq 3\epsilon_1$, $m \leq 3\epsilon_1$ the integrand is a smooth function of $z = j\mu h$ and a distribution with respect to τ. Using condition (7.3.4) we then obtain that for $|\tau| \leq \epsilon_1$ the integrand is a smooth function of $(j\mu h, \tau)$ and a distribution with respect to (x_1, ξ_1). Therefore we can replace summation with respect to j by integration with respect to $z = j\mu h$ and we then obtain that the sum is equal to

$$\varkappa(\tau)\mu^{-2}h^{-1} + O(\mu^{-1}) \tag{7.4.74}$$

with $\varkappa(\tau)$ depending smoothly on τ.

(iii) Furthermore, we have terms with total contribution of the same order; they are associated with $\mathcal{U}_{j,j\pm 1}$ and they originate from

$$\pm ih\bar{G}^{\pm}(W - \bar{W}\cdot I)\bar{G}^{\pm}\delta(x-y)\delta(t)I$$

of the successive approximation series for $\mathcal{U}$. In this case the diagonal blocks of the operator $\tilde{\psi} = T^{*}\mathcal{F}^{*}\psi\mathcal{F}T$ are replaced by blocks $\tilde{\psi}_{j\pm 1,j}$ which are of the form

$$\mu^{-1}\psi'\bigl(\sqrt{2j\mu h}\,\Pi^{+} + \sqrt{2(j-1)\mu h}\,\Pi^{-}\bigr)$$

with the scalar operator ψ'. The same arguments as before again give expression (7.4.74).

(iv) Moreover, we have terms associated with $\mathcal{U}_{j,j}$ and originating from the lower order terms of $\tilde{\psi}_{j,j}$; these terms originate from the same terms of the successive approximation series for $\mathcal{U}$ as terms of type (i).

For $d = 2$ these terms have the same order of total contribution as (ii) and are equal to $\sum_j j\Theta^{(1)}_{(j,1)}(\tau)$ where $\Theta_{j,1}(\tau)$ are given by (6.4.29) with the function f independent of $j\mu h$, and the previous arguments give the final expression (7.4.74).

For $d = 3$ the essential terms of this type are associated with the difference between the operator $\tilde{\psi}$ and the same operator with $\mu^{-1}hD_2$ replaced by 0; we obtain 0 if the symbol of the operator Q is even with respect to ξ_2.

(v) Finally, we have terms with total contribution of order $O(\mu^{-1})$ for $d = 2$ and $O(h^{-1})$ for $d = 3$; therefore these terms should be treated only for $d = 3$ and $\delta' > 0$. These terms originate from the freezing of the coefficients and commutators. Simple calculations (see section 4.4) give 0 if the symbol of the operator Q is even with respect to ξ_2.

Let us apply standard Tauberian arguments. Moreover, let us prove that the term $\varkappa(\tau)\mu^{-2}$ in the final answer vanishes (we compare the final answer obtained here with the results of section 7.3; here one can use averaging with respect to τ). Consequently we obtain the following statements in which we can get rid of condition (7.4.71) because these statements were proven under condition (6.3.1) (in section 7.3 and this section above):

Theorem 7.4.17. *Let $d = 2, 3$, $\vartheta = 0$ and suppose conditions* (7.1.1)–(7.1.3), (7.2.1), (7.4.1) *and one of the conditions* (7.3.4), (7.3.5) *are fulfilled. Then estimate* $(6.4.55)_{(d)}$ *holds.*

Theorem 7.4.18. *Let $d = 3$, $\vartheta = 0$. Then theorem 7.4.11 remains true with condition* (6.3.1) *replaced by* (7.4.1).

In the proof of this theorem we use the same arguments as in the proof of theorem 7.4.11 but in the zone $\{|\xi_2| \geq \epsilon\}$ we use the arguments of this subsection instead of arguments of section 7.3.

The same arguments are applicable in the case when conditions (7.3.4)–(7.3.6) are skipped but there is mollification on τ; under condition (6.3.1) we use the results of section 7.3. So the following statement is proven:

Proposition 7.4.19. *If $d = 2, 3$ and conditions* (7.1.1)–(7.1.3), (7.2.1) *and* (7.4.1) *are fulfilled then the following estimates hold:*

$$(7.4.75)_{(2)} \quad \mathcal{R}_{III} = \Big| \iint \phi(\tau)\psi(x)\big(d_\tau e(x,x,\tau_1,\tau) - \partial_\tau S(x,\tau_1,\tau,\mu h)h^{-d}d\tau\big)dx\Big| \leq C \text{ for } d = 2,$$

$$(7.4.75)_{(3)} \qquad \mathcal{R}_{III} \leq Ch^{\delta'-2} \qquad \text{for } d = 3$$

with arbitrarily small exponent $\delta' > 0$.

This proposition yields the following

Corollary 7.4.20. *Statements 7.4.11–7.4.18 remain true when the condition* $\operatorname{supp}\phi \subset [-\epsilon_0, \epsilon_0]$ *is replaced by the condition* $\operatorname{supp}\phi \subset [-c, c]$.

Certain improvements and generalizations of proposition 7.4.20 will appear in sections 7.6 and 7.8.

7.5 The Case of a Very Strong Magnetic Field

In this section we consider the case when

$$(7.5.1) \qquad \mu \geq 1, \quad \mu h \geq \epsilon_0$$

forbidden for the Schrödinger operator A_S unless we replace it by $A_S - B$. This case is in fact rather simple, because we actually need to consider only one Landau level.

7.5.1 Additional reduction

Let us note that the semiclassical parameter for $d = 2$ is $\mu^{-1}h$ instead of h. Therefore in this case we can weaken the condition $h \in (0, 1]$. Namely, we replace it by the condition

$$h \in (0,1] \quad \text{for } d = 3; \qquad h \in (0, \mu^{1-\delta}] \quad \text{for } d = 2,\ \delta > 0. \tag{7.5.2}$$

In subsection 7.5.4 even this condition will be weakened for $d = 2$. Under conditions (7.5.1) and (7.5.2) equalities (7.2.31) and (7.2.32) are fulfilled modulo functions f and f' respectively such that

$$|F_{t\to h^{-1}\tau}\chi_{T,T'}(t)Q_x f| \le C'T\mu^{1-s}h^s, \tag{7.5.3}$$

$$|Q_x f'| \le C'\mu^{1-s}h^s \tag{7.5.4}$$

with the same restrictions on Q, T as in (6.2.51), (6.2.52) for $d = 3$ and under the condition $T \ge h\mu^{\delta'}$ with $\delta' > 0$ for $d = 2$. Moreover, (6.2.53), (6.2.55) hold and the left-hand expression in (6.2.56) does not exceed $C'T\mu^{1-s}h^s$. Finally, the norm of the left-hand expression of (6.2.54) does not exceed $C\mu^l$ with some exponent l; this rough estimate will be improved in what follows.

Let us introduce $\mathcal{U}_{jk}$ as we did it in section 7.4; then estimate (7.4.6) and equality (7.4.5) remain true. Moreover, the construction of $\mathcal{U}$ and the fact that we multiplied $\mathcal{U}$ by two $\mu^{-1}h$-pseudo-differential operators with symbols vanishing outside of a neighborhood of Λ^0 (one pseudo-differential operator acts with respect to x and another acts with respect to y) yield that all our estimates remain true if we replace $\mathcal{U}$ by its cutoff

$$\sum_{j,k\le N} \mathcal{U}_{jk}\Upsilon_j(x_d)\Upsilon_k(y_d) \tag{7.5.5}$$

with $N = N(s,\delta)$. We also denote this cutoff by $\mathcal{U}$. The error will be $O(\mu^{-s})$.

We need the following statement which is obvious by standard elliptic arguments:

Proposition 7.5.1. *Let us assume that $d = 3$ and that $Q = \mathrm{Op}(q)$ with $q \in S_{\mathbf{h}',\rho,1,K}(T^*\mathbb{R}^2)$ supported in $\{\rho \le |\xi_2| \le 2\rho\}$ with $\rho \ge C_0$. Then the estimate*

$$|F_{t\to h^{-1}\tau}\chi_{T,T'}(t)Q_x\mathcal{U}_{jk}| \le C'T\mu h^s\rho^{-s} \tag{7.5.6}$$

holds for $|\tau| \le c$, $T \ge h^{1-\delta}\rho^\delta$, $j,k \le N$.

Proposition 7.5.2. *There exists an operator $\mathcal{K} = (\mathcal{K}_{j,k})_{j,k\le N}$ such that in the sense of $\mathbf{h}'$-pseudo-differential operators in $\tilde{\Omega}$ (7.4.27) is fulfilled and*

$$(m+1)\mathcal{K}^*\mathcal{A}\mathcal{K} \equiv (m+1)(\mathcal{A}_0 + \mathcal{A}_1) + \mathcal{B}' \tag{7.5.7}$$

where $\mathcal{B}'$ is of block-diagonal type and

$$|\mathcal{D}^\alpha\mathcal{K}_{jk}| \le c_\alpha\bar{\nu}(\mu^{-1}h)^{\frac{1}{2}|j-k|}(\mu h)^{-\frac{1}{2}} \qquad (j \ne k), \tag{7.5.8$_1$}$$

$$|\mathcal{D}^\alpha(\mathcal{K}_{jj} - I)| \le c_\alpha\mu^{-2}\bar{\nu}^2, \tag{7.5.8$_2$}$$

$$|\mathcal{D}^\alpha\mathcal{B}'_{(j)}| \le c_\alpha\mu^{-2}\bar{\nu}^2. \tag{7.5.8$_3$}$$

Proof. Let us construct $\mathcal{K}$ by induction with respect to M assuming that at every step (7.5.7) holds with an additional term $\mathcal{R}$ in the right-hand expression such that

$$|\mathcal{D}^\alpha \mathcal{R}_{jk}| \le c_\alpha (\mu^{-1}h)^{\frac{1}{2}\max(|j-k|,M)}. \tag{7.5.10}$$

Then we set $\mathcal{K}_{M+1} = \mathcal{K}_M \exp i\mathcal{L}$ with the Hermitian operator $\mathcal{L} = (\mathcal{L}_{jk})_{j,k\le N}$ where $\mathcal{L}_{jj} = 0$ and the principal symbols of the out-of-diagonal blocks of $\mathcal{L}$ are defined by the equations

$$i(m+1)\bigl(\mathcal{A}^0_{(j)}\mathcal{L}^0_{jk} - \mathcal{L}^0_{jk}\mathcal{A}^0_{(j)}\bigr) = \mathcal{R}^0_{jk}. \tag{7.5.11}$$

Note that the distances between the spectra of the Hermitian matrices $(m+1)\mathcal{A}^0_{(j)}$ and $(m+1)\mathcal{A}^0_{(k)}$ with $j\ne k$ are not less then $\epsilon_1\bigl((m+1)\sqrt{\mu h}+\mu h\bigr)$ with a constant $\epsilon_1>0$. Then Appendix A yields that $\mathcal{L}^0_{jk}$ are defined uniquely and satisfy the inequalities

$$|\mathcal{D}^\alpha \mathcal{L}^0_{jk}| \le c_\alpha \bar{\nu}(\mu^{-1}h)^{\frac{1}{2}\max(|j-k|,M)}(\mu h)^{-1}. \tag{7.5.12}$$

All the assertions of our proposition follow from this construction. □

Let us now set $\mathcal{U}' = \mathcal{K}^*_x \mathcal{U}^t \mathcal{K}^*_y$; then in $\tilde{\Omega}$

$$(hD_t - \mathcal{B}_{(j)})\mathcal{U}'_{jk} \equiv 0, \tag{7.5.13}$$

$$\mathcal{U}'_{jk}|_{t=0} \equiv \delta_{jk}\delta(x'-y')I \tag{7.5.14}$$

where $\mathcal{B}_{(j)} = (m+1)(\mathcal{A}_{0(j)} + \mathcal{A}_{1(j)}) + \mathcal{B}'_{(j)}$.

Then standard elliptic arguments yield the following

Proposition 7.5.3. *If the support of the symbol of Q is small enough then there exists $j_0\in\mathbb{Z}^+$ such that*

$$|F_{t\to h^{-1}\tau}\chi_{T,T'}(t)Q_x\mathcal{U}'_{jk}| \le C'T\mu^{-s} \qquad \forall\tau: |\tau|\le\epsilon_0 \quad \forall T>\mu^{-\delta'}h \quad \forall j\ne j_0 \tag{7.5.15}$$

where $\delta'>0$. In particular, if $\mu h\ge C_0$ then $j_0=0$ and this estimate remains true with $Q=I$.

Recall that the eigenvalues of the symbol $\mathcal{A}_{0(j)}(x',\xi')$ are $\bar{W}\pm(m^2+2j\mu h b_0)^{\frac{1}{2}}$ for $d=2$ and $\bar{W}\pm\bigl(m^2+(2j\mu h+\xi_2^2)b_0\bigr)^{\frac{1}{2}}$ for $d=3$. For $j\ge1$ all these eigenvalues are disjoint (two eigenvalues with $j=0$ are not necessarily disjoint). Therefore the following statement holds:

Proposition 7.5.4. *For $\mu h\le C_0$, $j\ge 1$ there exist operators $\mathcal{T}_{(j)}$ satisfying* (7.4.21) *and*

$$\mathcal{T}_{(j)}\mathcal{B}_{(j)}\mathcal{T}_{(j)} \equiv \tilde{\mathcal{B}}_{(j)} = \begin{pmatrix}\mathcal{B}^+_{(j)} & 0\\ 0 & \mathcal{B}^-_{(j)}\end{pmatrix} \tag{7.5.16}$$

in $\tilde{\Omega}$ where the principal symbols of the $\mathbf{h}'$-pseudo-differential operators $\mathcal{B}^\pm_{(j)}$ are $(m+1)(\bar{W}\pm(m^2+2j\mu h b_0)^{\frac{1}{2}})I$ for $d=2$ and $(m+1)\bigl(\bar{W}\pm(m^2+(2j\mu h+\xi_2^2)b_0)^{\frac{1}{2}}\bigr)I$ for $d=3$ and the blocks are $\frac{D}{2}\times\frac{D}{2}$-dimensional.

Let us consider the case $j = 0$. For $D = 2$ the operator $\mathcal{B}_{(0)}$ is one-dimensional from the beginning and its principal symbol is equal to $(m+1)(\bar{W}+\varsigma m)$ for $d = 2$ and $\bar{W}+\varsigma\xi_2 B_0$ for $d = 3$, $D = 2$ (and $m = 0$). In the case $d = 3$, $D = 4$ the operator $\mathcal{B}_{(0)}$ is two-dimensional and the eigenvalues of its principal symbol are equal to $(m+1)(\bar{W} \pm (m^2 + \xi_2^2 b_0)^{\frac{1}{2}})$; these eigenvalues are disjoint if either $m \geq \epsilon_0$ or $|\xi_2| \geq \epsilon_0$ in $\tilde{\Omega}$ and in these two cases there exists an operator $\mathcal{T}_{(0)}$ satisfying (7.4.21), (7.5.16) with blocks which are now one-dimensional.

So in all cases excluding $d = 3$, $D = 4$, $j = 0$, $m \leq \epsilon_0$, $|\xi_2| \leq \epsilon_0$ on $\tilde{\Omega}$ our problem is reduced to the analysis of operators with one-dimensional principal symbols and all the arguments of section 4.4 are applicable with the following modification: we have $(d-1)$ instead of d and there is an additional factor μ in the right-hand expressions of the estimates. Since the real semiclassical parameter is $\mu^{-1}h$ in the case $d = 2$, all the negligibility estimates have right-hand expression $C'\mu^{-s}T$ in this case. For $d = 3$ we have the real semiclassical parameter h and even $\frac{h}{\sqrt{\nu}}$ in the case when we treat the operator with scalar principal symbol and v_* satisfies the condition

$$|D^\alpha(v_* + 2j\mu h)| \leq c\nu \qquad \forall\alpha : |\alpha| \leq K$$

with $\nu \in [h^2, 1]$.

In this case the negligibility estimates have $C'\mu h^{-1}\bigl(\frac{h}{\sqrt{\nu}}\bigr)^s T$ as their right-hand expression.

Finally, let us note that standard elliptic arguments yield that we do not need to treat the case $j = 0$ for $|\tau| \leq \epsilon_0$, $d = 2$ and for $|\tau| \leq \epsilon_0$, $d = 3$ and symbol of Q supported in $\{|\xi_2| \leq \epsilon_0\}$ provided condition (7.3.24) is fulfilled. On the other hand, the opposite condition $(V^2 - m^2 \leq \epsilon_1)$ excludes the case $j \geq 1$ from the analysis.

7.5.2 Estimates: $d = 3$

It is easy to see that proposition 7.4.9 remains true with the following modification: we discuss smoothness with respect to $z = j\mu h$ neither here nor below and we do not replace summation by integration; the transition from $\mathcal{R}_4$ to $\mathcal{R}$ is now trivial. As a result we have the following statements:

Theorem 7.5.5. *Let $d = 3$, $\tau = 0$, $\phi \in C_0^K([-c, c])$ and let conditions (7.1.1)–(7.1.3), (7.2.1), (7.5.1), (7.5.2), (6.3.64)* be fulfilled with $\nu \in [h^2, 1]$. Then*

(i) Let either $m \geq 1$ or condition (7.3.24) be fulfilled. Then the estimate

$$\mathcal{R} \leq C\mu h^{-1-\delta}\nu^{\frac{\delta}{2}} + C' \tag{7.5.17}$$

holds with arbitrarily small $\delta > 0$. Moreover, under the condition

$$\min_{j\in\mathbb{Z}^+} |V_* - 2j\mu hF| + \sum_{\alpha:1\leq|\alpha|\leq L} |D^\alpha\bigl(\frac{V_*}{F}\bigr)| \geq \epsilon_0\nu \tag{7.5.18}$$

the estimate

$$\mathcal{R} \leq C\mu h^{-1} \tag{7.5.19}$$

holds.

(ii) If one of the conditions (7.3.4), (7.3.5) is fulfilled then estimate (7.5.19) holds.

Let us note that we cannot apply this statement in the most general case (with $m \leq \epsilon_0$) because there is a term with $j = 0$. Let us treat the case $j = 0$ more carefully.

Let us recall that for $j=0$ we have an operator with principal symbol $W+\varsigma\xi_2$ for $D=2$ and $W\cdot I+\sigma_1\xi_2+\sigma_0 m$ for $D=4$. In the first case we have a scalar operator of principal type and therefore the contribution of this term to the remainder estimate does not exceed $C\mu h^{-1}$. In the second case we can assume without loss of generality that either $m=1$ (a case which has been treated) or $\nu=1$; in section 7.8 we will show that the general case can be reduced to one of these (maybe with a small "loss of information").

Applying the arguments of subsection 4.5.4 (see theorems 4.5.16, 4.5.18)[28] one can easily prove the following

Theorem 7.5.6. *Let all the conditions of theorem 7.5.5 be fulfilled and let $m\le 1$. Then estimate (7.5.17) holds with $\nu=1$. Moreover, under condition (7.5.18) (with $\nu=1$) estimate (7.5.19) holds.*

7.5.3 Estimates: $d=2$

Let us treat the case $d=2$. In this case we obtain a scalar operator for $j=0$ as well. The theory of section 4.4 immediately yields the following two statements:

Theorem 7.5.7. *Let $d=2$, $\tau=0$ and let conditions (7.1.1)-(7.1.3), (7.2.1), (7.5.1) and (7.5.2) be fulfilled. Let $\phi\in C_0^K([-c,c])$. Then*

(i) Under the conditions

$$|\frac{V_*}{F}+2j\mu h|+|\nabla\frac{V_*}{F}|\le\epsilon_0\implies \operatorname{Hess}\frac{V_*}{F}\ge\epsilon_0\quad\forall j\ge 1 \tag{7.5.20$_+$}$$

$$|V_0|+|\nabla V_0|\le\epsilon_0\implies \operatorname{Hess}V_0\ge\epsilon_0 \tag{7.5.21$_+$}$$

with $V_0=(m+1)(V-\varsigma m)$ the estimate

$$\mathcal{R}\le C \tag{7.5.22}$$

holds.

(ii) Under the conditions

$$|\frac{V_*}{F}+2j\mu h|+|\nabla\frac{V_*}{F}|\le\epsilon_0\implies |\operatorname{Hess}\frac{V_*}{F}|\ge\epsilon_0\quad\forall j\ge 1 \tag{7.5.20}$$

$$|V_0|+|\nabla V_0|\le\epsilon_0\implies |\operatorname{Hess}V_0|\ge\epsilon_0 \tag{7.5.21}$$

the estimate

$$\mathcal{R}\le C|\log\hat\lambda|\le C\log\mu \tag{7.5.23}$$

holds where $\hat\lambda=\min_{j\in\mathbb{Z}^+,x\in\mathrm{M}_j}\hat\lambda_{j,x}$, $\hat\lambda_{j,x}=|V_*+2j\mu h|$ $(j\ge1)$, $\hat\lambda_{j,x}=|V-\varsigma m|$ $(j=0)$ and M_j is the set of saddle points of $v_{*j}=\frac{V_*}{F}$ for $j\ge1$ and $v_{*0}=V_0$ for $j=0$.

(iii) Finally, under the conditions

$$\min_{j\in\mathbb{Z}^+,j\ge1}|V_*-2j\mu hF|+\sum_{\alpha:1\le|\alpha|\le L}|D^\alpha(\frac{V_*}{F})|\ge\epsilon_0\nu, \tag{7.5.18$'$}$$

[28] We apply the second theorem in order to get rid of the logarithmic factor.

$$\sum_{\alpha:|\alpha|\le L} |D^\alpha V_0| \ge \epsilon_0 \tag{7.5.18$''$}$$

the estimate

$$\mathcal{R} \le C\mu^\delta \tag{7.5.24}$$

holds with arbitrarily small $\delta > 0$.

Theorem 7.5.8. *Let* $d = 2$, $\tau = 0$ *and let conditions* (7.1.1)–(7.1.3), (7.2.1), (7.5.1), (7.5.2) *and* $(7.2.25)_{1,2}$ *be fulfilled. Moreover, let*

$$|D^\alpha v_{*j}| \le c\nu \qquad \forall \alpha : 1 \le |\alpha| \le K \tag{7.5.25}$$

and $\bar\nu \ge \mu^{-l}$, $1 \ge \nu \ge C_0\bar\nu\mu^{-1}h$ *where* $v_{j*} = \frac{1}{F}(m^2 - V^2) + 2jmh$ *and* j *is defined by the inequality* $|v_{*j}| \le \epsilon_1$ *with arbitrarily small* $\epsilon_1 > 0$. *Then*

(i) If

$$|v_{*j}| + |\nabla v_{*j}| \le \epsilon_0\nu \implies \det \operatorname{Hess} v_{*j} \ge \epsilon_0\nu^2 \tag{7.5.26$_+$}$$

then estimate (7.5.22) *holds for* $\mathcal{R}_4$ *and*

$$\mathcal{R} \le C\frac{\bar\nu}{\nu}. \tag{7.5.27}$$

(ii) If

$$|v_{*j}| + |\nabla v_{*j}| \le \epsilon_0\nu \implies |\det \operatorname{Hess} v_{*j}| \ge \epsilon_0\nu^2 \tag{7.5.26}$$

then estimate (7.5.23) *holds for* $\mathcal{R}_4$ *and*

$$\mathcal{R} \le C\frac{\nu}{\bar\nu}\bigl(\log\frac{\mu\nu}{\bar\nu h} + 1\bigr). \tag{7.5.28}$$

(iii) If

$$\sum_{\alpha:|\alpha|\le L} |D^\alpha v_{*j}| \ge \epsilon_0\nu \tag{7.5.29}$$

then estimate (7.5.24) *holds for* $\mathcal{R}_4$.

7.5.4 The case $d = 2$ and very large h

Let us weaken condition (7.5.2) for $d = 2$. Namely, let us now assume that

$$h_1 = \mu^{-1}h \in (\mu^{-\delta}, 1] \tag{7.5.30}$$

with a small exponent $\delta > 0$ and let us treat h_1 as a small parameter. Now the Schwartz kernels of operators with operator norm in L^2 not exceeding $C'h_1^s$ where t is treated as parameter are considered negligible (before the restriction to $x = y$ and integration). Then the presence of a cutoff with respect to x_1 and h_1D_1 yields that $\mathcal{U}$ is equal to decomposition (7.5.5) modulo a negligible function. The first difficulty is that the remainder in the transformation of $\mathcal{A}^0$ is of order $\mu\bar\nu h_1^N$ where we assume that conditions $(7.2.25)_{1,2}$ are fulfilled. In order to make this remainder negligible (i.e., with operator norm less than $C'h_1^N$) we assume that

(7.5.31) Conditions $(7.2.25)_{1,2}$ are fulfilled and $\bar\nu \le c\mu^{-1}h_1^{-l}$.

Here we do not divide the right-hand expression of this inequality by $(m+1)$ because standard elliptic arguments yield that the components $\mathcal{U}_{jk}^{\zeta\zeta'}$ with $\zeta=\varsigma$ or (and) $\zeta'=\varsigma$ are Schwartz kernels of operators bounded even after being multiplied by $(m+1)$ ($(m+1)^2$ respectively) and cutting with respect to hD_t; i.e., after reduction is applied.

Then assuming that $\delta=\delta(s,l)>0$ in (7.5.30) is small enough standard elliptic arguments yield that the functions $T^{-1}F_{t\to h^{-1}\tau}\chi_{T,T'}(t)\mathcal{U}_{jk}$ are negligible for $T\ge\mu^{\delta'}h$, $j\ge 1$, $|\tau|\le c$ with $\delta'=\frac{1}{s}$. Therefore the same is true for $k\ge 1$. So only one essential component $\mathcal{U}_{00}$ remains and

$$(hD_t-(m+1)(\bar{W}-\varsigma m))\mathcal{U}_{00}\equiv 0, \tag{7.5.32$_1$}$$

$$\mathcal{U}_{00}|_{t=0}\equiv\delta(x_1-y_1) \tag{7.5.32$_2$}$$

where the first equality is fulfilled modulo functions negligible in the sense indicated above (in the same sense $\mathcal{U}_{jk}$ with $j+k\ge 1$ are negligible).

Therefore our routine arguments yield the following

Theorem 7.5.9. *Let $d=2$ and let conditions* (7.1.1)–(7.1.3), (7.2.1), (7.5.1), (7.5.30) *and* (7.5.31) *be fulfilled where $\delta>0$ in* (7.5.30) *is small enough. Let us assume that V_0 satisfies* (7.5.25) *with $\nu\in[h_1^l,1]$. Then all the statements of theorem 7.5.8 remain true and, moreover, under condition* (7.5.29) *the estimate*

$$\mathcal{R}_4\le Ch_1^{-\delta'} \tag{7.5.33}$$

holds with arbitrarily small $\delta'>0$.

7.6 Propagation of Singularities and Improved Asymptotics

We need to revise all the results of sections 7.2–7.5 in the manner of section 6.5.

7.6.1 Reduction

In the manner of theorem 6.5.1 we can prove

Theorem 7.6.1. *Let $d=3$, $X\supset B(0,1)$ and let conditions* (7.1.1)–(7.1.3) *and* (7.2.1) *be fulfilled. Let $\hat{x}\in B(0,\frac{1}{2})$ and for all $x\in B(\hat{x},\zeta^l)$ let the flow $\phi_t(x)$ be defined by formula* (6.5.1) *with $t\in[-T_1,T_2]$ where $\zeta\in[\mu^{\delta'},1]$, $\delta'>0$ is small enough, $0\le T_i\le\zeta^{-l}$ and $T_1T_2=0$. Let us assume that along this flow*

$$|D^\alpha\omega^{jk}|\le c\zeta^{-l},\qquad |D^\alpha F^j|\le c\zeta^{-l}. \tag{7.6.1$_{1,2}$}$$

Moreover, let us assume that conditions (6.5.4)–(6.5.8) *are fulfilled along this flow. Then one can perform all the constructions of section 7.2 in $\cup_t\phi_t(B(\hat{x},\epsilon\zeta^l))$ with t running over $[-T_1,T_2]$, $\epsilon=\epsilon(l,c,\epsilon_0)>0$, and all the symplectomorphisms and symbols satisfy the inequalities indicated in theorem 6.5.1.*

Moreover, we can define ϕ_t by the system (6.5.1)′ *under the additional assumption that conditions* (6.5.10)$_{1,2}$ *are fulfilled in $\cup_t\phi_t(B(\hat{x},\zeta^l))$.*

7.6.2 The case of a weak magnetic field

The statement concerning propagation of singularities (similar to theorem 6.5.2) remains true with the Hamiltonian (6.5.13) replaced by

$$H^{\pm}_{(\bar{x}_1,\bar{\xi}_1)r}(x_2,\xi_2) = (m+1)\Big(\bar{W} \pm \big(m^2 + b_0(\bar{x}_1,x_2,\bar{\xi}_1)\big)(r+\xi_2^2)^{\frac{1}{2}}\Big) \tag{7.6.2}$$

where we assume in addition that along trajectories

$$|\bar{W}| \geq \epsilon_0 \zeta^l. \tag{7.6.3}$$

We assume from the beginning that $|\tau| \leq h^{1-\delta''}$ with a small enough exponent $\delta'' > 0$ and so (7.6.3) is equivalent to a similar inequality for the subradical expression in (7.6.2). Surely, condition (6.3.8) should be replaced by a stronger condition (see remark to theorem 7.3.3). Moreover, arguments similar to those of section 6.5 yield that one can replace condition (6.5.7) by (6.5.7)′. Repeating the arguments which followed in section 6.5 one can replace condition (6.5.22) by (6.5.29)*.

As a result of this analysis we obtain the following

Theorem 7.6.2. [29] *Let $d = 3$ and suppose conditions (7.1.3), (7.2.1), (7.3.1), (6.3.1) and (7.3.24) are fulfilled. Let $\Xi_1, \dots, \Xi_n$ be closed subsets of $B(0,1)$ and $1 \leq T_1 \leq \dots \leq T_n \leq \mu^{\delta'}$ with small enough $\delta' = \delta'(\delta, l) > 0$. Let us assume that through every point x in the γ_i-neighborhood of Ξ_i (with $\gamma_i = (\mu^{\frac{1}{2}} h^{\frac{1}{2}-\delta} + \mu^{-1})\zeta_i j^{-L}$, $\zeta_i = T_i^{-1}$ and large enough $L = L(\delta, l)$) there passes a trajectory of the flow $\phi_t(x)$ given by the system*

$$\frac{dx_l}{dt} = (m+1)\frac{\sqrt{-v_*}}{FV} F^l \qquad (l = 1,2,3) \tag{7.6.4}$$

with $\phi_0(x) = x$, $\pm t \in [0, T_i]$ along which conditions (6.5.29),*

$$v_* \leq -\epsilon \zeta^2, \tag{7.6.5}$$

and conditions $(7.6.1)_{1,2}$, (6.5.4), (6.5.5), (6.5.8), (6.5.6)′, (6.5.7)′ are fulfilled with $\zeta = \zeta_i$. Moreover, along this trajectory let the following conditions be fulfilled:

$$|D^\alpha V| \leq c\zeta^{-l}(m+1)^{-1} \qquad \forall \alpha : 1 \leq |\alpha| \leq K, \tag{$7.6.1_3$}$$

$$|v_*| \leq c\zeta^{-l}. \tag{$7.6.1_4$}$$

Then

(i) For $\mu \leq h^{\delta - \frac{1}{3}}$ estimate (6.5.31) holds with $\mathcal{R}_1$ defined in section 7.3, $T_0 = 1$, $\Xi_0 = B(0,1) \cap \{v_* \leq 0\} \cap (\Xi_1 \cup \dots \cup \Xi_n)$.

(ii) Under condition (7.3.5) this estimate holds for $\mu \leq h^{\delta - 1}$.

(iii) Under condition (7.3.4) this estimate also holds for $\mu \leq h^{\delta-1}$ and in this case condition (7.3.24) is not necessary.

We have also used the arguments of remark 6.5.4 here.

---

[29] Compare with theorem 6.5.5.

7.6.3 The case of a strong magnetic field

Let us now assume that condition (7.4.1) is fulfilled. Let us improve the arguments of section 7.4. In order to do this let us introduce one-dimensional Hamiltonians (7.6.2) with $r = 2j\mu h$. Then propositions 6.5.5, 6.5.7 and remark 6.5.6 remain true and so propositions 6.5.8, 6.5.9 should be replaced by the following statements:

Proposition 7.6.3. *Let $d = 3$ and suppose that conditions* (7.1.1)–(7.1.3), (7.2.1), (7.3.1) *and* (6.3.1), (7.4.1), (7.3.24) *are fulfilled. Let $\Xi_1, \dots, \Xi_n$ be closed subsets of $B(0,1)$ and $1 \le T_1 \le \dots \le T_n \le \mu^{\delta'}$ with small enough $\delta' = \delta'(\delta, l) > 0$. Let us assume that through every point x in the γ_i-neighborhood of Ξ_i (with $\gamma_i = (\mu^{\frac{1}{2}} h^{\frac{1}{2}-\delta} + \mu^{-1})\zeta_i^{-L}$, $\zeta_i = T_i^{-1}$ and large enough $L = L(\delta, l)$) there passes a trajectory of the flow $\phi_t(x)$ given by system* (7.6.4) *with $\phi_0(x) = x$, $\pm t \in [0, T_i]$ along which conditions* (6.5.29)*, (7.6.5), $(7.6.1)_{1-3}$, (6.5.6)′, (6.5.7)′ *are fulfilled with $\zeta = \zeta_i$. Then*

(i) Estimate (6.5.31) *holds for $\mathcal{R}_4$ defined in section 7.4, $T_0 = 1$, $\Xi_0 = B(0,1) \cap \{v_* \le 0\} \cap (\Xi_1 \cup \dots \cup \Xi_n)$.*

(ii) For $\mathcal{R}$ estimate (6.5.40) *holds with R_6 and γ given by* $(6.5.38)_{1,2}$, (6.5.39) *and V_* instead of V. In particular, one can skip R_6 for $\mu \ge h^{-\frac{1}{3}-\delta}$.*

Let us now replace condition (6.3.1) by condition (7.4.1). The reader can easily prove that proposition 6.5.12 and the arguments which follow in subsection 6.5.3 remain true with the dynamical equations

$$\frac{dx_l}{dt} = \frac{1}{FT(\bar{x}, j)}(-V_* - 2j\mu h F)^{\frac{1}{2}} F^l \qquad (l = 1, 2, 3). \tag{7.6.6}$$

The same is true for $\epsilon \le \mu h \le c_0$. Finally, under conditions (7.5.1), (6.5.44)* and (6.5.45)* with $\nu \ge h^2$ one can improve the remainder estimate $O(\mu h^{-1})$ under conditions on the dynamical system (7.6.6). For $\mu h \ge C_0$ only $j = 0$ is important.

7.6.4 Propagation of singularities ($d = 2$)

In this case all the results for the Schrödinger operator remain true. Now under condition (7.3.24) the drift lines are defined by the equations

$$\frac{dx_1}{dt} = \frac{2f}{(\tau - V)\sqrt{g}} \partial_{x_2} v_*, \qquad \frac{dx_2}{dt} = -\frac{2f}{(\tau - V)\sqrt{g}} \partial_{x_1} v_* \tag{7.6.7}$$

(at the levels $v_* = 2j\mu h$) and for $j = 0$ the drift lines are defined by the equations

$$\frac{dx_1}{dt} = \frac{1}{f\sqrt{g}} \partial_{x_2} V, \qquad \frac{dx_2}{dt} = -\frac{1}{f\sqrt{g}} \partial_{x_1} \tag{7.6.8}$$

at the level $0 = V - \varsigma m$ where $f = \frac{1}{\sqrt{g}} F_{12}$ is the pseudo-scalar intensity of the magnetic field and $v_* = \frac{1}{f}(m^2 - (\tau - V)^2)$.

I leave the spectral asymptotics in this case to the reader (see theorem 6.5.15).

7.7 Lacunas in the Semiclassical Approximation to the Spectrum

7.7.1 Existence of spectral gaps

The majority of this section is devoted to the case $d = 2$ because for $d = 3$ there can exist only one spectral gap, namely $(-m + \sup V, m + \inf V)$ under the condition that the first number is less than the second. On the other hand, for $d=2$ other spectral gaps can appear. In this section we analyze the case $d=2$ in two different situations: under condition (7.4.1) and under the condition

$$h^{\delta-1}\le\mu; \tag{7.7.1}$$

later we join these cases. Let us start with the statements that gaps exist. By standard elliptic arguments the following statements are easily proven:

Proposition 7.7.1. *Let $d=2$ and suppose that conditions (7.1.1)–(7.1.3), (7.2.1), $(7.2.25)_{1-3}$, (7.4.1), (7.3.1) and $(6.3.10)_1$ are fulfilled. Then*

(i) Let condition (7.3.24) be fulfilled and for some $j\in\mathbb{Z}^+$ let

$$|2j\mu h+v_*|\ge\zeta^2\ge C_0\bar\nu\mu^{-2}\qquad\forall x\in B(0,1). \tag{7.7.2}$$

Then for $\mathcal{U}'''_{jk}$ constructed in section 7.4 the following estimate holds:

$$|F_{t\to h^{-1}\tau}\chi_T(t)Q_x\mathcal{U}'''_{jk}|\le C'\mu^{-s}h^s \tag{7.7.3}$$
$$\forall x,y\in\mathbb{R}^2\quad\forall T\ge h^{1-\delta}\zeta^{-2}\quad\forall k\in\mathbb{Z}^+\quad\forall\tau:|\tau|\le\epsilon\zeta^2.$$

In particular, if condition (7.7.2) is fulfilled for all $j\in\mathbb{Z}^+$ then estimate (7.7.3) holds for all j,k and hence the estimate holds for $\mathcal{U}$ instead of $\mathcal{U}'''$, and therefore the estimate

$$|F_{t\to h^{-1}\tau}\chi_T(t)U|\le C'T\mu^{-s}h^s \tag{7.7.4}$$

holds for the same T and τ as in (7.7.3) and for $x,y\in B(0,\frac{1}{2})$.

(ii) Let

$$v_*\ge\zeta^2\ge C_0\bar\nu\mu^{-2}\qquad\forall x\in B(0,1). \tag{7.7.5}$$

Then estimates (7.7.3), (7.7.4) hold for the x,y,T,τ indicated above and for all j,k.

Proposition 7.7.2. *Let $d=2$ and suppose that conditions (7.1.1)–(7.1.3), (7.2.1), $(7.2.25)_{1-3}$, (7.5.1), (7.5.2) and (7.3.1) are fulfilled with $\bar\nu\ge\mu^{-l}$. Then*

(i) For some $j\in\mathbb{Z}^+$ let

$$|2j\mu h+v_*|\ge\zeta^2\ge C_0\bar\nu\mu^{-1}h\qquad\forall x\in B(0,1). \tag{7.7.6}$$

Then for the $\mathcal{U}'_{jk}$ constructed in section 7.5 estimate (7.7.3) holds. In particular, if condition (7.7.6) is fulfilled for all $j\in\mathbb{Z}^+$ then estimate (7.7.3) holds for all j,k; hence the estimate holds for $\mathcal{U}$ instead of $\mathcal{U}'$, and therefore estimate (7.7.4) holds.

(ii) For some $j\in\mathbb{Z}^+$ let

$$|H_{\varkappa j}|\ge\zeta^2\ge C_0(m+1)^{-1}\bar\nu\mu^{-1}h\qquad\forall x\in B(0,1)\quad\forall\varkappa:(\varkappa,j)\in\mathbf{Z} \tag{7.7.7}$$

where $H_{\varkappa j}=V+\varkappa(2j\mu hF+m^2)^{\frac{1}{2}}$, $\varkappa=\pm$ and $\mathbf{Z}$ were introduced in section 7.4. Then estimate (7.7.3) holds. Moreover, if condition (7.7.7) is fulfilled for all j then estimate (7.7.3) holds for all j and therefore this estimate holds for $\mathcal{U}$ instead of $\mathcal{U}'$ and hence estimate (7.7.4) holds.

Let us discuss the case when conditions (7.5.1) and (7.5.2) are fulfilled and $\mu h\ge C_0$ more carefully. In this case one should check conditions of (7.7.2)-type only for $j=0$, $\varkappa=-\varsigma$ and in order to prove estimates (7.7.3) and (7.7.4) it is sufficient to check the microlocal hypoellipticity of the block $\mathcal{B}_{(0)}$ of the operator $\mathcal{B}=(m+1)\mathcal{K}^*\mathcal{A}\mathcal{K}$; this operator is of block-diagonal type. Let us note that the distance between eigenvalues of the symbol $(m+1)\mathcal{A}^0$ corresponding to $j=0$, $\varkappa=-\varsigma$ and eigenvalues corresponding to other $(j,\varkappa)$ surely is not less than $(m+1)\mu h(m+\sqrt{\mu h})^{-1}$ and that blocks $(m+1)\mathcal{A}_{0j}$ are uniformly $\mu^{-1}h$-pseudo-differential operators with an additional factor $\bar{\nu}(\mu^{-1}h)^{\frac{1}{2}}$ $(j\geq 1)$. Therefore the operator $\mathcal{M}=\mathcal{B}_{(0)}-\mathcal{A}_{00}$ is equal to $\rho\tilde{\mathcal{M}}$ where

(7.7.8) $$\rho=\bar{\nu}^2\mu^{-2}(1+\sqrt{\mu h}(m+1)^{-1})$$

and $\tilde{\mathcal{M}}$ is a $\mu^{-1}h$-pseudo-differential operator uniformly. Moreover,

$$\mathcal{A}_{00}=\mathcal{G}^*(V-\varsigma m)\mathcal{G}(m+1)$$

where $\mathcal{G}=T\mathcal{F}L$, L is the operator: $v(x_1)\to Lv=v(x_1)\Upsilon_0(x_2)e$ where e is a fixed element of $\operatorname{Ker}(\sigma_1-i\sigma_2)$ with $|e|=1$.

So we are interested in the microlocal hypoellipticity of the operator

$$(m+1)\mathcal{G}^*(V-\varsigma m)\mathcal{G}+\mathcal{M}.$$

Let us consider the map $\bar{\Psi}:\Lambda^0\ni(x_1,\xi_1)\to x\in B(0,1)$ introduced in section 6.2. It is easy to see that if $\psi=1$ in a neighborhood of $\bar{\Psi}(\operatorname{supp}q)$ and $Q=\operatorname{Op}(q)$ then $Q\mathcal{G}^*\psi\equiv Q\mathcal{G}^*$ and $\psi\mathcal{G}Q\equiv\mathcal{G}Q$. On the other hand, if $q=1$ in a neighborhood of $\bar{\Psi}^{-1}(\operatorname{supp}\psi)$ then $Q\mathcal{G}^*\psi\equiv\mathcal{G}^*\psi$ and $\psi\mathcal{G}Q\equiv\mathcal{G}Q$; here Q is a $\mu^{-1}h$-pseudo-differential operator with respect to x_1 and $\psi=\psi(x)$. This observation easily yields that $\mathcal{B}_{(0)}$ is a microlocally hypoelliptic $\mu^{-1}h$-pseudo-differential operator provided

(7.7.9) $$(m+1)|V-\varsigma m|\geq C_0\rho$$

(let us note that the left-hand expression equals $|H_{-\varsigma 0}|$). Moreover, it is easy to see that

(7.7.10) $$\mathcal{M}=\rho\Big(\sum_{p,r}f^{pr}\partial_p\tilde{V}\cdot\partial_r\tilde{V}+\sqrt{\mu^{-1}h}\sum_p f^p\partial_p\tilde{V}+\mu^{-1}hf\Big)\circ\bar{\Psi}$$

with uniformly smooth symbols f^{pr}, f^p, f and with $\tilde{V}=(m+1)\bar{\nu}^{-1}V$. Therefore condition (7.7.9) can be replaced by the condition

(7.7.11) $$(m+1)|V-\varsigma m|\geq C_0\rho\mu^{-1}h.$$

So we have proven the following

Proposition 7.7.3. *Let $d=2$ and suppose that conditions* (7.1.1)–(7.1.3), (7.2.1), $(7.2.25)_{1-3}$, (7.5.1), (7.5.2) *and* (7.3.1) *are fulfilled with $\bar{\nu}\geq\mu^{-l}$, $\mu h\geq C_0$. Moreover, let condition* (7.7.11) *be fulfilled. Then for all j,k estimate* (7.7.3) *holds and estimate* (7.7.4) *holds as well.*

In a similar way one can prove

Proposition 7.7.4. *Let $d=2$ and suppose that conditions* (7.1.1)–(7.1.3), (7.2.1), $(7.2.25)_{1-2}$, (7.5.1), (7.5.32), (7.5.35) *and* (7.3.1) *are fulfilled with*

(7.7.12) $$\nu\geq h_1^l,\qquad \bar{\nu}\leq\nu\mu^{-1}h_1^{-l}$$

where $\delta=\delta(l,s)>0$ in condition (7.5.31) *is small enough. Moreover, let*

(7.7.13) $$|H_{-\varsigma 0}|\geq\zeta^2\geq C_0\nu h_1\qquad\forall x\in B(0,1).$$

Then the estimate

$$|F_{t\to h^{-1}\tau}\chi_T(t)\Gamma' \mathcal{U}'_{jk}| \le C'\mu^{1-j-k}T\nu\zeta^{-2}h_1^s \qquad \forall x,y\in\mathbb{R}^2 \tag{7.7.14}$$

holds for $T\ge hh_1^{\delta'}$, $|\tau|\le\epsilon\zeta^2$. *Therefore the estimate*

$$|F_{t\to h^{-1}\tau}\chi_T(t)\Gamma\psi U|\le C'\mu T\nu\zeta^{-2}h_1^s \tag{7.7.15}$$

holds for the same T *and* τ.

Remark. *Surely, one can weaken condition (7.7.13) in the spirit of (7.7.9) or (7.7.11).* ◇

By standard elliptic arguments one can prove the following

Proposition 7.7.5. *Let* $d=3$ *and suppose that conditions* (7.1.1)–(7.1.3), (7.3.1), $(7.2.25)_3$ *are fulfilled with* $1\ge\nu\ge h^2$. *Moreover, let*

$$v_*\le-\epsilon\nu \qquad \text{in } B(0,1). \tag{7.7.16}$$

Then the estimate

$$|F_{t\to h^{-1}\tau}\chi_T(t)\Gamma\psi U|\le C'(\mu h^{-1}+h^{-2})\Bigl(\frac{h}{\sqrt{\nu}}\Bigr)^s \tag{7.7.17}$$

holds for the same $T\ge h^{1-\delta'}\nu^{\frac12\delta'-1}$ *and* $|\tau|\le\nu$.

By means of standard Tauberian arguments we then obtain the following

Theorem 7.7.6. *Suppose that conditions* (7.1.1)–(7.1.3), (7.2.1), (7.3.1) *and* $(7.2.25)_{1-3}$ *are fulfilled. Then*

(i) Let $d=2$, *let conditions* (7.4.1), $(6.3.10)_1$, (7.3.24) *be fulfilled and let condition* (7.7.2) *be fulfilled for all* $j\in\mathbb{Z}^+$. *Then the estimate*

$$|e(x,y,\tau_1,\tau_2)|\le C'\mu^{-s}h^s \qquad \forall x,y\in B(0,\frac12) \quad \forall\tau_1,\tau_2\in(-\epsilon\zeta^2,\epsilon\zeta^2) \tag{7.7.18}$$

holds.

(ii) Let $d=2$, *let conditions* (7.5.1), (7.5.2) *be fulfilled and suppose* $\bar\nu\ge\mu^{-l}$. *Moreover, let* ***either*** *condition* (7.7.6) *be fulfilled for all* $j\in\mathbb{Z}^+$ ***or*** *condition* (7.7.7) *be fulfilled for all* $(\varkappa,j)\in\mathbf{Z}$. *Then estimate* (7.7.18) *also holds.*

(iii) Let $d=2$, *let conditions* (7.5.1), (7.5.2) *be fulfilled and suppose* $\mu h\ge C_0$. *Moreover, let* ***either*** (7.7.9) ***or*** (7.7.11) *be fulfilled. Then estimate* (7.7.18) *also holds.*

(iv) Let $d=2$, *let conditions* (7.5.1), (7.5.32), (7.5.35), (7.7.12), (7.7.13) *be fulfilled (and in* (7.2.25) *only the first two conditions should be checked) and let* $\delta=\delta(l,s)>0$ *be small enough. Then the estimate*

$$|\Gamma\psi e(.,.,\tau_1,\tau_2)|\le C'\mu\nu\zeta^{-2}h_1^s \qquad \forall\tau_1,\tau_2\in(-\epsilon\zeta^2,\epsilon\zeta^2) \tag{7.7.19}$$

holds.

(v) Let $d=3$, *let condition* (7.3.59) *hold with* $\nu\in[h^2,1]$, $\bar\nu=1$ *in* (7.2.25) *and let condition* (7.7.16) *be fulfilled. Then the estimate*

$$|\Gamma\psi e(.,.,\tau_1,\tau_2)|\le C'(\mu h^{-1}+h^{-2})\Bigl(\frac{h}{\sqrt{\nu}}\Bigr)^s \qquad \forall\tau_1,\tau_2\in(-\epsilon\nu,\epsilon\nu) \tag{7.7.20}$$

holds.

7.7.2 Calculations in the spectral gaps

In what follows we assume that $d = 2$. In order to finish our analysis we should treat only mollified asymptotics. Theorem 4.3.1 immediately yields

Proposition 7.7.7. *Let $d = 2$ and suppose that conditions* (7.1.1)–(7.1.3), (7.2.1), (7.4.1), (7.3.1) *and* (7.3.24) *are fulfilled. Let $\phi \in C_0^K((-\epsilon,\epsilon))$, $L \in [h^{1-\delta'},1]$. Then*

$$(7.7.21)\quad |\int \phi(\frac{\tau}{L})\bigl(d_\tau(\Gamma\psi e)(.,\tau) - h^{-2}\int \psi d_\tau S(x,.,\tau,\mu h)dx - \sum_{\varkappa,n,r:n\le 2r,r\le N}\sum_j \Theta^{(n+1)}_{j\varkappa nr}(\tau)(\mu^{-1}h)^{r-1}d\tau\bigr)| \le C'h^s$$

where

$$(7.7.22)\qquad \Theta_{j\varkappa nr}(\tau) = \int \theta(\tau - H_{\varkappa j})f_{nr\varkappa}(j\mu h,\mu^{-1},\mu h,x_1,\xi_1)dx_1d\xi_1$$

and for $n = 0$

$$(7.7.23)\qquad f_{0r\varkappa} = \sum_{1\le|\alpha|\le K} F_{r\varkappa}(D^\alpha\psi)\circ\bar{\Psi}(j\mu h)^{\frac{|\alpha|}{2}}$$

and the coefficients $F_{r\varkappa}$ do not depend on ψ. Moreover, for $j\mu h \ge \epsilon_0$ all the functions are uniformly smooth.

Let us now consider the method of successive approximations applied without the reduction construction and without condition (7.3.24) and with the non-perturbed operator $\mathcal{A}_0^0 + \sigma_0 m + \bar{W}I$ where in $\bar{W}$ we set $x = y$ and skip all the terms with factor μ^{-2}. Let us note first of all that this construction is provided by the condition $T\mu^{-1}h^{-1} \le \mu^{-\delta'}$ with arbitrary small $\delta' > 0$ (under conditions (7.4.1) or (7.5.1)–(7.5.2)) and hence we obtain only the condition $L \ge \mu^{\delta''-1}$ on the mollification parameter where $\delta'' > 0$ is arbitrarily small. Furthermore, let us note that for $j > k \ge 1$ the blocks W_{jk} of the perturbed operator are

$$(7.7.24)\qquad \begin{aligned} W_{jk} = &\alpha^-_{(j-k)}(\mu^{-1}h)^{\frac{1}{2}(j-k)}\bigl(\frac{j!}{k!}\bigr)^{\frac{1}{2}}\Pi_- + \\ &\alpha^+_{(j-k)}(\mu^{-1}h)^{\frac{1}{2}(j-k)}\bigl(\frac{(j-1)!}{(k-1)!}\bigr)^{\frac{1}{2}}\Pi_+ \end{aligned}$$

where $\Pi_\pm = \frac{1}{2}(I \pm \sigma_0)$ and $\alpha^\pm_{(l)} = \alpha^\pm_{(l)}(x_1,h_1D_1,h_1,jh_1)$ are polynomials with respect to $z = jh_1$; the blocks with $k > j \ge 1$ can be obtained by Hermitian conjugation. The blocks of the perturbing operator with $j = k \ge 1$ have similar form but in them $\alpha^\pm_{(l)}(.,.,0,0) = 0$. Moreover, the blocks of the operator

$$\tilde{\psi} = T^*\mathcal{F}^*\psi\mathcal{F}T - \psi'(x_1,h_1D_1)I$$

where $\psi' = \psi\circ\bar{\Psi}$ have the same form and in this case $\alpha_{\dots}$ are expressed in terms of ψ in the following way: these coefficients (and therefore the coefficients $\tilde{\psi}_{jk}$ as well) are linear combinations of $(D^\beta\psi)\circ\bar{\Psi}$ with $1 \le |\beta| \le K$ and in $\tilde{\psi}_{jk}$ every derivative $D^\beta\psi$ surely has a factor $j^r h_1^{|\beta|}$ with $r \le |\beta|$. Finally, it is easy to see that in $\tilde{\psi}_{jk}$ as well as in W_{jk} the coefficients $\alpha^\pm_{(l)}$ satisfy the equality $\alpha^+_{(l)}(.,.,0,.) = \alpha^-_{(l)}(.,.,0,.)$.

On the other hand, let us note that

$$\left(\tau - \sigma_0 m - \sigma_1\sqrt{2j\mu h f}\right)^{-1} = \left(\tau + \sigma_0 m + \sigma_1\sqrt{2j\mu h f}\right)\left(\tau^2 - m^2 - 2j\mu h f^2\right)^{-1},$$

that $\sigma_1 \Pi_\pm = \Pi_\pm \sigma_1$ and that the trace of the product of any number of matrices σ_1 and σ_0 (taken in any order) vanishes unless there is an even number of σ_0 as well as even number of σ_1.

These arguments yield that modulo terms generating $O(\mu^{-s})$ after mollification with the indicated L, the expression $\mp i \operatorname{tr} F_{t\to h^{-1}\tau} \Gamma \psi U^\pm$ is (for $\pm \operatorname{Im}\tau < 0$) a sum of terms of the following two types:

(i) Terms of the form

$$\prod_{1\le\iota\le n+1} (\tau_1^2 - m^2 - 2j_\iota \mu h f^2)^{-1} \times (j\mu^{-1}h)^p (\mu^{-1}h)^q (j\mu h)^r (\mu h)^l \tau_1^\gamma G(j_2 - j_1, j_3 - j_1, \dots, j_n + 1 - j_1) \tag{7.7.25}$$

with $\tau_1 = \tau - \bar W$, $j_1 = j$, $p,q,r,l \in \frac{1}{2}\mathbb{Z}$, $\gamma, n \in \mathbb{Z}$, such that $p + r, q + l \in \mathbb{Z}$, $2(p+q) \ge \max(n, |j_1 - j_2| + \dots + |j_n - j_{n+1}|)$, $n + 1 \ge 2(r + l) + \gamma$ and with bounded coefficients $G(\dots)$;

(ii) Terms of the same form but with $j \le C$ and with G depending arbitrarily on j (in a non-polynomial way; these terms arise because in all the matrices $j, k \ge 0$ and blocks with $j = 0$ or $k = 0$ are special).

However, after we take the difference of the traces from the upper and lower complex half-planes these terms of type (ii) will produce 0 for $|\tau| \le \epsilon$ under conditions (7.3.24) and (7.4.1). Therefore taking the difference of the traces from the upper and lower complex half-planes we again obtain formula (7.7.21) but with a more explicit formula for $f_{nr\varkappa}$. In particular, let us calculate $f_{0r\varkappa}$; namely, $f_{0r\varkappa}$ are polynomials of $j\mu^{-1}h$, $\mu^{-1}h$, $j\mu h$ and $(m^2 + 2j\mu h f^2)^{-\frac{1}{2}}$ and all the monomials in these polynomials satisfy the following conditions:

$$1 + 2(\deg_1 + \deg_2) \ge \deg_5 \ge 2(\deg_3 + \deg_4)$$

where $\deg_k$ means the degree with respect to the k-th argument, $k = 1, \dots, 5$. Since these and the original formulas for $f_{\dots}$ should agree for $h^{\delta'} \le \mu h \le \epsilon$ we obtain the following statement:

Proposition 7.7.8. *In formula (7.7.22) the functions $f_{0r\varkappa}$ are polynomials with respect to $j\mu^{-1}h$, $\mu^{-1}h$ $j\mu h$, μh and $(m^2 + 2j\mu h f^2)^{-\frac{1}{2}}$ with the properties indicated above.*

Let us now calculate all the other terms in the decomposition without conditions (7.3.24) and (7.4.1). Moreover, without loss of generality one can assume that (7.5.1) holds. Let us temporarily assume that condition (7.5.2) also is fulfilled. It is easy to see that these additional terms are due to the non-standard form of the blocks $\mathcal{A}_0^0$, W_{0k}, W_{k0}, ψ_{00}, ψ_{0k} and ψ_{k0}. It is easy to see that these terms have the same form as above but with functions $f_{n\varkappa r}$ non-regularly depending on j. Namely, these functions are now $O((\mu^{-1}h)^{\max(j,1)})$.

However, in all cases the following remark is true:

Remark 7.7.9. *All terms with $n = 0$ which are different from terms contained in $\partial_\tau h^{-2} S(x, \tau, \mu h)$ have functions $f_{\dots}$ depending linearly on $D^\beta \psi \circ \bar\Psi$ with $1 \le |\beta| \le K$.* $\diamondsuit$

7.7.3 The main theorem

Now we can obtain the main statements of this section. We need the following two definitions:

Definition 7.7.10. Let $d = 2$ and let conditions (7.1.1)–(7.1.3), (7.2.1), (7.3.1) be fulfilled. Let us replace the potential V by the potential $V_\tau = V - \tau$ in the operator A. Then

(i) $\tau \in \mathcal{Z}_{L1}$ if conditions $(7.2.25)_{1-3}$, (7.4.1), $(6.3.10)_1$, (7.3.24) and (7.7.2) are fulfilled with $v_{\tau *} = (V-\tau)^2 - m^2$ instead of v_*.

(ii) $\tau \in \mathcal{Z}_{L2}$ if conditions $(7.2.25)_{1-3}$, (7.4.1), $(6.3.10)_1$ and (7.7.5) are fulfilled for all $j \in \mathbb{Z}^+$.

(iii) $\tau \in \mathcal{Z}_{L3}$ if conditions $(7.2.25)_{1-3}$, (7.5.1), (7.5.2) and (7.7.6) are fulfilled for all $j \in \mathbb{Z}^+$ and if $\bar\nu \geq \mu^{-l}$.

(iv) $\tau \in \mathcal{Z}_{L4}$ if conditions $(7.2.25)_{1-3}$, (7.5.1), (7.5.2) and (7.7.7) are fulfilled for all $(\varkappa, j) \in \mathbf{Z}$. ◊

Definition 7.7.11. Let $d = 2$ and let conditions (7.1.1)–(7.1.3), $(7.2.25)_{1-3}$, (7.3.1), (7.5.1), (7.5.31), (7.5.34), (7.7.12) and (7.7.13) be fulfilled with small enough $\delta > 0$ in (7.5.31). Then $\tau \in \tilde{\mathcal{Z}}_{L5}$. ◊

Now repeating arguments of theorems 6.6.6, 6.6.7 we easily obtain

Proposition 7.7.12. *Let $d = 2$ and let conditions* (7.1.1)–(7.1.3), (7.2.1) *and* (7.3.1) *be fulfilled. Then*

(i) If $\tau_0 \in \mathcal{Z}_L = \mathcal{Z}_{L1} \cup \cdots \cup \mathcal{Z}_{L4}$ then for the L indicated above the following estimate holds:

$$(7.7.26)\quad \Big|\int \phi\Big(\frac{\tau-\tau_0}{L}\Big)\Big(\Gamma\psi e(\tau,\tau_0) - h^{-2}\int S(x,\tau,\tau_0,\mu h)\psi dx - \sum_{\varkappa,n,r:n\leq 2r, r\leq N}\sum_j \Theta^{(n)}_{j,\varkappa n r}(\tau,\tau_0)(\mu^{-1}h)^r\Big)\Big| \leq C'\mu^{-s}h^s$$

where $\Theta^{(n)}_{j,\dots}(\tau,\tau_0) = \Theta^{(n)}_{j,\dots}(\tau_0) - \Theta^{(n)}_{j,\dots}(\tau)$ and $\Theta^{(n)}_{j,\dots}(\tau_0) = 0$ here for $r \geq 1$. Moreover, $\Theta_{j,\dots}$ are given by (7.7.22), (7.7.23) *and the functions $f\dots$ are uniformly bounded and smooth with respect to all their arguments for $\tau \in \mathcal{Z}_{L1}$.*

(ii) If $\tau_0 \in \tilde{\mathcal{Z}}_{L5}$ then the left-hand expression in (7.7.26) *does not exceed $C'\mu\nu\zeta^{-2}h_1^s$.*

Definition 7.7.13. Let $d = 2$ and let us assume that there exists $\eta \in (0,1)$ such that after multiplication by η^{-1} the operator A satisfies (7.1.1)–(7.1.3) with $h' = h\eta^{-1}$, $\mu' = \mu\eta^{-1}$, $m' = m\eta^{-1}$. Let us assume that after this procedure $\tau' = \tau\eta^{-1} \in \mathcal{Z}_{Lk}(A')$ where A' is a new operator. Then $\tau \in \mathcal{Z}'_{Lk} = \mathcal{Z}'_{lk}(A)$ for the original operator. Moreover, $\tilde{\mathcal{Z}}'_{L5}$ is defined in the same way. ◊

Proposition 7.7.12 then obviously yields

Corollary 7.7.14. *Estimate* (7.7.26) *with $L\eta$ instead of L holds for $\tau \in \mathcal{Z}'_L$.*

The same arguments yield the following

Theorem 7.7.15. *Let $d = 2$ and let conditions* (7.1.1)–(7.1.3), (7.2.1) *and* (7.3.1) *be fulfilled. Then for $\tau_1, \tau_2 \in \mathcal{Z}'_L$ the following estimate holds:*

$$(7.7.27)\quad \Big|\Gamma\psi e(\tau_1,\tau_2) - h^{-2}\int S(x,\tau_1,\tau_2,\mu h)\psi dx - \sum_{\varkappa,r\leq N}\sum_j \Theta_{j,\varkappa 0 r}(\tau_1,\tau_2)(\mu^{-1}h)^r\Big| \leq C'\mu^{-s}h^s.$$

In this theorem the values of the parameter η for τ_1 and τ_2 do not necessarily coincide.

Remark 7.7.16. *$\tau \in \tilde{\mathcal{Z}}'_{L5}$ can be defined similarly. The case when $\tau_1 \in \tilde{\mathcal{Z}}'_{L5}$ and $\tau \in \mathcal{Z}'_L$ or when $\tau_1, \tau_2 \in \tilde{\mathcal{Z}}'_{L5}$ can be treated using by theorem 7.7.15 and proposition 7.7.4.* ◊

7.8 Certain Generalizations

7.8.1 Getting rid of the condition "$v_* \le -\epsilon_0$"

Let us consider the violation of the condition $v_* \le -\epsilon_0$. Recall that for $d = 3$ we need this condition only for $m \le h^{-1+\delta}$ and for $d = 2$ we need it only for $\mu h \le \epsilon_0$. Let us first assume that $m \ge 1$. Then one can assume that

$$V = \pm m + \frac{1}{m} V'$$

with

$$|D^\alpha V'| \le c \qquad \forall \alpha : |\alpha| \le K.$$

Let us assume for the sake of definiteness that the sign is "+" here. Then the condition $v_* \le -\epsilon_0$ means exactly that $V' \ge \epsilon_0$ (with another constant $\epsilon_0 > 0$). Let us treat the following cases:

(a) $v_* \ge \epsilon_0 \nu$ in $B(0,1)$;
(b) $v_* \le -\epsilon_0 \nu$ in $B(0,1)$;
(c) $|v_*| \le c\nu$ in $B(0,1)$.

In cases (a) and (b) let us divide the operator A by $r = |v_*(0)|^{\frac{1}{2}}$ and let us introduce $h' = \frac{h}{r}$, $\mu' = \frac{\mu}{r}$, $m' = \frac{m}{r}$, $\nu' = \frac{\nu}{r^2}$. One can easily see that $\frac{1}{r}V = m' + \frac{1}{m'}V''$ with $|D^\alpha V''| \le c \quad \forall \alpha : |\alpha| \le K$. Therefore we are in the context of the cases

(a′) $v'_* \ge \epsilon_0$ in $B(0,1)$;
(b′) $v'_* \le -\epsilon_0$ in $B(0,1)$

respectively. Applying the arguments of subsection 6.7.1 we obtain that in cases (a) and (b) the condition $v_* \le -\epsilon_0$ (or $v_* \ge \epsilon_0$) in our theorems is not necessary.

Similarly, in case (c) we take $r = \sqrt{\nu}$ and apply the same arguments. In all cases $h' \le 1$ for $d = 3$ (because $\nu \ge h^2$); for $d = 2$ the condition ${\mu'}^{1-\delta} \ge h'$ follows from the same condition for μ, h and the restriction $\nu \ge \mu^{-l} h^l$; the same assertion is also true for condition (7.5.30).

Repeating the arguments used in the proof of proposition 6.7.1 and referring to theorem 7.5.5 for $\mu' h' \ge 1$ and with a partition of unity such that $\mu' h' \le 1$ we can easily prove

Proposition 7.8.1. *Theorems 7.4.11–7.4.14, 7.4.17, 7.4.18 remain true for $m \ge 1$ without condition* (7.3.24).

Now let $m \le 1$. Then the following cases should be considered (if necessary we apply an ϵ-admissible partition of unity):

(i) $r = |v_*(0)| + \nu + \mu h \le cm^2$. Division by m reduces this case to the case $h' = \frac{h}{m}$, $\mu' = \frac{\mu}{m}$, $m' = 1$, $\nu' = \frac{\nu}{m^2}$ which has been treated.

(ii) $|v_*(0)| \ge \epsilon_0 r \ge cm^2 + \mu h$. Similarly, division by $\sqrt{r}$ reduces this case to the case $h' = \frac{h}{\sqrt{r}}$, $\mu' = \frac{\mu}{\sqrt{r}}$, $\nu' = \frac{\nu}{r}$, $m' = \frac{m}{\sqrt{r}}$, $|v_*(0)| \ge \epsilon_0$.

(iii) $\mu h \ge \epsilon_0 r \ge m^2 + |v_*(0)| + \nu$. In this case we get $\mu' h' = 1$ and we do not need condition (7.3.24).

(iv) $|v_*(0)| \le \epsilon_0 \nu$, $m^2 + \mu h \le \epsilon_0 \nu$. In this case the reduction is not complete: we can only conclude that

$$|D^\alpha V| \le c\nu' \qquad \forall \alpha : |\alpha| \le K'$$

with $\nu' = \nu^{\frac{1}{2}-\delta'}$ with any $\delta' > 0$ and K' (and then $K = K(K', \delta')$ in the original conditions). Therefore one can either divide by ν' or change variables $x' = x\nu^{-\delta'}$ and divide by $\sqrt{\nu}$. In both cases some "loss of information" appears.

Let $\mu \le h^{-1+\delta}$ as before and let us now assume $m \le 1$. Let us introduce ρ by formula (6.7.1) and let us consider the zone $\{\rho \ge \bar\rho = m^{\frac12} + (\mu h)^{\frac14-\delta'}\}$. Let us consider a ρ-admissible partition of unity in this zone and let us divide the operator by ρ^2 on each element of the partition. On this element let us introduce $m' = m\rho^{-2}$, $h' = h\rho^{-3}$, $\mu' = \mu\rho^{-1}$, $\nu' = \nu\rho^{-3}$. Then for $d = 3$ the contribution of every element of the partition to the remainder estimate does not exceed $Ch^{-2}\rho^4$ and the total contribution does not exceed Ch^{-2}. So we have proven

Proposition 7.8.3. *Theorem 7.4.11 remains true without the conditions* (7.3.24) *and* $m \ge 1$.

The same arguments give us the following

Proposition 7.8.4. *Theorem 7.6.2 and proposition 7.6.3 remain true without the conditions* (7.3.24) *and* $m \ge 1$.

Let us treat now the case $d = 2$. The same arguments as before yield the following

Proposition 7.8.6. *Theorems 7.4.12–7.4.14, 7.4.17 and 7.4.18 remain true without condition* (7.3.24) *in all cases excluding* $m \le 1$, $1 \ge \nu \ge \epsilon_0 m^2$. *In this case these theorems remain true under the additional restriction*

$$|D^\alpha(\frac{V}{\sqrt{F}})| \le c\sqrt{\nu'} \qquad \forall \alpha : 1 \le |\alpha| \le K$$

fulfilled with $\nu' = \nu$. *This condition is fulfilled automatically with* $\nu' = \nu^{1-\delta'}$ *with arbitrary* $\delta' > 0$.

7.8.2 Getting rid of the condition "$F \ge \epsilon_0$"

Let us now assume that condition (6.2.1) is violated. Let conditions (6.7.3) and $(6.7.4)_m$ be fulfilled. Let us apply arguments of subsection 6.7.2 first assuming that $d = 3$. If $\mu h \le 1$ then $\mu' h' \le 1$ also and therefore we have the following

Proposition 7.8.7. *Let* $d = 3$, *let conditions* (7.1.1)–(7.1.3) *be fulfilled and let* $\mu h \le 1$. *Then under condition* $(6.7.4)_n$ *estimate* (6.7.5) *holds*[30] *and in the general case estimate* (6.7.8) *holds. Recall that the principal part is* $O(h^{-3})$ *and under an obvious condition it is* $\asymp h^{-3}$.

On the other hand if $\mu h > 1$ we need to consider the zone $\{\mu' h' \ge 1, h' \le 1\}$ (all other zones have been considered in subsection 6.7.2). Surely, we can always can include h in γ and keep $\gamma \ge h$ but then in the zone $\{\gamma \le C_0 h\}$ condition (6.2.1) is not necessarily fulfilled after rescaling. One can treat this zone by the variational method but here we simply avoid the trouble assuming that $\mu \le h^{-2}$ (and then this zone does not arise). Therefore we have the following

Proposition 7.8.8. *Let* $d = 3$, *let conditions* (7.1.1)–(7.1.3) *be fulfilled and let* $h^{-1} \le \mu \le h^{-2}$. *Then the estimate*

$$\mathcal{R} \le C\mu h^{-1-\delta} \tag{7.8.1}$$

holds with arbitrarily small $\delta > 0$. *Recall that the principal part is* $O(\mu h^{-2})$ *and under an obvious condition it is* $\asymp \mu h^{-2}$.

---

[30] With n instead of m here and below.

Remark 6.7.9 remains true. Now let $d = 2$ and let condition (7.3.24) be fulfilled. Moreover, let us assume that the condition

$$(7.8.2)_n \qquad \sum_{1\le|\alpha|\le n} |D^\alpha \frac{F}{V_*}| \ge \epsilon_0$$

is fulfilled (compare with $(6.7.10)_n$). Applying the arguments of the proof of theorem 6.7.10 one can easily prove

Theorem 7.8.9. *Let $d = 2$ and let conditions (7.1.1)–(7.1.3), (7.3.24) and $(7.8.2)_n$ be fulfilled. Then for $\mu h^2 \le 1$ estimate (6.7.11) holds. Recall that the principal part is $O(h^{-2} + \mu h^{-1})$ and under an obvious condition it is $\asymp h^{-2} + \mu h^{-1}$.*

Comments

The same case of a very strong magnetic field as in section 7.5 appears for the Pauli Hamiltonian $(\sigma \cdot (hD - \mathbf{V}(x)))^2 - V(x)$ where $\sigma = (\sigma_1, \sigma_2, \sigma_3)$ and $\sigma_1, \sigma_2, \sigma_3$ are Pauli matrices. [Lieb, Solovej, Ingvarson 1] is the only related paper of which I am aware.

Part III

Estimates of the Spectrum

8. Estimates of the Negative Spectrum

In this chapter we combine the results of Part II devoted to local semiclassical spectral asymptotics with variational estimates of Rozenblioum type (upper estimates for $\mathbf{N}^-(A)$) in order to derive upper and lower estimates for $\mathbf{N}^-(A) = \mathbf{N}(-\infty, A)$ where $\mathbf{N}^-(A)$ is the number of negative eigenvalues of the self-adjoint operator A (counting their multiplicities), provided $(-\infty, 0)$ does not intersect the essential spectrum of A and $\mathbf{N}^-(A) = \infty$ otherwise. Namely, the idea is that the domain consists of two overlapping zones: the regular zone (where local semiclassical spectral asymptotics are applied) and the singular zone (where formally variational estimates are applied). In the justification of this procedure (decoupling of asymptotics) the results of Chapters 2 and 3 concerning the finite speed of propagation of singularities in the intersection of these zones will play an important role.

In section 8.1 we formulate the main statements; in section 8.2 we prove them; in section 8.3 we formulate similar estimates for Riesz means and give a sketch of their proofs. In section 8.4 we transfer the results of section 8.1 to quadratic forms. In Appendix G a survey of variational estimates is given.

Later (in Part IV) we treat the case when the operator A depends on parameter(s); there the results of this chapter and Chapter 9 provide asymptotics of $\mathbf{N}^-(A)$ with respect to the parameters. Surely, the results of Part II provide some of the asymptotics of Part IV but without the additional analysis of this part we cannot cover many important cases.

8.1 Statement of the Basic Theorems

Let us divide our domain X into two overlapping subdomains X' and X'' in which specific conditions are assumed to be fulfilled. X' is the *zone of semiclassical asymptotics* or the *regular zone* and X'' is the *zone of variational estimates* or the *singular zone*.

Let us assume that

$$X \subset \mathbb{R}^d/(\bar\gamma_1\mathbb{Z} \times \bar\gamma_2\mathbb{Z} \times \cdots \times \bar\gamma_d\mathbb{Z}), \qquad 0 < \bar\gamma_j \le \infty \quad (j = 1, \dots, d) \tag{8.1.1}$$

(surely one can treat more general manifolds) and $X = X' \cup X''$. Let us assume that a function $\gamma = \gamma(x)$ is given on X' such that

$$0 < \gamma(x) \le \min_j \bar\gamma_j, \quad |\gamma(x) - \gamma(y)| \le |x - y| \qquad \forall x, y \in X'. \tag{8.1.2$_1$}$$

We say that $\gamma(x)$ is a *spatial scaling function*. We assume that a *momentum scaling function* $\rho(x)$ [1] is given on X' such that[2]

---

[1] In principle one can take this function also depending on ξ.

[2] In what follows functions ρ_k and ϱ_j will be introduced which should all satisfy condition $(8.1.2)_2$.

$$(8.1.2)_2 \quad \rho_k(x) > 0, \quad |\gamma(x) - \gamma(y)| \le \gamma(x) \implies \epsilon_0 \le \frac{\rho_k(x)}{\rho_k(y)} \le c, \qquad |\nabla \rho_k(x)|\gamma(x) \le c\rho_k(x).$$

Moreover, let $X_0' \subset X'$ and

$$(8.1.2)_3 \quad \rho(x)\gamma(x) \ge 1 \quad \forall x \in X_0' \quad \text{and} \qquad x \in X_0', y \in B(x, \gamma(x)) \cap X \implies y \in X',$$

$$X = X'' \cup X_0'. \tag{8.1.3}$$

8.1.1 General operators

Let us consider the operator A given on X' by the formula

$$A = \sum_{\alpha : |\alpha| \le m} a_\alpha(x) D^\alpha \tag{8.1.4}$$

with boundary conditions

$$B_j u = \sum_{\alpha : |\alpha| \le m_j} b_{j\alpha}(x) D^\alpha u = 0 \quad (j = 1, \dots, \mu) \qquad \text{on } \partial X \cap \overline{X_0'} \tag{8.1.5}$$

where a_α, $b_{j\alpha}$ are $D \times D$ matrices and $m_j \le m - 1$. That means that A is an operator in $L^2(X, \mathbb{H})$, $\mathbb{H} = \mathbb{C}^D$ [3] with domain $\mathrm{D}(A)$ but $Au(x)$ is given by (8.1.4) for $u \in \mathrm{D}(A)$, $x \in X'$. Moreover, it means that if $u \in H^m(X, \mathbb{H})$ is compactly supported in X' then $u \in \mathrm{D}(A)$ if and only if $Bu|_{\partial X} = 0$.

In our analysis we replace the operator A by the operator

$$\mathbf{A} = J^{-1} A \tag{8.1.6}$$

in $\mathcal{H}$ where

$(8.1.7)_1$ J is a Hermitian positive definite matrix depending on x

and here and below $\mathcal{H} = \mathcal{H}_J$ is a Hilbert space with inner product $(u, v)_{\mathcal{H}} = (Ju, v)$ and $(.,.)$ is the standard $L^2(X, \mathbb{H})$-inner product.

We assume that

$(8.1.7)_2$ $\mathbf{A}$ is a self-adjoint operator in $\mathcal{H}$.

Our goal is to derive upper and lower estimates for

$$\mathbf{N}^-(\mathbf{A}) = \int \operatorname{tr} e(x, x, -\infty, 0)\, dx \tag{8.1.8}$$

where $e(.,.,.,.)$ is the Schwartz kernel of the spectral projector of $\mathbf{A}$.

Let us note that $\mathbf{A}$ is unitarily equivalent to the operator $J^{-\frac{1}{2}} A J^{-\frac{1}{2}}$ in $\mathcal{H}_0 = L^2(X, \mathbb{H})$ and that $\mathbf{N}^-(\mathbf{A}) = \mathbf{N}^-(A)$. In what follows we use the notation $\mathbf{A}$ for the operator $J^{-\frac{1}{2}} A J^{-\frac{1}{2}}$ as well.[4]

[3] One can also treat the case when $\mathbb{H}$ is an infinite dimensional auxiliary Hilbert space.

[4] I hope that there will be no confusion.

Let us describe conditions on A, B and J. First of all, let us describe conditions in the regular zone X'. Namely, let us assume that in $X' \ni x$ the following conditions are fulfilled:

$$(8.1.9)_1 \qquad |D^\beta a_\alpha| \le c\varrho\rho^{-|\alpha|}\gamma^{-|\beta|} \qquad \forall\alpha : |\alpha| \le m,$$

$$(8.1.9)_2 \qquad |D^\beta b_{j\alpha}| \le c\varrho_j\rho^{-|\alpha|}\gamma^{-|\beta|} \qquad \forall j = 1,\dots,\mu \qquad \forall\alpha : |\alpha| \le m_j,$$

$$(8.1.9)_3 \qquad |D^\beta J| \le c\varrho_0\gamma^{-|\beta|}, \qquad |J^{-1}| \le c\varrho_0^{-1} \qquad \forall\beta : |\beta| \le K,$$

$(8.1.9)_4$ Either $B(y,\gamma(y)) \subset X$ or there exists a number $k = k(y) \in \{1,\dots,d\}$ and a real-valued function $\phi(x_{\hat{k}})$ such that

$$\partial X \cap B(y,\gamma(y)) = \{x_k = \phi(x_{\hat{k}})\} \cap B(y,\gamma(y))$$

and

$$|D^\beta \phi| \le c\gamma^{1-|\beta|} \qquad \forall\beta : 1 \le |\beta| \le K$$

where $x_{\hat{k}} = (x_1,\dots,x_{k-1},x_{k+1},\dots,x_d)$.

Moreover, let us assume that

$$(8.1.10)_1 \qquad \Big\langle \sum_{\alpha:|\alpha|=m} a_\alpha \xi^\alpha v, v \Big\rangle \ge \epsilon_0 \varrho\rho^{-m}|\xi|^m|v|^2 \qquad \forall v \in \mathbb{C}^D \quad \forall\xi \in \mathbb{R}^d$$

and that

$(8.1.10)_2$ If $\partial X \cap B(y,\gamma(y)) \ne \emptyset$ then after the boundary has been straightened in the coordinate system $\tilde{x}$: $\tilde{x}_{\hat{k}} = x_{\hat{k}}$, $\tilde{x}_k = x_k - \phi(x_{\hat{k}})$, after the operator A has been multiplied by $\varrho^{-1}\rho^m$ and the operators B_j have been multiplied by $\varrho_j^{-1}\rho^{m_j}$ $(j = 1,\dots,\mu)$ and all lower-order terms have been skipped, we obtain a boundary value problem which satisfies the Shapiro-Lopatinskii condition uniformly with respect to $\xi_{\hat{k}} \in \mathbb{R}^{d-1}$; moreover, this problem is semibounded from below and satisfies condition (5.1.11).

Under this conditions we know from sections 4.1 and 5.1 that

$$(8.1.11) \qquad |e(x,z,-\infty,0)| \le C\rho^d \qquad \forall x,z \in X \cap B(y,\frac{1}{2}\gamma(y)).$$

More precisely, these and all the other estimates (see below) follow from the estimates derived in Part II in the following way: we make the dilatation $x_{\text{new}} = \frac{1}{\gamma(y)}(x-y)$ (and we straighten the boundary unless $\partial X \cap B(y,\gamma(y)) = \emptyset$), we multiply A and $B_{(j)}$ by $\varrho(y)^{-1}$ and $\varrho_j(y)^{-1}$ respectively and we introduce an effective semiclassical parameter $h_{\text{eff}} = \frac{1}{\gamma(y)\rho(y)}$ and maybe some other effective parameters.

Moreover, if the condition

$$(8.1.12) \qquad \langle a(x,\xi)v, v\rangle \ge \epsilon_0\varrho|v|^2 \qquad \forall x \in B(y,\gamma(y)) \cap X \;\; \forall\xi \in \mathbb{R}^d \;\; \forall v \in \mathbb{C}^D$$

is fulfilled and if either $B(y,\gamma(y)) \subset X$ or the condition

$$(8.1.12)' \quad \langle a(x',D_1,\xi')w,w\rangle_+ \ge \epsilon_0\varrho|w|_+^2$$
$$\forall x' \in B(y,\gamma(y)) \cap \partial X \quad \forall\xi' \in \mathbb{R}^{d-1}$$
$$\forall w \in \mathcal{S}(\bar{R}^+,\mathbb{H}) : \; b_j(x',D_1,\xi')w|_{x_1=0} = 0 \quad (j = 1,\dots,d)$$

is fulfilled as well[5] then the estimate

$$|e(x,z,-\infty,0)| \le C\rho^{d-s}\gamma^{-s} \qquad \forall x,z \in X\cap B(y,\frac{1}{2}\gamma(y)) \tag{8.1.13}$$

holds with an arbitrarily large exponent s. Here and in what follows $a(x,\xi)=A^w(x,\xi)$, $\mathbf{a}(x,\xi)=\mathbf{A}^w(x,\xi)$ (or it can be replaced by $J^{-\frac12}a(x,\xi)J^{-\frac12}$ in what follows), $B_j(x,\xi)=b_j^w(x,\xi)$,[6] $\varsigma=\varrho\varrho_0^{-1}$.

In Chapters 4 and 5, under other additional conditions, intermediate estimates for

$$\mathcal{R}=|\int \psi(x)\bigl(\operatorname{tr} e(x,x,-\infty,0)-S(x,-\infty,0)\bigr)dx| \tag{8.1.14}$$

were derived where

$$S(x,\tau_1,\tau_2)=(2\pi)^{-d}\int \mathbf{n}(x,\xi,\tau_1,\tau_2)d\xi \tag{8.1.15}$$

here and in what follows, $\mathbf{n}(x,\xi,\tau_1,\tau_2)$ is the eigenvalue counting function for the operator $\mathbf{a}(x,\xi)$ in $\mathbb{H}$ and we assume that

$$|D^\beta\psi|\le c\gamma^{-|\beta|} \qquad \forall\beta: |\beta|\le K,\quad \operatorname{supp}\psi\subset B(y,\frac12\gamma(y)). \tag{8.1.16}$$

Moreover, under additional conditions of a global nature more precise estimates for

$$\mathcal{R}'=|\int \psi\bigl(\operatorname{tr} e(x,x,-\infty,0)-(S(x,-\infty,0)+S_1(x,-\infty,0)\bigr)dx - \int \psi S_2(x',-\infty,0)dx'| \tag{8.1.14$'$}$$

were derived where the additional terms are associated with the matrix character of $a(x,\xi)$ and with the boundary but their contributions surely do not exceed $C(\rho\gamma)^{d-1}$.

In particular we proved that

$$\mathcal{R}\le C(\rho\gamma)^{d-1} \tag{8.1.17}$$

provided one of the following conditions is fulfilled:

(8.1.18)$_1$ For every point $(x,\xi)\in\bigl(X\cap B(y,\gamma(y))\bigr)\times\mathbb{R}^d$ there exists a direction $\mathcal{T}$ such that $|\mathcal{T}_x|\le\gamma$, $|\mathcal{T}_\xi|\le\rho$ and the symbol $\varsigma^{-1}\mathbf{a}(x,\xi)$ is (uniformly) microhyperbolic at the point (x,ξ) at the energy level 0 in the direction $\mathcal{T}$ and, moreover, **either** $B(y,\gamma(y))\subset X$ **or** for every point $(x',\xi')\in\bigl(\partial X\cap B(y,\gamma(y))\bigr)\times\mathbb{R}^{d-1}$ there exists a multidirection $\mathcal{T}=(\mathcal{T}'_x,\mathcal{T}'_\xi,\nu_1,\dots,\nu_l)$ such that $|\mathcal{T}_x|\le\gamma$, $|\mathcal{T}'_\xi|\le\rho$, $|\nu_k|\le\rho$ $\forall k=1,\dots,l$ and the one-dimensional problem $(\varsigma^{-1}\mathbf{a}(x',D_1,\xi'),\varrho_1^{-1}b_1(x',D_1,\xi'),\dots,\varrho_\mu^{-1}b_\mu(x',D_1,\xi'))$ is (uniformly) microhyperbolic at the point (x',ξ') at the energy level 0 in the multidirection $\mathcal{T}$;

(8.1.18)$_2$ $X\supset B(y,\gamma(y))$ and a scalar function $\lambda(x,\xi)$ is given on $B(y,\gamma(y))\times\mathbb{R}^d$ such that

$$|D_x^\alpha D_\xi^\beta\lambda|\le c\varsigma\gamma^{-|\alpha|}\rho^{-|\beta|} \qquad \forall\alpha,\beta: |\alpha|\le K, |\beta|\le K \tag{8.1.19}$$

---

[5] Where we assume that the boundary is straightened and X coincides with $\{x, x_1\ge 0\}$ in the intersection with $B(y,\gamma(y))$.

[6] Surely, when replacing the operator $\mathbf{A}=J^{-1}A$ by the operator $\mathbf{A}=J^{-\frac12}AJ^{-\frac12}$ one should replace B_j by $B_jJ^{-\frac12}$.

and if $\tau \in (-\epsilon_0\varsigma, \epsilon_0\varsigma)$ is an eigenvalue of $\mathbf{a}_0(x,\xi)$ then $\tau = \lambda(x,\xi)$; moreover, $|\lambda| \le \epsilon_0\varsigma$, $|\tilde\nabla\lambda| \le \epsilon_0\varsigma \implies \widetilde{\mathrm{Hess}}\,\lambda$ has two eigenvalues of the same sign f_1, f_2 with $|f_j| \ge \epsilon_0\varsigma$ $\forall j$ where here and in what follows $\tilde\nabla$, $\tilde D$, $\widetilde{\mathrm{Hess}}$ mean ∇, D and Hess calculated *after* the change of variables $x_{\mathrm{new}} = \frac{x}{\gamma}$, $\xi_{\mathrm{new}} = \frac{\xi}{\rho}$ and a_0, $\mathbf{a}_0$ and b_{0j} are symbols such that the inequalities

$$|D_x^\alpha D_\xi^\beta \omega| \le c\varrho'\rho^{-|\beta|-1}\gamma^{-|\alpha|-1} \qquad \forall x \in B(y,\gamma(y)) \quad \forall \xi : |\xi| \le C_0\rho \tag{8.1.20}$$

hold for $\omega = a - a_0$, $\mathbf{a} - \mathbf{a}_0$, $b_j - b_{j0}$ with $\varrho' = \varsigma$, ϱ, ϱ_j respectively; it is natural to treat these symbols as principal symbols of the corresponding operators. Here and below C_0 is a large enough constant (such that $|a_0(x,\xi)v| \ge \epsilon_0\varrho|v|$ $\forall \xi : |\xi| \ge C_0\rho$ $\forall v$);

$(8.1.18)_3$ $X \supset B(y,\gamma(y))$ and on $B(y,\gamma(y)) \times \mathbb{R}^d$ a scalar function $\lambda(x,\xi)$ satisfying (8.1.19) is given such that if $\tau \in (-\epsilon_0\varsigma, \epsilon_0\varsigma)$ is an eigenvalue of $\mathbf{a}_0(x,\xi)$ then $\tau = \lambda(x,\xi)$; moreover, $|\lambda| \le \epsilon_0\varsigma$, $|\tilde\nabla\lambda| \le \epsilon_0\varsigma \implies \widetilde{\mathrm{Hess}}\,\lambda$ has two eigenvalues f_1, f_2 with $|f_j| \ge \epsilon_0$ $\forall j$ and

$$|D_x^\alpha D_\xi^\beta \omega| \le c\varrho'\rho^{-|\beta|-1-\sigma}\gamma^{-|\alpha|-1-\sigma} \qquad \forall x \in B(y,\gamma(y)) \quad \forall \xi : |\xi| \le C_0\rho \tag{8.1.20$'$}$$

with arbitrarily small $\sigma > 0$;

$(8.1.18)_4$ $X \supset B(y,\gamma(y))$ and on $B(y,\gamma(y)) \times \mathbb{R}^d$ a scalar function $\lambda(x,\xi)$ satisfying (8.1.19) is given such that if $\tau \in (-\epsilon_0\varsigma, \epsilon_0\varsigma)$ is an eigenvalue of $\mathbf{a}_0(x,\xi)$ then $\tau = \lambda(x,\xi)$; moreover, $|\lambda| \le \epsilon_0\varsigma$, $|\tilde\nabla\lambda| \le \epsilon_0\varsigma \implies \widetilde{\mathrm{Hess}}\,\lambda$ has an eigenvalue f_1 with $|f_1| \ge \epsilon_0\varsigma$ and after rescaling in appropriate coordinates z $\lambda = \pm z_1^2 + \lambda'(z')$ with condition $(4.4.47)_m$ fulfilled for $V = \lambda'$ and (8.1.20)$'$ holds with $m > 2, \sigma = \frac{2m-2}{m+2}$;

$(8.1.18)_5$ $d \ge 2$, $m = 2$,

$$\varsigma^{-1}\mathbf{a}_0(x,\xi) = \sum_{jk} g^{jk}(\xi_j - V_j)(\xi_k - V_k) + V \quad \text{in } B(y,\gamma(y)) \tag{8.1.21}$$

where $g^{jk} = g^{kj}$, V_j, V are real-valued functions such that

$$|D^\alpha g^{jk}| \le c\gamma^{-|\alpha|}, \quad |D^\alpha F_{jk}| \le c\rho\gamma^{-|\alpha|-1} \quad \forall j,k \tag{8.1.22$_{1-3}$}$$
$$|D^\alpha V| \le c\rho^2\gamma^{-|\alpha|} \quad \forall \alpha : |\alpha| \le K,$$

$$\sum_{jk} g^{jk}\xi_j\xi_k \ge \epsilon_0|\xi|^2 \qquad \forall x,\xi \tag{8.1.23}$$

and either $B(y,\gamma(y)) \subset X$ or $\mu = 1$ and $B = I$ or $\mu = 1$ and

$$b = -i\alpha(x')\sum_{jk}\nu_k g^{jk}(\xi_j - V_j) + \sum_j \ell^j(x')(\xi_j - V_j) + \beta(x') + \beta'(x') \tag{8.1.24}$$

where ν_k are components of the normal to ∂X, $\sum_{jk} g^{jk}\nu_j\nu_k = 1$, $\sum_j \ell^j\nu_j = 0$, α, β and ℓ^j are real-valued and either condition (5.1.57)[7] or condition (5.1.58) is fulfilled and β' is a lower order term in the sense of (8.1.20).

Finally, estimate (8.1.17) holds under the condition

[7] For $d = 2$ one should check condition (5.1.57)$'$ and reduce the operator $\varsigma^{-1}\mathbf{A}$ to the form (5.1.52).

$(8.1.18)_6 = (8.1.12)$;

in this case $S = 0$.

Summing with respect to a γ-admissible partition of unity in X_0' we obtain that if all the above conditions (providing estimate (8.1.17)) are fulfilled and if $\psi \geq 0$ supported in $\overline{X_0'}$ satisfies (8.1.16) then the estimates

(8.1.25)
$$\int \psi S(x,-\infty,0)dx - C\int_{X'_-} \wp_1 dx \leq \int \psi \operatorname{tr} e(x,x,-\infty,0)dx \leq \int \psi S(x,-\infty,0)dx + C\int_{X'_-} \wp_1 dx + C'\int_{X'_+} \rho^{d-s}\gamma^{-s}dx$$

hold where

(8.1.26)
$$\wp_1 = \rho^{d-1}\gamma^{-1},$$

X'_+ is the union of all the balls $B(x,\gamma(x))$ with $x \in X_0'$ in which conditions (8.1.12) and $(8.1.12)'$ are fulfilled and $X'_- = X' \setminus X'_+$.

Moreover, if the condition

$(8.1.18)_7$ $X \supset B(y,\gamma(y))$ and a scalar function $\lambda(x,\xi)$ satisfying (8.1.19) is given on $B(y,\gamma(y)) \times \mathbb{R}^d$ such that if $\tau \in (-\epsilon_0\varsigma, \epsilon_0\varsigma)$ is an eigenvalue of $\mathbf{a}_0(x,\xi)$ then $\tau = \lambda(x,\xi)$ and

(8.1.27)
$$\sum_{|\alpha|+|\beta|\leq m} \gamma^{|\alpha|}\rho^{|\beta|}|D_x^\alpha D_\xi^\beta \lambda| \geq \epsilon_0\varsigma$$

and $(8.1.20)'$ holds with large enough $\sigma = \sigma(m)$

is in the list of admissible conditions then the same estimate (8.1.25) holds where now

$(8.1.26)'$
$$\wp_1 = \rho^{d-1+\delta}\gamma^{-1+\delta}$$

with an arbitrarily small $\delta > 0$ in the union of balls where we appeal to condition $(8.1.18)_7$. Therefore the estimates

(8.1.28)
$$\int S(x,-\infty,0)dx - C\int_{X'_-} \wp_1 dx \leq \mathbf{N}^-(\mathbf{A}) \leq \int S(x,-\infty,0)dx + C\int_{X'_-} \wp_1 dx + C'\int_{X'} \rho^{d-s}\gamma^{-s}dx + \int \psi' \operatorname{tr} e(x,x,-\infty,0)dx$$

hold where ψ' is a function such that

(8.1.29) $0 \leq \psi' \leq 1$, $\psi' = 1$ in $X \setminus X'''$ and ψ' satisfies (8.1.16) in X' , $\psi' = 0$ in X_0' where X''' is the $\frac{1}{2}\gamma$-neighborhood of X_0'

and $\wp_1$ is introduced above. Thus, we have obtained a lower estimate for $\mathbf{N}^-(A)$ [8] and in order to obtain an upper estimate one should estimate the last term in the right-hand expression of (8.1.28).

[8] This estimate will be improved and modified under various conditions. Many improvements are left to the reader. In particular, I do not properly treat balls in which condition (8.1.12) and some additional conditions (microhyperbolicity of the boundary value problem, etc.) are fulfilled.

It appears that under appropriate conditions this term almost coincides with the term obtained in the following formal way: we replace X by X'' and apply a variational estimate of Rozenblioum type. It is more convenient not to write it explicitly. So, let us assume that

(8.1.30) There exist an operator $\mathbf{A}'$ in $L^2(X'', \mathbb{H})$ and boundary operators B'_j coinciding with $\mathbf{A}$ and B_j respectively on functions supported in $\overline{X''} \setminus \overline{X'_0}$ such that under the boundary conditions $B'_j u|_{\partial X''} = 0$ ($j = 1, \dots, \mu$) the operator $\mathbf{A}'$ is self-adjoint and the estimate

$$\mathbf{N}(\mathbf{A}', -\infty, t) \le R_0(t+1)^p \qquad \forall t \ge 0 \tag{8.1.31}$$

holds.

Finally, let us assume that

$$\varrho_0 \ge \epsilon_0 \varrho \rho^{-1} \gamma^{-1} \quad \text{in } X'_- \cap X'' \cap \{\rho\gamma \ge c\}, \tag*{(8.1.32)$_1$}$$

$$\epsilon_0 \varrho (\rho\gamma)^{-l} \le \varrho_0 \le c\varrho(\rho\gamma)^l \quad \text{in } X' \cap X''. \tag*{(8.1.32)$_2$}$$

The first condition is assumed to provide the following statement: *singularities (at a non-positive energy level) starting at X'_0 reach $X'' \setminus X'$ only after a time disjoint from* 0. In fact, in $X'_2 \cap \{\rho\gamma \le c\}$ the propagation of singularities is absent (because $h_{\text{eff}} = \frac{1}{\rho\gamma} \asymp 1$ and hence the notion of singularity is empty) and the results of Chapters 2 and 3 obviously yield that the spatial velocity of propagation of singularities for the operator $D_t - \mathbf{A}$ in $X'_1 \setminus X''$ does not exceed $C\varsigma\rho^{-1}$. On the other hand, the distance between $x \in X'_0$ and $y \in X'' \setminus X'$ is not less than $\epsilon_0 \gamma(x)$ and therefore the time is not less than $\epsilon_1 \varsigma \rho^{-1} \gamma^{-1}$.

Our first major result is the following

Theorem 8.1.1. *(i) Let conditions* (8.1.1)–(8.1.7), (8.1.9) *and* (8.1.10) *be fulfilled. Moreover, in every ball $B(x, \gamma(x))$ with $x \in X'_0$ suppose one of the conditions* (8.1.18)$_{1-7}$ *is fulfilled. Then the estimate*

$$\mathcal{N}^-_w(A) - CR_1 \le \mathbf{N}^-(A) \tag*{(8.1.33)$_1$}$$

holds with

$$\mathcal{N}^-_w(A) = \int_{X'} S(x, -\infty, 0)\,dx, \tag{8.1.34}$$

$$R_1 = \int_{X'} \wp_1\,dx \tag{8.1.35}$$

with $\delta \ge 0$ introduced above (one can vary δ from ball to ball).

(ii) Moreover, if conditions (8.1.30)–(8.1.32) *are fulfilled then the estimate*

$$\mathbf{N}^-(A) \le \mathcal{N}^-_w(A) + C(R_1 + R_2 + R_0) \tag*{(8.1.33)$_2$}$$

holds with

$$R_2 = \int_{X'} \rho^{d-s}\gamma^{-s}\,dx. \tag{8.1.36}$$

(iii) Both these estimates remain true if, in balls where one of conditions (8.1.18)$_{1,2,5}$ *is fulfilled with lower order terms estimate* (8.1.20)$'$, *we replace* $\wp_1$ *by*

$$\wp_1^* = \partial_\tau S(x,-\infty,\tau)\varrho(\rho\gamma)^{-1}\big|_{\tau=0} + \wp_1' \tag{8.1.26*}$$

where

$$\wp_1' = \rho^{d-1-\delta'}\gamma^{-1-\delta'}dx \tag{8.1.26''}$$

with a small exponent $\delta' > 0$.

8.1.2 Improved remainder estimates

The following theorem is based on the more precise estimates of Chapters 4 and 5; we replace $\wp_1$ (or $\wp_1^*$) by an expression with smaller integral. In order to do this we need to introduce new conditions. Let Λ_j $(j \ge 1)$ be closed subsets of the set

$$\Sigma = \{(x,\xi) \in X' \times \mathbb{R}^d,\ \det a(x,\xi) = 0\}.$$

Let us assume that in the $(\rho\gamma)^{\delta-\frac{1}{2}}(\gamma,\rho)$-neighborhood of Λ_j (with respect to (x,ξ)) a real-valued function $\lambda(x,\xi)$ is defined such that

(8.1.37) λ is an eigenvalue of $\mathbf{a}(x,\xi)$;

$$\operatorname{dist}(\lambda, \operatorname{Spec}\mathbf{a}\setminus\{\lambda\}) \ge (\rho\gamma)^{-\delta'}\varsigma \qquad \forall(x,\xi) \tag{8.1.38}$$

where here and below $\delta' > 0$ is a small enough exponent. Let $1 \le T_1 \le \dots \le T_n \le \dots$ be numbers such that

$$T_j \le (\rho\gamma)^{\delta'} \qquad \forall x \in \Pi_x\Lambda_j \tag{8.1.39}$$

and

(8.1.40) Through every point $(\bar{x},\bar{\xi})$ which belongs to the $(\rho\gamma)^{\delta-\frac{1}{2}}(\gamma,\rho)$-neighborhood of Λ_j there passes a Hamiltonian trajectory of the Hamiltonian

$$\lambda(x,\xi)\cdot\varsigma^{-1}\bar{\rho}\bar{\gamma}^{-1}$$

(where here and below the bar means that the corresponding function is calculated at the point $(\bar{x},\bar{\xi})$) along which either $t \in [0,T_j]$ or $t \in [-T_j,0]$ and $x(t) \in X'$, $x(0) = \bar{x}$, $\xi(0) = \bar{\xi}$,

$$\bar{\varrho}_*(\bar{\rho}\bar{\gamma})^{-\delta'} \le \varrho_* \le \bar{\varrho}_*(\bar{\rho}\bar{\gamma})^{\delta'},\quad \bar{\rho}(\bar{\rho}\bar{\gamma})^{-\delta'} \le \rho \le \bar{\rho}(\bar{\rho}\bar{\gamma})^{\delta'} \tag{8.1.41}$$
$$\bar{\gamma}(\bar{\rho}\bar{\gamma})^{-\delta'} \le \gamma \le \bar{\gamma}(\bar{\rho}\bar{\gamma})^{\delta'}$$

where the $*$ means any subscript, conditions (8.1.37) and (8.1.38) are fulfilled and

$$|D_{(x(0)\bar{\gamma}^{-1},\xi(0)\bar{\rho}^{-1})}(x(t)\bar{\gamma}^{-1},\xi(t)\bar{\rho}^{-1})| \le (\bar{\rho}\bar{\gamma})^{\delta'} \qquad \text{for } x(0)=\bar{x}, \xi(0)=\bar{\xi}, \tag{8.1.42}$$

$$\bar{\gamma}^{-1}|x(t)-\bar{x}| + \bar{\rho}^{-1}|\xi(t)-\bar{\xi}| \ge (\bar{\rho}\bar{\gamma})^{\frac{1}{2}-\delta}|t|. \tag{8.1.43}$$

Finally, we assume that on every segment of this trajectory one of the following two conditions is fulfilled:

$$\operatorname{dist}(x(t),\partial X) \ge \bar{\gamma}(\bar{\rho}\bar{\gamma})^{\frac{1}{2}-\delta} \tag{8.1.44$_1$}$$

or

$(8.1.44)_2$ In the $\bar{\gamma}(\bar{\rho}\bar{\gamma})^{-\delta'}$-neighborhood of $x(t)$, $\mathbf{a}(x,\xi)$ modulo lower order terms (in the sense of (8.1.20)) is the Schrödinger symbol (8.1.21) satisfying (8.1.22),

$$\sum_{jk} g^{jk}\eta_j\eta_k \ge \epsilon_0|\eta|^2(\bar{\rho}\bar{\gamma})^{-\delta'} \tag{8.1.23$'$}$$

and either $b = I$ or b is of the form (8.1.24) and condition (5.1.57) (or (5.1.57)$'$ for $d = 2$) is fulfilled **and**

$$|\sum_{jk} g^{jk}\nu_j(\xi_k - V_k)| \ge \epsilon_0\bar{\rho}(\bar{\rho}\bar{\gamma})^{-\delta'}; \tag{8.1.45}$$

then the trajectory reflects at ∂X; recall that $\nu = (\nu_1, \dots, \nu_d)$ is a unit normal to ∂X.

Theorem 8.1.2. *Let conditions (8.1.1)–(8.1.7), (8.1.9) and (8.1.10) be fulfilled. Moreover, in each ball $B(x,\gamma(x))$ with $x \in X_0'$ let one of the conditions $(8.1.18)_{1,2,5}$ with the lower order terms estimate (8.1.20)$'$ be fulfilled. Let Λ_j and T_j satisfy conditions (8.1.37)–(8.1.44). Then*

(i) The estimate

$$\mathcal{N}^-_{wp} - CR_3 \le \mathbf{N}^-(A) \tag{$(8.1.46)_1$}$$

holds with

$$\mathcal{N}^-_{wp} = \int_{X'} \bigl(S(x,-\infty,0) + S_1(x,-\infty,0)\bigr)dx + \int_{\partial X\cap \overline{X'}} S_2(x,-\infty,0)dx' \tag{8.1.47}$$

and

$$R_3 = \sum_{j\ge 1} \frac{1}{T_j}\int_{\Lambda_j} \varsigma(\rho\gamma)^{-1}dxd\xi : d\lambda + \partial_\tau\Bigl(\int_{\Omega_0} \varsigma\mathbf{n}(x,\xi,-\infty,\tau)(\rho\gamma)^{-1}dxd\xi\Bigr)\bigr|_{\tau=0} + R_1' \tag{8.1.48}$$

where Ω_0 is the $(\rho\gamma)^{\frac{1}{2}-\delta}(\gamma,\rho)$-neighborhood of $\Lambda_0 = \Sigma \setminus \cup_{j\ge 1}\Lambda_j$.

(ii) Moreover, if conditions (8.1.29)–(8.1.31) are fulfilled then the estimate

$$\mathbf{N}^-(A) \le \mathcal{N}^-_{wp} + C(R_3 + R_2 + R_0) \tag{$(8.1.46)_2$}$$

holds.

Remark 8.1.3. *Surely, we appeal to conditions $(8.1.18)_{3,4,7}$ in some balls in which case one can replace R_1 by R_3 with integration over Y_1' only and add the integral of $\wp_1$ over Y_2' where $Y_{1,2}'$ are domains in which the conditions of theorems 8.1.1 and 8.1.2 are assumed to be fulfilled. Similar remarks are valid in what follows.* ◊

A more complicated situation at the boundary leads us to weaker estimates. This fact is associated with other conditions on trajectories. In particular, $(\rho\gamma)^{-\delta'}$ should be replaced everywhere by an arbitrary small constant $\varepsilon > 0$ and one of the constants in the estimates depends on ε.

Let Λ_j ($j \ge 1$) be closed subsets of Σ as before and let $T_j \ge 1$ be numbers. Let us assume that a real-valued function λ satisfying (8.1.37) and

$$\operatorname{dist}(\lambda, \operatorname{Spec}\mathbf{a} \setminus \{\lambda\}) \ge \varepsilon\varrho \tag{8.1.38$'$}$$

is defined in the $(\rho\gamma)^{\delta-\frac{1}{2}}(\gamma,\rho)$-neighborhood of Λ_j. Let us assume that

$$T_j \le \varepsilon^{-1} \qquad \forall j \tag{8.1.39$'$}$$

and

(8.1.40)$'$ Through every point $(\bar{x},\bar{\xi})$ in the $(\rho\gamma)^{\delta-\frac{1}{2}}(\gamma,\rho)$-neighborhood of Λ_j there passes a Hamiltonian trajectory generated by the same Hamiltonian as before along which either $t\in[0,T_j]$ or $t\in[-T_j,0]$ and $x(t)\in X'$, $x(0)=\bar{x}$, $\xi(0)=\bar{\xi}$,

$$\bar{\varrho}_*\varepsilon \le \varrho_k \le \bar{\varrho}_*\varepsilon^{-1}, \quad \bar{\rho}\varepsilon \le \rho \le \bar{\rho}\varepsilon^{-1}, \quad \bar{\gamma}\varepsilon \le \gamma \le \bar{\gamma}\varepsilon^{-1}. \tag{8.1.41$'$}$$

Moreover, let us assume that conditions (8.1.37), (8.1.38)$'$ and

$$\bar{\gamma}^{-1}|x(t)-\bar{x}| + \bar{\rho}^{-1}|\xi(t)-\bar{\xi}| \ge \varepsilon|t|. \tag{8.1.43$'$}$$

are fulfilled along this trajectory. Finally, let us assume that on every segment of this trajectory either condition $(8.1.44)_1$ is fulfilled or

(8.1.45)$'$ In the $\varepsilon(\bar{\gamma},\bar{\rho})$-neighborhood of $(x(t),\xi(t))$ the coefficient of ζ^{mD} in the equation

$$\det(a(x,\xi+\nu\zeta)-\tau J(x))=0 \quad (\text{with } |\tau|\le\varepsilon\varsigma) \tag{8.1.49}$$

with respect to ζ has absolute value greater than ε, all the real roots of the equation are of constant multiplicity, the distances between the distinct roots are greater than $\varepsilon\bar{\rho}$, and finally by the boundary conditions one can express outgoing waves in terms of incoming waves with constant not exceeding ε^{-1}; here we have reflection of trajectories with branching at ∂X (if equation (8.1.49) has more than one real root corresponding to the outgoing wave) and we should consider *all possible trajectories with the given direction with respect to t and the previous conditions should be fulfilled on each trajectory.*

Theorem 8.1.4. *Let conditions* (8.1.1)–(8.1.7), (8.1.9) *and* (8.1.10) *be fulfilled. In each ball* $B(x,\gamma(x))$ *with* $x\in X'_0$ *let one of the conditions* $(8.1.18)_{1,2,5}$ *with lower order estimate*

$$|D_x^\alpha D_\xi^\beta \omega| \le \varepsilon c\varrho'\rho^{-|\beta|-1}\gamma^{-|\alpha|} \qquad \forall x\in B(y,\gamma(y)) \quad \forall\xi:|\xi|\le C_0\rho \tag{8.1.20$'''$}$$

be fulfilled. Let Λ_j *and* T_j *satisfy the modified conditions* (8.1.37)–(8.1.44)*. Then the estimate*

$$\mathcal{N}^-_{wp} - (CR_3 + C'R'_1) \le \mathbf{N}^-(A) \tag{$8.1.50_1$}$$

holds. Moreover, under conditions (8.1.29)–(8.1.31) *the estimate*

$$\mathbf{N}^-(A) \le \mathcal{N}^-_{wp} + C(R_3+R_0) + C'(R_2+R'_1) \tag{$8.1.50_2$}$$

also holds. In these estimates the constant C *does not depend on* ε *and the constant* C' *does depend on* ε.

8.1.3 The Schrödinger operator with strong magnetic field

Let us consider the Schrödinger operator with strong magnetic field. In this case we need to introduce additional scaling functions ρ_k. All these functions satisfy $(8.1.2)_{1-3}$. Moreover, let us assume that in $B(y,\gamma(y))\cap X$ with $y\in X'_0$ the conditions $(8.1.22)_{1,3}$ and

$$(8.1.22)_2^* \qquad |D^\beta F_{jk}| \le c\rho_1\gamma^{-|\beta|} \qquad \forall j,k \quad \forall \beta : |\beta| \le K$$

are fulfilled. Let condition (8.1.23) be fulfilled and

$$(8.1.2)_4 \qquad \rho_1 \ge \rho\gamma^{-1} \quad \text{in } X'.$$

Recall that F_{jk} and F are the tensor and scalar intensities of the magnetic field.

The zones $\{x \in X',\ \rho_1 \le 2c\rho\gamma^{-1}\}$ and $\{x \in X',\ \rho_1 \ge c\rho\gamma^{-1}\}$ are zones of normal and strong magnetic field respectively. In the first zone no new assumption is made. In the second zone we need a series of new assumptions. Let us first assume that

$$(8.1.51) \qquad F \ge \epsilon_0\rho_1 \qquad \forall x \in X' : \rho_1 \ge c\rho\gamma^{-1}$$

where F is the scalar intensity of the magnetic field.[9] Let us introduce $X_1' = \{x \in X',\ \rho_1 \le 2c\rho\gamma^{-1}, V \le \epsilon_0\rho^2\}$, $X_2' = \{x \in X', \rho_1 \ge c\rho\gamma^{-1}, V + F \le \epsilon_0\rho^2\}$ and $X_3' = \{x \in X', V + F \ge \epsilon_0\rho^2\}$. Let us assume that

$$(8.1.52)_1 \qquad B(x,\gamma(x)) \subset X \qquad \forall x \in X_2',$$

$$(8.1.52)_2 \qquad u|_{\partial X \cap B(x,\gamma(x))} = 0 \qquad \forall x \in X_3' \quad \forall u \in \mathrm{D}(\mathbf{A}).$$

Finally, let us assume that functions ϱ_2 and ϱ_3 are given in X_2' such that

$$(8.1.2)_5 \qquad \varrho_2 \ge \varrho_3,$$

$$(8.1.53)_1 \qquad |D^\beta(g^{jk}J^{-1})| \le c\varrho_2\gamma^{-|\beta|} \qquad \forall j,k,$$

$$(8.1.53)_2 \qquad |D^\beta(F_{jk}J^{-1})| \le c\rho_1\varrho_2\gamma^{-|\beta|} \qquad \forall j,k,$$

$$(8.1.53)_3 \qquad |D^\beta(VJ^{-1})| \le c\rho^2\varrho_3\gamma^{-|\beta|} \quad \forall \beta : 1 \le |\beta| \le K \quad \forall x \in B(y,\gamma(y)).$$

This means that in what follows $h_{\text{eff}} = (\rho\gamma)^{-1}$, $\mu_{\text{eff}} = \rho_1\gamma\rho^{-1}$, $\bar\nu_{\text{eff}} = \varrho_2$ and $\nu_{\text{eff}} = \varrho_3$. Let us set $\varrho = \rho^2$.

Let us define $S(x,-\infty,0)$ by the formulas $(6.1.7)_{(d)}$ with $\mu = h = 1$ for $d = 2,3$. For $F = 0$ we should take limits in these formulas and obtain Weylian expressions. Let us introduce

$$(8.1.54) \qquad \mathcal{N}^-(A) = \int_{X_1' \cap X_2'} S(x,-\infty,0)\,dx.$$

Let us consider a covering of X_2' by the balls $B(x^n,\gamma(x^n))$; we can assume that the multiplicity of the covering does not exceed an absolute constant. Let us define the remainder estimate R_4^n for every ball following the results of sections 6.3, 6.4, 6.7 and 6.6 (for $d = 2$); i.e., we should make a dilatation transforming the given ball into the unit ball, check the corresponding conditions for the operator $\mathbf{A} = J^{-\frac12}AJ^{-\frac12}$ where h, μ, $\bar\nu$ and ν were introduced above and define the remainder estimate as indicated in Chapter 6. Let us set

$$R_4 = \sum_n R_4^n$$

[9] Applying the results of section 6.8 one can get rid of this condition and condition $(8.1.52)_1$ below.

and

$$R_5 = \int_{X'} \rho_1^{-s}\gamma^{-d-2s}dx,$$

$$R_5 = \int_{X_2'\cap X''} \rho_1^{-1}\rho^2\gamma^{-2}dx \quad (\text{for } d=2), \qquad R_6 = 0 \quad (\text{for } d=3).$$

Let us introduce R_1 as before by formula (8.1.35) with the new definition of X_1'.

For upper estimates we need conditions of $(8.1.32)_1$-type:

$(8.1.55)_{(3)}$ $$\varrho_0 \ge \epsilon_0\rho\gamma^{-1} \quad \text{in } (X_1'\cup X_2')\cap X''\cap\{\rho\gamma\ge c\},$$

$(8.1.55)_{(2)1}$ $$\varrho_0 \ge \epsilon_0\rho\gamma^{-1} \quad \text{in } X_1'\cap X''\cap\{\rho\gamma\ge c\},$$

$(8.1.55)_{(2)2}$ $$\varrho_0 \ge c^{-1}\rho^2\rho_1^{-1}\gamma^{-2} \quad \text{in } X_2'\cap X''\cap\{\rho_1\gamma^2\ge c\}$$

for $d=2,3$ respectively. Instead of $(8.1.32)_2$ we assume that

(8.1.56) $$\epsilon_0\rho^2(\rho_1\gamma^2)^{-l} \le \varrho_0 \le c\rho^2(\rho_1\gamma^2)^{l} \quad \text{in } X'\cap X''.$$

Now we can formulate the corresponding theorem:

Theorem 8.1.5. *(i) Let $d=2,3$ and let A be a Schrödinger operator. Let us assume that conditions* (8.1.1), $(8.1.2)_{1-5}$, (8.1.3), (8.1.6), (8.1.7) *and* $(8.1.9)_{3,4}$ *are fulfilled and let us suppose that in every ball $B(x,\gamma(x))$ with $x\in X_1'$ either a Dirichletor Neumann boundary condition is given on ∂X.*[10] *Finally, let conditions* $(8.1.22)_{1,2}$, $(8.1.22)_2^*$, (8.1.51), $(8.1.52)_{1,2}$, $(8.1.53)_{1-3}$ *be fulfilled. Then the estimate*

$(8.1.57)_1$ $$\mathcal{N}^-(A) - C(R_1+R_4) \le \mathbf{N}^-(A)$$

holds.

(ii) Moreover, under conditions $(8.1.55)_{(d)}$, (8.1.56) *and* (8.1.32) *the estimate*

$(8.1.57)_2$ $$\mathbf{N}^-(A) \le \mathcal{N}^-(A) + C(R_1+R_4+R_6+R_0) + C'R_5$$

also holds.

(iii) Moreover, in these estimates we can replace R_1 by

$$R_1^* = \int_{\Sigma_1} dxd\xi : da + R_1', \qquad R_1' = \int_{X_1'} \rho^{d-1-\delta'}\gamma^{-1-\delta'}dx$$

with $\Sigma_1 = \{(x,\xi)\in X'\times\mathbb{R}^d, a(x,\xi) = g^{jk}(\xi_j-V_j)(\xi_k-V_k)+V=0\}$.

(iv) Finally, let Λ_j be closed subsets of Σ_1 and $1\le T_1\le\cdots\le T_n$ be numbers satisfying (8.1.38). *Let us assume that through every point of the $(\rho\gamma)^{-\frac12+\delta}(\gamma,\rho)$-neighborhood of Λ_j there passes a Hamiltonian trajectory generated by the Hamiltonian*

$$J^{-1}a\cdot\bar\varrho_0\bar\rho^{-1}\bar\gamma$$

with either $t\in[0,T_j]$ or $t\in[-T_j,0]$, along which conditions (8.1.40) *(for γ, ρ, ρ_1, ϱ_0),* (8.1.41) *and* (8.1.42) *are fulfilled and on every segment of which either condition* $(8.1.43)_1$ *or* $(8.1.423)_2$ *is fulfilled. Then in the above estimates one can replace R_1 by*

$$R_3 = \sum_{j\ge0}\frac{1}{T_j}\int_{\Lambda_j} dxd\xi : da + R_1'$$

with $\Lambda_0 = \Sigma_1\setminus(\cup_j\Lambda_j)$, $T_0=1$.

---

[10] I.e., either $B=I$ or $B=\sum_{j,k}g^{jk}\nu_j(D_k-V_k)+\beta$ with $|D^\alpha\beta|\le c\gamma^{-1-|\alpha|}\rho\ \forall\alpha:|\alpha|\le K$. This assumption is made in order to avoid the question of the positivity of the localized operator.

Remark 8.1.6. *(i) Surely one can improve certain items in R_4^n by applying the results of section 6.5. However these improvements are not difficult and have only a few applications. We leave them to the reader. Let us note only that system* (6.5.1)* *is transformed to*

$$\frac{dx_j}{dt} = \sqrt{-V}F^j(FJ)^{-1}\cdot\bar{\varrho}_0\bar{\gamma}\bar{\rho}^{-1} \qquad (j=1,2,3).$$

(ii) Under weak non-degeneracy conditions (for d = 2) one should apply non-effective asymptotics.

(iii) Applying the results of section 6.7 one can weaken the conditions of theorem 8.1.5. ◊

8.2 Proof of the Basic Theorems

Let us first note that lower estimates for $\mathbf{N}^-(\mathbf{A})$ were proven by the analysis of section 8.1: we made a partition of unity, dilatations and applied the results of Part II. Moreover, upper estimates for $\mathbf{N}^-(\mathbf{A})$ were proven in the special case when $X''=\emptyset$. In the general case we obtained an upper estimate for

$$\int_X \psi\,\mathrm{tr}\, e(x,x,\tau)dx$$

provided $\mathrm{supp}\,\psi\subset X_0'$, $\mathrm{supp}(1-\psi)\subset X''$ and

$$|D^\alpha\psi|\le c_1\gamma^{-|\alpha|}\qquad \forall\alpha:|\alpha|\le K\quad \text{in } X''\cap X_0'. \tag{8.2.1}$$

So, in order to obtain upper estimates for $\mathbf{N}^-(\mathbf{A})$ we need to prove upper estimates for

$$\int_X (1-\psi)\,\mathrm{tr}\, e(x,x,\tau)dx.$$

We will derive these estimates under the assumptions of theorem 8.1.1. All the other theorems are proven in the same way; moreover, for theorem 8.1.2 no special arguments are necessary. Let $U(x,y,t)$ be the Schwartz kernel of the operator $\exp it\mathbf{A}$; recall that $\mathbf{A}=J^{-\frac12}AJ^{-\frac12}$. We need some additional analysis in the zone $X'\cap X''\cap\{\rho\gamma\ge c\}$.

Proposition 8.2.1. *Suppose conditions* (8.1.1)–(8.1.7), (8.1.9), (8.1.10) *and* $(8.1.31)_{1,2}$ *are fulfilled. Let $\mathcal{X}_1$ be the $\epsilon\gamma$-neighborhood of $X''\setminus X'$ and $\mathcal{X}_2$ be the $\epsilon\gamma$-neighborhood of $\mathcal{X}_1$ where $\epsilon\in(0,\epsilon_1]$ and $T\in(0,T_0]$ are fixed; $\epsilon_1=\epsilon_0(d,c,\epsilon_0)>0$ and $T_0=T_0(d,c,\epsilon_0)>0$ are small enough constants. Then the following estimates hold:*

$$\|D_x^\alpha F_{t\to h^{-1}\tau}\chi_T(t)U^\pm\psi'(y)\|\le Ch^s(|\tau|+1)^{-s}\bar{\gamma}^{-\frac d2-|\alpha|}$$
$$\forall\tau\le c\quad \forall x\in B(\bar{x},\frac{\bar{\gamma}}{2})\cap X\setminus\mathcal{X}_2\quad\forall\alpha:|\alpha|\le s, \tag{8.2.2}$$

$$\|D_x^\alpha F_{t\to h^{-1}\tau}\chi_T(t)U^\pm\psi'(y)\|\le Ch^{-M}(|\tau|+1)^{M}\bar{\gamma}^{-\frac d2-|\alpha|}$$
$$\forall\tau\ge c\quad \forall x\in B(\bar{x},\frac{\bar{\gamma}}{2})\cap X\setminus\mathcal{X}_2\quad\forall\alpha:|\alpha|\le s. \tag{8.2.3}$$

Here L^2-norms with respect to y are calculated in the left-hand expression and here and below $\psi'\in L^\infty$ is an arbitrary function such that $|\psi'|\le1$, $\mathrm{supp}\,\psi'\subset\overline{\mathcal{X}_1}$, $M=M(d,m,|\alpha|)$, $\bar{x}\in X'\setminus\mathcal{X}_2$ is arbitrary, $h=\bar{\rho}^{-1}\bar{\gamma}^{-1}$.

Proof. Without loss of generality one can assume that $\bar{\gamma} = 1$ (otherwise we make a dilatation). Let us consider the operator $P = hD_t - \zeta\bar{\varsigma}^{-1}\mathbf{A}$ with $\zeta = \varsigma h$; recall that $\bar{\varsigma} = \bar{\varrho}_0^{-1}\bar{\varrho}$, $h = \bar{\rho}^{-1}\bar{\gamma}^{-1}$. Let $f \in L^2(X, \mathbb{H})$ be supported in $\overline{\mathcal{X}_1}$ and let us set

$$u(x,t) = \int U(x,y,t)(y)f(y)dy.$$

Then

$$\begin{aligned} Pu^{\pm} &= 0 \quad \text{in } B(\bar{x},\bar{\gamma}) \cap X \setminus \overline{\mathcal{X}_1} \quad \text{and} \\ B_j u^{\pm} &= 0 \quad \text{at } \partial X \cap B(\bar{x},\bar{\gamma}) \cap X \setminus \overline{\mathcal{X}_1} \quad (j = 1,\dots,\mu). \end{aligned} \tag{8.2.4}$$

Let us first assume that $\zeta \le c$. Then one can apply the theorem on the finite speed of propagation of singularities and therefore the following estimate holds:

$$\|F_{t\to h^{-1}\tau}\chi_T(t)u^{\pm}\| \le Ch^s\|f\| \qquad \forall x \in B(\bar{x}, \frac{\bar{\gamma}}{2}) \cap X \setminus \mathcal{X}_2 \quad \forall \tau \le c \tag{8.2.5}$$

where we have also used the estimate

$$\|u^{\pm}\|_{B(\bar{x},\bar{\gamma})} \le \|f\| \tag{8.2.6}$$

which follows from the fact that the operator $\exp it\mathbf{A}$ is unitary in $\mathcal{H}_0$.

Then let us apply the following *additional arguments:* the ellipticity of the problem $(\bar{\rho}^m\bar{\varsigma}^{-1}\mathbf{A},\ \ \bar{\rho}^{m_j}\bar{\varrho}_j^{-1}B_j\ \ j = 1,\dots,\mu)$ in the classical sense, the inequality $\zeta \ge h^l$ (which follows from $(8.1.31)_2$) and the standard embedding inequality. Then we obtain the estimate

$$|D^\alpha F_{t\to h^{-1}\tau}\chi_T(t)u^{\pm}| \le Ch^s\|f\| \qquad \forall x \in B(\bar{x}, \frac{\bar{\gamma}}{2}) \cap X \setminus \mathcal{X}_2 \quad \forall \alpha : |\alpha| \le s \quad \forall \tau \le c. \tag{8.2.7}$$

Moreover, the same *additional arguments* and the ellipticity of the h-differential problem $(P,\ \bar{\varrho}_j^{-1}B_j\ \ j = 1,\dots,\mu)$ easily yield the estimate

$$|D^\alpha F_{t\to h^{-1}\tau}\chi_T(t)u^{\pm}| \le Ch^s(|\tau|+1)^{-s}\|f\| \qquad \forall x \in B(\bar{x}, \frac{\bar{\gamma}}{2}) \cap X \setminus \mathcal{X}_2 \quad \forall \alpha : |\alpha| \le s \quad \forall \tau \le -C_0 \tag{8.2.8}$$

because now $\bar{\varsigma} \le ch^{-1}$. Thus, estimate (8.2.7) yields that estimate (8.2.8) holds for $\tau \le c$.

On the other hand, let $\zeta \ge c$. Then $B(\bar{x},\bar{\gamma}) \cap X'_- \cap X'' \cap \{\rho\gamma \ge c\} = \emptyset$ and the above *additional arguments* and the semiclassical elliptic arguments yield estimate (8.2.8) for all $\tau \le \epsilon_2\zeta$ with a small enough constant ϵ_2 because $\bar{\varsigma} \le ch^{-l}$. Therefore estimate (8.2.8) remains true without the condition $\zeta \le c$.

Finally, the *additional arguments* alone easily yield the estimate

$$|D^\alpha F_{t\to h^{-1}\tau}\chi_T(t)u^{\pm}| \le Ch^{-M}(|\tau|+1)^{M}\|f\| \qquad \forall x \in B(\bar{x}, \frac{\bar{\gamma}}{2}) \cap X \setminus \mathcal{X}_2 \quad \forall \alpha : |\alpha| \le s \quad \forall \tau. \tag{8.2.9}$$

Since $f \in L^2(X,\mathbb{H})$ is an arbitrary function supported in $\overline{\mathcal{X}_1}$, estimates (8.2.7) and (8.2.8) yield estimates (8.2.2) and (8.2.3). □

Remark 8.2.2. *The same proof gives that estimates (8.2.2) and (8.2.3) remain true for $U'^{\pm}(x,y,t)$ where $U'(x,y,t)$ is the Schwartz kernel of the operator $\exp it\mathbf{A}'$ for every operator $\mathbf{A}'$ satisfying condition (8.1.30) (with the corresponding boundary conditions) provided $x \in X'' \setminus \mathcal{X}_3$ where $\mathcal{X}_3$ is the $\epsilon\gamma$-neighborhood of X'_0.* ◊

Proposition 8.2.3. *Under the assumptions of proposition 8.2.1 the estimates*

$$(8.2.10)\quad |F_{t\to\tau}\chi_T(t)\int\psi'(x)\,\mathrm{tr}(U^{\pm}-U'^{\pm})(x,x,t)dx|\leq C(|\tau|+1)^{-s}\int_{X'\cap X''}\rho^{d-s}\gamma^{-s}dx$$

hold for all $\tau\leq\frac{1}{2}$.

Proof. Let ψ satisfy (8.2.1), $\psi=1$ in $\mathcal{X}_2$ and $\operatorname{supp}\psi\subset X''\setminus\mathcal{X}_3$. Let us consider

$$v^{\pm}(x,y,t)=\psi(x)(U^{\pm}-U'^{\pm})(x,y,t)\psi'(y).$$

Then

$$Pv^{\pm}=f^{\pm},\quad B_kv^{\pm}=g_k^{\pm}$$

with $P=D_t-\mathbf{A}$, $f^{\pm}$ and $g_k^{\pm}$ supported in $X''\setminus(\mathcal{X}_2\cup\mathcal{X}_3)$ and therefore by the Duhamel principle

$$(8.2.11)\quad v^{+}(x,y,t)=\int_0^t dt'\Bigl(\int U^{+}(x,z,t')f^{+}(z,y,t-t')dz+\sum_{1\leq k\leq\mu}\int_{\partial X}(\mathcal{B}_k(z,D_z)U^{+}(x,z,t'))g_k^{+}(z,y,t-t')dz'\Bigr)$$

where we have taken the sign to be "+" for definiteness and $\mathcal{B}_j$ are differential operators on z. Therefore $v^{+}=\sum_n v_n^{+}$ where v_n^{+} are given by formula (8.2.11) with additional factors $\psi_n^2(z)$ in the integrands where $\{\psi_n^2\}$ is a γ-admissible partition of unity in $X''\setminus(\mathcal{X}_2\cup\mathcal{X}_3)$.

Then proposition 8.2.1 and remark 8.2.3 yield that the L^2-norms with respect to y of $F_{t\to h^{-1}\tau}\chi_T(t)f^{+}\psi_n(z)$ and $F_{t\to h^{-1}\tau}\chi_T(t)g_k^{+}\psi_n(z)$ do not exceed

$$Ch^{-M}(|\tau|+1)^{M}\bar{\gamma}^{-\frac{d}{2}}$$

and

$$Ch^{-M}\bar{\varrho}_k(|\tau|+1)^{M}\bar{\gamma}^{-\frac{d}{2}}$$

respectively (for all $z\in\operatorname{supp}\psi_n$ where $h=\bar{\rho}^{-1}\bar{\gamma}^{-1}$ and $\bar{\rho},\bar{\gamma}$ are ρ,γ calculated at some point of $\operatorname{supp}\psi_n$; all of these and the following estimates hold for all n) and that the L^2-norms with respect to x of

$$\psi'(x)F_{t\to h^{-1}\tau}\chi_T(t)U^{+}(x,z,t)\psi_n(z)$$

and

$$\psi'(x)F_{t\to h^{-1}\tau}\chi_T(t)\mathcal{B}_k(z,D_z)U^{+}(x,z,t)\psi_n(z)$$

do not exceed

$$Ch^{-M}(|\tau|+1)^{M}\bar{\gamma}^{-\frac{d}{2}}$$

and

$$Ch^{-M}\bar{\varrho}_k^{-1}(|\tau|+1)^M\bar{\gamma}^{1-\frac{d}{2}}$$

respectively (for all $z \in \operatorname{supp}\psi_n$ also); recall that

$$U(x,y,t) = U^\dagger(y,x,-t) \tag{8.2.12}$$

because $\mathbf{A}$ is a self-adjoint operator in $\mathcal{H}_0$.

Similarly, for $\tau \le \frac{1}{2}c$ these four estimates hold with the exponent M replaced by $-s$ in the right-hand expressions. Therefore if $w_n^+(x,y,t)$ is given by formula (8.2.11) with the additional factor $\psi_n^2(z)\chi_T(t')\chi_T(t-t')$ in the integrand then the estimates

$$|F_{t\to h^{-1}\tau}\operatorname{tr}\Gamma w_n^+\psi''| \le \begin{cases} C\bar{\gamma}^{-d}h^s(|\tau|+1)^{-s} & (\tau \le c) \\ C\bar{\gamma}^{-d}h^{-M}(|\tau|+1)^M & (\tau \ge c) \end{cases}$$

hold. Therefore

$$|F_{t\to h^{-1}\tau}\chi_{\frac{1}{2}T}(t)\operatorname{tr}\Gamma w_n^+\psi''(y)| \le C\bar{\gamma}^{-d}h^s(|\tau|+1)^{-s} \qquad \forall \tau \le \frac{1}{2}c \tag{8.2.13}$$

where ψ'' is a function of the same type as ψ'.

Let us replace χ by $\bar{\chi}$ equal to 1 on $\operatorname{supp}\chi$ in the previous analysis and definitions. Then $\chi_T(t)(v_n^+ - w_n^+) = 0$ and therefore one can replace w_n^+ by v_n^+ in the estimate (8.2.13). Surely one can replace $h^{-1}\tau$ by τ in this estimate. Replacing h by $\rho^{-1}\gamma^{-1}$ and summing with respect to n we obtain estimate (8.2.10) with $\psi''' = \psi'^2\psi''$ instead of ψ'. Since ψ', ψ'' are arbitrary functions satisfying the above conditions this estimate holds for ψ' instead of ψ'''. □

Remark 8.2.4. *In fact we have proven that the trace norm of the operator (from $\mathcal{H}_0$ to itself)*

$$F_{t\to\tau}\psi'\chi_T(t)\bigl(\exp it\mathbf{A} - \exp it\mathbf{A}'\bigr)\psi'$$

does not exceed $CR_1''(|\tau|+1)^{-s}$ for $\tau \le c$ where

$$R_1'' = \int_{X'\cap X''} \rho^{d-s}\gamma^{-s}\,dx.$$

◊

Proof of theorems 8.1.1–8.1.4. Let Π be an arbitrary n-dimensional self-adjoint projector in $\mathcal{H}_0$. Then remark 8.2.4 yields that

$$|F_{t\to\tau}\chi_T(t)\operatorname{Tr}\Pi\psi'\bigl(\exp it\mathbf{A} - \exp it\mathbf{A}'\bigr)\psi'| \le CR_1''(|\tau|+1)^{-s} \qquad \forall\tau \le c. \tag{8.2.14}$$

Since the trace norm of Π does not exceed Cn, (8.2.14) remains true with its left-hand expression replaced by

$$|\int \widehat{\chi_T}(\tau-\lambda)d_\lambda \operatorname{Tr}\Pi\psi'(E(\lambda)-E'(\lambda))\psi'|$$

and therefore this estimate yields that

$$|\int \widehat{\chi_T}(\tau-\lambda)d_\lambda \operatorname{Tr}\Pi\psi' E(\lambda)\psi'| \le$$
$$|\int \widehat{\chi_T}(\tau-\lambda)d_\lambda \operatorname{Tr}\Pi\psi' E'(\lambda)\psi'| + CR_1''(|\tau|+1)^{-s}.$$

Let us take a Hörmander function χ (and so χ_T is of the same type). Then standard Tauberian arguments yield that

$$\operatorname{Tr} \Pi\psi'\bigl(E(\tau)-E(\tau-1)\bigr)\psi' \le C\int \widehat{\chi_T}(\tau-\lambda)\operatorname{Tr}\Pi\psi' d_\lambda E'(\lambda)\psi' + CR_1''(|\tau|+1)^{-s}.$$

The right-hand expression of this estimate does not decrease if we replace Π by I and ψ' by 1. Let us note that $\operatorname{Tr} E'(\lambda)=\nu(\lambda)=\mathbf{N}^-_{\mathbf{A}'}(\lambda)$; therefore for $\tau\le c$

$$\operatorname{Tr} \Pi\psi'\bigl(E(\tau)-E(\tau-1)\bigr)\psi' \le C\int \widehat{\chi_T}(\tau-\lambda)d_\lambda\nu(\lambda) + CR_1''(|\tau|+1)^{-s}.$$

Since this estimate is true for all the self-adjoint finite-dimensional projectors in $\mathcal{H}_0$ it is true for Π replaced by I:

$$\operatorname{Tr} \psi'\bigl(E(\tau)-E(\tau-1)\bigr)\psi' \le C\int \widehat{\chi_T}(\tau-\lambda)d_\lambda\nu(\lambda) + CR_1''(|\tau|+1)^{-s}; \tag{8.2.15}$$

we used Π only to justify the Fourier transform and the replacement of the left-hand expression in (8.2.14). Taking $\tau=-j$ in (8.2.15) and summing over $j=0,1,2,\dots$ we obtain the estimate

$$\operatorname{Tr} \psi' E(0)\psi' \le C\sum_{j\ge 0}\int \widehat{\chi_T}(-j-\lambda)d_\lambda\nu(\lambda) + CR_1''. \tag{8.2.16}$$

Let us note that $\sum_{j\ge0}\widehat{\chi_T}(-j-\lambda)$ does not exceed C and $C(|\lambda|+1)^{-s}$ for $\lambda\ge0$ and $\lambda\le0$ respectively. Therefore the first term in the right-hand expression of (8.2.16) does not exceed

$$C\int_0^\infty (|\lambda|+1)^{-s}d_\lambda\nu(\lambda) + C\int_{-\infty}^0 d_\lambda\nu(\lambda) \le C\int_0^\infty (|\lambda|+1)^{-s}d_\lambda\nu(\lambda) + C\nu(0)$$

where we have integrated by parts. Then condition (8.1.31) yields that the first term in the right-hand expression in (8.2.16) does not exceed CR_0 and therefore

$$\operatorname{Tr} \psi' E(0)\psi' \le CR_0 + CR_1''. \tag{8.2.17}$$

Namely this estimate was lacking for the proof of theorem 8.1.1. □

Proof of theorem 8.1.5. This proof repeats the proof of the previous theorems with only one modification: we now use the results of Chapter 6 (instead of Chapters 4 and 5) and for $d=2$ in the domain $X''\cap X'\cap\{\rho_1\gamma^2\ge c\}$ the spatial velocity of propagation of singularities does not exceed $C\rho^2\varrho_0^{-1}\rho_1^{-1}\gamma^{-1}$ instead of $\rho\varrho_0^{-1}$ in the above proof.[11] □

8.3 Riesz Means Estimates

Let us consider

$$\mathbf{N}^-_\vartheta(A) = \int \tau_+^{\vartheta-1}\mathbf{N}^-(A-\tau)d\tau, \qquad \vartheta>0 \tag{8.3.1}$$

[11] Moreover, at energy levels close to 0 the velocity does not exceed $C\rho^2\varrho_3\varrho_0^{-1}\rho_1^{-1}\gamma^{-1}$; this fact can be used to weaken conditions $(8.1.55)_*$.

instead of $\mathbf{N}^-(A)$. Then one can apply local asymptotics for Riesz means. However, it would be more convenient for us to write $\mathbf{N}^-_\vartheta(A)$ in the form

$$\mathbf{N}^-_\vartheta(A) = \int \tau_+^{\vartheta-1}\mathbf{N}^-(J^{-\frac{1}{2}}(A-\tau)J^{-\frac{1}{2}})d\tau \tag{8.3.2}$$

with the same function J as before in order to get better estimates in the singular zone. In this case one can apply the results of section 4.8.

8.3.1 General operators

Let us consider the operator A given by (8.1.4) with boundary condition (8.1.5) and assume that conditions (8.1.6) and $(8.1.7)_{1,2}$ are fulfilled (in which case the same is true for $A-\tau$ with $\tau<0$). Moreover, let us assume that conditions $(8.1.9)_{1-4}$, $(8.1.10)_{1,2}$ are fulfilled. Let us replace the list of conditions $(8.1.18)_{1-6}$ by the list $(8.3.3)_{1-4}$ with $(8.3.3)_1 = (8.1.18)_1$, $(8.3.3)_3 = (8.1.18)_5$, $(8.3.4)_4 = (8.1.18)_6 = (8.1.12)$ and
$(8.3.3)_2$ $X \supset B(y,\gamma(y))$ and on $B(y,\gamma(y))\times\mathbb{R}^d$ a scalar function $\lambda(x,\xi)$ is given satisfying (8.1.19).
Then summing with respect to a partition of unity we obtain lower order estimates. Moreover, applying the decoupling arguments of section 8.2 we obtain upper estimates. Namely, we get the following

Theorem 8.3.1. *(i) Let conditions* (8.1.1)–(8.1.7), (8.1.9) *and* (8.1.10) *be fulfilled. Moreover, in every ball $B(x,\gamma(x))$ with $x\in X_0'$ let one of the conditions $(8.3.3)_{1-4}$ be fulfilled. Then the estimate*

$$N^-_{\vartheta w}(A) - CR_{1\vartheta} \le \mathbf{N}^-_\vartheta(A) \tag*{$(8.3.4)_1$}$$

holds with

$$N^-_{\vartheta w}(A) = \int_{X'} S_\vartheta(x,-\infty,0)dx + \int_{\partial X\cap\overline{X'}} S_{2\vartheta}(x,-\infty,0)dx', \tag{8.3.5}$$

$$R_1 = \int_{X'} \wp_1 dx, \tag{8.3.6}$$

$$\wp_1 = \varrho^\vartheta\rho^{d-1-\vartheta}\gamma^{-1-\vartheta} \tag{8.3.7}$$

where S_ϑ and $S_{2\vartheta}$ are the principal part and the principal boundary part of the local asymptotics for the Riesz means.[12][13]

(ii) Moreover, if conditions (8.1.30)–(8.1.32) *are fulfilled then the estimate*

$$\mathbf{N}^-_\vartheta(A) \le N^-_{\vartheta w}(A) + C(R_{1\vartheta}+R_{2\vartheta}+R_{0\vartheta}) \tag*{$(8.3.4)_2$}$$

holds with

$$R_{2\vartheta} = \int_{X'} \varrho^\vartheta\rho^{d-s}\gamma^{-s}dx \tag{8.3.8}$$

and $R_{0\vartheta}$ taken from the estimate

---

[12] We include *everything* in S_ϑ, $S_{2\vartheta}$. Surely, for odd integers ϑ one should include correction terms in the final expression.

[13] One can easily prove that application of (8.3.2) instead of (8.3.1) does not lead to a different local expression.

$$\int_{-\infty}^{0} \tau_-^{\vartheta} d_\tau \mathbf{N}(\mathbf{A}' - \tau J, -\infty, t) \le R_{0\vartheta}(t+1)^p \quad \forall t \ge 0 \tag{8.3.9}$$

which is due to (8.1.31).

(iii) Both these estimates remain true if, in balls where one of conditions $(8.3.3)_{1,3}$ *is fulfilled with lower order terms estimate* (8.1.20)′*, we replace* $\wp_1$ *by*

$$\wp_1^* = \partial_\tau S(x, -\infty, \tau) \varrho^{1+\vartheta} (\rho\gamma)^{-1-\vartheta} \big|_{\tau=0} + \wp_1' \tag{8.3.7*}$$

where

$$\wp_1' = \rho^{d-1-\vartheta-\delta'} \gamma^{-1-\vartheta-\delta'} dx \tag{8.3.7$''$}$$

with a sufficiently small exponent $\delta' > 0$.

8.3.2 Improved remainder estimates

The same arguments prove

Theorem 8.3.2. *(i) Under the assumptions of theorem 8.1.2(i) the estimate*

$$\mathcal{N}^-_{wp\vartheta} - C R_{3\vartheta} \le \mathbf{N}^-_\vartheta(A) \tag{8.3.10$_1$}$$

holds with

$$\mathcal{N}^-_{wp\vartheta} = \int_{X'} S_\vartheta(x, -\infty, 0) dx + \int_{\partial X \cap \overline{X'}} S_{2\vartheta}(x, -\infty, 0) dx' \tag{8.3.11}$$

and

$$\begin{aligned} R_{3\vartheta} = \sum_{j\ge 1} \frac{1}{T_j^{1+\vartheta}} \int_{\Lambda_j} \varsigma (\rho\gamma)^{-1-\vartheta} \varrho^{\vartheta} dx d\xi : d\lambda + \\ \partial_\tau \Big(\int_{\Omega_0} \varsigma \mathbf{n}(x, \xi, -\infty, \tau) (\rho\gamma)^{-1-\vartheta} \varrho^{\vartheta} dx d\xi \Big) \big|_{\tau=0} + R_{1\vartheta}'. \end{aligned} \tag{8.3.12}$$

(ii) Under the assumptions of theorem 8.1.2(ii) the estimate

$$\mathbf{N}^-_\vartheta(A) \le \mathcal{N}^-_{wp\vartheta} + C(R_{3\vartheta} + R_{2\vartheta} + R_{0\vartheta}) \tag{8.3.10$_2$}$$

holds.

(iii) Under the assumptions of theorem 8.1.4(i),(ii) the estimates

$$\mathcal{N}^-_{wp\vartheta} - (C R_{3\vartheta} + C' R_{1\vartheta}') \le \mathbf{N}^-_\vartheta(A) \tag{8.3.13$_1$}$$

and

$$\mathbf{N}^-_\vartheta(A) \le \mathcal{N}^-_{wp\vartheta} + C(R_{3\vartheta} + R_{0\vartheta}) + C'(R_{2\vartheta} + R_{1\vartheta}') \tag{8.3.13$_2$}$$

hold respectively. In these estimates the constant C *does not depend on* ϵ *and the constant* C' *does depend on* ε.

Surely, remark 8.1.3 also remains true.

8.3.3 The Schrödinger operator with strong magnetic field

The same arguments obviously yield

Theorem 8.3.3. *Under the assumptions of theorem 8.1.5(i),(ii) the estimates*

$$\mathcal{N}_\vartheta^-(A) - C(R_{1\vartheta} + R_{4\vartheta}) \le \mathbf{N}_\vartheta^-(A) \tag{8.3.14$_1$}$$

and

$$\mathbf{N}_\vartheta^-(A) \le \mathcal{N}_\vartheta^-(A) + C(R_{1\vartheta} + R_{4\vartheta} + R_{6\vartheta} + R_{0\vartheta}) + C' R_{5\vartheta} \tag{8.3.14$_2$}$$

hold respectively with the definitions of $\mathcal{N}_\vartheta^-$ and $R_{k\vartheta}$ for $k \ge 4$ which are obvious from Chapter 6.

These estimates remain true after a modification analogous to that of theorem 8.1.5(iii).

Under the assumptions of theorem 8.1.5(iv) the above estimates remain true after the obvious improvement.

Surely, remark 8.1.6 remains true.

8.4 Estimates for Quadratic Forms

Let us proceed to another version of the basic theorems of this chapter. These new theorems deal with quadratic forms instead of operators and $\mathbf{N}^-$ is now the maximal dimension of the negative subspace lying in $C_0^K(\mathbb{R}^d, \mathbb{H}) \cap \mathrm{D}(Q)$. We have two quadratic forms, namely $Q(u) = (Au, u)$ and $Q_0(u) = (Ju, u)$, but we are interested in the negative subspaces of $Q(u)$, and the answer cannot depend on $Q_0(u)$.[14] Let us introduce the space $\mathcal{H} = \mathcal{H}_{Q_0}$ which is the closure of $C_0^K(\mathbb{R}^d, \mathbb{H}) \cap \mathrm{D}(Q)$ in the norm $\|u\|_0 = Q_0(u)^{\frac{1}{2}}$. If the form $Q(u)$ generates a self-adjoint operator $\mathbf{A}$ in this space then

$$\mathbf{N}^-(\mathbf{A}) = \mathbf{N}^-(Q)$$

and under appropriate conditions one of the theorems of section 8.1 can be applied.

Let us discuss how the answer depends on J. Let us first note that if we do not consider trajectories then neither the condition nor the estimates depend on the choice of J outside of X''. Surely, if we were to assume that $\mathbf{A}$ has a scalar principal symbol (or something like this) then we should take J satisfying this condition; it is simpler to assume that A has a scalar principal symbol and take a scalar matrix J. We try to take J in X'' as small as possible (in order to decrease R_0 in estimate (8.1.31)) but we need to fulfill $(8.1.9)_3$ and either $(8.1.32)_1$ or $(8.1.55)_{(d)}$. If we replace the latter conditions by the stronger conditions

$$\varrho_0 \asymp \varrho_1 \rho^{-1} \gamma^{-1} \qquad \text{in } X'_+ \cap \{\rho\gamma \ge c\} \tag{8.4.1}$$

or

$$\varrho_0 \asymp \rho\gamma^{-1} \qquad \text{in } (X'_1 \cup X'_2) \cap \{\rho\gamma \ge c\}, \tag{8.4.2$_{(3)}$}$$

$$\varrho_0 \asymp \rho\gamma^{-1} \qquad \text{in } X'_1 \cap \{\rho\gamma \ge c\}, \tag{8.4.2$_{(2)1}$}$$

$$\varrho_0 \asymp \rho^2 \rho_1^{-1} \gamma^{-1} \qquad \text{in } X'_2 \cap \{\rho_1\gamma^2 \ge c\}, \tag{8.4.2$_{(2)2}$}$$

---

[14] Nevertheless the estimates we will derive can depend on J.

instead of $(8.1.32)_1$, $(8.1.55)_{(3)}$, $(8.1.55)_{(2)1}$ and $(8.1.55)_{(2)2}$ respectively (in particular, we expand these conditions to a larger domain) then in the equations of trajectories and in the conditions along trajectories all the quantities with bars will disappear, and therefore these equations and conditions will be more natural. Despite these simplifications the constant R_0 has not yet been calculated and whether the quadratic form $Q(u)$ generates a self-adjoint operator in $\mathcal{H}$ is not yet clear.

However, in the lower bounds of $\mathbf{N}^-(Q)$ the latter condition is not necessary. In fact, these estimates are non-trivial only if $\mathbf{N}^-(Q) < \infty$ but in this case the quadratic form Q is semibounded from below on $\mathcal{H}$ (we assume that $\mathrm{D}(Q)$ is a dense subset of $\mathcal{H}$) and hence Q generates a self-adjoint operator on $\mathcal{H}$ which is semibounded from below. Therefore the lower and the upper estimates are provided and in the upper estimates the constant R_0 is deduced from the quadratic form $Q_t(u) = Q(u) - tQ_0(u)$, i.e., condition (8.1.31) should be replaced by

$$\mathbf{N}^-(Q - tQ_0) \le R_0(t+1)^l \qquad \forall t \ge 0 \tag{8.4.3}$$

with Q here restricted to the domain X''.

Let us consider upper estimates. Namely, let us assume that Q is given by the formula

$$Q(u) = \int_X \sum_{|\alpha|,|\beta|\le p} \langle a_{\alpha\beta} D^\beta u, D^\alpha u\rangle dx + \int_{\partial X} \sum_{|\alpha|,|\beta|\le p-1} \langle b_{\alpha\beta} D^\beta u, D^\alpha u\rangle dx' \tag{8.4.4}$$

with coefficients satisfying

$$a_{\alpha\beta} = a^\dagger_{\beta\alpha}, \quad b_{\alpha\beta} = b^\dagger_{\beta\alpha} \qquad \forall \alpha,\beta, \tag{8.4.5}$$

$$\sum_{\alpha,\beta:|\alpha|=|\beta|=p} \langle a_{\alpha\beta}\xi^{\alpha+\beta} v, v\rangle \ge \epsilon_0 \varrho\rho^{-2p}|\xi|^{2p}|v|^2 \qquad \forall \xi\in\mathbb{R}^d \quad \forall v\in\mathbb{H} \tag{8.4.6}$$

and with boundary conditions $B_j u|_{\partial X} = 0$ $(j = 1,\dots,\mu')$ where $m = 2p$ and $m_j \le p-1$ for $j \le \mu'$, $m_j \ge p$ for $j \ge \mu'+1$ (in which case boundary conditions for A with $j \ge \mu'+1$ follow from a variational problem). Moreover, let us assume that

(8.4.7) $J, J^{-1} \in L^\infty_{\mathrm{loc}}(\mathcal{X})$ where $\mathcal{X} \supset \overline{X'}$ and $\mathcal{X}$ is dense in X.

Let us also assume that

(8.4.8) $u \in C_0^K(\mathcal{X},\mathbb{H})$ belongs to $\mathrm{D}(Q)$ if and only if $B_j u|_{\partial X} = 0$ $\forall j = 1,\dots,\mu'$ and that Q is originally given on these functions and is expanded to $\mathcal{H}$ (and to $\mathcal{H}_0$) as a closure.

Then the subspaces in which we test whether Q is negative definite are the same for the chosen J and for $J = I$. All these conditions should be assumed from the beginning which we will do in the statements of theorems. To proceed further, let us first note that under the assumptions of our theorems the contribution of every ball $B(x,\gamma(x))$ with $x \in X'$ to $\mathcal{N} + R$ is either larger than a constant ϵ_1 or is $O(h^s)$ (or $O(\mu^{-s}h^s)$ where h and μ mean h_{eff} and μ_{eff}. We will call these balls *balls of the first type* and *balls of the second type* respectively. Therefore *if the right-hand expression is finite then the number of balls of the first type is finite and* $\rho,\gamma,\dots,J$ *on the union of these balls are disjoint from* 0 *and* ∞.

Thus, in order to obtain theorems similar to the theorems of section 8.1 but with $\mathbf{N}^-(Q)$ instead of $\mathbf{N}^-(\mathbf{A})$ we need only prove that Q generates a self-adjoint operator in $\mathcal{H}$. Surely it is sufficient to prove that $\mathbf{N}^-(Q_T) < \infty$ for some $T > 0$ with $Q_T(u) =$

$Q(u)+T(Ju,u)$. Moreover, the estimate $N^-(Q) \leq M$ is due to the estimate $N^-(Q_t) \leq M$ for $Q_t(u) = Q(u) + t(J_1 u, u)$ and for all $t > 0$ and for some function J_1 of the same type as J (with another weight function ϱ_0' instead of ϱ_0 and with a non-essential constant c' instead of c). If Q_t generates a self-adjoint operator A_t then one can apply the theorems of section 8.1 to this form. Let us note that in balls of the first type for small enough $t > 0$ the same conditions are fulfilled[15] which were assumed to be fulfilled for $t = 0$. Estimates in balls of the second type and in the singular zone are obvious. Therefore we get the estimate $N^-(Q_t) \leq M + o(1)$ for $t > 0$ which yields the estimates in question. Thus, it is sufficient to assume that the form $Q_{t,T}(u) = Q(u) + t(J_1 u, u) + T(Ju, u)$ is non-negative-definite for every $t > 0$ and sufficiently large $T = T_t$. But this condition is fulfilled provided

(8.4.9) The quadratic form $Q_T(u) = Q(u) + (J_0 u, u)$ is non-negative-definite for some function J_0 of the same type as J (with another weight function ϱ_0' instead of ϱ_0 and with a non-essential constant c' instead of c).

In fact, under this condition one can pick $J_1 = J_0^2 J^{-1}$ and $T = t^{-1}$. Therefore, *under condition* (8.4.9) *one can pass from operators to quadratic forms.*

There are many situations in which condition (8.4.9) is obviously fulfilled. I would like to list some of them.

Let $\{\psi_n^2\}_{n\geq 0}$ be a partition of unity such that ψ_n satisfy (8.2.1) in X' uniformly on n, $\operatorname{supp}\psi_0 \subset X'' \setminus (\epsilon\gamma\text{-neighborhood of } X \setminus X'')$, $\operatorname{supp}\psi_n \subset X' \setminus (\epsilon\gamma\text{-neighborhood of } X \setminus X')$, $\operatorname{diam}\operatorname{supp}\psi_n \leq \epsilon_2\gamma$ for $n \geq 1$ and the multiplicity of this partition does not exceed $C_0 = C_0(d)$. Then

$$
\begin{aligned}
Q(u) = \sum_{n\geq 0}\Bigl(\int_X \psi_n^2 \sum \langle a_{\alpha\beta} D^\beta u, D^\alpha u\rangle dx +{}\\
\int_{\partial X} \psi_n^2 \sum \langle b_{\alpha\beta} D^\beta u, D^\alpha u\rangle dx\Bigr) ={}\\
\sum_{n\geq 0}\Bigl(Q(\psi_n u) - 2\operatorname{Re}\int_X \sum \langle a_{\alpha\beta}[D^\beta, \psi_n]u, D^\alpha \psi_n u\rangle dx +{}\\
\operatorname{Re}\int_X \sum \langle a_{\alpha\beta}[D^\beta, \psi_n]u, [D^\alpha, \psi_n]u\rangle dx -{}\\
2\operatorname{Re}\int_{\partial X} \sum \langle b_{\alpha\beta}[D^\beta, \psi_n]u, D^\alpha \psi_n u\rangle dx' +{}\\
\operatorname{Re}\int_{\partial X} \sum \langle b_{\alpha\beta}[D^\beta, \psi_n], [D^\alpha \psi_n]u\rangle dx'\Bigr)
\end{aligned}
\tag{8.4.10}
$$

where we sum on α, β inside. Therefore (8.4.5) yields that

$$
\begin{aligned}
Q(u) \geq \sum_{n\geq 0} Q(\psi_n u) - \sum_{|\alpha|\leq p-1} C \int_X \chi\varrho(\rho\gamma)^{-2}\rho^{-2|\alpha|}|D^\alpha u|^2 dx -{}\\
\sum_{|\alpha|\leq p-1} C \int_{\partial X} \chi\varrho\rho^{-2|\alpha|-1}|D^\alpha u|^2 dx' \geq \sum_{n\geq 0} \tilde{Q}(\psi_n u)
\end{aligned}
\tag{8.4.11}
$$

where

---

[15] With constants $2c$ and $\frac{1}{2}\epsilon_0$ instead of c and ϵ.

(8.4.12) $$\tilde{Q}(v) = Q(v) - \sum_{|\alpha|\le p-1} C \int_X \chi\varrho_1(\rho\gamma)^{-2}\rho^{-2|\alpha|}|D^\alpha v|^2 dx - \sum_{|\alpha|\le p-1} C \int_{\partial X} \chi\varrho\rho^{-2|\alpha|-1}|D^\alpha v|^2 dx',$$

χ is the characteristic function of $X\setminus(2\epsilon\gamma$-neighborhood of $X'\setminus X'')$ and we assume from the beginning that

$(8.4.13)_1$ $$|D^\sigma a_{\alpha\beta}| \le c\varrho\rho^{-|\alpha|-|\beta|}\gamma^{-|\sigma|},$$

$(8.4.13)_2$ $$|D^\sigma b_{\alpha\beta}| \le c\varrho\rho^{-|\alpha|-|\beta|}\gamma^{1-|\sigma|} \qquad \forall\sigma : |\sigma| \le K \quad \forall\alpha,\beta.$$

One can easily see that under conditions (8.4.7)

$$\tilde{Q}(u) + 2(J_0 u, u) \ge \frac{1}{2}Q(u) + (J_0 u, u)$$

for appropriate J_0.

Let us treat the case $n \ge 1$. The case $n = 0$ will be treated later. In the case of balls of the first type we consider quadratic forms without boundary conditions and one can see easily that for a large enough T the corresponding quadratic form is positive definite. Let us consider balls of the second type.

The following new difficulty arises: $\psi_n u$ does not necessarily satisfy the boundary conditions defining the domain of the quadratic form Q.

However, this difficulty does not arise provided one of the following conditions is fulfilled:

$(8.4.14)_1$ $$B(\bar{x},\gamma(\bar{x})) \subset X \quad \forall\bar{x} \in X';$$

$(8.4.14)_2$ $m_j \le \frac{m}{2} - 1 \implies m_j = 0$; in particular, if $m = 2$;

$(8.4.14)_3$ The boundary conditions $B_j u = 0$ $(j = 1,\dots,\mu')$ are the conditions $D_1^k u|_{\partial X} = 0$ $(k = 0,\dots,\bar{k})$ plus one condition of order $\bar{k}$.

Moreover, this difficulty does not arise in some other cases. Finally, one can avoid this difficulty provided

$(8.4.14)_4$ $\forall j$ with $m_j \le \frac{m}{2} - 1$ the operator B_j is a $D_j \times D$ matrix operator with $D_j \le D$ and the coefficient of $D_1^{m_j}$ has rank D_j at any point of ∂X (where in the local coordinates $X = \{x_1 > 0\}$).

In fact, it is easy to prove (we leave it for the reader) that under this condition there exists an operator $\mathcal{L}$ such that

$(8.4.15)_1$ $$\operatorname{supp} v \subset B\bigl(\bar{x},(1-2\epsilon)\gamma(\bar{x})\bigr) \implies \operatorname{supp}\mathcal{L}v \subset B\bigl(\bar{x},(1-\epsilon)\gamma(\bar{x})\bigr),$$

$(8.4.15)_2$ $$\sum_{|\alpha|\le q} \rho^{-|\alpha|}\|D^\alpha \mathcal{L}v\| \le C_q \sum_{|\alpha|\le q-1} \rho^{-|\alpha|-1}\gamma^{-1}\|D^\alpha v\|$$

(K also depends on q where $q \ge p$ is arbitrary),

$(8.4.15)_3$ $$B_j(I+\mathcal{L})v|_{\partial X} = 0 \qquad \forall j : m_j \le \frac{m}{2} - 1,$$

$$(I+\mathcal{L})v=0,\ \operatorname{supp} v\subset B\big(\bar{x},(1-\epsilon)\gamma(\bar{x})\big)\implies v=0. \tag{8.4.15$_4$}$$

A similar operator $\mathcal{L}$ is also defined for $n=0$ with

$$\operatorname{supp} v\subset \mathcal{Z}_k \implies \operatorname{supp}\mathcal{L}v\subset\mathcal{Z}_1,\qquad v=\mathcal{L}v\quad\text{in } \mathcal{Z}_4, \tag{8.4.15$)'_1$}$$

$$\sum_{|\alpha|\le q}\|\varrho_1^{\frac12}\rho^{-|\alpha|}D^\alpha\mathcal{L}v\|\le C_q\sum_{|\alpha|\le q-1}\|\varrho_1^{\frac12}\rho^{-|\alpha|-1}\gamma^{-1}D^\alpha v\| \tag{8.4.13$)'_2$}$$

$$B_j(I+\mathcal{L})v|_{\partial X}=0\quad\forall j: m_j\le\frac{m}{2}-1 \tag{8.4.15$)'_3$}$$
$$\text{provided } B_jv|_{\partial X\cap\mathcal{Z}_3}=0\quad\forall j: m_j\le\frac{m}{2}-1,$$

$$(I+\mathcal{L})v=0,\ \operatorname{supp} v\subset\mathcal{Z}_1\implies v=0 \tag{8.4.15$)'_4$}$$

where $\mathcal{Z}_k$ is $\bar{X}\setminus(k\epsilon\gamma\text{-neighborhood of } X\setminus X'_0)$ and $\epsilon>0$ is a small enough constant. This construction is also left to the reader. Finally, we obtain the estimate

$$Q(u)\ge\sum_{n\ge0}\tilde{Q}\big((I+\mathcal{L})\psi_n u\big); \tag{8.4.11$)'$}$$

and therefore in all the cases $(8.4.14)_{1-4}$ the estimate

$$\mathbf{N}^-(Q)\le\sum_{n\ge1}\mathbf{N}^-(\tilde{Q};B(x^n(1-\epsilon)\gamma^n)+\mathbf{N}^-(\tilde{Q},\overline{X''}) \tag{8.4.16}$$

holds where Z denotes the geometrical domain on which the quadratic form Q is defined (i.e., all the test functions are supported) in the definition of $\mathbf{N}^-(\tilde{Q},Z)$, $x^n\in\operatorname{supp}\psi_n$ and $\gamma^n=\gamma(x^n)$ $(n\ge1)$.

Let us note that under the above conditions the estimate

$$\mathbf{N}^-(\tilde{Q},B(x^n,(1-\epsilon)\gamma^n)\le M(\mathcal{N}_n^-+R_n)$$

holds where $\mathcal{N}_n^-$ and R_n are the contributions of the ball $B(x^n,\gamma^n)$ to the principal part and the remainder estimate of the corresponding estimate. To prove this one should consider balls of the first and second type separately. Recall that there is only a finite number of balls of the first type.

Therefore the total contributions of these elements is finite provided the right-hand expression in the upper estimate is finite. Condition (8.4.3) yields that the contribution of $n=0$ is finite as well. Therefore, *for appropriate J_0 the number $N^-(Q_1)$ is finite and therefore $Q_T(u)=Q(u)+T(J_0u,u)$ is positive definite for some T.*

Appendix G. Variational Estimates

G.1. Rozenblioum estimates

Let us consider the following two quadratic forms defined on scalar functions:

$$Q_1(u)=\sum_{|\alpha|=m}\big(a(x)D^\alpha u,D^\alpha u\big), \tag{G.1}$$

$$Q_0(u) = \sum_{|\alpha_1|\le m, |\alpha_2|\le m, |\alpha_1|+|\alpha_2|=2r} \big(V_{\alpha_1,\alpha_2}(x) D^{\alpha_1}u, D^{\alpha_2}u\big) \tag{G.2}$$

where

$$m > r \ge 0, \quad m \in \mathbb{Z}, r \in \frac{1}{2}\mathbb{Z}, \quad a \ge 0, \quad V_{\alpha_1,\alpha_2} = V^{\dagger}_{\alpha_2,\alpha_1}. \tag{G.3}$$

Let us assume that

$$a^{-1} \in L^p, \quad V_{\alpha_1,\alpha_2} \in L^q, \tag{G.4}$$

$$\frac{1}{p} + \frac{1}{q} = \frac{1}{\theta} = \frac{2}{d}(m-r), \tag{G.5}$$

$$m - r \le d, \; p \ge 1; \quad \theta > \frac{1}{2} \implies q > 1. \tag{G.6}$$

Moreover, let us assume that Q_0 contains only terms with

$$\begin{aligned} &2(m - |\alpha_i|) < d(1 + \frac{1}{p}) \quad \text{for } p > 1, \\ &m - |\alpha_i| \le d \qquad \text{for } p = 1. \end{aligned} \tag{G.7}$$

Then the following is estimate due to G. Rozenblioum [Rozenblioum 1]:

$$\mathbf{N}^-_{Q_1 - Q_0} \le C\|a^{-1}\|^{\theta}_{L^p} \cdot \|V\|^{\theta}_{L^q}, \tag{G.8}$$

where $\|V\| = \sum_{\alpha_1,\alpha_2} \|V_{\alpha_1,\alpha_2}\|$ and $C = C(d,m,r,p,q)$, and one can take C which does not depend on p,q if conditions (G.6), (G.7) are fulfilled uniformly.

The most well known case is $m = 1, r = 0$ when for $d \ge 3$ one can take $p = \infty, q = \frac{d}{2}$ (the Lieb-Cwikel-Rozenblioum estimate [Reed, Simon 1]) and for $d = 2$ one can take arbitrary $q > 1$ and $p = \frac{q}{q-1}$.

Other estimates of this type are discussed in [Birman, Solomyak 3].

Let us now consider these quadratic forms on the space $\mathcal{V} \subset C^m(\bar{X})$. Let us assume that $X \subset \cup_{j\in\mathcal{J}} \mathcal{X}_j$ where $\mathcal{X}_j$ are open domains such that for every j there exists a linear operator $\mathcal{L}_j : \mathcal{V} \to C_0^m(\mathcal{Y}_j)$ such that

$$\mathcal{L}_j u = u \qquad \text{in } \mathcal{X}_j \quad \forall u \in \mathcal{V}, \tag{G.9}$$

$$Q_{1j}(\mathcal{L}_j u) \le Q'_{1j}(u) + Q'_{2j}(u) \quad \forall u \in \mathcal{V} \qquad \forall j \tag{G.10}$$

where Q_{1j} and Q'_{1j} are forms of (G.1)-type with a replaced by $a_j \ge 0$, $a'_j \ge 0$ respectively and Q'_{2j} are forms of (G.2)-type. Then summing (G.10) over j we obtain from (G.9) that if the quadratic form $Q_1 - Q_0$ is negative definite on the subspace $\mathcal{W} \subset \mathcal{V}$ then on this subspace

$$\sum_j Q_{1j}(\mathcal{L}_j u) \le Q_1(u) + \sum_j Q'_{2j}(u) \le Q_0(u) + \sum_j Q'_{2j}(u) \le \sum_j Q'_0(\mathcal{L}_j u) \tag{G.11}$$

provided

$$\sum_j a'_j = a, \tag{G.12}$$

(G.13) The forms Q_0, Q'_{2j} contain only terms with $\alpha_1 = \alpha_2$, $|\alpha_1| = r$ and $V_{\alpha_1,\alpha_1} \geq 0$ (and the same is true for the coefficients of Q'_{2j}) and

(G.14)
$$Q'_0 = Q_0 + \sum_k Q'_{2k}.$$

Therefore $\mathbf{N}^-_{(Q_1-Q_0),\mathcal{V}}$ does not exceed the maximal dimension of the subspace $\mathcal{W} \subset \prod_j C_0^m(\mathcal{Y}_j)$ with the following property:

For every $w = (w_j)_{j\in\mathcal{J}} \in \mathcal{W}$ there exists $u \in V$ with $w_j = \mathcal{L}_j u \ \forall j$ on which the quadratic form $\sum_j (Q_{1j} - Q'_{0j})(w_j)$ is negative definite.

Dropping the condition $\mathcal{L}_j u = w_j$ on w_j we obtain that

(G.15)
$$\mathbf{N}^-_{(Q_1-Q_0),\mathcal{V}} \leq \sum_j \mathbf{N}^-_{(Q_{1j}-Q'_{0j}),\mathcal{W}_j}$$

with $\mathcal{W}_j = C_0^m(\mathcal{Y}_j)$. Now one can apply the Rozenblioum estimate (G.8) in order to estimate every term in the right-hand expression. One should take L^p- and L^q-norms on $\mathcal{Y}_j$ in these estimates.

In particular, let $\mathcal{V}$ be the space of all compactly supported C^m functions satisfying the Dirichlet condition of order m on $S \subset \partial X$. Let us assume that $\mathcal{Z} = \partial X \setminus S = \cup_{j\geq 1}\mathcal{Z}_j$, $X = \cup_{j\geq 0}\mathcal{X}_j$ with $\mathcal{Z}_j = \mathcal{Z} \cap \overline{\mathcal{X}_j}$ and $\mathcal{X}_0 = \{x \in X, \mathrm{dist}(x, \mathcal{Z}_j) > h_j/2 \ \forall j\}$ and let Γ_j be open cones with vertex 0 in $\mathbb{R}^d$ and with angle $\phi > 0$ and altitude h_j such that

(G.16) For every point $\bar{x} \in \mathcal{X}_j$ the cone $\bar{x} + \Gamma_j$ does not intersect $\mathcal{Z}$.

Then taking the Sobolev prolongation operators [Sobolev 1] associated with cones $\epsilon\Gamma_j$ with a small enough constant $\epsilon > 0$ we see that for $j \geq 1$

$$\sum_{|\alpha|=m} \|D^\alpha \mathcal{L}_j u\| \leq C \sum_{|\alpha|=m} \|D^\alpha u\|_{Y_j} + Ch_j^{-m}\|u\|_{Y_j}$$

where $Y_j = (\mathcal{X}_j + 2\epsilon\Gamma_j) \cap X$. Then under the condition

$(\mathrm{G}.17)_1$ $Y_j = (\mathcal{X}_j + 2\epsilon\Gamma_j) \cap X$ form no more than a c-multiple covering of $X \setminus \mathcal{X}_0$

one can take $Q_1 \asymp \|.\|^2_m$, $Q'_{1j} \asymp \|.\|^2_m$, $Q_{1j} \asymp \|.\|^2_m$ in the domains X, Y_j and $\mathbb{R}^d$ respectively and $Q'_{2j} = Ch_j^{-2m}\|.\|^2_{Y_j}$.

Moreover, let us assume that

$(\mathrm{G}.17)_2$ $Y'_j = \{x \in X, \mathrm{dist}(x, \mathcal{Z}_j) < h_j\}$ form no more than a c-multiple covering of $X \setminus \mathcal{X}_0$.

Then as $\mathcal{L}_0$ one can take the operator of multiplication by a function ψ such that $\psi = 1$ in $\mathcal{X}_0$, $\psi = 0$ in the ϵh_j-neighborhood of $\mathcal{Z}_j$ (for every $j \geq 1$) satisfying the inequalities $|D^\alpha\psi| \leq c\bar{\gamma}^{-|\alpha|} \quad \forall\alpha : |\alpha| \leq m \quad$ with $\quad \bar{\gamma} = \mathrm{dist}(x, \mathcal{Z})$.

In this case one can take $Q_1(u) \asymp \sum_{|\alpha|=m}(aD^\alpha u, D^\alpha u)$, Q'_{10} and Q_{10} given by the same formula in X, $\mathcal{X}_0$ and $\mathbb{R}^d$ respectively and $Q'_{20} = C\sum_j h_j^{-2m}\|.\|^2_{Y'_j}$ provided

(G.18)
$$\epsilon_0 \leq a \leq c \qquad \text{in } Y'_j \quad \forall j \geq 1.$$

So, we obtain that $Q'_{0j}(v) = Q_0(v) + (W_j v, v)$ with $W_j = h_j^{-2m}\chi_{Y'_j}$ for $j \geq 1$ and $W_0 = \sum_{j\geq 1} h_j^{-2m}\chi_{Y'_j}$ where χ_M is the characteristic function of the set M. In fact, without loss of generality one can assume that $\mathcal{X}_j$ are contained in the $\frac{2}{3}h_j$-neighborhoods of $\mathcal{Z}_j$ and therefore Y_j are contained in Y'_j.

Let us apply the inequality

$$\mathbf{N}^-_{Q_{(1)}+Q_{(2)}} \le \mathbf{N}^-_{Q_{(1)}} + \mathbf{N}^-_{Q_{(2)}} \tag{G.19}$$

with $Q_{(1)} = \frac{1}{2}Q_{1j} - Q_0$ and $Q_{(2)} = \frac{1}{2}Q_{1j} - Q'_{2j}$. Let us apply the Rozenblioum estimate to $\mathbf{N}^-_{Q_{(1)}}$. Moreover, let us apply the same estimate to $\mathbf{N}^-_{Q_{(2)}}$ with $p = \infty$, $\theta = q = \frac{d}{2m}$ provided $d > 2m$. Finally, after summation over j we obtain the estimate

$$\mathbf{N}^-_{Q_1 - Q_0} \le C\|a^{-1}\|^\theta_{L^p}\|V\|^\theta_{L^q} + C\sum_j h_j^{-d} \operatorname{mes} Y'_j. \tag{G.20}$$

For $d \le 2m$ let us assume that

(G.21) $j = 0, \dots, J < \infty$ in the covering in question.

In this case let us again apply (G.19) and take a covering of Y'_j by h_j-cubes of multiplicity less than C_0. Let us apply the estimate $\mathbf{N}^-_{Q_1 - 2Q_0} \le C_1$ in every cube. Then we again obtain estimate (G.20) with a constant C depending on J.

So the following statement has been proven:

Proposition G.1. *Under the above conditions the estimate* (G.20) *holds for* $d > 2m$. *For* $d \le 2m$ *this estimate holds under the additional condition* (G.21) *with* C *depending on* J.

In order to conclude this subsection let us assume that $a \ge \epsilon_0$ in X'' and let us estimate $\mathbf{N}^-_{Q_1 - Jt + J_3}$ in X'' with $J = J_1 + J_2$, $J_1 = \chi_{X' \cap X''}\rho^{-2s}\bar\gamma^{-2s-2m}$ ($s \in \mathbb{R}$), $J_2 \ge 0$, $J_3 \ge 0$ with a Dirichlet condition on $\partial X'' \cap X$. Let us apply (G.19). Then in order to estimate $\mathbf{N}^-_{Q_1 - 2J_1 t}$ one can use a covering of $X' \cap X''$ by balls $B(x, \bar\gamma(x))$ with multiplicity less than C_0. In every ball without boundary conditions (because ∂X is regular in the intersection with this ball) $\mathbf{N}^-_{Q_1 - 2J_1 t + J_3}$ does not exceed $C(t\rho^{-2s}\bar\gamma^{-2s})^{\frac{d}{2m}} + C_0$ in the general case and the last term is absent provided $x \in X'_+ = \{x \in X', J_3 \ge \epsilon\rho^{2m} \text{ in } B(x, \bar\gamma(x))\}$. Therefore their sum does not exceed

$$Ct^{\frac{d}{2m}} \int_{X' \cap X''} \rho^{-\frac{ds}{m}}\bar\gamma^{-\frac{ds}{m} - d}\,dx + C\int_{X'_- \cap X''} \bar\gamma^{-d}\,dx$$

where $X'_- = X' \setminus X'_+$. Moreover, the contribution of $X'' \setminus X'$ is equal to 0 (because the form $Q_1 - 2J_1 t + J_3$ is non-negative definite on functions supported there). So in this case we obtain the estimate

$$\mathbf{N}^-_{Q_1 - Jt + J_3} \le \mathbf{N}^-_{Q_1 - 2J_2 t + J_3} + Ct^{\frac{d}{2m}} \int_{X' \cap X''} \rho^{\frac{ds}{m}}\bar\gamma^{\frac{ds}{m} - d}\,dx + C\int_{X'_- \cap X''} \bar\gamma^{-d}\,dx. \tag{G.22}$$

G.2. Riesz means estimates

Let us suppose $r = 0$, $Q_0 = (Wu, u)$ with $W \ge 0$ and $\bar Q = \|u\|^2$. Let us consider

$$\mathbf{N}^-_\vartheta = \int_{-\infty}^0 t_-^\vartheta\, d_t \mathbf{N}^-_{Q_1 - Q_0 - t\bar Q} = \vartheta \int_{-\infty}^0 t_-^{\vartheta - 1}\mathbf{N}^-_{Q_1 - Q_0 - t\bar Q}\,dt. \tag{G.23}$$

Then, according to estimate (G.8)

$$\mathbf{N}_\vartheta^- \le C\|a^{-1}\|_{L^p}^\theta \int_0^\infty \|(V-t)_+\|_{L^q}^\theta t^{\vartheta-1}\,dt. \tag{G.24}$$

In the case $p=\infty$, $q=\theta$ [16] we immediately obtain the estimate

$$\mathbf{N}_\vartheta^- \le C\|a^{-1}\|_{L^\infty}^{\frac{d}{2m}} \int W^{\frac{d}{2m}+\vartheta}\,dx. \tag{G.25}$$

On the other hand, let $p<\infty$; then $q>\theta$. Applying the Hölder inequality to the right-hand expression of estimate (G.8) we get the estimate

$$\mathbf{N}_\vartheta^- \le C\|a^{-1}\|_{L^p}^\theta \Bigl(\int \phi(t)dt\Bigr)^{1-\frac{\theta}{q}} \Bigl(\iint t^{\frac{q}{\theta}(\vartheta-1)}(W-t)_+^q (\phi(t))^{\frac{1-q}{\theta}}\,dxdt\Bigr)^{\frac{\theta}{q}}$$

with arbitrary function $\phi(t)\ge 0$; here $t\in(0,\infty]$. Taking $\phi(t)=t^{-1+\delta}(t+\tau)^{-\delta-\delta'}$ with $\delta>0,\delta'>0$ and parameter $\tau\ge 0$ we then obtain that

$$\mathbf{N}_\vartheta^- \le C\|a^{-1}\|_{L^p}^\theta \Bigl(\bigl(\int W^{l_1}dx\bigr)^{\frac{\theta}{q}}\tau^{-\delta'(1-\frac{\theta}{q})} + \bigl(\int W^{l_2}dx\bigr)^{\frac{\theta}{q}}\tau^{\delta(1-\frac{\theta}{q})}\Bigr)$$

with $l_1=q(1+\frac{\vartheta}{\theta})+\delta'(\frac{q}{\theta}-1)$, $l_2=q(1+\frac{\vartheta}{\theta})-\delta(\frac{q}{\theta}-1)$. Minimizing on τ we obtain

$$\mathbf{N}_\vartheta^- \le C\|a^{-1}\|_{L^p}^\theta \bigl(\int W^{l_1}dx\bigr)^{\frac{\nu\theta}{q}} \bigl(\int W^{l_2}dx\bigr)^{\frac{(1-\nu)\theta}{q}} \tag{G.26}$$

with arbitrary $\nu\in(0,1)$, $\rho>0$ $l_1=q(1+\frac{\vartheta}{\theta})+(1-\nu)\rho$, $l_2=q(1+\frac{\vartheta}{\theta})-\nu\rho$ where we have taken $\rho=(\delta+\delta')(\frac{q}{\theta}-1)$, $\nu=\delta(\delta+\delta')^{-1}$.

In particular, for $d=2m$ we obtain

$$\mathbf{N}_\vartheta^- \le C_q\|a^{-1}\|_{L^\infty}(\operatorname{mes} X)^{1-\frac{1}{q}}\bigl(\int W^{(\vartheta+1)q}dx\bigr)^{\frac{1}{q}} \tag{G.27}$$

with arbitrarily chosen $q>1$.

Comments

The principal idea of this chapter was first used in the papers [Ivrii 11] and [Ivrii, Fedorova 1] where a second order operator was treated via the wave equation, and the finite spatial velocity of propagation of supports was used instead of the finite spatial velocity of propagation of singularities. The next paper [Ivrii 12] and all other papers using semiclassical asymptotics in the regular zone and variational estimates in the singular zone were based on the slightly more complicated (and essentially more general) proof presented here. However, this proof was not yet published.

In all my papers and in Part IV just the Rozenblioum spectral estimate [Rozenblioum 2] is used. Sometimes even the earlier Birman-Solomyak estimate [Birman, Solomyak 1–4] is sufficient. But I do not want to "close the door" to using other estimates and hence a more abstract form of the basic theorems is presented.

Note that in the special case of the Schrödinger operator the Rozenblioum estimate is the famous Lieb-Cwikel-Rozenblioum estimate, at least five different proofs of which are known [Conlon 1], [Cwikel 1], [Lieb 1], [Li, Yau 1], [Rozenblioum 2].

Surely, in applications in which an operator depends on parameters, the auxiliary functions $\rho\dots$ (and maybe γ) also depend on the parameters, and regular and singular zones, etc., also depend on the parameters.

[16] This is possible if and only if $d>2m$.

9. Estimates of the Spectrum in an Interval

In this chapter we obtain upper and lower estimates for $\mathbf{N}(A;\tau_1,\tau_2)$ with $-\infty<\tau_1<\tau_2<\infty$. These results are useful in the case when $\mathbf{N}^-(A;\tau_i)=\infty$ for $i=1,2$. Our analysis is very similar to the analysis of Chapter 8 but here an important role is played by the Birman-Schwinger principle. It is rather strange that this extremely powerful and simple tool is used much less than it should be in papers devoted to spectral asymptotics. Moreover, in this chapter we do not consider Riesz means-like mollifications.[1)]

9.1 Statement of the Basic Theorems

9.1.1 Preliminary remarks

First of all let us just mention a series of results which follows trivially from the results of the previous chapter applied to operators $A-\tau_1 I$ and $A-\tau_2 I$ with corresponding functions J_1 and J_2, and from the formula

$$\mathbf{N}(A;\tau_1,\tau_2)=\mathbf{N}^-(A-\tau_1 I)-\mathbf{N}^-(A-\tau_2 I). \tag{9.1.1}$$

These results do not work, however, in the numerous cases when there is an indefiniteness of the form "$\infty-\infty$" in the right-hand expression.

Moreover, let us mention some results which follow trivially from Chapters 4–6: if the operators $A-\tau_1 I$ and $A-\tau_2 I$ satisfy the conditions of the previous chapter with $J_1=J_2=I$ (this is essential only in the restrictions on trajectories) and with the auxiliary functions $\gamma^{(1)}$, $\rho^{(1)}$, $\varrho_1^{(1)}$, ... and $\gamma^{(2)}$, $\rho^{(2)}$, $\varrho_1^{(2)}$, ... respectively and if in both cases $X''_{(1)}=X''_{(2)}=\emptyset$ then under the additional restriction $\gamma^{(1)}=\gamma^{(2)}=\gamma$, $\rho^{(1)}=\rho^{(2)}=\rho$, ... we can apply a γ-admissible partition of unity for

$$\mathbf{N}(A;\tau_1,\tau_2)=\int \operatorname{tr} e(x,x,\tau_1,\tau_2)dx \tag{9.1.2}$$

and obtain the estimate

$$|\mathbf{N}(A;\tau_1,\tau_2)-\int_X S(x,\tau_1,\tau_2)dx-\int_{\partial X} S_1(x,\tau_1,\tau_2)dx'|\le C\bigl(R_{(1)}+R_{(2)}\bigr) \tag{9.1.3}$$

where $R_{(1)}$ and $R_{(2)}$ are remainder estimates for operators $A-\tau_1 I$ and $A-\tau_2 I$ respectively and we have included in S and S_1 all the interior and all the boundary integrand terms of the local semiclassical spectral asymptotics respectively.

[1)] Riesz means themselves are usually infinite.

Moreover, the restriction $\gamma^{(1)} = \gamma^{(2)} = \gamma$, $\rho^{(1)} = \rho^{(2)} = \rho$, ... is not necessary. In fact, let us take a γ-admissible partition of unity with $\gamma = \max(\gamma^{(1)}, \gamma^{(2)})$. Then for every element of this partition[2] we have an estimate like

$$(9.1.4)\quad |\frac{1}{L}\phi(\frac{\tau}{L})\Big(\int_X (\psi(x)\operatorname{tr} e(x,x,\tau,\tau_i) - \tilde{S}_\psi(x,\tau,\tau_i))\,dx - \int_{\partial X} \tilde{S}_{1\psi}(x,\tau,\tau_i)\,dx'\Big)\,d\tau| \le CR_{(i)}$$

where $\tilde{S}_\psi$ and $\tilde{S}_{1\psi}$ can also contain extra terms (if we compare them with ψS and ψS_1) and we have noted that $R_{(i)}$ is surely not less than the remainder estimate due to the remainder estimate $O(h^s)$ (or $O(\mu^{-s}h^s)$ for the two-dimensional Dirac operator with strong magnetic field) in local semiclassical spectral asymptotics. After adding these estimates with the subtraction of the estimated quantities and with the mutual reduction of the extra terms we obtain estimate (9.1.3). More detailed arguments I leave to the reader. I would only like mention that results of this type are not very exciting because in real situations the condition $X''_{(1)} = X''_{(2)} = \emptyset$ is not necessarily fulfilled and that in the conditions on Hamiltonian trajectories $J_1 = J_2 = I$ is not necessarily natural.

In order to avoid the first difficulty we can apply the arguments of section 8.2 (assuming that $X'_{(1)} = X'_{(2)} = X'$ and $X''_{(1)} = X''_{(2)} = X''$ in the upper estimates; one picks up $J_1 = J_2 = I$). Therefore the following statement holds:

Theorem 9.1.1. *Let the operators $A - \tau_i I$ satisfy the conditions of the corresponding theorems of section 8.1 in $X'_{(i)}$ with $J_i = \lambda I$ (where λ is a constant parameter) and with auxiliary functions $\gamma^{(i)}$, $\rho^{(i)}$, ... instead of γ, ρ, ... ; conditions depending on i should be fulfilled for $i = 1$ and $i = 2$. Let $\tau_1 < \tau_2$ and $X'_{(1)} = X'_{(2)} = X'$ Then*

(i) The estimate

$$(9.1.5)_1\qquad \mathbf{N}(A;\tau_1,\tau_2) \ge \mathcal{N}(A;\tau_1,\tau_2) - C\big(R_{*(1)} + R_{*(2)}\big)$$

holds where

$$(9.1.6)\qquad \mathcal{N}(A;\tau_1,\tau_2) = \int_{X'} S(x,\tau_1,\tau_2)\,dx + \int_{\partial X\cap X'} S_1(x,\tau_1,\tau_2)\,dx'$$

and $R_{*(1)}$ and $R_{*(2)}$ are taken from the lower estimates for $\mathbf{N}^-(A-\tau_1 I)$ and $\mathbf{N}^-(A-\tau_2 I)$ respectively, obtained in the theorem of section 8.1 in which the conditions were fulfilled.

(ii) Moreover, let $X''_{(1)} = X''_{(2)} = X''$ and let either conditions $(8.1.32)_{1,2}$ or (8.1.55), (8.1.56) be fulfilled in $X' \cap X''$ with $\varrho_0^{(i)} = \lambda$ and let condition (8.1.30) be fulfilled with inequality (8.1.31) replaced by the inequality

$$(9.1.7)\qquad \mathbf{N}(A'';\tau_1 - t\lambda, \tau_2 + t\lambda) \le R_0(1+t)^l \qquad \forall t \ge 0$$

with some (arbitrarily large) l (we denote the auxiliary operator coinciding with A in the $\epsilon\gamma$-vicinity of $X' \setminus X'$ by A'' instead of $\mathbf{A}''$ because $J = I$). Then the estimate

$$(9.1.5)_2\qquad \mathbf{N}(A;\tau_1,\tau_2) \le \mathcal{N}(A;\tau_1,\tau_2) + C(R^*_{(1)} + R^*_{(2)})$$

holds with $R^*_{(1)}$ and $R^*_{(2)}$ taken from the upper estimates for $\mathbf{N}^-(A-\tau_1)$ and $\mathbf{N}^-(A-\tau_2)$ respectively, and these estimates are not less than CR_0.

[2] If the condition $\gamma^{(i)} \ge \epsilon_0\gamma$ is violated on this element one can make an additional $\gamma^{(i)}$-admissible partition on this element, apply on every element of this subpartition results of Part II and then sum.

This theorem does not follow from the results of Chapter 8 because $\mathcal{N}(A;\tau_1,\tau_2)$, $R_{(1)}$ and $R_{(2)}$ can be finite and $\mathcal{N}^-(A-\tau_i I)$ can be infinite at the same time. Usually this occurs in the case when $\tau_1<\tau_2$ belong to the same gap in the semiclassical approximation to the spectrum for $|x|\ge C_0$; examples provided by the Dirac operator and by the Schrödinger and Dirac operators with constant magnetic field in dimension 2 will be given in Chapters 10 and 11 respectively.

9.1.2 The Birman-Schwinger principle

In order to obtain less trivial results let us apply the following identity which is trivial due to the *Birman-Schwinger principle:*

$$\text{(9.1.8)}\quad \mathbf{N}(A;\tau_1,\tau_2)=\mathbf{N}(A-tJ;\tau_1,\tau_2)+\\ \mathbf{N}(J^{-1}(A-\tau_1 I);0,t)-\mathbf{N}(J^{-1}(A-\tau_2 I);0,t)$$

provided $\tau_1<\tau_2$, $t>0$ and J is a positive definite operator such that $J^{-1}(A-\tau_i I)$ are self-adjoint operators in the space $\mathcal{H}=\mathcal{H}_J$. As we will prove in section 9.2 it is sufficient to assume that J is a symmetric, positive,[3] bounded and A-compact[4] operator. We will justify formula (9.1.8) in section 9.2. This formula yields the more general formula

$$\text{(9.1.10)}\quad \mathbf{N}(A;\tau_1,\tau_2)=\mathbf{N}(A-t_1J_1-\dots-t_nJ_n;\tau_1,\tau_2)-\\ \sum_{1\le k\le n}\mathbf{N}(J_k^{-1}(A-t_1J_1-\dots-t_{k-1}J_{k-1}-\tau I);0,t_k)\Bigr|_{\tau=\tau_1}^{\tau=\tau_2}$$

where $\mathbf{N}(B;0,t)=-\mathbf{N}(B;0,-t)$ for $t<0$ by a definition and we assume that

(9.1.11) All the operators J_k $(k=1,\dots,n)$ are symmetric, positive definite, bounded and A-compact.

Let us assume that

$$\text{(9.1.12)}\qquad \phi_k\in C_0^K(\mathbb{R})\qquad\text{and}\qquad \int_{\mathbb{R}}\phi_k(t)dt=1.$$

Then (9.1.10) yields the following final formula which will be used:

$$\text{(9.1.13)}\quad \mathbf{N}(A;\tau_1,\tau_2)=\\ \int_{\mathbb{R}^n}\mathbf{N}(A-t_1J_1-\dots-t_nJ_n;\tau_1,\tau_2)\phi_1(t_1)\dots\phi_n(t_n)dt_1\dots dt_n-\\ \sum_{1\le k\le n}\int_{\mathbb{R}^k}\mathbf{N}(J_k^{-1}(A-t_1J_1-\dots-t_{k-1}J_{k-1}-\tau I);0,t_k)\Bigr|_{\tau=\tau_1}^{\tau=\tau_2}\times\\ \phi_1(t_1)\dots\phi_k(t_k)dt_1\dots dt_k.$$

Let $\{\psi_j^2\}_{j\ge 0}$ be a partition of unity of the same type as in chapter 8; then the right-hand expression of (9.1.13) is a sum of terms of the type

[3] I.e., $(Jv,v)>0\quad\forall u\ne 0$.

[4] I.e., $J(A+i)^{-1}$ is a compact operator. This condition is stronger than the standard definition

$$\text{(9.1.9)}\qquad \|Jv\|\le\epsilon\|Av\|+C_\epsilon\|v\|\qquad\forall v\in \mathrm{D}(A)\quad\forall\epsilon>0$$

which will be also useful.

$$(9.1.14)_{n+1} \qquad \int_{\mathbb{R}^n} \operatorname{tr}(\Gamma\psi^2 e_{A-t_1J_1-\dots-t_nJ_n})(\tau_1,\tau_2)\phi_1(t_1)\dots\phi_n(t_n)dt_1\dots dt_n$$

and

$$(9.1.14)_k \qquad \mp\int_{\mathbb{R}^k} \operatorname{tr}(\Gamma\psi^2 e_{J_k^{-1}(A-t_1J_1-\dots-t_{k-1}J_{k-1}-\tau I)})(0,t_k)\times \phi_1(t_1)\dots\phi_n(t_k)dt_1\dots dt_k.$$

Here for the sake of simplicity we skip the subscript j of ψ and the subscript $i=1,2$ of τ in terms with $k\le n$; in these terms $\tau=\tau_{1,2}$ correspond to signs $\mp$. Surely we can take different partitions of unity for different $k=1,\dots,n+1$ and for different $i=1,2$ if $k\le n$.

Let us assume that J_k are Hermitian positive definite matrices (bounded and A-compact as operators) and that in $B(\bar{x},\bar{\gamma})$

$$(9.1.15) \qquad |D^\alpha J_k|\le \varrho'_k\gamma^{-|\alpha|}, \qquad |J_k^{-1}|\le \varrho_k'^{-1}$$

with appropriate ϱ'_k and $\gamma=\gamma^{(i)}$, $i=1,2$. Later ϱ'_k will be scaling functions. Let $h=\bar{\rho}^{-1}\bar{\gamma}^{-1}$ and let us assume for the sake of simplicity that $\bar{\gamma}=\gamma=1$, $\bar{\rho}=\rho^i$.

Let us note first of all that after rescaling we get the mollification parameter $L_{\text{eff}}=\varrho'_k\varrho_1^{-1}$ for t_k and our results on spectral asymptotics with mollification yield that we have a complete (i.e., modulo $O(h^s)$ with arbitrary s) asymptotics of $(9.1.14)_{n+1}$ provided

$$(9.1.16)_k \qquad \max_{r\le k}\varrho'_r\ge \varrho_1 h^{\frac{1}{2}-\delta}$$

with $k=n$ and arbitrarily small $\delta>0$; moreover, for $k\ge 2$ we have a complete asymptotics of $(9.1.14)_k$ provided condition $(9.1.16)_{k-1}$ is fulfilled; here no microhyperbolicity condition[5)] is necessary. Moreover, in the scalar case we can replace the exponent $\frac{1}{2}-\delta$ by $1-\delta$ in condition $(9.1.16)_k$.

Let us now assume that for $\tau=\tau_i$ $(i=1,2)$ microhyperbolicity condition $(8.1.18)_1$ is fulfilled for the original problem $(A-\tau_iI,B)$. Then modulo $O(h^{1-d})$ we have asymptotics of the term $(9.1.14)_1$. Moreover, if condition $(9.1.16)_1$ is violated then this microhyperbolicity condition is fulfilled for the problem $(A-t_1J_1-\tau_iI,B)$ $(i=1,2)$ and we have asymptotics of the term $(9.1.14)_2$ modulo $O(h^{1-d})$ also, etc. Finally, we obtain that under the microhyperbolicity condition $(8.1.18)_1$ we have asymptotics modulo $O(h^{1-d})$ of all the terms $(9.1.14)_k$ with $k=1,\dots,n$ (and we have no need to check conditions $(9.1.16)_j$) and, moreover, the same is true for the modified term with $k=n+1$ now given by

$$(9.1.14)'_{n+1} \qquad \frac{1}{L}\int_{\mathbb{R}^{n+1}} \operatorname{tr}(\Gamma\psi^2 e_{A-t_1-\dots-t_nJ_n})(\tau',\tau_i)\times \phi_0(\frac{1}{L}(\tau'-\tau_i))\phi_1(t_1)\dots\phi_n(t_n)d\tau' dt_1\dots dt_n$$

with $L\ge h^{\frac{1}{2}-\delta}$ (or $L\ge h^{1-\delta}$ in the scalar case).

The same assertion is true if the microhyperbolicity condition for $(A-\tau_iI,B)$ is replaced by the ellipticity condition $(8.1.18)_6$ for the operator $A-\tau_iI$ (recall that this condition does not include the assumption that the boundary value problem is elliptic (or microhyperbolic)). If we assume that the boundary value problem $(A-\tau_iI,B)$ is also

5) With respect to (x,ξ) because these problems are automatically microhyperbolic with respect to $t_1,\dots,t_k$ and condition $(9.1.16)_k$ means that there is sufficient mollification with respect to t_r with appropriate $r\le k-1$.

elliptic then we can obtain all the terms $(9.1.14)_k$ ($k = 1, \dots, n$) and $(9.1.14)'_{n+1}$ modulo $O(h^s)$.

Let us now treat the Dirac operator. Let us assume that the conditions

$$\gamma|\nabla V| + |V - \tau_i| \geq \epsilon_0 \min(\frac{1}{m}\rho^2, \rho), \tag{9.1.17}$$

$$|F_{jl}| \leq c\rho\gamma^{-1} \tag{9.1.18}$$

are fulfilled in the ball $B(\bar{x}, \gamma)$. Then all the terms $(9.1.14)_k$ ($k = 1, \dots, n$) and $(9.1.14)'_{k+1}$ can be treated in the same way (we now refer to the theory of the generalized Dirac operator; see section 4.5). Let us consider this ball $B(\bar{x}, \gamma)$ in the general case. Without loss of generality one can assume that $\gamma = 1$. Let us make a γ'-admissible partition of unity with $\gamma' = \epsilon_0(|V'| + |\nabla V'| + m') + h^{\frac{1}{2}}$ where $V' = Vh$, $m' = mh$; recall that $h = \rho^{-1}$.

Let us now consider the zone $\{\gamma \leq 3m'\}$ (if it is not empty) in which we make a γ''-admissible partition of unity with $\gamma'' = \epsilon_0(|W| + |\nabla W|^2)^{\frac{1}{2}} + h^{\frac{1}{2}}$, $W = V'^2 - m'^2$. Let us repeat all the analysis made above (in this section and in section 4.5) and separately treat the case when ϱ'_k is larger than some critical value (in which case there are appropriate mollifications in all terms beginning with the $(k+1)$-th) and the case in which ϱ'_k is smaller than this value (then conditions (9.1.17), (9.1.18) remain true for all the terms ending with the $(k+1)$-th provided $\gamma \geq 2h^{\frac{1}{2}}$; for $\gamma \leq 2h^{\frac{1}{2}}$ we apply rough estimates). We finally obtain that in dimension $d \geq 2$ we can calculate all the terms $(9.1.14)_k$ ($k = 1, \dots, n$) and $(9.1.14)'_{n+1}$ modulo $O(h^{1-d})$ even if condition (9.1.17) is violated.

On the other hand, if $d = 1$ then we obtain the remainder estimate $O(h^{-\frac{1}{2}})$ in the general case and $O(\log h)$ under the Hessian non-degeneracy condition

$$\sum_{\alpha:|\alpha|\leq 2} \gamma^\alpha |D^\alpha v| \geq \epsilon_0 \rho^2 \tag{9.1.19}$$

with $v = (V - \tau_i \pm m)(m + \rho)$. Here the source of trouble is the zone $\{\gamma' \leq 3m'\}$ in which both eigenvalues of the principal symbol are simple and disjoint. Launching the complete technique of section 4.5, we obtain the remainder estimate $O(h^{-\delta})$ in the general case and $O(\log h)$ or even $O(1)$ under the non-degeneracy condition

$$\sum_{\alpha:|\alpha|\leq M} \gamma^\alpha |D^\alpha v| \geq \epsilon_0 \rho^2. \tag{9.1.20}$$

The one-dimensional Dirac operator will also be treated below.

Finally, let us treat the arbitrary matrix operator and let us assume that (after normalization, i.e., multiplication by an appropriate number if necessary) this operator satisfies the following condition:

(9.1.21) The interval $[\tau_i - \epsilon_0, \tau_i + \epsilon_0]$ contains no eigenvalues of the principal symbol of A different from the eigenvalue $\lambda(x, \xi)$ of constant multiplicity.

Then this condition is also fulfilled for operator $J_*^{-1}A$ provided one of the following three conditions is fulfilled:

$(9.1.22)_1$ λ is a simple eigenvalue of the principal symbol of A;

$(9.1.22)_2$ J_* is scalar;

$(9.1.22)_3$ The restriction of J_* to $\mathrm{Ker}(a^0 - \lambda)$ is scalar (a^0 is the principal symbol of A).

Condition $(9.1.22)_3$ is more general than $(9.1.22)_{1,2}$ but it is difficult to fulfill $(9.1.22)_3$ outside of the context of $(9.1.22)_{1,2}$ because J_* does not depend on ξ and $(a^0-\lambda)$ depends on ξ.

Application of the matrix J_* which does not satisfy one of the conditions $(9.1.21)_{1-3}$ is possible only if before there was mollification with parameter $L \geq h^{\frac{1}{2}-\delta}$ (we assume that the normalizing factor equals 1).

Thus, let us assume that condition $(9.1.22)_3$ is fulfilled. Let us make a γ-admissible partition of unity with $\gamma = \epsilon_0(|\lambda-\tau_i|+|\nabla\lambda|^2)^{\frac{1}{2}}+h^{\frac{1}{2}}$. Then for $\gamma \geq 2h^{\frac{1}{2}}$ the microhyperbolicity condition is fulfilled for the original operator after the dilatation-multiplication procedure; therefore in this case our remainder estimate is justified. For $\gamma \leq h^{\frac{1}{2}-\delta'}$ the condition $L' \geq h'^{\frac{1}{2}-\delta'}$ is fulfilled where $\delta' = \delta'(\delta) > 0$ is small enough, h' and L' are effective values of h and L after the dilatation-multiplication procedure and we assume that $L \geq h^{1-\delta}$. So the condition $L \geq h^{\frac{1}{2}-\delta}$ can be replaced by $L \geq h^{1-\delta}$. Finally, we obtain that in the general case the microhyperbolicity condition can be replaced by $(8.1.18)_{2,3}$ and

$(9.1.23)_p$ For every $k=1,\dots,p$ the matrix J_k satisfies $(9.1.22)_3$ and there exists $k \leq p$ such that $\varrho'_k \geq \varrho h^{1-\delta}$

with some $p \leq n+1$.

Moreover, let conditions (9.1.21) and $(9.1.22)_1$ be fulfilled. Then $(9.1.23)_p$ is reduced to the condition

$(9.1.24)_p$ There exists $k \leq p$ such that $\varrho'_k \geq \varrho h^{1-\delta} = \varrho\rho^{\delta-1}\gamma^{\delta-1}$;

which is, in fact, no restriction for $p = n+1$.

One can easily see that under conditions (9.1.21) and $(9.1.22)_1$ one can launch all the arguments of section 4.4 and replace $(8.1.18)_{2,3}$ by $(8.1.18)_4$. Certain generalizations for multiple eigenvalues are possible as well.

In all these cases the contribution of the ball $B(x,\gamma(x))$ to the remainder estimate is $O(h_{\text{eff}}^{1-d})$. Furthermore, under conditions (9.1.21), $(9.1.22)_1$ and $(8.1.18)_7$ the contribution of the ball $B(x,\gamma(x))$ to the remainder estimate is $O(h_{\text{eff}}^{1-d-\delta})$ with arbitrarily small $\delta > 0$. Moreover, the same remains true under the condition

$(9.1.23)'_p$ For every $k=1,\dots,p$ matrix J_k satisfies $(9.1.22)_1$ (i.e., is scalar) and there exists $k \leq p$ such that $\varrho'_k \geq \varrho h^{1-\delta}$

with some $p \leq n+1$.

All these arguments remain true for the Schrödinger operator with strong magnetic field, and having a "scalar-like operator" is not necessary. The same is true for the Dirac operator with strong magnetic field under condition $(9.1.23)'_p$; moreover, for $d=2$ one can replace the inequality $\varrho'_k \geq \varrho h_{\text{eff}}^{1-\delta}$ by $\varrho'_k \geq \varrho\mu_{\text{eff}}^{\delta-1}h_{\text{eff}}^{1-\delta}$ or even $\varrho'_k \geq \varrho\nu_{\text{eff}}\mu_{\text{eff}}^{\delta-1}h_{\text{eff}}^{1-\delta}$. Recall that $\varrho = \min(\rho, \frac{1}{m}\rho^2)$ for the Dirac operator. We refer to these conditions as $(9.1.23)''_1$ and $(9.1.23)'''_1$ respectively.

Moreover, under condition $(9.1.23)_1$ or $(9.1.23)'_1$ more precise remainder estimates associated with conditions on Hamiltonian trajectories remain true for general operators. Furthermore, improved remainder estimates are true for three-dimensional Schrödinger and Dirac operators with strong magnetic field under conditions on magnetic lines. Finally, improved remainder estimates are true for two-dimensional Schrödinger and Dirac operators with strong magnetic field under conditions on drift lines; in this case one can use condition $(9.1.23)''_1$ (or $(9.1.23)'''_1$).

On the other hand, applying the results of sections 6.7 and 7.8 one can weaken the conditions on these operators.

Remark 9.1.2. *In the lacunary zone, restrictions of the type $\varrho'_k \le \eta$ are also admissible; here $\eta > 0$ is a small parameter such that these inequalities (fulfilled for all k) provide the condition*

(9.1.25) τ_i lies in the lacuna of the problem $(A - t'_1 J_1 - \cdots - t'_k J_k, B)$ for all $t'_k \in [0, t_k]$ (or $t'_k \in [t_k, 0]$ for $t_k < 0$) for all $k = 1, \dots, n$ and for all $t_k \in \operatorname{supp} \phi_k$. ◊

Let us now treat the terms of the derived asymptotics. For the sake of simplicity we do not write the mollification on $t_1, \dots, t_k$. Then for $k = 1, \dots, n$ we have

$$\int_0^{t_k} \operatorname{Tr} \operatorname{Res}_{\mathbb{R}} \bigl(z - J_k^{-1}(A - t_1 J_1 - \cdots - t_{k-1} J_{k-1} - \tau)\bigr)^{-1} \psi dz \tag{9.1.26$_k$}$$

with $\tau = \tau_1, \tau_2$ and for $k = n + 1$ we have

$$\int_{\tau_1}^{\tau_2} \operatorname{Tr} \operatorname{Res}_{\mathbb{R}} (z - A + t_1 J_1 + \cdots + t_n J_n)^{-1} \psi dz \tag{9.1.26$_{n+1}$}$$

where $\operatorname{Res}_{\mathbb{R}}$ is the difference of the limits from the upper and lower complex semiplanes on z, we skip numerical factors depending on d only and G^{-1} means the formal resolvent of G (in the context of the formal calculus of pseudo-differential operators) obtained according to the rules of Chapters 4–7. We assume for the moment that $\max_k \varrho'_k \le 1$. We can only treat the case $n = 1$: the general case can be reduced to this case if we replace A by $A - t_1 J_1 - \cdots - t_{k-1} J_{k-1}$ and sum with respect to k. Therefore we skip the subscript "1" of t and J. Then the term $(9.1.26)_1$ is replaced by

$$\int_0^{t} \operatorname{Tr} \operatorname{Res}_{\mathbb{R}} (Jz - A + \tau)^{-1} J \psi dz \tag{9.1.26$'_1$}$$

(it is easy to see that the equality $(z - J^{-1}A)^{-1} = (Jz - A)^{-1} J$ holds for formal resolvents also) and the term $(9.1.26)_2$ is replaced by

$$\int_{\tau_1}^{\tau_2} \operatorname{Tr} \operatorname{Res}_{\mathbb{R}} (z - A + tJ)^{-1} \psi dz. \tag{9.1.26$'_2$}$$

Let $\Theta(t)$ be the last expression and let $\Theta'_i(t)$ be expression $(9.1.26)'_1$ at $\tau = \tau_i$. Let us consider

$$\Theta(0) - \Theta(t) - \Theta'_2(t) + \Theta'_1(t).$$

For $t = 0$ this expression vanishes. Moreover, its derivative with respect to t is equal to

$$-\int_{\tau_1}^{\tau_2} \operatorname{Tr} \operatorname{Res}_{\mathbb{R}} (z - A + tJ)^{-2} J (z - A + tJ)^{-1} [A, \psi] dz$$

because of the chain of equalities

$$\begin{aligned}
\partial_t \operatorname{Tr} \operatorname{Res}_{\mathbb{R}} (z - A + tJ)^{-1} \psi &= \\
&- \operatorname{Tr} \operatorname{Res}_{\mathbb{R}} (z - A + tJ)^{-1} J (z - A + tJ)^{-1} \psi = \\
- \operatorname{Tr} \operatorname{Res}_{\mathbb{R}} (z - A + tJ)^{-2} J \psi &- \operatorname{Tr} \operatorname{Res}_{\mathbb{R}} (z - A + tJ)^{-2} J (z - A + tJ)^{-1} [A, \psi]
\end{aligned}$$

which follow from the properties of the trace and from the fact that J and ψ commute. The first term in the right-hand expression is equal to

$$\partial_z \operatorname{Tr} \operatorname{Res}_{\mathbb{R}} (z - A + tJ)^{-1} J\psi$$

and after integration on z it reduces with $\partial_t(-\Theta_2'(t) + \Theta_1'(t))$.

In the context of condition $(9.1.23)_1$ or $(9.1.23)_1'$ these calculations are justified by microhyperbolicity for small ϱ_1' and by microhyperbolicity and mollifications otherwise; justification in other contexts is made similarly. Finally, after adding all the terms but without summation on the partition of unity we obtain that the right-hand expression of (9.1.13) contains the asymptotics of $(\Gamma\psi e_A)(\tau_1,\tau_2)$ described in Chapters 4–7 plus extra terms of the form

$$\sum_{1\le r\le s} h^{r-d} \int (\mathcal{L}_r\psi)dx$$

plus a remainder estimate as was obtained in Chapters 4–7. Here $\mathcal{L}_r$ are differential operators such that $\mathcal{L}_r 1 = 0$. Let us note that the operators $\mathcal{L}_r$ depend uniquely on A, $J_1, \dots, J_n, \phi_1, \dots, \phi_n$ and hence after summation on the partition of unity all these extra terms reduce. Let us note that for two-dimensional Schrödinger and Dirac operators with strong magnetic field h should be replaced by $\mu^{-1}h$ and $\mathcal{L}_r$ also depend on μ^{-2}, here and above and below.

All these assertions have been proven for $\varrho_k' \le \varrho$; for $\varrho_k' \ge \varrho$ with some number $k = k_0$ one can replace in all the terms with $k \ge k_0 + 1$ h by $h' = h(\frac{\varrho}{\varrho_k'})^{\frac{1}{m}}$, etc., and therefore after summation on the partition of unity we obtain that only terms supported in $X'\cap X''$ are not completely reduced, and the remaining terms do not exceed $Ch^{1-d}(1+\frac{(\varrho_1'+\dots+\varrho_n')}{\varrho})^{\frac{(d-1)}{m}}$.

Finally, let us note that if in some ball condition (9.1.25) is fulfilled then all the operators $\mathcal{L}_r$ vanish in this ball.

Let us now treat the zone X''. First of all we must assume that in $X'\cap X''\cap\{\rho\gamma \ge c\}$ [6] either an ellipticity condition or the condition

$$\varrho_1' \ge \varrho h \tag{9.1.27}$$

was fulfilled in every ball.

Finally, let us introduce a condition similar to (8.1.30). We assume that

(9.1.28) There exist a self-adjoint operator $\tilde{A}$ and symmetric operators $\tilde{J}_k$ coinciding with A and J_k respectively in X'' and such that for all k the operators $\tilde{J}_k$ are $\tilde{A}$-compact and

$$\mathbf{N}\big(\tilde{J}_k^{-1}(\tilde{A} - \tau_i - t_1\tilde{J}_1 - \dots - t_{k-1}\tilde{J}_{k-1}); -t, t\big) \le R_0(|t|+1)^l \qquad \forall k \quad \forall t_j \in \operatorname{supp}\phi_j \quad (j=1,\dots,k-1) \quad \forall t \in \mathbb{R}. \tag{9.1.29}$$

Moreover, we assume that condition (9.1.29) is fulfilled for sequences $\tau_{i,n} \to \tau_i$ with $\tau_{i,n} \in [\tau_1,\tau_2] \setminus \operatorname{Spec}_{\mathrm{ess}}(A)$; this condition will be denoted by (9.1.29)*.

Finally, let us assume that

$$\int_{X''} \operatorname{tr} e_{A-t_1J_1-\dots-t_nJ_n}(x,x,\tau_1,\tau_2)dx \le R_0 \qquad \forall t_j \in \operatorname{supp}\phi_j \quad (j=1,\dots,n). \tag{9.1.30}$$

---

[6] In the general case; modifications will be given below for the Schrödinger and Dirac operators.

Instead of this condition we will usually use the condition

(9.1.31) $$\mathbf{N}(A - t_1 J_1 - \cdots - t_n J_n; \tau_1, \tau_2) \le R_0 \qquad \forall t_j \in \operatorname{supp} \phi_j \quad (j = 1, \dots, n).$$

This condition may seem to be too restrictive because it is associated with X instead of X'', but in fact this is not so.

9.1.3 Basic theorems (general operators)

Now we can formulate and prove our basic theorems. The proofs which can be done easily by the methods of section 8.2 we leave to the reader.

Theorem 9.1.3. *Let* X, X', X'', $\gamma^{(i)}$, $\rho^{(i)}$, $\varrho_k^{(i)}$, ϱ'_k *satisfy conditions* (8.1.1)–(8.1.3). *Let operators* A *and* B *be of the forms* (8.1.4) *and* (8.1.5) *respectively and for all* $i = 1, 2$ *and* k *let* $A - \tau_i I$, B *and* J_k *satisfy conditions* $(8.1.9)_{1-4}$, $(8.1.7)_1$ *with* γ, ρ, $\varrho_* \dots$ *and* ϱ_0 *replaced by* $\gamma^{(i)}$, $\rho^{(i)}$, $\varrho_*^{(i)}$ *and* ρ'_k *respectively. Let us assume that in* X'

$(9.1.32)_1$ *The operator* $(\rho^{(i)})^m (\varrho_1^{(i)})^{-1} A$ *is classically uniformly elliptic for* $i = 1, 2$;[7]

$(9.1.32)_2$ *The boundary value problem* $\bigl((\rho^{(i)})^m (\varrho_1^{(i)})^{-1} A, (\rho^{(i)})^{m_j} (\varrho_j^{(i)})^{-1} B_j, j = 1, \dots, \mu\bigr)$ *is classically uniformly elliptic for* $i = 1, 2$.

Moreover, let us assume that

(9.1.33) *For* $i = 1, 2$, *for every ball* $B(x, \gamma^{(i)}(x))$ *in* $X'_{(i)} \cap X''_{(i)} \cap \{\rho^{(i)}\gamma^{(i)} \ge c\}$, ***either*** *the boundary value problem* $\bigl((\varrho_1^{(i)})^{-1}(A - \tau_i), (\varrho_j^{(i)})^{-1} B_j, j = 1, \dots, \mu\bigr)$ *is elliptic (as an* $h^{(i)}$*-boundary value problem with* $h^{(i)} = \frac{1}{\rho^{(i)}\gamma^{(i)}}$*)* **or** *condition* $(9.1.24)_1$ *is fulfilled (i.e.,* $\varrho'_1 \ge \varrho_1^{(i)} h^{(i)\,1-\delta}$*).*

Moreover, let us assume that J_k *are bounded and* A*-compact operators. Finally, let us assume that in every ball* $B(x, \gamma^{(i)}(x))$ *the operator* $A - \tau_i I$ *satisfies one of conditions* $(8.1.18)_{1-7}$ *with* ϱ_0 *replaced by* ϱ'_k *(and under conditions* $(8.1.18)_{4,7}$ *one of the conditions* $(9.1.22)_{1,2}$ *is assumed to be fulfilled and under conditions* $(8.1.18)_{2,3,5}$ *one of the conditions* $(9.1.22)_{1-3}$ *is assumed to be fulfilled).*

Finally, let us assume that there exists an operator $\tilde{A}$ *coinciding with* A *in* X''*, self-adjoint and such that for all* k *the operators* J_k *are* $\tilde{A}$*-compact and conditions* $(9.1.29)^*$, (9.1.31) *are fulfilled.*

Then

(i) The following estimate holds:

$(9.1.34)_1$ $$\mathbf{N}(A; \tau_1, \tau_2) \ge \mathcal{N}_w(A; \tau_1, \tau_2) - C \sum_{i=1,2} \bigl(R_1^i + R_2^i + R_4^i\bigr) - C R_0$$

where $\mathcal{N}_w$ *is given by* (9.1.6) *and here one can take* S *given by* (8.1.15), $S_1 = 0$ *and for every* i

$(9.1.35)_1$ $$R_1 = \int_{X'_-} \wp_1 \, dx, \qquad R_2 = \int_{X'} \rho^{d-s} \gamma^{-s} \, dx,$$

---

[7] In which case $(\rho^{(1)})^m (\varrho_1^{(1)})^{-1} \asymp (\rho^{(2)})^m (\varrho_1^{(2)})^{-1}$ and a similar deduction holds for the next condition.

$$R_4 = \int_{X' \cap X''} \left(\varrho'_1 + \dots + \varrho'_n\right)^{\frac{(d-1)}{m}} \varrho^{-\frac{(d-1)}{m}} \rho^{d-1}\gamma^{-1}dx, \tag{9.1.35$_2$}$$

X'_- is the set of all the points of X' in which the problem $\left(\varrho_1^{-1}(A-\tau), \varrho_j^{-1}B_j,\ j = 1,\dots,\mu\right)$ is not elliptic as an h-problem (in these formulas the superscript "(i)" is skipped for the sake of notational simplicity, by footnote [7])),

$$\wp_1 = \rho^{d-1+\delta'}\gamma^{-1+\delta'}$$

with $\delta' = 0$ under conditions $(8.1.18)_{1,2,3,6}$ and arbitrarily small $\delta' > 0$ otherwise.

(ii) Moreover, in balls in which of the conditions $(8.1.18)_{1,2,5}$ is fulfilled one can replace R_1 here and below by

$$\partial_\tau \int_{X'_+} S(x,.,\tau)\varrho_1(\rho\gamma)^{-1}dx\big|_{\tau=\bar\tau} + R'_1, \tag{9.1.36}$$

$$R'_1 = \int_{X'_+} \rho^{d-1-\delta'}\gamma^{-1-\delta'}$$

where here and below $\bar\tau = \tau_i$.[8])

(iii) Moreover, in some balls let one of the conditions $(8.1.18)_{1,2}$ be fulfilled, let J_1 be a scalar matrix and with $J = J_1$ let conditions (9.1.21), $(9.1.16)_1$, (8.1.37)–$(8.1.45)_1$ be fulfilled. Then in these balls one can replace R_1 by

$$\sum_j \frac{1}{T_j}\int_{\Omega_j} \varrho_1\rho^{m-1}\varrho_1'^{-1}\gamma^{-1}dxd\xi : d\omega +$$

$$\partial_\tau \int_{\Omega_0} \varrho_1\rho^{-1}\gamma^{-1}\mathbf{n}(x,\xi,.,\tau)dxd\xi\big|_{\tau=\bar\tau} + R'_1.$$

The notation of theorem 8.1.3 is used here.[8])

(iv) Moreover, in situations (i)–(iii), if condition (9.1.30) is fulfilled then

$$\mathbf{N}(A;\tau_1,\tau_2) \le \mathcal{N}^w(A;\tau_1,\tau_2) + C\sum_{i=1,2}\left(R_1^i + R_2^i + R_4^i\right) + CR_0. \tag{9.1.34$_2$}$$

9.1.4 Basic theorems (Dirac operator)

In the case of the Dirac operator one can get effective expressions for R_0 and deduce the following statements:

Theorem 9.1.4. *Let X, X', X'', γ and ρ satisfy conditions (8.1.1)–(8.1.3). Let operator A be of the form (4.5.3),[9]) $d \ge 2$. For $x \in X'$ let the following conditions be fulfilled:*

$$|D^\alpha \omega^{jk}| \le c\gamma^{-|\alpha|}, \qquad |D^\alpha F_{jk}| \le c\rho\gamma^{-|\alpha|-1},$$
$$|D^\alpha V| \le c\min(\rho, \frac{1}{m}\rho^2)\gamma^{-|\alpha|} \qquad (\alpha \ne 0)\quad \forall\alpha : |\alpha| \le K, \tag{9.1.37$_{1-3}$}$$

---

8) In the case of condition $(8.1.18)_2$ the modifications follow from subsection 4.4.1. Surely, we now use $\mathcal{N}_{wp}$ instead of $\mathcal{N}_w$ where the subscript p means that S and S_1 are calculated more carefully.

9) With $J = I$; i.e., $A = A + W\sigma_0$. This remark remains true throughout this chapter.

(9.1.37)$_4$
$$(V-\tau_2-m)_+\le c\min(\rho,\frac{1}{m}\rho^2),$$
$$(V-\tau_1+m)_-\le c\min(\rho,\frac{1}{m}\rho^2),$$

(9.1.38)
$$|\det(\omega^{jk})|\ge\epsilon_0,$$

(9.1.39)
$$B\bigl(x,\gamma(x)\bigr)\subset X.$$

Then

(i) The estimate

(9.1.40)$_1$
$$\mathbf{N}(A,\tau_1,\tau_2)\ge\mathcal{N}^w(A;\tau_1,\tau_2)-C(R_1+R_2)$$

holds with R_1, R_2 *given by* (9.1.35) *where* $\wp_1=\rho^{d-1}\gamma^{-1}$ *and* $X'_-=\{x\in X'$ *and one of the conditions*

(9.1.41)$_{1,2}$
$$V-\tau_2-m\le-\epsilon_0\min\bigl(\rho,\frac{1}{m}\rho^2\bigr),$$
$$-V+\tau_1-m\le-\epsilon_0\min\bigl(\rho,\frac{1}{m}\rho^2\bigr)$$

is violated$\}$.

(ii) Moreover, for $d\ge3$ *this estimate holds with*

(9.1.42)
$$R_1=\sum_\pm\int_{\Sigma_\pm\cap T^*X'}dxd\xi:dH_\pm+R'_1$$

where R'_1 *is given by* (9.1.36),

(9.1.43)
$$H_\pm=V\pm\bigl(m^2+g^{jk}(\xi_j-V_j)(\xi_k-V_k)\bigr)^{\frac{1}{2}},$$

$\Sigma_\pm=\{H_\pm=0\}$. *Moreover, estimate* (9.1.40)$_1$ *with the* R_1 *indicated here remains true for* $d=2$, $V_j=0$. *Finally, under appropriate modifications this estimate remains true for* $d=2$ *in the general case.*

(iii) Assertion (iii) of theorem 9.1.3 holds with $\varrho'_1=1$. *Moreover, this assertion holds with* $\varrho'_1=\min(\rho,\frac{1}{m}\rho^2)$ *if terms* $-C(R_3+R_4)$ *are included in the right-hand expression, where*

(9.1.44)$_1$
$$R_3=\begin{cases}\displaystyle\int_{X'\cap X''}\tilde\rho^d dx+\int_{X'}\tilde\rho^d dx & \text{for } d\ge3,\\ \displaystyle C\int_{X''\cap X'}\rho^2dx+C_p\Bigl(\int_{X''}\tilde\rho^{2p}dx\Bigr)^{\frac{1}{p}}(\operatorname{mes}X'')^{1-\frac{1}{p}} & \\ & \text{for } d=2,\ p>1,\end{cases}$$

(9.1.44)$_2$
$$R_4=\mu^\delta\int\rho^d(\rho+m)^{-\delta}dx,$$
$$\mu=\Bigl(\int(\rho+m)^{\frac{d}{4}}\rho^d dx\Bigr)^{\frac{4}{d}}\Bigl(\int\rho^d dx\Bigr)^{-\frac{4}{d}}$$

with arbitrarily small $\delta>0$.[10] *and we assume that conditions* (9.1.37)$_4$, (9.1.38) *and*

[10] There will be some improvement for R_4 at the end of the next section.

(9.1.45) $$\bar{X}'' \cap \partial X = \emptyset,$$

(9.1.46)$_{1,2}$ $$|V_j| \le c\tilde{\rho}_j, \qquad |D_j\omega^{kl}| \le c\tilde{\rho} \quad \text{in } X''$$

are fulfilled in X''.

Moreover, in the case when X'' *contains few connected components one can calculate* R_3 *and* R_4 *for every component and then sum the results.*

(iv) The estimate

(9.1.40)$_2$ $$\mathbf{N}(A;\tau_1,\tau_2) \le \mathcal{N}^w(A;\tau_1,\tau_2) + C\bigl(R_1 + R_2 + R_3 + R_4 + R_5\bigr)$$

holds with $R_1,\dots,R_4$ *defined above (we assume that conditions* (9.1.37)$_4$*,* (9.1.38)*,* (9.1.45) *and* (9.1.46) *are fulfilled in* X''*) and*

(9.1.47) $$R_5 = \min\bigl(\frac{Mm^{\frac{d}{2}}}{|\tau_2-\tau_1|}, M^{\frac12}m^{\frac{d}{2}}(\operatorname{mes} X'')^{\frac12}\bigr),$$

$$M = \int_{X'\cap X''} \gamma^{-d}dx$$

and only those connected components of X'' *on which condition* $\rho \le cm$ *is violated are taken into account.*

(v) All the modifications of estimate (9.1.38)$_2$ *in the spirit of (i), (ii) are valid with* $\rho_1' = \min(\rho, \frac{1}{m}\rho^2)$*. Surely, in the right-hand expression of this estimate the term* $C(R_3 + R_4)$ *should be added.*

Theorem 9.1.5. *Let* $d = 2,3$ *and let* X, X', X'', γ *and* ρ, ϱ_1 *satisfy conditions* (8.1.1)–(8.1.3)*. Let the operator* A *be of the form* (4.5.3)*. For* $x \in X'$ *let conditions* (9.1.35)$_{1,3,4}$*,* (9.1.36)*,* (9.1.37) *be fulfilled and let condition* (9.1.35)$_2$ *be replaced by the condition*

(9.1.37)$_2^*$ $$|D^\alpha F_{jk}| \le c\rho_1\gamma^{-|\alpha|} \qquad \forall\alpha : |\alpha| \le K.$$

Let $X_1' = \{x \in X', \rho_1 \le 2c\rho\gamma^{-1}\}$*,* $X_2' = \{x \in X', \rho_1 \ge c\rho\gamma^{-1}\}$*. Let us assume that*

(9.1.48) $$F \ge \epsilon_0\rho_1 \qquad \text{in } X_2'.$$

Let us assume that functions ϱ_2*,* ϱ_3 *are given on* X_2' *satisfying* (8.1.2)$_2$ *and such that in every ball* $B(\bar{x},\gamma(\bar{x})) \subset X_2'$ *after the dilatation transforming this ball into* $B(0,1)$ *conditions* (6.2.48)*,* (6.3.58)*,* (6.4.40) *are fulfilled with* $\nu = \varrho_2$*,* $\bar{\nu} = \varrho_3$*,* $h = (\rho\gamma)^{-1}$*,* $\mu = \rho_1\gamma\rho^{-1}$*. For* $d = 2$ *one can replace condition* (8.1.2)$_3$ *in* X_2' *by the condition*

(9.1.49) $$\rho_1\gamma^2 \ge 1.$$

Then

(i) The estimate

(9.1.50)$_1$ $$\mathbf{N}(A;\tau_1,\tau_2) \ge \mathcal{N}(A;\tau_1,\tau_2) - C(R_1^* + R_7)$$

holds with

(9.1.51) $$\mathcal{N} = \int_{X'} h^{-d}S(x,\tau_1,\tau_2,1)dx,$$

$S(x,\tau_1,\tau_2,\mu h) = S(x,\tau,\mu h)|_{\tau=\tau_1}^{\tau=\tau_2}$*,* $R_1^* = R_1 + R_2 + R_6$*, where* R_1*,* R_2 *are defined by* (9.1.42)*,* (9.1.34) *with* X' *replaced by* X_1'*,* R_6 *is the sum over a partition of unity in* X_2'

of the remainder estimates described in the theorems of Chapter 7 with $\mu, h, \nu, \bar{\nu}$ indicated above, and

$$R_7 = \begin{cases} \int_{X_2' \cap X''} \rho_1^{-1}\rho^2\gamma^{-2}\,dx & \text{for } d=2 \\ 0 & \text{for } d=3. \end{cases} \tag{9.1.52}$$

(ii) The contribution of the zone X_1' to the estimate can be improved in the spirit of assertions (ii), (iii) of theorem 9.1.4 with $\varrho_1' = 1$.

(iii) Independently for $d=3$ the contribution of the zone X_2' can be improved under conditions on magnetic lines and for $d=2$ the contribution of the zone X_2' can be improved under conditions on the drift lines (see section 7.6).

(iv) Moreover, if conditions (9.1.37)$_4$, (9.1.38), (9.1.45) *and* (9.1.46) *are fulfilled in X'' and conditions* (8.1.55), (8.1.56) *are fulfilled in $X' \cap X''$ then the estimate*

$$\mathbf{N}(A;\tau_1,\tau_2) \le \mathcal{N}(A;\tau_1,\tau_2) + C\bigl(R_1^* + R_7 + R_3 + R_4 + R_5\bigr) \tag{9.1.50$_2$}$$

holds where R_3 and R_5 are defined by (9.1.44), (9.1.47) *but all the connected components of X'' are taken into account, $R_4 = \operatorname{mes} X''$. Moreover, assertions (ii), (iii) of this theorem remain true.*

Theorem 9.1.6. *(i) Assertions (i)–(iii) of theorem 9.1.5 remain true with $\varrho_1' = \min(\rho, \frac{1}{m}\rho^2)$ but with the additional term $-CR_4$ in the right-hand expression of the estimate* (9.1.45) *where R_4 is defined in theorem 9.1.4(iii).*

(ii) Estimate (9.1.50)$_2$ *holds without conditions* (8.1.55), (8.1.56) *and with R_{3-5} defined in theorem 9.1.4.*

Remark 9.1.7. *The method of this chapter and the previous chapters provide estimates for $\mathbf{N}(A;\tau_1,\tau_2)$ mollified with respect to τ_1 or (and) τ_2 with improved remainder estimate in X'.* ◊

9.2 Functional-Analytic Considerations and the Proof of Theorems 9.1.2 and 9.1.3

9.2.1 Self-adjointness of the Dirac operator

Let us begin with the self-adjointness of the Dirac operator. First of all let us prove

Theorem 9.2.1. *Let us assume that A is an operator of the form* (4.5.3) *with real-valued coefficients ω^{jl}, V_j and V. Let us assume that* (4.5.7) *holds and*

(9.2.1) $\omega^{jl} \in C(X)$, $D_j\omega^{jl}, V_j, V \in L^p_{\mathrm{loc}}(X)$ *with $p = d$ for $d \ge 3$ and with some $p > 2$ for $d = 2$.*

Then the operator A with domain

$$\mathrm{D}(A) = \{u \in L^2(X,\mathbb{H}) \cap H^1_{\mathrm{loc}}(X,\mathbb{H}) : Au \in L^2(X,\mathbb{H})\}$$

is self-adjoint in $L^2(X,\mathbb{H})$.

Proof. (i) Let us first note that if $u \in H^1_{\mathrm{loc}}$ then the embedding theorem yields that $u \in L^q_{\mathrm{loc}}$ and

$$\|u\|_{L^q(Y)} \le C_Y \|u\|_{H^1(Y)} \tag{9.2.2}$$

on every compact Y where here and below $q = \frac{d}{d-2}$ for $d \geq 3$ and $q = \frac{p}{p-2}$ with arbitrary $p > 2$ for $d = 2$. Therefore (9.2.1) yields that $V_j u, Vu, \omega^l u \in L^2_{\text{loc}}$ where $\omega^l = iD_j \omega^{jl}$.

(ii) Let us prove that the operator A with the indicated domain is closed. Let us first assume that $\max_j \bar{\gamma}_j < \infty$.[11] Then let us write V, ω^l, V_j in the form $V = V' + V''$ with $\|V''\|_{L^p} \leq \varepsilon$, $\|V'\|_{L^\infty} \leq C_\varepsilon$, etc., and let us write ω^{jl} in the form $\omega^{jl} = \omega^{jl\prime} + \omega^{jl\prime\prime}$ with $\|\omega^{jl\prime\prime}\|_{L^\infty} \leq \varepsilon$, $\|\omega^{jl\prime}\|_{C^2} \leq C_\varepsilon$ with arbitrarily small $\varepsilon > 0$. Then

$$\|Au - A_1 u\| \leq C_0 \varepsilon \|u\|_{H^1} + C'_\varepsilon \|u\| \qquad \forall u \in \mathrm{D}(A) \tag{9.2.3}$$

where $A_1 = \omega^{jl\prime} \sigma_l h D_j$ and C_0 does not depend on ε. Moreover, the properties of $\omega^{jl\prime}$ and σ_l yield that

$$\left| \|A_1 u\|^2 - \int g^{jk\prime} (D_k u)^\dagger (D_j u) dx \right| \leq C'_\varepsilon \varepsilon' \|u\|^2_{H^1} + C''_{\varepsilon\varepsilon'} \|u\|^2 \qquad \forall u \in \mathrm{D}(A)$$

where $g^{jk\prime} = \sum_l \omega^{jl\prime} \omega^{kl\prime}$ and therefore $|g^{jk} - g^{jk\prime}| \leq C_0 \epsilon$; here $\varepsilon' > 0$ is arbitrarily small and C'_ε does not depend on ε'. This inequality and (9.2.3) yield the inequality

$$\left| \|Au\|^2 - \int g^{jk} (D_k u)^\dagger (D_j u) dx \right| \leq C_0 \varepsilon \|u\|^2_{H^1} + C'_\varepsilon \|u\|^2 \qquad \forall u \in \mathrm{D}(A)$$

and then (4.5.7) yields the estimate

$$\|u\|_{H^1} \leq C\big(\|Au\| + \|u\|\big) \qquad \forall u \in \mathrm{D}(A). \tag{9.2.4}$$

This estimate immediately yields that the operator A is closed provided $\max_j \bar{\gamma}_j < \infty$.

Let us consider the case $\max_j \bar{\gamma}_j = \infty$. Let $\psi \in C^1_0(X)$ be a scalar function. Then $A\psi u = \psi A u + \psi' u$ with a matrix-valued function $\psi' \in C_0(X)$. Then the above arguments yield the estimate

$$\|\psi u\|_{H^1} \leq C\big(\|Au\| + \|u\|\big) \qquad \forall u \in \mathrm{D}(A) \tag{9.2.4$'$}$$

where C depends on ψ. But then A is again obviously closed.

(iii) Let us prove that the operator A is symmetric. This is obvious for $\max_j \bar{\gamma}_j < \infty$. For $\max_j \bar{\gamma}_j = \infty$ the equality

$$(Au, v) = (u, Av) \qquad u, v \in \mathrm{D}(A)$$

holds for compactly supported v. Let us replace v by $\psi(x' t^{-1}) v$ where $\psi \in C^1_0(X')$ equals 1 in a neighborhood of 0 and $X' \ni x'$ means all the coordinates x_j with $\bar{\gamma}_j = \infty$. For t tending to $+\infty$ we obtain the same equality for arbitrary $v \in \mathrm{D}(A)$.

(iv) Let us consider the problem

$$D_t u = Au, \qquad u|_{t=0} = f \tag{9.2.5}$$

and let us prove that for every $f \in \mathrm{D}(A)$ there exists a solution $u \in C(\mathbb{R}, L^2(X, \mathbb{H})) \cap L^\infty(\mathbb{R}, \mathrm{D}(A))$ of this problem.

Let us first assume that $\max_j \bar{\gamma}_j < \infty$. Then problem (9.2.5) is obviously well-posed provided $\omega^{jl}, V_j, V \in C^2$. In this case $\mathrm{D}(A) = H^1(X, \mathbb{H})$. Let $\omega^{jl}_\delta, V_{j\delta}, V_\delta$ be sequences of mollifications of ω^{jl}, V_j, V respectively with mollifying function $\psi_\delta(x) = \delta^{-d} \psi(x\delta^{-1})$ where $\psi \in C^\infty_0(\mathbb{R}^d)$ is a real-valued non-negative function with $\int \psi dx = 1$. Then for $\delta \to +0$, ω^{jl}_δ tends to ω^{jl} in C and ω^l_δ, $V_{j\delta}$, V_δ tend to ω^l, V_j, V respectively in L^p. Moreover, $\|V''_\delta\|_{L^p} \leq C_0 \varepsilon$ and $\|V'_\delta\|_{L^\infty} \leq C_\varepsilon$, etc., where V' and V'' are introduced in

11) Recall that $X = \mathbb{R}^d / (\bar{\gamma}_1 \mathbb{Z} \times \cdots \times \bar{\gamma}_d \mathbb{Z})$.

(ii). Let u_δ be a sequence of solutions of problem (9.2.5) with the operator A_δ instead of A. Then the arguments of (ii) and the equalities $\|u_\delta\| = \|f\|$, $\|A_\delta u_\delta\| = \|A_\delta f\|$ $\forall t$ yield that for all $\delta > 0$ small enough,

$$\|D_t u_\delta\| + \|u_\delta\|_{H^1} \le C(\|Af\| + \|f\|) \qquad \forall t \in \mathbb{R}$$

where the constant C does not depend on δ. Then standard compactness arguments yield that one can choose a subsequence u_δ tending to a function u in $C(\mathbb{R}, L^2(X, \mathbb{H}))$ which lies in the required solution space and, moreover, $A_\delta u_\delta \to Au$ in $\mathcal{D}'(\mathbb{R} \times X)$ and therefore u is a solution of (9.2.5). The symmetry of A yields that this solution is unique and

$$\|u\| = \|f\|, \; \|Au\| = \|Af\| \qquad \forall t \in \mathbb{R}. \tag{9.2.6}$$

Now let $\max_j \bar{\gamma}_j = \infty$. Let us first take f supported in the domain $|x'| \le R$. Let us replace $\bar{\gamma}_j = \infty$ by $\bar{\gamma}_j = CR^2$ and let us solve problem (9.2.5). Then (4.5.7) yields that the propagation velocity for operators A_δ does not exceed C_0 and therefore the same is true for the operator A. Then u is supported in $\{|x'| \le \frac{1}{2}CR^2\}$ for $|t| \le R$ and sufficiently large R; moreover, $u(.,t)$ does not depend on $R \ge |t|$. Letting R tend to $+\infty$ we derive that problem (9.2.5) has a solution in the required space for every compactly supported $f \in \mathrm{D}(A)$ and the propagation velocity does not exceed C_0 and estimate (9.2.6) holds. But then the compactness condition is not necessary.

Now theorem 9.2.1 is due to the following proposition. □

Proposition 9.2.2. *Let A be a symmetric closed operator with dense domain in the Hilbert space* $\mathbf{H}$ *and let $U(t)$ be a group of unitary operators in* $\mathbf{H}$ *such that for $f \in \mathrm{D}(A)$*
(a) $U(t)f \in \mathrm{D}(A)$ $\forall t \in \mathbb{R}$ and $\|AU(t)f\| \le M_f e^{ct}$ with some constant $M = M_f$;
(b) $U(t)f \to f$ as $t \to 0$;
(c) $D_t U(t)f = AU(t)f$ in the weak sense.
Then A is a self-adjoint generator of $U(t)$.

Proof. Let us set

$$R_\pm = \mp i \int_0^\infty U(\pm t) e^{-\lambda t} dt$$

with $\lambda > c$. Obviously, the operator norm of $R_\pm$ in $\mathbf{H}$ does not exceed λ^{-1} and for $f \in \mathrm{D}(A)$ $R_\pm f \in \mathrm{D}(A)$ and $(A \pm i\lambda I)R_\pm f = f$. Then $R_\pm f \in \mathrm{D}(A)$ $\forall f \in \mathbf{H}$ and both defect indices vanish; here we used that A is closed and $\mathrm{D}(A)$ is dense. Therefore A is self-adjoint. But then a generator of $U(t)$ coincides with A. □

Remark 9.2.3. *(i) Obviously, the symmetric operator A is defined under conditions weaker than in theorem 9.2.1. For example, one can take $\mathrm{D}(A) = C_0^1(X)$ provided ω^{jl}, ω^l, $V_j, V \in L^2_{\mathrm{loc}}(X)$. Moreover, if Y is a measure zero closed subset of X and $\omega^{jl}, \omega^l, V_j, V \in L^2_{\mathrm{loc}}(X \setminus Y)$ than one can take $\mathrm{D}(A) = C_0^1(X \setminus Y)$. Obviously, A is a symmetric operator with domain dense in $L^2(X, \mathbb{H})$.*

(ii) Moreover, if $V_j = 0$ and either d is odd or d is even and $m = 0$ then A has a self-adjoint extension. In fact, there exists a unitary matrix β such that $\beta^\dagger \bar{\sigma}_j \beta = \sigma_j$ with $j = 0, \dots, d$ in the first case and with $j = 1, \dots, d$ in the second case where the bar means complex conjugation. But then $\beta^\dagger \bar{A} \beta = A$ and therefore both defect indices of A coincide.

(iii) Similarly, if $V = 0$, $m = 0$, then $\beta^\dagger A \beta = -A$ with $\beta = \sigma_0$ and therefore both defect indices of A again coincide. On the other hand, if $V = 0$ and $m > 0$ then the inequality

$$\|Au\| \ge m\|u\| \qquad \forall u \in \mathrm{D}(A) \tag{9.2.7}$$

holds.[12)] *Therefore both defect indices of A coincide in this case also; moreover, there exists a self-adjoint extension of A satisfying (9.2.7).*

Thus, if $V = 0$ then there exists a self-adjoint extension of A satisfying (9.2.7).

(iv) For $d = 2$ let us consider an operator A of the form

$$A_1 = \sigma_1(D_r + \frac{i}{2}r^{-1}) + \sigma_2 r^{-1}(D_\phi + \alpha) + \sigma_0 m \tag{9.2.8}$$

in polar coordinates. Let us set $\mathrm{D}(A) = \{u \in H^1(\mathbb{R}^2, \mathbb{H}), r^{-1}u \in L^2(\mathbb{R}^2, \mathbb{H})\}$. *It is easy to see that A is self-adjoint for $\alpha \notin \mathbb{Z}$ and that*

$$q^\dagger(\sigma_1 \cos\phi - \sigma_2 \sin\phi)q = \sigma_1, \quad q^\dagger(\sigma_1 \sin\phi + \sigma_2 \cos\phi)q = \sigma_2, \quad q^\dagger \sigma_0 q = \sigma_0$$

for the unitary matrix-valued function $q = \mathrm{diag}(e^{i\zeta\phi}, 1)$ *and σ_0, σ_1, σ_2 given by (C5). Moreover,*

$$A_2 = q^\dagger A_1 q = \sigma_1(D_1 - \alpha x_2|x|^{-2}) + \sigma_2(D_2 + \alpha x_1|x|^{-2}) + \sigma_0 m. \tag{9.2.9}$$

Then for $\alpha \notin \mathbb{Z}$ the operator A_2 (with the same domain as A_1) is self-adjoint. Moreover, $F = 0$ for $x \neq 0$ but for $\alpha \notin \mathbb{Z}$ one can remove the vector potential by means the of a gradient transform only in a subdomain in $\mathbb{R}^2 \setminus 0$ which contains no closed contour going around 0. Further, one can easily see that $A_2(\alpha')$ and $A_2(\alpha)$ are unitarily equivalent if and only if $\alpha' - \alpha \in \mathbb{Z}$.

(v) Other examples of self-adjoint Dirac operators will be given in chapter 11. ◊

9.2.2 Abstract theory

Let us now consider the abstract theory.

Proposition 9.2.4. *Let A be a self-adjoint operator and J_k be symmetric operators in the separable Hilbert space* $\mathbf{H}$. *Let* $\mathrm{D}(J_k) \supset \mathrm{D}(A)$, *let the operator $J_1(A+iI)^{-1}$ be compact and suppose the A-bound of J_2 is less than 1. Then*

(i) The A-bound of J_1 equals 0 and therefore the operators $A + J_k$ are self-adjoint.

(ii) The operator $J_1(A + J_2 + iI)^{-1}$ is compact.

(iii) $\mathrm{Spec}_{\mathrm{ess}}(A + J_1) = \mathrm{Spec}_{\mathrm{ess}}(A)$.

Proof. The compactness of the operator $J_1(A + iI)^{-1}$ yields that for every $\epsilon > 0$ there exists a finite-dimensional operator $K = K_\epsilon$ such that $\|J_1(A + iI)^{-1} - K\| \leq \epsilon$. This operator is of the form

$$Ku = \sum_{1 \leq j \leq N} f_j \langle u, g_j \rangle$$

and without loss of generality one can assume that $g_j \in \mathrm{D}(A)$ (otherwise they can be replaced by nearby elements of $\mathrm{D}(A)$). But then $K(A + iI)$ is a bounded operator and therefore

$$\|J_1 u\| \leq \|K(A + iI)u\| + \epsilon\|(A + iI)u\| \leq \epsilon\|Au\| + C'_\epsilon\|u\| \qquad \forall u \in \mathrm{D}(A)$$

and therefore the A-bound of J_1 equals 0. Assertion (i) is proven.

In order to prove (ii) let us note that

$$J_1(A + iI)^{-1} - J_1(A + J_2 + iI)^{-1} = J_1(A + iI)^{-1} \cdot J_2(A + J_2 + iI)^{-1}$$

---

12) This is due to the equality $\|Au\|^2 = \|(A - m\sigma_0)u\|^2 + m^2\|u\|^2$.

with compact operator $J_1(A+iI)^{-1}$ and bounded operator $J_2(A+J_2+iI)^{-1}$ (because the A-bound of J_2 is less than 1).

Let us prove that if $\tau \in \mathrm{Spec}_{\mathrm{ess}}(A)$ then $\tau \in \mathrm{Spec}_{\mathrm{ess}}(A+J_1)$; this assertion and (i), (ii) with $J_2 = J_1$ yield the converse implication. If $\tau \in \mathrm{Spec}_{\mathrm{ess}}(A)$ then there exists a sequence $u_n \in \mathrm{D}(A)$ such that $\|u_n\| = 1$, $u_n \overset{w}{\to} 0$ and $\|(A-\tau I)u_n\| \to 0$. Let $v_n = (A+iI)u_n$. Then $1 \le \|v_n\| \le M$ obviously and $\langle v_n, f\rangle = \langle u_n, (A-iI)f\rangle \to 0$ for $f \in \mathrm{D}(A)$. Therefore $v_n \overset{w}{\to} 0$ and then $J_1 u_n = J_1(A+iI)^{-1}v_n \overset{s}{\to} 0$ because the operator $J_1(A+iI)^{-1}$ is compact. Hence $\|(A+J_1-\tau I)u_n\| \to 0$ and therefore $\tau \in \mathrm{Spec}_{\mathrm{ess}}(A+J_1)$. □

Proposition 9.2.5. *Let A be a self-adjoint operator and let J be a symmetric operator in the separable Hilbert space $\mathbf{H}$. Let $\mathrm{D}(J) \supset \mathrm{D}(A)$ and let J be A-bounded (i.e. $\|Ju\| \le C\|Au\| + C\|u\| \quad \forall u \in \mathrm{D}(A)$). Then*

(i) If $\mathrm{dist}(\tau_0, \mathrm{Spec}\, A) \ge \epsilon > 0$ then $\mathrm{dist}(\tau_0, \mathrm{Spec}(A+tJ)) \ge \frac{1}{2}\epsilon$ for $|t| \le \delta = \delta(\epsilon) > 0$.

(ii) If $\mathrm{dist}(\tau_0, \mathrm{Spec}\, A \setminus \tau_0) \ge \epsilon > 0$ then $\mathrm{Spec}(A+tJ) \cap [\tau_0 - \frac{1}{2}\epsilon, \tau_0 + \frac{1}{2}\epsilon] \subset [\tau_0 - C|t|, \tau_0 + C|t|]$ for $|t| \le \delta = \delta(\epsilon) > 0$.

(iii) Moreover, in the context of (ii) the spectral projector of the operator $A+tJ$ corresponding to the interval $[\tau_0 - \frac{1}{2}\epsilon, \tau_0 + \frac{1}{2}\epsilon]$ depends analytically on t.

(iv) In particular, if τ_0 is an eigenvalue of A with multiplicity $n < \infty$ then for $|t| \le \delta$ $\mathrm{Spec}(A+tJ) \cap [\tau_0 - \frac{1}{2}\epsilon, \tau_0 + \frac{1}{2}\epsilon] = \{\tau_1(t), \dots, \tau_n(t)\}$ where $\tau_k(t)$ are Lipschitz functions and the number of eigenvalues of $A+tJ$ in the interval $[\tau_0 - \frac{1}{2}\epsilon, \tau_0 + \frac{1}{2}\epsilon]$ (counting multiplicities) is equal to n.

Proof. Assertions (i) and (ii) follow from the inequality

$$\|(A-\tau I)u\| \ge \mathrm{dist}(\tau, \mathrm{Spec}\, A)\|u\| \qquad \forall u \in \mathrm{D}(A)$$

for self-adjoint operators. Rewriting the spectral projector in question in the form

$$\Pi(t) = \frac{1}{2\pi i}\int_\Gamma (z - A - tJ)^{-1} dz$$

where $\Gamma = \{z \in \mathbb{C}, |z - \tau_0| = \frac{3}{4}\epsilon\}$ is a counter-clockwise oriented contour we obtain (iii) and the assertion concerning the number of eigenvalues in (iv). Moreover, $\mathrm{Spec}_{\mathrm{ess}}(A+tJ) \cap [\tau_0 - \frac{1}{2}\epsilon, \tau_0 + \frac{1}{2}\epsilon] = \emptyset$. The Lipschitz property of τ_k follows from the arguments of the proof of (i), (ii). □

Proposition 9.2.6. *Let the conditions of proposition 9.2.5(iv) be fulfilled and let*

$$\langle Jv, v\rangle \ge \epsilon\|v\|^2 \qquad \forall u \in \mathrm{Ker}(A - \tau_0 I). \tag{9.2.10}$$

Then

$$\tau_k(t') \ge \tau_k(t) + \delta_1(t'-t) \qquad \forall k \quad \forall t, t' \in [-\delta, \delta] : t' > t \tag{9.2.11}$$

with $\delta_1 = \delta_1(\epsilon) > 0$.

Proof. Obviously it is sufficient to prove (9.2.11) under the additional condition $t' < t + \delta_2(t)$ with $\delta_2(t) > 0$. Let $t = \bar{t}$ and let $\bar{\tau}$ be an eigenvalue of $A + \bar{t}J$ of multiplicity r; let $\bar{\Pi}$ be the corresponding spectral projector. Let τ' be an eigenvalue of $A + t'J$ close to $\bar{\tau}$ and $|t' - \bar{t}|$ be small enough; then proposition 9.2.5(iii) yields that $\|(I - \bar{\Pi})u\| \le M|t' - \bar{t}| \cdot \|u\|$ with $M = M(\bar{t})$. We know that $|\tau' - \bar{\tau}| \le C|t' - \bar{t}|$. These inequalities, the equalities $(A + t'J - \tau' I)v = 0$, $(A + \bar{t}J - \bar{\tau}I)\bar{\Pi}v = 0$ and the A-boundedness of the operator J yield that $\|A(I - \bar{\Pi})v\| \le M|t' - \bar{t}| \cdot \|v\|$ for $v \in \mathrm{Ker}(A + t'J - \tau' I)$. Then the same inequality holds for J instead of A (we increase M if necessary). These inequalities and the equality $(A + \bar{t}J - \bar{\tau}I)\bar{\Pi} = 0$ yield that

$$\left|(t'-\bar{t})\langle Jv,v\rangle + (\tau'-\bar{\tau})\|v\|^2\right| = |\langle (A+\bar{t}J-\bar{\tau}I)v,v\rangle| \le M|t'-t|^2\|v\|^2.$$

On the other hand, inequality (9.2.10) remains true (with $\frac{1}{2}\epsilon$ instead of ϵ) for $u \in \operatorname{Ker}(A + t'J - \tau'I)$ also with $|t'| \le \delta$, $|\tau' - \bar{\tau}| \le \delta$. These inequalities and proposition 9.2.5(iii) yield that for $t' \in [\bar{t}, \bar{t}+\delta_2(\bar{t})]$ $\operatorname{Spec}(A+t'J)\cap[\tau_0-\delta_3, \tau_0+\delta_1(t'-\bar{t})] = \emptyset$ where $\delta_3 = \delta_3(\bar{t}) > 0$ and $\delta_1 > 0$ does not depend on $\bar{t}$. So (9.2.11) is proven for $t' < \bar{t} + \delta_2(\bar{t})$. □

Proposition 9.2.7 (The Birman-Schwinger Principle). *Let A be self-adjoint operator and let J be a symmetric operator in the separable Hilbert space $\mathbf{H}$. Let $\mathrm{D}(J) \supset \mathrm{D}(A)$ and $J(A+iI)^{-1}$ be a compact operator. Let us assume that $\bar{\tau}_1, \bar{\tau}_2 \in \mathbb{R}$, $\bar{\tau}_1 < \bar{\tau}_2$, $\bar{t} > 0$ and*

$$(9.2.12)\qquad (A+tJ-\bar{\tau}_i I)u = 0 \implies \varkappa_i\langle Ju,u\rangle \ge \epsilon\|u\|^2 \qquad \forall i=1,2 \quad \forall t\in[0,\bar{t}]$$

with $\varkappa_i = \pm 1$, $\epsilon > 0$. Then

(i) The following equality holds:

$$(9.2.13)\quad \mathbf{N}\big(A;[\bar{\tau}_1,\bar{\tau}_2]\big) = \mathbf{N}\big(A+\bar{t}J;[\bar{\tau}_1,\bar{\tau}_2]\big) + \varkappa_2\tilde{\mathbf{N}}\big((A-\bar{\tau}_2 I),-J;[0,\bar{t}]\big) - \varkappa_1\tilde{\mathbf{N}}\big((A-\bar{\tau}_1 I),-J;[0,\bar{t}]\big)$$

where $\tilde{\mathbf{N}}(B,-J;\mathcal{J})$ is the number of eigenvalues $t\in\mathcal{J}$ of the problem

$$(9.2.14)\qquad (B+tJ)u = 0 \qquad u\in \mathrm{D}(B)\setminus 0$$

counting their multiplicities;[13] *on the other hand, $\mathbf{N}(B,\mathcal{J})$ is the cardinality of the spectrum of B in $\mathcal{J}$ and $\mathbf{N}(B,\mathcal{J}) = \infty$ if $\operatorname{Spec}_{\mathrm{ess}}(B)\cap\mathcal{J}\ne\emptyset$.*

(ii) Moreover, (9.2.13) *remains true if one replaces the interval $\mathcal{J}$ in the left-hand and right-hand expressions:* one removes *$\bar{\tau}_1$ and (or) $\bar{\tau}_2$ from $\mathcal{J}$* and replaces $[0,\bar{t}]$ *by $(0,\bar{t})$ in the third and (or) second terms in the right-hand expression; however, in order to avoid complications with the domain of u, one should assume that $\bar{\tau}_i \notin \operatorname{Spec}_{\mathrm{ess}}(A)$.*

(iii) Moreover, $\tilde{\mathbf{N}}(A-\bar{\tau}_i I,-J;[0,\bar{t}]) < \infty$ provided $\bar{\tau}_i\notin\operatorname{Spec}_{\mathrm{ess}}(A)$.

Proof. If $\operatorname{Spec}_{\mathrm{ess}}(A)\cap[\bar{\tau}_1,\bar{\tau}_2]\ne\emptyset$ then $\operatorname{Spec}_{\mathrm{ess}}(A+tJ)\cap[\bar{\tau}_1,\bar{\tau}_2]\ne\emptyset$ because of proposition 9.2.4(iii) and both sides of (9.2.13) are infinite (it will follow from assertion (iii) that there is no indefiniteness in the right-hand expression). On the other hand, if $\operatorname{Spec}_{\mathrm{ess}}(A)\cap[\bar{\tau}_1,\bar{\tau}_2]=\emptyset$ then the previous propositions yield that

$$\operatorname{Spec}(A+tJ)\cap[\bar{\tau}_1,\bar{\tau}_2] = \{\tau_k(t), t\in[t'_k,t''_k], k=0,\dots,N\}$$

where $N\le\infty$, $0\le t'_k < t''_k \le \bar{t}$, $\tau_k(t)$ are Lipschitz functions on the intervals $(\tilde{t}'_k,\tilde{t}''_k)\supset[t'_k,t''_k]$ and $t\in(\tilde{t}'_k,\tilde{t}''_k)$, $\tau_k(t)\in[\bar{\tau}_1,\bar{\tau}_2]$ if and only if $t\in[t'_k,t''_k]$; moreover, $\varkappa_i\tau_k(t)$ is monotone increasing function of t in a neighborhood of $t_{ik}\in[t'_k,t''_k]$ with $\tau_k(t_{ik}) = \bar{\tau}_i$. Then the various terms in (9.2.13) are the numbers of graphs $\{\tau=\tau_k(t)\}$ intersecting respectively the upper, lower, right and left sides of the rectangle $\{0\le t\le\bar{t},\ \bar{\tau}_1\le\tau\le\bar{\tau}_1\}$ taken with the correct signs and (9.2.13), and the modifications in (ii) follow from obvious geometric arguments.

Let us prove (iii). If $\tilde{\mathbf{N}}(A-\bar{\tau}I,-J;[0,\bar{t}]) = \infty$ then there would exist sequences $t_n\in[0,\bar{t}]$ and $u_n\in\mathrm{D}(A)$ such that $(A-\bar{\tau}I+t_nJ)u_n = 0$ and $\langle Ju_n,u_m\rangle = \delta_{nm}$. Then $\|u_n\|\le C$ by (9.2.12) and then the equality $(A+t_nJ-\bar{\tau}I)u_n = 0$ and the fact that the A-bound of J is 0 imply that $\|Au_n\|\le C_1$. Then the compactness of the operator $J(A+iI)^{-1}$ yields that the sequence Ju_n is strongly convergent (we take a subsequence if necessary) but this contradicts the equality $\langle Ju_n,u_m\rangle = \delta_{nm}$. □

[13] I.e., the dimensions of the subspaces of $\mathbf{H}$ on which (9.2.14) holds with given t; in this definition neither continuous nor fancier types of spectrum are discussed, just the bare dimension of the null space.

Remark 9.2.8. *Let A be a self-adjoint operator and J be a symmetric operator in the separable Hilbert space* $\mathbf{H}$. *Let* $\mathrm{D}(J) \supset \mathrm{D}(A)$ *and* $J(A+iI)^{-1}$ *be a compact operator. Moreover, let us assume that* $J \geq 0$, $\operatorname{Ker} J = 0$ *and*

$$\|(A-\bar{\tau}_i I)u\| + \|Ju\| \geq \epsilon_1 \|u\| \qquad \forall u \in \mathrm{D}(A') \tag{9.2.15}$$

with $\epsilon_1 > 0$. Then (9.2.12) holds. In fact, otherwise there exist sequences $t_n \in [0, \bar{t}]$ and $u_n \in \mathrm{D}(A)$ such that $Au_n = -t_n J u_n$ (we assume for the sake of simplicity that $\bar{\tau}_i = 0$), $\|u_n\| = 1$, $\langle Ju_n, u_n\rangle \to 0$. Then the estimate $\|Ju\| \leq \delta\|Au\| + C_\delta\|u\|$ (with arbitrarily chosen $\delta > 0$) yields that $\|Au_n\| \leq C$ and since $\|u\|_n = 1$ we obtain that the sequence Ju_n is strongly convergent (we take a subsequence if necessary). Then the sequence Au_n is also strongly convergent. Then (9.2.15) yields that the sequence u_n is also strongly convergent (because a Cauchy sequence is a converging sequence and conversely). Let $u = \lim u_n$. Then $u \in \mathrm{D}(A)$, $\|u\| = 1$ and $\langle Ju, u\rangle = \lim\langle Ju_n, u_n\rangle = 0$; this contradicts the conditions $J \geq 0$ and $\operatorname{Ker} J = 0$. ◊

Proposition 9.2.9. *Let the conditions of remark 9.2.8 be fulfilled. Let us replace condition* (9.2.15) *by the condition*

$$\|A'u\| + \|J^{\frac{1}{2}}u\| \geq \epsilon_1\|u\| \qquad \forall u \in \mathrm{D}(A) \tag{9.2.15*}$$

with $A' = A - \bar{\tau}_i I$; one can easily see that this condition is stronger. Then for small enough $\delta > 0$ the operators $(A' \pm i\delta J)^{-1}$, $A'(A' \pm i\delta J)^{-1}$, $J(A' \pm i\delta J)^{-1}$ are bounded (their norms depend on δ) and therefore the operator $J(A' \pm i\delta J)^{-1}$ is compact. Moreover, one can take arbitrarily large $\delta > 0$ provided J is a bounded operator.

Proof. Without loss of generality one can assume that $\bar{\tau}_i = 0$. If $(A + i\delta J)u = f$ then

$$\|J^{\frac{1}{2}}u\|^2 \leq \delta^{-1}\|u\| \cdot \|f\|, \quad \|Au\| \leq \delta\|Ju\| + \|f\| \leq C\delta(\|Au\| + \|u\|) + \|f\|$$

and hence $\|u\| + \|Au\| \leq C_\delta\|f\|$. This inequality yields the boundedness of the indicated operators. Moreover, if J is a bounded operator then $\|Ju\| \leq C\|J^{\frac{1}{2}}u\|$ and therefore $\|Au\| \leq C\delta\|J^{\frac{1}{2}}u\| + \|f\|$; this inequality combined with the inequalities $\|J^{\frac{1}{2}}u\|^2 \leq \delta^{-1}\|u\| \cdot \|f\|$ and (9.2.15)* yield that $\|u\| + \|Au\| \leq C_\delta\|f\|$ for arbitrary $\delta > 0$. □

Proposition 9.2.10. *Let the conditions of remark 9.2.8 be fulfilled and let J be a bounded operator (one can easily check that conditions* (9.2.15) *and* (9.2.15)* *are equivalent in this case). Then*

(i) The operator $B = J^{-\frac{1}{2}}A'J^{-\frac{1}{2}}$ with domain $\mathrm{D}(B) = \{u \in \mathrm{D}(J^{-\frac{1}{2}}) : J^{-\frac{1}{2}}u \in$ $\mathrm{D}(A')$ and $A'J^{-\frac{1}{2}}u \in \mathrm{D}(J^{-\frac{1}{2}})\}$ *is self-adjoint; here and below $A' = A - \bar{\tau}_i I$.*

(ii) The operators $(B \pm iI)^{-1}$ are compact.

(iii) The eigenvalues of the operator $-B$ and of the problem

$$(A - \bar{\tau}_i I)u + tJu = 0 \qquad u \in \mathrm{D}(A) \tag{9.2.16}$$

coincide (counting multiplicities) and

$$\operatorname{Ker}(B + tJ) = J^{\frac{1}{2}} \operatorname{Ker}(A' + tJ). \tag{9.2.17}$$

Proof. The proof is obvious. □

9.2.3 Proof of theorem 9.1.3

Thus *theorem 9.1.3 has been proven under the additional condition* (9.2.15) *which should be checked for operators* $A'_k = A - \tau_i I - t_1 J_1 - \cdots - t_{k-1} J_{k-1}$ *and* $J = J_k$ *for every* $k = 1, \dots$. In fact, by the construction $J_k \ge 0$ are bounded[14] operators with $\operatorname{Ker} J_k = 0$ and therefore proposition 9.2.10 and (9.2.13) yield

$$(9.2.13)^* \quad \mathbf{N}(A; [\bar\tau_1, \bar\tau_2]) = \mathbf{N}(A + \bar t J; [\bar\tau_1, \bar\tau_2]) + \varkappa_2 \mathbf{N}(-J^{-\frac12}(A - \bar\tau_2 I)J^{-\frac12}; [0, \bar t]) - \varkappa_1 \mathbf{N}(J^{-\frac12}(A - \bar\tau_1 I)J^{-\frac12}; [0, \bar t]).$$

Then the arguments of section 9.1 give us the statement of theorem 9.1.3. One can easily check that condition (9.2.15) is surely fulfilled provided

$$(9.2.18) \qquad \tau_1, \tau_2 \notin \operatorname{Spec}_{\mathrm{ess}}(A).$$

Moreover, proposition 9.2.4(iii) yields that in this case condition (9.2.15) is fulfilled for A'_k (see above) and J_k also for every k and *theorem 9.1.3 is proven under the additional condition* (9.2.18).

Let us now consider the general case. Let us replace condition (9.2.18) by the condition

(9.2.19) There exist sequences $\tau_{1,n} \to \tau_1 + 0$ and $\tau_{2,n} \to \tau_2 - 0$ with $\tau_{1,n}, \tau_{2,n} \notin \operatorname{Spec}_{\mathrm{ess}}(A)$;

this condition is due to (9.1.5)*. Under this condition we can apply theorem 9.1.3 to intervals $(\tau_{1,n}, \tau_{2,n})$.

Let us note that $\operatorname{Spec} a^0(x,\xi) = \{\lambda_1(x,\xi), \dots, \lambda_D(x,\xi)\}$ with Lipschitz functions λ_j such that $|\lambda_j(x,\xi)| \to \infty$ as $|\xi| \to \infty$ and therefore

$$(9.2.20) \qquad x \in X'_{(1)+} \cap X'_{(2)+} \implies \operatorname{Spec} a^0(x,\xi) \cap [\tau_1, \tau_2] = \emptyset.$$

On the other hand, it is easy to check (e.g., see the next subsection) that if $R_{(1)1} + R_{(2)1} < \infty$ (otherwise the statement of theorem 9.1.3 is empty) then $X'_{(1)-}$ and $X'_{(2)-}$ are compact sets. Therefore all the conditions in $X'_{(i)-}$ (i.e., the microhyperbolicity condition, etc.,) remain true (with $\bar\gamma' = \frac12 \bar\gamma$ and the same ρ ...) for $\tau_{i,n}$ instead of τ_i provided n is large enough. Moreover, the ellipticity condition which defines $X'_{(i)+}$ remains true (with $\bar\gamma' = \frac12 \bar\gamma$ and $\rho' \ge \rho$ and appropriately changed other parameters) for τ'_i instead of τ_i provided $|\tau'_i - \tau_i|$ is small enough. Therefore every term of the remainder estimate in the zone X' for the interval $(\tau_{1,n}, \tau_{2,n})$ tend to the corresponding term of the remainder estimate for the interval (τ_1, τ_2) and $\mathcal{N}(\tau_{1,n}, \tau_{2,n})$ tends to $\mathcal{N}(\tau_1, \tau_2)$ (even if it is infinite). Moreover, we can take the same R_0 for the interval $(\tau_{1,n}, \tau_{2,n})$ as for the original interval (τ_1, τ_2) (see conditions (9.1.27)* and (9.1.28), (9.1.29)). *So theorem 9.1.3 has been proven.* □

9.2.4 Proof of theorems 9.1.4, 9.1.5 (beginning)

Let us return to the Dirac operator. In order to prove theorems 9.1.4 and 9.1.5 one should derive some estimates of (9.1.27)- or (9.1.29)-type and construct the admissible matrix functions J_k.

Let us first consider the Dirac operator in the context of theorem 9.1.4. Let us note first of all that if $R_1 < \infty$ (otherwise all the assertions of this theorem are void) then X'_- are compact sets and on these sets ρ and γ are bounded and disjoint from 0. In fact,

[14] It is easy to see that otherwise $R_1 = \infty$ and then the statement of the theorem is empty.

otherwise there exists a family of non-intersecting balls $B(x^n, \gamma(x^n))$ with $n = 1, 2, \dots$ lying in X'_+, the contribution of each of which to R_1 is larger than $\epsilon > 0$. On the other hand, the definition of X'_+ immediately yields that $\rho \le c'm$ on this set. So we can assume that

(9.2.21) $$\rho \le M,\ \gamma \ge \epsilon \qquad \text{at } X'_+,$$

(9.2.22) $$\varepsilon \le \rho^i \le M,\ \epsilon \le \gamma^i \le M \qquad \text{at } X'_-.$$

Now we need to prove the A-compactness property.

Proposition 9.2.11. *Let all the conditions of theorem 9.1.4 associated with the zone X' be fulfilled. Let us assume that the operator A is self-adjoint. Moreover, let $\rho \le M$ at X' and $J \in L^\infty(X)$ be supported in X'. Let*

(9.2.23) $$\forall \delta > 0 \qquad \operatorname{mes}\{x, |J(x)| > \delta\} < \infty.$$

Then the operator $J(A + iI)^{-1}$ is compact in $L^2(X, \mathbb{H})$.

Proof. The results of section 4.5 and the boundedness of ρ yield the estimate

(9.2.24) $$|e(x, x, 0, \tau)| \le M(|\tau|^d + 1) \qquad \forall x \in X' \quad \forall \tau \in \mathbb{R}$$

where the constant M is increased if necessary. Hence if $E(\tau)$ is the spectral projector of A and $J \in L^1 \cap L^\infty$ then $J^{\frac12}\bigl(E(\tau) - E(0)\bigr)J^{\frac12}$ is a trace operator with trace norm not exceeding $M(|\tau|^d + 1)$. This operator is of the form LL^* and therefore $L = J^{\frac12}\bigl(E(\tau) - E(0)\bigr)$ is a Hilbert-Schmidt operator with Hilbert-Schmidt norm not exceeding $M(|\tau|^d + 1)$. Then the equalities

$$(A + iI)^{-p} = \int_{-\infty}^{\infty} (\tau + i)^{-p} d_\tau E(\tau) = \int_{-\infty}^{\infty} p(\tau + i)^{-p-1}\bigl(E(\tau) - E(0)\bigr)d\tau$$

yield that for $p > d$ the operator $J(A + iI)^{-p}$ is also a Hilbert-Schmidt operator. Let us note that for $\varepsilon > 0$

$$J(A + iI)^{-1} = J(A + iI)^{-1}(I + i\varepsilon A)^{-d} + JB_\varepsilon$$

with

$$B_\varepsilon = (A + iI)^{-1}(I + i\varepsilon A)^{-d}((I + i\varepsilon A)^d - I);$$

then obviously the first term in the right-hand expression of the first equality is a Hilbert-Schmidt operator and B_ε is a product of $\varepsilon A(A + iI)^{-1}$ and an operator with uniformly bounded operator norm; hence the operator norm of B_ε does not exceed $C\varepsilon$. Therefore $J(A + iI)^{-1}$ is approximated in the operator norm by compact operators and therefore is also compact. *Thus, for $J \in L^1 \cap L^\infty$ the proposition is proven.*

In the general case, for every $\varepsilon > 0$ there exists a representation $J = J' + J''$ with $J' \in L^1 \cap L^\infty$ and $\|J''\|_{L^\infty} \le \varepsilon$. Then $J'(A + iI)^{-1}$ is a compact operator and $\|J''(A + iI)^{-1}\| \le \varepsilon$ and so $J(A + iI)^{-1}$ is approximated by compact operators and so is also compact. □

Proposition 9.2.12. *Let all the conditions of theorem 9.1.4 associated with the zone X'' be fulfilled. Let $R_3 < \infty$ and $\gamma \ge \epsilon$ on $X' \cap X''$. Let $J \in L^p(X)$ with $p = d$ for $d \ge 3$ and with $p > 2$ for $d = 2$.*[15] *Suppose* $\operatorname{supp} J \subset X''$ *and* $\operatorname{dist}(\operatorname{supp} J, \partial X'' \setminus X') \ge \epsilon$. *Then the operator $J(A + iI)^{-1}$ is compact.*

[15] Let us recall that for $d = 2$ the condition $R_3 < \infty$ also means that $\operatorname{mes} X'' < \infty$.

Proof. The conditions of the proposition yield that there exists a function ϕ supported in the $\frac{2}{3}\epsilon$-neighborhood of $\operatorname{supp} J$ and equal 1 in the $\frac{1}{3}\epsilon$-neighborhood of $\operatorname{supp} J$ such that $0 \le \phi \le 1$ and $|D^\alpha \phi| \le C \quad \forall \alpha : |\alpha| \le K$.

Let us consider the equation $(A + iI)v = f \in L^2(X, \mathbb{H})$ with v supported in X''. Let us apply the arguments used in the proof of theorem 9.2.1, and the condition $R_3 < \infty$. It is easy to prove that if $v \in H^1$ then

$$\|Dv\| \le M(\|f\| + \|v\|). \tag{9.2.25}$$

Let us consider $v \in H^1_{\text{loc}} \cap L^2(X, \mathbb{H})$ and take a function ψ with $|\psi| + |\nabla\psi| \le C$. Then $(A+iI)\psi v = \psi f + [A, \psi]v$ and (9.2.25) yields the estimate $\|D\psi v\| \le M(\|f\| + \|v\|)$. Since this estimate holds for every ψ above we obtain that $v \in H^1$ and estimate (9.2.25) holds. Similarly, if $u \in \mathrm{D}(A)$ and $(A + iI)u = f$ then

$$\|\phi u\|_{H^1} \le M(\|f\| + \|u\|) \le (M + 1)\|f\|$$

and therefore the operator $\phi(A+iI)^{-1} : L^2(X, \mathbb{H}) \to H^1(X, \mathbb{H})$ is bounded. On the other hand, it is well known that if $J \in L^p$ with p indicated above (and mes $X'' < \infty$ for $d = 2$) then the operator $J : H^1(X'', \mathbb{H}) \to L^2(X, \mathbb{H})$ is compact. □

This proposition obviously yields

Corollary 9.2.13. *The assertion of proposition 9.2.12 remains true if the condition $J \in L^p$ is replaced by condition* (9.2.23).

9.2.5 Proof of theorems 9.1.4, 9.1.5 (construction of J_k)

It is easy to see that there exists a function ζ' on X' such that $\zeta'(x) > 0$, $c^{-1} \le \zeta'(x)/\zeta'(y) \le c$ for $y \in B(x, \epsilon_1\gamma(x))$, $x \in X'''$ where X''' is a domain intermediate between X' and $X' \setminus X''$, $\zeta' \asymp \min(\rho, \frac{1}{m}\rho^2)$. Let us set $\zeta = \zeta'$ in X' and $\zeta = \eta\zeta''$ in $X'' \setminus X'$ where $\zeta'' \in \mathcal{S}(X)$ is a fixed function, $\zeta''(x) > 0$ and $\mathbb{R} \ni \eta \to +0$ in what follows. Let us note that $\zeta \in L^\infty$ provided $R_1 < \infty$.

Let us set $J_1 = \begin{pmatrix} \zeta I & 0 \\ 0 & \epsilon_2\zeta I \end{pmatrix}$, $J_2 = \begin{pmatrix} \epsilon_2\zeta I & 0 \\ 0 & \zeta I \end{pmatrix}$ and let us take ϕ_1 supported in $[c_0, 2c_0]$ and ϕ_2 supported in $[-2c_0, -c_0]$, where $\epsilon_2 > 0$ is small enough and c_0 is large enough, such that

$$\begin{aligned} m - V - \tau_2 + t_1 j + \epsilon_2 t_2 \zeta &\ge \frac{\epsilon}{2}\min(\rho, \frac{1}{m}\rho^2), \\ -m + V - \tau_1 + t_1\epsilon_2 j + t_2\zeta &\le -\frac{\epsilon}{2}\min(\rho, \frac{1}{m}\rho^2) \\ \text{in } X' \qquad &\forall t_k \in \operatorname{supp}\phi_k. \end{aligned} \tag{9.2.26}$$

Let us apply the arguments of section 9.1. Then in order to estimate $\mathbf{N}(A, \tau_1, \tau_2)$ from below we need to estimate R_0 in condition (9.1.27); the compactness of the operator $J(A + iI)^{-1}$ obviously follows from $R_1 < \infty$ and proposition 9.2.11. Further, in order to estimate R_0 we need to estimate $\mathbf{N}(\tilde{A} - \tau_i I, -\tilde{J}_1, (-t, t))$ and $\mathbf{N}(\tilde{A} + t\tilde{J}_1 - \tau_i I, -J_2, (-t, t))$ where proposition 9.2.10(iii) yields that we can replace $\tilde{\mathbf{N}}$ by $\mathbf{N}$. Recall that ˜ means that the corresponding operator is extended from X''.[16] Let us consider the first quantity; one can analyze the second quantity in the same way.

[16] Or rather from its connected component, and later we sum the results obtained for these components.

Let us assume that on $X'\cap X''$ (here and below one can replace X'' by a given connected component) $\rho\le c\bar\rho$ with a parameter $\bar\rho=\mathrm{const}$. Let us assume for the sake of simplicity that $\tau_i=0$ (otherwise we replace V by $V-\tau_i$).

We know that $\mathbf{N}(\tilde A,-\tilde J_1;(-t,t))$ for $t\ge 0$ is the maximal dimension of the subspace $\mathcal{L}'\subset C_0^1(X,\mathbb{H})$ on which the inequality

$$\|\tilde J_1^{-\frac12}\tilde A u\|^2\le t^2\|\tilde J_1^{\frac12}u\|^2$$

holds. Therefore we should estimate the maximal dimension of the subspace $\mathcal{L}\subset C_0^1(X,\mathbb{H})$ on which the inequality

$$\|\tilde\zeta^{-\frac12}\tilde A u\|^2\le C_0t^2\|\tilde\zeta^{\frac12}u\|^2 \tag{9.2.27}$$

holds where $\tilde\zeta$ is an admissible continuation of ζ from X^{IV} (which is a domain intermediate between X'' and $X''\setminus X''')$ to $\mathbb{R}^d$ and none of the constants below depend on η.

Let $\tilde A$ be an operator of (4.5.3) type:

$$\tilde A=\tilde A_0+\tilde V,\qquad \tilde A_0=\sum\sigma_l(\omega^{kl}\tilde P_j+\tilde P_k\omega^{kl})+\sigma_0(m+W) \tag{9.2.28}$$

where the coefficients ω^{kl} are the same as in A, $\tilde P_k=D_k-\tilde V_k$, $|\tilde V_k|\le c\bar\rho$, $|\tilde V|\le m+c\min(\rho,\frac1m\rho^2)$ and $\tilde V_k$, $\tilde V$ are supported in X'', $W\ge 0$, $|D^\alpha W|\le c\rho\gamma^{-|\alpha|}\quad\forall\alpha:|\alpha|\le K$, $W\asymp\rho$ for $\rho\ge cm$ outside of X'' and W vanishes in X^{IV} and for $\rho\le\epsilon_1 m$.

For notational simplicity in this subsection we skip the˜ over A,P,V, etc.

Let us consider a partition of unity $1=\sum_{\nu\ge0}\psi_\nu^2$ in X'' where ψ_ν are supported in the $\epsilon\gamma$-vicinity of X' and γ-admissible for $\nu\ge1$, and where ψ_0 is supported in X''. Then the left-hand expression in (9.2.27) is equal to

$$2\|\zeta^{-\frac12}Au\|^2\ge 2\sum_\nu\|\lambda_\nu\psi_\nu Au\|^2\ge$$
$$\sum_\nu]\|\lambda_\nu A\psi_\nu u\|^2-\sum_\nu\|\lambda_\nu[A,\psi_\nu]u\|^2$$

provided

$$\zeta\lambda_\nu^2\le1\qquad\text{on }\operatorname{supp}\psi_\nu. \tag{9.2.29}$$

One can take a partition of unity such that

$$|\nabla\psi_\nu|\le\varPhi(\alpha)\psi_\nu+\alpha\qquad\forall\alpha>0 \tag{9.2.30}$$

with some fixed function $\varPhi(\alpha)$ slowly increasing as $\alpha\to+0$ (more precisely see below) and then

$$\|[A,\psi_\nu]u\|^2\le C\|\varPhi(\alpha_\nu)\vartheta_\nu\psi_\nu u\|^2+C\|\alpha_\nu\vartheta_\nu u\|^2$$

with arbitrary α_ν and one can replace $\vartheta_\nu=[A,\psi_\nu]$ by $\vartheta=\chi_Y\gamma^{-1}$ where Y is the $\epsilon\gamma$-vicinity of X' intersected with X'' and χ_Y is its characteristic function. Therefore

$$\sum_\nu\|\lambda_\nu A\psi_\nu u\|^2\le 2\|\zeta^{-\frac12}Au\|^2+\sum_{\nu\ge1}C\big(\lambda_\nu(1+\alpha_\nu)+$$
$$\lambda_0\varPhi(\alpha_0)\big)^2\|\chi_Y\gamma^{-1}u_\nu\|^2+\lambda_0^2(1+\alpha_0)^2\|\chi_Y\gamma^{-1}u_0\|^2+\|\lambda\varPhi(\alpha)\chi_Y\gamma^{-1}u_0\|^2$$

provided

(9.2.31)
$$\lambda \ge \lambda_\nu, \alpha \le \alpha_\nu \quad \text{on } \operatorname{supp}\psi_\nu \quad (\nu \ge 1),$$
$$\operatorname{supp}\psi_\nu \cap \operatorname{supp}\psi_\mu \ne \emptyset \implies \lambda_\nu \le c\lambda_\mu, \alpha_\nu \le \alpha_\mu \quad (\nu,\mu \ge 1)$$

and $u_\nu = \psi_\nu u$.

It is easy to see that

$$A_0^2 = \sum P_l g^{lk} P_k + (m+W)^2 + \sum (\alpha^l P_l + P_l \alpha^l) + \alpha$$

where α_j, α are Hermitian matrices such that $|\alpha^j| \le C\rho$, $|\alpha| \le C\tilde\rho^2 + C\rho\gamma^{-1}\chi_Y$. Therefore

(9.2.31)
$$\|A_0 u\|^2 \ge$$
$$\epsilon_1 \sum \|P_j u\|^2 + \|(m+W)u\|^2 - C\|\tilde\rho u\|^2 - C\|\rho^{\frac12}\bar\gamma^{-\frac12}\chi_Y u\|^2$$

with $\tilde\rho$ supported in X'' and

(9.2.32)$'$ Estimate (9.2.31) remains true with $\tilde P_j$ replaced by D_j in the right-hand expression.

Further,

$$\|A_0\psi_\nu u\|^2 = \|A\psi_\nu u\|^2 + \|V\psi_\nu u\|^2 - 2\operatorname{Re}(\psi_\nu A u, \tilde V \psi_\nu u),$$
$$(\psi_\nu A u, V\psi_\nu u) \le 2\lambda_\nu^{-2}\|\varsigma^{-\frac12}\psi_\nu A u\| + 2\lambda_\nu^2 \|\varsigma^{\frac12} V\psi_\nu u\|^2.$$

Multiplying by λ_ν^2 and summing on ν we obtain that

(9.2.33) *The dimension of $\mathcal{L}$ does not exceed the sum of the maximal dimensions of the spaces $\mathcal{L}_\nu$ (supported in $\operatorname{supp}\psi_\nu$) on which the following estimates hold:*

$$\|\nabla u_\nu\|^2 + \|(m+W)u_\nu\|^2 \le C\bigl(1 + \Phi(\alpha_\nu) + \lambda_\nu^{-1}\lambda_0\alpha_0)\bigr)^2 \|\chi_Y\gamma^{-1}u_\nu\|^2 +$$
$$C\lambda_\nu^2\|\varsigma^{\frac12} V u_\nu\|^2 + C\|\tilde\rho u_\nu\|^2 + C\|\rho^{\frac12}\gamma^{-\frac12}u_\nu\|^2 + Ct^2\lambda_\nu^{-2}\|\varsigma^{\frac12}u_\nu\|^2$$

for $\nu \ge 1$ and

$$\|\nabla u_0\|^2 + \|(m+W)u_0\|^2 \le$$
$$C\bigl(1 + \Phi(\alpha_0^2)\bigr)\|\chi_Y\gamma^{-1}u_0\|^2 + \lambda_0^{-2}\|\lambda^2\Phi(\alpha)\chi_Y\gamma^{-1}u_0\|^2 +$$
$$C\|\lambda^2\varsigma^{\frac12}V^2 u_0\|^2 + C\|\tilde\rho u_0^2\|^2 + C\|\rho^{\frac12}\gamma^{-\frac12}\chi_Y u_0\|^2 + Ct^2\lambda_0^{-2}\|\varsigma^{\frac12}u_0\|^2$$

for $\nu = 0$. Applying the Rozenblioum variational estimate, picking η very small in the definition of ς and recalling that in Y $\gamma^{-1} \le c\rho$ and $\varsigma \le c\rho^2(m+\rho)^{-1}$ we obtain that the sum of these dimensions does not exceed $C(t^{2d}+1)(R_0' + R_0'')$ with

(9.2.34)
$$R_0' = \int \min\bigl(\Phi(\alpha) + \lambda^{-1}\lambda_0\alpha_0 + \lambda(m+\rho)^{\frac12} + \lambda^{-1}(m+\rho)^{-\frac12},$$
$$\Phi(\alpha_0) + \lambda_0^{-1}\lambda\alpha\bigr)^d \rho^d\,dx$$

(9.2.35)
$$R_0'' = \int_{X''} \tilde\rho^d\,dx + \int_Y \rho^d\,dx$$

for $d \ge 3$. The estimate for $d = 2$ will be provided later. Here and below all the integrals without indicated domain are taken over $X' \cap X''$.

Let us pick

$$\lambda=(m+\rho)^{-\frac{1}{2}}+\lambda_0^{\frac{1}{2}}(m+\rho)^{-\frac{1}{4}},\qquad \alpha=\lambda^{-1}\lambda_0,\quad ,\alpha_0=1,\quad \lambda_0=\mu^{-2},$$
$$\mu=\Bigl(\int(\rho+m)^{\frac{d}{4}}\rho^d dx\Bigr)^{\frac{4}{d}}\Bigl(\int\rho^d dx\Bigr)^{-\frac{4}{d}}$$

(close to the optimal choice). Then we get the same estimate with

$$R_0'=\int\Phi_1((\rho+m)^{\frac{1}{4}}\mu)\rho^d dx$$

with $\Phi_1(t)=\Phi(t(t+1)^{-1})$ which is the same type of function as Φ.

One can easily see that $\Phi(\alpha)=c\alpha^{-1}$ works for any choice of ψ_ν. Replacing ψ_ν by $\chi(\psi_\nu)(\sum\chi(\psi_\nu)^2)^{-\frac{1}{2}}$ with $\chi\in C^\infty$, $\chi(t)=0$ for $t<0$, $\chi>0$ for $t>0$ we get that $\Phi(\alpha)$ works provided

$$|\frac{d}{dt}\chi(t)|\le\alpha\chi(t)+\Phi(\alpha).$$

One can take $\Phi(\alpha)=\alpha^{-\delta}$ with arbitrarily small $\delta>0$ in which case $\Phi_1(\alpha)=\alpha^{-\delta_1}$,

$$R_0'=\mu^\delta\int\rho^d(\rho+m)^{-\delta}dx.$$

However, one can pick $\Phi_1(\alpha)=\log^{1+\delta}\alpha^{-1}$ or even $\Phi_1(\alpha)=\log\alpha^{-1}\cdot\log_{(2)}\alpha^{-1}\cdots\log_{(n)}\alpha^{-1}\cdot\log_{(n+1)}^{1+\delta}\alpha^{-1}$ with arbitrarily small $\delta>0$, arbitrary n and $\log_{(n)}$ denoting the n-tuple logarithm.

One can easily see that if $\rho\le m$ in $X'\cap X''$ then $R_0'\le CR_0''$.

For $d=2$ only $\nu=0$ is affected. However, applying the same arguments as in Appendix G we see that one need only replace R_0'' by

$$R_0'''=\Bigl(\int_{X''}\bar\rho^p dx\Bigr)^{\frac{1}{p}}(\operatorname{mes}X'')^{1-\frac{1}{p}}+\int_Y\rho^d dx \tag{9.2.35$'$}$$

with arbitrary $p>1$ and $C=C_p$.

Let us note that in the right-hand expressions of estimate (9.2.33) for balls in $X'_+\cap X''$ one can replace $C\rho^2$ by $-\epsilon\rho^2+C\gamma^{-2}$ and, moreover, from the very beginning one can take $\zeta=\rho\gamma^{-1}m^{-1}$ here and therefore one can get a better remainder estimate:

$$\begin{aligned}R_0'=\int_{Y\cap X''_-}&\bigl(\Phi(\alpha)+\lambda^{-1}\lambda_0\alpha_0+\lambda(m+\rho)^{\frac{1}{2}}+\lambda^{-1}(m+\rho)^{-\frac{1}{2}}+\\&\Phi(\alpha_0)+\lambda_0^{-1}\lambda\alpha\bigr)^d\rho^d dx+\\&\int_Y\bigl(\Phi(\alpha)+\lambda^{-1}\lambda_0\alpha_0+\lambda m^{\frac{1}{2}}+\lambda^{-1}m^{-\frac{1}{2}}+\\&\Phi(\alpha_0)+\lambda_0^{-1}\lambda\alpha\bigr)^d\rho^{d-s}\gamma^{-s}dx\end{aligned}\tag{9.2.34*}$$

with arbitrarily large s with all further adjustments.

Taking an appropriate prolongation of ρ from the $\frac{1}{2}\epsilon\bar\gamma$-neighborhood of X'' one can easily see that the integration in third term in the right-hand expression of (9.2.33) should be taken only over the $\epsilon\gamma$-neighborhood of X'' and if we treat the case when X'' is not connected then in the analysis associated with the given connected component of X'' we should replace X'' by the connected component everywhere. Thus, *in the general case in the final estimates in the first term of estimate* (9.2.35)$'$ *one should consider every connected component of* X'' *separately (and only components with* $\bar{j}\ge cm$ *are essential), calculate the first term for every such component and then sum the results.*

9.2.6 Proof of theorems 9.1.4, 9.1.5 (end)

Thus, the lower estimate for $\mathbf{N}$ is proven. In order to prove the upper estimate we need to obtain an upper estimate for $\mathbf{N}(B;\tau_1,\tau_2)$ where here and in what follows $B = A - t_1 J_1 - \dots - t_n J_n$.

So B is an operator of (4.5.3) type satisfying all the conditions of section 4.5 with the following modifications: in the conditions on $D^\alpha V$, $D^\alpha W$ with $|\alpha| \geq 1$ there is an additional factor $\bar{t}$ in the right-hand expression, the conditions on V (without derivatives) remain true and, moreover, the condition

$$|V| \leq \bigl(m - \bar{t}\min(\rho, \frac{1}{m}\rho^2)\bigr)_+ + \min(\rho, \frac{1}{m}\rho^2) \qquad \text{in } X^V \tag{9.2.36$_1$}$$

is fulfilled. Moreover, the conditions on W

$$\bar{t}\min(\rho, \frac{1}{m}\rho^2) > m \implies W \geq \bar{t}\min(\rho, \frac{1}{m}\rho^2) \qquad \text{in } X^V, \tag{9.2.36$_2$}$$

$$c\bar{t}\min(\rho, \frac{1}{m}\rho^2) \geq W \geq 0 \qquad \text{in } X \tag{9.2.36$_3$}$$

are fulfilled where $\bar{t}$ is arbitrarily large and c does not depend on $\bar{t}$, X^V is the $\epsilon\gamma$-neighborhood of $X' \setminus X''$. All these conditions are provided by an appropriate choice of $J_1, \dots, J_n$ and $\phi_1, \dots, \phi_n$.

Let us divide the connected components of X'' into two types. The components with $\rho \leq cm$ at every point are components of the first type. We could choose $J_1, \dots, J_n$ such that conditions $(9.2.36)_{1-3}$ are also fulfilled on components of the first type (with ρ replaced by $\tilde{\rho}$).

Clearly $\mathbf{N}(B;(\tau_1,\tau_2))$ is the maximal dimension of the subspace $\mathcal{L} \subset C_0^1(X,\mathbb{H})$ on which

$$\|(B - \bar{\tau}I)u\|^2 < \tau^2\|u\|^2 \qquad (u \neq 0) \tag{9.2.37}$$

where $\bar{\tau} = \frac{1}{2}(\tau_2 + \tau_1)$, $\tau = \frac{1}{2}(\tau_2 - \tau_1)$. Let $\{\psi_k^2\}_{k=1,2}$ be a partition of unity such that $|D^\alpha\psi_k| \leq c\gamma^{-|\alpha|}\ \forall\alpha : |\alpha| \leq k$, let ψ_1 be supported in X_1 (which is the union of X^V and all components of the first type) and let ψ_2 be supported in X_2 (which is the union of all components of the second type). Then

$$\sum_k \|(B - \bar{\tau}I)\psi_k u\|^2 = \|(B - \bar{\tau})u\|^2 + \sum_k \|[B,\psi_k]u\|^2,$$

$$\|[B,\psi_k]u\| \leq C_0\|\gamma^{-1}\chi_Y u\|$$

where χ_Y is the characteristic function of $Y = X' \cap X''$. Therefore $\dim\mathcal{L}$ does not exceed $\max\dim\mathcal{L}_1 + \max\dim\mathcal{L}_2$ where $\mathcal{L}_k$ are subspaces of $C_0^1(X_k,\mathbb{H})$ on which the inequality

$$\|(B - \bar{\tau}I)u\|^2 < \tau^2\|u\|^2 + C_0\|\gamma^{-1}\chi_Y u\|^2 \quad (u \neq 0) \tag{9.2.38}$$

is fulfilled. Obviously there exists an extension B' of B from X_1 to X such that conditions $(9.2.36)_{1-3}$ are fulfilled on X. Then $\dim\mathcal{L}_1$ does not exceed the cardinality of the spectrum of the problem

$$(B' - \bar{\tau})v + \lambda Jv = 0 \qquad v \in \mathrm{D}(B') \tag{9.2.39}$$

in the interval $(-1,1)$ where $J = (\tau^2 + c\gamma^{-2}\chi_Y)^{\frac{1}{2}}$ and $J \in L^\infty$ provided the right-hand expression of the estimate in question is finite. Dividing v into $\frac{D}{2}$-dimensional components v_1 and v_2 we can rewrite the operator $B' - \bar{\tau}I + \tau J$ in the form $\begin{pmatrix} \zeta_1 & T \\ T^* & -\zeta_2 \end{pmatrix}$ where conditions $(9.2.36)_{1-3}$ fulfilled in X yield that $\zeta_j > \epsilon_0(1-|\lambda|)$, $|\lambda| \le 1 \quad \forall j = 1,2$ provided $\bar{t}$ is large enough and hence $(-1,1)$ does not intersect the spectrum of the problem (9.2.39) and therefore $\dim \mathcal{L}_1 = 0$.

Let us estimate $\dim \mathcal{L}_2$. Let us note that the conditions on V can be rewritten in the form

$$(9.2.40)_{1,2} \qquad \begin{aligned} V - \bar{\tau} &\ge \tau - m - c\min(\rho, \frac{1}{m}\rho^2), \\ V - \bar{\tau} &\le m - \tau + c\min(\rho, \frac{1}{m}\rho^2). \end{aligned}$$

In particular, for $\tau \ge m$ we obtain that $|V - \tau| + |\tau - m| \le c\min(\rho, \frac{1}{m}\rho^2)$.

Let us consider the operator $B_0 = B - V \cdot I - \sigma_0 W$. Standard arguments easily yield the estimate

$$(9.2.41) \qquad \epsilon_1\|Du\|^2 + m^2\|u\|^2 - C\|\tilde{\rho}u\|^2 \le \|B_0 u\|^2.$$

On the other hand, on $\mathcal{L}_2$ the estimates

$$\begin{aligned} \|B_0 u\|^2 \le (1+\varepsilon^{-2})\|(B - \bar{\tau}I)u\|^2 + (1+\varepsilon^2)\|(V + \sigma_0 W - \bar{\tau})u\|^2 \le \\ \bigl((1+\varepsilon^{-2})\tau^2 + (1+\varepsilon^2)(m-\tau)^2\bigr)\|u\|^2 + \\ C(1+\varepsilon^{-2})\|\chi\gamma^{-1}u\|^2 + C(1+\varepsilon^2)\|\rho u\|^2 \end{aligned}$$

are fulfilled where we have applied $(9.2.40)_{1,2}$ and the restrictions on W. These inequalities, (9.2.41) and the condition $|V_j| \le c\tilde{\rho}$ on X_2 yield that on $\mathcal{L}_2$ the estimate

$$(9.2.42) \qquad \|Du\|^2 \le C(\mathcal{K}u, u)$$

is fulfilled with

$$\mathcal{K} = 9\bigl((m-\tau)\varepsilon - \varepsilon^{-1}\tau\bigr)^2 + \rho^2(1+\varepsilon^2) + \chi\gamma^{-2}(1+\varepsilon^{-2}) + \tilde{\rho}^2.$$

If $\tau \ge 2m$ then $(9.2.40)_{1,2}$ yield that $m + \tau \le c\rho$ in X and therefore for $\varepsilon = 1$ one can skip the first and the third terms in $\mathcal{K}$; then the Rozenblioum estimate yields that

$$(9.2.43) \qquad \dim \mathcal{L}_2 \le CR_0''$$

with

$$R_0'' = \int_{X_2} (\rho + \tilde{\rho})^d dx \qquad (d \ge 3),$$

$$R_0'' = C_p\Bigl(\int_{X_2} (\rho + \tilde{\rho})^{2p} dx\Bigr)^{\frac{1}{p}} (\operatorname{mes} X_2)^{1-\frac{1}{p}} \qquad (d = 2)$$

where $p > 1$ is arbitrary.

Let us consider the case $\tau \le 2m$. One can assume without loss of generality that $\tau_2 - \tau_1 \le \frac{m}{2}$; otherwise one can divide the interval (τ_1, τ_2) into fewer than 10 intervals (τ_1', τ_2') with $\frac{m}{4} \le |\tau_1' - \tau_2'| \le \frac{m}{2}$. Moreover, if $\tau_2 - \tau_1 \ge \frac{m}{4}$ then the value of $\varepsilon > 0$ for which the first term in $\mathcal{K}$ vanishes is disjoint from 0 and ∞ and in this case estimate (9.2.40) remains true.

Let us now consider the case $\tau \le \frac{m}{4}$. Then the Rozenblioum estimate yields

$$\dim \mathcal{L}_2 \le C\Big(\big((m-\tau)\varepsilon - \varepsilon^{-1}\tau\big)^2 + a^2\varepsilon^2 + b^2\varepsilon^{-2}\Big)^{\frac{d}{2}} (\operatorname{mes} X_2) + CR_0'' \tag{9.2.44}$$

with

$$a = \begin{cases} \Big(\dfrac{1}{\operatorname{mes} X_2}\displaystyle\int_{X_2} \rho^d\,dx\Big)^{\frac{1}{d}} & (d \ge 3), \\ \Big(\dfrac{1}{\operatorname{mes} X_2}\displaystyle\int_{X_2} \rho^{2p}\,dx\Big)^{\frac{1}{2p}} & (d = 2, p > 1), \end{cases} \tag{9.2.45}$$

$$b = \Big(\frac{1}{\operatorname{mes} X_2}\int_{X_2} \chi\gamma^{-d}\Big)^{\frac{1}{d}} \qquad (d \ge 2) \tag{9.2.46}$$

(for $d = 2$ the procedure of covering of $X_1 \cap X_2$ by balls $B(x_n, \epsilon\bar\gamma(x_n))$ is applied; see above). Taking ε optimal we obtain from (9.2.41) the final estimate

$$\dim \mathcal{L}_2 \le CR_0'' + C\min(m^{\frac{d}{2}}b^{\frac{d}{2}}, m^{\frac{d}{2}}\tau^{-\frac{d}{2}}b^d)\operatorname{mes} X_2. \tag{9.2.47}$$

For $\tau \ge \epsilon_0 m$ this estimate coincides with (9.2.43).

Theorem 9.1.4 has been proven. Theorem 9.1.5 can be proven in the same way and theorem 9.1.6 can be proven much more simply, with $J = I$ and without the Birman-Schwinger principle.

Comments

The method of this chapter was first applied in [Ivrii 14]. I am aware that certain versions of the Birman-Schwinger principle of the type applied here were also used in the papers of M. Sh. Birman and M. Z. Solomyak.

Part IV

Asymptotics of Spectra

10. Weylian Asymptotics of Spectra

This chapter is devoted to applications of the results of Chapters 4, 5, 8 and 9 to eigenvalue asymptotics. In section 10.1 we treat semiclassical asymptotics. In sections 10.2–10.4 we consider asymptotics of eigenvalues tending to infinity: in section 10.2 the domain is bounded "in principal" (i.e., exits to infinity are thin enough) and the coefficients are weakly singular; in section 10.3 the coefficients are strongly singular and the coercivity condition is fulfilled but the domain is still bounded "in principal;" in section 10.4 the domain is "very unbounded" and the spectrum is discrete only because the lower order terms are singular. In section 10.5 we treat asymptotics of eigenvalues tending to the boundary of the essential spectrum. In particular, in section 10.4 we treat the Schrödinger operator with potential V tending to $+\infty$ at infinity and with the spectral parameter τ tending to $+\infty$, and in section 10.5 we treat the Schrödinger and Dirac operators with potentials tending to 0 at infinity and the spectral parameter τ tending to -0 and $M-0$ or $-M+0$ respectively (M is the mass in the case of the Dirac operator). Section 10.6 is devoted to multiparametrical asymptotics; for example, there is the semiclassical parameter $h \to +0$ and the spectral parameter τ tending either to $+\infty$ or $-\infty$ (for the Schrödinger and Dirac operators in bounded domains or for the Schrödinger operator in unbounded domains), or to -0, $M-0$ or $-M+0$ (for the Schrödinger and Dirac operators in unbounded domains), or to $\inf V + 0$, $\inf V + M + 0$, $\sup V - M - 0$ (for the Schrödinger and Dirac operators) where for the Schrödinger operator $\inf V$ is not necessarily finite. Moreover, the case of a vector semiclassical parameter $\mathbf{h} = (h_1, \dots, h_d)$ is treated.

For semibounded operators, Riesz means are also treated.

10.1 Semiclassical Eigenvalue Asymptotics

In this section the operator is fixed and only the semiclassical parameter $h \to +0$ varies.

10.1.1 Semibounded operators

Let us first apply the results of Chapter 8 obtained starting from the results of Chapters 4, 5. This means that $\tau_1 = -\infty$ and without loss of generality one can assume that $\tau_2 = 0$. We treat the operator

$$A_h = \sum_{|\alpha|,|\beta|\le m} (hD)^\alpha a_{\alpha,\beta} (hD)^\beta \tag{10.1.1}$$

with coefficients satisfying

$$a_{\alpha,\beta} = a^\dagger_{\beta,\alpha}, \tag{10.1.2}$$

(10.1.3) $$|D^\sigma a_{\alpha,\beta}| \le c\varrho\rho^{-|\alpha|-|\beta|}\gamma^{-|\sigma|} \qquad \forall\alpha,\beta: |\alpha|\le m, |\beta|\le m\ \forall\sigma: |\sigma|\le K$$

and

(10.1.4) $$\sum_{|\alpha|=|\beta|=m} \langle a_{\alpha,\beta}v_\alpha, v_\beta\rangle \ge \epsilon_0\varrho\rho^{-2m}\sum_{|\alpha|=m}|v_\alpha|^2$$
$$\forall v=(v_\alpha)_{|\alpha|=m} \quad \forall x: \rho\gamma\le h_0$$

where a function γ satisfying $(8.1.2)_1$ is given in X such that mes $X_s=0$ for $X_s=\{x,\gamma(x)=0\}$, functions ρ, ϱ satisfying $(8.1.2)_2$ are given in $X_r=X\setminus X_s$, and $h_0=\mathrm{const}>0$. The sets X_r and X_s are called *regular* and *singular sets* respectively.[1)]

For the Schrödinger operator

(Σ0) $$A=\sum_{j,k\le d}(hD_j-V_j)g^{jk}(hD_k-V_k)+V$$

with real-valued functions $g^{jk}=g^{kj}$, V_j and V we always take $\varrho=\rho^2$ and obtain

$(\Sigma 1)_{1-3}$ $$|D^\alpha g^{jk}|\le c\gamma^{-|\alpha|},\quad |D^\alpha V_j|\le c\rho\gamma^{-|\alpha|},\quad |D^\alpha V|\le c\rho^2\gamma^{-|\alpha|}$$

and

(Σ2) $$\epsilon_0|\xi|^2\le\sum_{j,k}g^{jk}\xi_j\xi_k\le c|\xi|^2 \qquad \forall\xi\in\mathbb{R}^d.$$

Moreover, we assume that conditions $(8.1.9)_4$ (the boundary regularity condition) and $(8.1.10)_1$ (the operator ellipticity condition) with m replaced by $2m$ [2)] are fulfilled in X_r. For the sake of simplicity we assume that Dirichlet boundary conditions are given in which case condition $(8.1.10)_2$ (the ellipticity condition for the boundary value problem) is fulfilled automatically.

Then one can apply theorem 8.1.1 with $X'=X'_h=\{x\in X,\rho\gamma>h\}$ and $X''=X''_h=\{x\in X,\rho\gamma\le 2h\}$ provided $h\le h_0$. Here conditions (10.1.2)–(10.1.4) yield the following estimate on X'':

$$(A_hu,u)\ge$$
$$\epsilon_0\int\varrho\Bigl(h^{2m}\rho^{-2m}\sum_{|\alpha|=m}|D^\alpha u|^2-C_0\sum_{|\alpha|\le m-1}h^{2|\alpha|}\rho^{-2|\alpha|}|D^\alpha u|^2\Bigr)dx$$

for all compactly supported functions $u\in C^{2m}(\bar X'',\mathbb{H})$ satisfying the Dirichlet boundary condition. Therefore the Birman-Solomyak-Rozenblioum estimate (see Appendix G) yields that the contribution of the zone X'' to the upper remainder estimate in theorem 8.1.1 does not exceed $CR''h^{-d}$ with

(10.1.5) $$R''=C\Bigl(\int_{X''}\varrho^{-p}\rho^{2mp}dx\Bigr)^{\frac{d}{2mp}}\Bigl(\int_{X''}\varrho^q dx\Bigr)^{\frac{d}{2mq}}\Bigl(\mathrm{mes}\ X''\Bigr)^{1-\frac{d}{2m}(\frac1p+\frac1q)}.$$

where p,q are arbitrarily chosen numbers such that

(10.1.6) $$\frac1p+\frac1q\le\frac{2m}{d},\qquad p\ge1, q\ge1$$

[1)] These sets do not coincide with the *singular* and *regular zones* $X'=\{x,\rho\gamma\le 2h\}$ and $X''=\{x,\rho\gamma\ge h\}$ respectively.

[2)] This modified condition $(8.1.10)_1$ is assumed to be fulfilled below.

and

$$\text{(10.1.7)} \qquad \begin{aligned} &\text{either} \quad \frac{1}{p}+\frac{1}{q}<\frac{2m}{d}, \\ &\text{or} \quad \frac{1}{p}+\frac{1}{q}=\frac{2m}{d}, d\geq m \text{ and } d>m \implies q>1 \\ &\qquad\qquad \text{and } p>1 \implies 2m<\frac{d}{p}(p+1). \end{aligned}$$

At the same time the contribution of the zone X' to the remainder estimate in the lower and upper estimates does not exceed $Ch^{1-d}R_1 + Ch^{s-d}R_2$ where

$$\text{(10.1.8)}_1 \qquad R_1 = \int_{X'_-} \rho^{d-1}\gamma^{-1}dx,$$

$$\text{(10.1.8)}_2 \qquad R_2 = \int_{X'} \rho^{d-s}\gamma^{-s}dx$$

with arbitrarily chosen s and we assume that in every ball $B\bigl(y,\gamma(y)\bigr)$ with $y\in X'_r$ one of the conditions $(8.1.18)_{1-6}$ is fulfilled.

Moreover, the principal term of the upper and lower estimates is $h^{-d}\mathcal{N}^w$ with

$$\text{(10.1.9)} \qquad \mathcal{N}^w = (2\pi)^{-d}\int_{T^*X'} \mathbf{n}(x,\xi,-\infty,0)dxd\xi$$

where $\mathbf{n}(x,\xi,.,.)$ is the eigenvalue counting function for the symbol

$$a(x,\xi)=\sum_{\alpha,\beta} a_{\alpha,\beta}\xi^{\alpha+\beta};$$

recall that X' depends on h.

So the following statement is proven:

Theorem 10.1.1. *Let A be a self-adjoint operator (with Dirichlet boundary conditions) of the form (10.1.1) in $L^2(X,\mathbb{H})$ and let conditions (10.1.2)–(10.1.4), $(8.1.2)_{1,2}$, $(8.1.9)_4$ and $(8.1.10)_1$ be fulfilled. Moreover, let us assume that, in every ball $B\bigl(y,\gamma(y)\bigr)$ with $y\in X'_r$, one of conditions $(8.1.18)_{1-6}$ is fulfilled. Then the estimates*

$$\text{(10.1.10)} \quad h^{-d}\mathcal{N}^w - Ch^{1-d}R_1 \leq \mathbf{N}^-(A_h) \leq h^{-d}\mathcal{N}^w + Ch^{1-d}R_1 + Ch^{s-d}R_2 + CR''h^{-d}$$

hold with $\mathcal{N}^w$, R_1, R_2 and R'' given by (10.1.9), $(10.1.8)_{1,2}$ and (10.1.5) respectively.

Corollary 10.1.2. *(i) For $\mathcal{N}^w$ in formula (10.1.9) let us replace the domain X' by the domain X_r (or X); then the error does not exceed $C\int_{X''}\rho^d dx \leq CR''$ by the Hölder inequality. Therefore*

$$\text{(10.1.11)} \qquad |\mathbf{N}^-(A_h) - \mathcal{N}^w h^{-d}| \leq Ch^{1-d}R_1 + Ch^{s-d}R_2 + Ch^{-d}R''$$

where $\mathcal{N}^w$ no longer depends on h.

(ii) This estimate yields that if

$$\text{(10.1.12)}_{1,2} \qquad \begin{aligned} &\rho^{d-s}\gamma^{-s}\in L^1(\{X,\rho\gamma\geq h_1\}), \\ &\rho^{d-1}\gamma^{-1}\in L^1(\{X_{r+},\rho\gamma\geq h_1\})\cap L^1(\{X_r,\rho\gamma\leq h_1\}) \end{aligned}$$

and

$$R'' = O(h) \qquad \text{as } h \to +0, \tag{10.1.13}$$

then

$$\mathbf{N}^-(A_h) = h^{-d}\mathcal{N}^w + O(h^{1-d}). \tag{10.1.14}$$

These asymptotics will be improved and generalized below, where other asymptotics due to the results of Chapter 8 will be derived.

(iii) Let us assume that

$$\varrho_1 \asymp \rho^k \quad \text{in } X_r \tag{10.1.15}$$

and that one of the following conditions are fulfilled:

$$(10.1.16)_{1-3} \qquad \begin{aligned} &d > 2m, k \in [0, 2m];\\ &m < d \le 2m, k \in [2m-d, d);\\ &d = m, k = d.\end{aligned}$$

Then one can choose p, q satisfying conditions (10.1.6) with equality and (10.1.7) and such that

$$R'' \asymp \int_{X''} \rho^d dx \le 2h \int_{X''} \rho^{d-1}\gamma^{-1} dx.$$

This inequality and $(10.1.12)_2$ yield that $R'' = o(h)$ and therefore condition (10.1.13) is fulfilled. In particular, for $d > 2m$ one can take $k = 2m$.

Remark 10.1.3. *Some of the conditions of theorem 10.1.1 and corollary 10.1.2 are too restrictive but we can weaken them.*

(i) First of all, one can assume that $X = X_r \cup X_s \cup Y$ where the domains $(X_r \cup X_s)$ and Y are properly overlapping (with respect to the function γ), and that in $(X_r \cup X_s)$ all the previous conditions are fulfilled and we assume that in Y the inequality

$$\sum_{\alpha,\beta} \langle a_{\alpha,\beta} v_\alpha, v_\beta\rangle \ge \epsilon_0 \sum_{|\alpha|=m} \varrho_5 |v_\alpha|^2 - c \sum_{|\alpha|\le m-1} \varrho_5 \varrho_6^{2|\alpha|} |v_\alpha|^2 \tag{10.1.17}$$

holds with appropriate measurable functions ϱ_5 and ϱ_6. Then one can take $X' = \{x \in X_r, \rho\gamma \ge h\}$ and $X'' = \{x \in X_r, \rho\gamma \le 2h\} \cup X_s \cup Y$ in which case R'' is given by the formula

$$(10.1.5)' \qquad \begin{aligned} R'' = C\Bigl(&\int_{X_r\cap\{\rho\gamma\le 2h\}} \rho^{-p}\rho^{2mp} dx + \int_Y \varrho_5^{-p} dx\Bigr)^{\frac{d}{2mp}} \times\\ \Bigl(&\int_{X_r\cap\{\rho\gamma\le 2h\}} \varrho^q dx + \int_Y \rho_5^q \varrho_6^{2mq} dx\Bigr)^{\frac{d}{2mq}} \times\\ &\bigl(\operatorname{mes}(X_r \cap \{\rho\gamma \le 2h\} \cup Y)\bigr)^{1-\frac{d}{2m}(\frac{1}{p}+\frac{1}{q})}.\end{aligned}$$

(ii) Moreover, one can assume that the Dirichlet boundary condition is given only on $\partial X \cap (\bar{X}_r \cap \{\rho\gamma \le h_0\} \cup \bar{Y})$. Moreover, let us assume that $Y = \emptyset$ and that for all $u \in \mathrm{D}(A)$ supported in $\bar{X}''$

$$(Au, u) \ge \epsilon_0 h^{2m} \sum_{|\alpha|=m} \|\varrho_1^{\frac{1}{2}} \rho^{-m} u\|^2 - c \sum_{|\alpha|\le m-1} h^{2|\alpha|} \|\varrho_1^{\frac{1}{2}} \rho^{-|\alpha|} u\|^2. \tag{10.1.18}$$

Finally, let us assume that the boundary condition which defines $\mathrm{D}(A)$ *is the Dirichlet boundary condition on* $\partial X \cap \bar{X}'' \setminus Z$ *where* Z *lies in the open set* Ω *and* $\varrho \asymp \rho^{2m}$ *on* $\Omega \cap \bar{X}''$ *and a function* γ_1 *is given satisfying* $(8.1.2)_1$*; let* Ω *contain a* γ_1*-neighborhood of* Z *and at every point of* $\Omega \cap \bar{X}''$ *the cone condition (with fixed angle and with height* γ_1*; see condition (G.16)) be fulfilled. Then the above estimates remain true with*

$$(10.1.5)'' \qquad R'' = \Big(\int_{X''} \rho^{2mq} dx + \int_{X''\cap\Omega} \gamma_1^{-2mq} dx\Big)^{\frac{1}{2mq}} (\text{mes } X'')^{1-\frac{d}{2mq}}.$$

A similar estimate is also possible for $Y \neq \emptyset$.

(iii) All these results remain true if we replace hD_j *by* $hD_j - V_j$ *in the operator* A *where* V_j *are real-valued functions supported in* X' *such that*

$$(10.1.19) \qquad |D^\alpha (D_j V_k - D_k V_j)| \le c\rho\gamma^{-|\alpha|-1} \qquad \forall \alpha : |\alpha| \le K.$$

(For the Schrödinger operator this condition is denoted by $(\Sigma 1)'_2$*.) In fact, after an appropriate gradient transform in every ball* $B(x, \epsilon\gamma(x))$ *lying in* X' *the condition*

$$(10.1.19)^* \qquad |D^\alpha V_j| \le c\rho\gamma^{-|\alpha|} \qquad \forall \alpha : |\alpha| \le K$$

is fulfilled. It is easy to check that conditions $(8.1.18)_{1-6}$ *remain true after the replacement* $hD_j \to hD_j - V_j$ *with* V_j *satisfying* $(10.1.19)^*$.

(iv) If the operator $A - \tau I$ *satisfies all the conditions of theorem 10.1.1 uniformly with respect to* $\tau \in [\tau_1, 0]$ *with* $\tau_1 < 0$ *then all the estimates and asymptotics of theorem 10.1.1 and corollary 10.1.2 remain true for this operator uniformly with respect to* $\tau \in [\tau_1, 0]$*. In particular, for the Schrödinger operator with* $d \ge 2$ *the condition of this statement is fulfilled automatically for* $\tau_1 = -\infty$.

(v) Including condition $(8.1.18)_7$ *in the list of admissible conditions we obtain upper and lower estimates with obvious modifications.*

(vi) Let us consider the Riesz means

$$(10.1.20) \qquad \mathbf{N}^-_\vartheta(A) = \int \tau_-^{\vartheta-1} \mathbf{N}^-(A - \tau I) d\tau$$

of order $\vartheta > 0$ *(i.e., we calculate them for the spectral parameter* $\tau = 0$*). Let us include in* $\mathcal{N}^w$ *(depending on* h*) a sufficient number of terms (see sections 4.3, 4.4 and footnote* [7] *below for modifications of these terms). Let us replace the list of conditions* $(8.1.18)_{1-6}$ *by the list* $(8.3.3)_{1-4}$*. Then the remainder estimates in the local semiclassical spectral asymptotics zone* X' *yield that* R_1 *and* R_2 *should be replaced by*

$$(10.1.8)'_1 \qquad R_{1\vartheta} = h^\vartheta \int_{X'_-} \varrho^\vartheta \rho^{d-\vartheta-1} \gamma^{-\vartheta-1} dx,$$

$$(10.1.8)'_2 \qquad R_2^\vartheta = \int_{X'} \varrho^\vartheta \rho^{d-s} \gamma^{-s} dx,$$

and the second factor in the right-hand expression of (10.1.5) should be given in the form

$$\int_0^\infty \tau^{\vartheta-1} \Big(\int_{X''} (\varrho_1 - \tau)_+^q dx\Big)^{\frac{d}{2mq}} d\tau.$$

In particular, if one can take $q = \frac{d}{2m}$ *and* $p = \infty$ *then*

$$R''_\vartheta = C \int_{X''} \varrho^{\frac{d}{2m}+\vartheta} dx \times \sup_{X''} (\varrho^{-\frac{d}{2m}} \rho^d)$$

and for the Schrödinger operator with $d = 2$ *one can take* R''_ϑ *according to Appendix G.*
In particular, replacing conditions $(10.1.12)_{1,2}$ *by the conditions*

$$(10.1.12)'_{1,2} \qquad \begin{aligned} &\varrho_1^\vartheta \rho^{d-s}\gamma^{-s} \in L^1(\{X, \rho\gamma \ge h_1\}),\\ &\varrho_1^\vartheta \rho^{d-1-\vartheta}\gamma^{-1-\vartheta} \in L^1(\{X_{r+}, \rho\gamma \ge h_1\}) \cap L^1(\{X_r, \rho\gamma \le h_1\}) \end{aligned}$$

and replacing condition (10.1.13) by the condition

$$(10.1.13)' \qquad R''_\vartheta = O(h^{1+\vartheta})$$

we obtain the asymptotics

$$(10.1.14)' \qquad \mathbf{N}^-_\vartheta(A) = h^{-d}\mathcal{N}^w + O(h^{1+\vartheta-d}).$$

In particular, in the context of (10.1.15)–(10.1.16) condition $(10.1.14)'$ *follows from* $(10.1.12)'_2$. ◊

Let us discuss the improvement of estimates $(10.1.14)'$ (and (10.1.14) which is treated as special case of $(10.1.14)'$ with $\vartheta = 0$).

Remark 10.1.4. *(i) If in the list of admissible conditions* $(8.1.18)_{1-6}$ *variant* $(8.1.18)_6$ *is rejected*[3] *then one can replace the first part of condition* $(10.1.12)'_2$[4] *by the two conditions*

$$(10.1.12)''_{2-1} \qquad \varrho^\vartheta \rho^{d-\vartheta-\sigma-1}\gamma^{-\vartheta-\sigma-1} \in L^1(X'_- \cap \{\rho\gamma \ge h_1\}),$$

$$(10.1.12)''_{2-2} \qquad \Big(\frac{\varrho}{\rho\gamma}\Big)^\vartheta \in L^1(T^*X \cap \Sigma_0 \cap \{\rho\gamma \ge h_1\})$$

where $\sigma = \sigma(d, \vartheta) > 0$ *is small enough and a natural density* $d\mu_\tau$ *is given on* $\Sigma_\tau = \{(x,\xi) : \tau$ *is an eigenvalue of* $a(x,\xi)\}$ *(see Chapter 4). In the general case this modification is possible in the union of balls* $B(x, \gamma(x))$ *in which we do not refer to condition* $(8.1.18)_6$.
(ii) Moreover, let us assume in addition that

(Ψ) *There exists a set* $\Lambda_0 \subset \Sigma_0$ *of measure 0 such that through every point of* $\Sigma_0 \setminus \Lambda_0$ *there passes an infinitely long trajectory of the Hamiltonian* $\varrho_1^{-1}\rho\gamma\lambda$ *lying in* T^*X_r *in a neighborhood of which* $\lambda = \lambda(x,\xi)$ *is an eigenvalue of constant multiplicity of the symbol* $a(x,\xi)$ *satisfying either the condition of simple reflection (see theorems 5.3.4 and 5.3.5) or the more general condition of theorem 5.3.5 (see also subsection 3.2.4)*[5] *and non-periodic.*

Then if

$$(10.1.13)'' \qquad R''_\vartheta = o(h^{1+\vartheta}) \quad \text{as } h \to +0$$

then

$$(10.1.21)' \qquad \mathbf{N}^-_\vartheta(A) = h^{-d}\mathcal{N}^w_{\vartheta+} + o(h^{1+\vartheta-d}).$$

Here the additional subscript "+" means that for $\vartheta \in \mathbb{Z}$ *the next term should be included in* $\mathcal{N}^w_\vartheta$ *and asymptotics (10.1.21) are these asymptotics with* $\vartheta = 0$.
(iii) Under more restrictive conditions along trajectories one can derive a better remainder estimate, which we do not do in this section. ◊

---

[3] In the case of Dirichlet boundary conditions this variant is excluded automatically.

[4] In fact, this condition consists of two parts because there is an intersection in its right-hand expression.

[5] In this case all the branches of this trajectory passing from this point in the chosen direction (in t) should be considered.

Let us consider a series of examples.

Example 10.1.5. *Let $d \geq 1$, let X be a bounded domain with $\partial X \in C^K$, let the coefficients of A and B belong to $C^K(\bar{X})$, $K = K(d, \vartheta)$ and let the operator (A, B) be elliptic. Moreover, let one of the conditions $(8.1.18)_{1-6}$ be fulfilled in every small ball $B(x, \epsilon)$. Then asymptotics (10.1.14)′ hold. Moreover, if $d \geq 2$, condition $(8.1.18)_6$ is rejected from the list of admissible conditions, and the standard condition (Ψ) on the Hamiltonian flow is fulfilled, then asymptotics (10.1.21)′ hold. Moreover, under more restrictive conditions on the Hamiltonian flow asymptotics with more precise remainder estimates hold (see theorems 4.3.13 and 5.3.4).* ◊

The conditions of our theorems are local and all our analysis is local (until the end of the subsection). So we assume that outside of the singularities indicated below everything is well (in the sense of example 10.1.5).

Example 10.1.6 (Conical singularity). *Let $d \geq 2$ and in a neighborhood of 0 let the domain X coincide with the cone $\{(r, \theta) \in \mathbb{R}^+ \times U\}$ where (r, θ) are spherical coordinates, U is a domain on the sphere $\mathbb{S}^{d-1}$ with boundary $\partial U \in C^K$ and let the conditions of theorem 10.1.1 be fulfilled with $\varrho = \rho^{2m}$, $\rho = r^\mu$, $\gamma = \epsilon r$ (here and below $\epsilon > 0$ is a small enough constant). Then ρ^d and (or) $\rho^{d-1}\gamma^{-1}$ belong to $L^1(X)$ if and only if $\mu > -1$. Moreover, let us assume that $\mu > \max(-1, -\frac{d}{2m})$.*[6] *Then asymptotics (10.1.14) hold and under the standard condition (Ψ) on the Hamiltonian flow asymptotics (10.1.21) also hold. Moreover, $\varrho^\vartheta \rho^{d-1-\vartheta}\gamma^{-1-\vartheta} \in L^1(X)$ if and only if*

$$(\mu + 1 - 2\mu m)\vartheta < (d-1)(\mu+1) \tag{10.1.22}$$

and if this condition is fulfilled and $\mu > -\frac{d}{2m}$ then asymptotics (10.1.14)′ hold for given ϑ. Moreover, under the standard condition (Ψ) on the Hamiltonian flow asymptotics (10.1.21)′ hold for given ϑ.

Furthermore, in the case

$$(\mu + 1 - 2\mu m)\vartheta = (d-1)(\mu+1) \tag{10.1.22$'$}$$

we obtain the asymptotics

$$\mathbf{N}^-_\vartheta(A) = h^{-d}\mathcal{N}^w_\vartheta + O(h^{1-d+\vartheta}\log h) \tag{10.1.23}$$

(we use theorem 10.1.1 instead of corollary 10.1.2).[7]

Finally, in the case

$$(\mu + 1 - 2\mu m)\vartheta > (d-1)(\mu+1), \quad 2\mu m\vartheta + d(\mu+1) > 0 \tag{10.1.22$''$}$$

we obtain the asymptotics

$$\mathbf{N}^-_\vartheta(A) = h^{-d}\mathcal{N}^w_\vartheta + O(h^p) \tag{10.1.24}$$

with $p = 2\mu m\vartheta(\mu+1)^{-1} \in (-d, 1 + \vartheta - d)$ where only terms with the degree of h less than $p + d$ are included in $\mathcal{N}^w_\vartheta$. ◊

---

[6] There are examples which show that otherwise the operator A is not necessarily semibounded.

[7] Here and below $\mathcal{N}^w_\vartheta$ is given by the previous formula but in general we only integrate over the domain $X'_h = \{x \in X, |x| \geq \epsilon h^{1/(\mu+1)}\}$. Moreover, one can skip all the terms with $n - d > p$ in the local semiclassical spectral asymptotics and in the terms with $n - d < p$ integrate over X; $p = 1 - d + \vartheta$ under condition (10.1.22)′.

Remark 10.1.7. *(i) One can easily see that in the domain* $X'' = X''_h = \{x \in X, r \le \epsilon h^{\frac{1}{\mu+1}}\}$ *the inequality*

$$(A_h u, u) \ge \epsilon_0 \sum_{|\alpha|=m} h^{2m} \|D^\alpha u\|^2 \qquad \forall u \in C^m(\bar{X}) : \operatorname{supp} u \subset B(0, \epsilon h^{\frac{1}{\mu+1}}) \tag{10.1.25}$$

holds. This inequality follows from the inequalities

$$h^k \sum_{|\alpha|=k} \|D^\alpha u\| \ge \epsilon_0 \epsilon^{-\delta} \|r^{\mu k} u\| \qquad \forall k = 1, \dots, m \tag{10.1.26}$$

for the same u *as in (10.1.25). Here* $\delta > 0$ *is a small enough exponent and* ϵ_0 *does not depend on* ϵ. *In fact, for* $h = \epsilon = 1$ *estimate (10.1.25) follows easily from the condition* $\mu > \max(-1, -\frac{d}{2m})$, *and the estimate in the general case follows from the estimate in this special case by means of dilatations.*

Therefore one can take $\rho = 0$, $\varrho = r^{2\mu m}$ *in* X''_h *and this simplifies the estimate of* R''. *Moreover, we obtain that for*

$$m\vartheta + d(\mu + 1) = 0 \tag{10.1.22$'''$}$$

asymptotics (10.1.24) hold with $p = -d$ *and with* $\mathcal{N}^w_\vartheta$ *given only by the integration of the* $n = 0$ *term over the domain* X'_h. *Here* $\mathcal{N}^w_\vartheta = O(\log h)$; *moreover, if*

$$\inf_{v:|v|=1, \xi \in \mathbb{R}^d} \langle a(x,\xi) v, v \rangle \le -\epsilon_0 \rho^{2m} \tag{10.1.27}$$

$\forall x \in \Gamma \cap \{|x| \le \epsilon\}$ *for some non-empty cone* Γ *with vertex in* 0 *then* $\mathcal{N}^w_\vartheta \asymp \log h$.

(ii) Moreover, let us note that under condition (10.1.22)$'$ *the main contribution to the remainder estimate in (10.1.24) is given by the zone* $\{X, ch^{\frac{1}{\mu+1}} \le |x| \le \varepsilon_h\}$ *with* $\varepsilon_h = o(1)$ *for* $h \to +0$ *(the contribution of the other zones is "o" with respect to the contribution of this zone). In order to improve the remainder estimate let us assume that the operator* A *tends at* 0 *to the quasi-homogeneous operator* A^0, *i.e., that*

$$D^\alpha (a_\beta - a^0_\beta) = o(|x|^{\mu|\beta| - |\alpha|}) \qquad \forall \beta : |\beta| \le 2m \quad \forall \alpha : |\alpha| \le K \tag{10.1.28}$$

for $|x| \to 0$ *where* $a^0_\beta \in C^K(\mathbb{R}^+ \times \bar{U})$ *are positively homogeneous of degrees* $\mu|\beta|$. *Moreover, let us assume that the symbol*

$$a^0(x,\xi) = \sum_\beta a^0_\beta(x) \xi^\beta \tag{10.1.29}$$

satisfies the standard condition (Ψ) *on Hamiltonian trajectories and* 0 *is a simple eigenvalue of* $a^0(x,\xi)$ *along trajectories starting from* $\Sigma^0_0 \setminus \Lambda_0$. *Here and below* Σ^0_T *is constructed for* a^0, *etc. Then the asymptotics*

$$\mathbf{N}^-_\vartheta(A) = h^{-d} \mathcal{N}^w_\vartheta + o(h^{1-d+\vartheta} \log h) \tag{10.1.30}$$

hold. Here in the term with $n = -d + 1 + \vartheta$ *we integrate over the domain* X'_h *and the term is* $O(h^{1+\vartheta} \log h)$; *in all the other terms we can integrate over* X. ◊

Example 10.1.8. *(i) Let us consider a more general case. Let* X *be a bounded domain locally coinciding with* $\mathbb{R}^{d'} \times \mathbb{R}^+ \times U$ *where spherical coordinates* (r, θ) *are introduced in* $\mathbb{R}^{d''}$ $(d'' = d - d')$, *and where* $U \subset \mathbb{S}^{d''-1}$, $\partial U \in C^K$, $0 \le d' \le d - 1$. *Let the conditions* *of theorem 10.1.1 be fulfilled with* $\varrho = \rho^{2m}$, $\rho = r^{\mu}$, $\gamma = \epsilon r$, $\mu > \max(-1, -\frac{d''}{2m})$. *Then for*

$$(\mu + 1 - 2\mu m)\vartheta < (\mu + 1)(d - 1) - d' \tag{10.1.31}$$

asymptotics (10.1.14)′ *hold. Moreover, under the standard condition* (Ψ) *on the Hamiltonian flow asymptotics* (10.1.21)′ *holds.*

On the other hand, for

$$(\mu + 1 - 2\mu m)\vartheta = (\mu + 1)(d - 1) - d' \tag{10.1.31$'$}$$

asymptotics (10.1.23) *hold and for*

$$(\mu + 1 - 2\mu m)\vartheta > (\mu + 1)(d - 1) - d', \quad 2\mu m\vartheta + d\mu + d'' \geq 0 \tag{10.1.31$''$}$$

asymptotics (10.1.24) *hold with* $\mathrm{p} = (2\mu m\vartheta - d')(\mu + 1)^{-1}$ *where for*

$$2\mu m\vartheta + d\mu + d'' = 0 \tag{10.1.31$'''$}$$

only one term in $\mathcal{N}^w_\vartheta$ *remains and integration is taken over* X'_h. *In this case* $\mathcal{N}^w_\vartheta = O(\log h)$ *and* $\mathcal{N}^w_\vartheta \asymp \log h$ *under condition (10.1.27) fulfilled in* $V \times \Gamma \cap \{x, |x''| \leq \epsilon\}$ *where* V *is a neighborhood of* 0 *in* $\mathbb{R}^{d'}$ *and* Γ *is a non-empty cone in* $\mathbb{R}^{d''}$ *with vertex at* 0.

Let us note that in this example (in contrast to example 10.1.6) the variants are possible even for $\vartheta = 0$.

(ii) Let $\vartheta = 0$ *and suppose condition* $(8.1.18)_7$ *is added to the list of admissible conditions. Then asymptotics (10.1.14) and (10.1.23) are replaced by asymptotics (10.1.24) with* $\mathrm{p} = -d+1-\delta$ *with arbitrarily small* $\delta > 0$ *while asymptotics (10.1.24) with* $\mathrm{p} < -d+1$ *remain true.*

(iii) Moreover, we can treat the case when the singularity is more complicated. Namely, for the Schrödinger operator we can consider positively homogeneous g^{jk}, V_j, V *which are not necessarily smooth on* $\mathbb{S}^{d-1}$ *as well as* ∂U. *We assume that on* $\mathbb{S}^{d-1}$ *all the conditions are fulfilled with* $\tilde\rho$ *and* $\tilde\gamma$ *(instead of* ρ *and* γ*). Then we can take* $\rho = \tilde\rho(\theta) r^{\mu}$ *and* $\gamma = \epsilon\tilde\gamma(\theta) r$ *with a small enough constant* $\epsilon > 0$. *Let us note that* $\rho^{2\vartheta+d} \in L^1$ *if and only if* $\tilde\rho^{2\vartheta+d} \in L^1(U)$ *and* $(2\vartheta+d)\mu+d > 0$; *moreover, let us note that* $\rho^{\vartheta+d-1}\gamma^{-\vartheta-1} \in L^1$ *if and only if* $\tilde\rho^{\vartheta d-1}\tilde\gamma^{-\vartheta-1} \in L^1(U)$ *and* $(\mu - 1)\vartheta + (\mu + 1)(d - 1) > 0$. *Therefore a more complicated structure is possible: one can treat a hierarchy of singularities: "vertices of singularities," "edges of singularities," "faces of singularities," etc.*

(iv) I leave to the reader the treatment of the case in which

$$\gamma = \epsilon \min(|x_{(1)}|, \dots, |x_{(l)}|), \qquad \rho = \prod_{S \subset \{1,\dots,l\}} \zeta_S^{\mu_S}, \qquad \zeta_S = \sum_{j \in S} |x_{(j)}|$$

where $x = (x_{(1)}, \dots, x_{(l)}) \in \mathbb{R}^{d_1} \times \dots \times \mathbb{R}^{d_l}$ *is an appropriate partition.* ◊

Example 10.1.9. *(i) Let* $0 \in \bar X$ *and all the conditions be fulfilled with* $\gamma = \epsilon r$, $\rho = \rho^{\mu}|\log r|^{-\nu}$, $r = |x|$ *where we assume that*

$$\begin{gathered} \mu > -\frac{2m}{d},\ \mu \geq -1,\ \vartheta \geq 0,\ \nu > 0, \\ 2m\mu\vartheta + (d - \vartheta - 1)(\mu + 1) = 0. \end{gathered} \tag{10.1.32}$$

Then repeating the arguments of the previous examples with

$$X'' = X''_h = \{x \in X, |x| \leq \epsilon h^{\frac{1}{\mu+1}} |\log h|^{\frac{\nu}{\mu+1}}\} \qquad \text{for } \mu > -1$$

and

$$X''_h = \{x \in X, \log|x| \le -Ch^{-\frac{1}{\nu}}\} \qquad \text{for } \mu = -1$$

and with corresponding X' we obtain that under the condition

$$(10.1.33) \qquad (2m\vartheta + d - \vartheta - 1)\nu > 1$$

asymptotics (10.1.14)$'$ *hold; moreover, under the standard condition* (Ψ) *on the Hamiltonian flow asymptotics* (10.1.21)$'$ *hold. On the other hand, under the condition*

$$(10.1.33)' \qquad (2m\vartheta + d - \vartheta - 1)\nu = 1$$

asymptotics (10.1.23) hold. Finally, if

$$(10.1.33)'' \qquad (2m\vartheta + d - \vartheta - 1)\nu < 1$$

and if either

$$(10.1.34)_1 \qquad \mu > -1$$

or

$$(10.1.34)_2 \qquad \mu = -1 (\implies \vartheta = 0),\ \nu \ge \frac{1}{d}$$

then asymptotics (10.1.24) hold with $p = 2m\vartheta - \vartheta - \frac{1}{\nu}$. Here footnote [7] *remains true. For $\mu = -1$, $\vartheta = 0$, $\nu = \frac{1}{d}$ we see that $\mathcal{N}^w = O(\log h)$ and $\mathcal{N}^w \asymp \log h$ provided condition (10.1.27) is fulfilled in some non-empty cone with vertex at 0 (for $|x| \le \epsilon$).*

(iii) Moreover, in the first critical case (10.1.33)$'$ *one can improve asymptotics (10.1.23). We do not do a careful analysis of the Hamiltonian trajectories here but refer to the arguments of the proofs of example 10.4.12 (see below). Let us assume that the coefficients a_β of A tend to functions $a^0_\beta\big|\log|x|\big|^{-\nu|\beta|}$ in the sense that*

$$(10.1.35) \qquad D^\sigma\big(a_\beta\big|\log|x|\big|^{\nu|\beta|} - a^0_\beta\big) = o(|x|^{\mu|\beta|-|\sigma|}) \qquad \forall\sigma : |\beta| \le 2m\ \forall\sigma : |\sigma| \le 1$$

where a^0_β are functions belonging to $C^K(\mathbb{R}^+ \times \bar{U})$ which are positively homogeneous of degree $\mu|\beta|$. Moreover, let us assume that the quasi-homogeneous symbol (10.1.29) satisfies the standard condition (Ψ) *on the Hamiltonian trajectories and 0 is a simple eigenvalue of $a^0(x,\xi)$ along trajectories starting from $\Sigma^0_0 \setminus \Lambda_0$. Then one can easily prove that asymptotics (10.1.30) hold with all the explanations made in remark 10.1.7.*

(iii) For $\nu > 0$ satisfying the critical condition (10.1.29)$'$ *one can treat the case in which $\rho = r^\mu|\log r|^{-\nu}\big|\log|\log r|\big|^{-\nu'}$ and analyze all the possible cases. In this case the conditions on ν are replaced by similar conditions on ν'. One can continue this process of complication.*

(iv) On the other hand, one can treat the case in which $\rho = r^\mu|\log r|^{-\nu}$ even if equality in (10.1.28) is violated, i.e., μ is not the critical value. In this case for $2m\mu\vartheta + (d-\vartheta-1)(\mu+1) > 0$ asymptotics (10.1.14)$'$ *(and even* (10.1.21)$'$ *under the standard condition* (Ψ) *on the Hamiltonian flow) hold and for $2m\mu\vartheta + (d-\vartheta-1)(\mu+1) < 0$ asymptotics (10.1.24) hold with remainder estimate $O(h^p|\log h|^q)$ instead of $O(h^p)$. The calculation of q is left to the reader.*

(v) In all the above statements the condition $\nu > 0$ is not necessary if $\mu > -1$.

(vi) Statement (ii) of example 10.1.8 remains true (with obvious modifications). Different combinations of (iii) and (iv) are possible. ◊

Example 10.1.10 (Quasi-conical singularity). *Consider the Schrödinger operator with functions g^{jk}, V_j and V which are L-quasi-homogeneous of degrees 0, μ and 2μ respectively, smooth away from 0. In a neighborhood of zero let X be an L-quasi-conical domain and let ∂X be smooth away from 0. Here $L = (L_1, \dots, L_d)$ is an exponent vector with $L_j > 0$ $\forall j$ and $\max_j L_j = 1$. Let us assume that either a Dirichlet or some other appropriate boundary condition is given and in the second case let us assume that the domain X satisfies the standard cone condition.*

In this case $\gamma = \epsilon[x]_L$, $\rho = \gamma^{2m\mu}$ with $[x]_L = \sum_j |x_j|^{\frac{1}{L_j}}$. Then repeating the arguments of example 10.1.6 and assuming that

$$\mu > \max\bigl(-1, -\frac{L^*}{2m}\bigr), \qquad L^* = L_1 + \dots + L_d \tag{10.1.36}$$

we obtain that for

$$(\mu + 1 - 2\mu m)\vartheta < (d-1)\mu + L^* \tag{10.1.37}$$

asymptotics (10.1.14) *hold and under the standard condition* (Ψ) *on the Hamiltonian flow asymptotics* (10.1.21)$'$ *hold. On the other hand, for*

$$(\mu + 1 - 2\mu m)\vartheta = (d-1)\mu + L^* \tag{10.1.32$'$}$$

asymptotics (10.1.23) hold and for

$$(\mu + 1 - 2\mu m)\vartheta > (d-1)\mu + L^*, \quad 2m\mu\vartheta + d\mu + L^* > 0 \tag{10.1.32$''$}$$

asymptotics (10.1.24) hold with $p = (2m\mu\vartheta + L^* - d)(\mu+1)^{-1}$.

However I have failed to apply more advanced arguments of remark 10.1.7 and we have failed to investigate the case $2m\mu\vartheta + d\mu + L^* = 0$. ◇

I leave to the reader the investigation of the case $\rho = r^{\mu}$, $\gamma = \epsilon r^{1+\nu}$ with $\nu > 0$ arising in the analysis of the Schrödinger operator with potential $V = \rho^{2\mu} \sin r^{-\nu}$ and other examples of this type.

Let us treat exits to infinity. We assume that

(10.1.38) Either $X'' = \emptyset$ or a Dirichlet boundary condition is given on $\partial X \cap \{|x| \ge c\}$.

Example 10.1.11 (Exit to infinity). *(i) In a neighborhood of infinity let X coincide with $\{x, x_1 > 0, x' = F(x_1)\Omega\}$ where $\Omega \subset \mathbb{R}^{d-1}$ is a bounded domain with C^K boundary, $x' = (x_2, \dots, x_d)$, $t\Omega = \{tx', x' \in \Omega\}$ and*

$$F(x_1) \asymp x_1^{\beta}, \qquad \beta \le 1, \tag{10.1.39}$$

$$|D^\alpha F| \le cF^{1-|\alpha|} \qquad \forall \alpha : |\alpha| \le K. \tag{10.1.40}$$

For $\beta = 1$ we have a conical exit, for $\beta \in (0,1)$ we have a slowly expanding quasi-conical exit, for $\beta = 0$ we have a cylindrical exit and for $\beta < 0$ we have a constricting exit which can be corrugated along the x_1 axis.

(ii) Similarly, let us consider the case in which X coincides with $\{x, x' = F(|x''|)\Omega\}$ in a neighborhood of infinity where $x = (x'', x') \in \mathbb{R}^{d''} \times \mathbb{R}^{d'}$, $0 < d' < d = d' + d''$ and the function F satisfies conditions (10.1.39) and (10.1.40) (I leave to the reader the treatment of the case of more general dependence of F on x'').

Then geometry yields that $\gamma \le \epsilon F(|x''|)$. Moreover, let us assume that the other conditions are fulfilled with $\gamma = \epsilon F(|x''|)$ and $\rho = |x''|^{\mu}$.

If $\mu+\beta\ge 0$ one can take $X'=X$ and $X''=\emptyset$. Then under the condition

$$2m\mu\vartheta + d\mu + d'\beta + d'' < 0 \tag{10.1.41}$$

(which provides that $\mathcal{N}^w_\vartheta$ converges) asymptotics (10.1.15) hold. Moreover, asymptotics (10.1.15)′ hold under the standard condition (Ψ) on the Hamiltonian flow. One can fulfill this condition only for $\vartheta>0$.

Let $\mu+\beta<0$. Then one can take $X'=X\cap\{r\le Ch^{\frac{1}{\mu+\beta}}\}$ and $X''=X\cap\{r\ge \frac{1}{2}Ch^{\frac{1}{\mu+\beta}}\}$ where $r=|x''|$. Let us assume that a Dirichlet condition is given on $\partial X\cap\{|x''|\ge C\}$. Then the inequality

$$h^m\sum_{\mu:|\mu|=m}\|D'^{\mu}u\|_{(x'')}\ge \epsilon_0F(r)^{-m}\|u\|_{(x'')}$$

(where $\|.\|_{(x'')}$ means the L^2-norm over sections of X at points x'') yields that under the condition

$$(2m\mu-\beta-\mu)\vartheta+(d-1)\mu-\beta+d'\beta+d''<0 \tag{10.1.42}$$

asymptotics (10.1.14)′ *hold and under the standard condition* (Ψ) *on the Hamiltonian flow asymptotics* (10.1.21)′ *hold. On the other hand, for*

$$(2m\mu-\beta-\mu)\vartheta+(d-1)\mu-\beta+d'\beta+d''=0 \tag{10.1.42$'$}$$

asymptotics (10.1.23) hold and for

$$(2m\mu-\beta-\mu)\vartheta+(d-1)\mu-\beta+d'\beta+d''>0,\qquad 2m\mu\vartheta+d\mu+d'\beta+d''\le 0 \tag{10.1.42$''$}$$

asymptotics (10.1.24) hold with $p=(2m\mu\vartheta+d''-d''\beta)(\mu+\beta)^{-1}$; here in the case of equality in the last condition the integral in $\mathcal{N}^w_\vartheta$ is calculated over the domain X'_h and $\mathcal{N}^w_\vartheta=O(\log h)$. Moreover, let condition (10.1.27) be fulfilled for $|x''|\ge c, x''\in\Gamma$ and $x'\in F(|x''|)\Omega'$ where $\Gamma\subset\mathbb{R}^{d''}$ is a non-empty cone and $\Omega'\subset\Omega$ is a non-empty domain. Then $\mathcal{N}^w_\vartheta\asymp\log h$. ◊

Example 10.1.12 (Conical exit to infinity). *(i) In a neighborhood of infinity let the domain X coincide with the cone $\{(r,\theta)\in\mathbb{R}^+\times U\}$ and let the conditions be fulfilled with $\gamma=\epsilon r$, $\rho=\rho^\mu$. Then if*

$$\mu\ge -1,\qquad 2m\mu\vartheta+(\mu+1)d<0$$

then asymptotics (10.1.14)′ *hold and under the standard condition* (Ψ) *on the Hamiltonian flow asymptotics* (10.1.21)′ *hold.*

Let us treat the case $\mu<\min(-1,-\frac{2m}{d})$; then estimates (10.1.25) and (10.1.26) hold for u supported in $\mathbb{R}^d\setminus B(0,\epsilon^{-1}h^{\frac{1}{\mu+1}})$ and therefore under condition (10.1.22)$^\#$ *asymptotics* (10.1.21)′ *hold and under the standard condition* (Ψ) *on the Hamiltonian flow asymptotics* (10.1.21)′ *hold. Here the superscript "#" means that the condition is replaced by its opposite (i.e., "$>$" by "$<$" and v.v., and "$=$" remains), $|x|\to 0$ is replaced by $|x|\to\infty$ and $|x|\le\epsilon$ is replaced by $|x|\ge C$. Moreover, under condition* (10.1.22)$'^\#$ *we obtain asymptotics (10.1.23) and under condition* (10.1.22)$''^\#\#$ *we obtain asymptotics (10.1.24) with p given by the same formula as before. Finally, under condition* (10.1.22)$'''$ *asymptotics (10.1.24) remain true with the same explanations as before. Moreover, under conditions* (10.1.22)$'^\#$, (10.1.28)$^\#$ *and the standard condition* (Ψ) *on the Hamiltonian flow we obtain asymptotics (10.1.30).*

(ii) Let $\gamma = \epsilon r$, $\rho = r^{\mu} \log^{-\nu} r$ where $\mu \le -1$, $\mu < -\frac{2m}{d}$ satisfies (10.1.22)′. Then all the results of example 10.1.9 remain true and the inequalities on ν do not change and in the stabilization conditions and in (10.1.27) $|x| \to 0$ is replaced by $|x| \to \infty$ and $|x| \le \epsilon$ is replaced by $|x| \ge C$.

Moreover, statements (ii) and (iii) of this example remain true with the inequalities on μ changed to their opposites. ◇

Example 10.1.13 (Quasi-conical exit to infinity). *In a neighborhood of infinity let X be an L-quasi-conical domain and $\partial X \in C^K$. Then in a neighborhood of infinity one can take $\gamma = \epsilon [x]_{\delta L}^{\delta}$ where we assume that $L = (L_1, \dots, L_d)$, $L_j \ge 1\ \forall j$ and $\delta \le \min_j L_j^{-1}$. Let us assume that $\rho = \gamma^{\mu}$ and $\varrho = \gamma^{2\mu m}$. In this case all the statements of example 10.1.10 remain true with the inequalities on μ replaced by their opposites.* ◇

I leave to the reader the case in which $X = \mathbb{R}^{d'} \times \mathbb{T}^{d''}$ in a neighborhood of infinity and $\gamma = 1$.

10.1.2 Non-semibounded operators

Let us now consider operators which are non-semibounded. In this case $\vartheta = 0$ unless otherwise indicated. Now m is the order of the operator.

Theorem 10.1.1 and corollary 10.1.2, as well as example 10.1.5, remain true in this case with obvious modifications. Moreover, one can treat the Dirac operator in $X = \mathbb{R}^d$ assuming that for $|x| \ge c$

(10.1.43) $$V + M - \tau_2 \ge \epsilon_0 \rho^2, \qquad M - V + \tau_1 \ge \epsilon_0 \rho^2$$

and

$(10.1.44)_{1,2}$ $$\rho\gamma \ge \epsilon_0, \qquad \int \rho^{d-s} \gamma^{-s} dx < \infty.$$

Recall that M is the mass. Moreover, one can include a neighborhood of infinity in X'' and skip condition $(10.1.44)_2$ and treat the case $X = \mathbb{R}^{d'} \times \mathbb{T}^{d''}$ with $d' > 0, d'' > 0$. Moreover, one can treat the *generalized Dirac* operator

(10.1.45) $$A = \begin{pmatrix} V_1 & \mathcal{L} \\ \mathcal{L}^* & V_2 \end{pmatrix}$$

where V_k are matrix-valued functions, $\mathcal{L}$ is a uniformly elliptic operator of order m and for $|x| \ge c$

(10.1.46) $$V_2 - \tau_2 \ge \epsilon_0 \rho^{2m}, \qquad V_1 - \tau_1 \le -\epsilon_0 (\rho + 1)^{2m}, \qquad \rho\gamma \ge \epsilon_0.$$

Moreover, we assume that the coefficients of $\mathcal{L}$ satisfy the conditions

(10.1.47) $$|D^{\alpha} b_{\beta}| \le c\rho^{m-|\beta|} \gamma^{-|\alpha|} \qquad \forall \beta : |\beta| \le m \quad \forall \alpha : |\alpha| \le K$$

and

(10.1.48) $$|D^{\alpha} V_2| \le c\rho^{2m} \gamma^{-|\alpha|}, \qquad |D^{\alpha} V_1| \le c(\rho + 1)^{2m} \gamma^{-|\alpha|} \qquad \forall \alpha : 1 \le |\alpha| \le K.$$

All the examples with singularities at 0 remain true for $X = \mathbb{R}^d$ if in the inequalities on exponents we replace $2m$ by m; however, we do not modify the inequalities $\mu > -\frac{d}{2m}$ and $\mu > -\frac{d''}{2m}$.

Moreover, all examples with singularities at infinity remain true without modification of the inequalities on exponents provided in a neighborhood of infinity

$$V_1 - \tau_1 \le -\epsilon_0, \qquad V_2 - \tau_2 \le -\epsilon_0 \rho^{2m}. \tag{10.1.49}$$

Recall that in these examples $\rho \le c$.

Let us note that under the conditions (in X)

$$V_1 - \tau_1 \le -\epsilon_0, \qquad V_2 - \tau_1 \ge \epsilon_0, \qquad \rho \le c \tag{10.1.50}$$

one can treat

$$\mathbf{N}^-_\vartheta(A) = \int_{\tau_1}^{\tau_2} (\tau_2 - \tau)_+^{\vartheta - 1} d\tau \tag{10.1.51}$$

with $\vartheta > 0$. Then example 10.1.5 and all the examples with singularities at infinity remain true.

10.2 Asymptotics of Large Eigenvalues: the Case of Weakly Singular Potentials

10.2.1 Getting started

In this and the following two sections we consider the eigenvalue asymptotics of an operator A tending to $+\infty$. In fact, one can treat the more general problem

$$(A - \lambda J)u = 0 \qquad u \in \mathrm{D}(A) \tag{10.2.1}$$

where A and J are operators of orders $2m$ and $n < 2m$ respectively ($m \in \frac{1}{2}\mathbb{Z}$). Recall that in this problem one should assume that A is a self-adjoint operator, J is a symmetric operator with $\mathrm{D}(J) \supset \mathrm{D}(A)$, J is A-compact (this condition will be fulfilled automatically in the context of the theorems below) and one of the operators A and J is positive definite. Surely, the case $J = I$ is the most natural and interesting from the point of view of physical applications; in the next sections that case will be treated but in this section we consider the more general case as well. In this section we assume that A is semibounded from below; at the end of the section $J = I$ and A is not necessarily semibounded.

If the operator J is positive definite then the Birman-Schwinger principle yields that the number of eigenvalues of the problem (10.2.1) less than τ is equal to $\mathbf{N}^-_\tau$ which is the number of negative eigenvalues of the operator $A - \tau J$. On the other hand, if the operator A is positive definite then the Birman-Schwinger principle yields that the number of eigenvalues $\lambda \in (0, \tau)$ of the problem (10.2.1) (with $\tau > 0$) is equal to $\mathbf{N}^-_\tau - \mathbf{N}^-_{+0}$ where $\mathbf{N}^-_\tau$ is the number of negative eigenvalues of the operator $A - \tau J$ and $\mathbf{N}^-_{+0} = \lim_{\tau \to +0} \mathbf{N}^-_\tau$. Therefore in the both cases we should consider the asymptotics of $\mathbf{N}^-_\tau$ as $\tau \to +\infty$.

Let us assume that A is given by (10.1.1) with $h = 1$, $m \in \mathbb{Z}$ and that conditions (10.1.2) and (10.1.3) are fulfilled and

$$J = \sum_{|\alpha|, |\beta| \le m, |\alpha| + |\beta| \le n} D^\alpha j_{\alpha,\beta} D^\beta, \tag{10.2.2}$$

$$(10.2.3)_{1,2} \qquad |D^\sigma a_{\alpha,\beta}| \le c\varrho\rho^{-|\alpha| - |\beta|}\gamma^{-|\sigma|}, \qquad |D^\sigma j_{\alpha,\beta}| \le c\varrho_0\rho^{-|\alpha| - |\beta|}\gamma^{-|\sigma|} \qquad \forall \sigma : |\sigma| \le K.$$

Moreover, we assume that the boundary satisfies condition $(8.1.9)_4$ and that the coefficients of the boundary operator $B=(B_1,\dots,B_l)$ satisfy condition $(8.1.9)_2$. We also assume that the operator $\varrho^{-1}\rho^{2m}A$ is uniformly elliptic and that the boundary value problem $(\varrho^{-1}\rho^{2m}A, \varrho_{(1)}^{-1}B_1,\dots,\varrho_{(l)}^{-1}B_l)$ is uniformly elliptic as well.

One can easily see that in the zone $\{x\in X, \tilde\rho\gamma\ge 1\}$ the operator $A-\tau J$ satisfies all the conditions of section 8.1 (perhaps excluding conditions of (8.1.18)-type) with functions $\tilde\gamma=\gamma$, $\tilde\rho=\rho\bigl(1+\frac{\tau\varrho_0}{\varrho}\bigr)^{\frac{1}{2m-n}}$, $\tilde\varrho=\varrho\bigl(1+\frac{\tau\varrho_0}{\varrho}\bigr)^{\frac{2m}{2m-n}}$ and $\tilde\varrho_{(j)}=\varrho_{(j)}\bigl(\frac{\tilde\rho}{\rho}\bigr)^{-m_j}$ instead of γ, ρ, ϱ and $\varrho_{(j)}$ respectively.[8)]

Let us treat condition (8.1.18) in the zone $X_1'=\{x\in X, \tau\varrho_0\ge C\varrho\bigl(1+(\rho\gamma)^{\frac{n-2m}{2m}}\bigr)\}$ in which $\tilde\rho\gamma\ge C'$ and $\tilde\rho\ge C'\rho$ provided $C=C(C')$. We easily see that for large enough C' in this zone one can replace the complete symbols of the operators A, B and J by their senior symbols[9)] in the verification of condition (8.1.18) and after dilatation on x and ξ (i.e., dilatation on x and multiplication of the operators by appropriate factors) we easily see that *one needs to check this condition with* $\gamma\partial_x$ *instead of* ∂_x *and with* ∂_ξ *preserved for the symbol* $\bigl(\tilde a(x,\xi)-\tilde\jmath(x,\xi),\tilde b(x,\xi)\bigr)$ where the tilde means that one needs to multiply a, $\jmath$ and b_k by ϱ^{-1}, ϱ_0^{-1} and $\varrho_{(k)}^{-1}$ respectively at point x; these factors are "well-differentiable" (in the corresponding scale) on x.

Let us first check the conditions which are not associated with $\tilde b$. For $|\xi|\ge\epsilon$ with an arbitrarily small constant $\epsilon>0$ the ellipticity of $\tilde a(x,\xi)$, the homogeneity of $\tilde a(x,\xi)$ and $\tilde\jmath(x,\xi)$ of degrees $2m$ and n with respect to ξ and the Euler identity yield that microhyperbolicity condition $(8.1.18)_1$ is fulfilled with $\mathcal{T}=\langle\xi,\partial_\xi\rangle$. Therefore *the remaining check should be done for* $\xi=0$ *only*. For general matrix operators one needs to check condition $(8.1.18)_1$ and since $\tilde a$ with its first order derivatives vanishes at $\xi=0$ this condition should be checked for $\tilde\jmath(x,\xi)$. The three following cases are possible:

(i) $n=0$; then the microhyperbolicity condition means exactly that

$(10.2.4)_1$ $n=0$ and for every $x\in X'$ there exists $\mathcal{T}\in T_xX$ such that $|\mathcal{T}|=1$ and

$$\gamma\langle\mathcal{T}\jmath(x)v,v\rangle\ge\epsilon_0\varrho_0|v|^2-c\varrho_0^{-1}|\jmath(x)v|^2\qquad\forall v\in\mathbb{H}$$

where $\jmath(x)=J(x)$ and we returned to $\jmath$ from $\tilde\jmath$. In particular, this condition is fulfilled automatically if

$(10.2.4)_2$ $n=0$ and $\quad|\jmath(x)v|\ge\epsilon\varrho_0|v|\qquad\forall x\in X'\quad\forall v\in\mathbb{H}$.

(ii) $n=1$; in this case $\tilde\jmath(x,0)=0$ and one should differentiate on ξ only; then microhyperbolicity condition means exactly that

$(10.2.4)_3$ $n=1$ and for every $x\in X'$ there exists $\xi\in T_x^*X$ such that $|\xi|=1$ and

$$\langle\jmath(x,\xi)v,v\rangle\ge\epsilon_0\varrho_0\rho^{-1}|v|^2\qquad\forall v\in\mathbb{H}$$

where $\jmath(x,\xi)$ is the senior symbol of $J(x,D_x)$.

(iii) $n\ge 2$; in this case $\tilde\jmath$ and its first order derivatives vanish at $\xi=0$ and the microhyperbolicity condition is surely violated.

For scalar symbols a and $\jmath$ [10)] the following additional possibilities arise:

(iv) $m=1$, $d\ge 2$; then condition $(8.1.18)_2$ holds automatically; so we assume that

---

8) We use the notation $\varrho_{(k)}$ instead of the ϱ_k of section 8.1 in order to avoid misunderstanding.

9) I.e., the higher degree homogeneous symbols.

10) It is sufficient to assume that either the eigenvalues of $a(x,\xi)-\jmath(x,\xi)$ close to 0 are simple in a neighborhood of $\{\xi=0\}$ or these eigenvalues are of constant multiplicity there and $\rho\gamma\le C$

$(10.2.4)_4$ $\qquad m=1,\ d\geq 2$ and a and j are scalar symbols.

(v) $m=1,\ d=1,\ n=0$; in this case condition $(8.1.18)_3$ is fulfilled if and only if

$(10.2.4)_5$ $m=1,\ d=1,\ n=0$ and for every $x\in X'$

$$\sum_{\alpha:|\alpha|\leq p}\gamma^{|\alpha|}|D^\alpha j|\geq \epsilon_0\varrho_0 \tag{10.2.5}$$

with appropriate p.

(vi) $m=1,\ d=1,\ n=1$; in this case condition $(8.1.18)_2$ is equivalent to condition $(10.2.4)_3$ which is equivalent to the condition

$(10.2.4)_6$ $m=1,\ d=1,\ n=1$ and $|j(x)|\geq \epsilon_0\varrho_0\rho^{-1}$ for every $x\in X'$

where now $J=\frac{1}{2}(D_xj+jD_x)+j'$; moreover, condition $(8.1.18)_3$ (without specification of the sign) is equivalent to the condition

$(10.2.4)_7$ $m=1,\ d=1,\ n=1$ and $|j(x)|+\gamma|\partial_xj(x)|\geq \epsilon_0\varrho_0\rho^{-1}$ for every $x\in X'$.

Finally, condition $(8.1.18)_4$ is equivalent to

$$\sum_{\alpha:|\alpha|\leq p}\gamma^{|\alpha}|D^\alpha j|\geq \epsilon_0\varrho_0\rho^{-1}. \tag{10.2.6}$$

(vii) If $m\geq 2$ then the first and the second order derivatives of a vanish at $\xi=0$ and we obtain the following cases:

$(10.2.4)_8$ $m\geq 2,\ d\geq 2,\ n=0$ and for every $x\in X'$ the inequality $|j(x)|+\gamma|\nabla j(x)|\leq \epsilon_0\varrho_0$ yields that $\operatorname{Hess} j(x)$ has two eigenvalues with absolute values greater than $\epsilon_0\varrho_0\gamma^{-2}$.

Then condition $(8.1.18)_3$ is fulfilled. Moreover, if these eigenvalues are of the same sign then condition $(8.1.18)_2$ is fulfilled.

$(10.2.4)_9$ $m\geq 2,\ d\geq 2,\ n=0$ and for every $x\in X'$ (10.2.5) holds with $p=2$; then condition $(8.1.18)_4$ is fulfilled;

$(10.2.4)_{10}$ $m\geq 2, n=1$ and for every $x\in X'$ the inequality $|j(x)|+\gamma|(Dj)(x)|\geq \epsilon_0\varrho_0\rho^{-1}$ holds

where here $J=\frac{1}{2}(\langle j,D\rangle+\langle D,j\rangle)+j'$, $j=j(x)$ is a vector field and (Dj) is the matrix of its derivatives, $j'=j'(x)$. Then condition $(8.1.18)_3$ is fulfilled. On the other hand, condition $(8.1.18)_4$ is not more general in this case.

$(10.2.4)_{11}$ $m\geq 2,\ d\geq 2,\ n=2$ and $\sum_{p,q}|j_{pq}|\geq \epsilon_0\varrho_0\rho^{-2}$ for every $x\in X'$ where j_{pq} are the coëfficients of the senior symbol of J.

Then condition $(8.1.18)_4$ is fulfilled. Moreover, if the matrix (j_{pq}) has two eigenvalues with absolute values greater than $\epsilon_0\varrho_0\rho^{-2}$ then condition $(8.1.18)_3$ is fulfilled. For $m\geq 2$,

---

in the latter case. For $n\geq 1$ that automatically yields that $a-j$ is a scalar symbol and the homogeneity of a and j yield that a and j are scalar symbols. For $n=0$ this conjecture is not necessary. In this case in the appropriate local coordinates $j=\begin{pmatrix} j'' & 0\\ 0 & j'I\end{pmatrix}$, $a=\begin{pmatrix} * & *\\ & a'I\end{pmatrix}$ where j'' is a non-degenerate matrix and j' and a' are scalar symbols and under the above condition all the following conditions should be checked for symbols a' and j'.

$d = 1$, $n = 2$ condition $(8.1.18)_4$ is surely violated. Let us note that for $n \geq 3$ condition $(8.1.18)_4$ is surely violated as well. On the other hand, condition $(8.1.18)_7$ is fulfilled automatically. Moreover, the following natural condition is assumed to be fulfilled:

(10.2.7) $j(x,\xi)$ (the principal symbol of J) is not a non-positive definite matrix.

(otherwise the principal part of the asymptotics vanishes. I leave the treatment of this case to the reader).

Let us treat the conditions associated with the boundary value problem. Recall that this condition for the operator $\ell = \tilde{a} - \tilde{j}$ is of the form

$$\text{(10.2.8)}\quad \operatorname{Re}\langle \mathcal{T}\ell v, v\rangle_+ + \operatorname{Re}\langle \mathcal{T}\tilde{b}v, \mathcal{B}v\rangle_0 \geq \epsilon_0 |v|_+^2 \qquad \forall v \in \mathcal{S}'(\mathbb{R}^+, \mathbb{H}) : \ell v = 0 \text{ and } bv|_{x_1=0} = 0$$

where we use the notation of Chapter 5 and make a dilatation; here $\ell = \ell(z, D_1)$, $b = b(z, D_1)$ and $\mathcal{B} = \mathcal{B}(z, D_1)$ is defined by the condition

$$\text{(10.2.9)}\qquad \langle \ell u, v\rangle_+ + \langle \tilde{b}u, \mathcal{B}v\rangle_0 = \langle u, \ell v\rangle_+ + \langle \mathcal{B}u, \tilde{b}v\rangle_0 \qquad \forall u, v \in \mathcal{S}(\mathbb{R}^+, \mathbb{H})$$

where $z = (x', \xi')$. Let us note that *condition* (10.2.8) *remains true after replacing* $D_1 \to \theta(z)D_1$ *with* $\theta \asymp 1$. In fact, it is obvious for $\theta = \text{const}$ and therefore we need check this statement only for θ equal to 1 at the point in question. In this case the additional term

$$\text{(10.2.10)}\qquad (\mathcal{T}\theta)(z) \times \operatorname{Re}\bigl(\langle \ell' v, v\rangle_+ + \langle \tilde{b}' v, \mathcal{B}v\rangle_0\bigr)$$

in the left-hand expression of (10.2.8) appears where the prime means derivative with respect to θ. Let us introduce $v_\theta(x_1) = v(\theta^{-1}x_1)$; then $\ell_\theta v_\theta = 0$ and $\tilde{b}_\theta v_\theta|_{x_1=0} = 0$. Therefore treating θ as an independent variable, differentiating and taking $\theta = 1$ we obtain that

$$\text{(10.2.11)}\qquad \ell' v + \ell v' = 0, \qquad \tilde{b}' v + \tilde{b} v' = 0$$

where the prime on v means derivative on θ also and therefore $v' = -x_1\partial_1 v$. Then (10.2.11), (10.2.9) and the equalities $\ell v = 0$, $\tilde{b}v|_{x_1=0} = 0$ obviously yield that the second factor in (10.2.8) vanishes. *QED*

Let us consider $|\xi'| \geq \epsilon_0$ and let us introduce $\mathcal{T} = \langle \xi', \partial_{x'}\rangle$, $\theta = |\xi'|$. Then the operators $\tilde{a}(x', \xi', \theta D_1)$ and $\tilde{j}(x', \xi', \theta D_1)$ are positive homogeneous of degrees $2m$ and n on ξ' respectively. Moreover, the operators $\tilde{b}_k(x', \xi', \theta D_1)$ are positive homogeneous of degrees m_k on ξ' and $b_k v = 0$. Therefore the Euler identity yields that the left-hand expression of (10.2.8) is equal to

$$2m\langle \tilde{a}v, v\rangle_+ - n\langle \tilde{j}v, v\rangle_+.$$

Since $(\tilde{a} - \tilde{j})v = 0$ we obtain that this expression is equal to

$$(2m - n)\langle \tilde{a}v, v\rangle_+ = (2m - n)\langle \tilde{j}v, v\rangle_+.$$

Therefore condition (10.2.8) is fulfilled provided either

$$\text{(10.2.12)}\qquad \langle \tilde{a}v, v\rangle_+ \geq \epsilon_0 |v|_+^2 \qquad \forall v \in \mathcal{S}(\mathbb{R}^+, \mathbb{H}) : \ \tilde{b}v = 0 \text{ at } x_1 = 0$$

or the same condition is fulfilled for $\tilde{j}$.

If A is a semibounded operator (under the boundary condition $Bv = 0$) then the ellipticity of the boundary value problem (A, B) yields that condition (10.2.12) is fulfilled for $\tilde{a}$. *In what follows when we consider non-semibounded operators A we assume that* $J = I$ and then condition (10.2.12) for $\tilde{\jmath}$ is fulfilled automatically.

Let us consider the case $\xi' = 0$. Let us prove that *in this case the kernel of the problem*

$$\ell v = 0, \qquad \tilde{b} v = 0 \quad \text{at } x_1 = 0, \qquad v \in \mathcal{S}(\mathbb{R}^+, \mathbb{H}) \tag{10.2.13}$$

is trivial. Then condition (10.2.8) will be automatically fulfilled. Let us note that $\tilde{a} = a_0 D_1^{2m}$, $\tilde{\jmath} = j_0 D_1^n$, $\tilde{b}_k = b_{0k} D_1^{m_k}$ where a_0, j_0, b_{0k} are matrices (depending on z). Let us replace D_1 by θD_1 and let us introduce $v_\theta(x_1) = v(\theta^{-1} x_1)$ with an arbitrary parameter θ. The above arguments yield the equality

$$\langle \ell' v, v\rangle_+ + \langle \tilde{b}' v, \mathcal{B} v\rangle_0 = 0$$

for any solution of the problem (10.2.13). Let us note that in our case $\tilde{b}' v = 0$ at $x_1 = 0$. Then the equalities $\ell v = 0$, $\ell = \tilde{a} - \tilde{\jmath}$ and $\ell' = 2m\tilde{a} - n\tilde{\jmath}$ obviously yield that

$$\langle \tilde{a} v, v\rangle_+ = 0, \qquad \langle \tilde{\jmath} v, v\rangle_+ = 0.$$

The first equality is impossible for $v \neq 0$ if A is a semibounded elliptic operator and the second equality is impossible for $v \neq 0$ if $J = I$.

So, if (a) one of the conditions $(10.2.4)_{1-3,6}$ is fulfilled, or (b) one of the conditions $(10.2.4)_{4,5,7-11}$ is fulfilled and $B(x, \gamma(x)) \cap \partial X = \emptyset$ (we mark this assumption by "#" in the corresponding condition), or (c)

$(10.2.4)_{12}$ $d \geq 2$ and A is the Schrödinger operator with a Dirichlet or Neumann boundary condition[11)] on $\partial X \cap B(x, \gamma(x))$,

then the contribution of $B(x, \gamma(x))$ to the principal part of the asymptotics is $O(\tau^{\frac{d}{2m-n}} \varrho_*^d \gamma^d)$ and its contribution to the remainder estimate is $O(\tau^{\frac{d-1}{2m-n}} \rho_*^{d-1} \gamma^{d-1})$ where $\rho_* = \rho\bigl(\frac{\varrho_0}{\varrho}\bigr)^{\frac{1}{2m-n}}$. Moreover, if condition (10.2.7) is fulfilled in this ball then its contribution to the principal part is exactly of the indicated order.

In the general case under the condition

$(10.2.4)_{13}$ A and J are scalar operators

(we refer to condition $(8.1.18)_7$) the contribution of $B(x, \gamma(x))$ to the remainder estimate is $O\bigl((\tau^{\frac{d-1}{2m-n}} \rho_*^{d-1} \gamma^{d-1})^{1+\delta}\bigr)$ with an arbitrarily small exponent $\delta > 0$.

We have not yet taken into account that the singularity is "weak." This condition arises from the following estimates. Variational arguments yield that the contribution of $X'' = \{x \in X, \tau \varrho_0 \leq C\varrho(1 + \rho\gamma)^{\frac{n-2m}{2m}}\}$ to the remainder estimate does not exceed

$$CR'' + C\tau^{\frac{d}{2m-n}} R'''$$

where R'' is given by (10.1.5) with exponents satisfying (10.1.6) and (10.1.7) and

$$R''' = \Bigl(\int_{X''} \varrho^{-p'} \rho^{2mp'}\, dx\Bigr)^{\frac{d}{(2m-n)p'}} \times \Bigl(\int_{X''} \varrho_0^{q'} \rho^{-nq'}\, dx\Bigr)^{\frac{d}{(2m-n)q'}} \times \bigl(\operatorname{mes} X''\bigr)^{1-\frac{d}{2mq'}(\frac{1}{p'}+\frac{1}{q'})} \tag{10.2.14}$$

with exponents p' and q' satisfying conditions $(10.1.6)^*$ and $(10.1.7)^*$ where the star means that we replace $2m$, p and q by $2m-n$, p' and q' respectively. We assume that

---

11) Or a Robin boundary condition with appropriate restrictions on lower order coefficients.

a Dirichlet boundary condition is given on $\partial X \cap \bar{X}''$. Otherwise we should assume that a cone condition is fulfilled (with constant angle and height of the cone), $\varrho \asymp \rho^{2m}$ and mes $X'' < \infty$.

One can easily see that if we replace integration over X' by integration over X in $\mathcal{N}^w$ then the error again does not exceed $CR'' + C\tau^{\frac{d}{(2m-n)}} R'''$.

Let us note that if $R'' < \infty$ then $\rho \in L^d(X)$ and therefore

$$\rho\gamma \le C, \quad \rho\gamma \to 0 \text{ as either } |x| \to \infty \text{ or } \gamma \to 0. \tag{10.2.15}$$

This justifies footnote [10] and proves that the remainder estimate remains true if we drop the lower degree terms in A and J in the formula for $\mathcal{N}^w$. After this modification

$$\mathcal{N}^w(\tau) = \varkappa_0 \tau^{\frac{d}{2m-n}}. \tag{10.2.16}$$

Later we improve the remainder estimate under conditions on the Hamiltonian trajectories in X'. Resulting from this analysis we obtain the following statements:

Theorem 10.2.1. *Let $X = X_r \cup X_s$, mes $X_s = 0$ and in X_r let all the conditions (8.1.1)–(8.1.7), (8.1.9) (excluding the restriction $\rho\gamma \ge 1$) and (8.1.10) be fulfilled. Let us assume that on $\partial X \cap \{\rho\gamma \le \epsilon_0\}$* ***either*** *an m-th order Dirichlet boundary condition is given* ***or*** *the cone condition is fulfilled (with constant angle and height of the cone), $\varrho \asymp \rho^{2m}$ and mes $X'' < \infty$. Moreover, let us assume that in every ball $B(x, \gamma(x))$ with $X \in X_r$ one of the conditions $(10.2.4)_{1-3,6,12}$ or $(10.2.4)^{\#}_{4,5,7-11}$ is fulfilled.*[12] *Finally, let us assume that $R'' < \infty$.*

Then

(i) The asymptotics

$$\mathbf{N}(0,\tau) = \varkappa_0 \tau^{\frac{d}{2m-n}} + O\bigl(\tau^{\frac{d}{2m-n}} R''' + \tau^{\frac{d-1}{2m-n}} R_{10}\bigr) \tag{10.2.17}$$

hold as $\tau \to +\infty$ where $R'' = R''(\tau)$ and $R''' = R'''(\tau)$ are defined by (10.1.5) and (10.2.14),

$$\varkappa_0 = (2\pi)^{-d} \int \mathbf{n}(x,\xi;0,1)\,dx d\xi, \tag{10.2.18}$$

$\mathbf{n}(x,\xi;0,1)$ is the eigenvalue counting function for the symbol $j^{-1}(x,\xi)a(x,\xi)$ on the interval $(0,1)$, and

$$R_{10} = \int_{X'} \rho_*^{d-1}\gamma^{-1}\,dx \tag{10.2.19}$$

also depends on τ.

(ii) In particular, if

$$\rho_*^d, \rho_*^{d-1}\gamma^{-1} \in L^1(X) \tag{10.2.20}$$

and

$$R'''(\tau) = O(\tau^{-\frac{1}{2m-n}}) \qquad \text{as } \tau \to +\infty \tag{10.2.21}$$

then the asymptotics

$$\mathbf{N}(0,\tau) = \varkappa_0 \tau^{\frac{d}{2m-n}} + O(\tau^{\frac{d-1}{2m-n}}) \tag{10.2.22}$$

[12] Recall that the superscript "#" means that $\partial X \cap B(x,\gamma(x)) = \emptyset$.

hold as $\tau \to +\infty$.

Moreover, in the case $\varrho = \rho^{2m}$ *and* $2m - n < d$ *one can set* $q' = \frac{d}{2m-n}$ *and then* (10.2.20) *yields* (10.2.21) *and* (10.2.20) *and condition* $\operatorname{mes} X < \infty$ *yield* $R'' < \infty$.

(iii) This theorem also remains true for compact C^K *manifolds.*

(iv) Moreover, if condition $(10.2.4)^{\#}_{13}$ *is added to the list of admissible conditions then the asymptotics*

$$\mathbf{N}(0,\tau) = \varkappa_0 \tau^{\frac{d}{2m-n}} + O\bigl(\tau^{\frac{d}{2m-n}} R''' + \tau^{\frac{d-1+\delta}{2m-n}} R'_{10}\bigr) \tag{10.2.23}$$

hold with an arbitrarily small exponent $\delta > 0$ *and with*

$$R'_{10} = \int_{X'} \rho_*^{d-1+\delta} \gamma^{-1+\delta}\, dx. \tag{10.2.19$'$}$$

Let us treat Riesz means under the natural condition

$$J = I \qquad (\implies n = 0,\ \varrho_0 = 1). \tag{10.2.24}$$

In this case condition $(10.2.4)_2$ is fulfilled automatically, the contribution of the ball $B(x, \gamma(x))$ to the principal part of the asymptotics is $\asymp \tau^{\vartheta + \frac{d}{2m}} \rho_*^d \gamma^d$, and the contribution to the remainder estimate is $O\bigl(\tau^{\vartheta(1-\frac{1}{2m}) + \frac{1}{2m}(d-1)} \rho_*^{d-\vartheta-1} \gamma^{d-\vartheta-1}\bigr)$ with $\rho_* = \varrho^{-\frac{1}{2m}} \rho$.

Moreover, the contribution of the zone X'' to the remainder estimate does not exceed $CR''_\vartheta + C\tau^{\vartheta + \frac{d}{2m}} R'''_\vartheta$ where

$$R''_\vartheta = \int_0^\infty t^{\vartheta-1} \Bigl(\int_{X''} (\varrho - t)_+^q\, dx\Bigr)^{\frac{d}{2mq}} dt \times \Bigl(\int_{X''} \varrho^{-p} \rho^{2mp}\, dx\Bigr)^{\frac{d}{2mp}} \times (\operatorname{mes} X'')^{1 - \frac{d}{2m}(\frac{1}{p} + \frac{1}{q})}, \tag{10.2.25}$$

$$R'''_\vartheta = \Bigl(\int_{X''} \varrho^{-p} \rho^{2mp}\, dx\Bigr)^{\frac{d}{2mp}} (\operatorname{mes} X'')^{1 - \frac{d}{2mp}} \tag{10.2.26}$$

and exponents p and q satisfy conditions (10.1.6)–(10.1.7).

As a result of this analysis we obtain the following

Theorem 10.2.2. *Let* $X = X_r \cup X_s$, $\operatorname{mes} X_s = 0$ *and let all the conditions* (8.1.1)–(8.1.7), (8.1.9) *(excluding the restriction* $\rho\gamma \geq 1$*) and* (8.1.10) *be fulfilled in* X_r. *Let us assume that on* $\partial X \cap \{\rho\gamma \leq \epsilon_0\}$ ***either*** *an m-th order Dirichlet boundary condition is given* **or** *the cone condition is fulfilled (with constant angle and height of the cone),* $\varrho \asymp \rho^{2m}$ *and* $\operatorname{mes} X'' < \infty$. *Moreover, let us assume that* $J = I$ *and* $R''_\vartheta < \infty$. *Then*

(i) The asymptotics

$$\mathbf{N}_\vartheta(\tau) = \mathcal{N}^{\mathsf{w}}_\vartheta(\tau) + O\bigl(R'''_\vartheta \tau^{\frac{d}{2m} + \vartheta} + R_{\vartheta,10} \tau^{\frac{d-1}{2m} + \vartheta(1 - \frac{1}{2m})}\bigr) \tag{10.2.27}$$

hold as $\tau \to +\infty$ *where* $R''_\vartheta = R''_\vartheta(\tau)$ *and* $R'''_\vartheta = R'''_\vartheta(\tau)$ *are given by* (10.2.25) *and* (10.2.26) *and*

$$R_{\vartheta,10} = \int_{X'} \rho_*^{d-\vartheta-1} \gamma^{-\vartheta-1}\, dx. \tag{10.2.28}$$

(ii) In particular, if

$$\rho_*^d,\ \rho_*^{d-\vartheta-1} \gamma^{-\vartheta-1} \in L^1(X) \tag{10.2.29}$$

then one can replace $\mathcal{N}^{w}_{\vartheta}(\tau)$ *by*

$$\sum_{k<\vartheta+1} \varkappa_k \tau^{\frac{1}{2m}(d-k)+\vartheta}$$

with constant coefficients $\varkappa_k$. *Moreover, if*

$$R'''_{\vartheta} = O\bigl(\tau^{-\frac{1}{2m}(\vartheta+1)}\bigr) \qquad \text{as } \tau\to+\infty \tag{10.2.30}$$

then the asymptotics

$$\mathbf{N}_{\vartheta}(\tau) = \sum_{k<\vartheta+1} \varkappa_k \tau^{\frac{1}{2m}(d-k)+\vartheta} + O\bigl(\tau^{\frac{1}{2m}(d-\vartheta-1)+\vartheta}\bigr) \tag{10.2.31}$$

hold as $\tau\to+\infty$.

Moreover, if $\varrho=\rho^{2m}$ *then* $R'''\asymp \operatorname{mes} X''$ *and* (10.2.29) *yields* (10.2.30).

Let us improve the results of these statements:

Theorem 10.2.3. *(i) Let the assumptions of theorem 10.2.1 be fulfilled. Moreover, let us assume that conditions* (10.2.23) *and*

$$R'''(\tau) = o(\tau^{-\frac{1}{2m-n}}) \qquad \text{as } \tau\to+\infty \tag{10.2.25*}$$

are fulfilled. Let us assume that the standard condition ($\Psi 0$) *on the Hamiltonian trajectories of the symbol* $a(x,\xi)-j(x,\xi)$ *is fulfilled at the energy level* 0 *in* X_τ *(here and below the "0" in condition* (Ψ) *means that only senior symbols are considered). Then the asymptotics*

$$\mathbf{N}(0,\tau) = \varkappa_0\tau^{\frac{d}{2m-n}} + \varkappa_1\tau^{\frac{d-1}{2m-n}} + o\bigl(\tau^{\frac{d-1}{2m-n}}\bigr) \tag{10.2.26*}$$

hold as $\tau\to+\infty$.

(ii) Let the assumptions of theorem 10.2.2 and condition (10.2.29) *be fulfilled and suppose*

$$R'''_{\vartheta} = o\bigl(\tau^{-\frac{1}{2m}(\vartheta+1)}\bigr) \qquad \text{as } \tau\to+\infty. \tag{10.2.30*}$$

Let us assume that the standard condition ($\Psi 0$) *on the Hamiltonian trajectories of the symbol* $a(x,\xi)-1$ *is fulfilled at the energy level* 0 *in* X_τ. *Then the asymptotics*

$$\mathbf{N}_{\vartheta}(\tau) = \sum_{k\le\vartheta+1} \varkappa_k \tau^{\frac{1}{2m}(d-k)+\vartheta} + o\bigl(\tau^{\frac{1}{2m}(d-\vartheta-1)+\vartheta}\bigr) \tag{10.2.31*}$$

hold as $\tau\to+\infty$.

(iii) In the context of (i) the assumptions $2m-n<d$, $\varrho=\rho^{2m}$ *and* (10.2.29) *yield* (10.2.25)*.

10.2.2 Examples

More precise asymptotics under more restrictive conditions on the Hamiltonian flow will be treated below. Let us consider some examples. Let us start with examples in which $J=I$, $\varrho=\rho=1$. Then condition (10.2.28) means that

$$\operatorname{mes} X<\infty, \qquad \gamma^{-1-\vartheta}\in L^1(X); \tag{10.2.32$'$}$$

the absence of the prime below means that $\vartheta=0$. *In the first examples the coefficients of* A *are regular and* γ *is completely defined by the geometry of the domain* X.

Example 10.2.4 (Quasi-polygonal domains). *(i) Let X be a bounded domain with ∂X contained in the union of a finite number of C^K hypersurfaces $\tilde{S}_l$ (with $l = 1, \dots, L$). Let $S_l = \tilde{S}_l \cap \partial X$ and δS_l be the relative boundary of S_l in $\tilde{S}_l$. Let*

$$\mathcal{K} = \bigcup_{l \neq k} (S_l \cap S_k) \cup \bigcup_l \delta S_k$$

be the singular set of ∂X.

Obviously,

$$\gamma(x) \asymp \min\Bigl(\min_{l \neq k}\bigl(\gamma_l(x) + \gamma_k(x)\bigr), \operatorname{dist}(x, \mathcal{K})\Bigr) \tag{10.2.33}$$

where $\gamma_l(x) = \operatorname{dist}(x, S_l)$.

Then condition (10.2.32)′ should be verified for $\gamma_{lk}(x) = \gamma_l(x) + \gamma_k(x)$ for all $l \neq k$ and for $\gamma'(x) = \operatorname{dist}(x, \mathcal{K})$ separately.

First of all, if the hypersurfaces $\tilde{S}_l$ and $\tilde{S}_k$ are transversal at $S_k \cap S_l$ then condition (10.2.32)′ is fulfilled for $\gamma = \gamma_{kl}$ and $\vartheta < \vartheta^* (= 1)$. Moreover, if $\mathcal{K}$ is contained in the union of a finite number of C^K manifolds of codimension 2 then condition (10.2.32)′ is fulfilled for $\gamma = \gamma'$ for $\vartheta < \vartheta^* = 1$ as well. Moreover, the above conditions yield the cone condition. Therefore for appropriate boundary conditions[13] asymptotics (10.2.31) hold for $\vartheta < \vartheta^*$. Moreover, under the standard condition ($\Psi 0$) on the Hamiltonian trajectories of the symbol $a(x, \xi)$ at the energy level 1 asymptotics (10.2.31) hold. Moreover, our analysis yields that for $\vartheta = \vartheta^*$ the asymptotics*

$$\mathbf{N}_\vartheta(\tau) = \mathcal{N}^{\mathrm{w}}_\vartheta(\tau) + O\bigl(\tau^{\frac{1}{2m}(d-\vartheta-1)+\vartheta} \log \tau\bigr) \tag{10.2.34}$$

hold as $\tau \to +\infty$ and for $\vartheta > \vartheta^*$ the asymptotics

$$\mathbf{N}_\vartheta(\tau) = \mathcal{N}^{\mathrm{w}}_\vartheta(\tau) + O\bigl(\tau^{\frac{1}{2m}(d-\vartheta^*-1)+\vartheta}\bigr) \tag{10.2.35}$$

hold as $\tau \to +\infty$.

(ii) Let us drop the transversality condition. After a fine enough covering we obtain that without loss of generality one can assume $\tilde{S}_k = \{x_1 = 0\}$ and $\tilde{S}_l = \{x_1 = h(x')\}$ with a smooth function $h(x')$ and with $x' = (x_2, \dots, x_d)$. Obviously, $\gamma_{kl} \geq \epsilon |h(x')|$ in the domain $Y' = \{0 < x_1 < h(x')\} \cup \{0 > x_1 > h(x')\}$ and therefore $\gamma_{kl}^{-1} \in L^1(Y')$. Let $Y'' = X \setminus Y'$. We consider the part of Y'' where $\gamma_k \geq \gamma_l$ (the other part is treated in the same way). Let us consider the ball $B\bigl(x, \gamma'(x)\bigr)$. This ball either does not intersect S_l (and then it also does not intersect S_k) or intersects S_l; in the second case S_l divides it into two connected components and the point x and S_k are in different components. In our analysis of local semiclassical spectral asymptotics in Chapter 5, only the connected component of $B\bigl(x, \gamma(x)\bigr)$ containing the point x was essential; so one can take $\gamma = \gamma'$ in this case. Therefore in the context of this example condition (10.2.31) needs to be checked only for $\gamma = \gamma'$.

(iii) In the series of examples $\gamma(x) \asymp \operatorname{dist}(x, \mathcal{K})$ where $\mathcal{K}$ is the union of a finite number of C^K manifolds of codimension $\delta \geq 2$. Namely, we consider domains with conical singularities of the boundary, domains the boundaries of which contain components of codimension $\delta \in [2, 4m-1]$.[14] Then all the statements of (i) remain true with $\vartheta^* = \delta - 1$. ◊

---

[13] Such as Dirichlet, Neumann and Robin boundary conditions, or boundary conditions which are smooth, elliptic, regular (see condition (5.1.12)) on $S_l \setminus \mathcal{K}$ and such that A is self-adjoint and $(Au, u) \geq \epsilon_0 \|u\|_m^2 - c\|u\|_0^2$.

[14] For operators of order $2m$ a boundary operator of degree less than $2m - \frac{1}{2}\delta$ can be considered on a δ-codimensional submanifold.

Example 10.2.5 (Cusps and horns). *(i) Let X' be a domain in $\mathbb{R}^{d'}$ satisfying condition* $(8.1.9)_4$ *with γ replaced by $\gamma' = \gamma'(x')$ such that $\gamma' \le c$ and $\gamma'^{1-\vartheta} \in L^1(X')$. Let $X = \{x = (x', x''), x'' \in \mathbb{R}^{d''}, x' \in F(x'')X'\}$* [15] *where the function F satisfies the condition*

$$(10.2.36) \qquad |D^\alpha F| \le cF^{1-|\alpha|} \qquad \forall \alpha : |\alpha| \le K.$$

Then for the domain X the function $\gamma(x) = \epsilon\gamma'(x'F^{-1}(x''))F(x'')$ is admissible and in this case condition (10.2.32)′ is equivalent to the condition

$$(10.2.37)' \qquad F^{d'-\vartheta-1} \in L^1;$$

in particular, (10.2.32) is equivalent to the condition "the $(d-1)$-dimensional measure of ∂X is finite."

(ii) This example includes cusps (exits to infinity): if $F \asymp |x''|^\nu$ with $\nu < 0$ for $|x''| \ge 1$ then the following cases are possible:[16]

(a) For $(d' - 1 - \vartheta)\nu + d'' < 0$, asymptotics (10.2.31) hold; moreover, under the standard condition ($\Psi 0$) on Hamiltonian trajectories, asymptotics (10.2.31) hold;*

(b) For $(d' - 1 - \vartheta)\nu + d'' = 0$, asymptotics with the remainder estimate $o\big(\tau^{\frac{1}{2m}(d-1-\vartheta)+\vartheta}\log\tau\big)$ hold;

(c) For $(d' - 1 - \vartheta)\nu + d'' > 0$, $d'\nu + d'' < 0$, asymptotics with the remainder estimate $O\big(\tau^{\frac{1}{2m}(d-1-\vartheta^*)+\vartheta}\big)$ hold where the critical value $\vartheta^* = d' - 1 - d''\nu^{-1}$ is defined by (b). Moreover, if $\vartheta^* \in \mathbb{Z}^+$ then the last term of these asymptotics is of the form $\varkappa_{\vartheta,\vartheta^*+1}(\tau)\tau^{\frac{1}{2m}(d-1-\vartheta^*)+\vartheta}$ with $\varkappa_{\vartheta,\vartheta^*+1}(\tau) = O(\log\tau)$ while all other terms in this case and all terms in the other cases are of the form $\varkappa_{\vartheta,n}\tau^{\frac{1}{2m}(d-n)+\vartheta}$;

(d) If $d'\nu + d'' = 0$ then the previous statement remains true with only the term $\varkappa_{\vartheta,0}(\tau)\tau^{\frac{d}{2m}+\vartheta}$ in the principal part with $\varkappa_{\vartheta,0}(\tau) \asymp \log\tau$.

We treat this example in section 12.1 again and obtain non-Weylian asymptotics with a more precise remainder estimate in cases (b)–(d) and $d'\nu + d'' > 0$.

(iii) The family of domains described in (i) again covers conical singularities as well as sharp quasi-conical singularities directed outside (i.e., singularities of the type $X = \{x = (x', x''), |x'| \le |x''|^\nu\}$ with $\nu \ge 1$; for $d'' = 1$ one can take only half of this singularity (lying in $\{x, x_1 > 0\}$ where here $x'' = (x_1)$ for $d'' = 1$)). The classification coincides with that in (ii) but in all the inequalities on the exponents "$<$" should be replaced by "$>$" and v.v., and case (d) is impossible.

(iv) If the domain X described in (i) is contained in the ball $B(0, R)$ then $\gamma(x) = \max\big(\gamma_{\rm old}(x)), \operatorname{dist}(x, \partial X)\big)$ works for the domain $\complement X$ where $\gamma_{\rm old}$ is defined as γ in (i).

One can easily see that the classification given in (ii) remains true. ◊

Example 10.2.6. *Let X' and γ' be as in example 10.2.5, let $0 \in X'$ and let $X = \{x = (x', x''), x'' \in B(0, R), x' \in f(|x''|)X'\}$ where $f(0) = 0$ and f is a monotone decreasing function such that the inverse function F satisfies condition (10.2.36) in dimension 1. Then for $x'' \in B(0, \frac{1}{2}R)$ one can take $\gamma(x) = \epsilon|x|\gamma'(x'f^{-1}(|x''|))$. Then condition (10.2.32)′ fulfilled if and only if*

$$(10.2.38) \qquad f^{d'}(r)r^{-\vartheta-1+d''-1} \in L^1([0, 1]).$$

This example covers various blunt quasi-conical singularities directed outward of the type $X = \{x = (x', x''), |x'| \le |x''|^\nu\}$ with $\nu \in (0, 1]$.

(ii) One can replace X by $B(0, R) \setminus X$ with the γ defined above. Again condition (10.2.38) is crucial. This example covers various blunt quasi-conical singularities directed inward. ◊

---

[15] For $d'' = 1$ one can take only one semiaxis, i.e., replace $\mathbb{R}$ by $\mathbb{R}^\pm$.

[16] In this example we assume that a Dirichlet boundary condition is given on ∂X.

Example 10.2.7. *(i) Let* $X = \{x = (x_1, x_2, x''), 0 < x_1 < 1,\ -1 < x_2 < f(x_1), |x''| < 1\}$ *where* $f(x_1)$ *is the jump function given by* $f(x_1) = h_i = \text{const} \geq 0$ *for* $x_1 \in I_i$ *and* $f(x_1) = 0$ *for* $x_1 \in I_\infty$ *where* $I_i \cap I_j = \emptyset$ *for* $i \neq j$, I_i *with* $i < \infty$ *are open intervals of lengths* l_i, $I_\infty = [0,1] \setminus \cup_{i<\infty} I_i$ *and* $\operatorname{mes} I_\infty = 0$. *Then the subdomain* $X \cap \{x_2 > 0\}$ *consists of separate "boxes" in each of which one should take* γ *as in example 10.2.5(i). Then for* $\vartheta \in [0,1)$ *the contribution of the i-th box to* $\int \gamma^{-1-\vartheta} dx$ *is of the order* $h_i l_i^{-\vartheta} + l_i^{1-\vartheta}$. *On the other hand, on* $X \cap \{x_2 \leq 0\}$ *one can take* $\gamma(x) = \frac{1}{2}(-x_2 + \delta(x_1))$ *where* $\delta(x_1) = \operatorname{dist}(x_1, I_\infty)$. *Therefore the contribution of this subdomain is of the order*

$$\int |\log \delta(x_1)| dx_1 \asymp \sum_i -l_i \log l_i \qquad \text{for } \vartheta = 0,$$

$$\int \delta(x_1)^{-\vartheta} dx_1 \asymp \sum_i l_i^{1-\vartheta} \qquad \text{for } \vartheta > 0.$$

Therefore condition (10.2.32) (i.e., with $\vartheta = 0$*) means that*

$$\sum_i h_i < \infty, \quad \sum_i -l_i \log l_i < \infty$$

and for $\vartheta > 0$ *condition* (10.2.32)$'$ *means that*

$$\sum_i h_i l_i^{-\vartheta} < \infty, \qquad \sum_i l_i^{1-\vartheta} < \infty.$$

(ii) Recall that under the cone condition, the Dirichlet boundary value problem can be replaced by an arbitrary regular boundary valuc problcm. Rccall that this condition does not mean that the cone Γ *lies completely in* X*; we assume that the intersection of* Γ *with* X *is connected and a Dirichlet condition is given on* $\Gamma \cap \partial X$. ◇

Let us discuss mixed boundary value problems assuming the cone condition is fulfilled in the indicated form.

Example 10.2.8. *Let the geometry of the domain (and the singularities of the coefficients of the operator A) give us the scaling function* $\gamma = \gamma_0$. *Then in order to take into account the singularities of the boundary value problem we should replace* γ *by* $\min(\gamma, \gamma'')$ *where* γ'' *corresponds to the singularities of the boundary value problem. If* $\bar\gamma = \gamma(\bar x) \leq 2\operatorname{dist}(x, \partial X)$ *then one can take* $\gamma'' \asymp \gamma$ *in* $B(\bar x, \frac{1}{3}\bar\gamma)$. *Otherwise one can straighten* ∂X *in* $B(\bar x, \frac{2}{3}\bar\gamma)$ *by means of a map which is uniformly smooth in the* γ*-scale with Jacobian disjoint from 0. Therefore without loss of generality one can assume that* $X \cap B(\bar x, \bar\gamma) = \{x_1 > 0\} \cap B(\bar x, \bar\gamma)$ *and then* $\gamma''(x) \asymp x_1 + \gamma''(x')$ *in* $X \cap B(\bar x, \frac{1}{2}\bar\gamma)$ *where here* γ'' *is given only on* ∂X *in order to take into account the singularities of the boundary value problem. Therefore if* $\gamma_0^{-1-\vartheta} \in L^1(X)$ *then* $\gamma''^{-1-\vartheta} \in L^1(X)$ *if and only if* $\log \gamma'' \in L^1(\partial X_r)$ *for* $\vartheta = 0$, $\gamma''^{-\vartheta} \in L^1(\partial X_r)$ *for* $\vartheta > 0$ *where the standard surface measure is used on* ∂X_r. *In particular, this condition is fulfilled in* $\vartheta \in [0, \delta)$ *and* $\gamma'' \geq \epsilon_0 \operatorname{dist}(x, \mathcal{K})$ *where* $\mathcal{K}$ *is a subset of* ∂X *with the following property:*

For every $y \in X_r$ $\mathcal{K} \cap B(y, \gamma(y))$ *is contained in* $\bigcup_{1 \leq l \leq L} \mathcal{K}_l$ *where* $\mathcal{K}_l$ *(depending on* y *also) are uniformly* C^K *submanifolds*[17] *of* ∂X_r *of codimension* δ.

This analysis provides us immediately with a series of examples for the Laplacian when on part of the boundary a Dirichlet condition is given and on the remaining part a Neumann (or Robin) boundary condition is given and $\vartheta < 1$.

[17] In the γ-scale.

Let us consider a boundary value problem of variational origin. Let us assume that the boundary terms of the functional $\boldsymbol{\Phi}(u)$ [18] *are regular in the sense that after the dilatation* $x_{\text{new}} = (x_{\text{old}} - \bar{x})\gamma(\bar{x})^{-1}$ *and multiplication by* $\gamma(\bar{x})^{2m-d}$ *we obtain a functional the coefficients of which (and all their derivatives of order* $\le K$*) are bounded in* $B(0,1)$*. The same should be true for the boundary conditions which define* $\mathrm{D}(\boldsymbol{\Phi})$ *but the* l*-th condition has its own weight* $\varrho_{(l)}$ *for multiplication and is given on the set* $\Xi_l \subset \partial X$ $(l = 1,\dots,L)$*. Let us assume that the cone condition is fulfilled in the version described above. Then* $\mathcal{K}$ *is the union of the relative (in* ∂X*) boundaries of* Ξ_l*. However, the possibly irregular structure of* $\mathcal{K}$ *prevents us from immediately applying the above arguments. Nevertheless, for* $\vartheta = 0$ *we avoid this difficulty by means of variational arguments. Let us consider a covering of* $\partial \Xi_l$ *of finite multiplicity by balls* $B\bigl(x^k, \epsilon\gamma(x^k)\bigr)$ *lying in the* $2\epsilon\gamma$*-neighborhood of* $\partial\Xi_l$*; let* $\omega_{l,\epsilon}$ *be the union of these balls. Let us replace* Ξ_l *by* $\Xi'_l = \Xi_l \cap \omega_{l,\epsilon}$ *or by* $\Xi''_l = \Xi_l \setminus \omega_{l,\epsilon}$*. We do the same with the set* Ξ_0 *where the Dirichlet boundary condition is given. In the first case (when all the sets* Ξ_l *expand) the domain of* $\boldsymbol{\Phi}$ *shrinks and* $\mathbf{N}^-(\tau)$ *decreases, and in the second case (when all the sets* Ξ_l *shrink) the domain of* $\boldsymbol{\Phi}$ *expands and* $\mathbf{N}^-(\tau)$ *increases. Moreover, the cone condition remains true in both cases for small enough* $\epsilon > 0$*. Moreover, in both cases the required regularity conditions are now fulfilled (because* $\partial\Xi'_l$ *and* $\partial\Xi''_l$ *intersected with* ∂X_r *lie in the union of spheres* $S\bigl(x^k, \epsilon\gamma(x^k)\bigr)$ *and every ball* $B\bigl(\bar{x}, \gamma(\bar{x})\bigr)$ *intersects no more than* $C = C_\epsilon$ *of these spheres. Therefore*

$$\mathbf{N}'(\tau) \le \mathbf{N}(\tau) \le \mathbf{N}''(\tau) \tag{10.2.39}$$

and $\mathbf{N}'(\tau) = \varkappa_0 \tau^{\frac{d}{2m}} + O(\tau^{\frac{1}{2m}(d-1)})$ *and the same is true for* $\mathbf{N}''$ *with the same coefficient* $\varkappa_0$ *(because it does not depend on the boundary conditions). Therefore the same is true for the problem in question:*

$$\mathbf{N}(\tau) = \varkappa_0 \tau^{\frac{d}{2m}} + O\bigl(\tau^{\frac{1}{2m}(d-1)}\bigr) \qquad \text{as } \tau \to +\infty. \tag{10.2.40}$$

Moreover, under the standard condition $(\Psi 0)$ *on Hamiltonian trajectories*[19] *the asymptotics*

$$\mathbf{N}'(\tau) = \varkappa_0 \tau^{\frac{d}{2m}} + \varkappa'_1 \tau^{\frac{1}{2m}(d-1)} + o\bigl(\tau^{\frac{1}{2m}(d-1)}\bigr)$$

hold and the same is true for $\mathbf{N}''(\tau)$*. Here* $\varkappa'_1$ *is a monotone non-increasing and* $\varkappa''_1$ *is a monotone non-decreasing function of* $\epsilon \in (0, \epsilon_0)$ *and* $\varkappa'_1 \le \varkappa''_1$*. Let us recall that the coefficient* $\varkappa_1$ *consists of two terms; one of them does not depend on boundary conditions and the other is an integral over* ∂X_r *of some function of the principal coefficients of the operator and the boundary operators. Therefore, assuming that*

$$\operatorname{mes}_{\partial X} \partial\Xi_l = 0 \qquad \forall l = 0,\dots,L \tag{10.2.41}$$

we obtain that the limits of $\varkappa'_1$ *and* $\varkappa''_1$ *as* $\epsilon \to +\infty$ *coincide and therefore the asymptotics*

$$\mathbf{N}(\tau) = \varkappa_0 \tau^{\frac{d}{2m}} + \varkappa_1 \tau^{\frac{1}{2m}(d-1)} + o\bigl(\tau^{\frac{1}{2m}(d-1)}\bigr) \qquad \text{as } \tau \to +\infty \tag{10.2.42}$$

hold. ◊

Remark 10.2.9. *Let us consider all cases in which the previous theory yields asymptotics of* $\mathbf{N}_\vartheta(\tau)$ *for* $\vartheta \le \vartheta^*$ *with remainder estimate* $O\bigl(\tau^{\frac{1}{2m}(d-1-\vartheta)+\vartheta}\bigr)$ *and under the*

---

[18] Where $\boldsymbol{\Phi}(u) = \boldsymbol{\Phi}_{\text{int}}(u) + \boldsymbol{\Phi}_{\text{bound}}(u) = \int_X \sum_{\alpha,\beta:|\alpha|\le m,|\beta|\le m} a_{\alpha,\beta}\langle a_{\alpha,\beta} D^\alpha u, D^\beta u\rangle dx + \int_{\partial X} \sum_{\alpha,\beta:|\alpha|\le m-1,|\beta|\le m-1} a_{\alpha,\beta}\langle b_{\alpha,\beta} D^\alpha u, D^\beta u\rangle dS$.

[19] This condition is not associated with $\partial\Xi'_l$ or $\partial\Xi''_l$ because the $2d$-dimensional measures of points lying on trajectories issuing from these sets obviously vanish.

additional condition ($\Psi 0$) *on Hamiltonian trajectories the theory yields asymptotics with the remainder estimate* $o\big(\tau^{\frac{1}{2m}(d-1-\vartheta)+\vartheta}\big)$. *Then it easily follows from the theory of Chapter 8 that for* $\vartheta > \vartheta^*$ *asymptotics with remainder estimate* $o\big(\tau^{\frac{1}{2m}(d-1-\vartheta^*)+\vartheta}\big)$ *hold with no condition on Hamiltonian trajectories. Moreover, a similar deduction holds with an additional factor* $\log^\nu \tau$ *in the remainder estimate,* $\nu > 0$. *This remark is applicable to the previous example with* $\vartheta^* = 0$ *under condition (10.2.41). This remark is also applicable in a different form in the previous and next sections.* ◊

Let us treat the case in which the boundary is irregular.

Example 10.2.10. *Let us assume that the operator* A *satisfies our conditions with* $\gamma =$ const > 0. *Then one can take* $\gamma(x) = \operatorname{dist}(x, \partial X)$ *in which case the remainder estimate in the asymptotics of* $\mathbf{N}_\vartheta(\tau)$ *is "O" of*

$$\tau^{\frac{1}{2m}(d-1-\vartheta)+\vartheta}\int_{r\geq\tau^{\frac{1}{2m}}} r^{-1-\vartheta}d\mu(r) + \tau^{\frac{d}{2m}+\vartheta}\mu(\tau^{-\frac{1}{2m}}) \asymp$$

$$\tau^{\frac{1}{2m}(d-1-\vartheta)+\vartheta}\int_{r\geq\tau^{\frac{1}{2m}}} r^{-2-\vartheta}\mu(r)dr + \tau^{\frac{d}{2m}+\vartheta}\mu(\tau^{-\frac{1}{2m}})$$

where $\mu(r) = \operatorname{mes}(\{x \in X, \operatorname{dist}(x, \partial X) \leq r\})$ *and we assume that* mes $X < \infty$. *In particular, if* $\mu(r) = O(r^\delta)$ *as* $r \to +0$ *then for* $\delta > 1+\vartheta$ *we obtain the remainder estimate* $O\big(\tau^{\frac{1}{2m}(d-1-\vartheta)+\vartheta}\big)$; *moreover, under the standard condition* ($\Psi 0$) *on the Hamiltonian trajectories we obtain the remainder estimate* $o\big(\tau^{\frac{1}{2m}(d-1-\vartheta)+\vartheta}\big)$. *On the other hand, we obtain the remainder estimate* $O\big(\tau^{\frac{1}{2m}(d-1-\vartheta)+\vartheta}\log\tau\big)$ *for* $\delta = 1+\vartheta$ *and we obtain the remainder estimate* $O\big(\tau^{\frac{1}{2m}(d-\delta)+\vartheta}\big)$ *for* $\delta < 1+\vartheta$. *For* $\vartheta = 0, \delta = 1$ *and for* $\vartheta = 0, \delta \in (0,1)$ *we obtain the results of the papers [Brüning 1] and [Fleckinger, Lapidus 1] respectively. Finally, in these two cases under the condition* $\mu(r) = o(r^\delta)$ *as* $r \to +0$ *we obtain "o" in the remainder estimates.* ◊

10.2.3 The strong non-periodicity condition. I

Let us consider examples in which restrictive conditions on the Hamiltonian flow imply asymptotics with highly accurate remainder estimates. Let us assume for the sake of simplicity that $\vartheta = 0$. Here we do not treat smooth closed manifolds without boundary; the reader can find all those results in [Volovoy 1,2] and in the earlier papers [Bérard 1], [Randoll 1]. For the Laplace-Beltrami operator these statements yield the following

Example 10.2.11. *(i) For the Laplace-Beltrami operator on a compact closed Riemannian* C^K *manifold of non-positive sectional curvature (for* $d \geq 3$*) or without conjugate points (for* $d = 2$*) the asymptotics of* $\mathbf{N}(\tau)$ *with remainder estimate* $O\big(\tau^{\frac{1}{2}(d-1)}\log^{-1}\tau\big)$ *hold.*

(ii) For a certain class of surfaces of rotation and for a certain class of graded Lie algebras the asymptotics of $\mathbf{N}(\tau)$ *with the remainder estimate* $O\big(\tau^{\frac{1}{2}(d-1-\delta)}\big)$ *hold with* $\delta > 0$. ◊

I would like to note that in the context of our book the results on Hamiltonian flows obtained in these papers yield asymptotics for $\mathbf{N}_\vartheta(\tau)$ with remainder estimates $O\big(\tau^{\frac{1}{2}(d+\vartheta-1)}\log^{-\vartheta-1}\tau\big)$, $O\big(\tau^{\frac{1}{2}(d+\vartheta-1-\delta)}\big)$ respectively.

Let us consider the Laplace(-Beltrami) operator in compact domains or on compact manifolds with singularities. The following statement is necessary:

Lemma 10.2.12. *Let* X *be a bounded domain in* $\mathbb{R}^d$ *which is a Riemannian manifold with boundary with metric* $g(x,\xi)$ *satisfying conditions*

$$g^{jk} = g^{kj}, \tag{10.2.43}$$

$$\epsilon_0|\xi|^2 \le g(x,\xi) = \sum_{jk} g^{jk}\xi_j\xi_k \le c|\xi|^2, \tag{10.2.44}$$

$$|D^\alpha g^{jk}| \le c\gamma^{-|\alpha|} \qquad \forall \alpha : |\alpha| \le K \quad \forall j,k \tag{10.2.45}$$

and with scaling function $\gamma(x)$ such that $\gamma^{-1} \in L^1(X)$. Then there exist a set of measure 0 of the first Baire category $\Lambda \subset S^*\bar{X}$ and a continuous Hamiltonian flow Φ_t on $S^*\bar{X} \setminus \Lambda$ with reflection at the boundary such that all its trajectories lie in $S^*\bar{X}_r$, $\bar{X}_r = \{x \in \bar{X}, \gamma(x) > 0\}$ and are transversal to ∂X. This flow conserves the natural measure $dxd\xi : d\sqrt{g}$ on S^*X.

Proof. The proof coincides with the arguments of Chapter 6 in the book [Cornfeld, Sinai, Fomin 1]. □

Let us note that if we are interested only in trajectories of length not exceeding T then in the statement of this lemma one can replace Λ by a closed nowhere dense subset Λ_T. However, $S^*\bar{X} \setminus \Lambda_T$ is not invariant with respect to the flow Φ_t.

Let $J_t(x,\xi)$ be the Jacobi matrix of Φ_t calculated at the point (x,ξ); then $\det J_t(x,\xi) = 1$. We need the following statements concerning the Hamiltonian flows on various manifolds:

Lemma 10.2.13 (Polyhedra). *Let X be a compact polyhedral domain in $\mathbb{R}^d$ with $g(x,\xi) = |\xi|^2$. Then*

$$|J_t(x,\xi)| \le C(|t|+1) \tag{10.2.46}$$

uniformly with respect to $(x,\xi) \in S^*\bar{X} \setminus \Xi$.

Proof. We know that for a polyhedral domain $\gamma^{-1} \in L^1$ and therefore lemma 10.2.12 is applicable. Note that Φ_t consists of linear movements $(x,\xi) \to (x + t_j\xi, \xi)$ and reflections from various faces; the reflection from the face Z is $(x,\xi) \to (\Pi_Z x + y_Z, \Pi_Z\xi)$ with $y_Z \in \mathbb{R}^d$, $\Pi_Z \in \mathcal{G}$ where $\mathcal{G}$ is a group of reflections in $\mathbb{R}^d$ (i.e., linear orthogonal maps Π such that $\Pi^2 = I$). Therefore if $(\bar{x},\bar{\xi}) \notin \Xi_{\bar{t}}$ and $(\bar{x},\bar{\xi}), \Phi_{\bar{t}}(\bar{x},\bar{\xi}) \notin S^*\bar{X}|_{\partial X}$ then in a neighborhood of $(\bar{x},\bar{\xi},\bar{t})$

$$\Phi_t(x,\xi) = (\bar{\Pi}(x+\xi t) + \bar{y}, \bar{\Pi}\xi)$$

with $\bar{y} \in \mathbb{R}^d$ and $\bar{\Pi} \in \mathcal{G}$ depending on $(\bar{x},\bar{\xi},\bar{t})$. This immediately yields (10.2.46). □

Let us treat different classes of strictly convex compact Riemannian manifolds with boundary. The *convexity* means that if $\phi \in C^K$, $\phi > 0$ on X, $\phi = 0$ and $\nabla\phi \ne 0$ on ∂X then

$$\phi(x) = 0,\ \{g,\phi\}(x,\xi) = 0 \implies \{g,\{g,\phi\}\}(x,\xi) < 0 \qquad \forall (x,\xi) \in S^*\bar{X}. \tag{10.2.47}$$

Recall that $\{.,.\}$ are Poisson brackets. In this case $\Xi = \bigl\{(x,\xi), \phi = \{g,\phi\} = 0\bigr\} \subset S^*\bar{X}$ is a manifold of codimension 2. Let

$$\delta(x,\xi) = \frac{1}{2}\{g,\phi\}^2 - \{g,\{g,\phi\}\}\phi$$

in a small enough neighborhood of Ξ (then $\delta(x,\xi) = 0 \iff (x,\xi) \in \Xi$) and $\delta \asymp 1$ outside of this neighborhood.

Lemma 10.2.14. *Under the assumptions made above the following estimates hold:*

$$C^{-1}e^{-C|t|} \le \frac{\delta(\Phi_t(x,\xi))}{\delta(x,\xi)} \le Ce^{C|t|}, \tag{10.2.48}$$

$$|J_t(x,\xi)| \le Ce^{C|t|}\delta(x,\xi)^{-M} \tag{10.2.49}$$

for all $(x,\xi) \in S^*\bar{X} \setminus \Xi$, $t \in \mathbb{R}$ where C does not depend on (x,ξ,t).

Proof. Without loss of generality one can assume that $t > 0$ and $\delta(x,\xi)$ is a smooth function; then calculating $\{g,\delta\}$ we see that along Hamiltonian trajectories $|\frac{d}{dt}\delta| \le C\delta$ and δ does not change at the moment of reflection at ∂X. This yields (10.2.48).

The forthcoming arguments are similar to the arguments from dynamical systems theory used in sections 3.3 and 5.3. I give only a brief sketch, leaving the detailed proofs to the reader. Let $\Sigma = S^*X$ and $\Sigma' = \Sigma|_{\partial X}$. Let us introduce the function $\eta = \{g,\phi\} = \pm(2\delta)^{\frac{1}{2}}$ and the density $d\nu = |\eta| dx d\xi : dg^{\frac{1}{2}}$ (which does not depend on the choice of ϕ) in Σ'. Let us consider the map $\Psi : \Sigma' \to \Sigma'$ transforming the point $z' \in \Sigma'$ to the point $\iota\Phi_{t_1(z')}(z')$ where $t_1(z') > 0$ is the time of the first return of the trajectory passing through z' to ∂X and $\iota : \Sigma' \to \Sigma'$ is the reflection map; if a trajectory never hits ∂X again then $t_1(z') = +\infty$ and $\Psi(z')$ is not defined. One can easily see that (10.2.49) holds if the trajectory $\Phi_{t'}(z)$ does not hit ∂X for $0 \le t' \le t$ or $0 \ge t' \ge t$. Moreover, one can easily see that

$$t_1(z') \asymp |\eta(z')|, \tag{10.2.50}$$

and $D\Psi(z') = I + O\bigl(t_1(z')\bigr)$ in the local coordinates provided $|\eta(z')| \le \epsilon_1$ (where $\epsilon_1 > 0$ is small enough) and therefore

$$|D\Psi^n(z')| \le e^{Ct_n(z')} \tag{10.2.51}$$

for $n = 1$ under the same assumption. Inverting time in our arguments if necessary we obtain that $\delta(z') \asymp \delta(\Psi(z'))$ and therefore (10.2.51) holds without the additional assumption that $\eta(z')$ is small.

Let us consider the map Ψ^n. Then (10.2.48) and (10.2.50) yield that

$$n \le Ce^{Ct_n}|\eta(z')|^{-1} \tag{10.2.52}$$

with an appropriate constant C and that (10.2.51) holds for arbitrary $n \in \mathbb{Z}^+$. Here $t_n(z') > 0$ is the time of the n-th return of the broken Hamiltonian trajectory to ∂X (it is possible that $t_n(z') = +\infty$ in which case $\Psi^n(z')$ is not defined).

Now let us consider an arbitrary point $z \in \Sigma$ and let us rewrite $\Phi_t(z)$ in the form

$$\Phi_t(z) = \Phi_{t-t_{n+1}(z)} \circ \Psi^n \circ \Phi_{t_1(z)}(z) \tag{10.2.53}$$

where $t_j(z) > 0$ is the time of the j-th return of the broken Hamiltonian trajectory to ∂X and we assume that $t_{n+1}(z) \le t < t_{n+2}(z)$ and $n \ge 0$. One can easily see that

$$\begin{aligned} &0 \le t_1(z) \le C\delta^{\frac{1}{2}}(z) \Longleftarrow \delta(z) \le \epsilon, \\ &0 \le t - t_{n+1} \le C\delta^{\frac{1}{2}}(\Phi_t(z)) \Longleftarrow \delta(\Phi_t(z)) \le \epsilon \end{aligned} \tag{10.2.54}$$

and

$$z' = \Phi_{t_1(z)} \in \Sigma' \implies |\eta(z')| \asymp \delta^{\frac{1}{2}}(z). \tag{10.2.55}$$

One can easily see that for $z' \in \Sigma'$ the inequality $|dt_1(z')| \le Ce^{Ct_1(z')}|dz'|$ holds and therefore

$$|dt_n(z')| \le C|dt_{n-1}(z')| + Ce^{C(t_n - t_{n+1})}|d\Psi^{n-1}(z')|$$

and finally we obtain by induction (using the above inequalities) that

$$|dt_n(z')| \le Ce^{Ct_n(z')}\delta(z')^{-\frac{1}{2}}|dz'|. \tag{10.2.56}$$

One can easily see that $|dt_1(z)| \le Ce^{Ct_1(z)}\delta(z)^{-\frac{1}{2}}|dz|$ for $z \in \Sigma$. These two inequalities yield that

$$|dt_n(z)| \le Ce^{Ct_n(z)}\delta(z)^{-1}|dz| \tag{10.2.57}$$

for all $n \in \mathbb{Z}^+$. The above inequalities yield (10.2.49). □

Lemma 10.2.15 (Elliptical domains). *(i) Let X be the interior of an ellipse in $\mathbb{R}^2$ or, more generally, a domain lying between two confocal ellipses where the standard Euclidean metric is given on $\mathbb{R}^2$. Then there exists a continuous non-negative function $\delta(z)$ on Σ such that*

$$\delta(\Phi_t(z)) = \delta(z), \tag{10.2.58}$$

$$|D\Phi_t(z)| \le C\delta^{-L}(z)(|t|+1)^{L'} \qquad \forall z \in \Sigma \setminus \Xi \quad \forall t \in \mathbb{R} \tag{10.2.59}$$

and

$$\mathrm{mes}_\Sigma(\{z : \delta(z) < h\}) = O(h^\sigma) \qquad \text{as } h \to +0 \tag{10.2.60}$$

with $\sigma > 0$ and $L' = 1$. Here mes_Σ is the standard measure on Σ.

(ii) The same is true for a ball in $\mathbb{R}^d$ and for a domain lying between two balls with the same center (where the standard Euclidean metric is given on $\mathbb{R}^d$).

Proof. (i) Let us consider an elliptical domain or a domain lying between two ellipses with the same foci; let Y_2 be the outer and the Y_1 inner ellipse. In the first case let us define Ξ as before. In the second case let us introduce Ξ_2 and Ξ_1' in the same way as before for Y_2 and Y_1 respectively and let us set

$$\Xi_1 = \bigl\{(x + t\xi, \xi) : (x,\xi) \in \Xi_1', t \in \mathbb{R}, x + t\xi \in \bar{X}\bigr\}$$

(thus, the definitions of Ξ_1 and Ξ_2 are different). Then $\Xi = \Xi_2 \cup \Xi_1$ is the same as in lemma 10.2.12 and $\Xi_2 \cap \Xi_1 = \emptyset$.

It is well known (see, e.g., [Babich, Buldyrev 1]) that in domains of this type every billiard belongs to one of the following three classes:

(a) All the straight lines containing at least one segment of this billiard do not pass between foci and are tangent to an ellipse Y which has the same foci as Y_1 and Y_2 and is smaller than Y_2 (although not necessarily larger than Y_1).

(b) All the straight lines containing at least one segment of this billiard pass between foci and are tangent to the same hyperbola Y which has the same foci as Y_1 and Y_2.

(c) All the straight lines containing at least one segment of this billiard pass through the foci (either by turns or simultaneously).

Then $\beta \overset{\text{def}}{=} +b^2, -b^2, 0$ in cases (a), (b), (c) respectively is an invariant of the Hamiltonian flow. Here a_k and b_k are semiaxes of Y_k ($k = 1, 2$) directed along the axes x_1 and x_2 respectively, $a_k > b_k$ and a and b are the same quantities for Y. Direct calculation shows that

(10.2.61) $$\beta = b_k^2 - a_k^2 b_k^2\Big(\frac{x_1\xi_1}{a_k^2} + \frac{x_2\xi_2}{b_k^2}\Big)^2 \qquad \text{for } x \in Y_k,\ |\xi| = 1.$$

Let us introduce $\Sigma_2' = \Sigma|_{Y_2}$ and the map $\Psi : \Sigma_2' \to \Sigma_2'$ in the same way as in the previous lemma where $t_1(z')$ is the time of the first return to Y_2 (maybe after one reflection from Y_1). Then

(10.2.62) $$t_1(z') \asymp \delta_2(z')^{\frac{1}{2}} = \Big|\frac{x_1\xi_1}{a_2^2} + \frac{x_2\xi_2}{b_2^2}\Big|.$$

Here the right-hand expression is an invariant of the map Ψ and therefore in what follows

(10.2.63) $$n \le C(|t_n(z')| + 1)\delta_2(z')^{-\frac{1}{2}}$$

with the obvious definition of $t_n(z')$. Let us introduce the function $\eta = \frac{x_1\xi_1}{a_2^2} + \frac{x_2\xi_2}{b_2^2}$ on Σ_2'. Then

$$|\nabla_{\Sigma'}\eta| \asymp (|x_1| + |\xi_1|)\cdot(|x_2| + |\xi_2|) = \delta_3(z').$$

In fact, let us introduce coordinates (u,v) : $x_1 = a_2\cos u$, $x_2 = b_2 \sin u$, $\xi_1 = \cos v$, $\xi_2 = \sin v$ on Σ_2'; in what follows $z' = (u,v)$. Then the previous relation is trivial.

Let us calculate the density $d\nu : d\eta$ on the lines $\{\eta = \text{const}\}$ by different methods where $d\nu$ is the density on Σ_2' corresponding to $\phi = \frac{1}{2}\big(1 - \frac{x_1^2}{a_2^2} - \frac{x_2^2}{b_2^2}\big)$. Then we obtain the equality $|\frac{du}{U}| = |\frac{dv}{V}|$ under the link condition $\eta = \text{const}$ where $U = U(z')$, $V = V(z')$ and

(10.2.64) $$|U| + |V| \asymp \delta_3.$$

Taking into account that $d\nu$ and $|\eta|$ are invariants of Ψ we see that $|d\nu : d\eta|$ is also an invariant and therefore the middle equality in this chain holds:

$$\Big|\frac{du}{U}\Big| = \Big|\frac{dv}{V}\Big| = \Big|\frac{du'}{U'}\Big| = \Big|\frac{dv'}{V'}\Big|$$

under the link condition $d\eta = 0$ where $(u',v') = \Psi(u,v)$ and $U' = U(u',v')$, $V' = V(u',v')$. One can easily see that $\delta_3(u,v) \asymp \delta_3(u',v')$.

Starting with this chain of equalities one can easily prove that without the link condition the following equivalence chain holds:

$$\Big|\frac{du}{U}\Big| \equiv \Big|\frac{dv}{V}\Big| \equiv \Big|\frac{du'}{U'}\Big| \equiv \Big|\frac{dv'}{V'}\Big|$$

modulo terms with absolute values no greater than $C\big(\delta_3^{-L} + \delta_1^{-L}|d\eta|\big)$ where $\delta_1(z') = \operatorname{dist}(z', \Xi) \asymp \min\big(|\beta - b_1^2|, |\beta - b_2^2|\big)$. Here δ_3 is not Ψ-invariant but $\delta_3 \ge \epsilon \min\big(|\beta|, |b + a_2^2 - b_2^2|\big)$ and β is invariant. Therefore for (u',v') the previous chain of equivalences holds modulo terms with absolute values not exceeding $Cn\delta^{-L}|d\eta|$ with

$$\delta = \min\big(|\beta|, |\beta + a_2^2 - b_2^2|, |\beta - b_1^2|, |\beta - b_2|^2\big).$$

Increasing L if necessary we obtain the inequality $|D\Psi^n| \le Cn\delta^{-L}$. The remainder of the proof repeats the final arguments of the proof of the previous lemma with (10.2.49) replaced by (10.2.63). This yields (10.2.58). The proof of (10.2.59) is trivial.

(ii) If $a_2 = b_2$ and $a_1 = b_1$ then the proof is trivial, and from the beginning we use $\delta(z)$ instead of $\delta_3(z)$. The case of a ball or of a ball layer can be reduced to case $d = 2$ because every billiard lies in some two-dimensional plane passing through origin and the Hamiltonian flow commutes with the map $(x,\xi) \to (Gx, G\xi)$ for every orthogonal matrix G. Therefore one can consider only billiards and their variations in the plane (x_1, x_2). Singularities related to this reduction near the manifold $\{(x,\xi), x = t\xi, t \in \mathbb{R}\} = \{(x,\xi), \beta = 0\}$ are taken into account because $|\beta| \ge \epsilon\delta$. □

Let us discuss the intermediate results obtained. In order to do this we need the following

Definition 10.2.16. Let X be a compact γ-admissible Riemannian manifold with boundary where γ is an admissible scaling function such that $\gamma^{-1} \in L^1(X)$. We say that a *power (exponential) geodesic flow with reflections* is given on X if for every $T \geq 0$ there exists a closed set $\Xi_T \subset \Sigma$ with $\operatorname{mes}_\Sigma \Xi_T \leq CT^{-1}$ such that through every point $z \in \Sigma \setminus \Xi_T$ there passes a geodesic trajectory $\Phi_t(z) = \bigl(x(t), \xi(t)\bigr)$ with reflections from ∂X with $t \in [0, T]$ or $t \in [-T, 0]$, along which $\gamma(x(t)) \geq \varepsilon(T)$, the angle with ∂X is not less than $\varepsilon(T)$ at every point where the distance from the boundary is less than $\varepsilon(T)$, and $|D\Phi| \leq \varepsilon(T)^{-1}$ where $\varepsilon(T) = T^{-L}$ ($\varepsilon(T) = e^{-CT}$ respectively) with large enough C, L. ◊

Thus,

(i) The condition of the power geodesic flow with reflections is fulfilled for Lie groups, example 4.2.11(ii), for polyhedral domains (including those with cuts which are γ-admissible with scaling function γ such that $\gamma^{-1-\sigma} \in L^1(X)$ with $\sigma > 0$). One can easily see that in this case the measure of the set $\bigl\{z, \exists t \in [0, T] : \gamma(\Phi_t(z)) \leq \varepsilon\bigr\}$ does not exceed $C\varepsilon^\sigma T$ and the measure of the set Q does not exceed $C\varepsilon T$ where $Q = \bigl\{z, \exists t \in [0, T]$ such that the distance from x to ∂X and the angle between $\frac{dx}{dt}$ and $T\partial X$ do not exceed $\epsilon\varepsilon$, and $\gamma(\Phi_t(z)) \geq \varepsilon\bigr\}$.

Moreover, this condition is fulfilled for balls and concentric ball layers in $\mathbb{R}^d$ and for elliptic domains and domains lying between two confocal ellipses in $\mathbb{R}^2$. Finally, this condition is obviously fulfilled for manifolds with periodic geodesic flow (such manifolds can be either smooth or γ-admissible with $\gamma^{-1-\sigma} \in L^1(X)$ for arbitrarily small $\sigma > 0$).

(ii) The condition of exponential geodesic flow with reflections is fulfilled for strictly convex Riemannian manifolds with boundary. It is trivially fulfilled for all compact Riemannian manifolds without boundary.

(iii) Moreover, one can construct new manifolds of the corresponding type by the following methods:

(a) Taking a Cartesian product;

(b) Taking a factor manifold of X with respect to a finite group of its isometries;

(c) Considering reflections, i.e., metric preserving diffeomorphisms $S : \bar{X} \to \bar{X}$ such that $S^2 = Id$ and $S(x) = x \implies dS(x) \neq dx$ (X can have cuts so S does not necessarily preserve X) and removing from X some appropriate subset K of the set of fixed points of S. The condition on K is $\delta^{-1-\sigma} \in L^1(X)$ for $\delta(x) = \operatorname{dist}(x, K)$.

For mixed boundary conditions (Dirichlet condition on Y_- and Neumann condition on Y_+) the condition associated with the boundary value problem is surely fulfilled provided

$$\operatorname{mes}_{\partial X}\bigl(\{x \in \partial X, \operatorname{dist}(x, \bar{Y}_+ \cap \bar{Y}_-) < \varepsilon(T)\}\bigr) = O(T^{-1}) \tag{10.2.65}$$

with $\varepsilon(T)$ defined above. In this case let us consider the set of all points of $\Sigma \setminus \Xi_T$ through which there pass trajectories $\Phi_t(z)$ hitting the $\varepsilon(T)$-neighborhood of $\bar{Y}_+ \cap \bar{Y}_-$. One can easily see that the measure of this set does not exceed CT^{-1} provided C, L in the definition of $\varepsilon(T)$ are large enough.

We have a non-trivial *non-periodicity* condition. In order to take it into account we introduce

Definition 10.2.17. In the context of definition 10.2.16 we say that there is a *non-periodic power (or exponential) geodesic flow* if in addition to the conditions of definition 10.2.16

$$\operatorname{mes}\bigl\{z \in \Sigma \setminus \Xi_T, \exists t \in (0, T) : |t|^{-1}\operatorname{dist}(z, \Phi_t(z)) \leq \varepsilon(T)\bigr\} = O(T^{-1}) \tag{10.2.66}$$

where we increase C and L in the definition of $\varepsilon(T)$ if necessary. ◊

Remark 10.2.18. *(i) Let us note that* $\operatorname{dist}(z,\Phi_t(z)) \asymp |t|$ *for* $|t| \le \epsilon\gamma(x)$, $z = (x,\xi)$. *Therefore the non-periodicity condition should be checked only for* $|t| \ge \epsilon\varepsilon(T)$.

(ii) Let us consider the Cartesian product X *of two manifolds* X' *and* X'' *with dimensions* ≥ 1. *Let us note that* $\Phi_t(x',x'',\xi',\xi'') = (\Phi'_{t|\xi'|}(x',\xi'),\Phi''_{t|\xi''|}(x',\xi''))$ *for* $(x',x'',\xi',\xi'') \in S^*X$. *Then, if the condition* $\operatorname{dist}(z',\Phi_t(z')) \ge \varepsilon|t|$ *is fulfilled for all* $t \in (0,T)$ *for a point* $z' = (x',\xi'_0) \in S^*X'$ *then the same condition with* ε^2 *instead of* ε *is fulfilled for a point* $(x',x'',\xi',\xi'') \in S^*X$ *with* $\xi' = |\xi'|\xi'_0$ *and* $|\xi'| \ge \varepsilon$. *The measure of the set of points with* $|\xi'| \le \varepsilon$ *is surely* $O(\varepsilon)$. *Therefore one should treat only points* $(x',x'',\xi',\xi'') \in S^*X$ *such that the non-periodicity condition is violated at both points* $z' = (x',\xi'|\xi'|^{-1}) \in S^*X'$ *and* $z'' = (x'',\xi''|\xi''|^{-1}) \in S^*X''$. *Let us set* $T' = \min(t \ge \epsilon\varepsilon : \operatorname{dist}(z',\Phi'_t(z')) \le \varepsilon)$ *and let* T'' *be defined in a similar way. We assume that* $|\xi'| \ge \varepsilon$ *and* $|\xi''| \ge \varepsilon$. *Then taking into account the condition on the Jacobi matrix we obtain that* $\operatorname{dist}(\Phi'_t(z'),z') \ge \varepsilon^2$ *provided* $|\frac{t}{T'} - m| \ge \varepsilon\ \forall m \in \mathbb{Z}$ *and* $|t| \le T$ *(where* $\varepsilon = \varepsilon(T)$*. Therefore one needs to estimate the measure of the set*

$$\mathcal{Q} = \{\phi \in \mathbb{S}^1 : \exists m',m'' \in \mathbb{Z} : \\ |m'| \le \varepsilon^{-\sigma}, |m''| \le \varepsilon^{-\sigma}, |m'T'\cos\phi + m''T''\sin\phi| \le \varepsilon\}$$

where ϕ *is associated with* ξ *by the relations* $|\xi'| = \cos\phi$, $|\xi''| = \sin\phi$ *and one can take* $\sigma > 0$ *arbitrarily small. However,* $\operatorname{mes}\mathcal{Q} = O(\varepsilon^\sigma)$ *obviously.*

Therefore for Cartesian products the non-periodicity condition always is fulfilled.

(iii) The above arguments also yield that $\operatorname{dist}(G^*z,\Phi_t(z)) \le \varepsilon^2$ *implies that* $\operatorname{dist}(G^{*k}z,\Phi_{kt}(z)) \le \varepsilon$ *for constant* k *and the isometry* $G^* : S^*\bar{X} \to S^*\bar{X}$ *induced by the isometry* $G : \bar{X} \to \bar{X}$. *Therefore definition 10.2.17 is stable under the above constructions (b),(c) of new manifolds.* ◊

Lemma 10.2.19. *Let* X *be a polyhedral domain with cuts; assume that* X *is* γ*-admissible with* $\gamma^{-1-\sigma} \in L^1$ *with* $\sigma > 0$. *Or let* X *be a ball or a concentric ball layer in* $\mathbb{R}^d$. *Or let* X *be a domain in* $\mathbb{R}^2$ *bounded by one or two confocal ellipses. Then there is a non-periodic power geodesic flow.*

Proof. (i) Let X be a polyhedral domain with cuts. Then the measure of the set $\{z = (x,\xi) \in \Sigma, \operatorname{dist}(x,\partial X) \le \varepsilon\}$ is less than $C\varepsilon$. Let us cover the set $\{z = (x,\xi) \in \Sigma, \operatorname{dist}(x,\partial X) \ge \varepsilon\}$ with no more than $C\varepsilon^{1-4d}$ balls of radius $O(\varepsilon^2)$ and let us divide the time interval $[0,T]$ into no more than $O(T\varepsilon^{-2})$ intervals of length $O(\varepsilon^2)$. Let $\varepsilon = O(T^{-L})$ with a large enough L. Then if the distance between the Φ_t-image of some ball and this ball is less than ε^2 for some t from the above subinterval then for all points (x,ξ) of this ball and for all t of this subinterval the geodesic flow is given by the formula

$$\Phi_t(x,\xi) = (y + \Pi x + t\Pi\xi, \Pi\xi)$$

with constant y and Π such that $\operatorname{dist}(\xi,\Pi\xi) \le \varepsilon$ for all ξ in question. Taking into account that for every x the $\mathbb{S}^{d-1}$-measure of the set

$$\{\xi \in \mathbb{S}^{d-1}, |y + \Pi x - x + t\Pi\xi| \le \varepsilon^N \quad \text{for some } t \text{ with } |t| \ge \varepsilon\}$$

does not exceed $O(\varepsilon^N)$ we obtain after summation over all the balls and subintervals that (10.2.66) holds.

(ii) Let X be a domain bounded by one or two confocal ellipses (balls and concentric ball layers are treated in a similar way). Then (10.2.66) is equivalent to

$$\operatorname{mes}_{\Sigma'_2}\{z' \in \Sigma'_2 : \exists k = 1,\dots,n : \operatorname{dist}(z',\Psi^k z') \le \varepsilon = n^{-M}\} = \\ O(n^{-\sigma}) \quad \text{as } n \to +\infty \tag{10.2.67}$$

for appropriate M and $\sigma > 0$.

Let us first consider the zone $\{(x,\xi),\beta>0\}$. Let us consider the lines $\{\eta=\text{const}\}$. These lines are analytically parameterized by a cyclic parameter $u\in\mathbb{R}/2\pi\mathbb{Z}$ (see the proof of lemma 10.2.15).[20] Let $w\in\mathbb{R}/\mathbb{Z}$ be a new cyclic coordinate: $w=w_\eta(u)=\nu_\eta((0,u))m_\eta^{-1}$ for $u\in[0,2\pi]$ and then by periodicity. Here $m_\eta=\nu_\eta([0,2\pi])$. One can easily see that $m_\eta\asymp\log|\beta|$ as $\beta\to+0$ for $a_2>b_2$ and $m_\eta\asymp\beta^{-1}$ as $\beta\to+0$ for $a_2=b_2$.

For every fixed η the map Ψ_η of this line is analytic and preserves the measure $dw_\eta=m_\eta^{-1}d\nu_\eta$ and the orientation[21] and therefore Ψ_η is a shift $\Psi_\eta: w_\eta\to w_\eta+f(\eta)$. Then in the proof of (10.2.67) for $\Sigma_2'\cap\{\beta>0\}$ one should prove that

$$Cn\,\mathrm{mes}\,\{\eta:\exists k,l=1,\dots,n:|f(\eta)-\frac{k}{l}|\le\varepsilon=n^{-L}\}=O(n^{-\sigma})\tag{10.2.68}$$

$$\text{as } n\to+\infty$$

for admissible L and $\sigma>0$; here mes means Lebesgue measure.

Obviously, $f(\eta)$ is a continuous function. One can easily prove that $\epsilon\beta^p\le|\frac{df}{d\eta}|\le c\beta^{-p}$ for $|\eta|\le a_2-\epsilon_0$, i.e., for $\beta\in(0,\beta_0)$ with small enough $\epsilon_0>0,\beta_0>0$. Recall that $\beta=b_2^2-a_2^2b_2^2\eta^2$. Therefore (10.2.68) holds for the η indicated here.

On the other hand, Ψ is an analytic map for $\beta\ne b_1^2$ (the case when rays are tangent to the inner ellipse Y_1). Therefore the function $f(\eta)$ is analytic for $|\eta|<\eta_1=a_2^{-1}b_2^{-1}(b_2^2-b_1^2)^{\frac12}$ and for $\eta_1<|\eta|\le a_2^{-1}$. Obviously, $f(\eta)$ is not constant in each of these three intervals. One can easily see that $f_\pm(\zeta)=f(\bar\eta\pm\zeta^2)$ is an analytic function for $|\zeta|\le\epsilon_0$ and $\bar\eta=+\eta_1,-\eta_1$. But one can easily see that (10.2.65) holds for every C^K function on the closed segment provided

$$\sum_{1\le j\le K-1}|f^{(j)}|\ge\epsilon_0.\tag{10.2.69}$$

Therefore (10.2.68) holds.

(iii) Let us treat the zone $\{(x,\xi):0>\beta>b_2^2-a_2^2\}$. In this case every line $\{\eta=\text{const}\}$ consists of two connected components and every component is equivalent to $\mathbb{S}^1$ and the x-projection maps every component to one of two segments on the ellipse Y_2 describing by the inequality $x_2^2\ge-\beta b_2^2(a_2^2-b_2^2)^{-1}$. The inverse map is two-valued. Let us introduce on this connected component the cyclic coordinate u' which can be done in the following way: let us consider the confocal hyperbola described by the equation $\frac{x_1^2}{a'^2}-\frac{x_2^2}{b'^2}$ with $a'^2+b'^2=a_2^2-b_2^2$. This hyperbola intersects Y_2 in four points $(\bar x_1,\bar x_2)$ with $\bar x_1^2=a_2^2(a_2^2-b_2^2+\beta)(a_2^2-b_2^2)^{-1}$, $\bar x_2^2=-b_2^2\beta(a_2^2-b_2^2)^{-1}$. We introduce the parameter $u'\in\mathbb{S}^1$ on this hyperbola: $x_1=a'\sec u'$, $x_2=b'\tan u'$. Through every point of the hyperbola corresponding to this parameter there passes a tangent straight line hitting the corresponding segment of the ellipse. Then (x_1,x_2) is an intersection point and (ξ_1,ξ_2) is a unit vector along this tangent line, directed inside and outside for $\eta<0$ and $\eta>0$ respectively.

Let us introduce the parameter $w'\in\mathbb{R}/\mathbb{Z}$ in the same way as before. Again we take $dw'=m_\eta^{-1}d\nu_\eta$ with $m_\eta\asymp\log|\beta|$ as $\beta\to-0$ and $m_\eta\asymp1$ as $\beta\to b_2^2-a_2^2$. Moreover, in the latter case m_η is an analytic function of $\zeta'=(\beta+a_2^2-b_2^2)^{\frac12}$. One can easily see that if there is no inner ellipse Y_2 then Ψ transforms one connected component of every line $\{\eta=\text{const}\}$ to another and therefore for $\beta<-n^{-\sigma}$ the inequality $\mathrm{dist}(z',\Psi^k(z'))<\epsilon$

[20] More precisely, this line consists of two connected components. Ψ transforms every component into itself and the x-projection is an analytic diffeomorphism of this component to the ellipse Y_2. Every component is parameterized by u. Permutation of these components is equivalent to replacing Ψ by Ψ^{-1}.

[21] Moreover, it is well known [Cornfeld, Sinai, Fomin 1] that for generic η, Ψ_η is an ergodic map.

is impossible for odd k and the measure of the set $\{|\beta| \le n^{-\sigma}\}$ is $O(n^{-\delta})$. On the other hand, if there is an inner ellipse Y_1 then the map Ψ preserves both connected components of $\eta = \text{const}$ but changes their orientations, i.e., the map is of the form $w_\eta \to \frac{1}{2}f(\eta) - w_\eta$ and therefore for odd k the $d\nu_\eta$-measure of points with $\operatorname{dist}(z', \Psi^k z') \le \varepsilon$ is $O(n^p\varepsilon)$. Therefore *in both cases only even k should be treated.* On the other hand, on every connected component Ψ^2 is of the form $w_\eta \to w_\eta + f(\eta)$ where the function f is again continuous and $\epsilon\beta^p \le |\frac{df}{d\eta}| \le c\beta^{-p}$ for $|\eta| \ge a_2^{-1} + \epsilon_0$, i.e., $\beta \in (-\beta_0, 0)$ with small enough ϵ_0 and β_0. This function is analytic for $a_2^{-1} < |\eta| < b_2^{-1}$, i.e., for $\beta \in (0, b_2^2 - a_2^2)$, and the functions $f(\pm(b_2^{-1} - \zeta^2))$ are analytic with respect to ζ with $|\zeta| \le \epsilon_0$. Therefore we obtain that (10.2.69) and (10.2.68) remain true in this case. □

10.2.4 The strong non-periodicity condition. II

Our theory now yields the following statement if we use the more precise estimates of section 5.3 on X' with $T = \epsilon \log\tau$, $T = \tau^\sigma$ in the exponential and power cases respectively where $\epsilon, \sigma > 0$ are small enough and $\varepsilon(T) = \tau^{-\sigma'}$ with a small enough $\sigma' > 0$ depending on d, σ:

Theorem 10.2.20. *Let A be the Laplace-Beltrami operator on a γ-admissible manifold X with $\gamma^{-1-\theta} \in L^1(X)$, $\theta > 0$. Moreover, lower order terms satisfying the conditions of theorem 10.2.1 can be included in A. In the case of mixed boundary conditions we assume that either X is a smooth manifold with boundary or a domain in $\mathbb{R}^d$ and the cone condition is fulfilled. Then*

(i) If the geodesic flow on X is of power non-periodic type and if condition (10.2.65) *is fulfilled (in the case of mixed boundary conditions) then the asymptotics*

$$\mathbf{N}(\tau) = \varkappa_0\tau^{\frac{d}{2}} + \varkappa_1\tau^{\frac{d-1}{2}} + O(\tau^{\frac{d-1}{2}-\sigma}) \tag{10.2.70}$$

hold as $\tau \to +\infty$ with $\sigma > 0$.

(ii) If the geodesic flow is of exponential non-periodic type and if condition (10.2.65) *is fulfilled (in the case of mixed boundary conditions) then the asymptotics*

$$\mathbf{N}(\tau) = \varkappa_0\tau^{\frac{d}{2}} + \varkappa_1\tau^{\frac{d-1}{2}} + O(\tau^{\frac{d-1}{2}}\log^{-1}\tau), \tag{10.2.71}$$

hold as $\tau \to +\infty$.

(iii) Moreover, if $\gamma^{-1-\theta} \in L^1(X)$ with $\theta > \vartheta$ and if $\gamma_1^{-1-\theta} \in L^1(X)$ in the case of mixed boundary conditions where γ_1 is the distance to the transition set then the asymptotics

$$\mathbf{N}_\vartheta = \sum_{n\le 1+\vartheta} \varkappa_n\tau^{\frac{d-n}{2}+\vartheta} + O(\tau^{\frac{d-1+\vartheta}{2}-\sigma}), \tag{10.2.72}$$

$$\mathbf{N}_\vartheta = \sum_{n\le 1+\vartheta} \varkappa_n\tau^{\frac{d-n}{2}+\vartheta} + O(\tau^{\frac{d-1+\vartheta}{2}}\log^{-1-\vartheta}\tau) \tag{10.2.73}$$

as $\tau \to +\infty$ hold in the power and exponential cases respectively.

10.2.5 The periodicity condition

Let us now consider a closed manifold X and a scalar operator A on X with periodic Hamiltonian trajectories of the principal symbol. As before we assume that the manifold and operator are smooth enough; furthermore, as before we divide the operator A

by τ, introduce $h=\tau^{-\frac{1}{2m}}$ and obtain an h-pseudo-differential operator. Then if all the Hamiltonian trajectories at the energy level $\Sigma=\{a(x,\xi)=1\}$ are periodic with period T_0 then every trajectory is T-periodic with $T(x,\xi)=T_0 a^{\frac{1}{2m}-1}(x,\xi)$ and then the Hamiltonian flow of the symbol $a^{\frac{1}{2m}}$ is $2mT_0$-periodic. *Without loss of generality we assume that* $2mT_0=1$. According to section 4.7 we must replace the operator A by the operator $\bar{A}=(\tau^{-1}A)^{\frac{1}{2m}}$ (one should use this formula only in a neighborhood of Σ). Then conditions (4.7.2), (4.7.3) and (4.7.6) are fulfilled with $\Omega_0\supset\Sigma$. Calculating the action by formula (4.7.5) we obtain that $\alpha_1=0$ because the symbol $\bar{a}$ is positively homogeneous of degree 1 on ξ. In what follows all the objects related to the operator $\bar{A}$ are denoted by the same letters as the corresponding objects related to A but equipped with a bar. Then according to (4.7.8)

(10.2.74) $$\ell_0(x,\xi)=\frac{1}{2mT_0}\int_0^{T_0}(a^s\circ\Phi_t)(x,\xi)dt-\frac{\pi\alpha_0}{2}\qquad\text{on }\Sigma$$

and is positively homogeneous of degree 0 on ξ in a neighborhood of Σ. Here α_0 is the Maslov index. Then theorem 4.7.1 yields

Theorem 10.2.21. *Let X be a closed compact C^K manifold and A an elliptic self-adjoint differential operator*[22] *of order $2m$. Let the Hamiltonian flow of $a(x,\xi)$ be T_0-periodic on Σ.*

Let $\mathcal{L}$ be the image of the map $\Sigma\ni(x,\xi)\to\ell(x,\xi)\xrightarrow{\iota}\mathbb{R}/\mu\mathbb{Z}$ where $\mu=\frac{\pi}{mT_0}$ and $\iota:\mathbb{R}\to\mathbb{R}/\mu\mathbb{Z}$ is a natural map. Assume $\mathcal{L}$ does not coincide with $\mathbb{R}/\mu\mathbb{Z}$; then $\iota^{-1}\mathcal{L}$ is a union of intervals $\{I_n\}$ of length $O(1)$ such that $\omega\in I_n\implies|\omega|\asymp n$ for $|n|\ge1$. It follows that

(10.2.75) $$\operatorname{Spec}(A^{\frac{1}{2m}})\subset\bigcup_{n\ge1}J_n\cup[-C,C]$$

where J_n are the $Cn^{\delta-1}$-neighborhoods of I_n, $\delta>0$ is arbitrarily small and $C=C_\delta$; without loss of generality we assume that $A\ge I$.

Moreover, for $\ell_0=\text{const}$ *one can take* $\delta=0$.

Theorem 4.7.11 yields the following

Theorem 10.2.22. *Let the conditions of theorem 10.2.21 be fulfilled and let $\omega_0\in\mathcal{L}$,*

(10.2.76) $$(x,\xi)\in\Sigma,\ell_0(x,\xi)=\omega_0\implies\text{either }d_\Sigma\ell_0(x,\xi)\ne0\\ \text{or }\operatorname{rank}\operatorname{Hess}_\Sigma\ell(x,\xi)\ge r.$$

Let us assume that either the Hamiltonian flow Φ_t has no subperiodic trajectory on Σ or

(10.2.77) $$\operatorname{dist}\bigl(\Phi_t(x,\xi),(x,\xi)\bigr)\ge t(T_0-t)\rho(x,\xi)\qquad\forall t\in(0,T_0)\ (x,\xi)\in\Sigma$$

where

(10.2.78) $$|\nabla\rho|\le c\quad\text{and}\quad\operatorname{mes}_\Sigma\bigl\{(x,\xi),\rho(x,\xi)\le h\bigr\}=O(h^q)\quad\text{as }h\to+0.$$

Then for $\tau\to+\infty$ such that $|\iota\tau^{\frac{1}{2m}}-\omega_0|\le\epsilon_0$ (where $\epsilon_0>0$ is small enough) the asymptotics

(10.2.79) $$\mathrm{N}(\tau)=\varkappa_0\tau^{\frac{d}{2m}}+\bigl(\varkappa_1+F(\tau^{\frac{1}{2m}})\bigr)\tau^{\frac{d-1}{2m}}+O\bigl(\tau^{\frac{d-1-p}{2m}}\bigr)$$

[22] Or a classical pseudo-differential operator. In all cases we assume that $a\ge0$.

hold with

$$F(z) = (2mT_0)^{-1}(2\pi)^{-d}\int_{\Sigma} v\Bigl(2mT_0\bigl(z - \ell(x,\xi)\bigr)\Bigr)dxd\xi : da, \tag{10.2.80}$$

$p = \min(1, \frac{r}{2} - \delta, \frac{q}{2} - \delta)$ *with an arbitrary small exponent* $\delta > 0$. *Here and below* $v(z)$ *is a* 2π*-periodic function on* $\mathbb{R}$ *equal to* $\pi - z$ *on* $[0, 2\pi)$.

Remark 10.2.23. *(i) One can easily show that for generic differential and pseudo-differential symbols* a^s *condition (10.2.76) is fulfilled with* $r = 2d - 2$ *for all* ω_0. *In fact the function* $\ell_0(x,\xi)$ *is given on* $\Sigma' = \Sigma/\Phi_t$ *(the factor of* Σ *with respect to the Hamiltonian flow) which is a* $2d-2$*-dimensional manifold* rather than on Σ.

(ii) More precise asymptotics are possible as well (see theorem 4.7.26). ◊

Surely, theorem 10.2.22 is applicable to a Laplace-Beltrami operator on a Riemannian manifold with periodic geodesic flow which is perturbed by a first-order operator. For a generic vector $(V_1, \dots, V_d)$ the field corresponding to the perturbation condition (10.2.76) is fulfilled with $r = 2d - 2$.

In particular, for the Laplace-Beltrami operator on the standard sphere $\mathbb{S}^d$ $\alpha_0 \equiv 2(d-1) \pmod{4\mathbb{Z}}$ and $T_0 = \pi$. Let us consider this operator with perturbation more carefully. Let $j : \mathbb{S}^d \to \mathbb{S}^d$ be the antipodal map $x \to -x$ and let $j^* : S^*\mathbb{S}^d \to S^*\mathbb{S}^d$ be the adjoint map $(x,\xi) \to (-x,-\xi)$. Then (10.2.74) yields that ℓ_0 depends only on the even part of a^s: $a^s_{\mathsf{even}} = \frac{1}{2}(a^s + a^s \circ j^*)$. Let us now assume that

$$a^s = -a^s \circ j^*. \tag{10.2.81}$$

Then $\ell_0 = 0$ and $\mathcal{M}$ is a (-1)-order operator (i.e., $\eta = h = \tau^{-\frac{1}{2}}$: see section 4.7). Let us calculate the principal symbol ℓ_{-1} of the operator $\mathcal{M}$. It will be more convenient for us to set $\bar{A} = h\Bigl(\bigl(A + \frac{1}{4}(d-1)^2\bigr)^{\frac{1}{2}} - \frac{d-1}{2}\Bigr)$ here instead of $\bar{A} = A^{\frac{1}{2}}$. Let $\bar{A}_0 = h\Bigl(\bigl(A_0 + \frac{1}{4}(d-1)^2\bigr)^{\frac{1}{2}} - \frac{d-1}{2}\Bigr)$ be the corresponding unperturbed operator where A_0 is the unperturbed Laplace-Beltrami operator. Let $U(t) = \exp(ih^{-1}t\bar{A})$ and $U_0(t) = \exp(ih^{-1}t\bar{A}_0)$. Then

$$U_0(2\pi) = I, \qquad U_0(\pi) = J : Jv = v \circ j. \tag{10.2.82}$$

So we want to calculate $U(2\pi)$ modulo operators of order (-2) because near Σ the orders in the sense of standard pseudo-differential operators and in the sense of h-pseudo-differential operators coincide. Let us calculate $U(\pi)J$ modulo operators of order (-1) first. To do it we consider $V(t) = U(t)U_0(-t)$ which is a standard (h-)pseudo-differential operator. Then

$$\frac{d}{dt}V(t) = ih^{-1}U(t)(\bar{A} - \bar{A}_0)U_0(-t) = iV(t)U_0(t)BU_0(-t)$$

where $B = h^{-1}(\bar{A} - \bar{A}_0)$ is an operator of order 0. Considering principal symbols we obtain that $\partial_t v = iv(b \circ \Phi_t)$. Since $v|_{t=0} = 1$ we obtain that

$$v(t) = \exp(-i\int_0^t b \circ \Phi_{t'}dt')$$

where $b = \frac{1}{2}a^s a^{-\frac{1}{2}}$ is the principal symbol of B. Therefore (10.2.82) yields that

(10.2.83) $U(\pi)J$ is an h-pseudo-differential operator with principal symbol $\exp(i\zeta \cdot \int_0^\pi b \circ \Phi_t dt)$

with $\zeta = 1$.

Let us consider the operator $\bar{A}_\zeta = \bar{A}_0 + \zeta hB$ and let us consider the corresponding unitary group $U_\zeta(t)$. Differentiating the equations

$$(hD_t - \bar{A}_\zeta)U_\zeta(t) = 0, \qquad U_\zeta(0) = I$$

with respect to ζ we obtain that

$$(hD_t - \bar{A}_\zeta)U'_\zeta(t) = hBU_\zeta, \qquad U'_\zeta(0) = 0$$

where the prime means the derivative with respect to ζ. Then

$$U'_\zeta(t) = i\int_0^t U_\zeta(t-t')BU_\zeta(t')dt'.$$

Let us set $t = 2\pi$. Let us divide the interval $[0, 2\pi]$ of integration into subintervals $[0, \pi]$ and $[\pi, 2\pi]$ and replace t' by t in the first subinterval and by $t+\pi$ in the second one. We obtain that

$$U'_\zeta(2\pi) = iU_\zeta(2\pi)\int_0^\pi U_\zeta(-t)\tilde{B}U_\zeta(t)dt \tag{10.2.84}$$

with $\tilde{B} = B + U_\zeta(-\pi)BU_\zeta(\pi)$. We know that $U_\zeta(\pi) = V_\zeta J$ and $U_\zeta(-\pi) = JV_\zeta^*$ where V_ζ is a unitary operator. Therefore

$$\tilde{B} = B + JBJ + JV_\zeta^*[B, V_\zeta]J.$$

Therefore (10.2.81) yields that $\tilde{B}$ is an operator of order (-1) with principal symbol

$$\tilde{b} = 2b_{\mathrm{even}} + \zeta\{b, \int_0^\pi b\circ\Phi_{t'}dt'\}$$

where b_{even} is the even (with respect to j^*) part of b^s and the second term is also even. Therefore (10.2.84) yields that $-iU_\zeta(-2\pi)U'_\zeta(2\pi)$ is an operator of order (-1) with principal symbol

$$\int_0^\pi \tilde{b}\circ\Phi_{-t}dt = \int_0^{2\pi} b^s\circ\Phi_t dt + \zeta\int_0^\pi\int_0^\pi \{b\circ\Phi_t, b\circ\Phi_{t+t'}\}dtdt'$$

where we changed variables in the integration in the obvious way. Integrating over $\zeta \in [0,1]$ we obtain that $U(2\pi) - I$ *is an operator of order* (-1) *with principal symbol* $2\pi i\ell_{-1}$ *where*

$$\ell_{-1} = \frac{1}{2\pi}\Bigl(\int_0^{2\pi} b^s\Phi_t dt + \int_0^\pi\int_0^\pi \{b\circ\Phi_t, b\circ\Phi_{t+t'}\}dtdt'\Bigr). \tag{10.2.85}$$

Now theorems 4.7.1 and 4.7.11 yield the following

Theorem 10.2.24. *Let $A = \Delta + A' + A''$ where Δ is the (positive) Laplace-Beltrami operator on the standard sphere $\mathbb{S}^d$, $A' = A'(x,D)$ and $A'' = A''(x,D)$ are symmetric operators of first and zero order respectively and A' anti-commutes with J (in which case a^s satisfies (10.2.81)). Let $\ell_{-1}(x,\xi)$ be given by (10.2.85) with $b = \frac{1}{2}a'a^{-\frac{1}{2}}$, $b^s = \frac{1}{2}a''a^{-\frac{1}{2}} - \frac{1}{8}a'^2a^{-\frac{3}{2}}$ where a' and a'' are the principal symbols of A' and A'' respectively. Let $\mathcal{L}' = [\lambda_1, \lambda_2]$ be the image of the map $\Sigma \ni (x,\xi) \to \ell_{-1}(x,\xi) \in \mathbb{R}$. Let $f(\tau) = (\tau + \frac{1}{4}(d-1)^2)^{\frac{1}{2}} - \frac{1}{2}(d-1)$. Then* *(i)* $f(\operatorname{Spec} A) \subset \bigcup_{n\geq 1} J'_n \cup [-C, C]$ *where*

$$J'_n = \left[n + \lambda_1 n^{-1} - Cn^{\delta-2}, n + \lambda_2 n^{-1} + Cn^{\delta-2}\right],$$

$C = C(\delta)$ is large enough, $\delta > 0$ is arbitrarily small and for $\lambda_1 = \lambda_2$ one can take $\delta = 0$.

(ii) Let $\omega_1 \in \mathcal{L}'$ and condition (10.2.76) be fulfilled with ω_1 and ℓ_{-1} instead of ω_0 and ℓ_0. Then for $\tau \to +\infty$ such that $f(\tau) \subset \bigcup_{n\geq 1}[n + (\omega_1 - \epsilon)\frac{1}{n}, n + (\omega_1 + \epsilon)\frac{1}{n}]$ with a sufficiently small constant $\epsilon > 0$ the asymptotics

$$\mathbf{N}(\tau) = \varkappa_0 \tau^{\frac{d}{2}} + \varkappa_1 \tau^{\frac{d-1}{2}} + F(f(\tau), \tau)\tau^{\frac{d-1}{2}} + O(\tau^{\frac{d-1-p}{2}}) \tag{10.2.86}$$

hold with $p = \min(1, \frac{r}{2} - \delta)$ with an arbitrarily small exponent $\delta > 0$ and with

$$F(z, \tau) = 2(2\pi)^{-d-1} \int_\Sigma \Upsilon\left(2\pi\left(z - \tau^{-\frac{1}{2}}\ell_{-1}(x, \xi)\right)\right)dxd\xi : da. \tag{10.2.87}$$

Remark 10.2.25. *One can easily prove that for an arbitrary fixed symbol a'' and for a generic differential symbol (i.e., vector field) as well as for a generic pseudo-differential symbol a' satisfying (10.2.81) the modified[23] condition (10.2.76) is fulfilled with $r = 2d-2$ for all $\omega_1 \in \mathbb{R}$. Moreover, the same is true for an arbitrary fixed a'' satisfying (10.2.81) and for a generic differential symbol (i.e., potential) as well as for a generic pseudo-differential symbol a''.* ◊

Let us now assume that

$$a' = 0, \quad a'' \circ j^* = -a''. \tag{10.2.88}$$

Then $\ell_{-1} = 0$ and the operator $U(2\pi) - I$ is an operator of order ≤ -2 (we will prove that its order is (-3)) and our goal is to calculate its principal symbol. Now the calculations are simpler because the perturbation is weaker. Let $\bar{A}$ and $\bar{A}_0$ be defined as above. Then $\bar{A} = \bar{A}_0 + B' + B''$ [24] where B' is an operator of order (-1) with principal symbol $b' = -\frac{1}{2}a^{-\frac{1}{2}}a''$ anti-commuting with J and B'' is an operator of order (-3) with principal symbol $b'' = -\frac{1}{8}a^{-\frac{3}{2}}a''^2$.[25] Then the equalities

$$(D_t - \bar{A}_0)U(t) = BU(t), \qquad U(0) = I$$

with $B = B' + B''$ yield that

$$U(t) = U_0(t) + i\int_0^t U(t - t')BU(t')dt'$$

(this is similar to the calculation of V). After iteration we obtain $U(t)$ as an asymptotic series of operators with order decreasing by step 1. Let us set $t = 2\pi$. We then obtain that modulo operators of order (-4)

$$U(t) \equiv I + i\int_0^{2\pi} B(t_1)dt_1 - \iint_{\Pi_2} B(t_1)B(t_2)dt_1dt_2 - i\iiint_{\Pi_3} B(t_1)B(t_2)B(t_3)dt_1dt_2dt_3 \tag{10.2.89}$$

[23] I.e., with ω_1, ℓ_{-1} instead of ω_0, ℓ_0.

[24] For the sake of simplicity I do not write the factor $\tau^{-\frac{1}{2}}$ which in any case is reduced in the final formula.

[25] In contrast with the above case we did not include a lower order operator commuting with J in A because we are mainly interested in differential operators. This generalization is left to the reader.

with $B(t) = U_0(-t)BU_0(t)$ where $\Pi_k = \{t = (t_1, \dots, t_k) \in \mathbb{R}^k, 0 \le t_1 \le \dots \le t_k \le 2\pi\}$. Moreover, in the third and fourth terms we can replace B by B'. Dividing the interval of integration in the second term into subintervals $[0,\pi]$ and $[\pi, 2\pi]$ and replacing t_1 by t and $t+\pi$ on these subintervals respectively we obtain that this term is a pseudo-differential operator with principal symbol $i\int_0^{2\pi} b'' \circ \Phi_t dt$.

The third term in the right-hand expression is equal to

$$-\frac{1}{2}\iint_{\Pi_2} (B(t_1)B(t_2) + B(t_2)B(t_1))dt_1dt_2 - \frac{1}{2}\iint_{\Pi_2} [B(t_1), B(t_2)]dt_1dt_2.$$

The first term here equals half of $-B(t_1)B(t_2)$ integrated over the square $(t_1,t_2) \in [0, 2\pi]^2$; therefore it is equal to $-\frac{1}{2}(\int_0^{2\pi} B(t)dt)^2$ and therefore is negligible. The second term is an operator of order (-3) and after simple calculations we obtain that its principal symbol is

$$i\int_0^{\pi}\int_0^{\pi} \{b \circ \Phi_t, b \circ \Phi_{t+t'}\}dtdt'.$$

Finally, the third term in (10.2.89) is obviously an operator of order (-3). Making permutations of (t_1,t_2,t_3) and summing we obtain that modulo an operator of order (-4) this term is one-sixth of $B(t_1)B(t_2)B(t_3)$ integrated over the cube $[0,2\pi]^3$ and is therefore negligible.

Thus, $U(2\pi) - I$ is an operator of order (-3) with principal symbol $2\pi i\ell_{-3}(x,\xi)$ where ℓ_{-3} is given by (10.2.85) with $b' = \frac{1}{2}a''a^{-\frac{1}{2}}$ and $b'' = -\frac{1}{8}a''^2a^{-\frac{3}{2}}$ instead of b and b^s.

Therefore theorems 4.7.1 and 4.7.3 yield the following

Theorem 10.2.26. *Suppose the conditions of theorem 10.2.24 and condition* (10.2.88) *are fulfilled. Let* $\mathcal{L}'' = [\lambda_1, \lambda_2]$ *be the image of the map* $\Sigma \ni (x,\xi) \to \ell_{-3}(x,\xi) \in \mathbb{R}$. *Then*

(i) $f(\operatorname{Spec} A) \subset \bigcup_{n\ge1} J''_n \cup [-C, C]$ *where*

$$J''_n = [n + \lambda_1 n^{-3} - Cn^{\delta-4}, n + \lambda_2 n^{-3} + Cn^{\delta-4}],$$

$C = C(\delta)$ *is small enough,* $\delta > 0$ *is arbitrarily small and for* $\lambda_1 = \lambda_2$ *one can take* $\delta = 0$.

(ii) Let $\omega_2 \in \mathcal{L}''$ *and condition* (10.2.76) *be fulfilled with* ω_2 *and* ℓ_{-3} *instead of* ω_0 *and* ℓ_0. *Then for* $\tau \to +\infty$ *such that* $f(\tau) \subset \bigcup_{n\ge1}[n + (\omega_2 - \epsilon)n^{-3}, n + (\omega_2 + \epsilon)n^{-3}]$ *with a sufficiently small constant* $\epsilon > 0$ *asymptotics* (10.2.86) *hold with the same* p *as above and with*

$$F(z,\tau) = 2(2\pi)^{-d-1}\int_{\Sigma} \Upsilon\Big(2\pi\big(z - \tau^{-\frac{3}{4}}\ell_{-3}(x,\xi)\big)\Big)dxd\xi : da. \tag{10.2.90}$$

Remark 10.2.27. *(i) One can easily prove that for a generic differential symbol (i.e., potential) as well as for a generic pseudo-differential symbol* a'' *satisfying (10.2.88) the modified*[26] *condition (10.2.76) is fulfilled with* $r = 2d - 2$ *for all* $\omega_2 \in \mathbb{R}$.

(ii) Thus, in all the above cases for generic perturbations the remainder estimate is $O(\tau^{\frac{d-2}{2}})$ *for* $d \ge 3$ *and* $O(\tau^{\frac{1}{2}(d-2+\delta)})$ *with an arbitrarily small exponent* $\delta > 0$ *for* $d = 2$. *Applying the more precise theorem 4.7.21 we obtain that for* $d = 2$ *the remainder estimate is* $O(1)$ *if the principal symbol* ℓ *has no saddle points at the corresponding level and* $O(\log\tau)$ *otherwise. Moreover, one can get the remainder estimate* $O(1)$ *in this case as well by including in the principal symbol the next symbol too (as we have seen it does not necessarily vanish even if the principal symbol vanishes identically).*

(iii) I leave Riesz means asymptotics to the reader. Recall that the best possible remainder estimate is $O(\tau^{\frac{1}{2}(d-1-n\vartheta)})$ *where* $n = 0, 1, 3$ *is the order of* B *taken with the opposite sign.* ◇

[26] I.e., with ω_2, ℓ_{-3} instead of ω_0, ℓ_0.

10.2.6 Periodic singular case

Let us now consider the singular case. Dividing Σ into the zones $\{\zeta \geq \frac{1}{2}\tau^{-\delta}\}$ and $\{\zeta \leq \tau^{-\delta}\}$ (where the function ζ is introduced below), applying theorem 4.7.11 in the first zone with remark 4.7.20 and applying standard results we obtain

Theorem 10.2.28. *Let X be a γ-admissible compact closed Riemannian manifold, A a scalar elliptic γ-admissible operator of order $2m$ with lower order terms satisfying the standard conditions of this section. Let us assume that $\gamma^{-1-\theta} \in L^1(X)$ with $\theta > 0$. Moreover, let us assume that the Hamiltonian flow of $a(x,\xi)$ is T_0-periodic on Σ. Furthermore, let us assume that a γ-admissible bounded function ζ is given on Σ such that*

$$(10.2.91)_{1,2} \qquad |J_t(z)| \leq C\zeta^{-n}(z), \quad \gamma(\Phi_t(z)) \geq \epsilon\zeta^n(z) \qquad \forall z \in \Sigma \quad \forall t.$$

Finally, let us assume that conditions (10.2.76) *with $r \geq 1$ (for some fixed ω_0),* (10.2.77) *and* (10.2.78) *with $q > 0$ are fulfilled. Then for $\tau \to +\infty$ such that $|\iota\tau^{\frac{1}{2m}} - \omega_0| \leq \epsilon$ asymptotics* (10.2.79) *hold with $p > 0$.*

As an example one can consider the Zoll-Tanner manifolds $[0, 2\pi] \times \mathbb{S}^{d-1} \ni (x_1, x')$ with the metric $\bigl(1 + \varphi(\cos x_1)\bigr)^{-2}\xi_1^2 + |\xi'|^2 \sin^{-2} x_1$ where $\varphi \in C^K((-1,1)) \cap C^1([-1,1])$ is an odd function such that $|\varphi(t)| < 1$ for $t \in [-1,1]$, $\varphi(1) = 0$. Then one can take A equal to the Laplace-Beltrami operator with lower order terms satisfying the standard conditions of this section and providing condition (10.2.68) with $p \geq 1$. The reader can find the methods for checking condition (10.2.91) in subsection 10.4.3 and choose a function ζ.

Repeating the arguments of theorems 10.2.24 and 10.2.28 we obtain the following

Theorem 10.2.29. *Let A be a Schrödinger operator on $\mathbb{S}^d$ with the standard metric and let the lower order terms satisfy the standard conditions (i.e., they are γ-admissible with $\gamma^{-1-\theta} \in L^1(X)$). Let the function ζ satisfy* $(10.2.91)_2$ *and* (10.2.78) *with $q > 0$.*

(i) Let us assume that condition (10.2.80) *and the modified condition* (10.2.76) *with $r \geq 1$ are fulfilled. Then the asymptotics* (10.2.86)–(10.2.87) *hold with $p > 0$.*

(ii) Let us assume that condition (10.2.88) *and the modified condition* (10.2.76) *with $r \geq 1$ are fulfilled. Then the asymptotics* (10.2.86), (10.2.90) *hold with $p > 0$.*

Remark 10.2.30. *Let us consider the Laplace-Beltrami operator on $\mathbb{S}^d$ perturbed by a potential $V \in L^p$ such that $\|V\|_{L_p} < \epsilon$ with $p > d = 2$ or $p = d \geq 3$. Then standard functional-analytic arguments yield that the cluster character of the spectrum remains true. The size of a cluster is $O(\epsilon\tau^{\frac{d}{2p}})$ and the distance between clusters is $\asymp \tau^{\frac{1}{2}}$; here $\epsilon = \|V\|_{L^p}$ is assumed to be small only for $p = d$.* ◊

10.2.7 The indefinite case

In this subsection we discuss the case when the principal symbol is neither positive nor negative. We assume that the lower order terms are appropriate: conditions on these terms will be discussed in the next subsection.

All the results of subsections 10.2.1–10.2.4 remain true for operators with elliptic principal symbols neither positive nor negative. In this case $\mathbf{N}(\tau)$ is the number of eigenvalues in the interval $(0, \tau)$ for $\tau > 0$ or in $(\tau, 0)$ for $\tau < 0$ and we consider the asymptotics for $\tau \to \pm\infty$. Surely, for $\tau \to +\infty$ and for $\tau \to -\infty$ we obtain different asymptotics and the conditions on Hamiltonian trajectories are different. Moreover, all the results for $\vartheta > 0$ remain true if we set

$$\mathbf{N}_\vartheta(\tau) = \pm \int_0^\tau |\tau'|^{\vartheta-1} N(\tau') d\tau'$$

for $\pm\tau > 0$. Furthermore, the results of subsections 10.2.5 and 10.2.6 remain true. In this case (instead of the assumption that $A(x,D)$ is a scalar operator) we assume that $a(x,\xi)$ has only one positive (or negative) eigenvalue $\lambda(x,\xi)$ which is simple provided we treat asymptotics as $\tau \to +\infty$ ($\tau \to -\infty$ respectively). In this case all the conditions of these subsections are assumed to be fulfilled for the Hamiltonian flow generated by $\lambda(x,\xi)$. Moreover, to obtain conditions on lower order terms one should find pseudo-differential operator Q such that $Q^*Q \equiv QQ^* \equiv I$ and $Q^*AQ \equiv \begin{pmatrix} A_1 & 0 \\ 0 & A_2 \end{pmatrix}$ where A_1 is a scalar operator with principal symbol $\lambda(x,\xi)$ and A_2 an operator with principal symbol of the opposite sign. Then all the conditions should be imposed on the lower order terms of A_1.

Moreover, we can treat the Dirac operator; nevertheless the eigenvalues of its principal symbol are simple only for $D = 2 \implies d = 1,2,3$ (for $d = 3$ one has a principally massless particle). However, one needs to check that all the symbols associated with A_1 and taken into account are scalar.

In order to prove the above statements one applies the results of Chapters 4 and 5 to the interval $(\frac{1}{2}\tau, \tau)$ (we consider the case $\tau > 0$ for the sake of simplicity). Then the routine remainder estimate $O(\tau^\nu)$ or $O(\tau^\nu \log^\sigma \tau)$ is obtained by standard methods. Let us replace τ by $2^{-j}\tau$ (with $j = 0,\dots,n$ such that $\tau \asymp 2^n$) and sum all these asymptotics; then we obtain asymptotics for the number of eigenvalues lying in (τ_1, τ) with the same remainder estimate for $\nu > 0$ and with the remainder estimate $O(\log^{\sigma+1}\tau)$ for $\nu = 0, \sigma \geq 0$. On the other hand it is easy to prove that $\mathbf{N}(0,\tau_1) < \infty$.

Moreover, in order to improve the final remainder estimate in the case when it grows slowly, one can apply the following arguments to obtain the remainder estimate $O(\tau^{-s})$ with arbitrarily chosen s for mollifications of $\mathbf{N}(\tau', \tau'')$ (the number of eigenvalues in the interval (τ', τ'')) with respect to τ' and τ'' with mollifying functions $\tau^{-1}\phi_1(\frac{\tau'}{\tau})$ and $\tau^{-1}\phi_2(\frac{\tau''}{\tau})$ where ϕ_1 and ϕ_2 are supported near $\frac{1}{2}$ and 1 respectively. Replacing τ by $2^{-j}\tau$ and making a partition of unity one obtains the remainder estimate $O(1)$ for $\mathbf{N}(0,\tau')$ mollified near $\frac{1}{2}\tau$. On the other hand, for $N(\tau',\tau)$ (even without mollification) we have the appropriate remainder estimate provided $\tau' \asymp \tau$.

For $\vartheta > 0$ one can apply analysis of the same type.

10.2.8 Conditions on lower order terms

Let us discuss conditions on the lower order terms. Let us note first that all that (ρ,γ)-admissible lower order terms work provided

(10.2.92) $$\rho^{d+2m\vartheta} \in L^1(X),$$

(10.2.93) $$\rho^{2m} \in L^1(X),$$

(10.2.93)$_\#$ $$\rho^{2m} \in L^2(X),$$

where (10.2.93) and (10.2.93)$_\#$ are conditions in the semibounded and non-semibounded cases respectively. Here $2m$ is the order of the operator and $m \in \frac{1}{2}\mathbb{Z}$.

We can give a series of admissible lower order terms for all the types of operators treated above. However, it is very simple in the context of (ρ,γ)-admissibility. We leave this analysis to the reader and discuss how one can weaken the conditions.

Example 10.2.31. *Let $\gamma = \epsilon|x|$ and $\rho = |x|^\mu$ (singularity at the origin). Then the conditions (10.2.92), (10.2.93) and (10.2.93)$_\#$ mean respectively that $\mu > -d(d+2m\vartheta)^{-1}$, $\mu > -\frac{d}{2m}$ and $\mu > -\frac{d}{4m}$. Let μ^* be the maximal forbidden value. Then for $\mu = \mu^*$*

we can treat $\rho = |x|^{\mu}\big|\log|x|\big|^{\nu}$ *with appropriate* ν, *etc. However, the arguments associated with estimates (10.1.25)–(10.1.26) give that one can take* $\mu > \max(-1, -\frac{d}{2m})$, $\mu > \max(-1, -\frac{d}{4m})$ *in the semibounded and non-semibounded cases respectively. Moreover, for* $-1 > -\frac{d}{2m}$, $-1 > -\frac{d}{4m}$ *in these cases one can even take* $\mu = -1$ *assuming that in a neighborhood of* 0

$$|a_{\alpha',\alpha''}| \le \varepsilon\rho^{2m-|\alpha'|-|\alpha''|} \qquad \forall\alpha',\alpha'' : |\alpha'| \le m, |\alpha''| \le m, |\alpha'| + |\alpha''| \ge 1, \tag{10.2.94}$$

$$|a_\alpha| \le \varepsilon\rho^{2m-|\alpha|} \qquad \forall\alpha : 1 \le |\alpha| \le 2m \tag{10.2.94$_\#$}$$

respectively with small enough $\varepsilon = \text{const} > 0$.

(ii) Let us assume that in a neighborhood of 0 *the domain* X *coincides with the cone* $\{(r,\theta) \in \mathbb{R}^+ \times U\}$ *where* $\partial U \in C^K$. *Moreover, let us assume that a Dirichlet boundary condition is given on part of the boundary* $\{(r,\theta) : r \le r_0, \theta \in U_1 \subset \partial U\}$ *and that* $r_0 > 0$ *and* $U_1 \ne \emptyset$ *contains the* ∂U*-neighborhood of some point. Then one can reject the condition* $\mu > -\frac{d}{2m}$ *in the semibounded case.*

(iii) Moreover, for the Schrödinger operator with $d \ge 3$ *and with the Dirichlet boundary condition or with* $d \ge 2$ *and a domain and boundary conditions of the indicated type, variational arguments yield that one can assume that in a neighborhood of* 0

$$V \ge -\varepsilon\rho^2, \qquad |V_j| \le \varepsilon\rho \tag{10.2.95}$$

with $\rho = |x|^{-1}$.

(iv) Let us treat the Schrödinger operator with $\mu = -1$, $d \ge 3$ *assuming that either a Dirichlet boundary condition is given on* ∂X *or the domain and boundary condition are of the type described in (ii). Then*

$$(Au,u) \ge \sum_j \epsilon_0\|(D_j - V_j)u\|^2 + (Vu,u).$$

Let us introduce spherical coordinates (r,θ) *and remove the radial component of the magnetic potential by means of a gradient transform. We obtain that for functions supported in the vicinity of* 0

$$(Au,u) \ge \epsilon_0\|D_r u\|^2 + \epsilon_0\sum_j \|r^{-1}\mathcal{L}_j u\|^2 + (Vu,u) - \varepsilon\|r^{-1}u\|^2$$

where $\mathcal{L}_j$ *are first-order operators with bounded coefficients on* $\mathbb{S}^{d-1}$ *and* $\varepsilon > 0$ *is arbitrarily small (the vicinity depends on* ε*). Let us consider*

$$\|D_r u\|^2 = \int\int_0^{r_0} r^{d-1}|D_r u|^2\,dr\,d\theta.$$

In order to do this we treat $\|r^\nu D_r u\|_1^2$ *where* $\|.\|_1$ *and* $(.,.)_1$ *mean the* L^2*-norm and inner product on* $[0,r_0]$, $\nu = \frac{1}{2}(d-1)$. *Substituting* $u = r^{-\nu}v$ *we obtain that*

$$\|r^\nu\partial_r u\|_1^2 = \|(\partial_r - \nu r^{-1})v\|_1^2 = \|\partial_r v\|_1^2 + \nu^2\|r^{-1}v\|_1^2 - \\ 2\nu\,\mathrm{Re}(\partial_r v, r^{-1}v)_1 = \|\partial_r v\|_1^2 - (\nu^2-\nu)\|r^{-1}v\|_1^2.$$

Let $d > 3$; *then* $\nu^2 > \nu$ *and we conclude that in this case* $\|r^\nu\partial_r\|_1^2 \ge \epsilon_1\|r^{\nu-1}u\|_1^2$. *On the other hand, let* $d = 3$. *Then* $\nu = 1$ *in which case* $\|r^\nu\partial_r u\|^2 \ge \|\partial_r v\|^2$ *and then*

$$\|r^{\nu}\partial_r u\|_1^2 \geq \epsilon_1 \|\nu r^{-1} v\|_1^2 = \epsilon_1 \|r^{\nu-1} u\|_1^2.$$

Therefore for $d \geq 3$ condition $(10.2.95)_2$ is not necessary.

On the other hand, for $d = 2$, $\nu = \frac{1}{2}$ and $\nu^2 - \nu = -\frac{1}{4}$. In this case only one operator $\mathcal{L}_1 = D_\theta - w(r,\theta)$ is present and if for $\theta = \theta_1$ a Dirichlet boundary condition is given then $\|\mathcal{L}_1 u\|_2 \geq \epsilon_1 \|u\|_2$ where $\|.\|_2$ is the L^2-norm with respect to θ. On the other hand, if $U = \{\theta, \theta_1 < \theta < \theta_2\}$ and if for $\theta = \theta_1$ and for $\theta = \theta_2$ a Neumann boundary condition is given then this estimate fails. However, if X is a complete neighborhood of 0 then this estimate remains true provided

$$|\int_0^{2\pi} w(\theta, r)d\theta - 2\pi m| \geq \epsilon_0 \qquad \forall m \in \mathbb{Z} \quad \forall r \in (0, r_0). \tag{10.2.96}$$

Conversely, if (10.2.96) fails for every $\epsilon_0 > 0$ then the previous inequality also fails.

Thus, if $d = 2$ and either $U = \{\theta_1 < \theta < \theta_2\}$ and for θ_1 or for θ_2 a Dirichlet boundary condition is given, or X contains a complete neighborhood of 0 and (10.2.96) holds then one can reject condition $(10.2.95)_2$.

(v) All these arguments remain true in the case $\gamma = \epsilon|x''|$, $\rho = |x''|^\mu$ with d replaced by d'' in the conditions. ◊

Example 10.2.32. *Using the arguments of remark 10.1.7 we see that in the context of example 10.2.5 one can assume that $\rho F(|x''|) \leq \epsilon$ (with small enough constant $\epsilon > 0$) instead of assumption (10.2.93).* ◊

In conclusion let us note that certain generalizations can be obtained from functional-analytic arguments. For example, for the Schrödinger operator asymptotics with remainder estimate $O(\tau^{\frac{1}{2}(d-1)})$ remain true if one replaces the potentials V and V_j by $V + V'$ and $V_j + V'_j$ with $V' \in L^d(X)$, $V'_j \in L^\infty(X)$. Moreover, asymptotics with remainder estimate $O(\tau^{\frac{1}{2}(d-1-\sigma)})$ remain true provided $V'_j = 0$, $V' \in L^{d(1-\sigma)^{-1}}(X)$. Therefore approximating $V' \in L^d(X)$ by the potentials $V'_\varepsilon \in L^\infty(X)$ we obtain that the remainder estimate $o(\tau^{\frac{1}{2}(d-1)})$ remains true. Moreover, the approximation arguments yield that the remainder estimate $O(\tau^{\frac{1}{2}(d-1)} \log^{-\nu} \tau)$ remains true provided $V' \log^\nu \in L^d(X)$. On the other hand, remainder estimates weaker than $O(\tau^{\frac{1}{2}(d-1)})$ remain true under weaker conditions. However, our weakly singular potentials (i.e., the potentials which we considered in this section) do not necessarily belong to the indicated classes. Thus, a combination of the two methods is the most effective.

10.3 Asymptotics of Large Eigenvalues: the Case of Strongly Singular Potentials

10.3.1 Getting started

In this section, under the assumptions of the previous section concerning the domain X and the principal symbol, we consider operators with lower order terms essential from the point of view of the domain of the operator and asymptotics.[27] We start with operators with positive principal symbol. In this case the coercivity condition is

$$\langle A(x,\xi)v, v\rangle \geq \epsilon_0(|\xi| + \rho)^{2m}|v|^2 \qquad \forall \xi \quad \forall v, \tag{10.3.1}$$

[27] Thus, it would be better to speak about senior symbols and junior terms.

$$\sum_{\alpha,\beta:|\alpha|\le m,|\beta|\le m} \langle a_{\alpha,\beta}v_\alpha, v_\beta\rangle \ge \epsilon_0 \sum_{\alpha:|\alpha|\le m} |v_\alpha|2\rho^{2m-2|\alpha|} \qquad \forall V=(v_\alpha)_{\{|\alpha|\le m\}} \tag{10.3.2}$$

where $\epsilon_0>0$,

$$A(x,\xi)=\sum_{\alpha,\beta:|\alpha|\le m,|\beta|\le m} a_{\alpha,\beta}\xi^{\alpha+\beta}$$

is the complete (symmetric) symbol of an operator A of the form (10.1.1) with $h=1$ and the first condition is assumed to be fulfilled for all $x\in X$ while the second condition (which is surely stronger) is assumed to be fulfilled only in $\{x\in X, \rho\gamma\le\epsilon_0, \gamma\le\epsilon_0\}$. However, in certain cases this condition is assumed to be fulfilled in X. We have assumed that

$$\varrho=\rho^{2m} \tag{10.3.3}$$

for the sake of simplicity. The reader can consider the more general case. In what follows we treat lower order terms which are sums of coercive and admissible indefinite (lower order) terms. For the Schrödinger operator conditions (10.3.1) and (10.3.2) coincide with the condition

$$V\ge\epsilon_0\rho^2 \qquad \forall x \tag{Σ3}$$

provided conditions (Σ0)–(Σ2) are fulfilled.

As before we set $\tilde\gamma=\gamma$ and $\tilde\rho=\rho+\tau^{\frac12}$ and consider the domain

$$X'_-=\{x\in X, \tau^{\frac{1}{2m}}\gamma\ge 1, \exists\xi:\exists v\ne 0 \quad \langle A(x,\xi)v,v\rangle\le(1+\epsilon_1)\tau|v|^2\}$$

with an arbitrarily small constant $\epsilon_1>0$. Then condition (10.3.1) yields that $\rho\le C\tau^{\frac{1}{2m}}$ in X'_-. As before we apply the results of Chapters 4 and 5 in X'_- to obtain local semiclassical spectral asymptotics in the ball $B(x,\gamma(x))$ with remainder estimate $O(\tau^{\frac{1}{2m}(d-1)}\gamma^{d-1})$ provided $\rho\le\epsilon_2\tau^{\frac{1}{2m}}$. Moreover, under a condition on the Hamiltonian flow this remainder estimate can be improved. On the other hand, for balls with $\rho\ge\epsilon_2\tau^{\frac{1}{2m}}$ the remainder estimate (as well as the principal part of the asymptotics) is $O(\tau^{\frac{d}{2m}}\gamma^d)$ in the general case and $O(\tau^{\frac{1}{2m}(d-1+\delta)}\gamma^{d-1+\delta})$ for scalar operators. In order to obtain the remainder estimate $O(\tau^{\frac{1}{2m}(d-1)}\gamma^{d-1})$ in local spectral asymptotics in this zone one should check microhyperbolicity or another (8.1.18)$_*$-type condition there. In particular, for the Schrödinger operator with $d\ge 2$ we obtain this remainder estimate without any additional conditions.

On the other hand, in the zone

$$X'_+=\{x\in X, \rho\gamma\ge\epsilon_1, \forall\xi \quad \forall v \quad \langle A(x,\xi)v,v\rangle\ge(1+\epsilon_1)\tau|v|^2\}$$

we obtain the remainder estimate[2)] $O(\rho^{-s}\gamma^{-s})$ with arbitrarily chosen s provided $B(x,\gamma(x))$ does not intersect ∂X, or an m-th order Dirichlet boundary condition is given on ∂X. Moreover, one can obtain the same answer provided

$$(Au,u)\ge\sum_{\alpha,\beta:|\alpha|,|\beta|\le m}(a_{\alpha,\beta}D^\alpha u, D^\beta v) \quad \forall u\in \mathrm{D}(A).^{28)} \tag{10.3.4}$$

Let us note that $X'_+\cup X'_-\supset\{x\in X, \tau^{\frac{1}{2m}}\gamma\ge C_0\}$.

[2)] The principal part vanishes here.

[28)] This inequality is a restriction on the boundary conditions.

Let us assume that condition (10.3.2) is fulfilled in the domain $\mathcal{X} \supset \{x \in X, \rho\gamma \le \epsilon_1, \gamma \le \epsilon_1\}$. Then for u supported in $\mathcal{X}$ the estimate

$$((A - \tau I)u, u) \ge -C\tau \int_{X''_-} |u|^2 dx$$

holds with $X''_- = \mathcal{X} \setminus X''_+$,

$$X''_+ = \{x \in X, \sum_{\alpha,\beta:|\alpha|\le m, |\beta|\le m} \langle a_{\alpha,\beta}(x)v_\alpha, v_\beta\rangle \ge \tau \sum_{\alpha:|\alpha|\le m} |v_\alpha|^2 \quad \forall v_\alpha\}.$$

Surely $X''_+ \supset \{x \in X, \rho \ge C\tau^{\frac{1}{2m}}\}$. Therefore the contribution of $\mathcal{X}$ to the remainder estimate does not exceed $C\tau^{\frac{d}{2m}} \operatorname{mes} X''_-$ provided either a Dirichlet boundary condition or any condition providing (10.3.4) on $\partial X \cap \mathcal{X}$ is given; in the latter case we assume that the cone condition is fulfilled. Let us note that $\mathcal{X}$ and X'_+ overlap and therefore we can subtract from X'_+ and add to $\mathcal{X}$ the zone $\tilde{X} = \{x \in X, \operatorname{dist}(x, X \setminus \mathcal{X}) \ge \epsilon_1\gamma\}$.

So we obtain the estimate

$$|\mathbf{N}(\tau) - \mathcal{N}^w(\tau)| \le C(R_1(\tau) + R_2(\tau) + R_3(\tau)) \tag{10.3.5}$$

with

$$\mathcal{N}^w(\tau) = (2\pi)^{-d} \iint_{X' \times \mathbf{R}^d} \mathbf{n}(x, \xi; \tau) dx d\xi \tag{10.3.6}$$

where $\mathbf{n}(x, \xi; \tau)$ is now the eigenvalue counting function for the complete symbol $A(x, \xi)$,

$$R_1(\tau) = \int_{X'_-} \tau^{\frac{d-1}{2m}} \gamma^{-1} dx, \tag{10.3.7$_1$}$$

$$R_2(\tau) = \tau^{\frac{d}{2m}} \operatorname{mes}(X''_- \cup \tilde{X}'_-), \tag{10.3.7$_2$}$$

$$R_3(\tau) = \int_{X'_+ \setminus \tilde{X}} \rho^{d-s} \gamma^{-s} dx \tag{10.3.7$_3$}$$

where $\tilde{X}'_- = \emptyset$ for the Schrödinger operator with $d \ge 2$ and $\tilde{X}'_- = \{x \in X'_-, \rho \ge \epsilon_2 \tau^{\frac{1}{2m}}\}$ in the general case. This zone can be reduced (or even voided) under microhyperbolicity or other (8.1.18)$_*$-type conditions.

Similarly, for $\vartheta > 0$ the estimate

$$|\mathbf{N}_\vartheta(\tau) - \mathcal{N}^w_\vartheta(\tau)| \le C(R_{1\vartheta}(\tau) + R_{2\vartheta}(\tau) + R_3(\tau)) \tag{10.3.5$'$}$$

holds with $\mathbf{N}_\vartheta = \tau_+^{\vartheta-1} * \mathbf{N}$, $\mathcal{N}^w_\vartheta = \tau_+^{\vartheta-1} * \tilde{\mathcal{N}}^w$, $\tilde{\mathcal{N}}^w$ containing lower order terms as well,

$$R_{1\vartheta}(\tau) = \int_{X'_-} \tau^{\frac{1}{2m}(d-1-\vartheta)+\vartheta} \gamma^{-1-\vartheta} dx, \tag{10.3.7$'_1$}$$

$$R_{2\vartheta}(\tau) = \tau^{\frac{d}{2m}+\vartheta} \operatorname{mes}(X''_- \cup \tilde{X}'_-). \tag{10.3.7$'_2$}$$

In particular, if $\gamma^{-1-\vartheta} \in L^1(X)$, $\rho^{d-s}\gamma^{-s} \in L^1(X \setminus \tilde{\mathcal{X}})$ and

$$\operatorname{mes} \tilde{X}'_- = O(\tau^{-\frac{1}{2m}(\vartheta+1)}) \qquad \text{as } \tau \to +\infty \tag{10.3.8$'$}$$

then

$$\mathbf{N}_\vartheta(\tau) = \mathcal{N}^w_\vartheta(\tau) + O\big(\tau^{\frac{1}{2m}(d-1-\vartheta)+\vartheta}\big). \tag{10.3.9$'$}$$

For the Schrödinger operator with $d \ge 2$ and other scalar operators with $\vartheta > 0$ condition (10.3.8)$'$ is fulfilled automatically, but in the general case this condition is very unpleasant and it is best to check the microhyperbolicity condition for $A - \tau I$ in $\tilde{X}'_-$ and then reduce this zone.

Let us now note that in the general case the principal parts of the asymptotics $\mathcal{N}^w$ and $\mathcal{N}^w_\vartheta$ cannot be decomposed into a power series of $\tau^{\frac{1}{2m}}$ as in the previous section. Usually this decomposition leads to much worse remainder estimate than the original one.

10.3.2 Improved remainder estimates

Let us discuss an improvement of the remainder estimate in the zone $X'_- \cap \{\rho \le C\tau^{\frac{1}{2m}}\}$; thus we consider the term $R_1(\tau)$ or $R_{1\vartheta}(\tau)$. We assume that $\gamma^{-1-\vartheta} \in L^1(X)$ (in which case $R_{1\vartheta} \asymp \tau^{\frac{1}{2m}(d-1-\vartheta)+\vartheta}$), $\vartheta \ge 0$ and that $\rho^{d-s}\gamma^{-s} \in L^1(X \setminus \mathcal{X})$ for some s. Then

(i) If the standard condition $(\Psi 0)$ on Hamiltonian trajectories is fulfilled in X_τ for the symbol $a(x,\xi) - 1$ at the energy level 0 and if

$$\operatorname{mes} \tilde{X}'_- = o\big(\tau^{\frac{1}{2m}(-\vartheta+1)}\big) \qquad \text{as } \tau \to +\infty \tag{10.3.10$'$}$$

then the asymptotics

$$\mathbf{N}_\vartheta(\tau) = \mathcal{N}^w_\vartheta + o\big(\tau^{\frac{1}{2m}(d-1-\vartheta)+\vartheta}\big) \tag{10.3.11$'$}$$

hold as $\tau \to +\infty$. In particular, for $\vartheta = 0$ the asymptotics

$$\mathbf{N}(\tau) = \mathcal{N}^w(\tau) + \varkappa_1 \tau^{\frac{d-1}{2m}} + o\big(\tau^{\frac{d-1}{2m}}\big) \tag{10.3.11}$$

hold.

(ii) All the examples of subsection 10.2.2 remain true. However, the principal part of these asymptotics should be given by the formulas of the present section.

(iii) In the context of subsection 10.2.3 one can take the remainder estimate $O\big(\tau^{\frac{1}{2m}(d-\vartheta-1)+\vartheta}\log^{-\vartheta-1}\tau\big)$ for the exponential non-periodic Hamiltonian flow and $O\big(\tau^{\frac{1}{2m}(d-\vartheta-1)+\vartheta-\delta}\big)$ for the power non-periodic Hamiltonian flow in asymptotics (10.3.11)$'$, (10.3.11) provided

$$\operatorname{mes} \tilde{X}'_- = O\big(\tau^{-\frac{1}{2m}(\vartheta+1)}\log^{-\vartheta-1}\tau\big) \qquad \text{as } \tau \to +\infty, \tag{10.3.12}$$

$$\operatorname{mes} \tilde{X}'_- = O\big(\tau^{-\frac{1}{2m}(\vartheta+1)-\delta}\big) \qquad \text{as } \tau \to +\infty \tag{10.3.13}$$

respectively.

(iv) In the context of subsection 10.2.6 under condition (10.3.13) the asymptotics

$$\mathbf{N}(\tau) = \mathcal{N}^w(\tau) + \tau^{\frac{d-1}{2m}}\big(\varkappa_1 + F(\tau^{\frac{1}{2m}})\big) + O\big(\tau^{\frac{d-1}{2m}-\delta}\big) \tag{10.3.14}$$

hold as $\tau \to +\infty$.

10.3.3 Examples

Let us consider some examples.

Example 10.3.1 (Power singularities). *(i) Let all the conditions be fulfilled with* $\gamma = |x''|$, $\rho \asymp |x''|^{\mu}$, $\mu \le -1$ *(here one can take* $\mathcal{X} = \emptyset$ *for* $\mu < -1$ *and* $\mathcal{X} = \{x \in X, |x''| \le \epsilon_0\}$ *for* $\mu = -1$*). Then in the general case one should take* $\tilde{X}'_- = \{\epsilon\tau \le |x''|^{2m\mu} \le C\tau\}$ *and then*

(a) For $d'' > \vartheta + 1$, $\mu \ge -d''(\vartheta+1)^{-1}$ *asymptotics* (10.3.9)′ *hold. Moreover, for* $\mu > -d''(\vartheta+1)^{-1}$ *and the standard condition* (Ψ0) *on the Hamiltonian trajectories asymptotics* (10.3.11)′ *hold. Furthermore, for* $\vartheta = 0$ *and* $\mu > -d''$ *one can replace* $\mathcal{N}^w(\tau)$ *by* $\varkappa_0\tau^{\frac{d}{2m}} + \varkappa_1\tau^{\frac{d-1}{2m}}$ *with the standard coefficients* $\varkappa_0, \varkappa_1$ *and with the same remainder estimate as above. On the other hand, for* $\mu = -d''$, $\mathcal{N}^w(\tau) = \varkappa_0\tau^{\frac{d}{2m}} + O\big(\tau^{\frac{d-1}{2m}}\log\tau\big)$ *in the general case and* $\mathcal{N}^w(\tau) = \varkappa_0\tau^{\frac{d}{2m}} + O\big(\tau^{\frac{d-1}{2m}}\big)$ *in the case of a scalar principal symbol.*

(b) For $d'' = \vartheta + 1$, $\mu = -1$ *the asymptotics*

$$\mathbf{N}_\vartheta(\tau) = \mathcal{N}^w_\vartheta(\tau) + O\big(\tau^{\frac{1}{2m}(d-1-\vartheta)+\vartheta}\log\tau\big) \tag{10.3.15}$$

hold and for $\vartheta = 0$, $d'' = 1$, $\mu = -1$ *one can replace* $\mathcal{N}^w(\tau)$ *by* $\varkappa_0\tau^{\frac{d}{2m}}$.

(c) If $\mu < -d''(\vartheta+1)^{-1}$ *then the asymptotics*

$$\mathbf{N}_\vartheta(\tau) = \mathcal{N}^w_\vartheta(\tau) + O\big(\tau^{\frac{d}{2m}+\frac{d''}{2m\mu}+\vartheta}\big) \tag{10.3.16}$$

hold. Let us note that $\mu \ge -d''(\vartheta+1)^{-1}$, $\mu \le -1$ *yield* $d'' \le \vartheta + 1$ *and therefore only the listed cases are possible.*

(ii) For the Schrödinger operator with $d \ge 2$ *and other scalar operators with* $\vartheta > 0$ *one can take* $\tilde{X}'_- = \emptyset$. *Moreover, one can easily see that if* X *is a conic with respect to the* x'' *domain in a neighborhood of* $\{x'' = 0\}$ *and*

$$D^\alpha(a_{\beta',\beta''} - a^0_{\beta',\beta''}) = o(\rho^{2m-|\beta'|-|\beta''|}\gamma^{-|\alpha|}) \qquad \forall \beta',\beta'' \quad \forall\alpha : |\alpha| \le 1 \tag{10.3.17}$$

as $|x''| \to 0$ *with functions* $a^0_{\beta',\beta''}$ *which are positively homogeneous of degrees* $(2m - |\beta'| - |\beta''|)\mu$ *then checking the microhyperbolicity condition with*

$$\mathcal{T} = \langle\xi, \partial_\xi\rangle + \mu^{-1}\langle x'', \partial_{x''}\rangle$$

we see that we can take $\tilde{X}'_- = \emptyset$. *In both cases we obtain that*

(a) For $d'' > \vartheta + 1$ *asymptotics* (10.3.9)′ *hold. Moreover, under the standard condition* (Ψ0) *on Hamiltonian trajectories asymptotics* (10.3.11)′ *hold.*

(b) For $d'' = \vartheta + 1$ *asymptotics* (10.3.15) *hold and for* $d'' < \vartheta + 1$ *asymptotics* (10.3.16) *hold.*

Let us note that in both cases for $\mu \le -d''$, *even for the Schrödinger operator, it is possible that*

$$\mathcal{N}^w(\tau) - \varkappa_0\tau^{\frac{d}{2m}} - \varkappa_1\tau^{\frac{d-1}{2m}} \asymp \tau^{\frac{d}{2m}+\frac{d''}{2m\mu}}$$

and therefore for $\mu < -d''$ *we cannot replace* $\mathcal{N}^w(\tau)$ *by* $\varkappa_0\tau^{\frac{d}{2m}}$ *and preserve the remainder estimate. Moreover, for* $\mu = -d''$ *we cannot replace* $\mathcal{N}^w(\tau)$ *by* $\varkappa_0\tau^{\frac{d}{2m}} + \varkappa_1\tau^{\frac{d-1}{2m}}$ *and preserve the remainder estimate* $o\big(\tau^{\frac{d-1}{2m}}\big)$. ◊

Example 10.3.2 (Power singularities again). *Let us consider the Schrödinger operator with* $d \ge 2$ [29] *and let us assume that in a neighborhood of* $\{x'' = 0\}$ *the domain* X *coincides with* $\{x = (x', r, \theta) \in \mathbb{R}^{d''} \times \mathbb{R}^+ \times U\}$ *where* $U \subset \mathbb{S}^{d''-1}$ *is a* $\tilde{\gamma}$*-admissible*

[29] One can consider general scalar operators for $\vartheta > 0$.

domain with $\tilde{\gamma} = \tilde{\gamma}(x',\theta)$. *Let us assume that* $g^{jk} = g^{jk}(x',\theta)$, $V_j = v_j(x',\theta)|x''|^{\mu}$, $V = v(x',\theta)|x''|^{2\mu}$ *and* $\mu \le -1$. *Let us assume that* $v \ge 0$ *and the functions* g^{jk}, v_i *and* v *on* $\mathbb{R}^{d'} \times \mathbb{S}^{d''-1}$ *satisfy the standard conditions* $(\Sigma 0)$–$(\Sigma 3)$ *with* $\tilde{\rho} = \sqrt{v}$ *and with* $\tilde{\gamma}$ *instead of* ρ *and* γ. *Then*

(i) *Under the additional condition* $|v_j| \le c\tilde{\rho}$ *one can take* $\mathcal{X} = \{x, |x''| \le \epsilon\}$, $\rho = \tilde{\rho}|x''|^{\mu}$, $\gamma = \epsilon_1\tilde{\gamma}|x''|$ *with small enough* $\epsilon > 0$, $\epsilon_1 > 0$. *Then we obtain that under the condition* $\tilde{\gamma}^{-1-\vartheta} \in L^1(\mathbb{R}^{d'} \times U)$, *asymptotics* (10.3.9)′ *hold for* $d'' > \vartheta + 1$. *Moreover, under the standard condition* $(\Psi 0)$ *on Hamiltonian trajectories asymptotics* (10.3.11)′ *hold. On the other hand, under the condition* $\tilde{\gamma}^{-1-\vartheta} \in L^1(\mathbb{R}^{d'} \times U)$ *for* $d'' = \vartheta + 1$, $d'' < \vartheta + 1$ *we obtain asymptotics (10.3.15), (10.3.16) respectively.*

(ii) *Rejecting the condition* $|v_j| \le c\tilde{\rho}$ *we should take* $\mathcal{X}'' = \emptyset$. *In this case we should assume that* $\mu < -1$ *and* $\tilde{\rho}\tilde{\gamma} \ge \epsilon_0 > 0$. *These conditions forbid the degeneration of* v *but permit certain singularities. Then all the statements of the previous assertion remain true.*

(iii) *Moreover, let us assume that* $\mu < -1$, $\tilde{\rho}\tilde{\gamma} \ge \epsilon_0 > 0$ *and* $V_j = v_j(x',\theta)|x''|^{\mu'}$ *with* $\mu' < \mu$ *(the case* $\mu' > \mu$ *is trivial). In this case one can take* $\rho = \tilde{\rho}|x'|^{\mu}$, $\gamma = \epsilon_1\tilde{\gamma}|x''|^{-\mu'+\mu+1}$. *Then* $\rho\gamma \ge \epsilon_2|x''|^{-\sigma}$ *with* $\sigma > 0$ *provided* $2\mu + 1 < \mu'$. *Then, assuming that* $\tilde{\gamma}^{-1-\vartheta} \in L^1(\mathbb{R}^{d'} \times U)$ *we obtain asymptotics* (10.3.9)′ *for* $d'' > (1+\vartheta)(1+\mu-\mu')^{-1}$. *Moreover, under the standard condition* $(\Psi 0)$ *on Hamiltonian trajectories we obtain asymptotics* (10.3.11)′. *On the other hand, for* $d'' = (1+\vartheta)(1+\mu-\mu')^{-1}$ *we obtain asymptotics* (10.3.15) *and for* $d'' < (1+\vartheta)(1+\mu-\mu')^{-1}$ *we obtain the asymptotics*

$$\bar{N}_{\vartheta}(\tau) = \mathcal{N}^{w}_{\vartheta}(\tau) + O\bigl(\tau^{\frac{d-\eta}{2}+\vartheta}\bigr) \tag{10.3.18}$$

with $\eta = \frac{1}{\mu}\bigl((1+\vartheta)(1+2\mu-\mu') - d''\bigr) \in (0, 1+\vartheta)$.

Here we cannot consider the case $\mu \le 2\mu + 1$ *because the magnetic field is too strong near the origin. This case is left to the next chapter.* ◊

Let us note that every time assertions with large enough s are used, K (in the conditions on coefficients) depends on μ, μ', etc. (via s).

Example 10.3.3 (Exponential singularities). *Let* X, g^{jk}, v_j *and* v *be as in assertions (i) or (ii) of the previous example. Let* $V_j = v_j \exp(|x''|^{-\nu})$, $V = v\exp(2|x''|^{-\nu})$ *with* $\nu > 0$. *Then all the conditions will be fulfilled with* $\rho = \tilde{\rho}\exp(|x''|^{-\nu})$, $\gamma = \epsilon_1\tilde{\gamma}|x''|^{\nu+1}$. *We obtain that for* $(1+\nu)(1+\vartheta) < d''$, *asymptotics* (10.3.9)′ *hold and under the additional condition* $(\Psi 0)$ *on Hamiltonian trajectories asymptotics* (10.3.11)′ *hold. Moreover, for* $(1+\nu)(1+\vartheta) = d''$ *the asymptotics*

$$\mathbf{N}_{\vartheta}(\tau) = \mathcal{N}_{\vartheta}(\tau) + O\bigl(\tau^{\frac{d-1+\vartheta}{2}} \log\log\tau\bigr) \tag{10.3.19}$$

hold, and for $(1+\nu)(1+\vartheta) > d''$ *the asymptotics*

$$\mathbf{N}_{\vartheta}(\tau) = \mathcal{N}_{\vartheta}(\tau) + O\bigl(\tau^{\frac{d-1+\vartheta}{2}} \log^{\sigma}\tau\bigr) \tag{10.3.20}$$

hold with $\sigma = \frac{1}{\nu}\bigl((1+\nu)(1+\vartheta) - d''\bigr)$. ◊

I leave it to the reader to consider examples in the spirit of example 10.3.2(iii) with exponential singularities, i.e., with $V_j = v_j|x''|^{-l}\exp(|x''|^{-\nu})$ or with $V_j = v_j\exp(k|x''|^{-\nu})$, $k \in (1,2)$. I also leave the estimate $\mathcal{N}^{w}(\tau) - \varkappa_0\tau^{\frac{d}{2}}$ from above and below to the reader. Moreover, I leave quasi-conical domains and domains with cusps for potentials which grow at 0 or infinity to the reader. Finally, I also leave general scalar operators with $\vartheta > 0$ to the reader.

10.3.4 Indefinite perturbations

Now we consider the case when a strongly singular coercive potential is perturbed by a less singular indefinite potential. We assume that conditions (10.1.3)–(10.1.4) are fulfilled with $\rho+\rho'$ instead of ρ and with $\varrho=(\rho+\rho')^{2m}$ where ρ and ρ' are γ-admissible weight functions. We replace conditions (10.3.1) and (10.3.2) by the conditions

$$\langle A(x,\xi)v,v\rangle \ge \epsilon_0\big((|\xi|+\rho)^{2m} - C{\rho'}^{2m}\big)|v|^2 \qquad \forall\xi\ \ \forall v, \tag{10.3.1$'$}$$

$$\sum_{\alpha,\beta:|\alpha|\le m,|\beta|\le m} \langle a_{\alpha,\beta}(x)v_\alpha, v_\beta\rangle \ge \epsilon_0 \sum_{\alpha:|\alpha|\le m} \rho^{2m-2|\alpha|}|v_\alpha|^2 - C\sum_{\alpha:|\alpha|\le m-1} {\rho'}^{2m-2|\alpha|}|v_\alpha|^2 \qquad \forall V=(v_\alpha)_{\{|\alpha|\le m\}} \tag{10.3.2$'$}$$

respectively.

Then the arguments of this and previous sections show that all the estimates of this section remain true with $\tilde{X}'_-$ defined by ρ if to their right-hand expressions we add

$$R'(\tau) = C\Big(\int_{X^{IV}} ({\rho'}^{2m})_+^q\,dx\Big)^{\frac{d}{2mq}} \times (\operatorname{mes} X^{IV})^{1-\frac{d}{2mq}} \tag{10.3.21}$$

for $\vartheta=0$, and

$$R'_\vartheta(\tau) = C\int_0^\infty t^{\vartheta-1}\Big(\int_{X^{IV}} ({\rho'}^{2m}-t)_+^q\,dx\Big)^{\frac{d}{2mq}}\,dt \times (\operatorname{mes} X^{IV})^{1-\frac{d}{2mq}} \tag{10.3.22}$$

for $\vartheta>0$ where $q\ge\frac{d}{2m}$, $q\ge 1$ and one of these inequalities is strict, and $X^{IV}=\{x\in X, \rho\le C\rho'\}$.

As an example we consider a power singularity under the assumptions of example 10.3.2(i) with potentials V, V_j replaced by $V+V'$, $V_j+V'_j$ respectively where $V'=w(x',\theta)|x''|^{2\nu}$, $V'_j=w_j(x',\theta)|x''|^{\nu}$, w and w_j are functions on $\mathbb{S}^{d''-1}\times\mathbb{R}^{d'}$ satisfying conditions $(\Sigma 0)$–$(\Sigma 3)$ with $\tilde\rho'$, $\tilde\gamma$. Here $\tilde\rho'$ is a $\tilde\gamma$-admissible weight function on $\mathbb{S}^{d''-1}\times\mathbb{R}^{d'}$ and $\nu>\mu$, $\mu\le -1$, $\nu<0$. Let us set $\gamma=\epsilon_1\tilde\gamma|x''|$, $\rho=\tilde\rho|x''|^\mu$, $\rho'=\tilde\rho'|x''|^\nu$ with $\tilde\rho=\sqrt{v}$. We obtain that a term should be added to the right-hand expressions of our estimates, and that this term is uniformly bounded provided one of the following conditions is fulfilled:

$$\iint_{\{\tilde\rho'^2\ge\epsilon v\}} \tilde\rho'^{-a} v^{\frac{b}{2}}\,d\theta dx' < \infty \qquad \text{if } b<0,\ d\ge 3 \tag{10.3.23$_1$}$$

where here and below $a=(d''+d\mu+2\vartheta\mu)(\nu-\mu)^{-1}$, $b=(d''+d\nu+2\vartheta\nu)(\nu-\mu)^{-1}$,

$$\iint_{\{\tilde\rho'^2\ge\epsilon v\}} \tilde\rho'^{d+2\vartheta}\Big(\log\frac{\tilde\rho'^2}{v}\Big)_+\,d\theta dx' < \infty \qquad \text{if } b=0,\ d\ge 3 \tag{10.3.23$_2$}$$

$$\iint_{\{\tilde\rho'^2\ge\epsilon v\}} \tilde\rho'^{d+2\vartheta}\,d\theta dx' < \infty \qquad \text{if } b>0,\ d\ge 3, \tag{10.3.23$_3$}$$

$$\iint_{\{\tilde\rho'^2\ge\epsilon v\}} \tilde\rho'^{-a-\delta\bar a} v^{\frac{b}{2}+\delta}\,d\theta dx' < \infty \qquad \text{if } b\le 0,\ d=2, \tag{10.3.23$_4$}$$

$$\iint_{\{\tilde\rho'^2\ge\epsilon v\}} \tilde\rho'^{d+2\vartheta+\delta}\,d\theta dx' < \infty \qquad \text{if } b>0,\ d=2, \tag{10.3.23$_5$}$$

where $\delta > 0$ is arbitrarily small, $\bar{a} = \mu(\nu-\mu)^{-1}$, $\bar{b} = \nu(\nu-\mu)^{-1}$.

I leave the analysis of the other examples to the reader.

10.3.5 Operators with indefinite principal symbols

I leave the treatment of operators with indefinite principal symbols to the reader. In this case one should replace conditions (10.3.1), (10.3.2) by

$$|A(x,\xi)v| \geq \epsilon_0(|\xi|+\rho)^{2m}|v| \qquad \forall \xi \quad \forall v, \tag{10.3.1$''$}$$

$$|\sum_{\alpha:|\alpha|\leq 2m} a_\alpha(x)v_\alpha| \geq \epsilon_0 \sum_{\alpha:|\alpha|\leq 2m} \rho^{2m-|\alpha|}|v_\alpha| \qquad \forall V = (v_\alpha)_{\{\alpha:|\alpha|\leq 2m\}} \tag{10.3.2$''$}$$

respectively. Moreover, one should replace conditions (10.3.1)$'$, (10.3.2)$'$ by

$$|A(x,\xi)v| \geq \epsilon_0\big((|\xi|+\rho)^{2m} - C\rho'^{2m}\big)|v| \qquad \forall \xi \quad \forall v, \tag{10.3.1$'''$}$$

$$|\sum_{\alpha:|\alpha|\leq 2m} a_\alpha(x)v_\alpha| \geq \epsilon_0 \sum_{\alpha:|\alpha|\leq 2m} \rho^{2m-|\alpha|}|v_\alpha| - C\sum_{\alpha:|\alpha|\leq 2m-1} \rho'^{2m}|v_\alpha| \qquad \forall V = (v_\alpha)_{\{\alpha:|\alpha|\leq 2m\}}. \tag{10.3.2$'''$}$$

respectively.

I do not know examples of this type which are interesting from the physical point of view.

10.4 Asymptotics of Large Eigenvalues: the Case of Essentially Unbounded Domains

Essentially unbounded domains are domains containing large (e.g., conical) exits to infinity. In domains of this type the spectrum of the operator in question is discrete only because the operator contains lower degree terms which are coercive with the function ρ tending to ∞ at infinity (in almost all directions). Therefore even the order of the asymptotics of $\mathbf{N}(\tau)$ depends on lower degree terms (which are not now "lower order terms" from our point of view). We treat only operators with positive definite senior symbols and leave the indefinite case to the reader.

10.4.1 Standard results

We begin with potentials of power growth at infinity. As the first example we use only the results of Chapters 4 and 5 to obtain the following well known assertion:

Theorem 10.4.1. *(i) Let X be a connected unbounded domain in $\mathbb{R}^d$ and let conditions (10.1.1)–(10.1.3), (10.3.1) and (10.3.3) be fulfilled with $\gamma = \epsilon_1\langle x\rangle$, $\rho = \langle x\rangle^\mu$, $\mu > 0$ where here and below $\langle x\rangle = (|x|^2+1)^{\frac{1}{2}}$.*

(i) Let us assume that for every $\tau > c$ one of the conditions $(8.1.18)_{1-12}$ is fulfilled for the symbol $\big(A(x,\xi)\tau^{-1}-1\big)$ uniformly with respect to τ, x, ξ (this assumption is only restrictive for $|x| \geq C_0$). Then the asymptotics

(10.4.1) $$\mathbf{N}(\tau) = \mathcal{N}^w(\tau) + O\bigl(\tau^{l(d-1)}\bigr)$$

hold as $\tau \to +\infty$ *where* $\mathcal{N}^w(\tau)$ *is given by* (10.3.6) *and* $\mathcal{N}^w \asymp \tau^{ld}$, $l = \frac{1}{2m\mu}(\mu+1)$.

(ii) Let us assume that for every $\tau > c$ *one of the conditions* $(8.1.18)_{1-13}$ *is fulfilled for the symbol* $\bigl(A(x,\xi)\tau^{-1} - 1\bigr)$ *uniformly with respect to* τ, x, ξ. *Then for* $\vartheta > 0$ *the asymptotics*

(10.4.1)′ $$\mathbf{N}_\vartheta(\tau) = \mathcal{N}^w_\vartheta(\tau) + O\bigl(R_\vartheta(\tau)\bigr)$$

hold as $\tau \to +\infty$ *with*

(10.4.2) $$R_\vartheta(\tau) = \begin{cases} \tau^{l(d-1)+\vartheta(1-l)} & \text{for } \vartheta < d-1, \\ \tau^{d-1}\log\tau & \text{for } \vartheta = d-1, \\ \tau^{\frac{1}{2m}(d-1-\vartheta)+\vartheta} & \text{for } \vartheta > d-1 \end{cases}$$

where $\mathcal{N}^w_\vartheta(\tau) \asymp \tau^{ld+\vartheta}$.

(iii) Moreover, let $X \cap \{x \ge c_1\} = X^0 \cap \{x \ge c_1\}$ *where* X^0 *is a conical (with respect to* x*) domain with* $\partial X^0 \setminus 0 \in C^K$ *and let the stabilization conditions*

(10.4.3) $$D^\alpha(a_\beta - a^0_\beta) = o\bigl(|x|^{(2m-|\beta|)\mu-|\alpha|}\bigr) \qquad \forall\beta : |\beta| \le 2m \quad \forall\alpha : |\alpha| \le 1$$

be fulfilled as $|x| \to \infty$ *with functions* $a^0_\beta \in C^K(\bar{X}^0 \setminus 0)$ *which are positive homogeneous of degrees* $(2m - |\beta|)\mu$ *with respect to* x. *Let us assume that at the energy level* $\tau = 1$ *the quasi-homogeneous symbol*

$$A^0(x,\xi) = \sum_{|\beta|\le 2m} a^0_\beta(x)\xi^\beta$$

satisfies the standard condition (Ψ) *on the Hamiltonian trajectories in* T^*X^0 *(and that these permitted trajectories never hit* $0 \times \mathbb{R}^d$*). Then the asymptotics*

(10.4.4) $$\mathbf{N}(\tau) = \mathcal{N}^w(\tau) + \bigl(\varkappa_1 + o(1)\bigr)\tau^{l(d-1)}$$

hold as $\tau \to +\infty$ *where, for the Schrödinger operator,*

$(10.4.5)_1$ $$\varkappa_1 = \mp\frac{1}{4}(2\pi)^{1-d}\omega_{d-1}\int_{\partial X^0} (1-V^0)_+^{\frac{d-1}{2}}\, dS^0$$

for Dirichlet and Neumann boundary conditions respectively, dS^0 *is the Riemannian density on* ∂X^0 *generated by the induced Riemannian metric. On the other hand, if only stabilization condition* (10.4.3) *is fulfilled then the microhyperbolicity condition is automatically fulfilled for the symbol* $\bigl(A(x,\xi)\tau^{-1} - 1\bigr)$ *and* $\mathcal{N}^w(\tau) = \varkappa_0\tau^{ld}(1+o(1))$ *as* $\tau \to +\infty$ *where, for the Schrödinger operator,*

$(10.4.5)_0$ $$\varkappa_0 = (2\pi)^{-d}\omega_d\int_{X^0} (1-V^0)_+^{\frac{d}{2}}\sqrt{g^0}\,dx$$

with $g^0 = \det(g^{jk0})^{-1}$.

Moreover, for $\vartheta < d-1$ *the asymptotics*

(10.4.4)′ $$\mathbf{N}_\vartheta(\tau) = \mathcal{N}^w_\vartheta(\tau) + o\bigl(R_\vartheta(\tau)\bigr)$$

hold with $R_\vartheta(\tau)$ *given by* (10.4.2)′.

(iv) On the other hand, if $\vartheta > d-1$, and the standard condition ($\Psi 0$) on Hamiltonian trajectories is fulfilled at the energy level $\tau = 1$ for the symbol $a(x,\xi) = \sum_{|\beta|=2m} a_\beta(x)\xi^\beta$ in the domain X, then asymptotics (10.4.4)′ also hold as $\tau \to +\infty$.

(v) Finally, if $\vartheta = d-1$, and the standard condition ($\Psi 0$) on Hamiltonian trajectories is fulfilled at the energy level $\tau = 1$ for the symbol $a^0(x,\xi) = \sum_{|\beta|=2m} a^0_\beta(x)\xi^\beta$ in the domain X^0 (and these trajectories never hit $0 \times \mathbb{R}^d$) then asymptotics (10.4.4)′ also hold as $\tau \to +\infty$

Note that in each of the cases (iii)–(v) we have different remainder estimates and different Hamiltonian trajectories.

Proof. Assertions (i)–(ii) obviously follow from the theorems of sections 4.3 and 5.3 (see the introduction). Let us prove (iii). We apply the theorems of sections 4.3, 4.4 and 5.3 and we examine the behavior of the billiards generated by the symbol $A(x,\xi)$ on $\Sigma_\tau = \{(x,\xi), \tau \in \operatorname{Spec} A(x,\xi)\}$. We take $L = 1$ and

$$\Lambda_1 = \Lambda_{1,\tau,\epsilon,T} = \Bigl\{(x,\xi) \in \Sigma_\tau, |x| \ge \epsilon\tau^{\frac{1}{2m\mu}},\quad |x\tau^{-\frac{1}{2m\mu}} - y| + |\xi\tau^{-\frac{1}{2m}} - \eta| \ge \epsilon \quad \forall (y,\eta) \in \Lambda^0_T\Bigr\}$$

where $\Lambda^0_T = \Sigma^0 \setminus \Lambda^1_T$, $\Sigma^0 = \{(y,\eta), 1 \in \operatorname{Spec} A^0(y,\eta)\}$ and through every point of Λ^1_T there passes a non-closed Hamiltonian billiard (a Hamiltonian trajectory with reflections at ∂X^0) of the symbol $|x|\bigl(A^0(x,\xi) - 1\bigr)$ of length $c_0 T$ along which all our conditions are fulfilled. By assumption mes $\Lambda^0_T = 0$ and Λ^0_T is a closed subset of $\Sigma^0 \setminus \{x = 0\}$ for every T. The remaining arguments repeat those in the proof of theorem 10.1.1: we check that the conditions on billiards generated by $A^0(x,\xi)$ and $A(x,\xi)$ coincide provided $\tau \ge C_{\varepsilon,T}$. Surely, it is important that *for $\vartheta \in [0, d-1)$ the principal contribution to the remainder estimate obtained in (ii) is given by the zone* $\{\varepsilon \le |x|\tau^{-\frac{1}{2m\mu}} \le C_0\}$ (with arbitrarily small fixed $\epsilon > 0$) in the following sense: the contribution of the zone $\{|x| \le \varepsilon\tau^{\frac{1}{2m\mu}}\}$ does not exceed $\varepsilon_1 R_\vartheta(\tau)$ with $\varepsilon_1 = \varepsilon_1(\varepsilon)$ tending to 0 as $\varepsilon \to 0$. On the other hand, for $\vartheta > d-1$ the principal contribution to the remainder estimate in (i) is given by the zone $\{x \in X, |x| \le R\}$ (with arbitrarily large fixed R) in the sense that the contribution of the remaining part does not exceed $\varepsilon_1 R_\vartheta(\tau)$ with $\varepsilon_1 = \varepsilon_1(R)$ tending to 0 as $R \to +\infty$. Therefore, in (iv) one should treat Hamiltonian billiards in this zone, in which one can replace $A(x,\xi)$ by $a(x,\xi)$ for large τ. Finally, for $\vartheta = d-1$ the principal contribution to the remainder estimate in (i) is given by zone $\{R \le |x| \le \varepsilon\tau^{\frac{1}{2m\mu}}\}$ for every fixed ε and R. Therefore, in (iv) one should treat Hamiltonian billiards in this zone, in which one can replace $A(x,\xi)$ by $a^0(x,\xi)$ for large τ, $|x|$. □

Let us consider scalar operators. Noting that the microhyperbolicity condition is automatically fulfilled for $|x| \le \epsilon\tau^{\frac{1}{2m\mu}}$ and condition $(8.1.18)_{13}$ is fulfilled automatically everywhere, we get the following

Theorem 10.4.2. *(i) Let X be a connected unbounded domain in $\mathbb{R}^d$ and let conditions (10.1.1)–(10.1.3), (10.3.1), (10.3.3) be fulfilled with $\gamma = \epsilon_1\langle x\rangle$, $\rho = \langle x\rangle^\mu$, $\mu > 0$ where $\langle x\rangle = (|x|^2+1)^{\frac{1}{2}}$ here and below. Moreover, let A be a scalar operator. Then*

(i) The asymptotics

$$\mathbf{N}(\tau) = \mathcal{N}^w_\vartheta(\tau) + O\bigl(\tau^{l(d-1)+\delta}\bigr) \tag{10.4.1″}$$

hold as $\tau \to +\infty$ with an arbitrarily small exponent $\delta > 0$ and the asymptotics (10.4.2)″ hold for $\vartheta > 0$.

(ii) Moreover, for $\vartheta > d-1$, $\vartheta = d-1$ respectively statements (iv),(v) of the previous theorem remain true.

10.4.2 Strongly non-periodic Hamiltonian flow

In what follows we consider homogeneous coefficients with singularities and (what is more essential here) with degenerations. However, we now improve the remainder estimates under stronger conditions on Hamiltonian trajectories.

Theorem 10.4.3. *Let the conditions of theorem 10.4.1 be fulfilled. Let A be a scalar operator and assume that either $X^0 = \mathbb{R}^d$ or A is the Schrödinger operator. Let the strong stabilization condition*

$$(10.4.3)^* \qquad D^\alpha(a_\beta - a^0_\beta) = O\bigl(|x|^{2m-|\beta|)\mu-|\alpha|-\delta}\bigr) \qquad \forall\beta : |\beta| \le 2m \quad \forall\alpha : |\alpha| \le 1$$

be fulfilled as $|x| \to \infty$ with some $\delta > 0$. Let us assume that the Hamiltonian flow (with reflections at ∂X^0) is of non-periodic power type[30] *and that along trajectories $(x(t), \xi(t))$ with $(x(0), \xi(0)) \in \Lambda^0_{\varepsilon,T} \subset \Sigma_1$, $|x(0)| = 1$, $\varepsilon = T^{-M}$, the condition $|x(t)| \ge T^{-M_1}$ is fulfilled. Then the asymptotics*

$$(10.4.6) \qquad \mathbf{N}(\tau) = \mathcal{N}^w(\tau) + \bigl(\varkappa_1 + O(\tau^{-\delta'})\bigr)\tau^{l(d-1)}$$

hold as $\tau \to +\infty$. Moreover, for $0 < \vartheta \le d-1-\delta$ the asymptotics

$$(10.4.6)' \qquad \mathbf{N}_\vartheta(\tau) = \mathcal{N}^w_\vartheta(\tau) + O\bigl(\tau^{l(d-1)+\vartheta(1-l)-\delta'}\bigr)$$

hold as $\tau \to +\infty$ with $\delta' = \delta'(d, m, \delta, L) > 0$.

Example 10.4.4. *(i) The condition of theorem 10.4.3 on the Hamiltonian flow is fulfilled for the Schrödinger operator with $X^0 = \mathbb{R}^d$ and $A^0(x,\xi) = |\xi|^2 + |x|^{2\mu}$, $0 < \mu \ne 1$.*

(ii) Therefore the statement of theorem 10.4.3 is true for this operator (provided all the other conditions of this theorem, including (10.4.3)*, *are fulfilled) and for the 2m-order operator with $A^0(x,\xi) = (|\xi|^2 + |x|^{2\mu})^m$, and one can replace $X^0 = \mathbb{R}^d$ by the X^0 obtained after (a sequence of) the following transformations:*

(a) We take the factor manifold of X^0 with respect to a finite rotation group (with center at the origin),

(b) We consider the reflection $S : \bar{X}^0 \to \bar{X}^0$ with respect to a hyperplane and remove from X^0 some subset $\mathcal{K}$ of this plane provided $\tilde{\gamma}^{-1-\vartheta} \in L^1(\mathbb{S}^{d-1})$ with $\tilde{\gamma} = \operatorname{dist}(x, \mathcal{K})$. ◊

Proof. Obviously it is sufficient to prove (i). Moreover, the potential $V^0 = |x|^{2\mu}$ is spherically symmetric and therefore angular momentum is constant along every Hamiltonian trajectory and therefore the x-projection of this trajectory lies in some 2-dimensional plane. Therefore it is sufficient to treat only the case $d = 2$. We introduce polar coordinates (r, θ) and use the fact that the angular momentum $M = \dot{\theta} r^2$ is constant where $\dot{\theta} = \frac{d\theta}{dt}$, etc. Therefore the energy is equal to $E = \frac{1}{2}\dot{r}^2 + \frac{1}{2}M^2 r^{-2} + r^{2\mu}$ and is also constant. Then a trajectory with $|M| \ge \varepsilon$ lies permanently in the zone $\{(\theta, r), r \ge M\sqrt{\frac{2}{E}}\}$. One can easily see that

$$|M| \le M_0(E) = (2\mu)^{\frac{1}{2}}\bigl(\frac{E}{\mu+1}\bigr)^{\frac{1}{2\mu}(\mu+1)}$$

and every trajectory oscillates between the largest and the smallest values $r = r_2$ and $r = r_1$ respectively where $r_1 < r_2$ [31] are roots of the equation $\frac{1}{2}M^2 r^{-2} + r^{2\mu} = E$ and the period of oscillation is

[30] I know of no example when the flow is of non-periodic exponential type but is not of non-periodic power type. I leave to the reader to formulate the statement in the case of non-periodic exponential type.

[31] And $r_1 = r_2$ if and only if $|M| = M_0(E)$.

$$T = T(E, M) = 2\int_{r_1}^{r_2} \left(2E - 2r^{2\mu} - M^2 r^{-2}\right)^{-\frac{1}{2}} dr$$

while the increment of θ during a period is

$$\theta_\Pi = 2\int_{r_1}^{r_2} Mr^{-2}\left(2E - 2r^{2\mu} - M^2 r^{-2}\right)^{-\frac{1}{2}} dr.$$

One can easily see that if $\varepsilon \le |M| \le M_0(E) - \varepsilon$ and $E \asymp 1$ then

$$|r_2 - r_1| \ge \epsilon\varepsilon^p, \quad T \ge \epsilon\varepsilon^p, \quad \theta_\Pi \ge \epsilon\varepsilon^p$$

and the functions $f = r_1, r_2, T, \theta_\Pi$ satisfy the inequalities

$$|D^\beta_{E,M} f| \le C_{|\beta|}\varepsilon^{-p|\beta|} \qquad \forall\beta$$

with appropriate constants $p, \epsilon > 0$ and $C_{|\beta|}$; $\varepsilon > 0$ is a small parameter here. On the other hand, the $dxd\xi : dA^0$-measure of the set of points $\{(r,\theta) : E = 1, |M| \le \varepsilon \text{ or} E = 1, |M| \ge M_0(E) - \varepsilon\}$ is less than $C\varepsilon^\delta$ with $\delta > 0$. These assertions yield that the Hamiltonian flow is of power type. Let us prove that this flow is of non-periodic power type. The arguments above yield that it is sufficient to prove that for $k, n \in \mathbb{Z}^+, m \le \varepsilon^{-1}, n \le \varepsilon^{-1}$ the measure of the set of points (x,ξ) with $E(x,\xi) = 1$ such that

$$|\theta_\Pi(1, M(x,\xi)) - 2\pi\frac{k}{n}| \le \varepsilon^L \tag{10.4.7}$$

does not exceed $O(\varepsilon^{\frac{L}{q}})$ where L is arbitrarily large. Obviously, $\theta_\Pi(1, M)$ is an analytic function of $M \in (0, M_0(1))$ and $\theta_\Pi(1, M_0(1) \;\; \eta^2)$ is an analytic function of $\eta \in (-\epsilon_0, \epsilon_0)$. Moreover, it is well known (see, e.g., [Arnold 2]) that for $\mu > 0, \mu \ne 1$, $\theta_\Pi(1, M)$ is not constant. Therefore the measure of points (x,ξ) with $E(x,\xi) = 1$ satisfying (10.4.7) and $|M(x,\xi)| \ge \epsilon_1$ is $O(\varepsilon^{\frac{L}{q}})$ as $\varepsilon \to +0$. Here $\epsilon_1 > 0$ is an arbitrary constant. On the other hand, one can easily see that θ_Π is continuous at $[0, M_0(1)]$ and $\theta_\Pi(+0) = \pi$. Therefore in the zone $\{(x,\xi), 0 < M(x,\xi) \le \epsilon\epsilon_0\}$ inequality (10.4.7) can be fulfilled only with $k \asymp n$. Moreover, one can easily see that for $M \in (0, \epsilon_0)$, $\theta_\Pi(1, M) = \mathcal{F}(M, M^{2\mu})$ where $\mathcal{F}$ is an analytic function of two variables in a neighborhood of $(0,0)$. Moreover, $\theta_\Pi(1,M) \ne \text{const}$ and therefore $|\nabla_M \theta_\Pi(1, M)| \ge \epsilon_1 M^p$ for $M \in (0, \epsilon_1)$ with appropriate $\epsilon_1 > 0$ and p and therefore the measure of the set of points with $E(x,\xi) = 1$ and $M(x,\xi) \in (0, \epsilon_1)$ satisfying (10.4.7) is $O(\varepsilon^{\frac{L}{p}})$. □

Remark 10.4.5. *(i) For $\vartheta > d - 1 + \delta$ one can easily improve the remainder estimate $O(\tau^{\frac{1}{2m}(d-1\vartheta)+\vartheta})$ to $O(\tau^{\frac{1}{2m}(d-1\vartheta)+\vartheta-\delta'})$ under a corresponding condition on the Hamiltonian flow of $a(x,\xi)$. In order to reach uniformity one can assume that stabilization condition (10.4.3)* is fulfilled only for terms with $|\alpha| = 2m$. In this case only these terms should be scalar but if some junior terms are not scalar one should check that a $(8.1.18)_*$-type condition is fulfilled.*

(ii) Moreover, one can consider the case $\vartheta = d - 1$. It is easy to replace the factor $\log\tau$ by $\log^{1-\delta'}\tau$ but better improvements are possible only under much more complicated and restrictive conditions. ◊

10.4.3 Periodic Hamiltonian flow

Let us assume that $X^0 = \mathbb{R}^d$,

(10.4.8) A is a scalar operator, the Hamiltonian trajectories of

$A^0(x,\xi)$ are periodic with period $T(A^0(x,\xi))$ and

$$A = A^{0w}(x,D) + A'^{w}(x,D), \tag{10.4.9}$$

$$D^\beta(a'_\alpha - a^1_\alpha) = O\bigl(|x|^{(2m-|\alpha|)\mu-\delta_1-\delta'-|\beta|}\bigr) \qquad \forall\alpha: |\alpha|\le 2m \quad \forall\beta: |\beta|\le K \qquad \text{as } |x|\to\infty \tag{10.4.10}$$

where $a^1_\alpha(x)$ are positively homogeneous of degrees $(2m-|\alpha|)\mu-\delta_1$ on x and are smooth away from 0, $0<\delta_1<2(\mu+1)$, $\delta'>0$.

Let us introduce the function

$$b(x,\xi) = \frac{l}{T_0}\int_0^{T_0} A^1\circ\Phi_t dt, \qquad A^1 = \sum_{|\alpha|\le 2m} a^1_\alpha(x)\xi^\alpha \tag{10.4.11}$$

on the surface $\Sigma^0 = \bigl\{(x,\xi), A^0(x,\xi)=1\bigr\}$; recall that $l = \frac{1}{2m\mu}(\mu+1)$ and $T_0 = T(1)$. It follows from the quasi-homogeneity of A^0 with respect to (x,ξ) that the action vanishes.[32)]

Then the results of section 4.6 immediately yield the following theorem:

Theorem 10.4.6. *Let $X^0=\mathbb{R}^d$ and let conditions (10.4.8)–(10.4.10) be fulfilled. Let us assume that*

$$\operatorname{mes}_\Sigma\bigl\{(x,\xi)\in\Sigma, |b(x,\xi)-\omega|\le\varepsilon\bigr\} = O(\varepsilon^\delta) \qquad \text{as } \varepsilon\to+0 \tag{10.4.12}$$

with $\delta>0$ and

$$\operatorname{dist}(\Phi_t(x,\xi),(x,\xi)) \ge t(T_0-t)\gamma(x,\xi) \qquad \forall(x,\xi)\in\Sigma \quad \forall t\in(0,T_0) \tag{10.4.13}$$

with $\gamma\ge 0$ such that $\operatorname{mes}\bigl\{(x,\xi),\gamma(x,\xi)\le\varepsilon\bigr\} = O(\varepsilon^\delta)$ as $\varepsilon\to+0$.

Then

(i) If $\delta_1<\mu+1$ and condition (10.4.12) is fulfilled for every $\omega\in\mathbb{R}$ then the asymptotics

$$\mathbf{N}(\tau) = \mathcal{N}^w(\tau) + O\bigl(\tau^{l(d-1)-\delta''}\bigr) \tag{10.4.14}$$

hold as $\tau\to+\infty$ with $\delta''>0$.

(ii) If $\delta_1\in[\mu+1,2\mu+2)$ then for $\tau\to+\infty$ such that $\tau^l\in\bigcup_{n\in\mathbb{Z}^+}J_n$ the asymptotics

$$\mathbf{N}(\tau) = \mathcal{N}^w(\tau) + F(\tau)\tau^{l(d-1)} + O\bigl(\tau^{l(d-1)-\delta''}\bigr) \tag{10.4.15}$$

hold with $\delta''>0$ where

---

32) In fact, the action is equal to

$$\frac{1}{T_0}\int_0^{T_0}\bigl(k\langle\xi,A^0_\xi\rangle + (1-k)\langle x,A^0_x\rangle - A\bigr)\circ\Phi_t dt$$

with an arbitrary constant k. Let us take $k=\mu(\mu+1)^{-1}$. Then the Euler identity yields that the integrand vanishes for $l=1$. On the other hand, we must assume that $l=1$: otherwise we should replace A by A^l in order to obtain $T=T_0$ for all (x,ξ).

$$J_n = \big[\nu(n+\frac{1}{4}\omega_0) + (\omega-\epsilon_1)(\nu n)^{l-\frac{\delta_1}{2m\mu}}, \nu(n+\frac{1}{4}\omega_0) + (\omega+\epsilon_1)(\nu n)^{l-\frac{\delta_1}{2m\mu}}\big],$$

ω_0 is the Maslov index of the trajectory $\Phi_t(x,\xi)$ with $t\in(0,T_0)$, $\nu = \frac{2\pi l}{T_0}$ and $\epsilon_1>0$ is small enough,

$$F(\tau) = (2\pi)^{-d} l T_0^{-1} \int_\Sigma v\Big(T_0 l^{-1}\big((\tau^l - \frac{1}{4}\pi\omega_0) - \tau^{l-\frac{\delta_1}{2m\mu}} b\big)\Big) dxd\xi : dA^0. \tag{10.4.16}$$

Let us consider the perturbed harmonic oscillator $A_0 = \frac{1}{2}(|D|^2+|x|^2)$ more carefully. In this case $\mu = l = m = 1$, $T_0 = 2\pi$, $\omega_0 \equiv 2d \pmod{4\mathbb{Z}}$ and the x-projections of the Hamiltonian trajectories are ellipses centered at the origin. Moreover, $\operatorname{Spec} A_0$ is $\{\tau_n = \frac{d}{2}+n, n\in\mathbb{Z}^+\}$. The multiplicity of τ_n is $r(n)$ where $r(.)$ is a polynomial of degree $d-1$. Functional-analytic arguments yield that if the perturbing operator is a potential V with $\|V\|_{L^\infty}\le\epsilon$ then the spectrum consists of clusters of length $\le\epsilon$.

It is easy to show that for generic functions a^1_α ($|\alpha|\le k$) which are positively homogeneous of degree $(2-|\alpha|-\delta_1)$ condition (10.4.12) is fulfilled for every ω in which case the theory of section 4.6 yields that asymptotics (10.4.14) hold for $0<\delta_1<2$ and asymptotics (10.4.15)–(10.4.16) hold for $\delta_1\ge 2$ where $\delta_1>0$ and $k=0,1,2$ are now arbitrary. In particular, one can take $k=0$ and a perturbing potential $V' = v_0 r^{2\mu-\delta_1}$ with arbitrary $v_0\ne 0$ and $0<\delta_1\ne 2$.

Let us consider the case $b\equiv 0$. Then we have two possibilities. The first (trivial) possibility is to assume that $A = A^{0w}+A^{1w}+A''^w$ where A'' (instead of A') satisfies condition (10.4.10) with a^2_α,δ_2 instead of a^1_α,δ_1 respectively, $\delta_2\in(\delta_1,2\delta_1)$. Then one can construct the symbol $b(x,\xi)$ by formula (10.4.11) with A^2 instead of A^1, assume that condition (10.4.12) is fulfilled and obtain asymptotics (10.4.14) or (10.4.15)–(10.4.16) with δ_2 instead of δ_1. However, for $\delta_2 = 2\delta_1$ the situation is more complicated because in this case b contains a term which is quadratic with respect to A^1.

So, let us assume that

$$A^1\circ j^* = -A^1 \tag{10.4.17}$$

where j^* is the antipodal map $(x,\xi)\to(-x,-\xi)$. Repeating the arguments of the proof of theorem 10.2.24 we again obtain formula (10.2.83):

$$b = \frac{1}{2\pi}\Big(\int_0^{2\pi} A^2\circ\Phi_t dt + \int_0^\pi\int_0^\pi \{A^1\circ\Phi_t, A^1\circ\Phi_{t+t'}\}dtdt'\Big) \tag{10.4.18}$$

is a positive homogeneous symbol of degree $(2-2\delta_1)$ with respect to (x,ξ).

Therefore we have proven

Theorem 10.4.7. *Let $X^0 = \mathbb{R}^d$, $A = A^{0w}+A^{1w}+A''^w$ where A'' satisfies (10.4.10) with a^2_α and $\delta_2>0$ instead of a^1_α and δ_1. Here a^1_α and a^2_α are functions smooth for $x\ne 0$ and positively homogeneous of degrees $2-|\alpha|-\delta_1$, $2-|\alpha|-\delta_2$, and $\delta_2 = 2\delta_1$. Let condition (10.4.17) be fulfilled and let the symbol b be defined by (10.4.18). Then*

(i) If $\delta_2<2$ and condition (10.4.12) is fulfilled for every $\omega\in\mathbb{R}$ then asymptotics (10.4.14) hold for $\tau\to+\infty$.

(ii) If $\delta_2\ge 2$ and condition (10.4.12) is fulfilled for some $\omega\in\mathbb{R}$ then asymptotics (10.4.15)–(10.4.16) hold for $\tau\to+\infty$ and belonging to the intervals indicated in theorem 10.4.6(ii) with $\omega_0 = 2(d-1)$, $\nu=1$. In these asymptotics as well as in the definition of the intervals J_n one should replace δ_1 with δ_2.

One can easily prove that for generic positively homogeneous functions a^2_α ($|\alpha|\le k$) with $k=0,1,2$ condition (10.4.14) is fulfilled for all $\omega\in\mathbb{R}$.

I leave the treatment of Riesz means with $\vartheta<d-1$ to the reader.

10.4.4 Singularities and degenerations

Let us consider operators with homogeneous terms with singularities and (what is now more interesting) degenerations. In order to avoid conditions of type (8.1.18)$_*$ we only consider the Schrödinger operator. By the same method as in theorem 10.4.1 but also with analysis in the zone X'' the following assertion can be proven:

Theorem 10.4.8. *(i) Let $d \geq 3$, $X = \{x = (r, \theta) \in \mathbb{R}^+ \times U\}$ where $U \subset \mathbb{S}^{d-1}$. Let $g^{jk} = g^{jk}(\theta)$, $V_j = v_j(\theta) r^\mu$, $V = v(\theta) r^{2\mu}$ with $\mu > 0$. Let $v \geq 0$ and let U, g^{jk}, v_j, v and $\tilde{\rho} = \sqrt{v}$, $\tilde{\gamma} = \tilde{\gamma}(\theta)$ satisfy conditions (Σ0)–(Σ3). Finally, let $v^{-\frac{d}{2\mu}}, v^{-\frac{d-1}{2\mu}}\tilde{\gamma}^{-1} \in L^1(U)$. Then asymptotics (10.4.1)–(10.4.2) hold with $l = \frac{1}{2\mu}(\mu+1)$ and $\mathcal{N}^w(\tau) = \varkappa_0 \tau^{ld}$ with the coefficient $\varkappa_0$ given by (10.4.5)$_0$.*

Moreover, if $0 < \vartheta < d-1$ and $v^{-\frac{1}{2\mu}(d-1-\vartheta)}\tilde{\gamma}^{-1-\vartheta} \in L^1(U)$ then asymptotics (10.4.1)$'$–(10.4.2)$'$ hold.

(ii) Let us assume that there exists a set $\Xi \subset \Sigma = \{(x,\xi), a(x,\xi) = 1\}$ such that $\operatorname{mes}_\Sigma \Xi = 0$ and through every point of $\Sigma \setminus \Xi$ there passes an infinitely long non-periodic Hamiltonian billiard generated by the Hamiltonian

$$H^0 = \tilde{\gamma}|x|\big(g^{jk}(\xi_j - V_j)(\xi_k - V_k) + V - 1\big). \tag{10.4.19}$$

Then asymptotics (10.4.4) hold with the coefficient $\varkappa_1$ given by (10.4.5)$_1$.

Moreover, if $0 < \vartheta < d-1$ and $v^{-\frac{1}{2\mu}(d-1-\vartheta)}\tilde{\gamma}^{-1-\vartheta} \in L^1(U)$ then asymptotics (10.4.4)$'$–(10.4.2)$'$ hold.

I leave the proof of this theorem to the reader. I indicate only that $\rho = \tilde{\rho} r^\mu$, $\gamma = \epsilon\tilde{\gamma} r$ with small enough $\epsilon > 0$ and that we use spherical coordinates and the change of variable $r = \tau^{\frac{1}{2\mu}} r'$ in the calculation of $\mathcal{N}^w(\tau)$. It should be noted that if we reject the regularizing factor $\tilde{\gamma} r$ in H^0 (i.e., replace H^0 by H^0_{st}) then the conditions of the theorem do not change because the set of the points through which pass Hamiltonian billiards generated by H^0_{st} and reaching $\{\tilde{\gamma} r = 0\}$ in a finite time is a set of measure 0. This remark follows from the conditions of theorem 10.4.8(i).

Example 10.4.9. *Let $X = \mathbb{R}^d$, $d \geq 3$, $V \asymp |x''|^{2\nu}|x|^{2\mu-2\nu}$, $x = (x'; x'') = (x_1, \dots, x_{d'}; x_{d'+1} \dots, x_d)$, $d'' = d - d'$, $0 < d'' < d$ (in which case one should try $\rho = |x''|^\nu |x|^{\mu-\nu}$, $\gamma = \frac{1}{2}|x''|$). Let us assume that the conditions on g^{jk}, V_j, V are fulfilled with the indicated γ and ρ. Then*

(i) $v^{-\frac{d}{2\mu}}$ and $v^{-\frac{1}{2\mu}(d-1-\vartheta)}\tilde{\gamma}^{-1-\vartheta}$ belong to $L^1(\mathbb{S}^{d-1})$ if and only if $\nu < \mu\frac{d''}{d}$ and $\nu < \mu(d''-1-\vartheta)(d-1-\vartheta)^{-1}$ respectively. Therefore for $\nu < \mu(d''-1)(d-1)^{-1}$ asymptotics (10.4.1)$'$–(10.4.2)[33] hold, and for $0 < \vartheta < d''-1$, $\nu < \mu(d''-1-\vartheta)(d-1-\vartheta)^{-1}$ asymptotics (10.4.1)$'$–(10.4.2)$'$[34] hold.

(ii) In order to treat the case $\nu \in [\nu_1^*, \nu_0^*] = [\mu(d''-1)(d-1)^{-1}, \mu d'' d^{-1}]$ recall that the original remainder estimate was

$$R(\tau) = \int_{\{V \leq C\tau, \sqrt{\tau}\gamma \geq 1\}} \tau^{\frac{d-1}{2}}\gamma^{-1}\,dx + \int_{\{V \leq C\tau, \sqrt{\tau}\gamma \leq 1\}} \tau^{\frac{d}{2}}\,dx.$$

Introducing spherical coordinates and integrating over r we obtain the remainder estimate

$$R(\tau) = \tau^{l(d-1)} \int_{\{v \leq C\tau^{\mu+1}\tilde{\gamma}^{2\mu}\}} v^{-\frac{d-1}{2\mu}}\tilde{\gamma}^{-1}\,d\theta + \tau^{ld}\int_{\{v \geq C\tau^{\mu+1}\tilde{\gamma}^{2\mu}\}} v^{-\frac{d}{2\mu}}\,d\theta.$$

---

[33] As well as asymptotics (10.4.4)–(10.4.2) under the standard condition on Hamiltonian trajectories.

[34] As well as asymptotics (10.4.4)$'$–(10.4.2) under the standard condition on Hamiltonian trajectories. These footnotes remain true below.

We can replace v by $|x''|^{2\nu}|x|^{2\mu-2\nu}$ here in which case we obtain the asymptotics

$$\mathbf{N}(\tau) = \mathcal{N}^{\mathrm{w}}(\tau) + O\big(\tau^{l(d-1)}\log\tau\big) \tag{10.4.20}$$

as $\tau \to +\infty$ provided $\nu = \nu_1^*$ and the asymptotics

$$\mathbf{N}(\tau) = \mathcal{N}^{\mathrm{w}}(\tau) + O\big(\tau^{l(d-\eta)}\big) \tag{10.4.21}$$

as $\tau \to +\infty$ provided $\nu \in (\nu_1^*, \nu_0^*)$ and $\eta = (-\nu d + \mu d'')(\mu-\nu)^{-1} \in (0,1)$. Here one can replace $\mathcal{N}^{\mathrm{w}}(\tau)$ by $\varkappa_0\tau^{ld}$ with $\varkappa_0$ given by $(10.4.5)_0$.

(iii)′ Similarly, for $\nu = \nu_{\vartheta+1}^* = \mu(d''-\vartheta-1)(d-\vartheta-1)^{-1}$ the asymptotics

$$\mathbf{N}_\vartheta(\tau) = \mathcal{N}_\vartheta^{\mathrm{w}}(\tau) + O\big(\tau^{l(d-1-\vartheta)+\vartheta}\log\tau\big) \tag{10.4.20$'$}$$

hold as $\tau \to +\infty$ and for $\nu \in (\nu_{\vartheta+1}^*, \nu_0^*)$ the asymptotics

$$\mathbf{N}_\vartheta(\tau) = \mathcal{N}_\vartheta^{\mathrm{w}}(\tau) + O\big(\tau^{l(d-\eta)+\vartheta}\big) \tag{10.4.21$'$}$$

hold as $\tau \to +\infty$ with $\eta = \big((d''-\vartheta)\mu - (d-\vartheta)\nu\big)(\mu-\nu)^{-1} \in (0, \vartheta+1)$.

(iv) Finally, for $\nu = \nu_0^*$ the asymptotics

$$\mathbf{N}(\tau) = \mathcal{N}^{\mathrm{w}}(\tau) + O(\tau^{ld}) \tag{10.4.22}$$

hold as $\tau \to +\infty$ where

$$\mathcal{N}^{\mathrm{w}}(\tau) = (2\pi)^{-d}\omega_d \int_{\gamma \geq \epsilon\tau^{-\frac{1}{2}}} (\tau - V)_+^{\frac{d}{2}}\sqrt{g}\,dx \tag{10.4.23}$$

and $\mathcal{N}^{\mathrm{w}}(\tau) \asymp \tau^{ld}\log\tau$.[35] *The results for $\vartheta > 0$ obviously follow from this statement. In order to prove this statement one need only note that in the domain $X'' = \{x, |x''| < \epsilon\tau^{-\frac{1}{2}}\}$ one can take $\tilde{\rho} = 0$ in the inequality*

$$((Au - \tau), u) \geq \epsilon\|u\|_m^2 - C\|\tilde{\rho}^m u\|^2 \tag{10.4.24}$$

for all $u \in \mathrm{D}(A)$ supported in $\overline{X''}$. Moreover, this yields that

(v) All the results of (i)–(iii) remain true for $d = 2$. ◊

It is interesting to perturb the potentials treated above by potentials which have less "average" increment at infinity. Namely, we can take perturbing potentials satisfying condition $(\Sigma 1)_{2,3}$ with the same γ as above and with $\rho' = (|x|+1)^{\mu'}$, $0 < \mu' < \mu$ instead of ρ. Then two different situations may arise:

(a) The perturbing potential is also coercive (i.e., satisfies $(\Sigma 3)$ with ρ replaced by ρ') in which case the perturbing potential "helps" to the unperturbed. In this case the above analysis can be applied with no change.

(b) The perturbing potential is indefinite.

In the first case we leave the detailed analysis to the reader and formulate the following statements:

Theorem 10.4.10 (Coercive perturbations). *(i) Let $d \geq 3$ and X, g^{jk}, V_j^0 and V^0 be as in theorem 10.4.8. Let $V_j = V_j^0 + V_j'$, $V = V^0 + V'$ where V_j' and V' satisfy $(\Sigma 1)_{2,3}$ and $(\Sigma 3)$ with $\rho' = \langle x\rangle^{\mu'}$, $0 < \mu' < \mu$ instead of ρ and with $\gamma = \tilde{\gamma}|x|$. Then asymptotics (10.4.1)–(10.4.2) hold with $\mathcal{N}^{\mathrm{w}}(\tau) = \varkappa_0\tau^{ld}(1+o(1))$ as $\tau \to +\infty$.*

(ii) Moreover, if the Hamiltonian H^0 given by (10.4.19) satisfies the conditions of statement (ii) of theorem 10.4.8 then the asymptotics (10.4.4)–(10.4.2) hold.

(iii) Finally, if $0 < \vartheta < d-1$ and $v^{-\frac{1}{2\mu}(d-1-\vartheta)}\tilde{\gamma}^{-1-\mu} \in L^1(U)$ then in the context of (i),(ii) asymptotics (10.4.1)′–(10.4.2)′, (10.4.4)′–(10.4.2)′ hold respectively.

---

[35] Moreover, under stabilization condition (10.4.3), $\mathcal{N}^{\mathrm{w}}(\tau) = (\bar{\varkappa}+o(1))\tau^{ld}\log\tau$. This statement remains true below.

Example 10.4.11. *Let X, g^{jk}, V_j^0 and V^0 be as in example 10.4.9. Let $V_j = V_j^0 + V'_j$, $V = V^0 + V'$ where V'_j and V' satisfy $(\Sigma 1)_{2,3}$ and $(\Sigma 3)$ with ρ replaced by $\rho' = \langle x\rangle^{\mu'}$, $0 < \mu' < \mu$, and $\gamma = \epsilon |x''|$. Then*

(i) If $\nu < \mu(d''-1-\vartheta)(d-1-\vartheta)^{-1}$, $0 \le \vartheta < d''-1$ then asymptotics (10.4.1)′–(10.4.2) [34] hold.

(ii) If $\nu = \mu(d''-1-\vartheta)(d-1-\vartheta)^{-1}$, $0 \le \vartheta < d''-1$ then asymptotics (10.4.20) hold.

(iii) If $\mu(d''-1-\vartheta)(d-1-\vartheta)^{-1} < \nu < \mu d'' d^{-1})$, $0 \le \vartheta < d-1$ then asymptotics (10.4.21)′ hold with η given in example 10.4.9 provided $\mu' \le \mu'_* = (\mu-\nu)(\nu+1)^{-1}$. On the other hand, for $\mu' > \mu'_*$ the remainder estimate is $O(\tau^q)$ with

$$q = \frac{1}{2}(d-\vartheta-1) + \frac{1}{2\mu'\nu}\left(d'\nu - (\mu-\nu+\mu')(d''-\vartheta-1)\right)$$

for $\vartheta < d''-1$ and with $q = \frac{d'}{2} + \frac{d'}{2\mu'} + \vartheta$ for $\vartheta > d''-1$. For $\vartheta = d''-1$ the remainder estimate is $O(\tau^q \log \tau)$ with the same q.

(iv) All these results remain true for $\nu = \mu\frac{d''}{d}$ and $\vartheta = 0$ with $\mathcal{N}^w(\tau) \asymp \tau^{lq}\log\tau$.

(v) Moreover, for $\nu > \mu\frac{d''}{d}$, $\vartheta = 0$ and $\mu' > \mu'_*$ these results remain true with $\mathcal{N}^w(\tau) \asymp \tau^p$ with $p = \frac{1}{2}(d + \frac{d}{\mu'} - \frac{d''\mu}{\mu'\nu} + \frac{d''}{\nu})$. Then $p - q = \frac{1}{2\nu}(\nu + 1 - \frac{1}{\mu'}(\mu-\nu)) > 0$.◊

Let us now consider indefinite perturbations. In this case the perturbation cannot be too strong.

Theorem 10.4.12 (Indefinite perturbations). *(i) Let $d \ge 3$ and let X, g^{jk}, V_j^0 and V^0 be as in theorem 10.4.8. Let $V_j = V_j^0 + V'_j$, $V = V^0 + V'$ where V'_j and V' satisfy $(\Sigma 1)_{2,3}$ with ρ replaced by $\rho' = \langle x\rangle^{\mu'}$, $0 < \mu' < \mu$ and $\gamma = \tilde\gamma|x|$. Then, if $v^{-nd-\mu'\vartheta(\mu-\mu')^{-1}}$ and $v^{-n(d-\vartheta-1)-\mu'\vartheta(\mu-\mu')^{-1}}\tilde\gamma^{-\vartheta-1}$ belong to $L^1(U)$ with $n = \frac{1}{2}(\mu'+1)(\mu-\mu')^{-1}$, $0 \le \vartheta < d-1$ then asymptotics (10.4.1)′–(10.4.2)′ hold and for $\vartheta = 0$ $\mathcal{N}^w(\tau) = \varkappa_0\tau^{ld} + o(\tau^{l(d-1)})$.*

(ii) Moreover, if the Hamiltonian H^0 given by (10.4.19) satisfies the conditions of theorem 10.4.8(ii), then asymptotics (10.4.4)′–(10.4.2)′ hold.

Proof. The proof is similar to the proof of theorem 10.4.8 but the domain $\{V \le \tau\}$ consists of two overlapping zones $\{V^0 \le C\tau, \rho' \le C\tau^{\frac{1}{2}}\}$ and $\{V^0 \le C\rho'^2, \tau \le C\rho'^2\}$. Then $\tilde\rho(\tau) \asymp \tau^{\frac{1}{2}}$ and $\tilde\rho(\tau) \asymp \rho'$ in the first and second zones respectively. □

Example 10.4.13. *Let X, g^{jk}, V_j^0 and V^0 be as in example 10.4.9. Let $V_j = V_j^0 + V'_j$, $V = V^0 + V'$ where V'_j and V' satisfy $(\Sigma 1)_{2,3}$ with ρ replaced by $\rho' = \langle x\rangle^{\mu'}$, $0 < \mu' < \mu$ and $\gamma = \epsilon|x''|$. Let $\mu > \mu' + \nu$ (otherwise the potential V can be semibounded from above or even tend to $-\infty$ in the pipe $\{|x''| \le \epsilon\}$). Moreover, let us assume that*

$$\nu < \frac{\mu - \mu'}{\mu' + 1}. \tag{10.4.25}$$

Then considering the zone

$$X'' = \left\{x : |x''| \le z(x,\tau) = \epsilon\min\left(\tau^{-\frac{1}{2}}, |x|^{-(\mu-\nu)(\nu+1)^{-1}}\right)\right\}$$

we see that we can take $\tilde\rho(\tau) = (\rho' - \epsilon_1 z^{-1})_+$ in condition (10.4.24) where $\epsilon_1 > 0$ does not depend on ϵ. Then (10.4.24) yields that $\tilde\rho(\tau) = 0$ in X'' for $\tau \ge \epsilon^{-1}$ and small enough ϵ. On the other hand, (10.4.25) yields that $(\tau^{\frac{1}{2}} + \rho)\gamma \ge \epsilon_2 > 0$, $\rho' \le C(\tau^{\frac{1}{2}} + \rho)^{1-\delta}$ outside of X'' where $\delta > 0$, $\rho = |x''|^\nu |x|^{\mu\mu-\nu}$, $\gamma = \epsilon_3|x''|$. Using these assertions one can easily prove that under condition (10.4.25) all the statements of example 10.4.9 remain true.◊

I leave the analysis of the Schrödinger operator with coefficients satisfying conditions (Σ0)–(Σ3) with $\rho = \langle x\rangle^{\mu}$, $\gamma = \epsilon\langle x\rangle^{1-\delta}$, $\mu > 0$, $\delta > 0$, $\mu + 1 > \delta$ to the reader. In this case the asymptotics $\mathbf{N}_{\vartheta}(\tau) = \mathcal{N}_{\vartheta}^{w}(\tau) + O(\tau^{q})$ appear with $\mathcal{N}^{w}(\tau) \asymp \tau^{p}$ in domains conical at infinity. On the other hand, in domains of type (10.4.32) (see below) similar asymptotics appear with other exponents p and q. In this example the magnetic field can grow strongly enough at infinity.

10.4.5 Quasi-homogeneous potentials

Let us now consider the quasi-homogeneous case.

Example 10.4.14. *(i) Let X be an L-quasi-conical domain in $\mathbb{R}^d$ and let g^{jk}, V_j, V be L-quasi-homogeneous on x of degrees 0, μ and 2μ respectively. Here $\mu > 0$ and $L = (l_1, \dots, l_d)$ with $l_j = 1$ for $1 \le j \le d'$ and $l_j > 1$ for $d' + 1 \le j \le d$, $0 < d' < d$. Let us assume that $\partial X \setminus 0 \in C^K$ and $g^{jk}, V_j, V \in C^K(\bar{X} \setminus 0)$ satisfy (Σ0) and $V(x) > 0$ for $x \ne 0$. Then all of the conditions (Σ0)–(Σ3) are fulfilled with $\rho = [x]_L^{\mu}$, $\gamma = \epsilon[x]_L$ for*

$$[x]_L \stackrel{\mathrm{def}}{=} \Bigl(\sum_j |x|^{\frac{\bar{l}}{l_j}}\Bigr)^{\frac{1}{\bar{l}}} \ge 1;$$

for $[x]_L < 1$ one should take $\gamma = [x]_L^{\bar{l}}$; $\bar{l} = \max_j l_j$ here. Then the asymptotics

$$\mathbf{N}(\tau) = \varkappa_0 \tau^{p} + O(\tau^{q}) \tag{10.4.26}$$

hold with $p = \frac{d}{2} + \frac{l}{2\mu}$, $q = \frac{1}{2}(d-1) + \frac{1}{2\mu}(l-1)$ with $l = \sum_j l_j$ and with the coefficient

$$\varkappa_0 = (2\pi)^{-d}\omega_d \int (1 - V)_+^{\frac{d}{2}} \sqrt{g}\, dx. \tag{10.4.27}$$

I leave the analysis for $\vartheta > 0$ to the reader.

(ii) In order to improve these asymptotics let us consider the Hamiltonian billiards generated by the Hamiltonian

$$H = [x]_L^{1-\mu}\bigl(g^{jk}(\xi_j - V_j)(\xi_k - V_k) + V - \tau\bigr). \tag{10.4.28}$$

Obviously, the contribution of the zones $\{x \in X, |x_i| \le \epsilon\tau^{\frac{l_j}{2\mu}}\}$ to the remainder estimate does not exceed $\epsilon_1\tau^{q}$ with $\epsilon_1 \to 0$ as $\epsilon \to +0$ ($i = 1, \dots, d$). So we consider billiards starting from these zones.

One can easily see that along the Hamiltonian trajectories

$$\frac{d\xi_i}{dt} = o(\tau^{\frac{1}{2}}), \qquad \frac{dx_i}{dt} = o(\tau^{\frac{l_i}{2\mu}})$$

as $\tau \to +\infty$ for $i \ge d' + 1$ provided $[x]_L \ge \epsilon\tau^{\frac{1}{2\mu}}$. Let us fix an arbitrarily large T and arbitrarily small $\epsilon > 0$ and treat billiards of length T.

Let us assume that $x' \ne 0$, $\nu' \ne 0$ on $\partial X \setminus 0$ where $x' = (x_1, \dots, x_{d'})$ and $\nu' = (\nu_1, \dots, \nu_{d'})$, $\nu = (\nu_1, \dots, \nu_d) = (\nu', \nu'')$ is a unit normal to ∂X. Then in the indicated zone $|\nu''| = o(1)$ and therefore the increment of ξ_i when a billiard reflects from ∂X is $o(\tau^{\frac{1}{2}})$ for $i \ge d' + 1$. So $\tau^{-\frac{1}{2}}\xi_i$ and $\tau^{-\frac{l_i}{2\mu}}x_i$ with $l \ge d' + 1$ are almost constant along billiards (if the length and the number of reflections are bounded). Therefore if $(\bar{x}', \bar{\xi}')$ is a good starting point for a billiard generated by the Hamiltonian $H_{(\bar{x}'', \bar{\xi}'')} = H(x', \bar{x}'', \xi', \bar{\xi}'')$ in the domain $X_{(\bar{x}'', \bar{\xi}'')} = \{x' : (x', \bar{x}'') \in X\}$ then $(\bar{x}, \bar{\xi})$ is a good starting point for the *original Hamiltonian. Therefore the phase measure of points which cannot be starting points does not exceed* $\epsilon_2\tau^q$ *with arbitrarily small* $\epsilon_2 > 0$. *Here* $\tau \ge \tau_0(\epsilon_2)$.

Let us treat the non-periodicity. Let us assume that

$$g^{jk} = 0, \quad V_k = 0 \quad \text{for } j \le d', k \ge d'+1. \tag{10.4.29}$$

Then one can easily show that $|x''(0) - x''(t)| \ge \varepsilon|t|\tau^{\frac{1}{2\mu}}$ *provided* $t \in [-T,T]$, $|\xi''| \ge C_{\epsilon,T}\varepsilon\tau^{\frac{1}{2}}$, $\tau \ge \tau_0(\epsilon,\varepsilon,T)$ *where* ϵ *is defined by the above conditions including the transversality condition. Therefore the non-periodicity condition is also fulfilled if we exclude a subset of measure* $\epsilon_2\tau^q$ *from the starting point set. Therefore the asymptotics*

$$\mathbf{N}(\tau) = \mathcal{N}^w(\tau) + \bigl(\varkappa_1 + o(1)\bigr)\tau^q \tag{10.4.30}$$

hold with the coefficient

$$\varkappa_1 = \mp\frac{1}{4}(2\pi)^{1-d}\omega_{d-1}\int_{\partial X}(1-V)_+^{\frac{d-1}{2}}\,dS'\sqrt{g''}dx'' \tag{10.4.31}$$

where dS' *and* $\sqrt{g''}dx''$ *are Riemannian densities on the projections* $(\partial X)_{x''} = \{x' : (x',x'') \in \partial X\}$ *and* $X_{x'} = \{x'' : (x',x'') \in X\}$ *generated by the metrics* $(g^{jk})_{j,k\le d'}$ *and* $(g^{jk})_{j,k\ge d'+1}$ *respectively (these metrics and densities depend on* x*). Here the sign "*$\mp$*" corresponds to Dirichlet and Neumann boundary condition respectively.* ◊

Surely, this assertion remains true in the case when the operator tends to an L-quasi-homogeneous operator or if we can use the ρ and γ indicated above (in the second case only assertion (i) remains true). In the first case $\mathcal{N}^w(\tau) = \bigl(\varkappa_0 + o(1)\bigr)\tau^p$ without exact equality.

10.4.6 Quickly growing potentials

Let us consider potentials with more than power growth at infinity:

Theorem 10.4.15. *(i) Let* $d \ge 2$ *and let* X *be a domain in* $\mathbb{R}^d$ *such that*

$$\operatorname{vol}\{x \in X, |x| \le r\} \asymp r^n \quad \text{as } r \to +\infty, \tag{10.4.32}$$

$d \ge n > 1-\mu$. *Let conditions* $(\Sigma 0)$*–*$(\Sigma 3)$ *be fulfilled with* $\gamma = \epsilon_1\langle x\rangle^{1-\mu}$, $\rho = \exp\langle x\rangle^\mu$, $\mu > 0$. *Then asymptotics* $(10.4.1)'$ *hold with principal part* $\mathcal{N}^w_\vartheta(\tau) \asymp \tau^{\frac{d}{2}+\vartheta}\log^{\frac{n}{\mu}}\tau$ *and remainder estimate* $R_\vartheta(\tau) = \tau^{\frac{1}{2}(d+\vartheta-1)}\omega_\vartheta(\tau)$,

$$\omega_\vartheta(\tau) = \begin{cases} \log^q\tau & \text{for } q \overset{\text{def}}{=} \dfrac{1}{\mu}\bigl(n-(1-\mu)(1+\vartheta)\bigr) > 0,\\ \log\log\tau & \text{for } q = 0,\\ 1 & \text{for } q < 0.\end{cases} \tag{10.4.33}$$

(ii) Moreover, let $q \ge 0$ *and for every fixed* $k > 0$ *let the condition*

$$\operatorname{vol}\{x \in X, |x| \le r, \operatorname{dist}(x,\partial X) \le k\gamma(x)\} = o(r^n) \quad \text{as } r \to +\infty \tag{10.4.34}$$

be fulfilled. Furthermore, let

$$|D^\alpha g^{jk}| = o(\gamma^{-1}) \quad \text{as } |x| \to +\infty \quad \forall j,k,\alpha : |\alpha| = 1, \tag{10.4.35$_1$}$$

$$|D^\alpha V_j| \le C\rho\gamma^{-|\alpha|} \quad \forall j,\alpha : |\alpha| \le 1. \tag{10.4.35$_2$}$$

Then asymptotics $(10.4.4)'$ *hold with* $R_\vartheta(\tau)$ *defined above and with* $\varkappa_1 = 0$ *for* $\vartheta = 0$.

Proof. Statement (i) is proven in the standard way. In order to prove (ii) let us note that if $q > 0$ then for $\tau \geq \tau_0(\varepsilon)$ the contributions of the zones $\{x \in X, |x| \leq \varepsilon \log^{\frac{1}{\mu}} \tau^{\frac{1}{2}}\}$ and $\{x \in X, |x| \geq (1-\varepsilon)\log^{\frac{1}{\mu}} \tau^{\frac{1}{2}}\}$ to the remainder estimate do not exceed $\varepsilon_1 R_\vartheta(\tau)$ with $\varepsilon_1 \to 0$ as $\varepsilon \to 0$. Moreover, condition (10.4.34) yields that the contribution of the zone $\{X, \mathrm{dist}(x, \partial X) \leq k\gamma(x)\}$ to the remainder estimate is $o(R_\vartheta(\tau))$ and therefore one need consider only billiards generated by the improved Hamiltonian and starting in the zones

$$X_{k,\varepsilon,\tau} = \{X \in X, \varepsilon \leq |x| \log^{-\frac{1}{\mu}} \tau^{\frac{1}{2}} \leq 1-\varepsilon, \mathrm{dist}(x, \partial X) \geq k(\log \tau^{\frac{1}{2}})^{\frac{1}{\mu}-1}\}.$$

Then for $t \in [-T, T]$ and $\tau \geq \tau_0(T, k, \varepsilon)$, $k \geq k_0(T, \varepsilon)$ these billiards lie in $X_{\frac{k}{2}, \frac{\varepsilon}{2}, \tau}$, do not hit ∂X, and along them

$$\tau^{-\frac{1}{2}} V_j \equiv \tau^{-1} V \equiv 0, \quad g^{jk} \equiv \mathrm{const}, \quad \gamma(\log \tau)^{1-\frac{1}{\mu}} \equiv \mathrm{const},$$
$$\tau^{-\frac{1}{2}} \xi_j \equiv \mathrm{const}, \quad \gamma^{-1} \frac{dx_j}{dt} \equiv 2\tau^{-\frac{1}{2}} g^{jk} \xi_k, \quad \tau^{-1} g^{jk} \xi_j \xi_k \equiv 1$$

modulo $o(1)$ as $\tau \to +\infty$. The remainder of the proof is similar to the proof of theorem 10.2.1(iii). The case $q = 0$ is treated in the same way with the removal of the zones $\{x \in X, |x| \leq R\}$, $\{x \in X, |x| \geq \epsilon \log^{\frac{1}{\mu}} \tau\}$ and $\{x \in X, \mathrm{dist}(x, \partial X) \leq k\gamma(x)\}$ with arbitrarily small $\epsilon > 0$ and arbitrarily large R. □

In the same way the following assertion is proven:

Theorem 10.4.16. *Let $d \geq 2$, $X = \mathbb{R}^+ \times X'$ where X' is a bounded domain in $\mathbb{R}^{d-1} \ni x'$ with C^K boundary. Let all the conditions $(\Sigma 0)$–$(\Sigma 3)$ be fulfilled with $\rho = \exp x_1$ and $\gamma = \epsilon_0 > 0$ and let*

$$g^{jk} - g^{jk0} = o(1), \qquad D_1 g^{jk} = o(1) \qquad \text{as } x_1 \to +\infty \tag{10.4.35}$$

where $g^{jk0} = g^{jk0}(x')$ and $g^{1k0} = 0$ for $k \geq 2$. Let condition (10.4.35) be fulfilled. Then for every $\vartheta > 0$ asymptotics (10.4.4)$'$ hold with $R_\vartheta(\tau) = \tau^{\frac{1}{2}(d+\vartheta-1)} \log \tau$. Moreover, for $\vartheta = 0$ the principal part is $\mathcal{N}^w(\tau) + \varkappa_1 \tau^{\frac{1}{2}(d-1)} \log \tau$ with coefficient

$$\varkappa_1 = \mp \frac{1}{8} (2\pi)^{1-d} \omega_{d-1} \int_{\partial X'} \frac{dS'}{\sqrt{g^{110}}} \tag{10.4.37$_1$}$$

where dS' is the Riemannian density generated by the metric $(g^{jk0})_{j,k \geq 2}$ on $\partial X'$. Moreover, $\mathcal{N}^w(\tau) = (\bar\varkappa_0 + o(1))\tau^{\frac{d}{2}} \log \tau$ with coefficient

$$\bar\varkappa_0 = \frac{1}{2} (2\pi)^{-d} \omega_d \int_{X'} \sqrt{g^{0\prime}} \frac{dx'}{\sqrt{g^{110}}} \tag{10.4.37$_2$}$$

where $\sqrt{g^{0\prime}} dx'$ is a Riemannian density on X' generated by the same metric.

10.4.7 Slowly growing potentials

Finally, let us consider potentials with slow growth at infinity. Namely, let $V = v(\theta) \log r$ in spherical coordinates (r, θ). We use $\tilde\rho(\tau) = (\tau - V)_+^{\frac{1}{2}}$ and $\tilde\gamma(\tau) = \epsilon_0 r \min(1, (\tau - V)_+ \log^{-1} r)$. Then we obtain

Theorem 10.4.17. *Let $d \ge 3$ and let $g^{jk} = g^{jk}(\theta)$, $V_j = 0$, $V = v(\theta)\log r$, $X = \{\theta \in U\}$ for $r \ge c$ where (r,θ) are spherical coordinates, $g^{jk}, v, \partial U \in C^K$ and (g^{jk}) is a positive definite matrix. Let $v_0 = \min_U v > 0$. Then*

(i) Asymptotics (10.4.1)′ hold with

$$\mathcal{N}^w_\vartheta \asymp \int_U \exp(\frac{\tau d}{v})d\theta,$$

$$R_\vartheta(\tau) = \tau^{1+\vartheta}\int_{U'} \exp(\frac{\tau(d-1-\vartheta)}{v})d\theta \qquad \text{for } \vartheta < d-1.$$

Moreover,

$$\mathcal{N}^w(\tau) = a\int_{U'} v^{\frac{d}{2}} \exp(\frac{\tau\, d}{v})\sqrt{g}d\theta$$

with $a = (2\pi)^{-d}\omega_d d^{-(1+\frac{d}{2})}\Gamma(1+\frac{d}{2}) = 2^{1-d}\pi^{-\frac{d}{2}}d^{-(1+\frac{d}{2})}$, $U' = \{\theta \in U, v \le v_0 + \varepsilon\}$ with arbitrarily small $\varepsilon > 0$.

(ii) In particular, if $v = \text{const} = v_0$ then the principal part is $\varkappa \exp(\frac{\tau d}{v_0})$ and the remainder estimate is $O(\tau^{1+\vartheta}\exp(\tau(d-\vartheta-1)v_0^{-1}))$.[36] *On the other hand, if $v(\theta) = v_0 \implies \theta \in \partial U, \nabla v(\theta) \ne 0$ and $\operatorname{rank}\operatorname{Hess} v|_{\partial U}(\theta) = d-2$ then the principal part is $\asymp \tau^{-\frac{d}{2}}\exp(\tau d v_0^{-1})$ and the remainder estimate is $O(\tau^{1+\vartheta-\frac{d}{2}}\exp(\tau(d-\vartheta-1)v_0^{-1}))$. In other cases the principal part of the asymptotics and the remainder estimate have intermediate orders.*

I leave the treatment of the case $\vartheta \ge d-1$ to the reader.[37] I leave the potential $V = v(\theta)\log^\mu r$ to the reader as well.

10.4.8 Non-Dirichlet boundary conditions

All our results have been obtained for Dirichlet boundary conditions. Surely, they remain true for other boundary conditions provided these conditions are regular in the obvious sense and the domain satisfies the cone condition. In particular, for the Schrödinger operator Neumann conditions (and Robin conditions provided the free term coefficient is regular in the obvious sense) are regular. Then we obtain that if the domain satisfies the cone condition and if the asymptotics do not distinguish Dirichlet and Neumann boundary conditions then the asymptotics remain true for mixed boundary condition when Dirichlet and Neumann boundary conditions are given on $Y_0 \subset \partial X$ and $Y = \partial X \setminus Y_0$ respectively.

Let us consider more accurate asymptotics distinguishing Dirichlet and Neumann the boundary conditions. Then, repeating arguments of subsection 10.2.2 we obtain that these asymptotics remain true for mixed boundary conditions with properly defined second coefficient provided either

(i) We are in the context of theorems 10.4.1(iii), 10.4.8(ii), 10.4.10(ii) and

$$\operatorname{mes}_{\partial X}\big(\bar{Y}_0 \cap \bar{Y}_1 \cap \{|x| \ge c\}\big)_{\text{cone}} = 0 \tag{10.4.38}$$

or

[36] One can take $\tilde\gamma(\tau) = \epsilon_0 r \min(1, (\tau - V)_+)$ in this case and reject the factor $\tau^{1+\vartheta}$.

[37] In this case one can take $\tilde\rho(\tau) = \tau^{\frac12} + \log^{\frac12} r$, $\tilde\gamma(\tau) = \epsilon_0 r$ and for $d \ge 2$ and smooth ∂X, g^{jk}, V_j and V we obtain the remainder estimate $R_\vartheta(\tau) = C\tau^d$ for $\vartheta = d-1$ and $R_\vartheta(\tau) = C\tau^{\frac12(d+\vartheta-1)}$ for $\vartheta > d-1$.

(ii) we are in the context of theorem 10.4.14(ii) and

$$\text{mes}_{\Pi'\partial X}\, \Pi'\big(\bar{Y}_0 \cap \bar{Y}_1 \cap \{|x| \ge c\}\big)_{\text{cone}} = 0 \tag{10.4.39}$$

where Π' is the projection $\mathbb{R}^d \ni x \to x' \in \mathbb{R}^{d'}$ or

(iii) We are in the context of theorem 10.4.16 and

$$\text{mes}_{\partial X'}\, \Pi'\big(\bar{Y}_0 \cap \bar{Y}_1 \cap \{|x| \ge c\}\big) = 0 \tag{10.4.40}$$

with the same Π'. Here M_{cone} is the conical hull of the set M.

Moreover, applying the arguments of example 10.2.8 we obtain that asymptotics (10.4.6)′ remain true in the context of theorem 10.4.4 if

$$\text{mes}_{\partial X}\{x \in \partial X, \text{dist}(x, Y) \le \varepsilon\} = O(\varepsilon^{\delta}) \qquad \text{as } \varepsilon \to +0 \tag{10.4.41}$$

with $\delta > 0$ and $Y = (\bar{Y}_0 \cap \bar{Y}_1 \cap \{|x| \ge c\})_{\text{cone}}$.

Finally, I would like to note that in almost all the above examples the remainder estimate is $O(\tau^{\frac{1}{2}(d+\vartheta-1)})$ for large enough ϑ (for the Schrödinger operator). In this case the principal contribution to the remainder estimate is provided by the zone $\{x \in X, |x| \le R\}$ with arbitrarily large R. In this case the standard condition $(\Psi 0)$ on the Hamiltonian $\sum_{j,k} g^{jk}\xi_j\xi_k$ at the energy level $\tau = 0$ provides the remainder estimate $o(\tau^{\frac{1}{2}(d+\vartheta-1)})$ (see, e.g., theorem 10.4.1(iii)). The situation in which the principal contributions to the principal part of the asymptotics and to the remainder estimate are provided by different zones is not very exotic.

10.5 Asymptotics of Eigenvalues Tending to −0

In this section we treat Schrödinger operators with potentials tending to 0 at infinity. Under reasonable conditions the spectrum is $[0, \infty)$ and the point -0 is the only possible limit point of the discrete spectrum. We obtain asymptotics of $\mathbf{N}(\tau)$ as $\tau \to -0$. In particular, we obtain criteria to determine when -0 is a limit point of the discrete spectrum. We treat more general higher-order operators and non-semibounded operators as well. In particular, we treat the Dirac operator with positive mass where a spectral gap appears.

10.5.1 Standard results

We start with semibounded operators. First of all the following statement holds:

Theorem 10.5.1 (compare with theorem 10.4.1). *Let X be a connected unbounded domain in $\mathbb{R}^d$ and let conditions* (10.1.1)–(10.1.3) *be fulfilled with $\gamma = \epsilon_1\langle x\rangle$, $\rho = \langle x\rangle^{\mu}$, $\mu \in (-1, 0)$. Let us assume that*

$$\langle a(x,\xi)v, v\rangle \ge \epsilon_0|\xi|^{2m}|v|^2 \qquad \forall \xi\ \forall v \tag{10.5.1}$$

where $a(x,\xi)$ is the senior symbol of A. Let us assume that for every $\tau \in (-\epsilon, 0]$ one of conditions (8.1.18)$_*$ *is fulfilled for the symbol $(A(x,\xi)\tau^{-1} - 1)$ uniformly with respect to (τ, x, ξ) (this condition is restrictive only for $|x| \ge C_0$). Then*

(i) The asymptotics

$$\mathbf{N}(\tau) = \mathcal{N}^{w}(\tau) + O(\tau^{l(d-1)}) \tag{10.5.2}$$

hold as $\tau \to -0$, *where* $\mathcal{N}^w(\tau)$ *is given by* (10.3.6) *and* $\mathcal{N}^w = O(\tau^{ld})$ *with* $l = \frac{1}{2m\mu}(\mu + 1) < 0$.

Moreover, for $0 < \vartheta \le \vartheta_0 = -ld$ *the asymptotics*

$$\mathbf{N}_\vartheta(\tau) = \mathcal{N}^w_\vartheta(\tau) + O\bigl(R_\vartheta(\tau)\bigr) \tag{10.5.2$'$}$$

hold as $\tau \to -0$ *with* $\mathcal{N}^w_\vartheta(\tau) = \tau_+^{\vartheta-1} * \tilde{\mathcal{N}}^w$, $\tilde{\mathcal{N}}^w$ *also containing lower order terms and*

$$R_\vartheta(\tau) = \begin{cases} |\tau|^{l(d-1)+\vartheta(1-l)} & \text{for } \vartheta < \vartheta_1, \\ \log|\tau| & \text{for } \vartheta = \vartheta_1, \\ 1 & \text{for } \vartheta > \vartheta_1 \end{cases} \tag{10.5.3}$$

where $\vartheta_1 = -l(d-1)(1-l)^{-1}$. *One can see easily that* $0 < \vartheta_1 < \vartheta_0$ *and that*

$$\mathcal{N}^w_\vartheta(\tau) = \begin{cases} O\bigl(|\tau|^{ld+\vartheta}\bigr) & \text{for } \vartheta < \vartheta_0 \\ O\bigl(\log|\tau|\bigr) & \text{for } \vartheta = \vartheta_0 \end{cases}$$

as $\tau \to -0$ *(for* $\vartheta > \vartheta_0$ *we have estimate* $\mathcal{N}^w_\vartheta = O(1)$*).*

Furthermore, if

$$\operatorname{mes}\bigl\{x \in X : |x| \le r \text{ and } \exists \xi \in \mathbb{R}^d : \operatorname{Spec} A(x,\xi) \cap (-\infty, -\epsilon_1 \rho^{2m}] \ne \emptyset\bigr\} \asymp r^d \qquad \text{as } r \to +\infty \tag{10.5.4}$$

for some $\epsilon_1 > 0$ *then*

$$\mathcal{N}^w_\vartheta(\tau) \asymp \begin{cases} |\tau|^{ld+\vartheta} & \text{for } \vartheta < \vartheta_0 \\ \log|\tau| & \text{for } \vartheta = \vartheta_0 \end{cases}$$

as $\tau \to -0$. *For the Schrödinger operator condition* (10.5.4) *is equivalent to*

$$\operatorname{mes}\bigl\{x \in X : |x| \le r, V \le -\epsilon_1 \rho^2\bigr\} \asymp r^d \qquad \text{as } r \to +\infty. \tag{10.5.4*}$$

(ii) Moreover, let us assume that $X \cap \{|x| \ge r_0\} = X^0 \cap \{|x| \ge r_0\}$ *where* X^0 *is a non-empty conical domain with* $\partial X^0 \setminus 0 \in C^K$ *and that stabilization conditions* (10.4.3) *are fulfilled. Let us assume that at the level* $\tau = -1$ *the quasi-homogeneous symbol*

$$A^0(x,\xi) = \sum_{|\beta| \le 2m} a^0_\beta(x)\xi^\beta$$

satisfies the standard condition (Ψ) *on Hamiltonian trajectories, and the trajectories never hit* $0 \times \mathbb{R}^d$. *Then the asymptotics*

$$\mathbf{N}(\tau) = \mathcal{N}^w(\tau) + \bigl(\varkappa_1 + o(1)\bigr)|\tau|^{l(d-1)} \tag{10.5.5}$$

hold as $\tau \to -0$. *In particular, for the Schrödinger operator*

$$\varkappa_1 = \mp\frac{1}{4}(2\pi)^{1-d}\omega_{d-1}\int_{\partial X^0} \bigl(-1 - V^0\bigr)_+^{\frac{d-1}{2}}\, dS^0 \tag{10.5.6$_1$}$$

for Dirichlet (Neumann) boundary conditions respectively where the notation of $(10.4.5)_1$ *is used.*

On the other hand, if only stabilization conditions (10.4.3) *(including the condition* $X \cap \{|x| \ge r_0\} = X^0 \cap \{|x| \ge r_0\}$*) are fulfilled then condition* $(8.1.18)_1$ *is automatically fulfilled and* $\mathcal{N}^w(\tau) = \varkappa_0|\tau|^{ld}\bigl(1 + o(1)\bigr)$ *as* $\tau \to -0$ *where for the Schrödinger operator*

$$(10.5.6)_0 \qquad \varkappa_0 = (2\pi)^{-d}\omega_d \int_{X^0} \left(-1 - V^0\right)_+^{\frac{d}{2}} \sqrt{g^0}\,dx$$

in the notation of $(10.4.5)_0$.

(iii) Moreover, in the context of (ii) for $0 < \vartheta < \vartheta_1$ *asymptotics* (10.4.4)′ *hold as* $\tau \to -0$.

(iv) Finally, let $\vartheta = \vartheta_1$*; the condition on the domain and the stabilization condition of (ii) remain true. Let the condition* (Ψ) *on Hamiltonian trajectories be fulfilled for the symbol* $A^0(x,\xi)$ *at the energy level* 0 *and let these trajectories never hit* $0 \times \mathbb{R}^d$*. Then asymptotics* (10.4.4)′ *also hold.*

Proof. The proof is similar to the proof of theorem 10.4.1 but now in the zone X'_-, $\tilde{\rho}_\tau \asymp \rho$ (instead of $\tilde{\rho}_\tau \asymp |\tau|^{\frac{1}{2m}}$) and $X'_- \subset \{\rho \geq \epsilon_2 |\tau|^{\frac{1}{2m}}\}$ with appropriate $\epsilon_2 > 0$. Moreover, every bounded domain gives a finite contribution to the remainder estimate (and therefore there is no assertion similar to theorem 10.4.1(iii)). Furthermore, the reduced Hamiltonian is now $|x|^{1-2m\mu+\mu} A^0(x,\xi)$.

I would like to emphasize that for $\vartheta > \vartheta_0$ we obtain the estimate $\mathbf{N}_\vartheta(-0) < \infty$ instead of asymptotics. This remark remains true in many examples below. Moreover, the remainder estimate is $o(|\tau|^{l(d-1)+\vartheta(1-l)})$ and $o(\log|\tau|)$ in asymptotics (10.4.4)′ for $\vartheta < \vartheta_1$ and $\vartheta = \vartheta_1$ respectively. Finally, the main contributions to the remainder estimate are made by the zones $\{x, |x| \geq \varepsilon |\tau|^{\frac{1}{2m}}\}$ and $\{x, R \leq |x| \leq \varepsilon|\tau|^{\frac{1}{2m}}\}$ for $\vartheta < \vartheta_1$ and $\vartheta = \vartheta_1$ respectively with arbitrarily small $\varepsilon > 0$ and arbitrarily large R. This difference leads to different (Ψ)-type conditions. □

For scalar operators we get the following

Theorem 10.5.2 (compare with theorem 10.4.2). *Let X be a connected unbounded domain in* $\mathbb{R}^d$ *and let conditions* (10.1.1)–(10.1.3) *be fulfilled with* $\gamma = \epsilon_1\langle x\rangle$, $\rho = \langle x\rangle^\mu$, $\mu \in (-1,0)$*. Let condition* (10.5.1) *be fulfilled and let A be a scalar operator. Then the asymptotics*

$$(10.5.2)'' \qquad \mathbf{N}(\tau) = \mathcal{N}^w(\tau) + O\left(\tau^{l(d-1)-\delta}\right)$$

hold as $\tau \to -0$, *and for* $\vartheta > 0$ *asymptotics* (10.5.2)′ *hold.*

10.5.2 Strongly non-periodic Hamiltonian flow

In what follows we consider homogeneous potentials with singularities. Now we obtain a more precise remainder estimate under more restrictive conditions on the Hamiltonian flow.

Theorem 10.5.3 (compare with theorem 10.4.3). *Let the conditions of theorem 10.5.1 be fulfilled and let A be a scalar operator. Moreover, let us assume that either* $X^0 = \mathbb{R}^d$ *or A is the Schrödinger operator. Let stabilization condition* (10.4.3)* *be fulfilled. Moreover, let us assume that the Hamiltonian flow (with reflections at* ∂X^0*) is of non-periodic power type*[38] *at negative energy levels and that along the trajectories* $(x(t),\xi(t))$ *with* $(x(0),\xi(0)) \in \Lambda^0_{\varepsilon,T} \subset \Sigma_{-1}$, $|x(0)| = 1$, $\varepsilon = T^{-M}$ *the condition* $|x(t)| \geq T^{-M_1}$ *is fulfilled. Then the asymptotics*

$$(10.5.7) \qquad \mathbf{N}(\tau) = \mathcal{N}^w(\tau) + \left(\varkappa_1 + O(|\tau|^{\delta'})\right)\tau^{l(d-1)}$$

hold as $\tau \to -0$ *with* $\delta' > 0$*. Moreover, for* $0 < \vartheta \leq \vartheta_1 - \delta$ *with* $\delta > 0$ *the asymptotics*

[38] Footnote [30] of the previous section remains true.

(10.5.7)′
$$\mathbf{N}_\vartheta(\tau) = \mathcal{N}^w_\vartheta(\tau) + O\big(|\tau|^{l(d-1)+\vartheta(1-l)+\delta'}\big)$$

hold as $\tau \to -0$ *with* $\delta' = \delta'(d,m,\delta,L) > 0$.

Moreover, let us consider the following

Example 10.5.4 (compare with example 10.4.4). *(i) The condition of theorem 10.5.3 on the Hamiltonian flow is fulfilled for the Schrödinger operator with* $X^0 = \mathbb{R}^d$ *and* $A^0(x,\xi) = |\xi|^2 - |x|^{2\mu}$, $\mu \in (-1,-\frac{1}{2}) \cup (-\frac{1}{2},0)$.

(ii) Therefore the statement of theorem 10.5.2 is true for this operator (provided all the other conditions of this theorem, including (10.4.3)*, *are fulfilled) and for any 2m-order operator with* $A^0(x,\xi) = (|\xi|^2 - |x|^{2\mu})^m$, *and one can replace* $X^0 = \mathbb{R}^d$ *by the* X^0 *obtained after (a sequence of) the transformations (a),(b) of example 10.4.4(ii).* ◊

10.5.3 Periodic Hamiltonian flow

As in the previous section let us consider the case of periodic Hamiltonian flow:

Theorem 10.5.5 (compare with theorem 10.4.6). *Let* $X^0 = \mathbb{R}^d$ *and conditions* (10.4.8)–(10.4.10), (10.4.13) *with* $\mu \in (-1,0)$ *be fulfilled in the domain of negative energies. Let us introduce the symbol*

(10.5.8)
$$b(x,\xi) = -\frac{l}{T_0}\int_0^{T_0} A^1 \circ \Phi_t dt$$

where T_0 *is the period at energy level* -1 *(compare with* (10.4.11)*; we have changed sign because* $A^0(x,\xi) = -1$ *now). Then*

(i) If $\delta_1 < (\mu+1)$ *and condition* (10.4.12) *is fulfilled for all* ω *then the asymptotics*

(10.5.9)
$$\mathbf{N}(\tau) = \mathcal{N}^w(\tau) + O\big(|\tau|^{l(d-1)+\delta''}\big)$$

hold as $\tau \to -0$ *with* $\delta'' > 0$.

(ii) If $\delta_1 \in [\mu+1, 2\mu+2]$ *then for* $\tau \to -0$ *such that* $|\tau|^l \in \bigcup_{n\in\mathbb{Z}^+} J_n$ *the asymptotics*

(10.5.10)
$$\mathbf{N}(\tau) = \mathcal{N}^w(\tau) + F(\tau)|\tau|^{l(d-1)} + O\big(|\tau|^{l(d-1)+\delta''}\big)$$

hold with $\delta'' > 0$ *and*

(10.5.11)
$$F(\tau) = (2\pi)^{-d} l T_0^{-1} \int_{\Sigma_{-1}} \upsilon\Big(T_0 l^{-1}\big((|\tau|^l - \frac{\pi}{4}\omega_0) - |\tau|^{l-\frac{\delta_1}{2m\mu}} b\big)\Big) dx d\xi : dA^0.$$

The intervals J_n *are defined in theorem 10.4.6(ii) with* $\nu = -\frac{2\pi l}{T_0}$, ω_0 *replaced by* $-\omega_0$ *where* ω_0 *is the Maslov index of the trajectory* Φ_t *with* $t \in (0,T_0)$.

Let us treat the perturbed Schrödinger operator with Coulomb potential more carefully. In this case $A_0 = |D|^2 - |x|^{-1}$ and $\mu = -\frac{1}{2}$, $l = -\frac{1}{2}$, $T_0 = \frac{\pi}{2}$, $\nu = 2$ and the Hamiltonian trajectories are Keplerian ellipses (with one focus located at the origin). Moreover, $\omega_0 \equiv 2(d-1) (\operatorname{mod} 4\mathbb{Z})$ and $\operatorname{Spec}(A_0) = \{\tau_n = -1(d+1+2n)^2,\ n \in \mathbb{Z}^+\}$ and the multiplicity of τ_n is $r(n)$ where $r(n)$ is a polynomial of degree $(d-1)$ in n. Let us introduce the function $f(\tau) = \frac{1}{2}(-\tau)^{-\frac{1}{2}} - \frac{1}{2}$.

One can easily see that for generic positively homogeneous functions a^1_β (with $|\beta| \le k$) of degree $(-1+2|\beta|-\delta_1)$ condition (10.4.12) is fulfilled for all $\omega \in \mathbb{R}$ and therefore asymptotics (10.5.9) hold for $0 < \delta_1 < \frac{1}{2}$ and asymptotics (10.5.10)–(10.5.11) hold for

$\delta_1 \ge \frac{1}{2}$, and now there is no restriction of the type "$\delta_1 \le 2\mu + 2$" on δ_1 and $k = 0, 1, 2$ is arbitrary. Moreover, one can take $V' = v_0|x|^{-1-\delta}$ with an arbitrary exponent $\delta > 0$ and an arbitrary constant $v_0 \ne 0$.

Let us assume that $b = 0$. There are two possibilities in this case: the trivial possibility is to take $\delta_2 \in (\delta_1, 2\delta_1)$ and the non-trivial one is to take $\delta_2 = 2\delta_1$. In the first case all the previous statements remain true with δ_1 replaced by δ_2 etc. In the second case we assume that condition (10.4.17) is fulfilled in which case the symbol $b(x,\xi)$ is given by formula (10.2.83) with b and b^s replaced by $-\frac{1}{2}A^1(-A^0)^{-\frac{1}{2}}$ and $-\frac{1}{2}A^2(-A^0)^{-\frac{1}{2}} + \frac{3}{4}(A^1)^2(-A^0)^{-\frac{3}{2}}$ respectively and with integration from 0 to π, $\frac{\pi}{2}$ instead of 2π, π respectively. Moreover, the numerical factor on every integral is $\frac{1}{\pi}$ instead of $\frac{1}{2\pi}$. Furthermore, one can move A^0 out of the Poisson brackets and then take $A^0 = -1$. Therefore

$$(10.5.12)\quad b(x,\xi) = \frac{1}{2\pi}\int_0^{\pi} \bigl(A^2 - \frac{3}{2}(A^1)^2\bigr) \circ \Phi_t dt + \frac{1}{2\pi}\int_0^{\frac{\pi}{2}}\int_0^{\frac{\pi}{2}} \{A^1 \circ \Phi_t, A^1 \circ \Phi_{t+t'}\} dt dt'.$$

Therefore the following assertion has been proven:

Theorem 10.5.6. *Let $x^0 = \mathbb{R}^d$, $A = A^{0w} + A^{1w} + A''^{w}$ where A'' satisfies condition (10.4.10) with a^2_β, δ_2 instead of a^1_β, δ_1, $\delta_2 = 2\delta_1$ and a^j_β are functions which are positively homogeneous of degrees $(-1+2|\beta|-\delta_j)$ with respect to x, smooth for $x \ne 0$. Let condition (10.4.17) be fulfilled and let us define the symbol b by (10.5.12). Then*

(i) If $\delta_2 < \frac{1}{2}$ and condition (10.4.12) is fulfilled for every $\omega \in \mathbb{R}$ then asymptotics (10.5.9) hold as $\tau \to -0$.

(ii) If $\delta_2 \ge \frac{1}{2}$ and condition (10.4.12) is fulfilled for some $\omega \in \mathbb{R}$ then asymptotics (10.5.10)–(10.5.11) hold as $\tau \to -0$ and $(-\tau)^{-\frac{1}{2}} \in \bigcup_{n\in\mathbb{Z}^+}$ where J_n are given in theorem 10.4.4(ii) with $\omega_0 = 2(d-1), \mu = 2$ and with δ_1 replaced by δ_2 everywhere.

One can easily prove that for fixed a^1_β with $|\beta| \le 2$ satisfying (10.4.17) and for generic a^2_β with $|\beta| \le k$ condition (10.4.14) is fulfilled for all $\omega \in \mathbb{R}$; here $k = 0, 1, 2$ is arbitrary.

10.5.4 Singularities

Let us consider homogeneous lower order terms with singularities (for the Schrödinger operator only):

Theorem 10.5.7 (compare with theorem 10.4.8). *(i) Let $d \ge 2$, $X = \{(r,\theta) \in \mathbb{R}^+ \times U\}$ where $U \subset \mathbb{S}^{d-1}$. Let $g^{jk} = g^{jk}(\theta)$, $V_j = v_j(\theta)r^\mu$, $V = v(\theta)r^{2\mu}$, $\mu \in (-1,0)$. Let $\tilde\rho = \tilde\rho(\theta)$ and $\tilde\gamma = \tilde\gamma(\theta)$ be functions on U and let conditions $(\Sigma 0)$–$(\Sigma 2)$ be fulfilled. Moreover, assume* **either** *$d \ge 3$* **or** *$\tilde\rho\tilde\gamma \ge \epsilon_0 > 0$. Finally, let $\tilde\rho^{-\frac{d}{\mu}}, \tilde\rho^{-\frac{1}{\mu}(d-1)}\tilde\gamma^{-1} \in L^1(U)$. Let $v(\bar\theta) < 0$ for some $\bar\theta \in U$. Then asymptotics (10.5.2)–(10.5.3) hold and $\mathcal{N}^w(\tau) = \varkappa_0|\tau|^{ld}$ with the coefficient $\varkappa_0$ given by $(10.5.6)_0$.*

(ii) Moreover, if $\vartheta \in (0,\vartheta_1)$, $\vartheta_1 = -l(d-1)(1-l)^{-1} = (1+\mu)(d-1)(1-\mu)^{-1}$ and if $\tilde\rho^{-\frac{1}{\mu}(d-1-\vartheta)}\tilde\gamma^{-1-\vartheta} \in L^1(U)$ then asymptotics (10.5.2)′–(10.5.3)′ hold.

(iii) Furthermore, let us assume that there exists a set $\Lambda \subset \Sigma = \Sigma_{-1}$ with $\mathrm{mes}_\Sigma(\Lambda) = 0$ such that through every point of $\Sigma\setminus\Lambda$ there passes an infinitely long non-periodic billiard generated by the Hamiltonian

$$H^0 = \tilde\gamma r^{1-\mu}(\tilde\rho + r^{-\mu})^{-1}\bigl(g^{jk}(\xi_j - V_j)(\xi_k - V_k) + V + 1\bigr)$$

which never hits $0 \times \mathbb{R}^d \cup \Lambda$. Then asymptotics (10.5.5) hold with the coefficient $\varkappa_0$ given by $(10.5.6)_1$.

Finally, let $\vartheta \in (0, \vartheta_1)$ *and* $\tilde{\rho}^{-\frac{1}{\mu}(d-1-\vartheta)}\tilde{\gamma}^{-1-\vartheta} \in L^1(U)$. *Then asymptotics* (10.5.5)′ *hold.*

Example 10.5.8 (compare with example 10.4.9). *Let* $x = \mathbb{R}^d$, $d \ge 2$, $\rho = |x''|^\nu |x|^{\mu-\nu}$, $\gamma = \frac{1}{2}|x''|$ *with* $\mu \in (-1, 0)$, $\nu \in (\mu, 0)$. *Then*

(i) $\tilde{\rho}^{-\frac{d}{\mu}} \in L^1(\mathbb{S}^{d-1})$ *and* $\tilde{\rho}^{-\frac{1}{\mu}(d-1-\vartheta)}\tilde{\gamma}^{-1-\vartheta} \in L^1(\mathbb{S}^{d-1})$ *if and only if* $\nu > \nu_0 = \mu d'' d^{-1}$ *and* $\nu > \nu_{\vartheta+1} = \mu(d''-1-\vartheta)(d-1-\vartheta)^{-1}$ *respectively. Therefore for* $\nu > \nu_1$ *asymptotics (10.5.2) hold and for* $\nu > \nu_{\vartheta+1}$, $\vartheta < \vartheta_1$ *asymptotics* (10.5.2)′ *hold.*[39]

(ii) The case $\nu \in [\nu_0, \nu_1)$ *is treated in the same way as case (ii) of example 10.4.9. We obtain that for* $\nu = \nu_1$ *the asymptotics*

$$\mathbf{N}(\tau) = \mathcal{N}^w(\tau) + O\big(|\tau|^{l(d-1)} \log|\tau|\big) \tag{10.5.13}$$

hold as $\tau \to -0$ *and for* $\nu \in (\nu_0, \nu_1)$ *the asymptotics*

$$\mathbf{N}(\tau) = \mathcal{N}^w(\tau) + O\big(|\tau|^q\big) \tag{10.5.14}$$

hold as $\tau \to -0$ *with* $q = -\frac{1}{2}(\mu+1)d'(\nu-\mu)^{-1} = l(d-\eta)$, $\eta = (\mu d'' - \nu d)(\mu-\nu)^{-1} \in (0,1)$.

In these asymptotics $\mathcal{N}^w(\tau) = \varkappa_0|\tau|^{ld}$.

(ii)′ Similarly, for $\nu = \nu_{\vartheta+1}$, $\vartheta \in (0, \vartheta_1)$ *the asymptotics*

$$\mathbf{N}_\vartheta(\tau) = \mathcal{N}^w_\vartheta(\tau) + O\big(|\tau|^{l(d-1-\vartheta)+\vartheta} \log|\tau|\big) \tag{10.5.13$'$}$$

hold as $\tau \to -0$ *and for* $\nu \in (\nu_0, \nu_{\vartheta+1})$ *the asymptotics*

$$\mathbf{N}_\vartheta(\tau) = \mathcal{N}^w_\vartheta(\tau) + O\big(|\tau|^{q+\vartheta}\big) \tag{10.5.14$'$}$$

hold as $\tau \to -0$ *with the same* q *as before.*

(iii) Finally, for $\nu = \mu d'' d^{-1}$ *the asymptotics*

$$\mathbf{N}(\tau) = \mathcal{N}^w(\tau) + O\big(|\tau|^{ld}\big) \tag{10.5.15}$$

hold as $\tau \to -0$ *with* $\mathcal{N}^w(\tau)$ *given by (10.4.23) and moreover* $\mathcal{N}^w(\tau) \asymp |\tau|^{ld}\log|\tau|$. *Moreover, footnote* [34] *of the previous section holds.* ◊

We now have no need of the coercivity condition and therefore theorems 10.4.10 and 10.4.12 and examples 10.4.11 and 10.4.11 are almost senseless. However, one can treat potentials of type $V^0 + V'$ where V^0 is a slowly decreasing and not very singular potential and V' decreases faster and is more singular, etc. I leave this task to the reader. Let us note that for $d \ge 3$ in the zone $\{V \ge 0\}$ no condition excluding $|V_j| \le c\sqrt{V}$ is necessary. I leave the treatment of examples with $\rho = \langle x\rangle^\mu$, $\gamma = \langle x\rangle^{1-\sigma}$ with $\mu < 0$, $\sigma \in (0, 1+\mu)$, where asymptotics with a worse remainder estimate may be obtained, to the reader. In this case the magnetic field can decrease very slowly at infinity.

10.5.5 Non-homogeneous operators

Let us consider non-homogeneous operators. Repeating the arguments of theorem 10.4.14 with obvious modifications we obtain

Theorem 10.5.9. *(i) Let* X *be an* L*-quasi-conical domain in* $\mathbb{R}^d$ *and let* g^{jk}, V_j, V *be positively* L*-quasi-homogeneous with respect to* x *of degrees* 0, μ, 2μ *respectively where* $\mu \in (-1, 0)$, $L = (l_1, \dots, l_d)$ *and* $l_j = 1$ *for* $1 \le j \le d'$ *and* $l_j > 1$ *for* $d'+1 \le j \le d$,

[39] And under a condition on the Hamiltonian flow asymptotics (10.5.5) and (10.5.5)′ (for $\nu > \nu_{\vartheta+1}$) hold.

$1 \le d' < d$. *Let us assume that* $g^{jk}, V_j, V \in C^K(\bar{X} \setminus 0)$ *satisfy condition* $(\Sigma 2)$ *and that* $V(\bar{\theta}) < 0$ *for some* $\bar{\theta} \in X$. *Then conditions* $(\Sigma 1)$ *are fulfilled with* $\rho = [x]_L^{\mu}$, $\gamma = \epsilon[x]_L$ *for* $[x]_L \ge 1$; *for* $[x]_L \le 1$ *one should take* $\gamma = \epsilon[x]_L^{\bar{l}}$. *Then the asymptotics*

$$\mathbf{N}(\tau) = \varkappa_0|\tau|^p + O(|\tau|^q) \tag{10.5.16}$$

hold as $\tau \to -0$ *with* $p = \frac{d}{2} + \frac{l}{2\mu}$, $q = \frac{1}{2}(d-1) + \frac{1}{2\mu}(l-1)$, $l = l_1 + \cdots + l_d$ *and*

$$\varkappa_0 = (2\pi)^{-d}\omega_d \int (-1 - V)_+^{\frac{d}{2}} \sqrt{g}dx. \tag{10.5.17$_0$}$$

(ii) Moreover, let us assume that $x' \ne 0$, $\nu' \ne 0$ *on* $\partial X \setminus 0$ *where* ν *is a unit normal to* ∂X *and* $x' = (x_1, \ldots, x_{d'})$, $\nu' = (\nu_1, \ldots, \nu_{d'})$. *Let us assume that* $g^{jk} = 0$, $V_k = 0$ *for* $j \le d'$, $k \ge d' + 1$. *Then the asymptotics*

$$\mathbf{N}(\tau) = \varkappa_0|\tau|^p + (\varkappa_1 + o(1))|\tau|^q \tag{10.5.18}$$

hold with

$$\varkappa_1 = \mp\frac{1}{4}(2\pi)^{1-d}\omega_{1-d} \int (-1 - V)_+^{\frac{d-1}{2}} dS' \sqrt{g''}dx'' \tag{10.5.17$_1$}$$

where dS' *and* $\sqrt{g''}dx''$ *are the Riemannian densities introduced in example 10.4.14.*

10.5.6 Quickly decreasing potentials

Let us consider potentials which decrease rather quickly.

Theorem 10.5.10. *(i) Let* X *be an unbounded domain in* $\mathbb{R}^d$ *and let conditions* (10.5.1)–(10.5.3) *be fulfilled with* $\gamma = \epsilon\langle x\rangle$, $\rho = \langle x\rangle^{-1}\log^{\nu}\langle x\rangle$ *where we now take* $\langle x\rangle = (|x|^2 + 4)^{\frac{1}{2}}$, $\nu > 0$. *Then the asymptotics*

$$\mathbf{N}(\tau) = \mathcal{N}^w(\tau) + O\big(|\log|\tau||^{\nu(d-1)+1}\big) \tag{10.5.19}$$

hold as $\tau \to -0$ *where* $\mathcal{N}^w(\tau) = O\big(|\log|\tau||^{\nu d+1}\big)$ *and under condition* (10.5.4)* $\mathcal{N}^w(\tau) \asymp |\log|\tau||^{\nu d+1}$.

(ii) Moreover, let $X \cap \{|x| \ge r_0\} = X^0 \cap \{|x| \ge r_0\}$ *where* X^0 *is a conical domain with* $\partial X^0 \setminus 0 \in C^K$ *and let stabilization conditions* (10.4.3) *be fulfilled for functions* $a_\beta \log^{-\nu(2m-|\beta|)}$ *where the functions* $a_\beta^0 \in C^K(\bar{X}^0 \setminus 0)$ *are positively homogeneous of degrees* $-(2m-|\beta|)$ *on* x. *Then the microhyperbolicity condition is fulfilled automatically. If the condition* (Ψ) *is fulfilled for the quasi-homogeneous symbol*

$$A^0(x,\xi) = |x|^{2m} \sum_{|\beta| \le 2m} a_\beta^0(x)\xi^\beta$$

at the energy level 0 *then the asymptotics*

$$\mathbf{N}(\tau) = \mathcal{N}^w(\tau) + (\varkappa_1 + o(1))|\log|\tau||^{\nu(d-1)+1} \tag{10.5.20}$$

hold where, for the Schrödinger operator,

$$\varkappa_1 = \mp 2^{-\nu(d-1)-3}(2\pi)^{1-d}\omega_{d-1} \int_{\partial U} (V^0)_-^{\frac{d-1}{2}} dS, \tag{10.5.21$_1$}$$

$U = X^0 \cap \mathbb{S}^{d-1}$ is a domain on $\mathbb{S}^{d-1}$, $dS = dS^0 : dr$ is the natural density on ∂U, dS^0 is the Riemannian density on ∂X^0 generated by the metric g^{jk0}. On the other hand, if only the stabilization condition is fulfilled then $\mathcal{N}^w(\tau) = \bigl(\varkappa_0 + o(1)\bigr)\bigl|\log|\tau|\bigr|^{\nu d+1}$ where, for the Schrödinger operator,

$$\varkappa_0 = 2^{-\nu d-1}(2\pi)^{-d}\omega_d \int_U (V^0)_-^{\frac{d}{2}} \sqrt{g^0}\, d\theta, \tag{10.5.21$_0$}$$

$d\theta = dx : dr$, $g^0 = \det(g^{jk0})^{-1}$ (compare with $(10.5.6)_{0,1}$).

Remark 10.5.11. *(i) One can prove a statement similar to theorem 10.5.7 with $V_j = v_j(\theta)r^{-1}\log^\nu r$, $V = v(\theta)r^{-2}\log^{2\nu} r$, $X = (1,\infty)\times U$. I leave the task to the reader, as well as a general analysis of the case $\rho = \langle x\rangle^\mu \log^\nu\langle x\rangle$, $\vartheta \ge 0$ with* ***either*** *$\mu > -d(d+2m\vartheta)^{-1}$* ***or*** *$\mu = -d(d+2m\vartheta)^{-1}$, $\nu \ge -1(d+2m\vartheta)^{-1}$. Under these conditions the principal part of the asymptotics is $\asymp I$ with either $I = |\tau|^p$ $(p<0)$ or $I = \bigl|\log|\tau|\bigr|^q$ $(q>0)$ or $I = \log\bigl|\log|\tau|\bigr|$ and the remainder estimate is either from the same list or $O(1)$.*

(ii) Moreover, I leave to the reader the treatment of the case of potentials decreasing even more quickly when $\rho = \langle x\rangle\bigl(\underbrace{\log\ldots\log}_{r \text{ times}}\langle x\rangle\bigr)^\nu$ when the principal part of the asymptotics and the remainder estimate are of the orders $\log|\tau|\zeta(\tau)^{\nu d}$ and $\log|\tau|\zeta(\tau)^{\nu(d-1)}$ respectively with $\zeta(\tau) = \underbrace{\log\ldots\log}_{r \text{ times}}|\tau|$.

(iii) One can easily see that for $\vartheta > 0$ we obtain $\mathbf{N}_\vartheta(-0) < \infty$. ◇

Proof of theorem 10.5.10. Assertion (i) follows immediately from our previous arguments; moreover, now $\rho\gamma = \log^\nu\langle x\rangle \to \infty$ as $|x| \to \infty$ and $X'' = \emptyset$. In order to prove (ii) let us note that the principal contribution to the remainder estimate is given by the zone $\{C \le |x| \le \epsilon\tau^{-\frac{1}{2m}}\}$ with arbitrary fixed C and $\epsilon > 0$. We need to examine the Hamiltonian trajectories starting in this zone and associated with the symbol

$$A_\tau(x,\xi) = \langle x\rangle^{2m}\log^{-\nu}\langle x\rangle\bigl(A(x,\xi)+\tau\bigr)$$

on $\Sigma_\tau = \{\det A_\tau(x,\xi) = 0\}$. Let $\Sigma^0 \supset \Lambda_T$ be a closed set of measure 0 such that through every point of $\Sigma^0 \setminus \Lambda_T$ there passes a non-periodic Hamiltonian trajectory (with reflections) of length $2T$ associated with the symbol A^0. Then one can take

$$\begin{aligned}\Lambda_{1,T,\epsilon,\varepsilon,\tau,C} = \bigl\{(x,\xi) \in \Sigma_\tau, C \le |x| \le \epsilon\tau^{-\frac{1}{2m}},&\\ \operatorname{dist}(\frac{x}{|x|},\partial X^0) > 0, \operatorname{dist}\bigl((\frac{x}{|x|},\xi|x|\log^{-\nu}|x|),\Lambda_T\bigr) > \varepsilon\bigr\}&\end{aligned}$$

with $C = C(\epsilon,\varepsilon,T)$. Moreover, one can see that $\log|x(t)|\log^{-1}|x(0)| = 1 + o(1)$ as $|x(0)| \to \infty$ uniformly with respect to τ and $t \in [-2T,2T]$ with arbitrary fixed T. I leave the details to the reader. □

10.5.7 Slowly decreasing potentials

Let us consider the slowly decreasing potential $V = v(\theta)\log^{-1} r$ in the spherical coordinate system (r,θ). Using $\tilde\rho_\tau = (\tau - V)_+^{\frac{1}{2}}$, $\tilde\gamma_\tau = \epsilon_0 r\min(1,(\tau - V)_+\log r)$ for $r \ge 2$ we obtain

Theorem 10.5.12. *Let $d \ge 3$ and $X \cap \{r \ge r_0\} = \{r \ge r_0, \theta \in U\}$. For $|x| \ge r_0$ let $g^{jk} = g^{jk}(\theta)$, $V_j = 0$, $V = v(\theta)\log^{-1} r$ where (r,θ) are spherical coordinates, $g^{jk}, v, \partial X \in C^K$ and the matrix (g^{jk}) is positive definite. Let $v_0 = \min_U v < 0$. Then*

(i) Asymptotics (10.5.2)′ *hold with*

$$\mathcal{N}^w_\vartheta(\tau) \asymp |\tau|^{d+2\vartheta} \int_U \exp(\frac{vd}{\tau})\, d\theta,$$

$$R_\vartheta(\tau) = |\tau|^{d-2} \int_U \exp(\frac{v}{\tau}(d-\vartheta-1))\, d\theta$$

for $\vartheta < d-1$. *Moreover, one can take*

$$\mathcal{N}^w(\tau) = a \int_{U'} |\tau|^d v_-^{-\frac{d}{2}} F_d(-\frac{\tau}{v_-}) \exp(-\frac{vd}{\tau}) \sqrt{g}\, d\theta$$

where $U' = U \cap \{v < v_0 + \varepsilon\}$ *with arbitrarily small fixed* $\varepsilon > 0$, $a = 2^{1-d}\pi^{-\frac{d}{2}} d^{-\frac{d}{2}+1}$,

$$F_d(s) = \frac{1}{\Gamma(\frac{d}{2}+1)} \int_0^{\frac{d}{s}-\varepsilon} t^{\frac{d}{2}} \bigl(1 - \frac{st}{d}\bigr)^{-\frac{d}{2}} e^{-t}\, dt = 1 + O(s) \tag{10.5.22}$$

as $s \to +0$.

(ii) In particular, for $v = v_0 = \mathrm{const}$ *the principal part of the asymptotics is* $\asymp |\tau|^{d+2\vartheta} \exp(\frac{v_0 d}{\tau})$ *and the remainder estimate is* $O(|\tau|^{d-2} \exp(\frac{v_0}{\tau}(d-1-\vartheta)))$. *Moreover, in this case one can take* $\tilde{\gamma}_\tau = \epsilon_0 r \min\bigl(1, (\tau - V)_+ \log^2 r\bigr)$ *and obtain a remainder estimate with the factor* $|\tau|^{d+\vartheta-1}$ *instead of* $|\tau|^{d-2}$. *On the other hand, if* $v = v(\theta) = v_0 \implies \theta \in \partial U, \nabla v(\theta) \neq 0$ *and* $\operatorname{rank} \operatorname{Hess} v|_{\partial U} = d-1$ *then the principal part is* $\asymp |\tau|^{\frac{3}{2}+2\vartheta} \exp(\frac{v_0 d}{\tau})$ *and the remainder estimate is* $O(|\tau|^{\frac{3}{2}d-3} \exp(\frac{v_0}{\tau}(d-1-\vartheta)))$. *In other cases the principal part and the remainder estimate are of intermediate orders.*

I leave the treatment of the case $\vartheta \geq d-1$ to the reader. In this case one can take $\tilde{\rho}_\tau = \log^{-\frac{1}{2}} r + |\tau|$ and $\tilde{\gamma}_\tau = \epsilon_0 r$ and for $d \geq 2$ one can obtain the remainder estimates $O(|\tau|^d)$ for $\vartheta = d-1$ and $O(1)$ for $\vartheta > d-1$. Moreover, I leave the treatment of the potential $V = v(\theta)\log^{-\nu} r$ with arbitrary $\nu > 0$ to the reader.

10.5.8 General boundary conditions

All the above results remain true for the Neumann boundary condition provided the domain satisfies the cone condition. Let us assume that this (cone) condition is fulfilled. Moreover, all the above results remain true for mixed boundary conditions (provided the asymptotics do not distinguish Dirichlet and Neumann boundary conditions) and also in the following cases:

(i) We are in the context of theorems 10.5.1(ii) and 10.5.7(iii), and condition (10.4.37) is fulfilled;

(ii) We are in the context of theorem 10.5.9(ii), and condition (10.4.38) is fulfilled.

Moreover, in the context of theorem 10.5.3 asymptotics (10.5.7)′ remain true if condition (10.4.40) is fulfilled.

10.5.9 Non-semibounded operators

Let us consider non-semibounded operators. We consider the Dirac operator with mass $M > 0$ (otherwise in the context of this section the spectrum coincides with $\mathbb{R}$). Moreover, let $V \to 0$, $V_j \to 0$ as $|x| \to \infty$. Then $\operatorname{Spec}_{\mathsf{ess}}(A) = (-\infty, M] \cup [M, \infty)$ and the spectral gap $(-M, M)$ contains either a finite or infinite number of eigenvalues. In the second case these eigenvalues tend to $-M+0$ and (or) to $M-0$ and we can consider the corresponding eigenvalue asymptotics. Let us consider $M-0$. In order to replace it by -0 we replace the operator A by $A - M \cdot I$. Let us fix $\tau_1 \in (-2M, 0)$ and let us consider

$$\mathbf{N}(\tau) \overset{\text{def}}{=} \mathbf{N}(\tau_1, \tau)$$

with $\tau \to -0$. Then

(i) Theorems 10.5.1, 10.5.9,[40] 10.5.10 and 10.5.3 with example 10.5.4 remain true and one can check conditions on the Hamiltonian flow for the corresponding Schrödinger operator with the symbol

$$H = g^{jk}(\xi_j - V_j)(\xi_k - V_k) + (V + M)^2 - M^2.$$

In this Hamiltonian one can drop the term V^2 which is negligible at infinity in which case one can consider the Hamiltonian

$$H = g^{jk}(\xi_j - V_j)(\xi_k - V_k) + 2MV.$$

(ii) Theorem 10.5.7 (and example 10.5.8(i) for $d \geq 3$) remain true. In order to prove this let us note that only the analysis in the zone $\{\tilde{\rho} r^{\mu} \geq 1\}$ should be changed. In this zone instead of $\rho = \tilde{\rho} r^{\mu}$ one should set $\rho = \tilde{\rho}^2 r^{2\mu}$. We include this zone in X''. In this zone $\rho^d \in L^1$ because of the condition $\tilde{\rho}^{-\frac{d}{2\mu}} \in L^1(U)$. Therefore the contribution of this zone to the asymptotics and to the remainder estimate is $O(1)$.

(iii) Theorem 10.5.12 also remains true. However, $\mathcal{N}^w(\tau)$ should be calculated for the Dirac operator, i.e., for the Schrödinger operator with the symbol

$$g^{jk}\xi_j\xi_k + (M - \tau - V)^2 - M^2$$

and with the spectral parameter equal to 0. In the proof one should replace $\rho = (\tau - V)_+^{\frac{1}{2}}$ by $\rho = (\tau - V + C_1 r^{-1}\log^{-2} r)_+^{\frac{1}{2}}$ with large enough C_1 in the zone $X'' \cap \{\tau - V \leq C|\tau|^{\frac{2}{3}} r^{-\frac{2}{3}}\}$. This does not increase the remainder estimate.

Let us consider theorems 10.5.5 and 10.5.6. For simplicity we assume that

$$|v| \leq 2M - \epsilon_0 \qquad \forall x. \tag{10.5.23}$$

The reader can easily reject this condition in what follows. Then for appropriate τ_1,

$$\mathbf{N}(\tau) = \mathbf{N}^-(\tau) - \mathbf{N}^-(\tau_1) \tag{10.5.24}$$

where $\mathbf{N}^-(\tau)$ is the number of the negative eigenvalues for the operator

$$A(\tau) = \mathcal{L}^*(2M - V + \tau)^{-1}\mathcal{L} + V - \tau$$

with

$$\mathcal{L} = \bigl(\omega^{jk}(D_k - V_k) - \frac{i}{2}\omega^j\bigr)\sigma'_j$$

where we use the Einstein summation convention and assume that

$$\sigma_0 = \begin{pmatrix} I & 0 \\ 0 & -1 \end{pmatrix}, \qquad \sigma_j = \begin{pmatrix} 0 & \sigma_j'^{*} \\ \sigma'_j & 0 \end{pmatrix} \ (j \geq 1)$$

and use the notation $\omega^j = \partial_k \omega^{jk}$.

If we could take $\rho = \langle x \rangle^{-\frac{1}{2}}$, $\gamma = \epsilon_0 \langle x \rangle$ for the original Dirac operator and if

$$D^\alpha(\omega^{jk} - \delta^{jk}) = O\bigl(|x|^{-\delta_0 - |\alpha|}\bigr), \tag{10.5.25$_1$}$$

---

[40] Now, however, $\mathcal{N}^w(\tau) = \varkappa_0|\tau|^p\bigl(1 + O(|\tau|^\delta)\bigr)$ with $\delta > 0$ instead of $\mathcal{N}^w(\tau) = \varkappa_0|\tau|^p$.

$$(10.5.25)_2 \qquad D^\alpha F_{jk} = O\big(|x|^{-\frac{3}{2}-\delta_3-|\alpha|}\big) \qquad \text{as } |x| \to \infty \quad \forall \alpha : |\alpha| \le K$$

(where $F_{jk} = i(\partial_j V_k - \partial_k V_j)$ is the tensor magnetic intensity) then the operator $A(\tau)$ is a scalar operator plus $B_j(D_j - V_j) + B_0$ where

$$(10.5.26) \qquad \begin{aligned} D^\alpha B_j &= O\big(|x|^{\max(-1-\delta_0,-2)-|\alpha|}\big),\\ D^\alpha B_0 &= O\big(|x|^{\max(-1-\delta_0,-2,-\frac{1}{2}-\delta_3)-|\alpha|-1}\big) \quad \text{as } |x| \to \infty \end{aligned}$$

and in the context of section 10.5 this error does not matter provided that under the assumptions of theorems 10.5.5 and 10.5.6

$$(10.5.27)_{1,2,3} \qquad \delta_\iota < \frac{3}{2}, \quad \delta_\iota < \frac{1}{2} + \delta_0, \quad \delta_\iota < \frac{1}{2} + \delta_3$$

respectively with $\iota = 1, 2$ (because the "order" of the operator $D_j - V_j$ is $-\frac{1}{2} + \delta'$ with arbitrarily small $\delta' > 0$ if we take the "order" of $|x|$ equal to 1). Then the operator $A(\tau)$ differs by negligible terms from the Schrödinger operator

$$B(\tau) = (D_j - V_j)g^{jk}(2M - V + \tau)^{-1}(D_k - V_k) + V - \tau$$

and moreover, for

$$(10.5.28) \qquad -2 < -1 - \delta_\iota \qquad (\iota = 1, 2)$$

one can replace this operator by the operator

$$\tilde{B}(\tau) = (2M)^{-1}(D_j - V_j)g^{jk}(D_k - V_k) + V - \tau.$$

Therefore, if conditions (10.5.25), (10.5.27) and (10.5.28) are fulfilled and if the conditions of some statement of theorems 10.5.5, 10.5.6 are fulfilled for the Schrödinger operator $\tilde{B}(\tau)$ (and then $\iota = 1, 2$ respectively) then the corresponding asymptotics hold for the Dirac operator. Namely, asymptotics (10.5.9) hold for $\delta_\iota < \frac{1}{2}$ (then condition (10.5.27) is fulfilled automatically) and (10.5.10) holds for $\delta_\iota \in (\frac{1}{2}, 1)$. In these asymptotics $\mathcal{N}^w(\tau)$ is calculated for the Dirac operator and $F(\tau)$ is calculated for the Schrödinger operator and then it is multiplied by $\frac{1}{2}D$. On the other hand, let condition (10.5.28) be violated but condition (10.5.27) be fulfilled. Then one should replace $B(\tau)$ by

$$\tilde{B}_1(\tau) = (2M)^{-1}(D_j - V_j)g^{jk}(D_k - V_k) + V - \tau - (2M)^{-1}(V - \tau)^2$$

and in the last term one can replace V by V^0 where V^0 is a smooth function equal to $-1(2M|x|)^{-1}$ for $|x| \ge c$. Changing variables $x \to \zeta x$ and multiplying by $2M\zeta^2$ with $\zeta = 1 - \frac{\tau}{M}$ we obtain, modulo negligible terms, the operator

$$\tilde{B}_2(\tau') = (D_j - V_j)g^{jk}(D_k - V_k) + W - \tau'$$

with the new potential $W = 2MV - (V^0)^2$ and with the modified spectral parameter $\tau' = \tau(1 - \frac{3\tau}{2M})$.

This yields that

If the Schrödinger operator $\tilde{B}_2(0)$ satisfies the conditions of theorem 10.5.5 or theorem 10.5.6 with $\delta_\iota \in [1, \frac{3}{2})$ and if conditions (10.5.25), $(10.5.27)_2$ are fulfilled then asymptotics (10.5.10) remain true where $\mathcal{N}^w(\tau)$ is calculated for the Dirac operator and $F(\tau')$ is calculated for the Schrödinger operator $\tilde{B}_2(\tau')$ and is then multiplied by $\frac{1}{2}D$, and $\tau' = \tau(1 - \frac{3\tau}{2M})$.

Let us weaken condition (10.5.27). For $d = 2$ one can assume without loss of generality that $D = 2$. Then $A(\tau)$ is surely a scalar operator but we have not taken into account some terms. Taking them into account and applying the above analysis we can reject condition (10.5.27) (in which case we have additional corrections for W and τ').

In the general case we assume that conditions (10.5.25), (10.5.26) are fulfilled. If $\delta_1 \geq 1$ then the non-scalar terms in $A(\tau)$ are

$$-\frac{i}{2}(2M)^{-3}r^{-3}\sum_{j\neq k}\sigma_j'^{*}\sigma_k'(x_jD_k - x_kD_j)$$

modulo operators of the form $b_j'D_j + b_0'$ where b_j', b_0' satisfy condition (10.5.25) with "−2" replaced by "−3." We used the skew-symmetry of the matrices $\sigma_j'^{*}\sigma_k'$ with $j\neq k$. This statement remains true for $\delta_2 \geq 1$ (i.e., for $\delta_1 \geq \frac{1}{2}$) in the context of theorem 10.5.5. However, in this case the operator $A(\tau)$ also contains terms of type $b_j''D_j + b_0''$ with b_j'', b_0'' satisfying condition (10.5.26) with "−2" replaced by "$-2-\delta_1$" and this operator skew-commutes with the antipodal map. In both cases these terms are negligible provided

$$\text{(10.5.27)}' \qquad \delta_\iota < 2, \quad \delta_\iota < \frac{1}{2} + \delta_3 \quad (\iota = 1,2).$$

However, the second inequality is not necessary: for $\iota = 1$ it is fulfilled automatically and for $\iota = 2$ it is associated with terms of the type

$$-i\,\mathrm{const}\sum_{l\neq k}\sigma_l'^{*}\sigma_k'F_{jk}.$$

However, in this case $V_l \circ j^* = V_l$ and therefore $F_{lk}\circ j^* = -F_{lk}$ and the same is true for this extra term. Then one should check the condition $-2-\delta_1 < -1 < -2\delta_1$ which follows from the first inequality.

Let us consider

$$\int_0^{T_0} r^{-3}(x_j\xi_k - x_k\xi_j)\circ\Phi_t dt$$

where Φ_t is the Hamiltonian flow generated by $|\xi|^2 - |x|^{-1}$. One can easily see that this integral is equal to $\pi\mu_{jk}\mu^{-3}$ where $\mu_{jk} = x_j\xi_k - x_k\xi_j$ are components of the angular momentum tensor and $\mu = (\sum_{j<k}\mu_{jk}^2)^{\frac{1}{2}}$. This calculation is easy for trajectories lying in the (x_1,x_2)-plane and then one can change coordinates. Then we obtain that the contribution of the non-scalar terms to the operator $U(\pi,\tau)$ [41] is the factor

$$I - i\pi\sigma_j'^{*}\sigma_k'\mu_{jk}\mu^{-3} + O\big(|\tau|^{\frac{3}{2}-\delta'}\big)$$

with arbitrarily small $\delta' > 0$. Applying corollary C.7 and calculating the eigenvalues of μ_{jk} we obtain that in order to take into account non-scalar terms for $d\geq 3$ we can treat the family of operators

$$\tilde{B}_2(\tau') + k\nu|\tau|^{\frac{3}{2}}\mathbf{M}^{-2}$$

with an appropriate constant k and with $\nu\in\{-1,0,1\}$ of the corresponding multiplicities where $\mathbf{M}^s$ is an operator with the symbol μ^s with $\mu = \mu(x,\xi)$ defined above. One should check conditions for every such operator, calculate and sum $\mathbf{N}_\nu(\tau)$ with the corresponding multiplicities. For $d = 2$ and $D = 2$ we have $\nu = 0,1$ and the operators

$$\tilde{B}_2(\tau') + 3\zeta\nu|\tau'|^{\frac{3}{2}}\mathcal{M}\mathbf{M}^{-3}$$

with $\mathcal{M} = x_1D_2 - x_2D_1$, $\zeta = i\sigma_0\sigma_1\sigma_2 = \pm 1$. The details are left to the reader.

[41] Here $U(\pi,\tau) = \exp\big(i\pi(-A(\tau))^{-\frac{1}{2}}\big)$ and π is the period of the corresponding Hamiltonian flow for $\tau = 0$.

10.6 Multiparametrical Asymptotics

In this section we consider operators depending on more than one parameter. First of all we consider semiclassical asymptotics in the case when the scalar semiclassical parameter h is replaced by a vector $\mathbf{h} = (h_1, \dots, h_d)$ with $h_j \to +0$. Next we assume that there are two parameters: the semiclassical parameter $h \to +0$ and the spectral parameter τ. Let us recall that each of the semiclassical asymptotics of section 10.1 were uniform with respect to the energy level τ lying in the interval in which conditions of the corresponding theorem were fulfilled[42] uniformly. Therefore there are two principally different situations in which non-trivial results are possible; I will explain them for the Schrödinger operator.

In the first situation τ increases and tends to a value τ^* such that the conditions of the corresponding theorem of section 10.1 are violated for $\tau = \tau^*$ and therefore the asymptotics are no longer uniform. There are two cases: $\tau \to +\infty$ as in sections 10.2–10.4 or $\tau \to \tau^* - 0$ with $\tau^* < \infty$ as in section 10.5 where in the second case $\tau^* = \lim_{|x|\to\infty} \inf V$. In the second case τ^* is the bottom of the essential spectrum and $\tau^* - 0$ is either a limit point of the discrete spectrum for every fixed $h > 0$ or there is only a finite discrete spectrum for every fixed $h > 0$ and in the second case the number of eigenvalues and (or) the remainder estimate near $\tau^* - 0$ are unusually large ($\gg \mathrm{const} \cdot h^{-d}$ and $\gg \mathrm{const} \cdot h^{1-d}$ respectively). In all these cases we obtain estimates for the number of eigenvalues and for the remainder which are *worse* than for "normal" τ.

In the second situation τ decreases to the special value $\tau^* = \inf V$ and we obtain estimates for the number of eigenvalues and for the remainder which are *better* than for "normal" τ. Again there are two different cases: either $\inf V$ is finite in which case $\tau \to \tau^* + 0$ or $\inf V = -\infty$ but $-\infty$ is not a limit point of the spectrum for every fixed $h > 0$. Then for $h \to +0$ the bottom eigenvalues of the operator "fall" to $-\infty$ and we consider this process. Only the last case is impossible for the Dirac operator.

10.6.1 Semiclassical asymptotics with a vector semiclassical parameter

Let us consider the Schrödinger operator

$$A = A_{\mathbf{h}} = P_j g^{jk} P_k + V, \qquad P_j = h_j - V_j \tag{10.6.1}$$

or the Dirac operator of the same type as before. Changing coordinates $x_{j\,\mathrm{new}} = x_{j\,\mathrm{old}} h_j^{-1}$ we obtain the operator

$$A' = A'_{\mathbf{h}} = (D_j - V_{j\mathbf{h}}) g^{jk}_{\mathbf{h}} (D_k - V_{k\mathbf{h}}) + V_{\mathbf{h}}$$

in the domain $X_{\mathbf{h}} = \{x, f_{\mathbf{h}}(x) \in X\}$ where X is the original domain and $v_{\mathbf{h}} = v \circ f_{\mathbf{h}}$ for every function v, $f_{\mathbf{h}}(x) = h \bullet x \stackrel{\mathrm{def}}{=} (h_1 x_1, \dots, h_d x_d)$, $\mathbf{h} = (h_1, \dots, h_d)$. One can easily see that if (some of) the conditions $(\Sigma 0)$–$(\Sigma 3)$ were fulfilled for $X, \rho, \gamma, g^{jk}, V_j, V$ then the same conditions remain true for $X_{\mathbf{h}}, \rho_{\mathbf{h}}, h^{-1}\gamma_{\mathbf{h}}, V_{j\mathbf{h}}, V_{\mathbf{h}}$ where $h = \max_j h_j$. Moreover, if the cone condition was fulfilled for the domain X then it remains true for $X_{\mathbf{h}}$ provided $h \le 1$. This immediately yields

Theorem 10.6.1. *(i) Let $d \ge 3$ and let conditions $(\Sigma 0)$–$(\Sigma 2)$ be fulfilled for $g^{jk}, V_j, V, \rho, \gamma$. Let $\gamma > 0$ a.e. and $V_-^{\frac{d}{2}}$, $\rho^{d-1}\gamma^{-1} \in L^1(X)$. Then the asymptotics*

$$\mathbf{N}^-(A_{\mathbf{h}}) = (h_1 \dots h_d)^{-1} (\mathcal{N}^w + O(h)) \tag{10.6.2}$$

[42] In this subsection for the sake of simplicity we consider only the Schrödinger and Dirac operators and we no not need to check $(8.1.18)_*$-type conditions.

hold as $h = \max_j h_j \to +0$ *where*

$$\mathcal{N}^w = (2\pi)^{-d}\omega_d \int V_-^{\frac{d}{2}} \sqrt{g} dx.$$

(ii) Moreover, let us assume that $g^{jk} = V_j = 0$ *for* $j \geq d'+1$, $k \leq d'$, $0 < d' < d$ *and that* $|\nu'| \geq \epsilon_0 > 0$ *on* $\partial X \cap \{V \leq 0\}$ *where* $\nu = (\nu_1, \dots, \nu_{d'}; \nu_{d'+1}, \dots, \nu_d) = (\nu'; \nu'')$ *is a unit normal to* ∂X*. Then the asymptotics*

$$\mathbf{N}^-(A_{\mathbf{h}}) = (h_1 \dots h_d)^{-1} (\mathcal{N}^w + \varkappa_1 h + o(h)) \tag{10.6.3}$$

hold as $h = \max_j h_j \to +0$, $h_j = o(h)$ $\forall j \geq d'+1$ *where*

$$\varkappa_1 = \mp \frac{1}{4} (2\pi)^{1-d} \omega_{d-1} \int_{\partial X} V_-^{\frac{d-1}{2}} dS' \sqrt{g''} dx'' \tag{10.6.4}$$

for Dirichlet and Neumann boundary conditions respectively and the Riemannian densities dS', $\sqrt{g''}dx''$ *were introduced in example 10.4.12(ii).*

(iii) Furthermore, for $\vartheta > 0$, $V_-^{\frac{d}{2}+\vartheta}, \rho^{d-1+\vartheta} \in L^1(X)$ *we have, in the context of (i),(ii), the asymptotics*

$$\mathbf{N}^-_\vartheta(A_{\mathbf{h}}) = (h_1 \dots h_d)^{-1} (\mathcal{N}^w_{\vartheta,\mathbf{h}} + O(h^{1+\vartheta})) \tag{10.6.2$'$}$$

and

$$\mathbf{N}^-_\vartheta(A_{\mathbf{h}}) = (h_1 \dots h_d)^{-1} (\mathcal{N}^w_{\vartheta,\mathbf{h}} + o(h^{1+\vartheta})) \tag{10.6.3$'$}$$

respectively.

(iv) The same results remain true for $d = 2$, $d' = d'' = 1$ *under the additional restrictions* $\operatorname{mes} X'' = \{X, \rho\gamma \leq h_0\} < \infty$ *for some* $h_0 > 0$ *and* $V_-^{p+\vartheta} \in L^1(X'')$ *for some* $p > 1$.

(v) Under the cone condition all these results remain true for Neumann boundary conditions and for the Dirac operator (with reasonable boundary conditions). In the second case we consider eigenvalues lying in the interval $[\tau_1, \tau_2]$ *with arbitrary fixed* τ_1, τ_2.

Proof. Statement (i) follows immediately from theorem 8.1.1. Statement (ii) follows from theorem 8.1.2 and analysis of Hamiltonian billiards. This analysis is based on the same arguments as in example 10.4.12(ii); namely ξ_j and $h_j x_j$ with $j \geq d'+1$ are "almost constant" along these billiards (which is not true for $j \leq d'$). The other statements are proven by a similar method. □

I leave the statements and proofs of assertions similar to examples 10.1.4–10.1.13 to the reader.

10.6.2 Asymptotics of eigenvalues tending to $+\infty$

Here and below we consider the Schrödinger operator A_h and consider the number of its eigenvalues less than τ. For the Dirac operator A_h (we use the same notation) we consider the number of eigenvalues lying in the interval $[\tau_1, \tau)$ with fixed τ_1.

Applying the arguments of sections 10.2–10.4 with $\bar{\rho}_{\tau,h} = h^{-1}(\rho + \sqrt{\tau})$, $\gamma_{\tau,h} = \gamma$ we obtain modifications of the statements of these sections. However, we cannot replace $\mathcal{N}^w(\tau)$ by $\varkappa_0 \tau^{\frac{d}{2m}}$ in these assertions. In particular, we obtain the following statements:

Theorem 10.6.2. [43] *(i) For the Schrödinger operator, let conditions* $(\Sigma 0)$–$(\Sigma 2)$ *be fulfilled with auxiliary functions* ρ, γ *and let* $\vartheta \ge 0$ *and* $\rho^{d+2\vartheta}, 1, \gamma^{-\vartheta-1} \in L^1(X)$. *For* $d = 2$ *let us in addition assume that* $\rho^{2p+2\vartheta} \in L^1(X)$ *with* $p > 1$. *Moreover, in the case of the Neumann condition let us in addition assume that the cone condition is also fulfilled.*[44] *Then the asymptotics*

$$\mathbf{N}^-_\vartheta(\tau, h) = \mathcal{N}^w_\vartheta(\tau, h) + O\big(\tau^{\frac{1}{2}(d+\vartheta-1)} h^{1-d+\vartheta}\big) \tag{10.6.4$'$}$$

hold as $\tau \to +\infty, h \to +0$. *In particular, for* $\vartheta = 0$

$$\mathcal{N}^w(\tau, h) = \mathcal{N}^w(\tau) h^{-d}.$$

(ii) Moreover, if the set of all points which are periodic with respect to the geodesic flow with reflections is of measure 0 *then the asymptotics*

$$\mathbf{N}^-_\vartheta(\tau, h) = \mathcal{N}^w_\vartheta(\tau, h) + o_\tau\big(\tau^{\frac{1}{2}(d+\vartheta-1)} h^{1-d+\vartheta}\big) \tag{10.6.5$'$}$$

hold as $\tau \to +\infty, h \to +0$ *where* $M' = o_\zeta(M)$ *means that* $\frac{M'}{M} \le \varepsilon(\zeta)$ *with* $\varepsilon(\zeta) = o(1)$ *as* ζ *tends to the described limit. In particular, for* $\vartheta = 0$ *the asymptotics*

$$\mathbf{N}^-(\tau, h) = \mathcal{N}^w(\tau) h^{-d} + \big(\varkappa_1 + o_\tau(1)\big) \tau^{\frac{1}{2}(d-1)} h^{1-d} \tag{10.6.5}$$

hold as $\tau \to +\infty, h \to +0$ *with the coefficient*

$$\varkappa_1 = \mp \frac{1}{4} (2\pi)^{1-d} \omega_{d-1} \operatorname{vol}'\big(\partial X \cap \{\gamma > 0\}\big)$$

for Dirichlet and Neumann boundary conditions respectively.

(iii) Furthermore, if $\gamma^{-1-\theta} \in L^1(X)$ *with* $\theta > \vartheta$ *and the geodesic flow is of non-periodic power type or of non-periodic exponential type then the remainder estimate in these asymptotics is* $O\big(\tau^{\frac{1}{2}(d+\vartheta-1-\delta')} h^{1-d+\vartheta}\big)$ *or* $O\big(\tau^{\frac{1}{2}(d+\vartheta-1)} h^{1-d+\vartheta} \log^{-1-\vartheta} \tau\big)$ *respectively.*

(iv) All these asymptotics remain true if we replace the condition $\rho^{d+2\vartheta} \in L^1(X)$ *by the coercivity condition* $(\Sigma 3)$.

(v) Statements (i)–(iii) remain true for $\vartheta = 0$ *for the Dirac operator.*[45] *In these assertions one should replace* $\mathbf{N}^-(\tau, h)$ *and* $\mathcal{N}^w(\tau_2)$ *by* $\mathbf{N}(\tau_1, \tau_2, h)$ *and* $\mathcal{N}^w(\tau_1, \tau_2)$ *respectively and take* $\tau = |\tau_1| + |\tau_2| \to \infty$. *One should assume that the boundary conditions are elliptic and the domain satisfies the cone condition.*

I leave it to the reader to consider examples 10.2.4–10.2.8, 10.2.31, 10.2.32, 10.3.1–10.3.3, remark 10.2.9 and transfer them into the context of the present two-parameter theory (automatically or with minor modifications).

Moreover, I leave it to the reader to modify theorems 10.2.20–10.2.22, 10.2.24 and 10.2.26–10.2.29. I would only like to note that one should assume that $h \ge \tau^{-n}$, and the principal part of the asymptotics is of the form $\mathcal{N}^w(\tau) h^{-d}$ and the second term is of the form

$$\big(\varkappa_1 + F(\tau h^{-m})\big) h^{1-d} \tau^{\frac{1}{2m}(d-1)}$$

and there is additional factor h^{1-d} in the remainder estimate.

Let us consider the statements of section 10.4. For the sake of simplicity I consider only the Schrödinger operator. Then the following statements hold:

---

43) I leave the modification of the complete set of theorems 10.2.1–10.2.3 to the reader.

44) I leave the analysis of mixed boundary conditions to the reader.

45) For $\vartheta > 0$ one should modify $\mathbf{N}_\vartheta(\tau_1, \tau_2, h)$ in the obvious manner.

Theorem 10.6.3. *(i) In the context of theorem 10.4.1(i) (with the additional assumption that A is the Schrödinger operator) the asymptotics*

$$\mathbf{N}_\vartheta(\tau,h) = \mathcal{N}_\vartheta^w(\tau,h) + O\bigl(R_\vartheta(\tau)h^{1-d+\vartheta}\bigr) \tag{10.6.6$'$}$$

hold as $\tau \to +\infty, h \to +0$ where $R_\vartheta(\tau)$ is defined by (10.4.2)$'$. *Moreover, under the additional assumptions of theorem 10.4.1(ii) the asymptotics*

$$\mathbf{N}_\vartheta(\tau,h) = \mathcal{N}_\vartheta^w(\tau,h) + o_\tau\bigl(R_\vartheta(\tau)h^{1-d+\vartheta}\bigr) \tag{10.6.7$'$}$$

hold as $\tau \to +\infty, h \to +0$. Finally, in the context of theorem 10.4.1(iii) the asymptotics

$$\mathbf{N}_\vartheta(\tau,h) = \mathcal{N}_\vartheta^w(\tau,h) + o_{\tau h^{-2}}\bigl(R_\vartheta(\tau)h^{1-d+\vartheta}\bigr) \tag{10.6.8$'$}$$

hold as $\tau \to +\infty, h \to +0$. Moreover, the principal part in asymptotics (10.6.6) *(i.e., in* (10.6.6)$'$ *with $\vartheta = 0$) is $\mathcal{N}^w(\tau)h^{-d}$ and in* (10.6.7)$'$, (10.6.8)$'$ *the principal part is $\mathcal{N}^w(\tau)h^{-d} + \varkappa_1\tau^{l(d-1)}h^{1-d}$ with the coefficient $\varkappa_1$ given by* $(10.4.5)_1$.

(ii) In the context of theorem 10.4.3 for $\vartheta \le d-1-\delta$ with $\delta > 0$ the asymptotics

$$\mathbf{N}_\vartheta(\tau,h) = \mathcal{N}_\vartheta^w(\tau,h) + O\bigl(\tau^{l(d-1)+\vartheta(1-l)-\delta'}h^{1-d}\bigr) \tag{10.6.9$'$}$$

hold as $\tau \to +\infty, h \to +0$. Here for $\vartheta = 0$ the principal part is the same as in (10.6.7)$'$. *In particular, example 10.4.4 applies.*

(iii) In the context of theorems 10.4.8, 10.4.10, 10.4.12 asymptotics (10.6.6)$'$ *remain true. Moreover, under the conditions of these theorems on the Hamiltonian trajectories with reflections asymptotics* (10.6.7)$'$ *remain true. I leave the analysis of examples 10.4.9, 10.4.11 and 10.4.13 to the reader.*

(iv) In the context of example 10.4.14(i),(ii) the asymptotics

$$\mathbf{N}(\tau,h) = \mathcal{N}^w(\tau)h^{-d} + O(\tau^q h^{1-d}), \tag{10.6.10}$$

$$\mathbf{N}(\tau,h) = \mathcal{N}^w(\tau)h^{-d} + \bigl(\varkappa_1 + o_\tau(1)\bigr)\tau^q h^{1-d}, \tag{10.6.11}$$

hold respectively with the coefficient $\varkappa_1$ given by (10.4.31).

(v) In the context of theorem 10.4.15(i), (ii) the asymptotics

$$\mathbf{N}_\vartheta(\tau,h) = \mathcal{N}_\vartheta^w(\tau)h^{-d} + O\bigl(\tau^{\frac{1}{2}(d+\vartheta-1)}h^{1-d+\vartheta}\omega(\tau)\bigr), \tag{10.6.12$'$}$$

$$\mathbf{N}_\vartheta(\tau,h) = \mathcal{N}_\vartheta^w(\tau)h^{-d} + o\bigl(\tau^{\frac{1}{2}(d+\vartheta-1)}h^{1-d+\vartheta}\omega(\tau)\bigr) \tag{10.6.13$'$}$$

hold respectively with

$$\mathcal{N}_\vartheta^w(\tau,h) \asymp \tau^{\frac{d}{2}+\vartheta}\log^{\frac{n}{\mu}}\tau h^{-d}$$

and $\omega(\tau)$ defined by (10.4.33). *For $\vartheta = 0$ we have $\mathcal{N}^w(\tau,h) = \mathcal{N}^w(\tau)h^{-d}$. Asymptotics* (10.6.13)$'$ *remain true in the context of theorem 10.4.16 and for $\vartheta = 0$*

$$\mathcal{N}_0^w(\tau) = \mathcal{N}^w(\tau)h^{-d} + \varkappa_1\tau^{\frac{1}{2}(d-1)}h^{1-d}\log\tau$$

where the coefficient $\varkappa_1$ is defined by $(10.4.37)_1$.

(vi) In the context of theorem 10.4.17(i) asymptotics with principal part

$$\mathcal{N}_\vartheta^w(\tau,h) \asymp \mathcal{N}^w(\tau)h^{-d}, \qquad \mathcal{N}_0^w(\tau,h) = \mathcal{N}^w(\tau)h^{-d}$$

and with remainder estimate $h^{1-d+\vartheta}R_\vartheta(\tau)$ hold where $\mathcal{N}^w(\tau)$ and $R_\vartheta(\tau)$ are defined in the theorem. Moreover, the improvements made in assertion (ii) and in footnotes [36),37)] *of section 10.4 remain true.*

I leave it to the reader to modify theorems 10.4.2, 10.4.6, and 10.4.7. I would like to note that the remarks made concerning the modification of theorems 10.2.20–10.2.29 also remain true in this case.

10.6.3 Asymptotics of eigenvalues tending to −0 and related asymptotics

Applying the arguments of section 10.5 with $\tilde{\rho}_{\tau,h} = h^{-1}(\rho + \sqrt{|\tau|})$, $\gamma_{\tau,h} = \gamma$ we obtain

Theorem 10.6.4. [46] *In the context of theorem 10.5.1(i) the asymptotics*

$$\mathbf{N}_\vartheta(\tau, h) = \mathcal{N}^w_\vartheta(\tau, h) + O\big(h^{1-d+\vartheta} R_\vartheta(\tau)\big) \tag{10.6.14}$$

hold as $\tau \to -0$, $h \to +0$ with $R_\vartheta(\tau)$ defined by (10.5.3)′. Moreover, in the context of theorem 10.5.1(ii) the asymptotics

$$\mathbf{N}_\vartheta(\tau, h) = \mathcal{N}^w_\vartheta(\tau, h) + o_\tau\big(h^{1-d+\vartheta} R_\vartheta(\tau)\big) \tag{10.6.15′}$$

hold as $\tau \to -0$, $h \to +0$. For $\vartheta = 0$ the principal parts of these asymptotics are $\mathcal{N}^w(\tau)h^{1-d}$ and $\mathcal{N}^w(\tau)h^{1-d} + \varkappa_1|\tau|^{l(d-1)}h^{1-d}$ respectively with the coefficient $\varkappa_1$ given by $(10.5.6)_1$.

(ii) In the context of theorem 10.5.3 with $\vartheta \le \vartheta_1 - \delta$, $\delta > 0$ the asymptotics

$$\mathbf{N}_\vartheta(\tau, h) = \mathcal{N}^w_\vartheta(\tau, h) + O(h^{1-d+\vartheta}|\tau|^{l(d-1)+\vartheta(1-l)+\delta'}) \tag{10.6.16′}$$

hold with $\delta' > 0$. In these asymptotics the principal part for $\vartheta = 0$ is $\mathcal{N}^w(\tau)h^{1-d} + \varkappa_1|\tau|^{l(d-1)}h^{1-d}$. In particular, example 10.5.3 applies.

(iii) In the context of theorem 10.5.7 asymptotics (10.6.14)′ remain true and under a condition on Hamiltonian trajectories with reflection asymptotics (10.6.15)′ also remain true. I leave the analysis of example 10.5.8 to the reader.

(iv) In the context of theorem 10.5.9(i),(ii) the asymptotics

$$\mathbf{N}(\tau, h) = \mathcal{N}^w(\tau)h^{-d} + O(|\tau|^q h^{1-d}), \tag{10.6.17}$$

$$\mathbf{N}(\tau, h) = \mathcal{N}^w(\tau)h^{-d} + \big(\varkappa_1 + o_\tau(1)\big)|\tau|^q h^{1-d} \tag{10.6.18}$$

hold with the coefficient $\varkappa_1$ given by $(10.5.17)_1$

(v) In the context of theorems 10.5.10(i),(ii) the asymptotics

$$\mathbf{N}(\tau, h) = \mathcal{N}^w(\tau)h^{-d} + O\big(\big|\log|\tau|\big|^{\nu(d-1)+1}h^{1-d}\big), \tag{10.6.19}$$

$$\mathbf{N}(\tau, h) = \mathcal{N}^w(\tau)h^{-d} + \big(\varkappa_1 + o_\tau(1)\big)\big|\log|\tau|\big|^{\nu(d-1)+1}h^{1-d} \tag{10.6.20}$$

hold with the coefficient $\varkappa_1$ given by $(10.5.21)_1$.

(vi) Moreover, assertion (v) remains true for $\nu = 0$ with $o_\tau(1)$ replaced by $o(1)$ in asymptotics (10.6.20).

(vii) In the context of theorem 10.5.11 asymptotics with principal part

$$\mathcal{N}^w_\vartheta(\tau, h) \asymp \mathcal{N}^w_{\vartheta,h}(\tau)h^{-d}, \qquad \mathcal{N}^w_0(\tau, h) = \mathcal{N}^w(\tau)h^{-d}$$

and remainder estimate $O\big(h^{1-d+\vartheta}R_\vartheta(\tau)\big)$ hold where $\mathcal{N}^w(\tau)$, $R_\vartheta(\tau)$ are defined in the theorem.

Statement (vi) has no analog in section 10.5 because for $\nu = 0$ for every small enough fixed $h > 0$ the eigenvalues of A_h tend to −0 but in (10.6.19) the principal part and the remainder estimate are of the same order $\log|\tau|$ and (10.6.19) is only an estimate, not asymptotics.

---

[46] Again we treat only the Schrödinger operator.

I leave the modification of theorems 10.5.2, 10.5.5 and 10.5.6 to the reader. I would like to note that the remark concerning the modification of theorems 10.2.20–10.2.29 remains true under the inequality $h \ge |\tau|^n$.

Let us consider the situation in which the conditions of theorem 10.5.10 are fulfilled with $\nu < 0$ instead of $\nu > 0$. In this case for every fixed $h > 0$ the discrete spectrum is finite. However, $V_-^{\frac{d}{2}} \notin L^1(X)$ for $\nu \ge -d^{-1}$ provided condition (10.5.4)* is fulfilled. In this case the above arguments give the following

Theorem 10.6.5. *(i) Under the assumptions of theorem 10.5.10(i) with* $\nu \in (-d^{-1}, 0)$ *asymptotics* (10.6.19) *hold for the Schrödinger operator for* $\tau \to -0$, $h = o(\bigl|\log|\tau|\bigr|^{\nu})$. *In these asymptotics* $\mathcal{N}^w(\tau) = O(\bigl|\log|\tau|\bigr|^{\nu d+1})$ *and under condition* (10.5.4)* $\mathcal{N}^w(\tau) \asymp \bigl|\log|\tau|\bigr|^{\nu d+1}$.

(ii) Under the assumptions of theorem 10.5.10 (i) with $\nu = -d^{-1}$, *asymptotics* (10.6.19) *hold for the Schrödinger operator for* $\tau \to -0$, $h = O(\bigl|\log|\tau|\bigr|^{\nu})$. *In these asymptotics* $\mathcal{N}^w(\tau) = O(\log\bigl|\log|\tau|\bigr|)$ *and under condition* (10.5.4)* $\mathcal{N}^w(\tau) \asymp \log\bigl|\log|\tau|\bigr|$.

(iii) Under the assumptions of theorem 10.5.10(ii) with $\nu \in [-d^{-1}, 0)$ *asymptotics* (10.6.20) *hold for the Schrödinger operator for* $\tau \to -0$, $h = o(\bigl|\log|\tau|\bigr|^{\nu})$. *In these asymptotics the coefficient* $\varkappa_1$ *is given by* $(10.5.21)_1$.

Let us consider the case $h \ge \epsilon_0 \bigl|\log|\tau|\bigr|^{\nu}$. Then the zone $X'' = \bigl\{(\rho + \sqrt{|\tau|})\gamma \le \epsilon h\bigr\}$ is not necessarily empty. Considering only the lower estimate we obtain that under condition (10.5.4)*

$$\mathbf{N}(\tau, h) \ge \epsilon_1 h^{\frac{1}{\nu}} \qquad \text{for } \nu \in (-\frac{1}{d}, 0), \tag{10.6.21}$$

$$\mathbf{N}(\tau, h) \ge \epsilon_1 h^{\frac{1}{\nu}} |\log h| \qquad \text{for } \nu = -\frac{1}{d} \tag{10.6.22}$$

with $\epsilon_1 > 0$.

On the other hand, for $d \ge 3$ one can apply a variational estimate in X'' (with $\rho = 0$ for $\log^{\nu} r \le \epsilon_2 h$. Then we obtain the estimate

$$\mathbf{N}(\tau, h) = O(h^{\frac{1}{\nu}}) \qquad \text{for } \nu \in (-\frac{1}{d}, 0), \tag{10.6.23}$$

and even the asymptotics

$$\mathbf{N}(\tau, h) = \bigl(\mathcal{N}_h^w(\tau) + O(1)\bigr)h^{-d} \qquad \text{for } \nu = -\frac{1}{d} \tag{10.6.24}$$

for $\tau \to -0$, $h \to +0$ with

$$\mathcal{N}_h^w(\tau) = (2\pi)^{-d}\omega_d \int_{\{|\log|x|| \le Ch^{-\frac{1}{\nu}}\}} (-\tau - V)_+^{\frac{d}{2}} \sqrt{g}\, dx \tag{10.6.25}$$

with an arbitrarily chosen constant C. Under condition (10.5.4)* this expression is $\asymp \min\bigl(\log\bigl|\log|\tau|\bigr|, |\log h|\bigr)$ for $\nu = -d^{-1}$. Let us recall that the asymptotics and estimates (10.6.21)–(10.6.24) were obtained under the restriction $h \ge \epsilon_0 \bigl|\log|\tau|\bigr|^{\nu}$.

Thus, if $d \ge 3$, $\nu \in (-d^{-1}, 0)$ then $\mathbf{N}(\tau, h) \asymp h^{\frac{1}{\nu}}$ for $h \ge \bigl|\log|\tau|\bigr|^{\nu}$ (under condition (10.5.4)*). One can easily obtain this result by purely variational arguments.

For $d \ge 3$, $\nu = -d^{-1}$ we combine assertion (ii) of theorem 10.6.5 and (10.6.24) to obtain

$$\mathbf{N}(\tau, h) = \mathcal{N}_h^w(\tau) h^{-d} + O\Bigl(\min\bigl(h^{-d}, h^{1-d}\bigl|\log|\tau|\bigr|^{\frac{1}{d}}\bigr)\Bigr) \tag{10.6.26}$$

with $\mathcal{N}_h^w(\tau)$ given by (10.6.25) and $\tau \to -0$, $h \to +0$.

I leave it to the reader to treat the case $\nu \in [-(d-1)^{-1}, -d^{-1})$. In this case the principal part is standard and $\asymp h^{-d}$ but the remainder estimate is worse than usual. In this case we also obtain asymptotics for $\tau = 0$ in the context of section 10.1.

All the results of this subsection remain true for the Dirac operator with $\tau_1^0 = -M + \lim_{|x|\to\infty} \sup V < \tau_2^0 = M + \lim_{|x|\to\infty} \inf V$.

One need only to replace $\mathbf{N}(\tau)$ by $\mathbf{N}(\tau_1, \tau)$ with fixed $\tau_1 \in (\tau_1^0, \tau_2^0)$ etc., and modify the formulas for $\mathcal{N}^w$, $\mathcal{N}_h^w$, $\varkappa_0$ and $\varkappa_1$; without loss of generality one can assume that $\tau_2^0 = 0$ and then $\tau \to -0$.

10.6.4 Asymptotics of eigenvalues near +0

Arguments of the same type as above permit us to treat the case $\tau \to +0$, $h \to +0$ and to obtain asymptotics of $\mathbf{N}(\tau, h)$ assuming that $\inf V = 0$ for the Schrödinger operator. Similar results may be obtained for the Dirac operator for $\mathbf{N}(\tau_1, \tau; h)$ provided $\tau_1^0 = -M + \sup V < \tau_1 \le \tau_2^0 = M + \inf V$; without loss of generality one can assume that $\tau_2^0 = 0$. We formulate our statements only for the Schrödinger operator (for the Dirac operator these statements remain true) and in the case of Neumann boundary conditions (and in the case of any admissible boundary condition for the Dirac operator) we assume that the domain satisfies the cone condition. I leave the treatment of general operators and the case $\vartheta > 0$ to the reader. It is easy to prove the following

Theorem 10.6.6. *(i) Let X be a bounded domain and let Y a C^1 manifold of codimension d'' in $\mathbb{R}^d$, $2 \le d'' \le d$ such that $\bar{X} \cap Y \ne \emptyset$. Let conditions $(\Sigma0)$–$(\Sigma3)$ be fulfilled with $\gamma = \epsilon r(x)$, $\rho(x) = r^\mu(x)$, $\epsilon > 0$, $\mu > 0$, $r(x) = \operatorname{dist}(x, Y)$. Then the asymptotics*

$$\mathbf{N}(\tau, h) = \mathcal{N}^w(\tau) h^{-d} + O\bigl(\tau^q h^{1-d}\bigr) \tag{10.6.27}$$

hold for $\tau \to +0$, $h \to +0$ with $\mathcal{N}^w(\tau) = O(\tau^p)$, $p = \frac{1}{2}d + \frac{1}{2\mu}d''$, $q = \frac{1}{2}(d-1) + \frac{1}{2\mu}(d''-1)$.[47] *Moreover, $\mathcal{N}^w(\tau) \asymp \tau^p$ provided*

$$\operatorname{vol}\bigl(X \cap \{r(x) < \bar{r}\}\bigr) \asymp r^n \quad \text{as } r \to 0 \tag{10.6.28}$$

with $n = d''$.

(ii) Furthermore, let us assume that $Y = \{0\}$, $X \cap \{|x| \le r_0\} = X^0 \cap \{|x| \le r_0\}$ where $r_0 > 0$ and X^0 is the non-empty open cone with $\partial X^0 \setminus 0 \in C^K$. Let stabilization conditions (10.4.3) be fulfilled for $|x| \to 0$ with functions g^{jk0}, V_j^0 and V^0 positively homogeneous of degrees 0, μ and 2μ respectively. Finally, let the standard condition (Ψ) be fulfilled for the domain X^0 and the Hamiltonian

$$H_{\mathrm{st}}^0 = g^{jk0}(\xi_j - V_j^0)(\xi_k - V_k^0) + V^0 - 1$$

at the energy level 0. Then the asymptotics

$$\mathbf{N}(\tau, h) = \mathcal{N}^w(\tau) h^{-d} + \bigl(\varkappa_1 + o(1)\bigr) \tau^{l(d-1)} h^{1-d} \tag{10.6.29}$$

hold as $\tau \to +0$, $h = o(\tau^l)$ with the coefficient $\varkappa_1$ given by $(10.4.5)_1$. Moreover, under the stabilization conditions alone, $\mathcal{N}^w = \bigl(\varkappa_0 + o(1)\bigr)\tau^{ld}$ with the coefficient $\varkappa_0$ given by $(10.4.5)_0$.

---

[47] Surely we obtain asymptotics only for $h = o(\tau^l)$ with $l = \frac{1}{2\mu}(\mu + 1)$, i.e., only if τ tends to $+0$ slowly enough. Otherwise we obtain only an upper estimate and for $h \ge \epsilon_0 h^l$ with a small enough constant ϵ_0 we obtain a lower estimate for $\mathbf{N}(\tau, h)$. This remark remains true for other asymptotics of this subsection.

(iii) Finally, let the stabilization conditions

$$D^\alpha(a_\beta - a^0_\beta) = O\bigl(|x|^{(2m-|\beta|)\mu-|\alpha|+\delta}\bigr) \qquad \forall\beta : |\beta| \le 2m \quad \forall\alpha : |\alpha| \le 1 \tag{10.6.30}$$

be fulfilled as $|x| \to 0$ with some $\delta > 0$. Let the Hamiltonian flow generated by H^0_{st} near 0 be of non-periodic power type. Then the asymptotics

$$\mathbf{N}(\tau, h) = \mathcal{N}^w(\tau)h^{-d} + \bigl(\varkappa_1 + O(h^{\delta'}\tau^{-l\delta'})\bigr)h^{1-d}\tau^{l(d-1)} \tag{10.6.31}$$

with $\delta' > 0$ hold for $\tau \to +0$, $h = o(\tau^l)$.

I leave it to the reader to generalize assertion (ii) to the case $d'' < d$. The arguments of theorem 10.4.14(ii) should be applied here. I also leave to the reader the case of periodic Hamiltonian flow. Finally, I leave to the reader the case $d'' = 1$ when the logarithmic factor appears in the remainder estimate.

Theorem 10.6.7. *Let X be a bounded domain, $0 \in X$ and $X \cap \{|x| \le r_0\} = X^0 \cap \{|x| \le r_0\}$ where X^0 is an L-quasi-conical domain in $\mathbb{R}^d$. Let g^{jk}, V_j and V be L-quasi-homogeneous with respect to x of degrees 0, μ, 2μ respectively with $\mu > 0$, $L = (l_1, \dots, l_d)$, $l_j = 1$ for $j \le d'$ and $l_j < 1$ for $j \ge d' + 1$, $0 < d' < d$. Let us assume that $\partial X \setminus 0 \in C^K$ and $g^{jk}, V_j, V \in C^K(\bar{X} \setminus 0)$ satisfy condition $(\Sigma 2)$.*[48] *Finally, let us assume that*

$$V(x) > 0 \qquad \forall x \in \bar{X} \setminus 0. \tag{10.6.32}$$

Then

(i) The asymptotics

$$\mathbf{N}(\tau, h) = \varkappa_0 \tau^p h^{-d} + O(\tau^q h^{1-d}) \tag{10.6.33}$$

hold for $\tau \to +0$, $h \to +0$ with $p = \frac{1}{2}d + \frac{1}{2\mu}l$, $q = \frac{1}{2}(d-1) + \frac{1}{2\mu}(l-1)$, $l = l_1 + \dots + l_d$ and with the coefficient $\varkappa_0$ given by (10.4.27) with X replaced by X^0.

(ii) Under the additional conditions of example 10.4.14(ii) the asymptotics

$$\mathbf{N}(\tau, h) = \varkappa_0 \tau^p h^{-d} + \bigl(\varkappa_1 + o(1)\bigr)\tau^q h^{1-d} \tag{10.6.34}$$

hold for $\tau \to -0$, $h = o(\tau^{\frac{1}{2} + \frac{1}{2\mu}})$ with the coefficient $\varkappa_1$ given by (10.4.30).

Theorem 10.6.8. *Let X be a bounded domain and let Y be a C^1 manifold of codimension d'' in $\mathbb{R}^d$, $2 \le d'' \le d$, $\bar{X} \cap Y \ne \emptyset$. Let conditions $(\Sigma 0)$–$(\Sigma 3)$ be fulfilled with $\gamma = \epsilon r(x)$, $\rho = \exp\bigl(-r(x)^\nu\bigr)$ with $\nu < 0$, $r(x) = \mathrm{dist}(x, Y)$. Let us assume that condition (10.6.28) is fulfilled with an exponent n which is now not necessarily equal to d''. Then*

(i) The asymptotics

$$\mathbf{N}(\tau, h) = \mathcal{N}^w(\tau)h^{-d} + O\bigl(\tau^{\frac{1}{2}(d-1)}|\log\tau|^{(\nu+n-1)\nu^{-1}}h^{1-d}\bigr) \tag{10.6.35}$$

hold for $\tau \to +0$, $h \to +0$ and $\mathcal{N}^w(\tau) \asymp \tau^{\frac{1}{2}d}|\log\tau|^{n\nu^{-1}}$.

(ii) Moreover, let $Y = \{0\}$, $n = d$. Let condition (10.6.34) be fulfilled for $r \to +0$ and every fixed k and let conditions $(10.4.35)_{1,2}$ be fulfilled for $|x| \to 0$. Then the asymptotics

$$\mathbf{N}(\tau, h) = \mathcal{N}^w(\tau)h^{-d} + o\bigl(\tau^{\frac{1}{2}(d-1)}|\log\tau|^{(\nu+n-1)\nu^{-1}}h^{1-d}\bigr) \tag{10.6.36}$$

hold for $\tau \to +0$, $h = o\bigl(\tau^{\frac{1}{2}}|\log\tau|^{(1-\nu)\nu^{-1}}\bigr)$.

---

[48] Then one can take $\gamma = \epsilon[x]_L$, $\rho = \epsilon[x]^\mu_L$.

Theorem 10.6.9. *Let $d \ge 3$, $X = (0, \frac{1}{2}) \times U$ where U is a domain on $\mathbb{S}^{d-1}$ with C^K boundary, $g^{jk} = g^{jk}(\theta) \in C^K(\bar{U})$ satisfy (Σ3). Let $V_j = 0$, $V = v(\theta)|\log r|^{-1}$ with $v \in C^K(\bar{U})$ and $\min_{\bar{U}} v = v_0 > 0$. Then the asymptotics*

$$\mathbf{N}(\tau,h) = \mathcal{N}^w(\tau)h^{-d} + O\Big(h^{1-d}\tau^{d-2}\int_U \exp\big(-v(d-1)\tau^{-1}\big)d\theta\Big) \tag{10.6.37}$$

hold as $\tau \to +0$, $h \to +0$ with

$$\mathcal{N}^w(\tau) = a\int_{U'} \tau^d v^{-\frac{d}{2}} F_d(\frac{\tau}{v})\exp(-vd\tau^{-1})\sqrt{g}d\theta \tag{10.6.38}$$

where $U' = U \cap \{v < v_0 + \varepsilon\}$, $\varepsilon > 0$ is arbitrarily small, and a and F_d are introduced in theorem 10.5.10 in which the orders of $\mathcal{N}^w(\tau)$ and the remainder estimate are also discussed.

10.6.5 Asymptotics of eigenvalues near $-\infty$

Let us consider asymptotics of $\mathbf{N}(\tau,h)$ for $\tau \to -\infty$, $h \to +0$. Recall that we consider the Schrödinger operator and now we assume that

$$\inf V = -\infty. \tag{10.6.39}$$

The results of this subsection surely make no sense for the Dirac operator. The same arguments as above give us the following

Theorem 10.6.10. *(i) Let X be a bounded domain in $\mathbb{R}^d$, $0 \in \bar{X}$ and let conditions (Σ0)–(Σ2) be fulfilled with $\gamma = \epsilon|x|$, $\rho = |x|^\mu$, $\mu \in (-1,0)$. Then the asymptotics*[49]

$$\mathbf{N}(\tau,h) = \mathcal{N}^w(\tau)h^{-d} + O\big(|\tau|^{l(d-1)}h^{1-d}\big) \tag{10.6.40}$$

hold for $\tau \to -\infty$, $h \to +0$ and $\mathcal{N}^w(\tau) = O(|\tau|^{ld})$. Moreover, $\mathcal{N}^w(\tau) \asymp |\tau|^{ld}$ provided

$$\operatorname{mes}\{x \in X, |x| \le r, V \le -\epsilon_0\rho^2(x)\} \asymp r^d \qquad \text{for } r \to +0. \tag{10.6.41}$$

(ii) Moreover, let $X \cap \{|x| \le r_0\} = X^0 \cap \{|x| \le r_0\}$ with some $r_0 > 0$ and an open non-empty cone X^0 such that $\partial X^0 \setminus 0 \in C^K$. Let stabilization conditions (10.4.3) be fulfilled for $|x| \to 0$ with functions $g^{jk0}, V^0_j, V^0 \in C^K(\bar{X}^0 \setminus 0)$. Let the standard condition (Ψ) on the Hamiltonian flow be fulfilled at the energy level 0 for the Hamiltonian

$$H^0(x,\xi) = |x|^{1-\mu}\big(g^{jk0}(\xi_j - V^0_j)(\xi_k - V^0_k) + V^0 + 1\big). \tag{10.6.42}$$

Then the asymptotics

$$\mathbf{N}(\tau,h) = \mathcal{N}^w(\tau)h^{-d} + \big(\varkappa_1 + o(1)\big)|\tau|^{l(d-1)}h^{1-d} \tag{10.6.43}$$

hold for $\tau \to -\infty$, $h = o(|\tau|^l)$ with the coefficient $\varkappa_1$ given by $(10.5.6)_1$.

(iii) Furthermore, let stabilization conditions (10.6.30) with $\delta > 0$ be fulfilled for $x \to 0$ and let the Hamiltonian flow corresponding to the Hamiltonian

$$H^0_{\mathrm{st}}(x,\xi) = g^{jk0}(\xi_j - V^0_j)(\xi_k - V^0_k) + V^0 + 1 \tag{10.6.43}$$

be of power non-periodic type near the energy level 0. Then the asymptotics

$$\mathbf{N}(\tau,h) = \mathcal{N}^w(\tau)h^{-d} + \big(\varkappa_1 + O(h^{\delta'}|\tau|^{-l\delta'})\big)|\tau|^{l(d-1)}h^{1-d} \tag{10.6.44}$$

with $\delta' > 0$ hold for $\tau \to -\infty$, $h \to +0$.

---

[49] See footnote [47].

I leave it to the reader to treat the case of a singularity at a manifold of codimension less than d.

Theorem 10.6.11. *Let the conditions of theorem 10.6.7 be fulfilled with $\mu \in (-1,0)$. Then the asymptotics*

$$\mathbf{N}(\tau,h) = \varkappa_0|\tau|^p h^{-d} + O\big(|\tau|^q h^{1-d}\big) \tag{10.6.45}$$

hold as $\tau \to -\infty$, $h \to +0$ with the same exponents p, q and coefficient $\varkappa_0$ as in theorem 10.6.7. Moreover, under the additional restrictions of theorem 10.6.7(ii) the asymptotics

$$\mathbf{N}(\tau,h) = \varkappa_0|\tau|^p h^{-d} + \big(\varkappa_1 + o(1)\big)|\tau|^q h^{1-d} \tag{10.6.46}$$

hold for $\tau \to -\infty$, $h = o(|\tau|^{\frac{1}{2\mu}(\mu+1)})$ with the coefficient $\varkappa_1$ given by (10.4.30).

Theorem 10.6.12. *Let X be a bounded domain in $\mathbb{R}^d$, $0 \in \bar{X}$ and let conditions $(\Sigma0)$–$(\Sigma2)$ be fulfilled with $\gamma = \epsilon|x|$, $\rho = |x|^{-1}\big(|\log|x|| + 1\big)^{\nu}$, $\nu \le -d^{-1}$. Then*

(i) For $\nu < -(d-1)^{-1}$ the asymptotics

$$\mathbf{N}(\tau,h) = \mathcal{N}^w(\tau)h^{-d} + O\big((\log|\tau|)^{\nu(d-1)+1}h^{1-d}\big) \tag{10.6.47}$$

hold for $\tau \to -\infty$, $h \to +0$.

(ii) For $\nu = -(d-1)^{-1}$ the asymptotics

$$\mathbf{N}(\tau,h) = \mathcal{N}^w(\tau)h^{-d} + O\big(\log(\log|\tau| \cdot h^{-\nu^{-1}})h^{1-d}\big) \tag{10.6.48}$$

hold for $\tau \to -\infty$, $h = o(\log^{\nu}|\tau|)$.

(iii) Moreover, let $X \cap \{|x| \le r_0\} = X^0 \cap \{|x| \le r_0\}$ with $r_0 > 0$ and a non-empty open cone X^0 with $\partial X^0 \setminus 0 \in C^K$. Let the functions g^{jk}, $V_j|\log|x||^{-\nu}$, $V|\log|x||^{-2\nu}$ satisfy stabilization conditions (10.4.3) for $|x| \to 0$ with $\mu = -1$ and with functions g^{jk0}, V_j^0, V^0 positively homogeneous of degrees 0, -1, -2 respectively. Finally, let condition (Ψ) be fulfilled with the Hamiltonian

$$H^0(x,\xi) = |x|^2\big(g^{jk0}(\xi_j - V_j^0)(\xi_k - V_k^0) + V^0\big) \tag{10.6.49}$$

at the energy level 0. Then in the context of (i) the asymptotics

$$\mathbf{N}(\tau,h) = \mathcal{N}^w(\tau)h^{-d} + \big(\varkappa_1 + o(1)\big)(\log|\tau|)^{\nu(d-1)+1}h^{1-d} \tag{10.6.50}$$

hold for $\tau \to -\infty$, $h = o(\log^{\nu}|\tau|)$ and in the context of (ii) the asymptotics

$$\mathbf{N}(\tau,h) = \mathcal{N}^w(\tau)h^{-d} + \big(\varkappa_1 + o(1)\big)\log(\log|\tau| \cdot h^{-\nu^{-1}})h^{1-d} \tag{10.6.51}$$

hold for $\tau \to -\infty$, $h = o(\log^{\nu}|\tau|)$.

(iv) For $\nu \in (-(d-1)^{-1}, -d^{-1})$ the asymptotics

$$\mathbf{N}(\tau,h) = \mathcal{N}^w(\tau)h^{-d} + O\big(h^{\nu^{-1}}\big) \tag{10.6.52}$$

hold for $\tau \to -\infty$, $h = o(\log^{\nu}|\tau|)$.

In all these asymptotics $\mathcal{N}^w(\tau) = O(\log^{\nu d+1}|\tau|)$ and $\mathcal{N}^w(\tau) \asymp \log^{\nu d+1}|\tau|$ under condition (10.6.40).

(v) Let $\nu = -d^{-1}$. Furthermore, suppose that **either** *$d \ge 3$* **or** *$0 \in \partial X$ and a Dirichlet condition is given on ∂X. Then the asymptotics*

$$\mathbf{N}(\tau,h) = \big(\mathcal{N}_h^w(\tau) + O(1)\big)h^{-d} \tag{10.6.53}$$

hold for $\tau \to -\infty$, $h = o(\log^{\nu}|\tau|)$ and $\mathcal{N}_h^w(\tau)$ is given by (10.6.25), $\mathcal{N}_h^w(\tau) = O\big(\log(\log|\tau| \cdot h^{\nu^{-1}})\big)$. Moreover, under condition (10.6.40) $\mathcal{N}_h^w(\tau) \asymp \log(\log|\tau| \cdot h^{\nu^{-1}})$.

Remark. *In the proof of the last assertion of this theorem we also use the arguments of remark 10.1.7(i).* ◊

Theorem 10.6.13. *Let $d \geq 3$, let X be a conical domain with $\partial X \setminus 0 \in C^K$, and let $g^{jk} = g^{jk}(\theta) \in C^K(\mathbb{S}^{d-1})$ satisfy condition ($\Sigma 3$). Let $V_j = 0$, $V = v(\theta)|\log r|$ with $v \in C^K(\mathbb{S}^{d-1})$ and $\min_{\bar{X}} v = v_0 < 0$. Then the asymptotics*

$$\mathbf{N}(\tau, h) = \mathcal{N}^w(\tau)h^{-d} + O\Bigl(h^{1-d}\tau \int_U \exp\bigl(-\tau(d-1)v^{-1}\bigr)d\theta\Bigr) \tag{10.6.54}$$

hold as $\tau \to -\infty$, $h \to +0$ with

$$\mathcal{N}^w(\tau) = a \int_U \exp(-\tau d v^{-1})\sqrt{g}d\theta \tag{10.6.55}$$

where $U = X \cap \mathbb{S}^{d-1}$ and the coefficient a is introduced in theorem 10.4.17. The orders of $\mathcal{N}^w(\tau)$ and the remainder estimate and more accurate results are discussed in assertion (ii) of this theorem.

Appendix H. Distributions of Spectra for Negative Order Operators

Let $A = A^w(x, D)$ be a classical pseudo-differential operator of order $m < 0$ on a compact closed manifold X. We are interested in the asymptotics of the eigenvalue counting function

$$\mathbf{N}^{\pm}(\tau) = \#\{\lambda > \tau,\ \pm\lambda \text{ is an eigenvalue of } A\} \tag{H.1}$$

and

$$\mathbf{N}^{\pm}_{\vartheta}(\tau) = \int (\lambda - \tau)^{\vartheta}_{+} d_\lambda \mathbf{N}^{\pm}(\lambda) \tag{H.2}$$

as $\tau \to +0$. First of all, the following theorem holds:

Theorem H.1. *(i) Let $a(x,\xi)$ be a microhyperbolic symbol at the energy level 0 (in a direction depending on the point). Then the asymptotics*

$$\mathbf{N}^{\pm}(\tau) = \varkappa^{\pm}\tau^{\frac{d}{m}} + O\bigl(\tau^{\frac{d-1}{m}}\bigr) \tag{H.3}$$

hold as $\tau \to +0$ with

$$\varkappa^{\pm} = (2\pi)^{-d} \int n^{\pm}(x,\xi;1)dxd\xi \tag{H.4}$$

where $n^{\pm}(x,\xi;\tau)$ is the number of eigenvalues of the symbol $\pm a(x,\xi)$ greater than τ and a is the principal symbol of A. Moreover, the asymptotics

$$\mathbf{N}^{\pm}_{\vartheta}(\tau) = \mathcal{N}^{\pm}_{\vartheta}(\tau) + O\bigl(R_\vartheta(\tau)\bigr) \tag{H.3$'$}$$

hold with

$$\mathcal{N}^{\pm}_{\vartheta}(\tau) = \sum_k \varkappa^{\pm}_{\vartheta k}\tau^{\frac{d-k}{m}} \tag{H.5}$$

and

$$R_\vartheta(\tau) = \begin{cases} \tau^{\frac{1}{m}(d-1-\vartheta)+\vartheta} & \text{for } \vartheta(1-m) < d-1 \\ |\log\tau| & \text{for } \vartheta(1-m) = d-1\,. \\ 1 & \text{for } \vartheta(1-m) > d-1 \end{cases} \tag{H.6}$$

(ii) Moreover, let $\vartheta(1-m) < d-1$ and let the standard condition ($\Psi 0$) be fulfilled for the Hamiltonian trajectories of the symbol $\pm a$ at the energy level 1. *Then the asymptotics*

$$\mathbf{N}^\pm(\tau) = \varkappa^\pm \tau^{\frac{d}{m}} + \bigl(\varkappa_1 + o(1)\bigr)\tau^{\frac{d-1}{m}} \tag{H.7}$$

and the asymptotics

$$\mathbf{N}^\pm_\vartheta(\tau) = \mathcal{N}^\pm_\vartheta(\tau) + o\bigl(R_\vartheta(\tau)\bigr) \tag{H.7$'$}$$

hold as $\tau \to +0$.

(iii) Finally, let $\vartheta(1-m) = d-1$ and the standard condition ($\Psi 0$) be fulfilled for the Hamiltonian trajectories of the symbol $\pm a$ at the energy level 0. *Then the asymptotics* (H.7)$'$ *hold.*

(iv) On the other hand, if the symbol $\pm a(x,\xi)$ is negative-definite for every (x,ξ) (which under the microhyperbolicity condition means precisely that $\varkappa^\pm = 0$) then

$$\mathbf{N}^\pm(\tau) = O(1). \tag{H.8}$$

I leave the proof to the reader and indicate only the dilatations which should be applied in this proof: namely, we apply the dilatation $(\epsilon, \epsilon r)$ with respect to (x,ξ) where $r = |\xi|$ and $h_{\text{eff}} = r + |\tau|^{\frac{1}{m}}$ afterwards.

Remark. *Asymptotics (H.3) and (H.3)$'$ are standard results under the condition*

$$|a(x,\xi)v| \ge \epsilon_0 |v|\cdot|\xi|^m \qquad \forall (x,\xi) \quad \forall v \in \mathbb{H} \tag{H.9}$$

and asymptotics (H.7) and (H.7)$'$ for $\vartheta(1-m) < d-1$ under this condition and condition ($\Psi 0$) at the energy level 1. ◊

Let us get rid of the microhyperbolicity condition. In the general case we can do nothing better than introduce the dilatations $(x_1, x', \xi_1, \xi') \to (\zeta^{-2}x_1, \zeta^{-1}x', r^{-1}\xi_1, \zeta^{-1}r^{-1}\xi')$ with $r \asymp |\xi|$ where the coordinate system is taken such that $\xi' = 0$ at some point of the element of the partition and $\zeta = \min_j(|\alpha_j| + \tau r^{-m} + r^{-\frac{1}{2}})$ for $\tau \le C r^m$ [50] and $\alpha_j = \alpha_j(x,\xi)$ are eigenvalues of $a(x,\xi|\xi|^{-1})$. Then $h_{\text{eff}} = r^{-1}\zeta^{-2}$ and the contribution of the zone $\{h_{\text{eff}} \le 1\}$ to the asymptotics does not exceed CR'_ϑ with

$$R'_\vartheta = \begin{cases} \tau^{l+\vartheta} & \text{for } \vartheta < -l \\ |\log\tau| & \text{for } \vartheta = -l \\ 1 & \text{for } \vartheta > -l, \end{cases} \qquad l = -\frac{2d}{1-2m}. \tag{H.10}$$

Then one can easily obtain the following

Theorem H.2. *(i) In the general case the following asymptotics hold:*

$$\mathbf{N}^\pm_\vartheta(\tau) = \mathcal{N}^\pm_\vartheta(\tau) + O\bigl(R_\vartheta(\tau) + R'_\vartheta(\tau)\bigr) \tag{H.11}$$

where $R_\vartheta, R'_\vartheta$ are given by (H.6), (H.10) *respectively.*

---

[50] Here and below we set $\zeta = 1$ for $\tau \ge r^m$.

(ii) Moreover, suppose $\vartheta < d(1-2m)^{-1} - 1 (\implies d > 1-2m \ \& \ \vartheta(1-m) < d-1)$ [51] *and let the standard condition* ($\Psi 0$) *be fulfilled for the Hamiltonian trajectories of the symbol* $\pm a$ *at the energy level* 1. *Then asymptotics* (H.7) *and* (H.7)$'$ *hold.*

(iii) On the other hand, if the symbol $\pm a(x,\xi)$ *is non-positive definite for every* $(x,\xi) (\Leftrightarrow \varkappa^{\pm} = 0)$ *then the estimate*

$$\mathbf{N}^{\pm}(\tau) = O(\tau^{l}) \tag{H.12}$$

holds.

Let us assume that

$$|D^{\alpha}_{(x,\xi)} a(x,\xi)| \le c\gamma^{\mu - |\alpha|} \qquad \forall \alpha : |\alpha| \le \mu \quad \forall (x,\xi) : |\xi| = 1, \tag{H.13}$$

$$|a(x,\xi)v| \ge \epsilon_0 \gamma^{\mu} |v| \quad \forall v \in \mathbb{H} \quad \forall (x,\xi) : |\xi| = 1, \tag{H.14}$$

where $\gamma(x,\xi) = \operatorname{dist}\big((x,\xi), \Lambda\big)$ and Λ is the union of a finite number of C^K manifolds of codimension $\ge p$. Then one can make the dilatations $(x,\xi) \to (\zeta^{-1}x, \zeta^{-1}r^{-1}\xi)$ with $\zeta = \epsilon(\gamma + \tau^{\frac{1}{\mu}} r^{-\frac{m}{\mu}} + r^{-\frac{1}{2}})$ and obtain the same results as before with $l = -(2d-p)(\mu - 2m)^{-1}$.

Let us now treat the scalar case. First of all, for $\vartheta > 0$ one can apply exactly the construction of the proof of theorem H.1 (no microhyperbolicity-type condition is necessary). For $\vartheta = 0$ one can apply the dilatations $(x,\xi) \to (\zeta^{-1}x, \zeta^{-1}r^{-1}\xi)$ with

$$\zeta = \epsilon\Big(\sum_{|\alpha| < M} |D^{\alpha} a|^{\frac{M}{M-|\alpha|}} + |\tau r^{-m}|\Big)^{\frac{1}{M}} + \epsilon r^{-\frac{1}{2}} \tag{H.15}$$

with $M = 2$. Then microhyperbolicity is fulfilled for $h_{\text{eff}} \le h_0$. After calculations one gets the remainder estimate $O(\tau^{\frac{d-1}{m}} + \tau^{l})$ with $l = -d(1-m)^{-1}$ which is good for $m < 1-d$. On the other hand, taking arbitrarily large M one can refer to condition $(8.1.18)_7$ and after easy calculations get the remainder estimate $O(\tau^{\frac{1}{m}(d-1+\delta)})$ with arbitrarily small $\delta > 0$. Therefore the following statement holds:

Theorem H.3. *Let* A *be a scalar operator. Then*

(i) For $\vartheta = 0$ *asymptotics* (H.3) *hold if* $m \le 1-d$ *and the asymptotics*

$$\mathbf{N}^{\pm}(\tau) = \varkappa^{\pm}\tau^{\frac{d}{m}} + O(\tau^{\frac{1}{m}(d-1+\delta)}) \tag{H.3$''$}$$

hold with arbitrarily small $\delta > 0$ *if* $m > 1-d$.

(ii) For $\vartheta > 0$ *asymptotics* (H.3)$'$ *hold.*

(iii) Suppose either $\vartheta > 0$ *or* $m \le 1-d$ *and let the standard condition* ($\Psi 0$) *be fulfilled for the Hamiltonian trajectories of the symbol* $\pm a$ *at the energy level* 1. *Then asymptotics* (H.7), (H.7)$'$ *hold.*

---

[51] I.e., $R_{\vartheta} \gg R'_{\vartheta}$.

11. Asymptotics of the Spectra of Schrödinger and Dirac Operators with Strong Magnetic Field

In this chapter we find eigenvalue asymptotics for two- and three-dimensional Schrödinger and Dirac operators in situations in which the role of the magnetic field is important. We saw in Chapters 6 and 7 that these four operators are essentially different. Moreover, it is convenient to separate the case of constant spectral parameter τ from the case in which τ tends to some specific limit τ^* in a given situation. Therefore this chapter is divided into sections in the following way: in sections 11.1 and 11.2, two- and three-dimensional (respectively) Schrödinger and Dirac operators are treated for fixed spectral parameter τ, and in sections 11.3–11.5, the same operators are treated for τ tending to a specific limit; the additional section appears because two-dimensional Schrödinger and Dirac operators are treated separately for the variable τ in section 11.3 and section 11.4 respectively.

The Schrödinger operator theory is more extensive than the Dirac operator theory: there are many more meaningful problems and questions for the Schrödinger operator than for the Dirac operator. Also, the two-dimensional theory is more extensive and provides more accurate remainder estimates than the three-dimensional theory. These circumstances are not due to technical difficulties but are instead due to the fact that the spectrum of the Schrödinger operator is discrete more often than the spectrum of the Dirac operator and that in dimension $d = 2$ the spectrum is discrete more often than in dimension $d = 3$.

Only the case $\vartheta = 0$ is treated in this chapter. Analysis of the case $\vartheta > 0$ is left to the reader.

Finally, in Appendix I the self-adjointness of the two-dimensional Dirac operator with a very singular magnetic field is proven and in Appendix J examples of operators satisfying the conditions of the statements of this chapter are provided in order to demonstrate that these statements are meaningful.

11.1 The Case $d = 2$ with Fixed Spectral Parameter

11.1.1 Getting started

We begin with an analysis of the Schrödinger operator

$$(\Sigma 0) \qquad A = P_j g^{jk} P_k + V, \qquad P_j = hD_j - \mu V_j$$

where we use the Einstein summation convention. Without loss of generality we can fix $\tau = 0$ and then in the important function $V_{\text{eff}} F_{\text{eff}}^{-1}$ the parameters μ and h enter as factors. Thus, we treat the operator $(\Sigma 0)(= (6.1.1))$ assuming that it is self-adjoint and that conditions $(8.1.53)_{1-3}$,[1] $(\Sigma 2)$ and $(8.1.2)_{1,2}$[2] are fulfilled and conditions $(8.1.2)_{3,5}$[3] are replaced by

[1] Inequalities in the terms of the weight functions γ, ρ, ρ_1, ϱ_2, ϱ_3 which I will review later.

[2] Inequalities on γ, ρ_l and ϱ_l.

[3] We no longer assume that $\rho_1 \geq \rho\gamma^{-1}$.

$$(11.1.1)_1 \qquad \rho\gamma \ge h \quad \forall x \in X'_0; \quad x \in X'_0, y \in B(x,\gamma(x)) \cap X \implies y \in X',$$

$$(11.1.1)_2 \qquad \rho_1\mu h \ge \rho \qquad \text{in } X'.$$

Recall that conditions $(\Sigma 0)$–$(\Sigma 3)$ were introduced in Chapter 10. Namely,

$$(\Sigma 2) \qquad \epsilon|\xi|^2 \le \sum_{j,k} g^{jk}\xi_j\xi_k \le c|\xi|^2 \qquad \forall \xi \in \mathbb{R}^d.$$

However, I now change condition $(\Sigma 1)_2$:

$$(\Sigma 1)_{1-3} \qquad |D^\alpha g^{jk}| \le c\gamma^{-|\alpha|}, \quad |D^\alpha F_{jk}| \le c\rho_1\gamma^{-|\alpha|}, \quad |D^\alpha V| \le c\rho^2\gamma^{-|\alpha|}.$$

Let us introduce the zones $X'_1 = \{x \in X', \rho_1\mu \le 2c\rho\gamma^{-1}\}$ and $X'_2 = \{x \in X', \rho_1\mu \ge c\rho\gamma^{-1}\}$ where the magnetic field is normal and where it is strong respectively. In what follows $h \to +0$ and X' tends to X. If μ is bounded then we can assume that X'_2 is fixed and the following conditions $(\Sigma 4)$, $(\Sigma 5)$ should be fulfilled only in X'_2; otherwise (if $\mu \to \infty$) X'_2 tends to X and we assume that this condition is fulfilled in X:

$$(\Sigma 4) \qquad |F| \ge \epsilon\rho_1 \qquad \text{in } X'_2$$

where F_{jk} and F are the tensor and scalar intensities of the magnetic field. Moreover, let us assume that

$$(11.1.2)_1 \qquad B(x,\gamma(x)) \subset X \qquad \forall x \in X'_{2-},$$

$$(11.1.2)_2 \qquad u|_{\partial X \cap B(x,\gamma(x))} = 0 \qquad \forall x \in X'_2 \quad \forall u \in \mathrm{D}(A)$$

where $X'_{2-} = \{x \in X'_2, V + \mu hF \le 2\epsilon\rho^2\}$ and $X'_{2+} = \{x \in X'_2, V + \mu hF \ge \epsilon\rho^2\}$ (see conditions $(8.1.52)_{1,2}$). Finally, let us assume that the following conditions are fulfilled in X'_{2-}:

$$(\Sigma 5)_1 \qquad |D^\alpha g^{jk}| \le c\varrho_2\gamma^{-|\alpha|}$$

$$(\Sigma 5)_2 \qquad |D^\alpha F_{jk}| \le c\varrho_2\rho_1\gamma^{-|\alpha|}$$

$$(\Sigma 5)_3 \qquad |D^\alpha \frac{V}{F}| \le c\varrho_3\rho^2\rho_1^{-1}\gamma^{-|\alpha|} \qquad \forall \alpha : 1 \le |\alpha| \le K;$$

recall that $\varrho_3 \le \varrho_2 \le 1$. I have now rewritten condition $(8.1.53)_{1-3}$ completely and the reader does not need to look at the original. Surely, if $\mu h \to 0$ then one should replace X'_{2-} by $X'_2 \cap X_-$ with $X_- = \{V \le 2\epsilon\rho^2\}$, but for μh disjoint from 0 the zone X'_{2-} is smaller. Finally, we should consider the singular zone $X'' = \{\rho\gamma \le 2h, \mu\rho_1\gamma^2 \le 2h\}$. In order to avoid this task we assume that

$$(11.1.3) \qquad \rho_1\gamma^2 + \rho\gamma \ge \epsilon.$$

Moreover, let the standard condition on the boundary be fulfilled:

(11.1.4) For every $y \in X$, $\partial X \cap B(y,\gamma(y)) = \{x_k = \phi_k(x_{\hat{k}})\}$ with

$$|D^\alpha \phi_k| \le c\gamma^{-|\alpha|}$$

and $k = k(y)$.

Then we obtain the following assertion from theorem 8.1.4:

Theorem 11.1.1. *Let $d = 2$ and let the Schrödinger operator A satisfy conditions (11.1.4), ($\Sigma1$), ($\Sigma2$), ($\Sigma4$) and ($\Sigma5$) in the corresponding regions where ρ, ρ_1, γ satisfy $(8.1.2)_{1-3}$, (11.1.2), (11.1.3) and $\varrho_2 = \varrho_3 = 1$.*[4] *Moreover, in X_2' let the condition*

$$(11.1.5)\quad \bigl|V + (2n+1)\mu h F\bigr| \le \epsilon\rho^2, \quad n \in \mathbb{Z}^+ \implies |\nabla v| \ge \epsilon\rho^2\rho_1^{-1}\gamma^{-1} \quad \text{or} \quad \det \operatorname{Hess} v \ge \epsilon\rho^4\rho_1^{-2}\gamma^{-4}$$

be fulfilled for $v = VF^{-1}$.[5] *Let $\rho_1 \ge \epsilon\rho\gamma^{-1}$ and*

$$(11.1.6)\qquad \rho^2\rho_1^{-1}\gamma^{-2} \in L^1\bigl(X \cap \{V + tF \le \epsilon\rho^2\}\bigr),$$

$$(11.1.7)\qquad \rho_1^{-s}\gamma^{-2s-d} \in L^1(X)$$

with some $s \ge 0$. Then for $h \to +0$, $1 \le \mu$ such that $\mu h \ge t$ the "standard" asymptotics

$$(11.1.8)\qquad \mathbf{N}^-(\mu, h) = \mathcal{N}^-(\mu, h) + O(\mu^{-1}h^{-1})$$

hold with

$$(11.1.9)\qquad \mathcal{N}^-(\mu, h) = h^{-d}\int S(x, \mu h)\,dx$$

and $S(x, \mu h)$ given by $(6.1.7)_{(2)}$.

Example 11.1.2. *(i) Let the conditions of theorem 11.1.1 be fulfilled in a bounded domain X, $0 \in \bar{X}$ with $\gamma = \epsilon_0|x|$, $\rho = |x|^m$, $\rho_1 = |x|^{m_1}$ and let $m_1 < \min(2m, -2)$. Then conditions (11.1.6), (11.1.7) are fulfilled automatically and asymptotics (11.1.8)–(11.1.9) hold for the indicated μ and h. Moreover,*

$$(11.1.10)_1\qquad \mathcal{N}^-(\mu, h) = O(h^{-2})$$

for $m > -1$,

$$(11.1.10)_2\qquad \mathcal{N}^-(\mu, h) = O\bigl(h^{-2}(|\log \mu h| + 1)\bigr)$$

for $m = -1$,

$$(11.1.10)_3\qquad \mathcal{N}^-(\mu, h) = O\bigl(h^{-2}(\mu h)^{(2m+2)(2m-m_1)^{-1}}\bigr)$$

for $m < -1$. Furthermore, if $\mu h \le t$ and

$$(11.1.11)\qquad \{X \setminus 0, V \le -tF - \epsilon\} \ne \emptyset$$

then

$$(11.1.12)_1\qquad \mathcal{N}^-(\mu, h) \asymp h^{-2}$$

for $m > -1$; on the other hand, if

$$(11.1.13)\qquad V \le -\epsilon\rho^2 \quad \text{in } \Gamma \cap \{|x| \le \epsilon\} \subset X$$

where Γ is an open non-empty sector (cone) with vertex at 0 then for $\mu h \le t$ with a small enough constant $t > 0$

---

[4] The case when $\varrho_3 < 1$ will be treated later.

[5] I impose this condition in order to avoid more complicated asymptotics.

$$\mathcal{N}^-(\mu,h) \asymp h^{-2}\bigl(|\log \mu h| + 1\bigr) \tag{11.1.12$_2$}$$

for $m = -1$ *and*

$$\mathcal{N}^-(\mu,h) \asymp h^{-2}(\mu h)^{(2m+2)(2m-m_1)^{-1}} \tag{11.1.12$_3$}$$

for $m < -1$.

(ii) Let the conditions of theorem 11.1.1 be fulfilled in the unbounded connected domain X *with* $\gamma = \epsilon_0\langle x\rangle$, $\rho = \langle x\rangle^m$, $\rho_1 = \langle x\rangle^{m_1}$ *and let* $m_1 > \max(2m,-2)$. *Then conditions (11.1.6) and (11.1.7) are fulfilled automatically and asymptotics (11.1.8)–(11.1.9) hold. Moreover, for* $m < -1$, $m = -1$, $m > -1$ *estimates* (11.1.10)$_1$, (11.1.10)$_2$ *and* (11.1.10)$_3$ *hold. If* $m < -1$ *and condition (11.1.11) is fulfilled then equivalence* (11.1.12)$_1$ *holds. On the other hand, if*

$$V \le -\epsilon\rho^2 \qquad \text{in } \varGamma \cap \{|x| \ge R\} \subset X \tag{11.1.13$^\#$}$$

where $\varGamma$ *is an open non-empty sector (cone) with vertex at* 0, *then for* $\mu h \le t$ *with a small enough constant* $t > 0$ *equivalences* (11.1.12)$_2$ *and* (11.1.12)$_3$ *hold for* $m = -1$, $m > -1$ *respectively.*

(iii) One can easily see that for $m > -1$ *in (i),* $m < -1$ *in (ii)*

$$\mathcal{N}^-(\mu,h) = \mathcal{N}^{w-}(\mu,h) + O\bigl(h^{-2}(\mu h)^{(2m+2)(2m-m_1)^{-1}}\bigr)$$

under condition (11.1.5) where $\mathcal{N}^{w-}$ *is the Weylian expression. Therefore for* $\mu \le h^p$ *with* $p = (2-m_1)(4m-m_1+2)^{-1}$ *the asymptotics remain true with* $\mathcal{N}^-$ *replaced by* $\mathcal{N}^{w-}$. *On the other hand, under condition (11.1.13) or* (11.1.13)$^\#$ *(respectively)* $\mathcal{N}^{w-} - \mathcal{N}^- \ge \epsilon h^{-2}(\mu h)^{(2m+2)(2m-m_1)^{-1}}$. *Thus, for* $\mu > h^p$ *one cannot replace* $\mathcal{N}^-$ *by* $\mathcal{N}^{w-}$ *and preserve the remainder estimates.* p *is not necessarily negative in our conditions! Therefore, due to the singularity, the magnetic field can be essential even for a fixed* μ. *In what follows I leave this type of analysis to the reader.* ◊

Example 11.1.3. *(i) Let the conditions of theorem 11.1.1 be fulfilled with a bounded domain* X, $0 \in \bar{X}$, $\gamma = \epsilon_0|x|^{1-\beta}$, $\rho = \exp a|x|^\alpha$, $\rho_1 = \exp b|x|^\beta$ *with either* $0 > a > \beta$, $a > 0$, $b > 0$ *or* $0 > \alpha = \beta$, $b > 2a > 0$. *Then conditions (11.1.6) and (11.1.7) are fulfilled automatically and for* $\mu \ge 1$, $h \to +0$ *asymptotics (11.1.8)–(11.1.9) hold. Moreover, under condition (11.1.13) for* $\alpha = \beta$, $\mu h < t$ *with a small enough constant* $t > 0$

$$\mathcal{N}^-(\mu,h) \asymp h^{-2}(\mu h)^{-2a(b-2a)^{-1}}|\log \mu h|^{\frac{2}{\alpha}-1}; \tag{11.1.14}$$

otherwise the left-hand expression is "O" only. I leave the calculations in the case $\alpha > \beta$ *to the reader.*

(ii) Let the conditions of theorem 11.1.1 be fulfilled with an unbounded connected domain X, $\gamma = \epsilon_0\langle x\rangle^{1-\beta}$, $\rho = \exp a\langle x\rangle^\alpha$, $\rho_1 = \exp b\langle x\rangle^\beta$ *where either* $0 < \alpha < \beta$, $a > 0$, $b > 0$ *or* $0 < \alpha = \beta$, $b > 2a > 0$. *Then conditions (11.1.6)–(11.1.7) are fulfilled automatically and asymptotics (11.1.8)–(11.1.9) hold. Moreover, under condition* (11.1.12)$^\#$ *for* $\alpha = \beta$, $\mu h < t$ *with a small enough constant* $t > 0$ *(11.1.14) holds; otherwise the left-hand expression is "O" only. I leave the calculations in the case* $\alpha < \beta$ *to the reader.* ◊

Example 11.1.4. *Let the conditions of theorem 11.1.1 be fulfilled with the bounded domain* X, $0 \in \bar{X}$, $\gamma = \epsilon_0|x_1|$, $\rho = |x_1|^m$, $\rho_1 = |x_1|^{m_1}$ *and* $m_1 < \min(2m-1,-2)$. *Then conditions (11.1.6)–(11.1.76) are fulfilled automatically and asymptotics (11.1.8)–(11.1.9) hold. Moreover, let* $V \le -\epsilon\rho^2$ *in the vicinity of* 0. *Then*

$$\mathcal{N}^-(\mu,h) \asymp h^{-2} \qquad \text{for } m > -\frac{1}{2}, \tag{11.1.15$_1$}$$

$$\mathcal{N}^-(\mu,h) \asymp h^{-2}\bigl(|\log \mu h| + 1\bigr) \qquad \text{for } m = -\frac{1}{2}, \tag{11.1.15$_2$}$$

$$\mathcal{N}^-(\mu,h) \asymp h^{-2}(\mu h)^{(2m+1)(2m-m_1)^{-1}} \qquad \text{for } m < -\frac{1}{2}; \tag{11.1.15$_3$}$$

otherwise the left-hand expression is "O" only. I leave the treatment of the case $\mu \le 1$ to the reader. ◊

Let us consider the case when $\varrho_3 < 1$, and let us modify theorem 11.1.1 in this case:

Theorem 11.1.1′. *Let $d = 2$ and let the Schrödinger operator A satisfy conditions (11.1.4), (Σ1), (Σ2), (Σ4) and (Σ5) in the corresponding regions where $\rho, \rho_1, \gamma, \varrho_3$ satisfy (8.1.2)$_{1-3}$, (11.1.2), (11.1.3) and $\varrho_2 = 1$. Moreover, in X_2' let condition (11.1.5) be fulfilled, $\rho_1 \ge \epsilon\rho\gamma^{-1}$ and in $X_2' \cap \{\varrho_3 \le \epsilon, V + tF \le \epsilon\rho^2\}$*

$$\varrho_3\mu^2\rho_1^2\gamma^2 \ge C\rho^2, \tag{11.1.16}$$

$$\bigl|V + (2n+1)\mu hF\bigr| \le \epsilon_0\varrho_3\rho^2, \quad n \in \mathbb{Z}^+ \implies |\nabla v| \ge \epsilon_0\varrho_3\rho^2\rho_1^{-1}\gamma^{-1}. \tag{11.1.17}$$

Let us assume that condition (11.1.7) is fulfilled and that

$$\varrho_3^{-1}\rho^2\rho_1^{-1}\gamma^{-2} \in L^1\bigl(X \cap \{V + tF \le \epsilon\varrho_3\rho^2\}\bigr) \tag{11.1.18}$$

for some $t \ge 0$. Then for $h \to +0$, $1 \le \mu$ such that $\mu h \ge t$, asymptotics (11.1.8)–(11.1.9) hold.

Example 11.1.5. *(i) Let the conditions of theorem 11.1.1′ be fulfilled with a bounded domain X, $0 \in \bar{X}$, $\gamma = \epsilon_0|x|$, $\rho = |x|^m\bigl(\bigl|\ln|x|\bigr| + 1\bigr)^\alpha$ and with $\rho_1 = |x|^{2m}\bigl(\bigl|\ln|x|\bigr| + 1\bigr)^\beta$, $\varrho_2 = 1$, $\varrho_3 = \bigl(\bigl|\ln|x|\bigr| + 1\bigr)^{-1}$. Let $m \le -1$, $\beta > 2\alpha$ and suppose either $m < -1$ or $\beta > 0$. Then condition (11.1.7) is fulfilled automatically. Moreover, for $t > 0$ conditions (11.1.16), (11.1.18) are fulfilled and hence asymptotics (11.1.8)–(11.1.9) hold for $\epsilon < \mu h \le c$. On the other hand, condition (11.1.18) is fulfilled for $\beta > 2\alpha + 2$ (in which case condition (11.1.16) is also fulfilled) and hence asymptotics (11.1.8)–(11.1.9) hold for $\mu h \le c$. It is easy to calculate that under condition (11.1.13) for $\mu h \le t$ with small enough $t > 0$*

$$\mathcal{N}^-(\mu,h) \asymp h^{-2}\bigl|\log(\mu h)\bigr| \qquad \text{for } m = -1, \alpha = -\frac{1}{2}, \tag{11.1.19$_1$}$$

$$\mathcal{N}^-(\mu,h) \asymp h^{-2}(\mu h)^{2\alpha+1)(\beta-2\alpha)^{-1}} \qquad \text{for } m = -1, \alpha > -\frac{1}{2}, \tag{11.1.19$_2$}$$

and

$$\epsilon h^{-2}\exp\bigl(\epsilon(\mu h)^{(2\alpha-\beta)^{-1}}\bigr) \le \mathcal{N}^-(\mu,h) \le C_1 h^{-2}\exp\bigl(C_1(\mu h)^{(2\alpha-\beta)^{-1}}\bigr) \tag{11.1.19$_3$}$$

for $m < -1$ where $C_1 > \epsilon > 0$ are constants. On the other hand, for $m = -1$, $\alpha < -\frac{1}{2}$ the equivalence $\mathcal{N}^-(\mu,h) \asymp h^{-2}$ holds.

(ii) Let the conditions of theorem 11.1.1′ be fulfilled with the unbounded connected domain X, $\gamma = \epsilon_0\langle x\rangle$, $\rho = \langle x\rangle^m\bigl(\log\langle x\rangle + 1\bigr)^\alpha$, $\rho_1 = \langle x\rangle^{2m}\bigl(\log\langle x\rangle + 1\bigr)^\beta$, $\varrho_2 = 1$, $\varrho_3 = \bigl(\log\langle x\rangle + 1\bigr)^{-1}$ and let $m \ge -1$, $\beta > 2\alpha$ and suppose either $m > -1$ or $\beta > 0$. Then all the statements of (i) remain true with the obvious modification: condition (11.1.13) should be replaced by (11.1.13)$^\#$ and asymptotics (11.1.19)$_3$ hold for $m < -1$.

Example 11.1.6. *(i) Let the conditions of theorem 11.1.1′ be fulfilled with a bounded domain X, $0 \in \bar{X}$, $\gamma = \epsilon_0|x|^{1-\beta}$, $\rho = |x|^m \exp|x|^\beta$, $\rho_1 = |x|^{m_1}\exp 2|x|^\beta$, $\varrho_2 = 1$, $\varrho_3 = |x|^{-\beta}$ and let $\beta < 0$, $m_1 < 2m$. Then conditions (11.1.7), (11.1.16) are fulfilled automatically. Moreover, for $t > 0$ condition (11.1.18) is fulfilled and asymptotics (11.1.8)–(11.1.9) hold. On the other hand, for $t = 0$ condition (11.1.18) is fulfilled for $m_1 < 2m + 3\beta$. So, for $m_1 < 2m + 3\beta$ asymptotics (11.1.8)–(11.1.9) hold and under condition (11.1.13)*

$$\epsilon h^{-2}\exp\bigl(\epsilon(\mu h)^{\beta(2m-m_1)^{-1}}\bigr) \le \mathcal{N}^-(\mu,h) \le C_1h^{-2}\exp\bigl(C_1(\mu h)^{\beta(2m-m_1)^{-1}}\bigr). \tag{11.1.20}$$

(ii) Let the conditions of theorem 11.1.1′ be fulfilled with an unbounded connected domain X, $\gamma = \epsilon_0\langle x\rangle^{1-\beta}$, $\rho = \langle x\rangle^m\exp\langle x\rangle^\beta$, $\rho_1 = \langle x\rangle^{m_1}\exp 2\langle x\rangle^\beta$, $\varrho_2 = 1$, $\varrho_3 = \langle x\rangle^{-\beta}$ where $\beta > 0$, $m_1 > 2m$. Then all the statements of (i) remain true with reversed inequalities for m_1, m, β and with condition (11.1.13) replaced by (11.1.13)$^\#$. ◊

11.1.2 Some improvements

Let us treat these two examples more carefully.

Theorem 11.1.7. *Let all the conditions of example 11.1.5(i) be fulfilled. Moreover, let us assume that either*

(a) $\beta > 2\alpha + 1$

or

(b) $\beta > 2\alpha$ and

$$-\langle x,\nabla\rangle v \ge \epsilon\varrho_3\rho^2\rho_1^{-1} \qquad \text{for } |x| \le \epsilon; \tag{11.1.21}$$

recall that $v = VF^{-1}$.

Moreover, let us assume that

(11.1.22) *Either $m < -1$ or $\beta > \frac{1}{3}(4\alpha+2)_+$.*

Then for $h \to +0$, $1 \le \mu = O(h^{-1})$ the asymptotics

$$\mathbf{N}^-(\mu,h) = \mathcal{N}^{-\prime}(\mu,h) + O(\mu^{-1}h^{-1}) \tag{11.1.8$'$}$$

hold where

$$\mathcal{N}^{-\prime}(\mu,h) = \int h^{-2}S'(x,\mu h)\,dx \tag{11.1.9$'$}$$

and S' is given by (6.4.89) *with $\psi = 1$. Moreover, condition* (11.1.13) *provides the equivalences* $(11.1.19)_{1-3}$ *for $\mathcal{N}^{-\prime}$ for corresponding m, α (see example 11.1.5(i)).*

(ii) Let all the conditions of example 11.1.5(ii) be fulfilled. Moreover, let us assume that either (a) holds or

(b′) $\beta > 2\alpha$ and

$$\langle x,\nabla\rangle v \ge \epsilon\varrho_3\rho^2\rho_1^{-1} \qquad \text{for } |x| \ge R. \tag{11.1.21$^\#$}$$

Moreover, let us assume that

(11.1.22)$^\#$ *Either $m > -1$ or $\beta > \frac{1}{3}(4\alpha+2)_+$.*

Then for $h \to +0$, $1 \le \mu = O(h^{-1})$ asymptotics (11.1.8)′–(11.1.9)′ *hold. Moreover, condition* (11.1.13)$^\#$ *provides equivalences* $(11.1.19)_{1-3}$ *for $\mathcal{N}^{-\prime}$ for the corresponding m, α (see example 11.1.5(ii)).*

Proof. Consider case (i); case (ii) is treated in the same way. Let us apply theorem 8.1.4 (see also theorem 6.4.18 which describes the remainder estimate in local semiclassical spectral asymptotics). Surely, condition (11.1.5) should be checked for $v^* = v + \mu^{-2}\omega_1$ instead of v; however, one can easily see that if either $m < -1$ or $\beta > (\alpha + \frac{1}{2})_+$ then one can check this condition for v. Furthermore, one can easily see that under condition (11.1.22) the contributions of the zones $\{\mu h \rho_1 \ge C_0\rho^2\}$ and $\{c_1\varrho_3\rho^2 \ge \mu h\rho_1\}$ to the remainder estimate are $O(\mu^{-1}h^{-1})$; here c_1 is an arbitrarily large constant. Moreover, all the contributions of the zone $\{c_1\varrho_3\rho^2 \le \mu h\rho_1 \le C_0\rho^2\}$ to the remainder estimate are $O(\mu^{-1}h^{-1})$ as well, excluding perhaps $C\gamma^{-2}$ integrated over this zone. Moreover, under condition (a) this term is also $O(\mu^{-1}h^{-1})$; thus, in case (a) the theorem is proven.

In the case (b) a more refined analysis is necessary. Let us note that in this case the zone $\{c_1\varrho_3\rho^2 \le \mu h\rho_1 \le C_0\rho^2\}$ consists of the "spectral strips"

$$\Pi_n = \{x, \left|v + (2n+1)\mu h\right| \le C_1\varrho_3\}$$

with $n = 0, \dots, n_0 = C_1(\mu h)^{-1(\beta - 2\alpha + 1)^{-1}}$ separated by the "lacunary strips." Condition (11.1.21) provides that for fixed n and $x|x|^{-1}$ (thus, only $|x|$ varies) on Π_n the inequality $\left|\mathrm{d}\log|x|\right| \le C_1$ holds where $\mathrm{d}f$ means the oscillation of f (on Π_n) and the constant C_1 does not depend on $n, x|x|^{-1}, \mu, h$. Therefore one can easily see that the contribution of every strip Π_n to the remainder estimate does not exceed C and their total contribution does not exceed $Cn_0 = O(\mu^{-1}h^{-1})$. The estimate of the contribution of the lacunary strips is obvious. □

The same arguments give us

Theorem 11.1.8. *(i) Let all the conditions of example 11.1.6(i) be fulfilled excluding condition $m_1 < 2m + 3\beta$. Let us assume that either*

(a) $m_1 < 2m + 2\beta$

or

(b) $m_1 < 2m + \beta$ and

(11.1.23) $$-\langle x, \nabla\rangle v \ge \epsilon|x|^{2m-m_1} \qquad \text{for } |x| \le \epsilon.$$

Then for $h \to +0$, $1 \le \mu = O(h^{-1})$ asymptotics (11.1.8)′–(11.1.9)′ hold. Moreover, under condition (11.1.13) estimates (11.1.20) hold for $\mathcal{N}^{-\prime}$.

(ii) Let all the conditions of example 11.1.6(ii) be fulfilled excluding condition $m_1 > 2m + 3\beta$. Let us assume that either

(a) $m_1 > 2m + 2\beta$

or

(b) $m_1 > 2m + \beta$ and

(11.1.23)# $$\langle x, \nabla\rangle v \ge \epsilon|x|^{2m-m_1} \qquad \text{for } |x| \ge R.$$

Then for $h \to +0$, $1 \le \mu = O(h^{-1})$ asymptotics (11.1.8)′–(11.1.9)′ hold. Moreover, under condition (11.1.12) estimates (11.1.20) hold for $\mathcal{N}^{-\prime}$.

Remark 11.1.9. *(i) The idea that the "spectral zone" is very narrow under certain conditions for $\varrho_2 < 1$ is also useful in other situations.*

(ii) The reader can prove some weaker remainder estimates here and in the examples above under weaker conditions. Moreover, under conditions of the type $\mu \ge \mu^*(h)$ these remainder estimates can be as good as $O(\mu^{-1}h^{-1})$. ◊

With certain restrictions one can replace $\mathcal{N}^{-\prime}$ by $\mathcal{N}^-$:

Corollary 11.1.10. *(i) In the context of theorem 11.1.8(i) let both conditions (a),(b) be fulfilled. Moreover, let*

$$|D^\sigma g^{jk}| \le c|x|^{-|\sigma|}, \qquad |D^\sigma \log F| \le c|x|^{\beta-|\sigma|} \qquad \forall \sigma : 1 \le |\sigma| \le 2. \tag{11.1.24}$$

Then for $h \to +0$, $1 \le \mu = O(h^{-1})$ asymptotics (11.1.8)–(11.1.9) *hold. Moreover, under condition* (11.1.13) *estimates* (11.1.20) *hold for $\mathcal{N}^-(\mu,h)$.*

(ii) In the context of theorem 11.1.8(ii) let both conditions (a), (b) be fulfilled. Let

$$|D^\sigma g^{jk}| \le c\langle x\rangle^{-|\sigma|}, \qquad |D^\sigma \log F| \le c\langle x\rangle^{\beta-|\sigma|} \qquad \forall \sigma : 1 \le |\sigma| \le 2. \tag{11.1.24$^\#$}$$

Then for $h \to +0$, $1 \le \mu = O(h^{-1})$ asymptotics (11.1.8)–(11.1.9) *hold. Moreover, under condition* (11.1.13)$^\#$ *estimates* (11.1.20) *hold for $\mathcal{N}^-$.*

Proof. Let us prove (i); (ii) can be proven in the same way. One can easily see that under condition (11.1.24)

$$|\omega_1| \le C_1 F^{-1}|x|^{\beta+4m-2m_1-2} = \omega_1^* \tag{11.1.25}$$

while the general theory yields only that

$$|\omega_1| \le C_1 F^{-1}|x|^{2\beta+4m-2m_1-2}.$$

Then the error $|\mathcal{N}^-(\mu,h) - \mathcal{N}^{-\prime}(\mu,h)|$ does not exceed

$$\begin{aligned} &C\mu^{-1}h^{-1}\int_{\{\rho_1\mu h \le \varrho_3\rho^2\}} \rho_1^2\rho^{-2}\varrho_3^{-1}\omega_1^*\,dx + \\ &Ch^{-2}\int_{\{\varrho_3\rho^2 \le \rho_1\mu h \le C_0\rho^2\}} \rho_1(\omega_1^*\mu^{-2} + \chi_\Pi)\,dx \end{aligned}$$

where χ_Π is the characteristic function of the set

$$\Pi = \bigcup_{n\in\mathbb{Z}^+} \{x, |v + (2n+1)\mu h| \le C_1\mu^{-2}\omega_1^*\}.$$

Applying estimate (11.1.25) and condition (11.1.23) one can prove that this error is $O(\mu^{-1}h^{-1})$. □

I leave it to the reader to formulate and prove a similar assertion for theorem 11.1.7.

11.1.3 Quickly growing μ

Let us consider the case when $\mu h \to +\infty$. Then the domain $\{V + \mu h F < 0\}$ shrinks (provided $F > 0$ in X) and in order to obtain $\mathcal{N}^-(\mu,h) \to +\infty$ one needs to assume that h tends to 0 quickly enough (i.e., μ grows slowly enough).

Theorem 11.1.11. *(i) Let X be a bounded domain, $0 \in \bar{X}$ and let conditions $(\Sigma 1)$, $(\Sigma 2)$, $(\Sigma 4)$ and $(\Sigma 5)$ be fulfilled in the corresponding regions with $\gamma = \epsilon_0|x|$, $\rho = |x|^m$, $\rho_1 = |x|^{m_1}$, $m_1 > 2m > -2$. In a neighborhood of 0 let*

$$|\nabla v| \ge \epsilon\rho^2\rho_1^{-1}\gamma^{-1}. \tag{11.1.26}$$

Then for $h \to +0$, $\mu h \to +\infty$, $\mu = o\bigl(h^{(m-m_1-1)(m+1)^{-1}}\bigr)$ the asymptotics

$$\mathbf{N}^-(\mu,h) = \mathcal{N}^-(\mu,h) + O\bigl(h^{-1}\mu^{-(m+1)(m_1+1-m)^{-1}}\bigr) \tag{11.1.27}$$

hold. Moreover, under condition (11.1.13)

$$\mathcal{N}^-(\mu, h) \asymp h^{-2}(\mu h)^{(2m+2)(2m-m_1)^{-1}}. \tag{11.1.28}$$

(ii) Let X be an unbounded connected domain and let conditions $(\Sigma 1)$, $(\Sigma 2)$, $(\Sigma 4)$ and $(\Sigma 5)$ be fulfilled in the corresponding regions with $\gamma = \epsilon_0\langle x\rangle$, $\rho = \langle x\rangle^m$, $\rho_1 = \langle x\rangle^{m_1}$, $m_1 < 2m < -2$. For $|x| \ge R$ let condition (11.1.26) *be fulfilled. Then for $h \to +0$, $\mu h \to +\infty$, $\mu = o\bigl(h^{(m-m_1-1)(m+1)^{-1}}\bigr)$ asymptotics* (11.1.27) *hold. Moreover, under condition* $(11.1.13)^{\#}$ *equivalence* (11.1.28) *holds.*

Proof. The proof follows easily from theorem 8.1.4. One should note that by means of an appropriate gradient transform one can obtain

$$|V_j| \le c\rho_1\gamma \tag{11.1.29}$$

for $|x| \le \epsilon$ and $|x| \ge c$ in (i), (ii) respectively; it is important that $m_1 > -2$ and $m_1 < -2$ in cases (i), (ii) respectively. Then in the zone $X'' = \{\rho_1\gamma^2 \le \mu^{-1}h, \gamma \le h\}$ one can take $W = (\rho_1\gamma^2\mu^2 + \rho^2)h^{-2}$, and for appropriate $p \in (1, \infty)$ the contribution of this zone to the remainder estimate is $O(R^*)$ with $R^* = h^{-2}r^{m+2} + h^{-2}\mu^2 r^{m+4}$ where $r = \max_{X''}|x|$, $r = \min_{X''}|x|$ in (i), (ii) respectively.

Moreover, there is a zone $X_2' = \{\rho\gamma^2 \ge h\mu^{-1}, \mu\rho_1\gamma \ge \rho\}$ of strong magnetic field; the contribution of the ball $B(x, \gamma(x)) \subset X_2'$ to the remainder estimate is $\rho^2(\rho_1\mu h)^{-1}$ if $\mu h\rho_1 \le C\rho^2$ and $\rho^2\gamma^2\bigl(\mu h\rho_1\gamma^2\rho^{-2}\bigr)^{-s}$ otherwise. Moreover, there is a zone $X_1' = X'\setminus X_2' = \{\rho\gamma \ge h, \mu\rho_1\gamma \le \rho\}$ of normal magnetic field; the contribution of the ball $B(x, \gamma(x)) \subset X_1'$ to the remainder estimate is $O(h^{-1}\rho\gamma)$. The total contribution of these zones to the remainder estimate is also $O(R^*)$. □

I leave it to the reader to prove similar assertions for a bounded domain X, $0 \in \bar{X}$ in the following cases:

(a) $\rho = |x|^{-1}\bigl(|\log|x|| + 1\bigr)^\alpha$, $\rho_1 = |x|^{m_1}$ with $m_1 > -2$, $\alpha < -\frac{1}{2}$;

(b) $\rho = |x|^m\bigl(|\log|x|| + 1\bigr)^\alpha$, $\rho_1 = |x|^{2m}\bigl(|\log|x|| + 1\bigr)^\beta$ with $m \ge -1$ and $\beta < 2\alpha$ provided either $m > -1$ or $\alpha < -\frac{1}{2}$;

in both cases $\gamma = \epsilon_0|x|$.

I leave the similar cases (a′),(b′) for an unbounded domain where $m_1 < -2$ in (a′) and $m \le -1$ in (b′) and either $m < -1$ or $\alpha < -\frac{1}{2}$ to the reader. I also leave it to the reader to treat the case

(c) $\gamma = \epsilon_0|x_1|$, $\rho = |x|^m$, $\rho_1 = |x_1|^{m_1}$ with $m_1 > 2m > -1$ in a neighborhood of 0. Moreover, one can treat the case $m_1 = 2m$ under the additional conditions $(\Sigma 5)_3$ with $\varrho_3 = \bigl(|\log|x|| + 1\bigr)^{-1}$ and (11.1.21).

11.1.4 The case $\mu \to 0$

To conclude the analysis of the Schrödinger operator we consider the case when $h \to +0$, $\mu = O(1)$.[6] Let us first consider the case when the conditions of theorem 11.1.11(i) are fulfilled with $m_1 < \min(2m, -2)$.[7] Let us apply theorem 8.1.4. In this case $X_2' = \{X, |x| \le \mu^{(m-m_1-1)^{-1}}\}$ and $X_1' = X \setminus X_2'$. Then assertion (i) of the theorem below holds; assertion (ii) is proven in the same way:

---

[6] A new issue is that μ is not necessarily disjoint from 0, so the really new case is when $\mu \to +0$.

[7] Surely now condition $(\Sigma 4)$ should be fulfilled only for $|x| \le \epsilon$ in (i) or for $|x| \ge C$ in (ii).

Theorem 11.1.12. *Let the conditions of theorem 11.1.11(i) be fulfilled with* $m_1 < \min(2m, -2)$. *Then for* $h \to +0$, $\mu = O(1)$ *the following asymptotics hold:*

$$\mathbf{N}^-(\mu, h) = \mathcal{N}^-(\mu, h) + O(h^{-1}) \tag{11.1.30$_1$}$$

for $m > -1$,

$$\mathbf{N}^-(\mu, h) = \mathcal{N}^-(\mu, h) + O(h^{-1}\log\mu) \tag{11.1.30$_2$}$$

for $m = -1$ *and*

$$\mathbf{N}^-(\mu, h) = \mathcal{N}^-(\mu, h) + O\big(h^{-1}\mu^{(m+1)(m-m_1-1)^{-1}}\big) \tag{11.1.30$_3$}$$

for $m < -1$.

Moreover, under condition (11.1.11) *for* $m > -1$ *equivalence* $(11.1.12)_1$ *holds and under condition* (11.1.13) *for* $m \le -1$ *equivalences* $(11.1.12)_{2,3}$ *hold.*

(ii) Let the conditions of theorem 11.1.11(ii) be fulfilled with $m_1 > \min(2m, -2)$. *Then for* $h \to +0$, $\mu = O(1)$ *asymptotics* $(11.1.30)_{1,2,3}$ *hold for* $m < -1$, $m = -1$ *and* $m > -1$ *respectively.*

Moreover, under condition (11.1.11) *for* $m < -1$ *equivalence* $(11.1.12)_1$ *holds and under condition* $(11.1.13)^{\#}$ *for* $m \ge -1$ *equivalences* $(11.1.12)_{2,3}$ *hold.*

(iii) Assertion (iii) of example 11.1.2 remains true.

Let us note that the contribution of the domain $\{|x| \le \varepsilon\}$ in (i) and $\{|x| \ge R\}$ in (ii) to the remainder estimate in asymptotics $(11.1.30)_1$ does not exceed $(\varepsilon' + o(1))h^{-1}$ with $\varepsilon' = \varepsilon'(\varepsilon) \to +0$ and $\varepsilon' = \varepsilon'(R) \to +0$ as $\varepsilon \to +0$ or $R \to +\infty$ respectively. Therefore applying Weylian asymptotics in this zone with more accurate remainder estimates we obtain the following

Theorem 11.1.13. *Let the conditions of theorem 11.1.12(i) with* $m > -1$, $m_1 < -2$ *or of theorem 11.1.12(ii) with* $m < -1$, $m_1 > -2$ *be fulfilled. Let the standard condition* (Ψ) *(see Chapter 10) be fulfilled for the Hamiltonian*

$$H(x,\xi) = g^{jk}(\xi_i - \mu V_j)(\xi_k - \mu V_k) + V \tag{11.1.31}$$

for fixed $\mu = \mu_0 \ge 0$. *Then for* $h \to +0$, $\mu \to \mu_0$ *the asymptotics*

$$\mathbf{N}^-(\mu, h) = \mathcal{N}^-(\mu, h) + \varkappa_1 h^{-1} + o(h^{-1}) \tag{11.1.32}$$

hold with the standard coefficient

$$\varkappa_1 = \mp\frac{1}{8\pi}\int_{\partial X} V_-^{\frac{1}{2}}\, dS \tag{11.1.33}$$

for Dirichlet and Neumann boundary conditions respectively where dS *is the Riemannian density on* ∂X *corresponding to the metric induced by the metric* (g^{jk}).[8)]

(ii) This assertion remains true for $m \le -1$ *provided* $\mu_0 > 0$.

Now let $m \le -1$ and $\mu \to +0$. Then the contribution of the zone $\{\mu\rho_1\gamma \ge \varepsilon\rho\}$ to the remainder estimate does not exceed $C_\varepsilon h^{-1}$. Moreover, the contribution of the ring $\{r \le |x| \le 2r\}$ to the remainder estimate does not exceed $Ch^{-1}r^{2m+2}$. Denoting $\eta_j = \xi_j - \mu V_j$, treating the dynamics (x, η) along Hamiltonian trajectories generated by the Hamiltonian (11.1.31) and applying more accurate estimates in the zone $\{\mu\rho_1\gamma \le \varepsilon\rho\}$ we obtain the following

8) Condition $(11.1.2)_1$ yields that $V_-|_{\partial X} = 0$ for $|x| \le \epsilon$ or $|x| \ge R$.

Theorem 11.1.14. *(i) Let the conditions of theorem 11.1.11(i) be fulfilled with $m = -1$, $m_1 < -2$ or let the conditions of theorem 11.1.11(ii) be fulfilled with $m = -1$, $m_1 > -2$. Moreover, let*

$$(11.1.34)_1 \qquad D^\sigma(g^{jk} - g^{jk0}) = o(|x|^{-|\sigma|}),$$

$$(11.1.34)_3 \qquad D^\sigma(V - V^0) = o(|x|^{2m-|\sigma|}) \qquad \forall \sigma : |\sigma| \le 1$$

as $|x| \to 0$ or $|x| \to \infty$ respectively where $g^{jk0}, V^0 \in C^K(\mathbb{R}^2 \setminus 0)$ are positively homogeneous functions of degrees 0 and $2m$ respectively. Let us assume that condition (Ψ) is fulfilled for the Hamiltonian

$$(11.1.35) \qquad H^0(x, \eta) = g^{jk0}\eta_j\eta_k + V_0$$

in $T^*\mathbb{R}^2$ at the energy level $\tau = 0$. Then for $h \to +0$, $\mu \to +0$ the asymptotics

$$(11.1.36) \qquad \mathbf{N}^-(\mu, h) = \mathcal{N}^-(\mu, h) + o(h^{-1}\log\mu)$$

hold.

(ii) Similarly, for $m < -1$ in case (i) and for $m > -1$ in case (ii) let us additionally assume that

$$(11.1.34)_2 \qquad D^\sigma(V_j - V_j^0) = o(|x|^{m_1+1-|\sigma|}) \qquad \forall \sigma : |\sigma| \le 1$$

as $|x| \to 0$ or $|x| \to \infty$ respectively where $V_j^0 \in C^K(\mathbb{R}^2 \setminus 0)$ are positively homogeneous functions of degree m_1+1. Let us assume that condition (Ψ) is fulfilled for the Hamiltonian

$$(11.1.37) \qquad H^0(x, \eta) = g^{jk0}(\eta_j - V_j^0)(\eta_k - V_k^0) + V^0$$

in $T^*\mathbb{R}^2$ at the energy level 0. Then for $h \to +0$, $\mu \to +0$ the asymptotics

$$(11.1.38) \qquad \mathbf{N}^-(\mu, h) = \mathcal{N}(\mu, h) + o\bigl(h^{-1}\mu^{-(m+1)(m_1+1-m)^{-1}}\bigr)$$

hold.

I leave it to the reader to formulate and prove statements similar to theorem 11.1.12 in the case of a bounded domain X, $0 \in \bar{X}$ and

(a) $\rho = |x|^{-1}\bigl(|\log|x|| + 1\bigr)^\alpha$, $\rho_1 = |x|^{m_1}$ with $m_1 < -2$;

(b) $\rho = |x|^m\bigl(|\log|x|| + 1\bigr)^\alpha$, $\rho_1 = |x|^{2m}\bigl(|\log|x|| + 1\bigr)^\beta$ with $m \le -2$, $\beta > 2\alpha$ such that either $m < -1$ or $\beta > 0$;

(c) $\gamma = \epsilon_0|x|^{1-\beta}$, $\rho = |x|^m \exp a|x|^\alpha$, $\rho = |x|^{m_1}\exp b|x|^\beta$ with $a \ge 0$, $b > 0$ and $\beta \le \alpha < 0$ such that $\beta < \alpha$ or $\beta = \alpha$, $b > 2a$ or $\beta = \alpha$, $b = 2a$, $m_1 < 2m$;

(d) $\gamma = \epsilon_0|x_1|$, $\rho = |x_1|^m$, $\rho_1 = |x_1|^{m_1}$ with $m_1 < \min(2m, -2)$.

I would like to note that one should take $\varrho_2 = 1$ in all these cases, $\varrho_3 = 1$ in (a), (d) and (c) with $\beta < \alpha$ or $\beta = \alpha$, $b > 2a$, $\varrho_3 = |x|^{-\beta}$ in case (c) with $\beta = \alpha$, $b = 2a$ and $\varrho_3 = \bigl(|\log|x|| + 1\bigr)^{-1}$ in case (b). Moreover, $\gamma = \epsilon_0|x|$ in cases (a),(b).

I leave the similar cases (a′)–(c′) in an unbounded domain to the reader.

11.1.5 Asymptotics for the Dirac operator

Let us now consider the Dirac operator

$$(11.1.39) \qquad \mathbf{A} = \frac{1}{2}\sigma_l(\omega^{jl}P_j + P_j\omega^{jl}) + \sigma_0 M + I \cdot V, \qquad P_j = hD_j - \mu V_j.$$

The theory of the Dirac operator is more complicated than the theory of the Schrödinger operator because it is different in the cases when V and F tend to 0 and (or) ∞; moreover, the value of ς is important (see Appendix C).

Without loss of generality one can assume that

$$\varsigma = 1 \tag{11.1.40}$$

and then the Landau levels (at the point x) are $V + \bigl(M^2 + 2j\mu hF\bigr)^{\frac{1}{2}}$ with $j = 1, 2, 3, \dots$ and $V - \bigl(M^2 + 2j\mu hF\bigr)^{\frac{1}{2}}$ with $j = 0, 1, 2, \dots$.

Moreover, we often *should prove* that the Dirac operator is essentially self-adjoint while the Schrödinger operator is *obviously* semibounded and therefore essentially self-adjoint. We do this in Appendix I.

So, we treat the operator given by $(\Sigma 0)$=(11.1.39) under the conditions

$$\begin{gathered} |D^\alpha \omega^{jk}| \le c\gamma^{-|\alpha|}, \qquad |D^\alpha F| \le c\rho_1\gamma^{-|\alpha|}, \\ |D^\alpha V| \le c\min\bigl(\rho, \frac{1}{M}\rho^2\bigr)\gamma^{-|\alpha|} \qquad (\alpha \ne 0) \quad \forall \alpha : |\alpha| \le K, \end{gathered} \tag*{$(11.1.41)_{1-3}$}$$

$$\begin{gathered} (V - \tau_2 - M)_+ \le c\min\bigl(\rho, \frac{1}{M}\rho^2\bigr), \\ (V - \tau_2 + M)_- \le c\min\bigl(\rho, \frac{1}{M}\rho^2\bigr) \end{gathered} \tag*{$(11.1.41)_4$}$$

(=$(9.1.35)_{1,3,4}$), $(9.1.35)_2$ = $(\Sigma 1)$, $(\Sigma 2)$ for $g^{jk} = \omega^{jl}\omega^{kr}\delta_{lr}$ (=(9.1.36)) and $(\Sigma 4)$. Here $(11.1.41)_4$ means the pair of conditions $(11.1.41)_{4\pm}$. Moreover, let condition (11.1.4) be fulfilled and

$$\bar{X}'' \cap \partial X = \emptyset, \tag{11.1.42}$$

$$|V_j| \le c\rho, \quad |D_j\omega^{kl}| \le c\rho \qquad \text{in } X'' \tag{11.1.43}$$

(see (9.1.43), (9.1.44)).

Theorem 11.1.15. [9] *(i) Let X be a compact manifold and*

(11.1.44) **Either $\partial X = \emptyset$ or $\mu = O(1)$** and $\partial X \cap X_2' = \emptyset$ (in what follows).

Let A be the Dirac operator satisfying conditions $(\Sigma 1)$, $(\Sigma 2)$, $(\Sigma 4)$, $(\Sigma 5)$ and (11.1.41)–(11.1.43) with $\gamma = \epsilon_0|x|$, $\rho = |x|^m$, $\rho_1 = |x|^{m_1}$, $m_1 < \min(2m, -2)$. Then

(a) Let $m < 0$ and let the conditions

$$|V| \ge \epsilon\rho \qquad \forall x : |x| \le \epsilon, \tag{11.1.45}$$

$$\begin{gathered} \bigl|V + \varsigma\bigl(M^2 + 2j\mu hF\bigr)^{\frac{1}{2}}\bigr| \le \epsilon\rho \implies \\ |\nabla v| \ge \epsilon\rho_1\rho^2\gamma^{-1} \qquad \text{or} \qquad \det \operatorname{Hess} v \ge \epsilon\rho_1^{-2}\rho^4\gamma^{-4} \qquad \text{or} \\ |\nabla V| \ge \epsilon\rho\gamma^{-1} \qquad \text{and} \qquad M \le \epsilon'\rho \quad \forall (j, \varsigma) \ne (0, 1) \end{gathered} \tag{11.1.46}$$

(with a small enough constant $\epsilon' = \varepsilon(c, \epsilon, \epsilon_0) > 0$, and this assumption is a part of condition (11.1.46))

---

[9] Compare with example 11.1.2.

be fulfilled. Recall that $v = F^{-1}\big((V - \tau_\iota)^2 - M^2\big)$ *and the conditions should be fulfilled for* $\iota = 1, 2$.

Then for $h \to +0$, $1 \le \mu = O(h^{-1})$ *asymptotics* (11.1.8)–(11.1.9) *hold with* S *given by* $(7.1.5)_1$ *with* $d = 2$. *Moreover, for* $m > -1$, $m = -1$, $m < -1$ *equivalences* $(11.1.12)_{1-3}$ *hold respectively.*

(b) Let $m = 0$ *and let conditions* (11.1.46) *and*

$$V - M \notin [\tau_1 - \epsilon, \tau_2 + \epsilon] \qquad \forall x : |x| \le \epsilon \tag{11.1.47}$$

be fulfilled. Then for $h \to +0$, $1 \le \mu = O(h^{-1})$ *asymptotics* (11.1.8)–(11.1.9) *hold and* $(11.1.10)_1$ *is also true. I leave it to the reader to clarify when* $(11.1.12)_1$ *is true here and below.*

(c) Let $m > 0$ *and let conditions* (11.1.46) *and*

$$V - M \notin [\tau_1 - \epsilon\varrho(x), \tau_2 + \epsilon\varrho(x)] \qquad \forall x : |x| \le \epsilon \tag{11.1.48}$$

be fulfilled with $\varrho = \min\big(\rho, \frac{1}{M}\rho^2\big)$. *Then (b) remains true.*[10)]

(ii) Let X *be an unbounded connected domain in* $\mathbb{R}^2$, *let* A *be a self-adjoint operator and let conditions* (11.1.41)–(11.1.45), $(\Sigma 1)$, $(\Sigma 2)$, $(\Sigma 4)$, (11.1.4) *be fulfilled with* $\gamma = \epsilon_0\langle x\rangle$, $\rho = \langle x\rangle^m$, $\rho_1 = \langle x\rangle^{m_1}$, $m_1 > \max(2m, -2)$. *Then*

(a) Let $m > 0$ *(and therefore* $m_1 > 0$*) and let conditions*

$$|V| \ge \epsilon\rho \qquad \forall x : |x| \ge c \tag{11.1.45$^\#$}$$

and (11.1.46) *be fulfilled. Then for* $h \to +0$, $1 \le \mu = O(h^{-1})$ *asymptotics* (11.1.8)–(11.1.9) *and* $(11.1.10)_3$ *hold.*

(b) Let $m = 0$, $m_1 > 0$ *and let conditions* (11.1.46) *and*

$$V - M \notin [\tau_1 - \epsilon, \tau_2 + \epsilon] \qquad \forall x : |x| \ge c \tag{11.1.47$^\#$}$$

be fulfilled. Then for $h \to +0$, $1 \le \mu = O(h^{-1})$ *asymptotics* (11.1.8)–(11.1.9) *hold and* $(11.1.10)_3$ *is also true. Moreover, if*

$$V + M \le \tau_2 - \epsilon \quad \text{or} \quad V - M \ge \tau_1 + \epsilon \quad \text{for } x \in \Gamma,\ |x| \ge c \tag{11.1.49$^\#$}$$

for some open non-empty cone Γ *then* $(11.1.12)_3$ *is true.*

On the other hand, if

$$V + M \ge \tau_2 + \epsilon \quad \text{or} \quad V - M \le \tau_1 - \epsilon \qquad \forall x : |x| \ge c \tag{11.1.50$^\#$}$$

then $(11.1.10)_1$ *is true.*

(c) Let $m < 0$, $m_1 > 0$ *and let conditions* (11.1.46) *and the condition*

$$V(\infty) + M = \tau_2, \qquad V(\infty) - M < \tau_1 \tag{11.1.51}$$

be fulfilled. Then asymptotics (11.1.8)–(11.1.9) *and* $(11.1.10)_{1-3}$ *(for* $m < -1$, $m = -1$ *and* $m > -1$ *respectively) remain true. Moreover, let*

$$V + M \le \tau_2 - \epsilon\varrho(x) \qquad \text{for } x \in \Gamma, |x| \ge c \tag{11.1.52$^\#$}$$

for some open non-empty cone Γ *where* $M > 0$ *and* $\varrho(x) = \langle x\rangle^{2m}$ *here. Then* $(11.1.12)_{2,3}$ *is true for* $m = -1$, $m > -1$ *respectively.*

Similarly, let us replace conditions (11.1.51), $(11.1.52)^\#$ *by*

---

10) This assertion is meaningful only for $V(0) - M = \tau_1$ or $V(0) - M = \tau_2$.

$$V(\infty)+M>\tau_2,\qquad V(\infty)-M=\tau_1 \tag{11.1.51$'$}$$

and

$$V-M\le \tau_1-\epsilon\varrho(x)\qquad \text{for } x\in\Gamma, |x|\ge c. \tag{11.1.52$^{\#\prime}$}$$

Then asymptotics (11.1.8)–(11.1.9) *and* $(11.1.12)_1$ *are true.*

(d) Let $m_1\le 0$ *(and then* $m<0$*),* $M>0$ *and*

$$V+M\ge\tau_2+\epsilon\varrho(x),\quad V-M\le\tau_1-\epsilon\varrho(x)\qquad \forall x:|x|\ge c. \tag{11.1.53$_{1,2}$}$$

Then asymptotics (11.1.8)–(11.1.9) *and* $(11.1.10)_1$ *remain true. On the other hand, if condition* $(11.1.53)_1$ *is violated then the same results as for the Schrödinger operator (in example 11.1.2(ii)) are true.*

I leave it to the reader to transfer examples 11.1.3, 11.1.4, 11.1.5, 11.1.6, theorems 11.1.7, 11.1.8 and corollary 11.1.10 to the Schrödinger operator.

11.1.6 The case $\mu h\to\infty$

Let us consider the case $\mu h\to\infty$ for the Dirac operator. There should be very essential differences between the two operators in this case because in regular situations $\mathbf{N}(\tau_1,\tau_2)=O(\mu^{-s})$ with arbitrarily large s for the Schrödinger operator, while

$$\mathbf{N}^-(\tau_1,\tau_2)=\mu h^{-1}\int S_0(x,\tau_1,\tau_2)dx+O(1) \tag{11.1.54}$$

for the Dirac operator where $S_0=\bigl(\theta(V-M-\tau_1)-\theta(V-M-\tau_2)\bigr)F\sqrt{g}$ provided $\varsigma=1$. Moreover, it is easy to prove that *in the context of theorems 11.1.1, 11.1.7, 11.1.8 and therefore in the context of examples 11.1.2–11.1.6 and corollary 11.1.10 the same asymptotics* (11.1.53) *hold under the additional condition* $\partial X=\emptyset$ *and under condition* (11.1.5) *checked on the whole manifold* X. Therefore we assume that X is a compact closed manifold when considering a singularity at 0 and $X=\mathbb{R}^2$ when considering a singularity at infinity.

It is also easy to prove that *theorem 11.1.11(i),(ii) also hold for the Dirac operator under the additional conditions* $m\ge 0$ *in case (i),* $m\le 0$ *in case (ii) and*

$$V-M\le\tau_1-\epsilon\rho^2\qquad\text{on } X, \tag{11.1.55}$$

$$V(0)+M=\tau_2, \tag{11.1.56}$$

$$V(\infty)+M=\tau_2 \tag{11.1.56$^{\#}$}$$

in (i),(ii) respectively; recall that X is a compact closed manifold in (i) and $X=\mathbb{R}^2$ in (ii). Moreover, (11.1.28) *holds provided*

$$V+M\le\tau_2-\epsilon\rho^2\qquad\text{for } x\in\Gamma, |x|\le\epsilon, \tag{11.1.57}$$

$$V+M\le\tau_2-\epsilon\rho^2\qquad\text{for } x\in\Gamma, |x|\ge c, \tag{11.1.57$^{\#}$}$$

in (i),(ii) respectively. On the other hand, if condition (11.1.55) *is fulfilled only for* $|x|\le\epsilon$, $|x|\ge c$ *in (i),(ii) respectively, then under the same remainder estimate* $\mathcal{N}^-(\mu,h)$ *is a sum of two terms; the first term is the principal part of asymptotics* (11.1.54) *and the second term is of order* $\mathcal{N}^-$ *for the Schrödinger operator. Moreover, if we replace the condition* $\mu=o\bigl(h^{(m-m_1-1)(m+1)^{-1}}\bigr)$ *(which was reasonable for the Schrödinger operator) with the opposite condition* $\mu\ge\epsilon h^{(m-m_1-1)(m+1)^{-1}}$ *then this additional term of the asymptotics is* $O(1)$ *and the asymptotics hold with* $\mathcal{N}$ *given by* (11.1.54).

I leave it to the reader to formulate and prove similar statements in cases (b), (a$'$), (b$'$) and (c) at the end of subsection 11.3.

11.1.7 The case $h \to +0$, $\mu = O(1)$

Let us consider the case $h \to +0$, $\mu = O(1)$ for the Dirac operator. I leave it to the reader to formulate and prove theorem 11.1.12(i),(ii) under the additional conditions (11.1.45), (11.1.47), (11.1.48) for $m < 0$, $m = 0$, $m > 0$ respectively in case (i) and under the additional conditions $(11.1.45)^{\#}$, $(11.1.47)^{\#}$, $(11.1.48)^{\#}$ for $m > 0$, $m = 0$, $m < 0$ respectively in case (ii). Moreover, I leave the same task with theorem 11.1.13 to the reader; surely two Hamiltonians

$$H_{\pm} = V \pm \left(g^{jk}(\xi_j - \mu V_j)(\xi_k - \mu V_k) + M^2\right)^{\frac{1}{2}} \tag{11.1.58}$$

should be treated and *each* of them should be treated at the energy levels τ_1 and τ_2.

Moreover, I leave theorem 11.1.14 to the reader; in case (i) of this theorem the two Hamiltonians

$$H^0_{\pm} = \pm(g^{jk0}\xi_j\xi_k)^{\frac{1}{2}} + V^0 \tag{11.1.59}$$

should be treated at the energy level 0 and in case (ii) under the assumption

$$V(\infty) + M = \tau_2, \qquad V(\infty) - M < \tau_1, \tag{11.1.60$_+$}$$

$$V(\infty) + M > \tau_2, \qquad V(\infty) - M = \tau_1, \tag{11.1.60$_-$}$$

only one Hamiltonian

$$\pm\frac{1}{2M} g^{jk0}\xi_j\xi_k + V^0 \tag{11.1.61}$$

should be treated again at the energy level 0 where V^0 is the principal homogeneous part of $V - V(\infty)$. In this theorem, it is very important to assume for the Dirac operator that condition (11.1.45) or $(11.1.45)^{\#}$ is fulfilled in cases (i),(ii) respectively.

Appendix I. On the Self-Adjointness of the Dirac Operator

The Dirac operators treated in section 11.1 are surely self-adjoint in the case of an exterior domain. However, the same fact should be proven for an interior domain with singular points.

Theorem I.1. *Let $X \subset \mathbb{R}^2/(\bar\gamma_1\mathbb{Z} \times \bar\gamma_2\mathbb{Z})$ be an open domain and let conditions* (11.1.4), $(\Sigma 1)$, $(\Sigma 2)$, $(\Sigma 4)$, (11.1.4) *and* (11.1.45) *be fulfilled. Let us assume that there exists a neighborhood of ∂X denoted by Y such that for every $r > 0$ the inequalities*

$$\rho_1 \geq \varepsilon\rho^2, \quad (\rho\gamma)^s \geq \varepsilon\rho_1, \quad \rho \geq 1, \tag{I.1}$$

$$|V + \sqrt{2j\mu h F}| \geq \varepsilon\sqrt{\rho_1} - \varepsilon^{-1} \qquad \forall j \in \mathbb{Z}^+ \setminus 0, \tag{I.2}$$

$$|V| \geq \varepsilon\rho - \varepsilon^{-1} \tag{I.3}$$

are fulfilled on $Y \cap X \cap \{|x| \leq r\}$ with appropriate $\varepsilon = \varepsilon(r) > 0$. Then for $\mu > 0$, $h > 0$ the operator A with domain $\mathrm{D}(A) = C^1_0(X, \mathbb{H})$ is essentially self-adjoint in $L^2(X, \mathbb{H})$.

Proof. Let us consider the adjoint operator A^*. This operator is given by the same formula with $\mathrm{D}(A^*) = \{u \in L^2, Au \in L^2\}$ with Au calculated as a distribution. We should prove that $\mathrm{Ker}(A^* \pm iI) = 0$ for both signs. So, here and below let $u \in L^2$ and $(A^* \pm iI) = 0$ for some sign.

The microlocal canonical form of section 7.1 yields the inequality

$$\text{(I.4)} \quad \|\rho v\| \le M\bigl(\|Av\| + \|v\| + \|\gamma^{-1}v\|\bigr)$$
$$\forall v \in C_0^1\bigl(B(y, \tfrac{1}{2}\gamma(y)\bigr) \qquad \forall y \in X_r = X \cap \{|x| \le r\}$$

with a constant $M = M(r)$; all the constants now depend on μ and h.

Let us prove that $(\rho\gamma)^n u \in L^2(X_r)$ for every $r > 0$ by induction on n. This is true for $n = 0$ by the assumption $u \in L^2$.

Let $(\rho\gamma)^n u \in L^2(X_r)$ for some n; then $v = \gamma(\rho\gamma)^n u$ also belongs to $L^2(X_r)$ and

$$Av = [A, \gamma(\rho\gamma)^n]u \mp iv.$$

One can easily see that $[A, \gamma(\rho\gamma)^n]$ is a matrix-valued function the matrix norm of which does not exceed $M(\rho\gamma)^n$. Therefore, taking a γ-admissible partition of unity in a neighborhood of X_r we see that $\sum_\nu \|A\psi_\nu v\|^2 < \infty$ and therefore (I.4) yields that $\sum_\nu \|\rho\psi_\nu v\|^2 < \infty$. Therefore $\rho v = (\rho\gamma)^{n+1} \in L^2(X_r)$ and the induction step is complete.

Thus we have proven that $(\rho\gamma)^n u \in L^2(X_r)$ for every $r > 0$. Then $\rho_1 u \in L^2(X_r)$ by $(\text{I.1})_2$. The ellipticity of A yields that $D_j u \in L^2(X_r)$. Therefore, if $\psi \in C_0^1(\mathbb{R}^2/(\bar\gamma_1\mathbb{Z} \times \bar\gamma_2\mathbb{Z}))$ then $\psi u \in \mathrm{D}(\bar A)$ where $\bar A$ is the closure of A. Moreover,

$$(\bar A \pm iI)\psi u = [\bar A, \psi]u.$$

Calculating the real part of the inner product with $i\psi u$ we obtain the inequality

$$\|\psi u\| \le \max|\nabla\psi^2| \cdot \|u\|^2.$$

Let us take $\psi = \psi^0(x/r)$ where $\psi^0 \in C_0^1$ is a fixed function equal to 1 in a neighborhood of 0. Then for $r \to +\infty$ we see that the left-hand expression of this inequality tends to $\|u\|^2$ and the right-hand expression tends to 0. Therefore $u = 0$. □

Theorem I.2. *Let all of the conditions of theorem I.1 excluding condition* (I.1) *be fulfilled with* $\gamma = \epsilon_0|x|$, $\rho = |x|^m$, $\rho_1 = |x|^{m_1}$, $m_1 \le 2m$, $m_1 < -2$. *Then the Dirac operator has a self-adjoint extension for* $\mu > 0$, $h > 0$.

Proof. For $m < -1$ condition (I.1) is also fulfilled and therefore the operator A is essentially self-adjoint. So, let us treat the case $m \ge -1$. Let us consider the Dirac operator A_S with the potential $V_\sigma = V + \sigma W$ where $\sigma > 0$, W is a potential which is regular away from $x = 0$, and $W = \pm|x|^{m'}$ in a neighborhood of $x = 0$ with $-1 > m' > \frac{1}{2}m_1$. This operator is essentially self-adjoint by theorem I.1. For $m < 0$ let us choose the sign of W coinciding with the sign of V on a neighborhood of 0 (condition (11.1.45) yields that this is possible). For $m \ge 0$ let us choose an appropriate interval $[\tau_1, \tau_2]$ with $\tau_1 < \tau_2$. Then applying the results of section 11.1 we see that the number of eigenvalues of the operator $\bar A_\sigma$ lying in the interval $[\tau_1, \tau_2]$ is bounded uniformly with respect to $\sigma > 0$. Then there exists a sequence $\sigma_k \to +0$ such that there exist $\tau_1' < \tau_2'$ which do not depend on k and such that $[\tau_1', \tau_2'] \cap \operatorname{Spec}(\bar A_\sigma) = \emptyset$ for $\sigma = \sigma_k$. Let $\tau = \frac{1}{2}(\tau' + \tau_2')$; then all the operators $(\bar A_\sigma - \tau)^{-1}$ are uniformly bounded and $\|A_\sigma u\| \ge \epsilon\|u\|$ $\forall u \in \mathrm{D}(A_\sigma)$ for $\sigma = \sigma_k$. Then the same is true for A. Therefore A has a self-adjoint extension $\tilde A$ satisfying the same estimate. □

11.2 The Case $d = 3$ with Fixed Spectral Parameter

In this section we transfer the results of the previous section to dimension $d = 3$. However, the remainder estimates are worse (does the reader recall that in regular situations, local semiclassical spectral asymptotics are worse?), certain improvements associated with magnetic lines are possible (for $d = 2$ one should use the drift lines instead) and, finally, for the Dirac operator the magnetic field does not extend the spectral gap, so a very singular magnetic field is now impossible for the Dirac operator.

11.2.1 Getting started

As before we start with the Schrödinger operator.

Theorem 11.2.1. [11] *(i) Let $d = 3$ and let the Schrödinger operator A satisfy conditions (11.1.4), ($\Sigma 1$), ($\Sigma 2$) and ($\Sigma 4$) in the corresponding regions where $\rho, \rho_1, \gamma, \varrho_3$ satisfy conditions $(8.1.2)_{1-3}$, (11.1.2), (11.1.3) and $\varrho_2 = 1$. Moreover, let the condition*

$$\sum_{|\alpha|\le m} \gamma^{|\alpha|}\bigl|D^\alpha\bigl(V + (2n+1)\mu hF\bigr)\bigr| \ge \epsilon_0\rho^2 \qquad \forall n \in \mathbb{Z}^+ \tag{11.2.1}$$

be fulfilled in X_2'. Let $\rho_1 \ge \epsilon_0\rho\gamma^{-1}$ and

$$\rho^2\gamma^{-1} \in L^1\bigl(X \cap \{V + tF \le \epsilon\rho^2\}\bigr), \tag{11.2.2}$$

$$\rho_1^{-s}\gamma^{-2s-d} \in L^1(X) \tag{11.2.3}$$

with some $t \ge 0$. Then for $h \to +0$, $1 \le \mu$ such that $\mu h \ge t$ the asymptotics

$$\mathbf{N}^-(\mu,h) = \mathcal{N}(\mu,h) + O(h^{-2}) \tag{11.2.4}$$

hold with $\mathcal{N}^-(\mu,h)$ given by (11.1.9) and $S(x,\mu h)$ given by $(6.1.7)_{(3)}$.

(ii) Moreover, let us assume that

(11.2.5) There exists a set $\Xi_0 \subset X$ of measure 0 such that through every point $\bar{x} \in X\setminus\Xi_0$ with $V(\bar{x}) < 0$ there passes a non-periodic magnetic line, infinitely long in some direction, along which $v(x) \le v(\bar{x})$.

Then for $h \to +0$, $\mu \to \infty$ such that $\mu h \ge t$ the asymptotics

$$\mathbf{N}^-(\mu,h) = \mathcal{N}^-(\mu,h) + o(h^{-2}) \tag{11.2.6}$$

hold.

Remark. *I leave it to the reader to obtain asymptotics with remainder estimate $o(\mu^{-1}h^{-1})$ in the two-dimensional case under the obvious condition on drift lines provided $\mu \to \infty$.* ◊

Theorem 11.2.2. *(i) Let X be a bounded domain, $0 \in \bar{X}$ and let the conditions of theorem 11.2.2 (excluding, perhaps, (11.2.2)) be fulfilled with $\gamma = \epsilon_0|x|$, $\rho = |x|^m$, $\rho_1 = |x|^{m_1}$, $m_1 < \min(2m, -2)$. Then*

(a) For $m > -1$ condition (11.2.2) is fulfilled and thus asymptotics (11.2.4)–(11.1.9) hold; moreover, under condition (11.1.13)

$$\mathcal{N}^-(\mu,h) = \mathcal{N}^{w-}(\mu,h) + O\bigl(h^{-3}(\mu h)^{(3m+3)(2m-m_1)^{-1}}\bigr); \tag{11.2.7}$$

therefore one can replace $\mathcal{N}^-(\mu,h)$ by $\mathcal{N}^{w-}(\mu,h)$ for μ tending slowly to infinity.

(b) For $m = -1$ the asymptotics

$$\mathbf{N}^-(\mu,h) = \mathcal{N}^-(\mu,h) + O\bigl(h^{-2}(|\log\mu h| + 1)\bigr) \tag{11.2.8$_2$}$$

and

$$\mathcal{N}^-(\mu,h) = O\bigl(h^{-3}(|\log\mu h| + 1)\bigr) \tag{11.2.9$_2$}$$

[11] Compare with theorems 11.1.1, 11.1.1′.

hold, and under condition (11.1.13) *for* $\mu h < t$ *with a small enough* $t > 0$ *the equivalence*

$$\mathcal{N}^-(\mu,h) \asymp h^{-3}(|\log \mu h| + 1) \tag{11.2.10$_2$}$$

holds.

(c) For $m < -1$ *the asymptotics*

$$\mathbf{N}^-(\mu,h) = \mathcal{N}^-(\mu,h) + O\bigl(h^{-2}(\mu h)^{2l}\bigr) \tag{11.2.8$_3$}$$

hold with $l = (m+1)(2m-m_1)^{-1}$ *and*

$$\mathcal{N}^-(\mu,h) = O\bigl(h^{-3}(\mu h)^{3l}\bigr) \tag{11.2.9$_3$}$$

and under condition (11.1.13) *for* $\mu h < t$ *with small enough* $t > 0$

$$\mathcal{N}^-(\mu,h) \asymp h^{-3}(\mu h)^{-3l}. \tag{11.2.10$_3$}$$

(ii) Let X *be an unbounded connected domain and let the conditions of theorem 11.2.1 (excluding, perhaps,* (11.2.2)*) be fulfilled with* $\gamma = \epsilon_0\langle x\rangle$, $\rho = \langle x\rangle^m$, $\rho_1 = \langle x\rangle^{m_1}$, $m_1 >$ $\min(2m,-2)$. *Then assertions (a), (b), (c) of (i) hold for* $m < -1$, $m = -1$ *and* $m > -1$ *respectively with* (11.1.13) *replaced by* $(11.1.13)^{\#}$.

I leave it to the reader to treat the following cases:
(a) X is a bounded domain, $0 \in \bar{X}$, $\gamma = \epsilon_0|x|$, $\rho = |x|^m\bigl(|\log|x|| + 1\bigr)^\alpha$, $\rho_1 = |x|^{m_1}\bigl(|\log|x|| + 1\bigr)^\beta$ with $-2 > 2m > m_1$ or
(b) X is an unbounded connected domain, etc.
Here I only treat some special cases:

Theorem 11.2.3. *(i) Let* X *be a bounded domain,* $0 \in \bar{X}$ *and let the conditions of theorem 11.2.1 (excluding, perhaps, condition* (11.2.2)*) be fulfilled with* $\gamma = \epsilon_0|x|$, $\rho = |x|^m\bigl(|\log|x|| + 1\bigr)^\alpha$, $\rho_1 = |x|^{m_1}\bigl(|\log|x|| + 1\bigr)^\beta$.
(i)$_1$ Let $m = -1$, $m_1 < -2$. *Then*

(a) For $\alpha < -\frac{1}{2}$ *condition* (11.2.2) *is fulfilled and* (11.2.4)–(11.1.9) *hold.*

(b) For $\alpha = -\frac{1}{2}$ *the asymptotics*

$$\mathbf{N}^-(\mu,h) = \mathcal{N}^-(\mu,h) + O\bigl(h^{-2}(\log|\log \mu h| + 1)\bigr) \tag{11.2.11$_2$}$$

hold.

(c) For $\alpha > -\frac{1}{2}$ *the asymptotics*

$$\mathbf{N}^-(\mu,h) = \mathcal{N}^-(\mu,h) + O\bigl(h^{-2}(|\log \mu h| + 1)^{2\alpha+1}\bigr) \tag{11.2.11$_3$}$$

hold.

Moreover, $\mathcal{N}^-(\mu,h) = O(h^{-3})$ *for* $\alpha < -\frac{1}{3}$ *and for* $\alpha = -\frac{1}{3}$ *and* $\alpha > -\frac{1}{3}$ *we have respectively*

$$\mathcal{N}^-(\mu,h) = O\bigl(h^{-3}(\log|\log \mu h| + 1)\bigr) \tag{11.2.12$_2$}$$

and

$$\mathcal{N}^-(\mu,h) = O\bigl(h^{-3}(|\log \mu h| + 1)^{3\alpha+1}\bigr) \tag{11.2.12$_3$}$$

and under condition (11.1.13) for $\mu h < t$ we obtain asymptotic equivalence relations.

(i)$_2$ Let $m = -1$, $m_1 = -2$ and $\beta > \max(2\alpha, 0)$. Then

(a) For $\alpha < -\frac{1}{2}$ condition (11.2.2) is fulfilled and asymptotics (11.2.4)–(11.1.9) hold.

(b) For $\alpha = -\frac{1}{2}$ the asymptotics

$$\mathbf{N}^-(\mu,h) = \mathcal{N}^-(\mu,h) + O\bigl(h^{-2}(|\log \mu h| + 1)\bigr) \tag{11.2.13$_2$}$$

hold.

(c) For $\alpha > -\frac{1}{2}$ the asymptotics

$$\mathbf{N}^-(\mu,h) = \mathcal{N}^-(\mu,h) + O\bigl(h^{-2}(\mu h)^{(2\alpha+1)(2\alpha-\beta)^{-1}}\bigr) \tag{11.2.13$_3$}$$

hold.

Moreover, $\mathcal{N}^-(\mu,h) = O(h^{-3})$ for $\alpha < -\frac{1}{3}$ and for $\alpha = -\frac{1}{3}$ and for $\alpha > -\frac{1}{3}$ we have respectively

$$\mathcal{N}^-(\mu,h) = O(h^{-3}(|\log \mu h| + 1)) \tag{11.2.14$_2$}$$

and

$$\mathcal{N}^-(\mu,h) = O\bigl(h^{-3}(\mu h)^{(3\alpha+1)(2\alpha-\beta)^{-1}}\bigr) \tag{11.2.14$_3$}$$

and under condition (11.1.13) for $\mu h < t$ we obtain asymptotic equivalence relations.

(ii) All the previous statements remain true in the case when X is an unbounded connected domain and $\gamma = \epsilon_0\langle x\rangle$, $\rho = \langle x\rangle^m \log^\alpha\langle x\rangle$, $\rho = \langle x\rangle^{m_1}\log^\beta\langle x\rangle$.

I leave it to the reader to treat the case $m_1 = 2m < -2$, $\beta > \max(2\alpha, 0)$ when $\mathcal{N}^-(\mu,h)$ as well as the remainder estimate grow exponentially as $\mu h \to 0$.

Theorem 11.2.4. *(i) Let X be a bounded domain, $0 \in \bar{X}$ and let the conditions of theorem 11.2.1 (perhaps excluding (11.2.2)) be fulfilled with $\gamma = \epsilon_0|x|^{1-\beta}$, $\rho = |x|^m$, $\rho_1 = \exp b|x|^\beta$, $b > 0$, $\beta < 0$. Then*

(a) For $2m+2+\beta > 0$ condition (11.2.2) is fulfilled and asymptotics (11.2.4)–(11.2.8) hold.

(b) For $2m+2+\beta = 0$ asymptotics (11.2.11)$_2$ hold.

(c) For $2m+2+\beta < 0$ the asymptotics

$$\mathbf{N}^-(\mu,h) = \mathcal{N}^-(\mu,h) + O\bigl(h^{-2}(|\log \mu h| + 1)^{(2m+2+\beta)\beta^{-1}}\bigr) \tag{11.2.15$_3$}$$

hold.

Moreover, for $m = -1$, $m < -1$ respectively (11.2.12)$_2$ and

$$\mathcal{N}^-(\mu,h) = O\bigl(h^{-3}(|\log \mu h| + 1)^{(3m+3)\beta^{-1}}\bigr) \tag{11.2.16$_3$}$$

hold and under condition (11.1.13) for $\mu h \le t$ we obtain asymptotic equivalence relations.

(ii) Let X be an unbounded connected domain and let the conditions of theorem 11.2.1 (perhaps excluding (11.2.2)) be fulfilled with $\gamma = \epsilon_0\langle x\rangle^{1-\beta}$, $\rho = \langle x\rangle^m$, $\rho = \exp b\langle x\rangle^\beta$, $b > 0$, $\beta > 0$. Then for $2m+2+\beta < 0$, $2m+2+\beta = 0$ and $2m+2+\beta > 0$ assertions (a), (b), (c) respectively of (i) hold, and for $m = -1$, $m > -1$ respectively (11.2.10)$_2$ and (11.2.14)$_3$ hold, and under condition (11.1.13)$^\#$ for $\mu h \le t$ we obtain asymptotic equivalence relations.

I leave it to the reader to treat the case $\gamma = \epsilon_0|x|^{1-\beta}$, $\rho = \exp a|x|^\alpha$, $\rho_1 = \exp b|x|^\beta$ where either $0 > \alpha > \beta$, $a > 0$, $b > 0$ or $0 = \alpha = \beta$, $0 < 2a < b$. In these cases different principal parts and remainder estimate arise. I leave it to the reader to treat similar cases in an unbounded domain.

Theorem 11.2.5. *Let X be a bounded domain, $0 \in \bar{X}$ and let all the conditions of theorem 11.2.1 (excluding, perhaps, (11.2.2)) be fulfilled with $\gamma = \epsilon_0|x|$, $\rho = |x'|^m$, $\rho_1 = |x'|^{m_1}$ where $x' = (x_1, \dots, x_k)$, $k = 1, 2$ and $m_1 < \min(2m, -2)$. Then*

(a) For $2m - 1 + k > 0$ condition (11.2.2) is fulfilled and asymptotics (11.2.4)–(11.1.9) hold.

(b) For $2m - 1 + k = 0$ asymptotics $(11.2.8)_2$ hold.

(c) For $2m - 1 + k < 0$ the asymptotics

$$\mathbf{N}^-(\mu, h) = \mathcal{N}^-(\mu, h) + O\big(h^{-2}(\mu h)^{(2m-1+k)(2m-m_1)^{-1}}\big) \tag{11.2.17$_3$}$$

hold.

Moreover, for $3m + k > 0$, $\mathcal{N}^-(\mu, h) = O(h^{-3})$ holds; for $3m + k = 0$, $\mathcal{N}^-(\mu, h) = O\big(h^{-3}(|\log \mu h| + 1)\big)$ holds and for $3m + k < 0$, $\mathcal{N}^-(\mu, h) = O\big(h^{-3}(\mu h)^{(3m+k)(2m-m_1)^{-1}}\big)$ holds. Moreover, if

$$V \le -\epsilon\rho^2 \qquad \text{for } x \in \{x' \in \Gamma', |x'| \le \epsilon, x'' \in B\} \subset X \tag{11.2.18}$$

where Γ' is a non-empty open cone in $\mathbb{R}^k$, $B \subset \mathbb{R}^{3-k}$ is a non-empty domain then for $\mu h \le t$ we have asymptotic equivalence relations instead of the last two estimates.

I leave it to the reader to obtain various results associated with additional factors of power of logarithm type in this case.

11.2.2 The case $\mu h \to \infty$

Let us now consider the case $\mu h \to \infty$.

Theorem 11.2.6. *In the context of theorem 11.1.11(i) or (ii)*[12] *the asymptotics*

$$\mathbf{N}^-(\mu, h) = \mathcal{N}^-(\mu, h) + O\big(h^{-2}(\mu h)^{2l}\big) \tag{11.2.19}$$

hold with $l = (m+1)(2m - m_1)^{-1}$ where

$$\mathcal{N}^-(\mu, h) = O\big(h^{-3}(\mu h)^{3l}\big) \tag{11.2.20}$$

and under condition (11.1.13) or (11.1.13)$^\#$ respectively we obtain an asymptotic equivalence relation instead of (11.2.20).

11.2.3 The case $\mu = O(1)$

Finally, let us treat the case $\mu = O(1)$. One can easily prove the following

Theorem 11.2.7. *(i) In the context of theorem 11.1.12(i) the asymptotics*

$$\mathbf{N}^-(\mu, h) = \mathcal{N}^-(\mu, h) + O(h^{-2}) \tag{11.2.21$_1$}$$

$$\mathbf{N}^-(\mu, h) = \mathcal{N}^-(\mu, h) + O(h^{-2} \log \mu h) \tag{11.2.21$_2$}$$

and

---

[12] Surely with $d = 3$ instead of $d = 2$.

$$\mathbf{N}^-(\mu,h) = \mathcal{N}(\mu,h) + O\big(h^{-2}(\mu h)^{2l}\big) \tag{11.2.21$_3$}$$

with $l = (m+1)(2m-m_1)^{-1}$ *hold for* $m > -1$, $m = -1$ *and* $m < -1$ *respectively. Moreover, in these cases* $\mathcal{N}^-(\mu,h) = O(h^{-3})$,

$$\mathcal{N}^-(\mu,h) = O(h^{-3}\log \mu h) \tag{11.2.22$_2$}$$

and

$$\mathcal{N}^-(\mu,h) = O\big(h^{-3}(\mu h)^{3l}\big) \tag{11.2.22$_3$}$$

respectively and under condition (11.1.13) *we obtain asymptotic equivalence relations instead of the last two estimates.*

(ii) In the context of theorem 11.1.12(ii) these results hold for $m < -1$, $m = -1$ *and* $m > -1$ *respectively.*

(iii) In the context of theorem 11.1.13(i) one can replace asymptotics $(11.2.21)_1$ *by*

$$\mathbf{N}^-(\mu,h) = \mathcal{N}^-(\mu,h) + \varkappa_1 h^{-2} + o(h^{-2}) \tag{11.2.23}$$

with the standard coefficient

$$\varkappa_1 = \mp\frac{1}{8\pi^2}\int_{\partial X} V_- dS \tag{11.2.24}$$

for Dirichlet and Neumann boundary conditions respectively.

I would like to note that there is a principal difference for $m \le -1$ in (i) and for $m \ge -1$ in (ii) between the two- and three-dimensional cases. Namely, for $d = 3$ the principal contribution to the remainder estimate is given by the zone of strong magnetic field $\{\mu h\rho_1 \le C_0\rho^2, \mu\rho_1\gamma \ge \rho\}$ but for $d = 2$ by the contributions of this zone and the zone of normal magnetic field $\{\mu\rho_1\gamma \le \rho\}$. Therefore theorems 11.1.13(ii), 11.1.14 have no analogs in the three-dimensional case. However one can prove theorems in which conditions $(11.1.34)_{1-3}$ are used for $\mu \to \mu_0$ with $\mu_0 > 0$ as well as for $\mu_0 = 0$ and conditions on the magnetic lines of the homogeneous limit operator are applied.

11.2.4 Some remarks

Let us note that applying theorem 6.4.14 one can modify certain of the above statements:

Remark 11.2.8. *(i) Theorems 11.2.2(i),(ii), 11.2.6(i),(ii) and 11.2.7(i),(ii) remain true if one replaces condition (11.2.1) by the condition that* v *is a positively homogeneous function for* $|x| \le \epsilon$ *and* $|x| \ge c$ *respectively.*[13)]

(ii) Theorem 11.2.5 remains true if one replaces condition (11.2.1) by the condition that v *is a positively homogeneous function of* x' *for* $|x'| \le \epsilon$.[13)] ◊

In the context of the improved remainder estimate one can easily prove

Theorem 11.2.9. *(i) In the context of theorems 11.2.2(i), 11.2.7(i) with* $m \le -1$ *and in the context of theorems 11.2.2(ii), 11.2.7(ii) with* $m \ge -1$ *and with* $\mu h \to +0$ *in all cases and in the context of theorem 11.2.6(i),(ii) let us assume that*

$$F^j - F^{0j} = o(|x|^{m_1}), \tag{11.2.25$_1$}$$

$$D^\sigma(v - v^0) = o(|x|^{2m-m_1-|\sigma|}) \qquad \forall\sigma : |\sigma| \le 1 \tag{11.2.25$_2$}$$

13) However, if condition (11.2.1) was assumed to be fulfilled for all $x \in X$ then it should be preserved for $|x| \ge \epsilon$ and $|x| \le c$ respectively.

for $|x| \to +0$, $|x| \to \infty$ in assertions (i),(ii) respectively where F^{0j} and v^0 are positively homogeneous functions of degrees $m-1$, $2m-m_1$ respectively. Let us assume that these functions satisfy condition (11.2.1). *Then in the corresponding asymptotics one can replace "O" by "o" in the remainder estimate.*

(ii) In the context of theorem 11.2.5 with $2m-1+k \le 0$ this assertion remains true provided

$$F^j - F^{0j} = o(|x'|^{m_1}), \tag{11.2.25$)'_1$}$$

$$D^\sigma(v - v^0) = o(|x'|^{2m-m_1-|\sigma|}) \qquad \forall \sigma : |\sigma| \le 1 \tag{11.2.25$)'_2$}$$

for $|x'| \to 0$ with functions F^{0j} and v^0 positively homogeneous with respect to x'.

(iii) In all the assertions one can replace condition (11.2.1) *by the condition that v is a positively homogeneous function (see remark 11.2.8 and footnote* [13]) *(in (ii) we should apply this footnote with $|x'|$ instead of $|x|$).*

I leave the proof to the reader.

11.2.5 The Dirac operator

Finally, let us discuss the Dirac operator. I have explained that we have only a few possibilities. One can easily prove theorem 11.2.6(i),(ii) for the Dirac operator under following additional conditions: one assumes that $m \ge 0$ and conditions (11.1.55), (11.1.56) and (11.1.57) are fulfilled in (i) and $m \le 0$, and conditions (11.1.55), (11.1.56)$^\#$ and (11.1.57)$^\#$ are fulfilled in (ii). The variations indicated in theorem 11.2.7 and remark 11.2.8 are possible. Surely, $\mathcal{N}^-(\mu, h)$ is now given by (7.1.5) with $d = 3$.

11.3 Asymptotics with the Spectral Parameter for Two-Dimensional Schrödinger Operators

The spectral parameter is the principal parameter in this section; moreover, it is the only parameter in many statements here. Recall that in this section we only consider the Schrödinger operator.

11.3.1 Asymptotics Weylian in principal

We start with asymptotics which are Weylian "in principal."

Theorem 11.3.1. *Let X be a bounded domain with C^K boundary, $0 \in X$ and let A satisfy conditions $(\Sigma 0)$, $(\Sigma 1)$, $(\Sigma 2)$, $(\Sigma 4)$*[14] *with $\gamma = \epsilon_0 |x|$, $\rho = |x|^m$, $\rho_1 = |x|^{m_1}$, $m_1 \le -2$, $m \le 0$. Suppose* **either** *$m_1 < -2m$* **or** *the coercivity condition*

$$V + F \ge \epsilon\rho^2 \qquad \text{for } |x| \le \epsilon, \tag{$\Sigma 3)'$}$$

$$V \ge \epsilon\rho^2 \qquad \text{for } |x| \le \epsilon, \tag{$\Sigma 3$}$$

is fulfilled for $m_1 < -2$, $m_1 = -2$ respectively.

Moreover, let one of the following three conditions be fulfilled:

(a) $m_1 < 2m$ and

$$|\nabla F| \ge \epsilon \rho_1 \gamma^{-1} \qquad \text{for } |x| \le \epsilon; \tag{11.3.1}$$

[14] Surely, the latter condition should be fulfilled only for $|x| \le \epsilon$.

(b) V and F are positively homogeneous functions of degrees $2m$ and m_1 respectively for $|x| \le \epsilon$;

(c) $m_1 = 2m$ and the inequalities (11.3.1) and

$$|\nabla V| \le \varepsilon |\nabla F| \qquad \text{for } |x| \le \epsilon \tag{11.3.2}$$

hold with small enough $\varepsilon = \varepsilon(c, \epsilon, \epsilon_0) > 0$.

Then
(i) For $\mu = h = 1$, $\tau \to +\infty$ the asymptotics

$$\mathbf{N}(\tau) = \mathcal{N}(\tau) + O(\tau^{\frac{1}{2}}) \tag{11.3.3}$$

hold with $\mathcal{N}(\tau)$ given by (11.1.9) with $S = S(x, \tau, \mu h)$ given by $(6.1.7)_{(2)}$.
Moreover,

$$\mathcal{N}(\tau) = \varkappa_0 \tau \bigl(1 + o(1)\bigr) \tag{11.3.4}$$

with the standard Weylian coefficient $\varkappa_0 = \frac{1}{4\pi} \operatorname{vol} X$ where $\operatorname{vol} X$ is the Riemannian two-dimensional volume of X corresponding to the metric (g^{jk}).
(ii) Moreover, for $m \ge -2$, $m_1 \ge -4$ the asymptotics

$$\mathbf{N}(\tau) = \varkappa_0 \tau + O(\tau^{\frac{1}{2}}) \tag{11.3.5}$$

hold.[15]
(iii) Moreover, for $m < -2$, $m_1 > \frac{3}{2}m - 1$[16] *the standard Weylian asymptotics*

$$\mathbf{N}(\tau) = \mathcal{N}^{w}(\tau) + O(\tau^{\frac{1}{2}}) \tag{11.3.6}$$

hold.
(iv) For $m_1 < \min(-4, 4 + 4m)$[16] *with condition (11.3.1) fulfilled the asymptotics*

$$\mathbf{N}(\tau) = \mathcal{N}^{M}(\tau) + O(\tau^{\frac{1}{2}}) \tag{11.3.7}$$

hold where here and below $\mathcal{N}^M$ is given by the same formula as $\mathcal{N}$ with $V = 0$.

I leave the proof to the reader; I only note that condition (6.3.44) for the operator $A - \tau$ is provided by condition (a), (b) or (c) above.

Remark 11.3.2. *(i) The main contribution to remainder estimate (11.3.3) is given by the zone $\{|x| \ge \varepsilon\}$ with arbitrarily small $\varepsilon > 0$. Therefore, under the additional condition $(\Psi 0)$ on the geodesic flow with reflections from ∂X the asymptotics*

$$\mathbf{N}(\tau) = \mathcal{N}(\tau) + \varkappa_1 \tau^{\frac{1}{2}} + o(\tau^{\frac{1}{2}}) \tag{11.3.8}$$

hold with the standard coefficient $\varkappa_1 = \mp\frac{1}{4} \operatorname{vol}' \partial X$ for Dirichlet and Neumann boundary conditions respectively where vol' means the one-dimensional geodesic volume (i.e., length). Moreover, under condition $(\Psi 0)$ in the context of theorem 11.3.1(ii)–(iv) asymptotics (11.3.5)–(11.3.7) remain true with "$O(\tau^{\frac{1}{2}})$" replaced by "$\varkappa_1 \tau^{\frac{1}{2}} + o(\tau^{\frac{1}{2}})$;" in (ii),(iii) one should take strict inequalities for m and m_1: "$m > -2, m_1 > -4$" and "$m_1 > \frac{3}{2}m - 1$" respectively.

---

[15] One can say that these asymptotics are Weylian; nevertheless the standard Weylian asymptotics (11.3.6) do not necessarily hold because $\mathcal{N}^w(\tau)$ does not necessarily converge.

[16] Then $m_1 > 2m$ and therefore condition $(\Sigma 3)$ should be fulfilled.

(ii) One can easily extend the list of statements. In particular, one can treat the case when μ and h vary; however, one should remember that if $m_1 < 2m$ and the coercivity condition $(\Sigma 3)'$ is violated then for $\mu h \to 0$ problems in the spirit of section 11.1 can arise. Moreover, for $\mu \to +\infty$ one should either assume that $\mu \le \tau^{\frac{1}{2}}$ or check the conditions associated with strong magnetic fields at every point of X (not only in a neighborhood of 0); in particular, in this case one should assume that X is a closed compact manifold. Furthermore, in order to avoid the "loss of order" of $\mathcal{N}(\tau)$ one should assume that $\mu h \le t\tau$ with small enough $t > 0$. ◊

Theorem 11.3.3. *Let X be a bounded domain with C^K boundary, $0 \in X$ and let A satisfy conditions $(\Sigma 0)$, $(\Sigma 1)$, $(\Sigma 2)$, $(\Sigma 4)$* [14] *with $\gamma = \epsilon_0 |x|^{1-\beta}$, $\rho = \exp a|x|^\beta$, $\rho_1 = \exp b|x|^\beta$, $\beta < 0$ and $0 < a < b$. Suppose* **either** *$b > 2a$* **or** *the coercivity condition $(\Sigma 3)'$ is fulfilled (for $|x| \le \epsilon$). Furthermore, let one of the following three conditions be fulfilled:*

(a) $b > 2a$ and condition (11.3.1) holds.

(b) $V = v(\theta) \exp a|x|^\beta$ and $F = f(\theta) \exp b|x|^\beta$ for $|x| \le \epsilon$ where $\theta = x|x|^{-1}$;

(c) $b = 2a$ and inequalities (11.3.1) and (11.3.2) are fulfilled with small enough $\varepsilon = \varepsilon(c, \epsilon, \epsilon_0) > 0$.

Then

(i) For $\mu = h = 1$, $\tau \to +\infty$ asymptotics (11.3.3) and (11.3.4) hold.

(ii) Moreover, if condition (b) is fulfilled and **either** *$b < \frac{3}{2}a$* **or** *$b = \frac{3}{2}a$ and $\beta \le 2$ then asymptotics (11.3.6) hold.*

(iii) On the other hand, if $b > 4a$ then asymptotics (11.3.7) hold.

(iv) Remark 11.3.2(i) remains true; surely one should also assume that the inequalities on a, α, β are strict for the improved asymptotics (11.3.6), (11.3.7).

Theorem 11.3.4. *Let X be a connected bounded domain with C^K boundary, let L be a closed curve lying in X and let conditions $(\Sigma 0)$, $(\Sigma 1)$, $(\Sigma 2)$, $(\Sigma 4)$*[17] *be fulfilled with $\gamma = \epsilon_0 \delta(x)$, $\rho = \delta(x)^m$, $\rho_1 = \delta(x)^{m_1}$ where $\delta(x) = \mathrm{dist}(x, L)$, $m \le 0$, $m_1 \le -2$. Moreover, let conditions $(\Sigma 3)'$, $(\Sigma 3)$ be fulfilled for $m_1 < -2$, $m_1 = -2$ respectively for $\delta(x) \le \epsilon$.*

Moreover, let one of the following three conditions be fulfilled:

(a) $m_1 < 2m$ and (11.3.1) holds for $\delta(x) \le \epsilon$;

(b) $V = v(x')\delta(x)^{2m}$ and $F = f(x')\delta(x)^{m_1}$ for $\delta(x) \le \epsilon$ where x' is a parameter on L;

(c) (11.3.1), (11.3.2) hold with small enough $\varepsilon = \varepsilon(c, \epsilon, \epsilon_0) > 0$ for $\delta(x) \le \epsilon$.

Then

(i) For $\mu = h = 1$, $\tau \to +\infty$ the asymptotics

$$\mathbf{N}(\tau) = \mathcal{N}(\tau) + O(\tau^{\frac{1}{2}} \log \tau) \tag{11.3.9}$$

and (11.3.4) hold.

(ii) Moreover, for $m_1 = -2$ the asymptotics

$$\mathbf{N}(\tau) = \varkappa\tau + O(\tau^{\frac{1}{2}} \log \tau) \tag{11.3.10}$$

hold.

(iii) On the other hand, for $m_1 \le 4m + 2$ the asymptotics

[17] Surely, condition $(\Sigma 4)$ is assumed to be fulfilled only for $\delta(x) < \epsilon$.

$$\mathbf{N}(\tau) = \mathcal{N}^M(\tau) + O\big(\tau^{\frac{1}{2}} \log \tau\big) \tag{11.3.11}$$

hold.

Let us consider **unbounded domains**:

Theorem 11.3.5. *Let X be a connected exterior domain*[18] *with C^K boundary. Let the Schrödinger operator A satisfy conditions $(\Sigma 0)$, $(\Sigma 1)$, $(\Sigma 2)$ with $\gamma = \epsilon_0\langle x\rangle$, $\rho = \langle x\rangle^m$, $\rho_1 = \langle x\rangle^{m_1}$, $m > 0$, $m_1 > m - 1$; moreover, let conditions $(\Sigma 4)$, $(\Sigma 3)$ be fulfilled for $|x| \geq c$. Let the stabilization conditions*

$$D^\sigma(V - V^0) = o(|x|^{2m-|\sigma|}), \tag{11.3.12$_1$}$$

$$D^\sigma(F - F^0) = o(|x|^{2m-|\sigma|}) \qquad \forall \sigma : |\sigma| \leq r \tag{11.3.12$_2$}$$

be fulfilled for $|x| \to \infty$ where $V^0, F^0 \in C^k(\mathbb{R}^2 \setminus 0)$ are positively homogeneous functions of degrees $2m$ and m_1 respectively. Then

(i) Let $m_1 > 0$ and $r = 1$ in stabilization conditions $(11.3.12)_{1,2}$. Then for $\mu = 1$, $\tau \to +\infty$, $h = O\big(\min(1, \tau^{\frac{1}{2}(2m-m_1)m^{-1}})\big)$ the asymptotics

$$\mathbf{N}(\tau, h) = \mathcal{N}(\tau, h) + O\big(h^{-1}\tau^{\frac{1}{2}(m_1+2)(m_1+1)^{-1}}\big) \tag{11.3.13}$$

and

$$\mathbf{N}(\tau, h) = \mathcal{N}^w(\tau, h) + O\big(h^{-1}\tau^{\frac{1}{2}(m_1+2)(m_1+1)^{-1}}\big) + O\big(\tau^{(m_1+1-m)m^{-1}}\big) \tag{11.3.14}$$

hold.

(ii) Let $m_1 < 0$ and $r = 2$ in stabilization conditions $(11.3.12)_{1,2}$. Let us assume that the function $(V^0)^{m_1}(F^0)^{-2m}$ has no degenerate critical point on the circle $\mathbb{S}^1$. Then for $\tau \to +\infty$, $\mu = 1$, $h = O(1)$ the asymptotics

$$\mathbf{N}(\tau, h) = \mathcal{N}(\tau, h) + O\Big(h^{-1}\tau^{\frac{1}{2}(2m-m_1)m^{-1}}\Big(\big(\log(h\tau^{3m_1-4m+2})\big)_+ + 1\Big)\Big) \tag{11.3.15}$$

and

$$\mathbf{N}(\tau, h) = \mathcal{N}^w(\tau, h) + O\Big(h^{-1}\tau^{\frac{1}{2}(2m-m_1)m^{-1}}\Big(\big(\log(h\tau^{3m_1-4m+2})\big)_+ + 1\Big)\Big) + O\big(\tau^{(m_1+1-m)m^{-1}}\big) \tag{11.3.16}$$

hold.

In both cases, under the additional condition $h = o\big(\tau^{\frac{1}{2}(2m-m_1)m^{-1}}\big)$ (which is a restriction only for $2m < m_1$),

$$\mathcal{N}(\tau, h) = \varkappa_0 h^{-2}\tau^{1+m^{-1}}\big(1 + o(1)\big) \tag{11.3.17}$$

with the standard coefficient $\varkappa_0 > 0$, and in the general case the Weylian formula gives the correct order of $\mathbf{N}$.

The case when $m_1 > 2m$ and h tends to 0 more slowly than is indicated above, is completely non-Weylian and will be treated below.

Let us note that for $m_1 > 0$ the main contribution to the remainder estimate (11.3.13) is given by the zone $\{\varepsilon \leq |x|\tau^{-\frac{1}{2}(m_1+1)^{-1}} \leq \varepsilon^{-1}\}$ with an arbitrarily small constant $\varepsilon > 0$. In this zone the magnetic field is normal (for every fixed $\varepsilon > 0$) and applying the improved Weylian asymptotics here we obtain the following

[18] I.e., a domain with compact complement $\complement X$ in $\mathbb{R}^d$.

Theorem 11.3.6. *Let $m_1 > 0$ and let the conditions of theorem 11.3.5 be fulfilled. Moreover, let the additional stabilization conditions*

$$D^\sigma(g^{jk} - g^{jk0}) = o(|x|^{-|\sigma|}), \tag{11.3.18$_1$}$$

$$D^\sigma(V_j - V_j^0) = o(|x|^{m_1+1-|\sigma|}) \qquad \forall\sigma : |\sigma| \le 1 \tag{11.3.18$_2$}$$

be fulfilled for $|x| \to \infty$ with the positively homogeneous functions $g^{jk0}, V_j^0 \in C^K(\mathbb{R}^2 \setminus 0)$ of degrees 0 and $m_1 + 1$ respectively. Let the standard condition (Ψ) be fulfilled for the Hamiltonian

$$H^0 = |x|\bigl(g^{jk0}(\xi_j - V_j^0)(\xi_k - V_k^0) - 1\bigr) \tag{11.3.19}$$

at energy level 0. Then for $\tau \to +\infty$, $\mu = 1$, $h = O\bigl(\min(1, \tau^{\frac{1}{2}(2m-m_1)m^{-1}})\bigr)$ the asymptotics

$$\mathbf{N}(\tau, h) = \mathcal{N}(\tau, h) + o\bigl(h^{-1}\tau^{\frac{1}{2}(m_1+2)(m_1+1)^{-1}}\bigr) \tag{11.3.20}$$

and

$$\mathbf{N}(\tau, h) = \mathcal{N}^w(\tau, h) + o\bigl(h^{-1}\tau^{\frac{1}{2}(m_1+2)(m_1+1)^{-1}}\bigr) + O\bigl(\tau^{(m_1+1-m)m^{-1}}\bigr) \tag{11.3.21}$$

hold.

Remark. *In the previous chapter we saw that it was already possible that the main contributions to the principal part and to the remainder estimate are made by different domains in Riesz means asymptotics. However, it is a new effect that the same is possible without averaging. In both cases this yields that one deduces improved asymptotics under a condition on the shortened Hamiltonian.* ◊

On the other hand, for $m_1 = 0$ one can apply the arguments associated with $\nu < 1$ and obtain the following

Theorem 11.3.7. *Let X be a connected exterior domain with C^K boundary. Let the Schrödinger operator A satisfy conditions ($\Sigma 0$), ($\Sigma 1$), ($\Sigma 2$) with $\gamma = \epsilon_0\langle x\rangle$, $\rho = \langle x\rangle^m$, $\rho_1 = 1$, $m \in (0, 1)$, $m_1 > -1$; moreover, let conditions ($\Sigma 4$), ($\Sigma 3$) be fulfilled for $|x| \ge c$. Furthermore, let stabilization conditions (11.3.12)$_1$ and*

$$D^\sigma(g^{jk} - \bar{g}^{jk}) = o(|x|^{2m-2-|\sigma|}), \tag{11.3.22$_1$}$$

$$D^\sigma(F - \bar{F}) = o(|x|^{2m-2-|\sigma|}) \qquad \forall\sigma : |\sigma| \le 1 \tag{11.3.22$_2$}$$

be fulfilled for $|x| \to \infty$ with $\bar{F} = \text{const} > 0$ and $\bar{g}^{jk} = \text{const}$.

Then for $\tau \to +\infty$, $\mu = 1$, $h = O(1)$ asymptotics (11.3.13) and (11.3.14) hold.

Let us now treat the case $\tau \to -0$ where 0 is the **bottom** of the essential spectrum:

Theorem 11.3.8. *Let X be a connected unbounded domain. Let the Schrödinger operator A satisfy conditions ($\Sigma 0$), ($\Sigma 1$), ($\Sigma 2$) with $\gamma = \epsilon_0\langle x\rangle$, $\rho = \langle x\rangle^m$, $\rho_1 = \langle x\rangle^{m_1}$, $m \in (-1, 0)$, $m_1 > m - 1$; moreover, let condition ($\Sigma 4$) be fulfilled for $|x| \ge c$. Let us assume that*

$$B\bigl(y, \gamma(y)\bigr) \subset X \qquad \forall y \in X : |y| \ge c,\ V \le \epsilon\rho^2. \tag{11.3.23}$$

Then

(i) Let $m_1 > 2m$ and let stabilization conditions (11.3.12)$_{1,2}$ be fulfilled with $r = 1$. Moreover, assume 0 is not a critical value of V^0. Then for $\tau \to -0$, $\mu = 1$, $h = O\bigl(\tau^{\frac{1}{2}(2m-m_1)m^{-1}}\bigr)$ the asymptotics

$$\mathbf{N}(\tau,h) = \mathcal{N}(\tau,h) + O(h^{-1}) \tag{11.3.24}$$

and

$$\mathbf{N}(\tau,h) = \mathcal{N}^w(\tau,h) + O(h^{-1}) + O\bigl(\tau^{(m_1+1-m)m^{-1}}\bigr) \tag{11.3.25}$$

hold.

(ii)[19]

In addition let the standard condition (Ψ) *on the Hamiltonian flow with reflections at the boundary for the standard Hamiltonian*

$$H_{\mathrm{st}}(x,\xi) = g^{jk}(\xi_j - V_j)(\xi_k - V_k) + V \tag{11.3.26}$$

be fulfilled in X *at the energy level* 0. *Then the asymptotics*

$$\mathbf{N}(\tau,h) = \mathcal{N}(\tau,h) + \varkappa_1 h^{-1} + o(h^{-1}) \tag{11.3.27}$$

and

$$\mathbf{N}(\tau,h) = \mathcal{N}^w(\tau,h) + \varkappa_1 h^{-1} + o(h^{-1}) + O\bigl(\tau^{(m_1+1-m)m^{-1}}\bigr) \tag{11.3.28}$$

hold with the standard coefficient $\varkappa_1$ *given by* (11.2.24).

(iii) Let $m_1 < 2m$ *and let stabilization conditions* $(11.3.12)_{1,2}$ *be fulfilled with* $r = 2$. *On the circle* $\mathbb{S}^1$ *let the function* $(V^0)^{m_1}(F^0)^{-2m}$ *have no degenerate critical point and assume* 0 *is not a critical value of* V^0. *Then for* $\tau \to -0$, $h = 1$, $\mu = O(1)$ *asymptotics* (11.3.13) *and* (11.3.14) *hold.*

Theorem 11.3.9. *Let* X *be a connected unbounded domain with* C^K *boundary. Let the Schrödinger operator* A *satisfy conditions* ($\Sigma 0$), ($\Sigma 1$), ($\Sigma 2$) *with* $\gamma = \epsilon_0\langle x\rangle$, $\rho = \langle x\rangle^{-1}\log^\alpha\langle x\rangle$, $\rho_1 = \langle x\rangle^{-2}\log^\beta\langle x\rangle$, $\alpha \ge -\frac12$, $\beta > \max(\alpha+\frac12, 2\alpha) \ge 0$. *Moreover, let conditions* (11.3.23), ($\Sigma 4$), ($\Sigma 5$) *be fulfilled for* $|x| \ge c$ *with* $\varrho_3 = \log^{-1}\langle x\rangle$. *Let us assume that*

$$\langle x,\nabla\rangle F = -2F\bigl(1+o(1)\bigr), \tag{11.3.29$_1$}$$

$$\langle x,\nabla\rangle \frac{V}{F} = (2\alpha-\beta)\frac{V}{F}\log^{-1}|x|\bigl(1+o(1)\bigr) \tag{11.3.29$_2$}$$

as $|x| \to \infty$.

Then

(i) For $\tau \to -0$, $\mu = 1$, $h = O(\log^{-\beta-1-\delta}|\tau|)$ *with arbitrarily small* $\delta > 0$ *the asymptotics* (11.3.24) *and*

$$\mathbf{N}(\tau,h) = \mathcal{N}^w(\tau,h) + O(h^{-1}) + O\bigl(\log^{2\beta-2\alpha}|\tau|\bigr) \tag{11.3.30}$$

hold.

(ii) Moreover, let the standard Hamiltonian (11.3.26) *satisfy condition* (Ψ) *at the energy level* 0. *Then the asymptotics* (11.3.27) *and*

$$\mathbf{N}(\tau,h) = \mathcal{N}^w(\tau,h) + \varkappa_1 h^{-1} + o(h^{-1}) + O(\log^{2\beta-2\alpha}|\tau|) \tag{11.3.31}$$

hold.

(iii) Moreover, for $\beta > 2\alpha+2$ *one can weaken the restriction on* h *in (i),(ii) and take* $h = O(\log^{2\alpha-\beta}|\tau|)$.

[19] We use the fact that the principal contribution to the remainder estimate is given by the zone $\{X, |x| \le \varepsilon^{-1}\}$ with an arbitrarily small constant $\varepsilon > 0$.

(iv) In all cases under condition (11.1.13)$^\#$

$$\mathcal{N}(\tau,h)\asymp\mathcal{N}^w(\tau,h)\asymp h^{-2}\log^{2\alpha+1}|\tau| \tag{11.3.32$_1$}$$

for $\alpha>-\frac{1}{2}$ *and*

$$\mathcal{N}(\tau,h)\asymp\mathcal{N}^w(\tau,h)\asymp h^{-2}\log\bigl|\log|\tau|\bigr| \tag{11.3.32$_2$}$$

for $\alpha=-\frac{1}{2}$.

Finally, let us apply theorem 6.4.14 and the "splitting zone into strips arguments" in the zone $\{C_1\bigl(\frac{\tau}{h}\bigr)^{(\beta+1)^{-1}}\le\log|x|\le C\tau^{\frac{1}{2}\alpha^{-1}}\}$ (this zone is empty for $h\le\tau^{\frac{1}{2}(2\alpha-\beta-1)\alpha^{-1}}$). These arguments were used in the proof of theorem 11.1.7 and corollary 11.1.10. We obtain the following statements:

Theorem 11.3.10. *Let X be a connected exterior domain with C^K boundary. Let the Schrödinger operator A satisfy conditions ($\Sigma0$), ($\Sigma1$), ($\Sigma2$) with $\gamma=\epsilon_0\langle x\rangle$, $\rho=\log^\alpha\langle x\rangle$, $\rho_1=\log^\beta\langle x\rangle$, $\alpha>0$, $\beta>0$. Moreover, for $|x|\ge c$ let conditions ($\Sigma3$), ($\Sigma4$), ($\Sigma5$) be fulfilled with $\varrho_3=\varrho_2=\log^{-1}\langle x\rangle$.*

Furthermore, let us assume that

$$\langle x,\nabla\rangle F=\beta F\log^{-1}|x|\bigl(1+o(1)\bigr), \tag{11.3.33$_1$}$$

$$\langle x,\nabla\rangle V=2\alpha V\log^{-1}|x|\bigl(1+o(1)\bigr) \tag{11.3.33$_2$}$$

as $|x|\to\infty$.

Finally, let $\beta\le\bar\beta\le\beta+1$ *and*

$$\kappa=O\bigl(|x|^{-2}\log^{-\bar\beta-1}|x|\bigr), \tag{11.3.34$_1$}$$

$$\Delta_gF=O\bigl(|x|^{-2}\log^{2\beta-\bar\beta-1}|x|\bigr), \tag{11.3.34$_2$}$$

$$\Delta_gV=O\bigl(|x|^{-2}\log^{2\alpha+\beta-\bar\beta-1}|x|\bigr) \tag{11.3.34$_3$}$$

as $|x|\to\infty$ where κ is the curvature corresponding to the metric $g^{jk}F^{-1}$ and $\Delta_g=g^{jk}\partial_j\partial_k$.[20]

Then

(i) For $\tau\to+\infty$, $\mu=1$, $h=O\bigl(\min(1,\tau^{\frac{1}{2}(2\alpha-\beta)\alpha^{-1}})\bigr)$ the asymptotics

$$\mathbf{N}(\tau,h)=\mathcal{N}(\tau,h)+O\bigl(h^{-1}\tau\log^{-\beta}\tau\bigr) \tag{11.3.35$_1$}$$

hold for $\bar\beta>1$, the asymptotics

$$\mathbf{N}(\tau,h)=\mathcal{N}(\tau,h)+O\bigl(h^{-1}\tau\bigr) \tag{11.3.35$_2$}$$

hold for $\bar\beta=1$ and the asymptotics

$$\mathbf{N}(\tau,h)=\mathcal{N}(\tau,h)+O\bigl(h^{-1}\tau^{\frac{1}{2}(2\alpha+1-\bar\beta)\alpha^{-1}}\bigr) \tag{11.3.35$_3$}$$

hold for $\bar\beta<1$.

(ii) Moreover, under the additional condition $h=O\bigl(\tau^{\frac{1}{2}(2\alpha-\beta-1)\alpha^{-1}-\delta}\bigr)$ with arbitrarily small $\delta>0$ asymptotics (11.3.35)$_1$ *also hold for $\bar\beta\le1$.*

(iii) Furthermore, in all these cases

$$\varepsilon h^{-2}\exp\bigl(\varepsilon|\tau|^{\frac{1}{2}\alpha^{-1}}\bigr)\le\mathcal{N}(\tau,h)\le Ch^{-2}\exp\bigl(C|\tau|^{\frac{1}{2}\alpha^{-1}}\bigr). \tag{11.3.36}$$

[20] For $\bar\beta=\beta$ these conditions are fulfilled automatically.

Theorem 11.3.11. *Let X be a connected unbounded domain with C^K boundary. Let the Schrödinger operator A satisfy conditions $(\Sigma 0)$, $(\Sigma 1)$, $(\Sigma 2)$ with $\gamma = \epsilon_0 \langle x \rangle$, $\rho = \log^\alpha \langle x \rangle$, $\rho_1 = \log^\beta \langle x \rangle$, $\alpha < 0$, $\beta > 2\alpha$. Moreover, for $|x| \ge c$ let conditions $(\Sigma 4)$, $(\Sigma 5)$ be fulfilled with $\varrho_3 = \varrho_2 = \log^{-1} \langle x \rangle$. Furthermore, let conditions (11.3.23) be fulfilled and conditions $(11.3.33)_{1,2}$, $(11.3.34)_{1-3}$ be fulfilled with $\beta \le \bar\beta \le \beta + 1$. Then*

(i) For $\tau \to -0$, $\mu = 1$, $h = O(\tau^{\frac12 (2\alpha - \beta)\alpha^{-1}})$ asymptotics (11.3.24) hold for $\bar\beta \ge 2\alpha + 1$ and for $\bar\beta < 2\alpha + 1$, $h = O(|\tau|^{\frac12 (2\alpha - \beta - 1)\alpha^{-1} + \delta})$ with arbitrarily small $\delta > 0$. On the other hand, asymptotics $(11.3.35)_3$ hold for $\bar\beta < 2\alpha + 1$, $h = O(|\tau|^{\frac12 (2\alpha - \beta - 1)\alpha^{-1}})$ and the asymptotics

$$\mathbf{N}(\tau, h) = \mathcal{N}(\tau, h) + O\bigl(h^{(\beta - \bar\beta + 2)(2\alpha - \beta - 1)^{-1}}\bigr) \tag{11.3.37}$$

hold for $\bar\beta < 2\alpha + 1$, $h \ge |\tau|^{\frac12 (2\alpha - \beta - 1)\alpha^{-1}}$.

(ii) Let one of $\bar\beta > 2\alpha + 1$ or $\bar\beta = 2\alpha + 1$ and $h = o(|\tau|^{\frac12 (2\alpha - \beta - 1)\alpha^{-1}})$ or $\bar\beta < 2\alpha + 1$ and $h = O(|\tau|^{\frac12 (2\alpha - \beta - 1)\alpha^{-1} + \delta})$ be fulfilled. Furthermore, let condition (Ψ) be fulfilled for the standard Hamiltonian (11.3.26) at the energy level 0. Then asymptotics (11.3.27) hold.

(iii) Moreover, the right-hand inequality (11.3.36) holds and under condition $(11.1.13)^\#$ the left-hand inequality (11.3.36) holds as well.

I leave it to the reader to formulate and prove statements similar to theorems 11.3.10, 11.3.11 and 11.3.13 (see below) in the case when $\rho = \log^\alpha \langle x \rangle$ and $\rho_1 = \langle x \rangle^{m_1}$ with $m_1 > 0$.

11.3.2 Asymptotics non-Weylian in principal

Let us consider asymptotics which are non-Weylian in principal. The following statement is proven by the same method as theorems 11.3.5, 11.3.6:

Theorem 11.3.12. *Let X be a connected exterior domain with C^K boundary. Let the Schrödinger operator A satisfy conditions $(\Sigma 0)$, $(\Sigma 1)$, $(\Sigma 2)$ with $\gamma = \epsilon_0 \langle x \rangle$, $\rho_1 = \langle x \rangle^m$, $\rho = \langle x \rangle^{m_1}$, $m_1 > 2m \ge 0$; moreover, let condition $(\Sigma 4)$ be fulfilled for $|x| \ge c$. Let the condition*

$$|\nabla F| \ge \epsilon \rho_1 \gamma^{-1}. \tag{11.3.38}$$

be fulfilled for $|x| \ge c$. Then

(i) For $\tau \to +\infty$, $\mu = 1$, $h = O(1)$ such that $h \ge C_0 \tau^{\frac12 (2m - m_1)m^{-1}}$ with a large enough constant $C_0 = C_0(c, \epsilon, \epsilon_0)$ asymptotics (11.3.13) hold and

$$\mathcal{N}(\tau, h) \asymp \tau^{(2+m_1)m_1^{-1}} h^{-(m_1+1)m_1^{-1}}. \tag{11.3.39}$$

(ii) Moreover, let stabilization conditions $(11.3.18)_{1,2}$ be fulfilled and let the Hamiltonian (11.3.19) satisfy the standard condition (Ψ) at the energy level 0. Then asymptotics (11.3.20) hold.

I leave it to the reader to estimate the error $\mathcal{N}(\tau, h) - \mathcal{N}^M(\tau, h)$.

The following statements are proven by the same method as theorem 11.3.10:

Theorem 11.3.13. *Let X be a connected exterior domain with C^K boundary. Let the Schrödinger operator A satisfy conditions $(\Sigma 0)$, $(\Sigma 1)$, $(\Sigma 2)$ with $\gamma = \epsilon_0 \langle x \rangle$, $\rho = \log^\alpha \langle x \rangle$, $\rho_1 = \log^\beta \langle x \rangle$, $\beta > 2\alpha \ge 0$; moreover, let conditions $(\Sigma 4)$, $(\Sigma 5)$ be fulfilled for $|x| \ge c$ with $\varrho_2 = \varrho_3 = \log^{-1} \langle x \rangle$. Let*

$$|\nabla F| \ge \epsilon \rho_1 \varrho_2 \gamma^{-1} \qquad \text{for } |x| \ge c \tag{11.3.40}$$

and let conditions $(11.3.34)_{1-3}$ *be fulfilled with* $\beta \le \bar\beta \le \beta+1$. *Then*

(i) For $\tau \to +\infty$, $\mu = 1$, $h = O(1)$ *such that* $h \ge C_0\tau^{\frac12(2\alpha-\beta)\alpha^{-1}}$ *asymptotics* $(11.3.35)_{1,2,3}$ *hold for* $\bar\beta > 1$, $\bar\beta = 1$ *and* $\bar\beta < 1$ *respectively.*

(ii) Moreover, in all cases

$$\varepsilon \exp\bigl(\varepsilon\tau^{\beta^{-1}}h^{-\beta^{-1}}\bigr) \le \mathcal{N}(\tau,h) \le C\exp\bigl(C\tau^{\beta^{-1}}h^{-\beta^{-1}}\bigr). \tag{11.3.41}$$

Theorem 11.3.14. *Let X be a connected exterior domain with* C^K *boundary. Let the Schrödinger operator A satisfy conditions* $(\Sigma0)$, $(\Sigma1)$, $(\Sigma2)$ *with* $\gamma = \epsilon_0\langle x\rangle^{1-\beta}$, $\rho = \exp a\langle x\rangle^\beta$, $\rho_1 = \exp b\langle x\rangle^\beta$, $\beta > 0$, $b > 2a > 0$; *moreover, let conditions* $(\Sigma4)$ *and* (11.3.38) *be fulfilled for* $|x| \ge c$. *Then for* $\tau \to +\infty$, $\mu = h = 1$ *the asymptotics*

$$\mathbf{N}(\tau) = \mathcal{N}(\tau) + O\bigl(\tau^{\frac12}\log^{(\beta+1)\beta^{-1}}\tau\bigr) \tag{11.3.42}$$

hold with

$$\mathcal{N}(\tau) \asymp \tau\log^{2\beta^{-1}}\tau. \tag{11.3.43}$$

One can easily prove the following statement with three variable parameters:

Theorem 11.3.15. *Let X be a connected domain with* $0 \in X$. *Let the Schrödinger operator A satisfy conditions* $(\Sigma0)$, $(\Sigma1)$, $(\Sigma2)$ *with* $\gamma = \epsilon_0|x|$, $\rho = |x|^m$, $\rho_1 = |x|^{m_1}$, $m > 0$, $m_1 > 0$; *moreover, let conditions* $(\Sigma4)$ *and* (11.3.38) *be fulfilled for* $|x| \le \epsilon$. *Let us assume that* μ, h [21] *are such that*

$$V + \mu hF \ge \epsilon(\rho^2 + \mu h\rho_1) \qquad \text{in } X. \tag{11.3.44}$$

(i) For $\tau \to +0$, $h > 0$, $\mu > 0$ *such that* $\tau^{\frac12(1-m_1)m^{-1}} \le \varepsilon\mu h$ *with a small enough constant* $\varepsilon = \varepsilon(c,\epsilon,\epsilon_0) > 0$ *the asymptotics*

$$\mathbf{N}(\tau,\mu,h) = \mathcal{N}(\tau,\mu,h) + O\bigl(h^{-1}\tau^{\frac12(m_1+2)(m_1+1)^{-1}}\mu^{-(m_1+1)^{-1}}\bigr) \tag{11.3.45}$$

hold provided $h^{2(m_1+1)}\mu^2 = O(\tau^{m_1+2})$ *with*

$$\mathcal{N}(\tau,\mu,h) \asymp \tau^{2m_1^{-1}}\mu^{-2m_1^{-1}}h^{-(2m_1+2)m_1^{-1}}. \tag{11.3.46}$$

(ii) Moreover, if stabilization conditions $(11.3.18)_{1,2}$ *are fulfilled and if the Hamiltonian* (11.3.19) *satisfies the standard condition* (Ψ) *at the energy level* 0 *then the asymptotics*

$$\mathbf{N}(\tau,\mu,h) = \mathcal{N}(\tau,\mu,h) + o\bigl(h^{-1}\tau^{\frac12(m+1)(m_1+2)^{-1}}\mu^{-(m_1+1)^{-1}}\bigr) \tag{11.3.47}$$

hold provided $h^{2(m_1+1)}\mu^2 = o(\tau^{m_1+2})$.

One can easily see that in the context of this theorem for given $\tau > 0$ neither μ nor h can be chosen completely arbitrarily. I leave it to the reader to prove statements similar to the other theorems above in the case when $\mu \to +\infty$ as well.

11.3.3 Operators with stabilization at infinity

Let us treat the case when F, g^{jk} and V tend to constants at infinity and are regular at all the points of the exterior domain X. Namely in this case the difference between two- and three-dimensional cases is the most fundamental because of the following statement which the reader can easily prove:

[21] Not necessarily fixed.

Theorem 11.3.16. *Let X be an exterior domain with C^K boundary. Let the Schrödinger operator A satisfy conditions ($\Sigma 0$), ($\Sigma 1$), ($\Sigma 2$) and ($\Sigma 4$), ($\Sigma 5$) for $|x| \ge c$ with $\rho = \rho_1 = 1$, γ, ϱ_2 and ϱ_3 satisfying standard conditions (8.1.2) and such that $\gamma \to \infty$, $\varrho_2 \to 0$, $\varrho_3 \to 0$ as $|x| \to \infty$. Then*

(i) For $d = 2$, $\operatorname{Spec}_{\mathrm{ess}}(A)$ is the set of limit points of the set $\{V + (2n+1)\mu h F, n \in \mathbb{Z}^+\}$ as $|x| \to \infty$.

(ii) For $d = 3$, $\operatorname{Spec}_{\mathrm{ess}}(A) = [\tau_*, \infty)$ with $\tau_* = \liminf_{|x|\to\infty}(V + \mu h F)$.

Then theorem 6.4.19 immediately yields the following statements:

Theorem 11.3.17. *Let X be a connected exterior domain with C^K boundary. Let the Schrödinger operator A satisfy conditions ($\Sigma 0$), ($\Sigma 1$), ($\Sigma 2$) with $\gamma = \epsilon_0 \langle x \rangle$, $\rho = \rho_1 = 1$; moreover, let conditions ($\Sigma 4$), ($\Sigma 5$) be fulfilled for $|x| \ge c$ with $\varrho_2 = \varrho_3 = \langle x \rangle^m$, $m < 0$. Moreover, let*

$$|V| + |F - \bar{F}| \le c\varrho_3 \tag{11.3.48}$$

with $\bar{F} = \mathrm{const} > 0$.

Let $n \in \mathbb{Z}^+$ be fixed, $\mu = h = 1$ and let the condition

$$\pm V_{(n)} \ge -\epsilon\varrho_3 \implies |\nabla V_{(n)}| \ge \epsilon\varrho_3\gamma^{-1} \qquad \text{for } |x| \ge c \tag{11.3.49$_{n,\pm}$}$$

be fulfilled with

$$V_{(n)} = V + (2n+1)(F - \bar{F}). \tag{11.3.50}$$

Then

(i) Let condition $(11.3.49)_{n,-}$ be fulfilled and let us fix $\tau_1 \in (\tau^*_{n-1}, \tau^*_n)$ for $n \ge 1$, $\tau_1 \in (-\infty, \tau^*_0)$ for $n = 0$ where $\tau^*_n = (2n+1)\bar{F}$. Then for $\tau_2 = \tau^*_n - \eta$ with $\eta \to +0$ the asymptotics

$$\mathbf{N}(\tau_1, \tau^*_n - \eta) = \mathcal{N}_{n,-}(\eta) + O(\log \eta) \tag{11.3.51$_{n,-}$}$$

hold. Let condition $(11.3.49)_{n,+}$ be fulfilled and let us fix $\tau_2 \in (\tau^*_n, \tau^*_{n+1})$. Then for $\tau_1 = \tau^*_n + \eta$ with $\eta \to +0$ the asymptotics

$$\mathbf{N}(\tau^*_n + \eta, \tau_2) = \mathcal{N}_{n,+}(\eta) + O(\log \eta) \tag{11.3.51$_{n,+}$}$$

hold; here and below $\mathcal{N}_{n,\pm}(\eta)$ are defined by the formula

$$\mathcal{N}_{n,\pm}(\eta) = \frac{1}{2\pi}\int_{\{\pm V_{(n)} > \eta\}} F\sqrt{g}\,dx \tag{11.3.52}$$

and in our conditions $\mathcal{N}_{n,\pm}(\eta) = O(\eta^{\frac{2}{m}})$. Moreover,

$$\mathcal{N}_{n,\pm}(\eta) \asymp \eta^{\frac{2}{m}} \tag{11.3.53$_{n,\pm}$}$$

provided

$$\pm V_{(n)} \ge \epsilon\varrho_3 \qquad \text{for } x \in \Gamma, |x| \ge c \tag{11.3.54$_{n,\pm}$}$$

where Γ is a non-empty open sector (cone) in $\mathbb{R}^d$.

(ii) Moreover, if

$$\pm V_{(n)} \ge \epsilon\varrho_3 \qquad \text{for } |x| \ge c \tag{11.3.55$_{n,\pm}$}$$

then[22] *the improved asymptotics* $(11.3.54)_{n,\pm}$ *hold:*

$$\mathbf{N}(\tau_1,\tau_2)=\mathcal{N}_{n,\pm}(\eta)+O(1). \tag{11.3.56$_{n,\pm}$}$$

Furthermore, in this case the points $\tau_n^*\mp 0$ *are not limit points of the discrete spectrum.*

Let us formulate similar statements in the case when $h\to +0$ and $\mu=h^{-1}$; in this case the obvious conditions in the bounded part of X should also be fulfilled.

Let us now treat the case when $\mu h=1$ is fixed but $h\to+0$ and $\mu\to\infty$ and τ is the same as above. In this case conditions $(\Sigma 4)$, $(\Sigma 5)$ should be assumed in the whole domain X and ∂X should be empty. First, we treat the case when τ_1 and τ_2 are fixed.

Theorem 11.3.18. *Let* $X=\mathbb{R}^2$ *and let the Schrödinger operator* A *satisfy conditions* $(\Sigma 0)$, $(\Sigma 1)$, $(\Sigma 2)$, $(\Sigma 4)$, $(\Sigma 5)$ *with* $\rho=\rho_1=1$, $\gamma\ge\epsilon_0\langle x\rangle^\sigma$, $\varrho_2=\varrho_3\ge\epsilon\langle x\rangle^{-s}$ *such that* $\varrho_2=o(1)$ *as* $|x|\to\infty$. *Let*

$$F=\bar F+o(1),\qquad V=o(1)\qquad \text{as } |x|\to\infty \tag{11.3.57}$$

with $\bar F=\text{const}>0$. *Let* $n\in\mathbb{Z}^+$, τ_1 *and* τ_2 *be fixed such that*

$$\begin{aligned}&\tau_n^*<\tau_1<\tau_2<\tau_{n+1}^* &&\text{for } n\ge 1;\\ -&\infty\le\tau_1<\tau_2<\tau_0^* &&\text{for } n=1.\end{aligned} \tag{11.3.58}$$

Then

(i) Let us assume that there do not exist $n'\in\mathbb{Z}^+$, $k=1,2$ *such that* 0 *is a degenerate critical value of* $\tau_k-V_{(n')}$. *Then for* $h\to+0$, $\mu=h^{-1}$ *the asymptotics*

$$\mathbf{N}(\tau_1,\tau_2,\mu,h)=\mathcal{N}(\tau_1,\tau_2,\mu,h)+O(\log h) \tag{11.3.59}$$

hold with

$$\mathcal{N}(\tau_1,\tau_2,\mu,h)=\frac{1}{2\pi h^2}\int \mathbf{n}(\tau_1,\tau_2,x)F\sqrt{g}\,dx, \tag{11.3.60}$$

$$\mathbf{n}(\tau_1,\tau_2,x)=\#\{n'\in\mathbb{Z}^+:\tau_1<V_{(n')}<\tau_2\}; \tag{11.3.61}$$

recall that $\#M$ *means the number of elements in the set* M.

(ii) Let us in addition assume that there do not exist $n'\in\mathbb{Z}^+$, $k=1,2$ *such that* 0 *is a minimax value of* $\tau_k-V_{(n')}$. *Then for* $h\to+0$, $\mu=h^{-1}$ *the asymptotics*

$$\mathbf{N}(\tau_1,\tau_2,\mu,h)=\mathcal{N}(\tau_1,\tau_2,\mu,h)+O(1) \tag{11.3.62}$$

hold.

(iii) On the other hand, if $\{x:\tau_1\le V_{(n')}\le\tau_2\}=\emptyset$ *for every* $n'\in\mathbb{Z}^+$ *then*

$$\mathbf{N}(\tau_1,\tau_2,\mu,h)=0$$

for sufficiently small $h>0$ *and* $\mu=h^{-1}$.

The following statement is a combination of the two preceding:

[22] In this case condition $(11.3.49)_{n,\pm}$ yields that

$$|\nabla V_{(n)}|\ge\epsilon\varrho_3\gamma^{-1}\qquad\text{for } |x|\ge c. \tag{11.3.49$^*_{n,\pm}$}$$

Theorem 11.3.19. *Let $X = \mathbb{R}^2$ and let the Schrödinger operator A satisfy conditions $(\Sigma 0)$, $(\Sigma 1)$, $(\Sigma 2)$, $(\Sigma 4)$, $(\Sigma 5)$ with $\rho = \rho_1 = 1$, $\gamma = \epsilon_0\langle x\rangle$, $\varrho_2 = \varrho_3 \geq \langle x\rangle^{-s}$. Moreover, let condition $(11.3.49)_{n,\mp}$ also be fulfilled. Let us fix $\bar\tau_1$ and $\bar\tau_2$ satisfying the following condition: there do not exist $n' \in \mathbb{Z}^+$, $k = 1,2$ such that $\bar\tau_k$ is a critical value of $V_{(n')}$. Then*

(i) Let condition $(11.3.49)_{n,-}$ be fulfilled, $\bar\tau_1 \in (\tau^*_{n-1}, \tau^*_n)$ for $n \geq 1$, $\bar\tau_1 \in [-\infty, \tau^*_0)$ for $n = 0$ and $\bar\tau_2 = \tau^*_n$. Then for $\tau_1 = \bar\tau_1$, $\tau_2 = \bar\tau_2 - \eta$ with $\eta \to +0$, $h \to +0$, $\mu = h^{-1}$ the asymptotics

$$\mathbf{N}(\tau_1,\tau_2,\mu,h) = \mathcal{N}(\tau_1,\tau_2,\mu,h) + O(\log h) + O(\log\eta) \tag{11.3.63}$$

hold with $\mathcal{N}$ given by (11.3.60)–(11.3.61).

On the other hand, let condition $(11.3.49)_{n,+}$ be fulfilled and $\bar\tau_1 = \tau^*_n$, $\bar\tau_2 \in (\tau^*_n, \tau^*_{n+1})$. Then the above asymptotics hold for $\tau_1 = \bar\tau_1 + \eta$, $\tau_2 = \bar\tau_2$, $\eta \to +0$, $h \to +0$, $\mu = h^{-1}$.

(ii) Let us in addition assume that there do not exist $n' \in \mathbb{Z}^+$, $k = 1,2$ such that $\bar\tau_k$ is a minimax value of $V_{(n')}$. Then one can replace $O(\log h)$ by $O(1)$ in asymptotics (11.3.63) in both cases described in (i).

(iii) On the other hand, let condition $(11.3.56)_{n,\pm}$ be fulfilled. Then one can replace $O(\log\eta)$ by $O(1)$ in asymptotics (11.3.63) in the corresponding case.

(iv) One can combine statements (ii) and (iii).

(v) Finally, if $\{x : \tau_1 \leq V_{(n')} \leq \tau_2\} = \emptyset$ for every $n' \in \mathbb{Z}^+$ in the context of (i) then

$$\mathbf{N}(\tau_1,\tau_2,\mu,h) = 0$$

for sufficiently small $h > 0$ and $\mu = h^{-1}$.

On the other hand, theorem 6.1.13 also yields the following

Theorem 11.3.20. *Let X be an exterior domain with C^K boundary. Let the Schrödinger operator A satisfy conditions $(\Sigma 0)$, $(\Sigma 1)$, $(\Sigma 2)$ with $\gamma = \epsilon_0\langle x\rangle$, $\rho = \rho_1 = 1$, $\varrho_3 = \varrho_4 = \tilde\varrho\log^{-1}\langle x\rangle$, $\tilde\varrho = \log^\alpha\langle x\rangle$, $\alpha < 0$. Let*

$$|V| + |F - \bar F| \leq c\tilde\varrho \tag{11.3.64}$$

with $\bar F = \mathrm{const} > 0$. Moreover, let

$$\pm V_{(n)} \geq -\epsilon\tilde\varrho \implies |\nabla V_{(n)}| \geq \epsilon\varrho_3\gamma^{-1} \qquad \text{for } |x| \geq c. \tag{11.3.65$_{n,\pm}$}$$

Then

(i) Let condition $(11.3.65)_{n,-}$ be fulfilled and let $\tau_1 = \bar\tau_1$ be fixed such that $\bar\tau_1 \in (\tau^*_{n-1}, \tau^*_n)$ for $n \geq 1$ and $\bar\tau_1 \in (-\infty, \tau^*_0)$ for $n = 0$; let $\bar\tau_2 = \tau^*_n$. Then for $\tau_2 = \bar\tau_2 - \eta$ with $\eta \to +0$ the asymptotics

$$\mathbf{N}(\bar\tau_1, \bar\tau_2 - \eta) = \mathcal{N}_{n,-}(\eta) + O(\eta^{\frac{1}{\alpha}}) \tag{11.3.66$_{n,-}$}$$

hold.

On the other hand, let condition $(11.3.65)_{n,+}$ be fulfilled. Then for fixed $\tau_2 = \bar\tau_2 \in (\tau^*_n, \tau^*_{n+1})$ and for $\tau_1 = \bar\tau_1 + \eta$, $\bar\tau_1 = \tau^*_n$ the asymptotics

$$\mathbf{N}(\bar\tau_1 + \eta, \bar\tau_2) = \mathcal{N}_{n,+}(\eta) + O(\eta^{\frac{1}{\alpha}}) \tag{11.3.66$_{n,+}$}$$

hold. In both cases $\mathcal{N}_{n,\pm}(\eta)$ is given by $(11.3.52)_{n,\pm}$ and does not exceed $C \cdot \exp(C\eta^{\frac{1}{\alpha}})$; on the other hand, if

$$\pm V_{(n)} \geq \epsilon\tilde\varrho \qquad \forall x \in \Gamma : |x| \geq c \tag{11.3.67}$$

where Γ is an open non-empty sector in $\mathbb{R}^2$ then $\mathcal{N}_{n,\pm}(\eta) \geq \varepsilon\exp(\varepsilon\eta^{\frac{1}{\alpha}})$.

I leave it to the reader to combine theorems 11.3.18 and 11.3.20. Moreover, I leave it to the reader to generalize theorems 11.3.17, 11.3.19 and 11.3.20 when ϱ_2 decays more slowly than ϱ_3. Instead, I give the following generalization of theorems 11.3.17(ii), 11.3.19(iii),(iv):

Theorem 11.3.21. *Let X be an exterior domain with C^K boundary. Let the Schrödinger operator A satisfy conditions $(\Sigma 0)$, $(\Sigma 1)$, $(\Sigma 2)$, $(\Sigma 5)$ with $\gamma = \epsilon_0\langle x\rangle$, $\rho = \rho_1 = 1$, $\varrho_3 = \varrho_2 \ge \epsilon\langle x\rangle^{-s}$; moreover, let condition $(\Sigma 4)$ be fulfilled for $|x| \ge c$. Let condition (11.3.64) be fulfilled with $\tilde\varrho \ge \varrho_3$. For some $n \in \mathbb{Z}^+$ let*

$$(11.3.68)_{n,\pm} \qquad \epsilon\tilde\varrho \le \pm V_{(n)} \le c\tilde\varrho \qquad \text{for } |x| \ge c,$$

$$(11.3.69)_n \qquad |\nabla V_{(n)}| \ge \epsilon\varrho_3\gamma^{-1} \qquad \text{for } |x| \ge c.$$

Finally, let us assume that $\tilde\varrho$ is a γ-admissible weight function, $\tilde\varrho = o(1)$ as $|x| \to \infty$ and

$$(11.3.70) \qquad -\langle x, \nabla\rangle\tilde\varrho \ge \epsilon\varrho_3 \qquad \text{for } |x| \ge c.$$

Then, under condition $(11.3.68)_{n,-}$ for fixed $\tau_1 = \bar\tau_1 \in (\tau^*_{n-1}, \tau^*_n)$ or $\tau_1 = \bar\tau_1 \in [-\infty, \tau^*_0)$ for $n \ge 1$, $n = 0$ respectively and for $\tau_2 = \bar\tau_2 - \eta$ with $\bar\tau_2 = \tau^*_n$, $\eta \to +0$ asymptotics $(11.3.56)_{n,-}$ hold. On the other hand, under condition $(11.3.68)_{n,+}$ for fixed $\tau_2 = \bar\tau_2 \in (\tau^*_n, \tau^*_{n+1})$ and for $\tau_1 = \bar\tau_1 + \eta$ with $\bar\tau_1 = \tau^*_n$, $\eta \to +0$ asymptotics $(11.3.56)_{n,+}$ hold.

Theorem 11.3.22. *Let $X = \mathbb{R}^2$ and let the Schrödinger operator A satisfy conditions $(\Sigma 0)$, $(\Sigma 1)$, $(\Sigma 2)$, $(\Sigma 4)$, $(\Sigma 5)$ with $\gamma = \epsilon_0\langle x\rangle$, $\rho = \rho_1 = 1$, $\varrho_3 = \varrho_4 \ge \epsilon\langle x\rangle^{-s}$. Moreover, let conditions (11.3.64), $(11.3.68)_{n,\pm}$ and $(11.3.69)_n$ be fulfilled and let $\tilde\varrho$ satisfy (11.3.70), $\tilde\varrho = o(1)$ as $|x| \to \infty$. Let $\bar\tau_1$ and $\bar\tau_2$ be fixed and assume there do not exist $n' \in \mathbb{Z}^+$, $k = 1,2$ such that $\bar\tau_k$ is a degenerate critical value of $V_{(n')}$. Then*

(i) Under condition $(11.3.68)_{n,-}$ for $\tau_1 = \bar\tau_1 \in (\tau^*_{n-1}, \tau^*_n)$, $\bar\tau_1 \in [-\infty, \tau^*_0)$ for $n \ge 1$, $n = 0$ respectively and for $\tau_2 = \bar\tau_2 - \eta$, $\bar\tau_2 = \tau^*_n$, $\eta \to +0$, $h \to +0$, $\mu = h^{-1}$ the asymptotics

$$(11.3.71) \qquad \mathbf{N}(\tau_1, \tau_2, \mu, h) = \mathcal{N}(\tau_1, \tau_2, \mu, h) + O(\log h)$$

hold where $\mathcal{N}$ is given by (11.3.60)–(11.3.61).

On the other hand, the same asymptotics hold under condition $(11.3.68)_{n,+}$ for $\tau_1 = \bar\tau_1 + \eta$, $\bar\tau_1 = \tau^*_n$, $\tau_2 = \bar\tau_2 \in (\tau^*_n, \tau^*_{n+1})$ with $\eta \to +O$, $h \to +0$, $\mu = h^{-1}$.

(ii) Moreover, assume there do not exist $n \in \mathbb{Z}^+$, $k = 1,2$ such that $\bar\tau_k$ is a minimax value of $V_{(n)}$. Then the asymptotics

$$(11.3.72) \qquad \mathbf{N}(\tau_1, \tau_2, \mu, h) = \mathcal{N}(\tau_1, \tau_2, \mu, h) + O(1)$$

hold.

Example 11.3.23. *(i) One can easily see that for $\tilde\varrho = \varrho_2 = \varrho_3 = \langle x\rangle^{2m}$ with $m < 0$, $\mathcal{N} \asymp \eta^{\frac{2}{m}} h^{-2}$.*

(ii) One can take $\tilde\varrho = \log^\alpha\langle x\rangle$, $\varrho_2 = \varrho_3 = \log^{\alpha-1}\langle x\rangle$ with $\alpha < 0$.

(iii) One can take $\tilde\varrho = \log^\alpha\log\langle x\rangle$, $\varrho_2 = \varrho_3 = \log^{\alpha-1}\log\langle x\rangle \cdot \log^{-1}\langle x\rangle$ with $\alpha < 0$; then

$$\varepsilon h^{-2}\exp\bigl(\varepsilon\exp(\varepsilon\eta^{\frac{1}{\alpha}})\bigr) \le \mathcal{N} \le Ch^{-2}\exp\bigl(C\exp(C\eta^{\frac{1}{\alpha}})\bigr).$$

(iv) Moreover, one can take

$$\tilde\varrho = \bigl(\underbrace{\log\ldots\log}_{r\ times}\langle x\rangle\bigr)^\alpha$$

with $\alpha < 0$ *and* $\varrho_2 = \varrho_3 = \partial_{|x|}\tilde{\varrho}$ *(for* $|x| > c$*). In this case* $\mathcal{N}$ *is estimated from above by*

$$Ch^{-2}\underbrace{\exp(C\ldots(\exp(C\eta^{\frac{1}{\alpha}})\ldots)}_{r\ \text{times}}$$

and is estimated from below by the same expression with C *replaced by* ε*. Surely, here we take* $\langle x\rangle = (|x|^2 + R^2)^{\frac{1}{2}}$ *with large enough* R *and assume that (11.3.67) is fulfilled.* ◇

The following statement, which is also due to theorem 6.4.19, is a good supplement to theorems 11.3.21 and 11.3.22:

Theorem 11.3.24. *(i) Let* X *be an exterior domain with* C^K *boundary. Let the Schrödinger operator* A *satisfy conditions* $(\Sigma0)$, $(\Sigma1)$, $(\Sigma2)$, $(\Sigma5)$, (11.3.64) *and* $(\Sigma4)$[23] *with* $\rho = \rho_1 = 1$, $\gamma \ge \epsilon_0\langle x\rangle^{\sigma}$, $\sigma > 0$, $\varrho_2 = \varrho_3 \ge \epsilon\langle x\rangle^{-s}$, $\tilde{\varrho} \ge \varrho_3$ *such that* $\tilde{\varrho} = o(1)$ *as* $|x| \to \infty$*. Let* $\bar{\tau} = \tau_n^*$, $h \in (0,1]$, $\mu = h^{-1}$ *and let* ε *be a small enough constant. Then, under condition* $(11.3.68)_{n,-}$ *the interval* $J_{n,-} = [\bar{\tau}, \bar{\tau} + \varepsilon]$ *contains no more than* Ch^{-2} *eigenvalues of the operator* $A = A_{\mu,h}$ *and under condition* $(11.3.68)_{n,+}$ *the interval* $J_{n,+} = [\bar{\tau} - \varepsilon, \bar{\tau}]$ *contains no more than* Ch^{-2} *eigenvalues of the operator* $A = A_{\mu,h}$ *(counting multiplicities) with* $C = C(c, \epsilon, \epsilon_0, s, \sigma, n)$*.*

(ii) Let us in addition assume that $X = \mathbb{R}^2$ *and condition* $(\Sigma4)$ *is fulfilled in* X*. Let condition* $(11.3.68)_{n,\pm}$ *be fulfilled and* $V_{(n')} \notin J_{n,\pm}$ $\forall n' \in \mathbb{Z}^+$ $\forall x \in X$*. Then for small enough* $h > 0$ *and* $\mu = h^{-1}$ *the interval* $J_{n,\pm}$ *contains no eigenvalue of* $A_{\mu,h}$*.*

Remark 11.3.25. *(i) Let us note that we have proven assertions (i),(ii) for the interval* $J'_{n,\pm} = J_{n,\pm}\setminus\bar{\tau}$*. However, these assertions remain true for the operator* $A'_{\mu,h} = A_{\mu,h} - \zeta W$ *with* $W \in \mathcal{S}(\mathbb{R}^2)$ *and small enough* $\zeta > 0$ *and therefore they also hold for the interval* $J_{n,\pm}$*.*

(ii) I would like to note that the results of this subsection as well as certain results below (where F *stabilizes at infinity and* $\mu = h^{-1}$*) can be proven in the following situation: instead of fixed* $\eta = 0$ *there is* $\eta' \to +0$ *in the restriction* $\mu h = 1 \pm \eta'$*; one can also treat combined situations.*

There is a deep but non-obvious connection between these results and results of A. Sobolev [1–4] and G. Raikov [1–6]; see also section 12.5. ◇

11.3.4 Singular case

The same arguments as in the previous subsection but with $W = C_0 h^{-2}$ for $|x| \le \varepsilon h$ provide the following

Theorem 11.3.26. *Let* $X = \mathbb{R}^2$ *and let the Schrödinger operator* A *satisfy conditions* $(\Sigma0)$, $(\Sigma1)$, $(\Sigma2)$, $(\Sigma4)$, $(\Sigma5)$ *with* $\rho = \rho_1 = 1$, $\gamma = \epsilon_0|x|$, $\varrho_2 = \varrho_3 = |x|^m$, $m > 0$*. Moreover, let* $V = O(|x|^{2m})$ *and* $F = \bar{F} + O(|x|^{2m})$ *as* $x \to 0$ *where* $\bar{F} = \text{const} > 0$*. Let* $V_{(k)}$ *be given by* (11.3.50) *and*

$(11.3.73)_n$ $$|V_{(n')}| \ge \epsilon \qquad \forall n' \in \mathbb{Z}\setminus n \qquad \forall x \in \mathbb{R}^2,$$

$(11.3.74)_{n,\pm}$ $$\pm V_n \ge \epsilon \min(1, |x|^{2m}) \qquad \forall x \in \mathbb{R}^2,$$

$(11.3.75)_n$ $$|\nabla V_{(n)}| \ge \epsilon |x|^{2m-1} \qquad \forall x : |x| \le \epsilon.$$

Finally, let $\bar{\tau} = \tau_n^*$ *and let* $\varepsilon > 0$ *be a small enough constant. Then under condition* $(11.3.74)_{n,+}$ *for fixed* $\tau_1 \in J_{n,-} = [\bar{\tau} - \varepsilon, \bar{\tau}]$ *and for* $\tau_2 = \bar{\tau} + \eta$ *with* $\eta \to +0$, $h = O(\eta^{\frac{1}{2m}})$, $\mu = h^{-1}$ *asymptotics* (11.3.73) *hold where* $\mathcal{N}$ *is given by* (11.3.60)–(11.3.61) *and in our assumptions*

[23] Surely, the latter condition should be fulfilled only for $|x| \ge c$.

(11.3.76) $$\mathcal{N} = \frac{1}{2\pi h^2} \int_{\{V_{(n)} \in (\tau_1, \tau_2)\}} F\sqrt{g}dx \asymp \eta^{\frac{1}{2m}} h^{-2}.$$

This statement remains true under condition $(11.3.74)_{n,-}$ *for fixed* $\tau_2 \in J_{n,+} = [\bar\tau, \bar\tau + \varepsilon]$ *and for* $\tau_1 = \bar\tau - \eta$ *with* $\eta \to +0$, $h = O(\eta^{1/2m})$, $\mu = h^{-1}$.

11.4 Asymptotics with the Spectral Parameter for Two-Dimensional Dirac Operators

11.4.1 Eigenvalues tending to $\pm M$. I

Let A be a Dirac operator given by $(\Sigma 0) = (11.1.39)$ with coefficients satisfying conditions $(\Sigma 1)_{1-4} = (11.1.41)_{1-4}$, $(\Sigma 2)$ for $g^{jk} = \omega^{jl}\omega^{kr}\delta_{lr}$, $(\Sigma 4)$ and $(\Sigma 5)$. Let us start with the case when the coefficients are regular inside the domain, the potential decreases at infinity and $\mu = 1$ is fixed. If $M > 0$ is also fixed then the interval $(-M, M)$ contains only the discrete spectrum of A and the asymptotics of eigenvalues tending to $-M + 0$ or $M - 0$ should be treated.

Theorem 11.4.1. [24] *Let* $X = \mathbb{R}^2$ *and let the Dirac operator* A *satisfy conditions* $(\Sigma 0)$, $(\Sigma 1)_{1-3}$, $(\Sigma 2)$ *and* $(\Sigma 4)$[25] *with* $M > 0$, $\gamma = \epsilon_0\langle x\rangle$, $\rho = \langle x\rangle^m$, $\rho_1 = \langle x\rangle^{m_1}$, $m < 0$, $m_1 > \max(-2, m-1)$. *Let us assume that*

(11.4.1) $$\rho_1\gamma \ge c\rho \implies F_{12}\det(\omega^{jk}) > 0, \qquad \varsigma = -1$$

where $\varsigma = \pm 1$ *characterizes the set of Pauli matrices*[26] *and* $F_{12} = \partial_{x_2}V_1 - \partial_{x_1}V_2$. *Moreover, let us assume that*

(11.4.2) $$\lim_{|x|\to\infty} V = 0$$

(this limit exists by condition $(\Sigma 1)_3$*). Then*

(i) Let $m_1 > 2m$ *and let the stabilization conditions*

$(11.4.3)_1$ $$D^\sigma(V - V^0) = o(|x|^{m' - |\sigma|}),$$

$(11.4.3)_2$ $$D^\sigma(F - F^0) = o(|x|^{m_1 - |\sigma|}) \qquad \forall\sigma : |\sigma| \le r$$

be fulfilled with $m' = 2m$, $r = 1$ *and positively homogeneous functions* $V^0, F^0 \in C^K(\mathbb{R}^2 \setminus 0)$ *of degrees* m' *and* m_1 *respectively. Moreover, let the condition*

(11.4.4) *0 is not a critical value of* V^0

be fulfilled. Then for fixed $\tau_1 \in (-M, M)$ *and for* $\tau_2 = M - \eta$ *with* $\eta \to +0$, $\mu = 1$, $h \in (0, 1]$ *the asymptotics*

(11.4.5) $$\mathbf{N}(\tau_1, \tau_2, \mu, h) = \mathcal{N}(\tau_1, \tau_2, \mu, h) + O(h^{-1}) + O(\log\eta)$$

hold with

(11.4.6) $$\mathcal{N}(\tau_1, \tau_2, \mu, h) = O\big(h^{-2}\mathcal{M}(\eta) + h^{-1}\eta^{\frac{1}{2m}(m_1+2)}\big)$$

---

[24] Compare with theorem 11.3.8.

[25] In subsections 11.4.1–11.4.5 this condition and condition (11.4.1) are assumed to be fulfilled only for $|x| \ge c$.

[26] Namely, $\sigma_0\sigma_1\sigma_2 = \varsigma iI$ (see Appendix C).

where $\mathcal{M}(\eta) = \eta^{\frac{1}{m}(m+1)}$ *for* $m > -1$, $\mathcal{M}(\eta) = \log \eta$ *for* $m = -1$ *and* $\mathcal{M}(\eta) = 1$ *for* $m < -1$.

In addition, let condition $(11.4.7)_+$ *be fulfilled where*

$$\pm V \le -\epsilon \min\bigl(\rho, \frac{1}{M}\rho^2\bigr) \qquad \text{for } x \in \Gamma, |x| \ge c \tag{11.4.7$_\pm$}$$

and Γ *is an open non-empty cone in* $\mathbb{R}^d$. *Then*

$$\mathcal{N}(\tau_1, \tau_2, \mu, h) \asymp h^{-2}\mathcal{M}(\eta) + h^{-1}\eta^{\frac{1}{2m}(m_1+2)}. \tag{11.4.6$'$}$$

(ii) Furthermore, let **both** *standard Hamiltonians*

$$H_{\mathrm{st},\pm} = \pm\bigl(g^{jk}(\xi_j - V_j)(\xi_k - V_k) + M^2\bigr)^{\frac{1}{2}} + V \tag{11.4.8}$$

satisfy the standard condition (Ψ) *at the energy levels* τ_1 *and* $\tau_2 = M$. *Then asymptotics* (11.4.5) *hold for* $h \to +0$ *with "*$O(h^{-1})$*" replaced by "*$o(h^{-1})$*."*

(iii) Moreover, let condition $(11.4.9)_+$ *be fulfilled for* $|x| \ge c$ *where*

$$\pm V \le -\epsilon \min\bigl(\rho, \frac{1}{M}\rho^2\bigr). \tag{11.4.9$_\pm$}$$

Then asymptotics (11.4.4) *hold with "*$O(\log \eta)$*" replaced by "*$O(1)$*."*

(iv) Both improvements (i),(ii) can be made simultaneously.

(v) Let $m_1 < 2m$ *(then* $m > -1$*) and let stabilization conditions* $(11.4.3)_{1,2}$ *be fulfilled with* $r = 2$. *Assume* 0 *is not a critical value of* V^0 *and let the function* $(V^0)^{m_1}(F^0)^{-2m}$ *have no degenerate critical point on the circle* $\mathbb{S}^1$. *Then for fixed* $\tau_1 \in (-M, M)$ *and for* $\tau_2 = M - \eta$ *with* $\eta \to +0$, $\mu = 1$, $h \in (0, 1]$ *the asymptotics*

$$\mathbf{N}(\tau_1, \tau_2, \mu, h) = \mathcal{N}(\tau_1, \tau_2, \mu, h) + O\Bigl(h^{-1}\eta^{\frac{1}{2m}(2m-m_1)}\bigl((\log h\eta^{\frac{1}{2m}(3m_1-4m+2)})_+ + 1\bigr)\Bigr) \tag{11.4.10}$$

hold. Moreover, the estimates of $\mathcal{N}$ *remain true.*

(vi) Finally, let $\varsigma = 1$ *(in contrast to the second part of condition* (11.4.1)*). Then all the assertions of theorem 11.3.8 remain true with* $\tau_1 \in (-M, M)$ *instead of* $\tau_1 = -\infty$, $\tau_2 = M - \eta$ *instead of* $\tau_2 = -\eta$. *Surely, the particular formula for* $\mathcal{N}$ *changes.*

I leave it to the reader to formulate and prove statements similar to theorems 11.3.9 and 11.3.10.

11.4.2 Eigenvalues tending to $\pm M$. II

Let us fix both parameters μ and h (we can assume that $\mu = h = 1$). We know that if $X = \mathbb{R}^2$ then under appropriate conditions on the coefficients, $\mathrm{Spec}_{\mathrm{ess}}(A) = (-\infty, -M] \cup [M, \infty)$ provided V and F tend to 0 at infinity. On the other hand, if F is disjoint from 0 and $\sqrt{F} \ge \epsilon|V|$ for $|x| \ge c$ [27] then $\mathrm{Spec}_{\mathrm{ess}}(A)$ is contained in[28] the union over $(\zeta, n) = \Upsilon = \{-1, 1\} \times \mathbb{Z}^+ \setminus (\varsigma, 0)$ of the limit points of the functions

$$V_{\zeta n} = \zeta\bigl(2\mu h n F + M^2\bigr)^{\frac{1}{2}} + V \tag{11.4.11}$$

as $|x| \to \infty$. In particular, $\mathrm{Spec}_{\mathrm{ess}}(A) = \emptyset$ provided both functions V and F tend to infinity at infinity.

Let us treat first the case when V decreases at infinity.

[27] The second restriction is important for V which is not bounded at infinity.

[28] Coincides with, in fact.

Theorem 11.4.2. [29] *Let $X = \mathbb{R}^2$ and let the Dirac operator A satisfy conditions $(\Sigma 0)$, $(\Sigma 1)_{1-3}$, $(\Sigma 2)$, $(\Sigma 4)$, (11.4.1) and (11.4.2) with $\gamma = \epsilon_0\langle x\rangle$, $\rho = \langle x\rangle^m$, $\rho_1 = \langle x\rangle^{m_1}$, $m < 0$, $m_1 > -2$, $m_1 \ge 2m$. Then*

$$\text{Spec}_{\text{ess}}(A) = (-,\infty - M] \cup [M,\infty) \qquad \text{for } m_1 < 0, \tag{11.4.12$_1$}$$

$$\text{Spec}_{\text{ess}}(A) \cup [-M-\varepsilon, M+\varepsilon] = \{M\} \qquad \text{for } m_1 = 0, \tag{11.4.12$_2$}$$

$$\text{Spec}_{\text{ess}}(A) = \{M\} \qquad \text{for } m_1 > 0, \tag{11.4.12$_3$}$$

provided $\varepsilon > 0$ is a small enough constant. Moreover, even for $m_1 < 0$ the point $-M+0$ is not a limit point of the discrete spectrum of the operator A provided **either** *$m_1 > 2m$* **or** *the condition*

$$(1-\epsilon)F \ge M|V| \qquad \text{if } M > 0, \tag{11.4.13$_1$}$$

$$\sqrt{(2-\epsilon)F} \ge |V| \qquad \text{if } M = 0 \tag{11.4.13$_2$}$$

holds for $|x| \ge c$.

Moreover, let us assume that for $|x| \ge c$

$$|\nabla V| \ge \epsilon\gamma^{-1}\min\bigl(\rho, \frac{1}{M}\rho^2\bigr), \tag{11.4.14}$$

Then

(i) Let us fix τ_1 such that $\tau_1 \in (-M, M)$ for $m_1 < 0$, $\tau_1 \in (-M-\varepsilon, M)$ for $m_1 = 0$ and $\tau_1 \in (-\infty, M)$ for $m_1 > 0$. Then for $\mu = h = 1$ and $\tau_2 = M - \eta$ with $\eta \to +0$ the asymptotics

$$\mathbf{N}(\tau_1,\tau_2) = \mathcal{N}(\tau_1,\tau_2) + O(\log\eta) \tag{11.4.15}$$

hold with

$$\mathcal{N}(\tau_1,\tau_2) = O\bigl(\eta^{\frac{1}{m'}(m_1+2)}\bigr) \tag{11.4.16}$$

where here and below (excluding some specially indicated cases)[30] *$m' = 2m$ for $M > 0$ and $m' = m$ for $M = 0$. Moreover, $\mathcal{N} \asymp \eta^{\frac{1}{m'}(m_1+2)}$ under condition $(11.4.7)_+$.*

(ii) Let $m_1 \ge 0$. Let us fix τ_2 such that $\tau_2 \in (M, M+\varepsilon)$ for $m_1 = 0$, $\tau_2 \in (M,\infty)$ for $m_1 > 0$. Then for $\mu = h = 1$, $\tau_1 = M + \eta$ with $\eta \to +0$ asymptotics (11.4.15) and estimate (11.4.16) remain true. Moreover, $\mathcal{N} \asymp \eta^{\frac{1}{m'}(m_1+2)}$ under condition $(11.4.7)_-$.

(iii) Furthermore, under the additional condition $(11.4.9)_\pm$ in (i),(ii) respectively the asymptotics

$$\mathbf{N}(\tau_1,\tau_2) = \mathcal{N}(\tau_1,\tau_2) + O(1) \tag{11.4.17}$$

hold. On the other hand, under this condition in (i),(ii) respectively $M \pm 0$ is not a limit point of the discrete spectrum.

I leave it to the reader to formulate and prove statements similar to theorems 11.3.18–11.3.20.

Theorem 11.4.3. [31] *Let $X = \mathbb{R}^2$ and let the Dirac operator A satisfy conditions $(\Sigma 0)$, $(\Sigma 1)_{1-3}$, $(\Sigma 2)$, $(\Sigma 4)$, (11.4.1) and (11.4.2) with $\gamma = \epsilon_0\langle x\rangle$, $\rho \to 0$ as $|x| \to \infty$, $\rho_1 = \langle x\rangle^{m_1}$, $m_1 \ge 0$. Moreover, for $|x| \ge c$ let*

---

[29] Compare with theorems 11.3.16, 11.3.17.

[30] The above definition of m' works well for decreasing potentials. For increasing potentials one should always set $m' = m$.

[31] Compare with theorem 11.3.21.

$$(11.4.18)_\pm \qquad \epsilon\tilde\rho \le \mp V \le c\tilde\rho,$$

$$(11.4.19) \qquad |D^\alpha V| \le c\tilde\rho\varrho_2\gamma^{-|\alpha|} \qquad \forall\alpha : |\alpha| \le K,$$

$$(11.4.20) \qquad |\nabla V| \ge \epsilon\tilde\rho\varrho_2\gamma^{-1},$$

$$(11.4.21) \qquad -\langle x, \nabla\rangle\tilde\rho \ge \epsilon\varrho_2\tilde\rho$$

where $\tilde\rho = \min\bigl(\rho, \frac{1}{M}\rho^2\bigr)$ *and we assume that*

$$(11.4.22) \qquad \rho \ge \epsilon\langle x\rangle^{-s}, \qquad \varrho_2 \ge \epsilon\langle x\rangle^{-\delta}$$

with some s *and with a small enough exponent* $\delta > 0$. *Then statements* $(11.4.12)_{2-3}$ *for* $\operatorname{Spec}_{\mathrm{ess}}(A)$ *remain true. Moreover,*

(i) *Let condition* $(11.4.18)_+$ *be fulfilled. Let us fix* τ_1 *such that* $\tau_1 \in (-M-\varepsilon, M)$ *for* $m_1 = 0$, $\tau_1 \in (-\infty, M)$ *for* $m_1 > 0$. *Then for* $\mu = h = 1$, $\tau_2 = M - \eta$ *with* $\eta \to +0$, *asymptotics* (11.4.17) *hold.*

(ii) *Let condition* $(11.4.18)_-$ *be fulfilled. Let us fix* τ_2 *such that* $\tau_2 \in (M, M+\varepsilon)$ *for* $m_1 = 0$, $\tau_2 \in (M, \infty)$ *for* $m_1 > 0$. *Then for* $\mu = h = 1$, $\tau_1 = M + \eta$ *with* $\eta \to +0$, *asymptotics* (11.4.17) *hold.*

(iii) *Under condition* $(11.4.18)_\pm$ *the point* $M \pm 0$ *is not a limit point of the discrete spectrum of* A.

I leave it to the reader to formulate and prove a statement similar to theorem 11.3.22 and construct examples similar to example 11.3.23.

Theorem 11.4.4. *Let* $X = \mathbb{R} \times \mathbb{R}/\bar\gamma$ *with* $\bar\gamma < \infty$ *and let the Dirac operator* A *satisfy conditions* $(\Sigma 0)$, $(\Sigma 1)_{1-3}$, $(\Sigma 2)$, $(\Sigma 4)$, (11.4.1), (11.4.2), (11.4.18)–(11.4.20) *with* $\gamma = \epsilon_0$, $\rho = \langle x_1\rangle^m$, $m < 0$, $\rho_1 = \exp b|x_1|$, $b > 0$, $\varrho_2 = \langle x_1\rangle^{-1}$. *Then* $\operatorname{Spec}_{\mathrm{ess}}(A) = \{M\}$ *and*

(i) *Under condition* $(11.4.18)_+$ *for fixed* $\tau_1 \in (-\infty, M)$ *and for* $\mu = h = 1$, $\tau_2 = M - \eta$ *with* $\eta \to +0$, *asymptotics* (11.4.17) *hold.*

(ii) *Under condition* $(11.4.18)_-$ *for fixed* $\tau_2 \in (M, \infty)$ *and for* $\mu = h = 1$, $\tau_1 = M + \eta$ *with* $\eta \to +0$, *asymptotics* (11.4.17) *hold.*

(iii) *Assertion (iii) of the previous theorem remains true.*

Moreover, in both cases $\mathcal{N} \asymp \exp(\eta^{\frac{1}{2m'}})$ *with the exponent* m' *indicated above.*

Theorem 11.4.5. *Let* $X = \mathbb{R} \times \mathbb{R}/\bar\gamma$ *with* $\bar\gamma < \infty$ *and let the Dirac operator* A *satisfy conditions* $(\Sigma 0)$, $(\Sigma 1)_{1-3}$, $(\Sigma 2)$, $(\Sigma 4)$, (11.4.1), (11.4.2), (11.4.14), $(11.4.9)_\pm$ *with* $\gamma = \epsilon_0$, $\rho = \exp a|x_1|$, $a < 0$, $\rho_1 = \exp b|x_1|$, $b > 0$. *Then* $\operatorname{Spec}_{\mathrm{ess}}(A) = \{M\}$ *and all the statements of the previous theorem remain true (with condition* $(11.4.18)_\pm$ *replaced by* $(11.4.9)_\pm$*). Moreover,* $\mathcal{N}(\tau_1, \tau_2) \asymp \eta^{\frac{b}{a'}}$ *where* $a' = 2a$ *for* $M > 0$ *and* $a' = a$ *for* $M = 0$.

I leave it to the reader to treat multiple exponents, slowly decreasing (as logarithm or as multiple logarithm) potentials V, the case $X = \mathbb{R}^2$, etc.

11.4.3 Eigenvalues tending to other points of the essential spectrum

Moreover, I leave it to the reader to treat the case $X = \mathbb{R}^2$, $V \to 0$, $F \to \bar F = \mathrm{const} > 0$ as $|x| \to \infty$ and to prove asymptotics of the eigenvalues of A tending to the points $\Delta_{\zeta n} \mp 0$ where $\Delta_{\zeta n} = \zeta(2\mu hn\bar F + M^2)^{\frac12}$ with $(\zeta, n) \in \{-1, 1\} \times \mathbb{N}$ together with $-\varsigma M$ are all the points of the essential spectrum. In this case all the statements of theorems 11.3.16–11.3.25 remain true. Surely, we now have limit points $\Delta_{\zeta n}$ instead of $(2n+1)\mu h\bar F$ and therefore other intervals in which we choose spectral parameters; finally, we have functions $V_{(\zeta n)} = V_{\zeta n} - \Delta_{\zeta n}$ instead of $V_{(n)}$ and other formulas for $\mathcal{N}$. Note that for $\mu = h = 1$ only one term in the formula for $\mathcal{N}$ is essential.

11.4.4 Eigenvalues tending to infinity. I

Let us now consider asymptotics of eigenvalues tending to infinity. I treat only the case of power growth of F and V and leave generalizations to the reader.

Theorem 11.4.6. *Let $X = \mathbb{R}^2$ and let the Dirac operator A satisfy conditions $(\Sigma 0)$, $(\Sigma 1)_{1-3}$, $(\Sigma 2)$, $(\Sigma 4)$ and (11.4.1) with $\gamma = \epsilon_0\langle x\rangle$, $\rho = \langle x\rangle^m$, $\rho_1 = \langle x\rangle^{m_1}$, $m_1 > 0$. Let*

$$|\nabla F| \ge \epsilon\rho_1\gamma^{-1} \tag{11.4.23}$$

for $|x| \ge c$ and let one of the following three conditions be fulfilled:

(a) $m \le 0$, $\tau_1 > \varsigma M + \lim_{|x|\to\infty}\sup V$;

(b) $m_1 \ge 2m > 0$ and for $|x| \ge c$ conditions $(11.4.9)_+$ and

$$|V| \le \sqrt{(2-\epsilon)F}, \tag{11.4.24}$$

$$|\nabla V| \le \varepsilon|\nabla F| \tag{11.4.25}$$

with $\epsilon > 0$ and with a small enough constant $\varepsilon = \varepsilon(c, \epsilon, \epsilon_0) > 0$;

(c) $m_1 = 2m > 0$ and for $|x| \ge c$ conditions $(11.4.9)_+$, (11.4.24) and stabilization conditions $(11.4.3)_{1,2}$ with $r = 1$ and $m' = m$ are fulfilled.

Then

(i) For $\mu = h = 1$, fixed τ_1 and for $\tau_2 \to +\infty$ the asymptotics

$$\mathbf{N}(\tau_1, \tau_2) = \mathcal{N}(\tau_1, \tau_2) + O\big(\tau_2^{(m_1+2)(m_1+1)^{-1}}\big) \tag{11.4.26}$$

hold with

$$\mathcal{N}(\tau_1, \tau_2) \asymp \tau_2^{2(m_1+2)m_1^{-1}}. \tag{11.4.27}$$

(ii) Moreover, let stabilization conditions $(11.1.34)_{1,3}$ be fulfilled and let the Hamiltonian

$$H^0(x,\xi) = g^{jk0}(\xi_j - V_j^0)(\xi_k - V_k^0)$$

satisfy the standard condition (Ψ) at the energy level $\tau = 1$. Then the asymptotics

$$\mathbf{N}(\tau_1, \tau_2) = \mathcal{N}(\tau_1, \tau_2) + o\big(\tau_2^{(m_1+2)(m_1+1)^{-1}}\big) \tag{11.4.28}$$

hold.

(iii) For fixed $\tau_2 < M\varsigma + \lim_{|x|\to\infty}\inf V$, $m = h = 1$ and $\tau_1 \to -\infty$, similar asymptotics hold under condition (a).

Theorem 11.4.7. *Let $X = \mathbb{R}^2$ and let the Dirac operator A satisfy conditions $(\Sigma 0)$, $(\Sigma 1)_{1-3}$, $(\Sigma 2)$, $(\Sigma 4)$ and (11.4.1) with $\gamma = \epsilon_0\langle x\rangle$, $\rho = \langle x\rangle^m$, $\rho_1 = \langle x\rangle^{m_1}$, $0 < 2m \le m_1$. Let conditions (11.4.23), $(11.4.9)_-$ be fulfilled and suppose* **either** *condition (11.4.25) with small enough $\varepsilon = \varepsilon(c, \epsilon, \epsilon_0) > 0$ is fulfilled* **or** *stabilization conditions $(11.4.3)_{1,2}$ with $r = 1$, $m' = m$ are fulfilled. Then statements (i),(ii) of the previous theorem hold with*

$$\mathcal{N}(\tau_1, \tau_2) \asymp \tau_2^{\frac{1}{m}(m_1+2)} \tag{11.4.29}$$

(i.e., for $m_1 > 2m$ the principal part now grows more quickly).

So, in the context of theorem 11.4.7 the essential spectrum of the operator A is empty and asymptotics (11.4.26) (and even (11.4.28) under the standard condition on the Hamiltonian flow) hold for eigenvalues tending to $+\infty$ and similar asymptotics hold for eigenvalues tending to $-\infty$ (however the orders of the principal parts do not necessarily coincide).

I leave it to the reader to weaken condition (11.4.25) for $m_1 = 2m$.

11.4.5 Multiparametrical asymptotics

Let us consider asymptotics with variable μ and h, $h \in (0,1)$, $\mu \to +\infty$ and $\mu h \to \omega > 0$. Without loss of generality one can assume that either $\omega = 1$ or $\omega = \infty$. *In this subsection we assume that conditions* ($\Sigma 4$) *and* (11.4.1) *are fulfilled for all* $x \in X$.

Theorem 11.4.8. [32] *Let $X = \mathbb{R}^2$ and let the Dirac operator A satisfy conditions ($\Sigma 0$), ($\Sigma 1$), ($\Sigma 2$), ($\Sigma 4$) and (11.4.1) with $\gamma = \epsilon_0 \langle x\rangle^\kappa$, $\rho = \langle x\rangle^m$, $\rho_1 = \langle x\rangle^{m_1}$, $\kappa \le 1$, $m_1 \ge 2m$, $m_1 > -2\kappa$. Let us fix $\tau_1 < \tau_2$ such that for every $(\zeta, n) \in \Upsilon$ the following conditions are fulfilled:*

(11.4.30) $\forall i = 1,2$ *τ_i is not a degenerate critical value of $V_{\zeta n}$;*

(11.4.31) *For $|x| \ge c$ one of the following inequalities is fulfilled:*

$$V_{\zeta n} - \tau_1 \le -\epsilon \min\bigl(\rho, \frac{1}{2M}\rho^2\bigr), \tag{11.4.32$_1$}$$

$$V_{\zeta n} - \tau_2 \ge \epsilon \min\bigl(\rho, \frac{1}{2M}\rho^2\bigr). \tag{11.4.32$_2$}$$

Then

(i) For fixed τ_1 and τ_2 and for $\mu \to \infty$, $\mu h \to 1$ asymptotics (11.3.59) *hold.*

(ii) Moreover, for every $(\zeta, n) \in \Upsilon$ let the condition

(11.4.33) $\forall i = 1,2$ *τ_i is not a minimax value of $V_{\zeta n}$*

be fulfilled. Then asymptotics (11.3.62) *hold.*

(iii) Finally, if for every $(\zeta, n) \in \Upsilon$ one of the inequalities $(11.4.32)_{1,2}$ is fulfilled for every $x \in X$ then $\mathbf{N}(\tau_1, \tau_2, \mu, h) = 0$ provided $\mu \ge \mu_0$ and $|\mu h - 1| \le \varepsilon'$ with sufficiently large μ_0 and small $\varepsilon' > 0$.

Theorem 11.4.9. *Let $X = \mathbb{R}^2$ and let the Dirac operator A satisfy conditions ($\Sigma 0$), $(\Sigma 1)_{1-3}$, ($\Sigma 2$), ($\Sigma 4$) and (11.4.1) with $\gamma = \epsilon_0 \langle x\rangle^\kappa$, $\rho = \langle x\rangle^m$, $\rho_1 = \langle x\rangle^{m_1}$, $\kappa \ge 1$, $m_1 \ge 2m$, $m_1 > -2\kappa$. Let us fix τ_1 and τ_2, $\tau_1 < \tau_2$ such that for $(\zeta, n) = (-\varsigma, 0)$ conditions* (11.4.30) *and* (11.4.31) *are fulfilled. Then*

(i) For $\mu \to \infty$, $h \in (0,1)$ and $\mu h \to \infty$ the asymptotics

$$\mathbf{N}(\tau_1, \tau_2, \mu, h) = \mathcal{N}(\tau_1, \tau_2, \mu, h) + O(\log \mu) \tag{11.4.34}$$

hold with

$$\mathcal{N}(\tau_1, \tau_2, \mu, h) = \frac{D\mu}{4\pi h} \int_{\{\tau_1 < V - \varsigma M < \tau_2\}} F\sqrt{g}\,dx. \tag{11.4.35}$$

(ii) Moreover, for $(\zeta, n) = (-\varsigma, 0)$ let condition (11.4.33) *be fulfilled. Then asymptotics* (11.3.62) *hold.*

---

[32] Compare with theorem 11.3.18.

(iii) Finally, for $(\zeta, n) = (-\varsigma, 0)$ *let one of the inequalities* $(11.4.32)_{1,2}$ *be fulfilled for all* $x \in X$. *Then* $\mathbf{N}(\tau_1, \tau_2, \mu, h) = 0$ *provided* $\mu h \geq \omega_0$, $h \in (0,1)$ *with sufficiently large* ω_0.

Theorem 11.4.10. *(i) Let the conditions of theorem 11.4.2(i) be fulfilled. In addition, for* $\tau = \tau_1$ *and for* $\tau = M$ *let condition* (11.4.30) *be fulfilled for all* $(\zeta, n) \in \Upsilon$. *Then for fixed* τ_1, $\mu \to +\infty$, $\mu h \to 1$, $\tau_2 = M - \eta$ *with* $\eta \to +0$, *asymptotics* (11.3.63) *hold.*

(ii) Let the conditions of theorem 11.4.2(ii) be fulfilled. In addition, for $\tau = \tau_2$ *and for* $\tau = M$ *let condition* (11.4.30) *be fulfilled for all* $(\zeta, n) \in \Upsilon$. *Then for fixed* τ_2, $\mu \to +\infty$, $\mu h \to 1$, $\tau_1 = M + \eta$ *with* $\eta \to +0$, *asymptotics* (11.3.63) *hold.*

(iii) If in statement (i) ((ii) respectively) conditions $(11.4.9)_+$ *(*$(11.4.9)_-$ *respectively) are fulfilled then one can replace* $O(\log \eta)$ *by* $O(1)$ *in these asymptotics.*

(iv) If in statement (i) ((ii) respectively) condition (11.4.33) *is fulfilled for all* $(\zeta, n) \in \Upsilon$ *for* $\tau = \tau_1$ *and* $\tau = M$ *(for* $\tau = \tau_2$ *and* $\tau = M$ *respectively) then one can replace* $O(\log h)$ *by* $O(1)$ *in these asymptotics.*

(v) Both improvements (iii),(iv) can be done at the same time.

(vi) In statement (i) ((ii) respectively) for $\tau = \tau_1$ *and* $\tau = M$ *(for* $\tau = \tau_2$ *and* $\tau = M$ *respectively) for all* $(\zeta, n) \in \Upsilon$ *one of inequalities* $(11.4.32)_{1,2}$ *be fulfilled for all* $x \in X$. *Then* $\operatorname{Spec}(A) \cap [\tau_1, M) = \emptyset$ *(*$\operatorname{Spec}(A) \cap (M, \tau_2] = \emptyset$ *respectively) provided* $\mu \geq \mu_0$ *and* $|\mu h - 1| \leq \varepsilon'$ *with sufficiently large* μ_0 *and small* $\varepsilon' > 0$.

(vii) In statements (i),(ii) $\mathcal{N} = O(h^{-2}\eta^{\frac{1}{m'}(m_1+2)})$ *and* $\mathcal{N} \asymp h^{-2}\eta^{\frac{1}{m'}(m_1+2)}$ *under condition* $(11.4.7)_\pm$.

Remark 11.4.11. *(i) The same type of statements also hold for* $\mu h \to \infty$ *after the following modifications: in the conditions, "for all* $(\zeta, n) \in \Upsilon$*" should be replaced with "for* $(\zeta, n) = (-\varsigma, 0)$*" (condition (11.4.1) is assumed to be fulfilled) and in the remainder estimates* $O(\log h)$ *should be replaced with* $O(\log \mu)$. *With these modifications* $\mathcal{N}(\tau_1, \tau_2, \mu, h)$ *is given by (11.4.35) and* $\mathcal{N} = O(\mu h^{-1}\eta^{\frac{1}{m'}(m_1+2)})$ *and* $\mathcal{N} \asymp \mu h^{-1}\eta^{\frac{1}{m'}(m_1+2)}$ *under condition* $(11.4.7)_\pm$.

(ii) Moreover, for $\mu h \to 1$ *and for* $\mu h \to \infty$ *statements similar to theorems 11.4.3–11.4.5 hold. Namely, condition (11.4.30) should be added to their hypotheses for corresponding values of* τ_1 *and* τ_2 *(one of these values is* M, $\varsigma = -1$*) and for corresponding values of* (ζ, n). *These conditions provide asymptotics (11.4.34) with the order of* $\mathcal{N}$ *given in the corresponding theorem (with the additional factor* $(\mu h + 1)h^{-2}$*). Furthermore, under condition (11.4.33) (for corresponding* τ_1, τ_2 *and* (ζ, n)*) one can replace* $O(\log \mu)$ *(or* $O(\log h)$*) by* $O(1)$ *in the remainder estimate; statement (vi) of theorem 11.4.10 also remains true (with the obvious modification for* $\mu h \to \infty$*).*

(iii) Finally, statements similar to theorems 11.3.19, 11.3.22 and 11.3.24 also remain true. Now $\mu \to \infty$, $h = \mu^{-1}$ *and we are only interested in the points of the essential spectrum different from* $-\varsigma M$ *(i.e., we are interested in the points* $\Delta_{\zeta,n}$ *with* $n \neq 0$; *see subsection 11.4.4).*

The exact statements are all left to the reader. ◊

11.4.6 Eigenvalues tending to infinity. II

Recall that there exist self-adjoint Dirac operators with potentials with strong singularities (see Appendix I). In this subsection I treat their spectral asymptotics.

Theorem 11.4.12. *Let* $X = \mathbb{R}^2/(\bar{\gamma}_1\mathbb{Z} \times \bar{\gamma}_2\mathbb{Z}) \setminus 0$, $\bar{\gamma}_1 < \infty$, $\bar{\gamma}_2 < \infty$ *and let the Dirac operator* A *satisfy conditions* $(\Sigma 0)$, $(\Sigma 1)_{1-3}$, $(\Sigma 2)$, $(\Sigma 4)$ *and* $(11.4.1)$[33] *with* $\gamma = \epsilon|x|$, $\rho = |x|^m$, $\rho_1 = |x|^{m_1}$, $m_1 < -2$, $m_1 \leq 2m$. *Moreover, let condition* $(11.4.23)^\#$ *be*

[33] In this subsection conditions $(\Sigma 4)$ and $(11.4.1)$ are assumed to be fulfilled only for $|x| \leq \epsilon$.

fulfilled (for $|x| \le \epsilon$) and let one of the following three conditions be fulfilled (compare with theorem 11.4.6):

(a) $m \ge 0$, $\tau_1 > M\varsigma + \lim_{\gamma(x)\to 0}\sup V$;

(b) $m < 0$ *and conditions* $(11.4.9)^{\#}_{+}, (11.4.24)^{\#}, (11.4.25)^{\#}$ *are fulfilled with a small enough constant* $\varepsilon = \varepsilon(c, \epsilon, \epsilon_0) > 0$;

(c) $m_1 = 2m < -2$ *and conditions* $(11.4.9)^{\#}_{+}$, $(11.4.24)^{\#}$ *and stabilization conditions* $(11.4.3)^{\#}_{1,2}$ *with* $m' = m$, $r = 1$ *are fulfilled;*

here and below $^{\#}$ *means the following modification:* $|x| \ge c$ *is replaced by* $\gamma(x) \le \epsilon$ *and* $|x| \to \infty$ *is replaced by* $\gamma(x) \to 0$. *Then*

(i) For $\mu = h = 1$, *fixed* τ_1 *and for* $\tau_2 \to +\infty$ *the asymptotics*

$$\mathbf{N}(\tau_1, \tau_2) = \mathcal{N}(\tau_1, \tau_2) + O(\tau_2) \tag{11.4.36}$$

hold with

$$\mathcal{N}(\tau_1, \tau_2) = \varkappa_0 \tau_2^2 (1 + o(1)) \tag{11.4.37}$$

where $\varkappa_0 = \frac{1}{4\pi}\operatorname{vol} X$ *is the standard coefficient.*

(ii) Moreover, let the standard condition $(\Psi 0)$ *be fulfilled for the geodesic flow. Then the asymptotics*

$$\mathbf{N}(\tau_1, \tau_2) = \mathcal{N}(\tau_1, \tau_2) + o(\tau_2) \tag{11.4.38}$$

hold.

Theorem 11.4.13. *Let* X, γ, ρ *and* ρ_1 *be as in the previous theorem, let* $m < 0$ *and let the Dirac operator* A *satisfy conditions* $(\Sigma 0)$, $(\Sigma 1)_{1-3}$, $(\Sigma 2)$, $(\Sigma 4)$, $(11.4.9)^{\#}_{-}$ *and* $(11.4.23)^{\#}$, $(11.4.24)^{\#}$. *Moreover, let one of the following two conditions be fulfilled:*

(b) Inequality $(11.4.25)^{\#}$ *holds with a small enough constant* $\varepsilon = \varepsilon(c, \epsilon, \epsilon_0) > 0$;

(c) $m_1 = 2m < -2m$ *and stabilization conditions* $(11.4.3)^{*}_{1,2}$ *are fulfilled with* $r = 1$, $m' = m$.

Then both statements (i),(ii) of theorem 11.4.12 remain true; however, (11.4.37) *now holds only for* $m_1 + 2 \ge 2m$; *for* $m_1 + 2 \le 2m$ *it should be replaced by*

$$\mathcal{N}(\tau_1, \tau_2) \asymp \tau_2^{\frac{1}{m}(m_1+2)}. \tag{11.4.39}$$

Theorem 11.4.14. *Let* $X \subset \mathbb{R}^2/(\bar\gamma_1\mathbb{Z} \times \bar\gamma_2\mathbb{Z})$ *be a bounded domain with non-empty boundary* ∂X *which consists of transversally intersecting* C^1 *curves. Let the Dirac operator* A *satisfy conditions* $(\Sigma 0)$, $(\Sigma 1)_{1-3}$, $(\Sigma 2)$, $(\Sigma 4)$, (11.4.1) *and* $(11.4.23)^{\#}$ *with* $\gamma(x) = \epsilon_0 \operatorname{dist}(x, \partial X)$, $\rho = \gamma^m$, $\rho_1 = \gamma^{m_1}$, $m_1 < -2$, $m_1 \le 2m$. *Let one of the conditions (a),(b) of theorem 11.4.12 be fulfilled. Then for* $\mu = h = 1$, *fixed* τ_1 *and* $\tau_2 \to \infty$ *the asymptotics*

$$\mathbf{N}(\tau_1, \tau_2) = \mathcal{N}(\tau_1, \tau_2) + O(\tau_2 \log \tau_2) \tag{11.4.40}$$

and (11.4.37) *hold.*

Theorem 11.4.15. *Let X, γ, ρ and ρ_1 be the same as in the previous theorem, $m < 0$ and let the Dirac operator A satisfy conditions* ($\Sigma 0$), ($\Sigma 1)_{1-3}$, ($\Sigma 2$), ($\Sigma 4$), (11.4.9)$^{\#}_{-}$ *and* (11.4.23)$^{\#}$, (11.4.24)$^{\#}$, (11.4.25)$^{\#}$ *with a small enough constant* $\varepsilon = \varepsilon(c, \epsilon, \epsilon_0) > 0$. *Then for $\mu = h = 1$, fixed τ_1 and $\tau_2 \to \infty$ asymptotics* (11.4.39) *hold; however,* (11.4.37) *holds only for $m_1 + 1 > 2m$; for $m_1 + 1 \le 2m$*

$$\mathcal{N}(\tau_1, \tau_2) \asymp \tau^{\frac{2}{m}(m_1+1)}. \tag{11.4.41}$$

I leave it to the reader to compare these statements with results for the Schrödinger operator (theorems 11.3.3 and 11.3.4). Moreover, I leave the treatment of other cases (exponential case, or case when $F = |x|^{-2}\bigl|\log|x|\bigr|^{\alpha}$ with $\alpha > 0$, etc) to the reader.

11.5 Asymptotics with the Spectral Parameter for Three-Dimensional Schrödinger and Dirac Operators

11.5.1 The Schrödinger operator in a bounded domain

Let us start with the Schrödinger operator. First of all one can easily prove

Theorem 11.5.1. [34] *Let X be a bounded domain with C^K boundary, $0 \in X$ and let the Schrödinger operator A satisfy conditions* ($\Sigma 0$), ($\Sigma 1$), ($\Sigma 2$), ($\Sigma 4$)[35] *with $\gamma = \epsilon_0|x|$, $\rho = |x|^m$, $\rho_1 = |x|^{m_1}$, $m_1 \le -2$, $m \le 0$. Suppose either $m_1 < -2m$ or the coercivity condition* ($\Sigma 3)'$, ($\Sigma 3$) *is fulfilled for $m_1 < -2$, $m_1 = -2$ respectively. Then*

(i) For $\mu = h = 1$ and $\tau \to +\infty$, the asymptotics

$$\mathbf{N}(\tau) = \mathcal{N}(\tau) + O(\tau) \tag{11.5.1}$$

hold with $\mathcal{N}$ given by (11.1.9) *with $S = S(x, \tau, \mu h)$ given by* $(6.1.7)_{(3)}$ *and*

$$\mathcal{N}(\tau) = \varkappa_0 \tau^{\frac{3}{2}}(1 + o(1)) \tag{11.5.2}$$

with the standard coefficient $\varkappa_0 = \frac{1}{6\pi^2}\operatorname{vol} X$.

(ii) Moreover, for $m \ge -3$, $m_1 \ge -6$ the asymptotics

$$\mathbf{N}(\tau) = \varkappa_0 \tau^{\frac{3}{2}} + O(\tau) \tag{11.5.3}$$

hold.

(iii) Let one of the following two conditions be fulfilled:

(e) $m < -3$, $m_1 \ge \frac{1}{3}(4m - 6)$,

(f) $m < -3$, $m_1 \ge \frac{3}{2}(m-1)$ *and V and F are positively homogeneous functions of degrees $2m$ and m_1 respectively for $|x| \le \epsilon$.*

Then the asymptotics

$$\mathbf{N}(\tau) = \mathcal{N}^{w}(\tau) + O(\tau) \tag{11.5.4}$$

hold.

(iv) Let $m_1 \le \min(-6, 6 + 4m)$ and (11.3.1) *be fulfilled. Then the asymptotics*

$$\mathbf{N}(\tau) = \mathcal{N}^{M}(\tau) + O(\tau) \tag{11.5.5}$$

hold.

---

[34] Compare with theorem 11.3.1.

[35] Surely, here and in theorem 11.5.3 condition ($\Sigma 4$) should be fulfilled only for $|x| \le \epsilon$. I sometimes do not make this or similar remarks below.

Remark 11.5.2. *(i) Statements (i)–(ii) of remark 11.3.2 remain true.*

(ii) One can treat the case when the singularity lies on a closed curve in X (as in theorem 11.3.4 with $|x'|$ instead of x_1 where $x' = (x_1, x_2)$) and prove asymptotics (11.5.1)–(11.5.5). Moreover, in the statements one should change only the inequalities on m and m_1 and positive homogeneity is assumed with respect to x' instead of x. Let us list the new inequalities (in the corresponding statements):
(ii) $m \ge -2$, $m \ge -4$;
(iiie) $m < -2$, $m_1 \ge \frac{4}{3}(m-1)$; *(iiif)* $m_1 \ge \frac{1}{2}(3m-2)$;
(iv) $m_1 \le \min(-4, 4+4m)$.

Theorem 11.5.3. [36] *Let X be a bounded domain with C^K boundary, $0 \in X$ and let the Schrödinger operator A satisfy conditions ($\Sigma 0$), ($\Sigma 1$), ($\Sigma 2$), ($\Sigma 4$) with $\gamma = \epsilon_0 |x|^{1-\beta}$, $\rho = \exp a|x|^{\beta}$, $\rho_1 = \exp b|x|^{\beta}$, $\beta < 0$ and $0 < a < b$. Suppose* **either** *$b > 2a$* **or** *the coercivity condition ($\Sigma 3$)′ is fulfilled (for $|x| \le \epsilon$). Moreover, let one of the following four conditions be fulfilled:*

(a) $b > 2a$ *and (11.3.1) holds.*

(b) $V = v(\theta)\exp a|x|^{\beta}$ *and* $F = f(\theta)\exp b|x|^{\beta}$ *for* $|x| \le \epsilon$ *where* $\theta = x|x|^{-1}$;

(c) $b = 2a$ *and inequalities (11.3.1) and (11.3.2) are fulfilled with small enough* $\varepsilon = \varepsilon(c, \epsilon, \epsilon_0) > 0$.

(d) $b < 2a$.

Then

(i) For $\mu = h = 1$, $\tau \to +\infty$ asymptotics (11.5.1) and (11.5.2) hold. Furthermore, without these conditions the asymptotics

$$\mathbf{N}(\tau) = \mathcal{N}(\tau) + O(\tau^{1+\delta}) \tag{11.5.6}$$

hold with an arbitrarily small exponent $\delta > 0$.

(ii) Moreover, let one of the following conditions be fulfilled:

(e) $b \le \frac{4}{3}a$;

(f) $b \le \frac{3}{2}a$ *and condition (b) is fulfilled*

Then asymptotics (11.5.4) hold.

(iii) On the other hand, if $b \ge 4a$ then asymptotics (11.5.5) hold.

(iv) Remark 11.3.2(iii) remains true; surely one should also assume that the inequalities on a, α, β are strict for the improved asymptotics (11.5.3), (11.5.4) and (11.5.5).

Theorem 11.5.4. [37] *Let X be a bounded domain with C^K boundary, let L be a closed C^K surface lying in $\bar{X}$ and let conditions ($\Sigma 0$), ($\Sigma 1$), ($\Sigma 2$), ($\Sigma 4$)[38] be fulfilled with $\gamma = \epsilon_0 \delta(x)$, $\rho = \delta(x)^m$, $\rho_1 = \delta(x)^{m_1}$ where $m_1 \le -2$, $m \le 0$ and $\delta(x) = \mathrm{dist}(x, L)$. Suppose* **either** *$m_1 < -2m$* **or** *the coercivity condition ($\Sigma 3$)′, ($\Sigma 3$) is fulfilled for $m_1 < -2$, $m_1 = -2$ respectively. Let one of the following additional conditions be fulfilled:*

(a) $m_1 < 2m$ *and (11.3.1) holds for* $\gamma(x) \le \epsilon$;

[36] Compare with theorem 11.3.3.
[37] Compare with theorem 11.3.4.
[38] Surely, condition ($\Sigma 4$) should be fulfilled only for $\gamma \le \epsilon$.

(b) V and F are positively homogeneous functions of degrees $2m$ and m_1 respectively on $x_1 = \delta(x)$ for $\gamma(x) \le \epsilon$;

(c) Inequalities (11.3.1), (11.3.2) hold for $\gamma(x) \le \epsilon$ with $\epsilon > 0$ and a small enough constant $\varepsilon = \varepsilon(c, \epsilon, \epsilon_0) > 0$;

(d) $m_1 > 2m$.

Then
(i) For $\mu = h = 1$ and $\tau \to +\infty$ the asymptotics

$$\mathbf{N}(\tau) = \mathcal{N}(\tau) + O(\tau \log \tau) \tag{11.5.7}$$

and (11.5.2) hold. Furthermore, without these conditions asymptotics (11.5.6) hold.
(ii) Moreover, for $m \ge -1$, $m_1 = -2$ the asymptotics

$$\mathbf{N}(\tau) = \varkappa_0 \tau^{\frac{3}{2}} + O(\tau \log \tau) \tag{11.5.8}$$

hold.
(iii) Let $m_1 < \min(-2, 2 + 4m)$ and (11.3.1) be fulfilled. Then the asymptotics

$$\mathbf{N}(\tau) = \mathcal{N}^M(\tau) + O(\tau \log \tau) \tag{11.5.9}$$

hold.

11.5.2 The Schrödinger operator in an unbounded domain

Let us now treat unbounded domains. In contrast to the two-dimensional case the magnetic field (without conditions on the magnetic lines) does not now permit us to improve the remainder estimate and therefore instead of $\mu = 1$ one can now assume that $\mu = O(1)$. Moreover, the main contribution to remainder estimate is now given by the large $|x|$ zone while in the two-dimensional case the main contribution is given by the intermediate $|x|$ zone (where the effective magnetic field was bounded); therefore now in order to improve the remainder estimate one should assume a condition on the magnetic flow instead of a condition on the Hamiltonian flow. Let us start with a case which is Weylian in principal:

Theorem 11.5.5. [39] *Let X be an exterior connected domain with C^K boundary and let the Schrödinger operator A satisfy conditions (Σ0), (Σ1), (Σ2) and (Σ4), (Σ3)[40] with $\gamma = \epsilon_0 \langle x \rangle$, $\rho = \langle x \rangle^m$, $\rho_1 = \langle x \rangle^{m_1}$, $m > 0$. Then*
(i) Let $\tau \to +\infty$, $h = O(1)$, $\mu h^{1-\delta} = O(\tau^{\frac{1}{2m}(2m - m_1)})$. Then the asymptotics

$$\mathbf{N}(\tau, \mu, h) = \mathcal{N}(\tau, \mu, h) + O(h^{-2-\delta} \tau^{2l+\delta}) \tag{11.5.10}$$

hold with $l = \frac{1}{2m}(m+1)$ and

$$\mathcal{N}(\tau, \mu, h) \asymp h^{-3} \tau^{3l}; \tag{11.5.11}$$

here and below the exponent $\delta > 0$ is arbitrarily small.
(ii) Moreover, in the context of (i) let $\mu h^{1-\delta} = O(\tau^{\frac{1}{2m}(2m - m_1)) - \delta})$ and suppose either $\mu h^{1-\delta} = O(\tau)$ or $\tau h^\delta \ge 1$. Then the asymptotics

$$\mathbf{N}(\tau, \mu, h) = \mathcal{N}(\tau, \mu, h) + O(h^{-2} \tau^{2l}) \tag{11.5.12}$$

hold.

[39] Compare with theorem 11.3.5.
[40] These two condition are assumed to be fulfilled here and below only for $|x| \ge c$.

(iii) Let stabilization conditions $(11.3.12)_{1,2}$ *be fulfilled with* $r = 1$ *for* $m_1 \geq 0$ *and with* $r = 2$ *for* $m_1 < 0$. *Moreover, let us assume that for* $m_1 = 0$ *either* $F = F^0$ *for* $|x| \geq c$ *or* F *stabilizes to* F^0 *very quickly at infinity. Then asymptotics (11.5.12), (11.5.11) hold for* $\tau \to +\infty$, $h = O(1)$, $\mu h = O\big(\tau^{\frac{1}{2m}(m_1-2m)}\big)$ *such that either* $\mu h^{1-\delta} = O(\tau)$ *or* $\tau h^\delta \geq 1$.

In both cases for $\mu h = o\big(\tau^{\frac{1}{2m}(m_1-2m)}\big)$

$$\mathcal{N}(\tau,\mu,h) = \mathcal{N}^w(\tau,h)\big(1+o(1)\big). \tag{11.5.13}$$

(iv) For $\tau \to +\infty$, $h = O(1)$, $\mu^3 h = O(\tau^{\min(2,\frac{1}{2m}(4m-3m_1-2))})$ *the asymptotics*

$$\mathbf{N}(\tau,\mu,h) = \mathcal{N}^w(\tau,h) + O(h^{-2}\tau^{2l}) \tag{11.5.14}$$

hold.

(v) In the context of (iii) asymptotics (11.5.12) hold for $\tau \to +\infty$, $h = O(1)$, $\mu^2 h = O(\tau^{\min(\frac{3}{2},\frac{1}{2m}(3m-2m_1-1))})$.

Theorem 11.5.6. [41] *Let* X *be an unbounded connected domain and let the Schrödinger operator* A *satisfy conditions* $(\Sigma 0)$, $(\Sigma 1)$, $(\Sigma 2)$, *and* $(\Sigma 4)$, *(11.3.23) with* $\gamma = \epsilon_0\langle x\rangle$, $\rho = \langle x\rangle^m$, $\rho_1 = \langle x\rangle^{m_1}$, $m \in (-1,0)$. *Then*

(i) Let $\tau \to -0$, $h = O(1)$, $\mu h = O\big(\tau^{\frac{1}{2m}(2m-m_1)}\big)$. *Then the asymptotics*

$$\mathbf{N}(\tau,\mu,h) = \mathcal{N}(\tau,\mu,h) + O\big(h^{-2-\delta}|\tau|^{2l-\delta}\big) \tag{11.5.10$'$}$$

hold with $\mathcal{N}(\tau,\mu,h) = O(h^{-3}\tau^{3l})$ *and with the same exponent* l *as before. Recall that the exponent* $\delta > 0$ *is arbitrarily small.*

Moreover, let $\mu h = o\big(\tau^{\frac{1}{2m}(m_1-2m)}\big)$ *and let condition* $(11.1.13)^\#$ *be fulfilled. Then equivalence relation (11.5.11) holds.*

(ii) Moreover, in the context of (i) let $\mu h^{1-\delta} = O\big(\tau^{\frac{1}{2m}(2m-m_1))-\delta}\big)$ *and suppose either* $\mu h^{1-\delta} = O(1)$ *or* $|\tau| \leq h^\delta$ *with an arbitrarily small exponent* $\delta > 0$. *Then asymptotics (11.5.12) hold.*

(iii) Let us assume that one of the following conditions is fulfilled:

(a) $m_1 \neq 2m$, $V \leq -\epsilon\rho^2$ *and stabilization conditions* $(11.3.12)_{1,2}$ *are fulfilled with* $r = 1$ *for* $m_1 > 2m$ *and with* $r = 2$ *for* $m_1 < 2m$.

(b) Stabilization conditions $(11.3.12)_{1,2}$ *are fulfilled,* $m_1 \neq 2m$ *and*

$$\sum_{\alpha:|\alpha|\leq r} |x|^{|\alpha|}|D^\alpha V^0| \geq \epsilon\rho^2. \tag{11.5.16}$$

(c) Stabilization conditions $(11.3.12)_{1,2}$ *are fulfilled,* $m_1 = 2m$ *and*

$$\sum_{\alpha:1\leq|\alpha|\leq r} |x|^{|\alpha|}\Big|D^\alpha \frac{V^0}{F^0}\Big| \geq \epsilon\rho^2\rho_1^{-1} \qquad \forall x : V^0 \leq \epsilon\rho^2. \tag{11.5.16}$$

Then asymptotics (11.5.12) hold for $\tau \to -0$, $h = O(1)$, $\mu h = O\big(\tau^{\frac{1}{2m}(m_1-2m)}\big)$ *such that either* $\mu h^{1-\delta} = O(\tau)$ *or* $\tau h^\delta \geq 1$.

(iv) For $\tau \to -0$, $h = O(1)$, $\mu^3 h = O\big(\min(1,\tau^{\frac{1}{2m}(4m-3m_1-2)})\big)$ *asymptotics (11.5.13) hold.*

(v) Let (a) be fulfilled or let one of the following conditions be fulfilled:

(d) Stabilization conditions $(11.3.12)_{1,2}$ *are fulfilled with* $r = 2$, $m_1 \neq 2m$ *and* 0 *is not a degenerate critical value of* V^0.

---

[41] Compare with theorem 11.3.8.

(e) Stabilization conditions $(11.3.12)_{1,2}$ *are fulfilled with* $r=2$, $m_1=2m$ *and there are no non-positive degenerate critical values of* V^0.

Then for $\tau\to -0$, $h=O(1)$, $\mu^2h=O\bigl(\min(1,\tau^{\frac{1}{2m}(3m-2m_1-1)})\bigr)$ *asymptotics* (11.5.13) *hold.*

Theorem 11.5.7. [42] *Let X be an unbounded connected domain and let the Schrödinger operator A satisfy conditions* $(\Sigma 0)$, $(\Sigma 1)$, $(\Sigma 2)$ *and* $(\Sigma 4)$, (11.3.23) *with* $\gamma=\epsilon_0\langle x\rangle$, $\rho=\langle x\rangle^{-1}\log^{\alpha}\langle x\rangle$, $\rho_1=\langle x\rangle^{-2}\log^{\beta}\langle x\rangle$, $\alpha>-\frac{1}{2}$. *Then*

(i) Let $\tau\to -0$, $h=O(1)$, $\mu h=O\bigl(|\log|\tau||^{2\alpha-\beta}\bigr)$. *Then the asymptotics*

$$\mathbf{N}(\tau,\mu,h)=\mathcal{N}(\tau,\mu,h)+O\bigl(h^{-2-\delta}|\log|\tau||^{2\alpha+1+\delta}\bigr) \tag{11.5.17}$$

hold with $\mathcal{N}(\tau,\mu,h)=O\bigl(h^{-3}|\log|\tau||^{3\alpha+1}\bigr)$.

Moreover, for $\alpha\ge -\frac{1}{3}$, $\mu h=o\bigl(|\log|\tau||^{2\alpha-\beta}\bigr)$ *under condition* $(11.1.13)^{\#}$

$$\mathcal{N}(\tau,\mu,h)\asymp h^{-31}|\log|\tau||^{3\alpha+1}. \tag{11.5.18}$$

(ii) Moreover, in the context of (i) let $\mu h^{1-\delta}=O\bigl(|\log|\tau||^{2\alpha-\beta-\delta}\bigr)$ *and suppose either* $\mu h^{1-\delta}=O(1)$ *or* $|\log|\tau||^{\delta}h\ge 1$. *Then the asymptotics*

$$\mathbf{N}(\tau,\mu,h)=\mathcal{N}(\tau,\mu,h)+O\bigl(h^{-2}|\log|\tau||^{2\alpha+1}\bigr) \tag{11.5.19}$$

hold.

(iii) For $|x|\ge c$ *let*

$$V=v(\theta)|x|^{-2}\log^{2\alpha}|x|, \tag{11.5.20$_1$}$$

$$F=f(\theta)|x|^{-2}\log^{\beta}|x|,\qquad \theta=x|x|^{-1} \tag{11.5.20$_2$}$$

where $v,f\in C^K$. *Moreover, let us assume that* **either** $\beta\ne 2\alpha$ *and*

$$\sum_{\alpha:|\alpha|\le r}|D^{\alpha}v|\ge\epsilon\qquad \forall x:|x|=1. \tag{11.5.21}$$

or $\beta=2\alpha$ *and*

$$\sum_{\alpha:|\alpha|\le r}|D^{\alpha}\frac{v}{f}|\ge\epsilon\qquad \forall x:|x|=1, v\le\epsilon. \tag{11.5.21$'$}$$

Then asymptotics (11.5.19) *hold for* $\tau\to -0$, $\mu=O(1)$, $h=O(1)$, $\mu h=O\bigl(|\log|\tau||^{2\alpha-\beta}\bigr)$, $h=O\bigl(|\log|\tau||^{\alpha}\bigr)$.

(iii)$'$ Let condition $(\Sigma 5)_3$ *be fulfilled with* $\varrho_3=\log^{-\delta}\langle x\rangle$. *Moreover, let stabilization conditions* $(11.3.29)_{1,2}$ *be fulfilled. Then asymptotics* (11.5.19) *hold under the same restrictions on* μ, h *as in (iii).*

(iv) For $\tau\to -0$, $h=O(1)$, $\mu^3h=O\bigl(\min(1,|\log|\tau||^{4\alpha-3\beta}),\ h=O(|\log|\tau||^{\alpha})\bigr)$ *the asymptotics*

$$\mathbf{N}(\tau,\mu,h)=\mathcal{N}^{w}(\tau,h)+O\bigl(h^{-2}|\log|\tau||^{\alpha+1}\bigr) \tag{11.5.22}$$

hold.

---

[42] Compare with theorem 11.3.9.

Remark 11.5.8. *(i) One can replace equalities* $(11.5.20)_{1,2}$ *by a quick enough stabilization condition as* $|x| \to \infty$*; this condition includes the condition that* r *is large enough.*

(ii) The list of statements can be extended (with other restrictions on μ*).*

(iii) In the proof of statements (v) of theorems 11.5.5–11.5.6 the estimate of γ_1 *from below is important (*γ_1 *is introduced in section 6.4). One can easily see that in the context of theorem 11.5.5(v) with* $m_1 > 0$ *and in the context of theorem 11.5.6(v) with* $m_1 > 2m$ *the inequality* $\gamma_1 \ge \epsilon_1 |x|$ *holds for* $|x| \ge c_0$*; moreover, in the context of theorem 11.5.5(v) with* $m_1 < 0$ *and in the context of theorem 11.5.6(v) with* $m_1 < 2m$ *the inequality* $\gamma_1 \ge \epsilon_1 \big| |x| - r_0(\theta, \tau) \big|$ *holds where* $r_0 \asymp \tau^{\frac{1}{2m}}$*. On the other hand, in the context of theorem 11.5.5(v) with* $m_1 = 0$ *the inequality* $\gamma_1 \ge \epsilon_1 |x|^{2m+1} \tau^{-1}$ *holds.*

(iv) I leave the cases $m_1 = 2m$ *and* $2\alpha = \beta$ *in theorem 11.5.6(v) to the reader.* ◇

The following statements are close to the statements above:

Theorem 11.5.9. *Let* $0 \in X$ *and let the Schrödinger operator* A *satisfy conditions* $(\Sigma0)$, $(\Sigma1)$, $(\Sigma2)$, $(\Sigma4)$ *and* $(\Sigma3)'$[43] *with* $\gamma = \epsilon_0 |x|$, $\rho = |x|^m$, $\rho_1 = |x|^{m_1}$, $m > 0$. *Then*

(i) Let $\tau \to +0$, $h = o(\tau^l)$ *(recall that* $l = \frac{1}{2m}(m+1)$*),* $\mu h = O\big(\tau^{\frac{1}{2m}(2m - m_1)}\big)$*. Then the asymptotics*

$$\mathbf{N}(\tau, \mu, h) = \mathcal{N}(\tau, \mu, h) + O\big(h^{-2-\delta} \tau^{2l}\big) \tag{11.5.23}$$

hold and (11.5.11) is fulfilled.

Moreover, for $\mu h = o\big(\tau^{\frac{1}{2m}(2m - 2m_1)}\big)$ *asymptotics (11.5.13) hold.*

(ii) Moreover, if $\mu h^{1-\delta} = O\big(\tau^{\frac{1}{2m}(2m - m_1)}\big)$ *then asymptotics (11.5.12) hold.*

(iii) Let stabilization conditions $(11.3.12)_{1,2}$ *be fulfilled with* $r = 1$ *for* $m_1 \ge 0$ *and with* $r = 2$ *for* $m_1 < 0$ *as* $x \to 0$*. Moreover, let us assume that for* $m_1 = 0$ *either* $F = F^0$ *for* $|x| \le \epsilon$ *or* F *stabilizes to* F^0 *very quickly at 0. Then asymptotics (11.5.12), (11.5.11) hold for* $\tau \to +0$, $h = O(1)$, $\mu h = O\big(\tau^{\frac{1}{2m}(m_1 - 2m)}\big)$.

(iv) Moreover, for $\tau \to +0$, $h = o(\tau^l)$ *such that* $\mu^3 h = O\big(\tau^{\frac{1}{2m}(4m - 3m_1 - 2)}\big)$*, asymptotics (11.5.14) hold.*

(v) In (iii), for $\tau \to +0$, $h = o(\tau^l)$, $\mu^2 h = O\big(\tau^{\frac{1}{2m}(3m - 2m_1 - 1)}\big)$*, asymptotics (11.5.12), (11.5.10) hold.*

Theorem 11.5.10. *Let* $0 \in \bar{X}$ *and let the Schrödinger operator* A *satisfy conditions* $(\Sigma0)$, $(\Sigma1)$, $(\Sigma2)$, $(\Sigma4)$ *and (11.3.23), (11.1.13) with* $\gamma = \epsilon_0 |x|$, $\rho = |x|^m$, $\rho_1 = |x|^{m_1}$, $m \in (-1, 0)$*. Then*

(i) Let $\tau \to -\infty$, $h = O(|\tau|^l)$ *such that* $\mu h = O\big(\tau^{\frac{1}{2m}(2m - m_1)}\big)$*. Then asymptotics (11.5.23) hold with* $\mathcal{N}(\tau, \mu, h) = O(h^{-3} |\tau|^{3l})$ *and with the same exponent* l *as before.*

Moreover, let $\mu h = o(\tau^{\frac{1}{2m}(m_1 - 2m)})$ *and let condition* $(11.1.13)^\#$ *be fulfilled. Then equivalence relation (11.5.11) holds.*

(ii) Moreover, in the context of (i) let $\mu h^{1-\delta} = O\big(|\tau|^{\frac{1}{2m}(2m - m_1)) - \delta}\big)$*. Then asymptotics (11.5.12) hold.*

(iii) Let one of the conditions $(a)^\#$*,* $(b)^\#$*,* $(c)^\#$ *of theorem 11.5.6 be fulfilled.*

Then asymptotics (11.5.12) hold for $\tau \to -\infty$, $h = O(|\tau|^l)$ *such that* $\mu h = O\big(|\tau|^{\frac{1}{2m}(m_1 - 2m)}\big)$.

(iv) For $\tau \to -\infty$, $h = O(|\tau|^l)$, $\mu^3 h = O\big(\min(1, \tau^{\frac{1}{2m}(4m - 3m_1 - 2)})\big)$ *asymptotics (11.5.13) hold.*

(v) Let one of the conditions $(a)^\#$*,* $(d)^\#$*,* $(e)^\#$ *of theorem 11.5.6 be fulfilled.*

Then asymptotics (11.5.13) hold for $\tau \to -0$, $h = O(|\tau|^l)$ *such that* $\mu^2 h = O\big(\min(1, |\tau|^{\frac{1}{2m}(3m - 2m_1 - 1)})\big)$.

---

[43] Surely, conditions $(\Sigma3)'$, $(\Sigma4)$ should be fulfilled here and in theorems 11.5.10 and 11.5.11 only for $|x| \le \epsilon$.

Theorem 11.5.11. *Let $0 \in \bar{X}$ and let the Schrödinger operator A satisfy conditions $(\Sigma 0)$, $(\Sigma 1)$, $(\Sigma 2)$, $(\Sigma 4)$, (11.3.23), (11.1.13) with $\gamma = \epsilon_0|x|$, $\rho = |x|^{-1}\bigl|\log|x|\bigr|^{\alpha}$, $\rho_1 = |x|^{-2}\bigl|\log|x|\bigr|^{\beta}$, $\alpha < -\frac{1}{2}$. Then*

(i) Let $\tau \to -\infty$, $h = o(\log^{\alpha}|\tau|)$, $\mu h = O\bigl(\log^{2\alpha-\beta}|\tau|\bigr)$. Then the asymptotics

$$\mathbf{N}(\tau,\mu,h) = \mathcal{N}(\tau,\mu,h) + O\bigl(h^{-2-\delta}\log^{2\alpha+1}|\tau|\bigr) \tag{11.5.24}$$

hold with $\mathcal{N}(\tau,\mu,h) = O\bigl(h^{-3}\log^{3\alpha+1}|\tau|\bigr)$.

Moreover, for $\alpha \le -\frac{1}{3}$, $\mu h = o\bigl(\log^{2\alpha-\beta}|\tau|\bigr)$ under condition (11.1.13)$^{\#}$ relation (11.5.18) holds.

(ii) Moreover, in the context of (i) let $\mu h^{1-\delta} = O\bigl(\log^{2\alpha-\beta}|\tau|\bigr)$. Then asymptotics (11.5.19) hold.

(iii) For $|x| \le \epsilon$ let conditions $(11.5.20)_{1,2}$ be fulfilled with $v, f \in C^K$. Moreover, let us assume that **either** *$\beta \ne 2\alpha$ and (11.5.21) is fulfilled* **or** *$\beta = 2\alpha$ and (11.5.21) is fulfilled.*

Then for $\tau \to -\infty$, $h = o\bigl(\log^{\alpha}|\tau|\bigr)$, $\mu h = O\bigl(\log^{2\alpha-\beta}|\tau|\bigr)$, asymptotics (11.5.19) hold.

(iii)$'$ Let condition $(\Sigma 5)_3$ be fulfilled with $\varrho_3 = \log^{-\delta}|x|$. Moreover, let stabilization conditions $(11.3.29)_{1,2}$ be fulfilled. Then asymptotics (11.5.19) hold under the same restrictions on μ, h as in (iii).

(iv) Let $\tau \to -\infty$, $h = o(\log^{\alpha}|\tau|)$, $\mu^3 h = O\bigl(\log^{4\alpha-3\beta}|\tau|\bigr)$. Then asymptotics (11.5.22) hold.

11.5.3 Asymptotics non-Weylian in principal

Let us proceed to asymptotics which are non-Weylian "in principal." The following statements have standard proofs.

Theorem 11.5.12. [44] *Let X be an exterior domain with C^K boundary and let the Schrödinger operator A satisfy conditions $(\Sigma 0)$, $(\Sigma 1)$, $(\Sigma 2)$, $(\Sigma 4)$ with $\gamma = c_0\langle x\rangle$, $\rho = \langle x\rangle^{m}$, $\rho_1 = \langle x\rangle^{m_1}$, $m \ge 0$, $m_1 > 0$. Then*

(i) For $\tau \to \infty$, $h \in [0,1]$, $\mu > 0$ such that $\frac{\tau}{\mu h} \to \infty$, $(\mu h)^{2m}\tau^{m_1-2m} \ge \epsilon$ the asymptotics

$$\mathbf{N}(\tau,\mu,h) = \mathcal{N}(\tau,\mu,h) + O(\lambda^{2+\delta}) \tag{11.5.25}$$

hold with $\lambda = \tau^{\frac{1}{m_1}+\frac{1}{2}}\mu^{-\frac{1}{m_1}}h^{-\frac{1}{m_1}-1}$ and

$$\mathcal{N}(\tau,\mu,h) \asymp \mathcal{N}^{M}(\tau,\mu,h) \asymp \lambda^{3}. \tag{11.5.26}$$

Moreover, for $(\mu h)^{2m}\tau^{m_1-2m} \to +\infty$ the asymptotics $\mathcal{N} \asymp \mathcal{N}^{M}\bigl(1+o(1)\bigr)$ hold.

(ii) Moreover, let

$$\sum_{\alpha:|\alpha|\le K} |D^{\alpha}F|\gamma^{|\alpha|} \ge \epsilon_0\rho_1. \tag{11.5.27}$$

Then for $\tau \to \infty$, $h \in [0,1]$, $\mu > 0$ such that $\frac{\tau}{\mu h} \to \infty$, $(\mu h)^{2m}\tau^{m_1-2m} \ge C_0$ with a sufficiently large constant $C_0 = C_0(c,\epsilon_0)$ the asymptotics

$$\mathbf{N}(\tau,\mu,h) = \mathcal{N}(\tau,\mu,h) + O(\lambda^{2}) \tag{11.5.28}$$

hold.

(iii) On the other hand, if condition (11.5.27) is fulfilled only for $|x| \ge c$ then asymptotics (11.5.28) hold under the additional restriction $\frac{\tau}{\mu h} \ge \lambda^{\delta}$.

(iv) Under the additional condition $\tau^{2m+1-\frac{1}{2}m_1} = O\bigl(h^{m_1+2m+1}\mu^{2m+1}\bigr)$ one can replace $\mathcal{N}$ by $\mathcal{N}^{M}$ in statements (i)–(iii).

[44] Compare with theorem 11.3.12.

I leave the most interesting case $\mu = 1$, $m_1 > 2m$, $h = 1$ to the reader.

Theorem 11.5.13. [45] *Let X be an exterior domain with C^K boundary and let the Schrödinger operator A satisfy conditions $(\Sigma 0), (\Sigma 1)$, $(\Sigma 2)$ and $(\Sigma 4)$ with $\gamma = \epsilon_0([x]_L + 1)$, $\rho = ([x]_L + 1)^m$, $\rho_1 = ([x]_L + 1)^{m_1}$ where $[x]_L = \sum_j |x|^{\frac{1}{l_j}}$, $L = (l_1, l_2, l_3)$, $\min_j l_j = 1$, $-m > 0$ is large enough and $m_1 > 0$. Moreover, let F be a positively L-quasi-homogeneous function of degree m_1 for $|x| \geq c$ (one can instead assume that F tends to such functions very quickly). Then for $\mu = h = 1$, $\tau \to +\infty$ the asymptotics*

$$\mathbf{N}(\tau, 1, 1) = \varkappa \tau^p + O(\tau^q) \tag{11.5.29}$$

hold with $\varkappa > 0$, $p = \frac{3}{2} + \frac{1}{m_1}\sum_j l_j$, $q = 1 + \frac{1}{m_1}(\sum_j l_j - 1)$.

Theorem 11.5.14. [46] *Let X be an exterior domain with C^K boundary and let the Schrödinger operator A satisfy conditions $(\Sigma 0)$, $(\Sigma 1)$, $(\Sigma 2)$, $(\Sigma 4)$ and (11.5.27) with $\gamma = \epsilon_0 \langle x\rangle^{1-\beta}$, $\rho = \exp a\langle x\rangle^\beta$, $\rho_1 = \exp b\langle x\rangle^\beta$, $\beta \geq 2a > 0$. Then*

(i) For $\mu = h = 1$, $\tau \to +\infty$ the asymptotics

$$\mathbf{N}(\tau, 1, 1) = \mathcal{N}(\tau, 1, 1) + O\big(\tau \log^{\frac{2}{\beta}} \tau\big), \tag{11.5.30}$$

$$\mathcal{N} = \mathcal{N}^M\big(1 + o(1)\big) \asymp \tau^{\frac{3}{2}} \log^{\frac{3}{\beta}} \tau \tag{11.5.31}$$

hold.

(ii) Moreover, if **either** $4a < b$ **or** $4a = b$ *and $\beta \geq 1$ then the asymptotics*

$$\mathbf{N}(\tau, 1, 1) = \mathcal{N}^M(\tau, 1, 1) + O\big(\tau \log^{\frac{2}{\beta}} \tau\big), \tag{11.5.32}$$

hold.

I leave it to the reader to generalize these two statements for variable μ and h.

Theorem 11.5.15. *Let $0 \in X$ and let the Schrödinger operator A satisfy conditions $(\Sigma 0)$, $(\Sigma 1)$, $(\Sigma 2)$, $(\Sigma 4)$ with $\gamma = \epsilon_0|x|$, $\rho = |x|^m$, $\rho_1 = |x|^{m_1}$, $m > 0$, $m_1 > 0$. Let*

$$V + h\mu F \geq \epsilon_0(\rho^2 + \mu h \rho_1) \qquad \text{in } X. \tag{11.5.33}$$

Then

(i) For $h \in (0, 1]$, $\tau \to +0$, $(\mu h)^{2m}\tau^{m_1 - 2m} \geq \epsilon_0$ asymptotics (11.5.25), (11.5.26) hold. Moreover, for $(\mu h)^{2m}\tau^{m_1-2m} \to \infty$ asymptotics $\mathcal{N} = \mathcal{N}^M\big(1 + o(1)\big)$ hold.

(ii) Moreover, let condition (11.5.25) be fulfilled. Then for $h \in (0, 1]$, $\tau \to +0$, $(\mu h)^{2m}\tau^{m_1-2m} \geq C_0$ with a large enough constant $C_0 = C_0(c, \epsilon_0)$ asymptotics (11.5.28), (11.5.26) hold.

(iii) Under the additional condition $\tau^{2m+1-\frac{1}{2}m_1} = O\big(h^{m_1+2m+1}\mu^{2m+1}\big)$ one can replace $\mathcal{N}$ by $\mathcal{N}^M$ in statements (i)–(iii).

11.5.4 Very strong magnetic field

Let us now treat the important special case of non-Weylian asymptotics when the formula

$$\mathbf{N} \sim \mathcal{N}_0 = \frac{1}{2\pi^2}\mu h^{-2} \int \big(\tau - V - \mu h F\big)_+^{\frac{1}{2}} F\sqrt{g}\,dx \tag{11.5.34}$$

applies (i.e., one takes only terms with $n = 0$); we assume that $\mu h = 1$. The above arguments provide the following series of statements:

---

[45] Compare with example 10.4.12.

[46] Compare with theorem 11.3.14.

Theorem 11.5.16. *Let X be an exterior domain with C^K boundary and let the Schrödinger operator A satisfy conditions $(\Sigma 0)$, $(\Sigma 1)$, $(\Sigma 2)$, $(\Sigma 4)$ and $(\Sigma 5)_3$ with $\gamma = \epsilon_0\langle x\rangle$, $\rho = \langle x\rangle^m$, $\rho_1 = \langle x\rangle^{m_1}$, $\varrho_3 = \langle x\rangle^{m_2-m_1}$, $m_1 = 2m$, $0 < m_2 < m_1$, $n = 0$.*[47] *Moreover, let*

$$(\Sigma 3)'' \qquad V + F \ge \epsilon\varrho_3\rho^2$$

for $|x| \ge c$ and let stabilization conditions $(11.3.12)_2$ *and*

$$D^\sigma(V + F - V^1) = o\bigl(\varrho_3\rho^2\gamma^{-|\sigma|}\bigr) \qquad \forall\sigma : |\sigma| \le 1 \tag{11.5.35}$$

be fulfilled for $|x| \to \infty$ where F^0 and V^1 are positively homogeneous functions of degrees m_1 and $2m_2$ respectively. Then

(i) For $\tau \to +\infty$, $h \in (0,1]$, $\mu = h^{-1}$ the asymptotics

$$\mathbf{N}(\tau,\mu,h) = \mathcal{N}(\tau,\mu,h) + O\bigl(h^{-2}\tau^{\frac{1}{m_2}(m+1)}h^{-\delta}\bigr) \tag{11.5.36}$$

hold. Moreover

$$\mathcal{N} \asymp \mathcal{N}_0 \asymp h^{-3}\tau^{\frac{1}{2m_2}(2m+3)+\frac{1}{2}}, \tag{11.5.37}$$

$$\mathcal{N} = \mathcal{N}_0 + O\bigl(h^{-3}\tau^{\frac{3}{2m_2}(m+1)}\bigr). \tag{11.5.38}$$

(ii) Moreover, under the additional restriction $\tau \ge h^\delta$ the asymptotics

$$\mathbf{N}(\tau,\mu,h) = \mathcal{N}(\tau,\mu,h) + O\bigl(h^{-2}\tau^{\frac{1}{m_2}(m+1)}\bigr) \tag{11.5.39}$$

hold.

(iii) Furthermore, let $X = \mathbb{R}^3$, let condition $(\Sigma 4)$ be fulfilled in X and let F satisfy (11.5.27). *Then asymptotics* (11.5.39) *hold for $\tau \to +\infty$, $h \in (0,1]$, $\mu = h^{-1}$.*

Therefore for $\mu = h^{-1} = O\bigl(\tau^{(m+1)(\frac{1}{m_2}-\frac{3}{2m})}\bigr)$, $3m_2 \le 2m$ asymptotics (11.5.39) *hold with $\mathcal{N}$ replaced by $\mathcal{N}_0$.*

Theorem 11.5.17. *Let X be an exterior domain with C^K boundary and let the Schrödinger operator A satisfy conditions $(\Sigma 0)$, $(\Sigma 1)$, $(\Sigma 2)$, $(\Sigma 4)$ and $(\Sigma 5)_3$ with $\gamma = \epsilon_0\langle x\rangle$, $\rho = \langle x\rangle^m$, $\rho_1 = \langle x\rangle^{m_1}$, $\varrho_3 = \langle x\rangle^{m_2-m_1}$, $m_1 = 2m$, $m_2 < m_1$, $m_1 \in (-1,0)$, $n = 0$. Moreover, let*

$$V + F \le -\epsilon_3\varrho\rho^2 \qquad \text{for } x \in \Gamma, |x| \ge c \tag{11.5.40}$$

where Γ is a non-empty open cone and let stabilization conditions $(11.3.12)_2$ *and* (11.5.35) *be fulfilled for $|x| \to \infty$. Then*

(i) For $\tau \to -0$, $h \in (0,1]$, $\mu = h^{-1}$ asymptotics (11.5.36) *hold.*

(ii) Under the additional restriction $\tau \ge h^\delta$ asymptotics (11.5.39) *hold.*

(iii) Furthermore, let $X = \mathbb{R}^3$, condition $(\Sigma 4)$ be fulfilled in X and

$$\sum_{|\alpha|\le K} |D^\alpha \frac{V}{F} + (2n+1)|\gamma^{|\alpha|} \ge \epsilon\rho_3\rho^2\rho^{-1} \tag{11.5.41}$$

for any $n \in \mathbb{Z}^+$. Then asymptotics (11.5.39) *hold for $\tau \to -0$, $h \in (0,1]$, $\mu = h^{-1}$.*

Moreover, (11.5.37) *holds and $\mathcal{N} - \mathcal{N}_0 = O(h^{-3})$. Therefore, for $\mu = h^{-1} = O\bigl(\tau^{\frac{1}{m_2}(m+1)}\bigr)$ asymptotics* (11.5.39) *hold with $\mathcal{N}$ replaced by $\mathcal{N}_0$.*

---

[47] Recall this condition:

$$(\Sigma 5)_3 \qquad |D^\alpha(V + (2n+1)F)| \le c\varrho_3\rho^2\gamma^{-|\alpha|} \qquad \forall\alpha : |\alpha| \le K.$$

Theorem 11.5.18. *Let the conditions of theorem 11.5.17 be fulfilled with* $\gamma = \epsilon_0\langle x\rangle$, $\rho = \langle x\rangle^m$, $\rho_1 = \langle x\rangle^{m_1}$, $\varrho_3 = \langle x\rangle^{-2m-2}\log^\alpha\langle x\rangle$, $m_1 = 2m > -2$, $\alpha > 0$. *Then*

(i) For $\tau \to -0$, $h \in (0,1]$, $\mu = h^{-1}$ *the asymptotics*

$$\mathbf{N}(\tau,\mu,h) = \mathcal{N}(\tau,\mu,h) + O\bigl(h^{-2-\delta}\tau^{-m-1}\bigl|\log|\tau|\bigr|^{2\alpha(m+1)}\bigr) \tag{11.5.42}$$

hold.

(ii) Moreover, in the context of theorem 11.5.17(ii),(iii) the asymptotics

$$\mathbf{N}(\tau,\mu,h) = \mathcal{N}(\tau,\mu,h) + O\bigl(h^{-2}\tau^{-m-1}\bigl|\log|\tau|\bigr|^{2\alpha(m+1)}\bigr) \tag{11.5.43}$$

hold.

(iii) Furthermore,

$$\mathcal{N} \asymp \mathcal{N}_0 \asymp h^{-3}\tau^{-m-1}\bigl|\log|\tau|\bigr|^{\alpha(2m+3)} \tag{11.5.44}$$

and $\mathcal{N} - \mathcal{N}_0 = O(h^{-3})$. *In particular, in statement (i) in asymptotics* (11.5.42) *one can replace* $\mathcal{N}$ *by* $\mathcal{N}_0$.

Theorem 11.5.19. *Let* X *be an exterior domain with* C^K *boundary and let the Schrödinger operator* A *satisfy conditions* $(\Sigma 0)$, $(\Sigma 1)$, $(\Sigma 2)$, $(\Sigma 4)$ *and* $(\Sigma 5)_3$ *with* $\gamma = \epsilon_0\langle x\rangle$, $\rho = \langle x\rangle^{-1}\log^{\frac{1}{2}\beta}\langle x\rangle$, $\rho_1 = \rho^2$, $\varrho_3 = \log^{2\alpha-\beta}\langle x\rangle$, $\beta > 2\alpha > 0$. *For* $|x| \geq c$ *let condition* (11.5.40) *be fulfilled; moreover, for* $|x| \geq c$ *let conditions* $(11.5.20)_2$ *and*

$$V + F = v(\theta)|x|^{-2}\log^{2\alpha}|x| \tag{$11.5.20_3$}$$

be fulfilled with v *satisfying* (11.5.21). *Then*

(i) For $\tau \to -0$, $h \in (0,1]$, $\mu = h^{-1}$ *the asymptotics*

$$\mathbf{N}(\tau,\mu,h) = \mathcal{N}(\tau,\mu,h) + O\bigl(h^{-2-\delta}\bigl|\log|\tau|\bigr|^{\beta+1}\bigr) \tag{11.5.45}$$

hold.

(ii) Under the additional assumption $h^\delta \geq \bigl|\log|\tau|\bigr|$ *the asymptotics*

$$\mathbf{N}(\tau,\mu,h) = \mathcal{N}(\tau,\mu,h) + O\bigl(h^{-2}\bigl|\log|\tau|\bigr|^{\beta+1}\bigr) \tag{11.5.46}$$

hold.

(iii) In the context of theorem 11.5.18(iii) asymptotics (11.5.46) *hold for* $\tau \to -0$, $h \in (0,1]$, $\mu = h^{-1}$.

(iv) Furthermore,

$$\mathcal{N} \asymp \mathcal{N}_0 \asymp h^{-3}\bigl|\log|\tau|\bigr|^{\alpha+\beta+1} \tag{11.5.47}$$

and $\mathcal{N} - \mathcal{N}_0 = O(h^{-3})$; *therefore for* $\mu = h^{-1} = O\bigl(\bigl|\log|\tau|\bigr|^{\beta+1}\bigr)$ *asymptotics* (11.5.46) *remain true with* $\mathcal{N}$ *replaced by* $\mathcal{N}_0$.

Remark 11.5.20. *(i) The most important case in theorem 11.5.18 is* $\mu = h = 1$ *and* $m = 0$.

(ii) Let us note that taking $\mu = h = 1$ *and formally setting* $m_2 = -1$ *or* $\alpha = 0$ *we obtain that the principal part of the asymptotics and the remainder estimate are of the same orders. However, in contrast to the Weylian case, these expressions are not* $O(1)$ *even formally. There is a principal difference between the present situation and the Weylian situation: now* $m_2 = -1$ *and* $\alpha = 0$ *are only the limits of the validity of the asymptotics derived here but they are no longer boundaries separating operators with*

discrete spectrum limit -0 *from operators with finite discrete spectrum. Now* -0 *is also (under some assumptions) the limit point of the discrete spectrum for* $m_2 < -1$ *but in this case quite different asymptotics hold: these asymptotics were derived for* $m_2 < -1, m_1 = 0$ *and a constant magnetic field by A. Sobolev [1–4] (see also [Raikov 1–5]) and for* $m_2 = -1$ *in very special case they were derived by G. Raikov [6]. We return to these asymptotics in the next chapter. The idea of the asymptotics for* $m_2 = -1$ *is the following: one should integrate the potential* V^1 *along magnetic lines and then the two-dimensional Schrödinger operator should be treated with the potential* $-\bar{F} - cW^2$ *where* W *is the integral above and* c *is an appropriate numerical constant; we assume that* $F = \bar{F}$ *for* $|x| \ge c$ *and* $\bar{F} = \mathrm{const} > 0$*. Then the asymptotics of section 11.4 should be applied to this new operator.* ◊

One can prove the following statements by the same methods:

Theorem 11.5.21. *Let* $0 \in X$ *and let the Schrödinger operator* A *satisfy conditions* $(\Sigma0)$, $(\Sigma1)$, $(\Sigma2)$, $(\Sigma4)$, $(\Sigma5)_3$ *and* $(\Sigma3)''$ *with* $\gamma = \epsilon_0|x|$, $\rho = |x|^m$, $\rho_1 = |x|^{m_1}$, $\varrho_3 = |x|^{2m_2-2m}$, $m_1 = 2m$, $m > -1$, $m_2 > \max(m, 0)$. *Let stabilization conditions* $(11.3.12)_2$ *and* (11.5.35) *be fulfilled for* $|x| \to 0$. *Then for* $\tau \to +0$, $h = o(\tau^{l'})$, $\mu = h^{-1}$, $l' = \frac{1}{2m_2}(m_2+1)$ *asymptotics* (11.5.39), (11.5.38) *hold with* $\mathcal{N} = \mathcal{N}_0$.

Theorem 11.5.22. *Let* $0 \in X$ *and let the Schrödinger operator* A *satisfy conditions* $(\Sigma0)$, $(\Sigma1)$, $(\Sigma2)$, $(\Sigma4)$, $(\Sigma5)_3$ *and* $(\Sigma3)''$ *with* $\gamma = \epsilon_0|x|$, $\rho = |x|^m$, $\rho_1 = |x|^{m_1}$, $\varrho_3 = |x|^{2m_2-2m}$, $m_1 = 2m$, $0 > m_2 > m > -1$. *Let stabilization conditions* $(11.3.12)_2$ *and* (11.5.27) *be fulfilled for* $|x| \to 0$. *Then for* $\tau \to -\infty$, $h = o(\tau^{l'})$, $\mu = h^{-1}$, $l' = \frac{1}{2m_2}(m_2+1)$ *asymptotics* (11.5.39), (11.5.38) *hold with* $\mathcal{N} = \mathcal{N}_0$.

I leave it to the reader to formulate and prove similar statements for $\rho_1 = \rho^2 = |x|^{-2}\bigl(|\log|x|| + 1\bigr)^{\beta}$, $\varrho_3\rho^2 = |x|^{2m_2}$, $\beta < -1$ in the following cases: $m_2 > 0$ and $\tau \to +0$; $m_2 \in (-1, 0)$ and $\tau \to -\infty$. In these cases under appropriate conditions $\mathcal{N} \asymp h^{-3}\tau^{l'}|\log|\tau||^{\beta}$ and the remainder estimate is $O\bigl(h^{-2}|\log|\tau||^{\beta+1}\bigr)$. I leave the case $\varrho_3\rho^2 = |x|^{-2}\langle\log|x|\rangle^{2\alpha}$ with $0 < \alpha < \frac{1}{2}\beta$ to the reader; under appropriate assumptions, asymptotics (11.5.46), (11.5.47) hold. In all these cases $\mathcal{N} = \mathcal{N}_0$. Moreover, I leave the cases when the potential well is a curve or a surface instead of a point to the reader.

11.5.5 The Dirac operator

Let us now consider the Dirac operator.

Theorem 11.5.23. [48] *Let* $X = \mathbb{R}^3$, $M > 0$ *and let the Dirac operator* A *satisfy conditions* $(\Sigma0)$, $(\Sigma1)_{1-3}$, $(\Sigma2)$, $(\Sigma4)$, (11.4.2) *with* $\gamma = \epsilon_0\langle x\rangle$, $\rho = \langle x\rangle^m$, $\rho_1 = \langle x\rangle^{m_1}$, $m \in (-1, 0)$. *Then*

(i) Let $\tau_1 \in (-M, M)$ *be fixed,* $\tau_2 = M - \eta$ *with* $\eta \to +0$ *and* $h = O(1)$, $\mu h = O\bigl(\eta^{\frac{1}{2m}(2m-m_1)}\bigr)$. *Then the asymptotics*

$$\mathbf{N}(\tau_1, \tau_2, \mu, h) = \mathcal{N}(\tau_1, \tau_2, \mu, h) + O\bigl(h^{-2-\delta}\eta^{2l-\delta}\bigr) \tag{11.5.48}$$

hold with $l = \frac{1}{2m}(m+1)$ *and* $\mathcal{N} = O(h^{-3}\eta^{3l})$. *Moreover,* $\mathcal{N} \asymp h^{-3}\eta^{3l}$ *under condition* $(11.4.7)_+$.

(i) Let* $\tau_1 \in (-M, M)$ *be fixed,* $\tau_2 = M - \eta$ *with* $\eta \to +0$ *and* $h = O(1)$, $\mu h \ge \epsilon\eta^{\frac{1}{2m}(2m-m_1)}$. *Then the asymptotics*

$$\mathbf{N}(\tau_1, \tau_2, \mu, h) = \mathcal{N}(\tau_1, \tau_2, \mu, h) + O\bigl(\mu h^{-1-\delta}\eta^{\frac{1}{2m}(m_1+2)-\delta}\bigr) \tag{11.5.49}$$

---

[48] Compare with theorems 11.4.1, 11.5.6.

hold and $\mathcal{N} = O(\mu h^{-2}\eta^{\frac{1}{2m}(m+m_1+3)})$. *Moreover,* $\mathcal{N} \asymp \mu h^{-2}\eta^{\frac{1}{2m}(m+m_1+3)}$ *under condition* $(11.4.7)_+$.

(ii) Moreover, in the context of (i) let $\mu h^{1-\delta} = O\big(\eta^{\frac{1}{2m}(2m-m_1))-\delta}\big)$ *and either* $\mu h^{1-\delta} = O(1)$ *or* $|\eta| \le h^\delta$ *with an arbitrarily small exponent* $\delta > 0$. *Then the asymptotics*

$$\mathbf{N}(\tau_1,\tau_2,\mu,h) = \mathcal{N}(\tau_1,\tau_2,\mu,h) + O(h^{-2}\eta^{2l}) \tag{11.5.50}$$

hold.

(iii) Let one of the conditions (a)–(c) of theorem 11.5.6(iii) be fulfilled. Let $\tau_1 \in (-M,M)$ *be fixed,* $\tau_2 = M - \eta$ *with* $\eta \to +0$ *and* $h = O(1)$, $\mu h = O\big(\eta^{\frac{1}{2m}(2m-m_1)}\big)$. *Then asymptotics* (11.5.50) *hold.*

(iii) Let one of the conditions (a)–(c) of theorem 11.5.6(iii) be fulfilled. Let* $\tau_1 \in (-M,M)$ *be fixed,* $\tau_2 = M - \eta$ *with* $\eta \to +0$ *and* $\mu = O(1)$, $h = O(1)$, $\mu h \ge \epsilon\eta^{\frac{1}{2m}(2m-m_1)}$ *and* $\eta \le h^\delta$. *Then the asymptotics*

$$\mathbf{N}(\tau_1,\tau_2,\mu,h) = \mathcal{N}(\tau_1,\tau_2,\mu,h) + O\big(\mu h^{-1}\eta^{\frac{1}{2m}(m_1+2)}\big) \tag{11.5.51}$$

hold.

(iv) In the context of (i) for $\mu^3 h = O\big(\min(1,\eta^{\frac{1}{2m}(4m-3m_1-2)})\big)$ *and in the context of (iii) for* $\mu^2 h = O\big(\min(1,\eta^{\frac{1}{2m}(3m-2m_1-1)})\big)$ *asymptotics* (11.5.50) *hold with* $\mathcal{N}$ *replaced by* $\mathcal{N}^w$.

In the conditions of the following theorem only $n=0$ should be treated:

Theorem 11.5.24. *Let* $X = \mathbb{R}^3$ *and let the Dirac operator* A *satisfy conditions* $(\Sigma 0)$, $(\Sigma 1)_{1-3}$, $(\Sigma 2)$, $(\Sigma 4)$ *and* (11.4.2) *with the same* ρ, ρ_1, γ *and* $M > 0$ *as before. Let* $m_1 \ge 2m$ *and*

$$|\nabla V| \ge \epsilon\rho^2\gamma^{-1} \qquad \text{for } |x| \ge c, \tag{11.5.52}$$

$$(1-\epsilon)F \ge M|V| \qquad \text{for } |x| \ge c. \tag{11.5.53}$$

Then for $\mu = h = 1$, *fixed* $\tau_1 \in (-M,M)$ *and* $\tau_2 = M - \eta$ *with* $\eta \to +0$ *the asymptotics*

$$\mathbf{N}(\tau_1,\tau_2) = \mathcal{N}(\tau_1,\tau_2) + O\big(\eta^{\frac{1}{2m}(m_1+2)}\big) \tag{11.5.54}$$

hold with

$$\mathcal{N} = O\big(\eta^{\frac{1}{2m}(m+m_1+3)}\big). \tag{11.5.55}$$

Moreover, under condition $(11.4.7)_+$ *we obtain an equivalence relation. On the other hand, under condition* $(11.4.9)_-$, $M-0$ *is not a limit point of the discrete spectrum.*

I leave it to the reader to prove statements similar to theorems 11.5.7, 11.5.9, 11.5.15, 11.5.18, 11.5.19 and 11.5.21. In these statements the principal parts and the remainder estimates consist of two terms. For example, for $\mu = h = 1$, $\gamma = \epsilon_0\langle x\rangle$, $\rho = \langle x\rangle^{-1}\log^\alpha\langle x\rangle$, $\rho_1 = \langle x\rangle^{-2}\log^\beta\langle x\rangle$, $\beta > 2\alpha$ the best possible remainder estimate is $O\big(|\log\eta|^{2\alpha} + |\log\eta|^{\beta+1}\big)$ and $\mathcal{N} = O\big(|\log\eta|^{3\alpha} + |\log\eta|^{\alpha+\beta+1}\big)$. On the other hand, for $\rho_1 = \langle x\rangle^{m_1}$ with $m_1 > -2$ we obtain the best possible remainder estimate $O\big(\eta^{-\frac{1}{2}(m_1+2)}|\log\eta|^{\frac{1}{2}(m_1+2)\alpha}\big)$ and $\mathcal{N} = O\big(\eta^{-\frac{1}{2}(m_1+2)}|\log\eta|^{\frac{1}{2}(m_1+4)\alpha}\big)$. Finally, the reader can easily generalize the conditions of all the above statements.

Appendix J. Examples of Vector Potentials

In this appendix we prove that the results of this chapter are meaningful. The only questionable part of their conditions is the existence of vector potentials with the given properties of the scalar intensities F and the doubts are only in the three-dimensional case. The construction of a conformal asymptotic Euclidean metric tensor[49] g^{jk} and the construction of a scalar potential V are obvious in all cases and are left to the reader. Analysis of the two-dimensional case is also obvious. In what follows $g^{jk} = \delta^{jk}$ in the definition of F^j and F.

Proposition J.1. *(i) For any $m \in \mathbb{R}$ there exist functions $V_j \in C^\infty(\mathbb{R}^3 \setminus 0)$ positively homogeneous of degree $m+1$ such that*

$$F \geq \epsilon |x|^m. \tag{J.1}$$

(ii) For any $m \geq -2$ there exist functions $V_j \in C^\infty(\mathbb{R}^3)$ such that

$$|D^\alpha V_j| \leq c_\alpha \langle x \rangle^{m+1-|\alpha|} \qquad \forall \alpha, \tag{J.2}$$

$$F \geq \epsilon \langle x \rangle^m \tag{J.3}$$

and V_j and F tend to positively homogeneous functions at infinity.

Proof. One can take

$$V_1 = ar^{m-1}x_1^2, \quad V_2 = x_3 r^m + ar^{m-1}x_1x_2, \quad V_3 = -x_2r^m + ar^{m-1}x_1x_3$$

where in (i), $r = |x|$ and in (ii), $r = \langle x \rangle$ and $a = 0$ for $m \neq -2$ and $a \neq 0$ for $m = -2$. □

Proposition J.2. *(i) For any $m \in \mathbb{R}$, $\beta \in \mathbb{R}$ there exist functions $v_j \in C^\infty(\mathbb{S}^2)$ such that for $V_j = v_j(\theta)|x|^{1+m}\bigl|\log|x|\bigr|^\beta$ with $\theta = x|x|^{-1}$ the following conditions are fulfilled for $|x| \leq \epsilon$:*

$$F = f(\theta)|x|^m \bigl|\log|x|\bigr|^\beta (1+\varphi); \qquad f(\theta) \geq \epsilon, \tag{J.4}$$

$$D^\alpha \varphi = O\bigl(|x|^{-|\alpha|} \log^{-1}|x|\bigr) \qquad \forall \alpha \tag{J.5}$$

as $|x| \to 0$.

(ii) For any $m \in \mathbb{R}$, $\beta \in \mathbb{R}$ there exist functions $V_j \in C^\infty(\mathbb{R}^3)$ such that

$$|D^\alpha V_j| \leq c_\alpha \langle x \rangle^{1+m-|\alpha|} \log^\beta \langle x \rangle \qquad \forall \alpha,$$

$$F \geq \epsilon \langle x \rangle^m \log^\beta \langle x \rangle$$

and for $|x| \geq c$ (J.4), (J.5) hold.

Proof. Let $\bar{V}_j$ be the potentials constructed in proposition J.1; one can set $V_j = \bar{V}_j \langle \log r \rangle^{\frac{1}{2}\beta}$. □

Proposition J.3. *(i) For any $\beta < 0$, $a > 0$, $m \in \mathbb{R}$ there exist potentials V_j such that*

$$|D^\alpha V_j| \leq c_\alpha \exp(a|x|^\beta)|x|^{(\beta-1)|\alpha|+m} \qquad \forall \alpha$$

and

$$F \geq \epsilon \exp(a|x|^\beta)|x|^{\beta-1+m} \qquad \text{for } |x| \leq \epsilon.$$

(ii) For any $\beta > 0$, $a > 0$, $m \in \mathbb{R}$ there exist potentials V_j such that

$$|D^\alpha V_j| \leq c_\alpha \exp(a\langle x \rangle^\beta)\langle x \rangle^{(\beta-1)|\alpha|+m} \qquad \forall \alpha$$

and

$$F \geq \epsilon \exp\bigl(a\langle x \rangle^\beta\bigr)\langle x \rangle^{\beta-1+m} \qquad \text{for } |x| \leq \epsilon.$$

---

[49] I.e., a tensor $g^{jk} = \delta^{jk}(1+\varphi)$ with $|1+\varphi| \geq \epsilon > 0$ such that $D^\alpha \varphi = o(\gamma^{-|\alpha|})$ as $|x| \to \infty$ or $|x| \to 0$ for all α.

Proof. One can set

$$V_1 = 0, \qquad V_2 = \exp(a\nu(x))\cos\bigl(b\psi(x)\bigr)r(x)^m,$$
$$V_3 = \exp(a\nu(x))\sin\bigl(b\psi(x)\bigr)r(x)^m$$

with the $r(x)$ introduced in proposition J.1, $\nu(x) = |x|^\beta$ in (i), $\nu(x) = a\langle x\rangle^\beta + O(1)$ such that $|D^\alpha\nu(x)| = O\bigl(r(x)^{\beta-|\alpha|}\bigr)$ $\forall\alpha$ in (ii),

$$\psi(x_1, x') = \int_0^{x_1} r(y_1, x')^{\beta-1}dy_1, \qquad x' = (x_2, x_3)$$

and b is large enough. □

Proposition J.4. *For any $L = (l_1, l_2, l_3)$ with $\min_j l_j = 1$ and $m \ne -1$ there exist potentials $V_j \in C^\infty(\mathbb{R}^3\setminus 0)$ positively L-quasi-homogeneous of degree $m+1$ and such that $F = F^0(1+\varphi)$ for $|x| \ge c$ where $F^0 > 0$ is positively L-quasi-homogeneous of degree m and*

$$D^\alpha\varphi = O([x]_L^{-\sigma-|\alpha|}) \qquad \text{as } [x]_L \to \infty \quad \forall\alpha$$

with $\sigma > 0$.

Proof. Let $l_1 = 1$; then one can set $V_1 = 0$, $V_2 = \omega^{m+1}$, $V_3 = x_1\omega^m$ where $\omega \in C^\infty(\mathbb{R}^3\setminus 0)$ is a positive function positively L-quasi-homogeneous of degree 1. □

Proposition J.5. *For any $m \ne -2$ there exist potentials $V_1, V_2, V_3 \in C^K(\Omega)$, $\Omega = (\mathbb{R}^2\setminus 0)\times\mathbb{R}/\mathbb{Z} \ni (x', x_3) = (x_1, x_2, x_3)$, positively homogeneous on x' of degrees $m+1$, $m+1$, $m+2$ respectively*[50] *such that $F^3 \ne 0$, integral curves of the vector field $\frac{1}{F^3}(F^1, F^2, F^3)$ are non-closed and along these curves $|x'| = \text{const}$. Moreover, all the functions depend only on x', and F and F^3 depend only on $|x'|$.*

Proof. One can set

$$V_1 = x_2|x'|^m, \quad V_2 = -x_1|x'|^m, \quad V_3 = a|x'|^{m+2}$$

with irrational $\frac{a}{\pi}$. □

Remark J.6. *In the case of codimension 1 (of the manifold L where the singularity or degeneration is placed) one can set $V_2 = v(x_1)$, $V_1 = V_3 = 0$ if $L = \{x_1 = 0\}$.* ◇

Comments

The number of papers devoted to asymptotics in which the magnetic field is important is not large (compared with the huge number of papers devoted to the opposite case; see Chapter 10). I would like to mention the papers of Y. Colin de Verdiere [7,8], A. Mohamed [4,5] and A. Mohamed-J. Nourrigat [1], G. Raikov [1–6] and H. Tamura [6–9] which are closely related to this chapter. There are also papers of G. Raikov [1–6], A. Sobolev [1–4] and S. Solnyshkin [1] with less obvious links to this chapter; these papers are closely related to section 12.5. The papers of S. Vugalter-G. M. Zhislin [2,5,6] deal with many particle Hamiltonians and some one-particle analysis is necessary which is closely related to this chapter.

[50] Then F^1, F^2, F^3 are positively homogeneous of degrees $m+1$, $m+1$, m respectively.

12. Miscellaneous Asymptotics

In this chapter I apply the results and methods of the previous chapters in order to derive asymptotics of the spectrum for the following problems: in section 12.1 I treat operators in domains with "thick" cusps, in section 12.2 I consider operators with potentials degenerate at infinity (such as the Schrödinger operator with a positively homogeneous potential $V^0 \ge 0$ with thick cusp $\{V^0 \le \tau\}$ for any fixed $\tau > 0$); section 12.3 is devoted to maximally hypoelliptic operators with symplectic degeneration variety Λ. Moreover, section 12.4 is devoted to asymptotics of Riesz means for operators with singularities at separate points and in section 12.5 I treat the same for the Schrödinger operator with strong magnetic field. Finally, in section 12.6 I consider the three-dimensional Schrödinger and Dirac operators with constant magnetic field and with electric potential quickly decreasing at infinity and treat the asymptotics of eigenvalues below the essential spectrum.

All these problems have certain common features: I do not apply the results of Chapters 8 and 9; the asymptotics either contain non-Weylian terms or are completely non-Weylian (otherwise they would follow from the results of Chapters 8 and 9); all the problems in question differ from problems which can be treated by means of results of Chapters 8 and 9 by some parameters (exponents, dimensions, orders of singularity, of degeneration, of Riesz means, etc.). Moreover, in almost all sections (excluding section 12.4) I apply the theory of Chapters 2 and 4 for operators with operator-valued symbols; namely, in sections 12.1, 12.2, after a coordinate transform we obtain cylindrical cusps and I treat operators given on the base of the cusp for which the auxiliary Hilbert space is L^2 over sections of the cusp. In section 12.3 I consider ΨDO with phase space Λ and with auxiliary Hilbert space $L^2(\mathbb{R}^{d'})$ with $d' = \frac{1}{2}\operatorname{codim}\Lambda$. Finally, in section 12.6 I direct the x_3-axis along the magnetic field and then treat operators on $\mathbb{R}^2 \ni x' = (x_1, x_2)$ with auxiliary Hilbert space $L^2(\mathbb{R})$ with $\mathbb{R} \ni x_3$. In sections 12.1–12.3 there are zones where I apply the standard Weylian asymptotics of Chapters 4 and 5, zones where we can obtain the same asymptotics with improved remainder estimates (these enhancements are due to results on the propagation of singularities for operators with operator-valued symbols; compare with the analysis in sections 6.3, 7.3) and zones where results of Chapter 4 are applicable; the second and third zones are overlapping, and one can combine the asymptotics obtained.

In sections 12.4 and 12.5, instead of operators with operator-valued symbols I treat model operators with positively homogeneous coefficients and apply the Weylian method to the difference of the spectral functions of the original and the model operators.

The results of section 12.1 were obtained (partly) in cooperation with my graduate student E. Filippov; the results of section 12.4 were obtained (partly) in cooperation with I. M. Sigal and the results of section 12.5 were obtained in cooperation with A. Sobolev.

12.1 Operators in Domains with Thick Cusps

12.1.1 Introduction

Let us consider a uniformly elliptic operator of order $2m$

$$A = A^w(x,D), \qquad A(x,\xi) = \sum_{|\alpha|\le 2m} a_\alpha(x)\xi^\alpha \tag{12.1.1}$$

where

(12.1.2) $a_\alpha(x)$ are Hermitian $D\times D-$matrix valued functions such that

$$|D^\beta a_\alpha| \le c\langle x\rangle^{-|\beta'|+\mu(2m-|\alpha|)+\mu|\beta''|} \qquad \forall\alpha: |\alpha|\le 2m\ \forall\beta: |\beta|\le K, \tag{12.1.3}$$

$$D^\beta a_\alpha = o\bigl(|x'|^{\mu(2m-|\alpha|)+\mu|\beta''|}\bigr) \qquad \text{as } |x'|\to\infty \\ \forall\alpha: |\alpha|\le 2m\ \forall\beta'': |\beta''|\le K: 2m-|\alpha|+|\beta''|>0, \tag{12.1.4}$$

$$\langle a(x,\xi)v,v\rangle \ge \epsilon|v|^2 \qquad \forall v\quad \forall(x,\xi) \tag{12.1.5}$$

where

$$a(x,\xi) = \sum_{|\alpha|=2m} a_\alpha(x)\xi^\alpha$$

is the principal symbol of A. We consider this operator in the domain $X\subset\mathbb{R}^d$ such that

$$X = X_0\cup X_1,\ \partial X\in C^K,\ X_0\Subset\mathbb{R}^d, \\ X_1 = \bigl\{x=(x',x''): x'\in Z=\mathbb{R}^{d'},\ x''\in\varrho(x')Y\bigr\},\quad Y\Subset\mathbb{R}^{d''},\ \partial Y\in C^K \tag{12.1.6}$$

with

$$\varrho(x')\asymp\langle x'\rangle^{-\mu},\quad |D^{\beta'}\varrho|\le c\varrho\langle x'\rangle^{-|\beta'|}\quad \forall\beta': |\beta'|\le K,\ \mu>0. \tag{12.1.7}$$

Moreover, for $d'=1$ one can take $Z=\mathbb{R}^\pm$ instead of $Z=\mathbb{R}$. One says that X_0 is the *compact part*of X, X_1 is the *cusp*, Z and Y are the *cusp base and cusp section* respectively and ϱ is the *cusp form defining function*. The same asymptotics as below in the case of a domain with more than one cusp (with different characteristics) surely follow from the analysis of this section; showing that is left to the reader.

Let us assume that

(12.1.8) A Dirichlet boundary condition (of order m) is given on ∂X.

Generalization to the cases of mixed boundary conditions which are Dirichlet for $|x'|\ge c$, or Dirichlet on the upper boundary of the cusp and Neumann on the lower boundary of the cusp for $|x'|\ge c$ and $d'=1$, is left to the reader, as are other natural generalizations associated with the boundary conditions.

Under these conditions A is a self-adjoint operator in $L^2(X,\mathbb{K})$, $\mathbb{K}=\mathbb{C}^D$ and has compact resolvent. Moreover, the following statement follows from Chapter 10:

Proposition 12.1.1. *Let conditions (12.1.1)–(12.1.8) be fulfilled. Then*

(i) If $\varrho^{d''-1} \in L^2(Z)$ ($\iff \mu > \mu_1 = d'(d''-1)^{-1}$) then the Weylian asymptotics

$$\mathbf{N}(\tau) = \mathcal{N}^w(\tau) + O(\tau^{\frac{1}{2m}(d-1)}) \tag{12.1.9$_1$}$$

hold as $\tau \to \infty$ with

$$\mathcal{N}^w(\tau) \asymp \tau^{\frac{d}{2m}}. \tag{12.1.10$_1$}$$

Moreover, under the standard condition ($\Psi 0$) on the Hamiltonian trajectories of $a(x,\xi)$ with reflections at ∂X (and perhaps with branching at the points of reflection for $m \geq 2$) the asymptotics

$$\mathbf{N}(\tau) = \mathcal{N}^w(\tau) + \varkappa_1 \tau^{\frac{1}{2m}(d-1)} + o(\tau^{\frac{1}{2m}(d-1)}) \tag{12.1.11$_1$}$$

hold. Here $\mathcal{N}^w$ is given by the standard Weylian formula, and moreover, one can replace it here by $\varkappa_0 \tau^{\frac{d}{2m}}$. Moreover, $\varkappa_0$ and $\varkappa_1$ are standard coefficients which are finite for $\mu > \mu_0 = d'd^{-1}$, $\mu > \mu_1$ respectively.

(ii) For $\mu = \mu_1$ the asymptotics

$$\mathbf{N}(\tau) = \mathcal{N}^w(\tau) + O(\tau^{\frac{1}{2m}(d-1)} \log \tau) \tag{12.1.9$_2$}$$

hold as $\tau \to \infty$ and for $\mu \in [\mu_0, \mu_1)$ the asymptotics

$$\mathbf{N}(\tau) = \mathcal{N}^w(\tau) + O(\tau^{\frac{1}{2m}(d-q)}) \tag{12.1.9$_3$}$$

hold as $\tau \to \infty$ with $q = d'\mu^{-1} + d'' \in [0,1)$; here $q = 0 \iff \mu = \mu_0$, $q = 1 \iff \mu = \mu_1$, and for $\mu > \mu_0$ (12.2.10)$_1$ holds, and for $\mu = \mu_0$

$$\mathcal{N}^w(\tau) \asymp \tau^{\frac{d}{2m}} \log \tau. \tag{12.1.10$_3$}$$

Remark 12.1.2. *(i) To be honest, our conditions here are more restrictive than necessary but the case $\mu > \mu_1$ is not our goal here. Our goal is to treat the case $\mu \leq \mu_1$ and to derive asymptotics with better remainder estimates. Namely, we obtain asymptotics with the remainder estimate $O(\tau^{\frac{1}{2m}(d-1)})$ and even $o(\tau^{\frac{1}{2m}(d-1)})$ under the standard condition ($\Psi 0$) for $d' = 1$ and for $d' \geq 2$, $\mu > \mu^*$ with $\mu^* < \mu_1$; otherwise we obtain asymptotics with a worse remainder estimate. Surely, one should replace $\mathcal{N}^w$ by another expression and some additional conditions appear.*

(ii) For Riesz means the crucial values in the Weylian asymptotics are $\mu_{\vartheta,1} = d'(d''-\vartheta-1)^{-1}$ and μ_0 respectively. For $\mu > \mu_{\vartheta,1}$ we get the remainder estimate $O(\tau^{\frac{1}{2m}(d-\vartheta-1)+\vartheta})$ or even "o" instead of "O" under the standard condition ($\Psi 0$). For $\mu = \mu_{\vartheta,1}$ the remainder estimate is $O(\tau^{\frac{1}{2m}(d-\vartheta-1)+\vartheta} \log \tau)$ and for $\mu_0 < \mu < \mu_{\vartheta,1}$ it is $O(\tau^{\frac{d'(\mu+1)}{2m\mu}+\vartheta})$. One can easily see that this exponent lies in $\left(\frac{1}{2m}(d-\vartheta-1)+\vartheta, \frac{d}{2m}+\vartheta\right)$. At the same time the principal Weylian part is of order $\tau^{\frac{d}{2m}+\vartheta}$ and $\tau^{\frac{d}{2m}+\vartheta} \log \tau$ for $\mu > \mu_0$ and $\mu = \mu_0$ respectively. $\diamondsuit$

12.1.2 Precise asymptotics

First of all let us divide $\mathbf{N}(\tau)$ into the sum

$$\mathbf{N}(\tau) = \mathbf{N}_0(\tau) + \mathbf{N}_1(\tau) \tag{12.1.12}$$

where

$$\mathbf{N}_j(\tau)=\int \psi_j(x)\,\mathrm{tr}\, e(x,x,\tau)dx, \tag{12.1.13}$$

$e(x,y,\tau)$ is the Schwartz kernel of the spectral projector of A and $\psi_0,\psi_1\in C^K$ are supported in $\{|x'|\le 3c\}$, $\{|x'|\ge 2c\}$ respectively, $\psi_0+\psi_1=1$. We assume that $X_0\subset\{|x'|\le\frac{3}{2}c\}$. Then *Weylian asymptotics hold for* $\mathbf{N}_0(\tau)$ *with the proper remainder estimate* $O\bigl(\tau^{\frac{1}{2m}(d-1)}\bigr)$ *(and even* $o\bigl(\tau^{\frac{1}{2m}(d-1)}\bigr)$ *under the standard condition* $(\Psi 0)$ *on the Hamiltonian billiards). So we need to derive asymptotics of* $\mathbf{N}_1(\tau)$. Let us introduce the coordinates (x',x'') with $\mathrm{x}''=x\varrho(x')^{-1}$. In these coordinates the cusp X_1 transforms to the cylinder $Z\times Y$ and the differential expression A transforms to a differential expression which is symmetric with weight $\varrho(x')^{d''}$ provided our test functions vanish for $|x'|\le\frac{1}{3}c$. In order to get rid of this weight we multiply the differential expression by $\varrho(x')^{\frac{1}{2}d''}$ and $\varrho(x')^{-\frac{1}{2}d''}$ on the left and right respectively, obtaining

$$\tilde{A}=\tilde{A}^w(x,D),\quad \tilde{A}(x,\xi)=\sum_{|\alpha|\le 2m} a_\alpha(\xi'-V')^{\alpha'}\xi''^{\alpha''}\varrho(x')^{-|\alpha''|} \tag{12.1.14}$$

$$V_k=\sum_{d'+1\le j\le\le d}\mathrm{x}_j(\partial_k\log\varrho)\xi_j\qquad k=1,\dots,d'$$

in the coordinates $x=(x',\mathrm{x}'')$. We treat this operator as an operator in $L^2(Z,\mathbb{H})$ with $\mathbb{H}=L^2(Y,\mathbb{K})$. Surely, we cannot discuss self-adjointness now because we cannot guarantee that our functions vanish for $|x'|\le\frac{5}{3}c$.

So, we are interested in the spectral projector of A corresponding to the spectral parameter τ. Let us consider the $\gamma(\bar{x}')$-neighborhood of the point $\bar{x}'$ with $\gamma(x')=\frac{1}{2}|x'|$ where we always assume that $|\bar{x}'|\ge c_1$ with a large enough constant c_1. Let us make a dilatation and the corresponding multiplication.

Let us first treat the case $\varrho(\bar{x}')\le c\tau^{-\frac{1}{2m}}$ (i.e., $|x'|\ge\epsilon_1^{-1}\tau^{\frac{1}{2\mu m}}$) where here and below $\epsilon_j>0$ are sufficiently small. Multiplying the operator by $\varrho^{2m}(\bar{x}')$ after dilatation we obtain an h-differential operator with $h=\varrho\gamma^{-1}<1$; we are interested in the operator's spectral projector corresponding to the spectral parameter $\tau\varrho(\bar{x}')^{2m}<\epsilon_2$. One can easily see that this operator is of the type considered in Chapter 4 provided $B=(L_{(m)\mathrm{D}})$, $B_1=(L_{(M)\mathrm{N}})^{mM^{-1}}$ where $L_M=(\Delta''+I)^M$ and the subscripts "D" and "N" mean m-th order Dirichlet and Neumann boundary conditions respectively (of the proper order), Δ'' is the positive Laplacian in Y and M is taken large in order to obtain large $l=l(M)$ (see Chapter 4 for B, B_1 and l). Then for $j\in[0,l]$ we see that $\mathbb{H}_j=\mathrm{D}(B_1^j)=H_{2mj}(Y)$ are Sobolev spaces and for $j\in[-l,0]$ $\mathbb{H}_j=B_1^{-j}(\mathbb{H})$ are the spaces dual to $H_{-2mj}(Y)$. In particular, one can easily check conditions (4.1.6), (4.1.7), $(4.1.11)_{1/2}$, $(4.1.12)_+$, (4.1.13), (4.1.15) and $(4.1.17)_+$.

In particular, we see that $\tilde{A}$ is elliptic at the energy levels $\le\epsilon_2$. One can check this easily by dropping lower order terms (with $|\alpha|<2m$) in $\tilde{A}$ and assuming that D'' and a_α commute and the energy level is non-positive. This procedure is justified by condition (12.1.4) and the appropriate choice of $\epsilon_2>0$. Since A is a self-adjoint operator, $\|\exp itA\|=1$ and $\|E(\tau)\|\le 1$. On the other hand, $\exp itA$ and $E(\tau)$ generate operator-valued operators in $B\bigl(\bar{x}',\gamma(\bar{x}')\bigr)$ via their Schwartz kernels and one can apply the theory of Chapter 4. This trick permits us to avoid the discussion of the self-adjointness of $\tilde{A}$, etc. Then the results of Chapter 4 immediately yield the following

Proposition 12.1.3. *Let conditions* (12.1.1)–(12.1.8) *be fulfilled and let* c_0 *be large enough. Then the estimate*

$$|e(x,y,\tau)|\le C\langle x'\rangle^{-s}\langle y'\rangle^{-s} \tag{12.1.15}$$

$$\forall\tau\ge 1\quad\forall x,y\in X:|x'|+|y'|\ge c_0\tau^{\frac{1}{2\mu m}}$$

holds.

Let us now treat the zone $\{c_0 \le |x'| \le c_0 \tau^{\frac{1}{2m\mu}}\}$. Here one should multiply the operator by τ^{-1} and so one should consider the operator $\tau^{-1}A$ at the energy level 1. One can easily check that we are again in the situation of Chapter 4 with $h = \tau^{-\frac{1}{2m}}\bar{\gamma}^{-1}$, $B = \tau^{-1}\bar{\varrho}^{-2m}(L_{(M)\mathrm{D}})^{mM^{-1}} + I$, $B_1 = \tau^{-1}\bar{\varrho}^{-2m}(L_{(M)\mathrm{N}})^{mM^{-1}} + I$ with the previous definition of $L_{(M)\mathrm{D}}$ and $L_{(M)\mathrm{N}}$; here and below $\bar{f} = f(\bar{x}')$. One can easily check that this operator $\tilde{A}$ (treated as an operator with operator-valued symbol) is elliptic at energy levels ≤ 1 for $|\xi'| \ge C_0$. Moreover, one can easily check that this operator is microhyperbolic at the energy level 1 in the direction $\mathcal{T}$ in each of the following cases:

(a) $m = 1$, $a_\alpha = o(1)$ for $|\alpha'| = |\alpha''| = 1$, $|\xi'| \ge \epsilon$, $\mathcal{T} = \langle \xi', \partial_{\xi'}\rangle$,

(b) $m = 1$, $|\xi'| \le \varepsilon$ with a small enough constant $\varepsilon > 0$, $|\ell\varrho| \ge \epsilon\varrho|x'|^{-1}$, $\mathcal{T} = \langle \ell, \partial_{x'}\rangle$ with an appropriate vector ℓ such that $|\ell| \le c$ and

(12.1.16) $$\ell a_\alpha = o\bigl(|x'|^{-1+(2m-|\alpha|\mu+|\beta''|\mu)}\bigr) \quad \text{as } |x'| \to \infty \quad \forall \alpha : |\alpha| < 2m$$

(here this condition should be checked for $\alpha' = 0$ only),

(c) $m \ge 1$, $\ell\varrho = \varrho\bigl(1 + o(1)\bigr)|x'|^{-1}$ as $|x'| \to \infty$ with an appropriate vector ℓ, $|\ell| \le c$, (12.1.16) holds and $\mathcal{T} = \langle \xi', \partial_{\xi'}\rangle + \langle \ell, \partial_{x'}\rangle$.

All the conditions are given in the original coordinates and in conditions (b), (c) one can take ℓ depending on x'.

Therefore, the microhyperbolicity condition is fulfilled for any ξ' provided one of the following conditions holds:

$(12.1.17)_1$ $m = 1$, $a_\alpha = o(1)$ for $|\alpha'| = |\alpha''| = 1$, $|\ell\varrho| \ge \epsilon\varrho|x|^{-1}$ and (12.1.16) holds for $\alpha' = 0$,

$(12.1.17)_2$ $m \ge 1$, $\ell\varrho = \varrho\bigl(1 + o(1)\bigr)|x'|^{-1}$ and (12.1.16) holds.

So, *suppose either* $(12.1.17)_1$ *or* $(12.1.17)_2$ *is fulfilled*. Then, applying the results of Chapter 4 we obtain that if $\psi = \psi(x')$ is a γ-admissible function supported in $\{|x' - \bar{x}'| \le \frac{1}{2}\bar{\gamma}\}$ then

(12.1.18) $$|B_1^j \int \psi(x')\bigl(e(x,y,\tau)_{x'=y'} - (2\pi)^{-d'}\int \mathbf{e}(x',\xi',\tau)d\xi'\bigr)dx' \cdot B_1^k| \le C\tau^{\frac{1}{2m}(d'-1)}\langle x'\rangle^{d'-1} \quad \forall j,k \le l$$

where the norm in the left-hand expression is the operator norm in $\mathbb{H}$ and $\mathbf{e}(x',\xi',\tau)$ is the spectral projector of operator $\mathbf{a}(x',\xi';\mathrm{x}'',\mathrm{D}'')$; this operator is an operator-valued symbol of $\tilde{A}(x,D)$ (where $\tilde{A}$ is the operator $\varrho^{\frac{1}{2}d''}A\varrho^{-\frac{1}{2}d''}$ in cylindrical coordinates (x',x'') (without division by τ)) so we obtain it by replacing D' with ξ' in $\tilde{A}(x,D)$; this operator acts in $\mathbb{H} = L^2(Y,\mathbb{K})$ and $\mathrm{D}(\mathcal{A}) = H_{2m}(Y,\mathbb{K}) \cap H^0_m(Y,\mathbb{K})$. Surely, this operator is self-adjoint. Recalling the definition of B_1 one can easily see that B_1^{-l} is a trace operator in $\mathbb{H}$ for $l > \frac{1}{2m}d''$ with trace norm less than $C\tau^{\frac{d''}{2m}}\varrho^{d''}$. Therefore we immediately obtain the following estimate from (12.1.18):

(12.1.19) $$|\int \psi(x')\,\mathrm{tr}\,e(x,x,\tau)dx - (2\pi)^{-d'}\iint \psi(x')\mathbf{n}(x',\xi',\tau)dx'd\xi'| \le C\tau^{\frac{1}{2m}(d-1)}\langle x'\rangle^{d'-1-\mu d''}$$

where $\mathbf{n}(x',\xi',\tau) = \operatorname{Tr}_{\mathbb{H}} \mathbf{e}(x',\xi',\tau)$ is the eigenvalue counting function of the operator $\mathbf{a}(x',\xi')$ in $\mathbb{H}$.

Summing over a partition of the unity in the zone $\{c_0 \le |x'| \le c_0\tau^{\frac{1}{2m\mu}}\}$ we obtain in the right-hand expression

$$R^*(\tau) = \begin{cases} \tau^{\frac{1}{2m}(d-1)} & \text{for } d' < \mu d'' + 1 \\ \tau^{\frac{1}{2m}(d-1)}\log\tau & \text{for } d' = \mu d'' + 1 \\ \tau^{\frac{1}{2m\mu}(d'-1)(\mu+1)} & \text{for } d' > \mu d'' + 1. \end{cases} \tag{12.1.20}$$

Then, taking into account results in the zones $\{|x'| \le 2c_0\}$ (Weylian asymptotics) and $\{|x'| \ge \frac{1}{2}c_0\tau^{\frac{1}{2m\mu}}\}$ (proposition 12.1.3) we obtain the following basic statement:

Theorem 12.1.4. *Let conditions* (12.1.1)–(12.1.8) *be fulfilled. Moreover, let one of conditions* $(12.1.17)_{1,2}$ *be fulfilled. Then the asymptotics*

$$\mathbf{N}(\tau) = \mathcal{N}(\tau) + O\bigl(R^*(\tau)\bigr) \tag{12.1.21}$$

hold as $\tau \to +\infty$ *where* $R^*(\tau)$ *is given by* (12.1.20) *and*

$$\begin{aligned}\mathcal{N}(\tau) = &(2\pi)^{-d}\iint \psi_0(x')n^{\mathrm{w}}(x,\xi,\tau)\,dxd\xi\, + \\ &(2\pi)^{-d'}\iint \bigl(1-\psi_0(x')\bigr)\mathbf{n}(x',\xi',\tau)\,dx'd\xi',\end{aligned} \tag{12.1.22}$$

$n^{\mathrm{w}}(x,\xi,\tau)$ *is the eigenvalue counting function for* $a(x,\xi)$, $\psi_0 \in C^K$ *is supported in* $\{|x'| \le 2c_0\}$ *and* $\psi_0 = 1$ *for* $|x'| \le c_0$. *Moreover,*

$$\mathcal{N}(\tau) \asymp \begin{cases} \tau^{\frac{d}{2m}} & \text{for } d' < \mu d'' \\ \tau^{\frac{d}{2m}}\log\tau & \text{for } d' = \mu d'' \\ \tau^{\frac{1}{2m\mu}d(\mu+1)} & \text{for } d' > \mu d''. \end{cases} \tag{12.1.23}$$

Moreover, if $d' < 1 + \mu d''$ then the main contribution to the remainder estimate is given by the zone $\{|x'| \le t\}$ with an arbitrarily large constant t (the contribution of the complement does not exceed $\tau^{\frac{1}{2m}(d-1)}t^{-\delta}$ with $\delta > 0$); on the other hand, in this zone under the standard condition $(\Psi 0)$ on Hamiltonian billiards we have the remainder estimate $o\bigl(\tau^{\frac{1}{2m}(d-1)}\bigr)$. Therefore the following statement holds:

Theorem 12.1.5. *Let conditions* (12.1.1)–(12.1.8) *be fulfilled. Moreover, let one of conditions* $(12.1.17)_{1,2}$ *be fulfilled. Finally, let* $d' < 1 + \mu d''$ *and let the standard condition* $(\Psi 0)$ *on Hamiltonian billiards be fulfilled. Then the asymptotics*

$$\mathbf{N}(\tau) = \mathcal{N}'(\tau) + o\bigl(\tau^{\frac{1}{2m}(d-1)}\bigr) \tag{12.1.24}$$

hold as $\tau \to +\infty$ *where*

$$\begin{aligned}\mathcal{N}(\tau) = &(2\pi)^{-d}\iint \psi_t(x')\bigl(n^{\mathrm{w}}(x,\xi,\tau) + \varkappa_1\tau^{\frac{d-1}{2m}}\bigr)\,dxd\xi\, + \\ &(2\pi)^{1-d}\int \psi_t(x')\tilde{\varkappa}_1\tau^{\frac{d-1}{2m}}\,dS\, + \\ &(2\pi)^{-d'}\iint \bigl(1-\psi_t(x')\bigr)\mathbf{n}(x',\xi',\tau)\,dx'd\xi',\end{aligned} \tag{12.1.25}$$

$(2\pi)^{-d}\int \psi\varkappa_1\tau^{\frac{1}{2m}(d-1)}dx$ and $(2\pi)^{1-d}\int_{\partial X}\psi\tilde{\varkappa}_1\tau^{\frac{1}{2m}(d-1)}dS$ are the standard second terms in the Weylian asymptotics, $\psi_t(x') = \psi_0(\frac{x'}{t})$ where $t = t(\tau)$ is an appropriate function such that $t(\tau) \to +\infty$ as $\tau \to +\infty$, and if $t(\tau)$ works then every function $t'(\tau) \le t(\tau)$ such that $t'(\tau) \to +\infty$ as $\tau \to +\infty$ also works.

On the other hand, if $d' > 1 + d''\mu$ then the main contribution to the remainder estimate in theorem 12.1.3 is provided by the zone $\{\varepsilon \le \tau^{-\frac{1}{2m\mu}}|x'| \le c_0\}$ with arbitrarily small $\varepsilon > 0$; in this zone $\mathbf{n}(x',\xi',\tau)$ does not exceed ν_ε. Applying theorem 4.3.6 and replacing $\mathbf{a}$ if necessary by another operator in $\mathbb{H}$ with eigenvalues asymptotically close to those of $\mathbf{a}$ we obtain the following statement:

Theorem 12.1.6. *Let conditions (12.1.1)–(12.1.8) be fulfilled. Moreover, let one of conditions $(12.1.17)_{1,2}$ be fulfilled and let $d' > 1 + \mu d''$. Let the stabilization conditions*

$$(12.1.26)_1 \qquad D^{\beta'}(a_\alpha - a^0_\alpha) = o(|x'|^{-|\beta'|}) \qquad \forall\alpha : |\alpha| = 2m,$$

$$(12.1.26)_2 \qquad D^{\beta'}(\varrho - \varrho^0) = o(|x'|^{-\mu-|\beta'|}) \qquad \forall\beta' : |\beta'| \le K$$

be fulfilled as $|x'| \to \infty$ where $a^0 = a^0(x')$, $\varrho^0(x')$ are positively homogeneous functions of x' of degrees 0, $-\mu$ respectively (surely, a^0_α can be matrix-valued). Let us consider the operator

$$(12.1.27) \qquad \mathbf{a}^0(x',\xi';\mathrm{D}'') = \sum_{|\alpha|=2m} a^0_\alpha \varrho^{0-|\alpha''|}\xi'^{\alpha'}\mathrm{D}''^{\alpha''}$$

in $\mathbb{H}$ and let $\zeta_j(x',\xi')$ be its eigenvalues.[1] *Let us assume that*

(12.1.28) There exist sets Λ_j of measure 0 on $\Sigma_j = \{(x',\xi'), \zeta_j(x',\xi') = 1\}$ such that for $(x',\xi') \notin \Lambda_j$ either $\zeta_j(x',\xi')$ does not belong to the interval $(1-\epsilon, 1+\epsilon)$ or has constant multiplicity and through each point $(x',\xi') \in \Sigma_j \setminus \Lambda_j$ there passes an infinitely long (in one direction) non-periodic Hamiltonian trajectory of ζ_j in a neighborhood of which x' does not vanish and ζ_j has constant multiplicity.

Then the asymptotics

$$(12.1.29) \qquad \mathbf{N}(\tau) = \mathcal{N}(\tau) + \bigl(\varkappa_1^* + o(1)\bigr)\tau^{\frac{1}{2m\mu}(d'-1)(\mu+1)}$$

hold as $\tau \to +\infty$ with an appropriate coefficient $\varkappa_1^*$ (given in theorem 4.3.6).

Remark 12.1.7. *(i) In theorem 12.1.6 one can assume that a^0 depend on x''; then one should assume that $a^0(x',\varrho^0(x')\mathrm{x}'')$ are positively homogeneous of degree 0 with respect to x'; then $\mathbf{a}^0 = \mathbf{a}^0(x',\xi';\mathrm{x}'',\mathrm{D}'')$ and one should take Weylian quantization in the right-hand expression of (12.1.27).*

(ii) One can also treat Riesz means. In this case

$$(12.1.30) \qquad R^*_\vartheta(\tau) = \begin{cases} \tau^{\frac{1}{2m}(d-1-\vartheta)+\vartheta} & \text{for } d' < \mu d'' + 1 + \vartheta \\ \tau^{\frac{1}{2m}(d-1-\vartheta)+\vartheta}\log\tau & \text{for } d' = \mu d'' + 1 + \vartheta \\ \tau^{\frac{1}{2m\mu}(d'-1-\vartheta)(\mu+1)+\vartheta} & \text{for } d' > \mu d'' + 1 + \vartheta. \end{cases}$$

and in the analogs of theorems 12.1.5 and 12.1.6 one should assume that $d' < \mu d'' + \vartheta + 1$, $d' > \mu d'' + \vartheta + 1$ respectively. ◊

[1] Then, obviously $\zeta_j(t^\mu x', t\xi') = t^{2m}\zeta_j(x',\xi')$.

12.1.3 Effective asymptotics. I

The main defect of theorems 12.4.3–12.4.6 is that the expression for $\mathcal{N}(\tau)$ is not effective and includes the eigenvalue counting function for the family of auxiliary operators $\mathbf{a}(x',\xi')$. I would like to derive a formula including the eigenvalue counting function for one auxiliary operator or for a family of auxiliary operators with a smaller number of parameters (under certain conditions). I am aware that for a thick cusp (i.e., with $\varrho^{d''-1-\vartheta} \notin L^1(Z)$) we cannot completely get rid auxiliary operators in the final answer. Moreover, in theorem 12.1.5 I would like to obtain a more precise description of $t(\tau)$.

Let us start with a more accurate Weylian analysis. Let $U(x,y,t)$ be the Schwartz kernel of the operator $\exp(it\tau^{-1}A)$. Then

$$\tau D_t U = A_x U, \qquad U|_{t=0} = \delta(x-y)I_{\mathbf{K}}. \tag{12.1.31}$$

Let us fix a point $\bar{x} \in X$ and consider its ϱ-neighborhood. Let us make the dilatation $x_{\text{new}} = (x_{\text{old}} - \bar{x})\bar{\varrho}^{-1}$ and divide A by τ; then we transform A into an h_1-differential operator A_1 with $h_1 = \bar{\varrho}^{-1}\tau^{-\frac{1}{2m}}$. Introducing a new time scale $t_1 = th_1$ we obtain from (12.1.31) the scaled problem

$$h_1 D_{t_1} U = A_{1x} U, \qquad U|_{t_1=0} = \bar{\varrho}^{-d}\delta(x-y)I \tag{12.1.31$_1$}$$

where we skip the subscript "new" on x and y. Then one can apply the results of section 4.2; moreover, at the energy level 1 the microhyperbolicity condition is fulfilled provided h_1 is small enough. Let $\bar{\chi} \in C_0^K([-1,1])$, $\bar{\chi} = 1$ on $[-\frac{1}{2},\frac{1}{2}]$, let ψ' be $\bar{\varrho}$-admissible and be supported in the $\bar{\varrho}$-neighborhood of $\bar{x}$, $h_1 \le \varepsilon$ (i.e., $\bar{\varrho} \ge c_0\tau^{-\frac{1}{2m}}$), and let $T_1 = \varepsilon h_1^{-1}$ where here and below $\varepsilon > 0$ is a small enough constant. We then obtain the following estimate:

$$\begin{aligned} &\Bigl|\int \bigl(F_{t\to\lambda}\bar{\chi}_{T_1}(t)\Gamma_x U - \sum_{0\le j\le J-1} \varkappa'_j(\lambda\tau)^{\frac{1}{2m}(d-j)-1}\tau\bigr)\psi'\,dx - {}\\ &\int_{\partial X}\sum_{1\le j\le J-1} \tilde{\varkappa}'_j(\lambda\tau)^{\frac{1}{2m}(d-j)-1}\tau\psi'\,dS\Bigr| \le C\tau^{\frac{1}{2m}(d-J)}\bar{\varrho}^{d-J}\end{aligned} \tag{12.1.32}$$

where $\lambda \in [\epsilon_0, c]$ here and below, dS is the standard Riemannian density on ∂X, $\varkappa'_j$, $\tilde{\varkappa}'_j$ are the standard Weylian coefficients and

$$|\varkappa'_j| \le C\bar{\varrho}^{-j}, \qquad |\tilde{\varkappa}'_j| \le C\bar{\varrho}^{1-j}. \tag{12.1.33}$$

Recall that one can obtain these coefficients in the following way:

$$\operatorname{Tr}_x \operatorname{Res}_{\mathbf{R}} \psi'(\tau - A)^{-1} \sim \int \sum_j \varkappa'_j \tau^{\frac{1}{2m}(d-j)-1}\,dx \tag{12.1.34$_0$}$$

and

$$\operatorname{Tr}\operatorname{Res}_{\mathbf{R}} \psi'\bigl((\tau - A)_D^{-1} - (\tau - A)^{-1}\bigr) \sim \sum_j \int_{\partial X} \psi'\tilde{\varkappa}'_j \tau^{\frac{1}{2m}(d-j)-1}\,dS \tag{12.1.34$_1$}$$

where $(\tau - A)^{-1}$ and $(\tau - A)_D^{-1}$ are parametrices of the pseudo-differential operator $(\tau - A)$ in terms of the formal calculus of pseudo-differential operators and the Dirichlet problem for the same operator in terms of the formal calculus of pseudo-differential operators (with the transmission property) and Green operators (see, e.g., [Boutet de Monvel 1]), and for the sake of simplicity tr is included in the operation Γ_x. Therefore under the same conditions for the $\bar{\gamma}$-admissible function $\psi = \psi(x')$ supported in the $\bar{\gamma}$-neighborhood of $\bar{x}'$ the estimate

(12.1.35)
$$\Big|\int \big(F_{t\to\lambda}\bar{\chi}_{T_1}(t)\Gamma_x U - \sum_{0\le j\le J-1} \varkappa'_j(\lambda\tau)^{\frac{1}{2m}(d-j)-1}\tau\big)\psi' dx - \int_{\partial X} \sum_{1\le j\le J-1} \tilde{\varkappa}'_j(\lambda\tau)^{\frac{1}{2m}(d-j)-1}\tau\psi' dS\Big| \le C\tau^{\frac{1}{2m}(d-J)}\bar{\varrho}^{d''-J}\bar{\gamma}^{d'}$$

holds.

Let us take another point of view. Namely, let us treat the $\bar{\gamma}$-neighborhood of $\bar{x}'$, make the dilatation $x'_{\text{new}} = (x'_{\text{old}} - \bar{x}')\bar{\gamma}^{-1}$ and divide A by τ; earlier we transformed X_1 to a cylinder in the routine way. Then A is transformed to a h_2-differential operator (with respect to x') A_2 with operator-valued symbol; surely $h_2 = \bar{\gamma}^{-1}\tau^{-\frac{1}{2m}} \ll 1$ here. Introducing the new time scale $t_2 = th_2\tau^{-1}$ we obtain from (12.1.31) another scaled problem

$$(12.1.31)_2 \qquad \tau D_{t_2}U = A_{2x}U, \qquad U|_{t_2=0} = \bar{\gamma}^{-d'}\delta(x'-y')I_{\mathbb{H}}.$$

Let one of conditions $(12.1.17)_{1,2}$ *be fulfilled.* Then A_2 is a microhyperbolic operator with operator-valued symbol. Let us apply the strongest version of theorem 2.1.15: namely, formula (4.2.28) yields that in the right-hand expressions in (2.1.35) and (2.1.36) one can add the factor $h = h_2$, i.e., one can take the right-hand expressions to be $h^{1-d-|\alpha|}\big(\frac{h}{T}\big)^s$. Therefore we obtain that if $\chi' \in C_0^K([-1,1])$ vanishes in a neighborhood of 0 and ψ is $\bar{\gamma}$-admissible and supported in the $\bar{\gamma}$-neighborhood of $\bar{x}'$ then

$$(12.1.36) \qquad |F_{t\to\lambda}\chi'_T\Gamma'\psi U| \le Ch_2^{-d'}T^{-s}$$

for any $T \in [\epsilon, T_2]$, $T_2 = \varepsilon h_2^{-1}$; here Γ' means that we take $x' = y'$ and integrate over x', and the operator norm in $\mathbb{H}$ is in the left-hand expression. Applying $(12.1.31)_2$ and the embedding theorem (or applying only formula (4.2.28)) we can take the trace norm in the left-hand expression with the additional factor $h_1^{-d''}$ in the right-hand expression; we still assume that $\bar{\varrho} \ge c_0\tau^{-\frac{1}{2m}}$. Taking an appropriate partition of unity in the interval $[T'_1, T'_2]$ we obtain the estimate

$$(12.1.37) \qquad |F_{t\to\lambda}(\bar{\chi}_{T'_2} - \bar{\chi}_{T'_1})\Gamma\psi U| \le Ch_2^{-d'}h_1^{-d''}T_1'^{-s}$$

for $\varepsilon \le T'_1 < T'_2 \le T_2$. Let us set $T'_j = T_j \ (= \varepsilon h_j^{-1})$. Then estimates (12.1.37) and (12.1.35) yield that estimate (12.1.35) *remains true with* T_1 *replaced by* T_2. Applying the routine Tauberian technique we then obtain the following estimate:

(12.1.38)
$$\Big|\Big(\int \psi\big(\operatorname{tr} e(x,x,\tau) - \sum_{0\le j\le J-1}\varkappa_j\tau^{\frac{1}{2m}(d-j)}\big)dx - \int_{\partial X}\psi\sum_{1\le j\le J-1}\tilde{\varkappa}_j\tau^{\frac{1}{2m}(d-j)}dS\Big)\Big|_{\tau=\tau'}^{\tau=\tau}\Big| \le C\bar{\varrho}^{d''}\bar{\gamma}^{d'-1}\tau^{\frac{1}{2m}(d-1)} + C\bar{\varrho}^{d''-J}\bar{\gamma}^{d'}\tau^{\frac{1}{2m}(d-J)}$$

provided $\tau \ge \epsilon\bar{\varrho}^{-2m}$, $\tau' \asymp \tau$ and $J \le d$. For $J > d$ one should replace the terms with $j = d$ by $\varkappa_d \log\tau$ and $\tilde{\varkappa}_d \log\tau$. Here and below the Weylian coefficients $\varkappa_j$ and $\tilde{\varkappa}_j$ satisfy (12.1.33).[2] Thus, *using propagation of singularities for operators with operator-valued symbols we have improved the Weylian asymptotics in the zone* $\{x : \varrho(x') \ge \tau^{-\frac{1}{2m}}\}$.

[2] A similar estimate also holds for Riesz means. Namely, the first term in the right-hand expression of (12.1.38) should be multiplied by $\tau^{\vartheta}\bar{\varrho}^{\vartheta}\bar{\gamma}^{-\vartheta}$ and the second term should be multiplied by τ^{ϑ}.

Let A^0 be another operator in the domain X^0 with the same cusp part X^1 also satisfying all the conditions above with the same exponents. Moreover, let us assume that the operators A and A^0 are close at infinity in the sense that

$$|D^\beta(a_\alpha - a^0_\alpha)| \le C\eta\langle x'\rangle^{-|\beta'|+\mu(2m-|\alpha|+|\beta''|)} \qquad \forall\alpha: |\alpha|\le 2m \quad \forall\beta: |\beta|<K \tag{12.1.39}$$

with

$$\eta = \eta(x') \asymp \langle x'\rangle^{-\nu},\ \nu>0 \tag{12.1.40}$$

(in most of the arguments below one need only assume that η is a γ-admissible weight function and $\eta = o(1)$ as $|x'|\to\infty$). All the objects associated with A^0 have an additional superscript "0." Let us introduce the formal differential expression $A^\theta = A^0 + \theta(A - A^0)$ with $\theta\in[0,1]$; then $A^1 = A$. We also assign Dirichlet boundary conditions to A^θ. Below all the objects associated with A^θ have the additional superscript "θ." Then applying estimate (12.1.38) to both operators we obtain the estimate

$$\begin{aligned}\Big|\Big(\int\Big(\psi\big(e(x,x,\tau)-e^0(x,x,\tau)\big)\\ -\sum_{0\le j\le J-1}\big(\varkappa_j-\varkappa^0_j\big)\tau^{\frac{1}{2m}(d-j)}\Big)dx -\\ \int_{\partial X}\sum_{1\le j\le J-1}\big(\bar\varkappa_j-\bar\varkappa^0_j\big)\tau^{\frac{1}{2m}(d-j)}dS\Big)\Big|_{\tau=\tau'}^{\tau=\tau}\Big| \le\\ C\bar\varrho^{d''}\bar\gamma^{d'-1}\tau^{\frac{1}{2m}(d-1)} + C\bar\varrho^{d''-J}\bar\gamma^{d'}\tau^{\frac{1}{2m}(d-J)}\end{aligned} \tag{12.1.41}$$

provided $J\le d$; otherwise terms with $j=d$ contain the additional factor $\log\tau$. Moreover, differentiating representations $(12.1.34)_{0,1}$ on θ one can easily see that $|\partial_\theta\varkappa^\theta_j|\le c_1\bar\eta\bar\varrho^{-j}$ and $|\partial_\theta\bar\varkappa^\theta_j|\le c_1\bar\eta\bar\varrho^{1-j}$ with the same notation as before (we first obtain these estimates for the coefficients $\varkappa'^{\theta}_j$, $\bar\varkappa'^{\theta}_j$). Integrating on θ from 0 to 1 and passing from ψ' to ψ we obtain that

$$|\varkappa_j-\varkappa^0_j|\le c_1\bar\eta\bar\varrho^{1-j}, \qquad |\bar\varkappa_j-\bar\varkappa^0_j|\le c_1\bar\eta\bar\varrho^{1-j} \tag{12.1.42}$$

These estimates and (12.1.41) yield that

$$\begin{aligned}\Big|\Big(\int\Big(\psi\big(e(x,x,\tau)-e^0(x,x,\tau)\big) -\\ \sum_{0\le j\le J-1}\big(\varkappa_j-\varkappa^0_j\big)\tau^{\frac{1}{2m}(d-j)}\Big)dx -\\ \int_{\partial X}\psi\sum_{1\le j\le J-1}\big(\bar\varkappa_j-\bar\varkappa^0_j\big)\tau^{\frac{1}{2m}(d-j)}dS\Big)\Big|_{\tau=\tau'}^{\tau=\tau}\Big| \le\\ C\bar\varrho^{d''}\bar\gamma^{d'-1}\tau^{\frac{1}{2m}(d-1)} + C\bar\varrho^{d''-s}\bar\gamma^{d'}\tau^{\frac{1}{2m}(d-s)} + C\bar\eta\bar\varrho^{d''-J}\bar\gamma^{d'}\tau^{\frac{1}{2m}(d-J)}.\end{aligned} \tag{12.1.43}$$

with arbitrarily large s (K surely depends on s).

This estimate is accurate enough to obtain the final theorems under certain restrictions on ν from below (I leave this task to the reader). However, in order to prove these theorems under weaker conditions one needs a more refined analysis. Namely, we want to reject the second term in the right-hand expression of (12.1.43) which is not nice for h_1 close to 1 (i.e., for $\bar\gamma$ close to $\tau^{\frac{1}{2m\mu}}$). Let us assume that $X = X^0$; one can arrange this by replacing both X and X^0 with $X^1 \cap \{|x'| \ge c_0\}$ (or $X^1 \cap \{x_1 \ge c_0\}$ in the case $d' = 1$) and taking Dirichlet boundary conditions on ∂X^1; surely the operators remain self-adjoint and all the functions $e^\theta(x,y,\tau)$ and $U^\theta(x,y,t)$ do not lose meaning. Then the routine Tauberian arguments yield that in the context of theorems 12.1.4–12.1.6 the errors in $\int \psi e^\theta(x,x,\tau)dx$ (due to changing the operators) do not exceed the remainder estimates in the asymptotics of these terms in the corresponding theorem, $\theta = 0, 1$.

Under the assumption $X = X^0$ the operator A^θ is self-adjoint and all the functions $e^\theta(x,y,\tau)$ and $U^\theta(x,y,t)$ do not lose meaning for $\theta \in [0,1]$. Then

$$\partial_\theta \exp\bigl(it\tau^{-1}A^\theta\bigr) = i\tau^{-1}\int_0^t U \exp\bigl(it'\tau^{-1}A^\theta\bigr)\cdot\bigl(A^1 - A\bigr)\cdot\exp\bigl(i\tau^{-1}(t-t')A^\theta\bigr)dt'$$

and one can easily prove that for $\lambda \in [\epsilon, c]$

$$\bigl|F_{t\to\lambda}\bar\chi_T(t)\partial_\theta U^\theta(x,y,t)\bigr| \le C\bar\eta h_1^{-p}T \tag{12.1.44}$$

for any $T > 0$ with large enough p. One can improve this estimate and calculate p but this improvement gives no advantage below. Then, applying routine arguments associated with the successive approximation method and increasing T_1 from $h_1^{\frac{1}{2}+\delta}$ (*only here*, the subscript 1 on T means that we use the t_1-scale) to ε in sections 5.2, 5.3 one can see that

$$\Bigl|\partial_\theta\Bigl(\int \bigl(F_{t\to\lambda}\bar\chi_T(t)\Gamma_x U - \sum_{0\le j\le J-1} \varkappa'_j(\lambda\tau)^{\frac{1}{2m}(d-j)-1}\tau\bigr)\psi' dx - \int_{\partial X}\sum_{1\le j\le J-1} \tilde\varkappa'_j(\lambda\tau)^{\frac{1}{2m}(d-j)-1}\tau\psi' dS\Bigr)\Bigr| \le C\bar\eta\tau^{\frac{1}{2m}(p-J)}\bar\varrho^{p-J} \tag{12.1.45}$$

where $T = T_1 = \varepsilon h_1^{-1}$ and p is large enough.[3] Therefore this estimate holds with ψ instead of ψ' and with the additional factor $h_2^{-d'}h_1^{d'}$ in the right-hand expression. Moreover, applying arguments associated with operators with operator-valued symbols one can easily obtain from (12.1.44) that

$$\bigl|F_{t\to\lambda}\chi'_T(t)\Gamma\psi\partial_\theta U^\theta\bigr| \le C\bar\eta h_2^{-d'}h_1^{-p}T^{-s} \tag{12.1.46}$$

for $T \in [T_1, T_2]$ and therefore

$$\bigl|F_{t\to\lambda}\bigl(\bar\chi_{T'_2} - \bar\chi_{T'_1}\bigr)\Gamma\psi\partial_\theta U^\theta\bigr| \le C\bar\eta h_2^{-d'}h_1^{-p}T_1'^{-s} \tag{12.1.47}$$

with the same $T_1, T_2, T'_1, T'_2, \chi'$ and $\bar\chi$ as in (12.1.36) and (12.1.37). Therefore (12.1.45) holds with $T = T_2$ instead of $T = T_1$. Integrating over θ from 0 to 1 we see that (12.1.45) holds with the derivative on θ replaced by the difference operator.

Let us apply the Tauberian technique. Surely, $\Gamma\psi\bigl(e(.,.,\tau) - e^0(.,.,\tau)\bigr)$ is not a monotone function of τ and one should apply the estimate

$$\bigl|\Gamma\psi\bigl(e^\theta(.,.,\tau) - e^\theta(.,.,\tau')\bigr)\bigr| \le \bar\gamma^{d'}\bar\varrho^{d''}\tau^{\frac{d}{2m}}h_2 = \bar\gamma^{d'-1}\bar\varrho^{d''}\tau^{\frac{1}{2m}(d-1)}$$

for $|\tau' - \tau| \le h_2$. This estimate was obtained by the Tauberian method (see section 4.3) for functions $\Gamma\psi e^\theta(.,.,\tau)$ monotone in τ. This gives the improved estimate (12.1.43)

---

[3] However, considering this estimate with large $J' = J + p$ instead of J and applying estimates (12.1.33) one can reach $p = d$.

(12.1.48)
$$
\begin{aligned}
&|\Bigl(\int\Bigl(\bigl(e(x,x,\tau)-e^0(x,x,\tau)\bigr)\\
&-\sum_{0\le j\le J-1}\bigl(\varkappa_j-\varkappa_j^0\bigr)\tau^{\frac{1}{2m}(d-j)}\Bigr)\psi' dx-\\
&\int_{\partial X}\sum_{1\le j\le J-1}\bigl(\tilde\varkappa_j-\tilde\varkappa_j^0\bigr)\tau^{\frac{1}{2m}(d-j)}\psi' dS\Bigr)\Big|_{\tau=\tau'}^{\tau=\tau}|\le\\
&C\bar\varrho^{d''}\bar\gamma^{d'-1}\tau^{\frac{1}{2m}(d-1)}+C\bar\eta\bar\varrho^{d''-J}\bar\gamma^{d'}\tau^{\frac{1}{2m}(d-J)}.
\end{aligned}
$$

Recall that $\tau'\in[\epsilon\tau,\tau]$, $\tau\bar\varrho^{2m}\ge\epsilon$ here and terms with $j=d$ contain the factor $\log\tau$.[4)]

So, the following statement has been proven:

Proposition 12.1.8. *(i) In the context of theorem 12.1.4 estimate* (12.1.38) *holds.*

(ii) Moreover, let both operators A and A^0 satisfy the conditions of theorem 12.1.4 and let (12.1.39) *be fulfilled (in $X^1\cap|x'-\bar x'|\le\bar\gamma$). Then estimate* (12.1.48) *holds.*

Recall that ψ is γ-admissible and supported in the $\bar\gamma$-neighborhood of $\bar x'$. Let us replace τ and τ' by $2^{-n}\tau$ and $2^{-n-1}\tau$ and sum over $n=0,\dots,n_0=c_0[\log(\tau\bar\varrho^{2m})]+c_1$. Applying proposition 12.1.3 we obtain

Corollary 12.1.9. *In the context of theorem 12.1.4 the estimate*

$(12.1.49)_1$
$$
\begin{aligned}
\mathcal{R}_1=&|\int\psi\Bigl(\operatorname{tr} e(x,x,\tau)-\sum_{0\le j\le J-1}\varkappa_j\tau^{\frac{1}{2m}(d-j)}\Bigr)dx-\\
&\int_{\partial X}\psi\sum_{1\le j\le J-1}\tilde\varkappa_j\tau^{\frac{1}{2m}(d-j)}\bigr)dS|\le\\
&C\bar\varrho^{d''}\bar\gamma^{d'-1}\tau^{\frac{1}{2m}(d-1)}+C\bar\varrho^{d''-J}\bar\gamma^{d'}\tau^{\frac{1}{2m}(d-J)}
\end{aligned}
$$

holds for $J<d$; for $J=d$ the estimate

$(12.1.49)_2$
$$
\mathcal{R}_1\le C\bar\varrho^{d''}\bar\gamma^{d'-1}\tau^{\frac{1}{2m}(d-1)}+C\bar\varrho^{-d'}\bar\gamma^{d'}\log(\tau\bar\varrho^{2m})
$$

holds and for $J=d+1$ the estimate

$(12.1.49)_3$
$$
\mathcal{R}_1\le C\bar\varrho^{d''}\bar\gamma^{d'-1}\tau^{\frac{1}{2m}(d-1)}+C\bar\varrho^{-d'}\bar\gamma^{d'}
$$

holds, and in the last case the terms with $j=d$ are $\varkappa_d\log(\tau\bar\varrho^2m)$ and $\tilde\varkappa_d\log(\tau\bar\varrho^2m)$.

(ii) In the context of proposition 12.1.8(ii) the estimates

$(12.1.50)_1$
$$
\begin{aligned}
\mathcal{R}_2=&|\Bigl(\int\psi\Bigl(\operatorname{tr} e^\theta(x,x,\tau)-\sum_{0\le j\le J-1}\varkappa_j^\theta\tau^{\frac{1}{2m}(d-j)}\Bigr)-\\
&\int_{\partial X}\psi\sum_{1\le j\le J-1}\tilde\varkappa_j^\theta\tau^{\frac{1}{2m}(d-j)}dS\Bigr)\Big|_{\theta=0}^{\theta=1}dx-|\le\\
&C\bar\varrho^{d''}\bar\gamma^{d'-1}\tau^{\frac{1}{2m}(d-1)}+C\bar\eta\bar\varrho^{d''-J}\bar\gamma^{d'}\tau^{\frac{1}{2m}(d-J)},
\end{aligned}
$$

$(12.1.50)_2$
$$
\mathcal{R}_2\le C\bar\varrho^{d''}\bar\gamma^{d'-1}\tau^{\frac{1}{2m}(d-1)}+C\bar\eta\bar\varrho^{-d'}\bar\gamma^{d'}\log(\tau\bar\varrho^{2m})
$$

and

---

4) For Riesz means both terms in the right-hand expression should be modified in the way indicated in footnote 2).

$$\mathcal{R}_2 \le C\bar{\varrho}^{d''}\bar{\gamma}^{d'-1}\tau^{\frac{1}{2m}(d-1)} + C\bar{\eta}\bar{\varrho}^{-d'}\bar{\gamma}^{d'} \tag{12.1.50$_3$}$$

hold for $J<d$, $J=d$, $J=d+1$ respectively (I remind the reader that terms with $j=d$ contain a logarithmic factor).

Then after summation over a γ-admissible partition of unity in Z we can obtain miscellaneous formulas (later under certain conditions we will calculate the terms associated with the operator A^0):

(i) Let us sum the estimates for $\mathcal{R}_2$ on the domain $\{t_1 \le |x'| \le c_0\tau^{\frac{1}{2m\mu}}\}$ and apply proposition 12.1.2 in the domain $\{|x'|\le 2t_1\}$. Then we derive the formula

$$\begin{aligned}\mathbf{N}(\tau) = &\sum_{j=0,1}\int \phi(\frac{x'}{t_1})\varkappa_j(x)\tau^{\frac{1}{2m}(d-j)}dx +\\ &\int \phi(\frac{x'}{t_1})\tilde{\varkappa}_1(x)\tau^{\frac{1}{2m}(d-1)}dS +\\ &\sum_{j=0,1}\int \bigl(1-\phi(\frac{x'}{t_1})\bigr)\bigl(\varkappa_j(x)-\varkappa_j^0(x)\bigr)\tau^{\frac{1}{2m}(d-j)}dx +\\ &\int \phi(\frac{x'}{t_1})\bigl(\tilde{\varkappa}_1(x)-\tilde{\varkappa}_1^0(x)\bigr)\tau^{\frac{1}{2m}(d-1)}dS +\\ &\int \bigl(1-\phi(\frac{x'}{t_1})\bigr)\operatorname{tr} e^0(x,x,\tau)dx +\\ &O\bigl(R^*(\tau)+\tau^{\frac{1}{2m\mu}(d'(\mu+1)-\nu)}\bigr);\end{aligned} \tag{12.1.51}$$

we have taken $J=2$ because we gain nothing by increasing J and so we have dropped all the terms with $j\ge 2$. Here $R^*(\tau)$ is introduced by (12.1.20) and here and below $\phi\in C_0^K(\mathbb{R}^{d'})$, $\phi=1$ in a neighborhood of the origin.

Here t_1 is either a large enough constant or a slowly increasing function of τ. Moreover, one can reject terms with $j=1$ in the third and fourth items in the right-hand expression in all cases excluding $d'=(d''-1)\mu+\nu$, $\nu<\mu+1$ when the remainder estimate is $O(\tau^{\frac{1}{2m}(d-1)})$, and the contribution of this term is $O\bigl(\tau^{\frac{1}{2m}(d-1)}\log\tau\bigr)$ (and this estimate is the best possible). Moreover, for $t_1=\text{const}$ one can reject terms with $j=1$ in the first and second items in the right hand expression. In any case, one can effectively calculate all terms with $j=0,1$.

(ii) Moreover, let $d'<1+\mu d''$, $\nu>\mu(1-d'')+d'$. In this case the remainder estimate in (12.1.51) is $O\bigl(\tau^{\frac{1}{2m}(d-1)}\bigr)$. Moreover, the contribution of the zone $\{t_1\le|x'|\}$ to the remainder estimate is $o\bigl(\tau^{\frac{1}{2m}(d-1)}\bigr)$ for $t_1=t_1(\tau)\to+\infty$ as $\tau\to+\infty$. Then, applying the improved Weylian estimate in the zone $\{|x'|\le t_1\}$ under the standard condition ($\Psi 0$) on Hamiltonian billiards we obtain the asymptotics

$$\begin{aligned}\mathbf{N}(\tau) = &\sum_{j=0,1}\int \phi(\frac{x'}{t_1})\varkappa_j(x)\tau^{\frac{1}{2m}(d-j)}dx +\\ &\int \phi(\frac{x'}{t_1})\tilde{\varkappa}_1(x)\tau^{\frac{1}{2m}(d-1)}dS +\\ &\int \bigl(1-\phi(\frac{x'}{t_1})\bigr)\bigl(\varkappa_0(x)-\varkappa_0^0(x)\bigr)\tau^{\frac{d}{2m}}dx +\\ &\int \bigl(1-\phi(\frac{x'}{t_1})\bigr)\operatorname{tr} e^0(x,x,\tau)dx +\\ &o\bigl(\tau^{\frac{1}{2m}(d-1)}\bigr)\end{aligned} \tag{12.1.52}$$

where we have dropped terms with $j=1$ in the third and fourth items in the right-hand expression of (12.1.52).

(iii) Let $d' > 1+\mu d''$, $\nu > \mu+1$; then the remainder estimate in (12.1.51) is $O\bigl(\tau^{\frac{1}{2m\mu}(d'-1)(\mu+1)}\bigr)$. Moreover, in this case one can drop all terms with $j=1$. Furthermore, the contribution of the zone $\{|x'|\le t_2\}$, $t_2=\varepsilon\tau^{\frac{1}{2m\mu}}$ to the remainder estimate is $\varepsilon'\tau^{\frac{1}{2m\mu}(d'-1)(\mu+1)}$ with $\varepsilon'=\varepsilon'(\varepsilon)\to 0$ as $\varepsilon\to 0$. On the other hand, for every fixed $\varepsilon>0$ in the zone $t_2(\tau)\le|x'|\le c_0\tau^{\frac{1}{2m\mu}}$ in the context of theorem 12.1.5 one can improve the remainder estimate. Namely, taking a standard partition of unity in (x',ξ')-space and applying standard Tauberian arguments we obtain that the contributions of both elements of this partition do not exceed $\varepsilon''\tau^{\frac{1}{2m\mu}(d'-1)(\mu+1)}$ with arbitrarily small $\varepsilon''>0$ because the support of the first element is small enough and in the second element one can take $T_2=(h_2\varepsilon''')^{-1}$ with arbitrarily small $\varepsilon'''>0$ in the Tauberian arguments. The only element of these arguments which is not completely standard is that we apply the propagation results derived from the theory of operators with operator-valued symbols to Weylian asymptotics. Then we obtain the asymptotics

$$
\begin{aligned}
\text{(12.1.53)}\qquad \mathbf{N}(\tau) = &\int \phi(\frac{x'}{t_1})\varkappa_j(x)\tau^{\frac{d}{2m}}\,dx + \\
&\int \bigl(1-\phi(\frac{x'}{t_1})\bigr)\bigl(\varkappa_0(x)-\varkappa_0^0(x)\bigr)\tau^{\frac{d}{2m}}\,dx + \\
&\int \bigl(1-\phi(\frac{x'}{t_1})\bigr)\operatorname{tr} e^0(x,x,\tau)dx + \\
&o\bigl(\tau^{\frac{1}{2m\mu}(d'-1)(\mu+1)}\bigr);
\end{aligned}
$$

here we have dropped all terms with $j=1$ in (12.1.51). Here as well as above t_1 is either a large enough constant or a slowly increasing function of τ.

(iv) If one replaces t_1 by $t_1'\ge t_1$ in (12.1.51)–(12.1.53) then the left-hand expression does not change, and in order to estimate the increment of the right-hand expression one can apply estimate (12.1.49) for $\mathcal{R}_1$. Then one can easily see that for $t_1'\le\tau^\delta$ with a small enough exponent $\delta>0$ this increment does not exceed the remainder estimate and therefore in (i)–(iii) one can formulate the condition of slow growth of t_1 more exactly: $t_1\le\tau^\delta$. I leave the details and the calculation of $\delta>0$ to the reader.

(v) One could apply formula (12.1.49) for $|x'|\le 2t_2$ and (12.1.50) for $t_2\le|x'|\le c_0\tau^{\frac{1}{2m\mu}}$ where t_2 is now less than $c_0\tau^{\frac{1}{2m\mu}}$. Therefore the following formula holds:

$$
\begin{aligned}
\text{(12.1.54)}\qquad \mathbf{N}(\tau) = &\sum_{0\le j\le J}\int \phi(\frac{x'}{t_2})\varkappa_j(x)\tau^{\frac{1}{2m}(d-j)}dx + \\
&\sum_{1\le j\le J}\int_{\partial X} \phi(\frac{x'}{t_2})\tilde\varkappa_j(x)\tau^{\frac{1}{2m}(d-j)}dx + \\
&\int \bigl(1-\phi(\frac{x'}{t_2})\bigr)e^0(x,x,\tau)dx + O(R^*+R_2)
\end{aligned}
$$

with some remainder estimate $R_2=R_{2,J}(\tau,t_2)$. Taking $t_2=t_{2,J}(\tau)$ as providing the best possible estimate one can obtain certain asymptotics, maybe with worse remainder estimates. Improvements in the spirit of (ii),(iii) are possible.

(vi) Let us pick two cases in which the formulas based on (12.1.50) and (12.1.49) are "well-corresponding" and in (12.1.54) we can take $J=2$ and $t_2=c_0\tau^{\frac{1}{2m\mu}}$ and the same remainder estimate as in formulas (12.1.51)–(12.1.53). This is the case when $\varkappa_j^0=\varkappa_j$ (j=0,1), $\tilde\varkappa_1^0=\tilde\varkappa_1$. And these equalities surely hold in the following cases:

(a) A and A^0 differ only in lower order terms. The only coefficient in the Weylian formula depending on lower order terms is $\varkappa_1$ but even this coefficient surely does not depend on lower order terms because A is a differential operator of even order; moreover, this coefficient vanishes provided the principal symbol a is not only Hermitian but also real-valued (in which case it is symmetric).

(b) The principal symbol of A can be obtained from the principal symbol of A^0 by the transformation $\xi' \to \xi' - T(x,\xi'')$ with scalar function $T(x,\xi'')$ linear in ξ''. In fact, under the completely Weylian procedure this transform surely preserves $\varkappa_0$, $\tilde{\varkappa}_1$ and some part of $\varkappa_1$ which does not depend on the lower order terms; however, the other part surely vanishes.

12.1.4 Effective asymptotics. II

Now let us consider the term

$$\int (1-\phi(\frac{x'}{t}))\,\mathrm{tr}\,e^0(x,x,\tau)dx \tag{12.1.55}$$

with $t \ge c_0$ and apply theorems 12.1.4–12.1.6. Then the remainder estimate is preserved. Moreover, applying theorem 12.1.5 one can replace the condition on billiards (which is now unnatural because it is associated with the operator A^0) with the condition "$t \to +\infty$ as $\tau \to +\infty$." Let us investigate when the formulas for (12.1.55) associated with A^0 are simpler than in the general case.

(i) After straightening the cusp let A^0 have the principal operator-valued symbol

$$\mathbf{a}^0(x',\xi') = \varrho(x')^{-2m}\bar{a}(\mathrm{x}'',\xi'\varrho(x');\mathrm{D}'') \tag{12.1.56}$$

in which case the answer is

$$(2\pi)^{-d'}\iint (1-\phi(\frac{x'}{t}))\bar{\mathbf{n}}(\xi',\tau\varrho(x')^{2m})\varrho^{d'}dx'd\xi'$$

where $\bar{\mathbf{n}}(\xi',\tau)$ is the eigenvalue counting function for the operator $\bar{a}(\xi',\mathrm{x}'',\mathrm{D}'')$ in $\mathbb{H} = L^2(Y,\mathbb{K})$. In this case the final answer uses the eigenvalue counting function for the operator family depending only on ξ'. Moreover, one can consider the case when $\bar{a}$ depends on $x'|x|^{-1}$.

(ii) Moreover, let the principal operator-valued symbol of A^0 (after straightening the cusp) be of the form

$$\mathbf{a}^0(x',\xi') = \varrho(x')^{-2m}F(x',\xi'\varrho(x');\Lambda(x')) \tag{12.1.57}$$

where $\Lambda = \Lambda(x';\mathrm{x}'',\mathrm{D}'')$ is a self-adjoint operator in $\mathbb{H}$ (an elliptic differential operator of order $2m$ with Dirichlet boundary conditions) and $F(x',\xi',\lambda)$ is a monotone increasing function of λ. Then the answer is

$$(2\pi)^{-d'}\iint \kappa(x',f(x',\xi',\tau\varrho(x')^{2m}))\varrho(x')^{-d'}dx'd\xi' \tag{12.1.58}$$

where $\kappa(x',\lambda)$ is the eigenvalue counting function for Λ and $\lambda = f(x',\xi',\tau)$ is solution of the equation $F(x',\xi',\lambda) = \tau$.

(iii) In particular, let $m = 1$ and let the principal operator-valued symbol of A^0 (after straightening the cusp) be of the form

$$\mathbf{a}^0(x',\xi') = b(x',\xi') + \varrho(x')^{-2}\Lambda(x';\mathrm{x}'',\mathrm{D}'') \tag{12.1.59}$$

where $b(x',\xi')=\sum_{jk}b_{jk}(x')\xi'_j\xi'_k$ is a positive scalar quadratic form in ξ'. Then the final answer is

$$\frac{1}{2}(2\pi)^{-d'}d'\omega_{d'}\iint \kappa(x',\lambda)\varrho(x')^{-d'}\beta(x')^{-\frac{1}{2}}\bigl(\tau\varrho(x')^2-\lambda\bigr)_+^{\frac{1}{2}(d'-2)}d\lambda dx' \tag{12.1.60}$$

where $\beta(x')$ is the determinant of the matrix of the quadratic form $b(x',.)$.

12.1.5 Effective asymptotics. III

Let us formulate the principal theorems proved in this section. They are due to results above. Moreover, we should assume that A^0 is a symmetric operator given in the cusp and we should not assume that A^0 is self-adjoint. Otherwise it could be replaced by $\frac{1}{2}(\mathcal{A}+\mathcal{A}^*)$ with $\mathcal{A}=\phi(\frac{x'}{t_0})A+\bigl(1-\phi(\frac{x'}{t_0})\bigr)A^0$. Then we obtain the following statements:

Theorem 12.1.10. *(i) Let conditions* (12.1.1)–(12.1.8), *condition* $(12.1.17)_1$ *or* $(12.1.17)_2$, *and conditions* (12.1.39), (12.1.40) *be fulfilled. Then the asymptotics*

$$\mathbf{N}(\tau)=\mathcal{N}_1(\tau)+O\bigl(R^*(\tau)+\tau^{\frac{1}{2m\mu}(d'(\mu+1)-\nu)}\bigr) \tag{12.1.61}$$

hold where $R^*(\tau)$ *is given by* (12.1.20) *and*

$$\begin{aligned}\mathcal{N}_1(\tau)=&\sum_{j=0,1}\int\phi(\frac{x'}{t_1})\varkappa_j(x)\tau^{\frac{1}{2m}(d-j)}dx+\\ &\int_{\partial X}\phi(\frac{x'}{t_1})\bar{\varkappa}_1(x)\tau^{\frac{1}{2m}(d-1)}dS+\\ &\sum_{j=0,1}\int\bigl(1-\phi(\frac{x'}{t_1})\bigr)\bigl(\varkappa_j(x)-\varkappa_j^0(x)\bigr)\tau^{\frac{1}{2m}(d-j)}dx+\\ &\int_{\partial X}\phi(\frac{x'}{t_1})\bigl(\bar{\varkappa}_1(x)-\bar{\varkappa}_1^0(x)\bigr)\tau^{\frac{1}{2m}(d-1)}dS+\\ &(2\pi)^{-d'}\iint\bigl(1-\phi(\frac{x'}{t_1})\bigr)\mathbf{n}^0(x',\xi',\tau)dx'd\xi'\end{aligned} \tag{12.1.62}$$

where $\mathbf{n}^0$ *is the eigenvalue counting function for* $\mathbf{a}^0(x',\xi';\mathrm{x}'',\mathrm{D}'')$ *in* $\mathbb{H}$, ϕ *is introduced above,* $c_0\le t_1\le\tau^\delta$ *with a small enough exponent* $\delta>0$. *Moreover,* (12.1.23) *remains true for* $\mathcal{N}_1$.

(ii) Moreover, let $d'<1+\mu d''$, $\nu>d'-\mu(d''-1)$.[5] *Let the standard condition* $(\Psi 0)$ *be fulfilled for the Hamiltonian billiards of* $a(x',\xi')$. *Then the asymptotics*

$$\mathbf{N}(\tau)=\mathcal{N}_1(\tau)+o\bigl(\tau^{\frac{d-1}{2m}}\bigr) \tag{12.1.63}$$

hold as $\tau\to+\infty$ *provided* $t_1\to+\infty$ *(we assume that* $t_1\le\tau^\delta$*).*

(iii) Let $d'>1+\mu d''$, $\nu>\mu+1$.[6] *Moreover, let stabilization conditions* $(12.1.26)_{1,2}$ *and condition* (12.1.28) *be fulfilled.*[7] *Then the asymptotics*

$$\mathbf{N}(\tau)=\mathcal{N}_1(\tau)+\varkappa_1^*\tau^{\frac{1}{2m\mu}(d'-1)(\mu+1)}+o\bigl(\tau^{\frac{1}{2m\mu}(d'-1)(\mu+1)}\bigr) \tag{12.1.64}$$

hold with the coefficient $\varkappa_1^*$ *given in theorem 4.3.6.*

---

5) Then the remainder estimate in (12.1.61) is $O\bigl(\tau^{\frac{1}{2m}(d-1)}\bigr)$.

6) Then the remainder estimate in (12.1.61) is $O\bigl(\tau^{\frac{1}{2m\mu}(d'-1)(\mu+1)}\bigr)$.

7) In these conditions a_α^0 should not necessarily coincide with the coefficients in (12.1.39) and in the definition of $\mathcal{N}_1$.

I leave it to the reader to obtain results in the spirit of (12.1.54) (it would be better to take $J = 2$ because we do not know nice formulas for $\varkappa_j$, $\tilde{\varkappa}_j$ with $j \geq 2$). I leave it to the reader to consider cases when ϱ and η are not purely power functions (for example, if these functions contain logarithmic factors), which is significant for the orders of the principal part of asymptotics and remainder estimate only for $d' \geq \mu d''$, $d' \geq 1 + \mu d''$ respectively. Moreover, I leave to the reader the case of a few different cusps (this problem is trivial even if the dimensions of these cusps (d') and the exponents of shrinking (μ) differ. I only note that in some situations the bounded part of the domain and even some cusps can be ignored in the final answer). I leave it to the reader to treat Riesz means. I indicate only that R^* in this case is defined by (12.1.30) and the additional error due to perturbation is $O\bigl(R^*(\tau) + \tau^{\frac{1}{2m\mu}(d'(\mu+1)-\nu)+\vartheta}\bigr)$ and so the remainder estimate is $O(R^*)$ provided $\nu \geq \min\bigl((\mu+1)(1+\vartheta), d' - d''\mu + (1+\vartheta)\mu\bigr)$.

Remark 12.1.11. *If we wish to improve the remainder estimate in the final theorem in the case when the stabilization exponent ν is not large enough then the following algorithm is suggested. In the domain $|x'| \leq t$ we apply the previous analysis. So we need to treat*

$$\int \psi(x')\,\mathrm{tr}\,\mathbf{e}^{\theta}(x',\xi',\tau)dx'd\xi'$$

with $\theta = 1$ and a γ-admissible function ψ supported in $B(\bar{x}',\bar{\gamma})$ (and later sum on a partition of unity in the zone $|x| \geq t_2$). This expression is smooth in θ because of the microhyperbolicity property. Then one can apply Taylor's formula with N terms. The remainder estimate is $O\bigl(\bar{\gamma}^{d'}\bar{\varrho}^{d''}\bar{\eta}^{N}\tau^{d'}\bigr)$ (for $|\bar{x}''| \leq \tau^{\frac{1}{2m\mu}}$) and after summation on the partition of unity we obtain $O\bigl(R^*(\tau) + R_3(t_2,\tau)\bigr)$ with R_3 depending on N and ν. So, the total remainder estimate is $O\bigl(R^*(\tau) + R_2(\tau,t_2) + R_3(\tau,t_2)\bigr)$ where R_2 is an upper estimate for $\mathcal{R}_2$. Picking $t_2 = t_2(\tau)$ as providing the best possible remainder estimate we obtain the final formula. However, this formula is rather sophisticated even in case (iii) of the previous subsection. Namely, the additional term with $n = 1$ in the Taylor expansion is

$$\mathrm{const}\iint \bigl(\tau\varrho(x')^{2m} - \lambda\bigr)^{\frac{d-2}{2}} d_{\lambda}\bigl(\mathrm{tr}\,\mathbf{a}'\mathbf{E}(\lambda)\bigr)dx'$$

provided $\mathbf{a} - \mathbf{a}^0 = \mathbf{a}'(x, D'')$ where $\mathbf{E}(\lambda)$ is the spectral projector for A in $\mathbb{H}$, and is slightly more complicated for more general perturbations. However, the next terms (with $n \geq 2$) are very sophisticated. I leave this task to the reader as well as the generalization to Riesz means. ◊

12.2 Operators with Potentials Degenerate at Infinity

12.2.1 Introduction

In this section we consider the Schrödinger operator A in $\mathbb{R}^d$ given by ($\Sigma 0$) such that

$$|D^{\alpha} g^{jk}| \leq c\langle x\rangle^{-|\alpha|}, \tag{12.2.1$_1$}$$

$$|D^{\alpha} V_j| \leq c\langle x\rangle^{q-|\alpha|}, \tag{12.2.1$_2$}$$

$$V = W_0 + W_1, \tag{12.2.2}$$

$$|D^{\alpha} W_0| \leq c\rho(x)^{(2n-|\alpha|)_+}\langle x\rangle^{2(m-n)-(2n-|\alpha|)_-}, \tag{12.2.1$_3$}$$

$$|D^{\alpha} W_1| \leq c\langle x\rangle^{2q-|\alpha|}, \tag{12.2.1$_4$}$$

$$W_0 \geq \epsilon_0\rho(x)^{2n}\langle x\rangle^{2(m-n)}, \tag{12.2.1$_5$}$$

(12.2.4) $\rho(x) = \operatorname{dist}(x, \Lambda)$ where Λ is a conical C^K manifold in $\mathbb{R}^d$,

$$q \geq 0, \quad n > 0, \quad m > n + q. \tag{12.2.5}$$

Surely, we assume that condition ($\Sigma 2$) is fulfilled. Condition (12.2.5) means exactly that the cusp $\{x, V(x) \leq \tau\}$ shrinks at infinity (on $|x|$) for any $\tau > 0$ and for any W_1 satisfying $(12.2.1)_4$.

First of all the standard theory of Chapter 10 yields the following

Proposition 12.2.1. *Let conditions ($\Sigma 0$), ($\Sigma 2$), (12.1.1)–(12.1.5) be fulfilled and at the point $\bar{x} \in \mathbb{R}^d$ let the conditions*

$$(12.2.6)_{1-3} \qquad \rho(x) \geq c_0 \langle x \rangle^{(q-m+n)n^{-1}}, \quad \rho(x) \geq c_0 \langle x \rangle^{(q-m)n^{-1}} \tau^{(2n)^{-1}}, \\ \rho(x) \geq \rho_0 = c_0 \langle x \rangle^{(q-m+n)(n+1)^{-1}}$$

be fulfilled with a large enough constant c_0. Then

$$|e(x, y, \tau)| \leq C \langle x \rangle^{(m-n)d(n+1)^{-1}} \rho(x)^{-s} \rho_0^s \qquad \forall x, y \in B(\bar{x}, \frac{1}{2}\rho(\bar{x})) \tag{12.2.7}$$

with an arbitrarily large exponent s. Moreover, if $U(x, y, t)$ is the Schwartz kernel of the operator $\tau^{-1} A$ then

$$\bigl| F_{t \to \lambda} \chi_{T, T'}(t) U(x, y, t) \bigr| \leq C \langle x \rangle^{(m-n)d(n+1)^{-1}} \rho(x)^{-s} \rho_0^s \\ \forall x, y \in B(\bar{x}, \frac{1}{2}\rho(\bar{x})) \quad \forall T' \in \mathbb{R}, T = \rho(\bar{x}) \quad \forall \lambda \leq 1 + \varepsilon \tag{12.2.8}$$

with a small enough constant $\varepsilon > 0$.

Remark 12.2.2. *For $qn + q + n \geq m$ ($qn + q + n \leq m$) the third (respectively the first) condition in (12.2.6) is due to the first (the third) one. Moreover, for $qn + q + n > m$ one can replace the right-hand expression in (12.2.7), (12.2.8) by $\tau^{-s} \langle x \rangle^{-s}$.* ◊

Let $\mathcal{Y}$ be the $\gamma(x)$-neighborhood of Λ with $\gamma(x) = \epsilon_1 \langle x \rangle$. Let us introduce the coordinates $x = (x', x'') \in \Lambda \times \mathbb{R}^p$, $p = \operatorname{codim} \Lambda$ in $\mathcal{Y}$; we treat only the domain $\{|x| \geq c_0\}$. Let us consider the equation

$$(\tau D_t - A_x) U = 0 \tag{12.2.9}$$

in this neighborhood and let us treat A as an operator on x' with operator-valued symbol in the auxiliary space $\mathbb{H} = L^2(\mathbb{R}^p)$. We consider only energy levels of A less than $(1+\epsilon)\tau$ in the zone $\{|x| \geq c_1 \tau^{\frac{1}{2m}}\}$. Then (12.2.8) yields that $U(x, y, t)$ is negligible near $\partial \mathcal{Y}$ and therefore the Dirichlet boundary condition is fulfilled on $\partial \mathcal{Y}$ modulo negligible functions.

Let $\phi_1^2 + \phi_2^2 = 1$ be a partition of unity with $\phi_1 \in C_0^K(\mathbb{R}^p)$ vanishing near the origin and $\phi_2 \in C^K(\mathbb{R}^p)$. Let $\tilde{\phi}_j(x'') = \phi_j(x'' \rho_1^{-1})$ where ρ_1 will be chosen later. Then

$$A = \tilde{\phi}_1 A \tilde{\phi}_1 + \tilde{\phi}_2 A \tilde{\phi}_2 + W_2 = \tilde{\phi}_1 (A + W_2) \tilde{\phi}_1 + \tilde{\phi}_2 (A + W_2) \tilde{\phi}_2$$

where $W_2 = \frac{1}{2} \sum_{j,k=1,2} \bigl[[A, \tilde{\phi}_j], \tilde{\phi}_j \bigr]$ is a function such that $|W_2| \leq c_1 \rho_1^{-2}$ and all the constants do not depend on the choice of ρ_1. Let us assume that

$$(12.2.10)_2 \qquad V - \tau + W_2 \geq \epsilon W_0 \quad \text{on } \operatorname{supp} \tilde{\phi}_2;$$

then

$$(12.2.10)_1 \qquad V - \tau + W_2 \geq W_1 - \tau \quad \text{on } \operatorname{supp} \tilde{\phi}_1.$$

Let $\mathbf{a} = \mathbf{a}(x', \xi')$ be the operator-valued symbol of A. Then under condition $(12.2.10)_{1,2}$,

$$(12.2.12)_1 \qquad \langle \mathbf{a}\tilde{\phi}_2 v, \tilde{\phi}_2 v\rangle \ge \epsilon_1\bigl(|x'|^2 + \tau\bigr)|\tilde{\phi}_2 v|^2 + \epsilon_1\langle W_0\tilde{\phi}_2 v, \tilde{\phi}_2 v\rangle$$

and

$$(12.2.12)_1 \qquad \langle \mathbf{a}\tilde{\phi}_1 v, \tilde{\phi}_1 v\rangle \ge \epsilon_1\bigl(|x'|^2 + \rho_1^{-2}\bigr)|\tilde{\phi}_1 v|^2 + \epsilon_1\langle (W_1 - \tau)\tilde{\phi}_1 v, \tilde{\phi}_1 v\rangle$$

where the norms and inner products are calculated in $\mathbb{H}$. Moreover, let us assume that

$$(12.2.12) \qquad \rho_1^{-2} \ge c_0\bigl(|x|^{2q} + \tau\bigr).$$

Then the conditions of lemma 4.1.5 are fulfilled. One can easily see that condition $(12.2.10)_2$ is equivalent to $(12.2.6)_{1-3}$ for $\rho = \rho_1$. On the other hand, conditions $(12.2.6)_{1-3}$ for $\rho = \rho_1$ and (12.2.12) are compatible if and only if

$$(12.2.13) \qquad m > q + n + qn$$

and $|x| \ge r_0(\tau) \stackrel{\mathrm{def}}{=} c_0\tau^{\frac{1}{2}(n+1)(m-n)^{-1}}$.

Then lemma 4.1.5 yields the following

Proposition 12.2.3. *Let conditions (Σ0), (Σ2), (12.2.1)–(12.2.5), (12.2.13) be fulfilled. Then the estimate*

$$(12.2.14) \qquad |e(x,y,\tau)| \le C\langle x\rangle^{-s}\langle y\rangle^{-s}\tau^{-s} \qquad \forall x, y : |x| \ge r_0(\tau), |y| \ge r_0(\tau)$$

holds.

Remark 12.2.4. *(i) Obviously, $(V-\tau)_+^{\frac{d}{2}} \in L^1$ provided*

$$(12.2.15) \qquad qdn + pq + dn < pm$$

(where $p = \operatorname{codim}\Lambda$) and this condition is necessary if

$$(12.2.16) \qquad W_1 \le -\epsilon'|x|^{2q} \qquad \text{for } |x| \ge c,\ \rho(x) \le \epsilon'$$

with an arbitrarily small constant $\epsilon' > 0$. Under this condition the principal part of the Weylian formula is $\asymp \tau^{\frac{1}{2m}d(m+1)}$. This condition is surely stronger than (12.2.13).

(ii) Let us reject condition (12.2.15) and let us integrate only over the domain $\{x, |x| \le r_0(\tau)\}$ in the Weylian formula by proposition 12.2.3. Then the principal part of the Weylian asymptotics (modified in this way) is

$$(12.2.17) \qquad \asymp \begin{cases} \tau^{\frac{1}{2}d(m+1)m^{-1}} & \text{for } mp > nd,\\ \tau^{\frac{1}{2}d(m+1)m^{-1}}\log\tau & \text{for } mp = nd,\\ \tau^{\frac{1}{2}(d-p)(m+1)(m-n)^{-1}} & \text{for } mp < nd.\end{cases}$$

(iii) Let us consider the remainder estimate in the Weylian formula. Under condition (12.2.13) we can integrate only over the domain $\{x, |x| \le r_0(\tau)\}$ because of proposition 12.2.3. Then the remainder estimate in the modified Weylian asymptotics is

$$(12.2.18) \qquad \asymp \begin{cases} \tau^{\frac{1}{2}(d-1)(m+1)m^{-1}} & \text{for } m(p-1) > n(d-1),\\ \tau^{\frac{1}{2}(d-1)(m+1)m^{-1}}\log\tau & \text{for } m(p-1) = n(d-1),\\ \tau^{\frac{1}{2}(d-p)(m+1)(m-n)^{-1}} & \text{for } m(p-1) = n(d-1).\end{cases}$$

(iv) Let $m(p-1) > n(d-1)$. Then under stabilization conditions on the positively homogeneous functions g^{jk0} and V^0 at infinity and standard restrictions on the Hamiltonian flow generated by the corresponding Hamiltonian H_0 one can obtain the remainder estimate $o(\tau^{\frac{1}{2}(d-1)(m+1)m^{-1}})$. In this remark the role of the "operator-valued theory" is very limited: it only provides a nice cutoff in the Weylian formulas.

(v) I leave it to the reader to calculate the principal part of the asymptotics and remainder estimate for Riesz means in this remark. ◊

12.2.2 Accurate asymptotics

So, our analysis should be done under conditions (12.2.13) and $m(p-1) \le n(d-1)$. Here we should treat the zone

$$\mathcal{X} = \bigl\{c_1\tau^{\frac{1}{2m}} \overset{\text{def}}{=} r_1(\tau) \le |x'| \le r_0(\tau), |x''| \le \varrho(x',\tau) \overset{\text{def}}{=} c_2|x'|^{\frac{1}{n}(m-n)}\tau^{\frac{1}{2n}+\delta}\bigr\}$$

with an arbitrarily small exponent $\delta > 0$. Really, estimate (12.2.7) and proposition 12.2.3 yield that for $|x'| \ge c_1\tau^{\frac{1}{2m}}$ outside of $\mathcal{X}$ the function $e(x,x,\tau)$ is negligible; on the other hand, the contributions of the zone $\{|x'| \le c_1\tau^{\frac{1}{2m}}\}$ to the principal part of the Weylian asymptotics and to the remainder estimate are $O\bigl(\tau^{\frac{1}{2m}d(m+1)}\bigr)$ and $O\bigl(\tau^{\frac{1}{2m}(d-1)(m+1)}\bigr)$ [8] respectively. Moreover, the same arguments yield that treating the zone $|x'| \le c_0 r_0(\tau)\tau^{-\delta'}$ with an arbitrarily small exponent $\delta' > 0$ one can set $\delta = 0$ in the definition of $\mathcal{X}$; in this case the constant c_2 should be large enough. We treat $\mathcal{X}$ as the cusp below.

Let us first consider the zone $\bigl\{|x'| \le c_0 r_0(\tau)\tau^{-\delta'}\bigr\}$ and pick $\delta = 0$ in the definition of ϱ and $\mathcal{X}$. Let us introduce coordinates $(x', \mathrm{x}'') = (x', x''\varrho(x',\tau)^{-1})$ in $\mathcal{X}$. In these coordinates $\mathcal{X}$ is straightened into the cylinder

$$\tilde{\mathcal{X}} = \bigl\{r_1(\tau) \le |x'| \le c_0 r_0(\tau), \mathrm{x}'' \in Y = B(0,1) \subset \mathbb{R}^p\bigr\}.$$

Let us treat the operator $\tau^{-1}A$ in this cylinder as an operator with operator-valued symbol in the auxiliary space $L^2(Y)$ with Dirichlet boundary condition on ∂Y (which is fulfilled modulo negligible functions and therefore can be applied in all the estimates and successive approximations provided we associate $e(x,y,\tau)$ and $U(x,y,t)$ with the original operator but not with the new one which is not even necessarily self-adjoint). Then one can take $B = L_{(2)\mathrm{D}} = \zeta\Delta''_{\mathrm{D}} + I$ and $B_1 = L^{\frac{1}{M}}_{(M)\mathrm{N}}$ where $L_{(M)\mathrm{N}} = (\zeta\Delta'' + I)^M$, Δ'' is the positive Laplacian in Y, $M \in \mathbb{Z}^+$ is large enough, the subscripts "D" and "N" mean the Dirichlet and Neumann boundary conditions of the corresponding order, $\zeta = \bar{\varrho}^{-1}\tau^{-1}$, $\bar{f} = f(\bar{x}')$, $\bar{x}'$ is a point, $\gamma(x') = \frac{1}{2}|x'|$ and we treat the $\bar{\gamma}$-neighborhood of $\bar{x}'$ in Λ.

On the other hand, in the zone $\bigl\{c_0 r_0(\tau)\tau^{-\delta'} \le |x'| \le c_0 r_0(\tau)\bigr\}$ we should pick $\delta > 0$ and set $B = L'_{(2)\mathrm{D}}\zeta\Delta_{\mathrm{D}} + v_\tau$, $B_1 = L^{\frac{1}{M}}_{(M)\mathrm{N}}$ where $L_{(M)*} = (\zeta\Delta'' + v_\tau)^M$ with the same M and ζ as before and with $v_\tau = \tau^{2n\delta}(|\mathrm{x}''| - 1)_+ + 1$ treated as an operator in $\mathbb{H}$.

The reader can easily check that the operator $\tau^{-1}A$ (considered as an operator with operator-valued symbol) is microhyperbolic (after the dilatation $x'_{\text{new}} = (x' - \bar{x}')\bar{\gamma}^{-1}$) at the energy level $\lambda = 1$ in the direction $\mathcal{T}$ provided one of the following two conditions is fulfilled:

(a) $|\xi'| \ge \varepsilon$, $\mathcal{T} = \langle \xi', \partial_{\xi'}\rangle$,

(b) $|\xi'| \le \varepsilon$, $\mathcal{T} = \langle x', \partial_{x'}\rangle$ and

---

[8] Moreover, under stabilization conditions and the standard condition (Ψ) on the Hamiltonian field the contribution of the zone $\{|x'| \le r_1(\tau)\}$ to the remainder estimate is $o\bigl(\tau^{\frac{1}{2m}(d-1)(m+1)}\bigr)$.

(12.2.19)
$$\langle x', \partial_{x'} \rangle V \geq \epsilon \rho^{2n}(x)|x|^{2(m-n)} - c|x|^{2q}$$
$$\text{for } |x'| \geq c_0,\ \rho(x) \leq \varrho_0(x', \tau)$$

with a sufficiently small constant $\varepsilon > 0$.

Let us apply theorem 4.3.3. Let us take the trace norm; one can easily see that for $l > \frac{1}{2}p$, $c_1\tau^{\frac{1}{2m}} = r_1(\tau) \leq |x'| \leq r_0(\tau)$ the trace norm of B_1^{-l} does not exceed $C\tau^{\frac{p}{2}}\varrho_0(x', \tau)^p$ with $\varrho_0(x', \tau) = |x'|^{-\frac{1}{n}(m-n)}\tau^{\frac{1}{2n}}$. Therefore we obtain the following estimate for a γ-admissible function ψ supported in $B(\bar{x}', \frac{1}{2}\bar{\gamma})$ with $\gamma(x') = |x'|$:

(12.2.20)
$$\Big|\int \psi(x')e(x, x, \tau)dx - (2\pi)^{p-d}\iint \psi(x')\mathbf{n}(x', \xi', \tau)dx'd\xi'\Big| \leq C\tau^{\frac{1}{2}(d-1)+\frac{p}{2n}}\bar{\gamma}^{d-1-m\frac{p}{n}}$$

where $\mathbf{n}(x', \xi', \tau)$ is the eigenvalue counting function of $\mathbf{a}(x', \xi') = \mathbf{a}(x', \xi'; \mathrm{x}'', \mathrm{D}'')$ which is the operator-valued symbol of A (after a coordinate transform). Let us sum over a γ-admissible partition of unity in the zone $\{r_1(\tau) \leq |x'| \leq r_0(\tau)\}$ and apply Weylian asymptotics in the zone $\{|x'| \leq t_1\tau^{\frac{1}{2m}}\}$. Then we obtain the following

Theorem 12.2.5. *Let conditions* ($\Sigma 0$), ($\Sigma 2$), (12.2.1)–(12.2.4), (12.2.13) *and* (12.2.19) *be fulfilled. Let $\phi \in C_0^K(\mathbb{R}^{d-p})$, $\phi = 1$ in a neighborhood of the origin. Then the asymptotics*

(12.2.21)
$$\mathbf{N}(\tau) = \mathcal{N}(\tau) = (2\pi)^{-d}\omega_d\int \phi(\frac{x'}{t_1})(\tau - V)_+^{\frac{d}{2}}\sqrt{g}dx + (2\pi)^{p-d}\iint (1 - \phi(\frac{x'}{t_1}))\mathbf{n}(\xi', \xi', \tau)dx'd\xi' + O(R(\tau))$$

hold with $c_0\tau^{\frac{1}{2m}} \leq t_1 \leq c_1\tau^{\frac{1}{2m}}$ [9] *and with*

(12.2.22)
$$R(\tau) = \begin{cases} \tau^{\frac{1}{2}(d-1)(m+1)m^{-1}} & \text{for } (d-1)n < pm \\ \tau^{\frac{1}{2}(d-1)(m+1)m^{-1}}\log\tau & \text{for } (d-1)n = pm \\ \tau^{\frac{1}{2}(d-p-1)(m+1)(m-n)^{-1}} & \text{for } (d-1)n > pm \end{cases}$$

and the order of $\mathcal{N}(\tau)$ is indicated in (12.2.17).

Let us note that for $(d-1)n < pm$ the main contribution to the remainder estimate is provided by the zone $|x'| \leq t\tau^{\frac{1}{2m}}$ with arbitrarily large t; then applying improved Weylian asymptotics we obtain

Theorem 12.2.6. *Let the conditions of theorem 12.2.5 be fulfilled with $(d-1)n < pm$. Moreover, let stabilization conditions* (10.4.3) *and condition* (Ψ) *on the Hamiltonian flow generated by the quasi-homogeneous part of Hamiltonian also be fulfilled. Then the asymptotics*

(12.2.23)
$$\mathbf{N}(\tau) = \mathcal{N}(\tau) + o(\tau^{\frac{1}{2m}(d-1)(m+1)})$$

hold as $\tau \to +\infty$ where $t_1 = t_1(\tau)$ in the definition (12.2.19) *of $\mathcal{N}(\tau)$ is now such that $t_1(\tau)\tau^{(2m)^{-1}}$ tends to $+\infty$ slowly enough.*

On the other hand, for $(d-1)n > pm$ the main condition on the remainder estimate is provided by the zone $\varepsilon r_0(\tau) \leq |x'| \leq r_0(\tau)$ with an arbitrarily small constant $\varepsilon > 0$.

---

9) One can assume that $t_1 = t_1(\tau)$ and $t_1(\tau)\tau^{-\frac{1}{2m}}$ is either bounded or tends slowly to $+\infty$.

In order to formulate a condition on Hamiltonian trajectories let us assume that the following stabilization conditions are fulfilled:

$$g^{jk}(x) = g^{jk0}(x') + o(1), \tag{12.2.24$_1$}$$

$$W_0(x) = W^0(x',x'') + o(|x'|^{2(m-n)(n+1)^{-1}}), \tag{12.2.24$_2$}$$
$$\text{as } |x'| \to +\infty,\ |x''| \le c_2|x'|^{-(m-n)(n+1)^{-1}}.$$

where

(12.2.25) $W^0(x)$ is positively homogeneous of degree $2m$ in x and $W_0 \ge \epsilon\rho(x)^{2n}\cdot|x|^{2m-2n}$ with $\epsilon > 0$. Let $\mathbf{a}^0(x',\xi')$ be the operator-valued symbol of the operator A^0 obtained from A by $g^{jk} \to g^{jk0}$, $V_j \to 0$, $V \to W^0$, i.e.,

$$\mathbf{a}^0(x',\xi') = \sum_{jk} P_j g^{jk0} P_k + W^0 \tag{12.2.26}$$

with $P_j = \xi_j$ for $j \le d-p$ and $P_j = D_j$ for $j \ge d-p+1$. Let us consider $\mathbf{a}^0$ as an operator in $\mathbb{H}^* = L^2(\mathbb{R}^p)$.[10] Then the eigenvalues of $\mathbf{a}^0$ are positively quasi-homogeneous on (x',ξ') and for $|x'| \to \infty$, $|\xi'| \to \infty$ the eigenvalues of $\mathbf{a}(x',\xi')$ tend to those of $\mathbf{a}^0(x',\xi')$. Then standard arguments and theorem 4.3.5 immediately yield the following

Theorem 12.2.7. *Let the conditions of theorem 12.2.5 be fulfilled and suppose $(d-1)n > pm$. Moreover, let stabilization conditions (12.2.24)$_{1,2}$, (12.2.25) be fulfilled. Finally, let the eigenvalues $\zeta_j(x',\xi')$ of the operator $\mathbf{a}^0(x',\xi')$ in $\mathbb{H}$ satisfy condition (12.1.28). Then the asymptotics*

$$\mathbf{N}(\tau) = \mathcal{N}(\tau) + \left(\varkappa_1^* + o(1)\right)\tau^{\frac{1}{2}(d-p-1)(m+1)(m-n)^{-1}} \tag{12.2.27}$$

hold as $\tau \to +\infty$ with the standard coefficient $\varkappa_1^*$ given in theorem 4.3.6.[11]

Theorems 12.2.5–12.2.7 are similar to theorems 12.1.4–12.1.6 and all the remarks after those theorems remain true for our new theorems. So our goal is to obtain more effective formulas and to obtain a more explicit upper estimate for $t_1(\tau)$.

Remark 12.2.8. *Let us assume that*

$$\begin{aligned} \bar V(x) &= V(x) && \text{for } |x''| \le c_2|x'|^{-(m-n)(n+1)^{-1}},\\ \bar V(x) &\ge \epsilon|x'|^{2(m-n)}|x''|^{2n} && \text{for } |x''| \ge c_2|x'|^{-(m-n)(n+1)^{-1}} \end{aligned} \tag{12.2.28}$$

with an arbitrarily small constant $\epsilon > 0$. Then in the definition of $\mathbf{n}(x',\xi',\tau)$ one can treat $\mathbf{a}(x',\xi')$ as an operator in $\mathbb{H}^* = L^2(\mathbb{R}^p)$ and so one can avoid straightening the cusp to a cylinder in the final formula. It makes all of the calculations simpler. ◇

12.2.3 Effective asymptotics

Repeating the arguments of the previous section one can easily prove the following estimate similar to (12.1.38):

---

[10] I.e., we have not yet made the change of coordinates. However, one can easily see that under condition (12.2.25) the eigenvalues of $\mathbf{a}^0$ defined in the previous way are asymptotically close to those under this definition. See also remark 12.2.8.

[11] Surely, in the definition of this coefficient one can replace $\mathbf{a}$ by $\mathbf{a}^0$.

$$(12.2.29)\qquad \Bigl|\int \psi(x')\bigl(\operatorname{tr} e(x,x,\tau)-\sum_{0\le j\le J-1}\varkappa_j(x,\tau)\bigr)_{\tau=\tau'}^{\tau=\tau}dx\Bigr|\le$$
$$C\bar\gamma^{d'-1-\frac{mp}{n}}\tau^{\frac{d-1}{2}+\frac{p}{2n}}+C\min\bigl(\tau^{\frac{d-J}{2}+\frac{p-J}{2n}}\bar\gamma^{d-J-\frac{m}{n}(p-J)},\bar\gamma^{(d-J)(m+1)}\bigr)$$

where $\psi=\psi(x')$ is γ-admissible and supported in $B(\bar x',\frac{1}{2}\bar\gamma)$, $r_1(\tau)\le|\bar x'|\le r_0(\tau)$, $\tau'\in[\epsilon\tau,\tau)$ and the Weylian coefficients $\varkappa_j$ satisfy the inequalities

$$(12.2.30)\qquad \Bigl|\int\psi(x')\varkappa_j(x,\tau)dx\Bigr|\le C\min\bigl(\tau^{\frac{d-j}{2}+\frac{p-j}{2n}}\bar\gamma^{d-j-\frac{m}{n}(p-j)},\bar\gamma^{(d-j)(m+1)}\bigr)$$

and

$$(12.2.31)\qquad \varkappa_0=(2\pi)^{-d}\omega_d(\tau-V)_+^{\frac{d}{2}}\sqrt{g},\qquad \varkappa_1=0.$$

Remark 12.2.9. *I would like to note that in the previous section under appropriate restrictions on the derivatives of a_α with respect to x'' one could modify the estimates for $\varkappa_j$ in this way. However, we could not modify the estimates for $\tilde\varkappa_j$ and (12.1.38) because the boundary is shrinking.* ◊

Let A^0 be the same type of operator and let the following conditions be fulfilled:

$$(12.2.32)_1\qquad D^\alpha(g^{jk}-g^{jk0})=O\bigl(|x'|^{-\nu-(\mu-|\alpha''|)_--|\alpha'|}|x''|^{(\mu-|\alpha''|)_+}\bigr),$$
$$(12.2.32)_2\qquad D^\alpha(V_j-V_j^0)=O\bigl(|x'|^{m-n-\nu-(\mu+n-|\alpha''|)_--|\alpha'|}|x''|^{(\mu+n-|\alpha''|)_+}\bigr),$$
$$(12.2.32)_3\qquad D^\alpha(V-V^0)=O\bigl(|x'|^{2m-2n-\nu-(\mu+2n-|\alpha''|)_--|\alpha'|}|x''|^{(\mu+2n-|\alpha''|)_+}\bigr)$$
$$\text{for } |x''|\le|x'|^{\delta-(m-n)(n+1)^{-1}}$$

with

$$(12.2.33)\qquad \nu\ge\mu\ge0,\qquad \mu+\nu>0.$$

Then the following estimate similar to $(12.1.49)_2$ can be proven in the same way:

$$(12.2.34)\qquad \Bigl|\int\Bigl(\bigl(e(x,x,\tau)-e^0(x,x,\tau)\bigr)-\sum_{0\le j\le J-1}\psi(x')\bigl(\varkappa_j-\varkappa_j^0\bigr)\Bigr)_{\tau=\tau'}^{\tau=\tau}dx\Bigr|\le$$
$$C\bar\gamma^{d-1-\frac{mp}{n}}\tau^{\frac{d-1}{2}+\frac{p}{2n}}+C\bar\gamma^{d-J+\mu-\nu-\frac{m}{n}(p-J+\mu)}\tau^{\frac{d-J}{2}+\frac{p-J+\mu}{2n}}$$

where

$$(12.2.35)\qquad \Bigl|\int\bigl(\varkappa_j-\varkappa_j^0\bigr)\psi dx\Bigr|\le C\bar\gamma^{d-j+\mu-\nu-\frac{m}{n}(p-j+\mu)}\tau^{\frac{d-j}{2}+\frac{1}{2n}(p-j+\mu)}.$$

Remark 12.2.10. *(i) The above arguments yield that for $g^{jk}=g^{jk0}$ one can weaken condition (12.2.33) and replace it by*

$$(12.2.33)'\qquad \mu+n\ge0,\quad \nu\ge0,\quad \mu+\nu+n>0$$

and for $g^{jk}=g^{jk0}$, $V_j=V_j^0$ one can replace it by

$$(12.2.33)''\qquad \mu+2n\ge0,\quad \nu\ge0,\quad \mu+2\nu+n>0.$$

*(ii) Moreover, one can easily see that if **either** $\mu=0$ **or** $g^{jk}=g^{jk0}$ and $\mu=-n$ **or** $g^{jk}=g^{jk0}$, $V_j=V_j^0$ and $\mu=-2n$ then one can replace the second term in the right-hand expression of (12.2.34) by*

$$C\bar\gamma^{(d-J)(m+1)-2m-\nu}\tau \text{ or } C\bar\gamma^{(d-J)(m+1)-m-n-\nu}\tau^{\frac{1}{2}} \text{ or } C\bar\gamma^{(d-J)(m+1)-2n-\nu}$$

respectively and one can also replace the right-hand expression of (12.2.35) by the above expression with j instead of J. ◊

Summing the right-hand expression of (12.2.34) with $J \geq 2$ over a γ-admissible partition of unity in the domain $r_1(\tau) \leq |x'| \leq r_0(\tau)$ we obtain the remainder estimate $O\bigl(R(\tau) + \tau^L\bigr)$ with

$$L = \frac{(d-\mu)}{2} + \frac{(d-\nu)(n+1) - p(m+1)}{2(m-n)}. \tag{12.2.36}$$

Moreover, one can drop all terms with $j \geq 2$ and the term with $j = 1$ vanishes. Then we obtain the following statement similar to theorem 12.1.9:

Theorem 12.2.11. *(i) Suppose conditions* ($\Sigma 0$), ($\Sigma 2$), (12.2.1)–(12.2.4), (12.2.13), (12.2.19) *and* $(12.2.32)_{1-3}$, (12.2.33) *are fulfilled. Then the asymptotics*

$$\mathbf{N}(\tau) = \tilde{\mathcal{N}}(\tau) + O\bigl(R(\tau) + \tau^L\bigr) \tag{12.2.37}$$

hold with $R(\tau)$, L *given by* (12.2.22) *and* (12.2.36) *and*

$$\begin{aligned}\tilde{\mathcal{N}}(\tau) = &\int \phi(\frac{x'}{t_1})\varkappa_0(x,\tau)dx + \\ &\int \bigl(1-\phi(\frac{x'}{t_1})\bigr)\bigl(\varkappa_0(x,\tau) - \varkappa_0^0(x,\tau)\bigr)dx + \\ (2\pi)^{p-d} &\iint \bigl(1-\phi(\frac{x'}{t_1})\bigr)\mathbf{n}^0(x',\xi',\tau)dx'd\xi'\end{aligned} \tag{12.2.38}$$

where $\mathbf{n}^0(x',\xi',\tau)$ *is the eigenvalue counting function for* $\mathbf{a}^0(x',\xi')$ *in* $\mathbb{H}$, *the function* ϕ *is introduced above and* $c_0 \leq t_1\tau^{-\frac{1}{2m}} \leq \tau^\delta$ *with a small enough exponent* $\delta > 0$.

(ii) Moreover, let $(d-1)n < pm$, $(d-1)(m+1) > 2mL$ *and let stabilization conditions* (10.4.3) *and the standard condition* (Ψ) *on Hamiltonian trajectories of the quasi-homogeneous Hamiltonian be fulfilled. Then the asymptotics*

$$\mathbf{N}(\tau) = \tilde{\mathcal{N}}(\tau) + o\bigl(\tau^{\frac{1}{2m}(d-1)(m+1)}\bigr) \tag{12.2.39}$$

hold.

(iii) On the other hand, let $(d-1)n > pm$, $(d-p-1)(m+1) > 2mL$ *and let stabilization conditions* $(12.2.25)_{1,2}$ *be fulfilled. Moreover, let the eigenvalues* $\zeta_j(x',\xi')$ *of the operator* (12.2.26) *satisfy condition* (12.1.28) *on Hamiltonian trajectories. Then the asymptotics*

$$\mathbf{N}(\tau) = \tilde{\mathcal{N}}(\tau) + \bigl(\varkappa_1^* + o(1)\bigr)\tau^{\frac{1}{2}(d-p-1)(m+1)(m-n)^{-1}}\bigr) \tag{12.2.40}$$

hold with the coefficient $\varkappa_1^*$ *given in theorem 4.3.6.*

(iv) Remark 12.2.10 remains true.

Remark 12.2.12. *(i) Recall that the operator* $\mathbf{a}^0$ *due to* (12.2.32) *and used in calculation of* $\mathbf{n}^0$ *and* $\varkappa_0^0$ *does not necessarily coincide with the operator denoted in the same way (i.e.,* $\mathbf{a}^0$*) and used for the calculation of* ζ_j *and* $\varkappa_1^*$.

(ii) One can replace the right-hand expressions of $(12.2.32)_{1-3}$ *by sums of the same type of terms with different* μ_k, ν_k. *Then one should set* $L = \max_k L_k$ *where* L_k *is calculated by (12.2.36) with* μ_k, ν_k *instead of* μ, ν.

(iii) I leave the trivial generalization to the case when X *is an exterior domain to the reader. Another trivial generalization is to replace a single degeneration manifold* Λ *by a union of disjoint manifolds* Λ_j *of different dimensions and with different exponents* n_j *and* μ_j, ν_j.

(iv) I leave the simple generalization to the case when X *is a conical domain for* $|x| \geq c$ *to the reader. In this case one should assume that either* $\Lambda \cap \partial X = \emptyset$ *or* $\Lambda \subset \partial X$;

if Λ is a union of disjoint connected manifolds Λ_j then each manifold should satisfy this condition.

(v) In the case when the limit operator $\mathbf{a}^0$ due to $(12.2.32)_{1-3}$ is

$$\mathbf{a}^0 = \sum_{j,k\le d-p} g^{jk0}(x')\xi_j\xi_k + \sum_{j,k\ge d-p+1} D_j g^{jk0}(x) D_k + W^0 \tag{12.2.41}$$

where W^0 satisfies (12.2.25) then

$$\int \mathbf{n}^0(x',\xi',\tau)d\xi' = \sqrt{g'}\sum_l \bigl(\tau - \omega_l(x')\bigr)^{\frac{d-p}{2}} \tag{12.2.42}$$

where $g' = \det(g^{jk0})^{-1}_{j,k=1,\dots,d-p}$, $\omega_l(x')$ are eigenvalues of the operator

$$\sum_{j,k>p} D_j g^{jk}(x) D_k + W_0(x)$$

in $\mathbb{R}^p$ (if $\Lambda\subset\partial X$ then here and elsewhere one should consider operators in the half-space). In the case when g^{jk0} are positively homogeneous of degree 0 on x these eigenvalues ω_l are positively homogeneous of degree $2(m-n)(n+1)^{-1}$ on x'. ◊

Surely, the reader can attack more general operators than the Schrödinger operator treated here.

12.3 Spectral Asymptotics for Maximally Hypoelliptic Operators

12.3.1 Maximally hypoelliptic operators

In this section we consider a class of pseudo-differential operators on a compact closed $C^{K'}$ manifold X degenerating on a symplectic manifold $\Lambda\subset T^*X\setminus 0$ conical in ξ. The same results hold with no change for operators acting in Hermitian bundles. Moreover, similar results hold after an obvious modification for differential operators on manifolds with boundary provided $\Lambda|_{\partial X}=\emptyset$. I leave these modification to the reader. So, let us assume that

(12.3.1) X is a compact $C^{K'}$ manifold, $\Lambda\subset T^*X\setminus 0$ is a symplectic manifold conic in ξ, $d=\dim X$, $2d'=\dim\Lambda$, $1\le d'<d$, $d''=d-d'$.

Let $\Psi^{m,\mu}(X,\Lambda,\mathbb{K})$ with $m\in\mathbb{R}$, $\mu\in\mathbb{Z}^+$ denote the class of pseudo-differential operators from $H^s(X,\mathbb{K})$ to $H^{s-m}(X,\mathbb{K})$ [12)] with symbols in every local chart of the form

$$a = \sum_{0\le n\le K-1} a_{\frac{n}{2}} + a'_{\frac{K}{2}} \tag{12.3.2}$$

where $a_{\frac{n}{2}}$ are $C^{K'-n}$ symbols positively homogeneous of degrees $m-\frac{n}{2}$ in ξ and such that

$$a^{(\alpha)}_{\frac{n}{2}(\beta)}(z) = 0 \qquad \forall z\in\Lambda\ \forall\alpha,\beta: |\alpha|+|\beta|+n<\mu \tag{12.3.3}$$

and the remainder $a'_{\frac{K}{2}}$ is assumed to be smooth enough and decrease quickly enough as $|\xi|\to\infty$. Here and below $K'>2K$ and K are large enough.

12) With $s\in(-\bar{s},\bar{s})$ and with $\bar{s}=\bar{s}(K)\to+\infty$ as $K\to+\infty$. This remark holds until the end of this section.

It is well known ([Boutet de Monvel 1]; see also section 1.2) that these classes are microlocally invariant. In an appropriate microlocal symplectic coordinate system on $T^*X\setminus 0$ we have $\Lambda=\{x''=\xi''=0\}$ where $x=(x';x'')=(x_1,\dots,x_{d'};x_{d'+1},\dots,x_d)$ and similar notation is applied to ξ. Then for the operator $A\in\Psi^{m,\mu}(X,\Lambda,\mathbb{K})$ and for $\mathrm{z}\in\Lambda$ let us introduce the operator

$$A_{\mathrm{z}}(x'',D'')=\sum_{|\alpha|+|\beta|+n=\mu}\frac{1}{\alpha!\beta!}a^{(\alpha)}_{\frac{n}{2}\,(\beta)}(\mathrm{z})x''^{\beta}D''^{\alpha} \tag{12.3.4}$$

acting in the space $L^2(\mathbb{R}^{d''},\mathbb{K})$.[13] This operator is called the *microlocalization* (of order μ) of the operator A at the point $\mathrm{z}\in\Lambda$. It is well known ([Boutet de Monvel 1]; see also section 1.2) that in another microlocal coordinate system with the same form of Λ we obtain the microlocalization $A'_{\mathrm{z}}=Q^*_{\mathrm{z}}A_{\mathrm{z}}Q_{\mathrm{z}}$ where Q_{z} is a unitary metaplectic [Leray 1] operator in $L^2(\mathbb{R}^{d''})$ also acting in $L^2(\mathbb{R}^{d''},\mathbb{K})=L^2(\mathbb{R}^{d''})\otimes\mathbb{K}$.

One says that the operator $A\in\Psi^{m,\mu}(X,\Lambda,\mathbb{K})$ is *maximally hypoelliptic* if and only if the following two conditions are fulfilled: the inequality

$$|a_0(x,\xi)v|\ge\epsilon_0\varrho^{\mu}(x,\xi)|\xi|^m|v|\qquad\forall(x,\xi)\in T^*X\quad\forall v\in\mathbb{K} \tag{12.3.5}$$

holds with a constant $\epsilon_0>0$ and with $\varrho(x,\xi)=\operatorname{dist}\bigl((x,\xi|\xi|^{-1}),\Lambda\bigr)$ and

(12.3.6) The operator A_{z} is injective in $\mathbb{H}=L^2(\mathbb{R}^{d''},\mathbb{K})$ at every point $\mathrm{z}\in\Lambda$. In this case

$$\|A_{\mathrm{z}}v\|\ge\epsilon_0\|v\|\qquad\forall v\in\mathbb{H}\quad\forall\mathrm{z}\in\Lambda\cap\{|\xi'|=1\}$$

with an appropriate constant $\epsilon_0>0$. Here and below $|.|$, $\|.\|$ and $\|.\|$ are the norms in $\mathbb{K}$, $\mathbb{H}$ and $L^2(X,\mathbb{K})$ respectively and $\langle.,.\rangle_{\mathbb{K}}$, $\langle.,.\rangle_{\mathbb{H}}$ and $(.,.)$ are the corresponding inner products.

It is well known [Boutet de Monvel 1] that every maximally hypoelliptic operator is microlocally hypoelliptic; more precisely, the operator $A\in\Psi^{m,\mu}(X,\Lambda,\mathbb{K})$ satisfies

$$\mathrm{WF}^{s+l}(u)\subset\mathrm{WF}^{s}(Au)\qquad\forall u\in\mathcal{D}'(X,\mathbb{K}) \tag{12.3.7}$$

with $l=m-\frac{\mu}{2}$ if and only if it is maximally hypoelliptic; moreover, in this case the resolvent of A is a bounded operator from $H^s(X,\mathbb{K})$ to $H^{s+l}(X,\mathbb{K})$.

In this section we treat a maximally hypoelliptic operator A with

$$l=m-\frac{\mu}{2}>0. \tag{12.3.8}$$

Then the resolvent of A is a compact operator in $L^2(X,\mathbb{K})$ and the spectrum of A is discrete with no accumulation point (provided A is self-adjoint).[14]

It is easy to prove that if the Weylian symbol of A is Hermitian at every point $z\in T^*X$ then the Weylian symbol of A_{z} is Hermitian at every point $z''\in T^*(\mathbb{R}^{d''})$ and therefore A_{z} is obviously symmetric in $\mathbb{H}$. Moreover, under condition (12.3.5) A_{z} is an elliptic operator and is therefore self-adjoint. On the other hand, it is well known that a symmetric maximally hypoelliptic operator satisfying (12.3.8) is self-adjoint. Moreover,

---

[13] A Hermitian bundle $\mathbb{K}$ should be replaced by its fiber $\mathbb{K}_{\mathrm{z}}$ over z in this definition.

[14] One can easily prove that this is wrong for $l\le 0$; moreover, for $l<0, m>0$ the continuous spectrum fills either the semiaxis or axis, while for $l=0, m>0$ and $l<0, m=0$ the spectrum is more complicated: there is an infinite number of accumulation points or points of the infinite multiplicity point spectrum (provided the operator is self-adjoint); for $m<0$ the structure of the spectrum is transparent.

a self-adjoint maximally hypoelliptic operator satisfying (12.3.8) is semibounded from above if and only if both of the following two conditions are fulfilled:

(12.3.9) $a_0(z)$ is a negative definite matrix in $\mathbb{K}$ at every point $z \in (T^*X \setminus 0) \setminus \Lambda$;

(12.3.10) A_z is a negative definite operator in $\mathbb{H}$ at every point $z \in \Lambda$.

12.3.2 Getting started

So, let $A \in \Psi^{m,\mu}(X, \Lambda, \mathbb{K})$ be a maximally hypoelliptic and self-adjoint operator and let (12.3.8) be fulfilled. Let $\mathbf{N}^{\pm}(\tau)$ be the number of eigenvalues of A lying in the intervals $[0, \tau)$ and $(-\tau, 0]$ respectively. We are interested in the asymptotics of $\mathbf{N}^{+}(\tau)$ as $\tau \to +\infty$. One can obtain the asymptotics of $\mathbf{N}^{-}(\tau)$ by replacing A by $-A$.

First of all let us rewrite $\mathbf{N}^{\pm}(\tau)$ in the form

$$\mathbf{N}^{\pm}(\tau) = \mathbf{N}_0^{\pm}(\tau) + \mathbf{N}_1^{\pm}(\tau), \qquad \mathbf{N}_j^{\pm}(\tau) = \operatorname{tr} \int (\psi_j e)(x, x,)dx \tag{12.3.11}$$

where $e(x, y, \tau)$ is the Schwartz kernel of the spectral projector of A and ψ_j are classical pseudo-differential operators of order 0,

$$\operatorname{supp} \psi_0 \cap \Lambda = \emptyset, \qquad \operatorname{supp} \psi_1 \subset \mathcal{Z} = \{\varrho(x, \xi) < \varrho_0\} \tag{12.3.12}$$

where $\varrho_0 > 0$ is an arbitrarily small constant.

Then for $\mathbf{N}_0^{\pm}(\tau)$ the standard Weylian asymptotics hold with principal part

$$\kappa_0 \tau^{\frac{1}{m}(d-1)} + \kappa_{\frac{1}{2}} \tau^{\frac{1}{m}(d-\frac{1}{2})} + \kappa_1 \tau^{\frac{1}{m}(d-1)}$$

and with remainder estimate $O\bigl(\tau^{\frac{1}{m}(d-1)}\bigr)$ and even $o\bigl(\tau^{\frac{1}{m}(d-1)}\bigr)$ under the standard condition ($\Psi 0$) on the Hamiltonian trajectories of a_0. Here $\kappa_{\frac{1}{2}} = 0$ provided $a_{\frac{1}{2}} = 0$. Therefore we should treat $\mathbf{N}_1^{\pm}(\tau)$. First of all we need the following

Proposition 12.3.1. *Let $A \in \Psi^{m,\mu}(X, \Lambda, \mathbb{K})$ be a self-adjoint operator.*

(i) Let $\phi \in C_0^{K'}(\mathbb{R}^{d''})$ be supported in $B(0, \epsilon)$ with a small enough constant $\epsilon > 0$. Then the estimate

$$\bigl|\bigl(\phi(\tau^{-\frac{1}{m}} D')\psi_1 e\bigr)(x, y, \tau', \tau)\bigr| \le C\tau^{-s} \tag{12.3.13}$$

holds with arbitrarily large s for $\tau > 0$, $\tau' \in [\epsilon_0 \tau, \tau]$.

(ii) Let A be maximally hypoelliptic and let $1 - \phi \in C_0^{K'}(\mathbb{R}^{d'})$; let ϕ be supported in $\{|\xi'| \ge C_0\}$ with a large enough constant C_0. Then the estimate

$$\bigl|\bigl(\phi(\tau^{-\frac{1}{l}} D')\psi_1 e\bigr)(x, y, \tau', \tau)\bigr| \le C\tau^{-s} \tag{12.3.14}$$

holds with arbitrarily large s for $\tau > 0$, $\tau' \in [\epsilon_0 \tau, \tau]$.

(iii) Let $\phi \in C_0^{K'}(\mathbb{R}^{d'})$, $0 \notin \operatorname{supp} \phi$ and $\varphi \in C_0^{K'}(\mathbb{R}^{d''})$. Let $r \ge \tau^{\delta}$, $\varrho_0 \ge \rho \ge r^{-\frac{1}{2}}$ and $\rho^{\mu} r^m \le \varepsilon \tau$ with an arbitrarily small exponent $\delta > 0$ and with a small enough constant $\varepsilon > 0$. Then the estimate

$$\bigl|\bigl(\varphi(\rho^{-1} x'', \rho^{-1} r^{-1} D'')\phi(r^{-1} D')\psi_1 e\bigr)(x, y, \tau', \tau)\bigr| \le C\tau^{-s} \tag{12.3.15}$$

holds with arbitrarily large s for $\tau > 0$, $\tau' \in [\epsilon_0 \tau, \tau]$.

Proof. Statement (i) follows from the ellipticity of $\tau - A$ in the zone $\{|\xi| \le \varepsilon\tau^{\frac{1}{m}}\}$ with a small enough constant $\varepsilon > 0$ and (iii) follows from the ellipticity of $\tau - A$ in the zone $\{\epsilon r \le |\xi'| \le cr, |x''| + |\xi''|r^{-1} \le c\rho\}$ with the indicated r and ρ. On the other hand, (ii) follows from the hypoellipticity of $\tau - A$ in the zone $\{|\xi'| \ge C_0\tau^{\frac{1}{l}}, \varrho(x,\xi) \le \varrho_0\}$. More precisely, the estimate

$$\|Au\|_s + \|u\|_s \ge \epsilon_0 \|A'u\|_s \tag{12.3.16}$$

holds where $\|.\|_s$ means the norm in the Sobolev space $H^s(X,\mathbb{K})$ and A' means any operator belonging to $\Psi^{m,\mu}(X,\Lambda,\mathbb{K})$ [Boutet de Monvel 1]. In particular,

$$\|Au\|_s + \|u\|_s \ge \epsilon_0 \|u\|_{s+l} \tag{12.3.17}$$

and estimate (12.3.16) yields (ii). □

Therefore, we need to treat the zone

$$\mathcal{U} = \{\varrho(x,\xi) \le \varrho_0, \epsilon_0\tau^{\frac{1}{m}} \stackrel{\text{def}}{=} r_0(\tau) \le |\xi'| \le r_1(\tau) \stackrel{\text{def}}{=} C_0\tau^{\frac{1}{l}},$$
$$\epsilon_0 \le \varrho(x,\xi)|\xi'|^{-\frac{m}{\mu}}\tau^{\frac{1}{\mu}}\}.$$

In this zone let us make a $(\epsilon_1, \epsilon_1|\xi'|; \epsilon_1\varrho(x,\xi), \epsilon_1\varrho(x,\xi)|\xi'|)$-admissible partition of unity on $(x',\xi';x'',\xi'')$ with a small enough constant $\epsilon_1 > 0$ and let us treat some element $\bar q$ of this partition; let $\bar r$ and $\bar\varrho$ be the values of $|\xi'|$ and ϱ calculated in $\bar z \in \bar q$; then

$$\epsilon_0\tau \le \bar r^m \bar\varrho\mu \le c_0\tau, \qquad r_0(\tau) \le \bar r \le r_1(\tau). \tag{12.3.18}$$

12.3.3 Weylian asymptotics

Let us start with the completely Weylian theory. Let us consider an $(\epsilon_1, \epsilon_1\varrho r; \epsilon_1\varrho^2 r, \epsilon_1\varrho r)$-admissible partition of unity on $(x',\xi';x'',\xi'')$. Then making the dilatation $x''_{\text{new}} = x''\bar\varrho^{-1}$ we transform A to an h-pseudo-differential operator $\bar r^m\bar\varrho^\mu\tilde A_h$ provided $h = \bar r^{-1}\bar\varrho^{-2} \le h_0$. Then the contribution of this element to the principal part of the asymptotics is 0 and its contribution to the remainder estimate is $O(h^s)$ with an arbitrarily large exponent s provided one of the following two conditions is fulfilled:

(i) a_0 is negative definite on $T^*X \setminus 0$;

(ii) $\bar r^m\bar\varrho^\mu \ge C_0\tau$.

Then we easily see that the total contribution of all these elements of the partition lying in $\mathcal{U}$ to the principal part of the Weylian asymptotics is 0 and to the remainder estimate is $O(\tau^{d'l^{-1}})$. Therefore, *if A is maximally hypoelliptic and a_0 is negative definite then the contribution of $\complement\mathcal{V}$ to the asymptotics is $O(\tau^{d'l^{-1}})$ where $\mathcal{V} = \{|\xi| \ge c_0, \varrho(x,\xi) \le \varrho_0, \varrho^2(x,\xi)|\xi'|^2 \le c_0\}$. In the general case, the contribution of the zone $\{|\xi| \ge c_0, |\xi'|^m\varrho(x,\xi)^\mu \ge C_0\tau\}$ to the asymptotics is $O(\tau^{d'l^{-1}})$ as well.* On the other hand, let us consider the zone $\mathcal{V}$. We should consider only elements with $r_0(\tau) \le |\xi'| \le r_1(\tau)$. Applying the previous arguments with ϱ replaced by $\rho(\xi') = |\xi'|^{-\frac{1}{2}}$ and $h = 1$ we see *that the total contribution of $\mathcal{V}$ to the asymptotics is $O(\tau^{d'l^{-1}})$ as well.*

So we obtain as a corollary the following rather well known estimate

$$\mathbf{N}^+(\tau) = O(\tau^{\frac{1}{l}d'}) \qquad \text{as } \tau \to +\infty \tag{12.3.19}$$

for maximally hypoelliptic operators with negative definite symbols.

Let us consider the general case and let us treat the elements of the above partition with $r\varrho^2 \ge C_0$, $r \ge c_0$, $r^m\varrho^\mu \asymp \tau$. Let us note that after the coordinate transform $x''_{\text{new}} =$

$x''\bar{\varrho}^{-1}$ and multiplication by τ^{-1} the operator A is microhyperbolic in the direction $\langle x, \partial_\xi\rangle$ provided C_0 is large enough. Then the contribution of each element to the principal part is

$$(2\pi)^{-d}\iint \bar{q}_1^w(x,\xi)n^+(x,\xi,\tau)dxd\xi$$

where $n^+(x,\xi,\tau)$ is the number of eigenvalues of the symbol $a_0(x,\xi)+a_{\frac{1}{2}}(x,\xi)$ in the interval $[0,\tau)$; one can introduce $n^-(x,\xi,\tau)$ and $n(x,\xi,\tau)$ in the same way. On the other hand, the contribution of this element to the remainder estimate is $O(h^{1-d}) = O(\bar{r}^{d-1}\bar{\varrho}^{2(d-1)})$. Then the contribution of the element of a $(\epsilon_1, \epsilon_1|\xi'|; \epsilon_1\varrho, \epsilon_1\varrho|\xi'|)$-admissible partition to the remainder estimate is

$$O(\bar{r}^{d-1}\bar{\varrho}^{2(d''-1)}) = O(\bar{r}^{d-1-2(d''-1)\frac{m}{\mu}}\tau^{\frac{2}{\mu}(d''-1)})$$

and after summation on this partition of unity we obtain the remainder estimate

$$R^w(\tau) = \begin{cases} O(\epsilon'\tau^{\frac{1}{m}(d-1)}) & \text{for } (d-1)l > d'm \\ O(\tau^{\frac{1}{m}(d-1)}\log\tau) & \text{for } (d-1)l = d'm \\ O(\tau^{\frac{1}{l}d'}) & \text{for } (d-1)l < d'm \end{cases} \tag{12.3.20}$$

where $\epsilon' = \epsilon'(\varrho_0) \to +0$ as $\varrho_0 \to +0$.

So (also taking into account the estimates for the other zones) we obtain

Theorem 12.3.2. *Let $A \in \Psi^{m,\mu}(X, \Lambda, \mathbb{K})$ be a maximally hypoelliptic (self-adjoint) operator, $l > 0$ and suppose a_0 is not a negative definite symbol. Then*

(i) For $(d-1)l > d'm$ the asymptotics

$$\mathbf{N}^+(\tau) = \mathcal{N}^{w+}(\tau) + O(\tau^{\frac{d-1}{m}}) \tag{12.3.21}$$

hold with

$$\mathcal{N}^{w+}(\tau) = (2\pi)^{-d}\iint n^+(x,\xi,\tau)dxd\xi. \tag{12.3.22}$$

(ii) Moreover, if $(d-1)l > d'm$ and if the symbol a_0 satisfies the standard condition $(\Psi 0)$ at the energy level $\tau = 1$ then the asymptotics

$$\mathbf{N}^+(\tau) = \mathcal{N}^{w+}(\tau) + \kappa_1'\tau^{\frac{d-1}{m}} + o(\tau^{\frac{d-1}{m}}) \tag{12.3.22}$$

hold with the standard coefficient κ_1'.

(iii) For $(d-1)l = d'm$ the asymptotics

$$\mathbf{N}^+(\tau) = \mathcal{N}^{w+}(\tau) + O(\tau^{\frac{d-1}{m}}\log\tau) \tag{12.3.23}$$

hold.

(iv) For $(d-1)l < d'm \le dl$ the asymptotics

$$\mathbf{N}^+(\tau) = \mathcal{N}^{w+}(\tau) + O(\tau^{\frac{d'}{l}}) \tag{12.3.24}$$

hold where for $d'm = dl$ we integrate only in the domain $\{|\xi'| \ge \tau^{\frac{1}{l}}, \varrho(x,\xi) \ge |\xi'|^{-\frac{1}{2}}\}$ in (12.3.22).

(v) For $dl < d'm$ the estimate

$$\mathbf{N}^+(\tau) = O\big(\tau^{\frac{d'}{m'l}}\big) \tag{12.3.25}$$

holds. Moreover, the same estimate holds if $a_0(x,\xi)$ is negative definite.

(vi) Moreover, in all these asymptotics one can replace $\mathcal{N}^{w+}(\tau)$ by

$$\mathcal{N}_1^{w+}(\tau) = \begin{cases} \kappa_0\tau^{\frac{1}{m}d} + \kappa_{\frac{1}{2}}\tau^{\frac{1}{m}(d-\frac{1}{2})} + \kappa_1''\tau^{\frac{1}{m}(d-1)} & \text{for } (d-1)l > d'm \\ \kappa_0\tau^{\frac{1}{m}d} + \kappa_{\frac{1}{2}}\tau^{\frac{1}{m}(d-\frac{1}{2})} & \text{for } (d-\frac{1}{2})l > d'm \ge (d-1)m \\ \kappa_0\tau^{\frac{1}{m}d} + \kappa'_{\frac{1}{2}}\tau^{\frac{1}{m}(d-\frac{1}{2})}\log\tau & \text{for } (d-\frac{1}{2})l = d'm \\ \kappa_0\tau^{\frac{1}{m}d} & \text{for } (d-\frac{1}{2})l < d'm < (d-1)m < dl \\ \kappa'_0\tau^{\frac{1}{m}d}\log\tau & \text{for } d'm < (d-1)m = dl \end{cases} \tag{12.3.26}$$

with positive constants κ_0 and κ'_0. Moreover, $\kappa_{\frac{1}{2}} = \kappa'_{\frac{1}{2}} = 0$ provided $a_{\frac{1}{2}} \equiv 0$.

Remark 12.3.3. *(i) Similarly, for Riesz means the remainder estimate is*

$$R^w(\tau) = \begin{cases} O\big(\epsilon'\tau^{\frac{1}{m}(d-1-\vartheta)+\vartheta}\big) & \text{for } (d-1-\vartheta)l > d'm \\ O\big(\tau^{\frac{1}{m}(d-1-\vartheta)+\vartheta}\log\tau\big) & \text{for } (d-1-\vartheta)l = d'm \\ O\big(\tau^{\frac{1}{l}d'+\vartheta}\big) & \text{for } (d-1-\vartheta)l < d'm \end{cases} \tag{12.3.20$'$}$$

while the principal part contains the terms $\kappa_j\tau^{\frac{1}{m}(d-j)+\vartheta}$ with $j \in \frac{1}{2}\mathbb{Z}$; for $(d-j)l = d'm$ one should replace this term by $\kappa'_j\tau^{\frac{1}{m}(d-j)+\vartheta}\log\tau$ while the remainder estimate is $O\big(\tau^{\frac{1}{m}(d-j)+\vartheta}\big)$ (if $j < 1+\vartheta$).

(ii) Moreover, let $k \notin \mathbb{Z}$ and $a_j \equiv 0$ for all $j \notin \mathbb{Z} : j \le k$. Then $\kappa_k = 0$. ◊

So, in what follows we only treat the case $(d-1)l \le d'm$ ($(d-1-\vartheta)l \le d'm$ if we are interested in Riesz means).

12.3.4 Accurate asymptotics

Let us treat x'' as an additional variable; however, we do not now introduce the special Hilbert space $\mathbb{H}$. Let us note that standard ellipticity arguments yield the following

Proposition 12.3.4. *Let $q' \in C_0^{K'}(\mathbb{R}^{2d'})$, $q'' \in C_0^{K'}(\mathbb{R}^{2d''})$ be fixed functions vanishing in neighborhoods of $\{\xi' = 0\}$, $\{x'' = \xi'' = 0\}$ respectively. Moreover, let $\rho > 0$ be a small enough constant and suppose $\bar{r} \ge c_\rho\tau^{\frac{1}{m}}$ where ρ and c_ρ depend on q', q'' and c_ρ is large enough. Then for the operators $Q' = q'(x', \bar{r}^{-1}D')$ and $Q'' = q''(\rho^{-1}x'', \rho^{-1}\bar{r}^{-1}D'')$ the estimate*

$$\big|F_{t\to\lambda}\chi_T(t)Q'_x\psi_1 Q''_x U\big| \le C\bar{h}^s \qquad \forall x,y \in X \quad \forall\lambda : |\lambda-\tau| \le \epsilon\tau \quad \forall T \in [\bar{r}^\delta\tau^{-1}, T_0\bar{r}\tau^{-1}] \tag{12.3.27}$$

holds with an arbitrarily small exponent $\delta > 0$, an arbitrarily large exponent s and with small enough constants $T_0 > 0$ and $\epsilon > 0$; here and below $U(x,y,t)$ is the Schwartz kernel of the operator $\exp(itA)$.

Recall that $\bar{h} = \bar{r}^{-1}$.

This proposition permits us to include the x''-mollifying operator $q(x'', \bar{r}^{-1}D'')$ with compactly supported q in our expressions whenever necessary.

Let us note that if $\phi = \phi(x', \bar{h}D')$ then in the ϵ-neighborhood of the point $(\bar{x}', \bar{\xi}')$ with $|\bar{\xi}'| \asymp r$ the operator $\bar{r}[A, \phi]$ belongs to $\boldsymbol{\Psi}^{m,\mu}(X, \Lambda, \mathbb{K})$ uniformly in $(\bar{x}', \bar{\xi}')$. Let us apply the standard energy estimate technique of Chapter 2 and take into account estimate (12.3.16) and proposition 12.3.3. Then one can easily prove the following statement (details are left to the reader):

Proposition 12.3.5. *Let $q_1', q_2' \in C_0^{K'}(\mathbb{R}^{2d'})$ be fixed functions and let q' vanish in a neighborhood of $\{\xi' = 0\}$, $q_2' = 1$ in a neighborhood of* $\operatorname{supp} q_1'$. *Then for the operators $Q_j' = q_j'(x', \bar{r}^{-1}D')$ the estimate*

$$(12.3.27) \quad \left|F_{t\to\lambda}\psi_1\chi_T(t)Q'_{1x}(I - Q'_{2x})U\right| \le C\bar{h}^s \qquad \forall x, y \in X \quad \forall \lambda : |\lambda - \tau| \le \epsilon\tau \quad \forall T \in [\bar{r}^{\delta}\tau^{-1}, T_0\bar{r}\tau^{-1}]$$

holds with an arbitrarily small exponent $\delta > 0$, an arbitrarily large exponent s and with small enough constants $T_0 > 0$ and $\epsilon > 0$.

This proposition shows us that singularities with respect to (x', ξ') of U propagate with velocity $(\bar{r}^{-1}\tau, \tau)$ along (x', ξ') (without rescaling on ξ') while in the zone which is semiclassical with respect to all the variables singularities propagate along (x'', ξ'') with velocity $\bar{\varrho}^{-1}(\bar{r}^{-1}\tau, \tau)$.

Let us consider the $(\epsilon, \epsilon\bar{r})$-neighborhood of $(\bar{x}', \bar{\xi}')$ with $|\bar{\xi}'| = \bar{r}$. Let $\mathbf{a} = A(\bar{x}', x'', \bar{\xi}', D'')$ be an operator acting on x'' only. I assert that *for the operator $\mathbf{a}' = \langle \bar{\xi}', \partial_{\xi'} \rangle \mathbf{a}$ the microhyperbolicity estimate*

$$(12.3.29) \qquad \operatorname{Re}\langle \mathbf{a}'v, v\rangle \ge \epsilon|\tau| \cdot \|v\|^2 - c_2|\tau|^{-1}\|(\tau - \mathbf{a})v\|^2 - C\bar{r}^{-s}\|w\|^2$$

holds for $v = \psi_1 w$ where norms and inner products are here considered only on x'', (x', ξ') belonging to this neighborhood, and $\tau > 0$ is such that $r_0(\tau) \le \bar{r} \le r_1(\tau)$; here $\epsilon > 0$ is a small enough constant and c_2 and C are large enough constants; none of the constants depend on $(\bar{x}', \bar{\xi}')$ and τ and s is an arbitrarily large exponent. In fact, it is sufficient to prove this estimate for $\xi' = \bar{\xi}'$. Let us apply Taylor's formula to A; we obtain that

$$\mathbf{a}(x', \xi'; x'', D'') \equiv \sum_{|\alpha|+|\beta|+n=\mu} a^{(\alpha)}_{\frac{n}{2}(\beta)}(x', \xi')\frac{1}{\alpha!\beta!}x''^{\beta}D''^{\alpha} \qquad (\operatorname{mod} \boldsymbol{\Psi}^{m,\mu+1})$$

where here and below we take $x'' = \xi'' = 0$ in $a^{(\alpha)}_{\frac{n}{2}(\beta)}$. Then for $\xi' = \bar{\xi}'$ we obtain the equality

$$\mathbf{a}'(x', \xi'; x'', D'') \equiv \sum_{|\alpha|+|\beta|+n=\mu} (m - n - |\alpha|)a^{(\alpha)}_{\frac{n}{2}(\beta)}(x', \xi')\frac{1}{\alpha!\beta!}x''^{\beta}D''\alpha \equiv l\mathbf{a} - i[L, \mathbf{a}] \qquad (\operatorname{mod} \boldsymbol{\Psi}^{m,\mu+1})$$

with $L = \frac{1}{2}\langle x'', D''\rangle - \frac{1}{4}id''$. Then estimate (12.3.27) together with the assumption that $l > 0$ yields estimate (12.3.29).

Again applying the analysis of Chapter 2 we obtain that *in the context of proposition 12.3.4 the estimate*

$$|F_{t\to\lambda}\Gamma'_x\psi_1\chi_T(t)Q'_{1x}U|\le C\bar{r}^{-s} \tag{12.3.30}$$

holds for x', y', τ and T indicated in this proposition provided χ vanishes in a neighborhood of 0; Γ' means that we set $y'=x'$ and then integrate over x'.

Then we want to apply the successive approximation method but we lack auxiliary spaces $\mathbb{H}$ and $\mathbb{H}_j$. In order to overcome this difficulty we treat two cases:

(i) $\bar{r}\le r_1(\tau)\tau^{-\delta}$,

(ii) $\bar{r}\ge r_0(\tau)\tau^{\delta}$

with $\delta>0$; these cases overlap and together cover the case $\bar{r}\in[r_0(\tau),r_1(\tau)]$.

In case (i) we can apply our standard technique with an unperturbed operator $\bar{A}=a(y,D_x)$ and we obtain the estimate

$$|F_{t\to\lambda}\bar{\chi}_T(t)\operatorname{tr}(\Gamma Q_xU)-\sum_{0\le j\le J-1}\varkappa'_{\frac{j}{2}}(\tau)|\le C\tau^{-s}\qquad\forall\lambda:|\lambda-\tau|\le\epsilon\tau \tag{12.3.31}$$

with sufficiently large $J=J(d,m,\mu,s,\delta,\delta')$ and with arbitrary $T\in[\tau^{-1}\bar{h}'^{-\delta'},T_0\tau^{-1}\bar{h}'^{-1}]$ where $T_0>0$ is a small enough constant, $\delta'>0$ and s are arbitrary and $\bar{h}'=\bar{r}^{-1}\bar{\varrho}^{-2}$. Here $\varkappa'_{\frac{j}{2}}$ are Weylian coefficients depending on Q and satisfying the estimate

$$|\varkappa'_{\frac{j}{2}}|\le C(\bar{r}\bar{\varrho}^2)^{d-\frac{j}{2}}\tau^{-1} \tag{12.3.32}$$

provided the symbol of Q is supported in the $\epsilon(1,\bar{\varrho}^2\bar{r},\bar{\varrho},\bar{\varrho}\bar{r})$-neighborhood of $(\bar{x}',\bar{\xi}',\bar{x}'',\bar{\xi}'')$, $|\bar{\xi}'|\asymp\bar{r}$; here and below we refer to the standard 1-symbol of Q instead of the $\bar{h}$-symbol. These estimates are due to the standard microhyperbolicity of a_0 in the direction $\langle\xi,\partial_\xi\rangle$ after the coordinate transform $x'_{\text{new}}=x'_{\text{old}}$, $\xi'_{\text{new}}=\xi'_{\text{old}}\bar{\varrho}^{-2}$, $x''_{\text{new}}=x''_{\text{old}}\bar{\varrho}^{-1}$, $\xi''_{\text{new}}=\xi''_{\text{old}}\bar{\varrho}^{-1}$. It should be noted that in case (i) we automatically obtain that only points with $\bar{h}'\le\tau^{-\delta''}$ with sufficiently small exponent $\delta''>0$ should be treated. Then for the operator Q with symbol supported in the $\epsilon(1,\bar{r},\bar{\varrho},\bar{\varrho}\bar{r})$-neighborhood of $(\bar{x}',\bar{\xi}',\bar{x}'',\bar{\xi}'')$ we obtain the same result with another estimate of these coefficients:

$$|\varkappa'_{\frac{j}{2}}|\le C\bar{r}^{d-\frac{j}{2}}\bar{\varrho}^{2(d''-\frac{j}{2})}\tau^{-1}. \tag{12.3.32$'$}$$

Moreover, (12.3.27) yields that the same estimates hold with Q replaced by ψ_1Q' where $Q'=q'(x',\bar{r}^{-1}D')$ and q' is supported in the ϵ-neighborhood of $(\bar{x}',\bar{\xi}'\bar{r}^{-1})$. However, estimate (12.3.30) then yields that estimate (12.3.31) holds for arbitrary $T\in[\tau^{-1}\bar{h}'^{-\delta'},T_0\tau^{-1}\bar{h}^{-1}]$. Applying the standard Tauberian technique we obtain that

$$|\Bigl(\int\operatorname{tr}(Q'_x\psi_1e)(x,x,\tau)dx-\sum_{0\le j\le J-1}\varkappa_{\frac{j}{2}}(\tau)\Bigr)_{\tau=\tau'}^{\tau=\tau}|\le \\ C\bar{r}^{d-1}\bar{\varrho}^{2d''}\asymp C\bar{r}^{d-1-2d''\frac{m}{\mu}}\tau^{\frac{2d''}{\mu}} \tag{12.3.33}$$

with $\tau'\in[\epsilon_0\tau,\tau]$, $\tau>0$ where $\varkappa_{\frac{j}{2}}(\tau)$ are standard Weylian coefficients equal to the integrals of $\varkappa'_{j/2}(\tau)$ on τ and therefore

$$|\varkappa_{\frac{j}{2}}(\tau)|\le C\bar{r}^{d-\frac{j}{2}}\bar{\varrho}^{2(d''-\frac{j}{2})}. \tag{12.3.34}$$

Let us discuss the coefficients $\varkappa_{\frac{j}{2}}(\tau)$. Recall that

$$\sum_{0\le j\le J-1}\varkappa'_{\frac{j}{2}}\sim(2\pi)^{-d-1}i^{-1}\operatorname{Res}_{\mathbb{R}}\int\bigl((\tau-A)^{-1}Q'\psi_1\bigr)_w(x,\xi)dxd\xi \tag{12.3.35}$$

where the subscript "w" means that we take the Weylian symbol of the corresponding operator and $(\tau - A)^{-1}$ is a parametrix of $(\tau - A)$ in the context of the formal calculus of pseudo-differential operators; we cut this symbol after we have taken a large enough number of $(m,1)$-quasi-homogeneous terms (on (τ,ξ)) of the symbol of $(\tau - A)^{-1}$ and derivatives of these terms on (x,ξ) and therefore derivatives of the symbol of $Q'\psi_1$ on (ξ, x). Let us note that any differentiation of ψ_1 gives us a term which does not exceed the right-hand expression in (12.3.33); the same is true for any differentiation on x' and therefore on ξ' (and *vice versa*) made in the calculation of $(\tau - A)^{-1}$ and of its product with Q'. Therefore estimate (12.3.33) remains true if, in the definition of these coefficients, we replace the right-hand expression of (12.3.35) by

$$(2\pi)^{-d-1}i^{-1}\operatorname{Res}_{\mathbb{R}} \int \left((\tau - \mathbf{a})^{-1}\right)_w (x,\xi) Q'(x',\xi')\psi_1(x,\xi)\,dxd\xi \tag{12.3.36}$$

where $(\tau-\mathbf{a})^{-1}$ is the parametrix of $(\tau-\mathbf{a})$ in the context of the formal calculus of pseudo-differential operators and the subscript "w" means the Weyl symbol of the corresponding operator; here $\mathbf{a} = A(x',\xi',x'',D'')$ but later we will change this definition.

Recall that $\left((\tau - \mathbf{a})^{-1}\right)_w$ is a distribution. However, if we use spherical coordinates $\xi = (r,\varphi)$ with $\varphi \in \mathbb{S}^{d-1}$ and integrate on r (multiplying by a finite function beforehand) we obtain a regular function of (τ, x, φ) provided $(x'',\varphi'') \ne 0$. Let us introduce the spherical coordinate system $(\zeta, z) \in \mathbb{R}^+ \times \mathbb{S}^{2d''-1}$ in $\mathbb{R}^{2d''} \ni (x'',\varphi'')$; then one can rewrite the term of $\left((\tau - \mathbf{a})^{-1}\right)_w$ which is $(m,1)$-quasi-homogeneous of order $-m-\frac{j}{2}$ in the form

$$\sum_{0\le k\le J-1} \tau^{-1} b_{jk}(\tau,r,z)\zeta^{k-j} + b'_j(\tau,r,z,\zeta)$$

where b'_j is a regular function. Then a similar representation holds for $\operatorname{Res}_{\mathbb{R}}\left((\tau - \mathbf{a})^{-1}\right)_w$. Let us integrate it against the measure $dx'dzd\varphi' \times \zeta^{2d''-1}d\zeta \times r^{d-1}dr$. Then after integration over τ from τ' to τ and summation over a partition of unity in the zone $\{r_0(\tau) \le r \le \eta\}$ we obtain the following

Proposition 12.3.6. *Let conditions* (12.3.1), (12.3.2), (12.3.5), (12.3.6) *and* (12.3.8) *be fulfilled and let*

$$Q' = q_2(x', \varepsilon_1\tau^{-\frac{1}{m}}D') - q_1(x',\eta^{-1}D')$$

with fixed smooth functions q_j supported in $\{|\xi'| \le c\}$ and equal to 1 for $|\xi'| \le \epsilon_0$. Then the estimate

$$\Bigl|\Bigl(\int \operatorname{tr}(Q'_x\psi_1 e)(x,x,\tau)\,dx - \mathcal{N}(\tau,\eta)\Bigr)_{\tau=\tau'}^{\tau=\tau}\Bigr| \le CR_1^*(\tau) \tag{12.3.37}$$

holds with

$$\mathcal{N}(\tau,\eta) = \mathcal{N}_0(\tau) + \mathcal{N}_1(\tau,\eta), \tag{12.3.38}$$

$$\mathcal{N}_0(\tau) = \sum_{j=0,1,2} \kappa_{\frac{j}{2}}\tau^{\frac{1}{m}(d-\frac{j}{2})} + \sum_{j,k=0,1,2:(d-\frac{j}{2})l=(d'-\frac{k}{2})m} \kappa'_{\frac{j}{2}}\tau^{\frac{1}{m}(d-\frac{j}{2})}\log\tau, \tag{12.3.39}$$

$$\mathcal{N}_1(\tau,\eta) = \sum_{j,k\le J:(d-\frac{j}{2})l\ne(d'-\frac{k}{2})m} \kappa_{jk}\tau^{\frac{1}{\mu}(2d''+k-j)}\eta^{\frac{2}{\mu}((d'-\frac{k}{2})m-(d-\frac{j}{2})l)} - \sum_{j,k=0,1,2:(d-\frac{j}{2})l=(d'-\frac{k}{2})m} \kappa'_{\frac{j}{2}} m\tau^{\frac{1}{m}(d-\frac{j}{2})}\log\eta, \tag{12.3.40}$$

$$R_1^*(\tau) = \begin{cases} \varepsilon'\tau^{\frac{1}{m}(d-1)} & \text{for } (d-1)l > (d'-1)m \\ C\tau^{\frac{1}{m}(d-1)}\log\tau & \text{for } (d-1)l = (d'-1)m \\ C\tau^{\frac{1}{l}(d'-1)-\delta''} & \text{for } (d-1)l < (d'-1)m \end{cases} \tag{12.3.41}$$

where the coefficients κ_j, κ'_j and κ_{jk} do not depend on τ and η, $J = J(d,m,\delta)$ is large enough, $\delta'' > 0$ and $\varepsilon' \to +0$ as $\varepsilon \to +0$.

Let us now treat case (ii). In this case in order to obtain the unperturbed operator $\bar{A}$ we freeze $x' = y'$ in the operator $A_{(x',D')}(x'',D'')$ where $A_{(x',\xi')}(x'',D'')$ is the microlocalization of $A(x,D)$ at the point (x',ξ') and we take Weyl quantization on x'. We now treat this operator $\bar{A}$, and A itself, as pseudo-differential operators on x' with operator-valued symbols in $\mathbb{H} = L^2(\mathbb{R}^{d''},\mathbb{K})$. Then all the constructions of the propagation of singularities of Chapter 2 are applicable including the proof of microhyperbolicity (I explained this proof above) as well as the successive approximation method. Moreover, no framing procedure is necessary now because we apply a pseudo-differential operator ψ_1 mollifying on x''. So, we obtain that if $Q' = q'(x',\bar{r}^{-1}D')$ with q' supported in the ϵ-neighborhood of $(\bar{x}',\bar{\xi}'\bar{r}^{-1})$ with $\bar{r} \asymp |\bar{\xi}'|$ and $r_1(\tau) \geq \bar{r} \geq r_0(\tau)\tau^{\delta}$ then for $T \in [\tau^{-1}\bar{r}^{\delta'}, T_0\tau^{-1}\bar{r}]$ with a sufficiently small constant $T_0 > 0$ the asymptotics

$$\bigl|F_{t\to\lambda}\chi_T(t)\Gamma\psi_1 Q'_x\bigl(U - \sum_{0\leq n\leq M} U_n\bigr)\bigr| \leq C\tau^{-s} \qquad \forall\lambda: |\lambda-\tau| \leq \epsilon\tau \tag{12.3.42}$$

hold where U_n are the terms of the successive approximation method, the number $M = M(d,m,\mu,\delta,s)$ is large enough and

$$\bigl|F_{t\to\lambda}\chi_T(t)\Gamma\psi_1 Q'_x U_n\bigr| \leq C\bar{r}^{d-2d''m\mu^{-1}-\nu(n)}\tau^{2d''\mu^{-1}-1} \tag{12.3.43}$$

for the same values of T,τ with $\nu(n) \to +\infty$ as $n \to +\infty$. Therefore

$$\bigl|F_{t\to\lambda}\chi_T(t)\Gamma\psi_1 Q'_x U\bigr| \leq C\bar{r}^{d-2d''m\mu^{-1}}\tau^{2d''\mu^{-1}-1} \tag{12.3.44}$$

for the same values of T,τ. I leave all the details to the reader.

Then applying the standard Tauberian technique we obtain the estimate

$$\begin{aligned} \Bigl|\int \operatorname{tr}(Q'_x\psi_1 e)(x,x,\tau',\tau)\,dx &- \\ \sum_{0\leq n\leq M}\int_{\tau'}^{\tau}\bigl(F_{t\to\lambda}\bar{\chi}_T(t)\Gamma\psi_1 Q'_x U_{n(q)}\bigr)\,d\lambda\Bigr| &\leq C\bar{r}^{d-1-2d''m\mu^{-1}}\tau^{2d'\mu^{-1}} \end{aligned} \tag{12.3.45}$$

for $\tau' \in [\epsilon_0\tau,\tau]$ where the conditions on $\bar{r}$ and T were given above, the subscript "(q)" has no meaning at this time and $\bar{\chi} \in C_0^K([-1,1])$ is equal to 1 on $[-\frac{1}{2},\frac{1}{2}]$.

Let us note that if we sum the right-hand expression of estimate (12.3.45) on a partition of unity on the interval for the $\bar{r}$ indicated above we then obtain that the remainder does not exceed $CR_2^*(\tau)$ with

$$R_2^*(\tau) = \begin{cases} \tau^{\frac{1}{m}(d-1)-\delta''} & \text{for } (d-1)l > (d'-1)m, \\ \tau^{\frac{1}{m}(d-1)-\delta''}\log\tau & \text{for } (d-1)l = (d'-1)m, \\ \varepsilon'\tau^{\frac{1}{l}(d'-1)} & \text{for } (d-1)l < (d'-1)m \end{cases} \tag{12.3.46}$$

where $\varepsilon' = 1$ at this moment. However, for $(d-1)l < (d'-1)m$ if we only sum over the partition in the zone $\{\varepsilon r_1(\tau) \geq \bar{r} \geq r_0\tau^{\delta}\}$ then $\varepsilon' = \varepsilon'(\varepsilon) \to +0$ as $\varepsilon \to +0$ and C does not depend on ε. On the other hand, for $\bar{r} \in [\varepsilon r_1, r_1(\tau)]$ the operator $A_{(x',\xi')}(x'', D'')$ has no more than $L(\varepsilon)$ eigenvalues in $[-2\tau, 2\tau]$ and the operator $A(x', \xi', x'', D'')$ is close to this operator in the natural sense. Therefore for $\bar{r}$ from this interval and under the standard condition (Ψ) for the Hamiltonian trajectories of positive eigenvalues $A_{(x',\xi')}(x'', D'')$ in $\mathbb{H}$ (we write this condition below) one can replace the right-hand expression in estimate (12.3.45) by $o\big(\tau^{\frac{1}{l}(d'-1)}\big)$; this and the arbitrary choice of $\varepsilon' > 0$ yield that for $(d-1)l < (d'-1)m$ under this standard condition (Ψ) we obtain the remainder estimate $o\big(\tau^{\frac{1}{l}(d'-1)}\big)$ even for the sum over the partition of unity in the zone $\{r_1(\tau) \geq \bar{r} \geq r_0\tau^{\delta}\}$.

Let us consider the terms $F_{t\to\lambda}\bar{\chi}_T(t)\Gamma\psi_1 Q'_x U_{n(q)}$ with $\lambda \in [\epsilon_0\tau, \tau]$ in (12.3.45). Then the standard arguments of section 4.3 associated with microhyperbolicity yield that one can replace $\bar{\chi}_T$ with 1 in these terms; the error caused by this operation is $O(\tau^{-s})$. Moreover, explicit calculation yields that one can replace ψ_1 by I in these terms. On the other hand, one can easily see that for $Q' = q(x', r_1(\tau)^{-1}D')$ with $q'(x', \xi')$ supported in $\{|\xi'| \geq \frac{1}{2}\}$ and equal to 1 for $|\xi'| \geq 1$ we obtain from proposition 12.3.1(ii) that the first term in the left-hand expression of (12.3.45) is $O(\tau^{-s})$ and all the other terms modified above (i.e., with $\psi_1 = I, \bar{\chi} = 1$) simply vanish (we can write them explicitly and note that the operator $A_{(x',\xi')}(x'', D'')$ has no eigenvalue in $[-2\tau, 2\tau]$. So for this operator Q' the modified formula (12.3.45) also holds. Therefore we obtain the following

Proposition 12.3.7. *(i) For $Q' = q'(x', \eta^{-1}D')$ with q' supported in $\{|\xi'| \geq \frac{1}{2}\}$ and equal to 1 for $|\xi'| \geq 1$ and for $\eta \in [r_0(\tau)\tau^{\delta}, r_1(\tau)]$, $\tau' \in [\epsilon_0\tau, \tau]$ the asymptotics*

$$\text{(12.3.47)} \qquad \int \operatorname{tr}(Q'_x\psi_1 e)(x, x, \tau', \tau)dx = \sum_{0\leq n\leq M} \int_{\tau'}^{\tau} \big(F_{t\to\lambda}\Gamma Q'_x U_{n(q)}\big)d\lambda + O\big(R_2^*(\tau)\big)$$

hold.

(ii) Moreover, let $(d-1)l < (d'-1)m$ and let the standard condition (Ψ) on the Hamiltonian trajectories of the positive eigenvalues of the operators $A_{(x',\xi')}(x'', D'')$ in $\mathbb{H}$ be fulfilled. Then asymptotics (12.3.47) *hold with "$O\big(R_2^*(\tau)\big)$" replaced by "$o\big(R_2^*(\tau)\big)$."*

This proposition is not our final goal. Let us consider all the terms in the expansion in the right-hand expression of (12.3.47). Let us first note that any differentiation on x' (which is accompanied by differentiation on ξ' and *vice versa*) decreases the order of the term in (12.3.45) by an additional factor $\bar{r}^{-1}$ and therefore asymptotics (12.3.45) remain true even if we skip all the terms containing differentiation on (x', ξ'). Moreover, these asymptotics with remainder estimate "$o\big(R_2^*(\tau)\big)$" are also preserved if we preserve terms with only one differentiation on x' (and one on ξ'). Furthermore, even in this case one can skip all terms with differentiation of the symbol of Q' provided $\eta = o\big(r_1(\tau)\big)$. In fact, such a term is supported in the zone $\bar{r} \asymp \eta$ and its order coincides with the order of the right-hand expression of (12.3.45), i.e., it is "$o\big(R_2^*(\tau)\big)$" provided $(d-1)l < (d'-1)m$ and $\eta = o\big(r_1(\tau)\big)$.

Therefore one should consider only terms of the form

$$\text{(12.3.48)} \qquad U_{n(q)} = F_{\lambda\to t}^{-1} \operatorname{Res}_{\mathbb{R}} \mathcal{F}_n(x', \xi', \lambda)Q(x', \xi')$$

where $\mathcal{F}_n(x', \xi', \lambda)$ is a product of $k+1$ copies of the operator $\big(\lambda - A_{(x',\xi')}(x'', D'')\big)^{-1}$ separated by k operators $B_j(x', \xi', x'', D'')$ of the form

$$B_j = \sum_{\alpha,\beta:|\alpha|+|\beta|\leq\mu+q} b_{j\alpha\beta}(x', \xi')x''^{\alpha}D''^{\beta}$$

where $b_{j\alpha\beta}$ are positively homogeneous of degrees $m+\frac{1}{2}(|\alpha|-|\beta|-\mu-q)$ on ξ' and $k\ge n$, $q=1,2,\dots$. Therefore, assigning the index q to the operator B_j we also assign the index q to $\mathcal{F}_n$ as the sum of all the indices of factors of this type in the product. Then (12.3.43) holds with $\nu(n)=\delta'' q$.

Let us introduce spherical coordinates $\xi'=(r,\varphi)$. Then after the coordinate transform $x''_{\text{new}}=x''r^{\frac{1}{2}}$, $D''_{\text{new}}=D''r^{-\frac{1}{2}}$ we rewrite the corresponding term in the decomposition of $F_{t\to\lambda}U_{n(q)}$ in the form

$$\int_0^\infty \operatorname{Res}_{\mathbb{R}}\mathcal{F}_n(\lambda r^{-l})Q(r)r^{d'-1-l-\frac{q}{2}}\,dr \tag{12.3.49}$$

where here and below $\mathcal{F}_n$ is now calculated for the operator $A_{(x',\varphi)}(x'',D'')$ with $\varphi\in\mathbb{S}^{d'-1}$ and we do not indicate the dependence on (x',φ) and integration on these variables. Moreover, one can assume that $Q(r)=\phi(r^{-l}\eta^l)$ with $\phi\in C_0^K([-1,1])$ is a fixed function equal to 1 in a neighborhood of 0. Then after substituting $\lambda=zr^l$ we rewrite (12.3.49) in the form

$$\frac{1}{l}\lambda^{\frac{1}{l}(d'-\frac{q}{2})-1}\int_0^\infty \operatorname{Res}_{\mathbb{R}}\mathcal{F}_n(z)\phi(z\eta^l\lambda^{-1})z^{-\frac{1}{l}(d'-\frac{q}{2})}\,dz. \tag{12.3.50}$$

For $\ell=\lambda\eta^{-l}$ let us consider the expression

$$\Phi_{n(q)}(\ell)=\int_0^\infty \operatorname{Res}_{\mathbb{R}}\mathcal{F}_n(z)\phi(\frac{z}{\ell})z^{\frac{1}{l}(\frac{q}{2}-d')}\,dz. \tag{12.3.51}$$

Proposition 12.3.8. *For $\ell\to+\infty$ the asymptotics*

$$\operatorname{Tr}\Phi_{n(q)}=\kappa_q+\kappa'_q\log\ell+\sum_{p:\frac{1}{l}(d'-\frac{q}{2})\ne\frac{2}{p}(d''-\frac{p}{2})}\kappa'_{qp}\ell^{\frac{2}{\mu}(d''-\frac{p}{2})-\frac{1}{l}(d'-\frac{q}{2})}+O(\ell^{-s}) \tag{12.3.52}$$

holds with arbitrarily large s. Here we sum over $p\in\mathbb{Z}^+$ such that $p\le J(d,m,\mu,s)$ and the term $\kappa'_q\log\ell$ arises if and only if there exists $p\in\mathbb{Z}^+$ with $\frac{1}{l}(d'-\frac{q}{2})=\frac{2}{\mu}(d''-\frac{p}{2})$; here we do not indicate the dependence on n.

Proof. Let us differentiate $\Phi_{(q)}(\ell)$:

$$\operatorname{Tr}\Phi'_{(q)}=\ell^{-\frac{1}{l}(d'-\frac{q}{2})}\int \operatorname{Tr}\operatorname{Res}_{\mathbb{R}}\mathcal{F}(z)\psi(\frac{z}{\ell})\frac{dz}{\ell}$$

where $\psi=z_+^{1-\frac{1}{l}(d'-\frac{q}{2})}\phi'(z)$ obviously belongs to $C_0^K([\epsilon_0,c_0])$; the prime here means the derivative.

One can easily see that modulo $O(\ell^{-s})$ the integral in the right-hand expression is equal to

$$\operatorname{tr}\psi(\ell^{-1}D_t)\bar\chi_T(t)\Gamma'\mathcal{U}_k|_{t=0} \tag{12.3.53}$$

with arbitrary $T\ge\ell^{-\delta-1}$ and an arbitrarily small exponent $\delta>0$; here $\Gamma'v=\int v(x'',x'',.)dx''$ and one defines the functions $\mathcal{U}_k$ from the recurrence relations

$$(D_t-\mathbf{a})\mathcal{U}_0=0,\qquad \mathcal{U}_0|_{t=0}=\delta(x''-y'')I, \tag{12.3.54$_0$}$$

$$(D_t-\mathbf{a})\mathcal{U}_k=B_k\mathcal{U}_{k-1},\qquad \mathcal{U}_k|_{t=0}=0\quad(k\ge1), \tag{12.3.54$_k$}$$

$\mathbf{a}$ here means microlocalization, $(k+1)$ is the number of factors $(\tau-\mathbf{a})^{-1}$ in the definition of $\mathcal{F}$ and B_j are the other factors in this definition; the operators $\mathbf{a}=\mathbf{a}(x'',D'')$ and $B_j=B_j(x'',D'')$ are calculated at (x',ξ'). However, one can apply the standard Weylian successive approximation method of Chapter 4 on x'' with semiclassical parameter $\ell^{-\mu^{-1}}$; therefore the reader can easily prove (there are no Tauberian arguments!) that (12.3.53) is equal to

$$\sum_{p\le J}\kappa''_{qp}\ell^{\frac{1}{\mu}(2d'-p)-1}$$

modulo $O(\ell^{-s})$. Multiplying by $\ell^{-\frac{1}{\mu}(d'-\frac{q}{2})}$ and integrating we obtain (12.3.52). □

Let us set $\eta\le\tau^{\frac{1}{l}(1-\delta)}$ (in which case $\ell\ge\tau^{\delta}$). Then (12.3.52) yields that the trace of (12.3.50) is equal to

$$\big(\kappa_q+\kappa'_q\log\lambda-\kappa'_q l\log\eta\big)\lambda^{\frac{1}{l}(d'-\frac{q}{2})-1}+\sum_{p:\frac{1}{l}(d'-\frac{q}{2})\ne\frac{2}{\mu}(d''-\frac{p}{2})}\kappa'_{qp}\lambda^{\frac{1}{\mu}(2d''-p)-1}\eta^{-(2d''-p)\frac{l}{\mu}+(d'-\frac{q}{2})}\tag{12.3.55}$$

modulo $O(\lambda^{-s})$. After integration over λ from τ' to τ we obtain terms of expansion (12.3.45). Therefore we have proven the following

Proposition 12.3.9. *(i) For $\eta\in[\tau^{\frac{1}{m}+\delta},\tau^{\frac{1}{l}-\delta}]$ the asymptotics*

$$\int \operatorname{tr}(Q'_x\psi_1 e)(x,x,\tau',\tau)\,dx=\bar{\mathcal{N}}(\tau,\eta)\Big|_{\tau=\tau'}^{\tau=\tau}+O\big(R_2^*(\tau)\big)\tag{12.3.56}$$

hold with

$$\bar{\mathcal{N}}(\tau,\eta)=\bar{\mathcal{N}}_0(\tau)+\bar{\mathcal{N}}_1(\tau,\eta),\tag{12.3.57}$$

$$\bar{\mathcal{N}}_0(\tau)=\sum_q\bar\kappa_q\tau^{(d'-\frac{q}{2})}+\bar\kappa_0\log\tau,\tag{12.3.58}$$

$$\begin{aligned}\bar{\mathcal{N}}_1(\tau,\eta)=&\bar\kappa'_0\big(\log^2\tau-2l\log\tau\log\eta\big)+\\&\sum_{q,p:\frac{1}{l}(d'-\frac{q}{2})=\frac{1}{m}(2d''-p)}\bar\kappa''_q\tau^{\frac{1}{l}(d'-\frac{q}{2})}(\log\tau-l\log\eta)+\\&\sum_{q\ne2d'}\bar\kappa'''_q\eta^{(d'-\frac{q}{2})}\log\tau+\\&\sum_{q,p:\frac{1}{l}(d'-\frac{q}{2})\ne\frac{1}{\mu}(2d''-p)}\bar\kappa'_{qp}\tau^{\frac{1}{\mu}(2d''-p)}\eta^{(d'-\frac{q}{2})-(2d''-p)\frac{l}{\mu}}\end{aligned}\tag{12.3.59}$$

and $R_2^*(\tau)$ given by (12.3.46). Here all the coefficients $\bar\kappa^*_{**}$ do not depend on τ,τ',η and we sum over $p,q\le J=J(d,m,\mu,\delta)$.

(ii) Moreover, for $(d-1)l<(d'-1)m$ and the standard condition (Ψ) on the Hamiltonian trajectories of the positive eigenvalues of $A_{(x',\xi')}$ in $\mathbb{H}$ the asymptotics

$$\int \operatorname{tr}(Q'_x\psi_1 e)(x,x,\tau',\tau)\,dx=\bar{\mathcal{N}}(\tau,\eta)\Big|_{\tau=\tau'}^{\tau=\tau}+o\big(\tau^{\frac{1}{l}(d'-1)}\big)\tag{12.3.60}$$

hold.

Let us combine propositions 12.3.6 and 12.3.9 and note that the final answer *should not* depend on the choice of η (lying in the given interval). Moreover, we know the Weylian asymptotics for $\mathbf{N}_0^+(\tau)$ in (12.3.11). Then we obtain the following statement for $\mathbf{N}(\tau',\tau)$ with $\tau'\in[\epsilon_0\tau,\tau]$ instead of $\mathbf{N}^+(\tau)$ and with $\mathcal{N}(\tau',\tau)=\mathcal{N}^+(\tau)|_{\tau=\tau'}^{\tau=\tau}$ instead of $\mathcal{N}^+(\tau)$:

Theorem 12.3.10. *Let $A\in\Psi^{m,\mu}(X,\Lambda,\mathbb{K})$ be a self-adjoint operator satisfying conditions* (12.3.5), (12.3.6) *and* (12.3.8). *Then*

(i) The asymptotics

$$\mathbf{N}^+(\tau)=\mathcal{N}^+(\tau)+O\big(R^*(\tau)\big) \tag{12.3.61}$$

hold with

$$R^*(\tau)=\begin{cases}\tau^{\frac{1}{m}(d-1)} & \text{for } (d-1)l>(d'-1)m\\ \tau^{\frac{1}{m}(d-1)}\log\tau & \text{for } (d-1)l=(d'-1)m\\ \tau^{\frac{1}{l}(d'-1)} & \text{for } (d-1)l>(d'-1)m\end{cases} \tag{12.3.62}$$

where

$$\mathcal{N}^+(\tau)=\sum_{p=0,1,2}\kappa_{\frac{p}{2}}\tau^{\frac{1}{m}(d-\frac{p}{2})}+\sum_{q=0,1,2}\bar{\kappa}_{\frac{q}{2}}\tau^{\frac{1}{l}(d'-\frac{q}{2})} \tag{12.3.63}$$

provided $(d-\frac{p}{2})l\neq(d'-\frac{q}{2})m$ for all p,q indicated above; on the other hand, if $(d-\frac{p}{2})l=(d'-\frac{q}{2})m$ for some p,q then one should replace the term $\kappa_{\frac{p}{2}}\tau^{\frac{1}{m}(d-\frac{p}{2})}$ by $\kappa_{\frac{p}{2}}\tau^{\frac{1}{l}(d-\frac{p}{2})}\log\tau$.[15)]

(ii) Moreover, let $(d-1)l>(d'-1)m$ and let condition $(\Psi 0)$ on the Hamiltonian trajectories of the positive eigenvalues of $a_0(x,\xi)$ be fulfilled. Then the asymptotics

$$\mathbf{N}^+(\tau)=\mathcal{N}^+(\tau)+o\big(\tau^{\frac{1}{m}(d-1)}\big) \tag{12.3.64}$$

hold.

(iii) On the other hand, let $(d-1)l<(d'-1)m$ and let condition (Ψ) on the Hamiltonian trajectories of the positive eigenvalues of the operator $A_{(x',\xi')}$ in $\mathbb{H}=L^2(\mathbb{R}^{d''},\mathbb{K})$ be fulfilled. Then the asymptotics

$$\mathbf{N}^+(\tau)=\mathcal{N}^+(\tau)+o\big(\tau^{\frac{1}{l}(d'-1)}\big) \tag{12.3.65}$$

hold.

Finally, let us replace τ and τ' by $2^{-n}\tau$ and $2^{-n-1}\tau$ and sum over $n=1,\dots,n_0\le[\log\tau]+1$. Then we obtain this theorem in the final form (i.e., with $\tau'=0$).

Remark 12.3.11. *(i) Moreover, one obtains similar statements for Riesz means with*

$$R^*_\vartheta(\tau)=\begin{cases}\tau^{\frac{1}{m}(d-1-\vartheta)+\vartheta} & \text{for } (d-1-\vartheta)l>(d'-1-\vartheta)m\\ \tau^{\frac{1}{m}(d-1-\vartheta)+\vartheta}\log\tau & \text{for } (d-1-\vartheta)l=(d'-1-\vartheta)m\\ \tau^{\frac{1}{l}(d'-1-\vartheta)+\vartheta} & \text{for } (d-1-\vartheta)l>(d'-1-\vartheta)m\end{cases} \tag{12.3.62$'$}$$

and with principal part

$$\mathcal{N}^+_\vartheta(\tau)=\sum_{p\le 2+2\vartheta}\kappa_{\frac{p}{2}}\tau^{\frac{1}{m}(d-\frac{p}{2})+\vartheta}+\sum_{q\le 2+2\vartheta}\bar{\kappa}_{\frac{q}{2}}\tau^{\frac{1}{l}(d'-\frac{q}{2})+\vartheta} \tag{12.3.63$'$}$$

---

15) One can easily see that there is usually at most one logarithmic term; there are two logarithmic terms if $m=2l$, $d=2d'$.

provided $(d-\frac{p}{2})l \neq (d'-\frac{q}{2})m$ *for all* p,q *indicated above; on the other hand, if* $(d-\frac{p}{2})l = (d'-\frac{q}{2})m$ *for some* p,q *then one should replace the term* $\kappa_{\frac{p}{2}}\tau^{\frac{1}{m}(d-\frac{p}{2})}$ *by* $\kappa_{\frac{p}{2}}\tau^{\frac{1}{m}(d-\frac{p}{2})}\log\tau$. *The coefficients* κ_*, $\bar{\kappa}_*$ *depend on* $\vartheta \geq 0$.

Surely, some terms are smaller than the remainder estimate and should be removed. One can see that for $\vartheta > \frac{2}{\mu}(ld-md')-1$ *we obtain a Weylian remainder estimate (but the asymptotics contain non-Weylian terms).*

(ii) Moreover, let $p \notin \mathbb{Z}$ *and* $a_j \equiv 0$ *for all* $j \notin \mathbb{Z} : j \leq k$. *Then* $\kappa_p = 0$ *which is also true in the case* $(d-p)l = (d-q)m$ *for some* $q \in \frac{1}{2}\mathbb{Z}$: *so the logarithmic term does not arise.* ◊

12.3.5 Operators with negative principal symbols

Let us consider the case when $a_0(x,\xi)$ is a negative definite symbol. However, we assume that $A_{(x',\xi')}$ is not a negative definite operator (otherwise A is semibounded from above). Our goal is to prove the following

Theorem 12.3.12. *Let the conditions of theorem 12.3.10 be fulfilled and let the condition*

(12.3.66) $a_0(x,\xi)$ is a negative definite symbol on $T^*X \setminus 0 \setminus \Lambda$

also be fulfilled.

Then

(i) The asymptotics

$$\mathbf{N}^+(\tau) = \mathcal{N}^+(\tau) + O\big(\tau^{\frac{1}{l}(d'-1)}\big) \tag{12.3.67}$$

hold with

$$\mathcal{N}^+(\tau) = \bar{\kappa}_0\tau^{\frac{1}{m}d'} + \bar{\kappa}_{\frac{1}{2}}\tau^{\frac{1}{l}(d'-\frac{1}{2})} + \bar{\kappa}_1\tau^{\frac{1}{l}(d'-1)}. \tag{12.3.68}$$

(ii) Moreover, let the standard condition (Ψ) *on the Hamiltonian trajectories of the positive eigenvalues of* $A_{(x',\xi')}$ *in* $\mathbb{H}$ *be fulfilled. Then the asymptotics*

$$\mathbf{N}^+(\tau) = \mathcal{N}^+(\tau) + o\big(\tau^{\frac{1}{l}(d'-1)}\big) \tag{12.3.69}$$

hold.

Proof. First of all, (12.3.42) and the explicit formula for U_n yield that for $\chi \in C_0^K(\mathbb{R})$ vanishing in a neighborhood of 0 one can replace (12.3.44) by the estimate

$$\big|F_{t\to\lambda}\chi_T(t)\Gamma\psi_1 Q'_x U\big| \leq C\bar{r}^{d-\frac{2}{\mu}d''m}\tau^{\frac{2}{\mu}d''-1-s}T^{-sl} \tag{12.3.70}$$

for $T \in [T_0\tau^{-1}, T_0\tau^{-1}\bar{r}]$, $\lambda \in [\epsilon_0\tau, c\tau]$. Therefore the same estimate holds for

$$\big|F_{t\to\lambda}\big(\bar{\chi}_{T'}(t) - \bar{\chi}_T(t)\big)\Gamma\psi_1 Q'_x U\big|$$

for $T, T' \in [T_0\tau^{-1}, T_0\tau^{-1}\bar{r}]$, $T < T'$ and $\bar{\chi} \in C_0^K(\mathbb{R})$ equal to 1 in a neighborhood of 0. On the other hand, for $T \in [T_0, \tau^{-1}\bar{h}'^{-\delta}, T_0\tau^{-1}\bar{h}'^{-1}]$ with $\bar{h}' = \bar{r}^{-\frac{2}{\mu}l}\tau^{\frac{2}{\mu}}$, the standard Weylian formula (which is derived by setting $\bar{A} = a_0(y,D)$) holds modulo terms which do not exceed the right-hand expression in (12.3.70). Condition (12.3.56) also yields that all the coefficients in this formula vanish. Taking $T = T_0\tau^{-1}\bar{h}'^{-1}$ we replace (12.3.44) by the estimate

$$|F_{t\to\lambda}\bar{\chi}_T(t)\Gamma\psi_1 Q'_x U| \le C\bar{r}^{d-\frac{2}{\mu}d''m+s}\tau^{\frac{2}{\mu}d''-1-s}T^{-sl} \tag{12.3.72}$$

with $T\in[T_0,\tau^{-1}\tau^{-1}\bar{h}'^{-\delta},T_0\tau^{-1}\bar{r}]$, $\lambda\in[\epsilon_0\tau,c\tau]$ and with another still arbitrarily large s. Here we have omitted the prime on T.

Therefore one can replace $\bar{\chi}$ with a Hörmander function χ and apply the routine Tauberian technique to obtain that the left hand expression in (12.3.45) now does not exceed

$$C\bar{r}^{d-1-\frac{2}{\mu}d''m+sl}\tau^{\frac{2}{\mu}d''-s}.$$

Therefore after summation over a partition of unity in the zone $\{r_0(\tau)\le|\xi'|\le\varepsilon r_1(\tau)\}$ we obtain the remainder estimate $\varepsilon'\tau^{\frac{1}{l}(d'-1)}$ with $\varepsilon'=\varepsilon'(\varepsilon)\to+0$ as $\varepsilon\to+0$. Therefore under condition (12.3.66) proposition 12.3.7 holds with the remainder estimate $O(\tau^{\frac{1}{l}(d'-1)})$ and even $o(\tau^{\frac{1}{l}(d'-1)})$ under the standard condition (Ψ) on the Hamiltonian trajectories of the positive eigenvalues of $A_{(x',\xi')}$. On the other hand, proposition 12.3.1 yields that the remainder estimate in proposition 12.3.6 is now $O(\tau^{-s})$. Therefore for $\mathbf{N}(\tau',\tau)$ with $\tau'\in[\epsilon_0,\tau]$ the remainder estimate is $O(\tau^{\frac{1}{l}(d'-1)})$.[16] This estimate yields the same remainder estimate for $\mathbf{N}^+(\tau)$ for $d'\ge 2$. For $d'=1$ standard arguments with τ, τ' replaced by $2^{-n}\tau$, $2^{-n-1}\tau$, etc., yield the remainder estimate $O(\log\tau)$ but one can also easily obtain the remainder estimate $O(1)$. In fact, one has the remainder estimate $O(1)$ for $\mathbf{N}(\tau',\tau)$ and the remainder estimate $O(\tau^{-s})$ for

$$\int\phi(\frac{\lambda}{\tau})d_\lambda\mathbf{N}^+(\lambda)d\lambda$$

with $\phi\in C_0^K([\epsilon_0,c])$ (this estimate is the routine estimate with mollification, the deduction of which requires neither the Tauberian technique nor condition (12.3.66)). Replacing $\phi(\lambda)$ by $\bar{\phi}(\tau)-\bar{\phi}(2\tau)$ with $\bar{\phi}\in C^K(\mathbb{R})$ supported in $[-\infty,1]$ and equal to 1 on $[-\infty,\frac{1}{2}]$ and replacing τ by $2^{-n}\tau$, etc., we obtain the remainder estimate $O(1)$ for

$$\int\bar{\phi}(\frac{\lambda}{\tau})d_\lambda\mathbf{N}^+(\lambda)d\lambda,$$

and this remainder estimate and the remainder estimate for $\mathbf{N}(\tau',\tau)$ obviously yield the remainder estimate $O(1)$ for $\mathbf{N}^+(\tau)$. So, both remainder estimates in theorem 12.3.12 have been proven.

Moreover, in proposition 12.3.6 we obtain $\mathcal{N}(\tau,\eta)=0$ because of condition (12.3.66) (recall that standard Weylian asymptotics were applied here) and in proposition 12.3.8 one should replace (12.3.52) by the asymptotics

$$\operatorname{Tr}\Phi_{n(q)}(\ell)=\kappa_q+O(\ell^{-s}). \tag{12.3.71}$$

Therefore expression (12.3.55) is reduced to

$$\kappa_q\tau^{\frac{1}{l}(d'-\frac{q}{2})-1}$$

and $\bar{\mathcal{N}}(\tau,\eta)=0$.

Therefore for $d'\ge 2$ we obtain expression (12.3.68) exactly for $\mathcal{N}^+(\tau)$, but for $d'=1$ an additional term $\bar{\kappa}'_1\log\tau$ arises (this term is less than the remainder estimate for $d'\ge 2$). However, explicit calculation proves that $\bar{\kappa}'_1=0$ for $d'=1$. I leave this to the reader; the same calculation was done in the analysis of non-semibounded boundary value problems for operators with positive principal symbols in [Ivrii 1] (the arguments apply with no modification). □

[16] And even $o(\tau^{\frac{1}{l}(d'-1)})$ under the standard condition (Ψ) on the Hamiltonian trajectories of the positive eigenvalues of $A_{(x',\xi')}$.

Remark 12.3.13. *The remainder estimate is* $O\bigl(\tau^{\frac{1}{l}(d-1-\vartheta)+\vartheta}\bigr)$ *for Riesz means, and even* $o\bigl(\tau^{\frac{1}{l}(d-1-\vartheta)+\vartheta}\bigr)$ *under the standard condition* (Ψ) *on the Hamiltonian trajectories of the positive eigenvalues of* $A_{(x',\xi')}$. ◊

12.3.6 Coefficients in spectral asymptotics

Let us discuss the coefficients κ_p and $\bar\kappa_q$ in the asymptotics of theorems 12.3.10, 12.3.12 and remarks 12.3.11, 12.3.13. First of all the following obvious remark holds (we discuss coefficients for $\vartheta=0$; the coefficients of $\tau^{r+\vartheta}$ for $\vartheta>0$ have the extra factor $r(r+\vartheta)^{-1}$ compared with the coefficient of τ^r for $\vartheta=0$).

Remark 12.3.14. *(i) If* $p\,(=0,\frac{1}{2},1,\dots)$ *is such that* $l(d-p)>d'm$ *then for* κ_p *the standard Weylian formula holds, the integral in which surely converges; in particular,*

$$\kappa_0=(2\pi)^{-d}\int n_0^+(x,\xi,1)\,dxd\xi \tag{12.3.73}$$

where $n_0^+(x,\xi,1)$ *is the number of eigenvalues of* $a_0(x,\xi)$ *lying in* $[0,1]$. *Therefore* $\kappa_0>0$ *provided* $a_0(x,\xi)$ *is not negative definite.*

(ii) On the other hand, for $p\,(=0,\frac{1}{2},1,\dots)$ *let the equality* $(d-p)l=d'm$ *hold. Then the integral providing the standard Weylian coefficient of* $\tau^{\frac{1}{m}(d-p)}$ *diverges logarithmically; namely, modulo convergent integrals we have*

$$\int \omega(x',\xi';\varphi)\,dx'd\xi'd\varphi\frac{d\zeta}{\zeta}$$

with $\varphi=\varrho^{-1}(x'',\xi''|\xi''|^{-1})\in\mathbb{S}^{2d''-1}$, $\varrho=\bigl(|x''|^2+|\xi''|^2|\xi'|^{-2}\bigr)^{\frac{1}{2}}$; *then the coefficient* $\bar\kappa_p$ *of* $\tau^{(\frac{1}{m}d-p)}\log\tau$ *is given by formula*

$$\bar\kappa_p=\frac{1}{2l}\int\omega(x',\xi',\varphi)\,dx'd\xi'd\varphi.$$

In particular, for $dl=d'm$ *the coefficient* $\bar\kappa_0$ *of* $\tau^{\frac{d}{m}}\log\tau$ *is*

$$\bar\kappa_0=\frac{1}{2l(2\pi)^d}\int_{\Lambda\times\mathbb{S}^{2d''-1}}\tilde n^+(x',\xi',\varphi,1)\,dx'd\xi'd\varphi \tag{12.3.74}$$

where $\tilde n^+(x',\xi',\varphi,1)$ *is the number of eigenvalues of the symbol*

$$\tilde a_0(x',\xi',\varphi)=\sum_{|\alpha|+|\beta|=\mu}\frac{1}{\alpha!\beta!}a_{0(\beta)}^{(\alpha)}(x',\xi')x''^{\beta}\xi''^{\alpha} \tag{12.3.75}$$

with $\varphi=\varrho^{-1}(x'',\xi''|\xi''|^{-1})(x,\xi)\in\mathbb{S}^{2d''-1}$.

(iii) Either let $q=(0,\frac{1}{2},1,\dots)$ *be such that* $(d'-q)m>dl$ *or let* $a_0(x,\xi)$ *be a negative definite symbol. Then the coefficient* $\bar\kappa_q$ *is given by a convergent integral*

$$\frac{1}{2d'-q}\int_0^\infty \operatorname{Tr}\operatorname{Res}_{\mathbb{R}}\mathcal{F}_{(q)}(z)z^{-\frac{1}{l}(d'-q)}dz$$

where $\mathcal{F}_{(q)}$ *is the sum of all the symbols with the given index* q. *In particular, if either* $d'm>dl$ *or* $a_0(x,\xi)$ *is a negative definite symbol then*

$$\bar\kappa_0=(2\pi)^{-d'}\int_\Lambda \mathbf{n}^+(x',\xi',1)\,dx'd\xi'=\frac{1}{(2\pi)^d d}\int_{\Lambda\cap S^*X}\omega(\zeta)\,d\zeta \tag{12.3.76}$$

where $\omega(z) = \sum_j \eta_j^{-\frac{d}{1}}$, η_j *are all the positive eigenvalues of* $A_{(x',\xi')}$ *and* $d\zeta$ *is the natural density on* $\Lambda \cap S^*X$ *provided a Riemannian metric is introduced on* X.

(iv) $a_{\frac{1}{2}} = 0$ *yields that* $\kappa_{\frac{1}{2}} = 0$. *See also remark 12.3.3(ii).*

(v) Moreover, let μ *be even. Let us again treat the zone* $\{|\xi'| \le r_1(\tau)\tau^{-\delta}\}$. *Let us consider a term in the expansion of* $\varkappa_{\frac{j}{2}}(x,\xi)$ *in terms which are homogeneous with respect to* $(x'', \xi''|\xi'|^{-1})$. *Obviously, if this term is odd with respect to* $(x'', \xi''|\xi'|^{-1})$ *then its integral vanishes. Therefore a term with a logarithmic factor appears only if the equality* $\left(d - \frac{j}{2}\right)l = \left(d' - \frac{k}{2}\right)m$ *holds with* $j \equiv k \,(\mathrm{mod}\, 2)$. *So, if* μ *is even then a term* $\kappa_{\frac{j}{2}}\tau^{\frac{1}{m}(d-\frac{j}{2})}\log\tau$ *appears only if* $\left(d - \frac{j}{2}\right)l = \left(d' - \frac{k}{2}\right)m$ *and* $j \equiv k \,(\mathrm{mod}\, 2)$.

(vi) Let $a_{\frac{1}{2}} \equiv 0$. *Then in the previous analysis one should only treat even* j *(and therefore only even* k*). On the other hand, analysis in the zone* $\{|\xi'| \ge r_0(\tau)\tau^{\delta}\}$ *proves that* $\bar\kappa_{\frac{q}{2}} = 0$ *for odd* q. *So, if* $a_{\frac{1}{2}} \equiv 0$ *and* μ *is even then in (12.3.63), (12.3.68) one should treat only* $p = 0, 2$ *and* $q = 0, 2$ *in the calculation of powers and in the criterion for terms with a logarithmic factor. Similarly, for* $\vartheta > 0$ *in this case (*$a_{\frac{n}{2}} \equiv 0$ *for odd* n *and* μ *is even) one should consider only even* p, q *in the criterion for terms with a logarithmic factor.*

(vii) On the other hand, one can easily see that for an operator with the symbol $\begin{pmatrix} \lambda & \mathcal{L} \\ \mathcal{L}^* & \lambda \end{pmatrix}$ *with* $\mathcal{L} \in \Psi^{1,1}$ *the previous statement for* $\bar\kappa_q$ *is wrong.*

(viii) Finally, if $a_{\frac{1}{2}} \equiv 0$, μ *and* m *are even and* a_0 *and* a_1 *are homogeneous (not only positive homogeneous) of degrees* m *and* $m-1$ *in* ξ *respectively then in (12.3.63) one should treat only* $p = 0$ *in the calculation of powers and in the criterion for terms with a logarithmic factor.* ◊

Remark 12.3.15. *(i) I leave it to the reader to consider the asymptotics of Riesz means.*

(ii) I leave the case of manifolds with boundary to the reader. In this case one should assume that every connected component of Λ *either does not intersect* $T^*X|_{\partial X}$ *or is contained in this manifold. In the second case one should consider* $A_{(x',\xi')}(x'', D'')$ *as an operator in the half-space of* $\mathbb{R}^{d''}$ *and the problem of boundary conditions arises. Moreover, there are more variants for the leading term of the asymptotics because we know that the condition that the operator is semibounded on a manifold with boundary consists of two conditions, and if the first condition is fulfilled and the second condition is violated then the first term on the asymptotics vanishes in the non-degenerate case.*

(iii) The other problem left to the reader is to return to sections 12.1, 12.2 and obtain (under appropriate conditions) the non-Weylian term in the asymptotics in a simpler form (as we did here, and in the form used here). ◊

12.4 Asymptotics of Riesz Means for Operators Singular at a Point

12.4.1 Introduction

In this section we consider the h-differential operator $A = A_h$ with Weyl symbol

(12.4.1) $$A(x,\xi) = \sum_{|\alpha| \le m} a_\alpha(x)\xi^\alpha,$$

Hermitian and such that in $B(0,1)$

(12.4.2) $$|D^\beta a_\alpha| \le c|x|^{\mu(m-|\alpha|)-|\beta|} \qquad \forall\alpha : |\alpha| \le m \quad \forall\beta : |\beta| \le K$$

with

$$\mu > \max\left(-1, -\frac{d}{m}\right). \tag{12.4.3}$$

Moreover, we assume that

$$\langle a(x,\xi)v, v\rangle \ge \epsilon_0|\xi|^m|v|^2 \qquad \forall (x,\xi)\in T^*B(0,1) \quad \forall v\in\mathbb{K} \tag{12.4.4}$$

where $a(x,\xi)=\sum_{|\alpha|=m} a_\alpha(x)\xi^\alpha$ is the senior symbol of A.

Let us consider the operator A in the domain $X\subset\mathbb{R}^d$ assuming that

$$B(0,1)\subset X, \quad d\ge 2 \tag{12.4.5}$$

and

(12.4.6) A is a self-adjoint operator with $\mathrm{D}(A)\supset C_0^m\big(B(0,1),\mathbb{K}\big)$.

We are interested in semiclassical asymptotics of the Riesz mean

$$\int \operatorname{tr}\psi(x)e_\vartheta(x,x,0)dx$$

where $\psi\in C_0^K\big(B(0,\frac{1}{2})\big)$ and $e_\vartheta(.,.,\tau)$ is the Riesz mean of order ϑ of the Schwartz kernel $e(.,.,\tau)$ of the spectral projector A, $\vartheta>0$:

$$e_\vartheta(.,.,\tau)=\int_0^\tau (\tau-\lambda)^\vartheta d_\lambda e(.,.,\lambda).$$

In section 10.1 we proved that under an appropriate condition associated with micro-hyperbolicity (namely, condition (12.4.21) below) the asymptotics

$$\int \operatorname{tr}\psi(x)e_\vartheta(x,x,0)dx=\sum_{0\le j\le J} h^{-d+j}\int \psi(x)\varkappa_{\vartheta,j}(x)dx+O\big(R_1(\tau)\big) \tag{12.4.7}$$

hold where

$$R_1(\tau)=\begin{cases} h^{-d+\vartheta+1} & \text{for } \vartheta<\bar\vartheta_0,\\ h^{-d+\vartheta+1}\log h & \text{for } \vartheta=\bar\vartheta_0,\\ h^l & \text{for } \vartheta>\bar\vartheta_0\end{cases} \tag{12.4.8}$$

$$\begin{aligned} &\bar\vartheta_0=\big((d-1)(\mu+1)+p\big)(-m\mu+\mu+1)^{-1}>p && \text{for } m\mu<\mu+1\\ &\bar\vartheta_0=+\infty && \text{for } m\mu\ge\mu+1,\\ &l=l_0=m\mu\vartheta(\mu+1)^{-1}, \qquad l_p=(m\mu\vartheta+p)(\mu+1)^{-1}, \end{aligned} \tag{12.4.9}$$

$\varkappa_{\vartheta,j}$ are standard Weylian coefficients and $J=[\vartheta]+1$ for $\vartheta\le\bar\vartheta$, $J=d+[l]$ for $\vartheta\ge\bar\vartheta$ excluding the case $\vartheta=\bar\vartheta\in\mathbb{N}$ when $J=\vartheta$. Moreover, for $\vartheta>\bar\vartheta$ and $l\in\mathbb{N}$ one should cut the corresponding term near the origin, i.e., replace $\psi(x)$ by $\psi(x)\phi(\frac{x}{r_0})$ in the term with $j=d+l$ of the right-hand expression of (12.4.7) where here and below $r_0=h^{(\mu+1)^{-1}}$, $1-\phi\in C_0^K(\mathbb{R}^d)$ and ϕ vanishes in a neighborhood of 0.

Moreover, if $\vartheta<\bar\vartheta_0$ and the standard condition (Ψ) is fulfilled at the energy level $\tau=0$ then one can replace $O(h^{-d+\vartheta+1})$ by $o(h^{-d+\vartheta+1})$ in asymptotics (12.4.7).

The principal part of the asymptotics is $O(h^{-d})$ for $m\mu\vartheta + p > -d(\mu+1)$ and $O(h^{-d}\log h)$ for $m\mu\vartheta + p = -d(\mu+1)$ and $O(h^{l})$ for $m\mu\vartheta + p < -d(\mu+1)$;[17] so in the last case we have an estimate instead of asymptotics.

All these statements remain true with $l = l_p$ for $p > 0$ provided

$$\operatorname{supp}\psi \subset B(0,\frac{1}{2}), \qquad |D^\alpha\psi| \le c|x|^{p-|\alpha|}$$

instead of the previous assumptions.

12.4.2 The Scott correction term

Our goal is to treat the case $\vartheta \ge \bar{\vartheta}$ and to improve this remainder estimate. This better estimate, obtained under appropriate conditions, requires an additional non-Weylian term ωh^{l} in the asymptotics. Physicists call this term the "Scott correction." Later we will obtain more than one such term for large ϑ. In this analysis, without loss of generality we can treat

$$\int x^\alpha\psi(x)\operatorname{tr} e_\vartheta(x,x,0)dx$$

with $|\alpha| = p$. Let us assume that

(12.4.10) $\qquad \bar{\vartheta} < +\infty \qquad (\iff m\mu < \mu+1),$

(12.4.11) $\quad |\alpha| = p,\ \psi \in C_0^K\big(B(0,\frac{1}{2}),\ \psi = 1 \text{ in } B(0,\frac{1}{3})\big).$

First of all we should treat the positive homogeneous case:

Proposition 12.4.1. *Let conditions* (12.4.2)–(12.4.5) *be fulfilled. Moreover, let $X = \mathbb{R}^d$, let A be a self-adjoint operator with $\mathrm{D}(A) \supset C_0^m(\mathbb{R}^d,\mathbb{K})$ and let us assume that*

(12.4.12) The coefficients a_β are positively homogeneous of degrees $(m-|\beta|)\mu$ in x.

Finally, let us assume that

(12.4.13) $A(x,\xi)$ is microhyperbolic in (x,ξ) at the energy level $\tau = 0$ for $x \ne 0$.[18]

Then for any $\beta \in \mathbb{N}^d$ and for $W \in C^K\big(\mathbb{R}^d\setminus 0, \operatorname{Hom}(\mathbb{K},\mathbb{K})\big)$ positively homogeneous of degree $p-\mu|\beta|$ in x the equality

$$\int W(x)\Big(\big((hD_x)^\beta e_\vartheta\big)(x,x,0) - \phi(\frac{x}{r_0})\sum_{0\le j\le J}\varkappa_{\vartheta,j}(x)h^{-d+j}\Big)dx = \omega h^{l} \tag{12.4.14}$$

holds with the coefficients $\varkappa_{\vartheta,j},\omega$ [19] *depending on β and a certain constant coefficient $\omega \in \operatorname{Hom}(\mathbb{K},\mathbb{K})$ which depends on W,β,ϕ* [20] *and $J \ge [l]+d$ and the integral in the left-hand expression converges at infinity.*[21] *Moreover, in terms with $j < l+d$ one can replace ϕ by 1;*[22] *in particular, if either $l \notin \mathbb{N}$ or $l \in \mathbb{N}$ but $\varkappa_{\vartheta,l+d} \equiv 0$ then one can set $J = [l]+d$ and replace ϕ by 1 in all the terms.*

---

[17] There are equivalence relations under obvious conditions.

[18] Surely, the constants in the microhyperbolicity condition depend on $|x|$.

[19] Here we do not take the trace and therefore the coefficients $\varkappa_*$ and ω are matrix-valued in this statement.

[20] recall that ϕ is a fixed function such that $1-\phi \in C_0^K(\mathbb{R}^d)$, $\phi = 1$ near 0.

[21] It surely converges at 0 due to the cutting function ϕ.

[22] Surely, the constant ω changes.

Proof. Let us first note that in the context of the proposition

(12.4.15) $\varkappa_{\vartheta,\beta,j}$ are positively homogeneous in x of degrees $d\mu + m\mu\vartheta - j(\mu+1) - |\beta|$.

Moreover, let us note that after the dilatation $x_{\text{new}} = x_{\text{old}}h^{-(\mu+1)^{-1}}$ the operator $A^w(x, hD)$ is transformed into the operator $h^{m\mu(\mu+1)^{-1}}A^w(x,D)$ and therefore $e_\vartheta(x,y,0;h) = h^{(m\mu\vartheta+d)(\mu+1)^{-1}}e_\vartheta(xh^{-(\mu+1)^{-1}}, yh^{-(\mu+1)^{-1}}, 0; 1)$ due to the definition of $e_\vartheta(.,.,.)$; here the fourth variable is the semiclassical parameter h. This equality and (12.4.15) yield that the left-hand expression in (12.4.14) is positively homogeneous of degree l on h provided it converges; therefore we should prove only that the integral in (12.4.14) converges at infinity.

However, theorem 4.3.3 yields that if $\psi' \in C_0^K(\mathbb{R}^d, \operatorname{Hom}(\mathbb{K},\mathbb{K}))$ vanishes in a neighborhood of 0 then the estimate

$$|\int \psi'(\frac{x}{r})\Big(((hD_x)^\beta e_\vartheta)(x,x,0) - \phi(\frac{x}{r_0})\sum_{0\le j\le J}\varkappa_{\vartheta,j}(x)h^{-d+j}\Big)dx| \le \\ Ch^{-d+J+1}r^{l(\mu+1)(l-J-1)}$$

holds uniformly on $r \ge r_0 = h^{(\mu+1)^{-1}}$; this estimate immediately yields that the integral in the left-hand expression of (12.4.14) converges for $J \ge [l] + d$. □

12.4.3 The first attempt

Our first goal is to improve the remainder estimate $O(h^l)$ for $\vartheta > \bar{\vartheta}$ in the general situation. To do this we will pass from one operator to another to reduce the problem. First of all the same analysis as in Chapter 8 proves

Proposition 12.4.2. *Let the operators A and A' satisfy conditions* (12.4.2)–(12.4.6) *and coincide in $B(0,1)$. Let $\psi \in C_0^K(B(0,\frac{1}{2}))$, $\psi \ge 0$. Moreover, let the estimate*

$$|\int \psi(x)e(x,x,\tau+h,\tau)dx| \le Ch^{1-d} \qquad \forall\tau : |\tau| \le \epsilon_0 \tag{12.4.16}$$

hold. Then this estimate holds for A' as well and, moreover, the estimate

$$|\int \psi(x)\big(e_\vartheta(x,x,0) - e'_\vartheta(x,x,0)\big)dx| \le Ch^{-d+\vartheta+1} \tag{12.4.17}$$

holds where here and below $e(.,.,.)$ with a superscript is associated with the operator A with the same superscript.

Therefore the remainder estimate (no better than $O(h^{-d+\vartheta+1})$) is preserved if we change the operator A outside of $B(0,\bar{r})$ with some fixed $\bar{r} > 0$. *Let us pick an operator A' coinciding with A in $B(0,\bar{r})$ and such that the negative spectrum of A' is discrete for* every $h \in (0,h_0]$ (we can even pick an operator with negative spectrum, finite for every $h \in (0,h_0]$). Let us now assume that $\vartheta \ge 1$. In fact the arguments below are really only good for $\vartheta = 1$ but I will later modify them to work in the case $\vartheta < 1$ and to provide a really good estimate for $\vartheta > 1$.

Proposition 12.4.3. *For $\eta \in [0,1]$ let the operator $A^\eta = A^0 + \eta B$ satisfy conditions* (12.4.2)–(12.4.6), *let the negative spectrum of A be discrete and let B be supported in $B(0,\frac{1}{2})$. Then the equality*

$$\text{(12.4.18)} \quad \int \operatorname{tr}\bigl(e^1_\vartheta(x,x,\tau) - e^0_\vartheta(x,x,\tau)\bigr)dx = \\ - c_{(\vartheta)} \int_0^1 d\eta \int \operatorname{tr}\bigl(B_x e^\eta_{\vartheta-1}\bigr)(x,x,\tau)dx$$

holds for $\tau \le 0$ *and* $\vartheta \ge 1$ *with* $c_{(\vartheta)} = \vartheta$ *for* $\vartheta > 1$, $c_{(1)} = 1$.

Proof. Let $E^\eta(\tau)$ be the spectral projector of A^η. Let us differentiate the equality

$$\partial_\tau \operatorname{Tr} E^\eta(\tau) = \operatorname{Res}_{\mathbb{R}} \bigl(\tau - A^\eta\bigr)^{-1},$$

which holds for $\tau < 0$ in the distribution sense, with respect to η. We obtain

$$\partial_\tau \partial_t \operatorname{Tr} E^\eta(\tau) = -\partial^2_\tau \operatorname{Tr} BE^\eta(\tau).$$

This equality and the semiboundedness of A^η from below yield that

$$\partial_\eta \operatorname{Tr} E^\eta(\tau) = -\partial_\tau \operatorname{Tr} BE^\eta(\tau).$$

Integrating over η from 0 to 1 and convolving with $\tau_+^{\vartheta-1}$ we obtain (12.4.18) for $\tau < 0$. Semicontinuity from the left also yields (12.4.18) for $\tau = 0$. □

Let us assume that the coefficients b_α of the operator $B = A - A^0$ satisfy the conditions

$$\text{(12.4.19)} \qquad |D^\beta b_\alpha| \le c|x|^{\mu(m-|\alpha|)+\nu-|\beta|} \qquad \forall \alpha : |\alpha| \le m \quad \forall \beta : |\beta| \le K$$

with $\nu > 0$ and

$$\text{(12.4.20)} \qquad \operatorname{supp} b_\alpha \subset B(0,r), \qquad 1 \ge r \ge r_0 = h^{(\mu+1)^{-1}}$$

where r is a parameter as well as h.

Let us apply the Weylian approximation to the right-hand expression; this means that we should replace the integrand by the convolution of

$$(2\pi)^{-d} \int \operatorname{Res}_{\mathbb{R}} \operatorname{tr}\bigl(B(\tau - A)^{-1}\bigr)_w(x,\xi)d\xi$$

with $\tau_+^{\vartheta-1}$; recall that the subscript w means the Weyl symbol and all calculations are done in the context of the formal calculus of pseudo-differential operators. In this situation let us reverse the deduction of (12.4.18). We then obtain the routine answer

$$\sum_{0 \le j \le [\vartheta]} h^{-d+j} \int \bigl(\varkappa^1_{\vartheta,j}(x) - \varkappa^0_{\vartheta,j}(x)\bigr)dx$$

where $\varkappa^\eta_{\vartheta,j}$ are the Weylian coefficients for the operator A^η. However, we gain a remainder estimate: calculating the remainder estimate[23)] in our Weylian approximation we obtain

[23)] More precisely, we apply the Weylian approximation only in the semiclassical zone, i.e., for $r \ge r_0$. In the ball $B(0,1)$ we apply a routine estimate of Chapter 8 and obtain that the contribution of this ball to the asymptotics does not exceed $h^{l'}$ with l' given by (12.4.24). Surely we only treated the case of a bounded function B in Chapter 8. However, it follows from (12.4.3) that on $C_0^m(B(0,2r_0),\mathbb{K})$ the operator A satisfies the inequalities $A \ge -C_0 h^l$, $A + C_0 h^l \ge h^{l-2mq(m+1)^{-1}} L^*L$ for any operator L of the same type as A but with m replaced by $\frac{q}{2}$, $q \le m$; then the routine estimates of Chapter 8 yield the estimate $\operatorname{Tr} ME(-\infty,0) \le h^{mq(m+1)^{-1}}$ for $M = L^*L$ with M supported in $B(0,\frac{3}{2}r_0)$ and therefore the same estimate holds for every operator M supported in $B(0,r^*)$ of the same type as A but with m replaced by q.

Proposition 12.4.4. *Let the operator $A^\eta = A+\eta B$ satisfy conditions (12.4.2)–(12.4.6), (12.4.19), (12.4.20) uniformly on $\eta \in [0,1]$. In addition, let the negative spectrum of A^η be discrete and let the condition*

(12.4.21) The symbol $A(xr, \xi r^\mu) r^{-m\mu}$ is microhyperbolic in (x,ξ) at the energy level 0 uniformly with respect to $x \in \mathbb{S}^{d-1}$, $\xi \in \mathbb{R}^d$, $r \in (0,1)$.

be fulfilled. Then the estimate

$$\Big| \int \operatorname{tr}\Big(\big(e^1_\vartheta(x,x,0) - e^0_\vartheta(x,x,0)\big) - \sum_{0\le j\le [\vartheta]+1} \big(\varkappa^1_{\vartheta,j}(x) - \varkappa^0_{\vartheta,j}(x)\big) \phi\big(\frac{x}{r_0}\big) h^{-d+j}\Big) dx \Big| \le CR^*_1 \tag{12.4.22}$$

holds with

$$R^*_1 = \begin{cases} h^{-d+\vartheta} r^{m\mu\vartheta + (d-\vartheta)(\mu+1)+\nu} & \text{for } \vartheta < \bar\vartheta_1, \\ h^{-d+\vartheta} |\log r| & \text{for } \vartheta = \bar\vartheta_1, \\ h^{l'} & \text{for } \vartheta > \bar\vartheta_1 \end{cases} \tag{12.4.23}$$

and

$$\begin{aligned} &\bar\vartheta_1 = \big(d(\mu+1)+\nu\big)(\mu+1-m\mu)^{-1} > \bar\vartheta_0, \\ &l' = l_\nu = (m\mu\vartheta + \nu)(\mu+1)^{-1} > l. \end{aligned} \tag{12.4.24}$$

Recall that $r_0 = h^{(\mu+1)^{-1}}$, ϕ is a fixed function with $1-\phi \in C^K_0(\mathbb{R}^d)$, and ϕ vanishes in a neighborhood of 0.

Moreover, for $\vartheta < \bar\vartheta_1$ one can replace ϕ by 1 in all the terms; one can also do this for $\vartheta = \bar\vartheta_1$, skipping terms with $j = \vartheta$; furthermore, for $\vartheta > \bar\vartheta_1$ one can skip terms with $j > l' + d$ and one can replace ϕ by 1 in terms with $j < l' + d$.

Furthermore, let $X = \mathbb{R}^d$ and let conditions (12.4.2), (12.4.4) be fulfilled in $\mathbb{R}^d$ instead of $B(0,1)$. Applying the arguments of Chapter 8 one can prove the estimate

$$\Big| \int \big(1 - \psi(\frac{x}{r_0})\big) \big(e^1_\vartheta(x,x,0) - e^0_\vartheta(x,x,0)\big) dx \Big| \le C h^{-d+\vartheta+1} \tag{12.4.25}$$

provided $\psi = 1$ in $B(0,\frac{1}{3})$ [24] and $r \le \epsilon r_1$ where $r_1 > 0$ is an arbitrarily small constant and $\epsilon > 0$ is sufficiently small.

Moreover, the local spectral asymptotics of Chapter 4 applied in the zone $\{1 \ge |x| \ge \frac{1}{4} r\}$ yield that if the microhyperbolicity condition (12.4.21) is fulfilled for the operator $A = A^0$ (in which case it is also fulfilled for $A = A^\eta$ for $r \le r_1$) then the estimate

$$\Big| \int \operatorname{tr}\big(\psi(\frac{x}{r_1}) - \psi(\frac{x}{r})\big) \big(e^\eta_\vartheta(x,x,0) - \sum_{0\le j\le[\vartheta]+1} \varkappa^\eta_{\vartheta,j} h^{-d+j}\big) dx \Big| \le CR^*_2 \tag{12.4.26}$$

holds with

$$R^*_2 = \begin{cases} h^{-d+\vartheta+1} r^{m\mu\vartheta + (d-\vartheta-1)(\mu+1)} & \text{for } \vartheta > \bar\vartheta, \\ h^{-d+\vartheta+1} \log r & \text{for } \vartheta = \bar\vartheta. \end{cases} \tag{12.4.27}$$

---

[24] This condition is also assumed to be fulfilled below.

Then (12.4.22)–(12.4.23), (12.4.25) and (12.4.26)–(12.4.27) yield the estimate

$$\text{(12.4.28)}\quad |\int \psi(\frac{x}{r})\,\mathrm{tr}\Big(e^1_\vartheta(x,x,0) - e^0_\vartheta(x,x,0) - \sum_{0\le j\le[\vartheta]+1}\big(\varkappa^1_{\vartheta,j}(x) - \varkappa^0_{\vartheta,j}(x)\big)\phi(\frac{x}{r_0})h^{-d+j}\Big)dx| \le CR^*_1 + CR^*_2$$

with R^*_1, R^*_2 given by (12.4.23) and (12.4.27). Moreover, proposition 12.4.2 yields that *one can reject the assumptions that* $X = \mathbb{R}^d$, *that condition* (12.4.2) *is fulfilled in* X *instead of* $B(0,1)$ *and that the negative spectrum of* A *is discrete.* In fact, it is easy to construct an appropriate extension of A to $\mathbb{R}^d$.

Moreover, let $\bar{A}^0$ coincide with A^0 in $B(0,\frac{1}{2}r)$ and be self-adjoint. Then the arguments of Chapter 8 easily yield that

$$|\int \psi(\frac{x}{r})\,\mathrm{tr}\big(e^0_\vartheta(x,x,0) - \bar{e}^0_\vartheta(x,x,0)\big)dx| \le CR^*_1.$$

Furthermore, $\varkappa^0_{\vartheta,j}(x) = \bar{\varkappa}^0_{\vartheta,j}(x)$ in a neighborhood of $\mathrm{supp}\,\psi(\frac{x}{r})$. Therefore, *one can eliminate condition* (12.4.20) *on the support of* B *in the deduction of estimate* (12.4.28).

So, *estimate* (12.4.28) *remains true under the following weakened assumptions:* A^0 *is an operator in* $\mathbb{R}^d$ *satisfying the conditions of proposition 12.4.1,* A^1 *is an operator in the domain* $X \supset B(0,1)$ *satisfying conditions* (12.4.6) *and* $b_\alpha = a_\alpha - a^0_\alpha$ *satisfy* (12.4.19); here and below we skip the superscript "1."

Again applying estimates (12.4.26)–(12.4.27) we obtain that estimate (12.4.28) remains true under weakened assumptions if we replace r by r_0 in its left-hand expression. Moreover, dividing the left-hand expression into two parts (associated with A, $\varkappa_{\vartheta,j}$ and with A^0, $\varkappa^0_{\vartheta,j}$) we see that in the second part one can replace r by arbitrary $r_2 > r$; moreover, one can even pick $r_2 = +\infty$, i.e., replace ψ by 1 here. Minimizing the right-hand expression by $r \in [r_0, \epsilon r_1]$ (the left-hand expression surely does not depend on r) we obtain the estimate

$$\text{(12.4.29)}\quad |\int \psi(\frac{x}{r_2})\,\mathrm{tr}\Big(e_\vartheta(,x,x,0) - \sum_{0\le j\le[\vartheta]+1}\varkappa_{\vartheta,j}\phi(\frac{x}{r_0})h^{-d+j}\Big)dx - \int \psi(\frac{x}{r_1})\,\mathrm{tr}\Big(e^0_\vartheta(,x,x,0) - \sum_{0\le j\le[\vartheta]+1}\varkappa^0_{\vartheta,j}\phi(\frac{x}{r_0})h^{-d+j}\Big)dx| \le CR^*$$

with

$$\text{(12.4.30)}\quad R^* = \begin{cases} h^{-d+\vartheta+1}|\log h| & \text{for } \vartheta = \bar{\vartheta}_0,\\ h^{l''} & \text{for } \bar{\vartheta}_0 < \vartheta < \bar{\vartheta}_1,\\ h^{l'}|\log h| & \text{for } \vartheta = \bar{\vartheta}_1,\\ h^{l'} & \text{for } \vartheta > \bar{\vartheta}_1 \end{cases}$$

where

$$\text{(12.4.31)}\quad l'' = \big(m\mu\vartheta - (d-\vartheta-1)\nu\big)(\mu+\nu+1)^{-1}.$$

One can easily see that $l < l'' < l'$ for $\bar{\vartheta}_0 < \vartheta \le \bar{\vartheta}_1$, $l'' = l = -d+\vartheta+1$ for $\vartheta = \bar{\vartheta}_0$, $l'' = l'$ for $\vartheta = \bar{\vartheta}_1$.

Let us pick $r_2 = +\infty$ and apply proposition 12.4.1 with $J = [\vartheta]+1$. We obtain the asymptotics

$$\int \psi(\frac{x}{r_0}) \operatorname{tr} e_\vartheta(x,x,0)dx = \sum_{0\le j\le [\vartheta]+1} h^{-d+j}\int \operatorname{tr} \varkappa_{\vartheta,j}(x)\phi(\frac{x}{r_0})dx + \omega h^{l'} + R^* \tag{12.4.32}$$

with R^* given by (12.4.30).

Therefore, we have proven the following

Theorem 12.4.5. *Let the operator A satisfy conditions (12.4.2)–(12.4.6). Furthermore, let $b_\alpha = a_\alpha - a^0_\alpha$ satisfy (12.4.19) with (matrix-valued) coefficients a^0_α which are positively homogeneous of degrees $(m-|\alpha|)\mu$. Let $\vartheta \ge \max(\bar\vartheta_0, 1)$. Then asymptotics (12.4.32) with Weylian coefficients $\varkappa_{\vartheta,j}(x)$ and some (non-Weylian) coefficient ω and with the remainder estimate given by (12.4.30) hold. Recall that $r_0 = h^{(\mu+1)^{-1}}$ and that in the terms with $j < \vartheta + 1$ one can replace ϕ by 1 (changing ω).*

Let us note that here we gain nothing for $\vartheta = \bar\vartheta_0$.

12.4.4 Adjustment for $\vartheta < 1$

Let us consider the case $\vartheta < 1$ when one should modify the above arguments. Let us pick $\tau = 0$. Moreover, let us consider a partition unity on $\mathbb{R}$ corresponding to the scaling function $\frac{1}{2}(|\tau| + L)$ where we choose L later. Then we obtain a sum of terms of the form

$$\vartheta\iint \lambda_-^{\vartheta-1}\varphi(\frac{\lambda}{\ell})\psi(x)e(x,x,\lambda)d\lambda dx$$

with either $\varphi = \varphi_0 \in C_0^K([-1,1])$ and $\ell = L$ or $\varphi = \varphi_1 \in C_0^K([\frac{1}{2},1])$ and $\ell \ge L$. We apply this decomposition only with $\psi(\frac{x}{r})$ instead of ψ (we will choose $r \in [r_0, 1]$ later). Let us apply the arguments of the previous subsection which are not associated with proposition 12.4.3 (and therefore do not require the assumption $\vartheta \ge 1$). In particular, let us apply estimates (12.4.25) and (12.4.26).[25] We then obtain that one needs to evaluate

$$\vartheta\iint \lambda_-^{\vartheta-1}\varphi_0(\frac{\lambda}{\ell})\psi(\frac{x}{r}) \operatorname{tr} e^\eta(x,x,\lambda)d\lambda dx \tag{12.4.33}$$

and

$$\vartheta\iint \lambda_-^{\vartheta-1}\varphi_1(\frac{\lambda}{\ell}) \operatorname{tr}\bigl(e(x,x,\lambda) - e^0(x,x,\lambda)\bigr)d\lambda dx \tag{12.4.34}$$

where $\eta = 0,1$, the error $O(R_2^*)$ $(= O\bigl(h^{-d+1+\vartheta} r^{m\mu\vartheta + (d-\vartheta-1)(\mu+1)}\bigr)$ in this case) is already accounted for, R_2^* is given by (12.4.27) and without loss of generality one can assume that conditions (12.4.2), (12.4.4) are fulfilled in $\mathbb{R}^d$.

Applying standard Weylian asymptotics to (12.4.33) one gets the remainder estimate $O(L^\vartheta h_{\text{eff}}^{1-d})$. One can apply the arguments of the proof of proposition 12.4.3 to (12.4.34) and represent it in the form

$$\int_0^1 dt \iint \bigl(\partial_\tau^2 \lambda_-^\vartheta \varphi_1(\frac{\lambda}{\ell})\bigr) \operatorname{tr}(B_x e^\eta)(x,x,\lambda)d\lambda dx. \tag{12.4.35}$$

Let us apply Weylian asymptotics to this expression. We obtain the remainder estimate $O(r^{m\mu+\nu}\ell^{\vartheta-1}h_{\text{eff}}^{1-d})$ provided

[25] In order to get these with the convolutions, including the cutoff function φ, one should apply theorem 4.3.6 instead of 4.3.14.

(12.4.36) $$m\mu+\nu+(d-1)(\mu+1)>0.$$

After summation on the partition of unity we obtain that the total contribution of the terms (12.4.34) to the remainder estimate is $O(r^{m\mu+\nu}L^{\vartheta-1}h_{\mathrm{eff}}^{1-d})$ while the contribution of (12.4.33) is $O(L^{\vartheta}h_{\mathrm{eff}}^{1-d})$. Minimizing the sum with respect to L (the optimal value is $\asymp r^{m\mu+\nu}$) we obtain

(12.4.37) $$O\bigl(h^{1-d}r^{(m\mu+1)\vartheta+(\mu+1)(d-1)}\bigr).$$

Recall that there is also an error $O\bigl(h^{-d+1+\vartheta}r^{m\mu\vartheta+(d-\vartheta-1)(\mu+1)}\bigr)$. Minimizing their sum over r (the optimal value is $\asymp r_3=h^{(\mu+\nu+1)^{-1}}$) we obtain the final remainder estimate $O\bigl(h^{l''}\bigr)$ where l'' is given by (12.4.31).

So, the following statement has been proven:

Theorem 12.4.5$'$. *Let all the conditions of theorem 12.4.5 be fulfilled excluding $\vartheta\ge 1$. Let $\bar\vartheta_0<\vartheta<1$ and condition* (12.4.36) *be fulfilled (which means exactly that $\bar\vartheta_1>1$). Then asymptotics* (12.4.32) *remain true with the Weylian coefficients $\varkappa_{\vartheta,j}(x)$ and some (non-Weylian) coefficient ω and with the remainder estimate given by* (12.4.30).

Remark. *Surely, even if condition (12.4.36) is violated one can get semiclassical asymptotics for (12.4.35) but the remainder estimate will be worse and the final remainder estimate will be also worse. However, this condition is only technical, and I am going to get rid of it without deterioration of the remainder estimate in the general approach.* ◊

12.4.5 The general case

Let us fix $r_3=h^{(1+\mu+\nu)^{-1}}$ and let us apply the same arguments as before in the zone $\{r_3\le|x|\le 1\}$. Then *the contribution of this zone to the remainder estimate is $O\bigl(R_2^*(r_3)\bigr)$ as before.*

Let us consider $B(0,r_3)$ and apply the same partition of unity here. Thus one should consider the same $\int\tau_-^{\vartheta}\,\mathrm{Tr}\,\psi d_\tau E(\tau)$ as before. The index of ψ is skipped. Let $r=\frac12(|x|+r_0)$ at some point of $\operatorname{supp}\psi$ and let us fix $L=L(r)=hr^{m\mu+\nu-\mu-1}$ for this element. Let us consider the same partition on $\mathbb{R}$ as before and let us replace $\mathrm{Tr}\,\psi d_\tau E(\tau)$ with its Weylian expression (with cutoff for $|x|\le r_0$). Then the contribution of the element with $\ell\le 2L$ to the remainder estimate does not exceed $CR(r)$ with

(12.4.38) $$R(r)=h^{-d+1}r^{(d-1)(\mu+1)}L^{\vartheta}=h^{-d+1+\vartheta}r^{(m\mu+\nu)\vartheta+(d-1-\vartheta)(\mu+1)}.$$

Then summing on the partition of unity on x one obtains CR^* with

(12.4.39) $$R^*=\begin{cases} h^{l''} & \text{for } (m\mu+\nu)\vartheta+(d-1)(\mu+1)>0\\ h^{l'}\log h & \text{for } (m\mu+\nu)\vartheta+(d-1)(\mu+1)=0\\ h^{l'} & \text{for } (m\mu+\nu)\vartheta+(d-1)(\mu+1)<0\end{cases}$$

where

(12.4.40) $$l''=\bigl((m\mu+\nu)\vartheta-(d-1)\nu\bigr)(\mu+\nu+1)^{-1},$$ $$l'=(m\mu+\nu)\vartheta(\mu+1)^{-1}.$$

Moreover, the contribution of $B(0,3r_0)$ does not exceed $Ch^{l''}$.

Let us consider elements with $\ell\ge L$. Then one can apply the same arguments as before with k differentiations instead of 1 and easily get the following generalization of (12.4.18):

(12.4.41)
$$\int \varphi(\tau)\tau_-^{\vartheta} d_\tau E^1(\tau) =$$
$$\sum_{0\le j\le k-1} \int \varphi_{\vartheta,j}(\lambda_1,\dots,\lambda_{j+1})E^0(\lambda_1)BE^0(\lambda_2)\dots BE^0(\lambda_{j+1})d\lambda_1\cdots d\lambda_{j+1} +$$
$$\int_0^1 (k+1)(1-\eta)^k dt \times$$
$$\int \varphi_{\vartheta,k+1}(\lambda_1,\dots,\lambda_{k+1})E^\eta(\lambda_1)BE^\eta(\lambda_2)\dots BE^\eta(\lambda_{k+1})d\lambda_1\cdots d\lambda_{k+1}$$

where $E^\eta(\lambda)$ is the spectral projector of A^η and

(12.4.42)
$$\varphi_{\vartheta,j} = (-1)^{j+1}\int \delta(\tau-\lambda_1)\cdots\delta(\tau-\lambda_{j+1})\Bigl(\partial_\tau^{j+1}\bigl(\tau_-^{\vartheta}\varphi(\tau)\bigr)\Bigr)d\tau,$$

understood as a distribution on $(\lambda_1,\dots,\lambda_j)$, is in fact a smooth function such that

(12.4.43)
$$\operatorname{supp}\varphi_{j,\vartheta}\subset[\frac{1}{2}\ell,2\ell]^j,\qquad |D_\lambda^\beta\varphi_{\vartheta,j}|\le c\ell^{\vartheta+1-2j-|\beta|}.$$

Let us multiply by ψ and take the trace. Our goal is to evaluate the result.

First of all I am going to evaluate the elements with $r\ge 2r_0$. For this goal let us return to Chapters 4 and 5. Under the standard assumptions of those chapters let us consider the following problems:

$$(12.4.44)_0\qquad (hD_t-A)U_0(t)=0,\qquad U_0(0)=I$$
$$(12.4.44)_j\qquad (hD_t-A)U_j(t)=B_jU_{j-1}(t),\qquad U_j(0)=0,\qquad j\ge 1$$

where the operators B_k are also regular h-ψdos of order 0. Then the results of Chapters 2 and 3 concerning the oscillatory front set of $U=U_0$ obviously remain true for U_j as well. In particular, under the microhyperbolicity condition for the operator A at the energy level 0 (with boundary conditions, if there is a boundary), $\varphi(hD_t)\chi_T'(t)\Gamma\psi U_j$ is negligible for $\varphi\in C_0^K$ supported in a small neighborhood of 0, $\chi'\in C_0^K([-1,-\frac{1}{2}]\cup[\frac{1}{2},1])$ and a sufficiently small constant $T>0$ (compare with proposition 2.1.14 and theorem 3.2.5). Applying rescaling-decomposition arguments one can get estimate (2.1.35) for $U=U_j$ as well (compare with theorems 2.1.15, 3.2.6). On the other hand, one can apply the successive approximation method in the interval $|t|\le h^{\frac{1}{2}+\delta}$. Let us note that for $B_1=B_2=\dots=B_j=B$,

$$\varphi(hD_t)\Gamma\operatorname{Tr}\psi U_j(t)\big|_{t=0}$$

is exactly the term to be evaluated (for $j=k+1$ one should integrate over the parameter η, and A depends on this parameter. Thus for $r\asymp 1$ the remainder estimate (after we replace the whole thing by its Weylian expression) would be $Ch^{1-d}\ell^{\vartheta}\bigl(1+\frac{\ell}{h}\bigr)^{-s}$ where we have taken into account condition (12.4.43) on φ.

Remark 12.4.6. *The procedure to get the Weylian approximation is obvious: one replaces $d_\tau E(\tau)$ by the formal parametrix $(\tau-A)^{-1}$ of the operator A (with the corresponding boundary conditions), then takes $A=A^\eta=A_0+\eta B$ and calculates the j-th derivative with respect to η. Afterward one applies* $\operatorname{Res}_{\mathbb{R}}$ *and formally calculates the trace of this operator.* ◊

Rescaling, we obtain that for $r\ge r_0$ the contribution to the remainder estimate is

$$C\ell^{\vartheta-j}r^{(m\mu+\nu)j}\bigl(\frac{h}{r^{\mu+1}}\bigr)^{1-d+\vartheta-j}\bigl(1+\frac{\ell r^{\mu+1-m\mu}}{h}\bigr)^{-s}.$$

For $j > \vartheta$ we obtain that the sum with respect to $\ell \geq L(r)$ does not exceed this expression with $\ell = L(r)$ which is equal to $CR(r)$ with $R(r)$ given by (12.4.38) and after summation over the partition on r we get CR^* with with R^* given by (12.4.39). *Let us pick* $k > \vartheta$.

In order to treat $j \leq \vartheta$ let us replace $\sum_{r \leq r_3} \sum_{\ell \geq L(r)}$ by $\sum_{\ell \geq L^*} \sum_{r \leq r(L)}$ where $L^* = L(r_0) = h^{m\mu(\mu+1)^{-1}}$ and $r(\ell) = \left(\frac{\ell}{h}\right)^{(m\mu+\nu-\mu-1)^{-1}}$. Let us now consider the sum on r only. For

$$f(j) \stackrel{\text{def}}{=} m\mu\vartheta + (\mu+1)(d-1-\vartheta) + (\mu+\nu+1)j > 0$$

we replace the corresponding term in the partition of the right-hand expressions of (12.4.41) by its Weylian approximations. Then the error will not exceed $CR'(\ell)$ with $R'(\ell) = R\big(r(\ell)\big)$. After summation on $\ell \geq L^*$ one again obtains CR^*.

For $f(j) = 0$ one gets the same answer with the additional factor $|\log h|$ in the right-hand expression. Thus, let us replace R^* by

$$R^{**} = \begin{cases} h^{l''} & \text{for } (m\mu+\nu)\vartheta + (d-1)(\mu+1) > 0, \\ & \quad \vartheta - (\mu+1)(d-1)(\mu+\nu+1)^{-1} \notin \mathbb{Z}^+ \\ h^{l''}|\log h| & \text{for } (m\mu+\nu)\vartheta + (d-1)(\mu+1) > 0, \\ & \quad \vartheta - (\mu+1)(d-1)(\mu+\nu+1)^{-1} \in \mathbb{Z}^+ \\ h^{l'}\log h & \text{for } (m\mu+\nu)\vartheta + (d-1)(\mu+1) = 0 \\ h^{l'} & \text{for } (m\mu+\nu)\vartheta + (d-1)(\mu+1) < 0. \end{cases} \tag{12.4.45}$$

Let us consider the terms with $f(j) < 0$. We replace them by Weylian expressions as well but treat them in a different way. First, we take the difference between the terms and their Weylian approximations on an element of the partition of unity, then sum the difference with respect to $r \leq r(\ell)$, then replace $\sum_{r \leq r(\ell)}$ by $\sum_{r<\infty} - \sum_{r>r(\ell)}$. For $r > r(\ell)$ let us apply the above remainder estimates and get the same answer as before. One can easily see that the contribution of $B(0,r_0) \times [0, 3L^*]$ to the original expression does not exceed $Ch^{l''}$. Later we will prove that the contribution of $B(0,2r_0)$ to terms with $f(j) > 0$ does not exceed $Ch^{l''}$ as well. Then modulo $O(R^{**})$ our original expression is equal to

$$\text{Weyl} + \sum_{j:f(j)<0} c_j \iint \tau_-^{\vartheta-1} \operatorname{tr}\big(\mathrm{K}_{(B(\tau-A^0)^{-1})^j} - \text{Weyl}_j\big)\,dx d\tau$$

where the first "Weyl" is some combination of the Weylian expressions and "Weyl_j" is the Weylian approximation to $\mathrm{K}_{(B(\tau-A^0)^{-1})^j}$ where K_T is the Schwartz kernel of T; c_j are some numerical coefficients. All the Weylian expressions contain the factor $\big(1 - \phi(\frac{x}{r_0})\big)$ cutting them in $B(0,r_0)$.

Now, let us assume for the sake of simplicity that

(12.4.46) The operator B is quasi-homogeneous; i.e., $B(\zeta x, \zeta^\mu \xi) = \zeta^{m\mu+\nu} B(x,\xi)$ $\forall x, \xi, \zeta > 0$.

(later we will get rid of this condition with the possible deterioration of the remainder estimate, and it is not necessary from the very beginning if $f(1) \geq 0$). Under this condition let us consider terms with $f(j) < 0$ and with integration on $x \in \mathbb{R}^d$. The same homogeneity arguments as in proposition 12.4.1 give us that these terms are equal to $\omega_j h^{(m\mu\vartheta+\nu j)(\mu+1)^{-1}}$ with unknown numerical coefficients ω_j. Moreover, $f(j) < 0$ means exactly that $(m\mu\vartheta + \nu j)(\mu+1)^{-1} < \min(l', l'')$.

Moreover, one can easily see that modulo $O(R^{**})$ the first "Weyl" is equal to

$$\sum_{p<d+\min(l',l'')} \kappa_{\vartheta,p} h^{-d+p} + \sum_{\substack{p<d+\min(l',l'')\\ j:-d+p=(m\mu\vartheta+\nu j)(\mu+1)^{-1}}} \kappa'_{\vartheta,p} h^{-d+p}\log h$$

plus the sum of the terms $c_{p,j}h^{-d+p+\nu j}r_3^{\alpha_{p,j}}$ with $\alpha_{p,j} = (\mu+1)(d-p-\nu j) \neq 0$ and $c_{p,j}h^{-d+p+\nu j}\log r_3$ with $\alpha_{p,j}=0$. Here $\kappa_{\vartheta,p}$ are standard Weylian coefficients (one can easily see that the integrals defining them converge (at 0) for $p < d + m\mu\vartheta(\mu+1)^{-1}$. Taking into account the fact that one can pick any $r_3 \asymp h^{(\mu+\nu+1)^{-1}}$ we see that all these coefficients $c_{p,j}$ should vanish. Therefore the final answer is

$$\sum_{p<d+\min(l',l'')} \kappa_{\vartheta,p} h^{-d+p} + \tag{12.4.47}$$

$$\sum_{p<d+\min(l',l''),j:-d+p=(m\mu\vartheta+\nu j)(\mu+1)^{-1}} \kappa'_{\vartheta,p} h^{-d+p}\log h$$

$$\sum_{j:(m\mu\vartheta+\nu j)(\mu+1)^{-1}<\min(l',l'')} \omega_j h^{(m\mu\vartheta+\nu j)(\mu+1)^{-1}}$$

where $\omega_j h^{(m\mu\vartheta+\nu j)(\mu+1)^{-1}}$ are (multiple) Scott correction terms. More precisely, defining

$$\bar\vartheta_J = \max\Bigl(\bigl((\mu+1)(d-1) + (\mu+\nu+1)J\bigr)(\mu+1-m\mu)^{-1}, j\Bigr) \tag{12.4.48}$$

we see that

For $\bar\vartheta_{J-1} < \vartheta \leq \bar\vartheta_J$ there are exactly J Scott correction terms $(j = 1,\dots.)$

Let us now consider an element of the partition supported in $B(0,2r_0)$. Let us replace every factor B by $B_1 = \bar\psi B$ where the cutoff function $\bar\psi$ is supported in $B(0,4r_0)$ and equal to 1 in $B(0,r_0)$. I will explain how to get rid of this cutoff function later. Let us note that for $\tau \leq 0$ the operator norm of $\bar\psi BE^\eta(\tau)$ does not exceed $r_0^{m\mu+\nu}$ while the trace norm of $\bar\psi E^\eta(\tau)$ does not exceed C. One can prove this easily for $h = r_0 = 1$ and then extend to the general case by rescaling (recall that $r_0^{\mu+1} = h$).[26] Then the trace norm of the operator

$$\int \varphi_{\vartheta,k+1}(\lambda_1,\dots,\lambda_j)\psi E^\eta(\lambda_1)B_1E^\eta(\lambda_2)\dots B_1E^\eta(\lambda_{k+1})d\lambda_1\cdots d\lambda_{k+1}$$

does not exceed $\ell^{\vartheta-k}r_0^{(m\mu+\nu)k}$ and after summation on the partition we get the same expression with $\ell = L = r_0^{m\mu+\nu}$ provided $k > \vartheta$. Namely, we get $r_0^{(m\mu+\nu)\vartheta}$.

Finally, we should consider similar expressions with some copies of B replaced by $\bar B$ and other copies replaced by $\psi' B$ where ψ' is supported outside of $B(0,3r_0)$. Taking partitions one can replace the i-th copy of ψ' by ψ_i supported in $\{\frac12 r_i \leq |x| \leq 2r_i\}$. Combining the above arguments one can estimate this term through $r_0^{(m\mu+\nu)\vartheta}\prod_i\bigl(\frac{r_0}{r_i}\bigr)^s$ and the total contribution will not exceed $r_0^{m\mu+\nu}$.

Thus, all the estimates are done. Therefore the first statement of the following theorem has been proven:

---

[26] One can add the factor $\bigl(|\tau|r_0^{-m\mu}+1\bigr)^{-s}$ with arbitrarily large s to the right side of these estimates but it is not important here.

Theorem 12.4.7. *Let the conditions of theorem 12.4.5 be fulfilled.*

(i) Moreover, let condition (12.4.46) *be fulfilled. Then the asymptotics*

$$\int \psi(x) e_\vartheta(x,x,\tau)\,dx = \sum_{p:p<d+\min(l',l'')} \kappa_{\vartheta,p} h^{-d+p} + \sum_{p<d+\min(l',l''),j:-d+p=(m\mu\vartheta+\nu j)(\mu+1)^{-1}} \kappa'_{\vartheta,p} h^{-d+p}\log h + \sum_{j:(m\mu\vartheta+\nu j)(\mu+1)^{-1}<\min(l',l'')} \omega_j h^{(m\mu\vartheta+\nu j)(\mu+1)^{-1}} + O\bigl(R^{**}(h)\bigr) \tag{12.4.49}$$

hold with $R^{**}(h)$ *given by* (12.4.45).

(ii) Moreover, for $\vartheta \le \bar\vartheta_1$ *one can reject condition* (12.4.46).

(iii) Let us assume that

$$\bigl|D^\beta(a_\alpha - a^0_\alpha - b^0_\alpha)\bigr| \le c|x|^{(m-|\alpha|)\mu+\nu'-|\beta|} \tag{12.4.50}$$

where a^0_α, b^0_α *are positively homogeneous of degrees* $(m-|\alpha|)\mu$, $(m-|\alpha|)\mu+\nu$ *respectively,* $\nu' > \nu$. *Then asymptotics* (12.4.49) *hold with* R^{**} *replaced by* $R^{**}+R^*$ *where* R^* *is given by* (12.4.30) *with* ν' *instead of* ν.

The proof of statement (iii) is simple: we first make the transition from A to A^0+B^0 (then the remainder estimate $O(R^*)$ arises) and then apply (i). I believe that it would not be difficult to generalize this theorem to the case when $D^\beta(b_\alpha - b^1_\alpha - \dots - b^s_\alpha) = O\bigl(|x|^{(m-\alpha)\mu+\nu_{s+1}-|\beta|}\bigr)$ and b^k_α are positively homogeneous of degrees $(m-|\alpha|)\mu+\nu_k$.

12.4.6 Propagation of singularities

Now our goal is to improve the remainder estimate, which requires some condition of a global nature on the symbol $A^0(x,\xi)$. To formulate the condition the reader must recall the definitions of generalized bicharacteristics and generalized δ-bicharacteristics (see definition 2.2.4); otherwise we should assume that the eigenvalues of $A^0(x,\xi)$ which are close to 0 have constant multiplicities. Moreover, the reader should remember quasi-conical sets and the quasi-conical hull of a set; these notions are associated with the dilatation family $\Phi_\zeta : (x,\xi) \to (\zeta x, \zeta^\mu \xi)$ with $\zeta > 0$.

Let us assume that

(T) There exist quasi-conical sets Λ^+ and Λ^- closed in $(\mathbb{R}^d\setminus 0)\times\mathbb{R}^d$ and numbers $T_0>0$, $\epsilon_1>0$ such that

$$\Lambda^+\cup\Lambda^- = \Sigma^0 = \bigl\{(x,\xi)\in(\mathbb{R}^d\setminus 0)\times\mathbb{R}^d : \det A^0(x,\xi)=0\bigr\} \tag{12.4.51}$$

and for every point $(\bar x,\bar\xi)$ belonging to the ϵ_1-neighborhood of $\Lambda^\pm$ intersected with $\bigl\{(x,\xi): |x|=1\bigr\}\cap\Sigma^0$ all the generalized bicharacteristics of the symbol $-A^0(x,\xi)$ passing through $(\bar x,\bar\xi)$ in the direction of $\pm t>0$ are infinitely long with respect to t, and along them

$$(x(t),\xi(t))\in\Lambda^\pm,\quad |x(t)|\ge 2|\bar x| \qquad \forall t: \pm t\ge T_0; \tag{12.4.52}$$

these conditions should be fulfilled for every sign "$\pm$" separately.

We discuss this condition below; here we make certain corollaries and remarks concerning it.

Remark 12.4.8. *(i) Condition (12.4.4) yields that* $\Sigma^0 \cap \{|x| = 1\}$ *is compact.*

(ii) For every generalized bicharacteristic $(x(t), \xi(t))$ *with* $(x(0), \xi(0)) \in \Lambda^{\pm}_{\epsilon_1}$ *for* $\pm t > 0$ *the inequalities*

$$\epsilon_0 \le |x(t)|\bigl(|x(0)| + |t|^{\theta}\bigr)^{-1} \le C_0 \tag{12.4.53}$$

hold with $\theta^{-1} = \theta' = 1 + \mu - m\mu > 0$ *(otherwise* $\bar{\vartheta}_0 = +\infty$*) and with appropriate constants* ϵ_0, C_0 *which do not depend on* t *and* $(x(0), \xi(0))$*; here and below* $\Lambda^{\pm}_{\varepsilon}$ *denotes the intersection of the quasi-conical hull of the* ε*-neighborhood of* $\Lambda^{\pm} \cap \{|x| + 1\}$ *and* Σ^0*.*

In fact, let us pick t_n*:* $t_0 = 0$, $t_{n+1} = t_n + T_0 r_n^{\theta'}$ *with* $r_n = |x(t_n)|$*. Then (12.4.52) and the quasi-homogeneity of* A^0 *and the quasi-conic nature of* Λ^+ *yield that* $2r_n \le r_{n+1} \le c_0 r_n$ *(we have taken* $(x(0), \xi(0)) \in \Lambda^+$ *for notational simplicity). Therefore* $r_n \to +\infty$ *as* $n \to +\infty$*. Then these inequalities and the definition of* t_{n+1} *obviously yield the inequality* $r_{n+1} \le c_1 t_{n+1}^{\theta}$ *(and hence* $t_n \to +\infty$ *as* $n \to +\infty$*) and the inequality* $t_{n+1} r_{n+1}^{-\theta'} \le c_2 + (1 - \epsilon) t_n r_n^{\theta'}$ *with* $\epsilon > 0$*. This inequality yields by induction that* $t_n r_n^{-\theta'} \le c_3$ *with a sufficiently large constant* c_3*. So, both inequalities (12.4.53) are fulfilled for* $t = t_n$ *with* $n = 0, 1, \dots$ *and therefore these inequalities are fulfilled for* $t \in (t_n, t_{n+1}]$ *as well.*

(iii) Let us replace the symbol $A^0(x, \xi)$ *with another symbol* $A^1(x, \xi)$ *also of the form (12.4.1) such that*

$$\bigl|D^{\beta}(a^1_{\alpha} - a^0_{\alpha})\bigr| \le \varepsilon |x|^{\mu(m - |\alpha|) - |\beta|} \qquad \forall \alpha : |\alpha| \le m \quad \forall \beta : |\beta| \le K \tag{12.4.54}$$

with a small enough constant $\varepsilon > 0$*. We do not assume that* $A^1(x, \xi)$ *is a quasi-homogeneous symbol. Then*

$$|\xi| \le c_3 |x|^{\mu} \qquad \text{on } \Sigma^j \qquad j = 0, 1 \tag{12.4.55}$$

and the microhyperbolicity of A^0 *at the energy level* 0 *yields that*

$$\forall (x, \xi) \in \Sigma^0 \quad \exists (x', \xi') \in \Sigma^1 : |x - x'| \le c_3 \varepsilon |x|,\ |\xi - \xi'| \le c_4 \varepsilon |x|^{\mu} \tag{12.4.56}$$

and vice versa *(one can permute* Σ^0 *and* Σ^1 *in this assertion), and* c_4 *does not depend on* ε*.*

Let us introduce the sets

$$\Lambda^{\pm}_1 = \bigl\{(x, \xi) \in \Sigma^1 :\ \exists (x', \xi') \in \Lambda^{\pm} : |x - x'| \le \epsilon |x|,\ |\xi - \xi'| \le \epsilon |x|^{\mu}\bigr\} \tag{12.4.57}$$

with sufficiently small $\epsilon > 0$*. Moreover, let us introduce the sets* $\Lambda^{\pm}_2$ *in the same way but with* ϵ *replaced* 2ϵ*. Then the properties of generalized bicharacteristics (see lemma 2.2.5) and condition (T) yield that if* $\epsilon > 0$ *is a small enough constant and if* $\varepsilon = \varepsilon(\epsilon) > 0$ *is also small enough then every generalized bicharacteristic* $(x(t), (\xi(t))$ *of the symbol* $A^1(x, \xi)$ *passing through the point* $(\bar{x}, \bar{\xi}) \in \Lambda^{\pm}_2$ *in the direction of positive* $\pm t$ *can be extended to the time interval* $\pm t \in [0, T_1 |\bar{x}|^{\theta'}]$*, and for* $\pm t \in [2T_0 |\bar{x}|^{\theta'}, T_1 |\bar{x}|^{\theta'}]$ *along this bicharacteristic*

$$(x(t), \xi(t)) \in \Lambda^{\pm}_1, \qquad |x(t)| \ge \frac{3}{2} |\bar{x}|; \tag{12.4.52$'$}$$

$T_1 > 2T_0$ *is arbitrarily large and* ϵ *and* ε *depend on* T_1*. Then the same arguments as in (i) yield that along these trajectories inequalities (12.4.53) are fulfilled with constant* c_0, ϵ_0 *which do not depend on* T_1*. Then the arbitrary choice of* T_1 *yields that this statement remains true even for* $T_1 = +\infty$*.*

(iv) The same analysis remains true if we assume that (12.4.54) is fulfilled in $B(0, r)$ *instead of* $\mathbb{R}^d$*; surely, we should treat the trajectory only as long as it remains in* $B(0, r)$*.*

In this case (12.4.53) yields that all the statements remain true in the time interval $\pm t \in [0, \epsilon_2 r^{\theta'}]$ *with a small enough constant* $\epsilon_2 > 0$ *and* $|\bar{x}| \le \frac{1}{2} r$.

(v) Therefore in the analysis of the trajectories at the energy level τ *instead of* $\tau = 0$ *all the statements (i)–(iii) remain true with the following modifications:*

(a) For $\mu < 0$ *one should replace* r *with* $\min(r, \epsilon_2 |\tau|^{(m\mu)^{-1}})$, *i.e., one should treat the starting points* $(\bar{x}, \bar{\xi})$ *with* $|\bar{x}| \le \min(r, \epsilon_2 |\tau|^{(m\mu)^{-1}})$ *and the time interval* $\pm t \in [0, \epsilon_2 \min(r^{\theta'}, |\tau|^{(m\mu\theta)^{-1}})]$.

(b) For $\mu > 0$ *one should leave* r *and the time interval* $\pm t \in [0, \epsilon_2 r^{\theta'}]$ *unchanged and one should treat the starting points* $(\bar{x}, \bar{\xi})$ *with* $c_2 |\tau|^{(m\mu)^{-1}} \le |\bar{x}| \le r$.

(c) For $\mu = 0$ *one should leave* r *and the time interval* $\pm t \in [0, \epsilon_2 r^{\theta'}]$ *and the condition* $|\bar{x}| \le r$ *unchanged and one should only assume that* $|\tau| \le \epsilon_2$ *where* $\epsilon_2 > 0$ *is a small enough constant.*

(vi) Let us consider a more complicated flow. Namely, let us assume that at the moments $\pm t_n = T_1^n |\bar{x}|^{\theta'}$ *with* $n = 1, 2, \dots$ *the point* $(x(t), \xi(t))$ *makes random "jumps;" namely it jumps to an arbitrary point in its* εr_n*-neighborhood intersected with the set* Ω; $\Omega = \Pi_{\tau_0}$ *for* $\mu \le 0$ *and* $\Omega = \Pi_{\tau_n}$ *with* $\tau_n = \left(\frac{r_n}{r_0}\right)^{m\mu} \tau_0$ *for* $\mu > 0$ *where* $\Pi_\zeta = \cup_{\tau : |\tau| \le \zeta} \Sigma^1_\tau$ *and we assume that* $\tau_0 > 0$. *So, we assume that the "jump" and the flow itself preserve the energy levels only approximately. This exotic flow will be useful later. I assure the reader that for sufficiently small* $\varepsilon > 0$ *and for sufficiently large* $T_1 > 0$, *statements (i)–(v) remain true.*

In fact, let $r_n = |x(\pm t_n \pm 0)|$. *Then statements (i)–(v) and the conditions on the jump at the end of a segment of the trajectory yield that if*

$$0 < \epsilon_0 \le r_n t_n^{-\theta} \le c_0 \tag{12.4.57$_n$}$$

then for sufficiently large $T_1 > 0$ *depending on* c_0, ϵ_0 *the estimates*

$$r_n + \epsilon_1 t_{n+1}^{\theta} \le r_{n+1} \le r_n + c_1 t_{n+1}^{\theta}$$

hold with c_1, ϵ_1 *which do not depend on* c_0, ϵ_0. *Then under an appropriate choice of the constants* c_0, $\epsilon_0 > 0$ *condition* (12.4.57)$_n$ *yields* (12.4.57)$_{n+1}$ *with the same constants* c_0, ϵ_0; *then induction on* n *yields (12.4.53). Moreover, the energy level after a jump remains admissible.* ◊

Let us prove some long-time propagation of singularities statements:

Proposition 12.4.9. *Let the operator* A *satisfy* (12.4.2)–(12.4.6). *Let* $b_\alpha = a_\alpha - a^0_\alpha$ *satisfy* (12.4.19) *and let* a^0_α *be positively homogeneous functions of degrees* $(m - |\alpha|)\mu$. *Let the symbol* A^0 *satisfy conditions* (12.4.13) *and* (T). *Let* $r \in [r^* h^{-\delta}, \varepsilon_0]$ *with* $r^* = h^{(\mu+1)^{-1}}$, *an arbitrarily small exponent* $\delta > 0$ *and a sufficiently small constant* $\varepsilon_0 > 0$. *Let* $h' = h r^{-\mu-1}$, $\gamma = h'^{\rho}$ *with an exponent* $\rho \in (0, \frac{1}{2})$. *Let* $\tau_0 = \epsilon_3 r^{m\mu} \gamma$ *for* $\mu < 0$ *and* $\tau_0 = \epsilon_3 r^{m\mu}$ *for* $\mu \ge 0$. *Let* $Q \in \Psi_{h, \gamma r^\mu, \gamma r}$ *be an operator with symbol supported in the intersection of the* $\varepsilon(r, r^\mu)$*-neighborhood (in* (x, ξ)*) of the set* $\Lambda^\pm \cap \{\epsilon_0 r \le |x| \le c_0 r\}$ *and* Π_{τ_0}. *Finally, let*

$$T'' = \begin{cases} \epsilon_3 \min(1, r^{\theta'} \gamma^{(m\mu\theta)^{-1}}) & \text{for } \mu < 0 \\ \epsilon_3 & \text{for } 0 \le \mu < (m-1)^{-1} \end{cases} \tag{12.4.59}$$

and $T' = c_3 r^{\theta'}$ *with small enough constants* $\epsilon_3 > 0$ *and* $\varepsilon > 0$ *and a large enough constant* c_3. *Let* $\bar{T} \in [T', T'']$.

Then there exists an operator $\bar{Q} = \bar{Q}_{\bar{T}} \in \Psi_{h, \bar{\gamma} \bar{r}^\mu, \bar{\gamma} \bar{r}}$ *with symbol supported in the* $\varepsilon(\bar{r}, \bar{r}^\mu)$*-neighborhood (in* (x, ξ)*) of the set* $\Lambda^\pm \cap \{\bar{\epsilon}_0 r \le |x| \le \bar{c}_0 r\}$ *with* $\bar{r} = \bar{T}^{\theta'}$, $\bar{h}' = h \bar{r}^{-\mu-1}$, $\bar{\gamma} = \bar{h}'^{\bar{\rho}}$ *such that the equality*

(12.4.60) $$U(\pm\bar{T})Q \equiv \bar{Q}U(\pm\bar{T})Q$$

holds for the operator $U(t) = \exp(ih^{-1}tA)$ modulo operators with operator norm less than Ch^s with arbitrarily large s. Moreover, the constants $\bar{\epsilon}_0$, $\bar{c}_0$ do not depend on $\varepsilon > 0$ (but the constant c_3 does) and one can choose the exponent $\bar{\rho} > \rho$ arbitrarily close to ρ, and the statement that the operator belongs to the operator class is uniform with respect to the parameters $h, r, \bar{T}$.

Proof. Let us consider the sign "+" for definiteness. Let us pick $Q_0 = Q$ and define moments t_n according remark 12.4.8(vi) with T_1 depending on ε. Let $r_n = t_n^\theta$. Let us introduce γ_n, h'_n with r replaced by r_n and with ρ replaced by $\bar{\rho}$ as was done above. We want to construct the operators $Q_n \in \boldsymbol{\Psi}_{h,\gamma_n r_n^\mu,\gamma_n r_n}$ with symbols supported in the $\varepsilon(r_n, r_n^\mu)$-neighborhoods of the sets $\Lambda \cap \{\bar{\epsilon}_0 r_n \le |x| \le \bar{c}_0 r_n\}$ intersected with $\varPi_{\tau_n}$ (where we define τ_n later) such that for all n with $t_n \le T''$ the equalities

(12.4.61) $$U(t_{n+1} - t_n)Q_n \equiv Q_{n+1}U(t_{n+1} - t_n)Q_n$$

are fulfilled, and all statements that operators belong to operator classes and all equivalence relations are uniform in h, r, n; here $t_0 = 0$.

Let us make the dilatation $x_{\text{new}} = x_{\text{old}} r_n^{-1}$, $\xi_{\text{new}} = \xi_{\text{old}} r_n^{-\mu}$, $t_{\text{new}} = t_{\text{old}} r_n^{-\theta'}$ and set $h_n = hr_n^{-\mu-1}$. Then we get a routine problem of propagation of singularities in bounded time in the domain $\{\epsilon \le |x_{\text{new}}| \le c, |\xi_{\text{new}}| \le c\}$. This problem was treated in sections 2.1, 2.2 and therefore theorem 2.2.6 and the remark 12.4.8 yield the following procedure to construct Q_n: let Q_n be constructed and let M_n be the support of its symbol. Let us consider all the generalized bicharacteristics passing through M_n at the moment $t = 0$; let M'_{n+1} be the set of all the ends of these trajectories at the moment $t = t_{n+1} - t_n$. Then we can choose Q_n from the class indicated above, with symbol equal to 1 in the $\frac{1}{2}\varepsilon(r_{n+1}, r_{n+1}^\mu)$-neighborhood of M'_{n+1} intersected with $\varPi_{\tau_{n+\frac{1}{2}}}$ and supported in the $\varepsilon(r_{n+1}, r_{n+1}^\mu)$-neighborhood of M'_{n+1} intersected with $\varPi_{\tau_{n+1}}$ where $\tau_k = \tau_0(1 + kh^{\delta'})$ with a small enough exponent $\delta' > 0$ for $\mu \le 0$ and $\tau_k = \epsilon' r_k^{m\mu}$ with a small enough constant $\epsilon' > 0$ for $\mu > 0$.[27] Obviously, $t_k \le T''$ yields $k \le C_0|\log h|$ and therefore $\tau_k \le 2\tau_0$ for $\mu \le 0$.

Then (12.4.61) yields that for n with $t_{n+1} \le T''$

$$U(t_{n+1})Q_0 \equiv Q_{n+1}U(t_{n+1} - t_n)Q_n(t_n - t_{n-1})Q_{n-1} \cdots Q_1 U(t_1)Q_0$$

and we immediately obtain (12.4.60) for $\bar{T} = t_{n+1}$. However, if $t_n < \bar{T} < t_{n+1}$ then one can always redefine t_n and take $t_{n+1} = \bar{T}$ in our analysis. □

So, in our analysis for $\mu \ge 0$ we can take small enough constants $\tau_0 > 0$ and $T'' > 0$. For $\mu < 0$ the situation is more complicated: we took the smallest $\tau_0 = r^{m\mu}\gamma$ admissible by the uncertainty principle (for coordinates and momenta) and the condition $T'' \le \epsilon_3 r^{\theta'} \gamma^{(m\mu\theta)^{-1}}$ means that $\tau_0 \le |x|^{m\mu}$ along the trajectory. In order to avoid this restriction we must be tricky.

Now we want to extend the time interval:

Proposition 12.4.10. *Let the conditions of proposition 12.4.9 be fulfilled and let $\varphi \in C_0^K(\mathbb{R}^d)$ be supported in $\{1 \le |x| \le 2\}$, $\chi \in C_0^K([\frac{1}{4}, 1])$.[28] Then for the Schwartz kernel $u(x, y, t)$ of the operator $U(t) = \exp(ith^{-1}A)$ the estimate*

---

[27] A neighborhood of $\varPi_{\tau_{n+\frac{1}{2}}}$ may be cut out by standard elliptic arguments.

[28] Then this proposition remains true for $\chi \in C_0^K([-1, -\frac{1}{4}] \cup [\frac{1}{4}, 1])$.

(12.4.62) $$|F_{t\to h^{-1}\tau}\operatorname{tr}\chi_T(t)\int\varphi(\frac{x}{r})u(x,x,t)dx|\le Ch^s$$

$$\forall\tau:|\tau|\le\frac{1}{2}\tau_0\quad\forall T\in[3T',T^*]$$

holds where $\tau^*=\min(\frac{1}{2}\tau_0,\epsilon')$, $T^*>0$ *and* $\epsilon'>0$ *are small enough constants,* $T'=c_1r^{\theta'}$, c_1 *is a large enough constant and* $\chi\in C_0^K([-1,1])$ *is arbitrary. Recall that* $r\in(r^*,\varepsilon_0)$.

Remark 12.4.11. *(i) Therefore one can pick* $\tau^*=\epsilon'$ *for* $\mu\le 0$ *and* $\tau^*=\epsilon' r^{m\mu}$ *for* $\mu>0$.

(ii) This proposition and corollary 12.4.12 (see below) remain true if one replaces $u(x,y,t)$ *by* $B_{j+1}u_j(x,y,t)$ *where* u_j *is a solution of the recurrence problem* $(12.4.44)_0$–$(12.4.44)_j$. ◊

Proof. Let us make a partition of unity in a neighborhood of $\operatorname{supp}\varphi_r$ ($\varphi_r(x)=\varphi(\frac{x}{r})$): $\varphi_r=Q^++Q^-+Q'$ where $Q^+,Q^-,Q'\in\boldsymbol{\Psi}_{h,\gamma r^\mu,\gamma r}$ with symbols supported in the $\varepsilon(r,r^\mu)$-neighborhoods of $\Lambda^\pm\cap\{\epsilon_0 r\le|x|\le c_0r\}$ intersected with Π_{τ_0} and in $\{\epsilon_0r\le|x|\le c_0r\}\setminus\Pi_{\frac{2}{3}\tau_0}$. Then

$$U(t)\varphi_r=U(t)Q^++U(t)Q^-+U(t)Q'$$

and standard elliptic arguments yield that for $V(t)=U(t)Q'$ the estimate

(12.4.63) $$\|F_{t\to h^{-1}\tau}\chi_T(t)V(t)\|\le Ch^s\qquad\forall T\in[3T',T^*]\quad\tau:\forall\tau:|\tau|\le\tau^*$$

holds where the norm in the left-hand expression is the operator norm and we use the definition of Q', τ_0, T' and the inequalities $r\ge h^{(\mu+1)^{-1}-\delta}$ and $\tau_0T\ge h^{1-\delta'}$ due to $\mu<(m-1)^{-1}$.

We use "$\sim$" to denote equivalence modulo operator-valued functions satisfying (12.4.63). So, we write

(12.4.64) $$U(\pm t)\varphi_r\sim U(\pm t)Q^++U(\pm t)Q^-.$$

Then applying $F_{t\to h^{-1}\tau}\chi_T(t)\operatorname{Tr}$ we obtain that for $T\in[3T',T'']$, both terms in the right-hand expression of (12.4.64) will give $O(h^s)$ in the final answer; this follows from proposition 12.4.9, and for the second term one should pass to the adjoint operator and then permute Q^{-*} and $U(-t)$ because of the presence of "Tr."

Thus, for $\mu\ge0$ the proposition is proven and for $\mu<0$ it is proven only under the additional restriction $T<T''$. Let $\mu<0$ and $T>T''$; then $T''<T^*$. Let us take $T_1=\epsilon T''$ with sufficiently small $\epsilon>0$ and let us construct the operators $\bar Q^+$ and $\bar Q^-$ associated with $Q=Q_0^+=Q^+$ and with $Q=Q_0^-=Q^-$ respectively according to proposition 12.4.9; $\bar T=T_1$. Then (12.4.60) yields that

$(12.4.65)_1$ $$U(\pm t)Q_0^\pm\sim U(\pm(t-T_1))\bar Q^\pm U(\pm T_1)Q_0^\pm.$$

Let us represent $\bar Q^\pm$ in the form $\bar Q^\pm=Q_1^\pm+Q_1^{\pm\prime}$ where $Q_*^*\in\boldsymbol{\Psi}_{h,\gamma_1r_1^\mu,\gamma_1r_1}$ and the symbols of $Q_1^\pm$ and $Q_1^{\pm\prime}$ are supported in the $\varepsilon(r_1,r_1^\mu)$-neighborhoods of $\Lambda^\pm\cap\{\bar\epsilon_1r_1\le|x|\le\bar c_1r_1\}$ intersected with Π_{τ_1} and in $\{\bar\epsilon_1r_1\le|x|\le\bar c_1r_1\}\setminus\Pi_{\frac{2}{3}\tau_1}$ respectively where $r_n=T_n^\theta$, $h'_n=hr_n^{-\mu-1}$, $\gamma_n=h_n'^{\rho_n}$, $\rho<\rho_1<\dots<\rho_n$ now and one can choose ρ_n arbitrarily close to ρ; now the subscript "1" replaces the "bar." Elliptic arguments again yield that $U(t-T_1)Q^{\pm\prime}\sim0$ and the presence of the factor $U(T_1)Q^\pm$ is not essential because here we do not apply "Tr." Therefore $(12.4.59)_1$ remains true with $\bar Q_1^\pm$ replaced by $Q_1^\pm$ where we replace τ_0 by τ_1 in the definition of "$\sim$." Repeating this process we obtain that

$$(12.4.65)_n \quad U(\pm t)Q_0^\pm \sim$$
$$U(\pm(t-T_n))Q_n^\pm U(\pm(T_n-T_{n-1}))Q_{n-1}^\pm \cdots Q_1^\pm U(\pm T_1)Q_0^\pm$$

where we replace τ_0 by τ_n in the definition of "$\sim$."

Therefore again applying proposition 12.4.9 we obtain that

$$\pm U(\pm t)Q_0^\pm \sim \bar\varphi_{r_n} U(t)Q_0^\pm$$

provided $T\in[\bar\epsilon_n, T_{n+1}]$, $\bar\varphi=1$ in $\{|x|\ge\epsilon_4\}$ with a sufficiently small constant $\epsilon_4>0$ and χ is supported in $[\frac14,1]$. Hence, applying the operator $F_{t\to h^{-1}\tau}\chi_T(t)\operatorname{Tr}$ we obtain the relations

$$F_{t\to h^{-1}\tau}\chi_T(t)\operatorname{Tr} U(t)Q^\pm = O(h^s)$$

where in the "$-$" case one should pass to the adjoint operator and move Q_0^{-*} to the right (and replace Q_0^{-*} by Q_0^- in the analysis).

Therefore, it remains to prove that in the case $\mu<0$ we can pass from $r_0\ge h^{(\mu+1)^{-1}-\delta}$ to $r_n>\varepsilon_0$ with a small enough constant ε_0 in a finite number of steps depending only on the exponents. But this assertion obviously follows from the equality

$$hr_n^{-\mu-1}=\mathrm{const}_n\bigl(hr_{n-1}^{-\mu-1}\bigr)^{1-\rho_{n-1}(m\mu)^{-1}}$$

which is due to (12.4.59) for $r_n<\epsilon$; here $1-\rho_{n-1}(m\mu)^{-1}>1-\rho(m\mu)^{-1}>1$ and $hr_0^{-\mu-1}\le h^{\delta''}$ with $\delta''>0$ and therefore for appropriate n the inequality $hr_n^{-\mu-1}\le Ch$ holds. □

12.4.7 Improved remainder estimates

Applying proposition 12.4.10 we obtain the following

Corollary 12.4.12. *(i) In the context of proposition 12.4.10 the estimate*

$$(12.4.66)\quad \Bigl|F_{t\to h^{-1}\tau}\bigl(\bar\chi_{T_2}(t)-\bar\chi_{T_1}(t)\bigr)\int\varphi(\frac{x}{r})\Gamma u(x,x,)dx\Bigr|\le Ch^s$$
$$\forall\tau:|\tau|\le\tau^*$$

holds with a sufficiently small constant $T_2>0$ and $T_1=C_0r^{\theta'}$. Recall that $\bar\chi\in C_0^K([-1,1])$ and $\bar\chi=1$ in $[-\frac12,\frac12]$.

(ii) Moreover, $T_1=\varepsilon r^{\theta'}$ works with an arbitrarily small constant $\varepsilon>0$. In fact, in the time interval $|t|\le T_1$ theorem 2.2.6 is applicable with no modification and condition (T) *obviously forbids periodic generalized trajectories for the symbol A^0 at the energy level* 0. *So these trajectories are also forbidden for A under the restrictions $|x|=r$, $T\le Lr^{\theta'}$, $r\le r_0$ with arbitrarily large L and sufficiently small $r_0=r_0(L)>0$; moreover, one can consider energy levels τ with $|\tau|\le\tau^*$.*

Now we can get very precise remainder estimates:

Theorem 12.4.13. *Let the conditions of theorem 12.4.5 and condition* (T) *be fulfilled.*

(i) Moreover, let condition (12.4.46) *be fulfilled. Then the following asymptotics hold:*

$$(12.4.67)\qquad \mathcal{R}\overset{\mathrm{def}}{=}\Bigl|\int\psi(x)e_\vartheta(x,x,\tau)dx-\sum_p\kappa_{\vartheta,p}h^{-d+p}-$$
$$\sum_{p,j:-d+p=(m\mu\vartheta+\nu j)(\mu+1)^{-1}}\kappa'_{\vartheta,p}h^{-d+p}\log h-$$
$$\sum_j\omega_j h^{(m\mu\vartheta+\nu j)(\mu+1)^{-1}}\Bigr|=O(R^\#)$$

with

(12.4.68) $$R^{\#} = h^{\min(-d+1+\vartheta, l')};$$

recall that $l' = (m\mu + \nu)\vartheta(\mu+1)^{-1}$. *Moreover, for* $\vartheta \le 1$ *condition* (12.4.46) *is not necessary unless* $\vartheta = 1$ *and* $(m\mu+\nu) = (-d+p)(\mu+1)$ *with* $p = 0, 1$.

(ii) Furthermore, let the conditions of (i) be fulfilled. Let

(12.4.69) $$\vartheta(m\mu + \nu - \mu - 1) + (d-1)(\mu+1) > 0$$

(then (i) yields the remainder estimate $O(h^{-d+1+\vartheta})$*) and let condition* (Ψ) *for the operator* A *in* $B(0,1)$ *be fulfilled. Then one can replace the remainder estimate* $O(h^{-d+\vartheta+1})$ *with* $o(h^{-d+\vartheta+1})$.

(iii) Replacing condition (12.4.46) *by* (12.4.50) *one should replace* $R^{\#}$ *by* $R^{\#} + R^*$ *with* R^* *given by* (12.4.39) *with* ν' *instead of* ν.

Proof. Let us first consider the zone $\{|x| \ge r_3 = r_0 h^{-\delta}\}$. Let us change the coordinate system $x' = xr^{-1}$ and set $h' = hr^{-\mu-1}$, $t' = tr^{-\theta}$. Then

$$F_{t\to h^{-1}\tau}\chi_T(t)\int \varphi(\frac{x}{r})u(x,x,t)dx$$

transforms to

$$r^{\theta} F_{t'\to h'^{-1}\tau'}\chi_{T'}(t)\int \varphi(x')u'(x',x',t')dx'$$

where $T' = Tr^{-\theta}$, $\tau' = \tau r^{-m\mu}$ (then $|\tau| \le \tau_0 \longrightarrow |\tau'| \le \epsilon'$), $u'(.,.,t')$ is the Schwartz kernel of the operator $\exp(ith'^{-1}A')$, $A' = r^{-m\mu}A$ in the new coordinate system and then for $T' = T_1 = \epsilon_0 r^{0'}$ with small enough constants ϵ_0, ϵ_1 the results of section 4.2 yield the standard Weylian formula modulo $O(h^s)$. Therefore this formula also holds for $T = T_2$.

Moreover, for sufficiently small $\rho > 0$ in the definition of γ and τ_0 we obtain that $\tau^* \ge h^{1-\sigma'}$ with $\sigma' = (\mu + 1 - m\mu)(\mu+1)^{-1} > 0$ and therefore taking $T = T_2$ and applying standard Tauberian arguments we obtain the estimate

(12.4.70) $$\Bigl|\int \varphi(\frac{x}{r})\bigl(e_\vartheta(x,x,0) - \sum_{0\le p\le \bar p} \varkappa_{\vartheta,p}(x)h^{-d+p}\bigr)dx\Bigr| \le$$
$$Ch^{-d+\vartheta+1}r^{m\mu\vartheta+(d-\vartheta-1)(\mu+1)}\bigl(\frac{T_1}{T_2}\bigr)^{\vartheta+1} = Ch^{-d+\vartheta+1}r^{d(\mu+1)-m\mu}.$$

Here the right-hand expression has an additional factor

$$\bigl(\frac{T_1}{T_2}\bigr)^{\vartheta+1} \asymp r^{(\vartheta+1)\theta'} = r^{(\vartheta+1)(\mu+1-m\mu)}$$

if we compare it with the right-hand expression obtained in subsection 12.4.1, $\bar p = \bar p(d,m,\mu,\vartheta,s,\delta)$ is large enough.

Since $d(\mu+1) - m\mu > 0$ we immediately obtain that *the contribution of the zone* $\{r_3 \le |x| \le \varepsilon\}$ *to the remainder estimate does not exceed* $\epsilon' h^{1-d+\vartheta}$ *with* $\varepsilon' \to 0$ *as* $\varepsilon \to 0$.

Let us apply the arguments of subsection 12.4.3 with $r_3 = h^{(\mu+1)^{-1}-\delta}$, $L = L(r^*) = h^{(m\mu+\nu)(\mu+1)^{-1}}$ and sufficiently small $\delta > 0$ depending on $d, m, \mu, \vartheta < \bar\vartheta_1, \nu$. *Let us first assume that condition* (12.4.69) *is fulfilled.* Then the terms with $m\mu\vartheta + \nu j > (-d+1+\vartheta)(\mu+1)$ are negligible (their total contribution is less than $\varepsilon' h^{-d+1+\vartheta}$).

So, we only need to consider terms with $m\mu\vartheta+\nu j\le(-d+1+\vartheta)(\mu+1) \implies j<\vartheta$. In order to handle these terms by the same method as before we should estimate

$$\iint \tau_-^{\vartheta-1}\operatorname{tr}\bigl(\mathrm{K}_{(B(\tau-A^0)^{-1})^j}-\mathrm{Weyl}_j\bigr)dxd\tau.$$

We do this exactly as before but due to condition (T) an extra factor $T^{-1-\vartheta+j}$ appears in the right-hand expression where one can take $T(x)\asymp h_{\mathrm{eff}}^{-s}$ with arbitrarily large s. One can easily see that for an appropriate s the corresponding integrals converge and are negligible.

Thus, under condition (12.4.69) statement (i) has been proven. Statement (ii) follows by standard arguments (recall that we have also proven that the contribution of $B(0,\varepsilon)$ to the remainder estimate does not exceed $\varepsilon' h^{1-d+\vartheta}$). The proof of (iii) is obvious.

Let us now assume that condition (12.4.69) *is violated.* Then applying the same arguments as before we obtain the required estimates for the contributions of $B(0,2r_0)$ and $B(0,1)\setminus B(0,r_3)$. Moreover, in $B(0,2r_3)\setminus B(0,r_1)$ the only source of trouble is the presence of Weylian approximation errors of the form $h^{l'}\left(\frac{r}{r_0}\right)^\zeta$ with exponents $\zeta\ge 0$. To handle them let us apply the above propagation results. In order to do this let us make the dilatation $x'=\gamma^{-1}x$, then one obtains $r'=\gamma^{-1}$ and $h'=h\gamma^{-1-\mu}$ instead of r and γ. To fulfill the conditions $1\ge r'\ge h'^{(\mu+1)^{-1}-\delta'}$ let us pick $\gamma=r\left(\frac{r}{r_0}\right)^s$ with large s. Then for arbitrarily large s and sufficiently small δ and δ' depending on s one can set $T=\left(\frac{r}{r_0}\right)^s$ which means that we should multiply the right-hand expression of the remainder estimate by T^{-1} and obtain $h^{l'}\left(\frac{r}{r_0}\right)^\zeta$ with $\zeta<0$. Then the contribution of such terms does not exceed $Ch^{l'}$. □

12.4.8 Final remarks

Let us discuss condition (T) and some generalizations of our results.

Remark 12.4.14. *(i) One can easily see that condition (T) is fulfilled for* $A^0=\left(|\xi|^2-|x|^{2\mu}\right)^{\frac{m}{2}}$ *with* $\mu\in\left(-1,0\right)\cup\left(0,(m-1)^{-1}\right)$ *(because we treat energy level* 0*) and therefore one can apply theorem 12.4.13. For* $\mu=-\frac{1}{2}$, $d=3$, $\vartheta=1$ *the numerical value of the coefficient* ω_0 *is well known (see [Ivrii, Sigal 1], for example).*

(ii) One can easily transfer theorems 12.4.5, 12.4.5′ and 12.4.7 to the case when X *is a conical domain (in a neighborhood of* 0*) and Dirichlet boundary condition are given on* ∂X*. Moreover, one could extend theorem 12.4.13 under nice propagation of singularities results. Namely, condition (T) has the following important corollaries:*

(12.4.71) For $t=T_0$ *and* $(x,\xi)\in\Lambda^\pm_{\epsilon_1}\cap\{|x|=1\}$, $K^\pm_t(x,\xi)\subset\Lambda^\pm\cap\{|x|\ge 2\}$ *where* $K^\pm_t$ *mean propagation of singularities "cones" at the energy level* 0 *for* A^0 *(see sections 2.2, 3.3). Moreover, this property is stable with respect to small perturbations of* A^0.

and

(12.4.72) $(x,\xi)\notin K^\pm_t(x,\xi)$ *for* $t>0$*. Moreover, this property is also stable.*

Exactly these conditions along with a stability condition apply to boundary-value problems. By the geometric interpretation of section 3.3 one can replace $K_t(x,\xi)$ *by a union of generalized bicharacteristics (in which case stability is achieved automatically).*

(iii) Let us consider $\mu=0$ *and* $A^0(x,\xi)$ *which is* $(1,0)$*-homogeneous of degree* m *in* (x,ξ)*, and the energy level* $\tau=1$ *instead of* $\tau=0$*. We obtain this problem when we treat the asymptotics of Riesz means when the spectral parameter* τ *tends to* $+\infty$*. In*

order to prove condition (T) let us note that one can take the function $\phi = \langle x,\xi\rangle$ in the analysis of propagation of singularities in section 3.1. For $x \in \partial X$ this function is constant on the layers $\{(x,\xi + n\lambda), \lambda \in \mathbb{R}\}$ where n is a normal to ∂X at x and we assume that X is a conical set; moreover, the microhyperbolicity conditions are fulfilled with $\mathcal{T} = H_\phi$. Therefore, for $(x,\xi) \in \Sigma_{1,f}$, $(y,\eta) \in K^\pm(x,\xi)$ (see the definitions of $\Sigma_{\tau,b}$, $\Sigma_{\tau,f}$ in Chapter 3) the inequality $\pm\langle y,\eta\rangle \ge \pm\langle x,\xi\rangle + \epsilon_0 t$ holds. Therefore condition (12.4.72) is fulfilled. Moreover, ellipticity yields that ξ is bounded on Σ_1 and $\langle x,\xi\rangle |x|^{-1}$ is bounded on $\Sigma_{1,b}$. Therefore picking $\Lambda^\pm = \{\pm\langle x,\xi\rangle \ge -\epsilon_1|x|\}$ with sufficiently small ϵ_1 we obtain $K_t^\pm(x,\xi) \subset \{(y,\eta), |y| \ge \epsilon(|x|+t)\}$ for $t \ge 0$ and therefore condition (12.4.71) is also fulfilled. Note that condition (12.4.69) is now "$(\nu-1)\vartheta > 1-d$." Moreover, the non-Weylian terms are $\omega_j\tau^{\vartheta-\frac{1}{m}\nu j}$ while the Weylian terms are $\kappa_p\tau^{\vartheta+\frac{1}{m}(d-p)}$ and a logarithmic factor appears as soon as these exponents coincide or one of them vanishes. For $\nu = 1$ condition (12.4.69) is fulfilled for all ϑ and there are many terms with logarithms.

(iv) All the arguments in (ii),(iii) apply to other boundary conditions provided that for fixed h the problem is elliptic in the classical sense and the operator is semibounded from below. $\Diamond$

Remark 12.4.15. *Let us assume that $X = \mathbb{R}^d$ (or X is a cone). Moreover, let stabilization condition (12.4.19) be fulfilled as $|x| \to \infty$ as well but with the operator A^∞ and exponents $\mu' \in (-1,0), \nu' < 0$ instead of A^0 and μ,ν. Moreover, let us assume that condition (T) is fulfilled for this operator A^∞ as well with the sets $\Lambda^\pm_\infty$ instead of $\Lambda^\pm$. Let us define the sets $\Lambda^\pm_1$ according to (12.4.57); we define these sets for $|x| \le \epsilon$ and $|x| \ge R$ with $\epsilon > 0$ very small and R very large. Finally, let us assume that there exists t_0 such that if $(x(0),\xi(0))$ lies in the (small) set $\Lambda^\pm_1 \cap \{|x| \le \epsilon\}$ then $(x(\pm t_0),\xi(\pm t_0)) \in \Lambda^\pm_1 \cap \{|x| \ge R\}$.*

Under these assumptions we control the propagation of singularities for very large T ($\asymp h_{\rm eff}^{-s}$ with an arbitrarily large s) and therefore in the context of theorem 12.4.13 and these assumptions the remainder estimate is $O(h^{l'})$ with $l' = (m\mu+\nu)\vartheta(\mu+1)^{-1}$ even if $l' > -d+1+\vartheta$. Moreover, for $(\mu+1)d + m\mu\vartheta < 0$ one can take $\psi = 1$.

As an example one can consider $X = \mathbb{R}^d$ and the Schrödinger operator with potential $W(|x|)$ where $W(r)$ is positively homogeneous of degrees 2μ and $2\mu'$ for $|x| \le \epsilon$ and $|x| \ge R$ respectively, $-1 < \mu' \le \mu < 0$, and $W(r) < 0$, $\frac{d}{dr}r^2W(r) < 0$ $\forall r$. Moreover, one can perturb this Schrödinger operator slightly and get rid of its spherical symmetry and homogeneity. $\Diamond$

12.5 Asymptotics of Riesz Means for Schrödinger Operators Singular at a Point with Strong Magnetic Field

In this section we consider Riesz means for the 3-dimensional modified Schrödinger operator

$$(\Sigma 0) \qquad A = \sum_{j,k} P_j g^{jk} P_k + V - \mu h F, \qquad P_j = hD_j - \mu V_j$$

where $F = (\sum_{j,k} g_{jk}F^jF^k)^{\frac{1}{2}}$ is the scalar intensity of the magnetic field and

$$(\Sigma 2) \qquad \epsilon_0|\xi|^2 \le \sum_{jk} g^{jk}\xi_j\xi_k \le c|\xi|^2 \qquad \forall \xi \in \mathbb{R}^3.$$

The additional term $-\mu hF$ provides some features of this operator similar to features of the Dirac operator. However, no additional difficulties appear because we also treated the case $\nu < 1$ in Chapter 6. I leave it to the reader to treat the case without this term.

12.5.1 The regular case

I would like to recall the results for the regular case. Namely, let us assume that

$$B(0,1) \subset X \tag{12.5.1}$$

and in $B(0,1)$ the following conditions are fulfilled:

$$|D^\sigma g^{jk}| \le c, \quad |D^\sigma V_j| \le c, \quad |D^\sigma V| \le c \quad \forall \sigma : |\sigma| \le K, \tag{12.5.2$_{1,2,3}$}$$

$$F \ge \epsilon_0. \tag{12.5.3}$$

Then in Chapter 6 the following statement was proven:

Theorem 12.5.1. *Let $d = 3$, let the operator A given by (Σ0) be self-adjoint and let conditions (Σ2), (12.5.1)–(12.5.3) be fulfilled. Moreover, let us assume that*

$$V \le \epsilon_0 \implies |\nabla \frac{V}{F}| \ge \epsilon_0 \tag{12.5.4}$$

in $B(0,1)$. Finally, let $\psi \in C_0^K\bigl(B(0,\frac{1}{2})\bigr)$. Then

$$|\int \psi(x)\bigl(e_\vartheta(x,x,\tau) - h^{-3}S_\vartheta(x,\mu h,\tau)\bigr)dx| \le Ch^{-2+\vartheta} + C\mu h^{-1+\vartheta} \qquad \forall \tau \le \frac{1}{2}\epsilon_0 \tag{12.5.5}$$

where

$$S(x,\mu h,\tau) = \frac{1}{4\pi^2}\sum_{n\in\mathbb{Z}^+} \bigl(\tau - V - 2n\mu hF\bigr)_+^{\frac{1}{2}} F\mu h\sqrt{g} \tag{12.5.6}$$

and for $\vartheta \in [0,1]$

$$S_\vartheta(x,\mu h,\tau) = \vartheta\tau_+^{\vartheta-1} * S = \text{const}\sum_{n\in\mathbb{Z}^+} \bigl(\tau - V - 2n\mu hF\bigr)_+^{\frac{1}{2}+\vartheta} F\mu h\sqrt{g}. \tag{12.5.7}$$

Remark 12.5.2. *(i) This theorem also remains true without certain conditions for $\mu \le 1$. This follows from routine semiclassical asymptotics.*

(ii) Condition (12.5.4) is too restrictive; a weaker condition was used in Chapter 6. In fact, no condition was necessary for $\vartheta > 0$. However, that is not very important in this section.

(iii) For $\vartheta > 1$ it is difficult to calculate $S_\vartheta(x,\mu h,\tau)$. ◇

12.5.2 The singular case. I

Now let us consider the operator (Σ0) in the case when there is a singularity at 0. Namely, we assume that

$$|D^\sigma g^{jk}| \le c|x|^{-|\sigma|}, \qquad |D^\sigma V_j| \le c|x|^{\beta-|\sigma|+1}, \qquad |D^\sigma V| \le c|x|^{2\alpha-|\sigma|} \qquad \forall \sigma : |\sigma| \le K, \tag{Σ1$_{1,2,3}$}$$

$$F \ge \epsilon_0|x|^\beta, \tag{Σ4}$$

$$V \le \epsilon_0 |x|^{2\alpha} \implies |\nabla \frac{V}{F}| \ge \epsilon_0 |x|^{2\alpha-\beta-1} \tag{12.5.8}$$

instead of (12.5.2), (12.5.3), (12.5.4) where we assume that $\alpha \in (-1,0]$, $\beta > 2\alpha$.

Our standard goal is to derive the asymptotics of

$$\int \psi(x) e_\vartheta(x,x,0)dx$$

but instead we derive the asymptotics of

$$\int \psi(x)\bigl(e_\vartheta(x,x,0) - e^0_\vartheta(x,x,0)\bigr)dx$$

where A^0 is the operator given by $(\Sigma 0)$ with $\mu = 0$ in $B(0,1)$. The corresponding asymptotics for this operator are due to section 12.4 under appropriate conditions. First of all we treat the general situation without propagation of singularities arguments.

Let us consider the operator A in $B(\bar{x}, \frac{1}{2}r)$ with $r = |\bar{x}|$. Then the routine dilatation-multiplication procedure gives us an operator with $h' = hr^{-\alpha-1}$ and $\mu' = \mu r^{\beta+1-\alpha}$ and one should compare these quantities with 1. For $h' \le 1$ one can apply the semiclassical approach to $\int \bar{\psi} e_\vartheta(x,x,0)dx$ which provides the remainder estimate

$$R_1 = r^{2\alpha\vartheta}(h'^{-1} + \mu')h'^{-1+\vartheta} = h^{-2+\vartheta} r^{\alpha\vartheta-\vartheta+2\alpha+2} + \mu h^{-1+\vartheta} r^{\alpha\vartheta-\vartheta+\beta+2}$$

where $\bar{\psi}$ is an element of the partition of unity.

On the other hand, let us apply semiclassical asymptotics to $\bar{\psi} B e^\eta_{\vartheta-1}(x,x,0)dx$ where $B = A - A^0$ and $A^\eta = \eta A^0 + (1-\eta)A$. Then the remainder estimate is $\mu' h'^{-1} R_1$ (provided $\mu' \le 1$). This is justified for $\vartheta \ge 1$; for $0 < \vartheta < 1$ one should apply the arguments of subsection 12.4.4 to get the remainder estimate $(\mu' h'^{-1})^\vartheta R_1$ (rigorous analysis is done below). Both of these remainder estimates are better than R_1 if $\mu' < h'$. So one should introduce r_3 with $\mu' = h'$ and apply the first and second approaches for $r > r_3$ and for $r < r_3$ respectively. So, $r_3 = \bigl(\frac{h}{\mu}\bigr)^{(\beta+2)^{-1}}$. However, one should remember the condition $h' \le 1$ which yields two different situations:

(I) The magnetic field is not very strong:

$$\mu^{\alpha+1} h^{\beta+1-\alpha} \le 1. \tag{12.5.9}$$

(II) The magnetic field is rather strong:

$$\mu^{\alpha+1} h^{\beta+1-\alpha} \ge 1. \tag{12.5.9$'$}$$

In the first case surely $h' \le 1$ for $r \ge r_3$. Now I want a better implementation of the above idea. Let us introduce $r_4 = \mu^{-(\beta+1-\alpha)^{-1}}$ from the condition $\mu' = 1$. Due to (12.5.9) $r_4 \ge r_3$. Let us apply straightforward semiclassical approximation *only* in the zone $\{r_4 \le |x| \le \varepsilon\}$. Then the contribution of this zone to the remainder estimate is $O(R_1^*)$ with

$$R_1^* = \begin{cases} h^{-2+\vartheta} + \mu h^{-1+\vartheta} & \text{for } \vartheta < \bar{\vartheta}_0, \\ h^{-2+\vartheta} \log \mu + \mu h^{-1+\vartheta} & \text{for } \vartheta = \bar{\vartheta}_0, \\ h^{-2+\vartheta} \mu^{(1-m)(\vartheta-\bar{\vartheta}_0)(\beta+1-\alpha)^{-1}} + \mu\mu h^{-1+\vartheta} & \text{for } \vartheta > \bar{\vartheta}_0 \end{cases} \tag{12.5.10}$$

with the same $\bar{\vartheta}_0 = (2\alpha+2)(1-\alpha)^{-1}$ as in section 12.4.

Let us now treat the contribution of $B(0, r_4)$. Making the dilatation $x_{\text{new}} = r_4^{-1} x$ we are in exactly the situation of section 12.4 with $h' = \mu^{(m+1)(n+1-m)^{-1}} \le 1$. Let us apply the results of this section. We should consider the general situation and the situation of condition (T).

Let us first consider the general situation:

(i) First of all, one can easily see that the contribution of $B(0,r_4)$ to the remainder estimate is $O(h^{-2+\vartheta})$ for $\vartheta < \bar{\vartheta}_0$.

(ii) On the other hand, for $\bar{\vartheta} \ge \vartheta \ge \bar{\vartheta}_0$ (with $\bar{\vartheta} = -(d-1)(\alpha+1)(\alpha+\beta+1)^{-1}$ for $\alpha+\beta+1<0$, $\bar{\vartheta}=\infty$ for $\alpha+\beta+1\ge 0$) the contribution of $B(0,r_4)$ to the remainder estimate is given by the first or second line of (12.4.45)[29] and is $O\bigl(r_3^{2\alpha\vartheta}(hr_3^{-\alpha-1})^{-2+\vartheta}\bigr)$ with the factor $\log h$ if $\vartheta - 2(\alpha+1)(\beta+2)^{-1} \in \mathbb{Z}^+$.

(iii) Finally, for $\vartheta > \bar{\vartheta}$ the contribution of $B(0,r_4)$ to the remainder estimate is given by the third line in (12.4.45)[29] and is $O\bigl(h^{(\alpha+\beta+1)\vartheta(\alpha+1)^{-1}}\mu^{\vartheta}\bigr)$.

On the other hand, *under condition* (T) there is no adjustment in cases (i),(iii) but in case (ii) the contribution of $B(0,r_4)$ is $O\bigl(r_4^{2\alpha\vartheta}(hr_4^{-\alpha-1})^{-2+\vartheta}\bigr)$.

However, to be able to enjoy these remainder estimates in cases (ii),(iii) one should assume that either $\vartheta \le \bar{\vartheta}_1$ or the homogeneity condition is fulfilled. For the sake of simplicity in the second case we assume that

(12.5.11) $g^{jk} = g^{0jk} + g^{1jk}$, $V = V^0 + V^1$ and g^{0jk}, g^{1jk}, V^0, V^1, V_j are C^K functions (outside of 0), positively homogeneous of degrees $0, \beta+1-\alpha, 2\alpha, \alpha+\beta+1, \beta+1$ respectively.

In the first case we only assume that

(12.5.12) $g^{jk} = g^{0jk} + g^{1jk}$, $V = V^0 + V^1$ and g^{0jk}, V^0, are C^K functions (outside of 0), positively homogeneous of degrees $0, 2\alpha$ respectively and

$$|D^\sigma g^{1jk}| \le c|x|^{\beta+1-\alpha-|\sigma|}, \qquad |D^\sigma V^1| \le c|x|^{\beta+1+\alpha-|\sigma|} \qquad \forall \sigma : |\sigma| \le K$$

in addition to (12.5.9). Then the terms generated by the singularity are

$$\sum_{j,k} \omega_{jk} h^{(2\alpha+(j+k)(\beta-\alpha+1))(\alpha+1)^{-1}} \mu^j$$

while the other terms are semiclassical (but not necessary Weylian). To avoid computational difficulties I will only consider the case $\vartheta \le 1$ in the final statement. The calculations for $\vartheta > 1$ are left to the reader.

Theorem 12.5.3. *(i) Let the operator A given by* ($\Sigma 0$) *be self-adjoint and let conditions* ($\Sigma 1$), ($\Sigma 2$), ($\Sigma 4$) *and* (12.5.1), (12.5.8), (12.5.12) *be fulfilled. Let* $1 \ge \vartheta \ge \bar{\vartheta}_0$. *Then for* $\mu^{\alpha+1} h^{\beta+1-\alpha} \le 1$ *the estimate*

$$\begin{aligned} \mathcal{R} = \Bigl| \int \psi(x)\bigl(e_\vartheta(x,x,0) - e^0_\vartheta(x,x,0) - {} \\ h^{-3} S_1(x,\mu h, 0) + h^{-3}\varkappa_{1,0}(x)\bigr)\,dx \Bigr| \le \\ C h^{-2+\vartheta}\bigl(\frac{h}{\mu}\bigr)^{(\alpha\vartheta-\vartheta+2\alpha+2)(\beta+2)^{-1}} + C\mu h^{-1+\vartheta} \end{aligned} \tag{12.5.13}$$

holds.

(ii) Moreover, under condition (T) *the estimate*

$$\mathcal{R} \le C h^{-2+\vartheta} \mu^{-(\alpha\vartheta-\vartheta+2\alpha+2)(\beta+1-\alpha)^{-1}} + C\mu h^{-1+\vartheta} \tag{12.5.14}$$

holds.

---

[29] With $hr_4^{-\alpha-1}$ instead of h and with the additional factor $r_4^{2\alpha\vartheta}$.

Remark 12.5.4. *Let us compare the remainder estimates with the Scott correction ωh^l with $l = 2\alpha\vartheta(\alpha+1)^{-1}$. It is easy to see that for $\mu^{\alpha+1}h^{\beta+1-\alpha} \le 1$ the first terms in these remainder estimates are surely less than the Scott term. Therefore the remainder estimate is less than the Scott term if and only if $\mu \le h^l$ which means exactly that $\mu \le h^{(\alpha\vartheta+\alpha+1-\vartheta)(\alpha+1)^{-1}}$ and the idea to treat $e_\vartheta - e_\vartheta^0$ instead of e_1 is reasonable only in this case.* ◊

12.5.3 The singular case. II

Let us consider the case $\mu^{\alpha+1}h^{\beta+1-\alpha} \ge 1$. In this case we should treat

$$\mathcal{R}_\vartheta = \Bigl|\int \psi(x)\bigl(e_\vartheta(x,x,0) - h^{-3}S_\vartheta(x,\mu h,0)\bigr)dx\Bigr|.$$

Let us pick $r_0 = h^{(\alpha+1)^{-1}}$. In the zone $\{|x| \ge r_0\}$ the routine semiclassical technique is applicable and the contribution of this zone to the remainder estimate does not exceed CR with

$$R = \begin{cases} \mu h^{-1+\vartheta} & \text{for } (3\alpha+1)\vartheta + \beta - 2\alpha > 0 \\ \mu h^{-1+\vartheta}\log\mu & \text{for } (3\alpha+1)\vartheta + \beta - 2\alpha = 0 \\ \mu h^{2\alpha\vartheta(\alpha+1)^{-1}} & \text{for } (3\alpha+1)\vartheta + \beta - 2\alpha < 0 \end{cases} \tag{12.5.15}$$

So we need to estimate the contribution of $B(0,r_0)$. The routine estimate due to Chapter 8 and standard variational estimates gives a very poor result (which the reader can derive if desired). So I want to improve the result under certain conditions. Namely, let us assume that conditions $(12.5.1)_{1,2}$ are fulfilled.

The rest of this section is devoted to the proof of

Proposition 12.5.5. *Let the operator A given by $(\Sigma 0)$ be self-adjoint and let conditions $(\Sigma 2)$, $(\Sigma 1)_3$, (12.5.1), $(12.5.2)_{1,2}$ and $(\Sigma 4)$ be fulfilled with $\beta = 0$. Let $h = 1$, $\mu \ge 1$. Then the operator A in $B(0,1)$ (with Dirichlet boundary conditions) is semibounded from below uniformly with respect to μ and the estimate*

$$\mathbf{N}(\tau) \le C\mu\tau^{\frac{1}{2}} + C\tau^{\frac{3}{2}} \tag{12.5.16}$$

holds for $\tau \ge 1$.

Then the standard arguments of Chapter 8 yield the following theorem (details are left to the reader).

Theorem 12.5.6. *Let the operator A given by $(\Sigma 0)$ be self-adjoint and let conditions $(\Sigma 2)$, $(\Sigma 1)_3$, (12.5.1), $(12.5.2)_{1,2}$ and $(\Sigma 4)$ be fulfilled with $n = 0$. Then for $\mu^{\alpha+1}h^{1-\alpha} \ge 1$ the estimate*

$$\mathcal{R}_\vartheta \le C\mu h^{-1+\vartheta} \tag{12.5.17}$$

holds.

Proof of proposition 12.5.5. Without loss of generality one can assume that $F^1 = F^2 = 0$ and $V_2 = V_3 = 0$, $g^{13} = g^{23} = 0$. Then $V_1 = V_1(x')$, $x' = (x_1,x_2)$. Surely in this case the standard variational estimates are much better than in the general case but we also need to improve them. Let $\mathbb{H} = L^2\bigl(B(0,1)\bigr)$ and on $\mathbb{K} \subset C_0^2\bigl(B(0,1)\bigr)$ let the operator $A-\tau I$ be negative definite. We should prove that $\dim\mathbb{K} = 0$ for $\tau \le -C_0$ and $\dim\mathbb{K} \le C\mu\tau^{\frac{1}{2}} + C\tau^{\frac{3}{2}}$ for $\tau \ge 1$. Let us consider

$$A - \tau I = g^{33} D_3^2 + \sum_{1 \le j,k \le 2} P_j g^{jk} P_k - \mu F + V - \tau I.$$

Let us note that the operator $A' = \sum_{1 \le j,k \le 2} P_j g^{jk} P_k - \mu F$ is semibounded from below on $\mathbb{H}$ and, moreover, that $A' + C_0 I \ge \epsilon_0 A''$ where $A'' = P_1^2 + P_2^2 - \mu F''$ and the scalar intensity F'' is calculated for the Euclidean metric; then the operator A'' does not contain x_3. On the other hand, let us note that $V \ge -c_0 |x_3|^{2m}$ and the quadratic form $\epsilon \| D_3^2 u\|^2 - c \| W u\|^2$ is semibounded from below on $L^2([-1,1])$ for every $\epsilon > 0$, $W = |x_3|^m$ since $m > -1$.

Therefore for appropriate $\varepsilon_0 > 0$ and C_0 the operator

$$\varepsilon_0 D_3^2 + A'' - \tau - C_0$$

should be negative definite on $\mathbb{K}$. One can replace the ball by the cylinder $\Omega \times [-1,1]$ where $\Omega \subset \mathbb{R}^2$ is a unit circle and separate variables x' and x_3. So, our original problem is reduced to a problem for the operator A'' in the circle (and for the operator $\epsilon_0 D_3^2$ on $[-1,1]$) and estimate (12.5.17) follows from the well known estimate

$$\mathbf{N}(\tau) \le C\mu + C\tau \tag{12.5.18}$$

for this operator for $\tau > 0$ (and the routine estimate for the one-dimensional operator). □

12.5.4 The case $d = 2$

Let us discuss the case $d = 2$. Not going into details of the regular case, recall that under the condition of subsection 12.5.1 the remainder estimate is $O(\mu^{-1-\vartheta} h^{-1+\vartheta})$. To attack the singular case *let us assume that conditions of subsection 12.5.2 are fulfilled (including condition* (12.5.9)). Then $r_4 \ge r_3$ where r_0, r_3, r_4 are introduced in subsection 12.5.2. There are again two possibilities:

(i) For $0 \le \vartheta \le \vartheta^* = (\beta+2)(\beta+2-2\alpha)^{-1}$ the contribution of $\{r_4 \le |x| \le r_1\}$ to the remainder estimate is $O(\mu^{-1-\vartheta} h^{-1+\vartheta})$ with the factor $\log \mu$ for $\vartheta = \vartheta^*$. However, since $\vartheta^* < \bar{\vartheta}_0$ we can apply standard semiclassical asymptotics and no Scott correction term appears. The final remainder estimate is $O(\mu^{-1-\vartheta} h^{-1+\vartheta})$ with the factor $\log \mu$ for $\vartheta = \vartheta^*$. For $\vartheta^* < \vartheta \le \bar{\vartheta}_0$ the best possible remainder estimate is $O(h^{-1-\vartheta} \mu^{(\alpha\vartheta + \beta + 1 - \vartheta)(\beta+1-\alpha)^{-1}})$ which is also due to $B(0, r_3)$ and there is still no Scott correction term.

(ii) For $\vartheta > \bar{\vartheta}_0$ it is not important which regular remainder estimate (namely, $O(\mu^{-1-\vartheta} h^{-1+\vartheta})$ or $O(h^{-1+\vartheta})$) was applied in the zone $\{r_4 \le |x| \le r_1\}$. So, the arguments of subsection 12.5.2 remain valid and for $\vartheta \ge \bar{\vartheta}_0$ we can obtain the same type of remainder estimate as before with the Scott correction term(s).

I have not succeeded in getting a variational estimate in the case $h = 1$, $\mu \ge 1$ and I have no extension of the results of subsection 12.5.3.

12.6 Asymptotics of Eigenvalues Tending to the Bottom of the Essential Spectrum for Three-Dimensional Schrödinger and Dirac Operators with Constant Magnetic Field and with Electric Potential Quickly Decreasing at Infinity

In this section we consider the Schrödinger and Dirac operators in $\mathbb{R}^3$ with constant non-zero magnetic field and with electric potential decreasing at infinity and we derive the asymptotics of the eigenvalues of these operators tending to the boundary of the essential spectrum. Directing the magnetic field along the axis x_3 we introduce the auxiliary space $\mathbb{H} = L^2(\mathbb{R})$ and apply semiclassical asymptotics with respect to $x' = (x_1, x_2)$. Moreover, certain generalizations are discussed.

12.6.1 Introduction

Namely, let us consider the Schrödinger operator

$$A_{\mathrm{S}} = \sum_{j,k} P_j g^{jk} P_k + V$$

and the Dirac operator

$$A_{\mathrm{D}} = \frac{1}{2}\sum_{jl} \sigma_l \bigl(\omega^{jl} P_j + P_j \omega^{jl}\bigr) + \sigma_0 M + V \cdot I$$

where $P_j = D_j - V_j$, $g^{jk} = g^{kj}$, V_j, V are smooth real-valued functions, $\sigma_0, \dots, \sigma_d$ are Dirac matrices, $g^{jk} = \omega^{jl}\omega^{kr}\delta_{lr}$ in the Dirac operator case and we use the Einstein summation convention. We assume that

$$(\Sigma 2) \qquad \epsilon_0 |\xi|^2 \le g^{jk}\xi_j\xi_k \le c|\xi|^2 \qquad \forall x, \xi \quad (\epsilon_0 > 0).$$

Let $F_{jk} = \partial_k V_j - \partial_j V_k$ be the tensor intensity of the magnetic field. We assume that $(F_{jk}) \ne 0$.

First of all in the case when g^{jk}, ω^{jl} and F_{jk} are constant and $V = 0$ the spectra of these operators are known: recall that for $d = 2$ these operators have pure point infinite multiplicity spectra and for $d = 3$ the spectra are absolutely continuous and $\mathrm{Spec}(A_{\mathrm{S}}) = [F, \infty)$, $\mathrm{Spec}(A_{\mathrm{D}}) = (-\infty, -M] \cup [M, \infty)$. Moreover, the spectra are also known in the multi-dimensional case (see, e.g., section 6.1); in particular, the spectra are pure point of infinite multiplicity if $\mathrm{rank}(F_{jk}) = d$ and they are absolutely continuous otherwise.

Let us perturb these operators by a potential V tending to 0 at infinity. Moreover, we can consider more general perturbations: we can assume that g^{jk}, ω^{jl} and F_{jk} tend to constants at infinity. Then for $d = 2$ every former eigenvalue $\bar{\lambda}$ (of infinite multiplicity) will be separated into a sequence of eigenvalues λ_n tending to $\bar{\lambda}$ and the question is to obtain the asymptotic distribution of λ_n. Asymptotics of the eigenvalue counting function with highly accurate remainder estimates were obtained in sections 11.3 and 11.4 in this case and it is not difficult to generalize those results to the case when $\mathrm{rank}(F_{jk}) = d$ and λ appears only once in the list of eigenvalues of the non-perturbed operator given by formula (6.1.8).

Let us consider the case $d = 3$. In this case there arises a sequence of eigenvalues tending to the bottom F of $\mathrm{Spec}_{\mathrm{ess}}(A_{\mathrm{S}})$ or to the boundaries $\pm M$ of $\mathrm{Spec}_{\mathrm{ess}}(A_{\mathrm{D}})$. So, the problem is to obtain the asymptotic distribution of these eigenvalues. In the case when V decreases slowly enough the asymptotics of the eigenvalue counting function with fairly accurate remainder estimates were obtained in section 11.5. In particular, under appropriate conditions in that section the case when V [30] decreases as $|x|^{2m}$ with $m \in (-1, 0)$ was treated and the asymptotics of $\mathbf{N}^-(\bar{F} - \eta)$ were derived for the Schrödinger operator and the asymptotics of $\mathbf{N}(M - \epsilon_1, m - \eta)$ and of $\mathbf{N}(-M + \eta, -M + \epsilon_1)$ were derived for the Dirac operator where $\mathbf{N}(\lambda_1, \lambda_2)$ is the number of eigenvalues lying

[30] For the Schrödinger operator one should replace V here and in what follows in this subsection with the effective potential $V^* = V + F - \bar{F}$, $\bar{F} = \lim_{|x|\to\infty} F$.

in the interval $[\lambda_1, \lambda_2)$ with $\lambda_2 > \lambda_1$, $\mathbf{N}^-(\lambda) = \mathbf{N}(-\infty, \lambda)$, $\eta \to +0$ and $\epsilon_1 > 0$ is a small enough constant. The principal parts of these asymptotics are $\asymp \eta^{\frac{1}{2m}(3+m)}$ and the remainder estimates are $O(\eta^{\frac{1}{m}})$. So, in the limit case $m = -1$ we *formally* obtain the estimates $\mathbf{N}^-(\bar{F} - \eta) = O(\eta^{-1})$ and $\mathbf{N}(M - \epsilon_1, M - \eta) = O(\eta^{-1})$. Compare with the classical case $F = 0$ when the principal parts of the asymptotics are $\asymp \eta^{\frac{d}{2m}(m+1)}$ and the remainder estimates are $O(\eta^{\frac{1}{2m}(d-1)(m+1)})$ and in the limit case $m = -1$ one *formally* has the estimate $O(1)$. This difference is associated with the following principal difference: *In the classical case $F = 0$ the exponent $m = -1$ is critical in the sense that for $m < -1$ there is only finite number of eigenvalues. In our case (under an appropriate condition on the sign of V or V^*) there is an infinite number of eigenvalues for every $m < 0$ and the exponent $m = -1$ is critical only in the sense that the asymptotics derived in sections 11.4 and 11.5 fail for $m \le -1$.* Thus the problem is to derive asymptotics in the case $m \le -1$. And all this section is devoted to asymptotics in the case $d = 3$, $m < -1$ and some generalizations. Asymptotics of this type were obtained few years ago by Alexander Sobolev [1–4]; see also the related papers of S. Solnyshkin [1] and the more advanced results of G. Raikov [2–6].[31] However they obtained no remainder estimate and so the problem was to derive asymptotics with accurate remainder estimates. For a few years I considered this problem a challenge because I was unable even to reprove the results of A. Sobolev and G. Raikov by my methods. Concluding this introductory part I would like to note that results of the type "$d = 2$, $m > -1$" can be generalized to the case of even $d > 2$ provided $|F_{jk}| \le c$ and $\operatorname{Spec}(F_{jk}) \cap (-\epsilon_2, \epsilon_2) = \emptyset$ for all x with some constant $\epsilon_2 > 0$. It is an open problem to overcome this condition.

To my great regret there were more misprints than pages (and even one real error) in [Ivrii 16] where these results were first partially announced.

12.6.2 The main theorem

So let us assume that $d = 3$. Moreover, for the Dirac operator let us assume that $M > 0$ (otherwise $\operatorname{Spec}_{\mathrm{ess}}(A) = \mathbb{R}$). Without loss of generality one can assume that $M = 1$ and that σ_j are 4×4 matrices (see Appendix C). For the Dirac operator we treat eigenvalues tending to $1 - 0$.

Let us introduce the vector intensity F^j of the magnetic field: $F^j = \frac{1}{2}\epsilon^{jkl} F_{kl}$ where ϵ^{jkl} is an absolutely skew-symmetric pseudo-tensor with $\epsilon^{123} = g^{-\frac{1}{2}}$, $g = \det(g^{jk})^{-1}$. Let us assume that

$$F^1 = F^2 = 0, \quad g^{13} = g^{23} = 0 \qquad \text{for } |x| \ge c \tag{12.6.1}$$

(under reasonable conditions we can arrange this by an appropriate change of variables). Without loss of generality one can assume that

$$V_3 = 0 \tag{12.6.2}$$

(one can always arrange this by an appropriate gradient transform; see, e.g., sections 6.1 and 7.1.)

Furthermore, let us assume that

$$(12.6.3)_{1-3} \qquad \begin{aligned} &\left|D^\alpha(g^{jk} - \delta_{jk})\right| \le c\langle x\rangle^{2m-|\alpha|},\\ &\left|D^\alpha(F^3 - \bar{F})\right| \le c\langle x\rangle^{2m-|\alpha|},\\ &\left|D^\alpha V^*\right| \le c\langle x\rangle^{2m-|\alpha|} \qquad \forall \alpha : |\alpha| \le K \end{aligned}$$

---

[31] They are all pupils of M. S. Birman.

where K is large enough,

$$m < -1, \qquad \bar{F} = \text{const} \neq 0 \tag{12.6.4}$$

(without loss of generality one can assume that $\bar{F} = 1$) and V^* is an effective potential:

$$V^* = V + F - \bar{F} \tag{12.6.5$_S$}$$

and

$$V^* = 2MV \tag{12.6.5$_D$}$$

for the Schrödinger and Dirac operators respectively. Later we consider more general potentials and weaken the condition $m < -1$.

Let us consider the auxiliary Schrödinger operator

$$\mathbf{a} = D_3 g^{33} D_3 + V^* \tag{12.6.6}$$

on $\mathbb{R} \ni x_3$ depending on $x' = (x_1, x_2)$. We should treat it in both the Schrödinger and Dirac operator cases.

Remark 12.6.1. *Under fairly general conditions (obviously fulfilled in this section)* $\operatorname{Spec}_{\mathbf{ess}}(\mathbf{a}) = \mathbb{R}^+$, *all the eigenvalues of* $\mathbf{a}$ *are negative and simple and if* $\mathbf{a}$ *depends properly on some parameters these eigenvalues are smooth with respect to these parameters (until they hit* -0*). Under fairly general conditions there is only a finite number of eigenvalues.* ◊

Proposition 12.6.2. *Let us consider the operator*

$$\mathbf{a}_c = D_t g_\varepsilon(t) D_t + \varepsilon^{-1} V_\varepsilon(t) \tag{12.6.7}$$

in $\mathbb{H} = L^2(\mathbb{R})$ *with* $D = D_t$, $V_\varepsilon(x) = V(\frac{x}{\varepsilon})$, *etc.*, $t \in \mathbb{R}$,

$$\epsilon_0 \le g \le c, \quad |V| \le \rho^2, \quad 0 \le \rho \le c, \quad \|\rho\|_{L^1} + \|\rho^2 t\|_{L^1} \le c. \tag{12.6.8}$$

Then

(i) The number of negative eigenvalues of the operator $\mathbf{a}$ *does not exceed* C_0 *for* $|\varepsilon| \le 1$.

(ii) The number of negative eigenvalues of the operator $\mathbf{a}$ *does not exceed* 1 *for* $|\varepsilon| \le \epsilon$ *with a small enough constant* $\epsilon > 0$.

(iii) Moreover, let us assume that

$$W = -\int_{-\infty}^{+\infty} V(t)dt \ge \epsilon_0. \tag{12.6.9}$$

Then for $\varepsilon \in (0, \epsilon]$ *there is exactly one negative eigenvalue* $\lambda(\varepsilon)$ *and*

$$-\epsilon_2 \ge \lambda \ge -c_1. \tag{12.6.10}$$

(iv) Furthermore, let us assume that (12.6.9) *holds and*

$$|g - 1| \le c\rho. \tag{12.6.11}$$

Then

$$|\lambda + \frac{1}{4} W^2| \le C_0 \varepsilon. \tag{12.6.12}$$

(v) Moreover, let us assume that (12.6.9) *holds and that* g *and* V *depend on the parameter* $z \in \Omega$ *and*

$$|D_z^\alpha g| \le c, \quad |D_z^\alpha V| \le c\rho^2 \qquad \forall \alpha : |\alpha| \le K. \tag{12.6.13}$$

Then

$$|D_z^\alpha \lambda| \le C_0; \tag{12.6.14}$$

moreover, under the condition

$$|D_z^\alpha g| \le c\rho \qquad \forall \alpha : |\alpha| \le K \tag{12.6.15}$$

we obtain that

$$|D_z^\alpha (\lambda + \frac{1}{4}W)| \le C_0\varepsilon. \tag{12.6.16}$$

(vi) Finally, let $v \in \mathbb{H}$, $|v| = 1$ *be an appropriate eigenfunction of* $\mathbf{a}$ *with eigenvalue* λ. *Then*

$$|D_z^\alpha v| \le C_0, \quad |D_z^\alpha D_t v| \le C_0, \quad |D_z^\alpha v|_\infty \le C_0, \tag{12.6.17}$$

where $|.|_p$ *means the* L^p*-norm and we skip* $p = 2$ *in the notation.*

This proposition will be proven later. Making the transformation $x_{3\text{new}} = x_{3\text{old}} \langle x' \rangle^{-2m-1}$ and applying the proposition with $\varepsilon = |x'|^{2m+2}$ we obtain the following

Corollary 12.6.3. *Let all the conditions* ($\Sigma 0$), (12.6.1)–(12.6.4) *be fulfilled and let us consider the auxiliary operator* $\mathbf{a}(x, D_3)$ *depending on* x' *in* $\mathbb{R} \ni x_3$ *given by* (12.6.6). *Then*

(i) The number of negative eigenvalues of $\mathbf{a}$ *is finite;*

(ii) For $|x'| > c_0$ *the number of negative eigenvalues of* $\mathbf{a}$ *does not exceed* 1;

(iii) If

$$W(x') = -\int_{-\infty}^{\infty} V^*(x', x_3)dx_3 \ge \epsilon_0 \langle x' \rangle^{1+2m} \tag{12.6.18}$$

with $\epsilon_0 > 0$ *then for* $|x'| \ge c_0$ *the operator* $\mathbf{a}$ *has exactly one negative eigenvalue* $\lambda(x')$ *and*

$$|D^\alpha \lambda| \le c\langle x' \rangle^{2+4m-|\alpha|} \qquad \forall \alpha : |\alpha| \le K - 2, \tag{12.6.19}$$

$$|D^\alpha (\lambda + \frac{1}{4}W(x')^2)| \le \langle x' \rangle^{4+6m-|\alpha|}. \tag{12.6.20}$$

Moreover, for a properly chosen corresponding eigenfunction $v(x) \in \mathbb{H}$ *with* $|v| = 1$

$$|D_{x'}^\alpha v| \le C_0 \langle x' \rangle^{-|\alpha|}. \tag{12.6.21}$$

The principal result of this section (generalized later) is the following

Theorem 12.6.4. *Let conditions* ($\Sigma 0$), (12.6.1)–(12.6.4) *be fulfilled. Moreover, let condition* (12.6.18) *be fulfilled for* $|x| \ge c$ *and let*

$$|\nabla W| \ge \epsilon_0 |x'|^{2m} \qquad \forall x' : |x'| \ge c. \tag{12.6.22}$$

Then the asymptotics

$$\nu(\eta) = (2\pi)^{-1}\bar{F} \int_{\{-\lambda(x') \ge \eta\}} dx' + O(1) \tag{12.6.23}$$

hold as $\eta \to +0$ *where* $\nu(\eta) = \mathbf{N}(1 - \eta)$ *and* $\nu(\eta) = \mathbf{N}(1 - \epsilon_2, 1 - \eta)$ *for the Schrödinger and Dirac operators respectively,* $\epsilon_2 > 0$ *is a small enough constant and*

$$\nu(\eta) \asymp \eta^{(1+2m)^{-1}}. \tag{12.6.24}$$

12.6.3 Proof of proposition 12.6.2

Let us start with

Proof of proposition 12.6.2. Statement (i) follows from the fact that the operator $\rho^s(D_3^2+1)^{-s}$ is compact in $\mathbb{H}$ for any $s>0$. In order to prove statement (ii) let us consider the quadratic form $Q(u)=\langle \mathbf{a}u,u\rangle$ on the subspace $\mathbb{H}_1=\{u\in\mathbb{H}, \int_{-\varepsilon}^{\varepsilon} u\,dt=0\}$ of codimension 1.[32] Obviously $|u(t)|\le 3(|t|+\varepsilon)^{\frac12}|D_tu|$ for $u\in\mathbb{H}_1$ and therefore $|\langle V_\varepsilon u,u\rangle|\le C_0\varepsilon^2|D_t^2u|^2$ because of (12.6.8). Then (12.6.8) yields that the quadratic form $Q(u)$ is positive definite on $\mathbb{H}_1$ for $|\varepsilon|\le\epsilon_2$ and therefore $\mathbf{a}$ has no more than one negative eigenvalue λ. Moreover, for arbitrary $u\in\mathbb{H}$ the inequality $|u(t)|\le\epsilon|D_tu|+C_\epsilon|u|$ with arbitrarily small $\epsilon>0$ yields that

$$|\langle V_\varepsilon u,u\rangle|\le c_0\epsilon\varepsilon|D_tu|^2+C_\epsilon\varepsilon|u|^2$$

and hence $Q(u)$ is uniformly semibounded from below and therefore

$$\lambda\ge -C_0. \tag{12.6.25}$$

Let v be the corresponding eigenfunction with $|v|=1$ (if there exists a negative eigenvalue). Then obviously $|v(t)|\le C_0$ and then $|D_tv(t)|\le C_0$ and hence $|v(t)-v(0)|\le C_0|t|$. Then (12.6.8), (12.6.11) yield that

$$|Q(v)-\bar{Q}(v)|\le C_1\varepsilon$$

for the quadratic form $\bar{Q}(u)=|D_tu|^2-W|u(0)|^2$. Therefore $\lambda\ge\bar\lambda-C_1\varepsilon$ where $\bar\lambda$ is the lower bound of $\bar{Q}(u)|u|^{-2}$ at $\mathbb{H}$.

One can apply the same arguments to the eigenvalues and eigenfunctions of $\bar{Q}$; as a result we obtain that $\bar\lambda\ge\lambda-C_1\varepsilon$. On the other hand, one can easily see that $\bar\lambda=-\frac14W^2$ if $W>0$ (otherwise $\bar{Q}$ is non-negative definite) and $\bar{v}(t)=\frac{W}{2}\exp(-\frac12W|x|)$ and hence we obtain that for $W>0$ there is a negative eigenvalue and (12.6.12) holds. Moreover, if (12.6.11) is violated then one can treat the quadratic form $C_0|D_t^2|+\varepsilon^{-1}(V_\varepsilon u,u)$ instead of the original one and since (12.6.10) holds for this form it remains true for the original one.

So all the statements excluding those associated with derivatives on z are proven.[33] The proof of (12.6.14), (12.6.17) is standard, due to K. O. Friedrichs [1]. Let these estimates be proven for $|\alpha|\le n$; then applying the operator ∂_z^α with $|\alpha|=n$ to the equation

$$(\mathbf{a}-\lambda)v=0 \tag{12.6.26}$$

we obtain

$$\sum_{\beta\le\alpha}\frac{\alpha!}{\beta!(\alpha-\beta)!}(\mathbf{a}-\lambda)^{(\alpha-\beta)}v^{(\beta)}=0 \tag{12.6.27}$$

with $u^{(\alpha)}=\partial_z^\alpha u$. Let us multiply this equation by v. Then terms with $v^{(\alpha)}$ disappear and we obtain terms with $|\beta|<n$

$$\langle g_\varepsilon^{(\alpha-\beta)}D_tv^{(\beta)},D_tv\rangle,\quad \varepsilon^{-1}\langle V_\varepsilon^{(\alpha-\beta}v^{(\beta)},v\rangle,\quad \lambda^{(\alpha-\beta)}\langle v^{(\beta)},v\rangle. \tag{12.6.28}$$

---

32) One can consider the subspace $\{u\in D(\mathbf{a}), u(0)=0\}$ as well.

33) One can easily prove that for $W<0$ and small enough ε the operator $\mathbf{a}$ is non-negative definite. I think that it would be nice to treat the case $W=0$. However I am not an expert here.

Terms of the first and second types are bounded because of $(12.6.17)_2$, $(12.6.17)_3$ respectively for $|\beta| < n$. Terms of the third type are bounded for $\beta \neq 0$ by (12.6.14), $(12.6.17)_1$. Therefore the remaining term

$$-\lambda^{(\alpha)}|v|^2$$

should also be bounded and (12.6.14) holds for $|\alpha| = n$. Let us consider equation (12.6.27); we now multiply it by $w = v^{(\alpha)}$. We obtain terms with $|\beta| \leq n$

$$\langle g_\varepsilon^{(\alpha-\beta)} v^{(\beta)}, w\rangle, \quad \varepsilon^{-1}\langle V_\varepsilon^{(\alpha-\beta} v^{(\beta)}, w\rangle, \quad \lambda^{(\alpha-\beta)}\langle v^{(\beta)}, w\rangle.$$

For $|\beta| < n$ terms of the first and second type do not exceed $C|D_t w|$ and $C|w|_\infty$ by $(12.6.17)_2$, $(12.6.17)_3$. Finally, terms of the third type for $|\beta| < n$ do not exceed $C\|w\|$ by (12.6.14), $(12.6.17)_1$. So,

$$|\langle(\mathbf{a} - \lambda)w, w\rangle| \leq C|D_t w| + C|w|$$

because $|w|_\infty \leq C|D_t w| + C|w|$. Taking $\lambda \leq -\epsilon_0$ into account we obtain from this inequality that

$$|D_t w|^2 + |w|^2 \leq C\varepsilon^{-1}|\langle V_\varepsilon w, w\rangle| + C. \tag{12.6.29}$$

Let us assume that $v(0) = 1$. Surely, we should reject the condition $|v| = 1$ but our above arguments yield that $|v| \asymp |v(0)|$. Then $w(0) = 0$ and $|w(t)| \leq |t|^{\frac{1}{2}}|D_t w|$ and therefore $|\langle V_\varepsilon w, w\rangle| \leq C\varepsilon^2 |D_t w|$ and therefore (12.6.29) yields $(12.6.17)_{1,2}$ for $|\alpha| = n$; $(12.6.17)_3$ follows from these estimates.

In order to prove (12.6.16) let us note that equation (12.6.27) and (12.6.17) yield that

$$|D_z^\alpha D_t v|_\infty \leq C_0 \tag{$(12.6.17)_4$}$$

and therefore under condition (12.6.15) terms of the first type in (12.6.28) do not exceed $C\varepsilon$. Moreover, $(12.6.17)_4$ yields that

$$|v(t) - v(0)| \leq C_0|t|, \qquad |D_z^\alpha v| \leq C_0|t| \quad \forall \alpha \neq 0$$

(because of condition $v(0) = 1$) and therefore terms of the second type in (12.6.28) do not exceed $C\varepsilon$ for $\beta \neq 0$. Moreover, the error does not exceed $C\varepsilon$ if we replace $v(t)$ by $v(0)$ in this term with $\beta = 0$; we then obtain $W^{(\alpha)}|v(0)|^2$ and under additional the restriction $W = \text{const}$ this term vanishes. Then induction on n yields that $|\lambda^{(\alpha)}| \leq C_0\varepsilon$ under this restriction. So under this restriction (12.6.16) holds. However one can reduce the general case to the case $W = 1$ by introducing $t' = tW^{-1}$ and multiplying $\mathbf{a}$ by W^2. □

Remark 12.6.5. *Applying the above results one can find v in the form*

$$v = \exp\Bigl(\int_0^t \phi_\varepsilon(t')dt'\Bigr)\bigl(1 + \varepsilon^2\psi_\varepsilon + \ldots\bigr)$$

where the number of terms depends on m and $\lambda = -\frac{1}{4}W^2 + \mu\varepsilon + \cdots$ with $\partial_t\phi = V$ and one can obtain $\mu \neq 0$ in the generic case; so estimates (12.6.12), (12.6.20) are the best possible estimates without this correction term. Therefore (12.6.23) remains true with $\lambda(x')$ replaced by $-\frac{1}{4}W(x')^2$ provided $m \leq -2$. For $m > -2$ this is correct with the remainder estimate $O\bigl(\eta^{(m+2)(2m+1)^{-1}}\bigr)$ coinciding with the principal part for $m = -1$ (in the context of remark 12.6.4). ◊

12.6.4 Proof of the main theorem. I

Let us begin the proof of theorem 12.6.4 for the Schrödinger operator. We know that

$$\nu(\eta) = \mathbf{N}^-(\bar{F} - \eta) = \mathbf{N}^-(A - \bar{F} + \eta) = \mathbf{N}^-(\mathbf{A}_\eta) = \operatorname{Tr} E_\eta(0) \tag{12.6.30}$$

where

$$\mathbf{A}_\eta = J^{-\frac{1}{2}}(A - \bar{F} + \eta)J^{-\frac{1}{2}}$$

is a self-adjoint operator in $L^2(\mathbb{R}^3)$ and $E_\eta(\tau)$ is the spectral projector of this operator; we set $J = \bigl(1 + |x'|^{2p} + |x_3|^{2p}\bigr)^{\frac{m}{p}}$ with large p.

We consider only $\eta > 0$ and for any fixed $\eta > 0$ and τ this projector is finite-dimensional and its Schwartz kernel belongs to $\mathcal{S}(\mathbb{R}^6)$ uniformly on $\tau \le \tau_0$. Let us note that

$$((A - \bar{F})v, v) \ge \|\rho D_3 v\|^2 - C_0\|\langle x\rangle^m v\|^2 \qquad \forall v \in C_0^2(\mathbb{R}^3) \tag{12.6.31}$$

where without loss of generality we assume that (12.6.1), (12.6.2) are fulfilled and $g^{33} = \rho^2$. In fact,

$$A = D_3\rho^2 D_3 + L_1^* L_1 + L_2^* L_2 + V$$

where L_j are first-order differential operators with real symbols and $[L_1, L_2] = -iF + i\alpha_1 L_1 + i\alpha_2 L_2$, α_j are real-valued and $|D^\sigma \alpha_j| \le c\langle x\rangle^{2m-|\sigma|}$ (due to conditions (12.6.3)–(12.6.4)). Therefore

$$A = D_3\rho^2 D_3 + L^* L + \bar{F} + \alpha_3$$

with $L = L_1 + iL_2$. This yields (12.6.31) with the additional term $\|Lv\|^2$ in the right-hand expression (we refer to this better inequality as (12.6.31)$'$).

This inequality immediately yields the estimates

$$\begin{aligned}&\|D_3 J^{-\frac{1}{2}} E_\eta(\tau)\| \le C, \qquad \|J^{-\frac{1}{2}} E_\eta(\tau)\| \le C\eta^{-\frac{1}{2}},\\ &\|LJ^{-\frac{1}{2}} E_\eta(\tau)\| \le C\end{aligned} \tag{12.6.32}$$

for operator norms where here and below $\tau \le \tau_0$. The last estimate and the estimate $\|E_\eta(\tau)\| \le 1$ obviously yield that

$$\|L^* E_\eta(\tau)\| \le C. \tag{12.6.33}$$

Moreover, (12.6.31) yields that $\mathbf{A}_\eta$ is semibounded from below uniformly on $\eta > 0$. Therefore $\|\mathbf{A}_\eta E_\eta(\tau)\| \le C$ and hence $\|(A - \bar{F} + \eta)v\|^2 \le C$ for $v = J^{-\frac{1}{2}} E_\eta(\tau) w$ with $\|w\| = 1$; recall that $\|.\|$ means the $L^2(\mathbb{R}^3)$-norm. One can easily see that

$$\begin{aligned}\bigl|((A - \bar{F})^2 v, v) - Q(v)\bigr| \le CQ(v)^{\frac{1}{2}}\bigl(\|D_3 v\| + \|Lv\| + \|J^{\frac{1}{2}} v\|\bigr) +&\\ C\bigl(\|D_3 v\| + \|Lv\| + \|J^{\frac{1}{2}} v\|\bigr)^2&\end{aligned}$$

for all v where

$$Q(v) = \|D_3\rho^2 D_3 v\|^2 + 2\|\rho D_3 L v\|^2 + \|L^* L v\|^2$$

and we used condition (12.6.3). Therefore

$$\|(A-\bar{F}+\eta)v\|^2 \ge \frac{1}{2}Q(v) - C\big(\|D_3 v\| + \|Lv\| + \|J^{\frac{1}{2}}v\|\big)^2.$$

Then (12.6.32), (12.6.33) yields that $Q(v) \le C\tau^2$ for the indicated v and hence

$$\|D_3^2 J^{-\frac{1}{2}} E_\eta(\tau)\| \le C, \qquad \|LD_3 J^{-\frac{1}{2}} E_\eta(\tau)\| \le C$$
$$\|L^* L J^{-\frac{1}{2}} E_\eta(\tau)\| \le C \tag{12.6.34}$$

and the order of L and D_3 is not essential in the second inequality because $[D_3, L] = \alpha_4 L + \alpha_5$ with the same type of α_j as before.

Moreover, $(\mathbf{A}_\eta \mathbf{A}_\eta v, \mathbf{A}_\eta v)$ does not exceed $C\tau^3$ and we obtain some inequalities. I write only one of them: $\|D_3 J^{-\frac{1}{2}} \mathbf{A}_\eta v\| \le C\tau^{\frac{3}{2}}$. One can skip $J^{-\frac{1}{2}}$ here and permute A and D_3 in these estimates. Then applying the same arguments we obtain that estimates (12.6.34) remain true with $J^{-\frac{1}{2}}E_\eta(\tau)$ replaced by $D_3 J^{-\frac{1}{2}} E_\eta(\tau)$.[34] Obviously inequalities $(12.6.32)_1$, $(12.6.34)_2$ and the modified $(12.6.34)_3$ yield that

$$\|D_{x'}^\alpha D_3 J^{-\frac{1}{2}} E_\eta(\tau) w\|_\Omega \le C$$

for all $|\alpha| \le 2$ and for all w with $\|w\| = 1$ where $\Omega = \Omega' \times \mathbb{R}$, $\Omega' = B(0,r) \subset \mathbb{R}^2$, r is an arbitrary constant (C depends on r) and $\|.\|_\Omega$ means the $L^2(\Omega)$-norm. Then the embedding theorem yields that

$$|D_3 J^{-\frac{1}{2}} E_\eta(\tau) w| \le C \qquad \forall x' \in \Omega'. \tag{12.6.35}$$

The simplified proof of proposition 12.6.2 yields the estimate

$$|\langle x_3\rangle^{-\delta-1} w| \le C|D_3 w| + C|\langle x_3\rangle^{-l} w| \tag{12.6.36}$$

for all $w \in \mathbb{H}$ with arbitrary $\delta > 0$ and l.

Then estimate (12.6.35) yields that

$$|\langle x\rangle^{-m-1-\delta} E_\eta(\tau) w| \le C \tag{12.6.37}$$

for all $x' \in \Omega'$ and for all $w \in L^2(\mathbb{R}^3)$ with $\|w\| = 1$; here we applied the inequality $|E_\eta(\tau)w| \le C\langle x\rangle^n$ with large enough n which obviously follows from (12.6.32)–(12.6.35) and the embedding theorem.

Inequalities (12.6.36), (12.6.37) and the embedding theorem again yield that $|E_\eta(\tau)w| \le C\langle x\rangle^{m+\frac{1}{2}+\delta}$ for all $x \in \Omega$ and all w with $\|w\| = 1$. Therefore we have an inequality for the Schwartz kernel: $\|e_\eta(x,.,0)\| \le C\langle x\rangle^{m+\frac{1}{2}+\delta}$ for the $L^2(\mathbb{R}^3)$-norm on y for $x \in \Omega$ and then the equalities $E_\eta = E_\eta^* = E_\eta^2$ yield that $|e_\eta(x,x,0)| \le C\langle x\rangle^{2m+1+2\delta}$ for $x \in \Omega$. Therefore we have proven

Proposition 12.6.6. *In the context of theorem 12.6.4 and the definitions of J and $e_\eta(x,y,\tau)$,*

$$\int_{|x'|\le r_0} e_\eta(x,x,\tau)\,dx = O(1)$$

for all $\eta > 0$, $\tau \le \tau_0$ and for any fixed r_0 and τ_0.

Recall that $e_\eta(x,y,\tau)$ is the Schwartz kernel of $E_\eta(\tau)$. So, we only need to treat the contribution of the zone $\{|x'| \ge r_0\}$.

---

[34] Surely one can continue this process.

12.6.5 Proof of the main theorem. II

Let us fix $\bar{x}' \in \mathbb{R}^2$ and consider $\bar{\psi}(x')$, $\bar{\psi} \in C_0^K\big(B(\bar{x}', \frac{1}{2}\bar{\gamma})\big)$ with $\bar{\gamma} = \frac{1}{2}\langle \bar{x}'\rangle$ such that $|D^\sigma \bar{\psi}| \le c\bar{\gamma}^{-\sigma}$ $\forall \sigma : |\sigma| \le K$. We want to derive asymptotics of

$$\int \bar{\psi} e_\eta(x,x,0)dx = \int \bar{\psi}\, \mathrm{Tr}_{\mathbb{H}}\, \mathbf{e}_\eta(x',x',0)dx' = \mathrm{Tr}\, \bar{\psi}E_\eta(0)$$

where $\mathbf{e}_\eta(x',y',\tau)$ is the family of operators in $\mathbb{H}$ with Schwartz kernel $e_\eta(x,y,\tau)$. Let us make the dilatation $x'_{\mathrm{new}} = (x'-\bar{x}')\bar{\gamma}^{-1}$, $x_{3\mathrm{new}} = x_3\bar{\gamma}^{-1}$. Then we obtain a LSSA problem for an operator with operator-valued symbol, with semiclassical parameter $h = \bar{\gamma}^{-1}$ and with magnetic field intensity parameter $\mu = \bar{\gamma}$. Moreover, without loss of generality one could assume that

$$V_1 = 0, \quad V_2 = x_2 + O\big(\langle x\rangle^{2m}\langle x'\rangle\big)$$

and then the dilatation transforms the operator A to the operator

$$h^2\Big(D_3 g^{33} D_3 + D_1^2 + \big(D_2 - h^{-2}x_2 + h^{-2m-2}A''\big)^2 + h^{-2m-2}A'\Big) \tag{12.6.38}$$

where $A' = a'(x, hD', h)$, $A'' = a''(x, hD', h)$ are operators with uniformly smooth symbols a', a'' decreasing as $\langle x_3\rangle^{2m}$ as $x_3 \to \infty$ (we consider them more carefully later).

Let $U(x,y,t)$ be the Schwartz kernel of the operator $\exp(i\mathrm{h}^{-1}t\mathbf{A}_\eta)$, $\mathrm{h} = h^2$. Later we rescale t. Then

$$(\mathrm{h}D_t - \mathbf{A}_\eta)U = 0, \qquad U|_{t=0} = \delta(x-y)I. \tag{12.6.39}$$

The analysis made in the previous subsection for arbitrary $\langle \bar{x}'\rangle = h^{-1}$ yields that the operator norm of $Z\bar{\psi}E_\eta(\tau)Z^*$ does not exceed C for the operators $Z = h^{m+1}D_3J^{-\frac{1}{2}}$, $Z = h^{m+1}LJ^{-\frac{1}{2}}$, $Z = h^{2m+2}D_3^2J^{-\frac{1}{2}}$, $Z = h^{2m+2}D_3LJ^{-\frac{1}{2}}$, $Z = h^{2m+2}L^*LJ^{-\frac{1}{2}}$ and $Z = \langle x\rangle^{-1-\delta}J^{-\frac{1}{2}}$, $Z = \langle x\rangle^{-1-\delta}L^*J^{-\frac{1}{2}}$ where here and below $\tau \le \tau_0$ and rescaling was done, $J = (h^2 + |x'|^{2p} + |x_3|^{2p})^{mp^{-1}}$, $L = L_1 + iL_2$, $L_1 = hD_1$, $L_2 = hD_2 - h^{-1}x_1 + O\big(h^{2m-1}\langle x_3\rangle^{2m}\big)$. Therefore the same is true for $L = L_1 + iL_2$, $L_1 = hD_1$, $L_2 = hD_2 - h^{-1}x_1$. Therefore the operator norm of $F_{t\to \mathrm{h}^{-1}\tau}\chi_T(t)Z\bar{\psi}UZ^*$ does not exceed CT for the operators Z listed above where $\chi \in C_0^K(\mathbb{R})$ is fixed and $T \ge T_0$ with constant $T_0 > 0$; s is arbitrarily large.

Then the standard elliptic theory and these estimates yield that U is negligible outside of the $h^{1-\delta'}$-neighborhood of $\Lambda = \{x_1 = \xi_2 - x_1 = 0\}$ with arbitrarily small $\delta' > 0$ where we assume that $\tau \le c_0$. More precisely, let $Q = q(hD_1, hD_2 - h^{-1}x_1)$ with q supported outside of this neighborhood and $q = 1$ outside of a larger neighborhood. Then the operator norm of the operator $F_{t\to \mathrm{h}^{-1}\tau}\chi_T(t)QZUZ^*$ does not exceed Ch^s with arbitrarily large s (the reader surely understands where U... denotes an operator and when it denotes the Schwartz kernel of an operator). Then Tauberian arguments yield that the operator norm of $QZE(\tau)Z^*$ does not exceed Ch^s and the same is true for the trace norm of $QE(\tau)$.

Let us replace J by $J' = (1 + |x_3|^{2p})^{mp^{-1}}$ and replace E and U by $E' = MEM^*$, $U' = MEM^*$, $M = J'^{\frac{1}{2}}J^{-\frac{1}{2}}$; in obvious situations we do not distinguish operators and their Schwartz kernels. Then for sufficiently large p all the above estimates remain true for the modified operators and we skip the prime.

Let us apply the transformation $\mathcal{T} = \mathcal{T}_0^{-1}\mathcal{T}_1\mathcal{T}_0$ where $\mathcal{T}_0 = F_{x_2\to \mathrm{h}^{-1}\xi_2}$ and $\mathcal{T}_1 v(x_1,\xi_2) = v(x_1 - \xi_2, \xi_2)$. Then instead of hD_1 and $hD_2 - h^{-1}x_1$ we obtain hD_1 and $-h^{-1}x_1$ respectively. Let Ψ be the corresponding linear symplectic transformation.

Let $\bar{U} = \mathcal{T}_x\psi' U^t \mathcal{T}_y$ where ψ' is supported in $B(0, 1-\epsilon)$ and equals 1 in $B(0, 1-2\epsilon)$. Let us decompose U in terms of the functions $\Upsilon_n(x_1) = h^{-\frac{1}{2}}\upsilon_n(x_1 h^{-1})$ and $\Upsilon_n(y_1)$:

(12.6.40) $$\bar{U}(x,y,t) = \sum_{l,n\in\mathbb{N}} \Upsilon_l(x_1)\Upsilon_n(y_1)U_{ln}(x_2,x_3,y_2,y_3,t).$$

We make the same decomposition for E. Then the above estimates yield that the operator norm of $F_{t\to h^{-1}\tau}\chi_T(t)ZU_{ln}Z$ does not exceed CT in the general case and does not exceed $Ch^s(l+1)^{-1}(n+1)^{-1}$ for $l+n \ge h^{-\delta'}$ with arbitrarily small $\delta' > 0$ where here and below Z is an operator from the list $Z = h^{m+1}D_3 J^{-\frac{1}{2}}$, $Z = h^{m+1}D_3^2 J^{-\frac{1}{2}}$ and $Z = \langle x\rangle^{-1-\delta} J^{-\frac{1}{2}}$.

Then

$$\operatorname{Tr}\psi E = \sum_{l,n\le N} h^{|m-n|}\operatorname{Tr}\psi_{ln}E_{ln} + O(h^s)$$

where we have the original expression in the left-hand side, $\psi_{ln} = \psi_{ln}(x_2, \mathrm{h}D_2, \mathrm{h})$, $\psi_{ll}(x_2,\xi_2,0) = \psi(x_2,\xi_2)$ and $N = h^{-\delta'}$. Moreover, $\operatorname{supp}\psi_{ln}(x_2,\xi_2,\mathrm{h}) \subset \operatorname{supp}\psi$ and one can replace $\psi_{ln} - \delta_{ln}\psi$ by a linear combination of the derivatives of ψ of non-zero order.

Then the Cauchy problem

$$(\mathrm{h}\bar{M}D_t - \mathbf{A}_\eta)U = 0, \qquad U\big|_{t=0} = \delta(x-y)\bar{M}^{-1}$$

holds with $\mathbf{A}_\eta = J^{-\frac{1}{2}}(A - \bar{F} + \eta)J^{-\frac{1}{2}}$ and with redefined J and $\bar{M} = (MM^*)^{-1}$ and therefore U_{mn} satisfy the equations

(12.6.41) $$(\mathrm{h}M_l D_t - A_l)U_{ln} = \sum_{p\ne m} h^{|l-p|}\big(\mathcal{K}_{pn}J^{-\frac{1}{2}}U_{pn} + \mathcal{K}'_{lp}\mathrm{h}D_t U_{pn}\big)$$

in the indicated zone where

$$M_l = M_l(x_2, \mathrm{h}D_2), \qquad A_l = h^{2m}J^{-\frac{1}{2}}(B + 2m\bar{F})J^{-\frac{1}{2}}$$

and in $B = \mathbf{a}^w(x_2, x_3, \mathrm{h}D_2, hD_3)$ (we made a dilatation),

$$\mathcal{K}_{ln} = \mathcal{K}^0_{ln} + \mathcal{K}^1_{ln}hD_3 + \mathcal{K}^2_{ln}(hD_3)^2$$

where $\mathcal{K}_{ln}h^{-\delta_1}\langle x_3\rangle\delta_2$ are h-pseudo-differential operators on x_2 uniformly in x_3, h; $\delta_{1,2} > 0$ are small enough; and $\mathcal{K}'_{ln}$ are operators of the same type with $\delta_1 = 0$ and large $\delta_2 > 0$. Similar equations are fulfilled in y as well.

Treating the equations for $l \ge 1$ we obtain (for $\tau < \tau_0$ with sufficiently small $\tau_0 > 0$) U_{ln} via U_{0n} and the dual equations yield U_{ln} via U_{l0} for $n \ge 1$. So we obtain U_{ln} via U_{00} and then routine Tauberian arguments yield that the trace norm of $\mathbf{e}_{\eta ln}(x', x', \tau)$ in $\mathbb{H}$ does not exceed $Ch^{(l+n)(1+\delta)-2}$ with small enough $\delta > 0$ and therefore

(12.6.42) $$\operatorname{Tr}\psi E = \operatorname{Tr}\psi_{00}E_{00} + O(h^\delta).$$

So we need to get $\operatorname{Tr}\psi_{00}E_{00}$ because after summation on a partition of unity $O(h^\delta)$ provides $O(1)$ in the final answer.

Let us write down the equation for U_{00}:

(12.6.43) $$M^0\mathrm{h}D_t U_{00} = h^{2+2m}J^{-\frac{1}{2}}(B_\varepsilon + h^{-2}\eta)J^{-\frac{1}{2}}U_{00}$$

where $M^0 = M^0(x_2, x_3, \mathrm{h}D_2, \mathrm{h})$ is an operator the symbol of which is positive and uniformly smooth and disjoint from 0, quickly tending to $M^0_\infty(x_2, \mathrm{h}, D_2, h_2)$ as $x_3 \to \infty$,

$J = \langle x_3\rangle^{2m}$, $\varepsilon = h^{-2m-2}$ and $B^0_\varepsilon = D_3 g^{33} D_3 + \varepsilon V$ where we rescaled on x beforehand and replaced x_1 by $\mathrm{h}D_2$ and B_ε is the same type of operator with coefficients also dependent on h (so B^0_ε is the principal symbol of B_ε).

Let us note that the operator $J^{-\frac12}\bigl(B^0_\varepsilon(x_2,\xi_2)+\zeta\bigr)J^{-\frac12}$ has discrete spectrum in $\mathbb{H}$ and all the eigenvalues of this operator excluding at most one are positive and uniformly disjoint from 0 and there is one (the lowest) eigenvalue $\mu = O(\varepsilon + \zeta^{\frac12})$ and $\epsilon_0\zeta^{\frac12} - c_0\varepsilon \le \mu \le c_0\zeta^{\frac12} - \epsilon_0\varepsilon$ (this is due to proof of proposition 12.6.2).

Therefore for $\zeta \ge \epsilon_0$ all the eigenvalues are positive and disjoint from 0 [35] and one can easily prove that the trace norm (in $\mathbb{H}$) of $F_{t\to \mathrm{h}^{-1}\tau}\chi_T(t)u$ is then $O(h^s)$ for $\tau \le \epsilon_1 \mathrm{h}$, $T\in(h^{-\delta},h^{-n})$. Therefore the contribution of this zone to the final answer is negligible.

On the other hand, for $\zeta \le \epsilon_0$ the lowest eigenvalue is disjoint from the others and there exists a symbol $q(x_2,\xi_2,h_2) : \mathbb{H} \to \mathbb{C}\oplus\mathbb{H}$ such that for the operator $Q = q(x_2,\mathrm{h}D_2,\mathrm{h})$ and for $U' = QU_{00}Q^*$ we obtain separate equations for all four blocks of $U' = \begin{pmatrix} U'_{00} & U'_{01}\\ U''_{10} & U'_{11}\end{pmatrix}$. Moreover, for the blocks U'_{10} and U'_{11} the equations are elliptic for $\tau \le \epsilon_1\mathrm{h}$ and for U'_{10} and U'_{11} this is true for the dual equations. Therefore $U' \equiv (u,0)$ with $u = U'_{00}$ and

$$F_{t\to\mathrm{h}^{-1}\tau}\chi_T(t)\,\mathrm{Tr}\,\psi U \equiv F_{t\to\mathrm{h}^{-1}\tau}\chi_T(t)\,\mathrm{Tr}\,\psi'' u$$

for $\tau\le\epsilon_0 h^2$, $T\in(h^{-\delta},h^{-n})$ where $\psi''=\psi''(x_2,\mathrm{h}D_2,h_2)$ and $\psi''(x_2,\xi_2,0)=\psi(x,\xi_2)$.

We have an equation for u:

$$(\mathrm{h}D_t - h^2\Lambda)u = 0$$

where Λ is an h-pseudo-differential operator with principal symbol μ. More precisely:

(i) Let $\zeta \ge C_0\varepsilon^2$. Then $\mu \asymp \zeta^{\frac12}$ and then $\zeta^{-\frac12}\mu \ge \epsilon_0$ and $\zeta^{-\frac12}\mu$ is the principal symbol of the operator $\zeta^{-\frac12}\Lambda$. Then rescaling on t ($t_{\mathrm{new}} = h^{-2}\zeta^{-\frac12}t$) and applying elliptic arguments we obtain that $F_{t\to\mathrm{h}^{-1}\tau}\chi_T u = O(h^s)$ for $\tau \le \epsilon_2\zeta^{\frac12}h^2$, $T\in[h^{-\delta}\zeta^{-\frac12},h^{-n}]$. Therefore the contribution of this zone to the final answer is negligible. One can easily see that this zone is $\{\langle x\rangle^{2m+2}\eta \ge C_0\}$ in the original coordinate system.

(ii) Let $\zeta\le\epsilon_2\varepsilon^2$. Then $\mu\asymp\varepsilon$ and then $\varepsilon^{-1}\mu\le-\epsilon_0$ and ε^{-1} is the principal symbol of the operator $\varepsilon^{-1}\Lambda$. Then rescaling on t ($t_{\mathrm{new}} = h^{-2}\varepsilon^{-1}t$) and applying elliptic arguments we obtain that $F_{t\to\mathrm{h}^{-1}\tau}\chi_T u = O(h^s)$ for $|\tau|\le\epsilon_2\varepsilon h^2$, $T\in[h^{-\delta}\varepsilon^{-1},h^{-n}]$. Therefore the contribution of this zone to the final remainder estimate is negligible. One can easily see that this zone is $\{\langle x\rangle^{2m+2}\eta\le\epsilon_0\}$ in the original coordinate system. Moreover, after simple calculations one can easily see that for admissible ψ supported in this zone and equal to 1 in the same zone with $\frac12\epsilon_0$ instead of ϵ_0 the final answer is

$$\int_{\{\mu\le0\}}\psi\,dx + O(1) \tag{12.6.44}$$

(in the original coordinate system); here the restriction $\mu\le0$ is fulfilled automatically.

(iii) Now let us consider the zone $\bigl\{\frac12\epsilon_0 \le \langle x\rangle^{2m+2}\eta \le 2C_0\bigr\}$. In this zone $\zeta\asymp\varepsilon^2$, $\mu\asymp\varepsilon$ and $\varepsilon^{-1}\mu$ is the principal symbol of the operator $\varepsilon^{-1}\Lambda$. Condition (12.6.22) means that this operator is microhyperbolic at the energy level $\tau=0$. So, rescaling on T and applying all the routine arguments we obtain that the contribution of this zone to the principal part of the final answer is again given by (12.6.44) and the contribution to the remainder estimate is also $O(1)$ for admissible ψ supported in the indicated zone.

Thus, the final answer is also given by (12.6.42) for $\psi=1$ and since $\mu<0 \iff \lambda < -\eta$ we obtain that theorem 12.6.4 is proven for the Schrödinger operator.

---

[35] We assume that h and therefore ε are sufficiently small.

12.6.6 Proof of the main theorem. III

The proof for the Dirac operator is essentially the same. The required number $\nu(\eta)$ equals $\mathbf{N}(\eta;-\epsilon_2,0)$ modulo $O(1)$ where $\mathbf{N}(\eta;\tau_1,\tau_2)$ is the number of eigenvalues of the problem

$$(A+1+\eta)v+\tau\mathbf{J}v=0$$

lying in the interval (τ_1,τ_2) and $\mathbf{J}=\frac{1}{2}(I+\sigma_0)J$. This problem is equivalent to the problem

$$(\mathcal{A}_\eta-\tau J)w=0,\qquad \mathcal{A}_\eta=\mathcal{L}^*(2-\eta-V)^{-1}\mathcal{L}+V+\eta$$

where we assume that

$$\sigma_0=\begin{pmatrix} I & 0\\ 0 & -I\end{pmatrix},\quad \sigma_j=\begin{pmatrix} 0 & \sigma'_j\\ \sigma'_j & 0\end{pmatrix},\quad \sigma'^*_j=\sigma'_j\ (j=1,2,3),$$

$\mathcal{L}=\sum_{1\le j\le 3}\frac{1}{2}(P_k\omega^{jk}+\omega^{jk}P_k)\sigma'_j$. One can easily transform $\mathcal{A}_\eta$ to the form $\sum_{j,k}P_jg'^{jk}P_k+i\frac{1}{4}[\sigma'_j,\sigma'_k]F_{jk}$ with $g'^{jk}=(2-\eta-V)^{-1}g^{jk}$ and under an appropriate transformation this operator is reduced to two scalar operators (corresponding to two different eigenvalues of the matrix $\frac{1}{2}[\sigma'_1,\sigma'_2]F$), and the number of eigenvalues belonging to the interval in question is unbounded for only one of these operators. So we have obtained a problem for some Schrödinger operator. Using corollary 12.6.3 one can easily see that the answers for this operator and for the operator in theorem 12.6.4 coincide modulo $O(1)$. I leave the details to the reader.

12.6.7 Generalizations. I

Remark 12.6.5. *(i) One can attack the problem under the conditions*

$$\begin{aligned}&D^\alpha(g^{jk}-\delta_{jk})=O\bigl(\varrho(x)\langle x\rangle^{-|\alpha|}\bigr),\\ &D^\alpha(F-\bar{F})=O\bigl(\varrho(x)\langle x\rangle^{-|\alpha|}\bigr),\qquad D^\alpha V=O\bigl(\varrho(x)\langle x\rangle^{-|\alpha|}\bigr)\end{aligned}\tag{12.6.45}$$

for the Schrödinger operator (for the Dirac operator one should modify these conditions in the obvious way) where

$$\varrho(x)=\langle x\rangle^{-2-2p}\langle x'\rangle^{2p},\qquad p>0.\tag{12.6.46}$$

In this case instead of one lowest eigenvalue disjoint from the others we get a few of them (recall that we consider the operator $J^{-\frac{1}{2}}\bigl(B^0(x_2,\xi_2)+\eta\bigr)J^{-\frac{1}{2}}$ *where* B^0 *arises from the reduction; one can easily prove that all its eigenvalues are simple and disjoint from the others and the number of negative eigenvalues does not exceed* C_0*) and one should replace condition (12.6.22) by the microhyperbolicity condition for this operator. Surely, if we replace* V *by* εV *with a sufficiently small constant* $\varepsilon>0$ *then we still have at most one negative eigenvalue. The remainder estimate is* $O(1)$ *if all the eigenvalues in question are disjoint from* 0 *for* $\eta=0$ *(and then they are disjoint from* 0 *unless* $|x'|^{-2}\asymp\eta$*). I leave the details to the reader.*

(ii) Let us consider an auxiliary operator with potential W *which is* $\asymp|x_1|^{-2}$ *at infinity. One can easily prove (see proposition 12.6.6 below) that if*

$$W\ge -\frac{1}{4}|x|^{-2}\qquad \forall x:|x|\ge C\tag{12.6.47}$$

then the number of negative eigenvalues is finite and there is no more than one negative eigenvalue if this inequality holds for all x*. Moreover, under the conditions*

(12.6.47)′ $$W \ge (\epsilon - \frac{1}{4})|x|^{-2} \qquad \forall x : |x| \ge C$$

and (12.6.18) with arbitrarily small $\epsilon > 0$ all the statements of proposition 12.6.2 remain true. Moreover, under condition (12.6.47)′

$$(\mathrm{a}v, v) \ge \frac{\epsilon}{2}|\langle x\rangle^{-1}v|^2 - C|\langle x\rangle^{-s}v|^2 \qquad \forall v$$

with arbitrarily large s.

Due to this all our analysis applies under condition (12.6.45) with

(12.6.48) $$\varrho(x) = \langle x\rangle^{-2}\langle x'\rangle^{2m+2}, \qquad m \le -1$$

provided

(12.6.49) $$V_{\mathsf{eff}} \ge (\epsilon - \frac{1}{4})|x|^{-2} \qquad \forall x : |x_1| \ge c|x'|$$

with arbitrarily small $\epsilon > 0$. The remainder estimate is the same $O(1)$ as above. The details are left to the reader. ◊

Proposition 12.6.6. *(i) Under condition* (12.6.47) *the operator $\mathcal{L}$ has a finite number of negative eigenvalues.*

(ii) Moreover, if this condition is fulfilled for all x then there is at most one negative eigenvalue.

(iii) On the other hand, if

(12.6.50) $$W \le -(\frac{1}{4} + \epsilon)|x|^{-2} \qquad \forall x : x \ge C$$

then there is an infinite number of negative eigenvalues.

Proof. To prove (i),(ii) one needs to prove the estimate

(12.6.51) $$|u'|^2 \ge \frac{1}{4}||x|^{-1}u|^2 \qquad \forall u : u(0) = 0$$

where $|u|$ is the $L^2(\mathbb{R}^+)$-norm. However, the left side is equal to

$$|x^{\frac{1}{2}}(ux^{-\frac{1}{2}})' + \frac{1}{2}x^{-1}u|^2 = |x^{\frac{1}{2}}(ux^{-\frac{1}{2}})'|^2 + \frac{1}{4}||x|^{-1}u|^2$$

provided $u = o(x^{\frac{1}{2}})$ as $x \to 0$.

To prove (iii) it is sufficient to prove that the inequality $|u'|^2 \le (\frac{1}{4} + \epsilon)||x|^{-1}u|^2$ is fulfilled on some subspace of $L^2([1,\infty))$ of infinite dimension. It is enough to prove that for any n this inequality is fulfilled on some function supported in $[L^n, L^{n+1}$ with sufficiently large L. By homogeneity it is sufficient to consider only $n = 0$. Substituting $u = x^{\frac{1}{2}}v$, $x = e^t$ we obtain that it is sufficient to fulfill the inequality $|v'|^2 \le \epsilon|v|^2$ with some v such that $v(0) = v(\log L) = 0$. But this is obvious provided L is large enough. □

12.6.8 Generalizations. II

In this subsection I would like to discuss *possible* generalizations. I believe that the following is true but I have failed to prove it:

Conjecture 12.6.7. *One can reject condition (12.6.18) and the similar conditions in (i),(ii) and obtain the remainder estimate $O(\log \eta)$ instead of $O(1)$ for the Schrödinger and Dirac operators. Surely the function λ is not regular but in fact we are interested in the lowest eigenvalue of the operator $J^{-\frac{1}{2}}(\mathbf{a}+\eta)J^{-\frac{1}{2}}$. This eigenvalue is disjoint from the others and is regular. Moreover, under the condition $\varepsilon \ll 1$ all the other eigenvalues are greater than $\epsilon_1 > 0$.* ◊

Now I would like to consider how to improve the remainder estimates due to section 11.4 under conditions (12.6.45) with

$$\varrho(x) = \langle x\rangle^{-2-2p}\langle x'\rangle^{2(p+m+2)}, \qquad p>0 \tag{12.6.52}$$

with $0 > m > -1$. As well, we assume the microhyperbolicity condition

$$|x'|\langle(\ell\mathbf{a})v, v\rangle \ge \epsilon_0|\varrho v|^2 - C|\varrho^{-1}\varrho^{-1}(\mathbf{a}+\eta)v|^2 \qquad \forall x' \quad \forall \eta \le 0 \forall v \quad \text{with } \ell = \ell(x', \eta) : |\ell| = 1. \tag{12.6.53}$$

Under condition (12.6.52) the number of eigenvalues of $\mathbf{a}$ less than $-\eta$ (or 0, it does not matter in this estimate) is $O(\langle x'\rangle^{m+1})$ and therefore the following seems to be true:

Conjecture 12.6.8. *Under the microhyperbolicity condition the contribution of the layer $\{r \le \langle x'\rangle \le 2r\}$ to the remainder estimate is of the same order. However, there are no negative eigenvalues for $|x'|^{2m} \le \eta$ and so one should apply the same remainder estimate with an additional factor $|x'|^{-2ms}\eta^{-s}$. After summation on a partition of unity we get the total remainder estimate $O(\eta^{-\frac{1}{2m}(m+1)})$. Namely, we get asymptotics with principal part*

$$(2\pi)^{-1}\int \mathbf{n}(x_2, \xi_2, \eta)dx_1 d\xi_1$$

where $\mathbf{n}(x_2, \xi_2, \eta)$ is the eigenvalue counting function for the operator $\mathcal{L}$. Applying one-dimensional asymptotics one gets exactly the results of section 11.4. ◊

Comments

There are few papers devoted to non-classical asymptotics in domains with thick cusps (sorry, I know only about the existence of these papers); however, there are either no accurate remainder estimates or even no remainder estimate at all.

The theory of operators with strong degeneration of the potential at infinity also is not very well developed. I am aware of the paper [Solomyak 1] in which there is also no remainder estimate.

The list of papers devoted to operators with degenerations is much larger. I cannot list them all. The following papers are devoted to operators of the type we consider: [Menikoff, Sjöstrand 1–3], [Iwasaki, Iwasaki 1], [Mohamed 1,2], [Lopanikov 1]. Remainder estimates were not obtained in them. A special case was treated in [Melrose 1] and the remainder estimate was accurate in the case when it was Weylian and was not accurate otherwise.

There are many papers dealing with the Scott correction term in mathematical physics. The reader can find a survey in [Ivrii, Sigal 1]. However, the method was completely different and only the spherically symmetric case was treated.

Finally, the list of papers connected with section 12.6 is not very extensive: [Solnyshkin 1], [Sobolev 1–4] and [Raikov 1–5]. The remainder estimate obtained here was also not obtained in those papers.

Recently the methods and results of sections 12.4 and 12.5 were applied to asymptotics of the ground state energy in multiparticle quantum theory [Ivrii 30], [Ivrii 31] and [Ivrii 32].

References

Agmon, S.

[1] Asymptotic formulas with remainder estimates for eigenvalues of elliptic operators. *Arch. Rat. Mech. Anal.*, 28:165–183 (1968).

Agranovich, M. A.

[1] Elliptic operators on closed manifolds. *Sovremennye problemy matematiki. Fundamental'nye napravleniya. VINITI (Moscow)*, 63:5–129 (1990). The English translation has been published by Springer-Verlag: Partial Differential Equations VI, EMS volume 63, (1994).

[2] Elliptic boundary value problems. *Sovremennye problemy matematiki. Fundamental'nye napravleniya. VINITI (Moscow)*, 63 (1993). The English translation has been published by Springer-Verlag: Partial Differential Equations VIII, EMS volume 79 (1997).

[3] *Elliptic Boundary Problems*, volume 79. Springer-Verlag (1997).

Andreev, A. S.

[1] On an estimate for the remainder in the spectral asymptotics of pseudo-differential operators of negative order. *Probl. Mat. Fiz.*, 11:31–46 (1986). In Russian.

Arnold, V. I.

[1] *Geometrical Methods in the Theory of Ordinary Differential Equations*, volume 250 of *Grundlehren ...* . Springer-Verlag (1983).

[2] *Mathematical Methods of Classical Mechanics*. Springer-Verlag (1990).

Avakumovič, V. G.

[1] Uber die eigenfunktionen auf geschlossen riemannschen mannigfaltigkeiten. *Math. Z.*, 65:324–344 (1956).

Avron, J., Herbst, I., and Simon, B.

[1] Schrödinger operators with magnetic fields I: General interactions. *Duke Math. J.*, 45:847–883 (1978).

Babich, V. M.

[1] Focusing problem and the asymptotics of the spectral function of the Laplace-Beltrami operator II. *J. Soviet Math.*, 19(4):1288–1322 (1982).

Babich, V. M. and Buldyrev, V. S.

[1] *Short-wavelength Diffraction Theory*, volume 4 of *Springer Series on Wave Phenomena*. Springer-Verlag, New York (1991).

Babich, V. M. and Kirpicnikova, N. Y.

[1] *The Boundary-layer Method in Diffraction Theory*. Number 3 in Springer Electrophysics Series. Springer-Verlag (1979).

Babich, V. M. and Levitan, B. M.

[1] Focusing problem and the asymptotics of the spectral function of the Laplace-Beltrami operator. *Soviet Math. Dokl.*, 230:1017–1020 (1978).

Beals, R.

[1] A general calculus of pseudo-differential operators. *Duke Math. J.*, 42:1–42 (1975).

Beals, R. and Fefferman, C.
[1] Spatially inhomogenous pseudo-differential operators. *Commun. Pure. Appl. Math.*, 27:1–24 (1974).

Bérard, P.
[1] On the wave equation on a compact manifold without conjugate points. *Math. Zeit.*, 155(2):249–273 (1977).
[2] Spectre et groupes cristallographiques I: Domains Euclidiens. *Inv. Math.*, 58(2):179–199 (1980).
[3] Spectre et groupes cristallographiques II: Domains sphériques. *Ann. Inst. Fourier*, 30(3):237–248 (1980).
[4] *Spectral Geometry: Direct and Inverse Problems.* Number 1207 in Lect. Notes Math. Springer-Verlag (1985).

Berezin, F. A. and Shubin, M. A.
[1] Symbols of operators and quantization. In *Hilbert Space Operators and Operator Algebras: Proc. Intern. Conf., Tihany, 1970*, number 5, pages 21–52. Coloq. Math. Soc. Ja. Bolyai, North-Holland (1972).
[2] *The Schrödinger Equation.* Kluwer Ac. Publ. (1991). Also published in Russian by Moscow University, 1983.

Besse, A.
[1] *Manifolds all of whose Geodesics are Closed.* Springer-Verlag (1978).

Birman, M. S.
[1] Discrete spectrum in the gaps of a continuous one for perturbations with large coupling constant. *Adv. Soviet Math.*, 7:57–74 (1991).

Birman, M. S., Karadzhov, G. E., and Solomyak, M. Z.
[1] Boundedness conditions and spectrum estimates for operators $b(X)a(D)$ and their analogs. *Adv. Soviet Math.*, 7:85–106 (1991).

Birman, M. S. and Raikov, G.
[1] Discrete spectrum in the gaps for perturbations of the magnetic Schrödinger operator. *Adv. Soviet Math.*, 7:75–84 (1991).

Birman, M. S. and Solomyak, M. Z.
[1] Spectral asymptotics of nonsmooth elliptic operators I. *Trans. Moscow Math. Soc.*, 27:1–32 (1972).
[2] Spectral asymptotics of nonsmooth elliptic operators II. *Trans. Moscow Math. Soc.*, 28:1–32 (1973).
[3] Asymptotic behaviour of spectrum of differential equations. *J. Soviet Math.*, 12:247–282 (1979).
[4] Quantitative analysis in Sobolev imbedding theorems and application to the spectral theory. *AMS Trans. Ser. 2*, 114 (1980).
[5] *Spectral Theory of Self-adjoint Operators in Hilbert Space.* D. Reidel (1987).
[6] The estimates for the number of negative bound states of the Schrödinger operator for large coupling constants. In *Proc. Conf. Inverse Problems, Varna, Sept. 1989*, volume 235 of *Pitman Res. Notes in Math. Sci.*, pages 49–57 (1991).
[7] *Negative Discrete Spectrum of the Schrödinger Operator with Large Coupling Constant: Qualitative Discussion*, volume 46 of *Operator Theory: Advances and Applications.* Birkhäuser (1990).
[8] Estimates for the number of negative eigenvalues of Schrödinger operator and its generalizations. *Adv. Soviet Math. AMS*, 7:1–55 (1991).
[9] Schrödinger operator: Estimates of the number of bound states as a function-theoretical problem. *AMS Transl. Ser. 2*, 150 (1991).

Blanshard, P., Stubbe, J., and Reznde, J.
[1] New estimates on the number of bound states of Schrödinger operator. *Lett. Math. Phys.*, 14:215–225 (1987).

Bony, J. M. and Lerner, N.

[1] Quantification asymptotique et microlocalizations d'ordre supérier I. *Ann. Sci. Ec. Norm. Sup., sér. IV*, 22:377–433 (1989).

Boutet de Monvel, L.

[1] Boundary problems for pseudo-differential operators. *Acta Math.*, 126(1–2):11–51 (1971).

Brossard, J. and Carmona, R.

[1] Can one hear the dimension of the fractal? *Commun. Math. Phys.*, 104:103–122 (1986).

Brüning, J.

[1] Zur Abschätzzung der Spectralfunktion elliptischer Operatoren. *Mat. Z.*, 137:75–87 (1974).

Buslaev, V. S.

[1] On the asymptotic behaviour of spectral characteristics of exterior problems of the Schrödinger operator. *Math. USSR Izv.*, 39:139–223 (1975).

Candelpergher, B. and Nosmas, J. C.

[1] Propriétes spectrales d'opérateurs differentiels asymptotiques autoadjoints. *Commun. Part. Diff. Eq.*, 9:137–168 (1984).

Carleman, T.

[1] Über die asymptotische Verteilung der Eigenwerte partieller Differentialgleichungen. *Ber. Sachs. Acad. Wiss. Leipzig*, 88:119–132 (1936).

[2] Propriétes asymptotiques des fonctions fondamentales des membranes vibrantes. In *C. R. 8-ème Congr. Math. Scand., Stockholm, 1934*, pages 34–44, Lund (1935).

Charbonnel, A. M.

[1] Spectre conjoint d'Opérateurs qui commutent. *Annales de Toulouse*, 5(5):109–147 (1983).

[2] Calcul fonctionnel a plusieurs variables pour des o.p.d. sur $\mathbb{R}^n$. *Israel J. Math.*, 45(1):69–89 (1983).

[3] Localization et développement asymptotique des élements du spectre conjoint d'opérateurs pseudo-différentiels qui commutent. *Integral Eq. Op. Th.*, 9:502–536 (1986).

[4] Comportement semi-classique du spectre conjoint d'opérateurs pseudodifferentiels qui commutent. *Asymp. Anal.*, 1:227–261 (1988).

Chazarain, J.

[1] Formule de Poisson pour les varietes riemanniennes. *Inv. Math*, 24:65–82 (1977).

[2] Spectre d'un hamiltonien quantique et méchanique classique. *Commun. Part. Diff. Eq.*, 5(6):595–611 (1980).

[3] *Sur le Comportement Semi-classique du Spectre et de l'Amplitude de Diffusion d'un Hamiltonien Quantique.* Singularities in Boundary Value Problems, NATO. D. Reidel.

Colin de Verdiere, Y.

[1] Spectre du laplacien et longeurs des géodésiques periodiques II. *Comp. Math.*, 27:159–184 (1973).

[2] Spectre conjoint d'opérateurs qui commutent. *Duke Math. J.*, 46:169–182 (1979).

[3] Spectre conjoint d'opérateurs qui commutent. *Mat. Z.*, 171:51–73 (1980).

[4] Sur les spectres des opérateurs elliptiques à bicaractéristiques toutes périodiques. *Comment. Math. Helv.*, 54(3):508–522 (1979).

[5] *Sur les Longuers des Trajectories Periodiques d'un Billard*, pages 122–139. Hermann, Paris.

[6] Ergodicité et fonctions propres du laplacien. *Commun. Math. Phys.*, 102:497–502 (1985).

[7] Comportement asymptotique du spectre de boitelles magnétiques. In *Travaux Conf. EDP Sant Jean de Monts*, volume 2 (1985).

[8] L'asymptotique du spectre des boitelles magnétiques. *Commun. Math. Phys.*, 105(2):327–335 (1986).

Conlon, J.

[1] A new proof of the Cwickel-Lieb-Rozenbljum bound. *Rocky Mountain J. Math.*, 15:117–122 (1985).

Cornfeld, I. P., Fomin, S. V., and Sinai, Y. G.
[1] *Ergodic Theory*, volume 245 of *Grundlehren* ... . Springer-Verlag (1982).

Courant, R.
[1] Über die Eigenwerte bei den Differentialgleichungen der mathematischen Physik. *Mat. Z.*, 7:1–57 (1920).

Cwikel, M.
[1] Weak type estimates for singular values and the number of bound states of Schrödinger operator. *Ann. Math*, 106:93–100 (1977).

Cycon, H. L., Froese, R. G., Kirsh, W., and Simon, B.
[1] *Schrödinger Operators with Applications to Quantum Mechanics and Global Geometry.* Texts and Monographs in Physics. Springer-Verlag.

Dauge, M. and Robert, D.
[1] Formule de Weyl pour une classe d'opérateurs pseudodifférentiels d'ordre négatif sur $L^2(\mathbb{R}^n)$. *Lect. Notes Math.*, 1256:91–122 (1987).

Dencker, N.
[1] On the propagation of polarization sets for systems of real principal type. *J. Funct. Anal.*, 46(3):351–372 (1982).
[2] The Weyl calculus with locally temperate metrics and weights. *Ark. Mat.*, 24(1):59–79 (1986).
[3] On the propagation of polarization in conical refraction. *Duke Math. J.*, 57(1):85–134 (1988).
[4] On the propagation of polarization in double refraction. *J. Func. Anal.*, 104:414–468 (1992).

Dimassi, M.
[1] Spectre discret des opérateurs périodiques perturbés par un opérateur différentiel. *Prépublications Math. de l'Université Paris-Nord*, 97(3) (1997).
[2] Trace asymptotics formula and some applications. *Prépublications Math. de l'Université Paris-Nord*, 97(11) (1997).

Dimassi, M. and Sjöstrand, J.
[1] Trace asymptotics via almost analytic extensions. *Ecole Polytechnique Preprints*, (N1119) (1995). 17pp.

Dubrovin, B. A., Fomenko, A. T., and Novikov, S. P.
[1] *Modern Geometry—Methods and Applications 1*, volume 92 of *Graduate Texts in Mathematics.* Springer-Verlag (1984).
[2] *Modern Geometry—Methods and Applications 2*, volume 104 of *Graduate Texts in Mathematics.* Springer-Verlag (1985).

Duistermaat, J. J. and Guillemin, V.
[1] The spectrum of positive elliptic operators and periodic bicharacteristics. *Invent. Math.*, 29(1):37–79 (1975).

Efremov, D. V. and Shubin, M. A.
[1] Spectral asymptotics for elliptic operators of Schrödinger type on hyperbolic space. *Trudy sem. Petrovskogo*, 15:3–32 (1991).

Egorov, Y. V. and Kondrat'ev, V. A.
[1] On the estimate of the negative spectrum of the Schrödinger operator. *Math. USSR Sbornik*, 62(4):551–566 (1989).
[2] On the negative spectrum of elliptic operator. *Math. USSR Sbornik*, 69(4):155–177 (1991).
[3] Estimates of the negative spectrum of an elliptic operator. *AMS Transl. Ser. 2*, 150 (1992).

Eskin, G.
[1] Parametrix and propagation of singularities for the interior mixed hyperbolic problem. *J. Anal. Math.*, 32:17–62 (1977).
[2] *General Initial-Boundary Problems for Second Order Hyperbolic Equations*, pages 19–54. D. Reidel Publ. Co., Dordrecht, Boston, London (1981).
[3] Initial boundary value problem for second order hyperbolic equations with general boundary conditions, I. *J. Anal. Math.*, 40:43–89 (1981).

Fedoryuk, M. V. and Maslov, V. P.
[1] *Semi-classical Approximation in Quantum Mechanics.* D. Reidel (1981).

Fefferman, C. L.
[1] The uncertainty principle. *Bull. Amer. Math. Soc.*, 9:129–206 (1983).

Fefferman, C. L., Ivrii, V., Seco, L. A., and Sigal, I. M.
[1] The energy asymptotics of large Coulomb systems. In *Proc. Conf. "N-body Quantum Mechanics"*, Aarhus, Denmark (1991).

Feigin, V. I.
[1] The asymptotic distribution of the eigenvalues of pseudodifferential operators in $\mathbb{R}^n$. *Math. USSR Sbornik*, 28:533–552 (1976).
[2] The asymptotic distribution of eigenvalues and a formula of Bohr-Sommerfeld type. *Math. USSR Sbornik*, 38(1):61–81 (1981).

Fleckinger, J. and Lapidus, M.
[1] Tambour fractal: vers une résolution de la conjecture de Weyl-Berry pour les valeurs propres du laplacien. *C. R. A. S. Paris, Sér. 1*, 306:171–175 (1988).

Fleckinger, J. and Metivier, G.
[1] Théorie spectrale des opérateurs uniforment elliptiques des quelques ouverts irréguliers. *C. R. A. S. Paris, Sér. A-B*, 276:A913–A916 (1973).

Fleckinger, J. and Vassiliev, D. G.
[1] Tambour fractal: Exemple d'une formule asymptotique à deux termes pour la "fonction de compatage". *C. R. Acad. Sci. Paris, Sér. 1*, 311:867–872 (1990).

Friedlander, F. G.
[1] Notes on closed billiard ball trajectories in polygonal domains i. *Commun. Part. Diff. Eq.*, 12 (1990).
[2] Notes on closed billiard ball trajectories in polygonal domains II. *Commun. Part. Diff. Eq.*, 16(6):1687–1694 (1991).

Friedrichs, K.
[1] *Perturbations of Spectra in Hilbert Space.* AMS, Providence, RI (1965).

Grubb, G.
[1] Spectral sypmptotics for Douglis-Nirenberg elliptic systems and pseudodifferential boundary value problems. *Commun. Part. Diff. Eq.*, 2(9):1071–1150 (1977).
[2] Remainder estimates for eigenvalues and kernels of pseudo-differential elliptic systems. *Math. Scand.*, 43:275–307 (1978).
[3] *Functional Calculus of Pseudo-differential Boundary Problems.* Birkhäuser (1986).

Guillemin, V.
[1] Lectures on spectral properties of elliptic operators. *Duke Math. J.*, 44(3):485–517 (1977).
[2] Some spectral results for the Laplace operator with potential on the n-sphere. *Adv. Math.*, 27:273–286 (1978).
[3] *Some Classical Theorems in Spectral Theory Revised.* Princeton University Press (1979).

Guillemin, V. and Melrose, R. B.
[1] The Poisson summation formula for manifolds with boundary. *Adv. Math.*, 32:204–232 (1979).

Guillemin, V. and Sternberg, S.
[1] On the spectra of commuting pseudo-differential operators. In *Proc. Park City Conf., 1977*, number 48 in Lect. Notes Pure Appl. Math., pages 149–165 (1979).
[2] *Geometric Asymptotics*. AMS Survey Publ., Providence, RI (1977).

Gureev, T. and Safarov, Y.
[1] Accurate asymptotics of the spectrum of the Laplace operator on manifold with periodic geodesics. *Trudy Leningradskogo Otdeleniya Mat. Inst. AN SSSR*, 179:36–53 (1988). English translation in Proceedings of the Steklov Institute of Math., 2, 1988.

Helffer, B.
[1] Théorie spectral pour des opérateurs globalement elliptiques. *Astérisque*, 112 (1984).
[2] *Introduction to the Semi-classical Analysis for the Schrödinger Operator and Applications.* Number 1336. Lect. Notes Math. (1984).

Helffer, B., Knauf, A., Siedentop, H., and Weikart, R.
[1] On the absence of a first order correction for the number of bound states of a Schrödinger operator with Coulomb singularity. *Commun. Part. Diff. Eq.*, ?(?):615–? (1992).

Helffer, B., Martinez, A., and Robert, D.
[1] Ergodicité et limite semi-classique. *Commun. Math. Phys.*, 109:313–326 (1987).

Helffer, B. and Mohamed, A.
[1] Sur le spectre essentiel des opérateurs de Schrödinger avec champ magnétique. *Ann. Inst. Fourier*, 38(2) (1988).

Helffer, B., Nourrigat, J., and Wang, X. P.
[1] Spectre essentiel pour l'équation de Dirac. *Ann. Sci. Ec. Norm. Sup., Sér. IV*, 22:515–533 (1989).

Helffer, B. and Parisse, B.
[1] Moyens de Riesz d'etats bornes et limite semi-classique en liasion avec la conjecture de Lieb-Thirring III. *Ec. Norm. Sup. Paris*, Prep. LMENS-90-12 (1990).

Helffer, B. and Robert, D.
[1] Comportement semi-classique du spectre des hamiltoniens quantiques elliptiques. *Ann. Inst. Fourier*, 31(3):169–223 (1981).
[2] Comportement semi-classique du spectre des hamiltoniens quantiques hypoelliptiques. *Ann. Ec. Norm. Sup. Pisa, Sér. IV*, 9(3) (1982).
[3] Propriétes asymptotiques du spectre d'opérateurs pseudo-differentiels sur $\mathbb{R}^n$. *Commun. Part. Diff. Eq.*, 7:795–882 (1982).
[4] Étude du spectre pour un opérateur globallement elliptique dont le présente des symétries I: Action des groupes finis. *Amer. J. Math.*, 106(5):1199–1236 (1984).
[5] Étude du spectre pour un opérateur globallement elliptique dont le présente des symétries II: Action des groupes compacts. *Amer. J. Math*, 108:978–1000 (1986).
[6] Asymptotique des niveaux d'énergie pour des hamiltoniens a un degré liberté. *Duke Math. J.*, 49(4):853–868 (1982).
[7] Puits de potentiel généralisés et asymptotique semi-classique. 41(3):291–331 (1984).
[8] Riesz means of bound states and semi-classical limit connected with a Lieb-Thirring conjecture I. *Asymp. Anal.*, 3(4):91–103 (1990).
[9] Riesz means of bound states and semi-classical limit connected with a Lieb-Thirring conjecture II. *Ann. de l'Inst. Henri Poincaé, sect. Phys. Theor.*, 53(2):139–147 (1990).

Helffer, B. and Sjöstrand, J.
[1] Multiple wells in the semiclassic limit I. *Commun. Part. Diff. Eq.*, 9(4):337–408 (1984).
[2] On diamagnetism and Haas-van Alphen effect. *Ann. Inst. Henri Poincaré*, 52(4):303–375 (1990).

Hörmander, L.
[1] *The Analysis of Linear Partial Differential Operators I–IV.* Springer-Verlag (1983, 1985).
[2] The spectral function of an elliptic operator. *Acta Math.*, 121:193–218 (1968).
[3] On the Riesz means of spectral functions and eigenfunction expansions for elliptic differential operators. In *Yeshiva Univ. Conf., November 1966*, volume 2 of *Ann. Sci. Conf. Proc.*, pages 155–202. Belfer Graduate School of Sci. (1969).
[4] On the asymptotic distribution of eigenvalues of p.d.o. in $\mathbb{R}^n$. *Ark. Math.*, 17(3):169–223 (1981).

Houakni, Z. E. and Helffer, B.
[1] Comportement semi-classique en présence de symétries. action d'une groupe de Lie compact. *Asymp. Anal.*, 5(2):91–114 (1991).

Ivrii, V.
[1] Wave fronts of solutions of symmetric pseudo-differential systems. *Siberian Math. J.*, 20(3):390–405 (1979).
[2] Wave fronts of solutions of boundary-value problems for symmetric hyperbolic systems I: Basic theorems. *Siberian Math. J.*, 20(4):516–523 (1979).
[3] Wave fronts of solutions of boundary value problems for symmetric hyperbolic systems II. systems with characteristics of constant multiplicity. *Sibirsk. Mat. Zhurn.*, 20(5):1022–1038 (1979).
[4] Wave fronts of solutions of boundary value problems for symmetric hyperbolic systems III. systems with characteristics of variable multiplicity. *Sibirsk. Mat. Zhurn.*, 21(1):74–81 (1980).
[5] Wave fronts of solutions of boundary value problems for a class of symmetric hyperbolic systems. *Sibirsk. Mat. Zhurn.*, 21(4):62–71 (1980).
[6] Second term of the spectral asymptotic expansion for the Laplace-Beltrami operator on manifold with boundary. *Funct. Anal. Appl.*, 14(2):98–106 (1980).
[7] Accurate spectral asymptotics for elliptic operators that act in vector bundles. *Funct. Anal. Appl.*, 16(2):101–108 (1982).
[8] *Precise Spectral Asymptotics for Elliptic Operators.* Number 1100 in Lect. Notes Math. Springer-Verlag (1984).
[9] Precise spectral asymptotics for elliptic operators on manifolds with boundary. *Siberian Math. J.*, 28(1):80–86 (1987).
[10] Linear hyperbolic equations. *Sovremennye Problemy Matematiki. Fundamental'nye Napravleniya. VINITI (Moscow)*, 33:157–247 (1988).
[11] Weyl's asymptotic formula for the Laplace-Beltrami operator in Riemann polyhedra and in domains with conical singularities of the boundary. *Soviet Math. Dokl.*, 38(1):35–38 (1986).
[12] Estimations pour le nombre de valeurs propres negatives de l'operateurs de Schrödinger avec potentiels singuliesrs. *C.R.A.S. Paris, Sér. 1*, 302(13, 14, 15):467–470, 491–494, 535–538 (1986).
[13] Estimates for the number of negative eigenvalues of the Schrödinger operator with singular potentials. In *Proc. Intern. Cong. Math., Berkeley*, pages 1084–1093 (1986).
[14] Estimates for the spectrum of Dirac operator. *Dokl. AN SSSR*, 297(6):1298–1302 (1987).
[15] Estimates for the number of negative eigenvalues of the schrödinger operator with a strong magnetic field. *Soviet Math. Dokl.*, 36(3):561–564 (1988).
[16] Sharp spectral asymptotics for the two-dimensional Schrödinger operator with a strong magnetic field. *Soviet Math. Dokl.*, 39(3):437–441 (1989).
[17] Precise spectral asymptotics for Dirac operator with the strong magnetic field. *Soviet Math. Dokl.*, 40(2):122–125 (1989).
[18] Eigenvalue asymptotics for two-dimensional Dirac operator with strong magnetic field. *Dokl. AN SSSR*, 311(5):1102–1105 (1990). In Russian.
[19] Semiclassical microlocal analysis and precise spectral asymptotics, preprint 1. *Ecole Polytechnique*, (Preprint M964.1190) (1989).
[20] Semiclassical microlocal analysis and precise spectral asymptotics, preprint 2. *Ecole Polytechnique*, (Preprint M969.0191) (1991).
[21] Semiclassical microlocal analysis and precise spectral asymptotics, preprint 3. *Ecole Polytechnique*, (Preprint M971.0291) (1991).
[22] Semiclassical microlocal analysis and precise spectral asymptotics, preprint 4. *Ecole Polytechnique*, (Preprint M974.0591) (1991).
[23] Semiclassical microlocal analysis and precise spectral asymptotics, preprint 5. *Ecole Polytechnique*, (Preprint M980.0691) (1991).

[24] Semiclassical microlocal analysis and precise spectral asymptotics, preprint 6. *Ecole Polytechnique*, (Preprint M1018.1091) (1991).
[25] Semiclassical microlocal analysis and precise spectral asymptotics, preprint 7. *Ecole Polytechnique*, (Preprint M1031) (1992).
[26] Semiclassical microlocal analysis and precise spectral asymptotics, preprint 8. *Ecole Polytechnique*, (Preprint M1034) (1992).
[27] Semiclassical microlocal analysis and precise spectral asymptotics, preprint 9. *Ecole Polytechnique*, (Preprint N1041) (1992).
[28] Eigenvalue asymptotics for the Schrödinger and Dirac operators with the constant magnetic field and with electric potential decreasing at infinity. In *Travaux Conf. EDP Sant Jean de Monts, Exp. 10* (1991).
[29] Semiclassical spectral asymptotics. *Astérisque*, 207:7–35 (1992). Summer School on Semiclassical Analysis, Nantes, June 1991.
[30] Asymptotics of the ground state energy of heavy molecules in the strong magnetic field I. *Russian Journal of Math. Phys.*, 4(1):29–74 (1996).
[31] Asymptotics of the ground state energy of heavy molecules in the strong magnetic field II. *Russian Journal of Math. Phys.* (1996). To appear. 40pp.
[32] Heavy molecules in the strong magnetic field. *Russian Journal of Math. Phys.*, 4(4):449–456 (1996).

Ivrii, V. and Fedorova, S.
[1] Dilitations and the asymptotics of the eigenvalues of spectral problems with singularities. *Funct. Anal. Appl.*, 20(4):277–281 (1986).

Ivrii, V., Filippov, E., and Kachalkina, A.
[1] Precise spectral asymptotics. In *Proc. Int. Conf. Integral Equations and Inverse Problems, Varna 1989*, pages 146–153. Longman (1990).

Ivrii, V. and Sigal, I. M.
[1] Asymptotics of the ground state energies of large Coulomb systems. *Ann. of Math.*, 138:243–335 (1993).

Kac, M.
[1] Can one hear the shape of a drum? *Amer. Math. Monthly*, 73:1–23 (1966).

Kolmogorov, A. N. and Fomin, S. V.
[1] *Introductory Real Analysis*. Selected Russian publications in the mathematical sciences. Prentice-Hall, Englewood Cliffs, N.J. (1970).

Kozhevnikov, A. N.
[1] Remainder estimates for eigenvalues and complex powers of the Douglis-Nirenberg elliptic systems. *Commun. Part. Diff. Eq.*, 6:1111–1136 (1981).
[2] Spectral problems for pseudo-differential systems elliptic in the Douglis-Nirenberg sense and their applications. *Math. USSR Sbornik*, 21:63–90 (1973).

Kozlov, V. A.
[1] Asymptotic behavior of the spectrum of nonsemibounded elliptic systems. *Probl. Mat. Analiza, Leningrad Univ.*, 7:70–83 (1979).
[2] Estimates of the remainder in formulas for the asymptotic behavior of the spectrum of linear operator bundles. *Probl. Mat. Analiza, Leningrad Univ.*, 9:34–56 (1984).

Lapidus, M. L.
[1] Fractal drum, inverse spectral problems for elliptic operators and a partial resolution of the Weyl-Berry conjecture. *Trans. Amer. Math. Soc.*, 325:465–529 (1991).

Laptev, A.
[1] Asymptotics of the negative discrete spectrum of class of Schrödinger operators with large coupling constant. *Linköping University Preprint* (1991).

Laptev, A. and Safarov, Y.
[1] Global parametrization of Lagrangian manifold and the Maslov factor. *Linköping University Preprint* (1989).

Laptev, A., Safarov, Y., and Vassiliev, D.
[1] On global representation of Lagrangian distributions and solutions of hyperbolic equations. *Linköping University Preprint* (1992).

Lazutkin, V. F.
[1] Asymptotics of the eigenvalues of the Laplacian and quasimodes. *Math. USSR Sbornik*, 7:439–466 (1973).
[2] Convex billiards and eigenfunctions of the Laplace operators. *Leningrad Univ.* (1981). In Russian.
[3] Semiclassical asymptotics of eigenfunctions. *Sovremennye Problemy Matematiki. Fundamental'nye Napravleniya. VINITI (Moscow)*, 34:135–174 (1987).

Leray, J.
[1] *Analyse Lagrangienne et Mecanique Quantique.* Univ. Louis Pasteur, Strasbourg (1978).

Levendorskii, S. Z.
[1] Asymptotic distribution of eigenvalues. *Math. USSR Izv.*, 21:119–160 (1983).
[2] The approximate spectral projection method. *Acta Appl. Math.*, 7:137–197 (1986).
[3] The approximate spectral projection method. *Math. USSR Izv.*, 27:451–502 (1986).
[4] Non-classical spectral asymptotics. *Russian Math. Surveys*, 43:149–192 (1988).
[5] Asymptotics of the spectra of operators elliptic in the Douglis-Nirenberg sense. *Trans. Moscow Math. Soc.*, 52:533–587 (1989).
[6] *Asymptotic Distribution of Eigenvalues of Differential Operators.* Kluwer Acad. Publ. (1990).

Levitan, B. M.
[1] On the asymptotic behaviour of the spectral function of the second order elliptic equation. *Izv. AN SSSR, Ser. Mat.*, 16(1):325–352 (1952). In Russian.
[2] Asymptotic behaviour of the spectral function of elliptic operator. *Russian Math. Surveys*, 26(6):165–232 (1971).

Levy-Bruhl, P.
[1] Spectre d'opérateurs avec potentiel et champ magnétique polynomiaux. *Preprint* (1988?).

Li, P. and Yau, S.-T.
[1] On the Schrödinger equation and the eigenvalue problem. *Comm. Math. Phys.*, 88:309–318 (1983).

Lieb, E. H.
[1] Bounds on the eigenvalues of the Laplace and Schrödinger operators. *Bull. of the AMS*, 82:751–753 (1976).

Lions, J. L. and Magenes, E.
[1] *Non-homogeneous Boundary Value Problems*, volume 181, 182, 183 of *Grundlehren der mathematischen Wissenschaften in Einzeldarstellungen.* Springer-Verlag, Berlin (1972–).

Malgrange, B.
[1] *Ideals of Differentiable Functions.* Tata Institute and Oxford University Press, Bombay (1966).

Markus, A. S. and Matsaev, V. I.
[1] Comparison theorems for spectra of linear operators and spectral asymptotics. *Trans. Moscow Math. Soc.*, 45:139–187 (1982, 1984).

Maslov, V. P.
[1] *Théorie des Perturbations et Méthods Asymptotiques.* Dunod, Paris (1972).
[2] Non-standard characteristics for asymptotical problems. *Uspechi Mat. Nauk*, 38(6):3–36 (1983).

Maz'ya, V. G.
[1] *Sobolev Spaces.* Springer-Verlag (1985).

Melrose, R.
[1] Microlocal parametrices for diffractive boundary value problems. *Duke Math. J.*, 42:605–635 (1975).
[2] Local Fourier-Airy operators. *Duke Math. J.*, 42:583–604 (1975).
[3] Airy operators. *Commun. Part. Differ. Equat.*, 3(1):1–76 (1978).
[4] Weyl's conjecture for manifolds with concave boundary. In *Proc. Symp. Pure Math.*, volume 36, Providence, RI. AMS (1980).
[5] Transformation of boundary problems. *Acta Math.*, 147:149–236 (1981).
[6] The trace of the wave group. *Contemp. Math.*, 27:127–161 (1984).
[7] Weyl asymptotics for the phase in obstacle scattering. *Commun. Part. Differ. Equat.*, 13(11):1441–1466 (1989).

Melrose, R. B. and Sjöstrand, J.
[1] Singularities of boundary value problems I. *Commun. Pure Appl. Math.*, 31:593–617 (1978).
[2] Singularities of boundary value problems II. *Commun. Pure Appl. Math.*, 35:129–168 (1982).

Menikoff, A. and Sjöstrand, J.
[1] On the eigenvalues of a class of hypoelliptic operators I. *Math. Ann.*, 235:55–85 (1978).
[2] *On the Eigenvalues of a Class of Hypoelliptic Operators II*, pages 201–247. Number 755 in Lect. Notes Math. Springer-Verlag.
[3] On the eigenvalues of a class of hypoelliptic operators III. *J. d'Anal. Math.*, 35:123–150 (1979).

Métivier, G. M.
[1] Fonction spectrale et valeurs propres d'une classe d'opérateurs non-elliptiques. *Commun. Part. Differ. Equat.*, 1:467–519 (1976).
[2] Valeurs propres de problèmes aux limites elliptiques irreguliers. *Bull. Soc. Math. France, Mém*, 51–52:125–219 (1977).
[3] Estimation du reste en théorie spectral. In *Rend. Semn. Mat. Univ. Torino, Facs. Speciale*, pages 157–180 (1983).

Mohamed, A.
[1] Etude spectrale d'opérateurs hypoelliptiques a caracteristiques multiples I. *Ann. Inst. Fourier*, 32(3):39–90 (1982).
[2] Etude spectral d'opérateurs hypoelliptiques a caracteristiques multiples II. *Commun. Part. Differ. Equat.*, 8:247–316 (1983).
[3] Comportement asymptotique, avec estimation du reste, des valeurs propres d'une class d'opd sur $\mathbb{R}^n$. *Math. Nachr.*, 140:127–186 (1989).
[4] Quelques remarques sur le spectre de l'opérateur de Schrödinger avec un champ magnetique. *Commun. Part. Differ. Equat.*, 13(11):1415–1430 (1989).
[5] Estimations semi-classique pour l'operateur de Schrödinger a potentiel de type coulombien et avec champ magnétique. *Asymp. Anal.*, 4(3):235–255 (1991).

Mohamed, A. and Nourrigat, J.
[1] Encadrement du $N(\lambda)$ pour un opérateur de Schrödinger avec un champ magnétique et un potentiel électrique. *J. Math. Pures Appl.*, 70:87–99 (1991).

Nosmas, J. C.
[1] Approximation semi-classique du spectre de systemes différentiels asymptotiques. *C.R.A.S. Paris, Sér. 1*, 295(3):253–256 (1982).

Paul, T. and Uribe, A.
[1] Sur la formula semi-classique des traces. *C.R.A.S. Paris, Sér. 1*, pages 217–222 (1991).

Petkov, V. and Popov, G.
[1] Asymptotic behaviour of the scattering phase for non-trapping obstacles. *Ann. Inst. Fourier*, 32:114–149 (1982).

Petkov, V. and Robert, D.
[1] Asymptotique semi-classique du spectre d'hamiltoniens quantiques et trajectories classiques periodiques. *Commun. Part. Differ. Equat.*, 10(4):365–390 (1985).

Lai, P. T.
[1] Meilleurs estimations asymptotiques des restes de la fonction spectrale et des valeurs propres relatifs au Laplacien. *Math. Scand.*, 48:5–31 (1981).

Popov, G.
[1] Spectral asymptotics for elliptic second order differential operators. *J. Math. Kyoto Univ.*, 25:659–681 (1985).

Popov, G. and Shubin, M. A.
[1] Asymptotic expansion of the spectral function for second order elliptic operators on $\mathbb{R}^n$. *Funct. Anal. Appl.*, 17(3):193–199 (1983).

Raikov, G.
[1] Spectral asymptotics for the Schrödinger operator with potential which steadies at infinity. *Commun. Math. Phys.*, 124:665–685 (1989).
[2] Eigenvalue asymptotics for the Schrödinger operator with homogeneous magnetic potential and decreasing electric potential I: Behavior near the essential spectrum tips. *Commun. Part. Differ. Equat.*, 15(3):407–434 (1990).
[3] Eigenvalue asymptotics for the Schrödinger operator with homogeneous magnetic potential and decreasing electric potential II: Strong electric field approximation. In *Proc. Int. Conf. Integral Equations and Inverse Problems, Varna 1989*, pages 220–224. Longman (1990).
[4] Strong electric field eigenvalue asymptotics for the Schrödinger operator with the electromagnetic potential. *Letters Math. Phys.*, 21:41–49 (1991).
[5] Semi-classical and weak-magnetic-field eigenvalue asymptotics for the Schrödinger operator with electromagnetic potential. *C.R.A.S. Bulg. AN*, 44(1) (1991).
[6] Border-line eigenvalue asymptotics for the Schrödinger operator with electromagnetic potential. *Int. Eq. Oper. Theorem.*, 14:875–888 (1991).

Randoll, B.
[1] The Riemann hypothesis for Selberg's zeta-function and asymptotic behavior of eigenvalues of the Laplace operator. *Trans. AMS*, 276:203–223 (1976).

Reed, M. and Simon, B.
[1] *The Methods of Modern Mathematical Physics*, volume I–IV. Academic Press, New York (1972, 1975, 1979, 1978).

Robert, D.
[1] Propriétés spectrales d'opérateurs pseudodifférentiels. *Commun. Part. Differ. Equat.*, 3(9):755–826 (1978).
[2] Comportement asymptotique des valeurs propres d'opérateurs du type Schrödinger à potentiel dégénére. *J. Math. Pures Appl.*, 61(3):275–300 (1982).
[3] Calcul fonctionnel sur les opérateurs admissibles et application. *J. Funct. Anal.*, 45(1):74–84 (1982).
[4] *Autour de l'Approximation Semi-classique.* Number 68 in Publ. Math. Birkhäuser (1987).

Robert, D. and Tamura, H.
[1] Asymptotic behavior of scattering amplitudes in semi-classical and low energy limits. *Ann. Inst. Fourier*, 39(1):155–192 (1989).

Rozenblioum, G. V.
[1] The distribution of the discrete spectrum for singular differential operators. *Dokl. Akad. Nauk SSSR*, 202:1012–1015 (1972). In Russian. No frills English translation in Soviet Math. Dokl. 13 (1972).
[2] Distribution of the discrete spectrum for singular differential operators. *Soviet Math. Dokl.*, 13(1):245–249 (1972).
[3] On the eigenvalues of the first boundary value problem in unbounded domains. *Math. USSR Sb.*, 89(2):234–247 (1972).
[4] Asymptotics of the eigenvalues of the Schrödinger operator. *Math. USSR Sb.*, 22:349–371 (1974).
[5] An asymptotics of the negative discrete spectrum of the Schrödinger operator. *Math. Notes Acad. Sci. USSR*, 21:222–227 (1977).

[6] *On the Asymptotics of the Eigenvalues of Some 2-dimensional Spectral Problems*, pages 183–203. Number 7 in Univ. Publ. Leningrad (1979).

Rozenblioum, G. V., Solomyak, M. Z., and Shubin, M. A.
[1] Spectral theory of differential operators. *Sovremennye problemy matematiki. Fundamental'nye napravleniya. VINITI, Moscow*, 64 (1990). Translation to English has been published by Springer-Verlag: Partial Differential Equations VII, EMS volume 34 (1994).

Safarov, Y.
[1] The asymptotics of the spectrum of transmission problems. *J. Soviet Math.*, 32:519–525 (1986).
[2] The asymptotics of the spectrum of boundary value problem with periodic billiard trajectories. *Funct. Anal. Appl.*, 21:337–339 (1987).
[3] On the second term of the spectral asymptotics of the transmission problem. *Acta Appl. Math.*, 10:101–130 (1987).
[4] On the riesz means of the counting eigenvalues function of elliptic operator. volume 163, pages 143–145 (1987). In Russian.
[5] The asymptotics of the spectrum of pseudodifferential operator with periodic bicharacteristics. In *LOMI Proceedings*, volume 152, pages 94–104 (1986). English translation in J. Soviet Math., 40, no. 5 (1988).
[6] The asymptotics of the spectral function of positive elliptic operator without non-trapping condition. *Funct. Anal. Appl.*, 22(3):213–223 (1988).
[7] The precise asymptotics of the spectrum of boundary value problem and periodic billiards. *Izv. AN SSSR*, 52(6):1230–1251 (1988).

Safarov, Y. G. and Vassiliev, D. G.
[1] Branching Hamiltonian billiards. *Soviet Math. Dokl.*, 38(1) (1989).
[2] Asymptotic distribution of eigenvalues of differential operators. *AMS Transl., Ser. 2*, 150 (1992).
[3] *The Asymptotic Distribution of Eigenvalues of Partial Differential Operators*, volume 155 of *Translations of Mathematical Monographs*. American Mathematical Society (1997). 354pp.

Schenk, D. and Shubin, M. A.
[1] Asymptotic expansion of the density of states and the spectral function of the Hill operator. *Mat. Sborn.*, 12(4):474–491 (1985).

Seeley, R.
[1] A sharp asymptotic estimate for the eigenvalues of the Laplacian in a domain of $\mathbf{R}^3$. *Advances in Math.*, 102(3):244–264 (1978).
[2] An estimate near the boundary for the spectral function of the Laplace operator. *Amer. J. Math.*, 102(3):869–902 (1980).

Shigikawa, I.
[1] Eigenvalue problems for the Schrödinger operator with magnetic field on a compact Riemannian manifold. *J. Funct. Anal.*, 87(75):92–127 (1989).

Shubin, M. A.
[1] Weyl's theorem for the Schrödinger operator with an almost periodic potential. *Trans. Moscow Math. Soc.*, 35:103–164 (1976).
[2] *Pseudodifferential Operators and Spectral Theory.* Nauka (1978). In English: Springer-Verlag (1987).
[3] Spectral theory and the index of elliptic operators with almost periodic potential. *Soviet Math. Surv.*, 34(2):95–135 (1979).

Shubin, M. A. and Tulovskii, V. A.
[1] On the asymptotic distribution of eigenvalues of p.d.o. in $\mathbb{R}^n$. *Math. USSR Sbornik*, 21:565–573 (1973).

Sjöstrand, J.
[1] Analytic singularities and microhyperbolic boundary value problems. *Math. Ann.*, 254:211–256 (1980).
[2] On the eigenvalues of a class of hypoelliptic operators IV. *Ann. Inst. Fourier*, 30(2) (1980).
[3] Singularités analytiques microlocales. *Astérisque*, 95 (1982).
[4] Propagation of analytic singularities for second order Dirichlet problems I. *Commun. Part. Differ. Equat.*, 5(1):41–94 (1980).
[5] Propagation of analytic singularities for second order Dirichlet problems II. *Commun. Part. Differ. Equat.*, 5(2):187–207 (1980).
[6] Propagation of analytic singularities for second order Dirichlet problems III. *Commun. Part. Differ. Equat.*, 6(5):499–567 (1981).

Sobolev, A. V.
[1] On the asymptotic for energy levels of a quantum particle in a homogeneous magnetic field, perturbed by a decreasing electric field 1. *J. Soviet Math.*, 35(1):2201–2211 (1986).
[2] On the asymptotic for energy levels of a quantum particle in a homogeneous magnetic field, perturbed by a decreasing electric field 2. *Probl. Math. Phys*, 11:232–248 (1986).
[3] Discrete spectrum asymptotics for the Schrödinger operator with electric and homogeneous magnetic field. *Notices LOMI sem.*, 182:131–141 (1990).
[4] On the asymptotics of discrete spectrum for the Schrödinger operator in electric and homogeneous magnetic field. *Operator Theory: Advances and Applications*, 46:27–31 (1990).
[5] Weyl asymptotics for the discrete spectrum of the perturbed Hill operator. *Adv. Soviet Math.*, 7:159–178 (1991).
[6] The quasi-classical asymptotics of local Riesz means for the Schrödinger operator in a strong homogeneous magnetic field. *Duke Math. J.*, 74(2):319–429 (1994).
[7] Quasi-classical asymptotics of local Riesz means for the Schrödinger operator in a moderate magnetic field. *Ann. Inst. H. Poincaré*, 62(4):325–359 (1995).
[8] *Discrete Spectrum Asymptotics for the Schrödinger Operator in a Moderate Magnetic Field*, volume 78, pages 357–367. Birkhäuser (1995).

Sobolev, S. L.
[1] *Some Applications of Functional Analysis in Mathematical Physics*, volume 90 of *Trans. of Math. Monographs*. AMS (1991).

Solnyshkin, S. N.
[1] Asymptotic of the energy of bound states of the Schrödinger operator in the presence of electric and homogeneous magnetic fields. *Select. Math. Sov.*, 5(3):297–306 (1986).

Solomyak, M. Z.
[1] Spectral asymptotics of Schrödinger operators with non-regular homogeneous potential. *Math. USSR Sbornik*, 55(1):9–38 (1986).

Solomyak, M. Z. and Vulis, L. I.
[1] Spectral asymptotics for second order degenerating elliptic operators. *Math. USSR Izv.*, 8(6):1343–1371 (1974).

Tamura, H.
[1] Asymptotic formulas with sharp remainder estimates for bound states of Schrödinger operators. *J. d'Anal. Math.*, 40:166–182 (1981).
[2] Asymptotic formulas with sharp remainder estimates for bound states of Schrödinger operators II. *J. d'Anal. Math.*, 41:85–108 (1982).
[3] Asymptotic formulas with sharp remainder estimates for eigenvalues of elliptic second order operators. *Duke Math. J.*, 49(1):87–119 (1982).
[4] Asymptotic formulas with remainder estimates for eigenvalues of Schrödinger operators. *Commun. Part. Differ. Equat.*, 7(1):1–54 (1982).
[5] The asymptotic formulas for the number of bound states in the strong coupling limit. *J. Math. Soc. Japan*, 36:355–374 (1984).
[6] Eigenvalue asymptotics below the bottom of essential spectrum for magnetic Schrödinger operators. In *Proc. Conf. on Spectral and Scattering Theory for Differential Operators, Fujisakara-so, 1986*, pages 198–205, Tokyo. Seizo Itô (1986).
[7] Asymptotic distribution of eigenvalues for Schrödinger operators with magnetic fields. *Nagoya Math. J.*, 105(10):49–69 (1987).

[8] Asymptotic distribution of eigenvalues for Schrödinger operators with homogeneous magnetics fields. *Osaka J. Math.*, 25:633–647 (1988).
[9] Asymptotic distribution of eigenvalues for Schrödinger operators with homogeneous magnetic fields II. *Osaka J. Math.*, 26:119–137 (1989).

Taylor, M. E.
[1] Grazing rays and reflection of singularities of solutions to wave equations. *Comm. Pure Appl. Math.*, 29:1–38 (1976).
[2] *Diffraction Effects in the Scattering of Waves*, pages 271–316. D. Reidel Publ. Co., Dordrecht, Boston, London (1981).
[3] *Pseudodifferential Operators.* Princeton University Press (1981).

Treves, F.
[1] *Introduction to Pseudo-differential and Fourier Integral Operators, 1, 2.* Plenum Press (1982).

Vainberg, B. R.
[1] A complete asymptotic expansion of the spectral function of second order elliptic operators in $\mathbb{R}^n$. *Math. USSR Sb.*, 51(1):191–206 (1986).

Vassiliev, D. G.
[1] Binomial asymptotics of spectrum of boundary value problems. *Funct. Anal. Appl.*, 17(4):309–311 (1983).
[2] Two-term asymptotics of spectrum of boundary value problems with inner reflection of general form. *Funct. Anal. Appl.*, 18(4):1–13 (1984).
[3] Asymptotics of the spectrum of pseudodifferential operators with small parameters. *Math. USSR Sbornik*, 49(1):61–72 (1984).
[4] Two-term asymptotics of the spectrum of a boundary value problem in the case of a piecewise smooth boundary. *Soviet Math. Dokl.*, 33(1):227–230 (1986).
[5] Two-term asymptotics of the spectrum of natural frequencies of a thin elastic shell. *Soviet Math. Dokl.*, 41(1):108–112 (1990).
[6] One can hear the dimension of a connected fractal in R^2. In *Proc. Int. Conf. Integral Equations and Inverse Problems, Varna 1989.* Longman (1990).

Volovoy, A.
[1] Improved two-term asymptotics for the eigenvalue distribution function of an elliptic operator on a compact manifold. *Commun. Part. Differ. Equat.*, 15(11):1509–1563 (1990).
[2] Verification of the Hamiltonian flow conditions associated with Weyl's conjecture. *Ann. Glob. Anal. Geom.*, 8(2):127–136 (1990).

Vorobets, Y. B., Galperin, G. A., and Stepin, A. M.
[1] Periodic billiard trajectories in polygons. *Uspechi Matem. Nauk*, (6):165–166 (1991).

Vugal'ter, S. A.
[1] On asymptotics of eigenvalues of many-particles Hamiltonians on subspaces of functions of the given symmetry. *Theor. Math. Physics*, 83(2):236–246 (1990).

Vugal'ter, S. A. and Zhislin, G. M.
[1] On the spectrum of Schrödinger operator of multiparticle systems with short-range potentials. *Trans. Moscow Math. Soc.*, 49:97–114 (1986).
[2] Asymptotics of the discrete spectrum of Hamiltonians of quantum systems with a homogeneous magnetic field. *Operator Theory: Advances and Applications*, 46:33–53 (1990).
[3] On precise asymptotics of the discrete spectrum of n-particle Schrödinger operator in the spaces with symmetry. *Soviet Math. Dokl.*, 312(2):339–342 (1990).
[4] On the asymptotics of the discrete spectrum of the given symmetry for many-particle Hamiltonians. *Trans. Moscow Math. Soc.*, 54:187–213 (1991). In Russian.
[5] On the discrete spectrum of the given symmetry for the Schrödinger operator of the n-particle system in the homogeneous magnetic field. *Funct. Anal. Appl.*, 25(4):83–86 (1991). In Russian; also in Soviet Math. Dokl., 317, no 6, 1365–1369 (1991).

Wakabayashi, S.
[1] Singularities of solutions of the Cauchy problem for symmetric hyperbolic systems. *Commun. Part. Differ. Equat.*, 9(12):1147–1177 (1984).

Wang, X. P.
[1] Asymptotic behavior of spectral means of pseudo-differential operators I. *J. Approx. Theor. Appl.*, 1(2):119–136 (1985).
[2] Asymptotic behavior of spectral means of pseudo-differential operators II. *J. Approx. Theor. Appl.*, 1(3):1–32 (1985).

Weinstein, A.
[1] Asymptotics of the eigenvalues; clusters for Laplacian plus a potential. *Duke Math. J.*, 44:883–892 (1977).

Weyl, H.
[1] Das asymptotische Verteilungsgesets linearen partiellen Differentialgleichungen. *Math. Ann.*, 71:441–479 (1912).
[2] Über die Abhängigkeit der Eigenschwingungen einer Membran von deren Begrenzung. *J. Für die Angew. Math.*, 141:1–11 (1912).

Widom, H.
[1] A complete symbolic calculus for pseudo-differential operators. *Bull. Sci. Math., Sér. 2*, 104:19–63 (1980).

Zelditch, S.
[1] Selberg trace formulae, Ψdo's and equidistribution theorems for closed geodesics and Laplace eigenfunctions. *Lect. Notes Math.*, 1256:467–479 (1987).

Index

Springer and the environment

At Springer we firmly believe that an international science publisher has a special obligation to the environment, and our corporate policies consistently reflect this conviction.
We also expect our business partners – paper mills, printers, packaging manufacturers, etc. – to commit themselves to using materials and production processes that do not harm the environment. The paper in this book is made from low- or no-chlorine pulp and is acid free, in conformance with international standards for paper permanency.

Printing: Mercedesdruck, Berlin
Binding: Buchbinderei Lüderitz & Bauer, Berlin

GPSR Compliance

The European Union's (EU) General Product Safety Regulation (GPSR) is a set of rules that requires consumer products to be safe and our obligations to ensure this.

If you have any concerns about our products, you can contact us on ProductSafety@springernature.com

In case Publisher is established outside the EU, the EU authorized representative is:

Springer Nature Customer Service Center GmbH
Europaplatz 3
69115 Heidelberg, Germany

Batch number: 10704021

Printed by Printforce, the Netherlands